AZEPINES—PART 1

This is the forty-third volume in the series

THE CHEMISTRY OF HETEROCYCLIC COMPOUNDS

THE CHEMISTRY OF HETEROCYCLIC COMPOUNDS

A SERIES OF MONOGRAPHS

ARNOLD WEISSBERGER AND EDWARD C. TAYLOR

Editors

AZEPINES
Part 1

Authors:

Burt Renfroe

CIBA–GEIGY CORPORATION
ARDSLEY, NEW YORK

Clinton Harrington

DOW CHEMICAL COMPANY
MIDLAND, MICHIGAN

George R. Proctor

UNIVERSITY OF STRATHCLYDE
GLASGOW, SCOTLAND

Editor:

Andre Rosowsky

DANA-FARBER CANCER INSTITUTE
BOSTON, MASSACHUSETTS

AN INTERSCIENCE® PUBLICATION

JOHN WILEY AND SONS
NEW YORK · CHICHESTER · BRISBANE · TORONTO · SINGAPORE

An Interscience® Publication

Library of Congress Cataloging in Publication Data:

Renfroe, Burt.
Azepines.

(Chemistry of heterocyclic compounds, ISSN 0069-3154 v. 43)
"An Interscience publication."
Bibliography: p.
Includes index.
1. Azepines. I. Harrington, Clinton. II. Proctor, George. III. Rosowsky, Andre. IV. Title. V. Series.

QD401.R43 1984 547'.593 83-3497
ISBN 0-471-01878-3 (v. 1)

Printed in the United States of America

10 9 8 7 6 5 4 3 2 1

The Chemistry of Heterocyclic Compounds

The chemistry of heterocyclic compounds is one of the most complex branches of organic chemistry. It is equally interesting for its theoretical implications, for the diversity of its synthetic procedures, and for the physiological and industrial significance of heterocyclic compounds.

A field of such importance and intrinsic difficulty should be made as readily accessible as possible, and the lack of a modern detailed and comprehensive presentation of heterocyclic chemistry is therefore keenly felt. It is the intention of the present series to fill this gap by expert presentations of the various branches of heterocyclic chemistry. The subdivisions have been designed to cover the field in its entirety by monographs which reflect the importance and the interrelations of the various compounds, and accommodate the specific interests of the authors.

In order to continue to make heterocyclic chemistry as readily accessible as possible new editions are planned for those areas where the respective volumes in the first edition have become obsolete by overwhelming progress. If, however, the changes are not too great so that the first editions can be brought up-to-date by supplementary volumes, supplements to the respective volumes will be published in the first edition.

ARNOLD WEISSBERGER

Research Laboratories
Eastman Kodak Company
Rochester, New York

EDWARD C. TAYLOR

Princeton University
Princeton, New Jersey

Preface

Heterocyclic systems with seven atoms, once considered chemical oddities, are today just as easily obtained as their five- and six-membered cousins, thanks to the very substantial advances in synthetic art that have been made in this field over the past 20 years. In a previous volume of the *Heterocyclic Compounds* series (Volume 26), seven-membered oxygen ring systems (oxepins) and sulfur ring systems (thiepins) were reviewed. The present two-part volume extends this coverage to the seven-membered nitrogen heterocycles (azepines).

As with the oxepins and thiepins, interest in the azepines encompasses a broad spectrum of theoretical and applied disciplines. At the theoretical level, chemists are fascinated by these compounds from the molecular orbital standpoint. Of particular importance in this regard are the azulenoid cyclopentazepine ring systems, the bridgehead nitrogen benz[*a*]azepinium cation, the bridged annulene system 11-azabicyclo[4.4.1]undeca-1,3,5,7,9-pentaene ("1,6-azirino[10]annulene"), and the still unknown cycl[4.4.3]azine. At the "applied" end of the spectrum, one hardly need be reminded of the enormous amount of research that has been conducted in pharmaceutical laboratories since the early 1960s on tranquilizers, antidepressants, and other psychotropic drugs of the benzodiazepine and dibenzazepine class. Probably more than any other factor, the vast commercial success of these medicinal agents and their benefit to society have caused the chemistry of condensed azepines to evolve into a major area of research in heterocyclic chemistry. One of the consequences of this effort has been the appearance, on the chemical scene, of a host of new ring systems which collectively form a dazzling array of structural types. It is the purpose of this two-part volume to give an account of the current state of knowledge concerning the synthesis, chemical reactions, and physical properties of some—though by no means all—of these systems, with particular emphasis on those facets that pertain to the "seven-memberedness" of the azepine ring.

Part 1 consists of two chapters. Chapter I (Renfroe and Harrington) is devoted to tricyclic systems containing an azepine ring along with two other rings that can be either carbocyclic or heterocyclic. This category includes no fewer than 150 different heterocylic systems, ranging from oxireno[*d*][2]benzazepines (a 3,6,7-system) to benzo[*e*]cyclooct[*b*]azepines (a 6,7,8-system). The largest single family, the dibenz[*b*,*f*]azepines, has as its most famous members the antidepressant imipramine and its congeners, such as carbamazepine. Other dibenzazepines of pharmaceutical interest are the dibenz[*c*,*e*]azepines, a number of which possess hypotensive activity. Certain tricyclic azepine ring systems are also noteworthy because they appear in natural products. Here may be cited certain alkaloids of the rheadan family (1,3-dioxolo[4,5-*h*][3]benzazepines) and at least three groups of indole alkaloids, the erythroidine and tuberostemonine family (azepino[3,2,1-*hi*]indoles), the naucladerine family (azepino[4,5-*b*]indoles), and the ergot family (azepino[5,4,3-*cd*]indoles). Additionally, fused tricylic azepine systems have been generated from tricyclic terpene ketones (e.g., santonin) via ring enlargement. The identification and characterization of the products from these reactions have played an important role in increasing our understanding of the mechanism of the Schmidt and Beckmann rearrangements and in providing model systems for the synthesis of ring-enlarged azasteroid analogues.

Chapter II (Proctor) presents a review of the chemistry of bicyclic, as opposed to tricyclic, azepine derivatives. Here again a substantial number of fused, bridged, and spiran systems are covered, with emphasis being placed again on the seven-membered ring. Among these compounds the 1-, 2-, and 3-benzazepines are of interest because of their potential psychopharmacologic activity. A novel class of antiviral agents also cited in this chapter consists of compounds belonging to the bridged 1,6-azirino [10]annulene ring system. The development of these compounds represents an intriguing blend of "pure" and "applied" heterocyclic chemistry.

Part 2 also consists of two chapters. Chapter I (Watthey and Stanton) covers the tricyclic diazepine systems, a number of which have attracted attention because of the pharmaceutical importance of some of their members. Here may be cited anxiolytic agents such as benzazepam, oxazolam, and triazolam; antidepressants such as dibenzopin; and antipsychotic agents such as clozapine. All these tricyclic azepine derivatives have contributed in a major way to the modern treatment of mental illness. Another example of a biologically important class of tricyclic azepine derivatives is provided by the anthramycin antibiotics (pyrrolo[2,1-*c*][1,4]benzodiazepines), which have shown impressive activity as antineoplastic agents.

The practical importance of azepine derivatives is by no means limited to medical applications. For example, among the ring systems discussed by Dr. Peet in Chapter II are the 1,2,5-triazepines, some of which have found use in the agricultural field as pesticidal plant protectants. Other triazepines, of the 1,3,5-type with *N*-nitro substituents, have been investigated as high explosives.

In summary, the two-part volume for which this Preface is written bears witness to the fact that seven-membered heterocyclic compounds are no longer the esoteric species they were once considered to be. Quite to the contrary, the pace of research and development in this area is accelerating, and there seems to be virtually no limit to the number of interesting ring systems that can be created in the laboratory by a combination of ingenuity and perseverance. The future should bring rich rewards not only in terms of new academic knowledge but also in terms of practical applications that will benefit us all, chemists and nonchemists alike. I wish to graciously thank the several authors who have joined me in preparing this review. Their thoroughness and limitless patience cannot be sufficiently praised. Thanks are due, as well, to Drs. Arnold Weissberger and Edward C. Taylor for their encouragement and support of the project, and to the capable staff at John Wiley & Sons for their expeditious handling of these chapters as they were received.

ANDRE ROSOWSKY

Boston, Massachusetts
January 1984

Contents

CHAPTER I

DIBENZAZEPINES AND OTHER TRICYCLIC AZEPINES

BURT RENFROE

CIBA–Geigy Corporation
Ardsley, New York

CLINTON HARRINGTON

Dow Chemical Company
Midland, Michigan

I. INTRODUCTION

This chapter reviews the synthesis and chemistry of the 148 tricyclic azepine ring systems that have appeared in the chemical literature through Volume 86 of *Chemical Abstracts*. Ring systems have been arranged in ascending order of ring size according to the rules of *Chemical Abstracts*.

The chemistry in this survey ranges from isolated investigations of purely academic interest in heterocyclic chemistry to numerous patents generated by industrial research organizations in their vigorous efforts to uncover new classes of pharmacologically active agents. The largest single system, consisting of the dibenz[*b,f*]azepines (Section VI. 39), reflects the especially intense interest of the pharmaceutical industry in structural modifications of the clinically important antidepressant imipramine. It must be noted that the patent literature does not always provide melting points or other physical data. In this review, only those compounds which were supported by a melting point or boiling point in a patent or journal article are tabulated or discussed in the text.

A common method of forming azepine rings that is reported frequently in the literature is the Beckmann or Schmidt-type ring expansion of a six-membered ketone to a seven-membered lactam. Unfortunately, the structures of the lactams generated in this way are not always firmly established, and the reader should bear this in mind.

A recent review of the reactions of acetylenecarboxylic esters and nitrogen heterocycles by Acheson and Elmore (Section VI.24) offers new structural assignments for several ring systems previously thought to contain azepine rings. The tricyclic systems affected, in addition to azepino[1,2-*a*]-quinolines are: azepino[1,2-*a*]- and [1,2-*c*]quinazolines (Sections VI.12 and VI.13), azepino[1,2-*a*]quinoxalines (Section VI.15), azepino[2,1-*b*]-benzoxazoles (Section IV.4), azepino[1,2-*a*]benzimidazoles (Section IV. 20), and—quite possibly—azepino[2,1-*a*]isoquinolines (Section VI.23), azepino[2,1-*b*]benzothiazoles (Section IV.7), and azepino[2,1-*b*]benzoselenazoles (Section IV.13).

II. 3,6,7-SYSTEMS

1. Oxireno[*d*][2]benzazepines

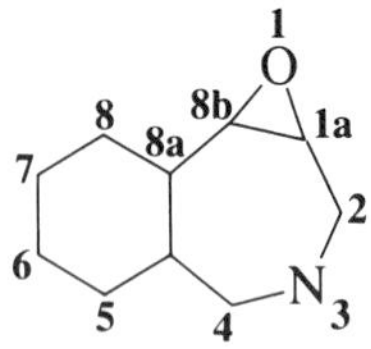

The only study of derivatives of the oxireno[*d*][2]benzazepine ring system was reported in 1974 by Hassner and Anderson (1), who found that rearrangement of the azirine–isobenzofuran adducts **1a** and **1b** on neutral alumina gave the 2H-oxireno[*d*]benzazepines **2a** and **2b.** Treatment of **2b** with HCl/HOAc under reflux (Eq. 1) gave the diketone **3,** and nmr spectra of **2a** and **2b** closely resembled those of the benzazepines **4a** and **4b**. These data were

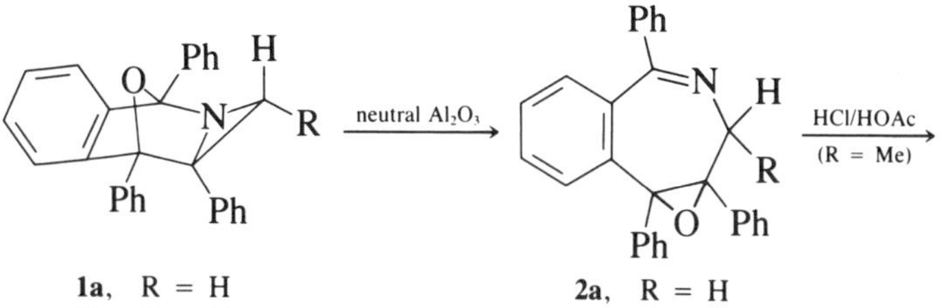

neutral Al_2O_3 → HCl/HOAc (R = Me) →

1a, R = H
1b, R = Me

2a, R = H
2b, R = Me

(1)

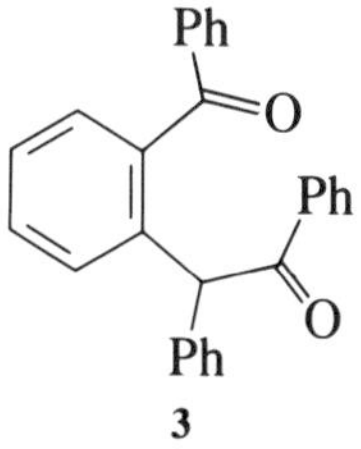

3

consistent with either the *N*-oxide **6,** arising via **5** (Eq. 2), or the oxide **2b,** arising by direct rearrangement of **1b** (Eq. 3). An authentic sample of **6,** prepared by oxidation of **4b** with *m*-chloroperbenzoic acid (Eq. 4), proved to be different from the rearrangement product of **1b.** Treatment of **4b** with excess *m*-chloroperbenzoic acid in refluxing chloroform gave a substance (**7**) that was identical to the product of room temperature oxidation of **2b.**

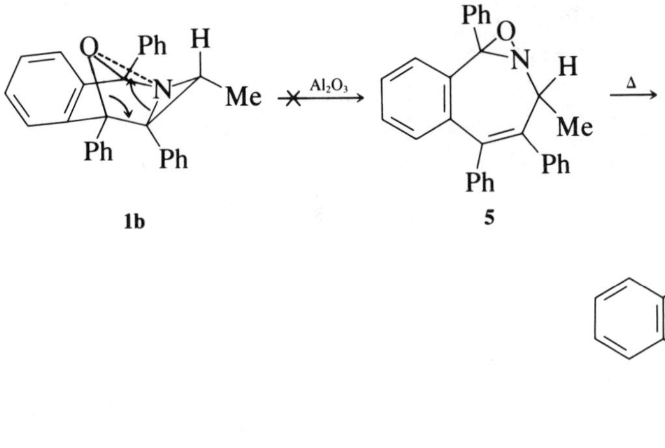

1b $\xrightarrow{Al_2O_3}$ (✕) **5** $\xrightarrow{\Delta}$ **6** (2)

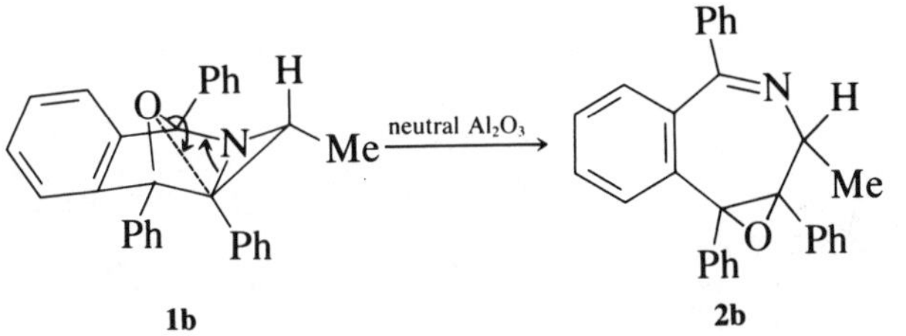

1b $\xrightarrow{\text{neutral } Al_2O_3}$ **2b** (3)

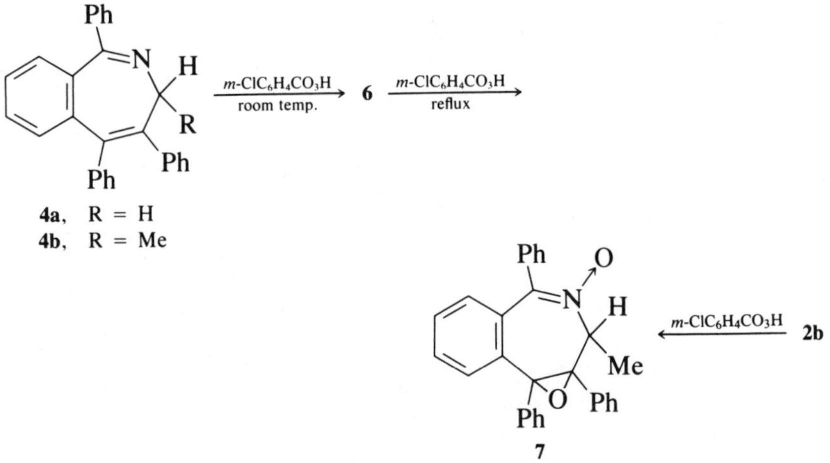

4a, R = H
4b, R = Me

4b $\xrightarrow[\text{room temp.}]{m\text{-}ClC_6H_4CO_3H}$ **6** $\xrightarrow[\text{reflux}]{m\text{-}ClC_6H_4CO_3H}$ **7** $\xleftarrow{m\text{-}ClC_6H_4CO_3H}$ **2b** (4)

The 2H-oxireno[*d*][2]benzazepines **2a** and **2b** were resistant to acid, base, and lithium aluminum hydride (cf. the resistance of tetraphenylethylene oxide to these reagents). However, prolonged treatment of **2a** with neutral alumina gave the pinacol rearrangement product **8a.** The structure was formulated as **8a** rather than the isomeric ketone **8b** on the basis of its infrared absorption at 1710 cm^{-1}.

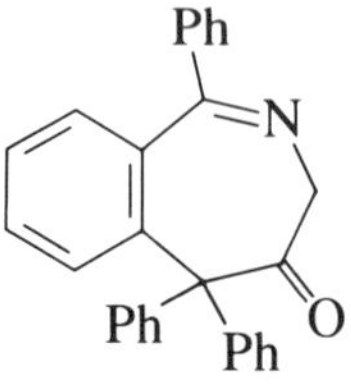

Ph N Ph Ph O **8a**   Ph N Ph O Ph **8b**

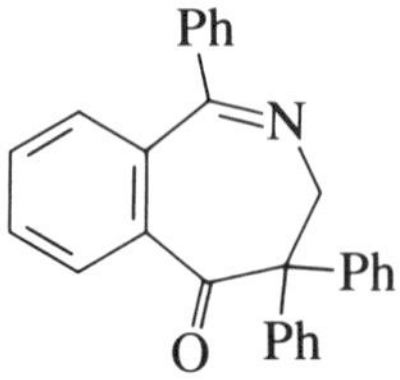

2. Benzo[*b*]cycloprop[*d*]azepines

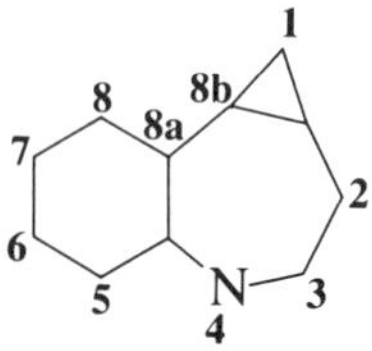

1 8 8b 8a 7 2 6 5 N 3 4

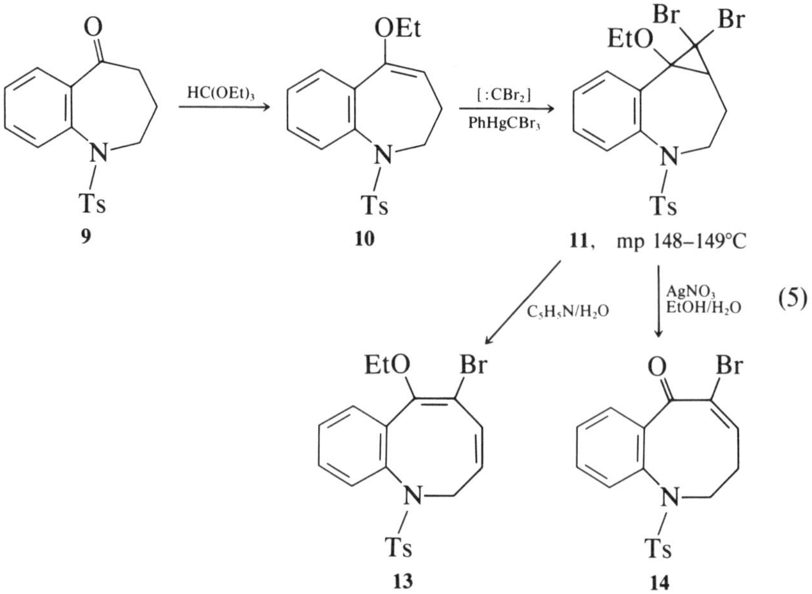

O N Ts **9** —HC(OEt)$_3$→ OEt N Ts **10** —[:CBr_2] PhHgCBr_3→ Br Br EtO N Ts **11**, mp 148–149°C

11 —C_5H_5N/H_2O→ EtO Br N Ts **13**

11 —$AgNO_3$, EtOH/H_2O→ O Br N Ts **14** (5)

Ross and Proctor (2) have generated an example of the benzo[*b*]-cycloprop[*d*]azepine ring system by addition of dibromocarbene to the enol ether **10,** as shown in Eq. 5. The dibromo adduct **11** (51–60% yield) can be ring-expanded with silver ion to the benzazocinone **14** (48% yield) or to the bromo ether **13** in refluxing aqueous pyridine (81% yield). The adduct **12,** prepared similarly from the corresponding olefin (Eq. 6), did not undergo ring expansion to a benzazocine (as did **11**) on treatment with silver ion. Since the isomeric adduct **16** (Section II.3) also did not undergo ring expansion, it is clear that the ether oxygen is required for this process.

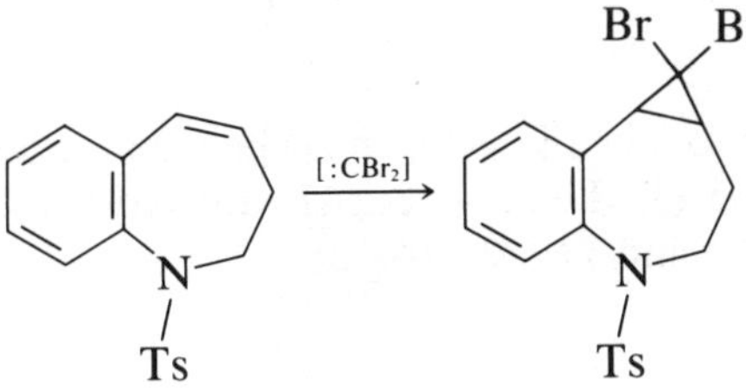

Br Br

[:CBr₂] → N Ts N Ts **12** (6)

3. Benzo[*d*]cycloprop[*b*]azepines

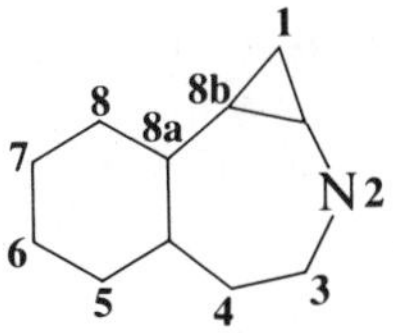

1 8 8b 8a 7 N2 6 5 4 3

A single example of the benzo[*d*]cycloprop[*b*]azepine ring system is found in the work of Proctor and Ross (2), who treated the *N*-tosylenamine **15** with dibromocarbene to form **16** in 40% yield (Eq. 7). In a similar manner to **12,** compound **16** failed to undergo ring expansion by reaction with silver ion (see Section II.2).

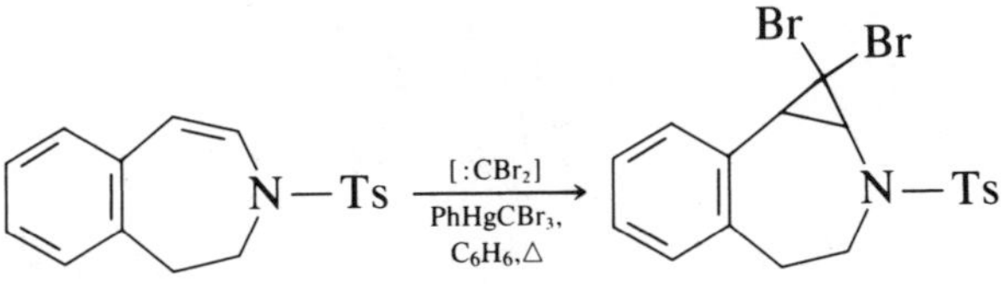

$$\textbf{15}\ \text{N—Ts} \xrightarrow[\text{PhHgCBr}_3,\ C_6H_6,\Delta]{[:CBr_2]} \textbf{16}\ \text{(Br, Br) N—Ts} \quad (7)$$

III. 5,5,7-SYSTEMS

1. Cyclopenta[*c*]tetrazolo[1,5-*a*]azepines

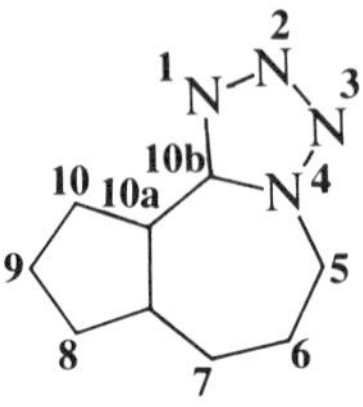

The only example to date of cyclopenta[*c*]tetrazolo[1,5-*a*]azepines is found in ferrocene chemistry (3). Treatment of **17** with hydrazoic acid (Eq. 8) yields a mixture of **18** (27%) and **19** (10–20%).

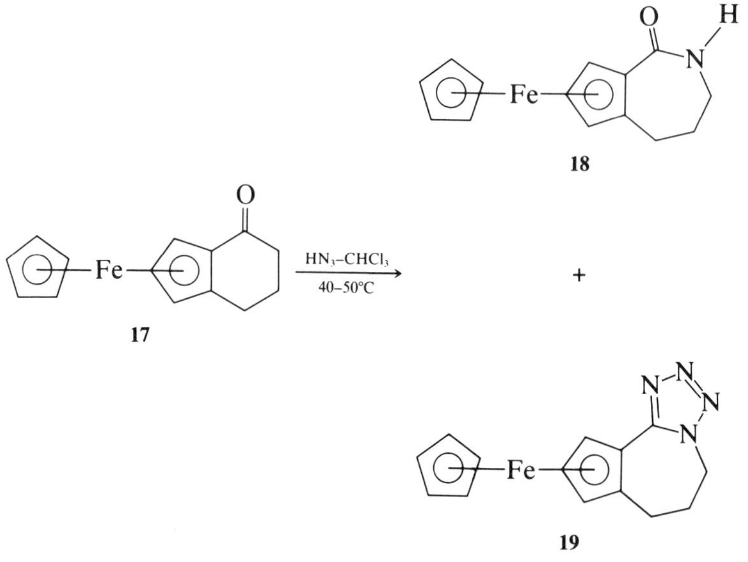

$$\mathbf{17} \xrightarrow[40\text{–}50^{\circ}\mathrm{C}]{HN_3\text{–}CHCl_3} \mathbf{18} + \mathbf{19} \qquad (8)$$

2. 2,2a,3-Trithia(2a-S^{IV})-4a-azacyclopent[*cd*]azulenes

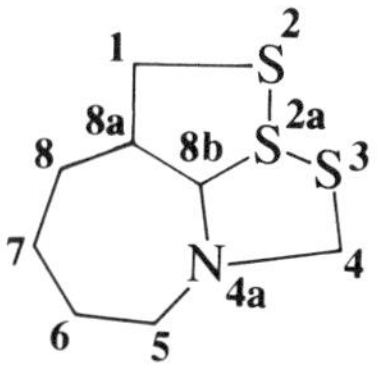

The sodium *tert*-amylate–catalyzed condensation of two equivalents of carbon disulfide with caprolactam, when quenched with excess methyl iodide, gave rise to the adduct **20.** Replacement of caprolactam with pyrrolidinone afforded the homolog **21.** When **20** and **21** were treated with phosphorous pentasulfide followed by perchloric acid (Eq. 9), the tricyclic products **22** and **23** were formed in 60 and 41% yields, respectively. Treatment of **22** and **23** with ethanolic ammonia produced **24** and **25** (Eq. 10a), but reaction with aniline (Fig. 1) took a different course; compound **22** afforded **26–28,** whereas **23** generated **29–31** (4).

In principle, reaction of **22** with ammonia might be expected to give **32** instead of its isomer **24,** and **23** could give rise to **33** and not **25** (Eq. 10b). Furthermore, **22** and **23** with aniline could give **34** and **35** instead of **26** and **29** (Fig. 1). In order to differentiate between these possibilities, the tricyclic analog **36** was prepared from **37** (Eq. 11). Loss of the ethyl signals in the nmr spectrum on reaction of **36** with ammonia was consistent with structure **25** as the reaction product. Likewise, **36** and *p*-toluidine produced **30** as the

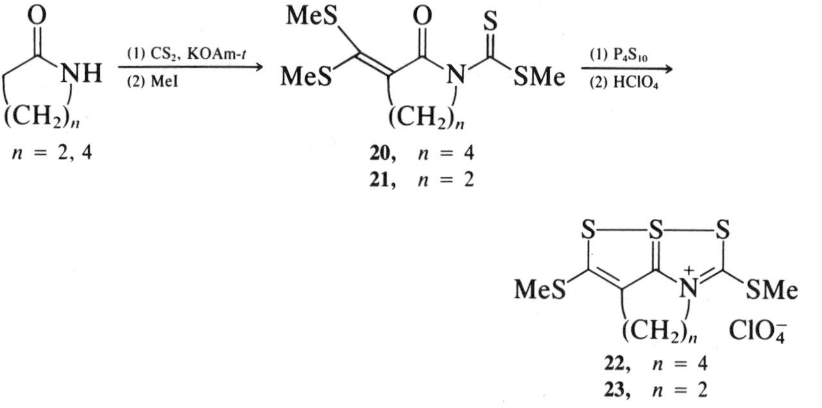

$n = 2, 4$ —(1) CS_2, KOAm-*t*; (2) MeI→ **20,** $n = 4$; **21,** $n = 2$ —(1) P_4S_{10}; (2) $HClO_4$→ **22,** $n = 4$; **23,** $n = 2$ (9)

(10a)

path *a* → **24,** $n = 4$ **25,** $n = 2$

path *b* → **32,** $n = 4$ **33,** $n = 2$

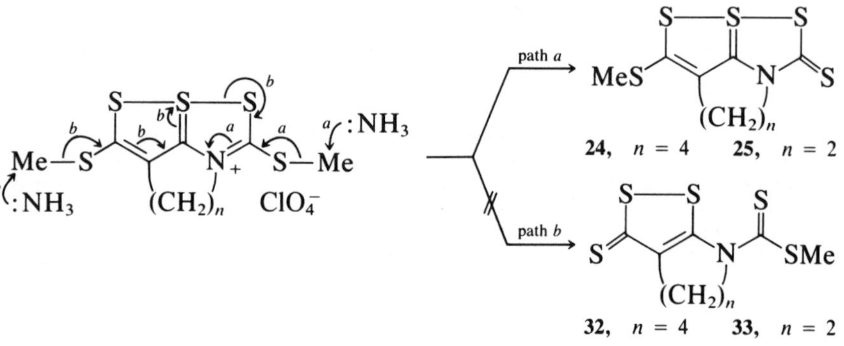

(10b)

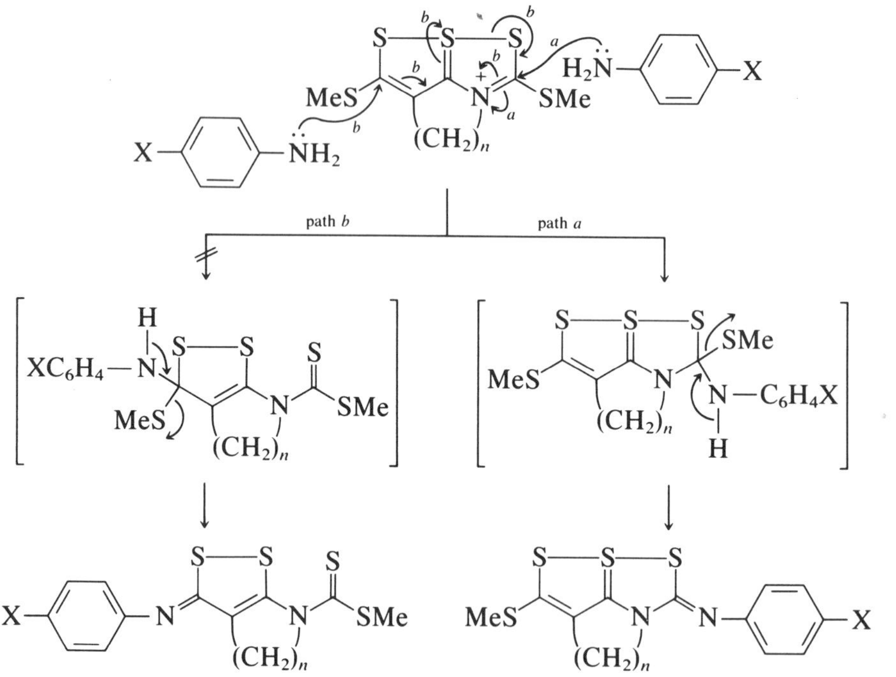

path b | path a

34, $n = 4$; X = H
35, $n = 2$; X = H

26, $n = 4$; X = H
27, $n = 4$; X = Me
28, $n = 4$; X = OMe
29, $n = 2$; X = H
30, $n = 2$; X = Me
31, $n = 2$; X = OMe

Figure 1

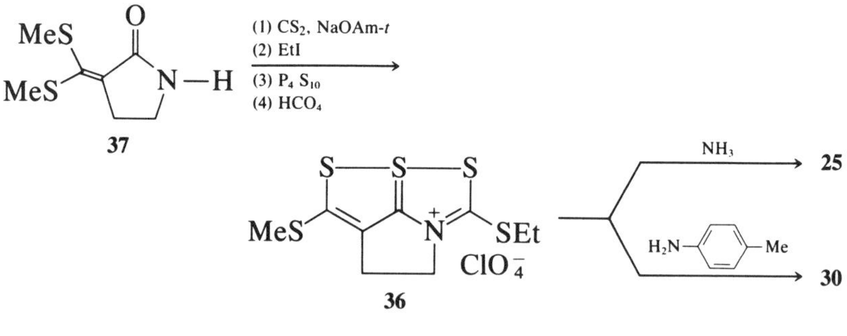

37 → (1) CS_2, NaOAm-*t* (2) EtI (3) P_4S_{10} (4) $HClO_4$ → 36 (ClO_4^-); 36 + NH_3 → 25; 36 + H_2N–C_6H_4–Me → 30 (11)

TABLE 1. 2,2a,3-TRITHIA(2a-S^{IV})-4a-AZACYCLOPENT[*cd*]AZULENES[a]

Product	mp (°C)	Spectra
22	190–193	pmr
24	159–162	pmr
26	147–149	
27	154–156	
28	168–170	

[a]Ref. 4.

sole product. These results strongly supported the assignment of the tricyclic structures **24** and **25** to the reaction products obtained from **22** and **23.** Table 1 summarizes the physical data for representatives of this system.

3. Dipyrazolo[4,3-*c*:3′,4′-*e*]azepines

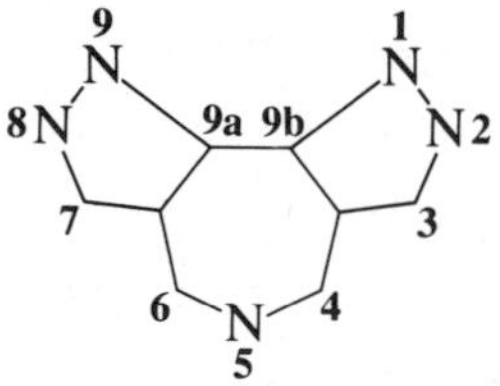

Two examples of the dipyrazolo[4,3-*c*:3′,4′-*e*]azepine system have been reported by Treibs and Lange (5), who prepared the dicyanoazepinone **38** in 63% yield. Treatment of **38** with hydrazine hydrate or phenylhydrazine in 50% aqueous HOAc gave **39a** and **39b,** respectively (Eq. 12).

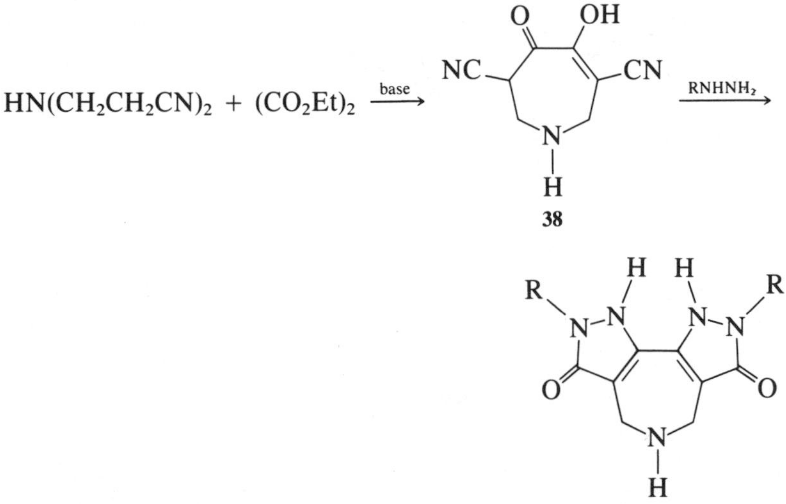

$$HN(CH_2CH_2CN)_2 + (CO_2Et)_2 \xrightarrow{\text{base}} \mathbf{38} \xrightarrow{RNHNH_2} \mathbf{39} \quad (12)$$

39a, R = H; mp 146–147°C
39b, R = Ph; mp 130–131°C

4. Thiazolo[2′,3′ :2,3]imidazo[1,5-*a*]azepines

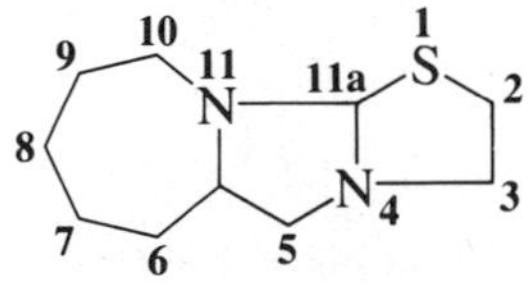

Only one example of the thiazolo[2′,3′:2,3]imidazo[1,5-*a*]azepine ring system has been described in the literature. Condensation of the imino ether of caprolactam **40** with nitromethane (Fig. 2) yielded the nitrovinyl derivative **41.** Catalytic reduction afforded the diamine **42,** which on reaction with carbon disulfide gave the imidazo[3,4-*a*]azepine thione **43.** Condensation of **43** with phenacyl bromide (Fig. 3) gave a single product (69% yield), which was formulated as the tricyclic structure **44** (6).

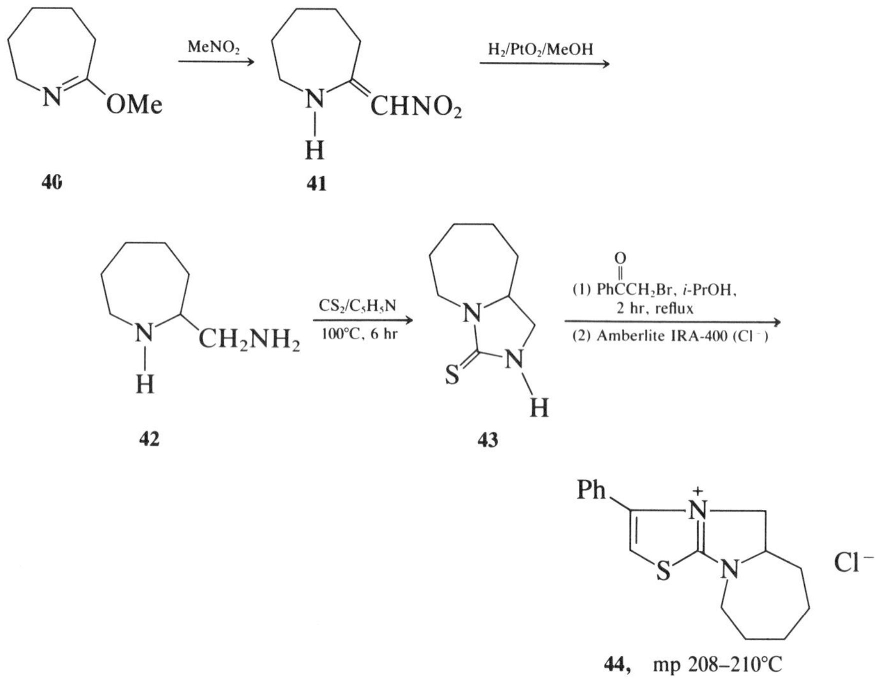

N OMe — $MeNO_2$ → N(H) $CHNO_2$ — $H_2/PtO_2/MeOH$ →

40 **41**

N(H) CH_2NH_2 — CS_2/C_5H_5N, 100°C, 6 hr → **43** — (1) $PhCOCH_2Br$, *i*-PrOH, 2 hr, reflux; (2) Amberlite IRA-400 (Cl^-) →

42 **43**

Cl^-

44, mp 208–210°C

Figure 2

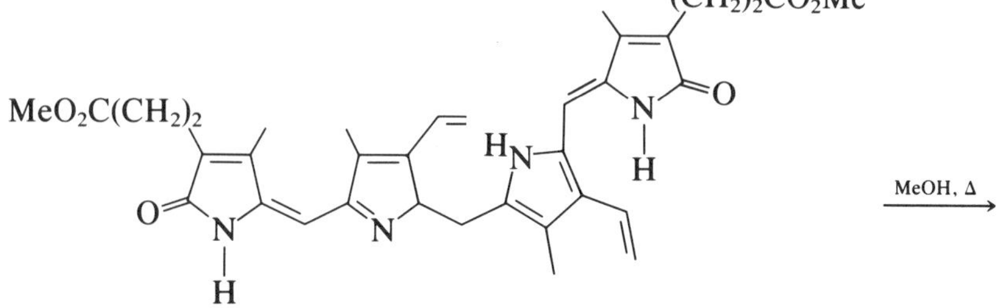

$(CH_2)_2CO_2Me$ $MeO_2C(CH_2)_2$ — MeOH, Δ →

45
Pterobiline, Biliverdine IV ν

Figure 3

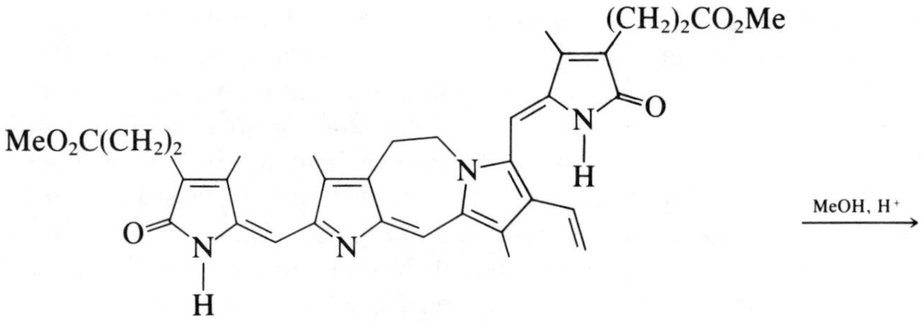

$(CH_2)_2CO_2Me$

$MeO_2C(CH_2)_2$

MeOH, H^+

46
Phorcabiline, Neobiliverdine ν

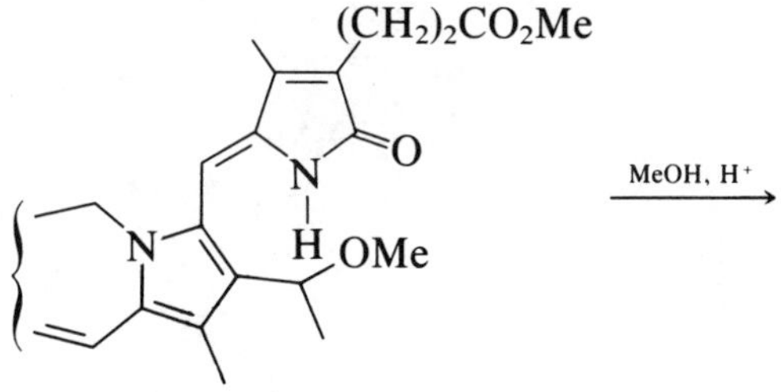

$(CH_2)_2CO_2Me$

H OMe

MeOH, H^+

47

CO_2Me

$(CH_2)_2$

$(CH_2)_2CO_2Me$

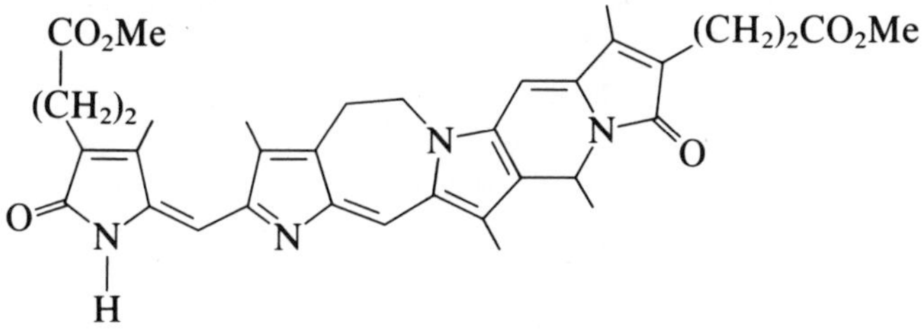

48
Figure 3 (*continued*)

5. Dipyrrolo[1,2-*a*:2′,3′-*d*]azepines

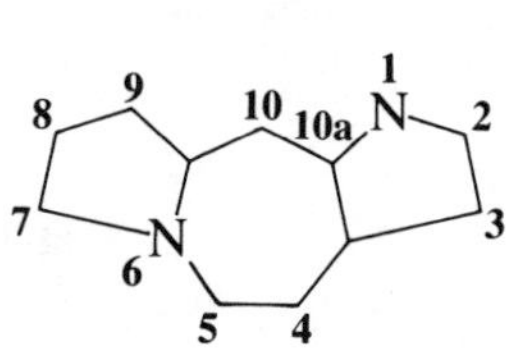

French workers have studied the structure and biosynthesis of the main bile pigments of *Lepidoptera* and—by employing field desorption mass spectroscopy and high-resolution nmr of the natural compounds and of degradation products obtained from thick-layer chromatography—have discovered two examples of the dipyrrolo[1,2-*a*:2′,3′-*d*]azepine ring system in phorcabiline (**46**) and neobiliverdine δ (**50**) (7). It was found that the pigment pterobiline (**45**) is converted to phorcabiline (**46**) *in vivo* and can also be cyclized to **46** on heating (Fig. 3). In similar fashion, biliverdine δ (**49**) is converted to neobiliverdine δ (**50**) (Fig. 4). Pigment **46** was distinguished from **50** by analysis of mass spectral and nmr patterns and by its conversion to **47** and **48** on treatment with methanol. Neobiliverdine δ, on chromic acid oxidation, yielded fragments **51, 52,** and **53,** which were isolated from tlc plates and identified by mass spectrometry. More recently, the conversions of **45** to **46** and of **49** to **50** have been carried out photochemically (8).

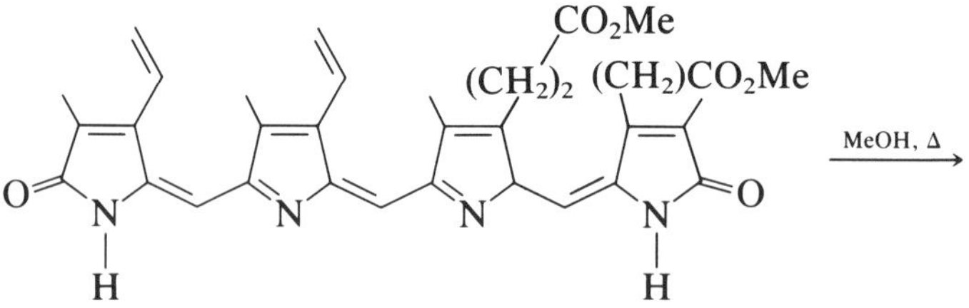

MeOH, Δ

49
Biliverdine IV δ

CrO_3

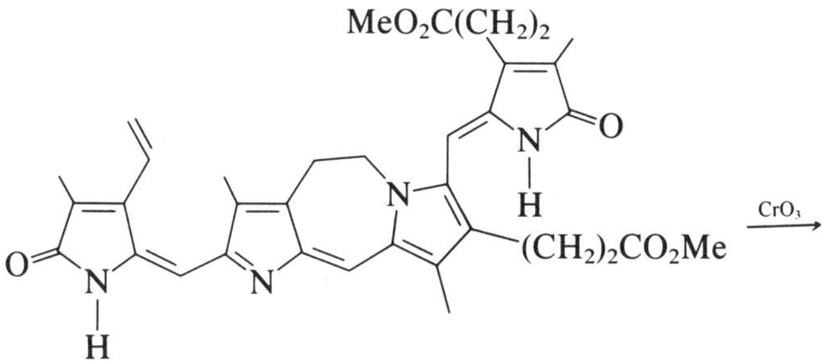

50
Neobiliverdine δ

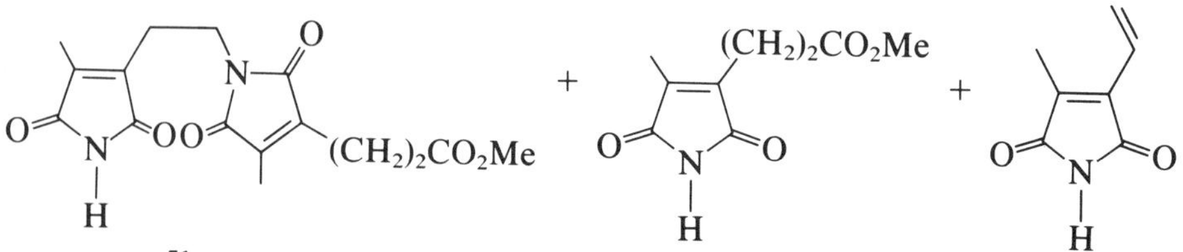

51 + **52** + **53**

Figure 4

6. Azepino[2,1,7-*cd*]pyrrolizines

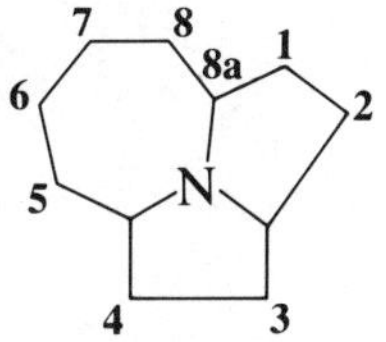

Several examples of the azepino[2.1.7-*cd*]pyrrolizine system, which is also referred to as cycl[4,2,2]azine, have been prepared (9) by the reaction of 3H-pyrrolizines with a five- or tenfold excess of an acetylene dicarboxylic acid ester (Fig. 5). Yields range from poor to moderate. The stereochemistry indicated in structure **55** was deduced from the coupling constants $J_{AB} = 2$ Hz and $J_{BC} = 12$ Hz, which are consistent with a *cis* relationship for H_A–H_B and a *trans* relationship for H_B–H_C. Catalytic reduction of **55a** at atmospheric pressure (Fig. 6) gave the dihydro derivative **57,** whereas dehydrogenation with dichlorodicyanoquinone (DDQ) yielded a dark red product that could be assigned structure **58** on the basis of its ultraviolet absorption and nmr spectral data. Hydrogenation of **58** led to **59,** an isomer of **55a.** All attempts at hydride ion abstraction from **58** to form the unknown 10π-electron cycl[4.2.2]azinium ion failed. In the preparation of **55d,** the yellow compound **56** was the major product (48% yield).

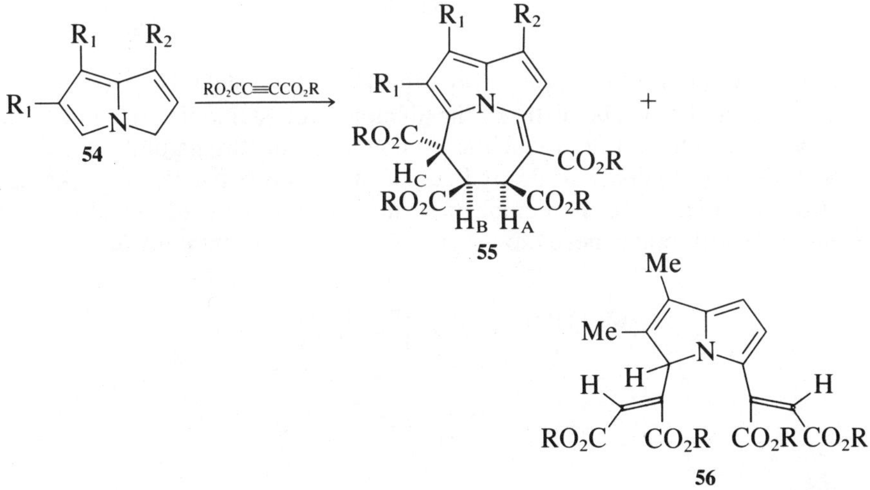

	R	R_1	R_2	Mp (°C)	Yield (%)	Color	Spectra
55a	Me	H	H	105–112	59	Red	uv, nmr
55b	Me	H	Me	153–161	52	Red	ir, ms
55c	Me	Me	H	151–152	19	Red	ms
55d	Et	H	Me	Oil	7	Red	ms

Figure 5

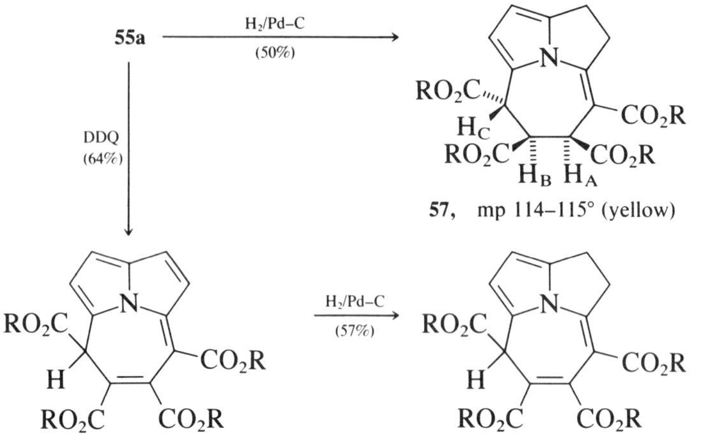

55a → (H2/Pd–C, 50%) → 57, mp 114–115° (yellow)

55a → (DDQ, 64%) → 58, mp 150°C (red)

58 → (H2/Pd–C, 57%) → 59, mp 112–113°C (colorless)

Figure 6

7. Furo[3,2-*c*]pyrrolo[1,2-*a*]azepines

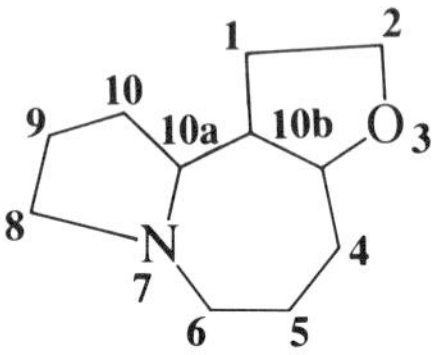

The only reported examples of the furo[3,2-*c*]pyrrolo[1,2-*a*]azepine system are found in the chemistry of the alkaloid protostemonine (**60**) (10). The (*E*) or (*Z*) configuration about the double bond in this natural product is unknown. The hydrochloride hydrate of **60** is converted via a two-phase hydrolysis (Fig. 7) to stemonine (**61**) and the furanone **62.** Oxidation of stemonine with manganese dioxide leads to dehydrostemonine (**63**).

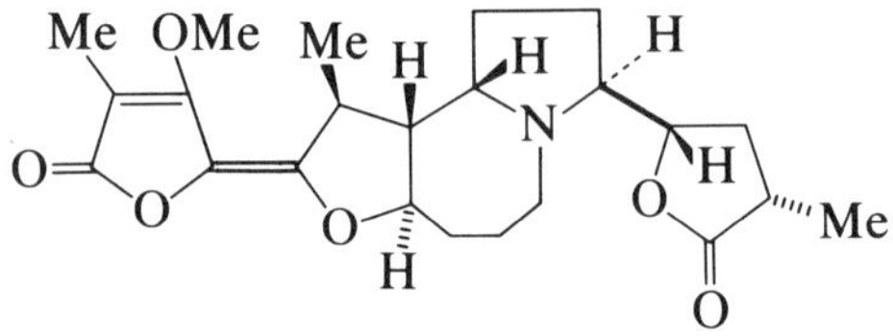

60

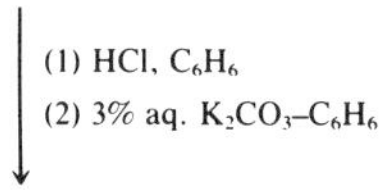

(1) HCl, C_6H_6
(2) 3% aq. K_2CO_3–C_6H_6

Figure 7

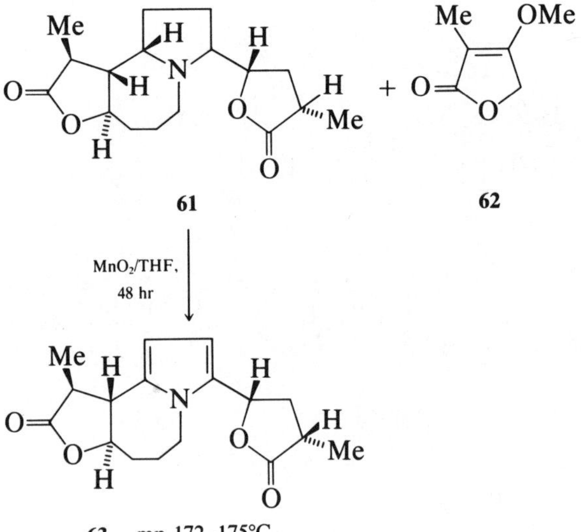

61 62

MnO_2/THF, 48 hr

63, mp 172–175°C

Figure 7 (*continued*)

8. Dithieno[2,3-*c*:3′,2′-*e*]azepines

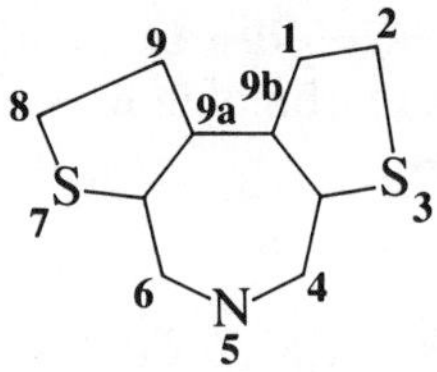

The dithienoazepines **65** (Eq. 13) were prepared in high yield from the dialdehyde **64** by condensation with amines in the presence of sodium dithionite (11). Initially the Schiff bases of **64** were isolated and reduced in a separate step, as was done in the preparation of 5H-dibenz[*c,e*]azepines (see Section VI. 40). Under these conditions, however, only a 7% yield of **65a** was obtained, along with a 74% yield of **66**. Similar observations were made in the two-stage synthesis of **65b**.

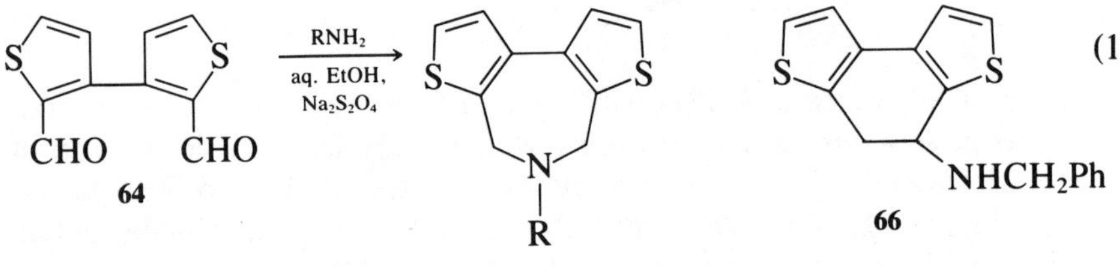

RNH_2, aq. EtOH, $Na_2S_2O_4$ (13)

64 65 66

	R	Yield (%)	mp (°C)	Spectra
65a	CH_2Ph	94	145–146	ir, uv,
65b	Ph	86	142–143	ms, nmr

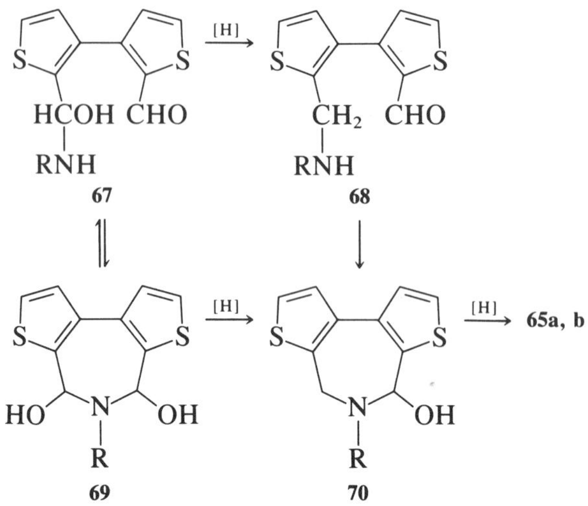

Figure 8

Because of the mild reaction conditions, the authors concluded that *bis*-Schiff bases of **64** were probably not involved in the formation of **65** and suggested the sequence **67**→**68**→**70**→**65** or **67**→**69**→**70**→**65** (Fig. 8). The benzyl-like hydroxyls of the carbinolamines **67** or **69** would be easily hydrogenolyzed in the final step.

IV. 5,6,7-SYSTEMS

1. Tetrazolo[5,1-*a*][2]benzazepines

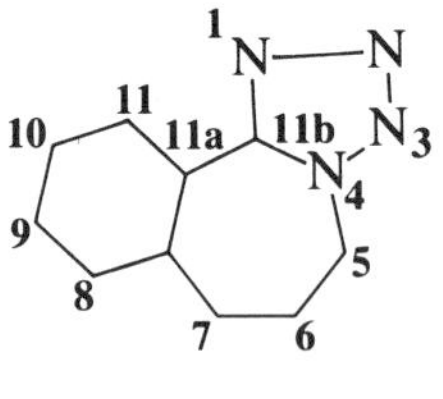

Treatment of α-tetralone (**71**) with sodium azide (Fig. 9) under Schmidt conditions gave a substance with the formula $C_{10}H_{10}N_4$ (12). Reduction of this material with lithium aluminum hydride in ether produced the tetrahydrobenzazepine **72.** Since the molecular formula ruled out simple lactam possibilities, the isomeric tetrazoles **73**, **74a**, and **74b** were considered as alternative structures for the Schmidt product. Since reduction of pentamethylenetetrazole with lithium aluminum hydride gave hexamethyleneimine, compound **72** could have arisen from either **73** or **74a.** However,

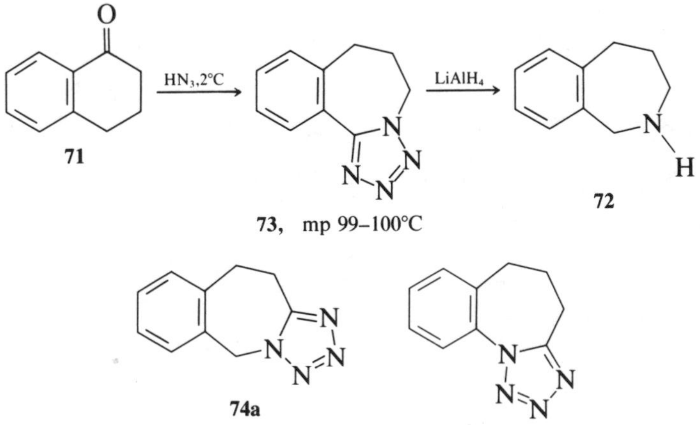

73, mp 99–100°C

Figure 9

the structure **73** was favored over the unlikely **74a** and **74b** on the basis of ultraviolet spectral studies, which indicated conjugation between the benzene and tetrazole rings.

2. Tetrazolo[5,1-*b*][3]benzazepines

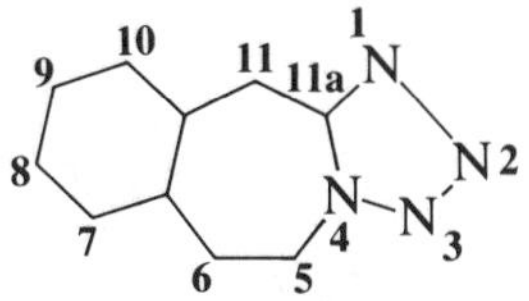

A single example of the tetrazolo[5,1-*b*][3]benzazepine system was produced by the action of two equivalents of hydrazoic acid on the diketone **75,** as shown in Fig. 10 (13). This study provides an interesting illustration of how the ring expansion of cyclic ketones to lactams in the Schmidt reaction can be altered by changing the nature of the acid or the stoichiometry of the reactants. Treatment of **75** with two equivalents of hydrazoic acid, generated *in situ* from sodium azide in hydrochloric acid, gave the tetrazolo[5,1-*b*][3]benzazepine derivative **76** (78% yield). In contrast, one equivalent of NaN_3 in HCl gave lactam **77** (47% yield), two equivalents of NaN_3 in sulfuric acid gave lactam **78** (74% yield), and three equivalents of NaN_3 in trichloroacetic acid gave lactam **79** (20% yield).

The infrared spectrum of **76** showed bands at 1639 and 1515 cm^{-1}, attributed to the C—C and C—N stretching frequencies, respectively, and the nmr spectrum contained a two-proton multiplet centered at δ 4.56. It had been reported that the nmr spectrum of the tetrazole system **80** exhibits a complex multiplet at δ 2.95, whereas that of **81** appears at δ 4.95. Similarly,

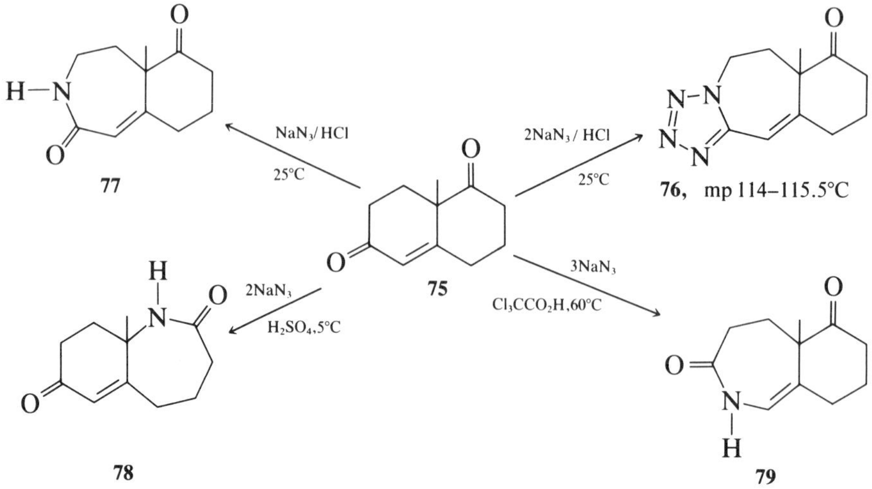

Figure 10

the methylene groups adjacent to the tetrazole ring in **82** were known to give different signals—at δ 4.48 and δ 3.02, respectively. Thus the triplet at δ 4.56 in the spectrum of **76** eliminated the tetrazole of lactam **79** as the structure of **76.**

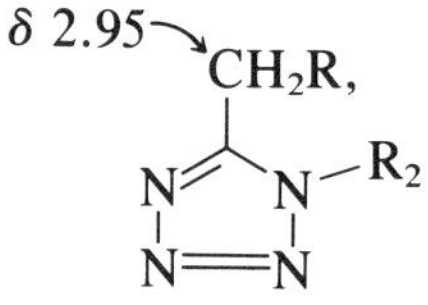

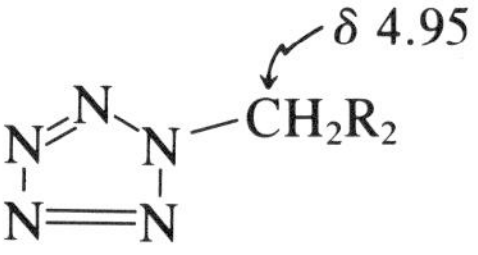

3. 1,2,4-Triazolo[3,4-*b*][3]benzazepines

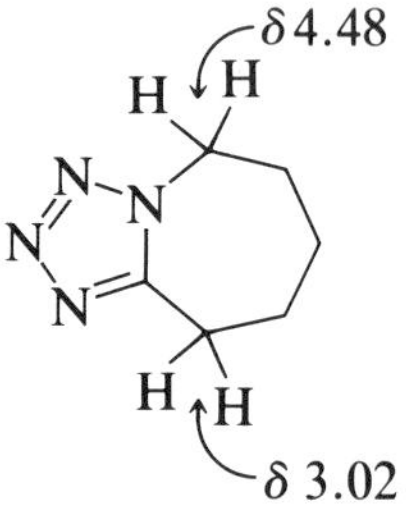

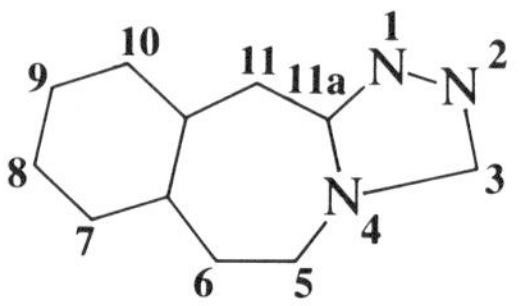

The only reported examples of the 1,2,4-triazolo[3,4-*b*]benzazepine ring system are found in the patent literature (14,15). Tricyclic products pos-

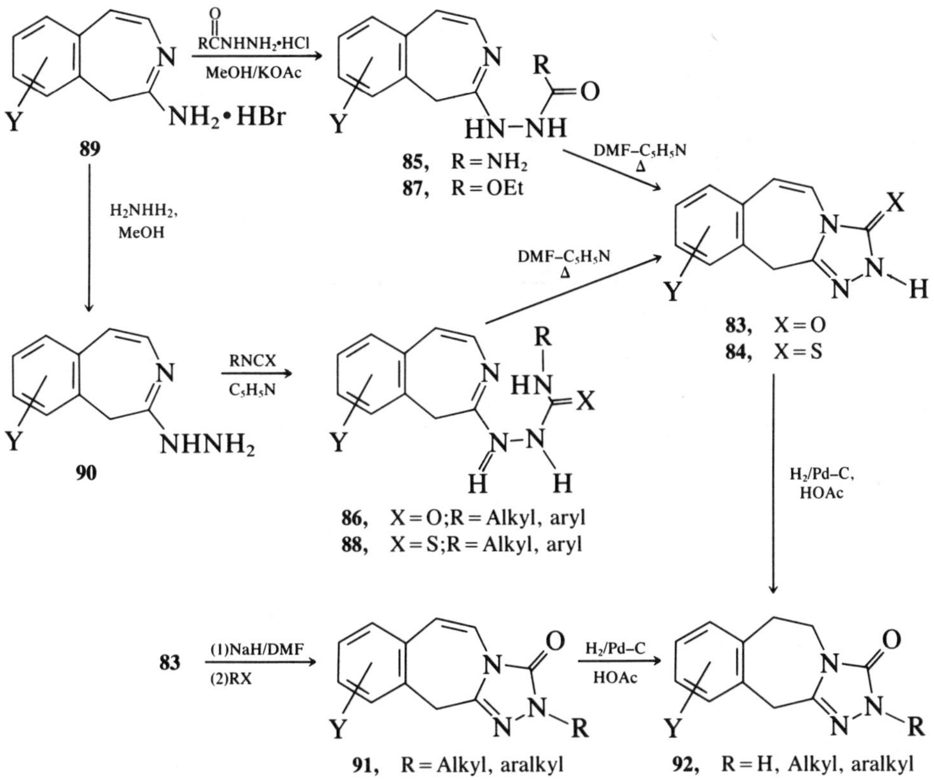

Figure 11

sessing the general structures **83** and **84** were obtained by heating the semicarbazides **85** and **86,** the carbazates **87,** or the thiosemicarbazides **88** (Fig. 11). In general, the cyclizations proceeded smoothly in a refluxing solution of pyridine–dimethylformamide. The requisite precursors **85–88** were prepared by two routes from the aminobenzazepine **89.** Compounds of general formula **85** and **87** were obtained by treating **89** directly with semicarbazide or ethyl carbazate in methanol solution. The preparation of **86** and **88** was achieved by treating **89** with anhydrous hydrazine and then condensing the resultant hydrazinobenzazepine **90** with an isocyanate or isothiocyanate, respectively.

The triazolobenzazepin-3-ones **83** were alkylated smoothly in dimethylformamide with formation of the 2-alkyltriazolobenzazepin-3-ones **91.** Both **83** and **91** underwent catalytic reduction to give the tetrahydrotriazolobenzazepin-1-ones **92.** Compounds of structures **84, 85, 91,** and **92** were claimed to exhibit analgetic and muscle-relaxing activity. A list of members of this ring system is given in Table 2.

TABLE 2. 1,2,4-TRIAZOLO[3,4-*b*][3]BENZAZEPINES

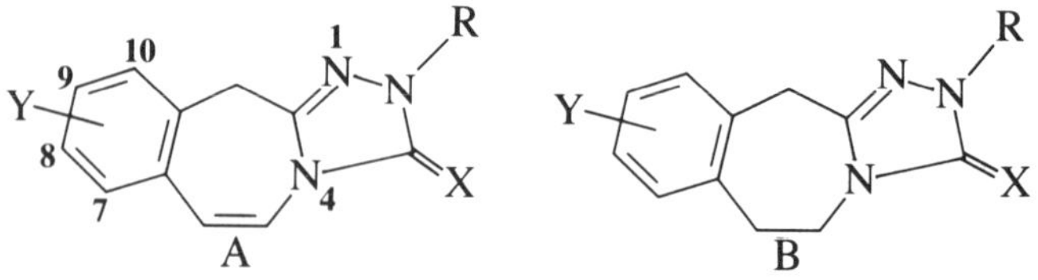

X	Y	R	mp (°C)	Refs.
		Structure A		
O	H	H	209–210	14,15
S	H	H	231	14,15
O	H	Me	123–124	15
O	H	Et	125–126	15
O	H	Ch_2Ph	133–134	15
O	H	$(CH_2)_2NMe_2$	205–207 (oxalate)	15
O	7-Cl	H	280–287	14
O	10-Cl	H	290–295	14
O	10-Cl	Me	199–200	15
O	7,8-Me_2	H	265–269	15
O	8,9-Me_2	H	265–269	14
O	8,9-$(MeO)_2$	H	273–274	14,15
O	8,9-$(MeO)_2$	Me	210–211	15
O	8,9-Me_2	Me	160–161	15
		Structure B		
O	H	H	253–261	15
O	H	Me	187	15
O	H	Et	122–123	15
O	8,9-Me_2	H	285–292	15
O	8,9-$(Me0)_2$	H	259–263	15

4. Azepino[2,1-*b*]benzoxazoles

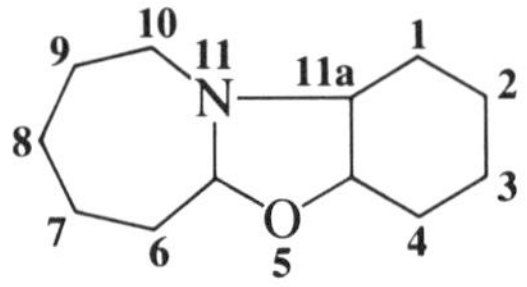

In the earliest reported synthesis of the azepino[2,1-*b*]benzoxazole system, 2-methylbenzoxazole (**93**) was condensed with dimethyl acetylenedicarboxylate (16) to give a low yield of a pale yellow product (mp 167–168°C), which was formulated as the azepino[2,1-*b*]benzoxazole **94** (Fig. 12). The structure of **94** was assigned on the basis of its nmr spectrum, which showed a pair of doublets at δ 5.45 and δ 5.75 (J = 5 cps) corresponding to

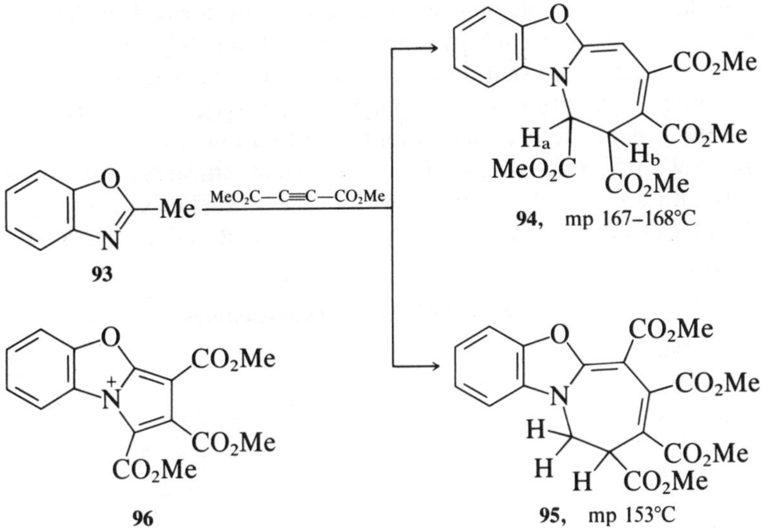

93

$MeO_2C—C≡C—CO_2Me$

94, mp 167–168°C

95, mp 153°C

96

Figure 12

the C_9 and C_{10} protons. However, repetition of this work (17) yielded a different product (mp 153°C, aq. MeOH), to which the isomeric structure **95** was assigned on the basis of the ABX pattern observed in the nmr spectrum for the C_9 and C_{10} protons. The mass spectrum showed the loss of the methyl acrylate characteristic of adducts of this type, with formation of the pyrrolobenzoxazole cation radical **96**. Since products **94** and **95** were isolated

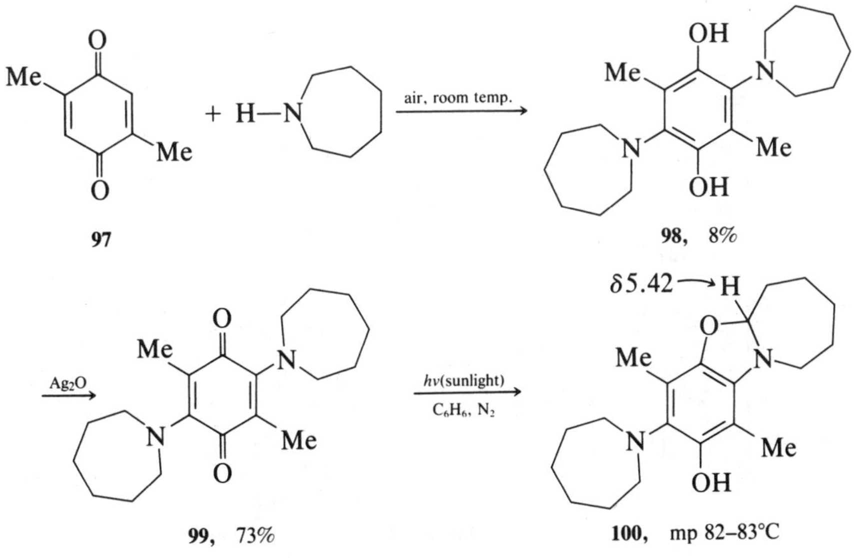

97

air, room temp.

98, 8%

Ag_2O

99, 73%

hν(sunlight) C_6H_6, N_2

δ5.42

100, mp 82–83°C

Figure 13

in low yield, it is possible that both were actually formed in the reaction, but that one was lost in each instance during chromatography.

A second mode of formation of this ring system (18) is shown in Fig. 13. Photochemical irradiation (using sunlight) of an oxygen-free solution of the quinone **99** in benzene produced rapid discoloration, with the formation of the azepino[2,1-*b*]benzoxazole **100** in 54% yield. Infrared, ultraviolet, and 100-mHz nmr spectral data confirmed the structure. The C_{5a} proton is seen as a doublet of doublets (J = 4.5 and 7.0 cps) centered at δ 5.42.

5. Oxazolo[2,3-*a*][2]benzazepines

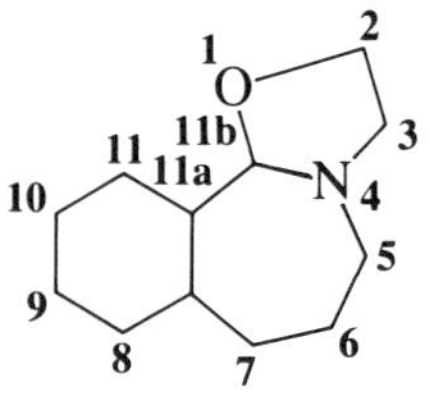

Single examples of both the oxazolo[2,3-*a*]benzazepine system and the corresponding thiazolo system (Section IV. 9) were prepared (19) by stirring the bromoaldehyde **102** at ambient temperature in dioxane with one equivalent of ethanolamine or mercaptoethylamine (Fig. 14). Physical constants for the products are listed in Table 3.

The intermediate Schiff base presumably quaternizes to give **103.** The free hydroxy or mercapto group then adds to the immonium bond, forming **104.**

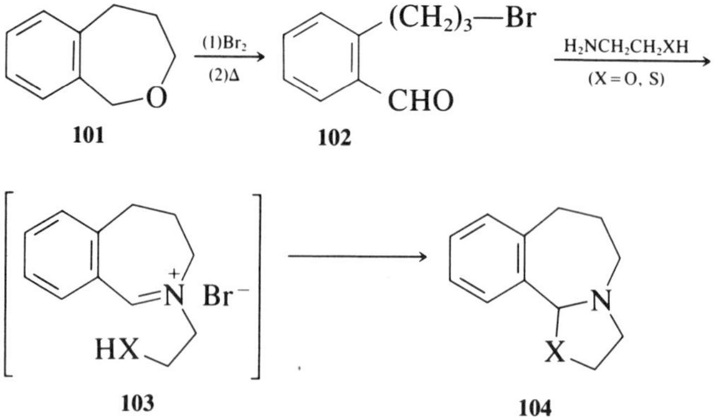

Figure 14

TABLE 3. PHYSICAL DATA FOR COMPOUND **104**

X	Yield (%)	bp (°C/torr)	mp (°C, HBr salts)	nmr (δ, C_{11b} proton)
O	85	90–92/0.1	110–112	5.37
S	70	138–140/0.2	214–216	5.45

6. Isothiazolo[5′,4′:4,5]pyrimido[1,2-*a*]azepines

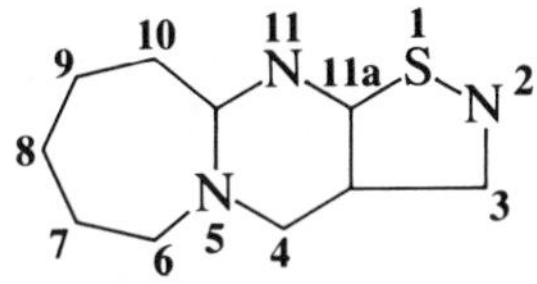

Condensation of 5-amino-3-methylisothiazole-4-carboxylic acid (**105**) with methyl caprolactim (**106**) gave a low yield of **107** when the reaction was run at 110°C in the absence of solvent (20). Since acids related to **105** were found to be prone to decarboxylation, the low yield of **107** was attributed to a competing thermal decarboxylation of **105.** To circumvent this problem, the condensation was repeated with **108,** the ethyl ester of **105.** When heated at 160°C for 30 min, **108** and **106** combined to produce **107** in 32% yield, a slight improvement over the original procedure (Eq. 14).

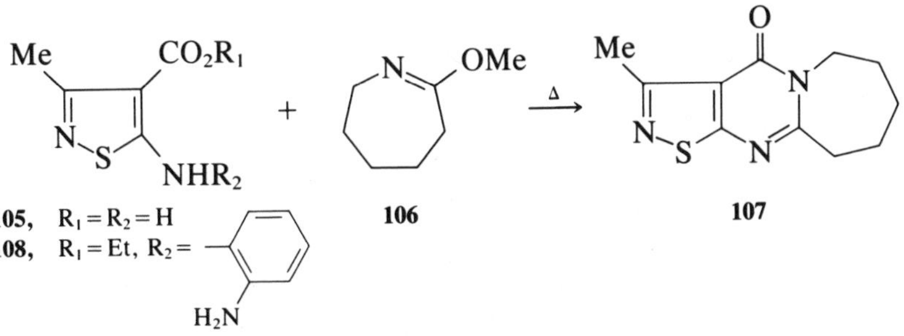

(14)

105, $R_1 = R_2 = H$
108, $R_1 = Et$, $R_2 =$ 2-aminophenyl

106

107

7. Azepino[2,1-*b*]benzothiazoles

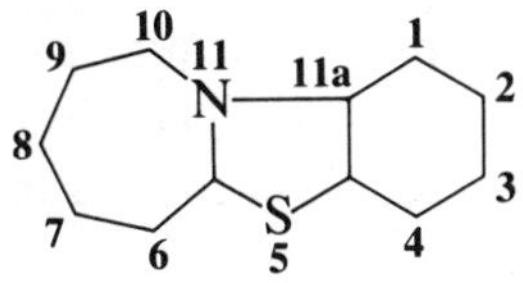

In continuing studies of the reaction of acetylenedicarboxylate esters with nitrogen-containing heterocycles, Acheson and co-workers prepared derivatives of the azepino[2,1-*b*]benzothiazole ring system from the dimethyl ester and 2-alkylthiazoles (21). These workers found that **109** (Fig. 15), as well as 2,4-dimethylthiazole and 2-methylbenzoselenazole, reacted with two moles of dimethylacetylenedicarboxylate to yield adducts possessing very similar ultraviolet, infrared, and nmr spectra. The most striking features of the nmr spectra were the disappearance of the 2-methyl group of the starting thiazole **109a** in **110a,** and the presence of a singlet at δ 1.88, indicating a methyl rather than ethyl group in **110b.** Nmr spectra of **110a** and **110b** also contain a typical AB pattern, about δ 6.0 and δ 5.5 (J = 6 Hz), corresponding to the MeO_2C—CHX—CHY—CO_2Me moiety. Additionally, the nmr spectrum

CO_2Me MeO_2C CO_2Me CO_2Me

MeO_2C—C≡C—CO_2Me, 2 moles

Raney Ni (R = H)

109a, R = Me
109b, R = Et

	R	Mp(°C)	Color	Spectra
110a	H	213	Yellow	ir,uv,nmr
110b	Me	186	Yellow	ir,uv,nmr,ms

111

Figure 15

of **110a** shows a vinylic proton at δ 5.2. The only structures consistent with these data are the ones shown in Fig. 15. When **110a** was treated with Raney nickel, **111** was produced in low yield. The formation of these adducts was rationalized as shown in Fig. 16. Mass spectra of these and related adducts have been published (26).

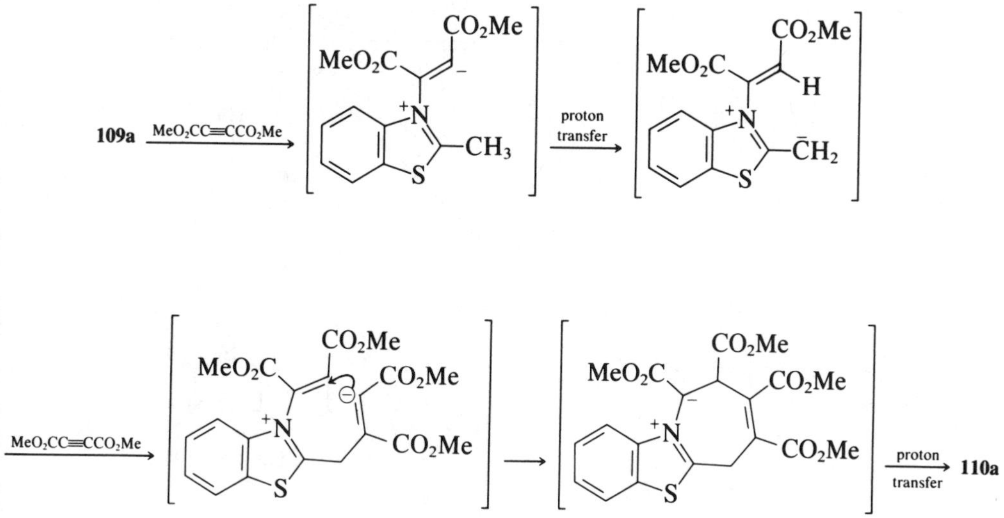

109a $\xrightarrow{MeO_2CC{\equiv}CCO_2Me}$ [intermediate] $\xrightarrow{\text{proton transfer}}$ [intermediate] $\xrightarrow{MeO_2CC{\equiv}CCO_2Me}$ [intermediate] $\longrightarrow$ [intermediate] $\xrightarrow{\text{proton transfer}}$ **110a**

Figure 16

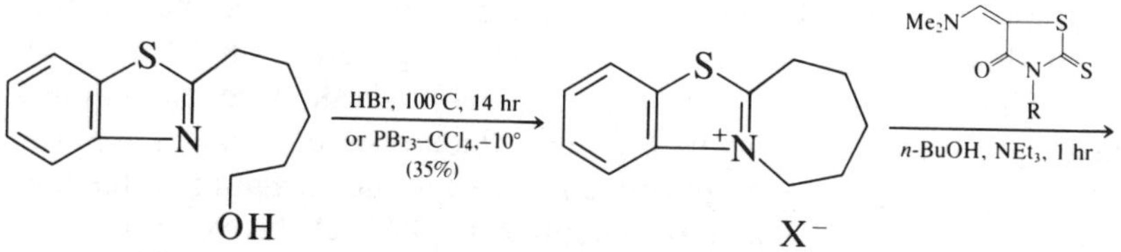

112 $\xrightarrow[\text{(35\%)}]{\text{HBr, 100°C, 14 hr or } PBr_3\text{–}CCl_4, -10°}$ **113** $\xrightarrow{n\text{-BuOH, } NEt_3\text{, 1 hr}}$

113a, X = Br; mp 216–217°C
113b, X = ClO_4; mp 166–167°C

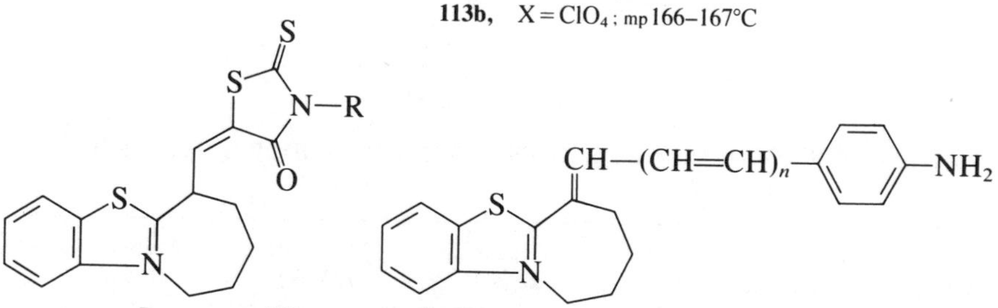

	R	mp(°C)	λmax(nm)
114a	Et	312	557
114b	Ph	290	557

	n	mp(°C)	λmax(nm)
115a	0	229	485
115b	1	248	528

Figure 17

113 $\underset{\text{HCl}}{\overset{\text{50\% aq.MeOH, } Et_3N, -25°C}{\rightleftharpoons}}$ 116

116 $\xrightarrow{CS_2,\ 48\ hr}$ 118

113 $\xrightarrow{X^- N_2^+ C_6H_4R}$ 117

(15)

117a, $R = NO_2$; mp 72°C
117b, $R = SO_3H$; mp 315°C

118, mp 104°C

In a search for new dyes, several azepino[2,1-*b*]benzothiazole derivatives, including the iminium salts **113** and **115** (Fig. 17) and the enamines **114** and **116–118** (eq. 15), were prepared (22–24).

Another source of the azepinobenzothiazole ring system (25) is the decomposition of 2-azidophenyl mesityl sulfide (**119**) under thermal or photolytic conditions (Table 4), which are known to generate singlet or triplet nitrenes. Three products are formed, presumably via the valence tautomer intermediates **120a–c** (shown in Fig. 18). Photolysis of 2-azido-5-chlorophenyl mesityl sulfide in acetophenone resulted in the formation of the corresponding disulfide (51%) and azepinothiazole (48%) at the expense of the thiazepine.

TABLE 4. PHOTOCHEMICAL DECOMPOSITION OF 2-AZIDOPHENYL MESITYL SULFIDE (**119**)

Exp.	Reaction Conditions (singlet or triplet)	Thiazepine **122** (%)	Disulfide **123** (%)	Azepinothiazole **121** (%)
1	Δ/Decalin/180° (s)	45	33	0
2	Δ/Decalin/150° (s)	34	33.5	Trace
3	Δ/PhBr/154° (t)	50.2	16.7	5.6
4	$h\nu$/CH_2Cl_2 (s)	0	14.2	4.4
5	$h\nu$/CH_2Cl_2/Pyrene (s)	2.2	16.7	10.8
6	$h\nu$/PhCOMe (t)	Trace	28.2	20

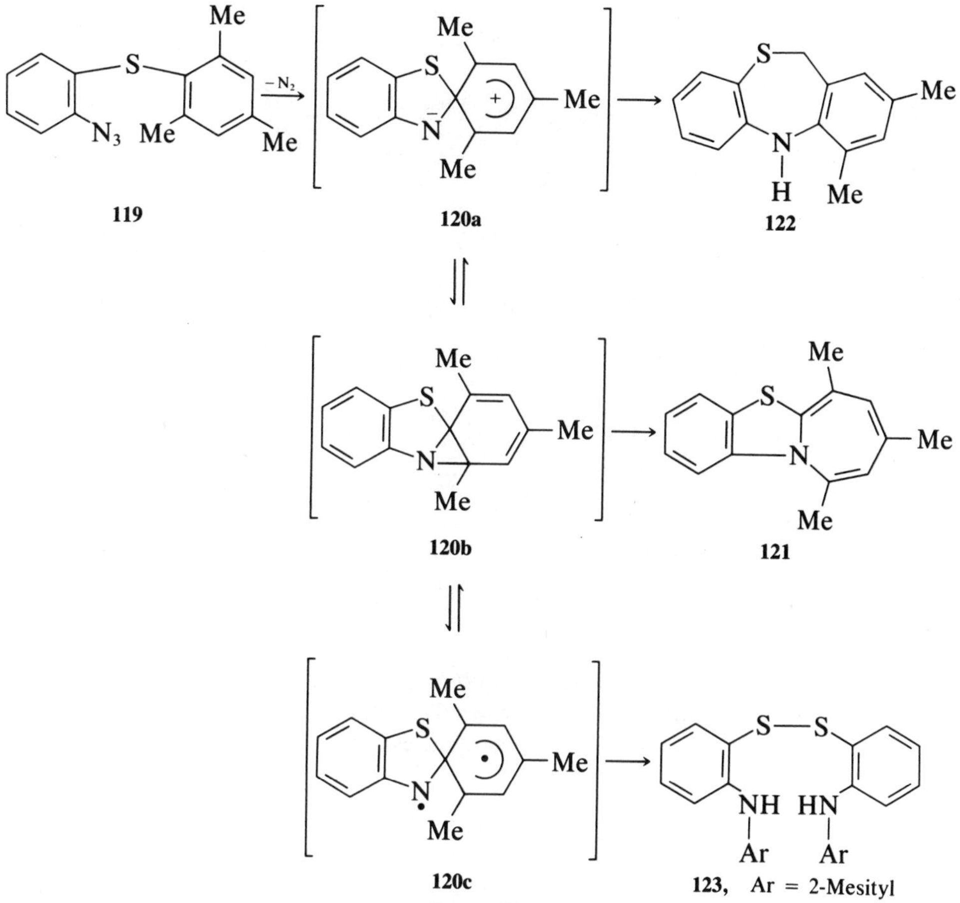

119 120a 122

120b 121

120c 123, Ar = 2-Mesityl

Figure 18

8. Isothiazolo[3,2-*b*][3]benzazepines

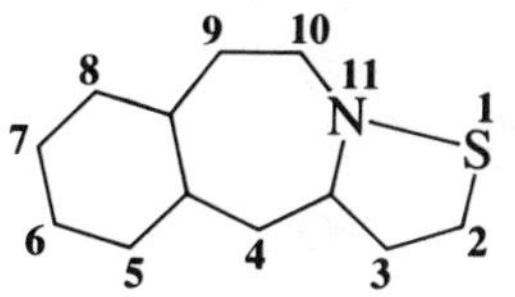

The preparation of several members of the isothiazolo[3,2-*b*][3]benzazepine ring system was achieved by the unusual annulation reaction shown in Fig. 19 (27). Treatment of the aldehyde **124** with cold methanolic potassium hydroxide afforded an unspecified yield of the alcohol **125,** which could be subsequently dehydrated in good yield to the yellow product **126.** Physical data in support of structure **126** included an infrared band at 1650 cm^{-1} and nmr signals corresponding to vinyl protons at δ 5.98 (singlet) and δ 6.85

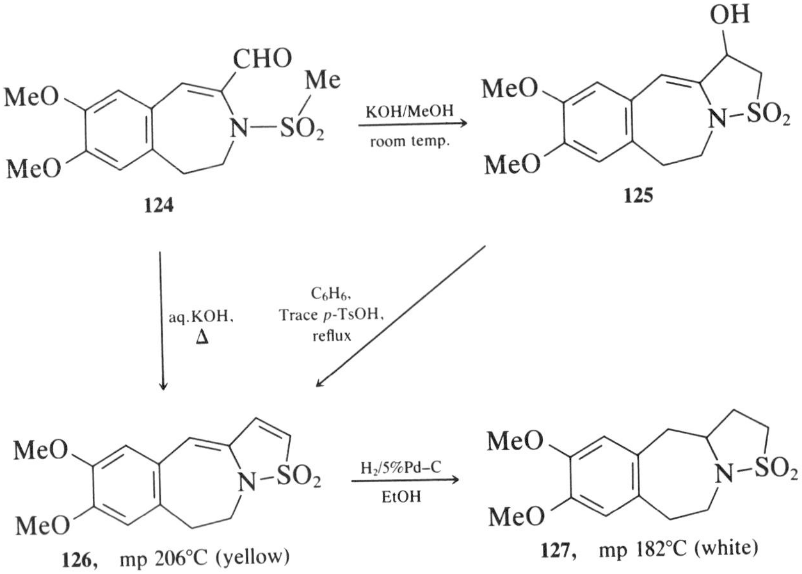

124 → (KOH/MeOH, room temp.) → 125

124 → (aq. KOH, Δ) → 126; 125 → (C_6H_6, Trace p-TsOH, reflux) → 126

126 → (H_2/5%Pd–C, EtOH) → 127

126, mp 206°C (yellow)

127, mp 182°C (white)

Figure 19

(doublet, $J = 6$ Hz). Hydrogenation of **126** gave a white product formulated as **127**, whose nmr spectrum showed no vinylic protons. The ethylenic aldehyde **124** was obtained via the synthesis shown in Eq. 16.

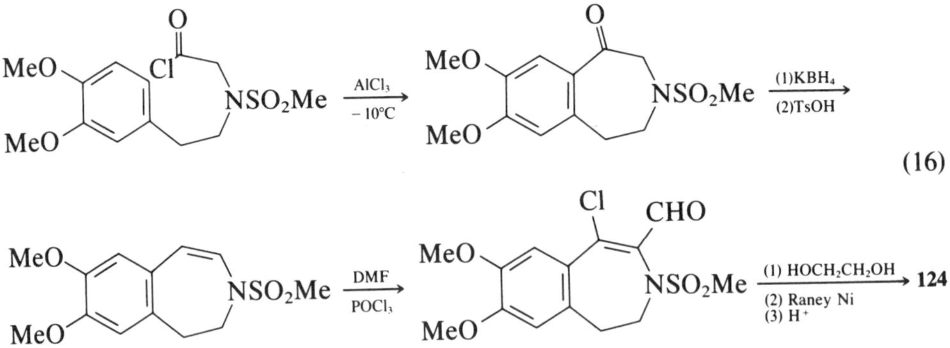

(AlCl$_3$, −10°C) → ((1) KBH_4, (2) TsOH) → (DMF, $POCl_3$) → ((1) $HOCH_2CH_2OH$, (2) Raney Ni, (3) H^+) → **124** (16)

9. Thiazolo[2,3-*a*][2]benzazepines

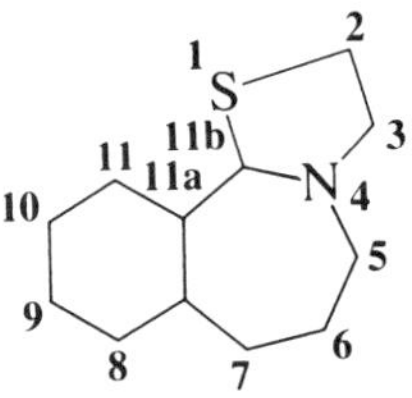

The single known example of the thiazolo[2,3-*a*][2]benzazepine ring system was discussed along with the oxygen analog in Section IV.5.

10. Thiazolo[2,3,-*b*][3]benzazepines

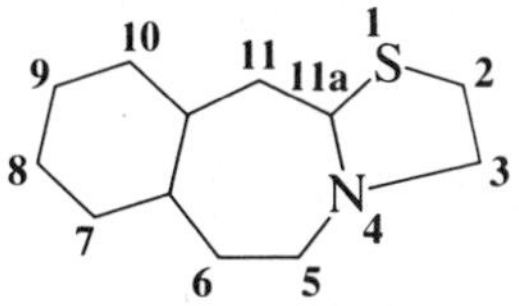

11. Thiazolo[3,2-*b*][2]benzazepines

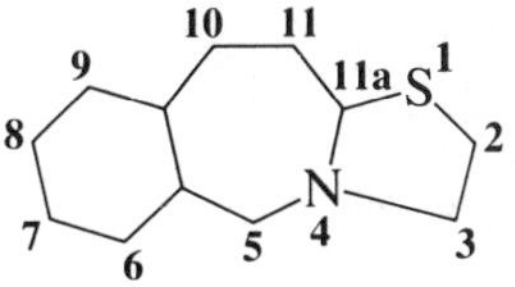

These two closely related isomeric ring systems have been investigated as part of a broad search for new cyanine dyes (28). As shown in Eq. 17, 2,3-disubstituted 10,11-dihydrothiazolo[3,2-*b*][2]benzazepinium perchlorates (**132**) can be prepared from 2-benzazepin-3-thione (**129**) by annulation with α-bromoketones. Likewise, 2-substituted derivatives (**133**) of the isomeric 5,6-dihydro[2,3-*b*][3] ring system can be prepared from 3-benzazepin-2-thione (**130**) (Eq. 18).

From these quaternary salts, listed in Table 5, the thiazolobenzazepine dyes in Table 6 were prepared by standard methods (such as heating **132** or **133** with 3-ethyl-2-(formylmethylene)benzothiazoline or its equivalent in acetic anhydride). The visible absorption maxima of each of these dyes are noted. Table 7 shows *bis*-cyanine dyes obtained from **132** or **133** by condensation with an orthoformate or malonaldehyde equivalent. Condensation of **132** or **133** with *p*-dimethylaminobenzaldehyde or *N*-methylrhodanine gave the dyes **134** and **135**. In each instance, there was a bathochromic shift of 50–60 nm in the λ_{max} relative to the unsubstituted open-chain thiazolocyanines.

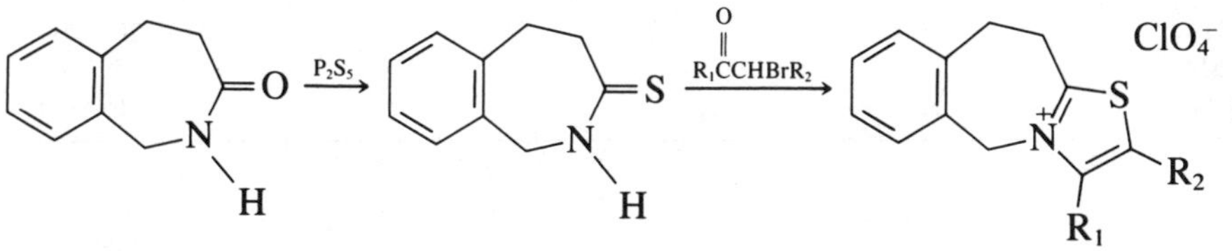

P_2S_5; $R_1COCHBrR_2$; ClO_4^- (17)

128 **129** **132**

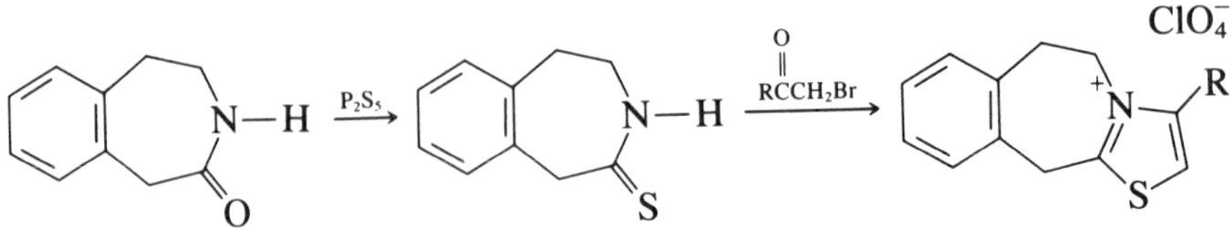

ClO_4^-

N—H $\xrightarrow{P_2S_5}$ N—H $\xrightarrow{RCOCH_2Br}$ R (18)

O S S

131 **130** **133**

Me

S

N S

O

O N

Me

Me S ClO_4^-

^+N CH NMe$_2$

$(CH_2)_m$ $(CH_2)_n$

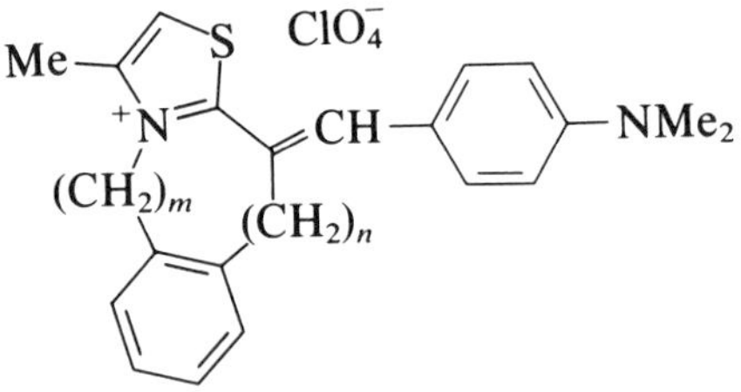

134a, $m=n=1$; mp 216°C (λ_{max}468 nm)
134b, $m=$ 2, $n=0$; mp 156°C (λ_{max}446 nm)

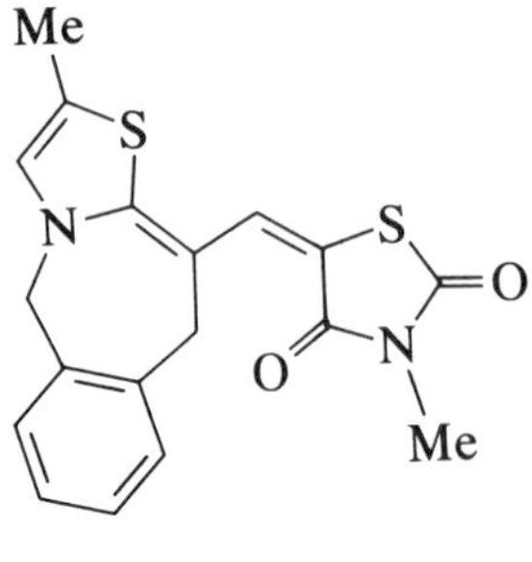

135, mp 215°C (λ_{max}568 nm)

TABLE 5. DERIVATIVES OF THIAZOLO[2,3-*b*][3]- AND [3,2-*b*][2]BENZAZEPINES

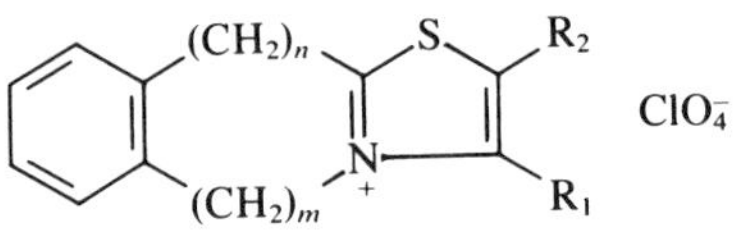

$(CH_2)_n$ S R_2

ClO_4^-

N

$(CH_2)_m$ + R_1

R_1	R_2	*m*	*n*	mp (°C)
		[3,2-b][2] series		
H	H	1	2	192–193
Me	H	1	2	235–236
Et	H	1	2	97–98
Ph	H	1	2	198
Ph	Me	1	2	210–211
Ph	Ph	1	2	212
		[2,3-b][3] series		
H	H	2	1	203–204
Me	H	2	1	159–160
Ph	H	2	1	169–170

TABLE 6. CYANINE DYES FROM THIAZOLO[2,3-*b*][2]- AND [3,2-*b*][2]BENZAZEPINES

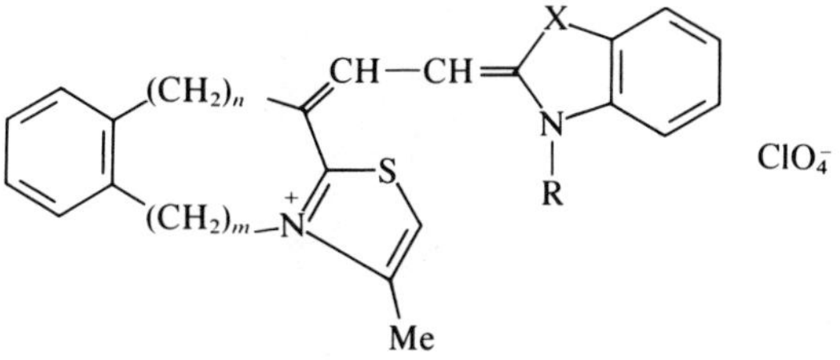

ClO_4^-

m	*n*	X	R	mp (°C)	λ_{max} (nm)
1	1	CMe_2	Me	234	534
1	1	S	Et	237	556
1	1	CH=CH	Et	243	600
2	0	CMe_2	Me	246	505
2	0	S	Et	243	565

TABLE 7. BIS-CYANINE DYES FROM THIAZOLO[2,3-*b*] [3]- AND [3,2-*b*] [2]BENZAZEPINES

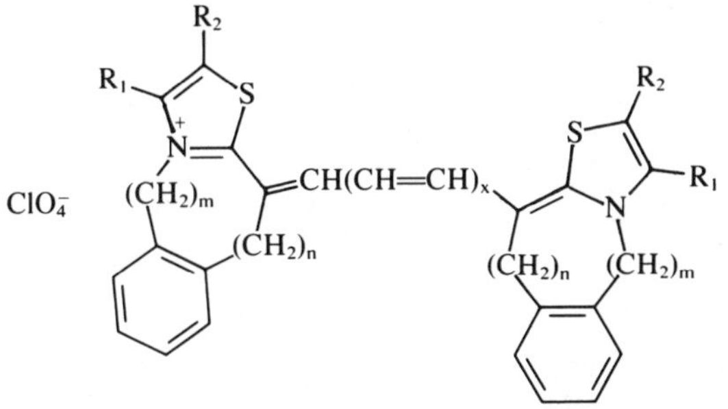

ClO_4^-

m	*n*	*x*	R_1	R_2	mp (°C)	λ_{max} (nm)
2	0	0	Me	H		580
2	0	0	Ph	H	174	610
1	1	0	H	H	140–141	599
1	1	0	Me	H	196	606
1	1	0	Ph	H	146	613
1	1	0	Ph	Me	174	620
1	1	0	Ph	Ph	212	643
1	1	1	Me	H	205	680
1	1	1	Ph	H	190	686

12. Thiazolo[3,2-*a*][1]benzazepines

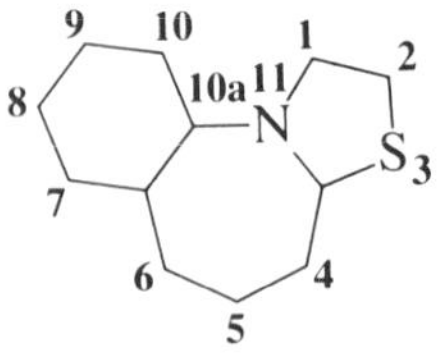

Dyes belonging to the thiazolo[3,2-*a*][1] benzazepine system, as well as the isomeric [2,3-*b*][3] and [3,2-*b*][2] ring systems, have been described (29). In a manner similar to the work on the latter two systems, 1-benzazepine-2-thione (**137**) was converted to the 4,6-dihydrothiazolo[3,2-*a*][1]benzazepinium perchlorate **138** by reaction with α-bromoacetone (Eq. 19). The cyanine dyes listed in Table 8 were then prepared by condensation with formaldehyde or aldehyde equivalents.

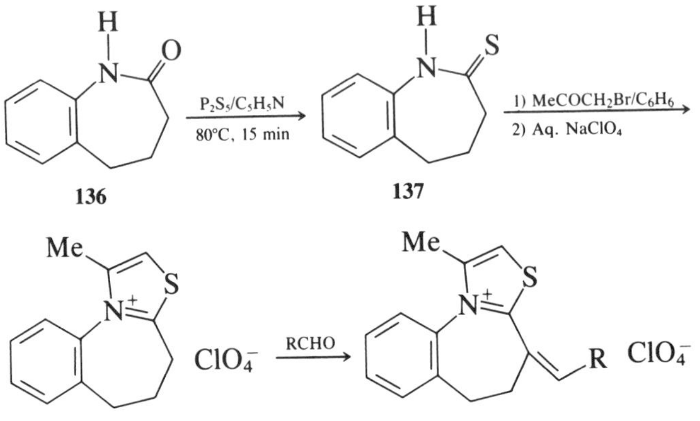

P_2S_5/C_5H_5N, 80°C, 15 min; 1) $MeCOCH_2Br/C_6H_6$ 2) Aq. $NaClO_4$; RCHO

136 **137** (19)

138, 74%; mp 195–196°C **139**

13. Azepino[2,1-*b*]benzoselenazoles

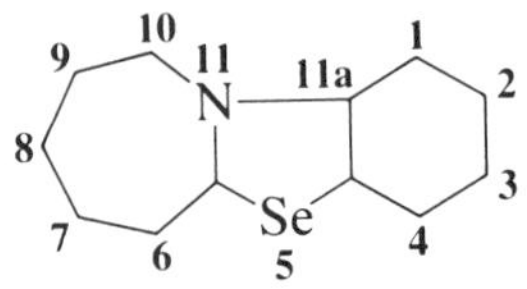

An example of the azepino[2,1-*b*]benzoselenazole system, **141,** has been prepared in modest yield (21), as shown in Eq. 20, by the addition of dimethyl acetylenedicarboxylate to 2-methylbenzoselenazole (**140**). The reaction of similar substrates with acetylene dicarboxylic esters has been a

TABLE 8. CYANINE DYE DERIVATIVES OF THIAZOLO[3,2-*a*][1]BENZAZEPINES

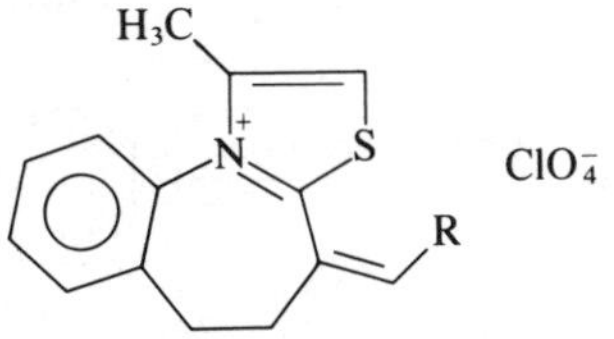

ClO_4^-

R	Yield (%)	mp (°C)	λ_{max} (nm)
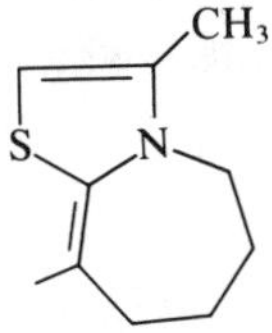 [structure: CH_3, S, N]	74	179	600
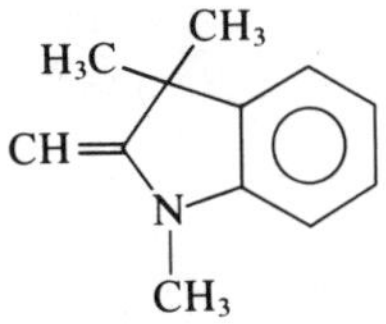 [structure: H_3C, CH_3, CH=, N, CH_3]	52	181	525
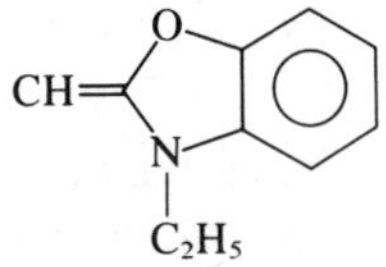 [structure: O, CH=, N, C_2H_5]	41	192	530
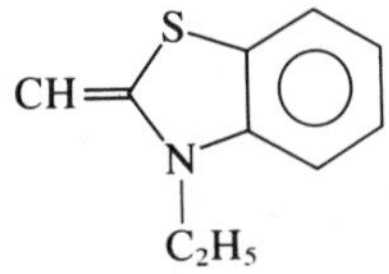 [structure: S, CH=, N, C_2H_5]	43	200	560
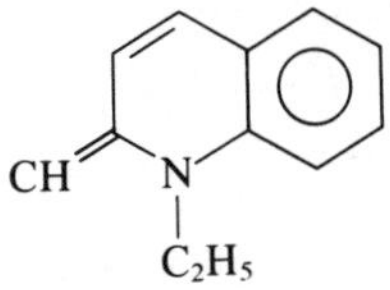 [structure: CH, N, C_2H_5]	64	185	602
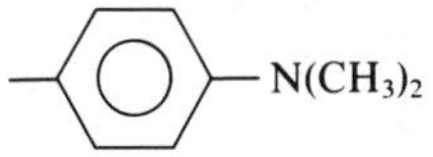 [structure: —C₆H₄—$N(CH_3)_2$]	61	164	444
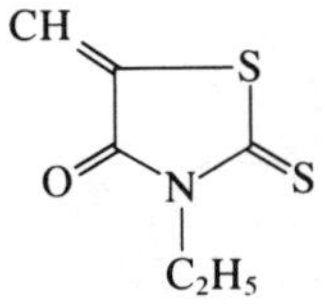 [structure: CH, S, O, N, S, C_2H_5]	79	215	574

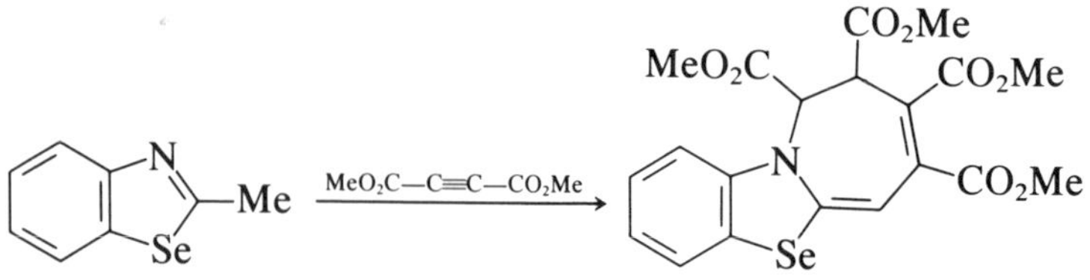
CO2Me
MeO2C
CO2Me
N
Me
MeO2C—C≡C—CO2Me
N
CO2Me
Se
Se

140 **141**, mp 214°C (yellow) (20)

source of other, related heterocycles. A suggested mechanism for this reaction is discussed in Section IV.7 on azepino[2,1-*b*]benzothiazoles, which are produced by the same route from 2-alkylbenzothiazoles.

14. Azepino[1,2-*e*]purines

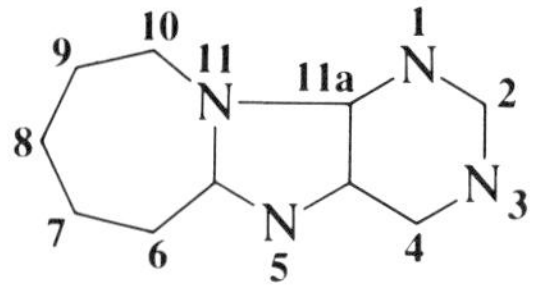
10
9
11
N
11a
1
N
2
8
N
3
7
6
N
5
4

The 4-oxo and 2-amino-4-oxo derivatives (**143** and **144**) of the azepino-[1,2-*e*]purine system have been prepared (30) via the reactions of **142** shown

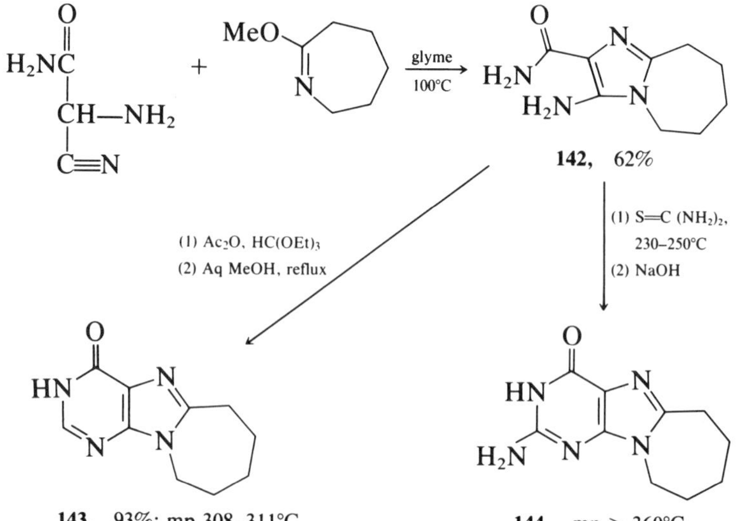
O
H2NC
CH—NH2
C≡N
+
MeO
N
glyme
100°C
O
H2N
N
H2N
N
142, 62%
(1) Ac2O, HC(OEt)3
(2) Aq MeOH, reflux
(1) S=C (NH2)2,
230–250°C
(2) NaOH
O
HN
N
N
N
O
HN
N
H2N
N
N
143, 93%; mp 308–311°C
144, mp > 360°C

Figure 20

TABLE 9. DERIVATIVES OF AZEPINO[1,2-*e*]PURINES[a]

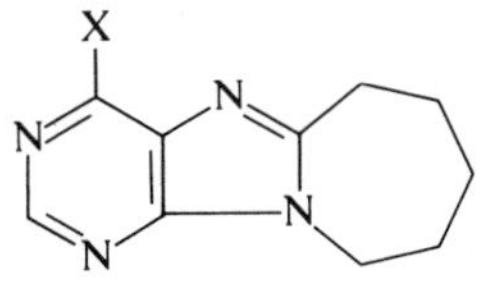

No.	X	Precursor	Reagent/Conditions	Yield (%)	mp (°C)
1	Cl	OH	$POCl_3/PhNMe_2$	80	141–143
2	SH	OH	P_2S_5, C_5H_5N	91	290–292
		Cl	$S{=}C(NH_2)_2$	96	
3	SCH_3	SH	$Me_2SO_4/NaOH$	88	139–141
4	H	SH	Raney Ni/H_2O, 100°C	81	98–100 (picrate, 166–169)
5	NH_2	Cl	$AgNH_3$/170°C, autoclave	93	227–229
6	Me_2N	Cl	Me_2NH/MeOH	90	142–144
7	$PhCH_2NH$	Cl	$PhCH_2NH_2$/glyme	74	153–155
8	2-Furylamino	Cl	Furfuryl amine/glyme	91	168–170
9	$(HOCH_2CH_2)_2N$	Cl	$(HOCH_2CH_2)_2NH$/glyme	89	177–179
10	$(ClCH_2CH_2)_2N$	No. 9	$SOCl_2/CHCl_3$	80	205–207
11	$HOCH_2CH_2NH$[b]	Cl	$HOCH_2CH_2NH_2$/glyme	92	127–128.5

[a]Ref. 30.
[b]On treatment with $SOCl_2$/MeOH, this substituent reacted with the adjacent nitrogen to form a new dihydroimidazole ring.

in Fig. 20. Compound **143** was converted in good yields (Table 9) to chloro, mercapto and amino derivatives by classical methods.

15. Azepino[2,1-*f*]purines

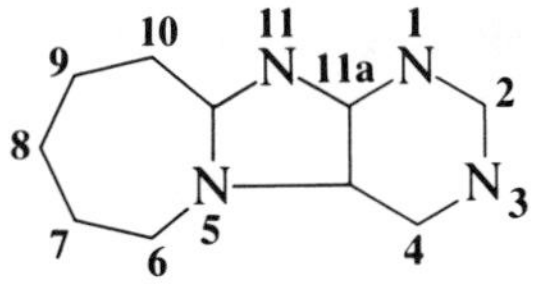

The single known example of the azepino[2,1-*f*]purine system to date was prepared (32–34) by the synthesis shown in Fig. 21 The 4-amino-5-nitroso-uracil (**146**) rearranges exothermally, by an unspecified mechanism, to the 8,8-pentamethylene-8H-xanthine (**147**) in good yield (31). Many examples of related xanthines were obtained via this efficient route. The 8,8-pentamethylenexanthine (**147**) in particular rearranges when heated above its melting point (250°C in an open flask or in refluxing DMF) to give 7,8-pentamethylenetheophylline (**148**).

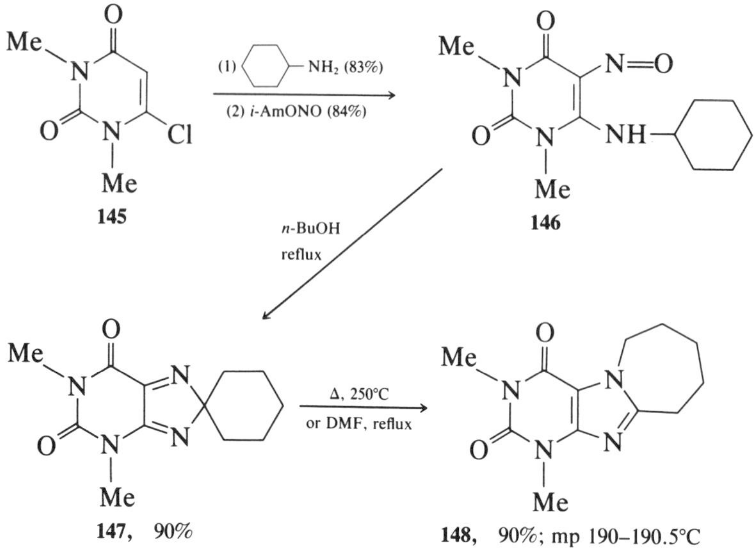

145

(1) cyclohexyl–NH_2 (83%)
(2) *i*-AmONO (84%)

146

n-BuOH reflux

147, 90%

Δ, 250°C or DMF, reflux

148, 90%; mp 190–190.5°C

Figure 21

16. Pyrazolo[3′,4′:4,5]pyrimido[1,2-*a*]azepines

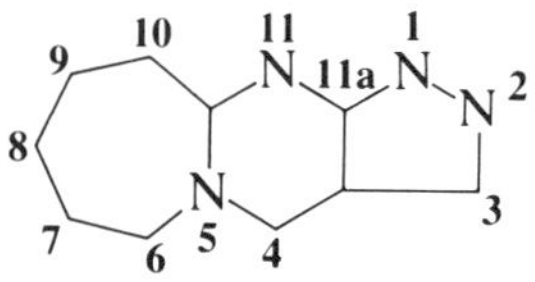

The two known examples of the pyrazolo[3′,4′;4,5]pyrimido[1,2-*a*]azepine system were prepared by condensation of the aminopyrazole esters **149** with caprolactim ethyl ether (**150**) to give **151** (35).

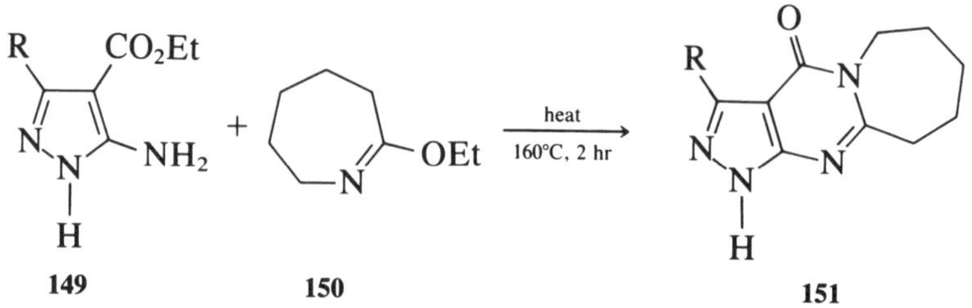

heat 160°C, 2 hr

149 **150** **151** (21)

	R	Yield (%)	mp (°C)	Spectra
151a	H	98	266	uv,ir,nmr
151b	NH_2	95	314–316	uv,ir,nmr

17. 2a,4,9a-Triazabenz[*c,d*]azulenes

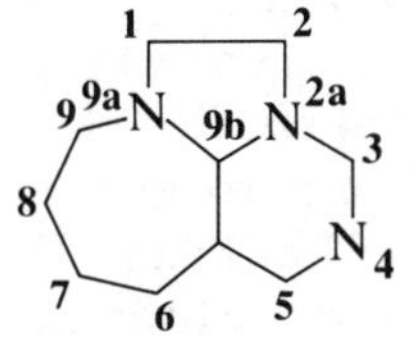

A patent (36) disclosing numerous annulated uracils as plant protective agents cites one example of the 2a,4,9a-triazabenz[*c,d*]azulene ring system (Eq. 22). Heating the bicyclic base **152** and iminodicarboxylic acid diethyl ester to 150–170°C leads to loss of ethanol with formation of the tricyclic product **153**. Physical data in support of the structure were not presented.

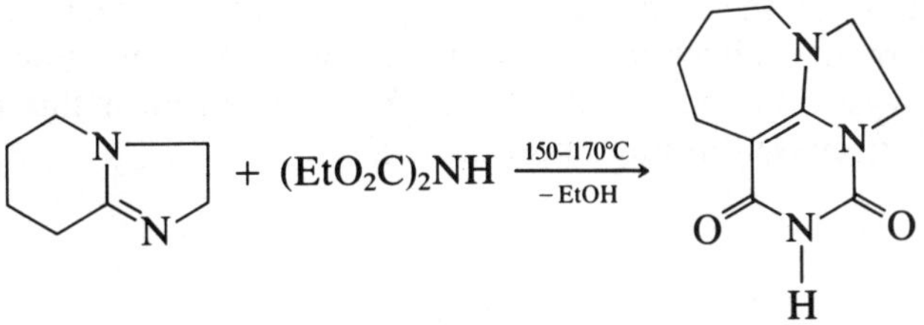

$$\mathbf{152} + (EtO_2C)_2NH \xrightarrow[-\,EtOH]{150\text{–}170°C} \mathbf{153} \qquad (22)$$

152 **153**, 63%; mp 273–274°C

18. Pyrido[3′,2′:4,5]imidazo[1,2-*a*]azepines

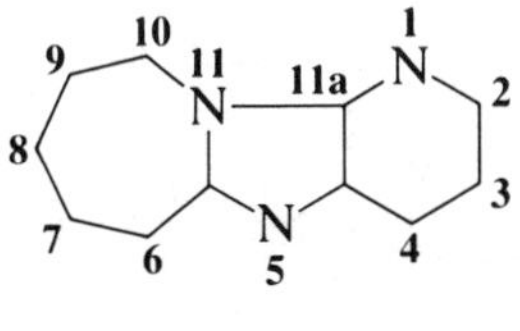

19. Pyrido[3′,4′:4,5]imidazo[1,2-*a*]azepines

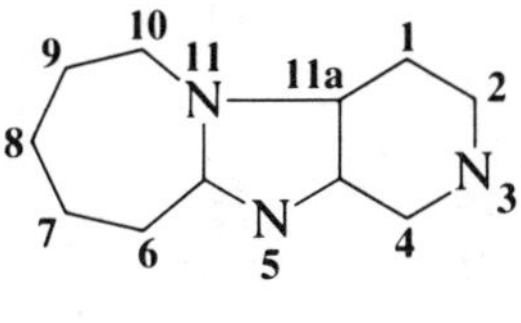

The 7,8,9,10-tetrahydro-6H derivatives of these two isomeric systems have been prepared by chemical transformations that are discussed more extensively in Section IV.20, which deals with azepino[1,2-*a*]benzimidazoles. The isomeric chloronitropyridines shown in Eqs. 23 and 24 were converted in good yields to the azides **154** and **156,** which were not characterized but were thermally decomposed in nitrobenzene to give the [3′,2′:4,5]isomer **155** and the [3′,4′:4,5]isomer **157** in 15 and 72% yield, respectively (37). The lower yield of **155** may have been due to an unstable diazonium intermediate.

Regardless of whether the reactive intermediate in the azide decomposition is an electrophilic $(R\ddot{N}:)$ or biradical $(R\dot{N}:)$ species, it probably combines with a methylene group of the hexahydroazepine to form an unstable dihydro intermediate such as **158** (Eq. 25). In accord with this mechanism, when the thermolysis of **156** was carried out in refluxing acetic anhydride a low yield of the *N*-acetyl derivative **159** was obtained. Treatment of the diamines formed by catalytic reduction (as shown in Eqs. 23 and 24) with performic acid afforded **155** and **157** in 15–20% yields (38), thereby providing an alternative route to these ring systems. Ultraviolet spectral data were given in support of the structures. It is also of interest that when 2-(*N*-hexahydroazepino)-3-nitropyridine was treated with titanous chloride, a 66% yield of **155** was obtained directly (39). The mechanism of this reductive cyclization is discussed in the following section.

(1) HN (hexahydroazepine); (2) H_2/Raney Ni; (3) HONO; (4) NaN_3 → [**154**] → $PhNO_2$, 170°C → **155,** 15%; mp 93°C (23)

(see Eq. 23) → [**156**] → Δ → **157,** 72%; mp 155°C (24)

156 → Δ → [**158**] → Ac_2O → **159,** mp 66°C (25)

20. Azepino[1,2-*a*]benzimidazoles

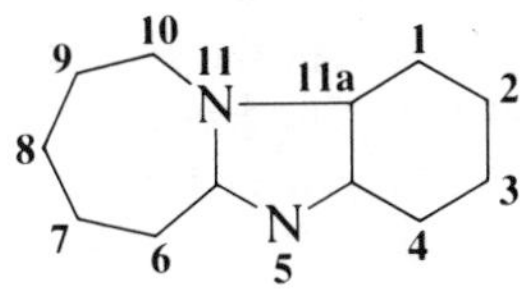

The first reported examples of the azepino[1,2-*a*]benzimidazole system were obtained in good yields (40) via the azide decomposition route shown in Eq. 26. No physical data were provided in this early work to support the structures of the products, but the hexahydroazepino[1,2-*a*]benzimidazole **162** (X, Y = H) could be compared directly with the known compound.

Since that initial report, several alternative syntheses have been reported for the preparation of this system that involve either oxidation or reduction of appropriate *ortho*-substituted *N*-arylhexahydroazepines.

A very effective procedure, similar to the azide route, employed trifluoroperacetic acid to oxidize the amine **163** (Eq. 27) to the hexahydro[1,2-*a*]benzimidazole **164** in 90% yield (41). Later, performic acid (98% formic acid plus 30% hydrogen peroxide) was reported to accomplish this same conversion in 80–90% yields (42,43). The hexahydro compound **164** was also obtained in good yield (44) via a straightforward annulative alkylation of the 2-bromoalkylbenzimidazole **165a** (Fig. 22). A related, patented process (45) claims the acid-catalyzed cyclization of **165b**, prepared from *o*-phenylene-

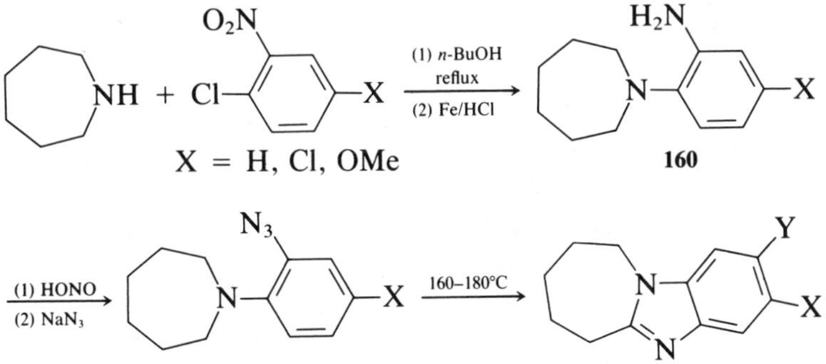

X = H, Cl, OMe; (1) *n*-BuOH reflux (2) Fe/HCl → **160**

(1) HONO (2) NaN_3 → **161** → 160–180°C → **162** (26)

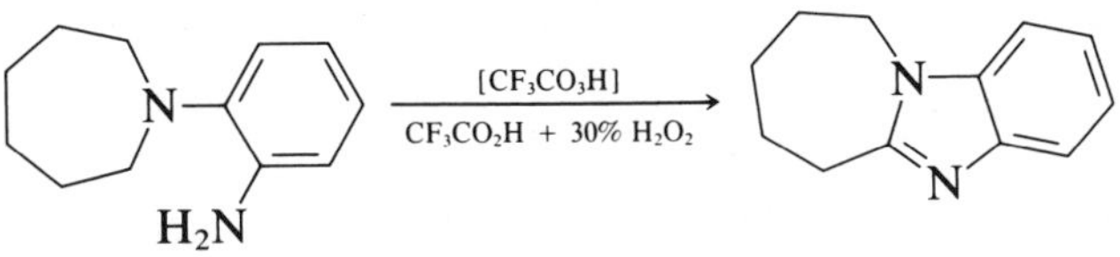

163 → $[CF_3CO_3H]$, CF_3CO_2H + 30% H_2O_2 → **164**, 90%; mp 124–125°C (27)

diamine and caprolactam, to **164** in 85% yield. Although this last method is satisfactory for the preparation of six- and seven-membered rings, it is less convenient than the peracid or azide routes

The acetamido derivatives **166** shown in Fig. 23 were prepared (45) by the performic acid oxidation route and could be converted into the corresponding azides **167** by diazotization of the amines followed by azide displacement. The azides were thermally stable up to 145–150°C and could be sublimed *in vacuo* at 100–130°C. Reaction of azide **167a** or **167b** with acetic acid in PPA (Fig 23) gave only one pentamethylene-bridged imidazo-[4,5-*g*]benzoxazole: **168** from **167a** and **169** from **167b.** The authors had previously developed this method of benzoxazole ring formation. Since one step in the process presumably involves nucleophilic substitution by an acyloxy moiety (derived from acetic acid) at a position adjacent to the azide group, the outcome of the oxazole formation is ambiguous when both *ortho* positions are available. On the other hand, cyclization might be selective if one *ortho* position were significantly more electron deficient than the other. This situation in fact exists in benzimidazoles, and is consistent with the regiospecific formation of benzoxazoles shown in Fig. 23. The structure of **168** or **169** is readily assigned, since the nmr spectrum of each compound contains only a single pair of aromatic *ortho* doublets.

Numerous papers on the synthesis and chemistry of this and related ring systems have originated from the research groups of Suschitzky and Meth-Cohn (46–57). In an initial report, an attempted preparation of the hexahydrophenazine **171** by reduction of the nitroamine **170** with iron or iron salts

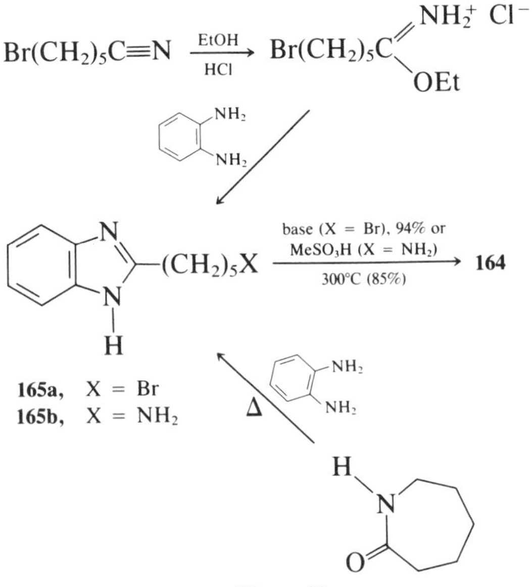

Figure 22

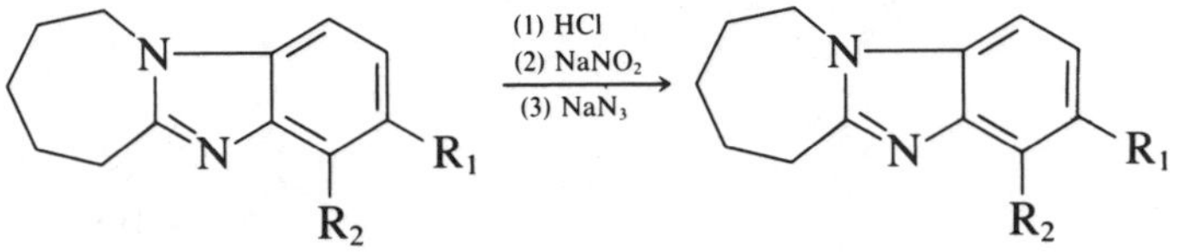

(1) HCl
(2) $NaNO_2$
(3) NaN_3

166 → 167

	R_1	R_2	mp (°C)	Yield (%)
166a	H	NHAc	222	66
166b	NHAc	H	252	58

	R_1	R_2	mp (°C)
167a	H	N_3	98
167b	N_3	H	108

167

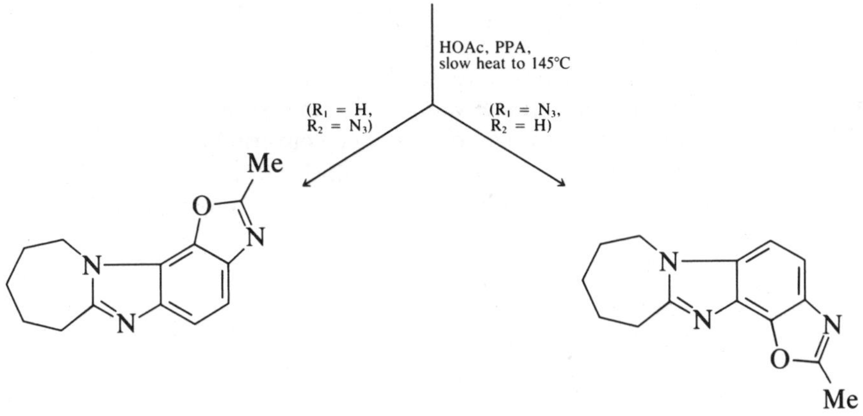

HOAc, PPA, slow heat to 145°C

(R_1 = H, R_2 = N_3) → **168**, 68%

(R_1 = N_3, R_2 = H) → **169**, 66%

Figure 23

(Eq. 28) gave a product having chemical and physical properties at variance with structure **171** (47). The product had an ultraviolet absorption spectrum similar to those of 2-alkylbenzimidazoles and proved to be identical with the product obtained via the decomposition of **161** (X = H). Yields of **164** from **170** were relatively poor: 21% with ferrous oxalate as the reductant and only 15% using iron. A mechanism was suggested that involved a cyclodehydration to an *N*-oxide of **164** followed by deoxygenation. Suschitsky has also reported the formation of **164** in unstated yield by reductive cyclization of aryldiazonium sulfonates (48) with sulfur dioxide.

A far superior method of synthesis of the hexahydroazepino[1,2-*a*]benzimidazole **164** is the reductive cyclization of *N*-(*o*-nitrophenyl)

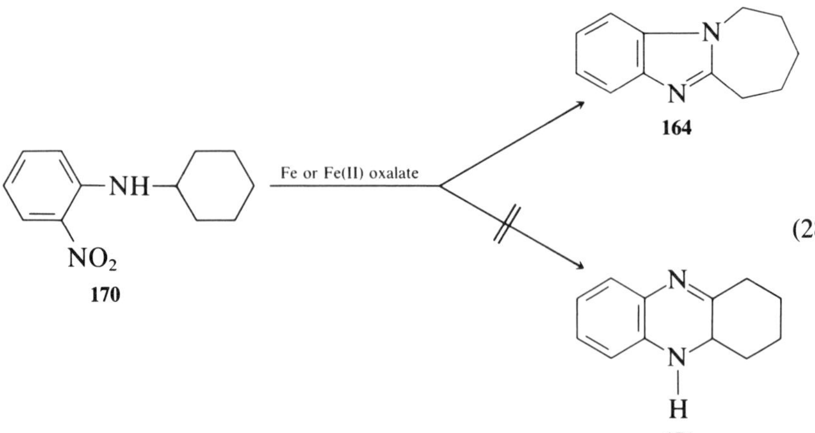

(28)

hexahydroazepine (**170**) under nitrogen at 80°C by adding $TiCl_3$ to a solution of **170** in HCl (Fig. 24). When two equivalents of $TiCl_3$ have been consumed, the product precipitates as a hydrochloride in quantitative yield. The mechanism for this reaction is believed to involve initial formation of the complex **172a,** followed by proton transfer and cyclization to the *N*-oxide **172d** via **172b** and **172c.** A second equivalent of reductant converts **172d** to the product.

A somewhat similar reduction using zinc chloride in acetic anhydride (Fig. 25) has been used to convert **170** to the 6-hydroxy compound **173** in 70%

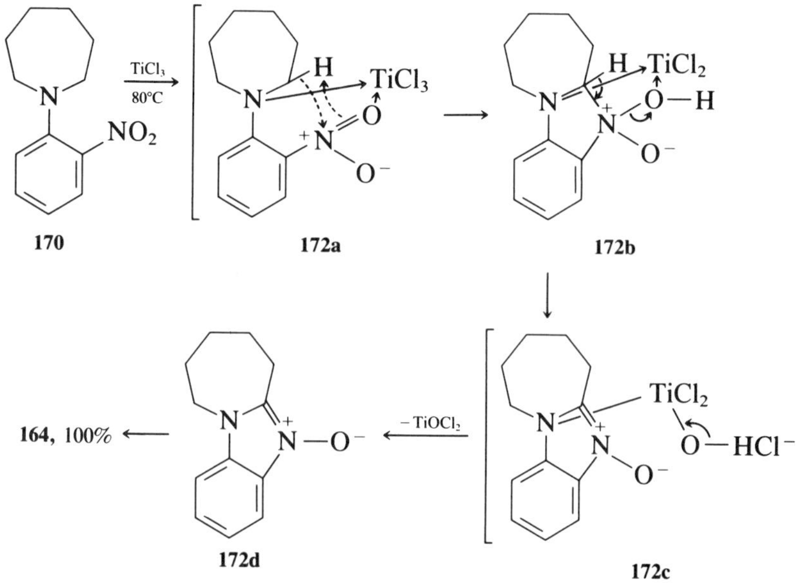

Figure 24

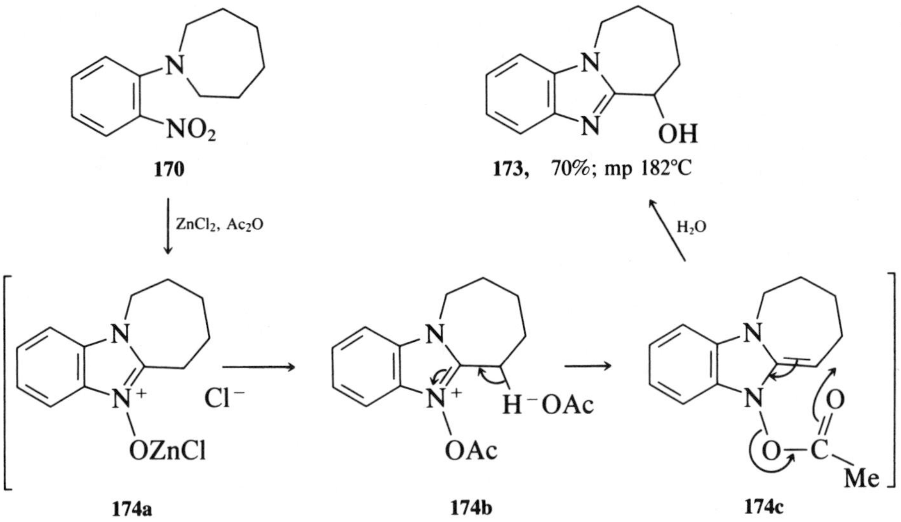

170

$ZnCl_2$, Ac_2O

173, 70%; mp 182°C

H_2O

174a

174b

174c

Figure 25

yield (50). The suggested mechanism is rather similar to the one shown in Fig. 24, except for the formation of the acetylated *N*-oxide **174b** from the initial zinc intermediate **174a.** The ensuing removal of a proton by acetate ion and rearrangement of **174c** to **173** are typical reactions of 2-alkylbenzimidazole-*N*-oxides.

Suschitzky and Meth-Cohn have studied the thermal and photochemical behavior of these same substrates in HCl solution (Fig. 26). When the nitroamines **175** were treated with hot 12 *N* HCl, the *N*-oxides **176** were isolated in good yield (51). The suggested pathway is indicated in Fig. 26. The results are seen in Table 10. Photocyclization of these nitroamines was generally free of side reactions when conducted in 1 *N* HCl in 10% aqueous methanol using a Hanovia 200-watt mercury lamp and pyrex filter (52,53). The results are shown in Table 11.

The *N*-oxide **176** was shown not to be a precursor to **177,** since it is stable under photolytic conditions. Whether the reaction yields **177** or **176** apparently depends on a combination of steric and electronic factors. Mechanisms involving photoexcitation of the nitro group to a diradical (which may attack the methylene group adjacent to the tertiary amine nitrogen to give **176** or react with the basic nitrogen itself to form **177**) are discussed in a full paper (52).

Suschitzky and Meth-Cohn have reported the nuclear substitution of benzimidazole *N*-oxides with concomitant loss of the *N*-oxide functionality (54, 55). Using the nitro *N*-oxide **176** (Fig. 27) as substrate, they obtained a mixture of the 1- and 4-substituted chloro derivatives (**178**), as shown in Table 12. The pathway for this transformation may be rationalized according to Fig. 27.

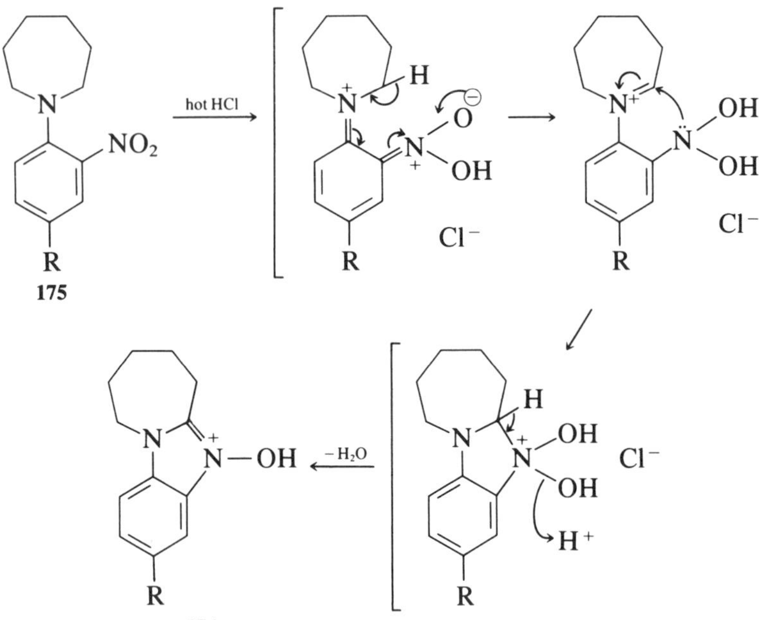

Figure 26

TABLE 10. HEXAHYDROAZEPINO[1,2-*a*]BENZIMIDAZOLES FROM ACID-CATALYZED CYCLIZATION OF **175**[a]

R	mp (°C)	Temperature (°C)	Reflux Time (hr)	Yield (%)	Refs.
H	212	110	40	74	51,52
NO_2	206 188	150	12	76	51,52
Cl	129	110	48	62	52

[a]Fig. 26.

TABLE 11. HEXAHYDROAZEPINO[1,2-*a*]BENZIMIDAZOLES FROM PHOTOLYSIS OF **175**[a]

R	Time (hr)	Unchanged **175** (%)	**177** (%)	**176** (%)	mp (°C)	Refs.
H	24	12	81	0	125	52,53
Cl	65	13	0	79	129	52,53

[a]Eq. 29.

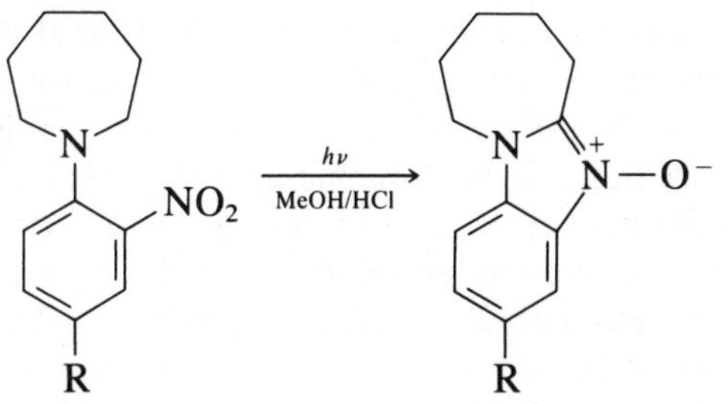

TABLE 12. REACTION PRODUCTS DERIVED FROM HEXAHYDROAZEPINO-[1,2-*a*]BENZIMIDAZOLE *N*-OXIDES AND REACTIVE HALIDES[a]

	178 (% Yield)		**179**	
Reagent	1-Cl	4-Cl	X	% Yield
$POCl_3$	35	65		
SO_2Cl_2	27	55		
$SOCl_2$	35	65		
AcCl			AcO	100
BzCl	50		BzO	50
$4\text{-}MeOC_6H_4COCl$	10		$4\text{-}MeOC_6H_4CO_2$	84
TsCl	19	27		
$POBr_3$			Deoxygenation only	

[a]Fig. 27.

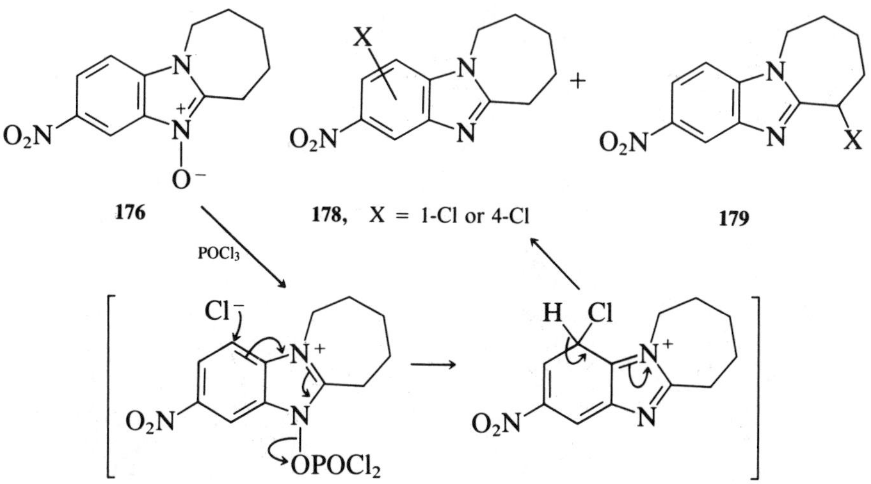

Figure 27

Since both nitrogen atoms are capable of bearing the formal charge, all the benzenoid positions are in principle capable of being substituted. A clear preference is seen, however, for the 1-position, which probably reflects both steric and electronic factors in its favor. Organic acid chlorides bring about a second type of substitution, well known in the pyridine series, to give products of type **179.** When the reagent is one that provides a good potential leaving group, and a strong nucleophile is added as well, the reaction represents a general synthesis of substituted benzimidazoles from their *N*-oxides. Thus, treatment of the *N*-oxide **176** (Fig. 28) with benzoyl chloride and potassium cyanide gave a mixture of the 4-cyano derivative **180a** and the α-substituted product **179** (X = BzO). When benzoyl chloride was replaced with tosyl chloride, only **180a** was isolated. With thiocyanate ion as the nucleophile, a higher yield of **180b** was obtained. With azide ion and tosyl chloride, two products were obtained once again. The major product was the 1-azido derivative **181,** and the minor product was identified as the furoxan **182.** Compound **182** is presumably formed from the initially produced 4-azido compound by cyclization with loss of nitrogen. This could be confirmed by formation of **182** from **183** on reaction with sodium azide. With oxidizable anions such as bromide, iodide and *n*-butylmercaptide, **176** underwent only deoxygenation.

The simultaneous action of an acyl halide and hydroxide ion on **176** was of particular interest. With benzoyl chloride, the α-benzoyloxy product **179** (X = BzO) was again obtained (62%) along with a minor product (33%).

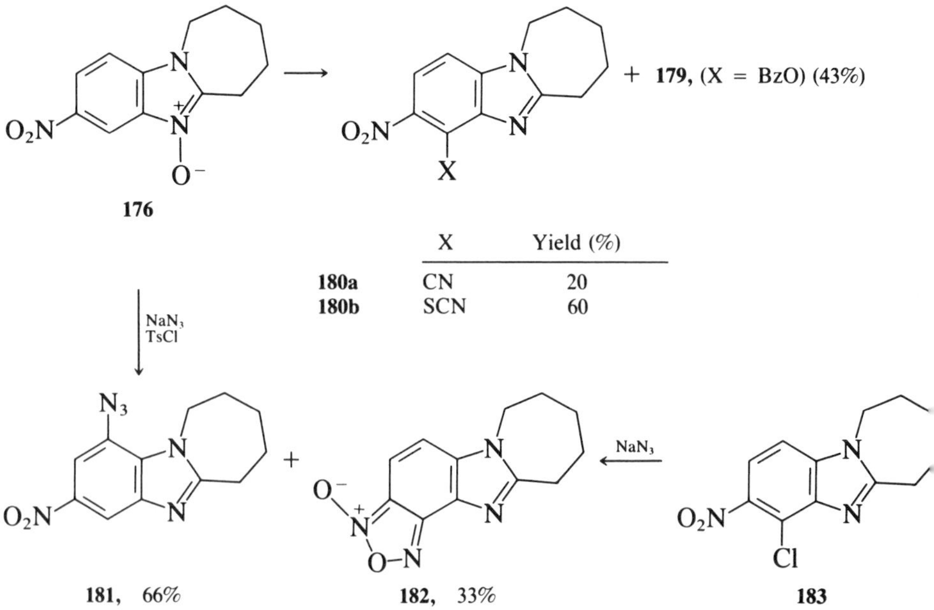

176 + **179,** (X = BzO) (43%)

	X	Yield (%)
180a	CN	20
180b	SCN	60

NaN_3, TsCl

181, 66% **182,** 33% **183**

Figure 28

The minor product formed exclusively when tosyl chloride was used with sodium hydroxide, and was shown by nmr and mass spectra to be **184** (Fig. 29). The structure of **184** was confirmed by alkaline hydrolysis, which gave the known *N,N′*-pentamethylene-*o*-phenylene diamine **185a,** and by $LiAlH_4$ reduction to the *N*-methyl analog **185b.** The considerable strain in **184** is seen in the uv and ir (C=O, 1750 cm^{-1}) spectra. The proposed pathway of formation is shown in Fig. 29.

Finally, the action of amines and tosyl chloride on **176** was investigated. *n*-Butylamine gave a mixture of **179** (X = *n*-BuNH, 60%) and **184.** The weaker base aniline gave **179** (X = NHPh, 90%) and the 2-anilino derivative **186.**

Similar chemistry was encountered upon ultraviolet irradiation of polymethylenebenzimidazole *N*-oxides **187** in methanol (56) (Eq. 30). The benzimidazolone **188** was obtained in modest yield along with the deoxygenated product **189** and, in the case of R = Cl, the solvent adduct **190.** The results are seen in Table 13. An oxaziridine similar to that in Fig. 29 was proposed as an intermediate leading to **188.**

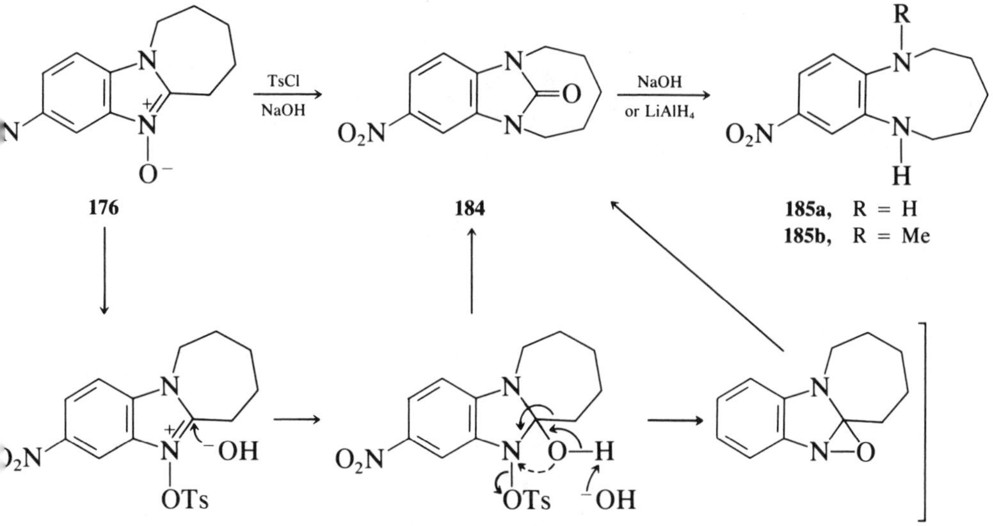

TsCl, NaOH

NaOH or $LiAlH_4$

176

184

185a, R = H
185b, R = Me

Figure 29

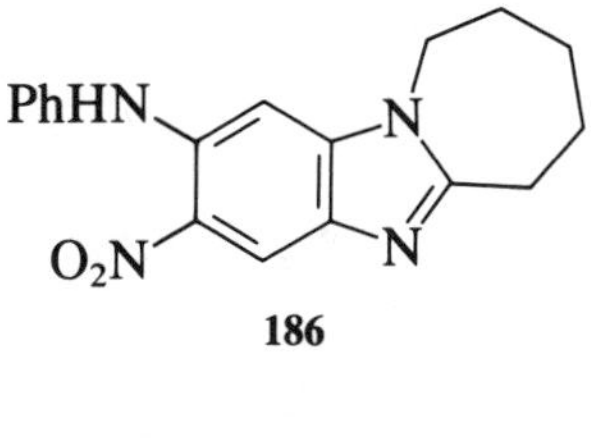

186

TABLE 13. PRODUCTS FROM IRRADIATION OF BENZIMIDAZOLE *N*-OXIDES **187**[a]

R	Product	Yield (%)
H	**187**	42
	188	25
	189	32
Cl	**187**	12
	188	44
	190	

[a]Eq. 30.

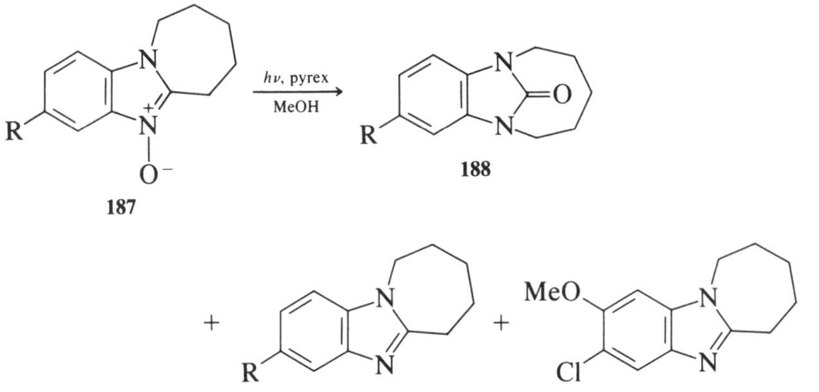

hν, pyrex / MeOH

187 → **188** + **189** + **190** (30)

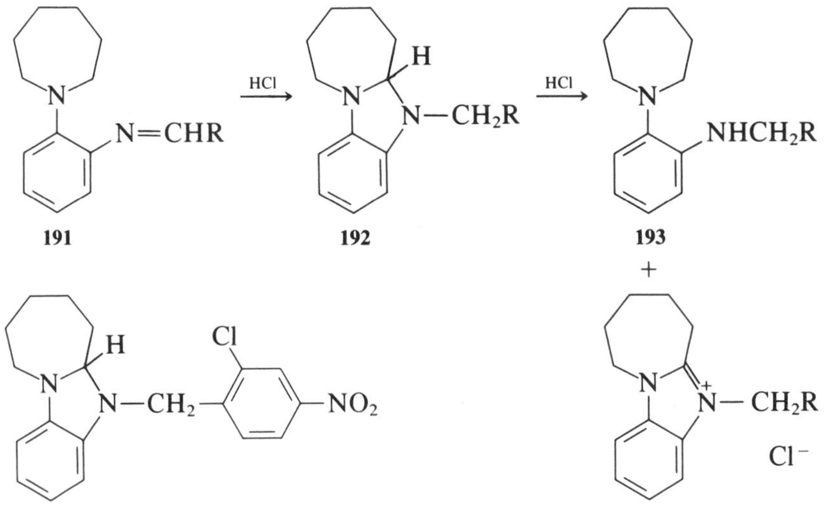

191 —HCl→ **192** —HCl→ **193** + **194**

195, mp 144°C

Figure 30

Anils derived from *o*-pyrrolidinoaniline and a wide variety of aldehydes were found to undergo a rapid acid-catalyzed cyclization in the cold to give dihydrobenzimidazoles (57). The corresponding reaction of the piperidine and hexahydroazepine analogs gave the benzimidazolium salts **194** (Fig. 30). The anil **191** was completely consumed when a 1:2 molar ratio of acid to anil was employed. However, instead of the dihydrobenzimidazoles **192,** equal yields of the amines **193** and the salt **194** were produced. In one case, red needles of the sparingly ethanol-soluble compound **195** (triplet at δ 5.12 corresponding to the bridgehead proton) could be isolated in 94% yield. Upon being warmed with acid, **195** underwent slow conversion to **193** and **194.** Table 14 lists the benzimidazolium salts obtained in this manner.

TABLE 14. HEXAHYDROAZEPINO[1,2-*a*]BENZIMDAZOLIUM SALTS FROM THE ACTION OF ACID ON ANILS **191**[a]

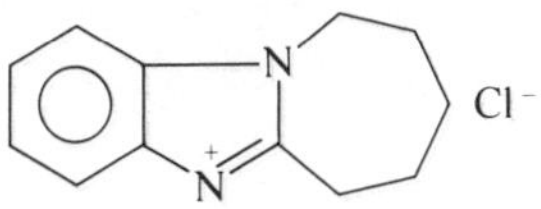

Cl⁻

R	mp (°C)	Yield (%)	Spectra
C_6H_5	288 (dec.)	84	nmr
o-ClC_6H_4	244–246	86	nmr
p-ClC_6H_4	252	90	nmr
2-Cl-5-$NO_2C_6H_3$	243 (dec.)	80	nmr

[a]Fig. 30.

It was previously reported (58) that simple 1,3-dialkyldihydrobenzimidazoles undergo disproportionation on treatment with acid. The mechanism proposed for this process (Eq. 31) involved hydride transfer and concomitant reductive ring cleavage to give **198** and **199.** This mechanism was confirmed by deuterium labeling studies.

In a similar transformation (59), the reaction of the amine **200** with alloxan (**201**) in ethanolic HCl (Eq. 32) led to the salt **202.** In this example, alloxan performs the role of both acid and oxidant to aromatize the dihydro intermediate.

A stable dihydroazepinobenzimidazole (**204**) was obtained in a 75% yield by thermal decomposition of the azide **203** (60). The closure to form **204** presumably occurs via nitrene insertion (Eq. 33).

The benzene ring in azepino[1,2-*a*]benzimidazoles has been the site of two annulation reactions. Thermal decomposition of the azides **205a** and **205b** in 2-ethoxyethanol (Fig. 31) gave the same yellow imidazo[4,5-*e*]benzofuroxan **206.** When the azides were decomposed in propionic acid or ethylene glycol the corresponding orange furazan **207** was obtained. This

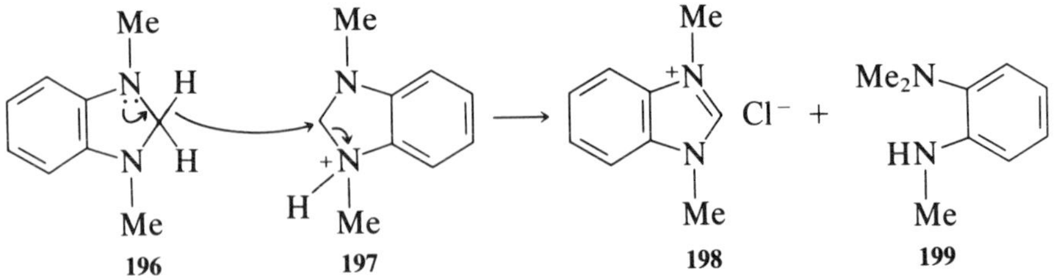

196 197 198 Cl^- + 199 (31)

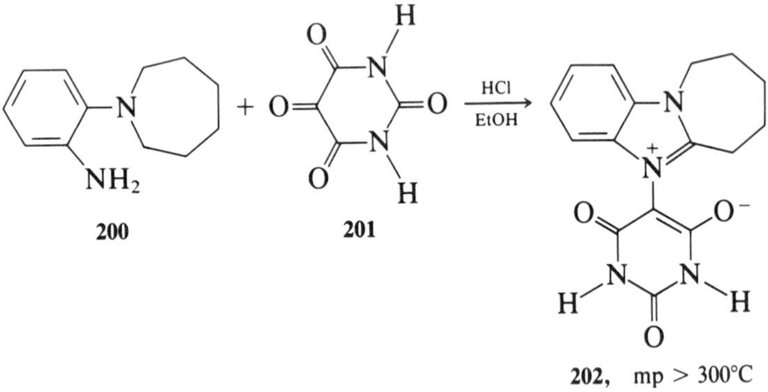

200 + 201 $\xrightarrow[\text{EtOH}]{\text{HCl}}$ 202, mp > 300°C (32)

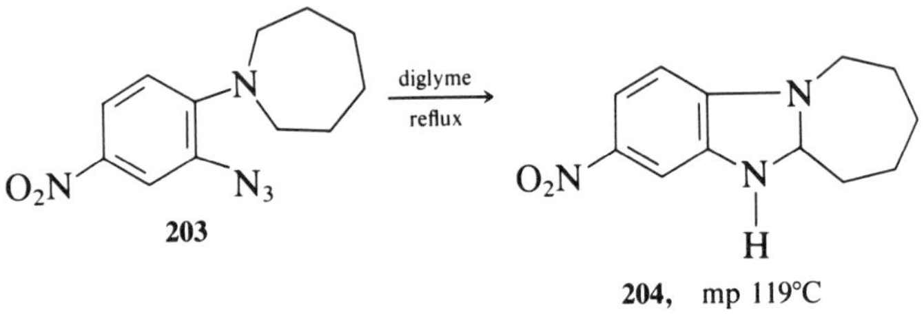

203 $\xrightarrow[\text{reflux}]{\text{diglyme}}$ 204, mp 119°C (33)

205a, $R_1 = N_3$, $R_2 = NO_2$
205b, $R_1 = NO_2$, $R_2 = N_3$

$EtOCH_2CH_2OH$, 133–135°C → **206**, 57%

205 → $CH_3CH_2CO_2H$ or $HOCH_2CH_2OH$, heat → **207**, 60%

206 → $HOCH_2CH_2OH$, heat → **207**

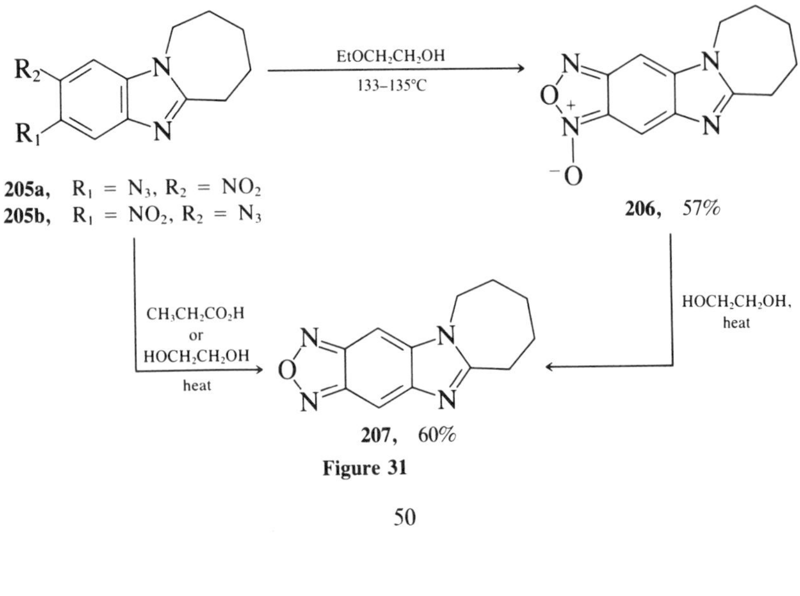

Figure 31

substance was also generated by thermal deoxygenation of **206** in refluxing ethylene glycol. Since Katritzky and co-workers (61) have demonstrated the tautomeric nature of benzofuroxans, the decomposition of the isomeric azides **205** was expected to yield the same product (**206**). Photolytic decomposition of the azides in glacial acetic acid also gave rise to **207**, but in only 32% yield.

Another interesting annulation reaction (62,70) was observed in this ring system when the amino acid **210** was diazotized and the resulting diazonium salt **211** was heated in ethylene glycol in the presence of tetracyclone (Fig. 32). Formation of the tetracyclic product **213** presumably involved a benzyne intermediate (**212**). Aprotic diazotization of **210** using isopentyl nitrite in THF–methylene chloride gave an excellent yield of the diazonium salt **211**

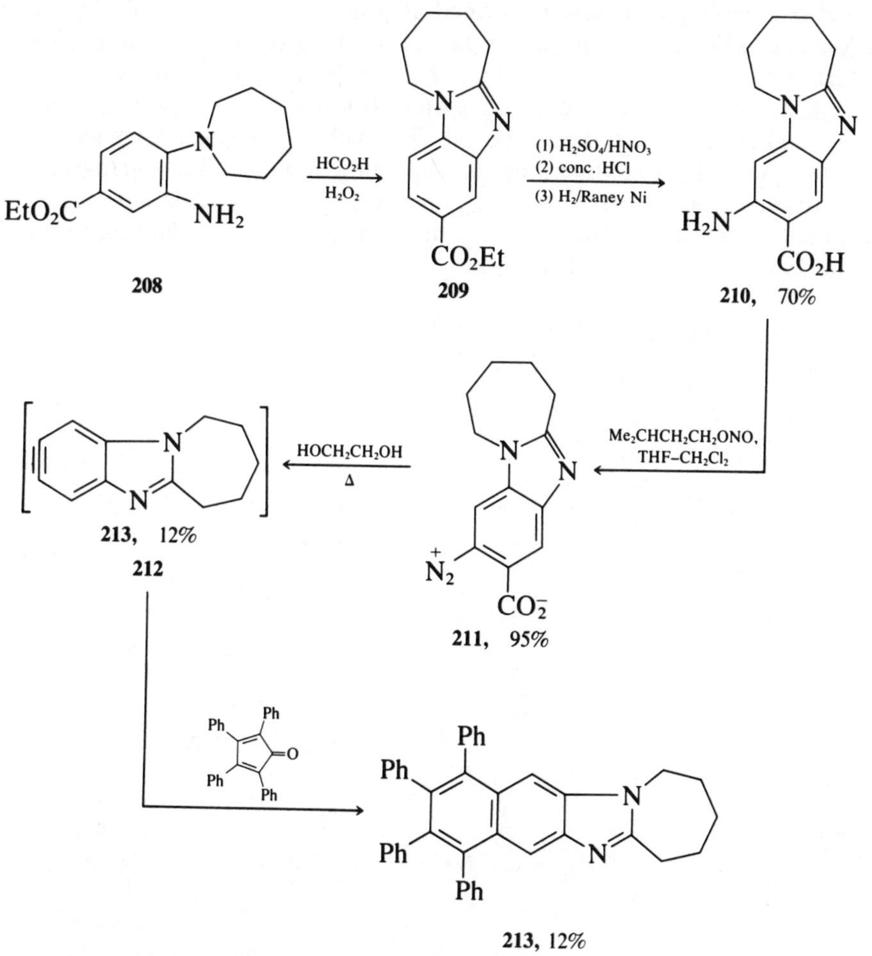

EtO_2C NH_2 N **208** → HCO_2H, H_2O_2 → CO_2Et **209** → (1) H_2SO_4/HNO_3 (2) conc. HCl (3) H_2/Raney Ni → H_2N CO_2H **210**, 70%

→ $Me_2CHCH_2CH_2ONO$, THF–CH_2Cl_2 → $\overset{+}{N_2}$ CO_2^- **211**, 95% → $HOCH_2CH_2OH$, Δ → **212** → (tetracyclone: Ph, Ph, Ph, Ph, O) → Ph Ph Ph Ph **213**, 12%

Figure 32

as the sole product even in the presence of tetracyclone. Thermal decomposition in the presence of tetracyclone produced a small amount of **213.** A similar reaction with anthracene failed.

Deoxygenation of the nitro compound **214** gave, in addition to **215,** an 11% yield of the 6H-azepino[1,2-*a*]benzimidazole **216** by nitrene insertion into the aromatic ring (63) (Eq. 34). The nmr spectrum was consistent for either a 6H- or 10H-structure, but the authors preferred the former since it was in accord with the presumed formation of an azanorcaradiene.

When the trimer **217** was heated to 200°C, a 35% yield of the azepino-[1,2-*a*]benzimidazol-10-one (**218**) was obtained (Eq. 35). The structure of the product was assigned on the basis of infrared (C═O, 1715 cm^{-1}) and nmr spectra (64).

The photolysis of substituted phenazine *N*-oxides has been investigated (65,66). As indicated in Fig. 33, a variety of azepino[1,2-*a*]benzimidazolones and oxepinoquinoxalines are formed as products. Photolysis of phenazine *N*-oxide (**219,** X = H) produces **221** (X = H). Since it had been proposed that either **223** or **224** could be an intermediate in the photolytic conversion of phenazine *N*-oxide to **221,** the photolysis of the substituted *N*-oxides **219** and **220** should allow these two possible pathways to be distinguished. If both **223** and **224** were intermediates, all four possible products **221a–d** could be formed from either **219** or **220,** and the ratios **221a/221b** and **221c/221d** should be the same. The data seen in Table 15, however, indicate that **223** and **224** are not intermediates.

Cl, NO_2, Me, N, Ac — **214** —($(EtO)_3P{=}O$, cumene, 150°C)→ Cl, N, Me, N — **215,** 20% (34)

+ Cl, N, N, Me — **216,** 11%; mp 148°C

[NH—C(=O)—$(CH_2)_4$—C(=O)—NH]$_3$ — **217** —(194–200°C)→ O, N, N — **218,** 35%; mp 51–52°C (35)

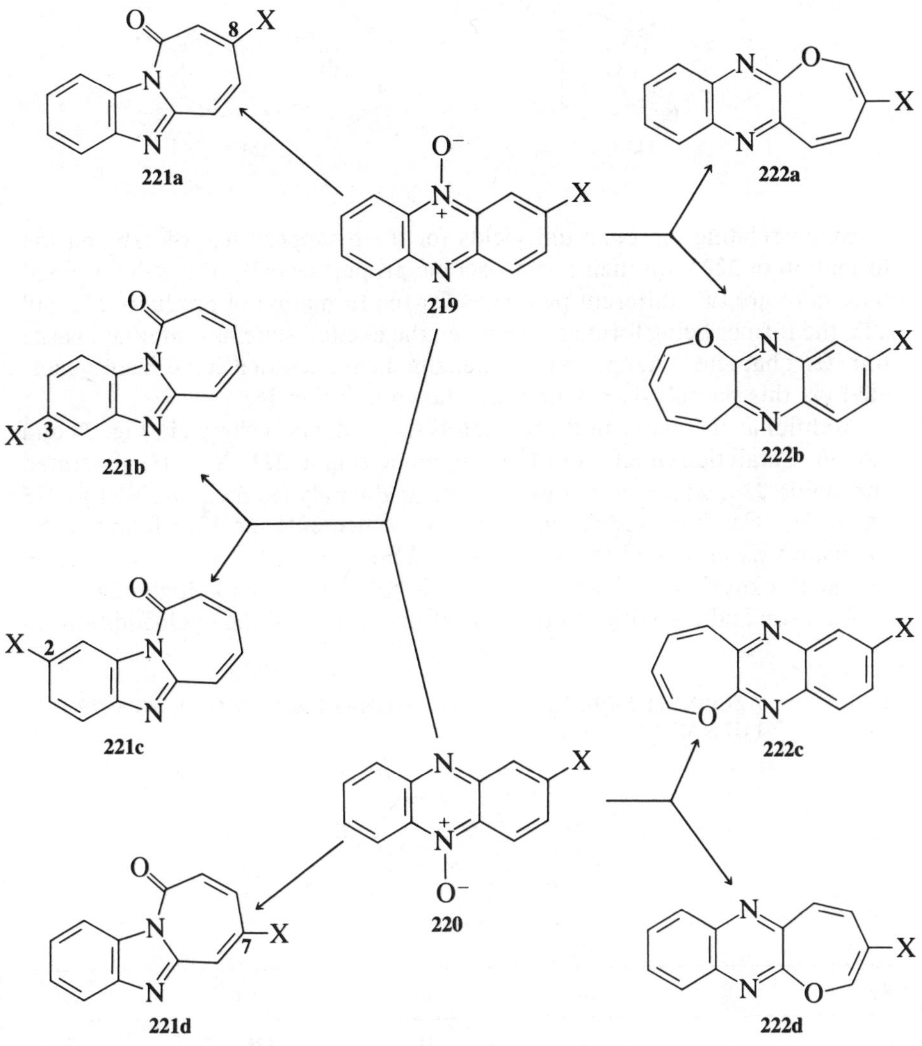

Figure 33

TABLE 15. FORMATION OF COMPOUNDS **221** (Fig. 33) BY PHOTOLYSIS OF PHENAZINE *N*-OXIDES IN ACETONITRILE

Substrate	Yield (%)[a]				Irradiation Time (min)	% Conversion
	221a	**221b**	**221c**	**221d**		
219, X = Cl	12.5	—	20	18.5	40	77
220, X = Cl	—	25	12	9.5	60	67
219, X = MeO	—	2	1	1	420	58
220, X = MeO	48	—	24	12	100	73

[a]Based on percentage of the starting material consumed.

223

224

By correlating the quantum yields for the disappearance of **219** and the formation of **220** with their spectroscopic properties (67), it was determined that there are two different pathways for the formation of products **221** and **222**, the former being formed only when the excited state has internal charge transfer character. Azepino[1,2-*a*]benzimidazolones that have been generated via this photolysis reaction are shown in Table 16.

Additional chemistry of these photolysis products is shown in Fig. 34 and Eq. 36. Catalytic reduction of the azepinone ring in **221** (X = H) generated the amide **226**, which was converted by acid-catalyzed methanolysis to **225** (X = H). Catalytic reduction of either isomer **221b** or **221c** followed by methanolysis produced the same ester (**225**, X = Cl or MeO). Reduction and methanolysis of **227** gave, after decarboxylation, the ketone **229**.

Acheson and co-workers (16,17,68,69) have studied the cycloaddition re-

TABLE 16. AZEPINO[1,2-*a*]BENZIMIDAZOL-10-ONES FROM PHOTOLYSIS OF PHENAZINE N-OXIDES[a]

R_1	R_2	R_3	R_4	mp (°C)	Refs.
H	H	H	H	176–177	65
Cl	H	H	H	129–130	66
H	Cl	H	H	138–139	66
H	H	Cl	H	176–177	66
H	H	H	Cl	159–161	66
MeO	H	H	H	146–147	66
H	H	MeO	H	153–154	66
H	H	H	MeO	147–148	66
Me	H	H	H	110–111	67
H	H	H	Me	113–116	67
CN	H	H	H	191–192	67
H	H	CN	H	238–239	67
H	H	H	CN	221–223	67

[a]Nmr, ir, uv, and mass spectra are reported for all compounds in the table.

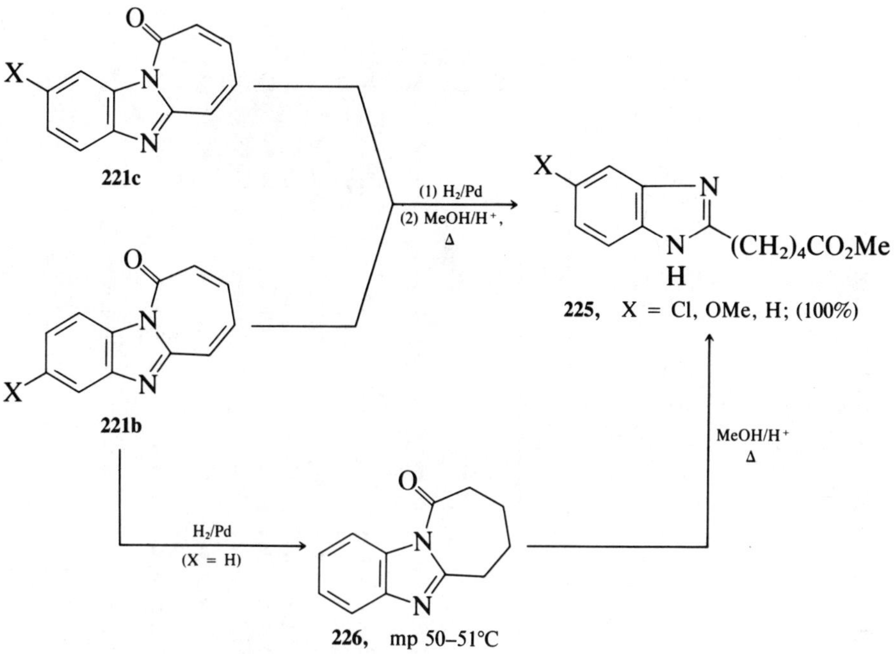

221c

(1) H_2/Pd
(2) MeOH/H^+, Δ

225, X = Cl, OMe, H; (100%)

$(CH_2)_4CO_2Me$

221b

H_2/Pd
(X = H)

MeOH/H^+
Δ

226, mp 50–51°C

Figure 34

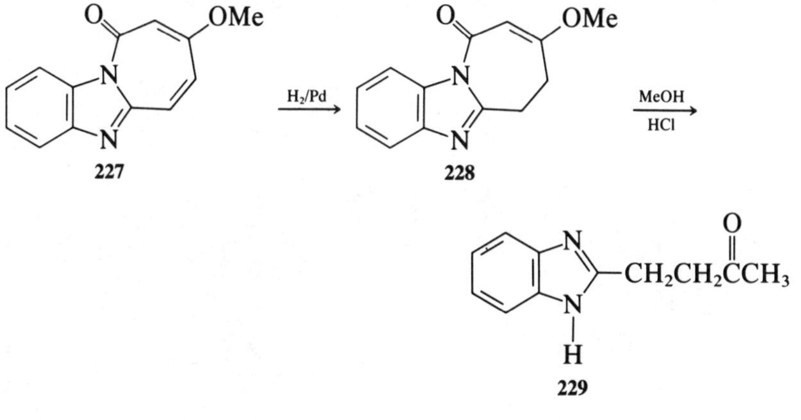

227 $\xrightarrow{H_2/Pd}$ 228 $\xrightarrow[HCl]{MeOH}$ 229 (36)

$CH_2CH_2CCH_3$

action of benzimidazoles with dialkyl acetylenedicarboxylates. Several products are usually produced in such a reaction, including examples of the azepino[1,2-*a*]benzimidazole system (see Table 17). The reaction does not appear to have preparative value, however. Extensive spectral data are reported for the compounds summarized in Table 17, including mass spectra (26). Tables 18, 19, and 20 list the azepino[1,2-*a*]benzimidazoles found in the literature.

TABLE 17. AZEPINO[1,2-*a*]BENZIMIDAZOLES FROM ACETYLENEDICARBOXYLIC ESTER CONDENSATIONS WITH BENZIMIDAZOLES

Type	R_1	R_2	X	Y	E	Conditions	Yield (%)	Color	mp (°C)	Refs.[a]
A	Me	H	CO_2Me	CO_2Me	CO_2Me	CH_3CN, reflux	13	Yellow	206–207	16
						THF, room temp.	29			
A	Et	H	CO_2Et	CO_2Et	CO_2Et	THF, room temp.	26	Yellow	204–205	16
B	Me	Ph	CO_2Me	H	CO_2Me	CH_3CN, reflux	<1	Green	157	68
A	Me	Ph	CO_2Me	CO_2Me	CO_2Me	CH_3CN, reflux	1	Red	210	17
A	Me	Ph	CO_2Et	CO_2Et	CO_2Et	CH_3CN, reflux	<1	Orange	105–110	17
B	Me	Ph	CO_2Et	H	CO_2Et	CH_3CN, reflux	<1	Green	120	17
A	H	CN	CO_2Me	CO_2Me	CO_2Me	CH_3CN, reflux	4	Yellow	253–255	17
A	Me	CO_2Me	CN	CO_2Me	CO_2Me	CH_3CN, reflux	4.7	Colorless	184–185	68
A	$CH{=}CHCO_2Me$	H	CO_2Me	CO_2Me	CO_2Me	CH_3CN, reflux	19	Colorless	180–181	68
A	$CH{=}CHCO_2Me$	H	CO_2Me	CO_2Me	CO_2Me	CH_3CN, reflux	4.6	Yellow	206–208	69
A	$C(CO_2Me){=}CHCO_2Me$	H	H	H	CO_2Me	CH_3CN, reflux	6.5	Yellow	180	69
A	$CH{=}CHCO_2Me$	Ph	CO_2Me	CO_2Me	CO_2Me	CH_3CN, reflux	3.5	Yellow	233–235	69

[a]Extensive spectral data (ir, uv, nmr) are reported in the references.

TABLE 18. HEXAHYDROAZEPINO[1,2-*a*]BENZIMIDAZOLES

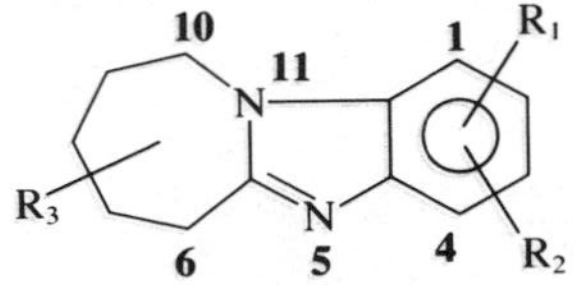

R_1	R_2	R_3	mp (°C) or bp (°C/torr)	Spectra	Refs.
H	H	H	125–126		40
			232–234 (MeI)		
			124–125	uv	41
			123.5–125.0	uv	44
			126		46
			124		49
			125		53
			212 (*N*-oxide)	nmr	51,52
2-NO_2	H	H	196–197		40
			199		43
2-NH_2	H	H	198–199		40
2-NHAc	H	H	222		43
2-N_3	H	H	98		45
3-Cl	H	H	107–109		40
			273–274 (MeI)		
			259–261 (HCl)		
			129 (*N*-oxide)	nmr	51,52,53

TABLE 18. *(Continued)*

R_1	R_2	R_3	mp (°C) or bp (°C/torr)	Spectra	Refs.
3-NO_2	H	H	174–175 177 206 (*N*-oxide)		40,55
3-NH_2	H	H	108.5–181.0		40
3-N_3	H	H	108		45
3-OH	H	H	295–296		40
3-NHAc	H	H	254–255 252		40
3-$NHCO_2Et$	H	H	238–240 270–272 (MeI)		40
3-NH(1-Hydroxy-2-naphthoyl)	H	H	310–311		40
3-NH—(pyrimidinium ring bearing NH_2, N, $\overset{+}{N}$—Me I^-, Me)	H	H	~220 (rapid heating) 335 (MeI)		40
3-*N*-Caprolactimido	H	H	345		40
3-$NHCONH_2$	H	H	335 316–318 (MeI)		40
1-Me	H	H	154–155/0.3		46
4-Me	H	H			
1-Me	3-Me	H	168–170/0.3		46

2-Me	4-Me	H			
H	H	8-*t*-Bu	143		46
H	H	10-Cyclohexyl	203–205/0.2		46
H	H	6-OH	182	nmr,uv	50
1-Cl	3-NO_2	H	136	nmr	55
1-N_3	3-NO_2	H	154	nmr	55
2-MeO	3-Cl	H	165	nmr	56
2-NO_2	3-CO_2H	H	165		62
2-NO_2	3-NHAc	H	190–191		40
			183		70
2-NHAc	3-NO_2	H	183	nmr	70
2-NO_2	3-NH_2	H	295–296		40
			273		70
2-NO_2	3-N_3	H	140		70
2-NH_2	3-NO_2	H	228	nmr	70
2-NH_2	3-NO_2	H	178 (dec.)	nmr	70
2-NHAc	3-NO_2	H	198	nmr	55
3-NO_2	4-Cl	H		nmr	55
3-NO_2	4-CN	H	205	nmr	55
3-NO_2	4-SCN	H	142		55
3-NO_2	4-NHBu-*n*	H	99	nmr	55
3-NO_2	4-SBu-*n*	H	71–72	nmr	55
3-NO_2	4-SO_2Bu-*n*	H	300	nmr	55
3-NO_2	H	6-AcO	204		55
3-NO_2	H	6-BzO	179	nmr	50
3-NO_2	H	6-O-*p*-C_6H_4OMe	155	nmr	55
2-NHPh	3-NO_2	H	198		55
3-NO_2	4-NHPh	H	150		55

TABLE 19. MISCELLANEOUS AZFPINO[1,2-*a*]BENZIMIDAZOLES

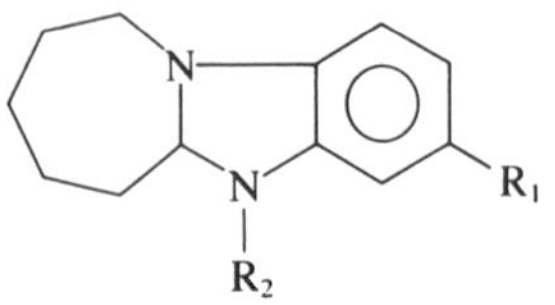

A 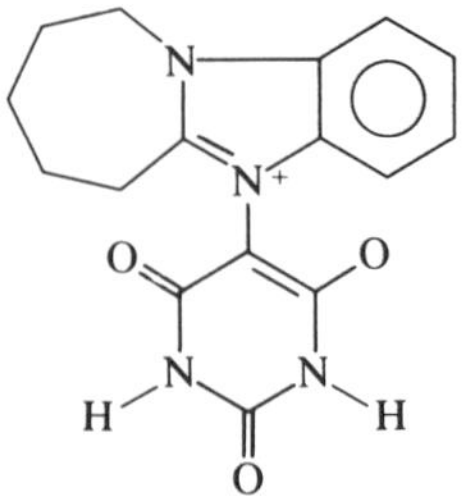 B 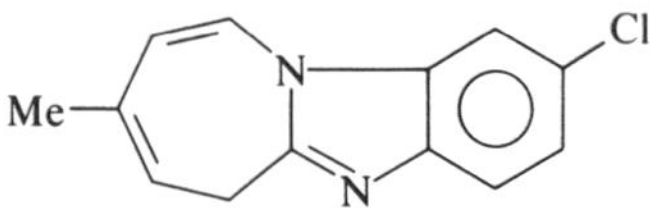 C

Compound	R_1	R_2	mp (°C)	Spectra	Refs.
A	H	2-Cl-5-NO_2-benzyl	144	nmr	57
	NO_2	H	119		60
B			>300	nmr	58
C			148	nmr,ir	63

TABLE 20. AZEPINO[1,2-*a*]BENZIMIDAZOL-10-ONES

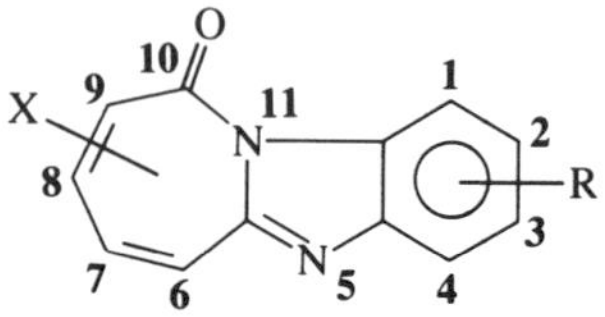

R	X	mp (°C)	Spectra	Refs.
H	H	176–177	ir,nmr,ms	65
H	H	50–51	ir	64,65
		6,7,8,9-Tetrahydro		
2-Cl	H	176–177	nmr,ir,uv	66
3-Cl	H	159–161	nmr,ir,uv	66
H	7-Cl	138–139	nmr,ir,uv	66
H	8-Cl	129–130	nmr,ir,uv	66
2-MeO	H	153–154	nmr,ir,uv	66
3-MeO	H	147–148	nmr,ir,uv	66
H	8-MeO	146–147	nmr,ir,uv	66
H	8-MeO	—[a]	nmr	66
		6,7-Dihydro		
3-Me	H	113–116	nmr,ir,uv	67
H	8-Me	110–111	nmr,ir,uv	67
2-CN	H	238–239	nmr,ir,uv	67
3-CN	H	221–223	nmr,ir,uv	67
H	8-CN	191–192	nmr,ir,uv	67

[a]No mp given, but structure is supported by nmr data and conversion to a known compound.

21. Imidazo[2,1-*a*][2]benzazepines

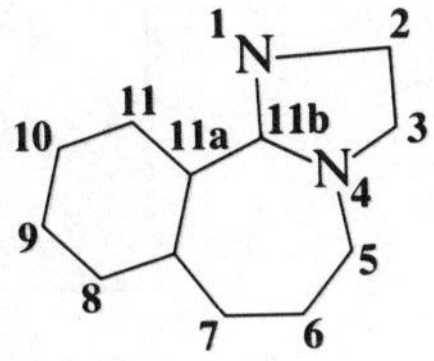

The only known example of the imidazo[2,1-*a*][2]benzazepine system is the 5H-6,7-dihydro derivative **234** (Fig. 35), which was described in a patent (71) devoted mostly to imidazoquinolines with cardiovascular activity. No yields or other physical data were provided.

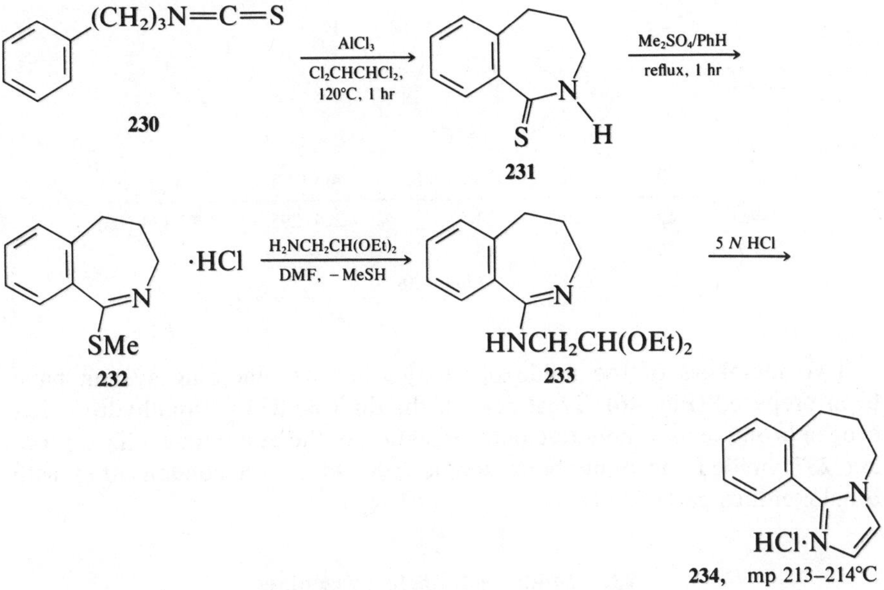

$$\text{230: } C_6H_5(CH_2)_3N{=}C{=}S \xrightarrow[\text{120°C, 1 hr}]{AlCl_3,\ Cl_2CHCHCl_2} \textbf{231} \xrightarrow[\text{reflux, 1 hr}]{Me_2SO_4/PhH} \textbf{232}\cdot HCl \xrightarrow[\text{DMF, } -MeSH]{H_2NCH_2CH(OEt)_2} \textbf{233} \xrightarrow{5\,N\,HCl} \textbf{234}$$

230; 231; 232 (SMe, ·HCl); 233 ($HNCH_2CH(OEt)_2$); 234 (HCl·N), mp 213–214°C

Figure 35

22. Imidazo[2,1-*b*][3]benzazepines

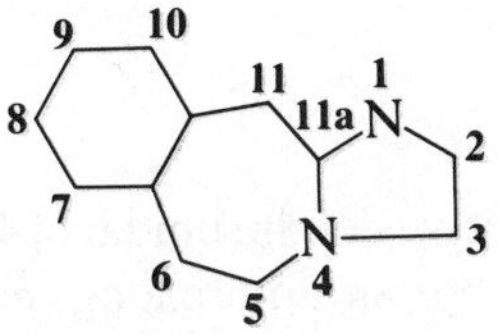

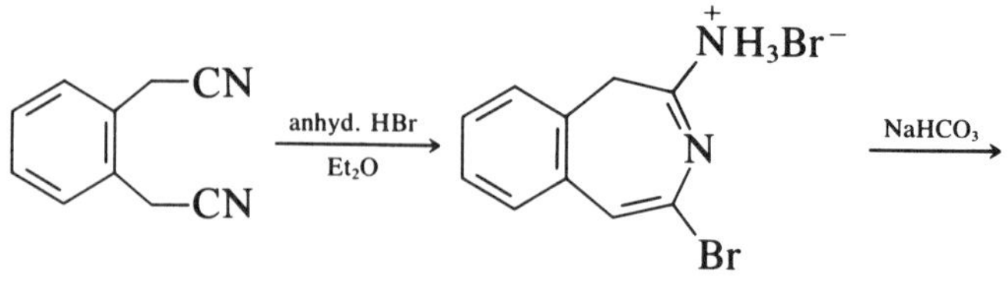

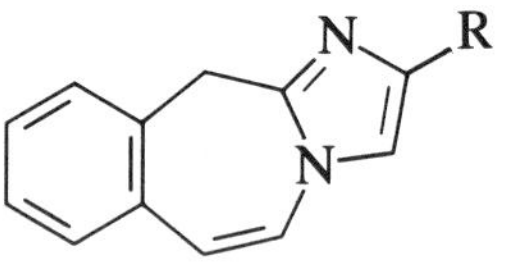

	R	X	Yield (%)	mp (°C)	λ_{max} (ϵ)
239a	Ph	Br	73.5	204–205	262 (28,600)
239b	Me	Cl	18	95–97	

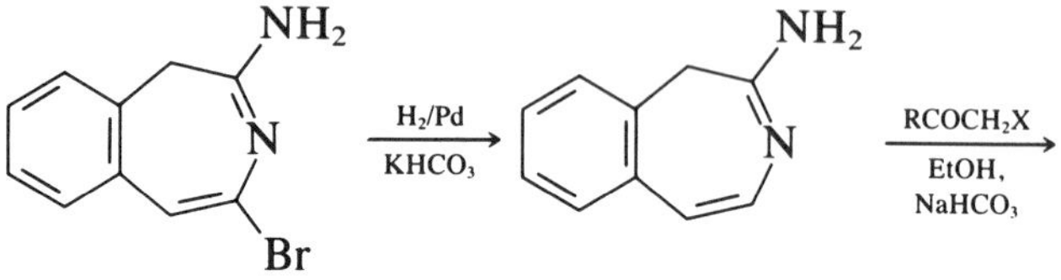

Figure 36

Two members of the imidazo[2,1-*b*][3]benzazepine ring system have been prepared (Fig. 36). Treatment of the dinitrile **235** with anhydrous hydrogen bromide and reductive debromination of the resultant cyclized product **237** yielded the aminobenzazepine **238**, which on condensation with α-haloketones gave **239** (237).

23. Imidazo[4,5-*jk*]benzazepines

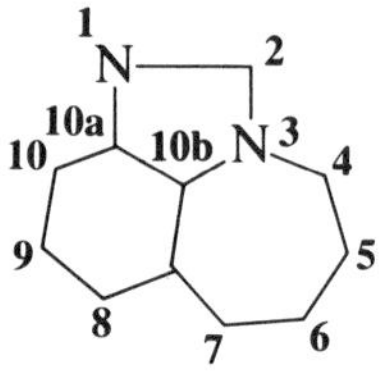

Several tetrahydro derivatives of the imidazo[4,5-*jk*]benzazepine ring system have been prepared by an interesting polyphosphoric acid (PPA) induced cyclization of the *N*-(2-acylaminophenyl)piperidines **240** (n = 3)

$$\text{240} \xrightarrow[150°C]{PPA} \text{241} \quad (37)$$

	n	R_2
241a	3	H
241b	5	Et

$$\text{244} \xrightarrow[(2)\ Ac_2O]{(1)\ H_2} \text{245} \xrightarrow{PPA} \text{246} \quad (38)$$

TABLE 21. DERIVATIVES OF 3,4,5,6-TETRAHYDROIMIDAZO-[4,5-*jk*]BENZAZEPINES

R_1	R_2	mp (°C)	Yield (%)	Spectra	Refs.
H	H	157/2[a]	58		
Me	H	117 163 (methiodide)	90	uv,nmr	72
Ph	H	130	55	uv	
Me	OH	196	20	ir	72
Me	Et	96	20	ir	73

[a]Boiling point (°C/torr).

(72,73). Structures were assigned to the products by analogy with a similar reaction in the imidazopyridine series. In the latter instances, the tricyclic product (**246**) was prepared independently from 4-methyl-8-nitroquinoline (**244**) (Eq. 38). Assignments were corroborated by infrared and nmr spectral evidence. When the heptamethylenimine derivative **240** ($n = 5$) was employed, cyclization was accompanied by ring contraction to give the imidazo[4,5-*jk*]benzazepine **241b** (Eq. 37). Products obtained via this reaction are listed in Table 21.

The mechanism suggested for this transformation is shown in Fig. 37. The spirobenzimidazole **248** forms as a result of attack by the basic nitrogen on the protonated acyl carbonyl, followed by loss of water. The piperidinium ring in this spiro intermediate opens with loss of a proton (or loss of HX if the ring has been opened by a counterion) to give the olefin **249.** Protonation of the olefin gives the more stable secondary carbonium ion, which cyclizes

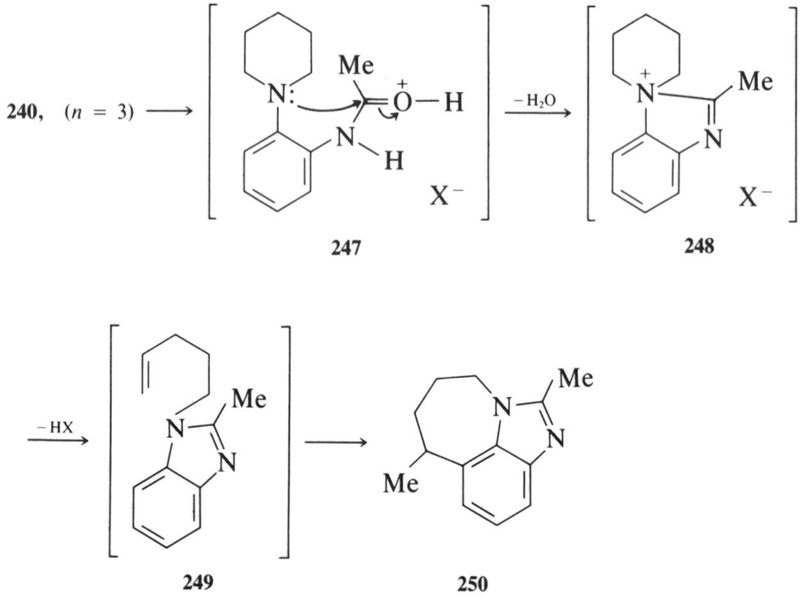

240, (n = 3) ⟶ [**247**] $\xrightarrow{-H_2O}$ [**248**] $\xrightarrow{-HX}$ [**249**] ⟶ **250**

Figure 37

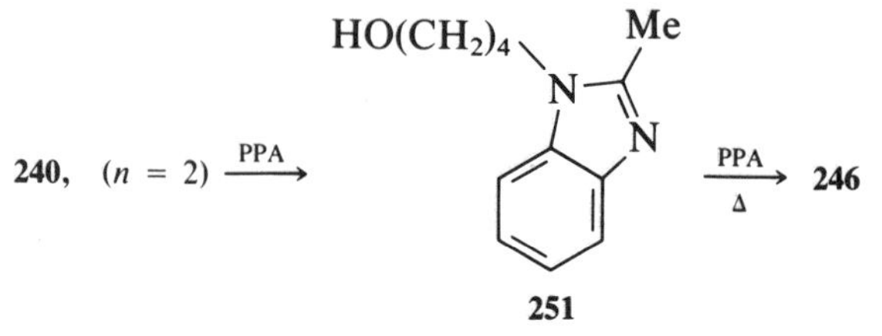

$$\mathbf{240},\ (n = 2) \xrightarrow{\text{PPA}} \mathbf{251}\ \xrightarrow[\Delta]{\text{PPA}} \mathbf{246} \qquad (39)$$

to the azepine **250** by electrophilic attack on the benzene ring. In support of this mechanism, the alcohol **251** was isolated in good yield (Eq. 39) from the cyclization of the pyrrolidino compound **240** ($n = 2$); on further heating in PPA, **251** gave **246.**

Oxidation of **241a** with potassium permanganate (Eq. 40) gave a benzylic alcohol (in 20–25% yield) that was assigned structure **242.** However, no firm physical data (nmr) were given in support of this assignment. The benzimidazole ring in **241a** was stable to boiling ethanolic sodium hydroxide and aqueous 30% sulfuric acid, but ring fission occurred rapidly when the methiodide derived from **241a** was heated in 2 N sodium hydroxide. The 1-acyl-1-benzazepine **243** was isolated in quantitative yield.

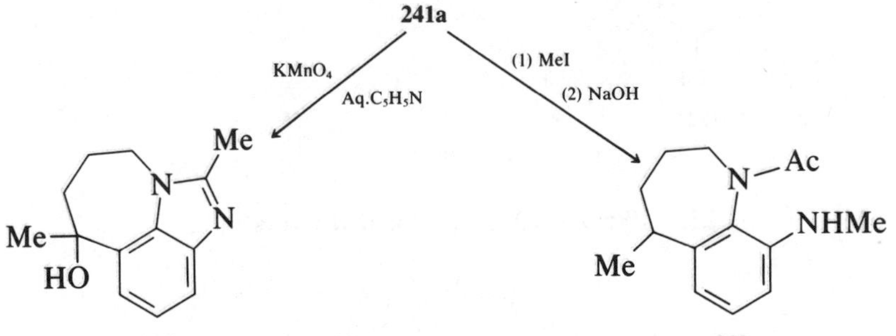

241a → ($KMnO_4$, Aq.C_5H_5N) → **242**; **241a** → ((1) MeI, (2) NaOH) → **243** (40)

24. Pyrazolo[3,4-*b*][1]benzazepines

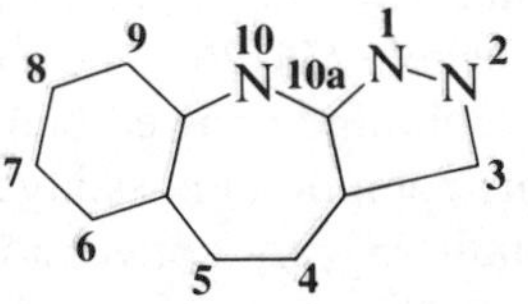

Conversion of 8-chlorotetrahydro-2H-1-benzazepin-2-(3H)-one (**252**) to the enaminoaldehyde **253** was accomplished in low yield (Fig. 38) via the Vilsmeier–Haack reaction (74). The structure of **253** was confirmed by the presence of an unsaturated carbonyl absorption in the infrared at 1640 cm^{-1} and an aldehydic proton signal at δ 7.37 in the nmr spectrum. Condensation of this crystalline substance with *p*-chlorophenylhydrazine gave a product that was tentatively assigned structure **254.** This assignment was based on the observation from the literature (75) that aryl hydrazines generally react with alkyl halides at their α-nitrogen. However, the alternate structure **255** could not be entirely ruled out.

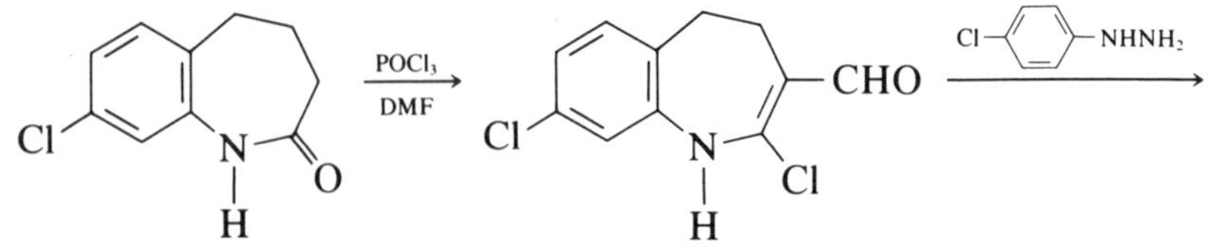

252

253, 23%

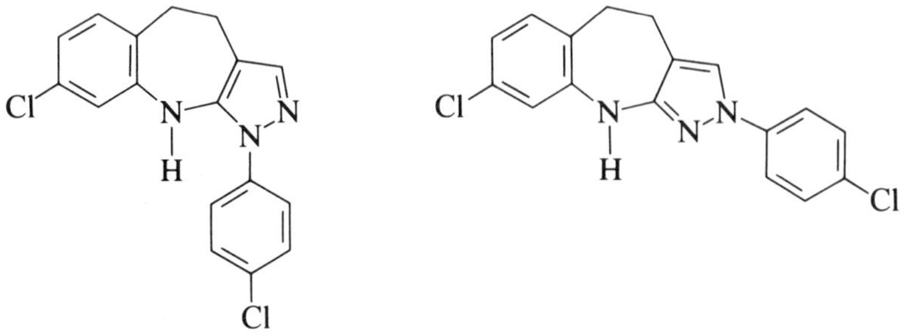

254, 24%; mp 190–192°C

255

Figure 38

25. Pyrazolo[3,4-*c*][1]benzazepines

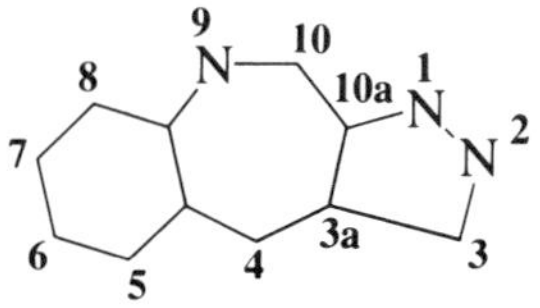

A single example of the pyrazolo[3,4-*c*][1]benzazepine ring system was generated as a minor product from the alkaline hydrogen peroxide oxidation (Fig. 39) of the pyrazolocarbazole **256** (76). The product, **258,** gave the correct mass spectral molecular weight and showed bands in the infrared spectrum at 1667 and 1626 cm^{-1}. This compound possibly arises from the major product, the expected ketolactam **257,** by conversion to **261** (not isolated) followed by lactamization (Eq. 41).

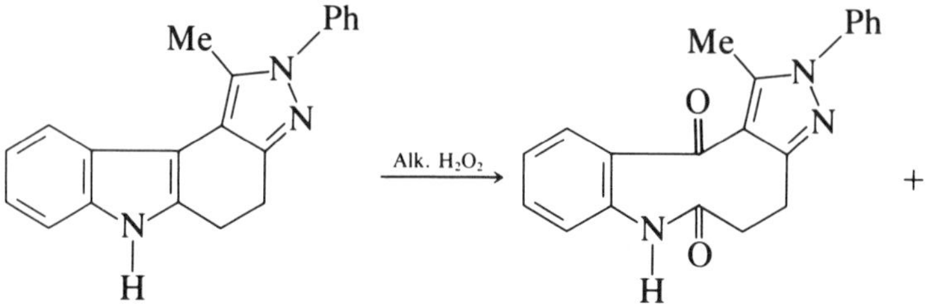

256

Alk. H_2O_2

257

Figure 39

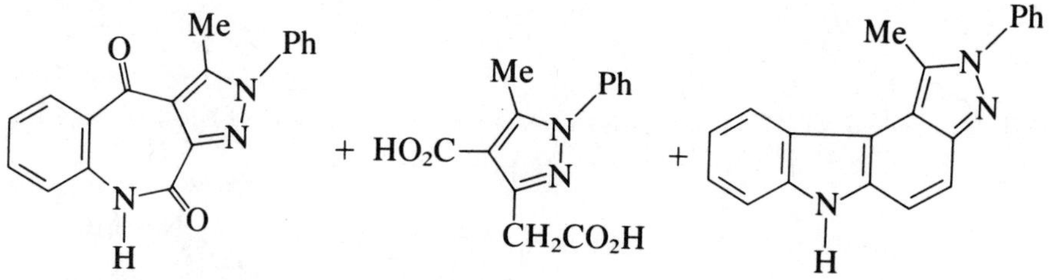

258, mp 308–310°C

259

260

Figure 39 (*continued*)

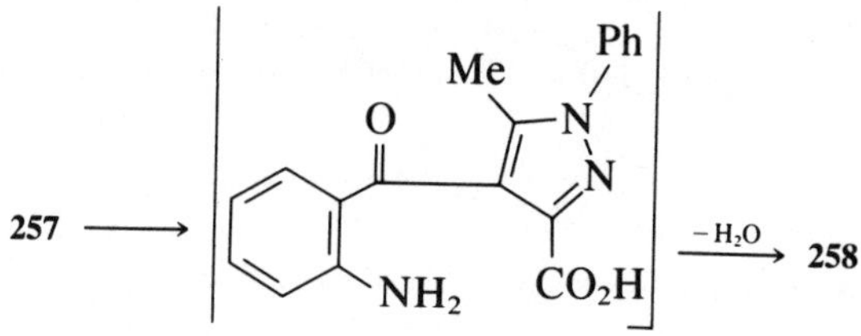

257 ⟶ 261 $\xrightarrow{-H_2O}$ 258 (41)

26. Pyrazolo[3,4-*c*][2]benzazepines

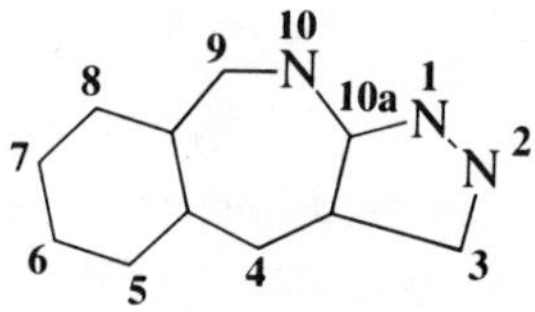

The pyrazolo[3,4-*c*][2]benzazepine ring system has been synthesized recently by a quite efficient route as shown in Fig. 40 (77). Molecular orbital calculations predicted that the maximum electron density of 5-aminopyrazoles (e.g., **263**) occurs at the 4-position, in agreement with experimentally observed electrophilic substitution at that site. If this reaction between 5-aminopyrazoles and *o*-phthalaldehydic acid (**262**) proceeds in the same manner as with indoles, the resulting products would be valuable intermediates in the synthesis of tricyclic systems. Indeed, the product of this condensation, **264**, is obtained in good yield and is easily converted to the pyra-

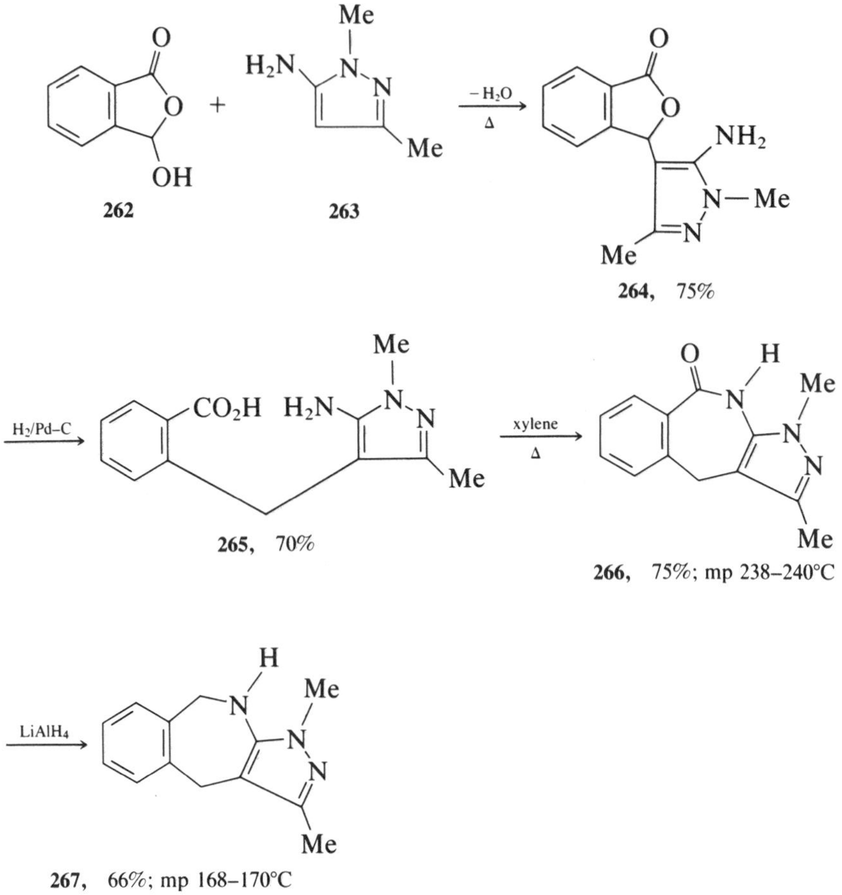

262 263 264, 75%

265, 70% 266, 75%; mp 238–240°C

267, 66%; mp 168–170°C

Figure 40

zolobenzazepin-9-one **266.** Infrared and nmr spectra support the structure: infrared bands at 3120 and 1655 cm^{-1}; two proton nmr singlet at δ 3.64 assignable to the methylene at position 4. Reduction to the basic amine **267** was accomplished with lithium aluminum hydride.

27. Pyrazolo[4,3-*d*][2]benzazepines

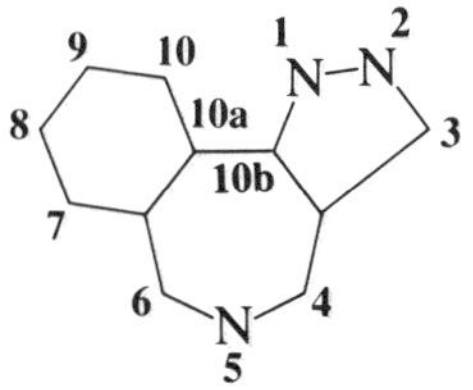

An extensive investigation of the synthesis of 6-aryl–substituted derivatives of the pyrazolo[4,3-*d*][2]benzazepine ring system has been carried out because of the interesting combination of anxiolytic and antidepressant properties of such compounds (78–80). Very creative synthetic chemistry has been employed to construct benzophenone and benzazepine precursors of the desired pyrazolo[4,3-*d*][2]benzazepines. Two routes to the protected benzophenone **268** are shown in Fig. 41, and the subsequent conversion of **268** to the 2-benzazepin-5-one **269** is outlined in Fig. 42. Condensation of **269** with DMF diethyl acetal yields the enamino ketone **270.** On reaction with hydrazines, compound **270** forms 1-substituted 1H,4H-pyrazolo[4,3-*d*]-[2]benzazepines (**272**), presumably via dehydration of the initial adduct **271.** With methylhydrazine, the intermediate **271** (R = CH_3, X = F) could be isolated (mp 114–115°C).

A second approach, which involves the construction of the seven-membered ring as the last step, is seen in Fig. 43. The protected benzophenone **268** is converted to the enamino ketone **274.** The key reaction is the transformation of the phenylthiomethyl side chain into a formyl group (**268→273**). The action of hydrazines on the enaminoketone moiety of **274** forms the pyrazole ring in **275.** On Vilsmeier–Haack formylation, followed by reductive amination and Schiff base formation, **275** is converted to the pyrazolo[4,3-*d*][2]benzazepine **276.**

Addition of hydrazines to enaminoketones such as **270** or **274** to give 1-substituted pyrazolo[4,3-*d*][2]benzazepines is regiospecific. However, experience showed that alkylation as well as acylation of the pyrazole ring in **277** gave the 2-substituted 2H,4H-products **278** (Eq. 42). Table 22 lists the pyrazolo[4,3-*d*][2]benzazepines generated in this study.

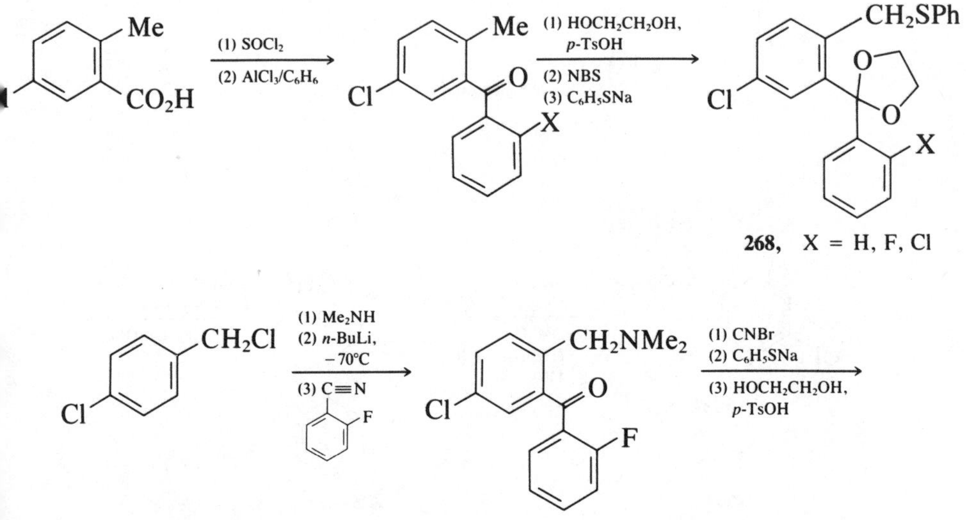

Me
CO₂H
(1) $SOCl_2$
(2) $AlCl_3/C_6H_6$
Cl
Me
O
X
(1) $HOCH_2CH_2OH$, *p*-TsOH
(2) NBS
(3) C_6H_5SNa
Cl
CH_2SPh
O
O
X

268, X = H, F, Cl

Cl
CH_2Cl
(1) Me_2NH
(2) *n*-BuLi, −70°C
(3) C≡N, F
Cl
CH_2NMe_2
O
F
(1) CNBr
(2) C_6H_5SNa
(3) $HOCH_2CH_2OH$, *p*-TsOH

Figure 41

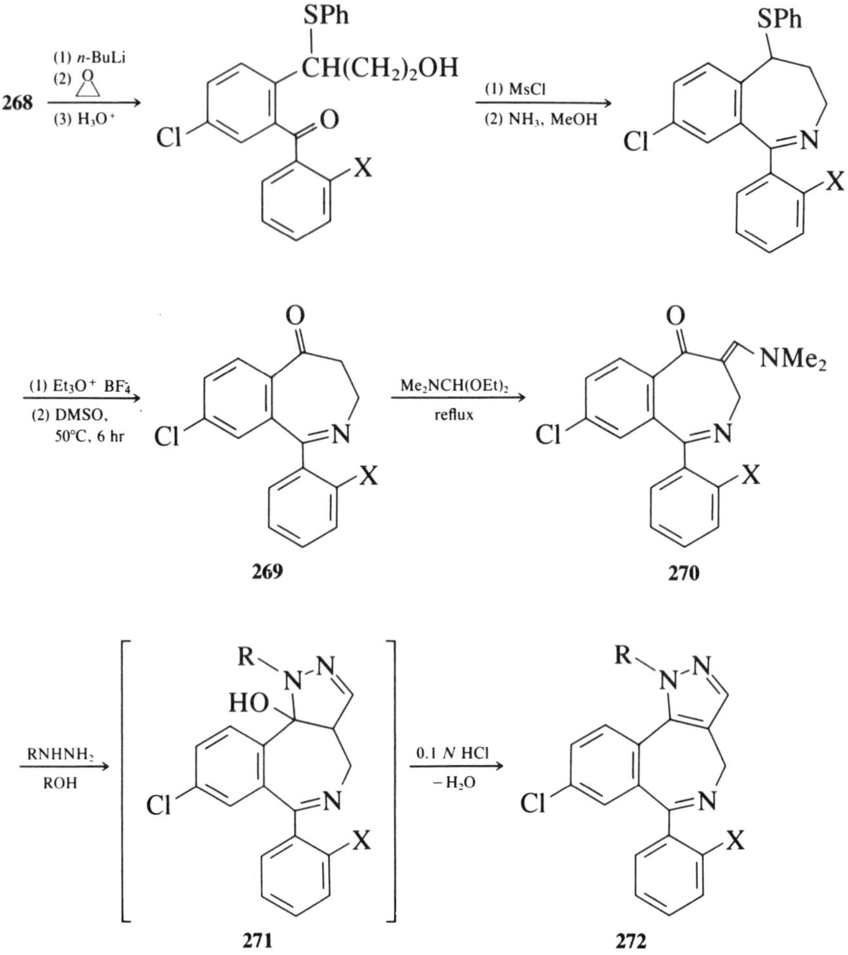

268 —(1) n-BuLi; (2) ethylene oxide; (3) H_3O^+→ SPh–CH$(CH_2)_2$OH, Cl, O, X —(1) MsCl; (2) NH_3, MeOH→ SPh, Cl, N, X

—(1) Et_3O^+ BF_4^-; (2) DMSO, 50°C, 6 hr→ O, Cl, N, X —$Me_2NCH(OEt)_2$, reflux→ O, NMe_2, Cl, N, X

269 270

—$RNHNH_2$, ROH→ [R, N–N, HO, Cl, N, X] —0.1 N HCl, $-H_2O$→ R, N–N, Cl, N, X

271 272

Figure 42

SPh, CH_2, Cl, O, O, F —(1) MeI, Δ, 5 days; (2) DMSO, $NaHCO_3$, 110°C→ CHO, Cl, O, O, F —(1) MeLi; (2) $Na_2Cr_2O_4$; (3) $Me_2NCH(OEt)_2$→

268 273

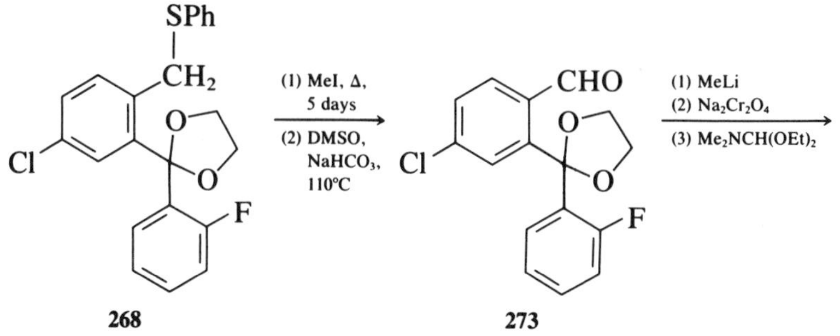

Figure 43

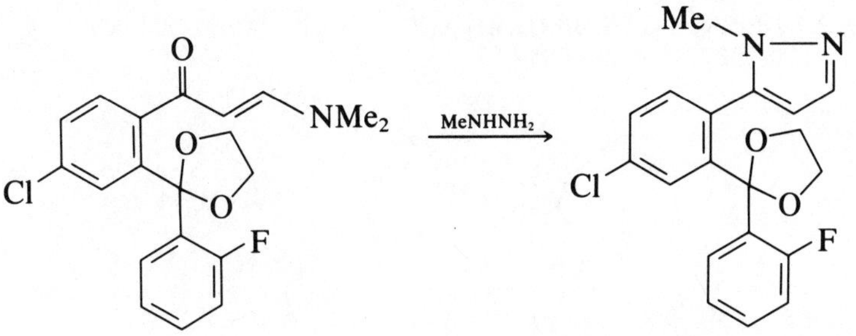

274 → ($MeNHNH_2$) → 275

(1) DMF/$POCl_3$; (2) NaOAc, H_2O → CHO intermediate → (1) $MeONH_2$; (2) $LiAlH_4$; (3) HCl, aq. dioxane; (4) NaOH, Et_2O; (5) H_3PO_4, EtOH → $NH_2 \cdot \frac{1}{3}H_3PO_4$ intermediate → (dil. base) → 276

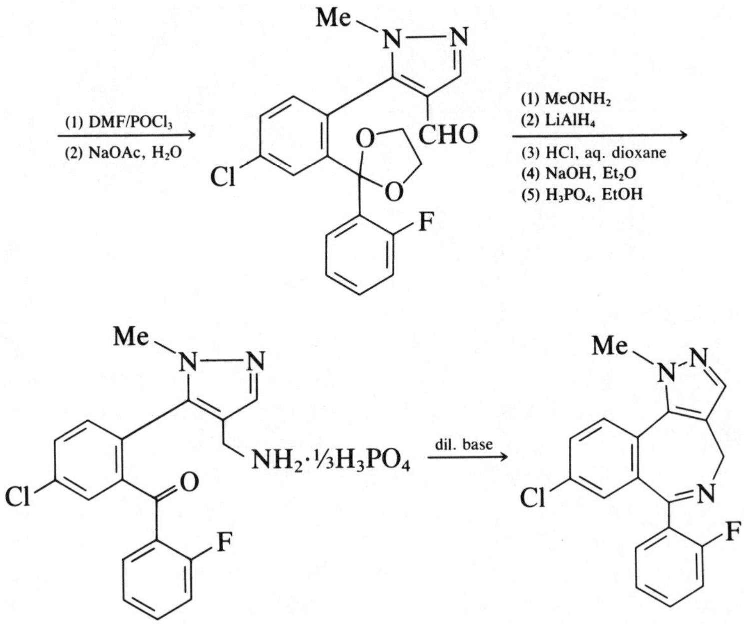

276, mp 128–130°C

Figure 43 (*continued*)

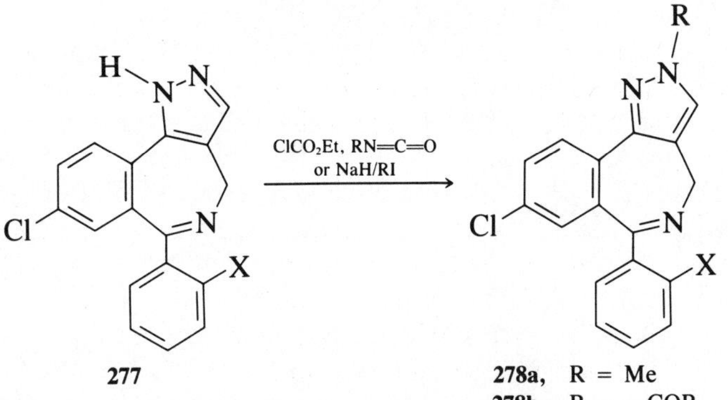

277 → ($ClCO_2Et$, RN=C=O or NaH/RI) → 278 (42)

278a, R = Me
278b, R = −COR

TABLE 22. DERIVATIVES OF 1H,4H- AND 2H,4H-PYRAZOLO[4,3-*d*]-[2]BENZAZEPINES

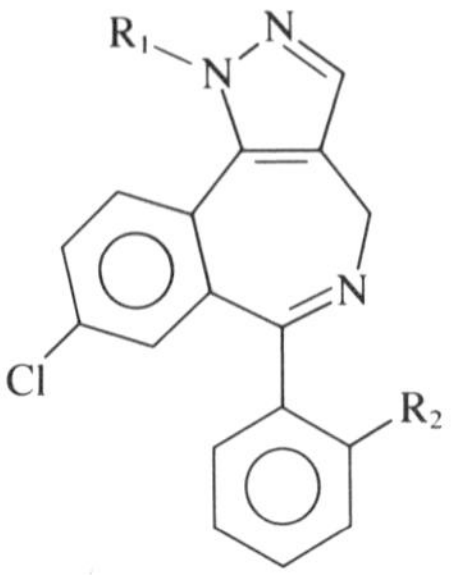

A (1H,4H) 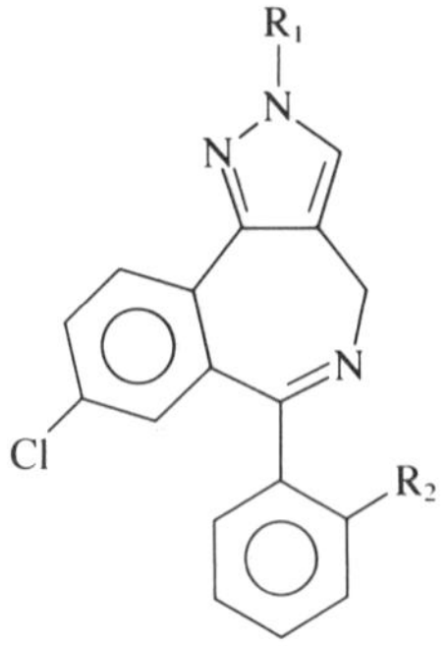 B (2H,4H)

R_1	R_2	Salt	mp (°C
		Type A	
H	H	2HCl	268–270
		CH_3SO_3H	221
Me	H	2HCl	255
H	F	2HCl	225 (dec.)
Me	F	—	128–130
		2HCl	190
n-Pr	F	2HCl	199
$(CH_2)_2NEt_2$	F	Cyclamate	105 (dec.)
$(CH_2)_2OH$	F	HCl	169
$(CH_2)_3NEt_2$	F	Cyclamate	110
H	Cl	HCl	220
CH_3	Cl	HCl	209–211
$PhCH_2$	Cl	HCl	218–221
$(CH_2)_2NEt_2$	Cl	2HCl	195
		Type B	
Me	F	—	139
CO_2Et	F	HCl	168
CONHMe	F	—	192–193
$CONMe_2$	F	—	170
CH_3	Cl	—	160–161
CO_2Et	Cl	—	151–152
CONHMe	Cl	—	192
$CONMe_2$	Cl	—	175–176
CONHMe	Cl	—	195

28. 1,3-Dioxolo[4,5-*h*][2]benzazepines

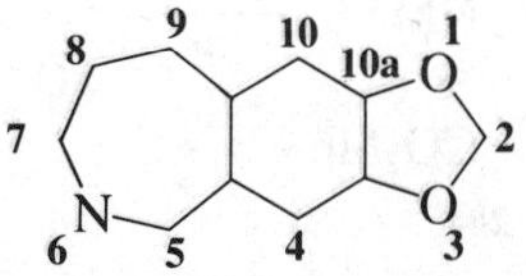

In a study of intramolecular Vilsmeier reactions, the cyclic aminal **280** was reported to form in good yield from the formamide **279**, but was not characterized (Fig. 44). This product was converted to the ether **281** and the hydroxamine **282**. Microanalytical data were given for these compounds, but additional supporting data were not provided (81).

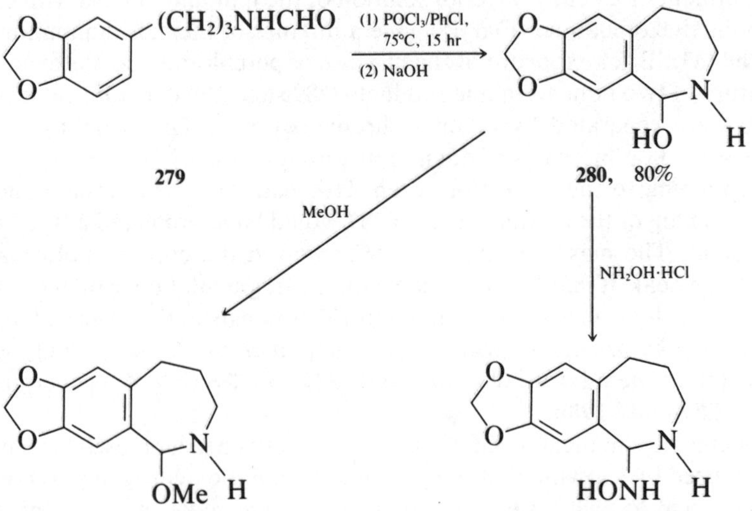

(1) $POCl_3$/PhCl, 75°C, 15 hr; (2) NaOH

279 → **280**, 80%

MeOH → **281**, 68%; mp 78–80°C

$NH_2OH{\cdot}HCl$ → **282**, mp 145–146°C

Figure 44

29. 1,3-Dioxolo[4,5-*h*][3]benzazepines

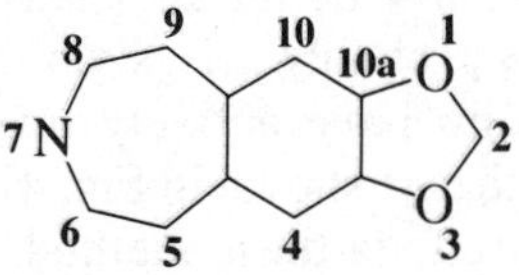

One of the earliest reports of the 1,3-dioxolo[4,5-*h*][3]benzazepine system was the condensation (82) of the phthalaldehyde **283** with the diester **284** to give the deeply colored dioxolo[4,5-*h*][3]benzazepine **285a** (Eq. 43). The

283 + **284** $\xrightarrow[\text{THF}]{\text{KOMe}}$ **285** (43)

	R	Yield (%)	mp (°C)	Color
285a	Me	20	167.5	Deep red
285b	H	17	>390	Dark red-brown

corresponding diacid **285b** was obtained on acidic workup of the mother liquors from the methyl ester. Apart from elemental analysis, no supporting evidence of structure was offered.

Treatment of a benzyl alcohol solution of the iminium ion **286** with excess ethanolic diazomethane (Fig. 45) gave a mixture of the aziridinium ions **287** and **288** (83). Brief exposure of the mixture of perchlorates to methanol gave a mixture of two noncrystalline products (**289a** and **290a**) in a 1:1 ratio (32%), which were separated by column chromatography. The structure of **289a** was assigned on the basis of the nmr absorption at δ 5.77 (*t*, 1H, $J = 5$ Hz), corresponding to the *X* portion of an *ABX* pattern. The alternate and less likely opening of the aziridine ring of **287** would have produced **291,** a known compound. The mass spectrum of **289a** showed the correct molecular ion and a base peak at *m*/*e* 177, explicable by the fragmentation sequence shown in Eq. 44. Dihydrobenzisopyrylium ion **297** is consistently observed in mass spectra of 3-benzazepines with a C_1 oxygen function. Similar solvolytic ring expansion of the mixture of compounds **287** and **288** (Fig. 45) generated the diacids **289b** and **290b.**

Conclusive chemical proof of the structure of benzazepines **289a** and **289b** was secured by showing that the product of catalytic hydrogenolysis of **289a** was identical to **298a,** prepared by an alternative route involving photocyclization of the *N*-chloroacetamide **299a** (Fig. 46). On the basis that lithium aluminum hydride reduction of simple aziridium salts leads to S_N2-like ring opening at the least substituted carbon, it was expected that reduction of the mixture of **287** and **288** would produce **292** and **293** (Fig. 45). Compound **293** was in fact isolated in 22% yield from this reaction and was shown not to be the positional isomer **294** by an alternative photochemical synthesis from **299b.** Compound **288** probably arises via the pathway seen in Eq. 45. Brossi and co-workers, who have actively investigated this ring system, developed the general syntheses shown in Fig. 47 (84,85,86) and Fig. 48 (87). Specific compounds prepared via these methods are listed in Tables 23 and 24.

The 1,3-dioxolo[4,5-*h*][3]benzazepine ring system occurs in Rheadan and related families of alkaloids, and examples have been reported by various groups as degradation products or synthetic intermediates. Rhoeadine (**300a**)

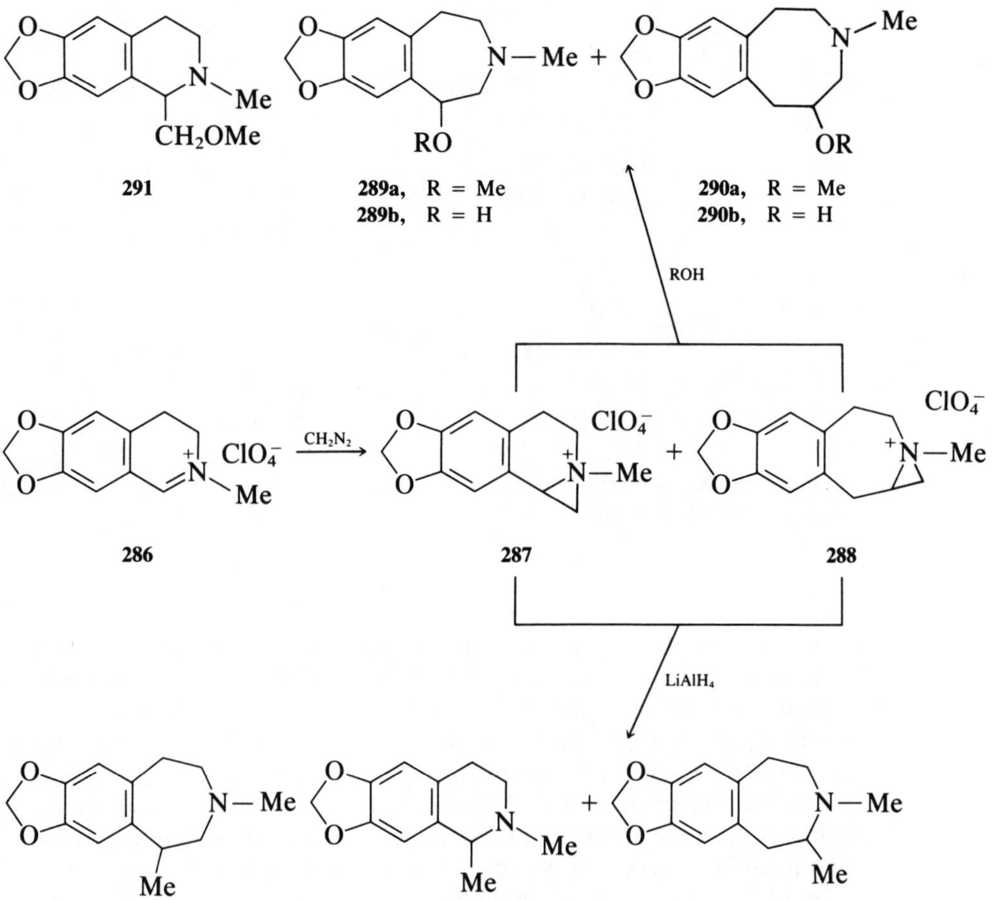

291

289a, R = Me
289b, R = H

290a, R = Me
290b, R = H

ROH

CH_2N_2

286

287

288

$LiAlH_4$

294

292

293

Figure 45

$-Me''$

$-CH_2{=}NMe$

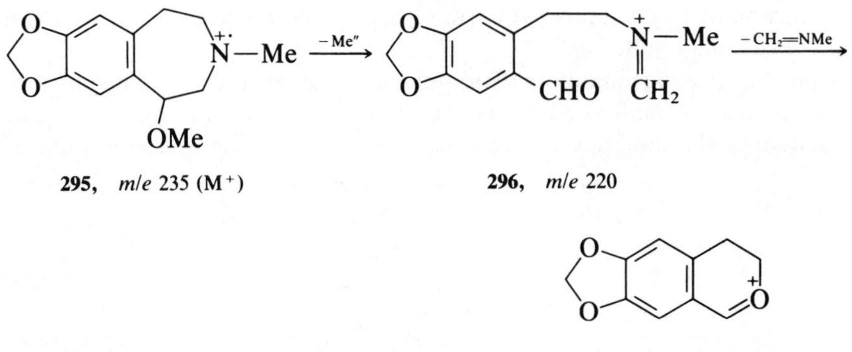

295, m/e 235 (M^+)

296, m/e 220

(44)

297, m/e 177

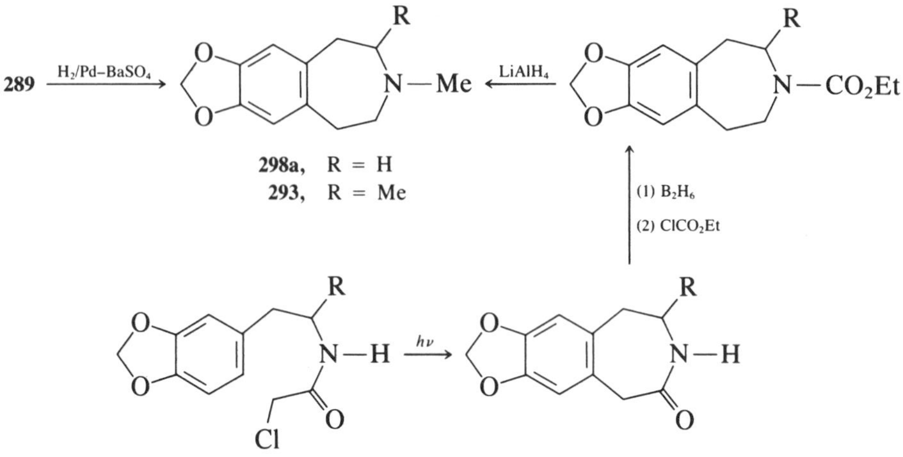

298a, R = H
293, R = Me

299a, R = H
299b, R = Me

Figure 46

is not reduced by lithium aluminum hydride or catalytic hydrogenolysis over platinum (Fig. 49), but rhoeagenine (**300b**) is reduced by these reagents to rhoeageninediol (**301a**), further characterized as its diacetate **301b** (88). Oxidation of either rhoeagenine or the diol gave oxyrhoeagenine (**302**). In a paper (88) dealing with the chiroptical properties and interrelationships in the rhoeadine (*cis* junction of B/D rings) and isorhoeadine (*trans*) alkaloids, hydrogenolysis of rhoeageninediol (**301a**) or the diacetate **301b** in neutral or alkaline media was reported to fail. However, Clemmensen reduction, catalytic hydrogenation in HOAc–perchloric acid, or reduction with aluminum chloride–lithium aluminum hydride converted **301a** into demethoxyrhoeadine (**303**). Hydrogenolysis of the diacetate **301b** gave a 2:1 mixture of **304** and **305.** The latter substance presumably arises via cyclization of the benzylic carbon of ring C onto the basic nitrogen with resulting demethylation of the quaternary salt.

Ring B of rhoeageninediol (**301a**) appears to adopt the pseudochair conformation, with axial hydroxyl and equatorial aryl moieties, respectively, as suggested by the coupling constant ($J_{1,2}$ = 2.0 Hz). The nmr spectrum of

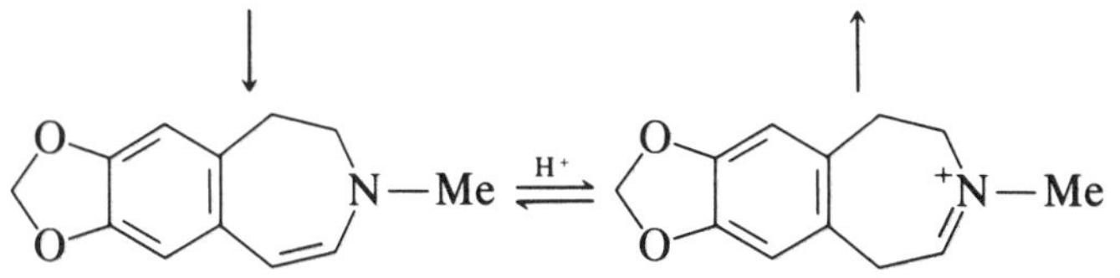

(45)

R
$CH_2CHNHAc$
CH_2O/HCl
(R = H, Me)
R
$CH_2CHNHAc$
CH_2Cl
NaCN
DMSO
R
$CH_2CHNHAc$
CH_2CN
EtOH/HCl
R
$CH_2CHNHAc$
CH_2CO_2Et
(1) NaOH
(2) HCl
R
CH_2CHNH_2
CH_2CO_2H
(1) PhCHO
(2) $NaBH_4$
R
$CH_2CHNHCH_2Ph$
CH_2CO_2H
R
$N-CH_2Ph$
O
B_2H_6
R
$N-CH_2Ph$

Figure 47

CH_2Cl CH_2Cl — NaCN, DMSO → CH_2CN CH_2CN — H_2/Raney Ni, 100°C/1000 psi → N—H

↓ Ac_2O, NaOAc

N—Ac — CH_2O, HCl → N—Ac $(CH_2Cl)_n$ (n = 1, 2) — H_2/Pd–c → N—Ac $(CH_3)_n$ (n = 1, 2)

$(CH_3)_n$ (n = 1, 2) — NaOH → N—H $(CH_3)_n$ (n = 1, 2)

$(CH_2Cl)_n$ (n = 1, 2) — AcOH, KOAc → N—Ac $(CH_2OAc)_n$ (n = 1, 2) — NaOH, EtOH → N—Ac $(CH_2OH)_n$ (n = 1, 2) — MnO_2 → N—Ac $(CHO)_n$ (n = 1, 2)

$(CH_2Cl)_n$ (n = 1, 2) — NaCN/DMSO → N—Ac $(CH_2CN)_n$ (n = 1, 2) — EtOH/HCl → N—Ac $(CH_2CO_2Et)_n$ (n = 1, 2) — NaOH → N—H $(CH_2CO_2H)_n$ (n = 1, 2)

Figure 48

TABLE 23. 1,3-DIOXOLO[4,5-*h*][3]BENZAZEPINES[a]

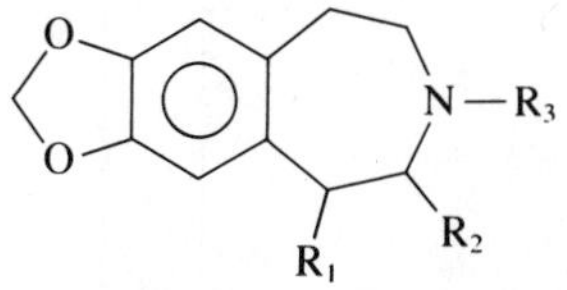

R_1	R_2	R_3	Salt	mp (°C)	Spectra	Refs.
MeO	H	Me	Methiodide	192–196	ms,nmr,uv	83
HO	H	Me	Methiodide	242–245	ms,nmr	83
H	H	Me	Methiodide	258–260	ms,nmr	83
			HCl	277	nmr	84,85,86
H	Me	Me	Methiodide	268–269.5	ms,nmr	83
H	H	CO_2Et		96–97	ir	83
H	Me	CO_2Et		135–140/0.15[b]	nmr	83
H	H	H		82–84	nmr	83
			HCl	287–287.5	nmr	84,85,86
H	Me	H			ir,nmr	83
			HCl	271.5–272.5	ir,nmr	84,85,86
H	H	CH_2Ph		98–100	nmr	84,85,86
			HCl	242–244		
H	H	Ac		178–180		87
H	H	Et		65–66		85,86
			HCl	268–269		
H	H	CH_2CH_2OH		112–113		85,86
			HCl	223–225		
H	H	$CH_2\ CH{=}CH_2$		83.5–85.5		85,86
			HCl	267.5–268.5		
H	H	$(CH_2)_3NMe_2$	2HCl	273–275		85,86
H	H	$(CH_2)_3CO_2H$	HCl	245–247		85,86
H	H	CH_2CO_2Et	HCl	89–90		85,86
H	H	$(CH_2)_3CO_2H$	HCl	245–247		85,86
H	H	CH_2CO_2Et	HCl	89–90		85,86
H	H	CH_2CO_2H	HCl	265–267		85,86
H	H	CH_2CONMe_2	HCl	238–240		85,86
H	H	$(CH_2)_3COC_6H_4F$-*p*	HCl	209–210		85,86
H	H	$COCH_2C_6H_4CF_3$-*m*		102–104		85,86
H	H	$COCH_2NHMe$	HCl	253–254		85,86
H	H	$COCH_2Cl$		125–126.5		85,86

[a]Figs. 45–47.
[b]Boiling point (°C/torr)

isorhoeageninediol (**306**) shows the same substitutents to be diequatorial ($J_{1,2}$ = 8.8 Hz). The fact that the chair conformation of ring B is the mirror image of the corresponding ring in rhoeageninediol explains the observed change of sign of the Cotton effect (89).

This could be further demonstrated as seen in Fig. 50. Reduction of rhoeagenine methiodide (**307**) with sodium borohydride gave the diol methiodide

TABLE 24. 1,3-DIOXOLO[4,5-*h*][3]BENZAZEPINES AND BENZAZEPINONES[a]

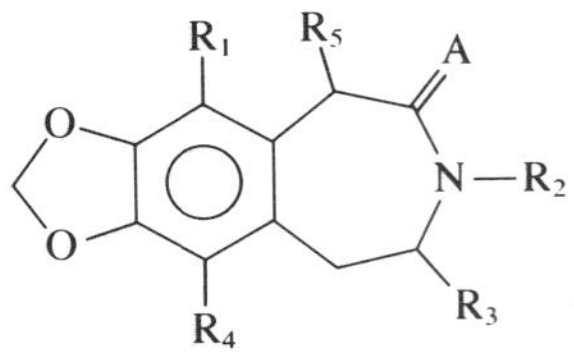

R_1	R_2	R_3	R_4	R_5	A	mp (°C)	Spectra	Refs.
H	H	H	H	H	O	235–237	ir	83
						236–237		85
H	H	Me	H	H	O	179–179.5	ir,nmr	83
						175–175.5	ir,nmr	84,85
H	CH_2Ph	H	H	H	O	141–142	ir,nmr	84
Br	H	H	H	H	H_2	137–139		85,86
H	H	H	H	Me	O	271.5–272.5		85,86
CH_2Cl	Ac	H	H	H	H_2	133–135	nmr	87
Me	Ac	H	H	H	H_2	103–104	nmr	87
CH_2OAc	Ac	H	H	H	H_2	167–168		87
CH_2OH	Ac	H	H	H	H_2	146–148		87
CHO	Ac	H	H	H	H_2	178–179	ir,nmr	87
CO_2H	Ac	H	H	H	H_2	204–206	nmr	87

CO_2H	H	H	H	H	H_2	273–275 (HCl)		87
CN	Ac	H	H	H	H_2	184–185		87
Me	H	H	H	H	H_2	264.5–266 (HCl)		87
CH_2OH	H	H	H	H	H_2	250–251 (HCl)		87
CH_2CO_2H	H	H	H	H	H_2	265–266 (HCl)		87
CH_2Cl	Ac	H	CH_2Cl	H	H_2	202–202.5	nmr	87
Me	Ac	H	Me	H	H_2	177–178	nmr	87
CH_2OAc	Ac	H	CH_2OAc	H	H_2	126–127		87
CH_2OH	Ac	H	CH_2OH	H	H_2	190.5–191.5		87
CH_2CN	Ac	H	CH_2CN	H	H_2	188.5–190		87
CH_2CO_2H	Ac	H	CH_2CO_2H	H	H_2	263–263.5		87
Me	H	H	Me	H	H_2	292–293 (HCl)	nmr	87
CH_2OH	H	H	CH_2OH	H	H_2	221–222		87
CH_2CO_2H	H	H	CH_2CO_2H	H	H_2	>330		87
Me	Me	H	Me	H	H_2	284.5–286 (HCl)		87
CH_2CO_2H	H	H	CH_2CO_2H	H	H_2	>300		87
Me	Me	H	Me	H	H_2	284.5–286 (HCl)		87
H	CH_2Ph	$C_6H_3(OCH_2O)$-3,4	H	MeO	H_2	141–142	ir,nmr,ms	97
H	Me	$C_6H_3(OCH_2O)$-3,4	H	MeO	H_2	189–190 (MeI)	ir,nmr,ms	97
H	Me	H	H	MeO	H,Ph	245		99
MeO	Me	H	H	MeO	H,Ph	207–209		99

[a]Figs. 48 and 49.

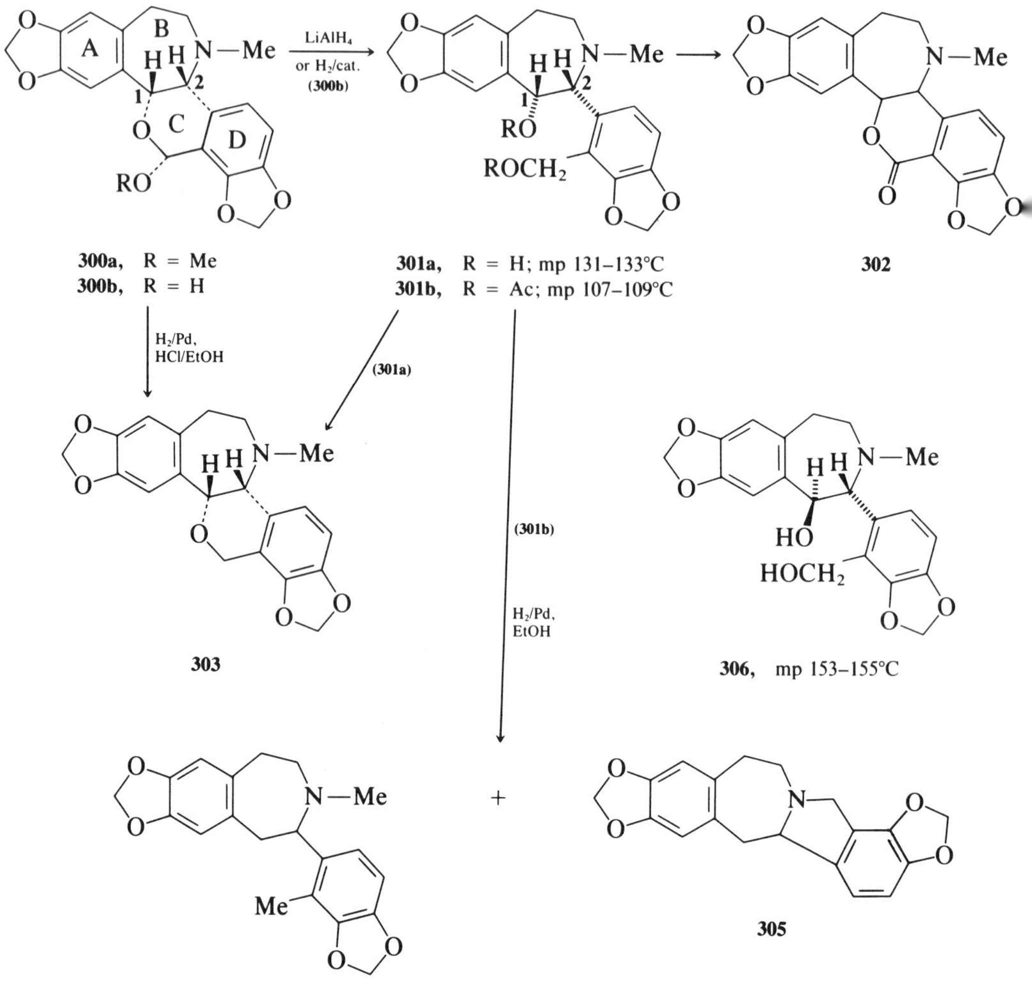
LiAlH4
or H2/cat.
(300b)
300a, R = Me
300b, R = H
301a, R = H; mp 131–133°C
301b, R = Ac; mp 107–109°C
302
H2/Pd,
HCl/EtOH
(301a)
(301b)
H2/Pd,
EtOH
303
306, mp 153–155°C
304, mp 153–154°C
305

Figure 49

308 quantitatively (90). Emde degradation of either **307** or **308** (Fig. 50) provided the optically active substance **309.** Emde degradation of isorhoeageninediol methiodide **310** (Eq. 46) gave optically active **311,** whose nmr spectrum was identical to that of **309** (91). The CD spectrum of **311,** however, was a mirror image of the spectrum of **309.** Since the absolute configuration of rhoeadine (**300a**) was known, these transformations were useful in assigning relative and absolute configurations in the isorhoeadine series.

Several syntheses of alkaloids of the rheadan and isorheadan classes that

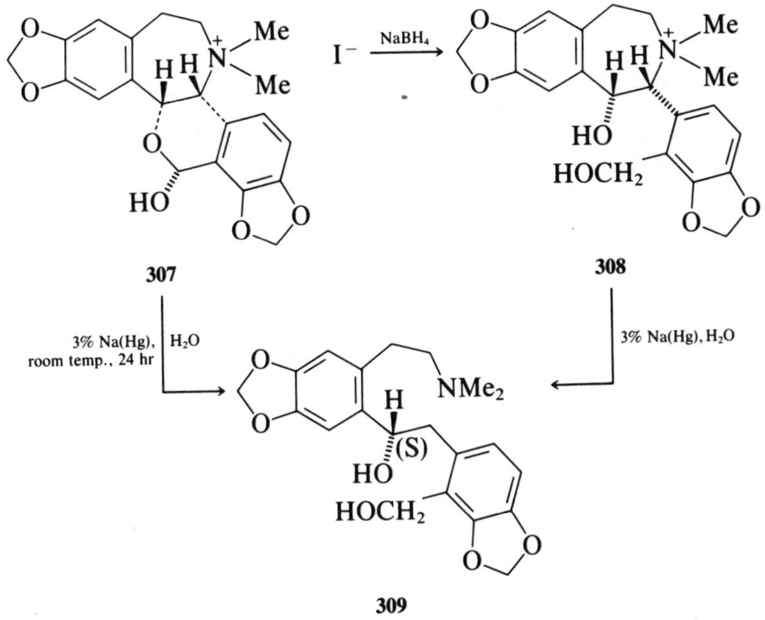

307 → 308 (NaBH$_4$)

307 → 309 (3% Na(Hg), H_2O, room temp., 24 hr)

308 → 309 (3% Na(Hg), H_2O)

Figure 50

involve representatives of the 1,3-dioxolo[4,5-*h*][3] ring system have been reported. The approaches devised by Brossi and co-workers are summarized in Figs. 51 and 52. Conversion of the opium alkaloid nornarceine (**312**) (92) and hydrastine (**313a**) (93) to representatives of the rheadan ring system (**317a** and **317b**) and of bicuculline (**313b**) (94) to the naturally occurring rhoeadine (**317c**) demonstrated the facile conversion of natural phthalide-isoquinolines to benzazepine alkaloids. Reduction of **314** gave *cis*-**315** as the major product. However, a small amount of the *trans* isomer was isolated as well (95), and both were converted to the *cis*-lactone **316** under acidic conditions. Another approach by Japanese workers in a somewhat related synthesis involved preparation of (±)-rhoeagenine diol (**301a**). Since **301a**

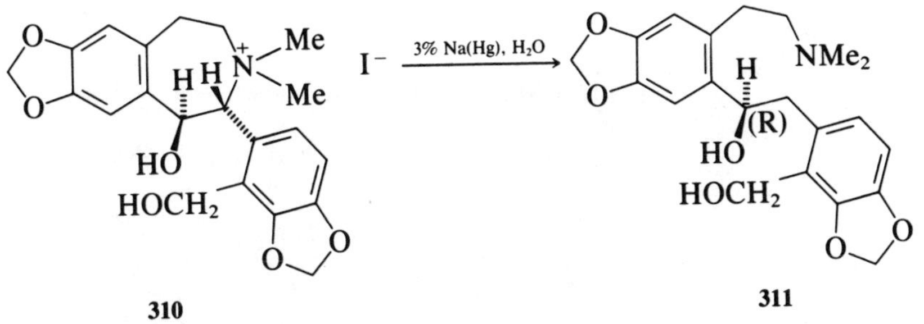

310 → **311** (3% Na(Hg), H_2O) (46)

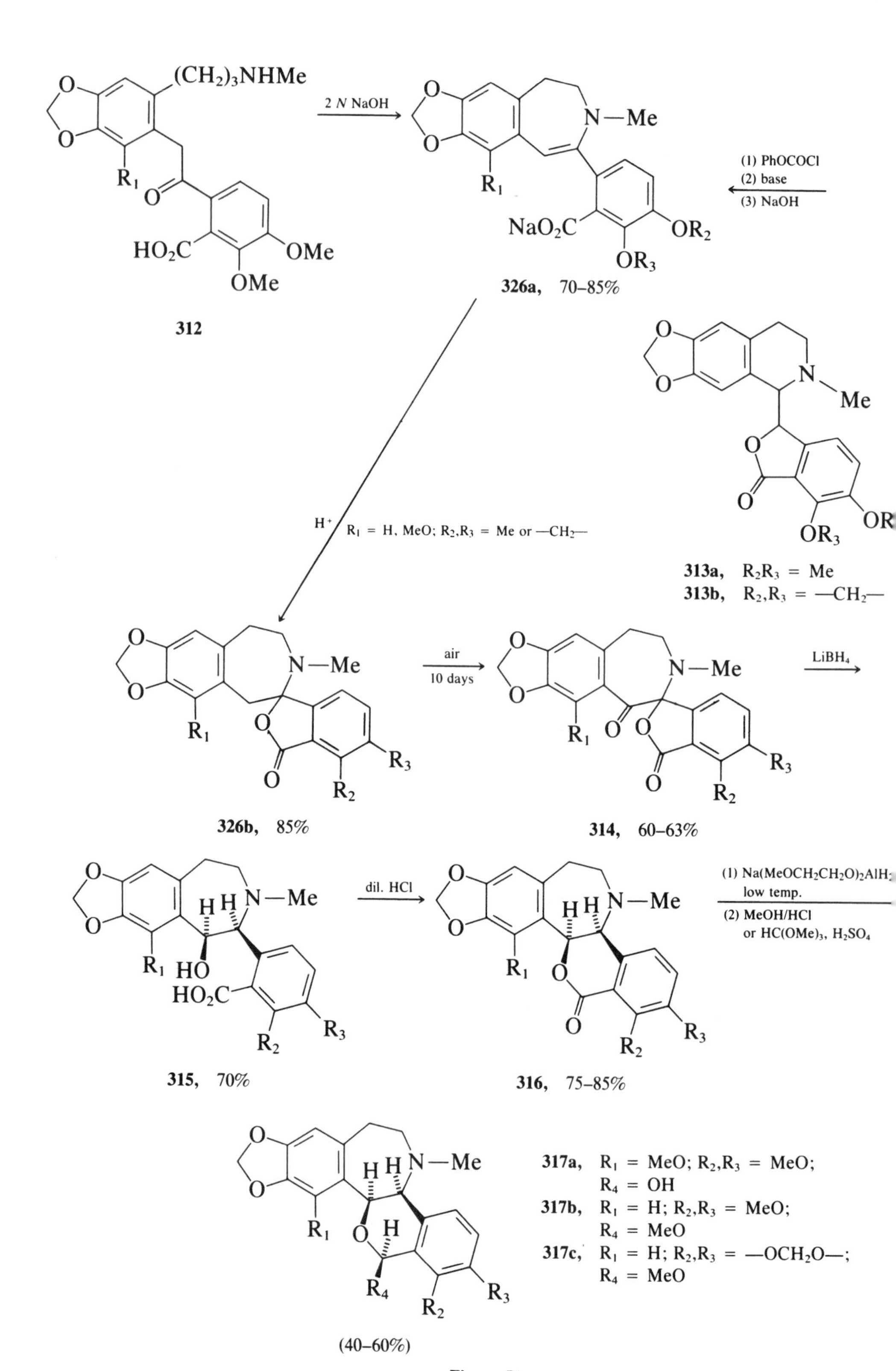

(CH2)3NHMe
R1
O
HO2C
OMe
OMe
312
2 N NaOH
N—Me
NaO2C
OR2
OR3
326a, 70–85%
(1) PhOCOCl
(2) base
(3) NaOH
Me
313a, R2R3 = Me
313b, R2,R3 = —CH2—
H+ R1 = H, MeO; R2,R3 = Me or —CH2—
326b, 85%
air
10 days
314, 60–63%
LiBH4
HO
315, 70%
dil. HCl
316, 75–85%
(1) Na(MeOCH2CH2O)2AlH2
low temp.
(2) MeOH/HCl
or HC(OMe)3, H2SO4
317a, R1 = MeO; R2,R3 = MeO;
R4 = OH
317b, R1 = H; R2,R3 = MeO;
R4 = MeO
317c, R1 = H; R2,R3 = —OCH2O—;
R4 = MeO
(40–60%)

Figure 51

had already been converted to rhoeadine, this represented an alternative total synthesis (96).

Another Japanese group (97) working on the synthesis of isopapaverine alkaloids gained entry to this ring system in a similar manner (Fig. 52). Conversion of the immonium salts **318a** and **318b** to the aziridinium salts **319a** and **319b** by diazomethane addition, followed by solvolytic ring opening, gave the intermediate 1,3-dioxolo[4,5-*h*][3]benzazepines **320a** and **320b** in low yields. The products could be isolated or converted *in situ* via the suggested intermediate **322** to alkaloids such as reframidine (**321a**).

The 10H-isoindolo[2,3-*c*]benzazepine **325a** (Schopf's base, R = H) and its methoxy analog **325b** could be prepared very conveniently from inter-

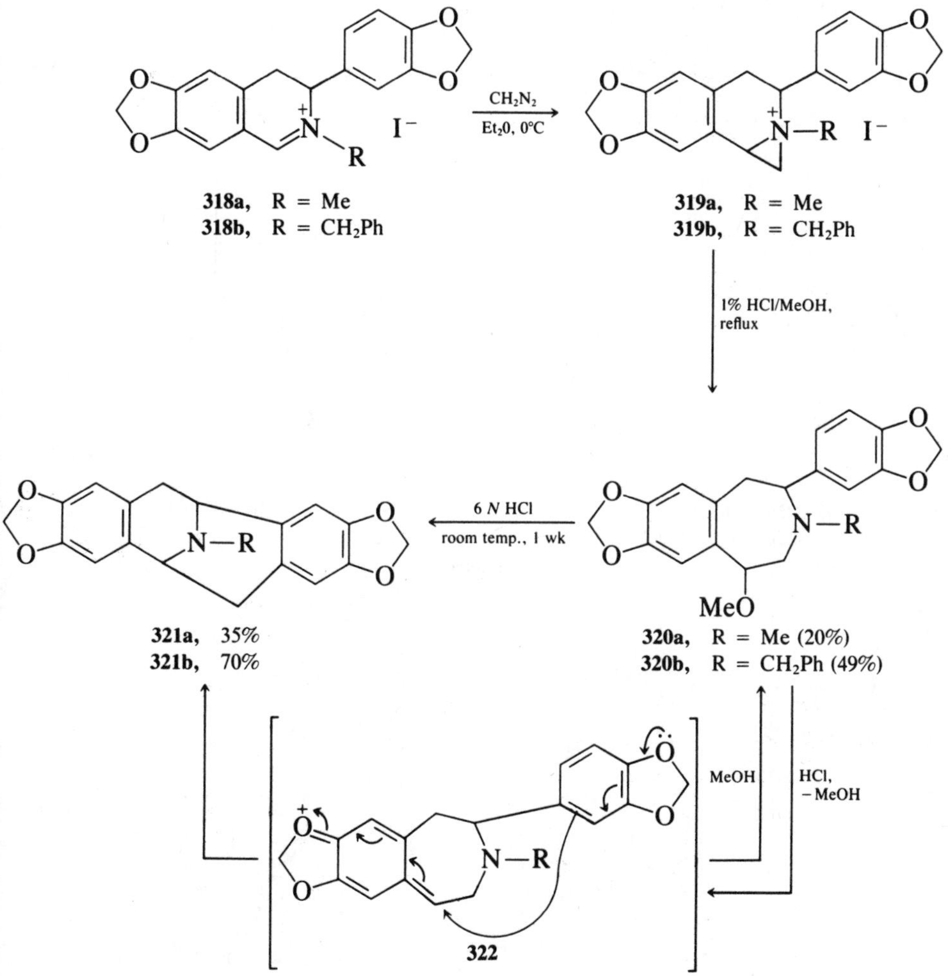

CH_2N_2, Et_2O, 0°C

318a, R = Me
318b, R = CH_2Ph

319a, R = Me
319b, R = CH_2Ph

1% HCl/MeOH, reflux

6 *N* HCl, room temp., 1 wk

321a, 35%
321b, 70%

320a, R = Me (20%)
320b, R = CH_2Ph (49%)

MeOH

HCl, −MeOH

322

Figure 52

mediates arising out of the phthalideisoquinoline–rheadan alkaloid series (98). Careful adjustment to pH 5 converted the sodium salt of **323** to the free acid, which was reduced to **324** with lithium borohydride (Fig. 53). More acidic conditions converted **323** to the spirolactones **326** (Fig. 51). Recently, diol **327** was converted in very low yield to the benzazepine **328** (99) (Eq. 47). Table 25 lists examples of this ring system that have been reported as part of the chemistry of the rheadan and isorheadan alkaloids.

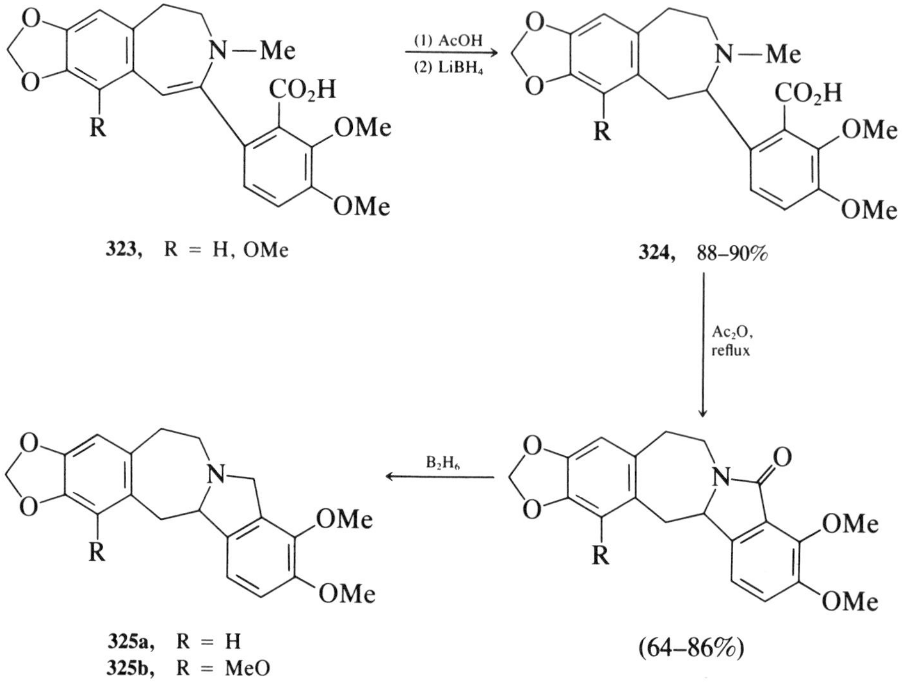

(1) AcOH, (2) $LiBH_4$

323, R = H, OMe

324, 88–90%

Ac_2O, reflux

B_2H_6

325a, R = H
325b, R = MeO

(64–86%)

Figure 53

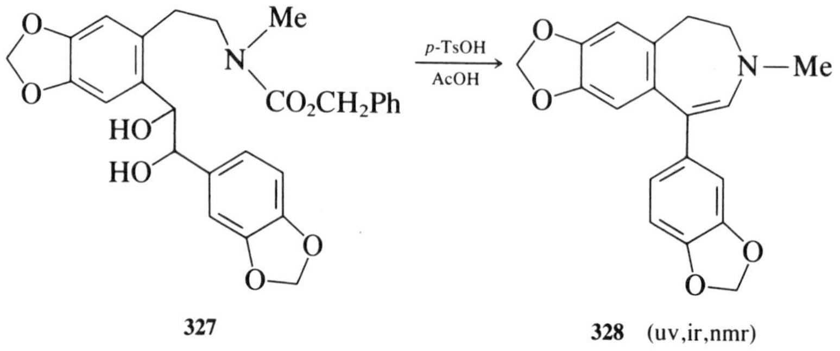

p-TsOH, AcOH

327

328 (uv,ir,nmr)

(47)

TABLE 25. 1,3-DIOXOLO[4,5-*h*][3]BENZAZEPINES REPORTED AS DEGRADATION PRODUCTS OR SYNTHETIC INTERMEDIATES IN RHEADAN AND ISORHEADAN ALKALOID CHEMISTRY

R_1	R_2	R_3	R_4	R_5	R_6	mp (°C)	Spectra	Refs.
MeO	CO_2Na	MeO	MeO	—[a]	H	190–195	uv,ir,nmr	92
MeO	C_2-Spirolactone	MeO	MeO	H	H	169–171		92
MeO	C_2-Spirolactone	MeO	MeO	—CO—		186–190	ir,nmr	92
MeO	CO_2H	MeO	MeO	H	OH	258–260	ir,nmr	92,95
H	CO_2Na	—OCH_2O—		—[a]	H	>300	uv	94
H	C_2-Spirolactone	—OCH_2O—		—CO—		195–197	uv,ir,nmr	94
H	CO_2H	MeO	MeO	OH	H	225	ir,nmr	95
H	CH_2OH	MeO	MeO	OH	H	161–162	nmr	95
H	CH_2OH	MeO	MeO	H	OH	164–166	uv,nmr	95
H	CO_2Na	MeO	MeO	—[a]	H	225–230	uv,ir	93
H	C_2-Spirolactone	MeO	MeO	—CO—		123–126	uv,ir,nmr	93
H	CH_2OH	—OCH_2O—		H	OH	131–133	ir,uv,nmr	88
							$[\alpha]_D$,CD	89
H	CH_2OH	—OCH_2O—		OH	H	153–155	CD	89
H	CH_2OAc	—OCH_2O—		H	OAc	107–109	ir,uv,nmr	88
							$[\alpha]_D$,CD	89
H	CO_2H	MeO	MeO	H	H	230–231	uv,ir,nmr	98
MeO	CO_2H	MeO	MeO	H	H	160–165	uv,ir,nmr	98
MeO	CO_2H	MeO	MeO	OH	H	173–176	ir,nmr	95
H	Me	—OCH_2O—		H	H	153–154	uv,nmr,ms	89

[a]Double bond between carbons 1 and 2.

30. [1,3]Oxazino[3,2-*a*]pyrrolo[2,3-*f*]azepines

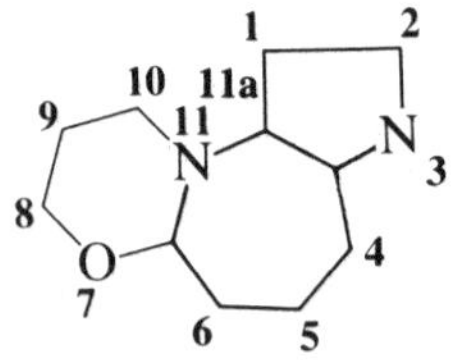

Within the scope of a program investigating the chemistry of pyrroloazepine **329** (Fig. 54), the oxazinopyrroloazepine **330** was formed in 69% yield from **329** and dimethylmalonyl chloride. Catalytic reduction of **330** gave rise to **331** in 81% yield. Together, **330** and **331** comprise the only reported members of the [1,3]oxazino[3,2-*a*]pyrrolo[2,3-*f*]azepine ring system. Thermolysis of **330** in methanol afforded the diester **332** (100).

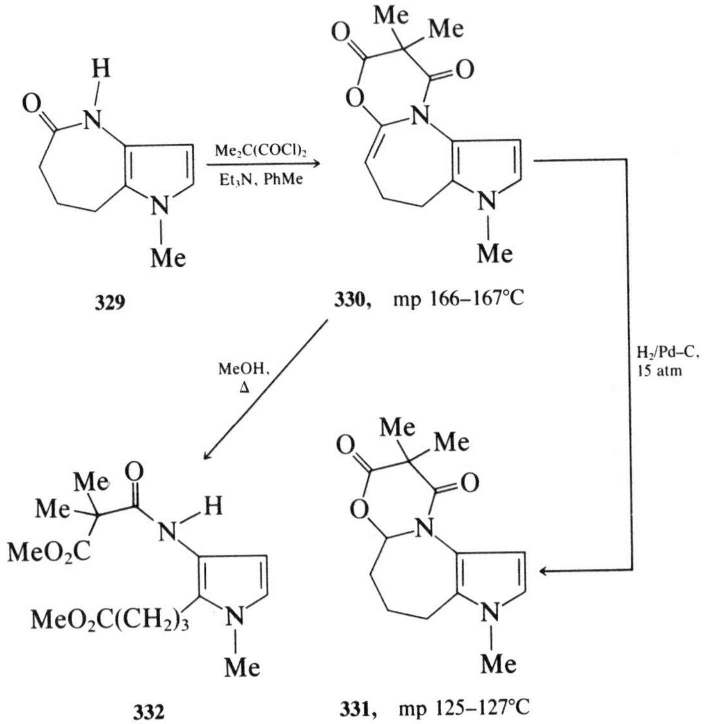

329 → ($Me_2C(COCl)_2$, Et_3N, PhMe) → **330**, mp 166–167°C

330 → (H_2/Pd–C, 15 atm) → **331**, mp 125–127°C

330 → (MeOH, Δ) → **332**

Figure 54

31. Azepino[2,1,7-*cd*]indolizines

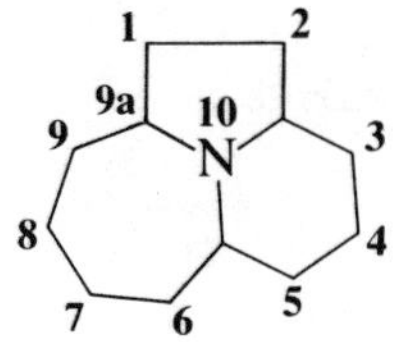

Tricyclic systems containing a completely conjugated perimeter of sp^2-hybridized carbon atoms held planar by a central nitrogen are called cyclazines (101,102). As originally defined by Boekelheide (101), azepino[2,1,7-*cd*]-indolizines would be classified as cycl[4.3.2]azines. Recently, Ceder and Beijer proposed a slight modification of the cyclazine nomenclature (103). The ring system would first be oriented and numbered according to IUPAC rules, with the bracket numerals then being assigned in the same order as the atoms from which the numerals were derived. This modification would name the title system cycl[2.3.4]azine. Because the literature to date has consistently utilized the Boekelheide nomenclature, we shall retain this system in the present review.

Phenalene (**333**) and the isoelectronic compound 2H-benz[*cd*]azulene (**334**) both possess a nonbonding molecular orbital, and therefore give rise to 12π-electron cations, 13π-electron radicals, and 14π-electron anions (104). Cycl[3.3.3]azine (**335**), which corresponds to the anion of **333**, was prepared and found to be paratropic (105,106). Cycl[4.3.2]azine **336** (see Fig. 55) was prepared in order to allow a comparison of tricyclic arenes and hetarenes with a 12π-electron periphery (107,108).

Vilsmeier–Haack formylation of azaazulenone **337** afforded aldehyde **338** in 70% yield, and Wittig olefination converted **338** to **339** in 50% yield (Fig. 55). Upon being heated for 2 hr in a mixture of pyridine and piperidine, **339** afforded the violet-black cycl[4.3.2]azine **336** in 50% yield. The pmr spectrum of **336** indicated that this molecule, like **335**, was highly paratropic. As expected for an antiaromatic structure, **336** behaved as though it consisted of a reactive butadiene moiety bridging an independent, stable indolizine ring system. For example, **336** reacted with *N*-phenylmaleamide to afford the cycloaddition product **340**. Cyclazine **336** also underwent intermolecular

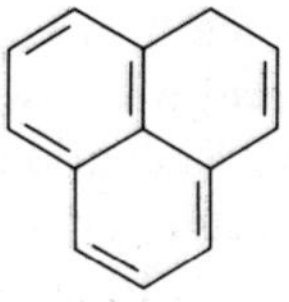

333 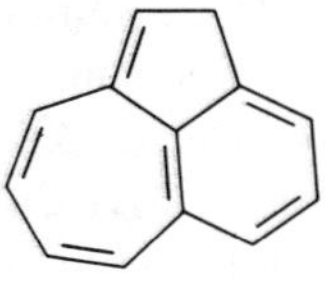 334 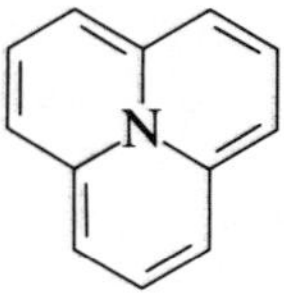 335

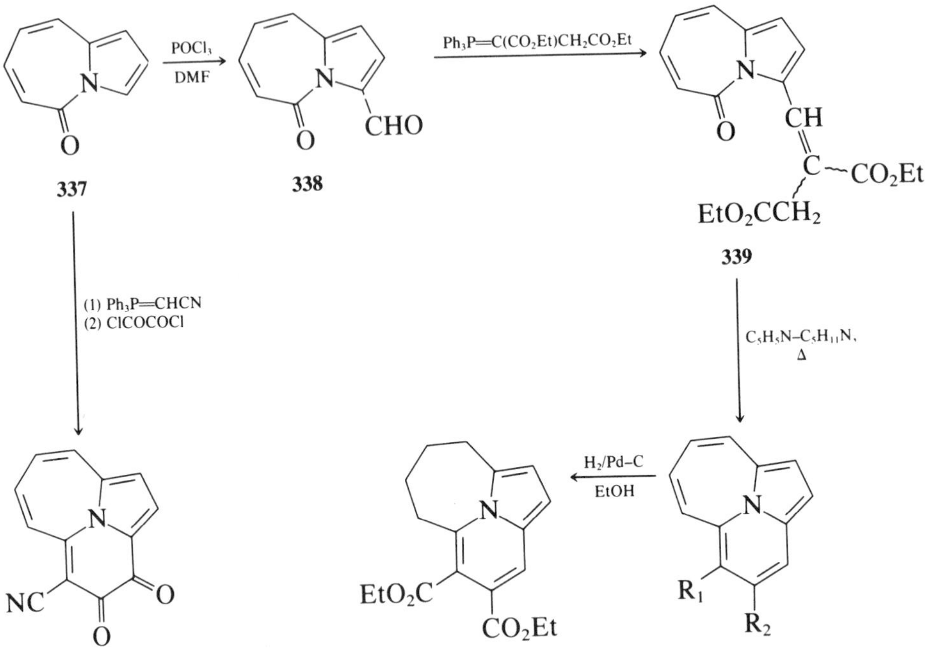

	R_1	R_2
336	CO_2Et	CO_2Et
343	CO_2Et	CO_2H
344	CO_2Et	CO_2Me

Figure 55

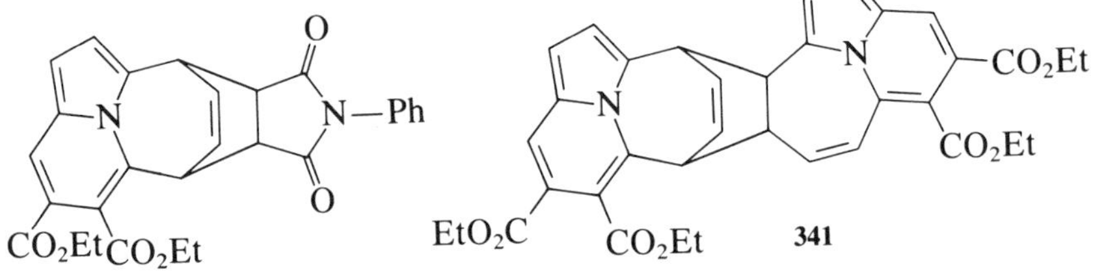

dimerization upon heating in toluene to give a 62% yield of a compound to which was assigned structure **341** (a definitive choice between **341** and its regioisomer could not be made). The same compound was also formed as a minor by-product in the conversion of **339** to **336.** Reduction of **336** over Pd/C-produced the tetrahydro derivative **342.** The ester groups of **336** were chemically and spectrally nonequivalent. For example, basic hydrolysis afforded the half-acid **343,** and transesterification in methanol afforded the mixed ester **344.**

TABLE 26. AZEPINO[2,1,7-*cd*]INDOLIZINES[a]

Structure	R_1	R_2	mp (°C)	bp (°C/torr)	Spectra
336	CO_2Et	CO_2Et	87		ir,nmr,ms,uv
342				125/0.2	ir,nmr,ms,uv
343	CO_2Et	CO_2H	>250		ir,nmr,ms
344	CO_2Et	CO_2Me	108–109		ir,nmr,ms
345			>300		ir,ms

[a]Fig. 55, Ref. 108.

Azaazulenone **337** was likewise a cycl[4.3.2]azine precursor. Thus, Wittig olefination of **337** with cyanomethylenetriphenylphosphorane followed by treatment with oxalyl chloride produced **345** in 10% yield (Fig. 55). The azepino[2,1,7-*cd*]indolizines discussed above are listed in Table 26.

32. Pyrido[3′,2′:4,5]pyrrolo[2,3-*c*]azepines

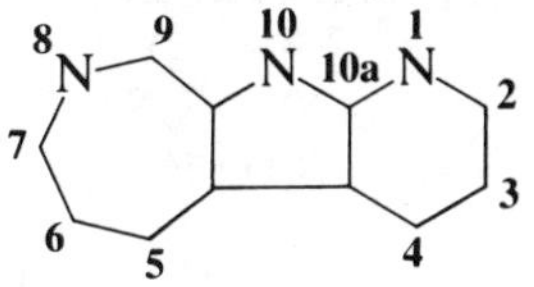

33. Pyrido[3′,4′:4,5]pyrrolo[2,3-*c*]azepines

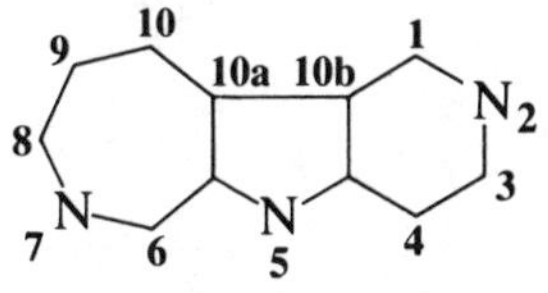

When azepinone **346** was fused with pyridyl hydrazine **347** and three moles of zinc chloride (Fig. 56), the azaindole derivative **348** was obtained in 67% yield (109). Presumably, **348** arose from a Fischer indole reaction via the intermediacy of the hydrazone **349.** Hydrazone **349** could be isolated from the reaction of **347** with azepinone **350,** and was shown to give a quantitative yield of the 9-azepinone **348** on Fischer cyclization (110). The lactam **348** was readily hydrolyzed to the corresponding amino acid **351** in 87% yield.

Compound **352** was obtained in 80% yield by Fischer cyclization of the isomeric hydrazone **353.** The latter is formed in 84% yield from pyridyl hydrazine **354** and azepinone **350.** Upon hydrolysis, lactam **352** afforded the amino acid **355** in quantitative yield (110).

346, R = MeO
350, R = Piperidino

347, X = CH, Z = N
354, X = N, Z = CH

349, X = CH, Z = N
353, X = N, Z = CH

3 eq. $ZnCl_2$, Δ

$(CH_2)_3NH_2$ CO_2H

351, X = CH, Z = N
355, X = N, X = CH

	X	Z	mp (°C)	Spectra
348	CH	N	300 (dec.)	ir,uv
352	N	CH	300 (dec.)	

Figure 56

34. Azepino[1,2-*a*]indoles

In 1967 Kaneko reported the formation of azepinoindole **358a** as a secondary product from the photolysis (Fig. 57) of 1,2,3,4-tetrahydroacridine *N*-oxide (**356a**) (111). He further observed that **358a** was formed when the primary photoadduct **357a** was chromatographed on silica gel (Fig. 58). The uv spectrum of **358a** was almost identical to that of *N*-acetyl-2-hydroxyindoline. Subsequent investigations revealed a dramatic solvent effect on the nature of the photolytic products (112). Specifically, when the acridine *N*-oxides **356a–e** were photolyzed in the aprotic solvent benzene, the major photoproducts were the benz[*d*]-1,3-oxazepines **357a–e,** respectively. The primary photoproducts were unstable to moisture and underwent solvolysis during chromatographic workup to afford azepinoindoles **358a–e** in 40–60% overall yield. Hydroxyindoles **358a–c** could be dehydrated thermally or by

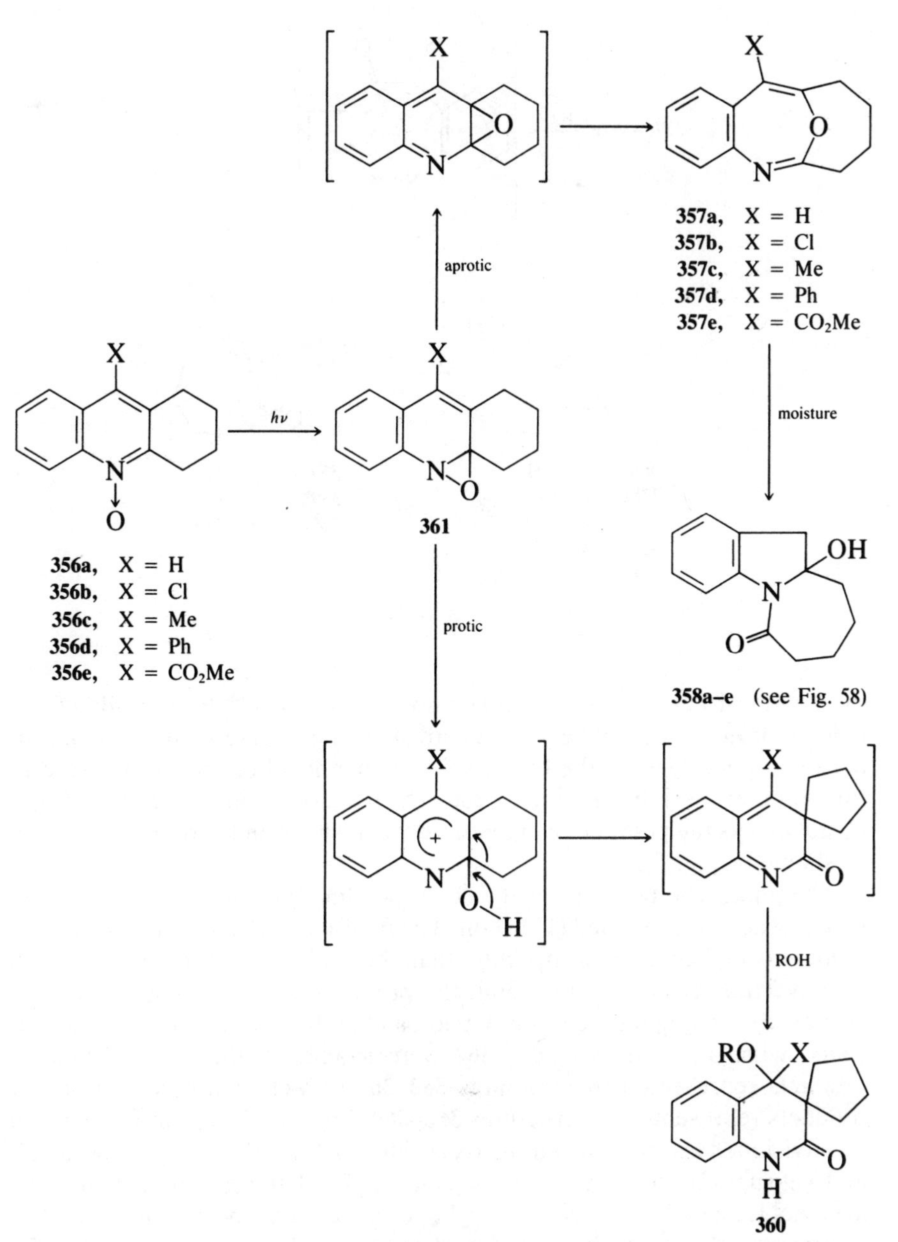
X
O
N
357a, X = H
357b, X = Cl
357c, X = Me
357d, X = Ph
357e, X = CO₂Me
aprotic
moisture
hν
N
O
361
OH
356a, X = H
356b, X = Cl
356c, X = Me
356d, X = Ph
356e, X = CO₂Me
protic
358a–e (see Fig. 58)
ROH
RO
H
360

Figure 57

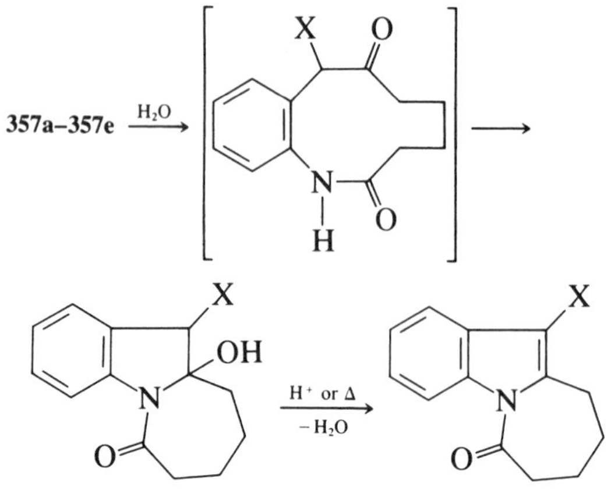

357a–357e $\xrightarrow{H_2O}$

X O

N H O

X OH N O

$\xrightarrow[-H_2O]{H^+ \text{ or } \Delta}$

X N O

358a, X = H
358b, X = Cl
358c, X = Me
358d, X = Ph
358e, X = CO_2Me

359a, X = H
359b, X = Cl
359c, X = Me

Figure 58

the action of acid in an appropriate solvent to give 85–95% yields of the indoles **359a–c,** respectively. In contrast to the course of the reaction in benzene, photolysis in the protic solvents methanol or ethanol resulted in formation of the spirocabostyrils **360.** The oxaziridine intermediate **361** was postulated as the point of divergence between protic and aprotic photolyses, as shown in Fig. 57.

The photolytic behavior of the corresponding fully aromatic acridine *N*-oxides was also examined (113–116). The products isolated were much more numerous and complex in structure than those obtained from the analogous tetrahydro derivatives. In general, the products could be divided into two groups, depending on the primary photoadduct from which they originated. Thus, when acridine *N*-oxides **362** were irradiated, there were obtained products represented by structures **363–365** as well as a second group of products represented by structures **366–369** (Fig. 59). The former group can be envisioned as arising from the oxaziridine **370** via thermally allowed 1,5- and 1,9-suprafacial oxygen rearrangements. The latter group, including the azepino[1,2-*a*]indoles **369,** is thought to come from the intermediate 1,2-oxazepines **371,** which have, in fact, been isolated and characterized (115).

The azepinoindoles **369** were characterized by their spectral properties and supported by their microanalytical data. More conclusive proof of structure was supplied by the reduction of **369a, 369c,** and **369g** to their tetrahydro derivatives **359a, 359c,** and **372** (Eq. 48). Compounds **359a** and **359c** were identical in all respects to those prepared earlier (Fig. 57).

Two features of the mechanism depicted in Fig. 59 merit comment. Com-

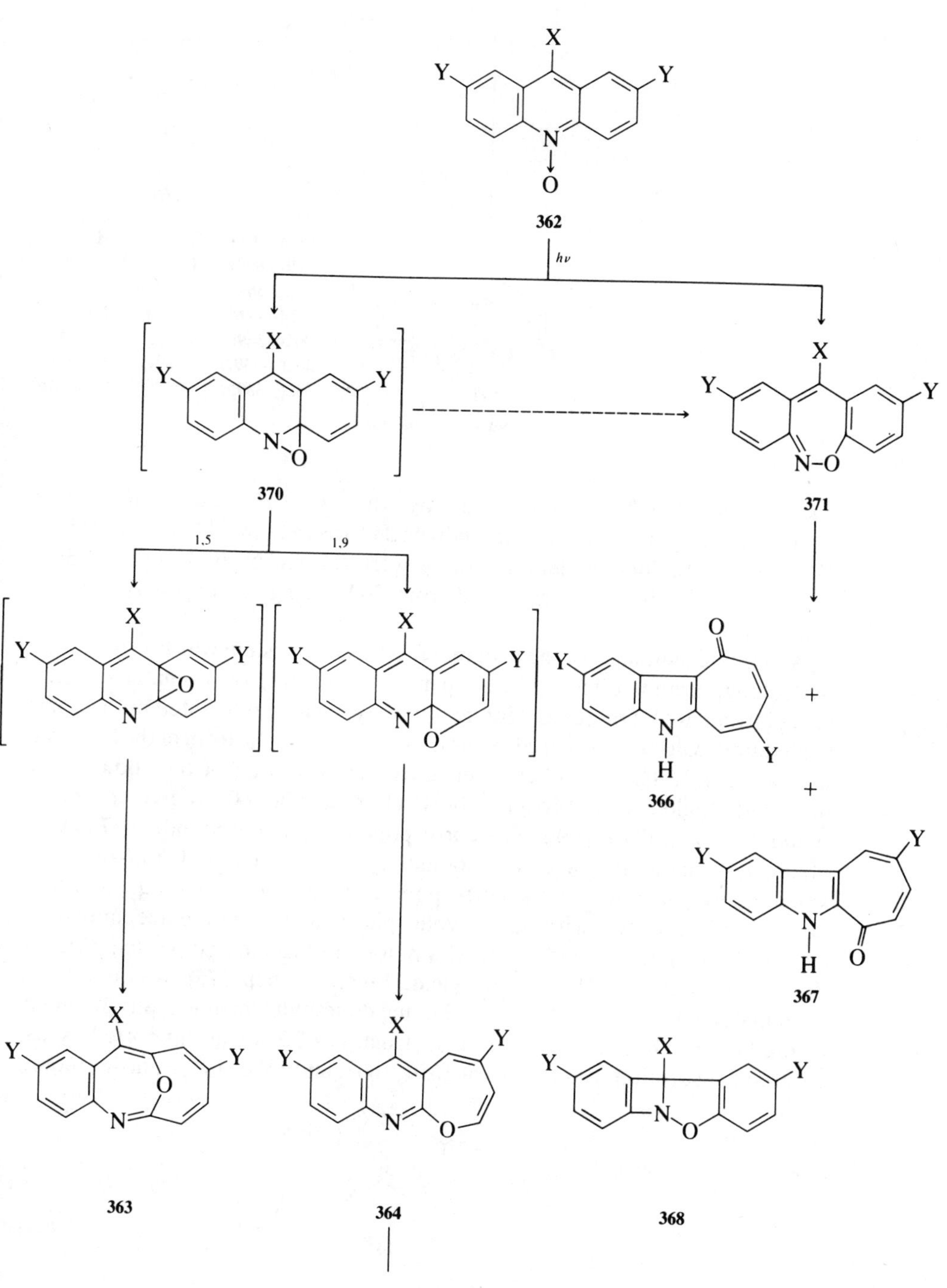
X
Y
Y
N
O
362
hν
X
Y
Y
N
O
370
X
Y
Y
N–O
371
1,5
1,9
X
Y
Y
O
N
X
Y
Y
N
O
O
Y
N
H
Y
366
+
+
Y
Y
N
H
O
367
X
Y
Y
O
N
363
X
Y
Y
N
O
364
X
Y
Y
N
O
368

Figure 59

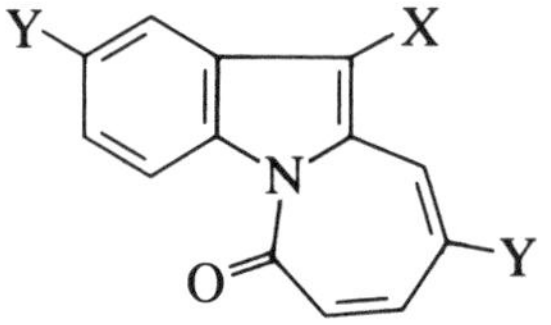

hν

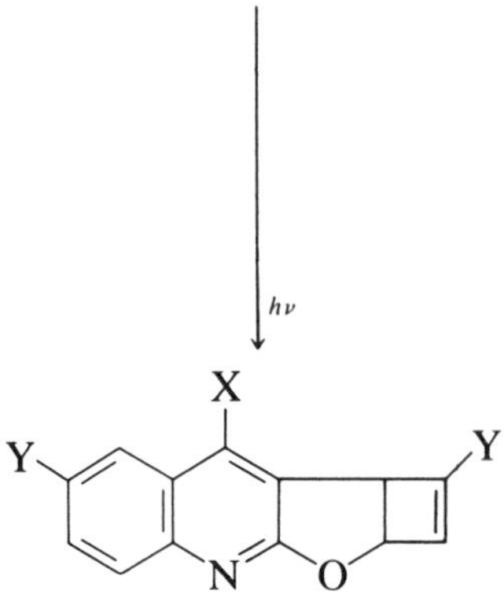

365

369

362a–369a,	X = Y = H
362b–369b,	X = H, Y = Me
362c–369c,	X = Me, Y = H
362d–369d,	X = Y = Me
362e–369e,	X = Et, Y = Me
362f –369f,	X = CN, Y = H
362g–369g,	X = CN, Y = Me

Figure 59 (*continued*)

pounds **366** and **367** can be formed only when X = H. Additionally, while it has been suggested that azepinoindoles **369** arise via the dibenzoxazepines **371,** the azepinoindoles **358** in Fig. 58 were formed by hydrolysis of **357.** The possibility that **369** originated from **363** during workup may also be considered.

A directed synthesis of the azepino[1,2-*a*]indole system was first achieved by Collington and Jones (117). As part of their study of aromatic systems isoelectronic with the quinolizinium ion, they became interested in "azonia-azulenium" salts and prepared the first such species in the form of the hydroxy derivative **373.** Their synthesis began by the alkylation of the sodium salt of 3-methylindole with 4-tosyloxybutyl chloride (Fig. 60) to give chlorobutylindole **374a** in 76% yield. The corresponding cyanobutyl indole **374b** was obtained in moderate yield with sodium cyanide in dimethyl sulfoxide. Alkaline hydrolysis afforded the corresponding acid **374c** in nearly quantitative yield, and cyclization with hot polyphosphoric acid gave the azepinoindole **375a** in 50% yield. Phenyltrimethyl ammonium bromide converted **375a** to the dibromoketone **375b** in 80% yield. Finally, when **375b** was dehydrobrominated with lithium chloride in boiling dimethylformamide, an 83% yield of the ketone **376** was obtained. The structure of **376** was supported by its carbonyl absorption frequencies at 1656, 1603, and 1570 cm^{-1}. Furthermore,

369a, 369c, and **369g** $\xrightarrow[\text{EtOH, 25°C}]{H_2/Pd\text{–}C}$ (48)

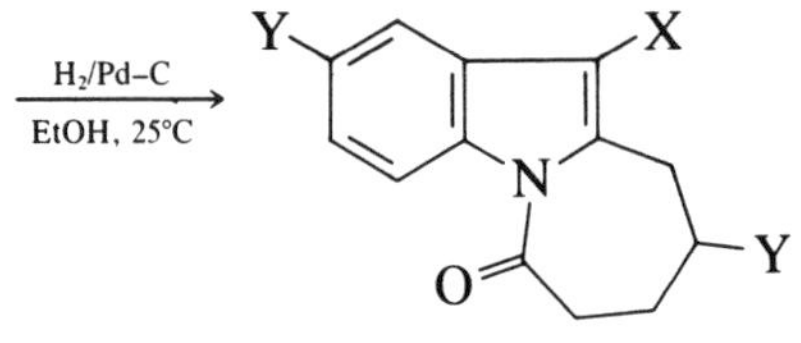

359a, X = Y = H
359c, X = Me, Y = H
372, X = CN, Y = Me

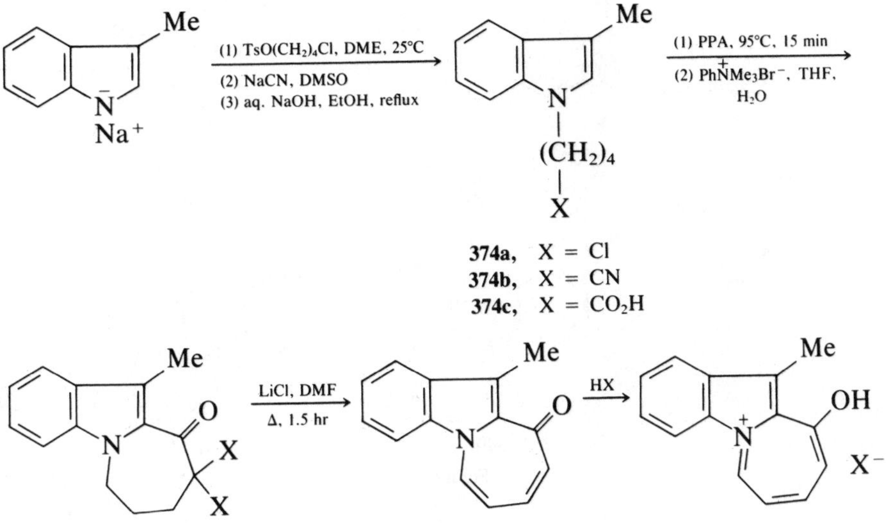

(1) $TsO(CH_2)_4Cl$, DME, 25°C
(2) NaCN, DMSO
(3) aq. NaOH, EtOH, reflux

(1) PPA, 95°C, 15 min
(2) $Ph\overset{+}{N}Me_3Br^-$, THF, H_2O

374a, X = Cl
374b, X = CN
374c, X = CO_2H

LiCl, DMF
Δ, 1.5 hr

HX

375a, X = H
375b, X = Br

376

373a, X = Br
373b, X = CF_3CO_2

Figure 60

microanalysis indicated the loss of two moles of hydrogen bromide. Solutions of **376** in trifluoroacetic or other strong acids were deep blue, a color often characteristic of the azulenes. Apparently protonation gave the "azonia-azulenium" ion **373,** and, in fact, satisfactory analyses were obtained for the bromide **373** (X = Br), a black solid. The salts were very unstable to protic media and, in alcoholic solution, were immediately reconverted to the ketone **376** (118).

With ketone **376** in hand, its chemistry was elaborated (119). Attempted reduction of **376** with either sodium borohydride or lithium aluminum hydride resulted in the formation of unstable products, which continued to show carbonyl absorption in the infrared. In contrast, reduction of the iron complex **377** (obtainable in 60% yield by the action of iron dodecacarbonyl on **376**) with either sodium borohydride in ethanol or lithium aluminum hydride in tetrahydrofuran (Eq 49) yielded mixtures composed of **376, 377,** and the reduction products **375a,** and **378–381.** The tetrahydroazepinoindoles **378**

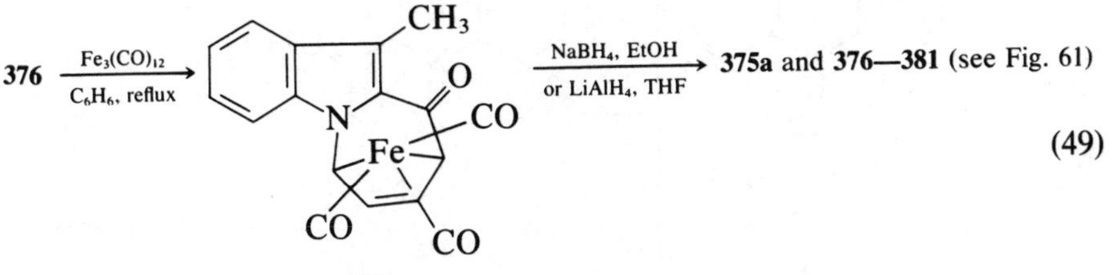

376 $\xrightarrow[C_6H_6,\ \text{reflux}]{Fe_3(CO)_{12}}$ **377** $\xrightarrow[\text{or } LiAlH_4,\ \text{THF}]{NaBH_4,\ \text{EtOH}}$ **375a** and **376—381** (see Fig. 61) (49)

377

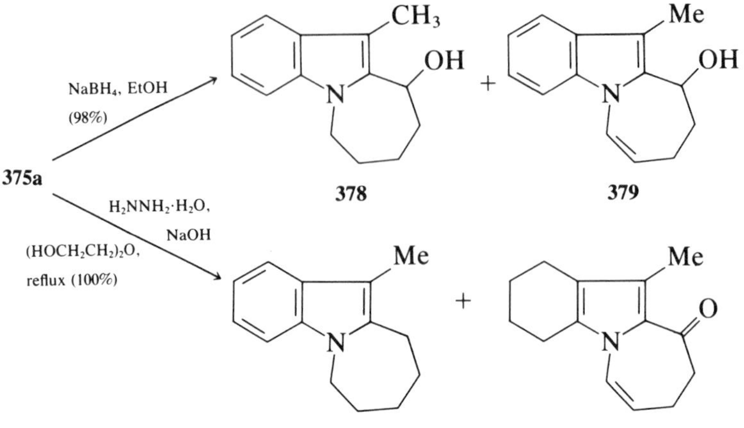

CH_3 OH N + Me OH N

$NaBH_4$, EtOH (98%)

375a

378 379

$H_2NNH_2 \cdot H_2O$, NaOH

$(HOCH_2CH_2)_2O$, reflux (100%)

Me N + Me O N

380 381

Figure 61

and **380** were prepared independently from **375a** by borohydride reduction and Huang–Minlon reduction, respectively (Fig. 61). Finally, it was observed that reaction of **376** with hydrazine hydrate in ethanol (Fig. 62) resulted in a single product, the pyrazoline derivative **382.** Surprisingly, other nitrogen nucleophiles did not react with **376.**

An alternative approach to azepino[1,2-*a*]indoles, first reported by Krbe-

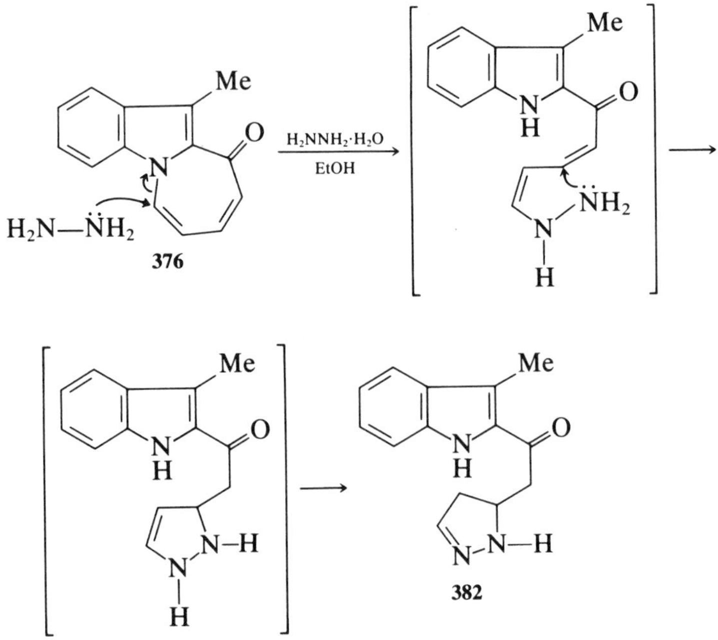

Me O N $H_2N—\ddot{N}H_2$ 376

$H_2NNH_2 \cdot H_2O$ EtOH

[Me O N H $\ddot{N}H_2$ N H] →

[Me O N H N–H N H] → Me O N H N-N–H 382

Figure 62

	R_1	R_2	R_3	R_4
383a	H	H	H	H
383b	ME	H	H	H
383c	H	H	MeO	H
383d	H	Me	H	H
383e	H	Me	Me	Me
383f	H	MeO	H	H
383g	H	MeO	MeO	MeO

	R_1	R_2	R_3	R_4
385a	H	H	H	H
385b	Me	H	H	H
385c	H	H	MeO	H
385d	H	Me	H	H
385e	H	H	H	Me
385f	H	Me	Me	Me
385g	H	H	H	MeO

chek (120), involves thermolysis of *o*-azidodiphenylmethanes (Eq. 50). The decomposition of **383a,** apparently via a nitrene intermediate, was reported to yield 11H-azepino[1,2-*a*]indole (**384**) in 66% yield. Hydrogenation of **384** gave the tetrahydro derivative **386a** in 80% yield. Subsequently, the structure assigned to **384** was challenged on the grounds that azepines generally tend to exist as 3H-tautomers whenever possible (119). When the thermolysis of **383a** was repeated (119), a product was formed with melting point, uv spectra, and elemental composition identical to those reported by Krbechek (120). However, the pmr spectrum revealed some differences, particularly a two-proton doublet ($J = 6$ Hz) at δ 3.55. This two-proton signal, which had originally been assigned to position 11 of **384,** appears more compatible with structure **385a.** Support for this reassignment was supplied by the subsequent thermoysis of the methyl derivative **383b** to the methylazepinoindole **385b.** The crucial methylene signal was again observed in **385b** as a doublet ($J = 6$ Hz) at δ 3.5. Furthermore, reduction of **385b** gave the tetrahydro derivative **386b,** which was identical to the material prepared earlier by the Huang–Minlon reduction of **375a.**

A more detailed investigation of the formation of azepinoindoles via nitrene insertion has been published (122). Whereas the major products from the thermolysis of the azides **383c–e** in trichlorobenzene at 180–200°C were azepino[1,2-*a*]indoles, **383f** and **383g** afforded acridines and acridanes as the major products. Decomposition of **383c** gave a 40% yield of the azepinoindole **385c** as the sole product. Thermolysis of **383d** produced a mixture from which **385d** was isolated in 27.5% yield, and although **385e** was never actually isolated in pure form, it was tentatively identified by its spectra. The *o*-(trimethylbenzyl)phenylazide **383e** decomposed to a mixture of the 6H-, 8H-, and 10H-azepinoindoles **387, 388,** and **385f,** respectively. These isomers were characterized most rigorously on the basis of their pmr spectra.

In contrast to the reasonably straightforward decomposition of **383c–e,**

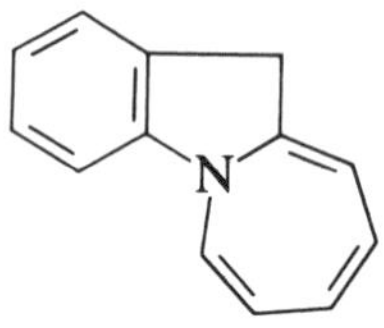

384

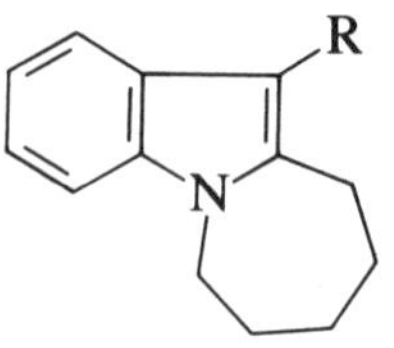

386a, R = H
386b, R = Me

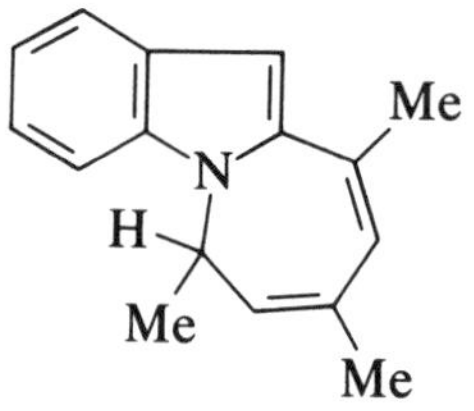

387

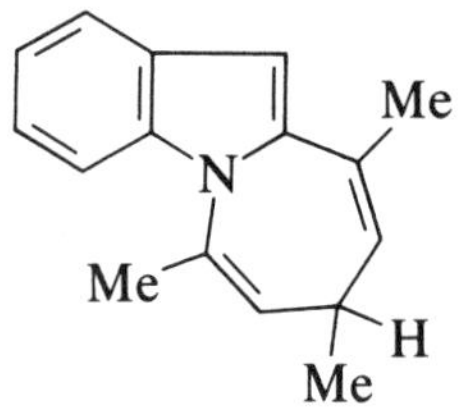

388

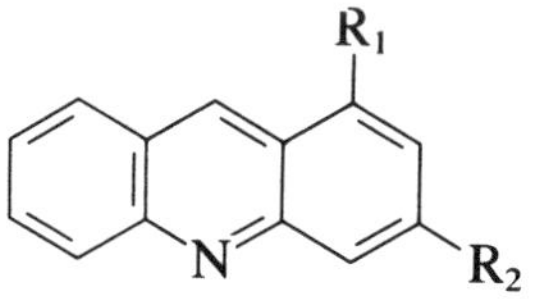

389a, $R_1 = R_2 = H$
389b, $R_1 = MeO, R_2 = H$
389c, $R_1 = R_2 = MeO$

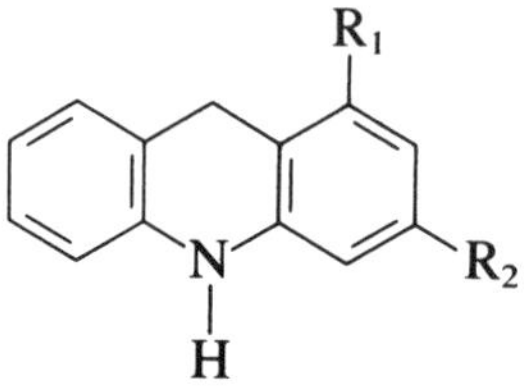

390a, $R_1 = R_2 = H$
390b, $R_1 = MeO, R_2 = H$
390c, $R_1 = R_2 = MeO$

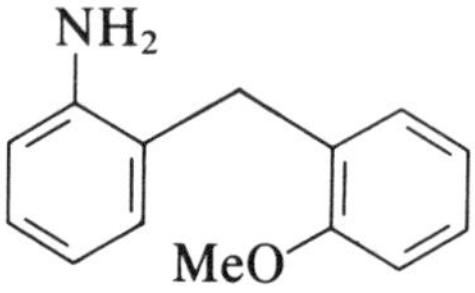

391

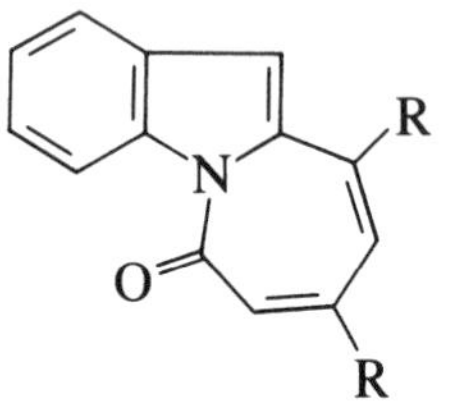

392a, R = H
392b, R = MeO

thermolysis of **383f** resulted in a mixture of 11 products (121,122). The products were predominantly the acridines **389a** and **389b** and the acridans **390a** and **390b.** Minor amounts of the aniline **391** and the azepinoindoles **385g** and **392a** were also obtained. The trimethoxyazide **383g** (Eq. 50) gave a less complex mixture, although the major products were acridine **389c** and acridane **390c.** A third compound isolated from **383g** was the azepinoindole **392b.**

On the basis of the results obtained from the azides **383a** and **383c–g,** a mechanism has been proposed to explain the observed products. The scheme, as originally suggested by Krbechek (120), involves the azanorcaradiene **393** as the principal intermediate, formed initially by nitrene insertion. The details are summarized in Fig. 63, which for simplicity contains only a single *o*-methoxy substituent.

The azepinoindole **385c,** the decomposition product of **383c,** was sensitive to acid (Fig. 64) and in the presence of HCl was rapidly hydrolyzed in 90%

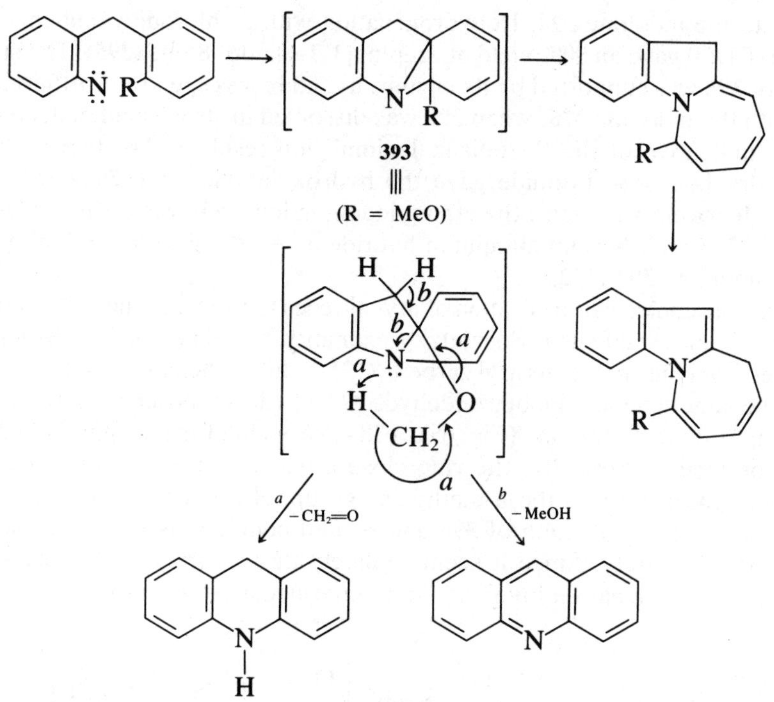

(R = MeO)

$-CH_2{=}O$

$-MeOH$

Figure 63

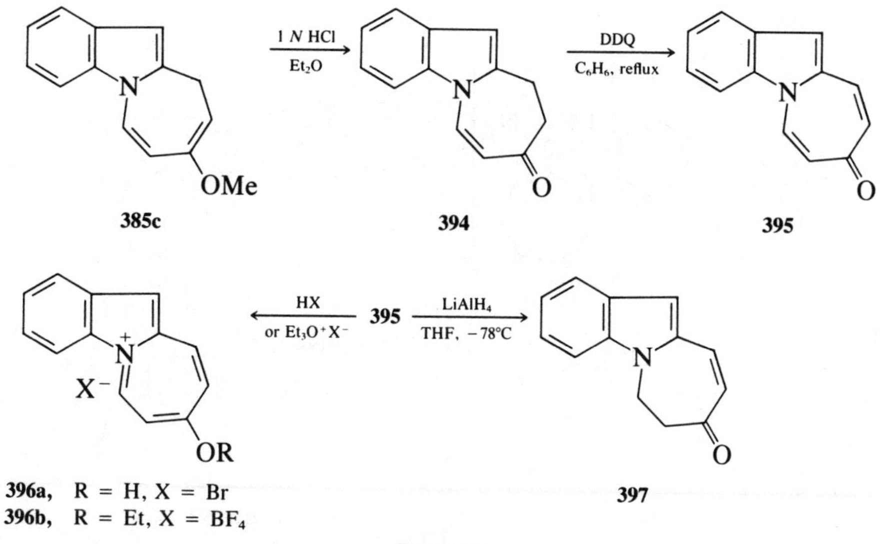

1 *N* HCl

Et_2O

DDQ

C_6H_6, reflux

OMe

385c

394

395

HX

or $Et_3O^+X^-$

395

$LiAlH_4$

THF, $-78°C$

OR

396a, R = H, X = Br
396b, R = Et, X = BF_4

397

Figure 64

yield to the indolone **394.** Dehydrogenation with dichlorodicyanobenzoquinone (DDQ) gave an 80% yield of azepino[1,2-*a*]indol-8-one (**395**). The structure of **395** was supported by its ir, nmr, uv, and mass spectra. Furthermore, as with the indolone **376,** when **395** was dissolved in strong acid, a deep blue color indicative of the "azoniaazulenium" ion resulted. Treatment of **395** with dry hydrogen bromide gave the hydroxyindolium ion **396a,** whereas with Meerwein's reagent, the ethoxyindolium ion **396b** was formed. Reduction of **395** with lithium aluminum hydride at −78°C gave the 6,7-dihydroazepinoindole **397** (122).

The intramolecular cyclization of aryl nitrenes to form azepino[1,2-*a*]indoles having been established, Suschitzky examined the generation of *N*-heterocycles from the analogous aryl carbene (123). Thus, when the tosylhydrazone of *o*-hexamethyleneiminobenzaldehyde **398** was decomposed in diglyme containing sodium methoxide (Fig. 65), a 38–45% yield of the azepinoindole **399** was obtained. Originally, the ring closure was ascribed to intramolecular carbene insertion into the α-methylene group. However, in the absence of a nitro group, cyclization of **398** and related compounds was repressed or eliminated entirely. Since it seems unlikely that a carbene is involved, an alternative explanation involving ionic intermediates was proposed.

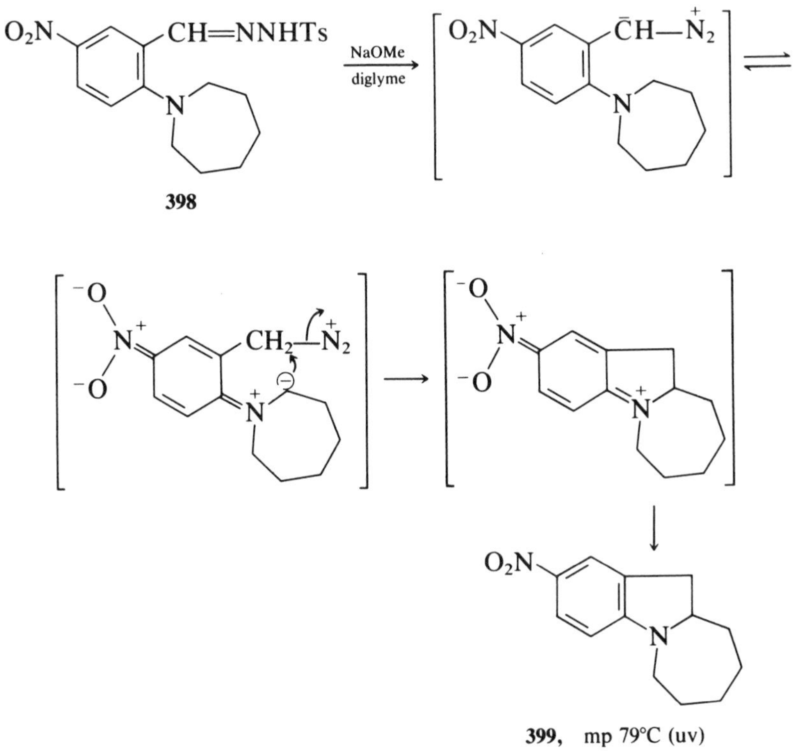

O_2N CH=NNHTs N NaOMe diglyme O_2N $\bar{C}H-\overset{+}{N}_2$ N

398

^-O $\overset{+}{N}$ ^-O $CH_2-\overset{+}{N}_2$ $\overset{+}{N}$ → ^-O $\overset{+}{N}$ ^-O $\overset{+}{N}$

O_2N N

399, mp 79°C (uv)

Figure 65

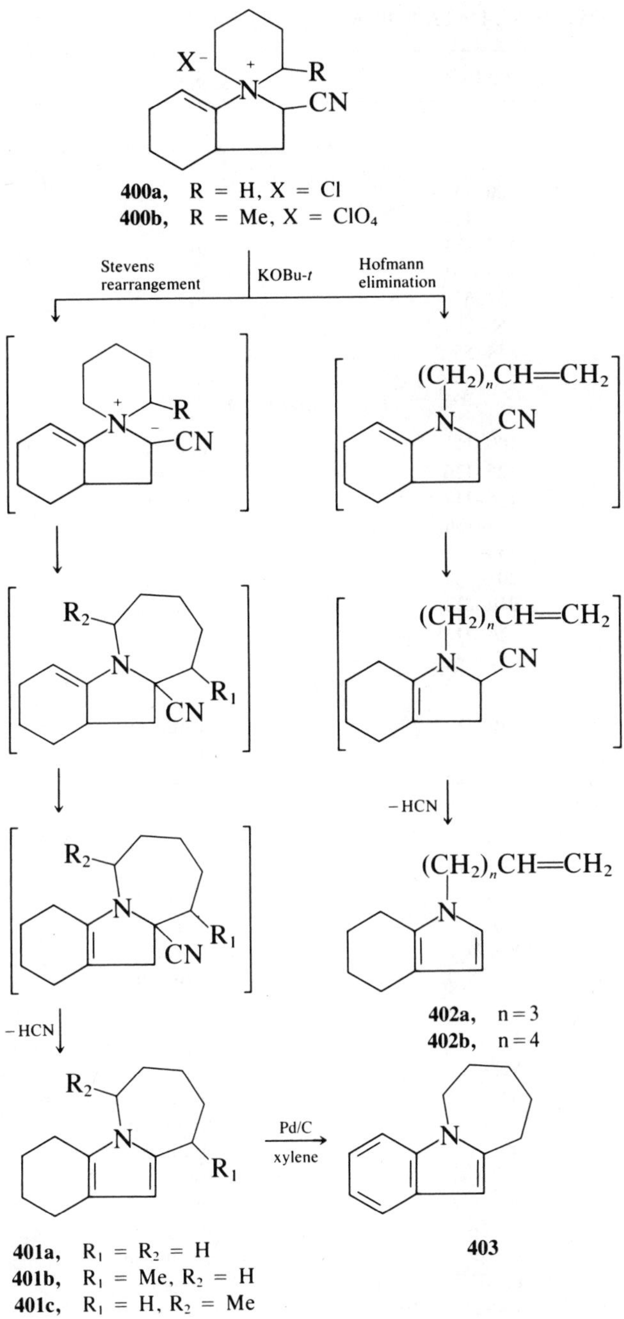

X^- R CN N

400a, R = H, X = Cl
400b, R = Me, X = ClO_4

Stevens rearrangement | KOBu-*t* | Hofmann elimination

R CN N | $(CH_2)_nCH{=}CH_2$ N CN

R_2 N R_1 CN | $(CH_2)_nCH{=}CH_2$ N CN

$-$HCN

R_2 N R_1 CN | $(CH_2)_nCH{=}CH_2$ N

402a, n = 3
402b, n = 4

$-$HCN

R_2 N R_1 — Pd/C, xylene → N

401a, $R_1 = R_2 = H$
401b, $R_1 = Me$, $R_2 = H$
401c, $R_1 = H$, $R_2 = Me$

403

Figure 66

TABLE 27. PHYSICAL DATA FOR AZEPINO[1,2-*a*]INDOLES

Structure	mp (°C)	Spectra	Refs.
		Fig. 58	
358a	150–155	uv	111,112
358c	122–124	uv	112
358d	157–158	uv	112
358e	159–161	uv	112
359a	54–57	uv	112
359b	68–70	uv	112
359c	54–55	uv	112
		Fig. 59 and Eq. 48	
369a	109–11	nmr,ir,uv	114,116
369b	125–126	nmr,ir,uv	114,116
369c	116–117	nmr,ir,uv	114,116
369d	134–136	nmr,ir,uv	116
369e	138–140	nmr,ir,uv	116
369f	201–204	nmr,ir,uv	113,116
369g	210–212	nmr,ir,uv	113,116
372	88–91	ms	116
		Fig. 60	
375a	70–71	ir,nmr,uv	117,118
375b	118–119	ir,nmr	117,118
376	124–125	ir,nmr,uv	117,118
		Eq. 49 and Fig. 61	
377	>196 (dec.)	ir,nmr,uv	119
378	67–75	ir,nmr,uv	119
379	Oil	nmr	118
380	94–95	nmr,uv	119
381	98–99	ir,nmr,uv	119
		Eq. 50	
385a	91.5	uv,nmr,ms	122
385b	47	uv,nmr,ms	122
385c	95.5	uv,nmr,ms	122
385d	59	uv,nmr,ms	122
386a	—	uv	120,122
386d	—	nmr	122
387	52–54	uv,nmr,ms	122
388	77–79	uv,nmr,ms	122
385f	60–62.5	uv,nmr	122
385g	150–200/4–10[a]	ms,nmr	122
392a	93–96	uv,nmr,ms,ir	122
392b	159–159.5	uv,nmr,ir	122

TABLE 27. (*Continued*)

Structure	mp (°C)	Spectra	Refs.
		Fig 64	
394	106.5–107.5	ir,nmr	122
395	128.5	ir,uv,nmr,ms	122
396a	140–145		122
396b	140–145	ir	122
397	109.5		
		Fig 65	
399	79	uv	123
		Fig. 66	
401a	56–57	nmr	124
401b	65–66	nmr	124
401c	—[b]		124
403	87	uv,nmr	124

[a]Boiling point (°C/torr).
[b]Not obtained free of **401b.**

Treatment of the *N,N*-dialkylhexahydroindolium salts **400a** and **400b** with potassium *t*-butoxide in ether resulted in competitive Stevens rearrangement and Hofmann elimination (Fig. 66). Compound **400a** gave a mixture of the Stevens product azepinoindole **401a,** and the Hofmann product **402a** in 57 and 9% yields, respectively. The related salt **400b** afforded the Stevens products **401b** (27% yield) and **401c** (1.5% yield) and the Hofmann product **402b** (22% yield). The predominance of **401b** as opposed to **401c** is consistent with the greater migratory aptitude of secondary over primary alkyl groups in the Stevens rearrangement. Azepinoindole **401a** was easily aromatized to the indole **403** by refluxing with 10% Pd/C in xylene.

A list of azepino[1,2-*a*]indoles and their physical data are given in Table 27.

35. Azepino[2,1-*i*]indoles

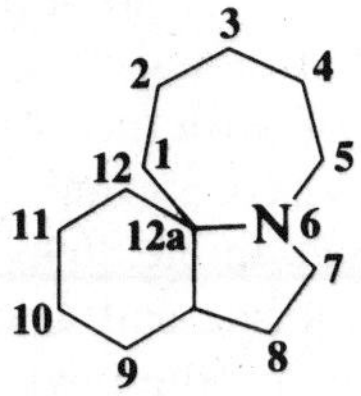

3
2 4
12 1 5
11 12a N6
7
10
9 8

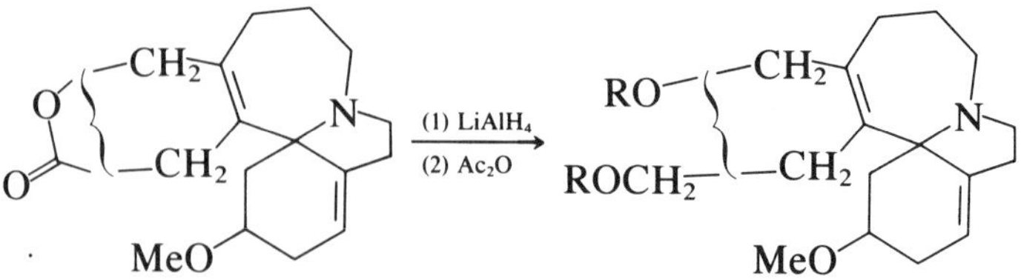

(1) $LiAlH_4$
(2) Ac_2O

(51)

404

405, R = H; mp 162°C (ir,ms)
406, R = Ac (ir,pmr,ms)

The only known examples of the azepino[2,1-*i*]indole system were prepared as degradation products during the structural elucidation of an alkaloid isolated from *Phelline billiardieri* (125). The alkaloid was tentatively given structure **404.** Lithium aluminum hydride reduction cleaved the lactone and gave the diol **405,** the first reported member of this ring system (Eq. 51). The diacetate **406** was also prepared. The position of the lactone ring in compounds **404–406** could not be unequivocally assigned.

36. Azepino[2,3-*b*]indoles

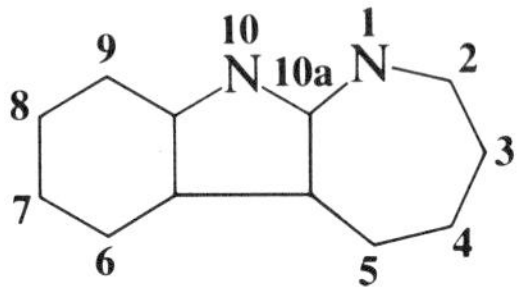

All of the published information about the azepino[2,3-*b*]indole system is due to Hester (126,127). Treatment of the tetrahydrocarbazolone **407a** with hydroxylamine in ethanol yielded a mixture of the *syn*- and *anti*-oximes **407c** and **407d,** respectively. The mixture was separable by chromatography, but both oximes gave a high yield of the same lactam (**408**) when subjected to Beckmann rearrangement in polyphosphoric acid. In contrast, when the *anti*-oxime **407d** was first converted to its *p*-toluenesulfonate ester (**407f**) and the rearrangement was carried out over neutral deactivated alumina, the reaction proceeded regiospecifically, albeit in modest yield, to afford the isomeric azepino[2,3-*b*]indole **409a.**

Lithium aluminum hydride reduction of **409a** and subsequent workup in the presence of oxygen gave a 71% yield of the alcohol **410a.** Structure **410a** was supported by its spectral data and by a series of chemical transformations. Ultimate confirmation of structure **410a** was provided by an X-ray crystallographic analysis of its hydrobromide. That **410a** was formed by air oxidation of the initially formed amine **409b** was strongly supported by the isolation of the stable hydrochloride salt **410b** and *N*-acetyl derivative **411a** on treatment of the reduction product with hydrogen chloride or acetic anhydride prior to exposure to oxygen. The assignment of the amidine tau-

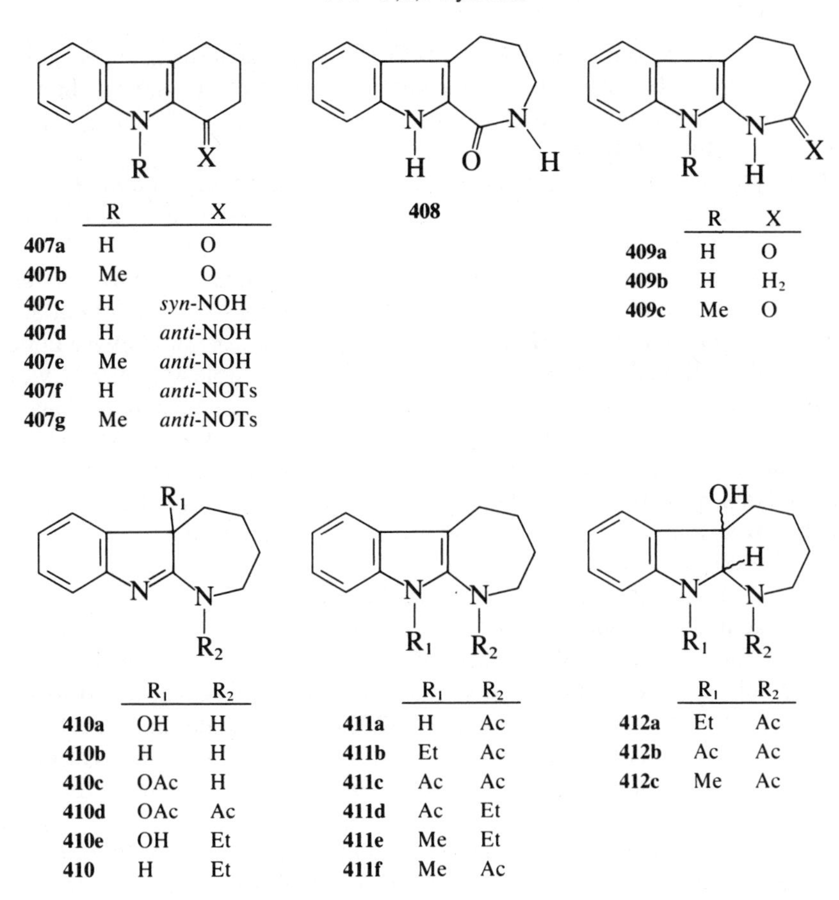

	R	X
407a	H	O
407b	Me	O
407c	H	*syn*-NOH
407d	H	*anti*-NOH
407e	Me	*anti*-NOH
407f	H	*anti*-NOTs
407g	Me	*anti*-NOTs

	R	X
409a	H	O
409b	H	H_2
409c	Me	O

	R_1	R_2
410a	OH	H
410b	H	H
410c	OAc	H
410d	OAc	Ac
410e	OH	Et
410	H	Et

	R_1	R_2
411a	H	Ac
411b	Et	Ac
411c	Ac	Ac
411d	Ac	Et
411e	Me	Et
411f	Me	Ac

	R_1	R_2
412a	Et	Ac
412b	Ac	Ac
412c	Me	Ac

tomer **410b** rather than the isomeric aminoindole **409b** to the hydrochloride salt was based on spectral analogy.

Reaction of **410a** with acetic anhydride gave chromatographically separable products **410c** and **410d** in 25 and 26% yields, respectively. The pmr spectrum of **410d** displayed two methyl singlets at δ 2.59 and δ 2.08, which were assigned to the amide and ester acetyl groups. The C_2 methylene of **410d** was observed as a strongly deshielded quartet a δ 4.95 ppm. The latter signal was confirmed by preparation of the C_2-dideuterio compound **415b** in which the signal at δ 4.95 was absent. Thus, lithium aluminum deuteride reduction of **409a** gave **415a,** which, upon acetylation, afforded **415b.** Basic hydrolysis of **410d** regenerated **410a,** whereas lithium aluminum hydride reduction of **410d** gave not only **410a** but also the *N*-ethylazepinoindolol **410e.** The latter compound was also obtained in 75% yield upon lithium aluminum hydride reduction of **411a** followed by workup in the presence of oxygen. Catalytic reduction of **410c** with a palladium catalyst in acetic anhydride gave **411a** as the only isolable product. Compound **411a** could arise by: (1) successive reduction of the amidine double bond, elimination of acetic acid to form **409b,** and *N*-acylation; (2) hydrogenolysis of the acetoxy moiety to give **410b,** with subsequent *N*-acylation. Regardless of the mechanism, the overall transformation of **410c** to **411a** provides strong chemical support for the structure of **410a.**

Catalytic hydrogenation of **410a** in acetic anhydride gave a complex mixture from which four crystalline products could be isolated by chromatography. The products, assigned structures **411b, 412a, 412b,** and **413,** were mainly characterized on the basis of spectra. Chemical support was also provided in some instances. For example, the alcohol **412a** underwent acid-catalyzed conversion to **411b** in 78% yield. The alcohol **412b** likewise underwent acid-catalyzed dehydration to give the diacetyl indole **411c** in 92% yield. Support for structure **411c** was derived from its spectral data and its cleavage to **411a** with sodium ethoxide in ethanol. Compound **413** was isomeric with **412a** and underwent dehydration to produce the noncrystalline indole **411d.** Ethanolysis of **411d** with sodium ethoxide in ethanol gave the crystalline alcohol **410e,** presumably via air oxidation of the initial product. Acidification of **411d** with hydrogen chloride, followed by crystallization from a mixture of methanol and ethyl acetate, gave **410f**·HCl.

The facile auto-oxidations encountered in this series prompted Hester to examine the analogous 10-alkylazepinoindoles. Alkylation of **407a** with dimethyl sulfate gave **407b,** which was in turn sequentially converted to **407e** and **407g.** Beckmann rearrangement of **407g** on neutral alumina deactivated with 0.4% water gave a 23% yield of **409c.** Lithium aluminum hydride reduction of **409c** followed by a workup in air gave a 71% yield of the auto-oxidation product **414a.**

Catalytic reduction of **414a** in acetic anhydride over 10% Pd/C gave a mixture of four compounds: **411e, 411f, 412c,** and **414b.** Compound **411e** was an oil that gave a crystalline hydrochloride. Of note was the fact that **411e,** a 1,10-dialkyl derivative of **409b,** was quite stable to air. Structure **411f** was supported by spectral data, as was structure **412c.** The *N*-methyl and *N*-acetyl moieties of **412c** were observed in the pmr spectrum as pairs of singlets, suggestive of a mixture of *cis* and *trans* isomers. Compound **412c** also underwent facile acid-catalyzed conversion to indole **411f.** Structure **414b** was supported by ir, pmr, uv, and mass spectral data. A mechanistic interpretation of the products obtained from the catalytic reduction of **410a** and **414a** has been proposed (127). A list of the known members of the azepino[2,3-*b*]indole ring system is given in Table 28.

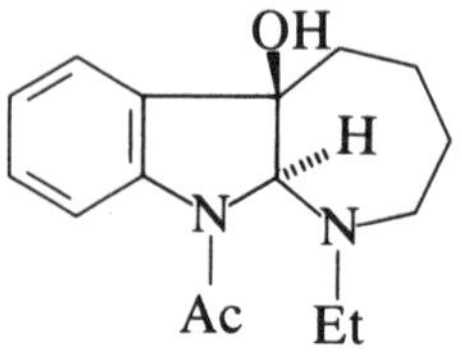

413

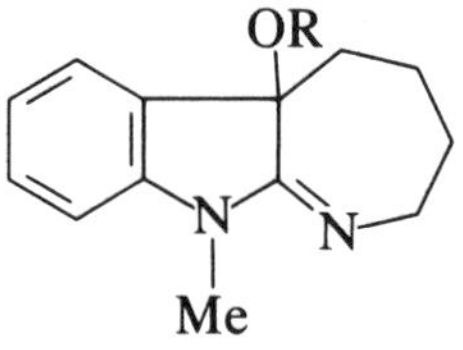

414a, R = H
414b, R = Ac

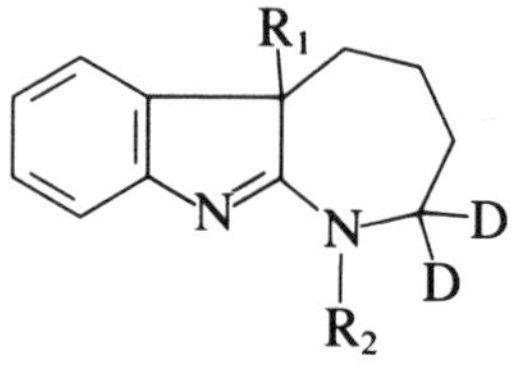

415a, R_1 = OH, R_2 = H
415b, R_1 = OAc, R_2 = Ac

TABLE 28. AZEPINO[2,3-*b*]INDOLES[a]

Compound	mp (°C)	Spectra
409a	205.5–206.5	ir,pmr,uv,ms
409c	193–194.5	ir,pmr,uv
410a	255–259.5	ir,pmr,uv,ms
410b	253.5–255.5	ir,pmr,uv
410c	176 (dec.)	ir,pmr,uv,ms
410d	127.5–129	ir,pmr,uv,ms
410e	251.5–252.5	ir,pmr,uv,ms
410f	226.5–228	ir,pmr,uv
410f (HCl)	229.5–230.5	ir,pmr,uv
410f (HBr)	205.5–206.5	
411a	193	ir,pmr,uv,ms
411b	140.5–141.5	ir,pmr,uv,ms
411c	113.5–115.5	ir,pmr,uv,ms
411d	Oil	
411e (HCl)	209–210 (dec.)	ir,pmr,uv,ms
411f	125–125.5	ir,pmr,uv
412a	111.5–112.5	ir,pmr,uv,ms
412b	200–201	ir,pmr,uv,ms
412c	141.5–142.5	ir,pmr,uv,ms
413	164–165	ir,pmr,uv,ms
414a	129–133	ir,pmr,uv
414a (HCl)	264.5–265	ir,pmr,uv,ms
414b	105–108	ir,pmr,uv,ms
415a	246–250 (dec.)	
415a (HCl)	231–232.5	ir,pmr,uv,ms
415b	120.5–121.5	ir,pmr,uv,ms

[a]Ref. 127.

37. Azepino[3,2-*b*]indoles

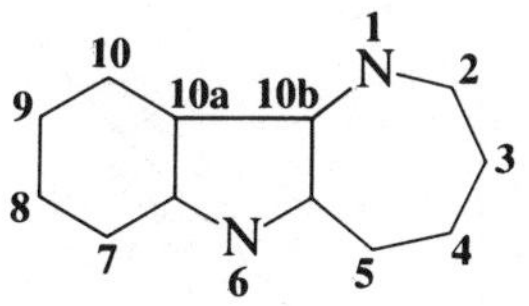

1 2 3 4 5 6 7 8 9 10 10a 10b N N

The Beckmann rearrangement of 1,2,3,4-tetrahydrocarbazol-4-one oxime (**416**) was studied as a possible route to the azepino[3,2-*b*]indole ring system (128,129). When ketone **417** was treated with hydroxylamine, a single sharp-melting oxime, which was assigned the *anti* configuration **416,** was obtained. Rearrangement of **416** with polyphosphoric acid gave a clean product that was not the azepino[3,2-*b*]indolone **418,** but instead the azepino[4,3-*b*]indolone **419** (see Section IV.41). If the oxime configuration were *anti* as assigned, the formation of **419** would suggest an acid-catalyzed isomerization of **416** to the *syn* configuration prior to rearrangement. When **416** was first converted

416, X = NOH (*anti*)
417, X = O
420, X = NOTs (*anti*)

418, R = H
427, R = Me

419

to its *p*-toluenesulfonate ester **420,** the Beckmann rearrangement proceeded stereospecifically on deactivated alumina to afford **418.**

Lithium aluminum hydride reduction of **418,** followed by careful workup under nitrogen, gave the amine **421** in 61% yield (Fig. 67). Compound **421** was very sensitive to auto-oxidation and, in ethyl acetate solution, was rapidly air-oxidized to **422.** The assignment of structure **422** was based on its uv spectrum, which was compatible with its pseudoindoxyl-type structure. Catalytic reduction of **422** with platinum oxide in acetic anhydride afforded the acetamide **423,** which was identical to the compound obtained from **421** and acetic anhydride. For pmr studies the deuterated *N*-acetyl

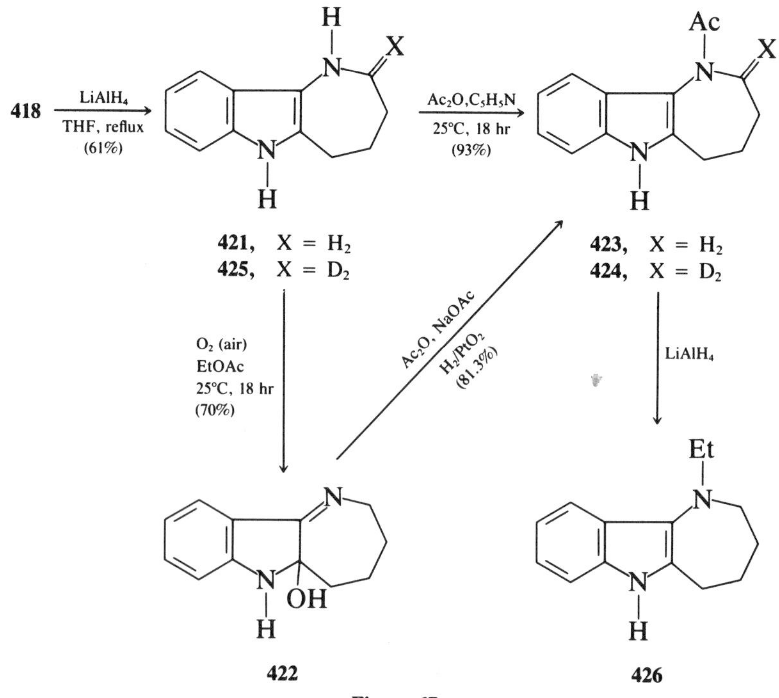

Figure 67

TABLE 29 AZEPINO[3,2-*b*]INDOLES[a]

Compound	mp (°C)	Spectra
418	243.5–246	ir,pmr,uv
421	175–178	pmr,uv
422	170.5–176 (dec.)	ir,uv,ms
423	217.5–218.5	ir,pmr,uv
424	217–218	pmr,uv,ms
426	128–129.5	pmr,uv

[a]Fig. 67, Ref. 128.

derivative **424** was prepared from the deuterated amine **425.** Lithium aluminum hydride reduction of **423** gave the corresponding amine **426** in good yield. Melting points and other data for members of this ring system are given in Table 29.

In a brief note (130), the rose bengal–sensitized photooxygenation of **427** was reported to produce the 10-membered lactam **428** and the 1,4-benzoxazine **429** in 50 and 25% yields, respectively.

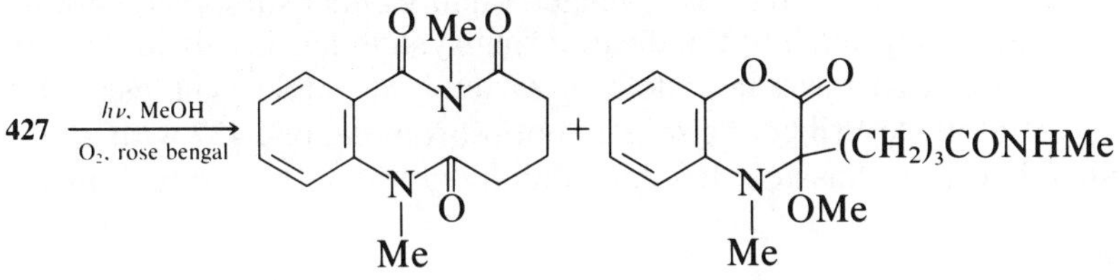

427 $\xrightarrow[O_2,\ \text{rose bengal}]{h\nu,\ \text{MeOH}}$ **428** + **429** $-(CH_2)_3CONHMe$, OMe (52)

38. Azepino[3,4-*b*]indoles

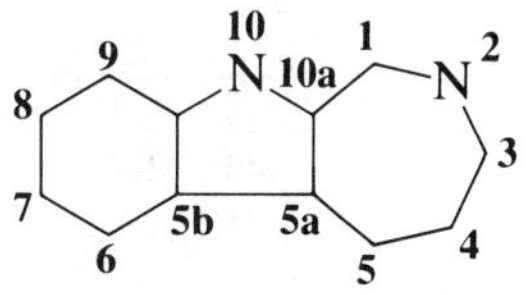

The chemical literature contains several examples of the azepino[3,4-*b*]indole system, and three distinctly different approaches have been used for its preparation. The first of these methods was reported in 1930 by Jackson and Manske (131) and represents the first reported example of the system. Methyl indolebutyrate **430** (Eq. 53) was converted sequentially to the hydrazide **431** and the carbonylazide **432.** When warmed in benzene, **432** rearranged to the corresponding isocyanate **433,** which when treated *in situ* with dry hydrogen chloride gave a 5% yield of the azepinoindole **435.** This low

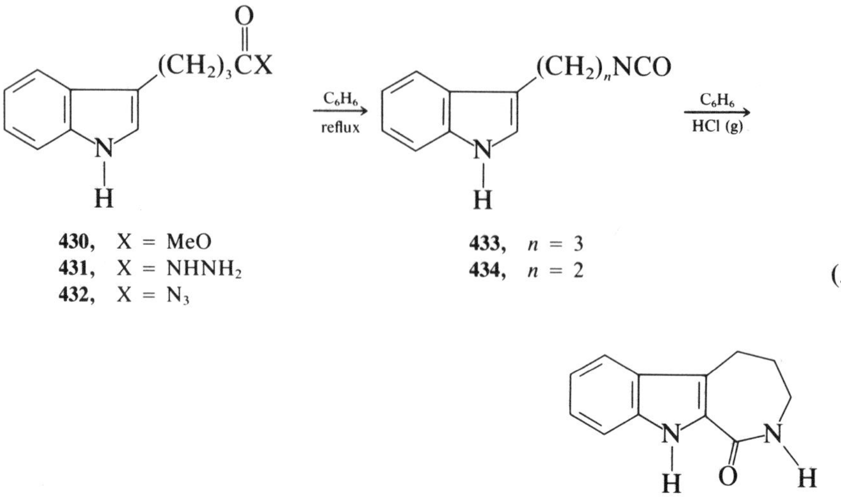

430, X = MeO
431, X = $NHNH_2$
432, X = N_3

433, n = 3
434, n = 2

(53)

435

yield must reflect an inherent difficulty in forming the seven-membered ring, since the next lower homolog (**434**) cyclized rapidly under similar conditions.

An alternative approach to the desired ring system has involved the conversion of azepines to azepinoindoles via standard indole syntheses. The first such example described the reaction of α-bromoketone **437** with aniline and *p*-anisidine in a classical Bischler indole synthesis to produce in low

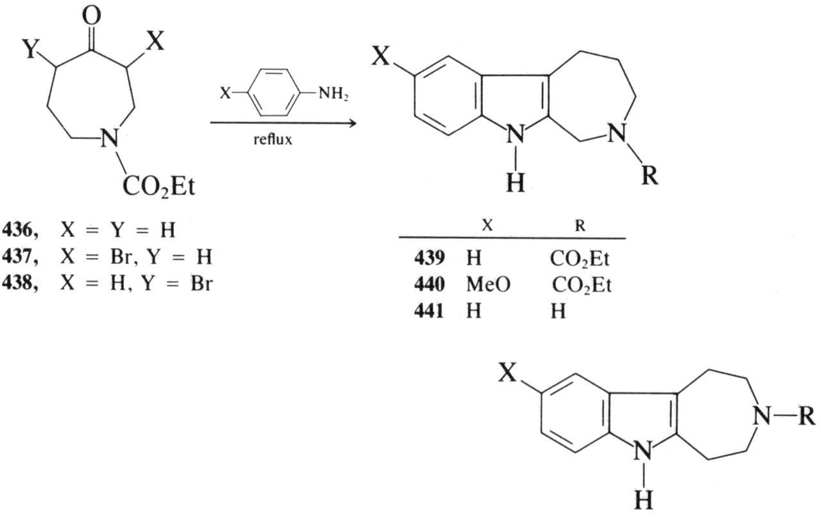

436, X = Y = H
437, X = Br, Y = H
438, X = H, Y = Br

	X	R
439	H	CO_2Et
440	MeO	CO_2Et
441	H	H

(54)

	X	R
442	H	CO_2Et
443	MeO	CO_2Et
444	H	H

yield the azepinoindoles **439** an **440,** respectively (Eq. 54). Acid hydrolysis of **439** gave the parent compound **441** (132). Bromination of **436** was reported to produce only **437** and none of the isomeric product **438.** However, **437** was not characterized, and the argument used in support of its preferential formation over **438** was based on an inappropriate model. Azepinones such as **436** are known to enolize in a manner more likely to yield **438** than **437** (see Section IV.43). Thus, in the absence of more definitive analytical data, structures **437,** and consequently **439–441** should be regarded with caution. Obviously if the bromoketone were actually **438,** the Bischler reaction would give the azepino[4,5-*b*]indoles **442–444** (see Section IV.43)

In a related sequence (Eq. 55), the Fischer indole synthesis was utilized to convert the caprolactam derivatives **447** and **448** to the azepinoindoles **435** and **449.** Caprolactam (**445**) was chlorinated with phosphoryl chloride or phosphorous pentachloride to yield the α,α-dichloro derivative **446,** which on further reaction with cyclic amines (e.g. piperidine, morpholine) or so-

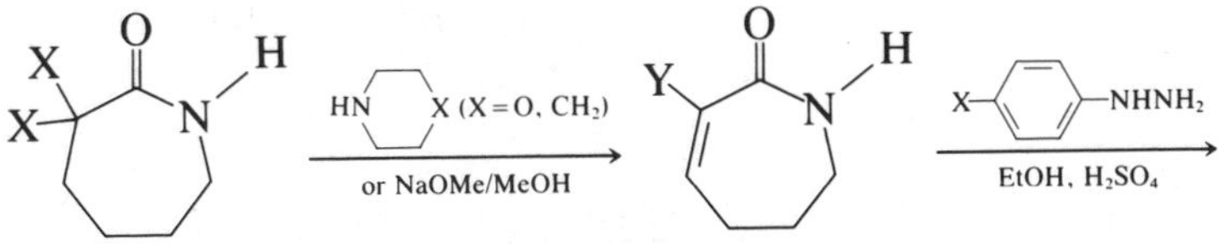

HN X (X = O, CH$_2$) or NaOMe/MeOH

X–C$_6$H$_4$–NHNH$_2$, EtOH, H_2SO_4

445, X = H
446, X = Cl

447, Y = N(cyclic amine)X
448, Y = OMe

(55)

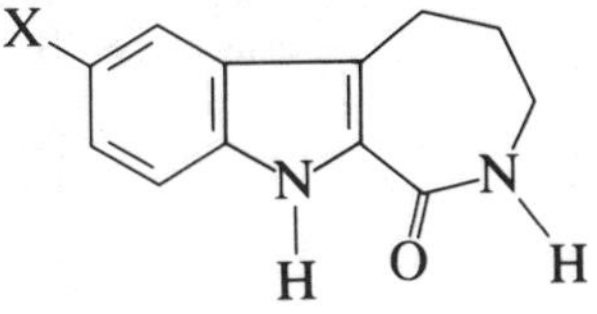

435, X = H
449, X = MeO

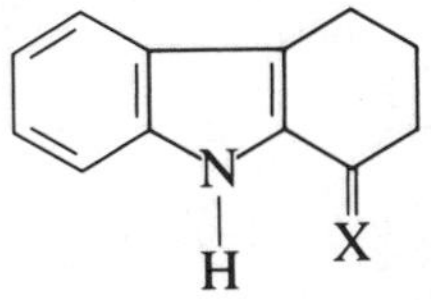

450a, X = O
450b, X = NOH (*syn*)
450c, X = NOH (*anti*)
450d, X = NOTs (*syn*)

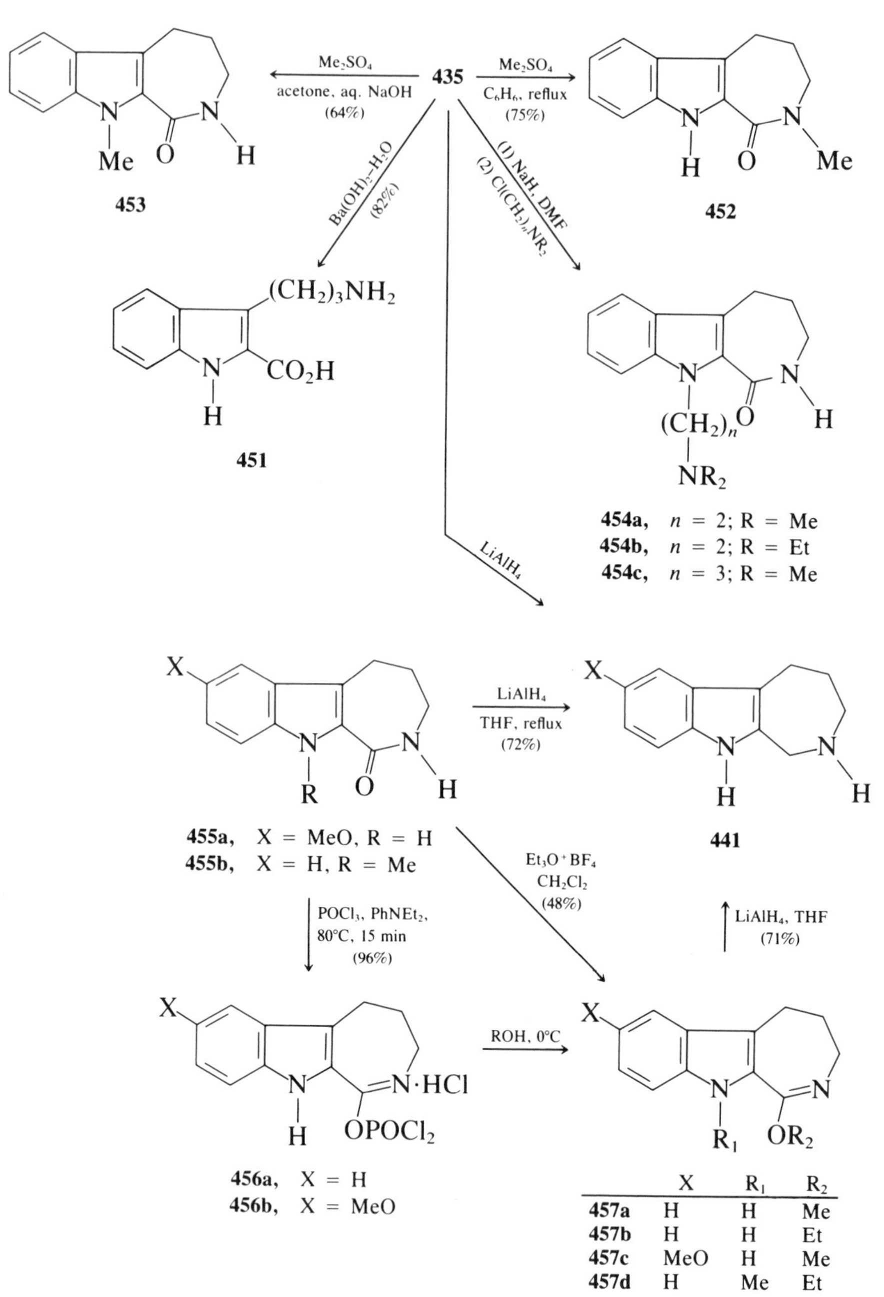
435
Me2SO4
acetone, aq. NaOH
(64%)
Me2SO4
C6H6, reflux
(75%)
Ba(OH)2–H2O
(82%)
(1) NaH, DMF
(2) Cl(CH2)nNR2
LiAlH4
453
452
(CH2)3NH2
CO2H
451
(CH2)n
NR2
454a, n = 2; R = Me
454b, n = 2; R = Et
454c, n = 3; R = Me
LiAlH4
THF, reflux
(72%)
455a, X = MeO, R = H
455b, X = H, R = Me
441
Et3O+BF4
CH2Cl2
(48%)
LiAlH4, THF
(71%)
POCl3, PhNEt2,
80°C, 15 min
(96%)
N·HCl
OPOCl2
ROH, 0°C
OR2
456a, X = H
456b, X = MeO
X R1 R2
457a H H Me
457b H H Et
457c MeO H Me
457d H Me Et

Figure 68

dium methoxide gave the caprolactam derivatives **447** and **448,** respectively. Condensation of **447** and/or **448** with arylhydrazines in acidic ethanol led to the indoles **435** and **449** in excellent yield (109,133,134).

The third synthetic entry into the ring system has been by ring expansion of the tetrahydrocarbazolone **450a.** For example, the Schmidt reaction, employing sodium azide in polyphosphoric acid, converted **450a** to **435** (127). Furthermore, when the Beckmann rearrangement was carried out in polyphosphoric acid, the same lactam **435** was obtained from either the *syn*-oxime **450b** or the *anti*-oxime **450c.** The *syn*-oxime **450b** was converted to its tosylate ester **450d,** which also rearranged in high yield to **435** (126,127,135).

The chemistry of azepinoindolones is characteristic of that normally associated with indoles and lactams (Fig. 68). Hydrolysis of the lactam ring converted **435** to the amino acid **451** (135), and lithium aluminum hydride readily reduced **435** to **441** in 72% yield (127). Alkylations of the azepinoindolones have been carried out and, in general, proceed cleanly. The position of alkylation can be controlled by an appropriate choice of reaction conditions. For example, alkylation of **435** with dimethyl sulfate in benzene gives the *N*-methylamide **452.** However, in the presence of aqueous hydroxide, **435** undergoes alkylation on the indole nitrogen with formation of **453** (138). Likewise, treatment of **435** with sodium hydride in dimethyl formamide followed by addition of aminoalkylhalides gives the *N*-alkylated products **454** (137,140). On the other hand, when the azepinoindolones **455** are treated with potent electrophiles such as Meerwein's reagent or phosphorous ox-

457a and/or **457b** $\xrightarrow[\text{EtOH}]{HNR_1R_2}$

456a $\xrightarrow[\text{DMF}]{HNR_1R_2}$

N—H, NR_1R_2 (56)

458a, R_1 = H, R_2 = $(CH_2)_2NEt_2$
458b, R_1,R_2 = $(CH_2)_5$
458c, R_1 = H, R_2 = Ph
458d, R_1 = H, R_2 = $CH_2C_6H_4(OMe)_2$-3,4
458e, R_1 = H, R_2 = NH_2

456a + (CO_2H, NH_2) $\xrightarrow[\text{reflux, 4 hr (96\%)}]{DMF}$ (57)

459

441 $\xrightarrow[\text{EtOH}]{\text{Raney Ni}}$ [indole with 3-$(CH_2)_3NEt_2$, 2-Me, N–H] **460** (58)

TABLE 30. AZEPINO[3,4-*b*]INDOLES[a]

Compound	mp (°C) or bp (°C/torr)	Spectra	Refs.
435	228–229	ir,uv	126,127
	220		131
	224–227	ir,uv	135
	224–225		134
439	156 (picrate)		
	(216/0.4)		132
440	164.5 (picrate)		
	220–228/0.5		132
441	254 (dec.; picrate)	ir	132
	212–214	uv,nmr	126,127
	182–186		
	280–282		139
449	190–191	ir	109,133,134
452	240–242	ir	138
453	126–128	ir	138
454a	122–123		
	253–254 (HCl)		137,140
454b	112–113		
	232–234 (HCl)		137,140
454c	170–171.5 (HCl)		137,140
456a	150–153	ir	138
456b			138
457a	100–102	ir	
	225–227 (HCl)		
	122–124 (bitartrate)		138
457b	122.5–124	ir,uv	126,127
	117–118		138
457c	142–143		138
457d	121–122	ir	138
458a	144.5–145.5	uv	126,127
458b	150–153 (HCl,$2H_2O$)	uv	126,127
	155–157 (HCl)		138
458c	164–165		138
458d	221–223 (HCl)		138
458e	177–178		
	235–238 (HCl)		138

[a]Eqs. 53–58, Fig. 68.

ychloride, attack occurs on the oxygen, with formation of the imino ethers **456** and **457** (126,127,138).

The imino ethers undergo several substitution reactions characteristic of this functionality, including reduction with lithium aluminum hydride (Fig. 68) (126,127,138,139), formation of the amidines **458** (Eq. 56), reaction of **456a** with anthranilic acid to afford the pentacyclic product **459** (139) (Eq. 57), and catalytic hydrogenolysis of **441** to the aminoalkyl indole **460** (138) (Eq. 58). Table 30 lists known members of this ring system.

39. Azepino[3,2,1-*hi*]indoles

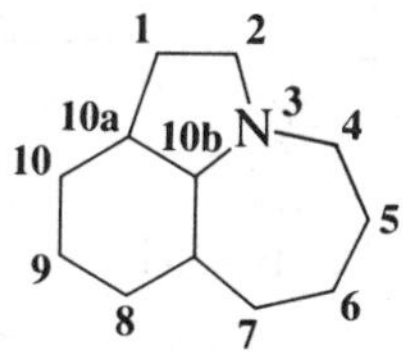

The azepino[3,2,1-*hi*]indole system first appeared in the chemical literature in an elegant series of papers by Boekelheide describing the structural determination of the alkaloids β-erythroidine (**461**) and apo-β-erythroidine (**462**) (Fig. 69) (141–147). When apo-β-erythroidine methiodide (**463a**) was passed over a basic ion-exchange resin, it was converted not to the expected methohydroxide **463b,** but to the zwitterionic azepinoindole **464** (142). When **464** was heated with methyl iodide, it reverted to the methiodide **463a.** Treatment of either **463a** or **464** with aqueous hydroxide effected a normal Hofmann elimination and gave the methine base **465** in 54% yield.

Additional degradative experiments helped to further confirm the structure assigned to apo-β-erythroidine as **462** (143). In order to more firmly establish the orientation of the lactone ring, experiments were directed toward the removal of the allylic oxygen. Lithium aluminum hydride reduction of **462** yielded the diol **466a,** which was subjected to hydrogenolysis over palladium–carbon (Fig. 70). This reduction unexpectedly gave only a trace of the desired deoxy compound **468** along with the major product dihydroapo-β-erythroidinol (**467**). Attempted removal of the allylic oxygen by reduction of **462** with alkaline Raney nickel gave only the hydroxy acid **469.** On heating to 115°C, **469** cyclized to dihydroapo-β-erythroidine (**470**).

The slow cleavage of the allylic carbon–oxygen bond prompted an attempt to convert **466a** to the dichloride **466b.** Unfortunately, treatment of **466a** with thionyl chloride gave instead of **466b** the cyclic ether **471** (Fig. 71). Accordingly, an alternate approach was developed that still allowed the definitive assignment of the allylic position. Treatment of β-erythroidinol **472** with thionyl chloride gave the corresponding dichloride **473** in 70% yield. Reduction of **473** with lithium aluminum hydride resulted in selective removal

MeO

O

O

N

461

N

O

O

462

MeI, EtOH

reflux, 3 hr

$\overset{+}{N}$

Me

X^-

O

O

463a, X = I

463b, X = OH

aq. KOH

25°C, 24 hr

MeI, EtOH

Amberlite IRA-400, H_2O

N

Me

CH=CH$_2$

O

O

465

aq. KOH

25°C, 24 hr

$\overset{+}{N}$

Me

CO_2^-

OH

464

Figure 69

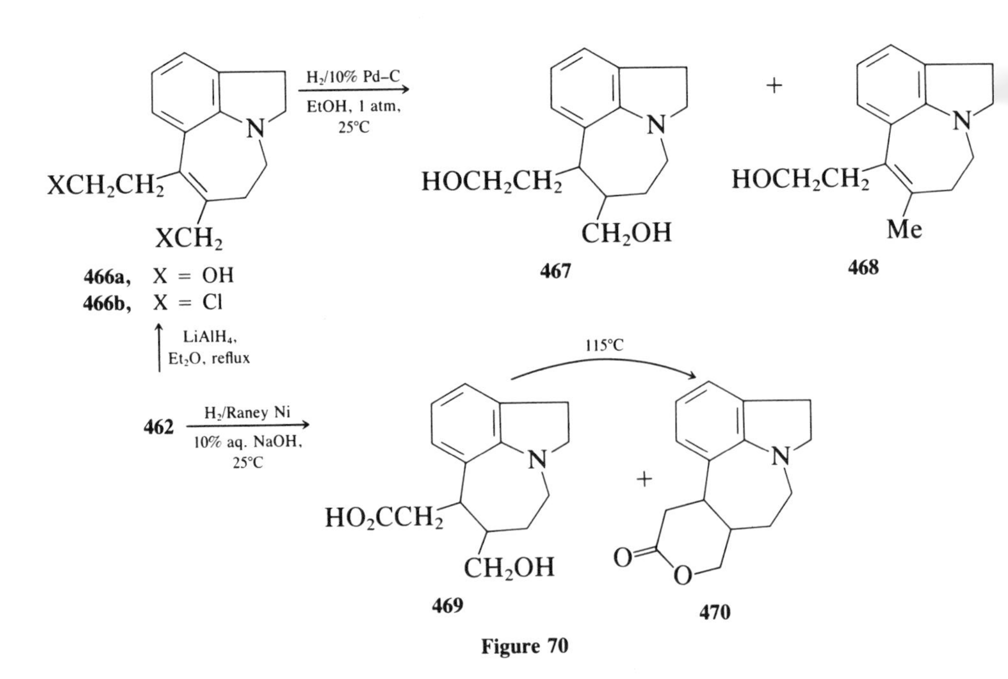

Figure 70

466a $\xrightarrow[0°C]{SOCl_2}$ **471** $\xleftarrow[105–120°C]{H_3PO_4}$ **472**

MeO
$HOCH_2CH_2$
$HOCH_2$
N

471

472

$\downarrow SOCl_2$

$\xleftarrow[110°C, 1 hr]{H_3PO_4}$ **474** $\xleftarrow[Et_2O–THF, reflux, 2 hr]{LiAlH_4}$ **473**

MeO
$ClCH_2CH_2$
Me
N

MeO
$ClCH_2CH_2$
$ClCH_2$
N·HCl

474

473

$ClCH_2CH_2$
N
Me

475

$\downarrow$ $LiAlH_4$, THF, reflux

476 $\xrightarrow[(2) Amberlite IRA-400]{(1) MeI, EtOH}$ **477** $\xrightarrow[\Delta]{H_2O}$ **478**

Et
Me
N

476

Et
Me
$\overset{+}{N}$
Me
X^-

477a, X = I
477b, X = OH

Et
Me
N
Me
$CH{=}CH_2$

478

H_2/PtO_2, EtOH

MeCEt (C=O) $\xleftarrow{O_3}$ **479**

Et
Me
Et
N
Me

479

Figure 71

of the allylic halogen to give **474**, and heating of **474** in phosphoric acid effected the apo-rearrangement to give chlorodesoxyapo-β-erythroidinol **475.** Prolonged reduction of **475** with lithium aluminum hydride in hot tetrahydrofuran removed the remaining chlorine to give the oily **476,** which was characterized as it crystalline methiodide **477a.** Basic ion-exchange resin converted **477a** to the methoxyhydroxide **477b,** which, when heated in aqueous solution, gave the diene **478.** Catalytic reduction of **478** led to **479,** which, on ozonolysis, yielded methyl ethyl ketone. The formation of methyl ethyl ketone instead of diethyl ketone confirmed the orientation of the lactone ring in **462.**

Passage of **462** over an alumina column induced a shift of the double bond into conjugation with the carbonyl group to give isoapo-β-erythroidine **480** (Fig. 72). Lithium aluminum hydride reduction afforded the diol **481,** which, on futher reduction over palladium, gave **469** as the main product (Fig. 72).

With the structure of apo-β-erythroidine (**462**) established, several groups turned their attention toward its total synthesis. In a classic Sandmeyer isatin synthesis (Fig. 72), treatment of benzazepine **482** with chloral hydrate and hydroxylamine led to the *N*-nitrosoacetyl derivative **483,** which in the presence of sulfuric acid yielded the isatin **484,** the first synthetic azepino-[3,2,1-*hi*]azepine derivative ever prepared (148). The same isatins could also be obtained with oxalyl chloride (see Section VII.3) (149).

The first successful total synthesis of **462** was described by Rapoport (150,151). Condensation of methyl 7-indolineacetate (**485**) with methyl β-bromopropionate (Fig. 73) gave the diester **486,** which on Dieckmann cy-

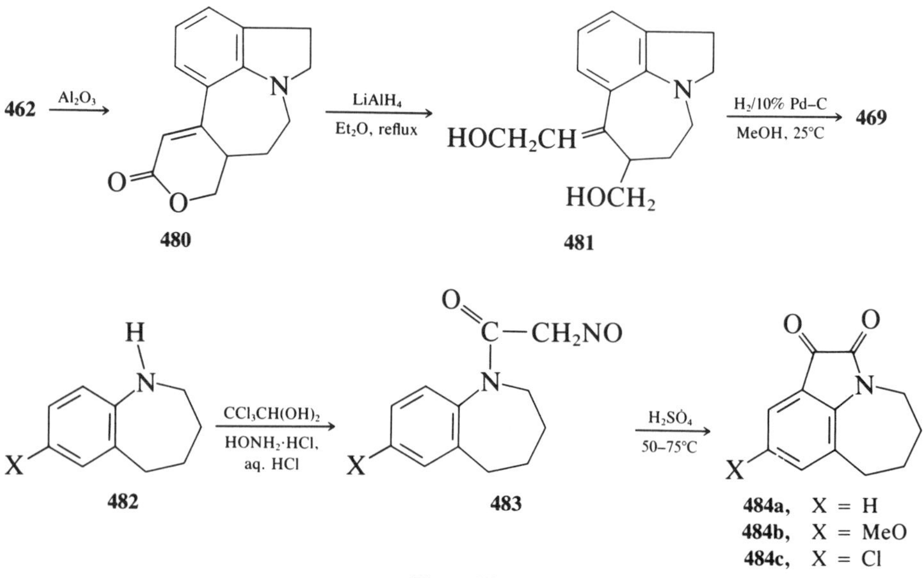

462 $\xrightarrow{Al_2O_3}$ **480** $\xrightarrow[Et_2O,\ reflux]{LiAlH_4}$ **481** (HOCH₂CH=, HOCH₂) $\xrightarrow[MeOH,\ 25°C]{H_2/10\%\ Pd\text{-}C}$ **469**

482 $\xrightarrow[HONH_2\cdot HCl,\ aq.\ HCl]{CCl_3CH(OH)_2}$ **483** (C(=O)–CH₂NO) $\xrightarrow[50\text{–}75°C]{H_2SO_4}$ **484**

484a, X = H
484b, X = MeO
484c, X = Cl

Figure 72

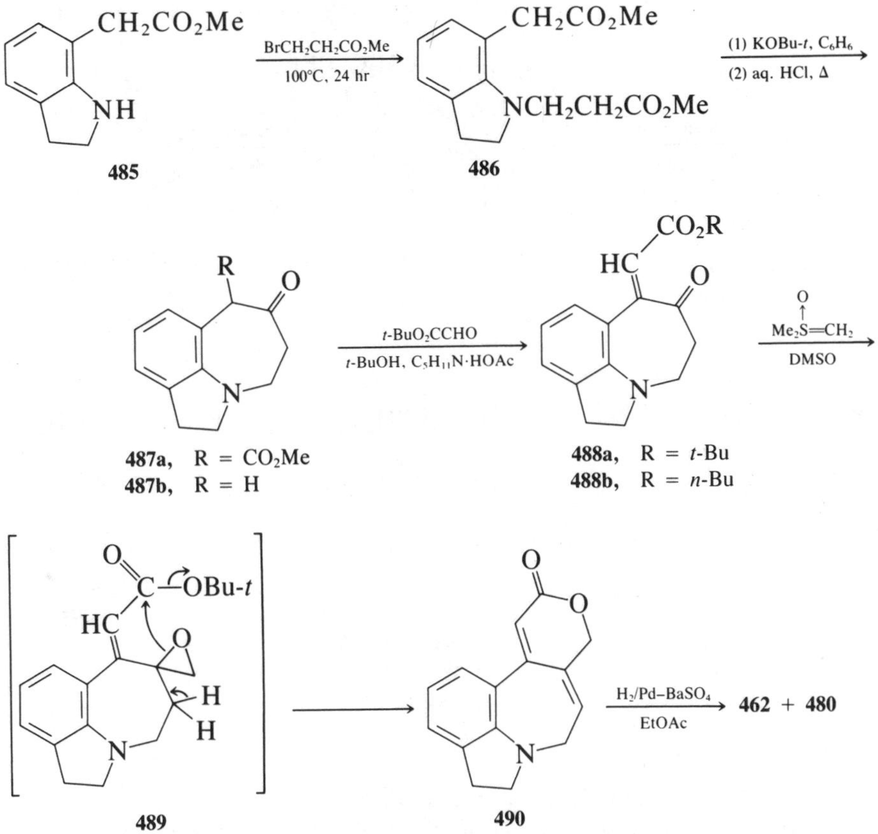

485 → ($BrCH_2CH_2CO_2Me$, 100°C, 24 hr) → 486 → ((1) KOBu-*t*, C_6H_6; (2) aq. HCl, Δ)

487a, R = CO_2Me
487b, R = H

→ (*t*-BuO_2CCHO; *t*-BuOH, $C_5H_{11}N\cdot HOAc$) →

488a, R = *t*-Bu
488b, R = *n*-Bu

→ ($Me_2S{=}CH_2$ (O), DMSO) →

489 → 490 → (H_2/Pd–$BaSO_4$, EtOAc) → **462** + **480**

Figure 73

clization with potassium *tert*-butoxide gave a 55% yield of the azepino-[3,2,1-*hi*]indole **487a.** Acid decarboxylation gave the azepinoindolone **487b,** and the latter was converted to the keto ester **488a** (59% yield) by condensation with *tert*-butyl glyoxylate. The reaction of **488a** with dimethyloxosulfonium methylide gave a 51% yield of a compound that was assigned structure **490,** presumably arising from the intermediate spiroepoxide **489.** Compound **490** was characterized on the basis of its elemental analysis and mass spectrum. In addition, the infrared spectrum contained a single carbonyl bond at 1723 cm^{-1}, and the ultraviolet spectrum showed evidence of extended conjugation. The final proof of structure for **490** came from its pmr spectrum. Catalytic reductions of **490,** using numerous catalysts and/or solvents, gave complex mixtures. With palladium on barium sulfate, **490** was converted to a mixture of apo-β-erythroidine (**462**) and isoapo-β-erythroidine (**480**) in yields of 14 and 20%, respectively.

Extensive work on the general chemistry of the azepino[3,2,1-*hi*]indole ring system has been carried out in connection with the total synthesis of

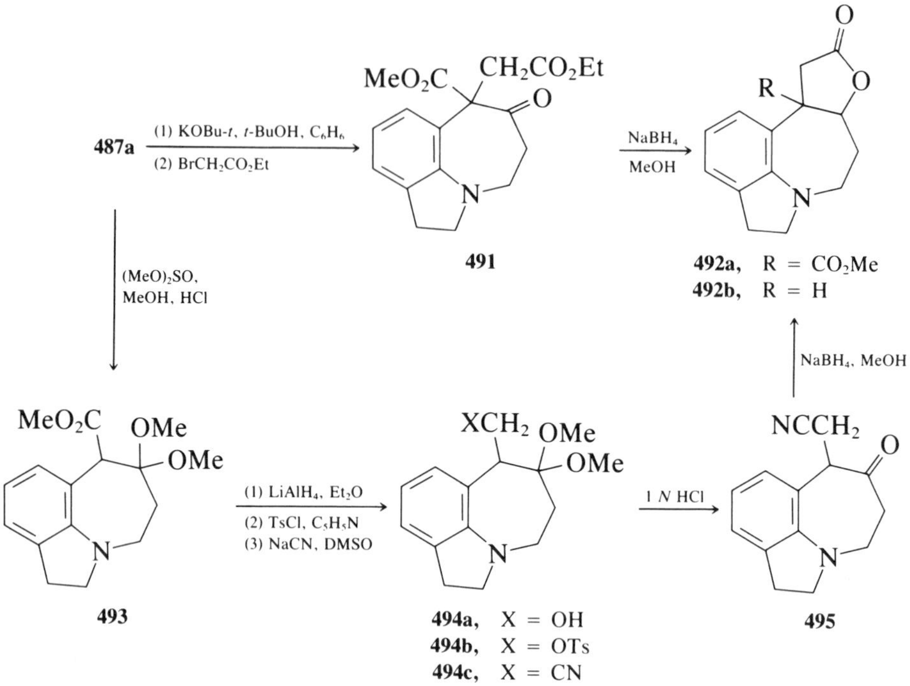

Figure 74

462 (150,151). For example, alkylation of **487a** with ethyl bromoacetate (Fig. 74) produced the keto diester **491** in 45% yield. Sodium borohydride reduction converted **491** to the lactam **492a.** The keto ester **487a** could also be converted in good yield to the ketal **493** with dimethyl sulfite. Successive reactions of **493** with lithium aluminum hydride, tosyl chloride, and sodium cyanide in dimethyl sulfoxide gave the alcohol **494a,** tosylate **494b,** and the nitrile **494c,** respectively. Acidic hydrolysis of **494c** gave the keto nitrile **495,** which, on sodium borohydride reduction, gave the lactone **492b.** Although lactone **492b** could also be obtained in low yield by decarboxylation of **492a,** this reaction gave primarily polymeric material.

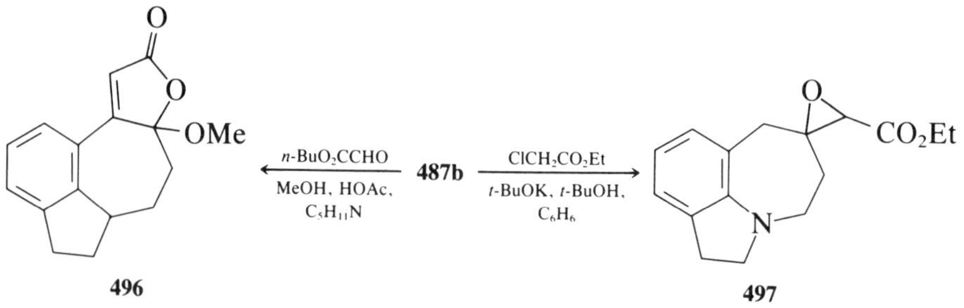

(59)

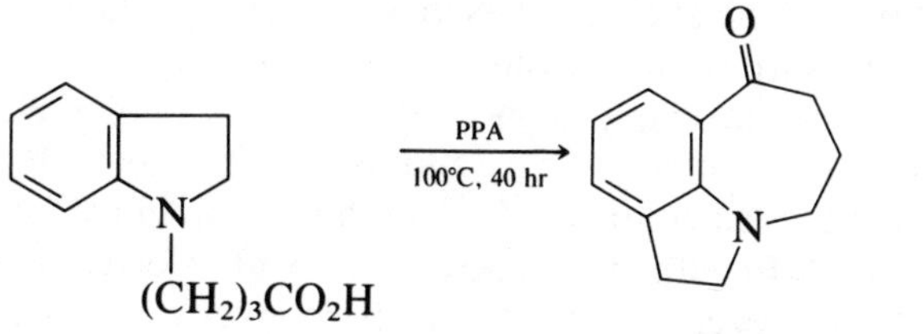

(CH$_2$)$_3$CO$_2$H

PPA
100°C, 40 hr

498

(60)

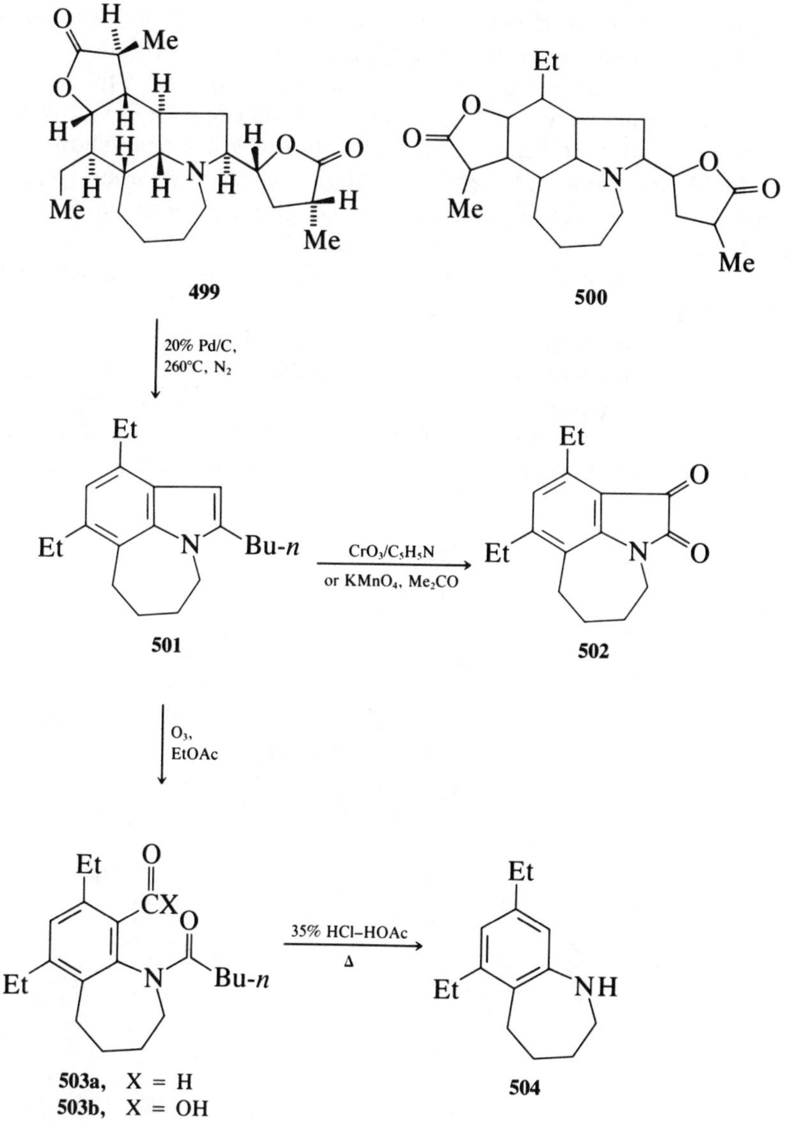

499

500

20% Pd/C, 260°C, N_2

501

CrO_3/C_5H_5N
or $KMnO_4$, Me_2CO

502

O_3, EtOAc

35% HCl–HOAc
Δ

503a, X = H
503b, X = OH

504

Figure 75

Condensation of the azepinone **487b** with *n*-butyl glyoxylate (Eq. 59) took a different course from the prviously described reaction with *tert*-butyl glyoxylate. Instead of the simple adduct **488b,** the lactol ether **496** was obtained. Compound **496** apparently arose by addition of the solvent methanol to **488b** with subsequent cyclization of the intermediate hemiketal. The Darzens condensation of **487b** with ethyl chloroacetate afforded the glycidic ester **497** in 59% yield. However, **497** could not be converted to the corresponding aldehyde. Cyclization of 1-indolinebutyric acid in polyphosphoric acid (Eq. 60) has been reported to give the 7-oxoazepino[3,2,1-*hi*]indole **498.**

Another alkaloid, tuberostemonine, the major alkaloid of *Stemona tuberosa,* has also been shown to contain the azepino[3,2,1-*hi*]indole system. A preliminary communication, published in 1961, established the structure of tuberostemonine as either **499** or its isomer **500** (Fig. 75) (152). Through much spectral and chemical evidence, structure **499** has been shown to be correct (585), and the stereochemistry and absolute configuration were subsequently determined by X-ray studies (153). Since a thorough discussion of the tuberostemonine work would be beyond the scope of this review (154), most of the discussion to follow will be restricted to the azepino[3,2,1-*hi*]-indole system.

In 1962, Uyeo identified the indole **501,** which was obtained in 70% yield from the dehydrogenation of **499** over 20% palladium on carbon (155). Oxidation of **501** with chromic acid–pyridine or potassium permanganate gave a low yield of a red-orange substance with the formula $C_{16}H_{19}NO_2$, which was tentatively identified as the isatin **502** from its infrared and ultraviolet

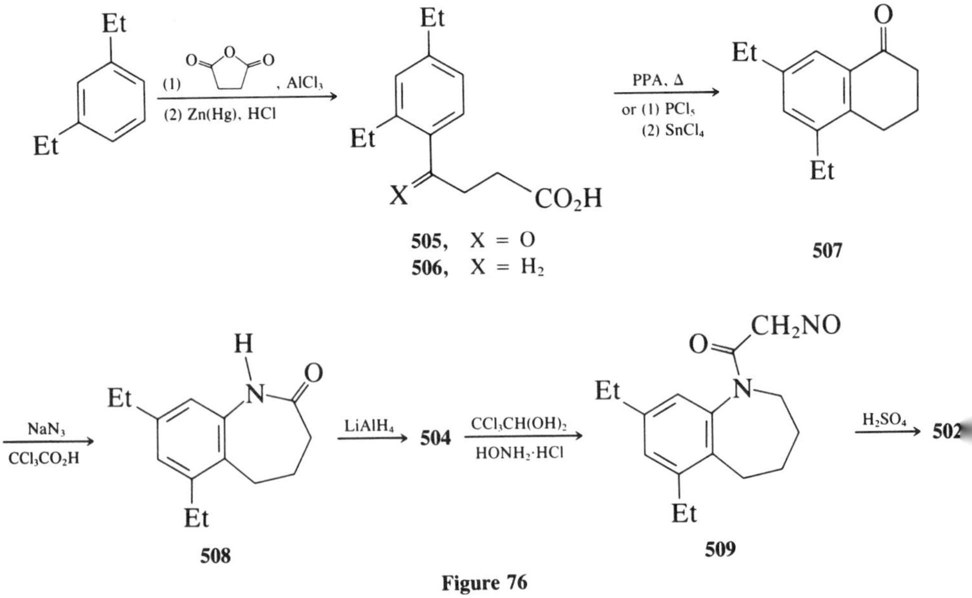

Figure 76

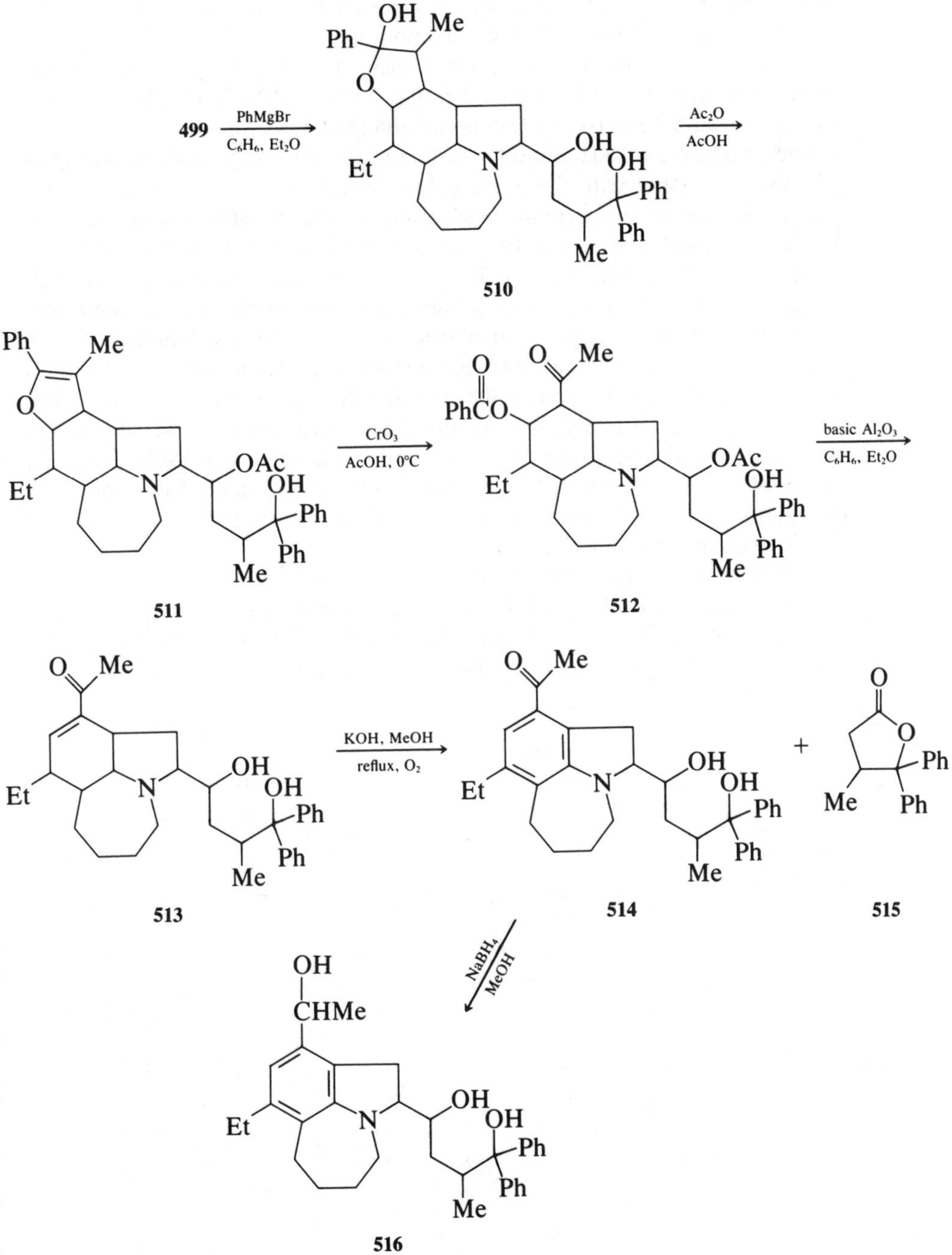

Figure 77

spectra. Ozonolysis of **501** gave the *N*-valeroylamino aldehyde **503a** and the corresponding acid **503b.** The latter compound was obtained by permanganate or nickel peroxide oxidation of **503a.** Hydrolysis and simultaneous decarboxylation was effected by heating **503b** in 35% hydrochloric acid–acetic acid, and gave rise to the amine **504** (155,156,157).

The structures of **502** and **504** were confirmed by independent synthesis (155,157). Together with the previously reported spectral and chemical evidence, the unambiguous chemical synthesis of **502** and **504** provided support for the structures assigned to the alkaloids **499** and **501.** Friedel–Crafts acylation of 1,3-diethylbenzene with succinic anhydride (Fig. 76) gave **505,** which was converted by Clemmensen reduction to the butyric acid **506.** Cyclization of **506** with polyphosphoric acid gave the tetralone **507,** which was in turn converted to the benzazepine **508** by a Schmidt reaction. Lithium aluminum hydride reduction of **508** afforded the amine **504,** which was identical to the product previously obtained by degradation of **501.** When **504** was treated with chloral hydrate and hydroxylamine, the *N*-nitrosoacetyl amine **509** was formed. In the presence of sulfuric acid, **509** underwent cyclization to **502,** which was identical in all respects to the product isolated above from natural sources.

Gotz and co-workers have published the results of their chemical studies of tuberostemonine (**499**) (158). Treatment of **499** with excess phenylmagnesium bromide (Fig. 77) afforded the triol **510,** which on treatment with acetic anhydride gave the crystalline anhydromonoacetate **511.** When **511**

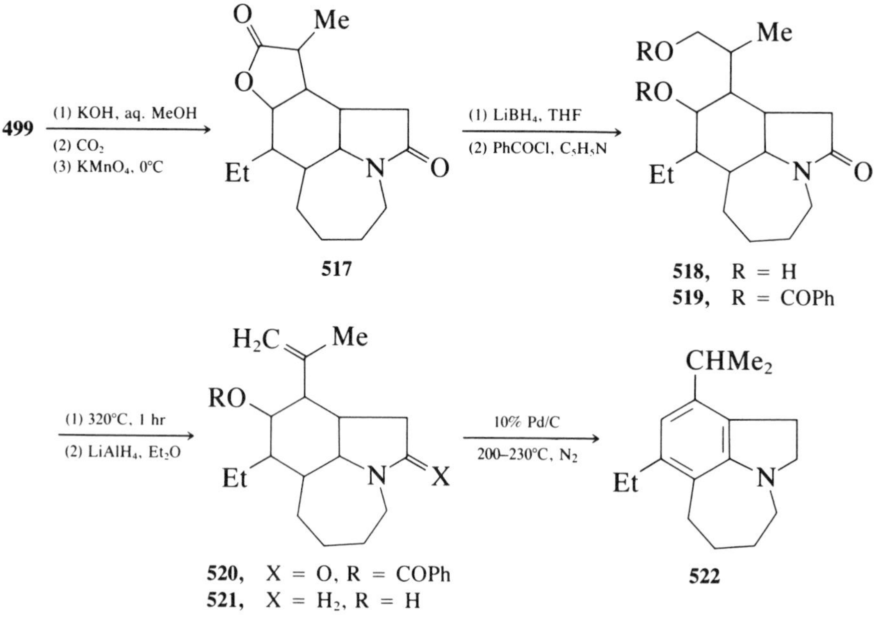

Figure 78

Pr-*i*

(1) $AlCl_3$, $C_2H_4Cl_2$
(2) Zn, HCl

X (CH$_2$)$_3$CO$_2$H

523, X = O
524, X = H_2

HF

525

MeMgI
THF

526

(1) Ac_2O, NaOAc, reflux
(2) TsOH, C_6H_6, reflux

527

(1) O_3, EtOAc, −78°C
(2) H_2/Pd–C
(3) $KMnO_4$, H_2O

528, R = Me
529, R = OH

(1) $NaBH_4$
(2) H_2/10% Pd–C, HOAc

530

(1) HF
(2) H_2NOH

531, X = O
532, X = NOH

(1) PPA, 125°C
(2) $LiAlH_4$, Et_2O

533, X = O
534, X = H_2

(1) ClC(O)C(O)Cl, Et_2O
(2) $AlCl_3$, CS_2

535

$Na_2S_2O_4$
H_2O

536

$LiAlH_4$, Et_2O

522

Figure 79

was oxidized with chromium trioxide in acetic acid, the methyl ketone **512** was obtained. Attempted recrystallization of **512** or passage through a basic alumina column effected elimination of benzoic acid and hydrolysis of the acetate ester, yielding the α,β-unsaturated ketone **513.** Heating **513** with methanolic potassium hydroxide under reflux in the presence of air gave the aromatic ketone **514** along with some unchanged starting material. Sodium borohydride reduction of the carbonyl group in **514** gave the triol **516.**

Permanganate oxidation of **499** (Fig. 78) produced a neutral compound (**517**), which, when reduced with lithium borohydride, gave the lactam diol **518;** the latter was converted to the dibenzoate **519.** Pyrolysis of **519** at 320–330°C gave a mixture in which the olefin **520** predominated. The product was reduced directly to **521** with lithium aluminum hydride. Without further purification, **521** was dehydrogenated by heating with palladium on carbon to give **522.** In subsequent work, **522** was synthesized by an unambiguous route and shown to be identical to the product obtained from **521** (159). Thus, the carbon skeleton of the lactam **517** was confirmed, and the structure of tuberostemonine (**499**) was rigorously established.

The indole **522** was prepared according to the sequence in Fig. 79 (159). Cumene and glutaric anhydride in the presence of aluminum chloride gave **523.** Clemmensen reduction of the keto acid **523** yielded **524,** which was cyclized to the benzosuberone **525** with hydrogen fluoride. Addition of methylmagnesium iodide to **525** yielded the alcohol **526,** which was dehydrated

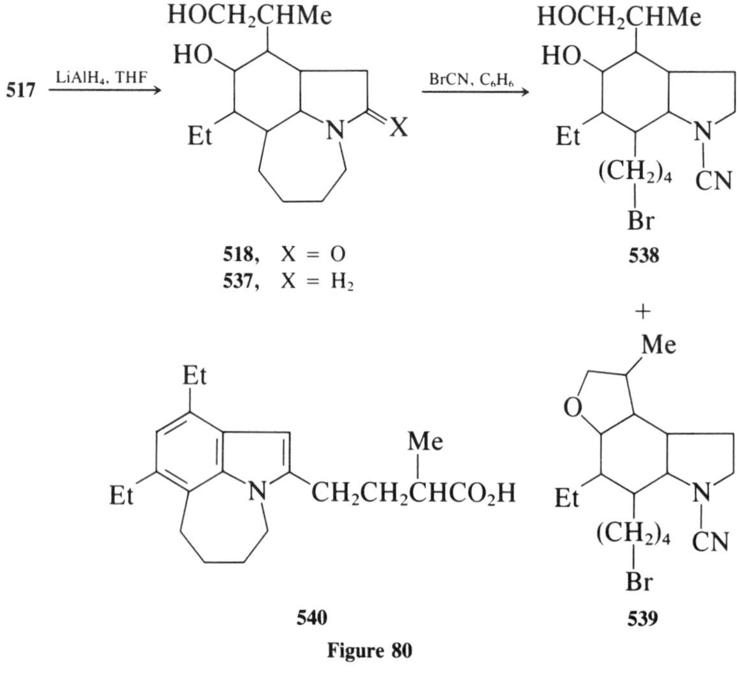

517 $\xrightarrow{\text{LiAlH}_4\text{, THF}}$ HOCH$_2$CHMe, HO, Et, N, X $\xrightarrow{\text{BrCN, C}_6\text{H}_6}$ HOCH$_2$CHMe, HO, Et, N, (CH$_2$)$_4$, CN, Br

518, X = O
537, X = H$_2$

538

+

Me, O, Et, N, (CH$_2$)$_4$, CN, Br

539

Et, Et, N, Me, CH$_2$CH$_2$CHCO$_2$H

540

Figure 80

with acetic anhydride–sodium acetate to a mixture of endocyclic and exocyclic olefins. Equilibration of the olefinic mixture with *p*-toluenesulfonic acid in benzene effected a complete conversion to the endocyclic olefin **527.** Ozonolysis followed by an oxidative workup gave the keto acid **528** and the diacid **529.** Reduction of the keto group in **528** with sodium borohydride followed by hydrogenolysis over palladium in acetic acid gave a good yield of the acid **530.** Hydrogen fluoride cyclized **530** to the tetralone **531** in excellent yield. On being heated in polyphosphoric acid at 125°C, the corresponding oxime **532** afforded the lactam **533,** which could be reduced with lithium aluminum hydride to the oily amine **534.** Treatment of **534** with oxalyl

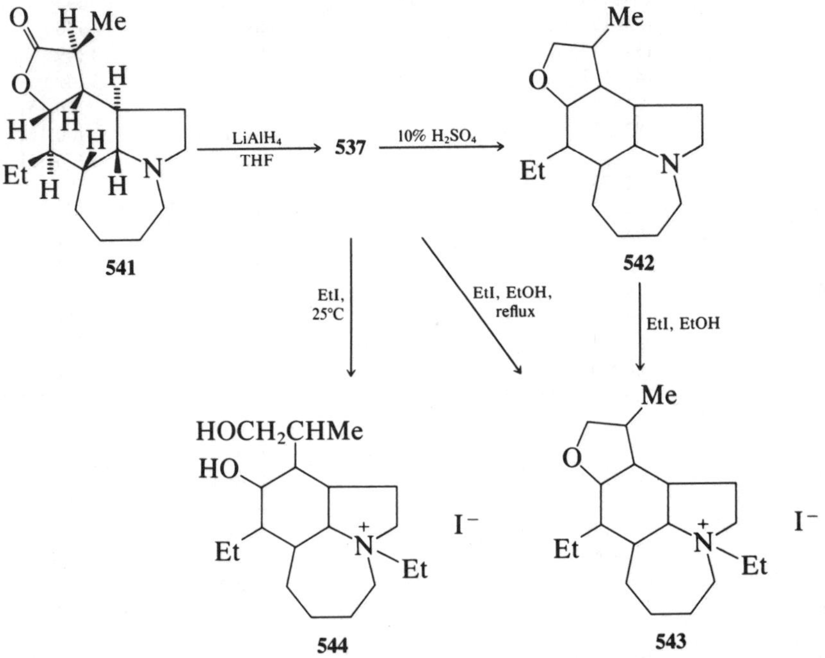

541 $\xrightarrow[\text{THF}]{LiAlH_4}$ **537** $\xrightarrow{10\%\ H_2SO_4}$ **542**

537 $\xrightarrow{\text{EtI, 25°C}}$ **544**

537 $\xrightarrow{\text{EtI, EtOH, reflux}}$ **543**

542 $\xrightarrow{\text{EtI, EtOH}}$ **543**

Figure 81

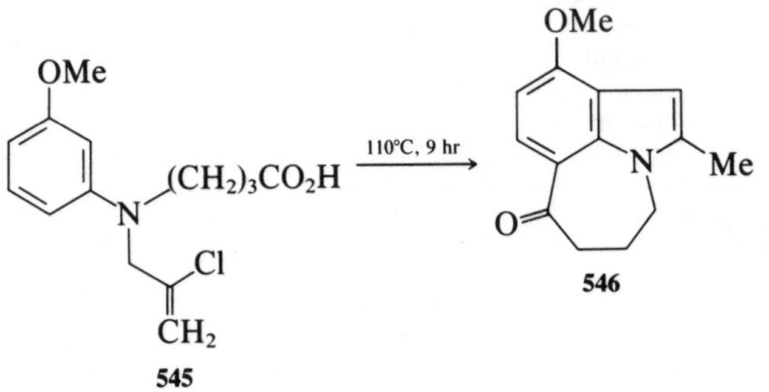

545 $\xrightarrow{110°C,\ 9\ hr}$ **546** (61)

TABLE 31. AZEPINO[3,2,1-*hi*]INDOLES[a]

Compound	mp (°C)	Salt	Spectra	Refs.
464	189.5–191.5			142
466a	112–112.5			143
466b	177–180	HCl		143
467	154–155			143
469	110–111			143
475	72–73			143
476	Colorless oil			143
477a	170–171			143
481	91–93			143
481	174–176	MeI		143
484a	155–155.5		uv	148
484b	134–139		ir	149
484c	145–147			149
487a	91–92		uv	150,151
487b	82–83		uv,ir	150,151
488a	—		ir,uv	150,151
491	72–73		uv	151
493	121–122		uv	151
494a	94–95		uv	151
494b	118–121		uv	151
494c	114–115			151
495	144–146			151
497	238–241			151
498	160/5.0[b]		uv	151
501	200–220/0.1[b]			155,156,157
502	110–112			155,156,157
512	120–130	AcOH	ir,uv,pmr	158
513	105–120		ir,uv,pmr	158
514	185–189		ir,uv,pmr	158
516	173–176		ir,uv,pmr	158
518	163–164		ir	158
519	169–172		ir,pmr	158
520	—		ir,pmr	158
521	—		ir	158
522	46–50; 49–53		ir,uv,nmr	158,159
535	79–82		ir,uv	159
536	149–162			159
537	208–214/0.075[b,c]		ir	161
	92–94			160
	165–170	HCl		161
540	96–98		ir,uv,nmr	162
544	204–205			160,161
546	145–146		ir,nmr,ms	163

[a]Figs. 71–81, Eqs. 59, 60.
[b]Boiling point (°C/torr).
[c]$[\alpha]_D^{30} = -26.39$.

chloride followed by aluminum chloride in carbon disulfide afforded the blood red isatin **535.** Reduction of **535** was effected in two steps. Treatment with sodium dithionite yielded the dioxindole **536,** and further reduction with lithium aluminum hydride converted **536** to the indole **522** in 47% yield. The product obtained via this route was indistinguishable from the "natural" indole.

Other aspects of the chemistry of azepino[3,2,1-*hi*]indoles are described in several more publications dealing with tuberostemonine. For example, whereas lithium borohydride reduction of the lactam **517** gave rise to the diol **518** (Fig. 78), lithium aluminum hydride reduction of **517** (Fig. 80) gave **518** and the fully reduced amine **537** (160,161). The amide group in **518** could be reduced on extended treatment with lithium aluminum hydride. Reaction of the amine **537** with cyanogen bromide gave rise to the cleavage products **538** and **539.** Earlier in this section, the dehydrogenation of tuberostemonine (**499**) was reported to give the indole **501** (Fig. 75). In a subsequent publication, Uyeo identified some of the minor components of that same reaction (162), one of which was the azepinoindole **540.**

In addition to tuberostemonine (**499**), several other alkaloids have been isolated from *Stemona tuberosa.* Uyeo reported the isolation and characterization of stenine (**541**) (160,153). Lithium aluminum hydride reduction of stenine (Fig. 81) gave the diol **537.** This diol was identical to the product obtained by lithium aluminum hydride reduction of the lactam **517,** a compound previously obtained from tuberostemonine by permanganate oxidation. Treatment of **537** with 10% sulfuric acid gave the cyclic ether **542.** Reaction of **537** with ethanolic ethyl iodide under reflux gave the ethiodide **543,** whereas at room temperature the ethiodide **544** was obtained.

In an isolated example of the preparation of new heterocycles by cyclization of 2-chloroallyl anilines (163), **545** was converted in low yield to the azepinoindole **546** (Eq. 61).

Melting points and other data for azepino[3,2,1-*hi*]indoles are given in Table 31.

40. Azepino[3,4,5-*cd*]indoles

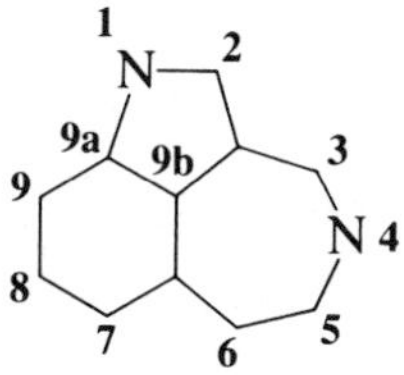

A series of patents by McMannus accounts for all but one of the reported examples of the azepino[3,4,5-*cd*]indole ring system (164–166). Condensation of nitromethane with the indole-4-carboxaldehyde **547** gave the nitro olefins **548,** which were reduced directly with lithium aluminum hydride to the

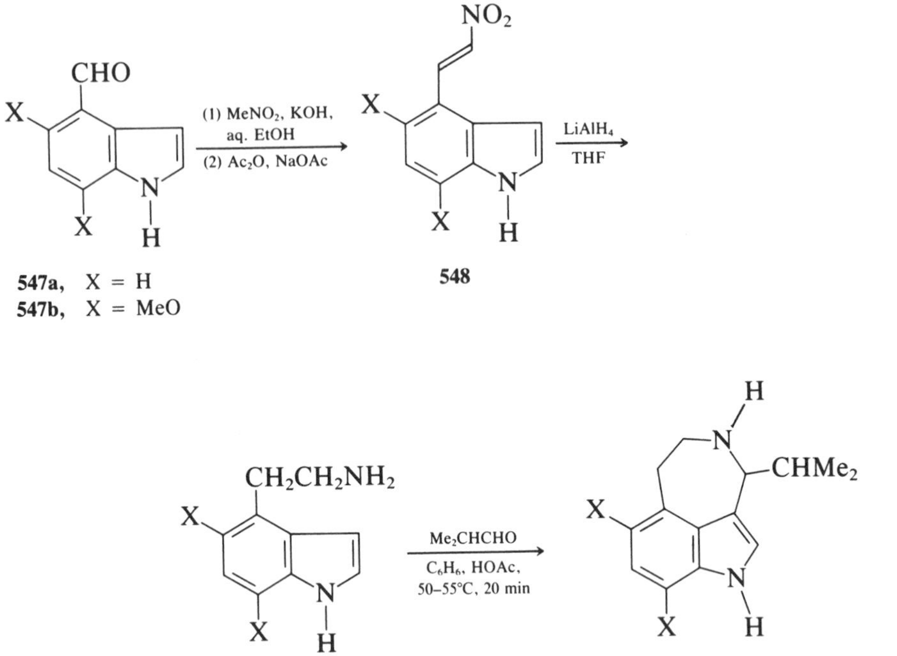

CHO

X

N

X H

(1) $MeNO_2$, KOH, aq. EtOH
(2) Ac_2O, NaOAc

NO_2

X

N

X H

$LiAlH_4$
THF

547a, X = H
547b, X = MeO

548

H

N

$CH_2CH_2NH_2$

$CHMe_2$

X

N

X H

Me_2CHCHO
C_6H_6, HOAc, 50–55°C, 20 min

X

N

X H

549

550a, X = H; mp 173.5–175°C
550b, X = MeO; mp 192°C (dec.)

Figure 82

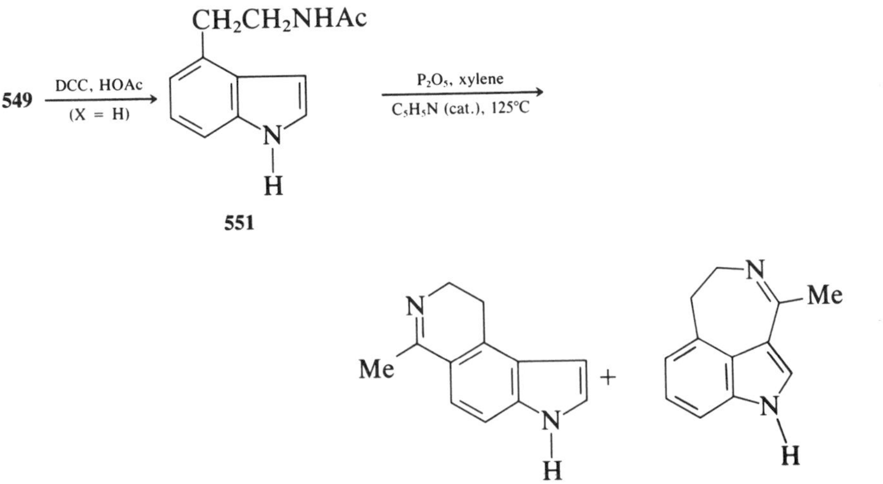

CH_2CH_2NHAc

549 DCC, HOAc (X = H)

N

H

P_2O_5, xylene
C_5H_5N (cat.), 125°C

551

N

N

Me

Me

+

N

N

H

H

552

553, mp 228–230°C (dec.) (ir,uv)

Figure 83

aminoethylindoles **549** (Fig. 82). Pictet–Spengler condensation of **549** with isobutyraldehyde gave rise to the azepinoindoles **550,** which were claimed to be orally active hypoglycemic agents.

In the only other reported example of this system, **549** (X = H) was acetylated with acetic acid and dicyclohexylcarbodiimide (Fig. 83). The resulting acetamide **551** was cyclized under Bischler–Napieralski conditions to give a mixture of **552** and **553** in 70 and 18% yields, respectively (167).

41. Azepino[4,3-*b*]indoles

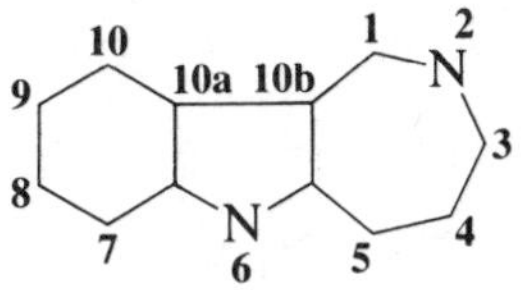

10 1 2 9 10a 10b N 3 8 N 4 7 6 5

The formation of **557** from the *anti*-oxime **554** was discussed briefly in Section IV.37. First reported by Teuber and co-workers (135), the reaction was shortly thereafter described in somewhat greater detail by Hester (128,129). Reaction of the ketotetrahydrocarbazole **555** with hydroxylamine gave rise to a single sharp-melting oxime **554,** which was assigned the *anti* configuration. Beckmann rearrangement in polyphosphoric acid (Eq. 62) converted **554** into **557** rather than the isomeric lactam **556.** The result can be accommodated by invoking an acid-catalyzed isomerization of **554** to **559** via the

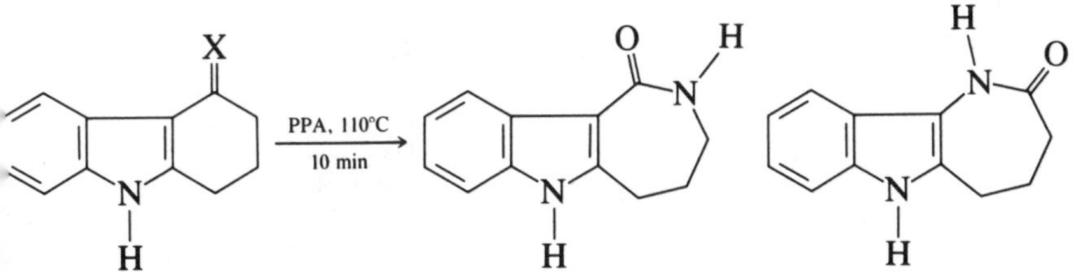

PPA, 110°C 10 min (62)

554, X = NOH
555, X = O

557

556

intermediate **558** (Eq. 62a). This isomerization may be energetically preferred over the sterically strained intermediate **560** which would be required for aryl migration.

The lactam **557** was stable to acidic or basic hydrolysis. On reaction with nitrous acid (Fig. 84), **557** afforded the *N*-nitroso lactam **561** (135). Lithium aluminum hydride in refluxing dioxane reduced **557** to the corresponding amine **562** in 70% yield. Formylation of **562** gave a 61% yield of **563a,** which could be reduced with lithium aluminum hydride in tetrahydrofuran to the tertiary amine **563b** in 84% yield. Azepino[4,3-*b*]indoles are listed in Table 32.

(62a)

Figure 84

TABLE 32. AZEPINO[4,3-*b*]INDOLES[a]

Compound	mp (°C)	Spectra	Refs.
557	219–220	uv,ir,nmr	128,129
	210	uv,ir,nmr	135
561	180	uv,ir,nmr	135
562	200–202.5	uv,nmr	128,129
563a	160–161	uv,ir	128,129
563b	168.5–170	uv	128,129

[a]Eq. 62, Fig. 84.

42. Azepino[4,3,2-*cd*]indoles

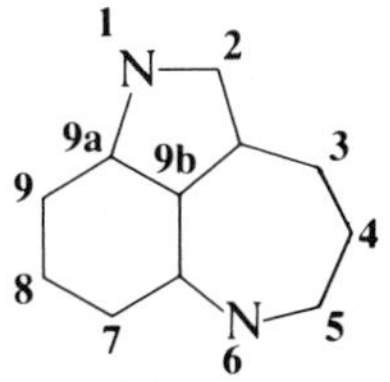

All of the published work on the azepino[4,3,2-*cd*]indole ring system has been conducted in a single laboratory (168–170). Condensation of 4-nitrogramine (**564a**) with diethyl malonate (Fig. 85) gave the expected adduct **564b** in 64% yield. Reduction of **564b** afforded, depending on the reaction conditions, a 41% yield of the azepinoindole **565** or a 92% yield of the amino malonate **564c.** Thermolysis of **564c** gave **565** in 72% yield. Mild alkaline hydrolysis of **565** gave the crude acid **566,** which upon heating to 170°C underwent smooth decarboxylation to the lactam **567** in 90% overall yield. Lithium aluminum hydride reduction of **567** afforded the corresponding amine **568** in 53% yield.

Either of the nitrogen atoms in **568** could be substituted, depending on the choice of reaction conditions (Fig. 86). Thus, formylation yielded the N_6-substituted product **569,** which could be subsequently reduced with lithium aluminum hydride to the tertiary amine **570.** On the other hand, alkylation with an alkyl halide and sodium hydride in dimethyl formamide yielded the N_1-substituted compounds **571.** Because of its apparent instability, **571a** could be isolated only as the N_6-acetyl derivative **572a.** Compounds **572b** and **572c** were also prepared.

The dihydrochloride salt of **571c** is an especially interesting molecule. Whereas most pmr spectra of the compounds in this series contain four-proton aromatic multiplets, the spectrum of **571c** in D_2O displays a two-proton aromatic doublet with peaks at 446 and 441 cps (recorded on a 60 MHz instrument). Furthermore, the exchangeable proton peak at 290 cps is

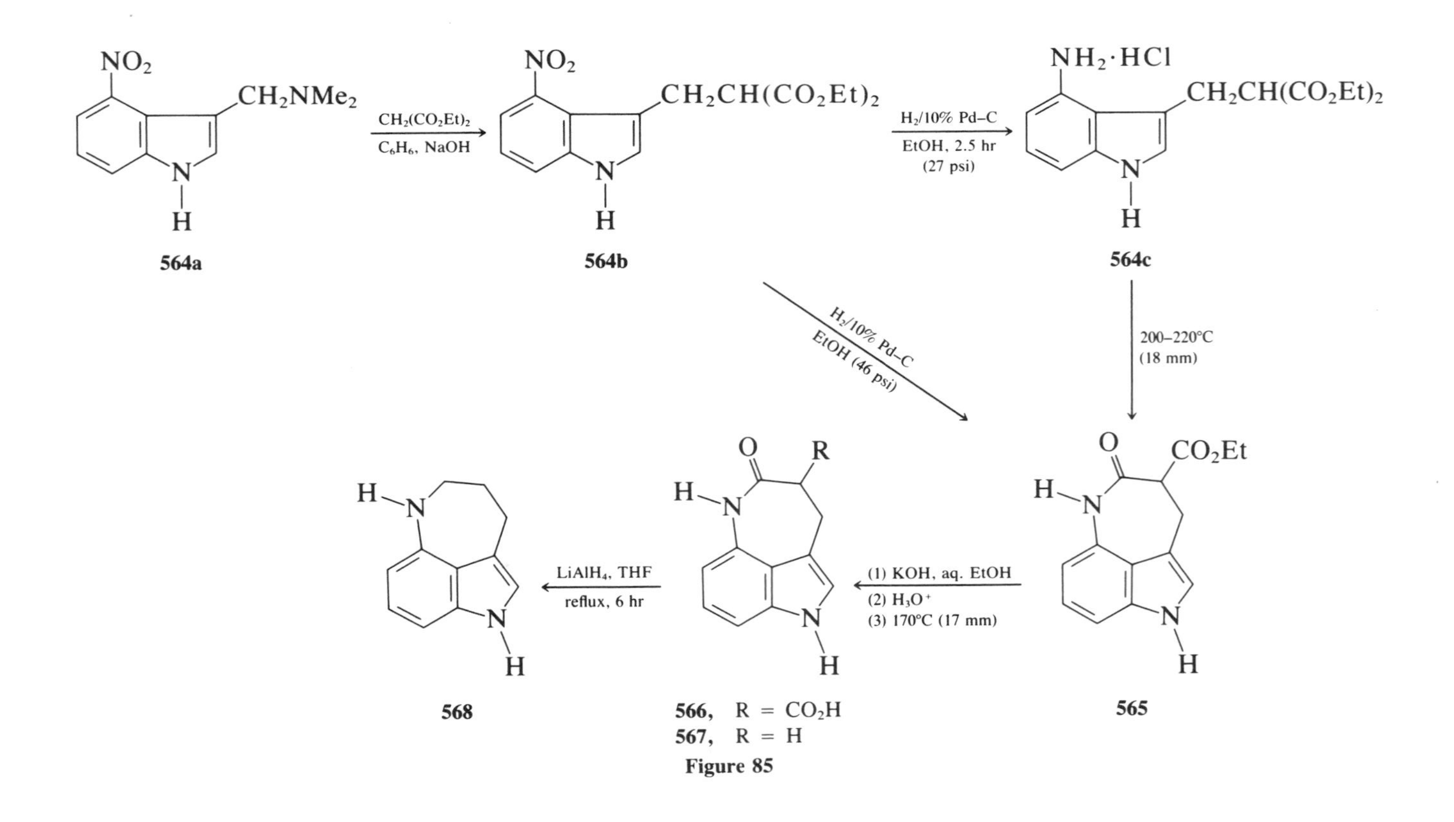

NO2
CH2NMe2
N
H
564a
CH2(CO2Et)2
C6H6, NaOH
NO2
CH2CH(CO2Et)2
N
H
564b
H2/10% Pd–C
EtOH, 2.5 hr
(27 psi)
NH2·HCl
CH2CH(CO2Et)2
N
H
564c
H2/10% Pd–C
EtOH (46 psi)
200–220°C
(18 mm)
O
CO2Et
H–N
N
H
565
(1) KOH, aq. EtOH
(2) H3O+
(3) 170°C (17 mm)
O
R
H–N
N
H
566, R = CO2H
567, R = H
LiAlH4, THF
reflux, 6 hr
H–N
N
H
568

Figure 85

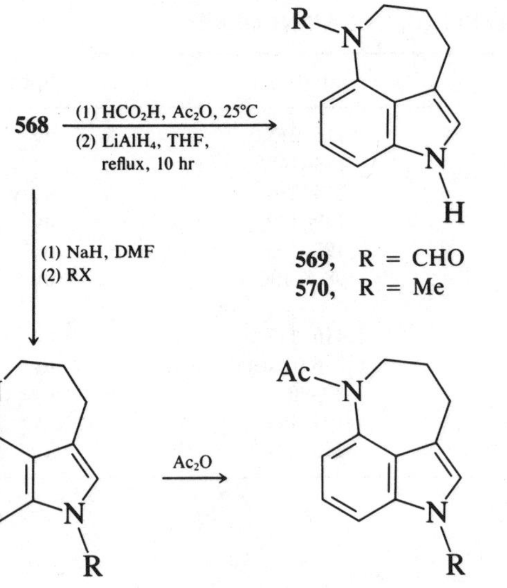

568 $\xrightarrow{\text{(1) HCO}_2\text{H, Ac}_2\text{O, 25°C; (2) LiAlH}_4\text{, THF, reflux, 10 hr}}$

(1) NaH, DMF
(2) RX

Ac₂O

569, R = CHO
570, R = Me

571a, R = Me
571b, R = $CH_2C_6H_4Cl$-*p*
571c, R = $(CH_2)_3NMe_2$

572a, R = Me
572b, R = H
572c, R = $(CH_2)_3NMe_2$

Figure 86

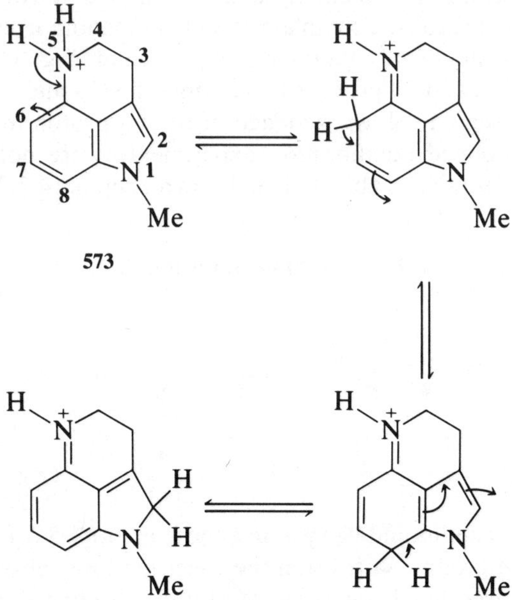

573

Figure 87

TABLE 33. AZEPINO[4,3,2-*cd*]INDOLES[a]

Compound	mp (°C)	Spectra	Refs.
565	211–213.5	uv	169,170
566	143–148		169,170
567	214–216	ir,uv	169,170
568	119–120	uv	168–170
569	192–193	uv	168–170
570	98.5–100.5	uv	168–170
571a			168–170
571b	116–117.5	uv	168–170
571c (2HCl)	276 (dec.)	uv	168–170
572a	89–90	uv	168–170
572b	197–199	uv	168–170
572c (HCl)	250–251	uv	168,170

[a]Figs. 85 and 86.

larger than expected, giving rise to the plausible rationale that two of the aromatic protons are exchangeable.

Because of the insolubility of **571c** in any other solvents, the pyrrolo[4,3,2-*de*]quinoline **573** was likwise examined. In d_6-DMSO, **573** showed a normal four-proton aromatic multiplet. However, in D_2O **573** displayed a two-proton aromatic doublet, and the exchangeable proton peak accounted for the remaining aromatic protons. To explain this phenomenon, it seems reasonable to suggest the existence of a tautomeric equilibrium (Fig. 87) in which rapid exchange of the C_6 and C_8 protons leaves only the C_2- and C_8-proton signals in the aromatic region. Slow exchange of the C_2 proton might also be expected, but prolonged equilibration experiments were not carried out to examine this possibility. Table 33 lists known azepino[4,3,2-*cd*]indoles.

43. Azepino[4,5-*b*]indoles

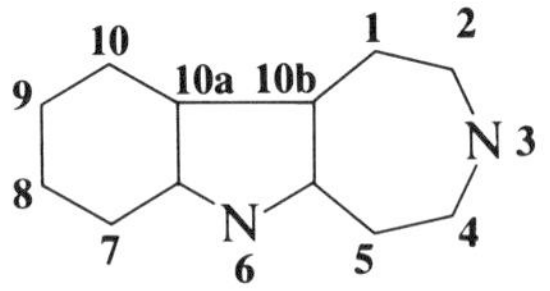

Of the various azepinoindole systems, the azepino[4,5-*b*]indoles comprise the largest single group. As is often the case, the large effort that has been devoted to this area has been prompted by the discovery of significant biological activity. The system first appeared in the literature in 1955 when the azepinoindole **578** (Fig. 88) was reported as an unexpected product of oxidation of the hydroquinonylpiperidine **574a.** Demethylation of the di-

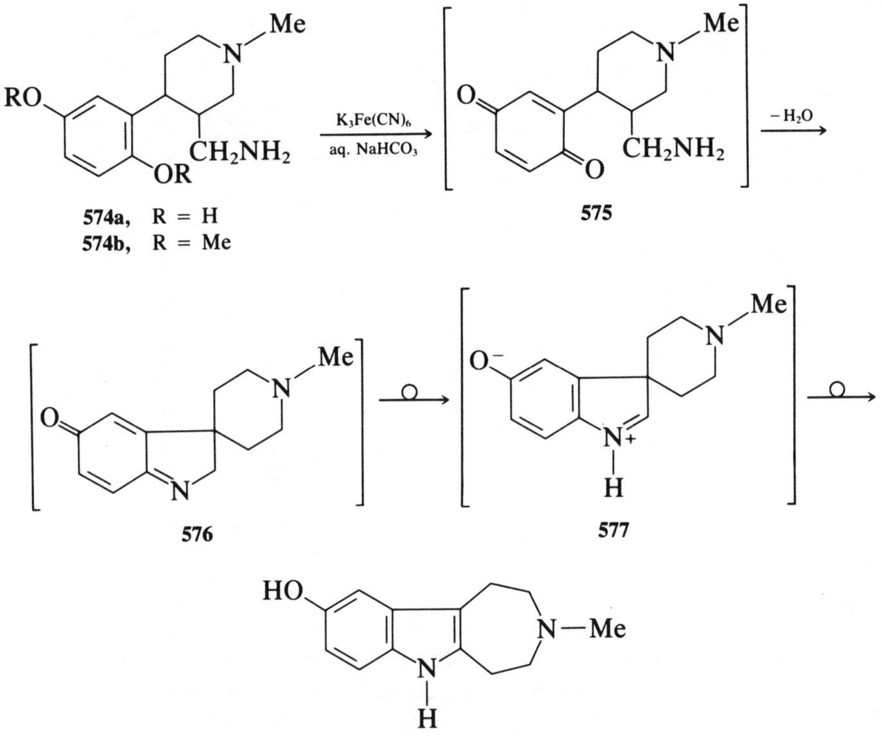

574a, R = H
574b, R = Me

$\xrightarrow[\text{aq. NaHCO}_3]{K_3Fe(CN)_6}$ **575** $\xrightarrow{-H_2O}$ **576** → **577** → **578**

Figure 88

methoxyarylpiperidine **574b** with hydrobromic acid gave **574a,** which on potassium ferricyanide oxidation afforded the azepinoindole **578** in 75% yield. Apparently the reaction proceeds via the *p*-quinone **575,** which spontaneously cyclizes to the quinone imine **576.** Intramolecular oxidation–reduction generates the intermediate **577,** which gives rise to the observed product **578** by the Wagner–Meerwein rearrangement (171).

Azepino[4,5-*b*]indoles were subsequently prepared by several other methods, but, because of a report in 1960 by Morosawa that 1-benzoyl-1-azacycloheptan-4-one (**582b**) did not undergo Fischer indole cyclization (132), 10 years elapsed before any more members of this ring system were described. In 1965 a new synthetic route to azepino[4,5-*b*]indoles, based on the Fischer cyclization, was described (172). This unique approach (Eq. 63) involved cyclization of the hydrazones **579a** and **579b** presumably via the spiroindolenines **580a** and **580b.**

Shortly thereafter, several publications described the successful Fischer cyclization of the *N*-alkyl derivative **582a** and the *N*-acyl derivative **582b** (Fig. 89) (173–177). Thus, **582a** and **582b** were converted to the arylhydrazones **583a** and **583b,** which were cyclized, either *in situ* or following iso-

579a, X = H
579b, X = MeO

580a, X = H
580b, X = MeO

(63)

581a, X = H
581b, X = MeO

lation, to the corresponding indoles **584a** and **584b.** One might reasonably have expected **582a** to give rise to a mixture of the azepino[4,5-*b*]indole **584a** and the azepino[4,3-*b*]indole **585a,** just as **582b** might be expected to generate a mixture of **584b** and **585b.** However, all published examples of this reaction have indicated exclusive formation of the [4,5-*b*] isomers. Consideration of the mechanism of the Fischer indole synthesis (178) suggests the isomeric intermediates **586** and **587,** which would lead to [4,5-*b*]- and [4,3-*b*] azepinoindoles, respectively. An examination of molecular models reveals very similar conformational preferences and nonbonded interactions in **586** and **587.** Since electronic arguments cannot explain the fact that both **583a** and **583b** give the same isomer, the selectivity of these reactions is difficult to explain.

The cyclization of the hydrazones **583** has been carried out under diverse reaction conditions, with yields of **584** varying widely as a function of both the conditions and the substituents R and Y. In some instances the hydrazones **583** have been isolated prior to cyclization, whereas in other reactions the hydrazones were generated and cyclized *in situ.* Conditions that have been reported for this reaction include the following: (1) **582a** and the appropriate arylhydrazine were heated under reflux in 3–10% ethanolic hydrogen chloride for 30–180 min (175); (2) **582a** and the arylhydrazine were heated under reflux in ethanol saturated with dry hydrogen chloride for 2 hr (173); (3) **582a** and the arylhydrazine were heated for 5 min in boiling water, dilute sulfuric acid was added, and the solution was refluxed for 1 hr (173); (4) the preformed hydrazone of **582a** was dissolved in acetic acid and treated at 25°C with boron trifluoride etherate, whereupon the mixture was slowly heated to 100°C, and finally heated for 30 min at 140°C (173); (5) the hydra-

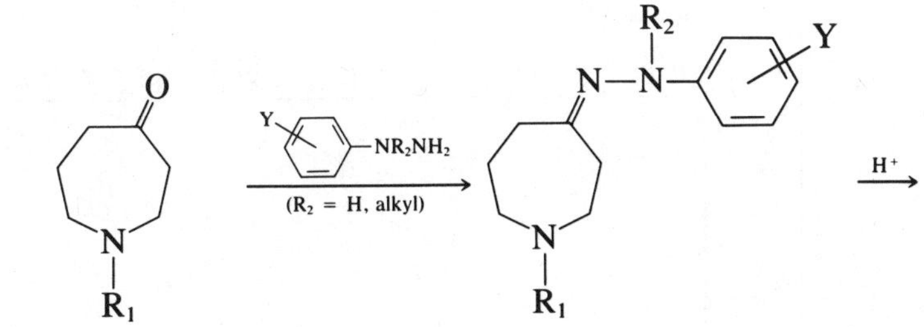

582a, R_1 = Alkyl
582b, R_1 = PhCO

$(R_2 = H, alkyl)$

H^+

583a, R_1 = Alkyl
583b, R_1 = PhCO

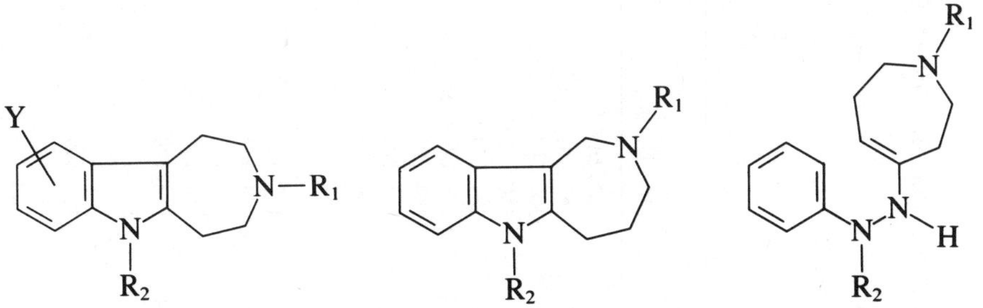

584a, R_1 = Alkyl
584b, R_1 = PhCO

585a, R_1 = Alkyl
585b, R_1 = PhCO

586

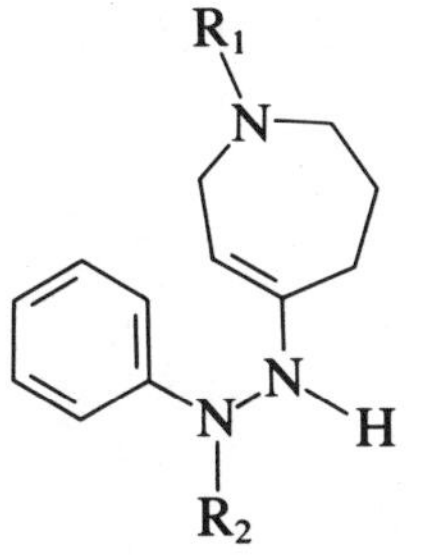

587

Figure 89

zone **583a** in polyphosphoric acid gave a spontaneous exothermic reaction (173); (6) the preformed hydrazone **583b** was heated at 80°C for 30–60 min in 88% formic acid (174). Generally, method 1 gave yields of 50–60%, whereas method 6 gave yields of about 40%, with some reactions giving yields below 10%. Methods 2–5 were described in the patent literature and specific yields were not given. Table 34 lists azepino[4,5-*b*]indoles that have been prepared by the Fischer indole synthesis.

Azepino[4,5-*b*]indolones have been prepared by several different methods. Compound **589** was reported to be the sole product formed in the

TABLE 34. 1,2,3,4,5,6-HEXAHYDROAZEPINO[4,5-*b*]INDOLES OBTAINED BY THE FISCHER INDOLE SYNTHESIS

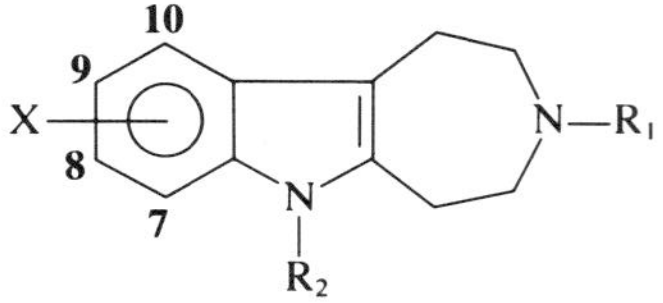

R_1	R_2	X	Yield (%)	mp (°C)	Salt	Refs.
Me	H	H	—	163–165		173
			58	162–162.5;		175
				200–201	CH_3I	
Me	H	9-Me	54	150.5–152		175
Me	H	9-MeO	52	231–231.5	HCl	175
Me	H	9-CO_2H	72	273–274 (dec.)		175
Me	H	9-OH	75	204–205;		171
				201–203	Dipicrate	
Et	H	H	—	106–108		173
Et	H	8-MeO	—	142		173
Et	H	8-Me	—	83–85		173
Et	H	10-Me	—	122–123		173
Et	H	8-Cl	—	134–135		173
Et	H	10-Cl	—	144–144.5		173
Et	H	9-Me	—	116–117		173
				178–179	Picrate	
Et	H	9-Cl	—	116–117		173
Et	H	9-MeO	—	137–138		173
$PhCH_2$	H	H	—	117–119		173
Ac	H	H	—	—		172
Ac	H	9-MeO	—	—		172

PhCO	H	H	36.9	169–170		174
PhCO	H	7-MeO	6.7	205–206		174
PCO	H	9-F	34.8	155–167		174
PhCO	H	7-Me	37.9	189–190		174
PhCO	H	9-Me	40.6	210–211		174
PhCO	H	10-MeO	6.8	264.5–266.5		174
PhCO	H	8-MeO	23.6	202–203.5		174
PhCO	H	9-MeO	9.4	126.5–127.5		174
Me	Me	H	94.5	231–233	HCl	175
Me	Me	H	—	237–238	CH_3I	175
Me	$PhCH_2$	H	—	201–202	HCl	173
Me	$PhCH_2$	8-Cl	—	122–124	HCl	133
Me	$PhCH_2$	9-Cl	—	248–249	HCl	173
Me	$PhCH_2$	10-Cl	—	223–225	HCl	173
Me	$PhCH_2$	9-CO_2Et	21	240–242	HCl	175
Et	$PhCH_2$	8-Me	—	201–203	HCl	173
Et	$PhCH_2$	9-Me	—	224–225	HCl	173
Et	$PhCH_2$	10-Me	—	232–235	HCl	173
Et	$PhCH_2$	8-Cl	—	220–222	HCl	173
Et	$PhCH_2$	9-Cl	—	240–242	HCl	173
Et	$PhCH_2$	10-Cl	—	248–251	HCl	173
Et	$PhCH_2$	8-MeO	—	215–217	HCl	173
Me	2-$ClC_6H_4CH_2$	H	—	228–229	HCl	173
Me	3-$MeOC_6H_4CH_2$	H	—	191–193	HCl	173
Et	4-$ClC_6H_4CH_2$	H	—	215–216	HCl	173
Me	$PhCH_2CH_2$	H	—	207–208	HCl	173
				231–232	Picrate	
Me	2-Picolyl	H	—	191–193	HCl	132
				178–180	Picrate	
H	$PhCH_2$	H	—	268–270	H_2SO_4;	173
				243–245	HCl	

Schmidt reaction of the tetrahydrocarbazalone **588** (Eq. 64) (135). Characterization of **589** was achieved primarily through spectral analysis, with the pmr spectrum being cited as definitive evidence. Apparently, none of the isomeric lactam **590** was formed. In the same publication the isomeric carbazalone **592** was likewise reported to give a single Schmidt lactam (Eq. 65), but in this instance the authors were unable to distinguish between the azepino[4,5-*b*]indole **593** and the [4,3-*b*] isomer **594.** Interestingly, while pmr data were used in order to characterize **589,** similar evidence was not reported for the latter lactam (**593** or **594**). Two subsequent publications (179,180) described the unequivocal synthesis of **593.** Since the melting point of **593** reported in these publications was different from that of the Schmidt lactam, it is possible that the latter compound actually has the structure **594.** Both Schmidt lactams, **589** and **594** (or **593**) were hydrolyzed with aqueous barium hydroxide to the corresponding amino acids **591** and **596** (or **595**), respectively.

An efficient synthesis of the 5-aryl–substituted azepino[4,5-*b*]indolones **598** was described by Freter (181), and the same methodology has since been described by Soviet authors (182). The indolones were prepared by intramolecular alkylation of the *N*-(α-chloroacetyl)tryptamines **597.** The reaction was carried out by briefly heating the amides **597** in a mixture of acetic acid-aqueous phosphoric acid (Eq. 66), and gave the indolones **598** in 50–70% yield. However, on extended heating in the same medium, the indolones were hydrolyzed with concomitant decarboxylation to the benzyl-

HN_3, $CHCl_3$, 0°C

588 → 589

aq. $Ba(OH)_2$ reflux

(64)

$CH_2CH_2NH_2$

CH_2CO_2H

590

591

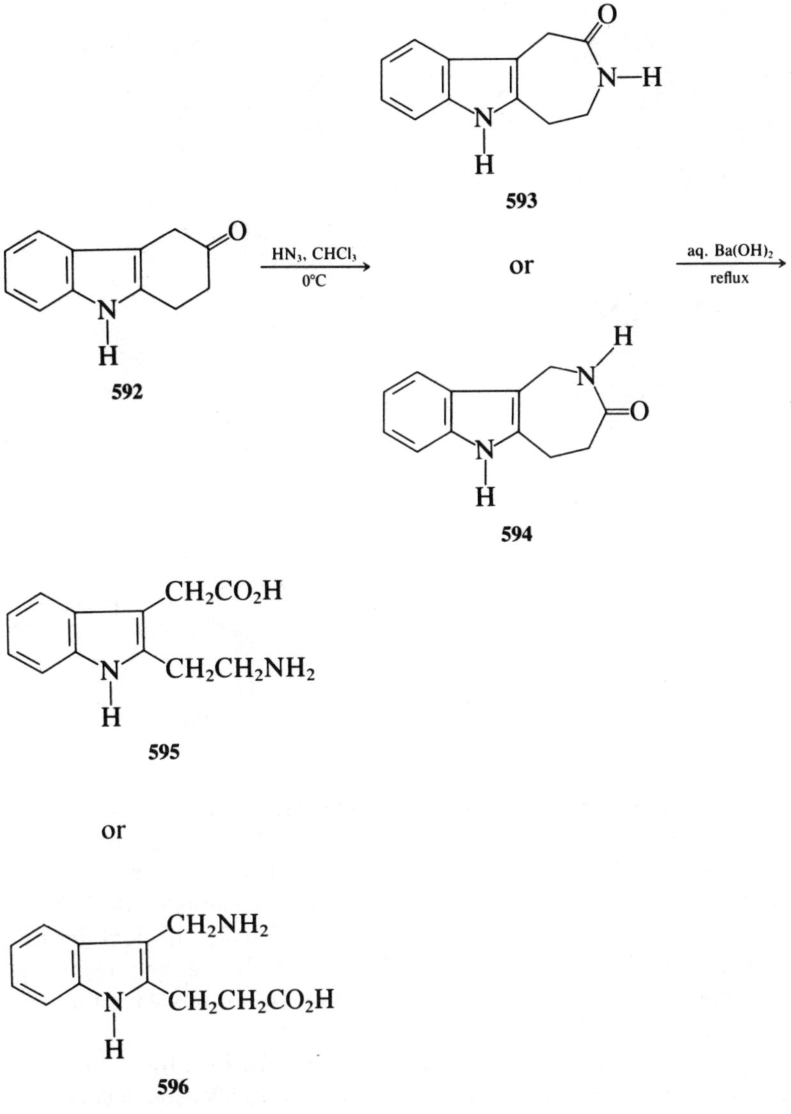

593

592 $\xrightarrow[0°C]{HN_3, CHCl_3}$ or $\xrightarrow[reflux]{aq. Ba(OH)_2}$

594

(65)

CH_2CO_2H

$CH_2CH_2NH_2$

595

or

CH_2NH_2

$CH_2CH_2CO_2H$

596

tryptamines **600** (181,182). Reduction of the lactams **598** with lithium aluminum hydride yields the corresponding amines **599** (181,183).

As was mentioned earlier, the azepinoindolone **593** has been prepared unequivocally by two independent groups (179,180). The method of preparation, the photocyclization of the *N*-chloroacetylaminoethylindole **601** (Eq. 67) is closely related to the synthesis of **598.** Irradiation of **601** in aqueous ethanolic base afforded an 85% yield of the lactam **593.** Similar photolysis of *N*-chloroacetyltryptamine (**602**) gave a 6:1 mixture (Eq. 68) of the azocinoindole **603** and the azepinoindole **589** (179).

(1) HOAc, aq. H_3PO_4, 1 hr, 120°C
(2) $LiAlH_4$, THF

HOAc, aq. H_3PO_4 reflux, 5 days

597

598, X = O
599, X = H_2

(66)

600

$h\nu$ (2537 Å)
50% aq. EtOH, Na_2CO_3 (or $NaHCO_3$)

$CH_2CH_2NHCOCH_2Cl$

601

593

(67)

A recent publication outlined a new method (Fig. 90) for the preparation of 4-oxo-3,4,5,6-tetrahydroazepino[4,5-*b*]indoles (**607**) starting from the 2-diazoacetylindoles **604** (184). Silver-catalyzed Wolff rearrangement of **604** in the presence of the aminoacetals **605** gave 50–70% yields of the indoleacetamides **606.** The acetamides could be cyclized using 15–17% hydrochloric acid at 70–80°C, polyphosphoric acid at 30–35°C, or even phosphoryl chloride. In practice, the first two sets of reaction conditions gave the best results.

One other route to azepino[4,5-*b*]indoles, described by French workers (Eq. 69), involves the ring expansion of the *β*-carbolines **608a** and **608b** (185). Reduction of these compounds with sodium borohydride in dioxane afforded the azepinoindoles **609a** and **609b** in 52 and 30% yields, respectively. Although formation of **609a** most likely occurs via reduction of the intermediate aziridinium ion **610,** this mechanism is less plausible for the formation of the sulfonamide **609b.** An alternative pathway would be the reductive cleavage of the C—N *β*-carboline bond, which is labile because of its benzylic relationship to the indole nucleus, as in **611.** Reductive cleavage to the sulfonamide anion **612** could be followed by tosylate displacement to give **609b.**

Alkylation of azepino[4,5-*b*]indoles has been thoroughly investigated and has been shown to occur at either the 3- or 6-position, depending on reaction

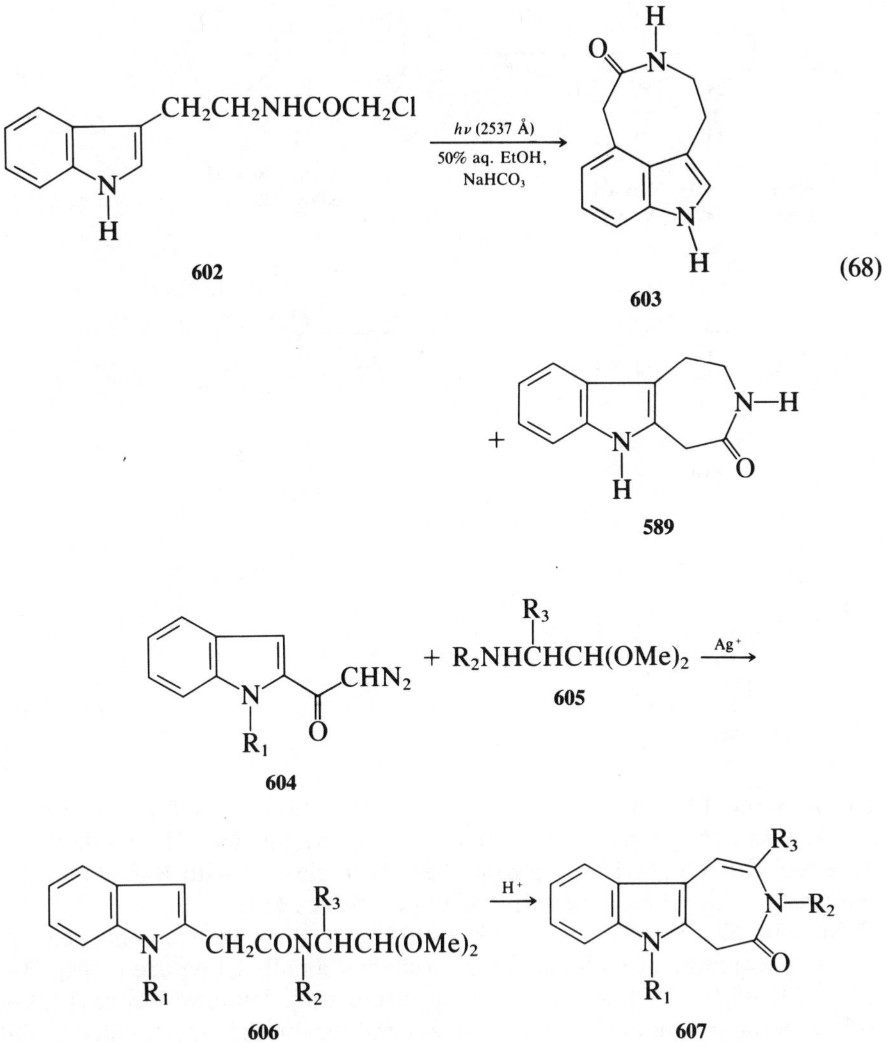

$CH_2CH_2NHCOCH_2Cl$

$h\nu$ (2537 Å)

50% aq. EtOH, $NaHCO_3$

602 603 (68)

+ 589

R_3

+ $R_2NHCHCH(OMe)_2$ $\xrightarrow{Ag^+}$

CHN_2

604 605

$CH_2CONCHCH(OMe)_2$ $\xrightarrow{H^+}$

606 607

Figure 90

conditions. In Fig. 91, some of these conditions are illustrated with the model azepino[4,5-*b*]indole **613.** Dimethyl and diethyl sulfate alkylate **613** exclusively at the 3-position to give **614a** and **614b,** respectively (173). The tertiary amine **614a** can also be obtained from the formate **614c** by lithium aluminum hydride reduction (173,174,186). Alkylation of azepino[4,5-*b*]indoles such as **613** with 4-chloro-*p*-fluorobutyrophenone gives exclusively the products of N_3-substitution, the teriary amines **615.** These adducts are of biological interest as potential neuroleptics (187,188). The tendency for preferential alkylation of the more basic nitrogen at the 3-position must be especially

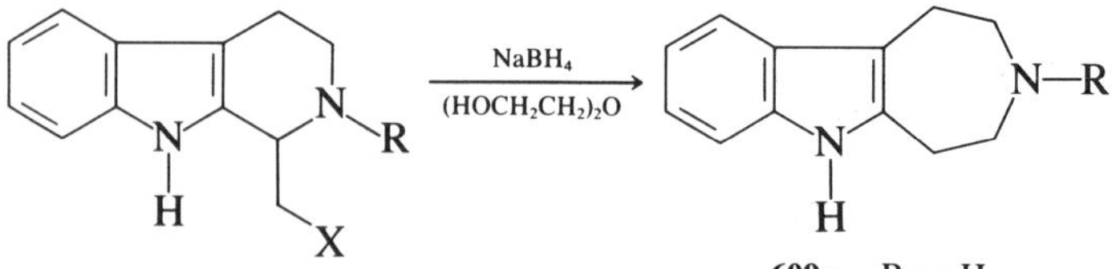

(69)

608a, R = H, X = Cl
608b, R = Ts, X = OTs

609a, R = H
609b, R = Ts

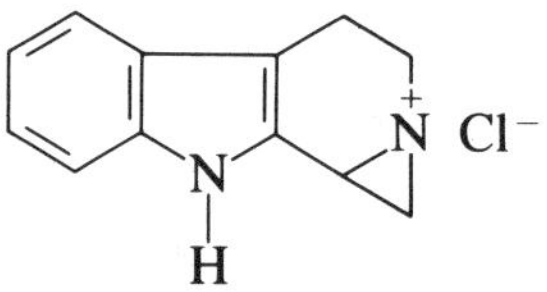

610

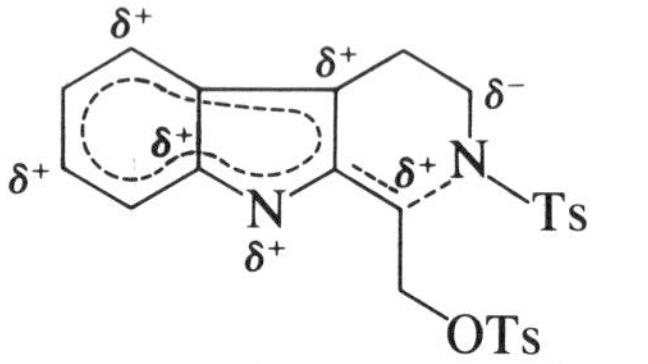

611

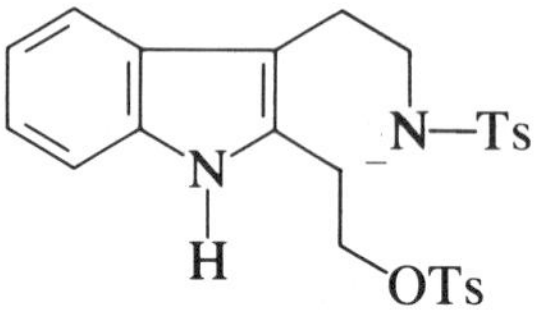

612

strong, since it has been found that **614a** and methyl iodide form the methiodide **616a,** to the virtual exclusion of the 6-methyl adduct (175). Methiodide **616a** and the 6-methyl analog **616b** were both cleaved with Raney nickel–aluminum alloy to give the tryptamine derivatives **617.**

Specific alkylation at the indole nitrogen (N_6) can be accomplished by sequential treatment with sodium hydride and an alkylating agent (Eq. 70) (173,174,186,187,189). However, in one instance alkylation with benzyl chloride generated a mixture of the 6-benzyl and the 3,6-dibenzyl adducts **618b** and **618c,** respectively (186). A second method of specific alkylation at N_6 has been to protect N_3 prior to alkylation (174,186,189,190). The benzoyl group has been a convenient protecting group because of its ease of subsequent removal. Thus the 3-benzoylazepinoindole **619** was converted to the 6-methyl derivative **620** by successive treatment with sodium hydride and methyl iodide in dimethylformamide (Fig. 92). Deprotection was accomplished stepwise by lithium aluminum hydride reduction to **621** and hydrogenolysis to **622.** The overall yield for the conversion of **619** to **622** was 77% (174).

Alkylation of 6-benzylazepino[4,5-*b*]indole (**623**) with dimethyl sulfate (Eq. 71) gave the corresponding 3-methyl compound **624.** The same alkylation could also be effected under Eschweiler–Clarke conditions as well as with

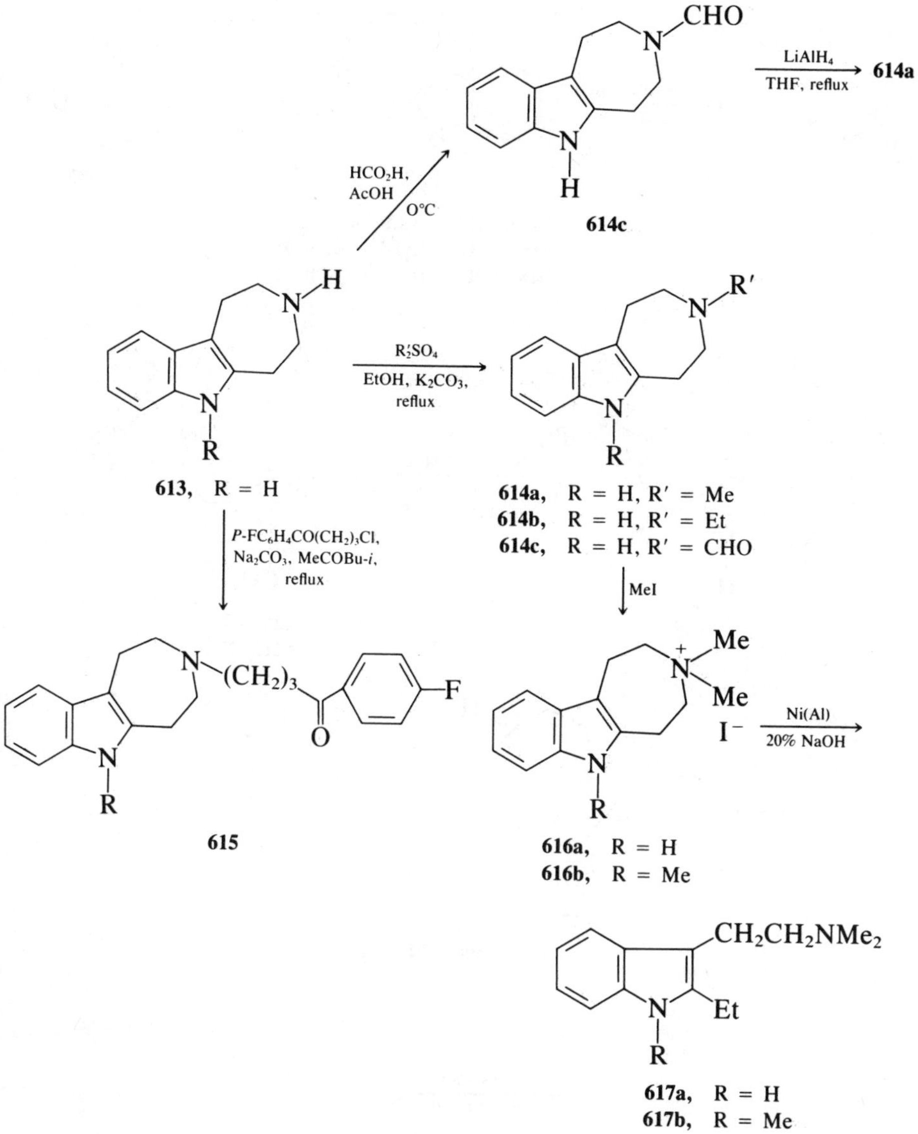

LiAlH$_4$, THF, reflux → **614a**

HCO$_2$H, AcOH, 0°C

614c

R'_2SO_4, EtOH, K_2CO_3, reflux

613, R = H

614a, R = H, R′ = Me
614b, R = H, R′ = Et
614c, R = H, R′ = CHO

MeI

P-FC$_6$H$_4$CO(CH$_2$)$_3$Cl, Na$_2$CO$_3$, MeCOBu-*i*, reflux

Ni(Al), 20% NaOH

615

616a, R = H
616b, R = Me

617a, R = H
617b, R = Me

Figure 91

formaldehyde–sodium borohydride (173). For other examples of 3,6-dialkyl-azepino[4,5-*b*]indoles prepared by these methods, the reader is referred to Tables 34–36.

Although the literature contains many examples of the alkylation of azepino[4,5-*b*]indoles, few dealkylations have been reported. As described above, benzyl groups are readily removed by hydrogenolysis (173,174,186,187,

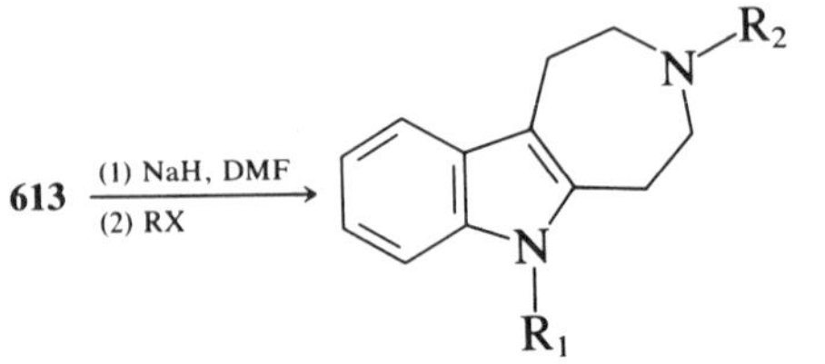

613 $\xrightarrow[\text{(2) RX}]{\text{(1) NaH, DMF}}$ (70)

618a, R_1 = Alkyl, R_2 = H
618b, R_1 = CH_2Ph, R_2 = H
618c, R_1 = R_2 = CH_2Ph

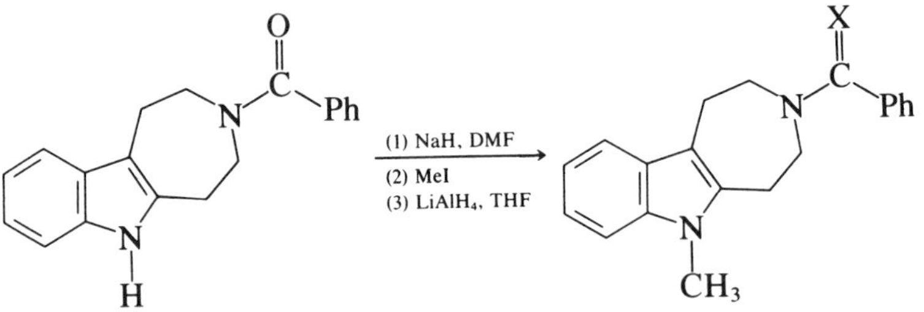

(1) NaH, DMF
(2) MeI
(3) $LiAlH_4$, THF

619

620, X = O
621, X = H_2

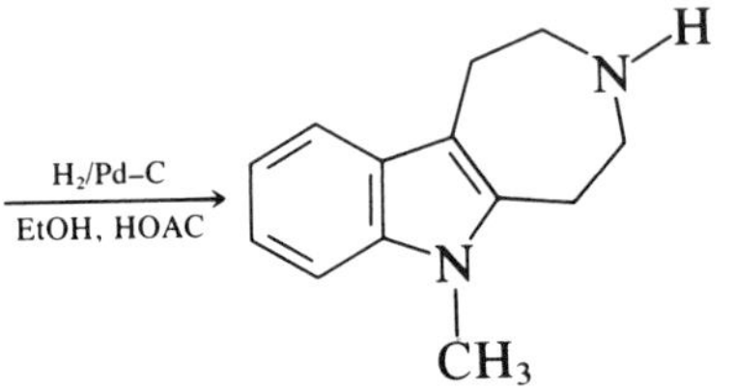

H_2/Pd–C
EtOH, HOAC

622

Figure 92

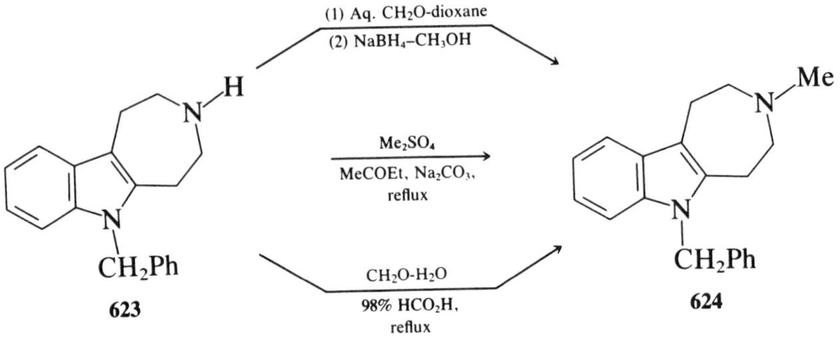

(1) Aq. CH_2O-dioxane
(2) $NaBH_4$–CH_3OH

Me_2SO_4
MeCOEt, Na_2CO_3,
reflux

CH_2O-H_2O
98% HCO_2H,
reflux

623

624

(71)

TABLE 35. 1,2,3,4,5,6-HEXAHYDROAZEPINO[4,5-*b*]INDOLES OBTAINED BY METHODS OTHER THAN THE FISCHER INDOLE SYNTHESIS

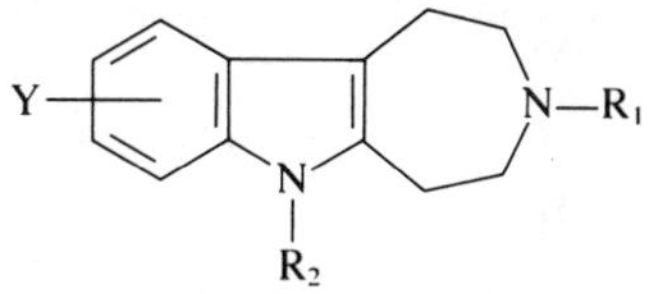

R_1	R_2	Y	mp (°C)	Other Data	Refs.
H	H	H	189–192		173
			190–193		174
			250.5–251	HCl,uv	
			164–165	Cyclamate	186
			247.5–248.5	HCl	
			185		185
			240–242	HCl	180
H	H	7-Me	271 (dec.)	HCl	174,176
H	H	9-Me	243.5–245		174,176
H	H	7-MeO	275–275.5		174,176
H	H	8-MeO	158–160.5		176
			276–276.5		174,186
H	H	9-MeO	174–176		176
			235–235.5	HCl	174,186
H	H	10-MeO	Oil		176
			236	HCl	174,18
H	H	9-F	179–180		174,176
Me	H	H	163–165		173
			165–166.5		174,186
			162		185
Et	H	H	106–108		173
$PhCH_2$	H	H	116–117	uv	174,176,186
CHO	H	H	220–221.5		174,186
CO_2Et	H	H	168–170		173
Tosyl	H	H	164		185
H	Me	H	214–215		174
			214–216	HCl	186
H	Et	H	253–254	HCl	174,176
H	$PhCH_2$	H	212–213	HCl	174,186
H	Me	9-MeO	272	HCl	176
H	Me	10-MeO	276.5–278	HCl	186,189
Me	H	9-OH	204–205		171
			201–203	Dipicrate, uv	
$PhCH_2$	H	7-Me	210.5–212	HCl	174,176,186
$PhCH_2$	H	9-Me	142.5–143.5		174,176,186
$PhCH_2$	H	7-MeO	247–248 (dec.)		174
			251–252.5		176,186
$PhCH_2$	H	8-MeO	146.5–147		174,176,186
$PhCH_2$	H	9-MeO	127.5–129.5		174,176,186
$PhCH_2$	H	10-MeO	163.5–164.5		174,176,186
$PhCH_2$	H	9-F	143–144		174,176,186
$PhCH_2$	H	8-CHO,9-MeO	207–208		194
$PhCH_2$	H	9-MeO,10-CHO	170–172		194
Me	Me	H	251–251.5	HCl	174,186
CHO	Me	H	125–128.5		174,186
Me	$PhCH_2$	H	201–202	HCl	173
			163–172 (0.02 mm)[a]		

TABLE 35. (*Continued*)

R_1	R_2	Y	mp (°C)	Other Data	Refs.
$PhCH_2$	$PhCH_2$	H	85.5–86.5		174,186
PhCO	Me	H	Oil		174
Me	 $(CH_2)_3N$ (piperazine ring) NCH_3	H	154–156	Maleate	175
Et	$PhCH_2$	H	199–202	Picrate	173
			223–228	HCl	
Me	Me	9-MeO	270 (dec.)		174
$PhCH_2$	Me	10-MeO	246–247.5	HCl	174,186
PhCO	Me	10-MeO	187.5–188		174,186

[a]Boiling point.

TABLE 36A. MISCELLANEOUS AZEPINO[4,5-*b*]INDOLES

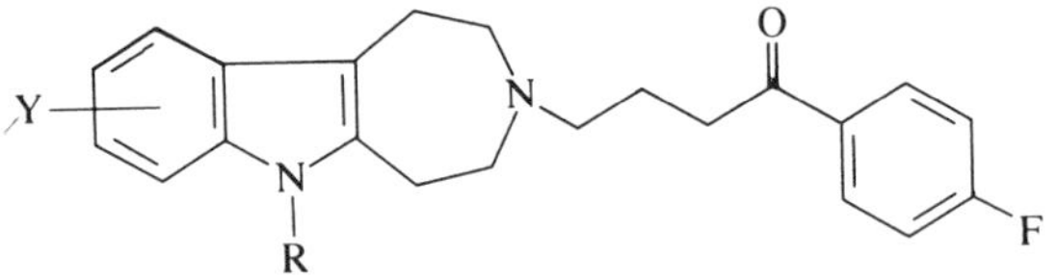

R	Y	mp (°C)	Salt	Refs.
H	H	101.5–102.5		176,187,188
Me	H	223–224		176,187
H	9-Me	103–104		176,187
H	8-MeO	137.5–138.5		176,187
H	9-MeO	171.5–172.5	Cyclamate	176,187
H	10-MeO	214.5–215.5	HCl	176,187
Me	10-MeO	87–89		187

TABLE 36B. MISCELLANEOUS AZEPINO[4,5-*b*]INDOLES

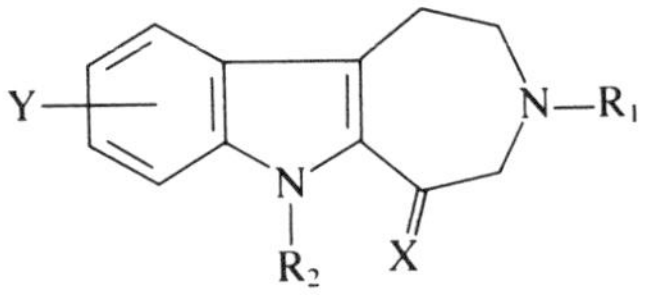

R_1	R_2	X	Y	mp (°C)	Salt	Refs.
Me	$PhCH_2$	H,AcO	H	110–112		193
Me	$PhCH_2$	H,OH	H	116–117		193
Et	$PhCH_2$	H,OH	H	103–105		193
Et	$PhCH_2$	H,OH	8-MeO	85–87		193
Me	$3\text{-}MeOC_6H_4CH_2$	H,OH	H	132–134		193
Me	$3\text{-}MeOC_6H_4CH_2$	H,OH	8-MeO	118–120		193
Me	$PhCH_2$	H,OH	9-Cl	134–135		193
Me	$PhCH_2$	O	H	203–205	H_2SO_4	193
Me	$3\text{-}MeC_6H_4CH_2$	O	H	170–172	H_2SO_4	193
Et	$PhCH_2$	O	9-Me	191–195	HCl	193

TABLE 36C. MISCELLANEOUS AZEPINO[4,5-*b*]INDOLES

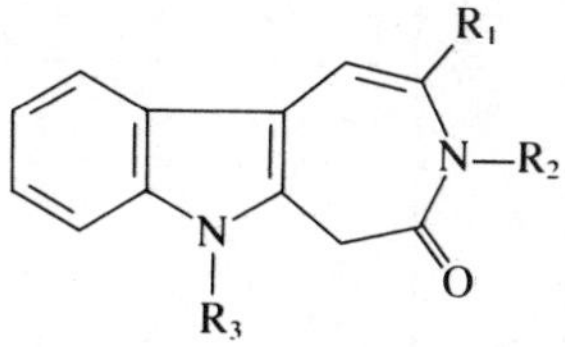

R_1	R_2	R_3	mp (°C)	Spectra	Refs.
Me	$PhCH_2$	H	180–181	ir,ms	184
H	$PhCH_2$	Me	119–120	ir,ms	184
H	$PhCH_2$	$PhCH_2$	163–164	ir	184
H	*i*-Pr	H	242–244	ir	184
Me	$PhCH_2$	Me	160–161	ir	184
Me	$PhCH_2$	$PhCH_2$	151–151.5	ir	184

TABLE 36D. MISCELLANEOUS AZEPINO[4,5-*b*]INDOLES

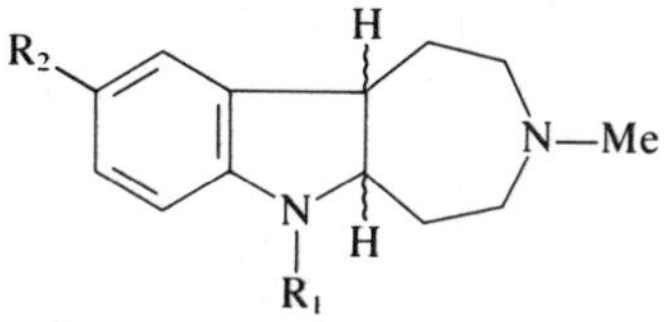

R_1	R_2	mp (°C)	Salt	Refs.
H	H	81–83		192
H	Me	58–60		191
		158–160 (2 mm)[a]		191
		235–237	2HCl	
NO	H	104–105.5		192
NH_2	H	73–75		192

[a]Boiling point.

189,190). Dealkylation of **625a** and **625b** with ethyl chloroformate (Eq. 72) gave the urethane **626,** which was hydrolyzed to **613** in unspecified yield with potassium hydroxide in diethylene glycol monomethyl ether (173).

Oxidation and reduction reactions of the azepino[4,5-*b*]indole ring system have been reported occasionally and are generally consistent with the known reactivity of substituted indoles. Reduction of **627a** and **627b** with zinc in aqueous ethanol–hydrochloric acid (Fig. 93) gave the corresponding indolines **628a** and **628b** in 90 and 80% yields, respectively (191,192). Nitrosation of **628a** and subsequent reduction with zinc–acetic acid gave sequentially the nitrosamine **629a** and the hydrazine **629b.** Compound **629b** underwent Fischer condensation with tetrahydro-4-thiopyranone to give the indole **630.**

Oxidation of 3,6-dialkylazepinoindoles **631** with mercuric acetate (Eq. 73)

TABLE 36E. MISCELLANEOUS AZEPINO[4,5-*b*]INDOLES

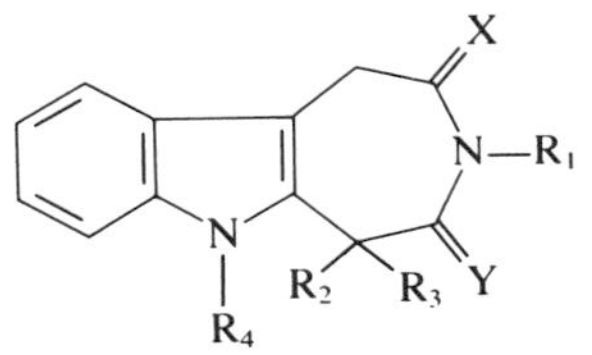

X	Y	R_1	R_2	R_3	R_4	mp (°C)	Other Data	Refs.
O	H_2	H	H	H	H	228–230	ir,pmr,ms	180
						230–232		179
						205–208 (?)		135
						245–247		
H_2	O	H	H	H	H	228–235	ir,pmr,uv	179
						261–263	pmr,ms	135
H_2	O	H	Ph	H	H	223		181
H_2	O	H	4-ClC_6H_4	H	H	225		181
H_2	O	H	4-BrC_6H_4	H	H	193–194		181
H_2	O	H	Ph	H	$PhCH_2$	87–88		182
H_2	O	H	$PhCH_2$	H	$PhCH_2$	134–136		182
H_2	O	H	$PhCOCH_2$	H	$PhCH_2$	100–102		182
H_2	O	H	Ph	Ph	$PhCH_2$	128–130		182
H_2	O	H	H	H	$PhCH_2$	109–111		182
H_2	O	$PhCH_2$	Ph	H	$PhCH_2$	96–97		182
H_2	O	$PhCH_2$	$PhCH_2$	H	$PhCH_2$	131–132		182
H_2	O	$PhCH_2$	$PhCOCH_2$	H	$PhCH_2$	102–103		182
H_2	O	$PhCH_2$	Ph	Ph	$PhCH_2$	75–76		182
H_2	O	$PhCH_2$	H	H	$PhCH_2$	117–119		182

H_2	H_2	H	Ph	H	$PhCH_2$	135–137	HCl	183
						76–77		
H_2	H_2	H	$PhCH_2$	H	$PhCH_2$	123–124	HCl	183
H_2	H_2	H	$PhCHOHCH_2$	H	$PhCH_2$	155–156	HCl	183
H_2	H_2	H	Ph	Ph	$PhCH_2$	106–107		183
						164–165	HCl	
H_2	H_2	H	H	H	$PhCH_2$	115–116		183
						173–175	HCl	
H_2	H_2	$PhCH_2$	Ph	H	$PhCH_2$	121–122	HCl	183
H_2	H_2	$PhCH_2$	$PhCH_2$	H	$PhCH_2$	149–151	HCl	183
H_2	H_2	$PhCH_2$	$PhCHOHCH_2$	H	$PhCH_2$	60–61	HCl	183
H_2	H_2	$PhCH_2$	Ph	Ph	$PhCH_2$	98–99	HCl	183
H_2	H_2	$PhCH_2$	H	H	$PhCH_2$	70–71	HCl	183
H_2	H_2	$CO(CH_2)_2Cl$	H	H	$PhCH_2$	131–132		183
H_2	H_2	$COCH_2Cl$	Ph	H	$PhCH_2$	172–173		183
H_2	H_2	$CO(CH_2)_2NEt_2$	H	H	$PhCH_2$	70–73	HCl	183
H_2	H_2	$COCH_2NEt_2$	Ph	H	$PhCH_2$	67–68	HCl	183
H_2	H_2	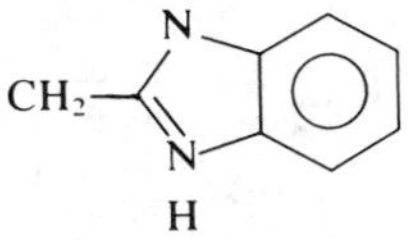 CH_2— N, N, H	Ph	H	$PhCH_2$	189–190	HCl	183
H_2	H_2	H	Ph	H	H	151–152		181,196
H_2	H_2	H	4-ClC_6H_4	H	H	259	HCl	181
H_2	H_2	H	4-BrC_6H_4	H	H	254–256	HCl	181
H_2	H_2	H	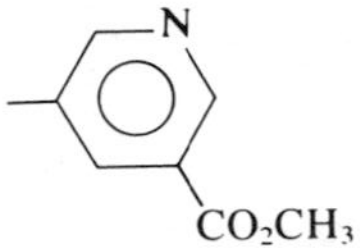 N, CO_2CH_3	H	H	102–124	ir,uv,pmr,ms	195
						94–108		196

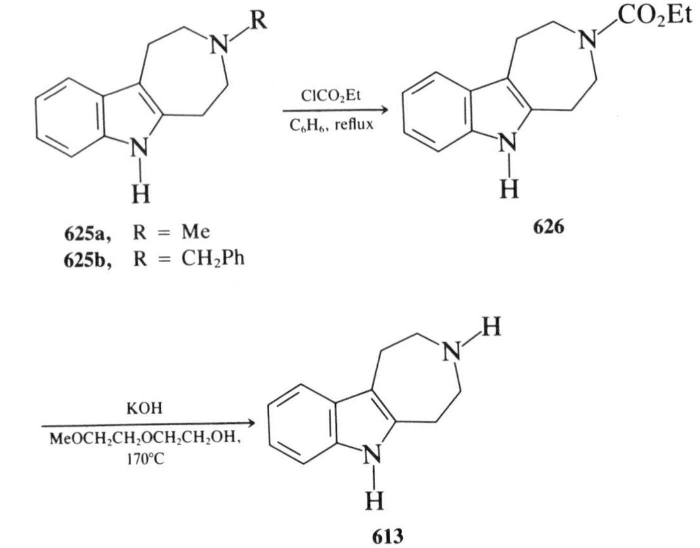

$ClCO_2Et$ / C_6H_6, reflux

625a, R = Me
625b, R = CH_2Ph

626 (72)

KOH / $MeOCH_2CH_2OCH_2CH_2OH$, 170°C

613

Zn, cat. $HgCl_2$ / conc. HCl, EtOH, reflux

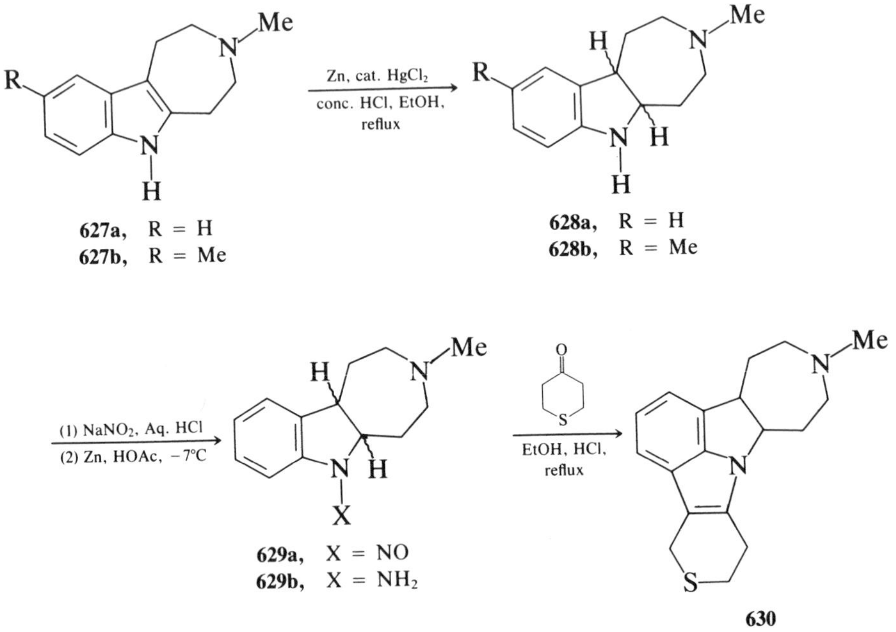

627a, R = H
627b, R = Me

628a, R = H
628b, R = Me

(1) $NaNO_2$, Aq. HCl / (2) Zn, HOAc, −7°C

629a, X = NO
629b, X = NH_2

EtOH, HCl, reflux

630

Figure 93

gave mixtures of the 5-hydroxy- and 5-acetoxyazepino idoles **632a** and **632b,** respectively (193). With one exception, these mixtures were not separated but were treated directly with alcoholic potassium hydroxide to obtain exclusively the alcohol **632b.** Chromic acid oxidation of **632b** in pyridine produced the azepino[4,5-*b*]indol-5-ones **633.** Oxidation of **631** was also attempted with periodic acid, but gave only small amount of **632b** in addition to much unchanged starting material.

A patent citation describes the only known electrophilic aromatic substitution reaction of azepino[4,5-*b*]indoles (194). Formylation of **634** with *s*-triazine in trifluoroacetic acid gave a mixture of the isomeric azepinoindole carboxaldehydes **635a** and **635b** (Eq. 74).

To date there is only one report of the azepino[4,5-*b*]indole system occurring naturally (195). Of six alkaloids isolated from the bark of *Nauclea diderrichii,* nauclederine was assigned structure **636.** The crystalline compound, which melted over a 22°C range, possessed the formula $C_{19}H_{19}N_3O_2$. The assignment of structure **636,** as opposed to the isomeric structures **637** and **638,** was made after careful consideration of mass spectral and other physical data and is supported by comparison of **636** with compounds **639a** and **639b,** which have been prepared unambiguously by Freter (181). Pmr and mass spectral fragmentation patterns of **639** showed a marked similarity to those of nauclederine.

Additional strong, though not unequivocal, support for the structure assigned to nauclederine was provided recently in a synthesis of **636** (196). The structural identity of synthetic **636** and naturally occurring nauclederine was judged by comparison of their spectral and chromatographic properties. The method of synthesis is outlined in Fig. 94. Reaction of the pyridyl epoxide **640** with tryptamine produced the corresponding amino alcohol **641** in 46% yield, and cyclization of **641** in polyphosphoric acid generated synthetic nauclederine (**636**) in 5.4% yield.

Unfortunately, this synthetic route does not represent unequivocal structure proof by itself. Acid catalyzed cyclization of **641** probably proceeds via an intermediate aziridinium ion that is capable of undergoing isomerization. Furthermore, the electrophilic aziridinium ion can certainly be attacked by the β-position of the indole nucleus (197) to produce a spiroindolenine intermediate. Subsequent Plancher rearrangement of the intermediate spiroindolenine can lead to several alternative structures.

While these potential mechanistic pitfalls could not be excluded *a priori,* the authors did provide circumstantial evidence in support of the structure they assigned to synthetic nauclederine. The unequivocal synthesis of **637** demonstrated its nonequivalence with nauclederine. More importantly, both the mono- and diadduct of tryptamine and styrene oxide, **642** and **643,** respectively, underwent cyclization in polyphosphoric acid to give the azepino[4,5-*b*]indole **639a.** Since **639a** had been previously prepared (181) by a completely unambiguous route, the mode of cyclization of **643** and—by extension—of **641** was confirmed. Tables 34, 35, and 36 list the known azepino[4,5-*b*]indoles.

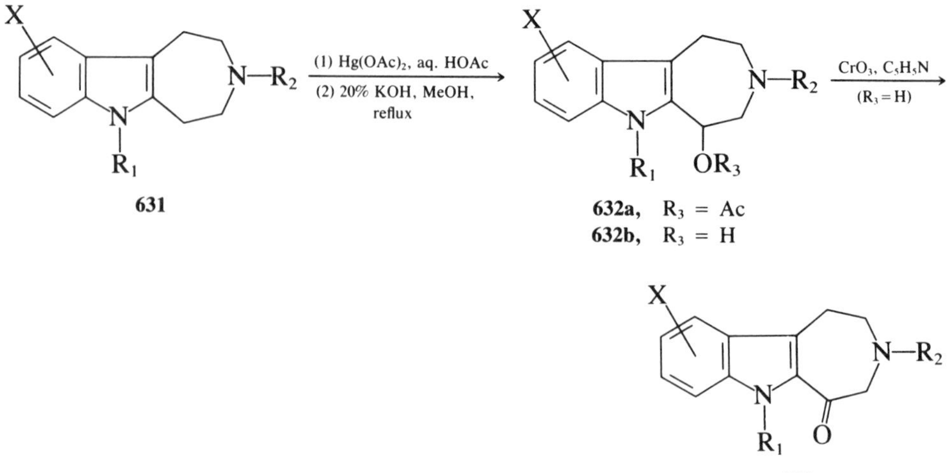

X

N—R_2

N

R_1

631

(1) $Hg(OAc)_2$, aq. HOAc
(2) 20% KOH, MeOH, reflux

X

N—R_2

N

R_1 OR_3

632a, R_3 = Ac
632b, R_3 = H

CrO_3, C_5H_5N
($R_3 = H$)

(73

X

N—R_2

N

R_1 O

633

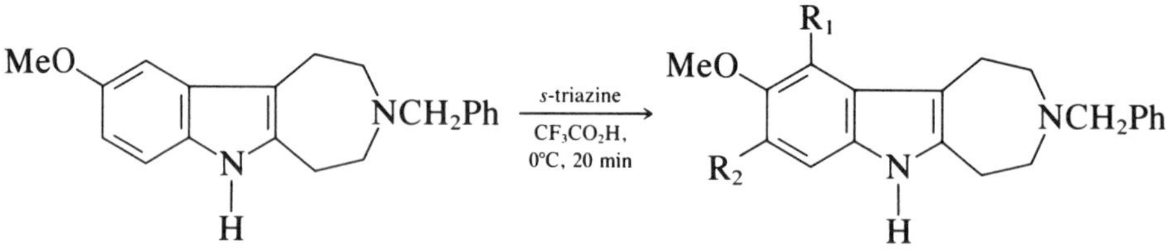

MeO

NCH_2Ph

N

H

634

s-triazine
CF_3CO_2H,
0°C, 20 min

R_1

MeO

NCH_2Ph

R_2

N

H

(74)

635a, R_1 = H, R_2 = CHO
635b, R_1 = CHO, R_2 = H

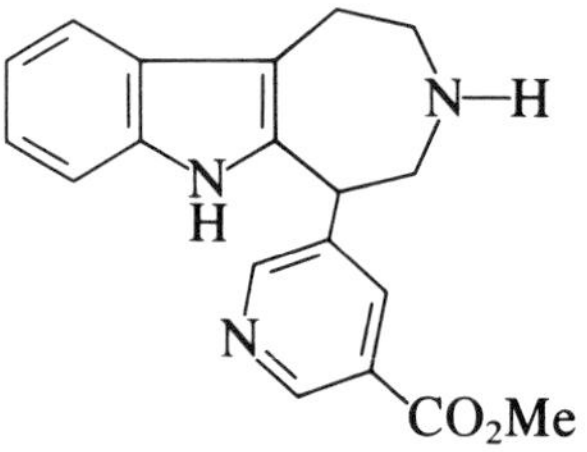

N—H

N
H

N

CO_2Me

636

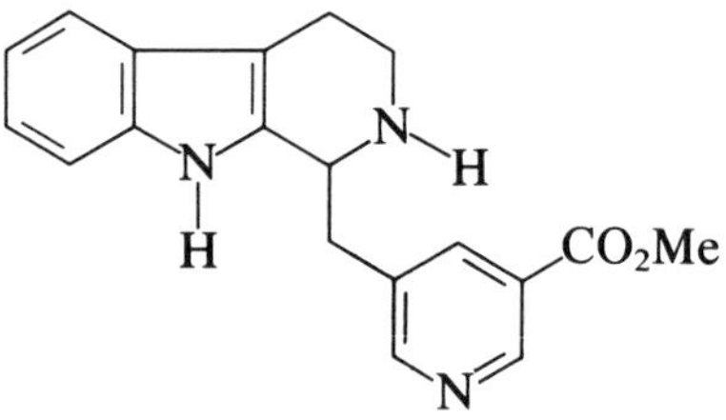

N

H

N
H

CO_2Me

N

637

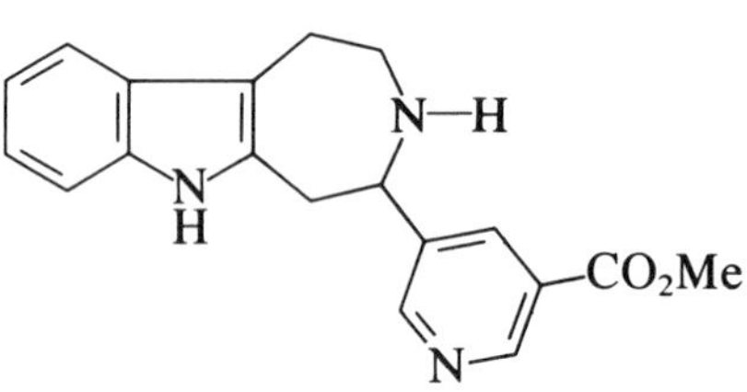

N—H

N
H

CO_2Me

N

638

N—H

N

H

X

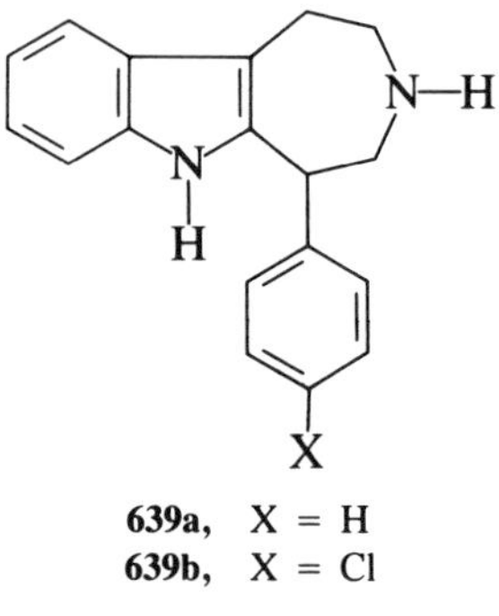

639a, X = H
639b, X = Cl

MeO_2C CH_2Br (pyridine) ← 48% HBr ← MeO_2C $\overset{O}{\|}CCHN_2$ (pyridine)

NaBH₄, MeOH

MeO_2C $\overset{OH}{|}CHCH_2Br$ (pyridine)

NaH, THF

MeO_2C $CH\!-\!CH_2$ (epoxide, pyridine) **640**

MeOH, reflux

636

PPA, 95°C, 8 hr

$CH_2CH_2NHCH_2\overset{OH}{|}CH$– (pyridine with CO_2Me) (indole, N–H)

641

$CH_2CH_2NH_2$ (indole, N–H)

PhCH—CH₂ (epoxide), 135°C, 3 hr

$CH_2CH_2N(CH_2\overset{OH}{|}CHPh)_2$ (indole, N–H)

643

$Ph\overset{OH}{|}CHCH_2Br$, MeOH, NaHCO₃, reflux

$CH_2CH_2NHCH_2\overset{OH}{|}CHPh$ (indole, N–H) **+643**

642 (55:45)

PPA, 85°C, 6.5 hr

639a

Figure 94

44. Azepino[5,4,3-*cd*]indoles

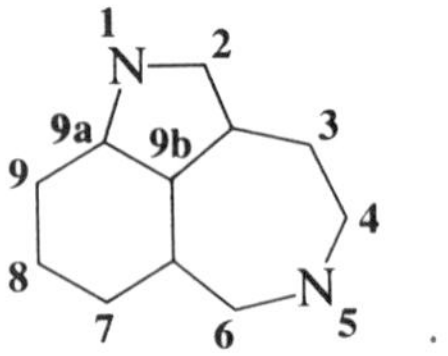

A preliminary report describing the preparation of several substituted tryptamines appeared in 1962 (198). Decarboxylation of 4-acetyl-2-carboxytryptamine (**644**) was found not to give the expected tryptamine **645**, but instead the azepinoindole **646**. The overall synthetic scheme, which was subsequently described in detail (199,200,207), is shown in Fig. 95.

Japp–Klingemann coupling of *m*-acetyl- and *m*-benzoylbenzenediazonium chloride with the piperidone carboxylic acid **647** gave the corresponding hydrazones **648a** and **648b** in 41 and 50% yields, respectively. Cyclization of the *m*-acylhydrazones **648** was accomplished with refluxing formic acid and gave mixtures of the 7-acyl-1,2,3,4-tetrahydro-1-oxo-β-carbolines **649** and the 5-acylcarbolines **650** in approximately 3:1 ratios. Basic hydrolysis of **650a** and **650b** gave the amino acids **644** and **651** in 93 and 88% yields, respectively. On refluxing in acetic acid–HCl, **644** and **651** gave the azepinoindoles **646** and **652** in 53 and 14% yields, respectively.

The problem with the sequence outlined above is that the Fischer cyclization gives a mixture of isomers, and the desired products **650a** and **650b** are the minor components of their respective mixtures. In order to block formation of the undesired isomers, 3-acetyl-6-chloroaniline was diazotized and allowed to react with **647** (Fig. 96) to give the hydrazone **653** in 75% yield. Fischer cyclization proceeded in 70% yield and gave the oxocarboline **654**. Alkaline hydrolysis converted **654** to **655** in 95% yield, and subsequent acid-catalyzed decarboxylation gave a 1:3 mixture of the azepinoindoles **656a** and **656b**. The latter reaction was very slow, in contrast to that of **653**, which was complete in 6 hours. The rate of decarboxylation is apparently decelerated by the electronegative affect of the chlorine substituent in **656b**. The acid **656b** was esterified to **656c** in nearly quantitative yield with ethanolic HCl.

Treatment of **656a** with acetic anhydride in pyridine (Fig. 97) gave an 80% yield of the enamide **657**, which on mild acid hydrolysis gave the diacetyltryptamine **658**. Alkaline hydrolysis of **658** resulted in the re-formation of **656a**. Alkylation of **656a** and **656c** with ethanolic methyl iodide (Fig. 98) afforded the quaternary salts **659** and **660** in 74 and 87% yields, respectively. Methylation of the indole nitrogen of **656a** to give **661** was accomplished in 56% yield with dimethyl carbonate in the presence of sodium hydride.

After fulfilling its function by directing the cyclization of hydrazone **653**, the chlorine was removed by catalytic hydrogenolysis. If the reduction of **656a** was interrupted after the uptake of 1 mole of hydrogen, the deschloro

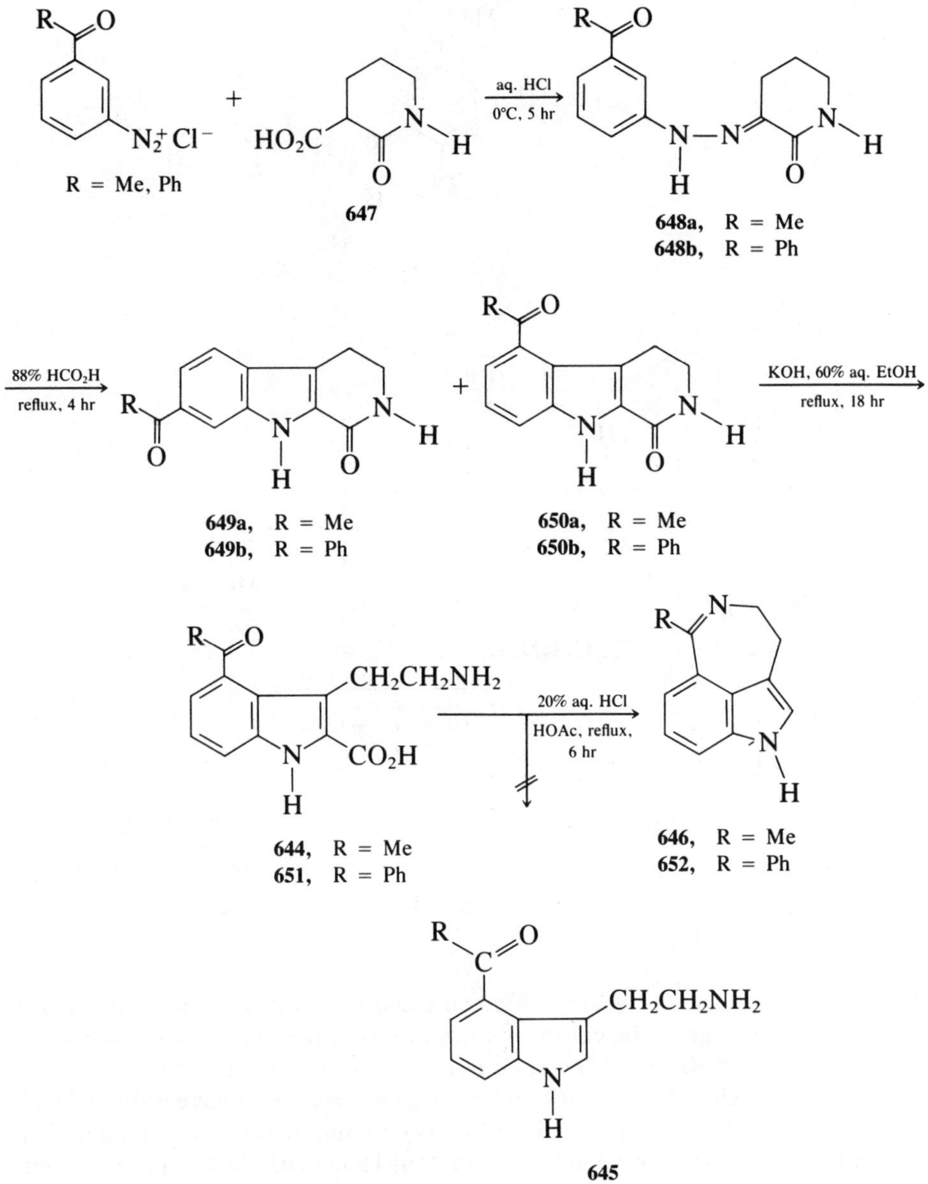
R = Me, Ph
647
aq. HCl
0°C, 5 hr
648a, R = Me
648b, R = Ph
88% HCO₂H
reflux, 4 hr
649a, R = Me
649b, R = Ph
650a, R = Me
650b, R = Ph
KOH, 60% aq. EtOH
reflux, 18 hr
644, R = Me
651, R = Ph
20% aq. HCl
HOAc, reflux,
6 hr
646, R = Me
652, R = Ph
645

Figure 95

compound **646** was isolated, but in only 11% yield. When reduction was allowed to proceed to completion, saturation of the imine bond yielded **662a** in 80% yield. In an analogous manner, catalytic reduction of **656c** gave **662b** in 90% yield, and reduction of the quaternary salt **659** gave the tertiary amine **663a**.

Whereas catalytic reduction using palladium–carbon led to a facile hy-

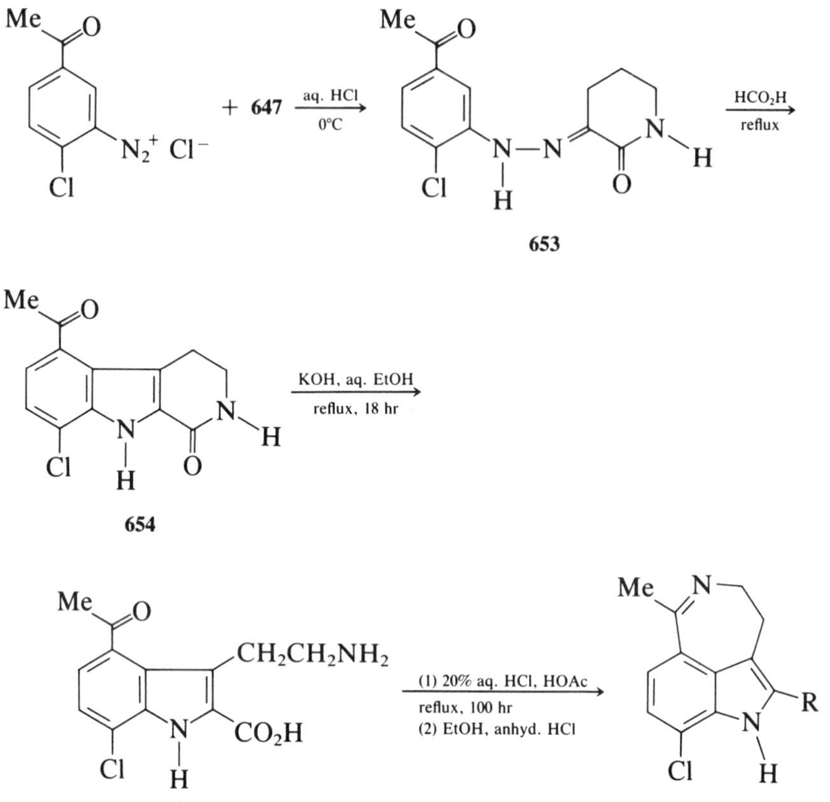

+ 647 aq. HCl, 0°C → 653 → HCO_2H, reflux → 654 → KOH, aq. EtOH, reflux, 18 hr → 655 → (1) 20% aq. HCl, HOAc, reflux, 100 hr (2) EtOH, anhyd. HCl → 656

656a, R = H
656b, R = CO_2H
656c, R = CO_2Et

Figure 96

drogenolysis, metal hydrides effected clean reductions of the imine moiety without cleavage of the chlorine bond. For example, potassium borohydride reduction of **656a** and **656c** gave **662c** and **662d** in 80 and 50% yields, respectively, while **659** and **660** under the same conditions gave **663b** and **663c** in 85 and 87% yields (Fig. 98). Lithium aluminum hydride reduction of **656c** and **660** gave the corresponding alcohols **664a** and **663d** in 43 and 55% yields, respectively. Surprisingly, acylation of **664a** gave only the *O*-acylated product **664b** and none of the product of *N*-acylation (Eq. 75). Lithium aluminum hydride reduction of the amide **665a,** which was prepared from **662b** and piperidine, gave the corresponding amine **665b** in only 23% yield (Eq. 76).

An alternative method of preparing azepino[5,4,3-*cd*]indole was reported by Evans and co-workers (201). The amines **666a** and **666b,** derived from the benz[*c,d*]indolones **666c** and **666d,** underwent Beckmann rearrangement to the azepinoindolones **667a** and **667b,** respectively, in the presence of

656a $\xrightarrow[100°C,\ 1\ hr]{Ac_2O,\ C_5H_5N}$ 657 $\xrightarrow[100°C,\ 3\ hr]{1.5\ N\ HCl}$

Ac, N, H_2C, N, Cl, H

657

Me, O, CH_2CH_2NHAc, N, Cl, H

658 $\xrightarrow{KOH,\ aq.\ EtOH}$ 656a

Figure 97

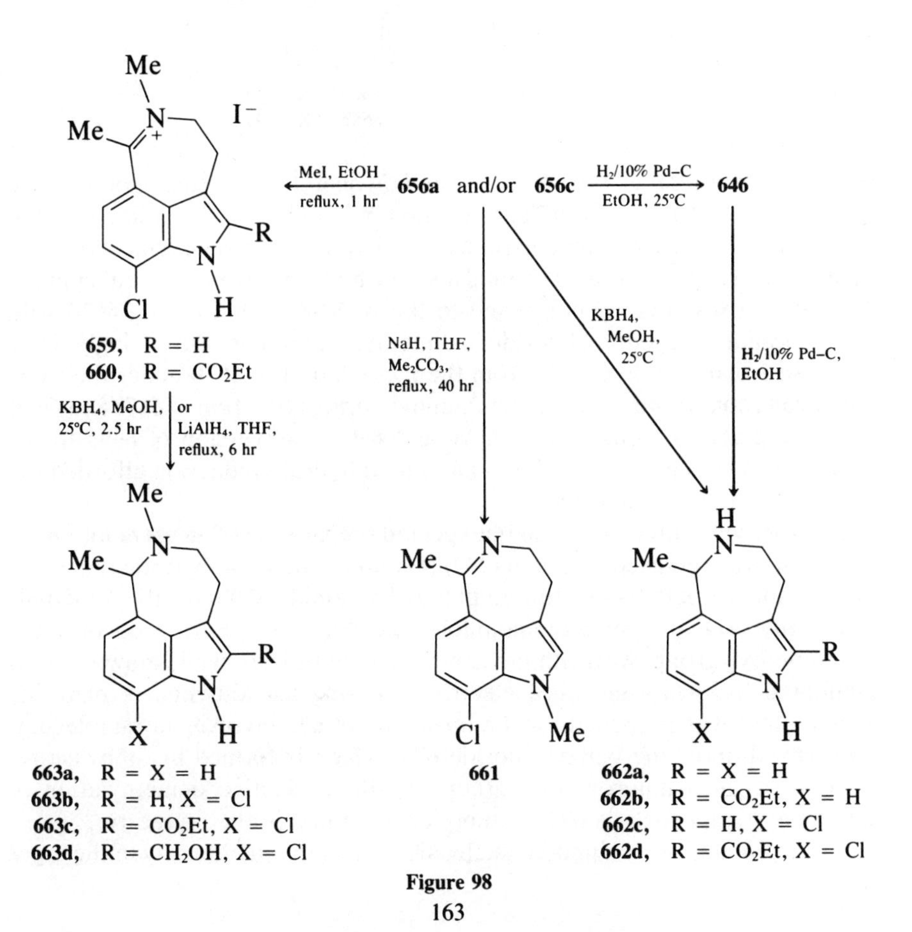

Figure 98

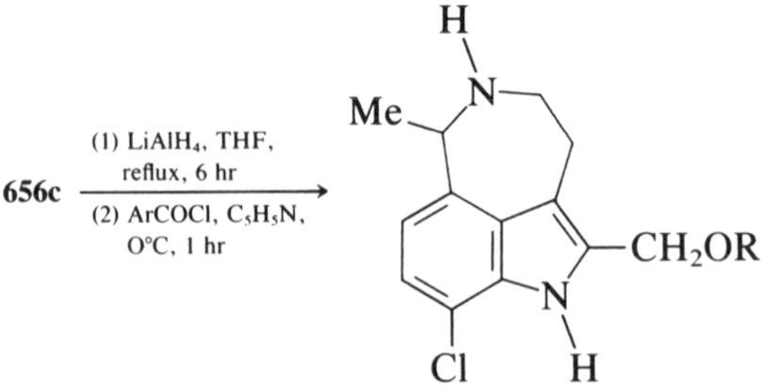

656c —(1) $LiAlH_4$, THF, reflux, 6 hr; (2) ArCOCl, C_5H_5N, 0°C, 1 hr→ (75)

664a, R = H
664b, R = $COC_6H_2(OMe)_3$-3,4,5

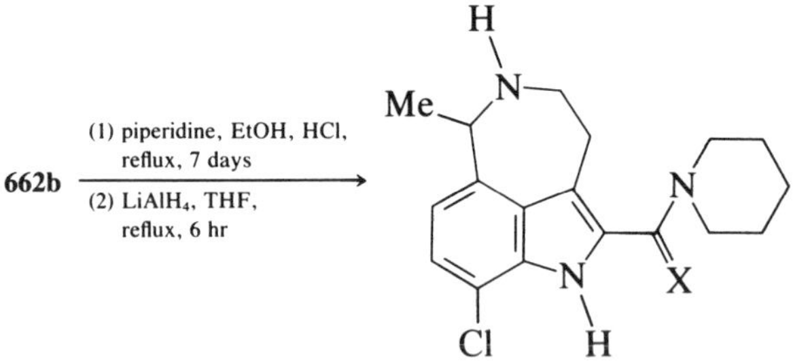

662b —(1) piperidine, EtOH, HCl, reflux, 7 days; (2) $LiAlH_4$, THF, reflux, 6 hr→ (76)

665a, X = O
665b, X = H_2

thionyl chloride (Fig. 99). *N*-Tosyl and *N*-acetyl protecting groups were necessary in order to prevent isomerization of 3,4-hydrobenz[*c,d*]indol-5(1H)-one (**666e**) to the more stable naphthalenoid system **668.** The *N*-acetyl group of **667b** was readily removed by mild alkaline hydrolysis to give the azepinone **667c,** which was readily converted to the *N,N'*-dimethyl analog **667d** with methyl iodide and sodium hydride in dimethyl formamide. Surprisingly, **667d** was also isolated in low yield from the methylation of **667b** under the same anhydrous conditions. Lithium aluminum hydride reduction of **667c** and **667d** gave the corresponding amines **669a** and **669b.** Conversion of **669a** to the urethane **669c** followed by lithium aluminum hydride reduction afforded the 5-methylazepinoindole **669d.**

Another very interesting and unexpected preparation of an azepino[5,4,3,-*cd*]indole was reported by Allen (202). Lithium aluminum hydride reduction of the trifluoromethyl–substituted indoleglyoxamide **670** gave the 4-methyltryptamine **671** and the azepinoindole **672** (Fig. 100). Reduction of a trifluoromethyl group with lithium aluminum hydride is well known and is facilitated by *ortho* and *para* electron-releasing substituents. A plausible scheme that was proposed for the formation of **672** involves intramolecular quaternization of the benzyl fluoride **673,** which is formed by stepwise reduction of the trifluoromethyl group of **670.** Reductive demethylation of quaternary salts such as **674** is amply documented in the literature.

The azepino[5,4,3-*cd*]indole nucleus has been found naturally in the ergot

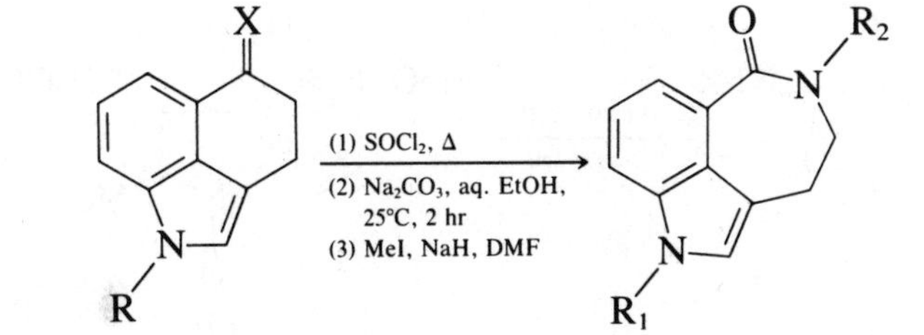

(1) $SOCl_2$, Δ
(2) Na_2CO_3, aq. EtOH, 25°C, 2 hr
(3) MeI, NaH, DMF

666a, R = Ts, X = NOH
666b, R = Ac, X = NOH
666c, R = Ts, X = O
666d, R = Ac, X = O
666e, R = H, X = O

667a, R_1 = Ts, R_2 = H
667b, R_1 = Ac, R_2 = H
667c, R_1 = R_2 = H
667d, R_1 = R_2 = Me

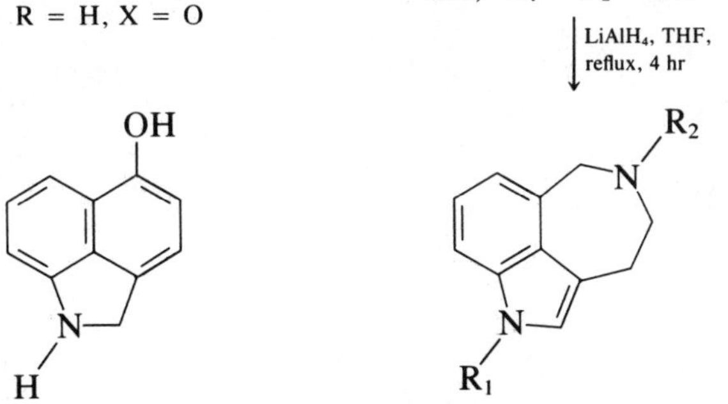

$LiAlH_4$, THF, reflux, 4 hr

668

669a, R_1 = R_2 = H
669b, R_1 = R_2 = Me
669c, R_1 = H, R_2 = CO_2Me
669d, R_1 = H, R_2 = Me

Figure 99

alkaloids (203–206). Roberts and Floss (206), in their studies of ergot alkaloid biosynthesis, isolated a 4-substituted indole amino acid which they called clavicipitic acid and to which they assigned structure **675.** The amino acid was obtained from submerged fungal cultures of a *Claviceps* species (strain SD-58) that had been inhibited with DL-ethionine. Subsequently, King and co-workers (203) isolated the same compound from a submerged fungal culture (with or without DL-ethionine) of *Claviceps fusiformis,* strain 139/2/1G, and reassigned the structure of clavicipitic acid as the azepino[5,4,3-*cd*]indole **676a.** They were able to chromatographically resolve the acid into two fractions (R_f = 0.24 and 0.29), both of which had ultraviolet and mass spectra identical with those of the original material. Circular dichroism measurements suggested a diastereomeric relationship between the two components. However, the diastereomers could not be interconverted by the action of ammonia or by heat.

Attempts to prepare a trimethyl derivative of **676a** by alkylation with diazomethane gave mixtures of mono- and dimethyl derivatives. The reaction of **676a** with acetic anhydride in methanol gave the *N*-acetylated methyl ester

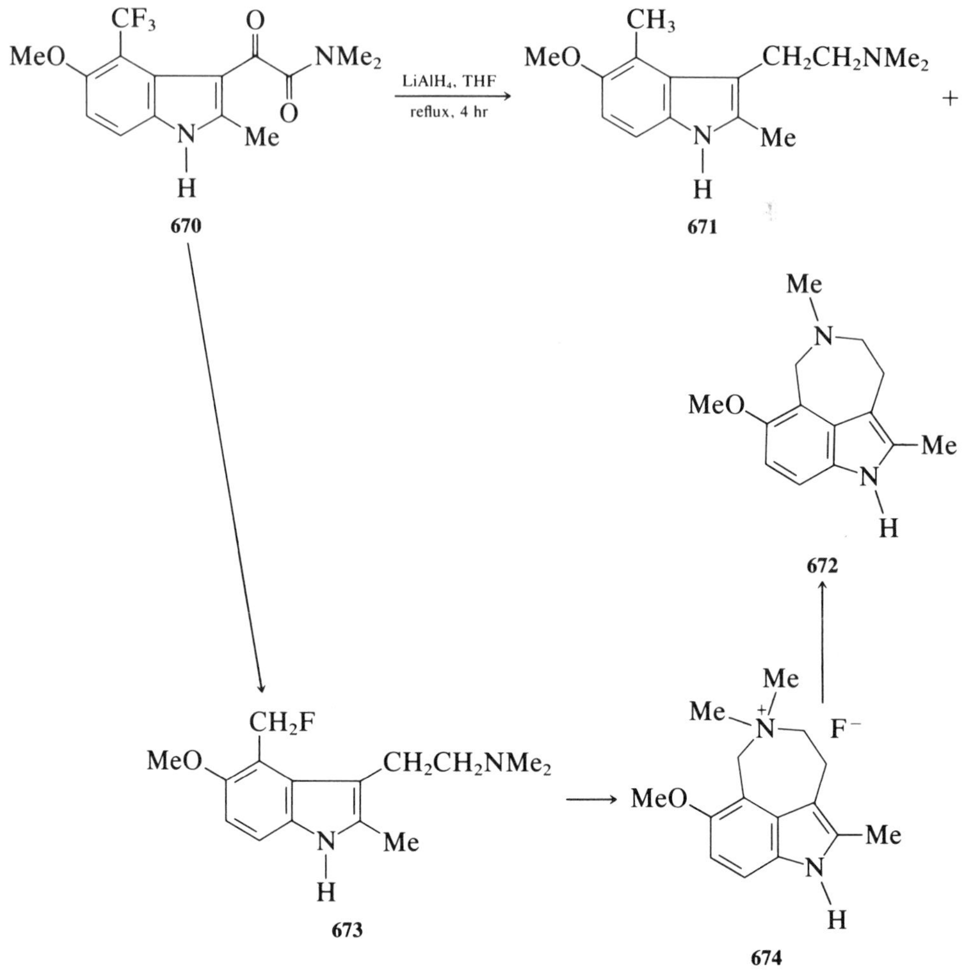

CF$_3$ O
MeO NMe$_2$
O
N Me
H
670

LiAlH$_4$, THF
reflux, 4 hr

CH$_3$
MeO CH$_2$CH$_2$NMe$_2$
N Me
H
671

+

Me
N
MeO
Me
N
H
672

CH$_2$F
MeO CH$_2$CH$_2$NMe$_2$
N Me
H
673

Me
Me $\overset{+}{N}$ F$^-$
MeO
Me
N
H
674

Figure 100

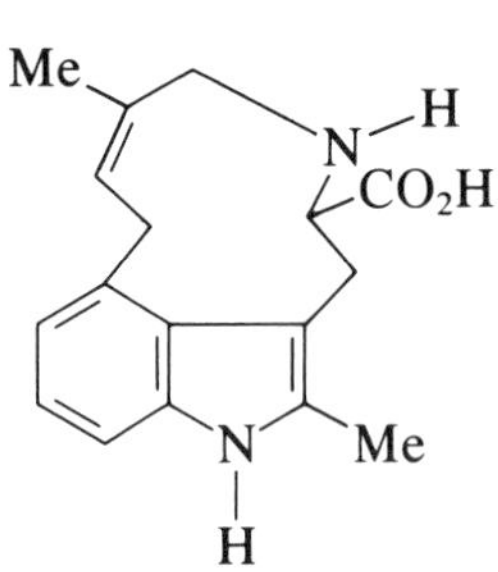

Me
N H
CO$_2$H
N Me
H
675

Me Me R$_3$ CO$_2$R$_2$
N H$_c$
H$_d$
H$_a$ H$_b$ H$_d$
N
R$_1$

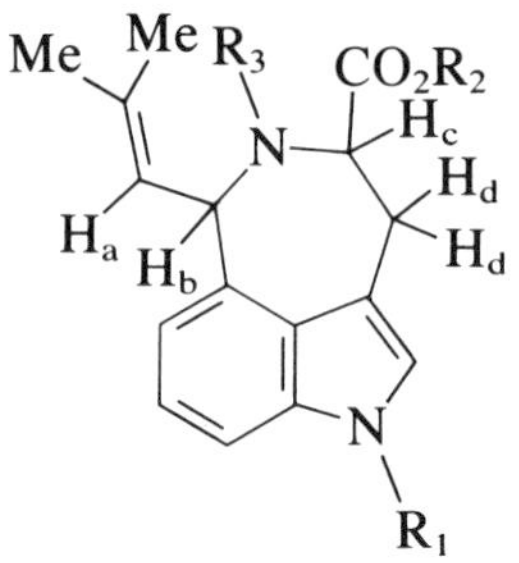

676a, $R_1 = R_2 = R_3 = H$
676b, $R_1 = H$, $R_2 = Me$, $R_3 = Ac$
676c, $R_1 = R_2 = Me$, $R_3 = Ac$

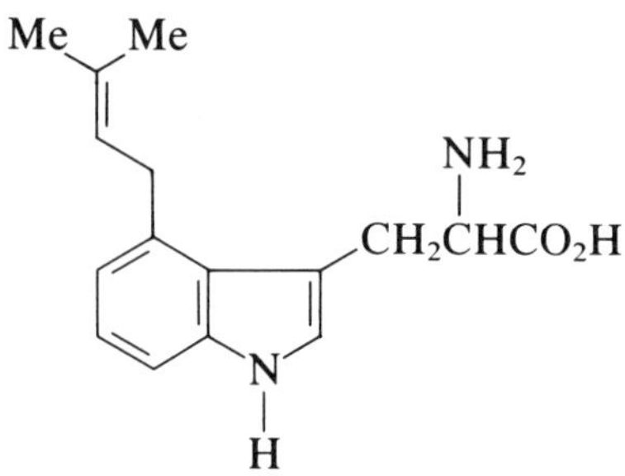

Me Me
NH$_2$
CH$_2$CHCO$_2$H
N
H
677

676b, which, in contrast to **676a,** was readily soluble in chloroform. Treatment of **676b** with methyl iodide–methylsulfinyl carbanion in dimethyl sulfoxide effected methylation of the indole nitrogen, giving **676c.** The structure of **676b,** and by analogy of **676a,** was based primarily on its 100-MHz pmr spectrum.

4-γ,γ-Dimethylallyltryptophan (**677**) has recently been converted to clav-

TABLE 37. AZEPINO[5,4,3-*cd*]INDOLES[d]

Compound	mp (°C)	Spectra	Refs.
646	274.5–279.5	uv	198–200
652	174–176	uv	199
656a	208–211	uv,ir	207
656b	360	uv,ir	207
656c	229–234	uv,ir	207
657	236–239	uv,ir	207
659	251–253	uv,ir	207
660	260–265	uv,ir	207
661	280–282 (HCl)	uv,ir	207
662a	200–205	uv,ir	207
	267–273 (HCl)		
662b	172–174	uv,ir	207
662c	152–154	uv,ir	207
662d	126–129	uv,ir	207
663a	207–210	uv,ir	207
663b	179–181	uv,ir	207
663c	99–101	uv,ir	207
663d	206–210	uv,ir	207
664a	215–218	uv,ir	207
664b	90–94 (0.5EtOH)	uv,ir	207
665a	221–223	uv,ir	207
665b	171–174	uv,ir	207
667a	192–194		201
667b	264–265	ir	201
667c	237–239		201
667d	129–132	ir	201
669a	231–233	ir	201
669b	45–50	—[a]	201
	250–252 (HCl)		
669c	210–212	ir	201
669d	195–197[b]		201
	252–257 (HCl)	—[c]	
672	213–215	uv,pmr	202
676a	262(dec.)	uv,ir,ms,pmr	203,206
676b	107–109	uv,ms,pmr	203
676c	—	ms	203

[a] pK_a = 7.82.
[b] pK_a = 7.92.
[c] pK_a = 8.12.
[d] Fig. 95-100; eq. 75,76

icipitic acid (**676a**) in cell-free *Claviceps* extracts (204,205). The tryptophan **677** is the first intermediate in clavine alkaloid biosynthesis, and its accumulation by ergot fungus has been previously demonstrated (see Refs. 204 and 206 and others cited therein). Conversion of **677** to clavicipitic acid required oxygen, but not cofactors, and was inhibited by *p*-hydroxymercuribenzoic acid and stimulated by diethyldithiocarbamate. Conversion was also favored at high pH. Table 37 lists the known azepino[5,4,3-*cd*]indoles.

45. Azepino[4,5-*e*]indoles

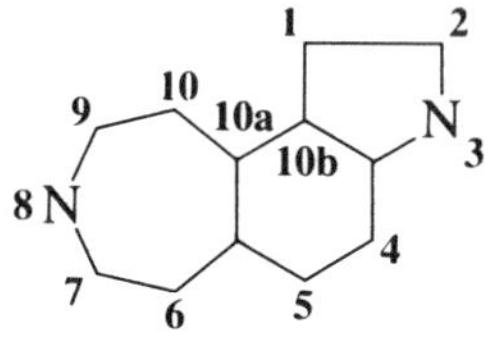

46. Azepino[4,5-*f*]indoles

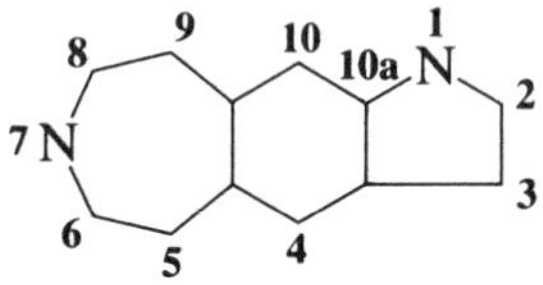

47. Azepino[4,5-*g*]indoles

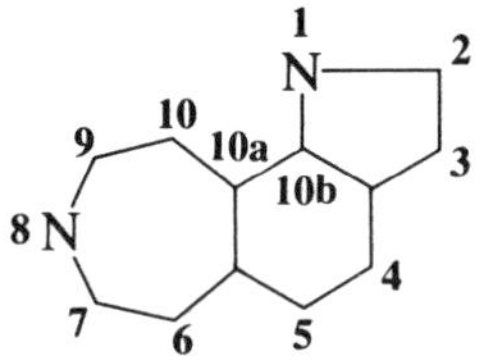

Photocyclization of the *N*-chloroacetylaminoethylindole **678** gave the isomeric azepino[4,5-*e*]indole **679** and the azepino[4,5-*f*]indole **680** in a 1:2 ratio (Eq. 77). Photocyclization of **681** gave only the azocinoindole **682** (Eq. 78) and none of the azepino[4,5-*e*]indole **683** (179). Photocyclization of **684** produced only the [4,5-*f*]azepinoindolone **685** and none of the [4,5-*g*] isomer **686** (Eq. 79). Analogous photocyclization of **687,** however, did provide the [4,5-*g*] indolone isomer **688** in unspecified yield (Eq. 80).

$ClCH_2CONH(CH_2)_2$ N Me

678

$h\nu$ (2537 Å)
50% aq. EtOH, $NaHCO_3$

H O N N Me

679, mp 187–189°C

(77)

+ H—N O N Me

680, mp 203–205°C

H N O O $(CH_2)_2NHCCH_2Cl$ N Me

$h\nu$

682 **681** **683**

(78)

686 $ClCH_2CONH(CH_2)_2$ N Me **684**

$h\nu$

(79)

O H—N N Me

685, mp 197–198°C

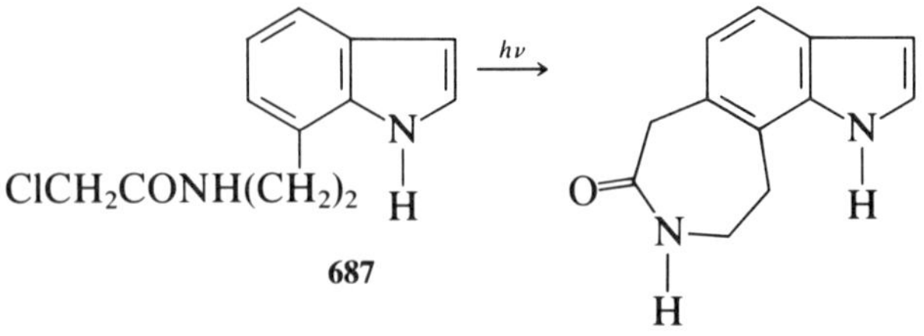

$ClCH_2CONH(CH_2)_2$ **687** $\xrightarrow{h\nu}$ **688,** mp 283–285°C (80)

48. Azepino[2,1-*a*]isoindoles

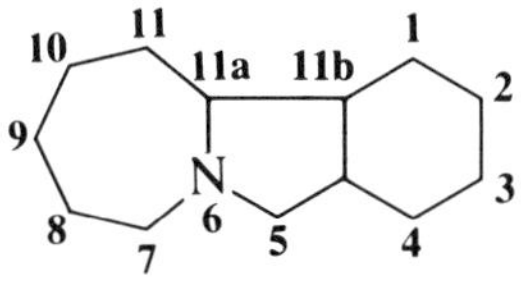

When the substituted phthalimide **689a** was photolyzed (Fig. 101), the Norrish type II photoproduct, azepinoisoindole **691a,** was formed in 12% yield (208). The isomeric compound **689b** failed to undergo cyclization. In an analogous manner, phthalimides **689c–f** underwent photocyclization to the azepinoisoindoles **691b–e** (209–212).

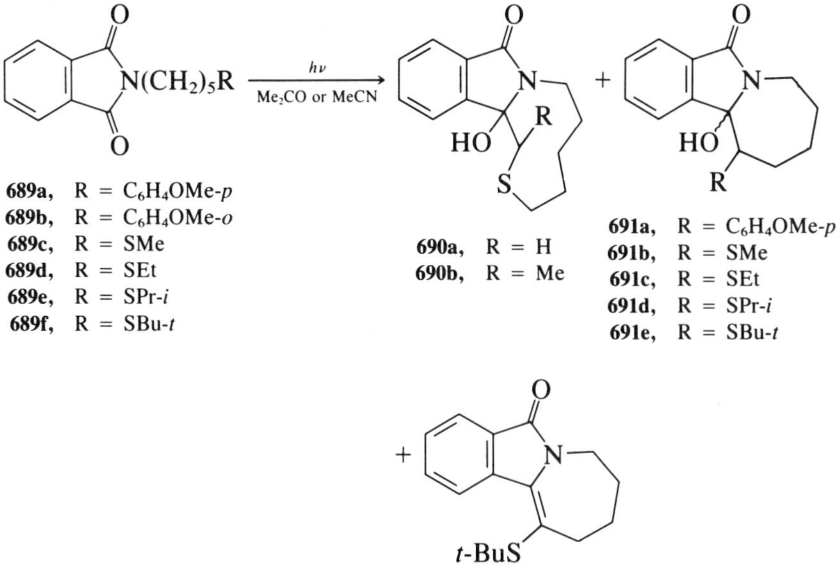

$N(CH_2)_5R$ $\xrightarrow[Me_2CO \text{ or } MeCN]{h\nu}$ + +

689a, R = C_6H_4OMe-*p*
689b, R = C_6H_4OMe-*o*
689c, R = SMe
689d, R = SEt
689e, R = SPr-*i*
689f, R = SBu-*t*

690a, R = H
690b, R = Me

691a, R = C_6H_4OMe-*p*
691b, R = SMe
691c, R = SEt
691d, R = SPr-*i*
691e, R = SBu-*t*

t-BuS **692**

Figure 101

TABLE 38. AZEPINO[2,1-*a*]ISOINDOLES AND AZATHIOCYCLOLS OBTAINED FROM PHOTOLYSIS OF PHTHALAMIDES

Starting Material	Product	Yield (%)	mp (°C)	Refs.
689a	**691a**	12	193–195	208
689c	**691b**	6	145–145.5	210–212
	690a	78.3	152–153	210–212
689d	**691c**	5	Oil	212
	690b	61	121–123	210–212
689e	**691d**	35	138–140	210,211
689f	**691e** (*trans*)	12.5	144–146	210,211
	691e (*cis*)	3.4	148–150	210,211
	692	26.6	165–167	210–212

Whereas **689e** and **689f** afforded only azepinoisoindoles, the main products from **689c** and **689d** were the azathiacyclols **690a** and **690b,** respectively, and the azepinoisoindoles **691b** and **691c** were only minor products. The stereochemistry of **691b–d** was not addressed, although the *cis* and *trans* isomers of **691e** were isolated and characterized. In addition to **691e,** the dehydration product **692** was also formed from **689f.** Yields and physical constants of the products formed in these reactions are given in Table 38.

49. Pyrrolo[1,2-*a*]benzazepines

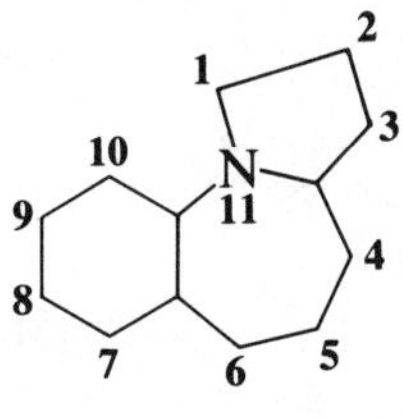

In 1938 Clemo prepared benzoquinolizidines **693a** and **694a** and subjected them to Wolff and Clemmensen reductions (213). The Clemmensen products did not correspond to the Wolff products **693b** and **694b.** The following year, Prelog demonstrated that quinolizidine **695a** afforded **696a,** and not **695a,** under Clemmensen conditions (214). The mechanism of formation of the rearranged Clemmensen product was deduced from the methylated quinolizidines **695c** and **695d,** which were prepared and reduced to afford the pyrroloazepines **696b** and **696c** (215,216). Unfortunately **696b** and **696c** could not be positively identified, and it remained for Leonard and Wildman to demonstrate unequivocally via independent synthesis that **696b** arose from **695c,** and **696c** from **695d** (217). On this basis it was concluded that Clem-

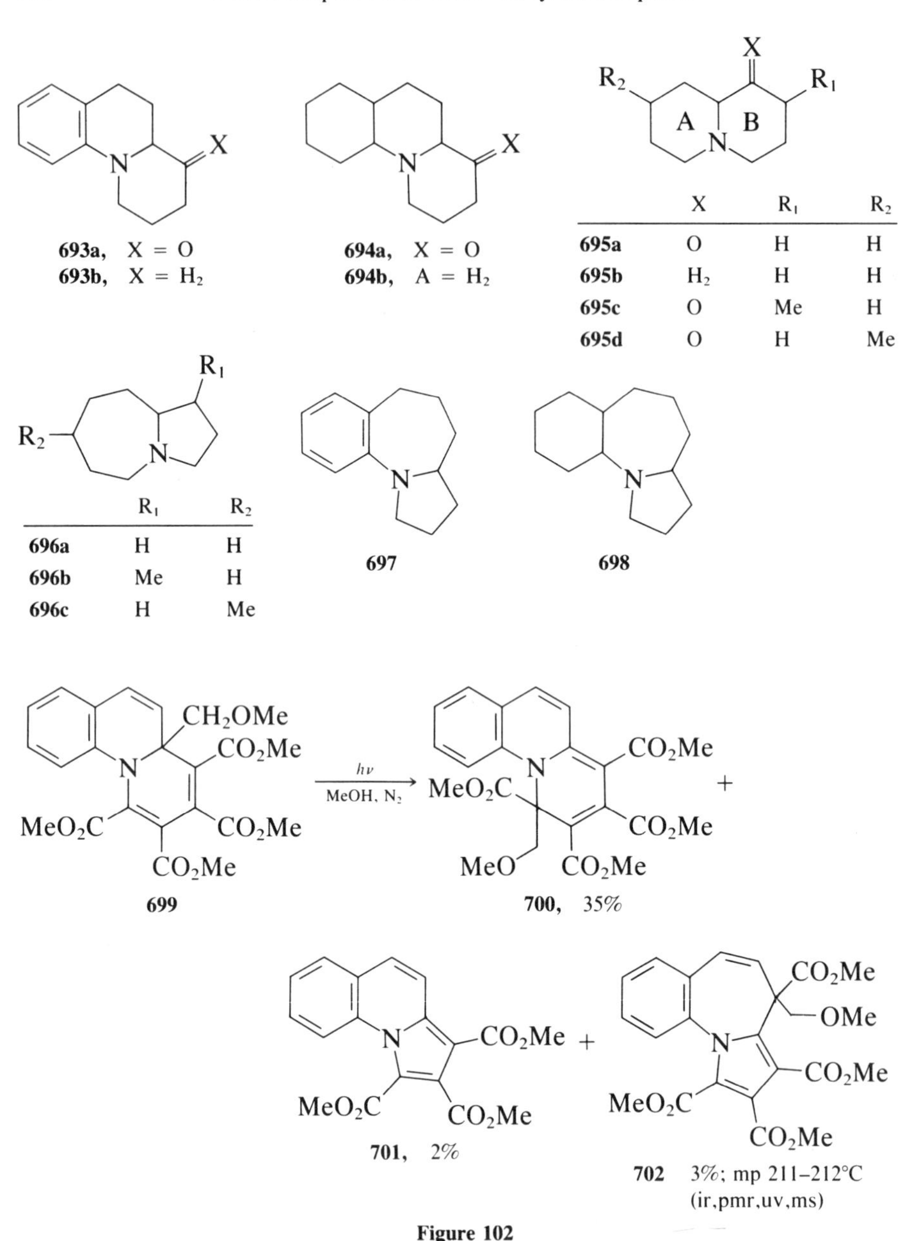

Figure 102

mensen reduction of the quinolizidones proceeded via contraction of the B ring, and that Clemo's early reductions of **693a** and **694a** must have produced the pyrrolo[1,2-*a*][1]benzazepines **697** and **698,** respectively.

More recently, the photolysis of **699** was reported to produce **702,** a red solid, as a minor product along with **700** and **701** (Fig. 102). Characterization of **702** was based on spectral arguments (218).

50. Pyrrolo[1,2-*b*][2]benzazepines

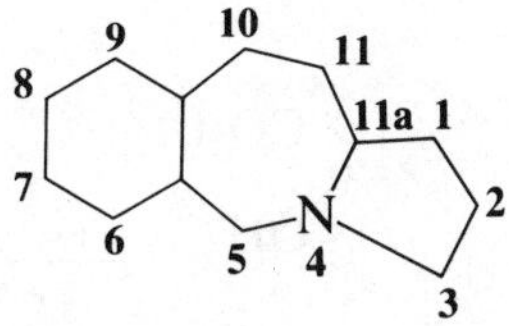

When the benzazabicyclo[4.4.0]decane **703** was electrolyzed in 30% sulfuric acid using a lead cathode, a three-component mixture was formed (Eq. 81). One of the components (41% yield) was the pyrrolobenzazepine **704.** The remaining two compounds (51% of the overall yield) were assigned structures **705** and **706.** The picrate of **704** was identical with the picrate of the product obtained in the rearrangement of **703** during Clemmensen reduction (see discussion in Section IV. 49) (219).

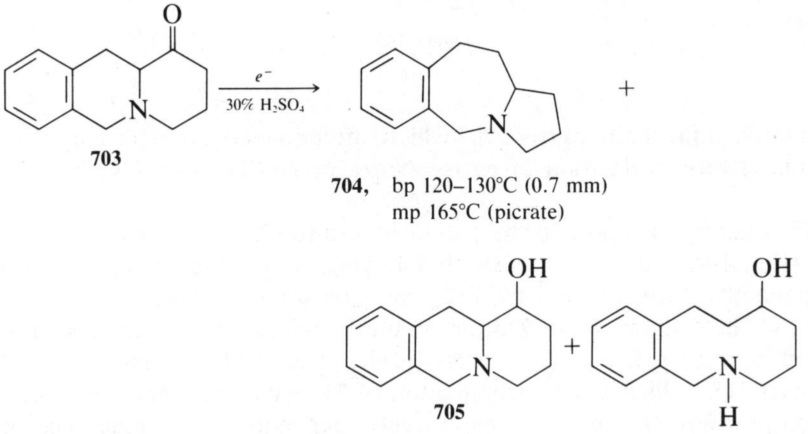

e^-, 30% H_2SO_4

703

704, bp 120–130°C (0.7 mm)
mp 165°C (picrate)

(81)

705

706

51. Pyrrolo[2,1-*b*][3]benzazepines

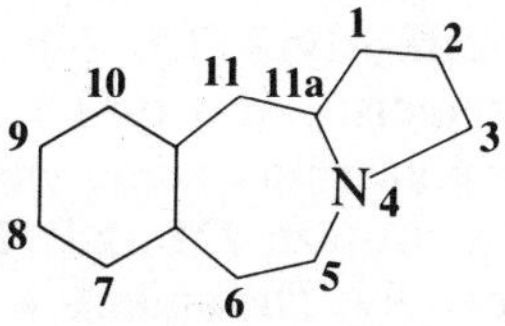

Huisgen and Laschtuvka (220) described the preparation of *N*-substituted pyrroles from 1,4-dicyano-2-butene (**707**). Condensation of **707** with diethyl oxalate gave a high yield of the *bis* adduct **708,** which on treatment with primary amines afforded *N*-substituted pyrroles. Thus treatment of **708** with

$NCCH_2CH{=}CHCH_2CN + EtO_2CCO_2Et \xrightarrow[EtOH]{NaOEt}$

707

NC, CO_2Et, OH, HO, EtO_2C, CN

708

$\xrightarrow[\substack{PhMe,\ \Delta,\\ -NCCH_2COCO_2Et}]{PhCH_2CH_2NH_2}$

CN, CO_2Et, N, CH_2CH_2Ph

709

(1) KOH, EtOH
(2) H_3O^-
(3) $SOCl_2$
(4) $AlCl_3$

N, O, CN

710, mp 123–124°C

Figure 103

2-phenethylamine led to pyrrole **709** which, after conversion to its acid chloride, underwent cyclization to pyrrolobenzazepine **710** in 76% yield (Fig. 103).

Subsequently, a report in the patent literature (221) enlarged this early work (Fig. 104). Alkylation of methyl-4-cyanopyrrole 2-carboxylate (**711**) with phenethyl bromide afforded **712,** which on saponification gave the free acid **718.** The sodium salt of **713,** on treatment with styrene oxide and subsequent acidic workup, gave the *N*-styryl pyrrole **714.** Reduction of **714** produced **715** in high yield. Bromination of **715** gave the 4-bromo- or 4,5-dibromopyrroles **716** and **717,** respectively, depending on whether one or two equivalents of bromine was used. The acids **715** and **718** were treated with thionyl chloride to obtain chlorides that underwent cyclization to the pyrrolobenzazepines **719** on heating in the presence of aluminum chloride.

Alternatively, substituted 6,11-dihydro-5H-pyrrolo[2,1-*b*][3]benzazepin-11-ones were prepared by cyclization of 2-(2-pyrrol-1-yl)ethylbenzoic acids. Nitration of 3,4-dihydroisocarbostyril (**720**, Fig. 105) afforded **721,** which in turn was reduced to the aminoisocarbostyril **722.** Diazotization of **722** and subsequent treatment with potassium iodide yielded **723.** Concentrated hydrochloric acid served to hydrolyze **721** and **723** with formation of amino acids **724a** and **724b,** respectively. The amines were converted to the desired pyrroles **725a** and **725b,** respectively, with 2,5-dimethoxytetrahydrofuran. Polyphosphate ester (PPE) treatment at room temperature caused cyclization of **725a** and **725b** to pyrrolobenzazepines **726a** and **726b** in high yield.

The parent 11H-pyrrolo[2,1-*b*][3]benzazepin-11-ones were obtained by

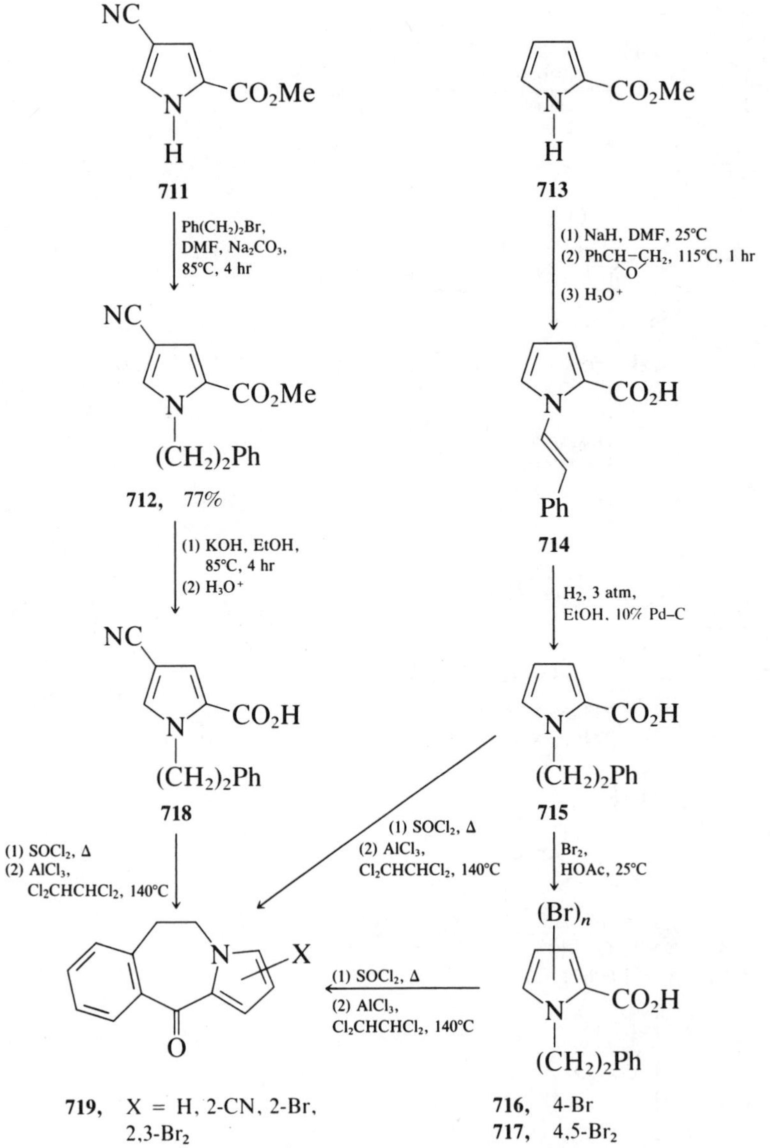

711 → [$Ph(CH_2)_2Br$, DMF, Na_2CO_3, 85°C, 4 hr] → 712, 77%

712 → [(1) KOH, EtOH, 85°C, 4 hr; (2) H_3O^+] → 718

718 → [(1) $SOCl_2$, Δ; (2) $AlCl_3$, $Cl_2CHCHCl_2$, 140°C] → 719

713 → [(1) NaH, DMF, 25°C; (2) PhCH–CH$_2$ (O), 115°C, 1 hr; (3) H_3O^+] → 714

714 → [H_2, 3 atm, EtOH, 10% Pd–C] → 715

715 → [(1) $SOCl_2$, Δ; (2) $AlCl_3$, $Cl_2CHCHCl_2$, 140°C] → 719

715 → [Br_2, HOAc, 25°C] → 716, 717

716, 717 → [(1) $SOCl_2$, Δ; (2) $AlCl_3$, $Cl_2CHCHCl_2$, 140°C] → 719

719, X = H, 2-CN, 2-Br, 2,3-Br$_2$

716, 4-Br
717, 4,5-Br$_2$

Figure 104

several different routes. As described above, reaction of the sodium salt of **711** with styrene oxide followed by acid treatment (Fig. 106) gave the *N*-(*trans*-1-styryl)pyrrole **727**. However, when **711** was reacted with styrene oxide in the presence of a catalytic amount of potassium *t*-butoxide, lactone **728** was obtained. Phosphorous oxychloride–phosphorous pentachloride converted **728** to **729**, which in turn was cyclized with aluminum chloride

O
N—H
720
HNO_3, H_2SO_4
30 min, 0°C

O_2N
721, 95%
H_2/Pd–C (1 atm)
MeOH
H_2N
722, 93%

conc. HCl,
145°C, 30 hr

(1) $NaNO_2$, aq. HCl
(2) KI, aq. Me_2CO,
50–60°C

MeO–(2,5-dimethoxytetrahydrofuran)–OMe
aq. HOAc,
55°C, 2 hr

X, CO_2H, $(CH_2)_2NH_2 \cdot HCl$
724a, X = NO_2
724b, X = I (80%)

conc. HCl
145°C, 30 hr

I
723, 73%

X, CO_2H, $(CH_2)_2$—N
725a, X = NO_2
725b, X = I (58%)

PPE
1.5 hr, 25°C

X, N, O
726a, X = NO_2
726b, X = I

Figure 105

711

PhCH—CH$_2$ (epoxide), DMF, cat. KOBut

728

(1) NaH, DMF, 25°C
(2) Ph CH—CH$_2$ (epoxide), 110°C, 6 hr
(3) H_3O^+

727

$POCl_3$–PCl_5, reflux, 36 hr

$PhCOCH_2Br$, DMF, K_2CO_3, 100°C, 30 min

(1) 6 *N* NaOH, EtOH
(2) $NaBH_4$, aq. EtOH
(3) $POCl_3$, PCl_5

729, 93%

734, 84%

(1) $SOCl_2$, reflux
(2) Cl_2, $CHCl_3$
(3) $AlCl_3$

$AlCl_3$, $Cl_2CHCHCl_2$, 140°C, 4–5 min

730

733

$CrCl_2$, aq. Me_2CO, 60°C (86%)

727

$h\nu$, MeCN

KOH, EtOH reflux, 15 min

$(CF_3CO)_2O$, $SnCl_4$

731

732

Figure 106

to **730** in fair yield. Dehydrochlorination produced 2-cyano-11H-pyrrolo[2,1-*b*][3]benzazepin-11-one (**731**). The styryl pyrrole **727** was converted to **731** in two ways. Photolysis of **727** gave *N*-(*cis*-1-styryl)pyrrole (**732**), which could be cyclized with trifluoroacetic anhydride to **731.** Alternatively, when **727** was treated sequentially with thionyl chloride, chlorine, and aluminum chloride, dichlorpyrrolobenzazepine **733** was formed. Dechlorination of **733** with chromous chloride produced **731.** Finally, **711** was converted to **729** via the phenacyl intermediate **734.**

The chemistry of the pyrrolobenzazepines was examined and is summarized in Eq. 82–84 and Fig. 107. Catalytic hydrogenolysis of 2,3-dibromo-6,11-dihydro-5H-pyrrolo[2,1-*b*][3]benzazepine (**735**) afforded **736** in good

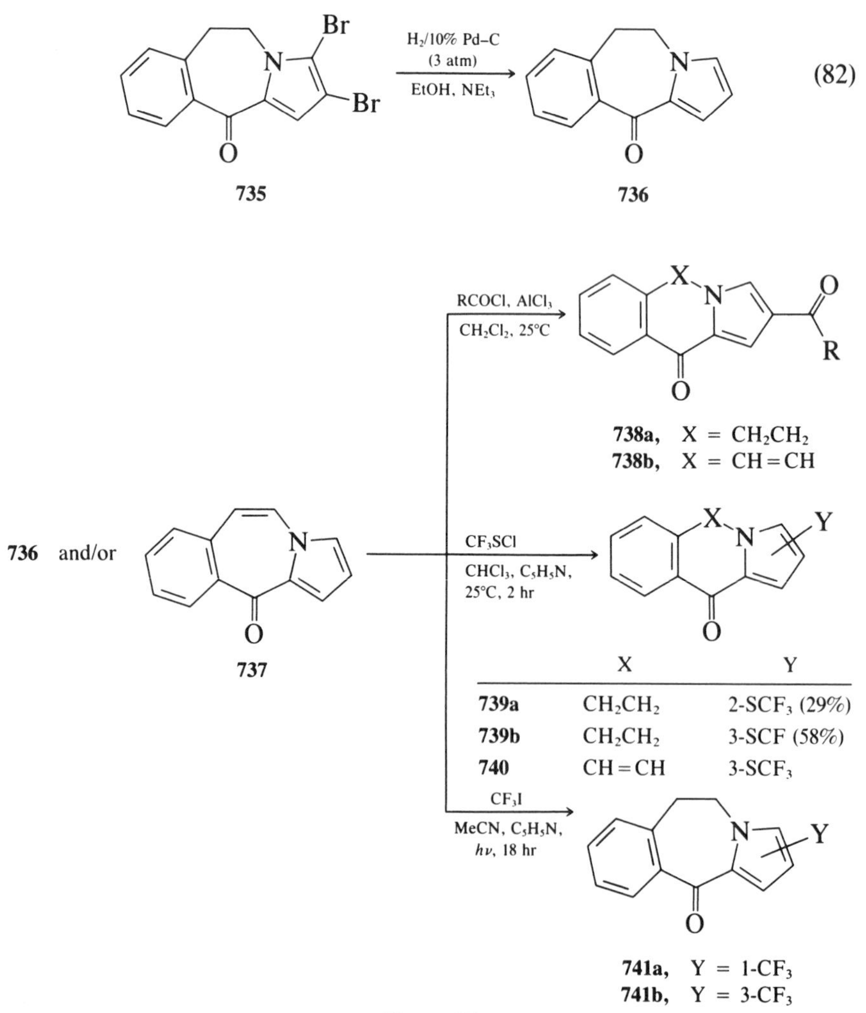

	X	Y
739a	CH_2CH_2	2-SCF_3 (29%)
739b	CH_2CH_2	3-SCF (58%)
740	CH=CH	3-SCF_3

Figure 107

yield (Eq. 82). The reactivity of **736** and the dehydro compound **737** toward electrophiles was examined (Fig. 107). For example, Friedel–Crafts acylation of **736** and **737** afforded exclusively the 2-isomers **738a** and **738b.** However, in the presence of excess acid chloride at elevated temperatures, with no aluminum chloride added, **737** afforded a mixture of the 2- and 3-isomers. Whereas treatment of **736** with trifluoromethanesulfenyl chloride produced a mixture of the isomers **739a** and **739b,** the same reaction with **737** gave only the 3-isomer **740.** Photolysis of a mixture of **736** and trifluoromethyl iodide afforded the 1- and 3-trifluoromethyl derivatives **741a** and **741b.** Iodoketone **726b** reacted with *bis*(trifluoromethylthio)mercury and copper in dimethylformamide to produce the keto sulfide **742** in 90% yield (Eq. 83). The nitroketone **726a** was reduced to **743a,** which could be converted by standard procedures to the mono- and dimethylamino ketones **743b** and **743c,** respectively. Similarly, the 2-cyano group of **719** (X = CN) was readily converted to the corresponding aldehyde, carboxylic acid, ester, amide, dimethylamide, and acid chloride by classical procedures. These derivatives can be found in Table 39.

11-Alkylidenepyrrolo[2,1-*b*][3]benzazepines such as **745** were prepared by reaction of pyrrolobenzazepinones with 3-(*N*,*N*-dimethylamino)-1-propylmagnesium chloride (Eq. 84) followed by dehydration of the resultant alcohols **744** (e.g., with hydrogen chloride). The exocyclic double bond in **745** was demonstrated to have both *cis* and *trans* stereochemistry. 11-Alkylidenepyrrole[2,1-*b*][3]benzazepines have been claimed to possess muscle relaxant properties (221).

$Hg(SCF_3)_2$, Cu, DMF, 100°C (X = I) → CF_3S ketone

726a, X = NO_2
726b, X = I

H_2/10% Pd–C, MeOH(X = NO_2) → R_1R_2N ketone

(83)

743a, $R_1 = R_2 = H$
743b, $R_1 = H$, $R_2 = Me$
743c, $R_1 = R_2 = Me$

TABLE 39A. PYRROLO[2,1-*b*][3]BENZAZEPINES[a]

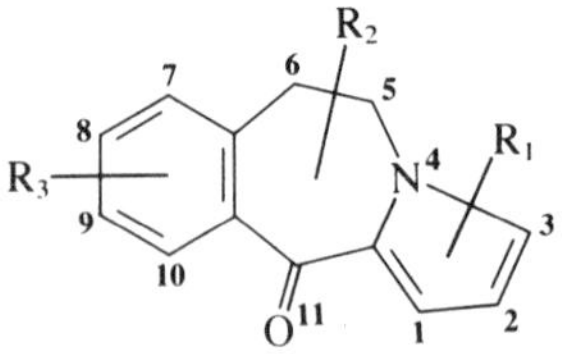

R_1	R_2	R_3	mp (°C)	Yield (%)
H	H	H	54–55	—
1-CN	H	H	123–124	76[b]
2-CN	H	H	146–147	90
2,3-Br_2	H	H	130–132	56
2-Br	H	H	101–103	—
2-Cl	H	H	100–105	—
3-CN	H	H	130–131	—
1,2,3-Br_3	H	H	150–155	—
2-CHO	H	H	135–136	56
2-CO_2H	H	H	287–290	89
2-COCl	H	H	147–148.5	89
2-CO_2Me	H	H	125–127	85
2-$CONH_2$	H	H	228	83
2-$CONMe_2$	H	H	148–149.5	84
2-SCF_3	H	H	77–78	29
3-SCF_3	H	H	95–95.5	58
1-CF_3	H	H	102–103	—
3-CF_3	H	H	90–93	—
2-Ac	H	H	160–161	—
2-COPr-*i*	H	H	124–126	—
2-COBu-*n*	H	H	137–138	85
2-SO_2NMe_2	H	H	134–137	—
H	H	9-I	120–122	—
H	H	9-SCF_3	81–83	90
H	H	9-NO_2	190–192	—
H	H	9-NH_2	166–167	—
H	H	9-NMe_2	Oil	—
2-CN	6-Cl	H	—	—
2-CN	5,6-Cl_2	H	213–223	33

[a]Ref. 221.
[b]Ref. 222.

TABLE 39B. PYRROLO[2,1-*b*][3]BENZAZEPINES[a]

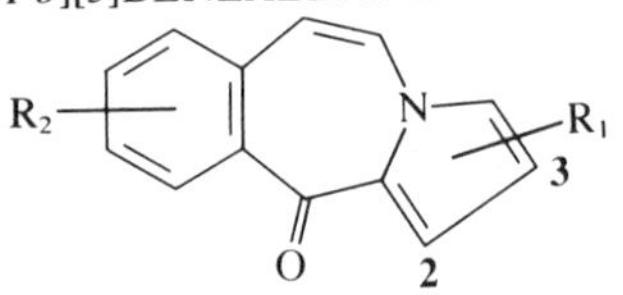

R_1	R_2	mp (°C)	Yield (%)
2-CN	H	197	86
3-SCF_3	H	137–137.5	

[a]Ref. 221.

TABLE 39C. PYRROLO[2,1-*b*][3]BENZAZEPINES[a]

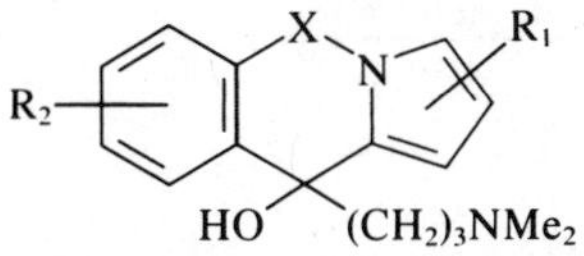

X	R_1	R_2	mp (°C)
CH_2CH_2	H	H	68–70
CH=CH	2-CN	H	123–126

[a]Ref. 221.

TABLE 39D. PYRROLO[2,1-*b*][3]BENZAZEPINES[a]

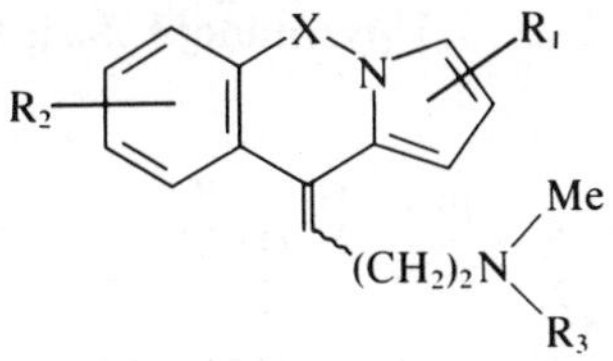

X	R_1	R_2	R_3	mp (°C)[b]
CH_2CH_2	H	H	Me	140–165
CH_2CH_2	2-SO_2NMe_2	H	Me	128–136
CH_2CH_2	2-CN	H	Me	211–215
CH_2CH_2	2-CO_2Me	H	Me	115 (dec.)
CH_2CH_2	2-Cl	H	Me	142–160
CH_2CH_2	2-CO_2H	H	Me	223–231
CH_2CH_2	2-$CONH_2$	H	Me	190 (dec.)
CH_2CH_2	2-SCF_3	H	Me	161 (dec.)
CH_2CH_2	3-SCF_3	H	Me	174 (dec.)
CH_2CH_2	2-$CONMe_2$	H	Me	167 (dec.)
CH_2CH_2	3-CF_3	H	Me	155–159
CH_2CH_2	2-COPr-*i*	H	Me	120–125
CH_2CH_2	2-COBu-*n*	H	Me	115–132 (dec.)
CH_2CH_2	2,3-Br_2	H	Me	180–185 (dec.)
CH_2CH_2	3-CN	H	Me	180–185 (dec.)
CH_2CH_2	H	9-SCF_3	Me	155–165 (dec.)
CH=CH	2-CN	H	Me	185–187
CH=CH	H	H	Me	100 (dec.)
CH=CH	H	9-SCF_3	Me	110 (dec.)
CH_2CH_2	2-CHO	H	Me	181 (dec.)
CH_2CH_2	2-CN	H	H	190 (dec.)
CH_2CH_2	H	H	Me (isomer 1)	188–189
CH_2CH_2	H	H	Me (isomer 2)	110–120
CH_2CH_2	2-CHO	H	Me (isomer 1)	104–105
CH_2CH_2	2-CHO	H	Me (isomer 2)	223

[a]Ref. 221.
[b]Oxalate.

$Me_2N(CH_2)_3MgCl$, THF

(84)

744 745

52. Furo[2′,3′:4,5]pyrimido[1,2-*a*]azepines

Heterocyclic β-enamino esters have been shown to react with imino ethers to form pyrimidines (222). When the reaction was extended to the β-enamino ester **746,** condensation with *O*-methyl caprolactim (**747**) gave the compound **748** in 57% yield (Eq. 85) (223).

160°C, 12 hr (85)

746 747 **748,** mp 148–150°C (pmr,uv)

53. Benzofuro[2,3-*b*]azepines

When *N*-methylcaprolactam diethylacetal (**749**) was condensed with *p*-benzoquinone (Eq. 86), a good yield of the condensed benzofuran **751** resulted. The reaction proceeded readily and was attributed to dissociation of **749** to the α-alkoxyenamine **750** (224).

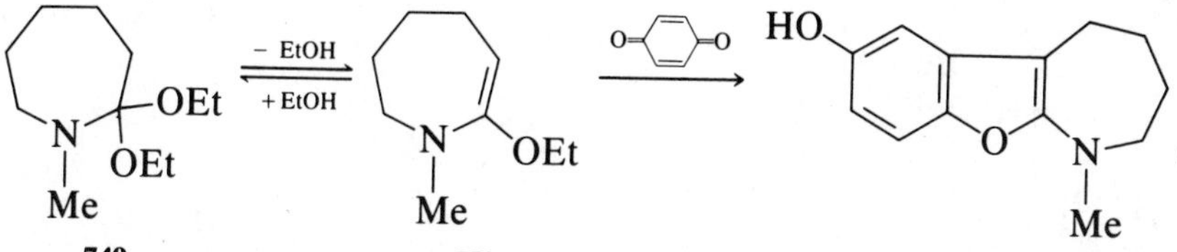

− EtOH / + EtOH

749 **750** **751**, mp 235–263°C (HCl) (86)

54. Benzofuro[2,3-*c*]azepines

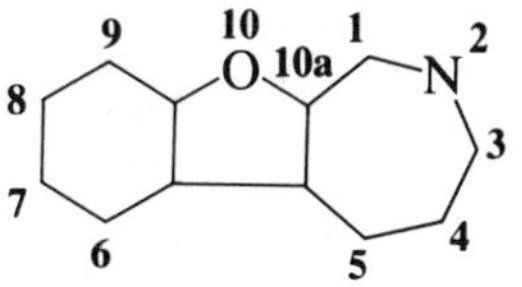

The oxime of 1,2,3,4-tetrahydro-4-oxodibenzofuran (**752**) underwent Beckmann ring expansion in hot polyphosphoric acid to give a high yield of the azepinone **753a** (Eq. 87). Lithium aluminum hydride reduction of lactam **753*a*** afforded 2,3,4,5-tetrahydro-1H-benzofuro[2,3-*c*]azepine (**753b**) (225).

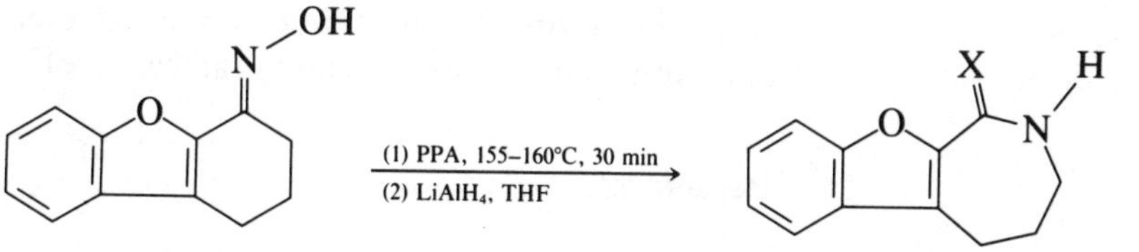

(1) PPA, 155–160°C, 30 min
(2) $LiAlH_4$, THF

752 (87)

753a, X= O; mp 247°C
753b, X = H_2; mp 300°C (HCl), mp 245°C (picrate)

55. Benzofuro[3,2-*b*]azepines

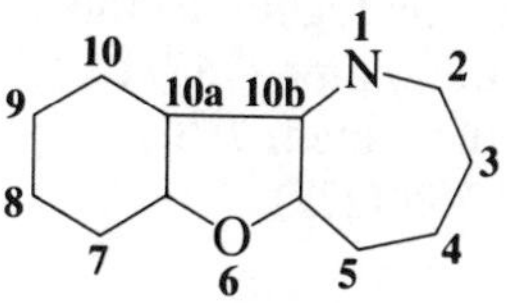

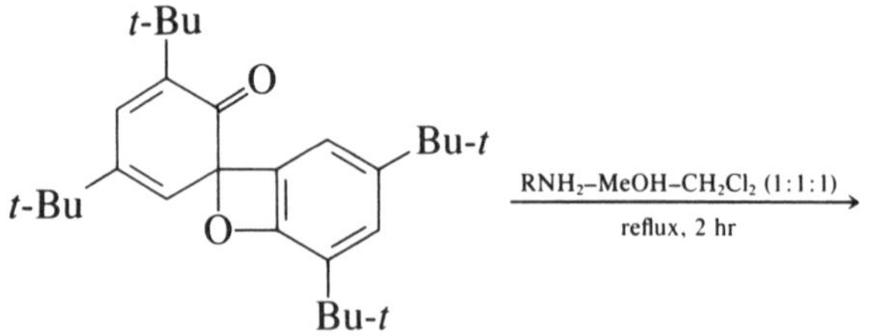

754

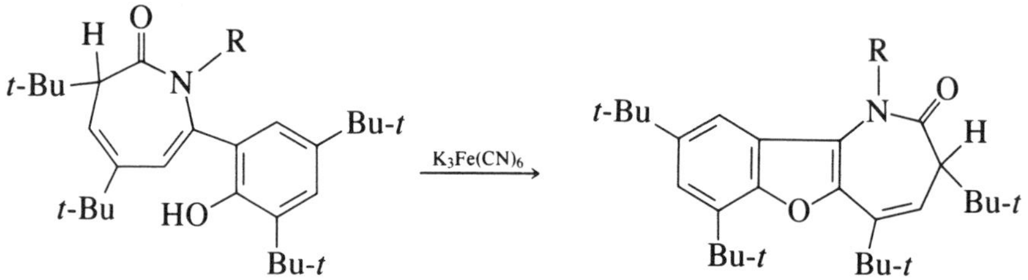

RNH_2–MeOH–CH_2Cl_2 (1:1:1), reflux, 2 hr

$K_3Fe(CN)_6$

755a, R = Me
755b, R = Pr-*n*
755c, R = *c*-C_6H_{11}

756a, R = Me (57%); mp 167–168°C
756b, R = *n*-Pr, (78%); mp 190–193°C
756c, R = *c*-C_6H_{11} (97%); mp 237–238°C

Figure 108

Refluxing a solution of the spiroquinol ether **754** in a 1:1:1 mixture of methylene chloride, methanol, and 40% aqueous primary amine for 2 hr afforded the azepinones **755a–c** (Fig. 108). When the azepinones were oxidized with potassium ferrocyanide, the benzofuro[3,2-*b*]azepines **756a–c** were produced in excellent yield. This type of intramolecular phenolic coupling was suggested to be the result of an electrophilic substitution on the azepine ring via disproportionation of an originally formed phenoxy radical (226).

56. Benzofuro[3,2-*c*]azepines

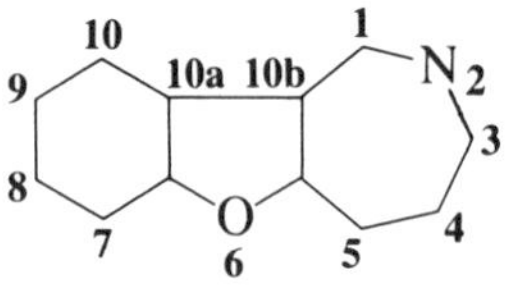

In a reaction completely analogous to the synthesis of benzofuro[2,3-*c*] azepines (Section IV. 54) the oxime (**757**) of 1,2,3,4-tetrahydro-1-oxodibenzofuran was converted to the lactam **758** with polyphosphoric acid (Eq. 88). Reduction of **758** afforded **759** (225).

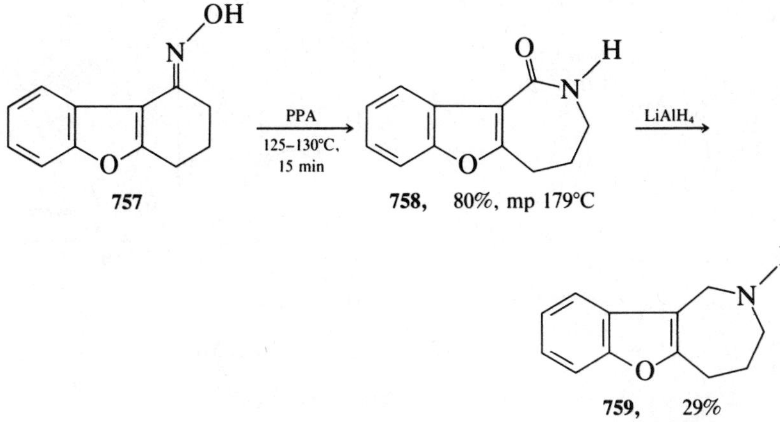

OH N PPA 125–130°C, 15 min O O N H LiAlH$_4$

757 **758**, 80%, mp 179°C

(88)

759, 29%
bp 133°C (1 mm)
mp 260°C (HCl)

57. Furo[2,3-*g*]benzazepines

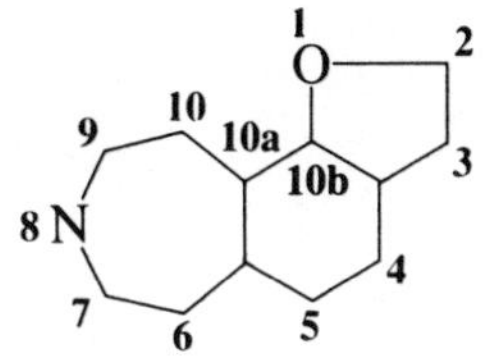

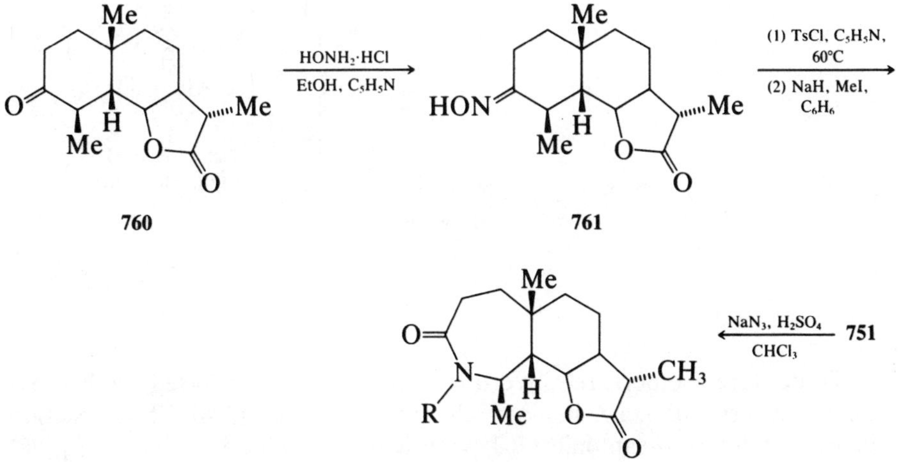

1 O 2 3 4 5 6 7 8 N 9 10 10a 10b

HONH$_2$·HCl EtOH, C_5H_5N (1) TsCl, C_5H_5N, 60°C (2) NaH, MeI, C_6H_6

760 **761** (89)

NaN$_3$, H$_2$SO$_4$ CHCl$_3$ **751**

762a, R = H
762b, R = Me

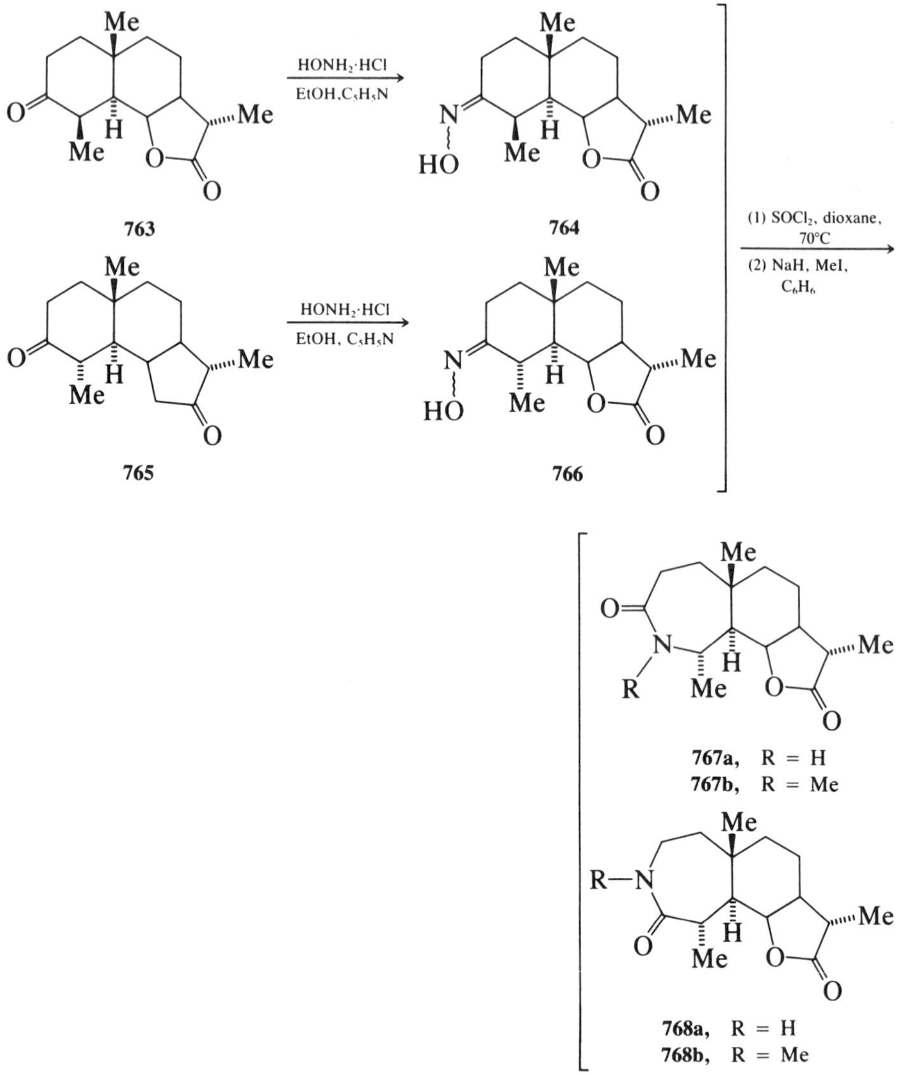

763 → ($HONH_2 \cdot HCl$, $EtOH, C_5H_5N$) → 764

765 → ($HONH_2 \cdot HCl$, $EtOH, C_5H_5N$) → 766

(1) $SOCl_2$, dioxane, 70°C; (2) NaH, MeI, C_6H_6 →

767a, R = H
767b, R = Me

768a, R = H
768b, R = Me

Figure 109

Three stereoisomeric tetrahydro-α-santonins were converted via their oximes to a series of aza-A-homotetrahydro-α-santonins (228). Thus, reaction of *cis*-tetrahydro-α-santonin (**760**) with hydroxylamine in ethanol (Eq. 89) afforded the corresponding oxime **761**, which on treatment with *p*-toluenesulfonyl chloride in pyridine rearranged in 76% yield to 4-aza-A-homo-*cis*-tetrahydro-α-santonin (**762a**). Since no other isomeric products were detected, the authors suggested that the oxime **761** must exist exclusively in the *anti* configuration. In contrast, *trans*-4β-tetrahydro-α-santonin (**763**) gave two oximes (**764**), in a 1:4 ratio (Fig. 109). Likewise, *trans*-4α-tetrahydro-α-santonin (**765**) gave two oximes **766**, but in a 5:1 ratio. On treatment with thionyl chloride in dioxane, each oxime mixture underwent Beckmann rearrangement to the lactams **767a** and **768a** in ratios of 2:3 and 2:1, respectively. The ketones **760** and **765** underwent the Schmidt reaction with hydrazoic acid in chloroform to give an 85% yield of **762a** and a 40% yield of **767a,** respectively. The Schmidt lactams were identical to the Beckmann lactams. *N*-Methylation of the lactams **762a, 767a** and **768a** with sodium hydride–methyl iodide in benzene gave lactams **762b, 767b,** and **768b,** respectively (227).

A detailed analysis of the circular dichroism (CD) and optical rotary dispersion (ORD) spectra of lactams **762, 767,** and **768** was performed (227) in order to elucidate the possible relationship between the conformation of a seven-membered lactam and the sign of the n–π^* Cotton effect. The data suggested that such a relationship does exist, and the authors proposed a lactam rule linking the sign of the n–π^* carbonyl effect with the conformation of the ring. According to this rule, lactams may be classified into two types on the basis of the octant rule, with seven-membered lactams of type *A* and type *B* displaying negative and positive Cotton effects, respectively. Table 40 lists available data on these lactams.

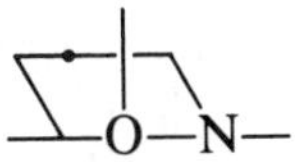

—O—N—   —N—O—

A 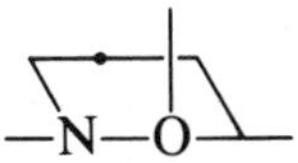   **B**

TABLE 40. FURO[2,3-*g*][3]BENZAZEPINES

Compound	mp (°C)	Spectra	Refs.
762a	222	$[\alpha]_D^{26}$ = +27.50° ir,nmr,ms	228
767a	214–218	$[\alpha]_D^{24}$ = −3.92° ir,nmr,ms	228
768a	211–213	$[\alpha]_D^{23}$ = +10.90° ir,nmr,ms	228
762b	113–114	ORD,CD, ir,nmr	227
767b	223–224	ORD,CD, ir,nmr	227
768b	207–209	ORD,CD, ir,nmr	227

[a]Eq. 89, Fig. 109.

58. Furo[3,2-*g*][1]benzazepines

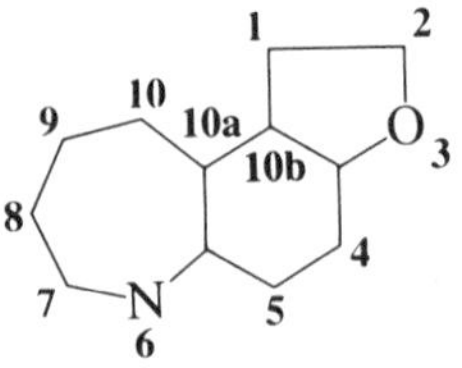

59. Furo[3,2-*g*][2]benzazepines

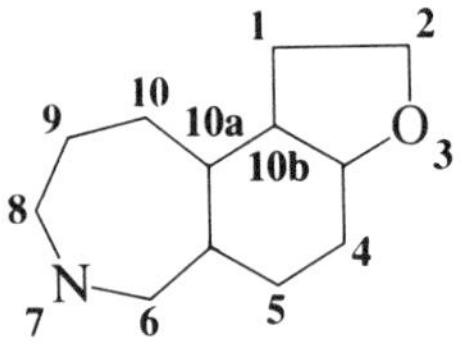

Succinoylation of trimethylbenzofuran **769a** (Fig. 110) resulted in an inseparable mixture of keto acids **769b** and **769c.** Esterification of the mixture produced a 65:35 mixture of **769d** and **769e,** which could be resolved by fractional crystallization. Saponification of **769e** gave pure keto acid **769c** which, on Huang–Minlon reduction, afforded **769f** in 40% yield. Cyclodehydration of **769f** proceeded in 90% yield with formation of **770a.** The corresponding oxime **770b** underwent Beckmann rearrangement in polyphosphoric acid to a 65:35 mixture of the [3,2-*g*][1] lactam **771a** and the [3,2-*g*][2] isomer **771b.** Each of the components of the isomeric mixtures **769b** and **769c, 769d** and **769e,** and **771a** and **771b** was unambiguously characterized by nmr spectrometry (225).

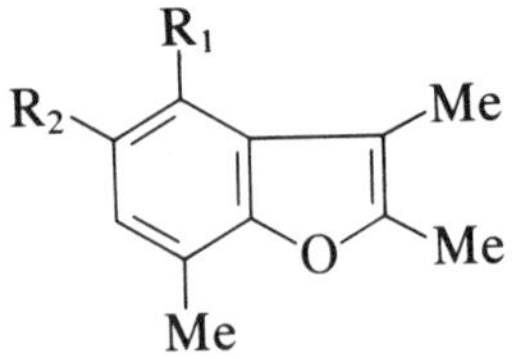

769a, $R_1 = R_2 = H$
769b, $R_1 = H, R_2 = CO(CH_2)_2CO_2H$
769c, $R_1 = CO(CH_2)_2CO_2H, R_2 = H$
769d, $R_1 = H, R_2 = CO(CH_2)_2CO_2Et$
769e, $R_1 = CO(CH_2)_2CO_2Et, R_2 = H$
769f, $R_1 = (CH_2)_3CO_2H, R_2 = H$

Figure 110

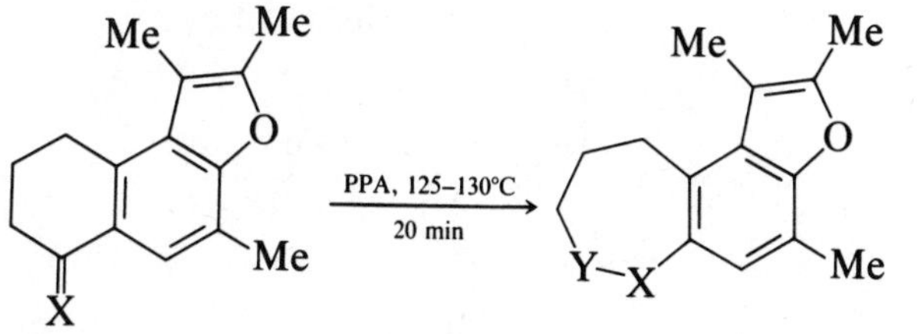

770a, X = O
770b, X = NOH

PPA, 125–130°C, 20 min

771a, X = NH, Y = CO; mp 241°C
771b, X = CO, Y = NH; mp 204°C

Figure 110 (*continued*)

60. Furo[3,2-*i*][1]benzazepines

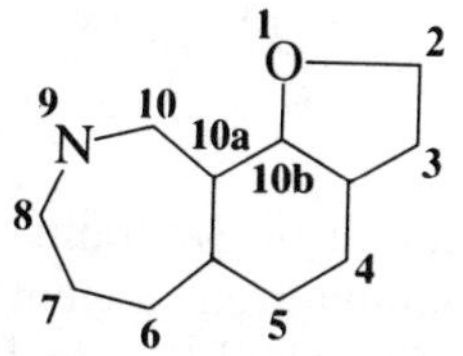

Trimethylbenzofuran **772a** (Eq. 90) was converted to ketone **773a** via a standard sequence. Thus, Friedel–Crafts acylation of **772a** with succinic anhydride gave 93% of the keto acid **772b**. Huang–Minlon reduction of **772b** afforded 89% of **772c**, which on cyclodehydration produced **773a**. The oxime **773b** underwent Beckmann rearrangement to **774** in 53% yield, and lithium aluminum hydride reduction of **774** gave **775** in 83% yield (225).

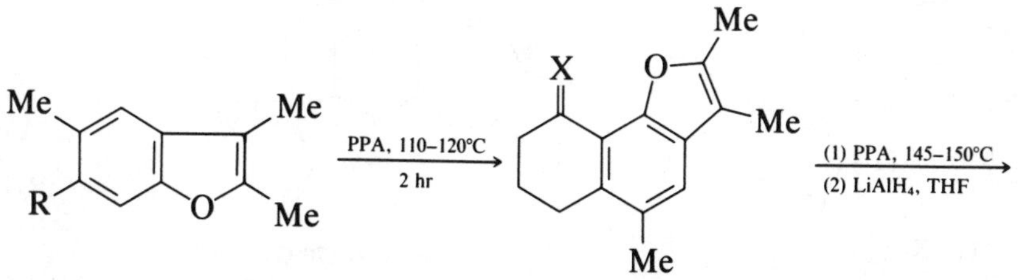

PPA, 110–120°C, 2 hr

(1) PPA, 145–150°C
(2) $LiAlH_4$, THF

(90)

772a, R = H
772b, R = $CO(CH_2)_2CO_2H$
772c, R = $(CH_2)_3CO_2H$

773a, X = O
773b, X = NOH

(*see next page*)

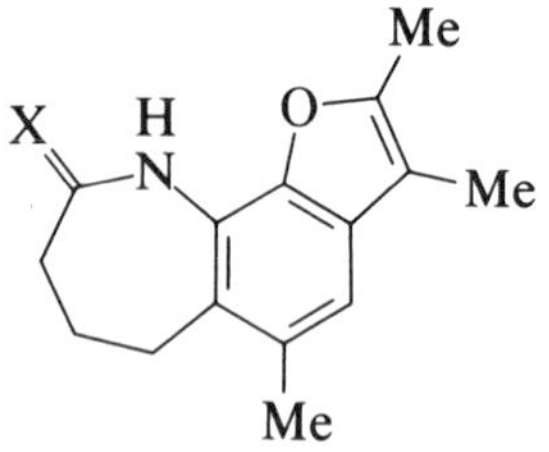

(90 *cont'd*)

774, X = O, mp 161°C
775, X = H_2, mp 108°C

61. Furo[3,2-*i*][2]benzazepines

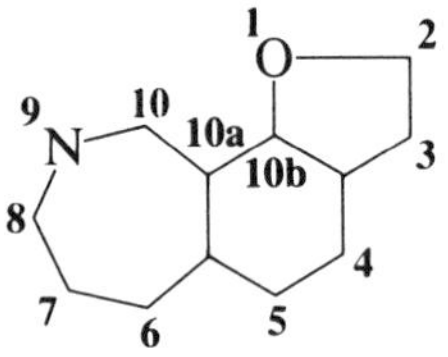

The conversion of *cis*-tetrahydro-α-santonin (**760**) to various ring-enlarged lactams via Beckmann and Schmidt rearrangement was described in Section IV. 57. A more recent note described the Beckmann rearrangement of the oxime **777** of α-santonin itself (**776**) (229). Treatment of **777** with phosphorous pentachloride in ether (Eq. 91) gave, after two recrystallizations, a 26% yield of the analytically pure lactam **778.** Lactam **778** was identified on the basis of its analytical and spectral data and by its hydrogenation products. Hydrogenation of **778** in acetone solution over 5% palladium on carbon gave a mixture of 4-aza-A-homo-*cis*-tetrahydro-α-santonin (**762a**) and its *trans* isomer **767a.**

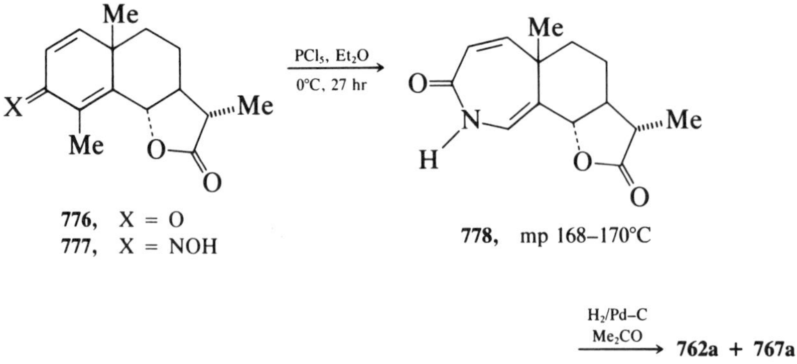

PCl_5, Et_2O, 0°C, 27 hr

776, X = O
777, X = NOH

778, mp 168–170°C

(91)

H_2/Pd–C, Me_2CO → **762a** + **767a**

62. Furo[3,4-*c*][1]benzazepines

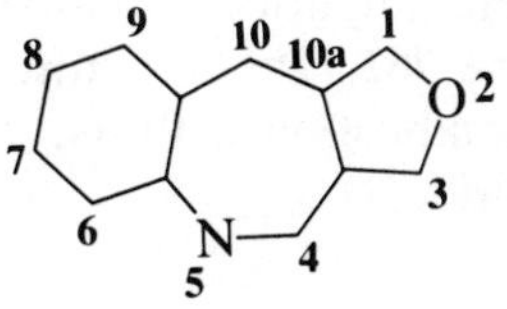

1
2
3
4
5
6
7
8
9
10
10a
O
N

Reaction of indole with dimethyl acetylenedicarboxylate in refluxing acetonitrile (Fig. 111) gave the benz[*b*]azepine **781a** (230,231), probably via the

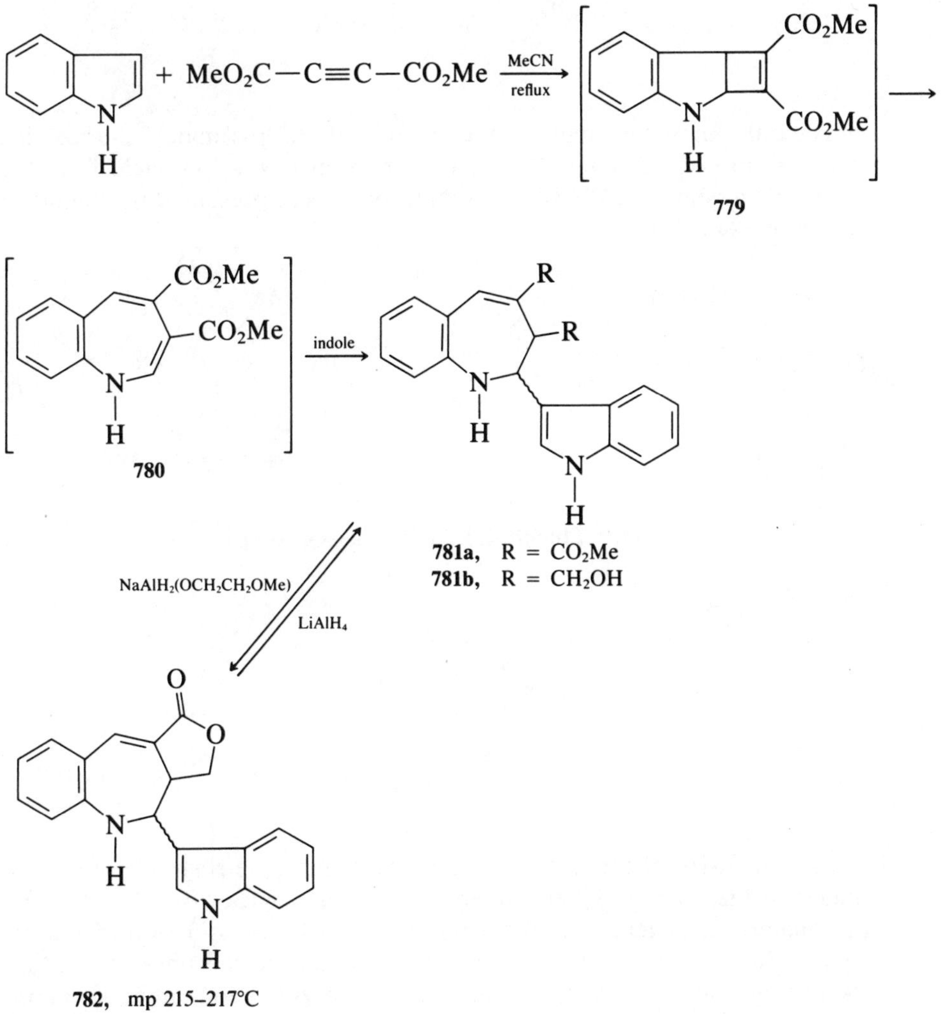

+ MeO2C—C≡C—CO2Me
MeCN
reflux
CO2Me
CO2Me
779
780
indole
781a, R = CO2Me
781b, R = CH2OH
NaAlH2(OCH2CH2OMe)
LiAlH4
782, mp 215–217°C

Figure 111

cycloaddition product **779** and the ring-opened benzazepine **780.** Although **781a** was stable to lithium aluminum hydride, reduction with sodium *bis*(2methoxyethoxy)aluminum hydride in refluxing benzene gave a mixture of the diol **781b** and lactone **782.** The structure of **782** was established on the basis of spectral data, combustion analysis, and further reduction to **781b** with lithium aluminum hydride.

63. Thieno[2′,3′:4,5]pyrimido[1,2-*a*]azepines

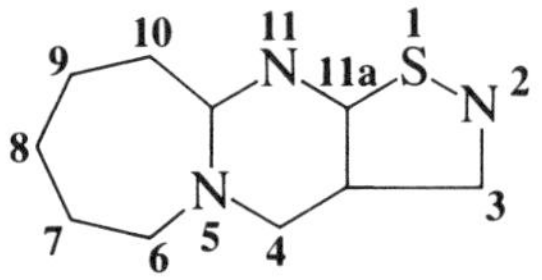

The only known example of the thieno[2′,3′:4,5]pyrimido[1,2-*a*]azepine ring system is compound **784,** which was prepared in low yield from the thiophene amino acid **783** (Eq. 92). Nmr data were presented in support of structure **784** (232).

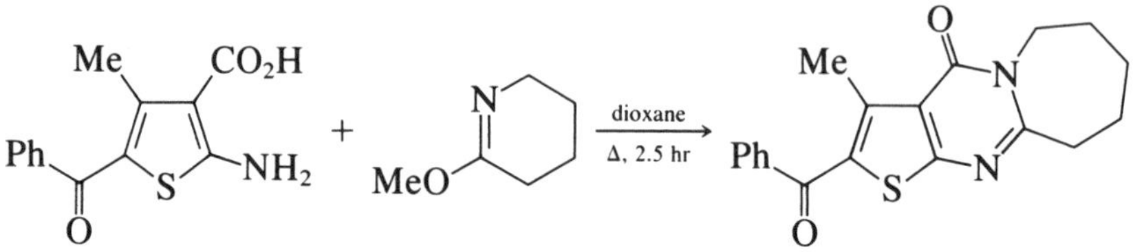

dioxane, Δ, 2.5 hr (92)

783 **784,** mp 133–135°C

64. Thieno[4,3,2-*ef*][1]benzazepines

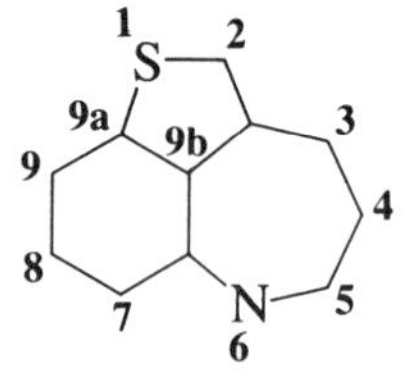

The chemistry of the thieno[4,3,2-*ef*][1]benzazepine ring system is very similar to that of the [1]benzothieno[2,3-*c*]azepines (Section IV. 65)(233). Beckmann rearrangement of the oxime **785** gave the 5-oxo-3,4-dihydro compound **786.** The structural assignment was based on the observation that alkaline hydrolysis yielded a product that, on diazotization, underwent coupling to β-naphthol. Lactam **786** was converted to the lactim ether **787,** from

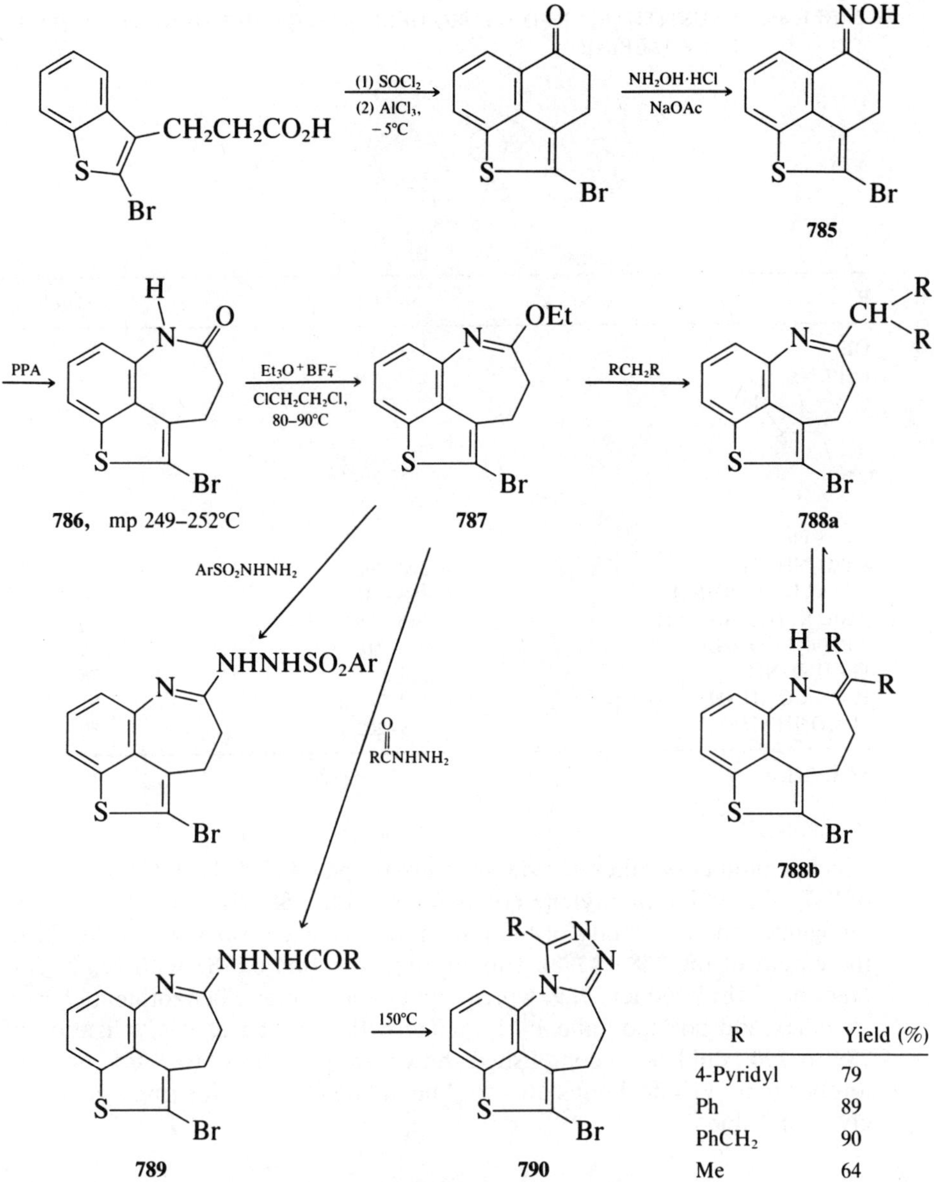

(1) $SOCl_2$ (2) $AlCl_3$, −5°C

$NH_2OH{\cdot}HCl$, NaOAc

785

PPA

786, mp 249–252°C

$Et_3O^+BF_4^-$, $ClCH_2CH_2Cl$, 80–90°C

787

RCH_2R

788a

788b

$ArSO_2NHNH_2$

$RCONHNH_2$

789

150°C

790

R	Yield (%)
4-Pyridyl	79
Ph	89
$PhCH_2$	90
Me	64

Figure 112

TABLE 41. 5-SUSTITUTED 3,4-DIHYDRO DERIVATIVES OF THIENO [4,3,2-*ef*] [1] BENZAZEPINE[a]

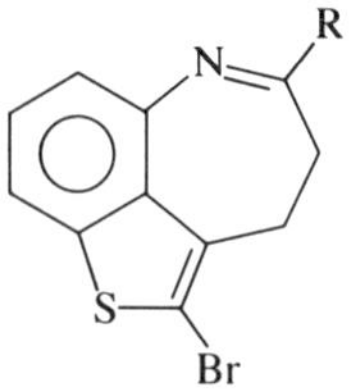

R	mp (°C)	Yield (%)
OEt	181–183	37
$CH(CN)_2$	257–259	12
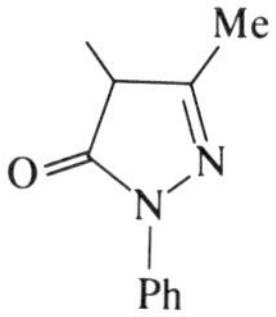 3-Methyl-5-oxo-1-phenyl-2-pyrazolin-4-yl (structure: Me, O, N, N, Ph)	234.5–237	58
$PhSO_2NHNH$	200–202	87
p-$MeC_6H_4SO_2NHNH$	199–201	96
p-$MeOC_6H_4SO_2NHNH$	186–188	90
4-Pyridyl-CONHNH	216–218.5	79
PhCONHNH	236.5	78
$PhCH_2CONHNH$	169.5–172.5	91
MeCONHNH	199–202	89

[a]Ref. 233.

which a number of other derivatives were prepared (Fig. 112). Condensation of **787** with active methylene compounds led to **788.** Although the authors recognized the possibility of tautomers, no evidence was given to establish the extent of the **788a**⇌**788b** equilibrium. Reaction of **787** with acylhydrazines gave the products **789,** which, unlike the isomeric benzothieno[2,3-*c*]-azepines, did not spontaneously cyclize with loss of water. Cyclization of **789** to **790** could be accomplished, however, by heating to 150°C in dichlorobenzene. Physical constants for known members of this ring system are given in Table 41.

65. [1]Benzothieno[2,3-*c*]azepines

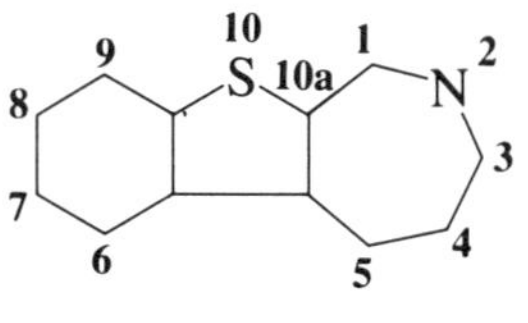

The tetrahydrodibenzothiophenone **791** (Fig. 113) was converted to a 1:1 mixture of oximes, which, when subjected to Beckmann rearrangement, formed the [1]benzothieno[2,3-*c*]azepin-1-one **792** in high yield (234). The assignment of structure of the product rests solely on the fact that hydrolysis gave an amino acid that, on diazotization, did not couple with β-naphthol. Although **792** arises from the presumably electronically less favored mode of ring expansion, more rigorous evidence is needed to rule out the alternate

H_2NOH; PPA or TSCl

N~OH

791

792, mp 198–199.5°C

$Et_3O^+BF^-_4$, CH_2Cl_2 20°C

OEt·HBF_4

793, mp 194°C

RCH_2R

794

$ArSO_2NHNH_2$

$RCNHNH_2$ (O above C)

$ArSO_2NHNH$

796

795

797

R	Yield (%)
4-Pyridyl	36
Ph	20
Me	43

Figure 113

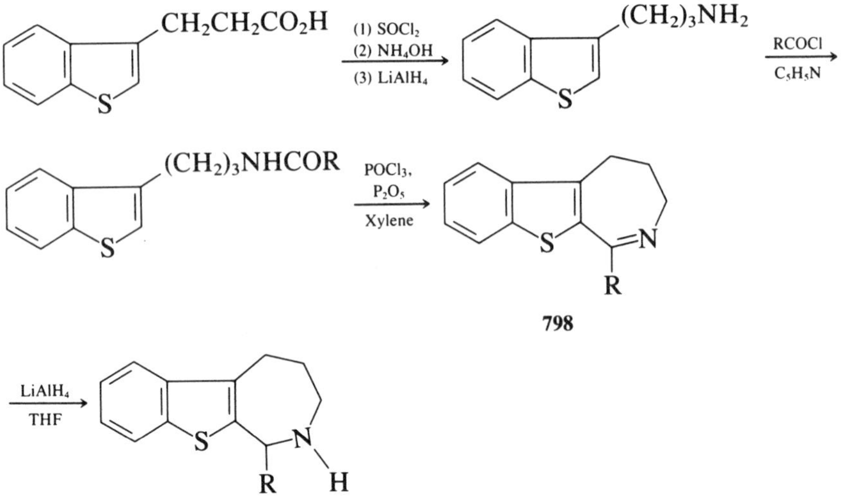

$CH_2CH_2CO_2H$ —(1) $SOCl_2$, (2) NH_4OH, (3) $LiAlH_4$→ $(CH_2)_3NH_2$ —$RCOCl$, C_5H_5N→

$(CH_2)_3NHCOR$ —$POCl_3$, P_2O_5, Xylene→ **798**

—$LiAlH_4$, THF→ **799**

Figure 114

TABLE 42. DERIVATIVES OF 1H-[1]BENZOTHIENO[2,3-*c*]AZEPINES

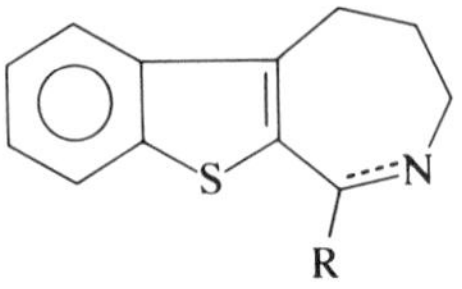

R	mp (°C)	Salt	Yield (%)	Refs.
		Dihydro		
OEt	33–36		33	234
	194	HBF_4		
$CH(CN)_2$	278.5		40	234
	220–223		43	234
Me, O, N, N, Ph	220–223		43	234
$PhSO_2NHNH_2$	156.5–158.5		51	234
p-$MeOC_6H_4SO_2NHNH_2$	204		43	234
Et	158–160	Fumarate ethanolate		235,236
Ph	Oil			235
		Tetrahydro		
Et	238–240	HCl		235
Ph	239–242	HCl		235

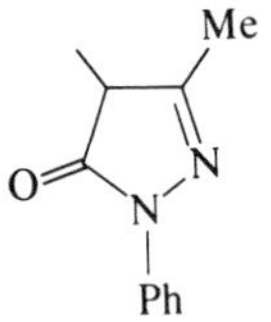

lactam structure. Conversion of **792** by Meerwein's reagent to the crystalline lactim ether **793** was accomplished in 37% yield. Further reaction of **793** with compounds containing an active methylene group generated adducts (**794**) that were shown, on the basis of infrared and nmr spectra, to be in equilibrium with the tautomers **795.** In the case of the malononitrile adduct, tautomer **795** predominated at room temperature and became the exclusive species at 100°C, as shown by the complete loss of the methine proton signal at δ 3.35 ppm (DMSO). Reaction of **793** with arylsulfonylhydrazines gave **796,** but with acylhydrazines the triazole products **797** formed spontaneously.

Several 1-substituted dihydro (**798**) and tetrahydro (**799**) derivatives of this system were prepared in unspecified yield as shown in Fig. 114 (235,236). These substances (Table 42) were claimed to have antihypertensive and tranquilizing effects.

66. [1]Benzothieno[3,2-*b*]azepines

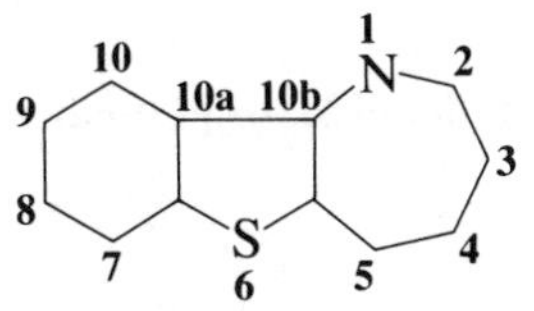

67. [1]Benzothieno[3,2-*c*]azepines

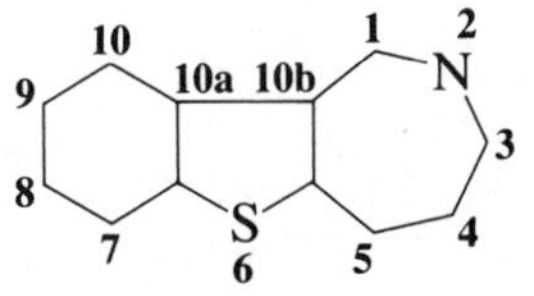

Derivatives of the isomeric [3,2-*b*] and [3,2-*c*] systems have been prepared as shown in Fig. 115 (238,239). Solvolysis of the oxime benzenesulfonate **801** gave only the [3,2-*b*] isomer **803** in 68% yield. Strong acid treatment of the oxime **800** gave only the [3,2-*c*] system **802** in low yield. Alkaline hydrolysis of **803** gave an amino acid that underwent diazo coupling with β-naphthol. Although this observation was offered as evidence that an aromatic amine had been formed in the hydrolysis of **803,** the same kind of hydrolysis–diazo coupling experiment was not reported for **802.** The lactams were said to be effective against hexanol-induced sleep and as central nervous system blocking agents (238).

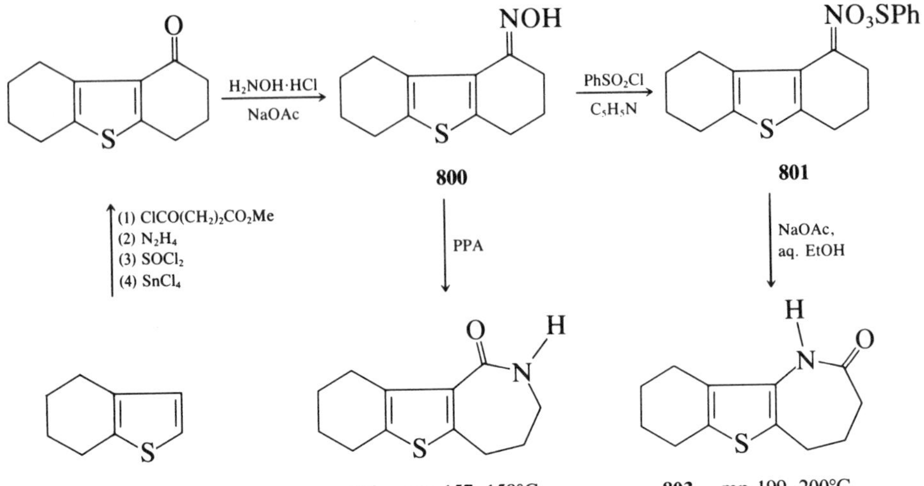

$H_2NOH{\cdot}HCl$, NaOAc; $PhSO_2Cl$, C_5H_5N; (1) $ClCO(CH_2)_2CO_2Me$ (2) N_2H_4 (3) $SOCl_2$ (4) $SnCl_4$; PPA; NaOAc, aq. EtOH

800, **801**

802, mp 157–158°C **803**, mp 199–200°C

Figure 115

68. Thieno[2,3-*c*][1]benzazepines

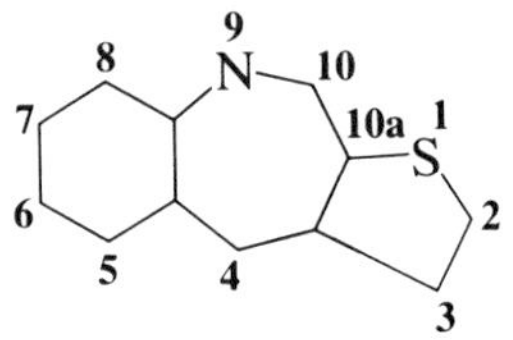

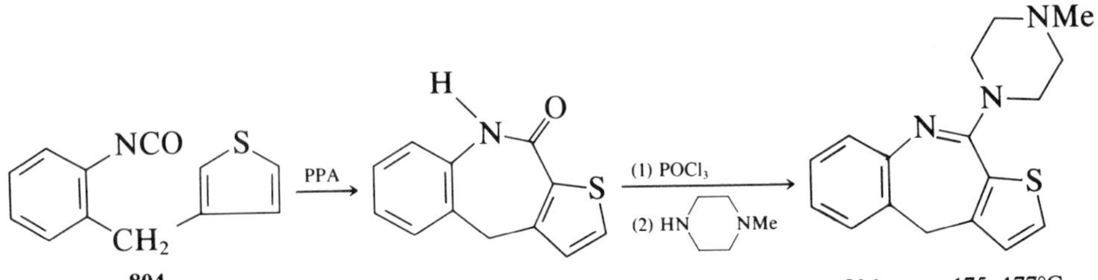

n-BuLi; (1) KOH/N_2H_4 (2) $COCl_2$; PPA; (1) $POCl_3$ (2) HN NMe

804

805, mp 236–238°C **806**, mp 175–177°C

Figure 116

A patent (240) has claimed the preparation of two examples of the thieno-[2,3-*c*][1]benzazepine system (Fig. 116): the lactam **805** (from the cyclization of the isocyanate **804**) and the piperazinyl amidine **806**. The patent claims antipsychotic activity, but no other data were provided.

69. Thieno[3,2-*b*][1]benzazepines

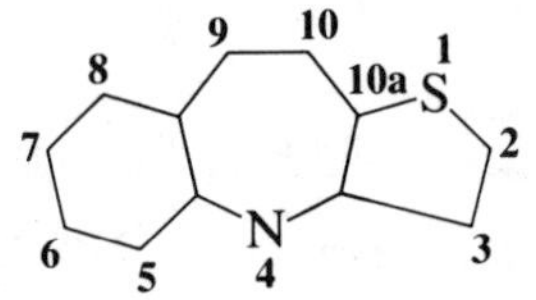

All reported examples of the thieno[3,2-*b*][1]benzazepine ring system are 4H-derivatives and are the outcome of an extensive search for new antidepressant drugs. Various chemical approaches to the synthesis of these structures were developed (241–244) and are summarized in Figs. 117–120 and Eqs. 93, 94. Specific compounds are listed in Tables 43 and 44, and biological data are given in Table 45.

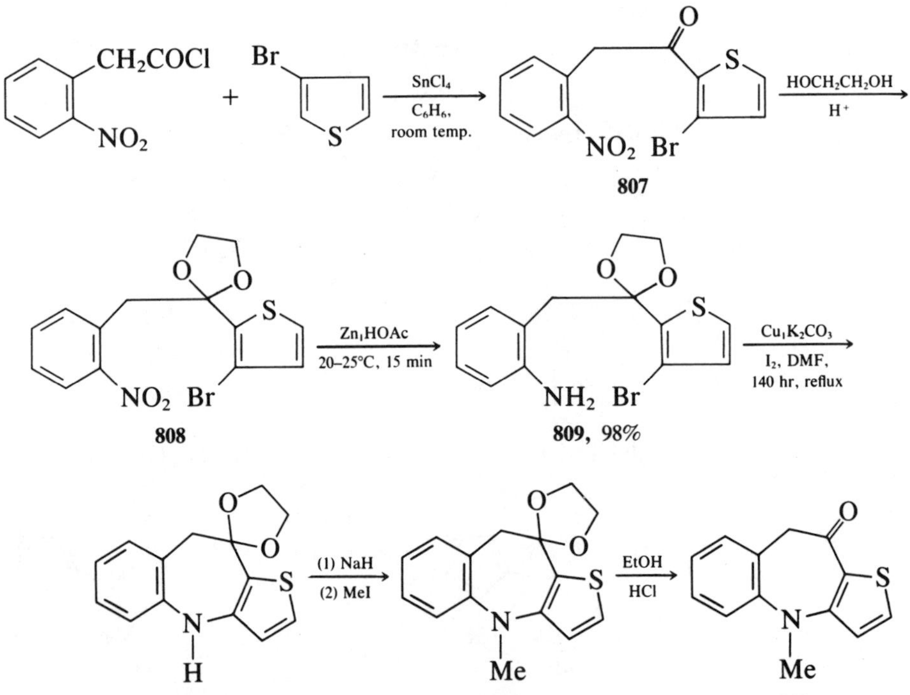

$SnCl_4$, C_6H_6, room temp.; $HOCH_2CH_2OH$, H^+; Zn, HOAc, 20–25°C, 15 min; Cu, K_2CO_3, I_2, DMF, 140 hr, reflux; (1) NaH (2) MeI; EtOH HCl

807, **808**, **809**, 98%, **810**, **811**, **812**

Figure 117

812 (1) $ZnCH_2CO_2Et$ (2) MeOH, 2 *N* HCl → H, CO_2Et, S, N, Me → KOH, EtOH → CH_2CO_2H, S, N, Me

HO–C_6H_4–NO_2, DCC, EtOAc → Activated *p*-nitrophenyl ester → $HNMe_2$, $CHCl_3$, room temp. → CH_2CONMe_2, S, N, Me

$LiAlH_4$ → $CH_2CH_2NMe_2$, S, N, Me

813

Figure 118

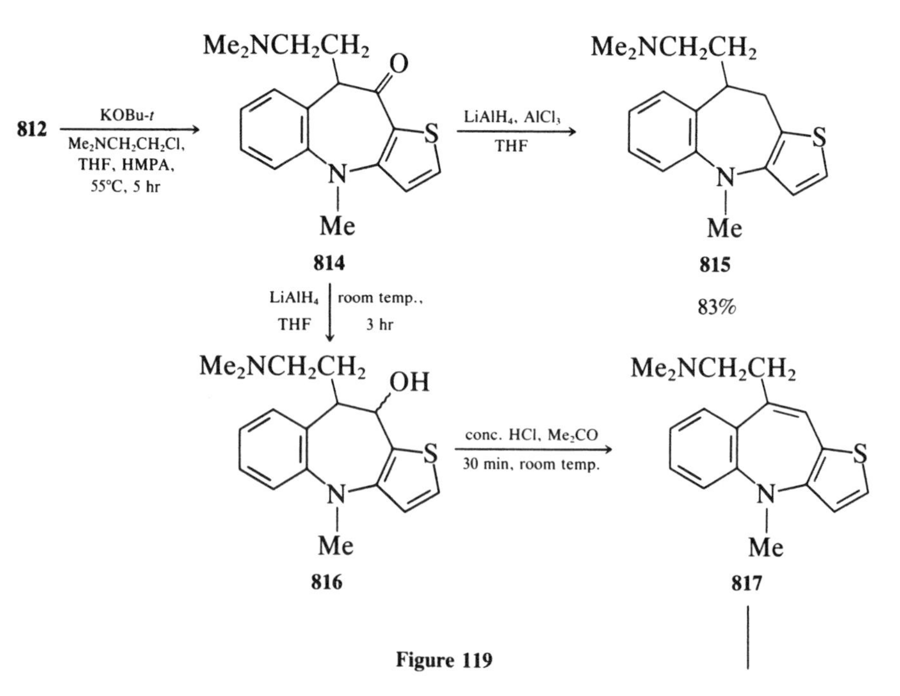

Figure 119

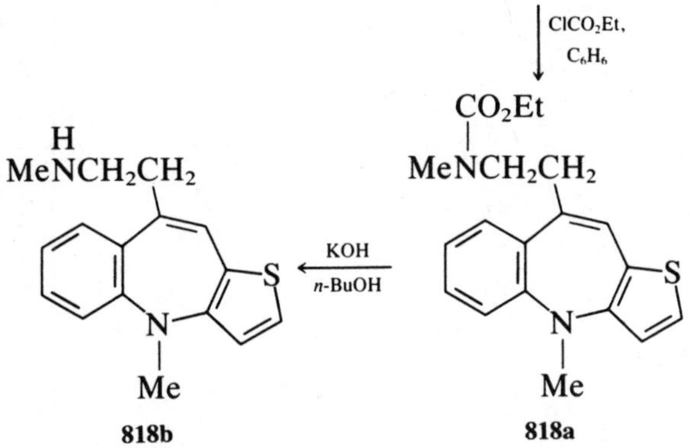

$ClCO_2Et$, C_6H_6

CO_2Et

MeNCH$_2$CH$_2$ / KOH, n-BuOH

H MeNCH$_2$CH$_2$

Me Me

818b **818a**

Figure 119 (*continued*)

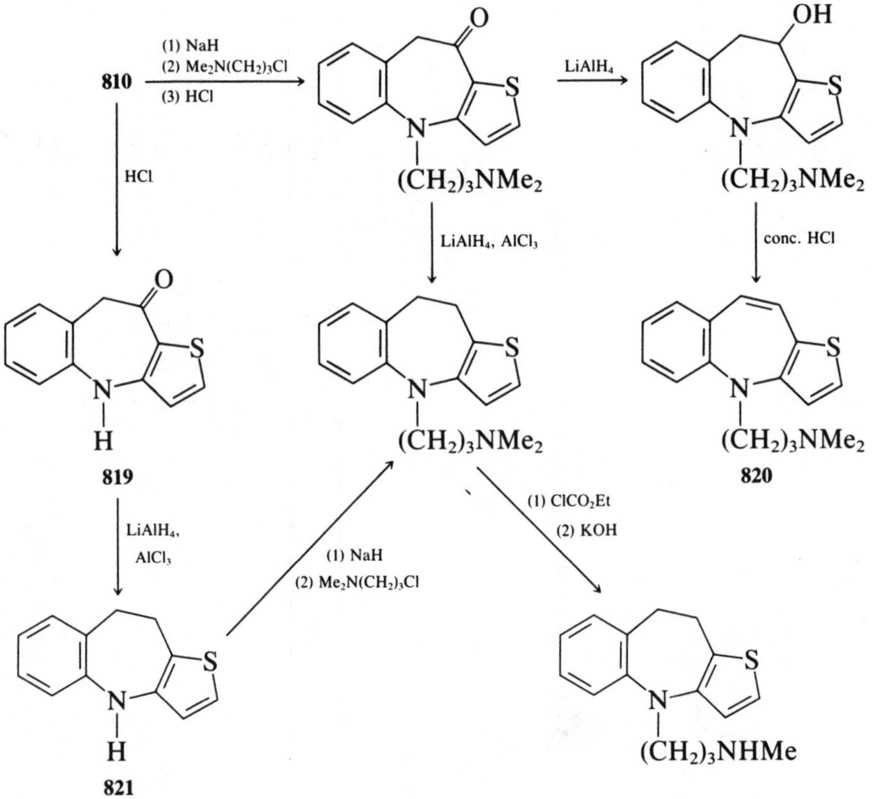

810 — (1) NaH (2) $Me_2N(CH_2)_3Cl$ (3) HCl → O, S, N, $(CH_2)_3NMe_2$ — $LiAlH_4$ → OH, S, N, $(CH_2)_3NMe_2$

HCl; $LiAlH_4$, $AlCl_3$; conc. HCl

O, S, N, H **819** — S, N, $(CH_2)_3NMe_2$ — S, N, $(CH_2)_3NMe_2$ **820**

$LiAlH_4$, $AlCl_3$; (1) NaH (2) $Me_2N(CH_2)_3Cl$; (1) $ClCO_2Et$ (2) KOH

S, N, H **821** — S, N, $(CH_2)_3NHMe$

Figure 120

TABLE 43. DERIVATIVES OF 4H-THIENO[3,2-*b*][1]BENZAZEPINE

R_1	R_2	R_3	mp (°C)	Salt	Spectra	Refs.
Me	H	CH_2CO_2Et	213			241
Me	H	CH_2CO_2Et (*exo*-methylene)	121			241
Me	H	$CH_2CO_2C_6H_4NO_2$-*p*	136			241
Me	H	CH_2CONMe_2	122			241
Me	H	$CH_2CONHMe$	149–150			241
Me	H	$(CH_2)_2NMe_2$	136	Maleate		241
Me	H	$(CH_2)_2NHMe$	186	Oxalate		241
Me	$(CH_2)_2NMe_2$	H	77		ir,uv	242
Me	$(CH_2)_3NMe_2$	H	173	Oxalate		242
Me	$(CH_2)_2NHMe$	H	158	Fumarate	ir,uv	242

$(CH_2)_3NMe_2$	H	H	166	Fumarate	uv	243
$(CH_2)_3NHMe$	H	H	115	Maleate		243
H	H	H	187		uv[a]	243
$ClCH_2CH_2CO$	H	H	130			245
$Me_2NCH_2CH_2CO$	H	H	193	Oxalate		245
$MeNHCH_2CH_2CO$	H	H	218	Oxalate		245
			9,10-Dihydro			
Me	$(CH_2)_2Me_2$	H	140	Fumarate		242
Me	$(CH_2)_2NMe_2$	OH			ir	242
$(CH_2)_3NMe_2$	H	OH	Amorphous solid		uv	243
$(CH_2)_3NMe_2$	H	H	Oil		uv	243
$(CH_2)_3NHMe$	H	H	217	HCl	uv,nmr	243
H	H	H	114			
$CH_2CHCH_3CH_2NMe_2$	H	H	210	HCl		243
$CH_2CH_2NMe_3$	H	H	218	HCl		243
CH_2CH_2NHMe	H	H	248	HCl		243
$ClCH_2CH_2CO$	H	H	108			245
$Me_2NCH_2CH_2CO$	H	H	102			245
$MeNHCH_2CH_2CO$	H	H	190	HCl		245

[a]Red solid.

TABLE 44. DERIVATIVES OF 4H-THIENO[3,2-*b*][1]BENZAZEPIN-10-ONE

R_1	R_2	mp (°C)	Salt	Spectra	Refs.
Me	H	148		ir,uv	242
Me	$(CH_2)_2NMe_2$	106		uv	242
Me	$(CH_2)_3NMe_2$	192	Oxalate		242
$(CH_2)_3NMe_2$	H	198	Fumarate	uv	243
H	H	195		ir,uv	243
		10, 10-Ethylene Ketals			
H	H	165			241
Me	H	125		uv,ir	242
$(CH_2)_3NMe_2$	H	180	Fumarate		244

819 —(1) $NaBH_4$; (2) Al_2O_3, C_6H_6, Δ→ **822** —$Me_2N(CH_2)_3Cl$, NaH→ **820** (93)

821 and **822** —$ClCH_2CH_2COCl$, C_6H_6→ [N-$COCH_2CH_2Cl$] —R-HNMe→ **823** [N-$COCH_2CH_2N(Me)R$] (94)

TABLE 45. BIOLOGICAL DATA OF THIENO[3,2-*b*][1]BENZAZEPINES FROM ANTIDEPRESSANT SCREENING MODELS

Test Compound	Antagonism to Reserpine-Induced Ptosis (rat)		Antagonism to Tetrabenazine-Induced Ptosis (rat)	
	Dose p.o. (mg/kg body wt.)	% Protection	Dose p.o. (mg/kg body wt.)	% Protection
4-(α-Dimethylamino propyl)-4H-thieno [3,2-*b*][1]benzazepine[a]	1	20	1	0
	2	49	2	25
	5	55	5	70
	10	83	10	51
9,10-Dihydro derivative[a]	1	0	1	13
	2	0	2	16
	5	69	5	50
	10	84	10	99
4-(α-Methylamino propyl)-9,10-dihydro-4H-thieno [3,2,-*b*][1]benzazepine[a]	1	7	1	42
	2	25	2	85
	5	32	5	98
	10	52	10	100
4-(α-Methylamino propyl)-4H-thieno [3,2-*b*][1]benzazepine[b]	1	0	1	50
	2	0	2	79
	5	16	5	48
	10	21	10	99

[a]Ref. 243.
[b]Ref. 244.

70. Thieno[3,2-*c*][1]benzazepines

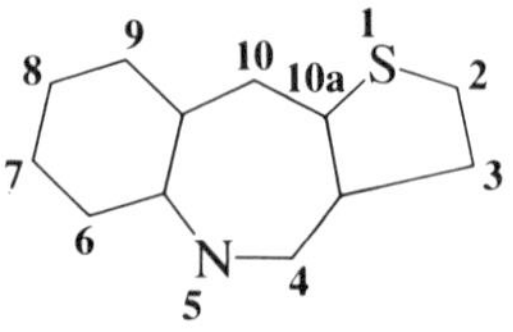

As in the related thienobenzazepine systems, neuroleptic activity has been claimed for the several known derivatives of the [3,2-*c*] system (246). Figure 121 shows the major synthetic route to the patented 4-piperazinyl derivatives **825** (Table 46A,B) via the thieno[3,2-*c*][1]benzazepin-4-ones **824**.

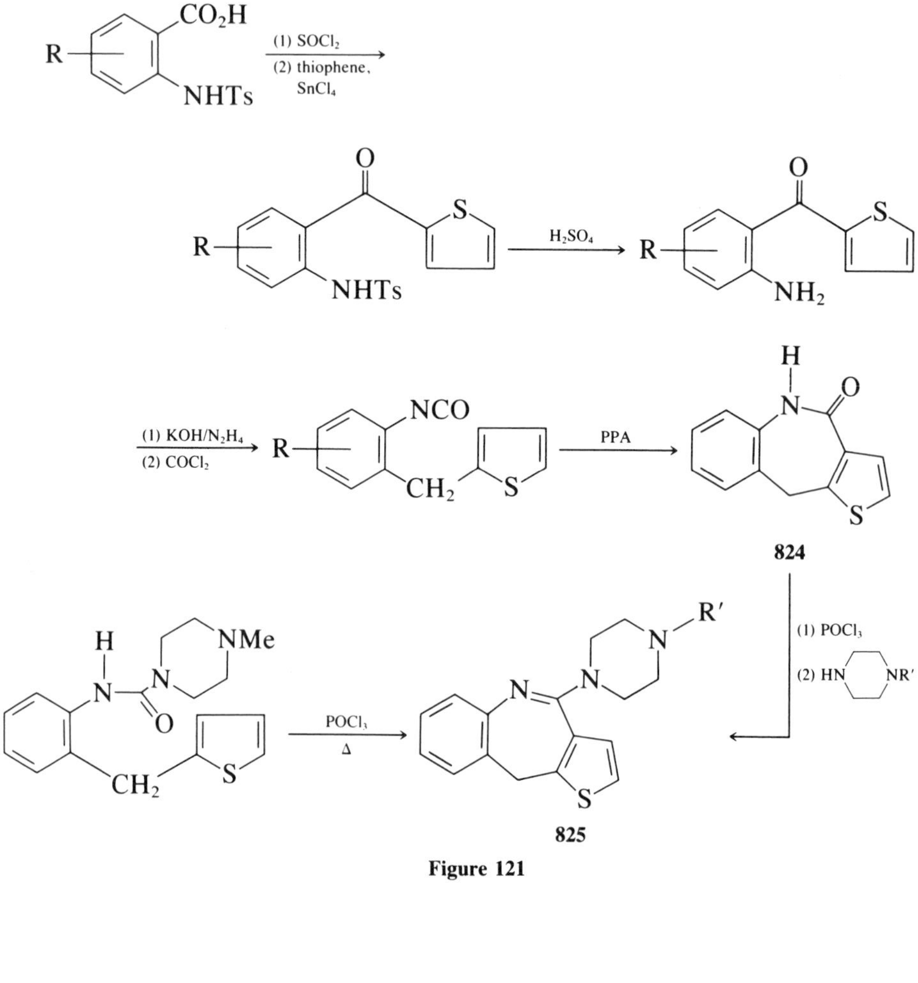

(1) $SOCl_2$ (2) thiophene, $SnCl_4$

H_2SO_4

(1) KOH/N_2H_4 (2) $COCl_2$

PPA

824

(1) $POCl_3$ (2) HN NR′

$POCl_3$ Δ

825

Figure 121

TABLE 46A. 4-(4-ALKYL-1-PIPERAZINYL)-10H-THIENO[3,2-*c*][1]BENZAZEPINES[a]

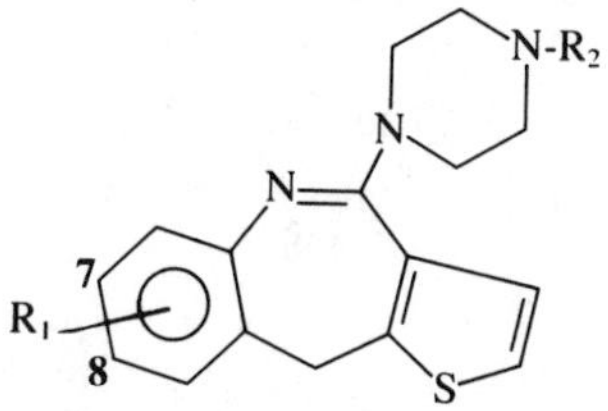

R_1	R_2	mp (°C)	Recrystallization Solvent
H	Me	145–147	Ether/petroleum ether
H	*t*-Bu	147–176 (?)	
7-Cl	H	155–157	Acetone
7-Cl	Me	184–185	Acetone, petroleum ether
7-Cl	CH_2CH_2OH	192–194	Ethyl acetate
8-Cl	H	80–100 (?)	Acetone/water
8-Cl	Me	193–195	Acetone/petroleum ether
8-Cl	CH_2CH_2OH	202–203	Ethyl acetate
8-Cl	CH_2CH_2OAc	185–189	Ether/petroleum ether
7-Me	Me	180–181	Acetone/petroleum ether
7-Me	CH_2CH_2OH	169–171	Acetone/petroleum ether

[a]Ref. 246.

TABLE 46B. 10H–4,5–DIHYDROTHIENO[3,2-*c*][1]BENZAZEPIN–4–ONES[a]

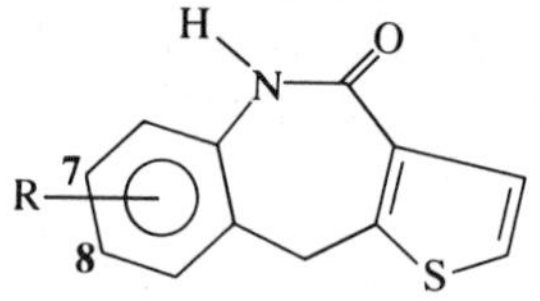

R	mp (°C)
H	225–226
7-Cl	264–266
8-Cl	280–281

[a]Ref. 246.

71. Benzo[*e*]cyclopent[*b*]azepines

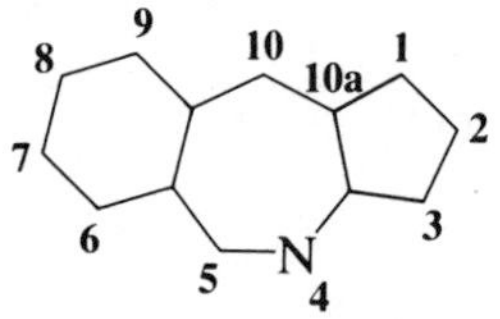

The only examples of the benzo[*e*]cyclopent[*b*]azepine system are discussed in Section VIII. 2.

72. Cyclopenta[*d*][1]benzazepines

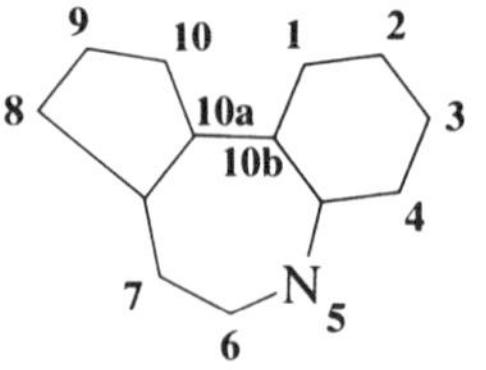

The condensation of anilines with cyclopentanone-2-acetic acid (**826a**) or the mixed anhydride **826b** in the presence of an acid catalyst (Fig. 122) produced the condensed benzazepines **827a–c** (247). In general, the best yields were obtained when the reactions were carried out using excess aniline as the solvent at reflux temperature in the presence of the corresponding aniline hydrochloride. Yields were always low, but were better than those obtained with polyphosphoric acid, Dowex 50, or ammonium chloride as the proton source. Compound **827a** was clearly different from its independently synthesized isomer **828.**

Lithium aluminum hydride reduction of lactam **827a** produced the corresponding amine **829a,** which could be acetylated with acetic anhydride–pyridine to **829b** or catalytically reduced to **830** (Fig. 123). In addition to **830,** catalytic reduction of **829a** produced as a by-product the perhydro compound **832.** Catalytic hydrogenation of **827a** produced the lactam **831,** which on lithium aluminum hydride reduction afforded **830.** Additional spectral data for **827a** and **830** have been published (248).

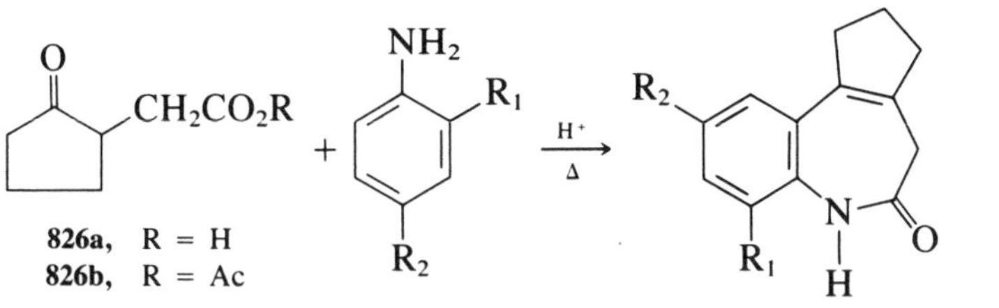

826a, R = H
826b, R = Ac

827a, $R_1 = R_2 =$ H; mp 196–197°C
827b, $R_1 =$ Me, $R_2 =$ H; mp 181–182°C
827c, $R_1 =$ H, $R_2 =$ Me; mp 201°C

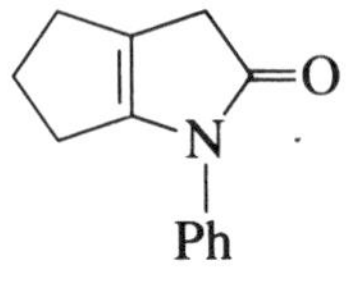

828

Figure 122

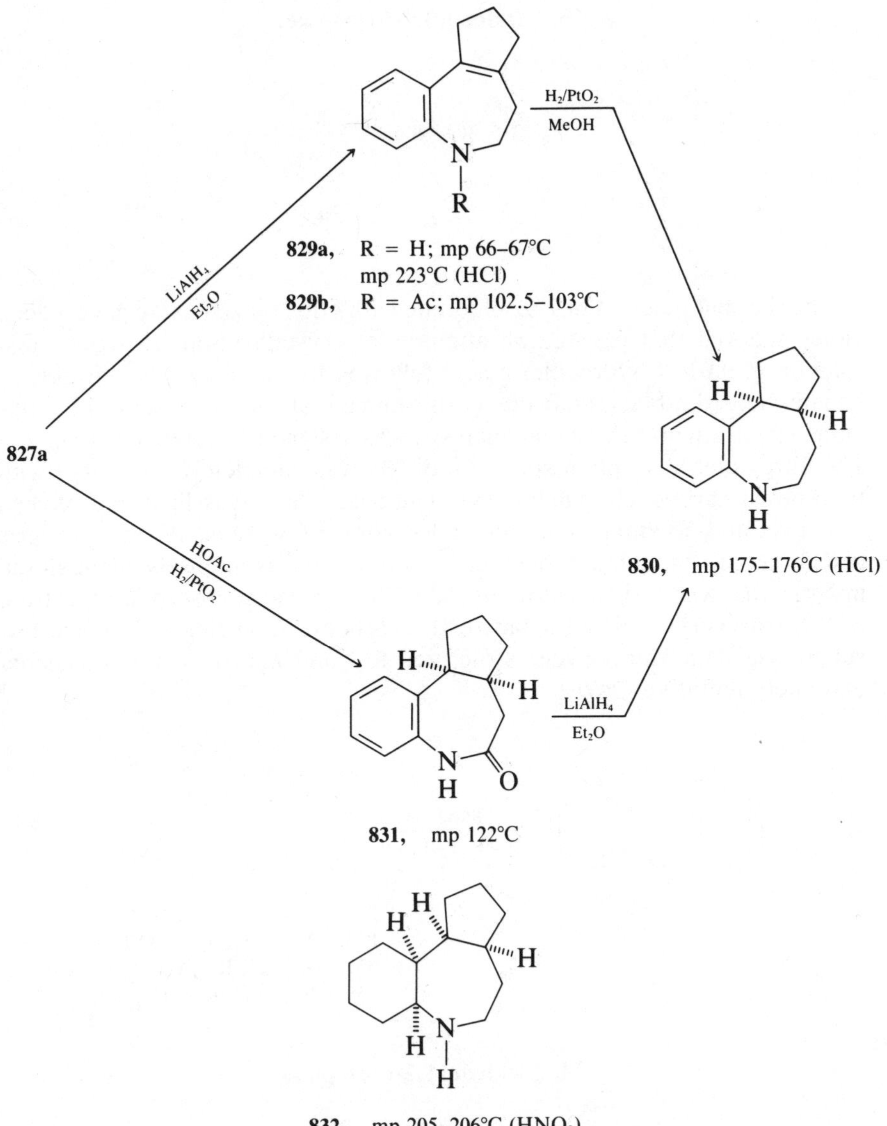
827a
LiAlH4
Et2O
829a, R = H; mp 66–67°C
mp 223°C (HCl)
829b, R = Ac; mp 102.5–103°C
N
R
H2/PtO2
MeOH
H
H
N
H
830, mp 175–176°C (HCl)
HOAc
H2/PtO2
H
H
N
H
O
LiAlH4
Et2O
831, mp 122°C
H
H
H
H
N
H
832, mp 205–206°C (HNO3)

Figure 123

73. Indeno[1,2-*b*]azepines

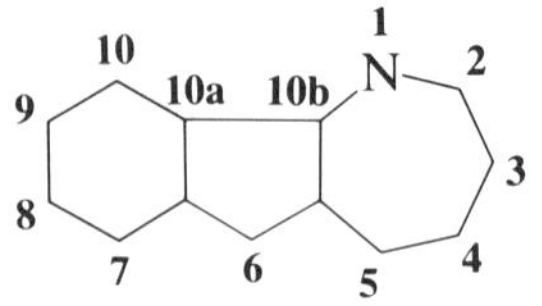

The thermal reaction of 2-azafluorene with ethyl diazoacetate gave a deep violet material that resisted all attempts at crystallization. However, dissolution in 0.01 *N* hydrochloric acid followed by neutralization with bicarbonate yielded a black amorphous substance that, on the basis of its ultraviolet spectrum and elemental analysis, was assigned structure **833** (Eq. 95). The ultraviolet absorption spectrum of **833** was completely consistent with the known carbocyclic analog **834.** The azaazulene was light and oxygen sensitive, and ethanolic solutions faded after a few days. Basic hydrolysis of **833** was accompanied by decarboxylation to give an intensely blue-colored material that was assigned structure **835.** The ultraviolet absorption spectrum of **835** was very similar to that of the carbocyclic analog **836.** Since few supporting data were given, structures **833** and **835** must be considered extremely tentative (249).

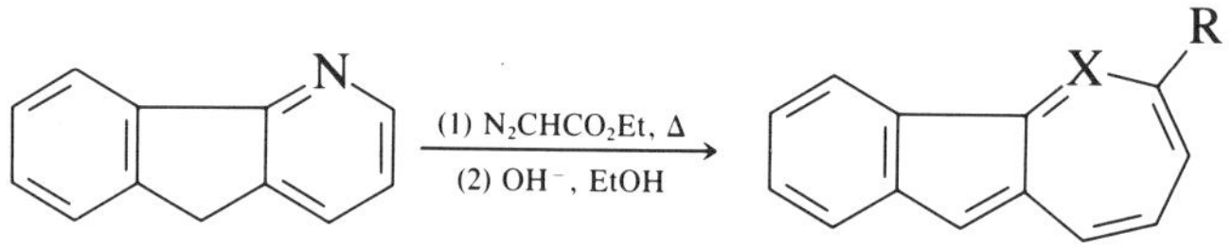

(1) N_2CHCO_2Et, Δ
(2) OH^-, EtOH

(95)

833, X = N, R = CO_2Et
834, X = CH, R = CO_2Et
835, X = N, R = H
836, X = CH, R = H

74. Indeno[1,2-*c*]azepines

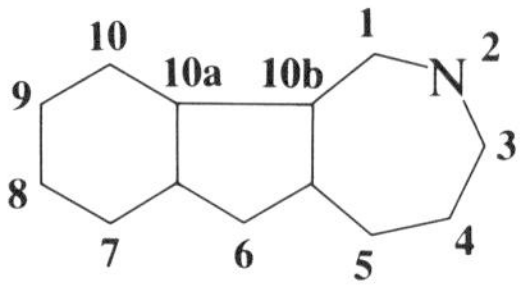

The *cis*- and *trans*-3-ketohexahydrofluorenes **837a** and **837b** were converted to their oximes **838a,b,** which on brief heating in polyphosphoric acid underwent regiospecific rearrangement to the indeno[1,2-*c*]azepine lactams

837a, X = O (*cis*-B/C)
837b, X = O (*trans*-B/C)
838a,b, X = NOH

PPA, 140°C, 10–15 min

839a,b, R = H
840a,b, R = Me
841a,b, R = Ac

$LiAlH_4$, dioxane, ether, anisole

842a,b, R = H
843a,b, R = Me
844a,b, R = Et

(96)

845, X = O
846, X = *syn*-NOH
847, X = *syn*-NOTs

aq. MeOH, reflux

848

(1) $LiAlH_4$ (2) Na, *n*-BuOH

849

(97)

839a,b (Eq. 96). The *N*-methyl derivatives **840a,b** and *N*-acetyl derivatives **841a,b** were obtained from the lactams by reaction with dimethyl sulfate and acetic anhydride, respectively. Lithium aluminum hydride reduction of the lactams **839a,b–841a,b** produced the amines **842a,b–844a,b,** respectively (250).

The structural assignment of **839a,b** was confirmed by independent synthesis of one member of the indeno[1,2-*d*]azepine family, **849,** the alternative rearrangement product from the oximes **838a,b.** The α,β-unsaturated ketone **845** was converted sequentially to the oxime **846,** the tosylate **847,** and the lactam **848** (Eq. 97). Reduction of **848** with lithium aluminum hydride followed by sodium in *n*-butanol gave **849.** The R_f value and melting point of **849** did not coincide with either **842a** or **842b.** While the stereochemistry of the ring junction in **849** was not determined, the authors assumed it to be *trans*. Physical constants for the known members of the indeno[1,2-*c*]azepine ring system are give in Table 47.

TABLE 47. INDENO[1,2-*c*]AZEPINES[a]

Compound	mp (°C)	Salt	Yield (%)
839a	146		92
839b	170		86
840a	117		73
840b	187		86
841a	94		92
841b	111		91
842a	144	HCl	82
842b	198	HCl	79
843a	123	HCl	90
843b	183	HCl	93
844a	138	HCl	92
844b	204	HCl	93
848	205		84
849	178	HCl	57

[a]Ref. 250.

75. Indeno[1,2-*d*]azepines

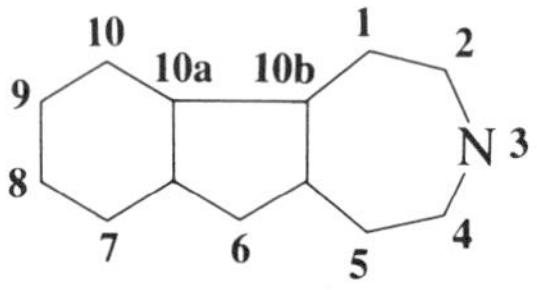

The only reported examples of the indeno[1,2-*d*]azepine system, compounds **848** and **849,** were discussed in Section IV. 74.

76. Indeno[1,7-*bc*]azepines

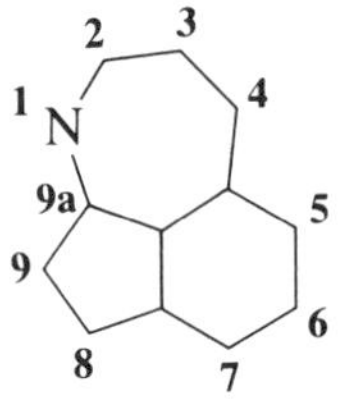

Heating 3,4-dimethoxyphenylpropionamide and cinnamic acid in polyphosphoric acid produced the indanone **850** (Fig. 124), which was cyclized to the indeno[1,7-*bc*]azepinone **851** with *p*-toluenesulfonic acid in refluxing

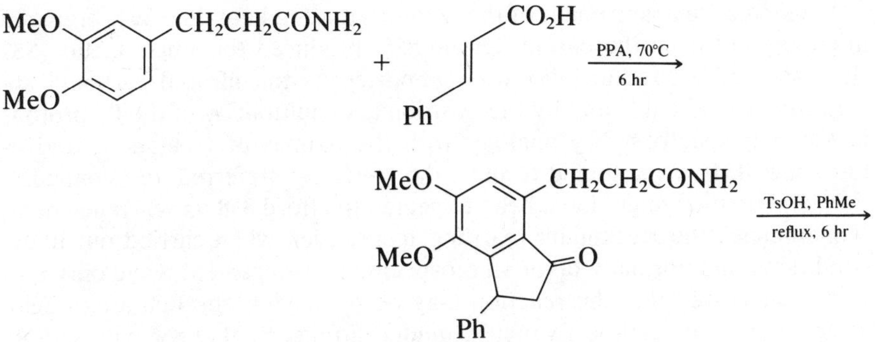

MeO, MeO, $CH_2CH_2CONH_2$ + CO_2H, Ph; PPA, 70°C, 6 hr

MeO, MeO, $CH_2CH_2CONH_2$, O, Ph; TsOH, PhMe, reflux, 6 hr

850, mp 194–200°C

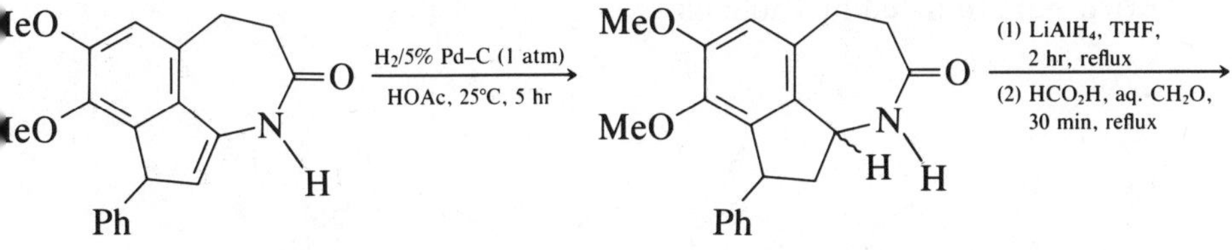

MeO, MeO, O, N, H, Ph; H_2/5% Pd–C (1 atm), HOAc, 25°C, 5 hr; MeO, MeO, O, N, H, H, Ph; (1) $LiAlH_4$, THF, 2 hr, reflux (2) HCO_2H, aq. CH_2O, 30 min, reflux

851, mp 203–205°C (dec.) **852**, mp 215–220°C

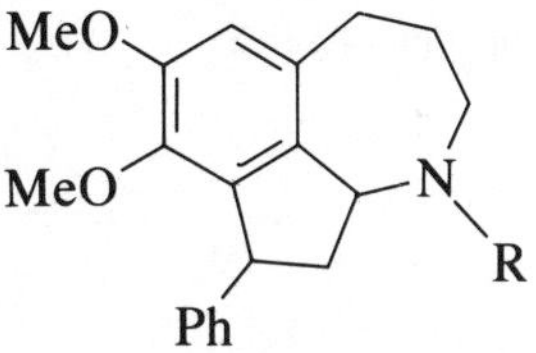

MeO, MeO, N, R, Ph

853a, R = H; mp 110–112°C
mp 267–270°C (HCl)
853b, R = Me; mp 261–263°C (naponate)

Figure 124

toluene. Catalytic reduction of **851** gave **852**, which on lithium aluminum hydride reduction afforded the indenoazepine **853a**. Eschweiler–Clarke methylation of **853a** gave the *N*-methyl derivative **853b** (251).

77. Indeno[2,1-*c*]azepines

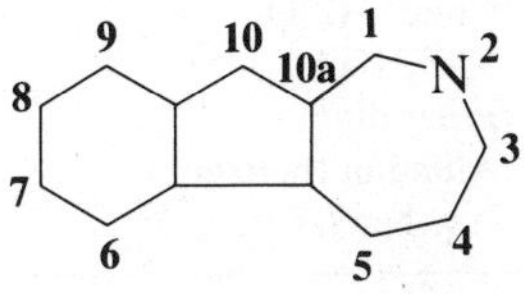

9, 10, 1, 10a, N, 2, 8, 3, 7, 6, 5, 4

Beckmann rearrangement of the α,β-unsaturated ketoxime **854,** prepared in good yield from the parent ketone **855,** produced the single lactam **857** (Eq. 98) (252). Structure **857** was supported by an infrared carbonyl absorption at 1645 cm^{-1} and by the position and multiplicity of the C_3 protons in the pmr spectrum. By analogy with the oximes of 1-indanone and α-tetralone, **854** was assumed to have the sterically preferred *anti* configuration, and as such might have been expected to afford **858** on rearrangement. Nonetheless, the Beckmann rearrangement, even when carried out under conditions that normally favor stereospecific rearrangement, gave only **857.** It was suggested that the reaction may be proceeding through a nonregioselective iminium cation. Even a Schmidt rearrangement of **855** with sodium azide in polyphosphoric acid, a reaction that might be expected to produce **858,** gave only **857.** The various conditions under which these reactions were carried out are listed in Table 48.

(98)

854, X = NOH
855, X = O
856, X = NOTs

857, mp 186.5–188°C

858

TABLE 48. REACTION CONDITIONS FOR THE PREPARATION OF **857**

Substrate	Reaction Conditions	Yield (%)
854	PPA, 160–170°C, 10 min	55
855	NaN_3–PPA, 100°C, 3 hr	65
854	PCl_5, Et_2O	31
854	Conc. H_2SO_4	27
856	50% HOAc	26
856	Piperidine	38
856	Alumina (benzene)	7
856	Aq. NaOH	21

78. Indeno[5,4-*c*]azepines

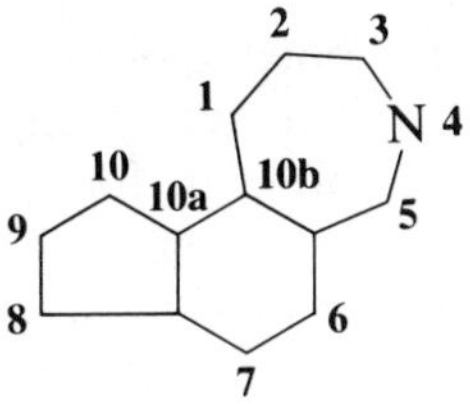

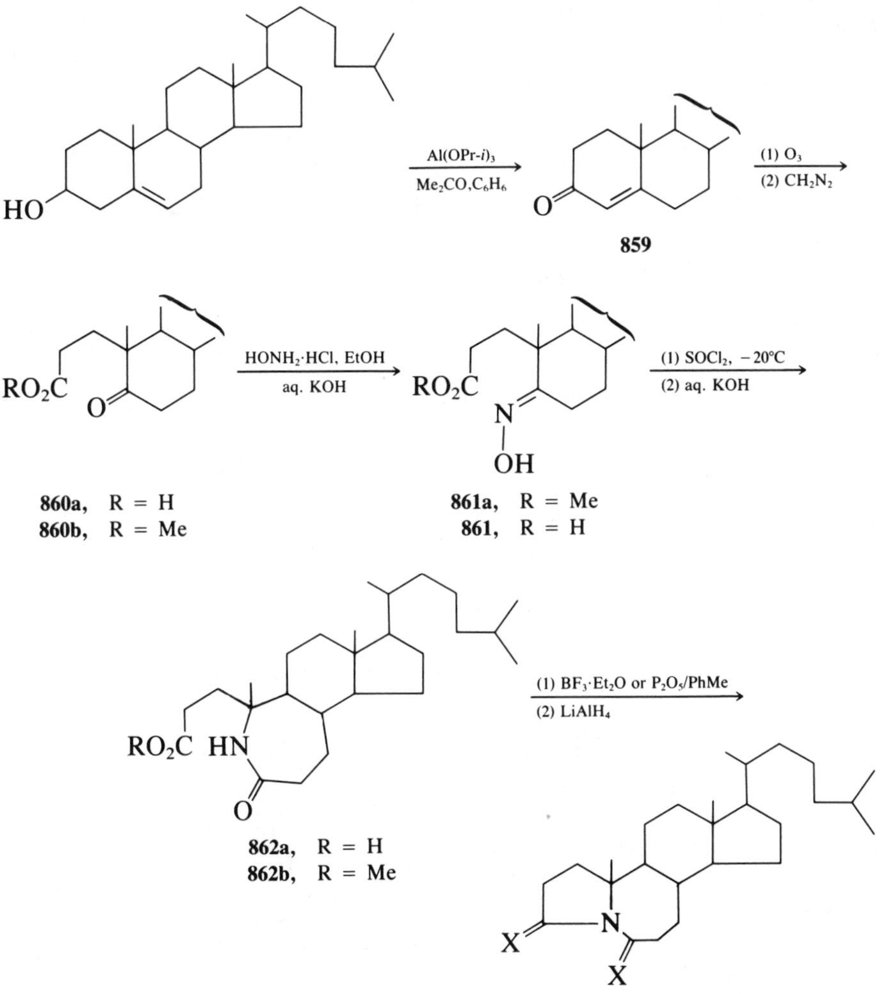

2 3 1 N 4 10 10a 10b 5 9 8 6 7

HO

$Al(OPr\text{-}i)_3$ / Me_2CO, C_6H_6

O

859

(1) O_3 (2) CH_2N_2

RO_2C O

HONH$_2$·HCl, EtOH / aq. KOH

RO_2C N OH

(1) $SOCl_2$, −20°C (2) aq. KOH

860a, R = H
860b, R = Me

861a, R = Me
861, R = H

RO_2C HN O

(1) $BF_3 \cdot Et_2O$ or P_2O_5/PhMe (2) $LiAlH_4$

862a, R = H
862b, R = Me

X N X

863a, X = O
863b, X = H_2

Figure 125

In the search for physiologically active analogs of steroid hormones, transformations of natural systems were carried out to replace carbon by nitrogen. The first reported aza analogs (Fig. 125) were derivatives of 5-azacholestane. Thus, Oppenauer oxidation of cholesterol yielded cholest-4-en-3-one (**859**) (253,254). The A ring of **859** was cleaved either by ozonolysis or by meta-periodate–permanganate oxidation with formation of 3,5-*seco*-4-nor-5-ketocholestan-3-oic acid (**860a**). Treatment of **860a** with diazomethane gave the corresponding ester **860b,** which with hydroxylamine gave the oxime **861a.** Alternatively, **861a** was obtained by treatment of keto acid **860a** with hydroxylamine and esterification of the resultant oxime **861b** with diazomethane. Beckmann rearrangement of **861a** with thionyl chloride at −20°C, followed by hydrolysis with aqueous potassium hydroxide, afforded 3,5-

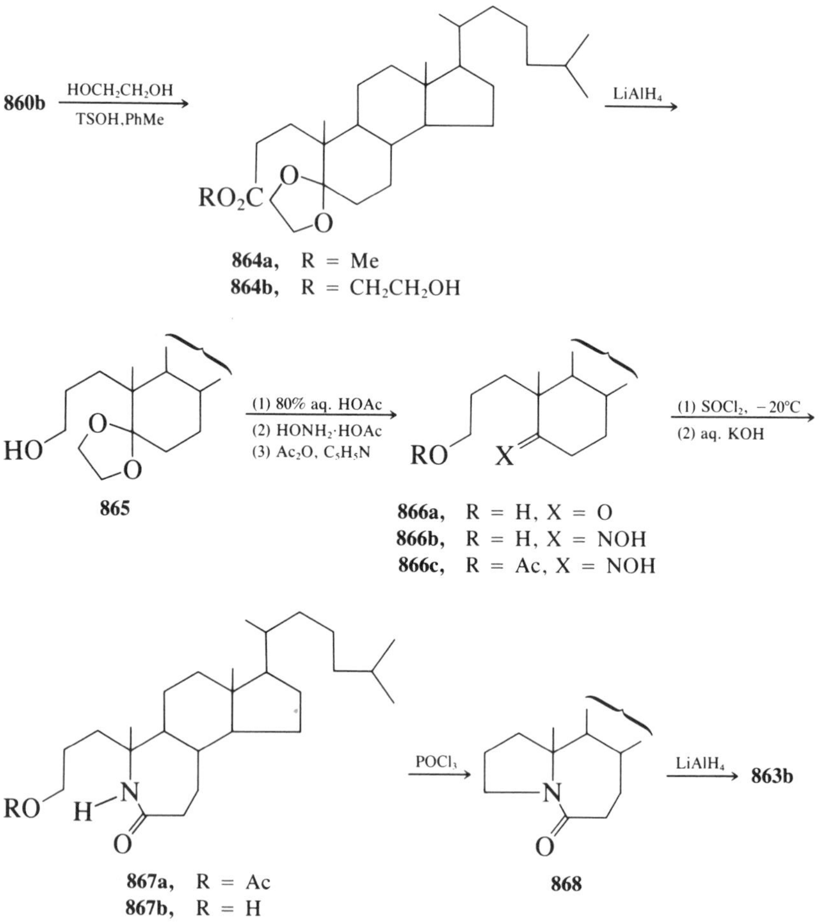

860b $\xrightarrow[\text{TSOH.PhMe}]{\text{HOCH}_2\text{CH}_2\text{OH}}$ $\xrightarrow{\text{LiAlH}_4}$

RO_2C O O

864a, R = Me
864b, R = CH_2CH_2OH

HO O O

865

$\xrightarrow{\text{(1) 80\% aq. HOAc}}$ (2) $HONH_2{\cdot}HOAc$ (3) Ac_2O, C_5H_5N

RO X

866a, R = H, X = O
866b, R = H, X = NOH
866c, R = Ac, X = NOH

$\xrightarrow{\text{(1) SOCl}_2,\ -20°\text{C}}$ (2) aq. KOH

RO H N O

867a, R = Ac
867b, R = H

$\xrightarrow{\text{POCl}_3}$

N O

868

$\xrightarrow{\text{LiAlH}_4}$ **863b**

Figure 126

seco-4-nor-5-azacholestan-5a-one-3-oic acid (**862a**), which after esterification yielded **862b.** When lactam **862a** was treated with ethereal boron trifluoride or phosphorous pentoxide, the A ring was regenerated, with formation of A-nor-B-homo-5-azacholestane-3,5a-dione (**863a**) in 20 and 70% yields, respectively. Lithium aluminum hydride reduction afforded the azasteroid analog **863b.**

Treatment of **860b** with ethylene glycol and *p*-toluenesulfonic acid in toluene (Fig. 126) produced the ketal esters **864a** and **864b.** Lithium aluminum hydride reduction of either ester gave **865,** which was hydrolyzed with aqueous acetic acid to give **866a.** Treatment of **866a** with hydroxylamine acetate afforded crystalline **866b,** which was acylated with acetic anhydride in pyridine to produce **866c.** Beckmann rearrangement of **866c,** as described for the formation of **862,** afforded cyclic lactam **867a** in 90% yield. Hydrolyis with alcoholic potassium hydroxide gave the indenoazepine **867b.** Phosphoryl chloride effected intramolecular cyclization of **867b** to A-nor-B-homo-5-azacholestan-5a-one (**868**). On reduction, **868** was converted to **863b.**

In a series of reactions completely analogous to those of cholest-4-en-3-one (**859**), shown in Fig. 125, testosterone (**869a**) has been converted to the indenoazepines **870a** and **870b** (Fig. 127). When **870b** was heated with acetic anhydride, intramolecular lactamization occurred, with production of 17β-acetoxy-A-nor-B-homo-5-azaandrostan-3,5a-dione (**871a**) in 70% yield (255).

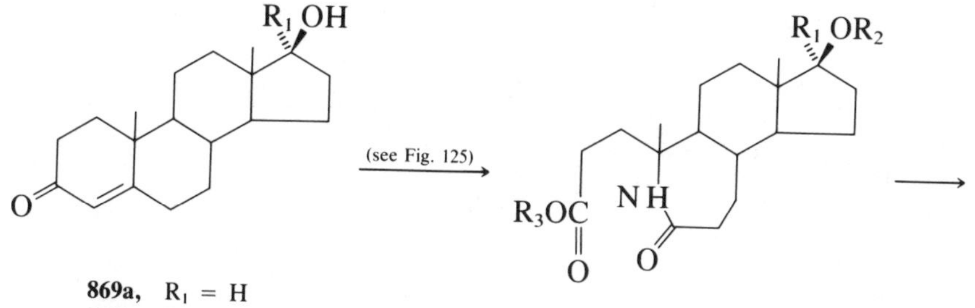

(see Fig. 125)

869a, R_1 = H
869b, R_1 = Me

	R_1	R_2	R_3
870a	H	Ac	Me
870b	H	H	H
870c	Me	H	Me
870d	Me	H	H
870e	Me	Ac	Me
870f	Me	Ac	H

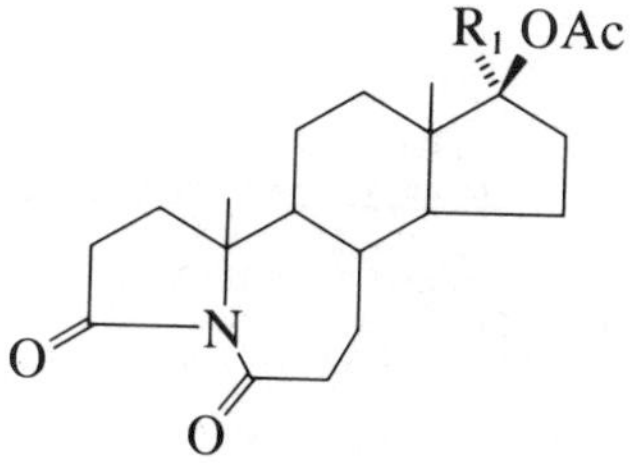

871a, R_1 = H
871b, R_2 = Me

Figure 127

OH

(1) $KMnO_4$, $NaIO_4$, K_2CO_3, aq. *t*-BuOH
(2) Ac_2O, C_5H_5N

872

OR

O O OH

873a, R = H
873b, R = Ac

(1) CH_2N_2, Et_2O
(2) $HONH_2{\cdot}HCl$, C_5H_5N

OAc

MeO_2C

X

(1) $SOCl_2$, dioxane, 10°C
(2) 25% aq. KOH, 80°C
(3) CH_2N_2, MeOH–Et_2O

874a, X = O
874b, X = NOH

OH

RO_2C

HN

O

+

OAc

Me

MeO_2C

NC

875a, R = H
875b, R = Me

876

(1) $LiAlH_4$ THF, reflux
(2) Ac_2O, C_5H_5N
(3) 10% K_2CO_3, aq. MeOH

OR_2

R_1O N

R_3

(1) $H_2Cr_2O_7$, Me_2CO
(2) $AgNO_3$, NaOH, aq. EtOH, 4 M
(3) CH_2N_2, Et_2O–MeOH
(4) 10% HCl, MeOH, Δ
(5) conc. HCl, aq. dioxane, Δ

877a, $R_1 = R_2 = R_3 =$ H
877b, $R_1 = R_2 = R_3 =$ Ac
877c, $R_1 = R_2 =$ H, $R_3 =$ Ac

Figure 128

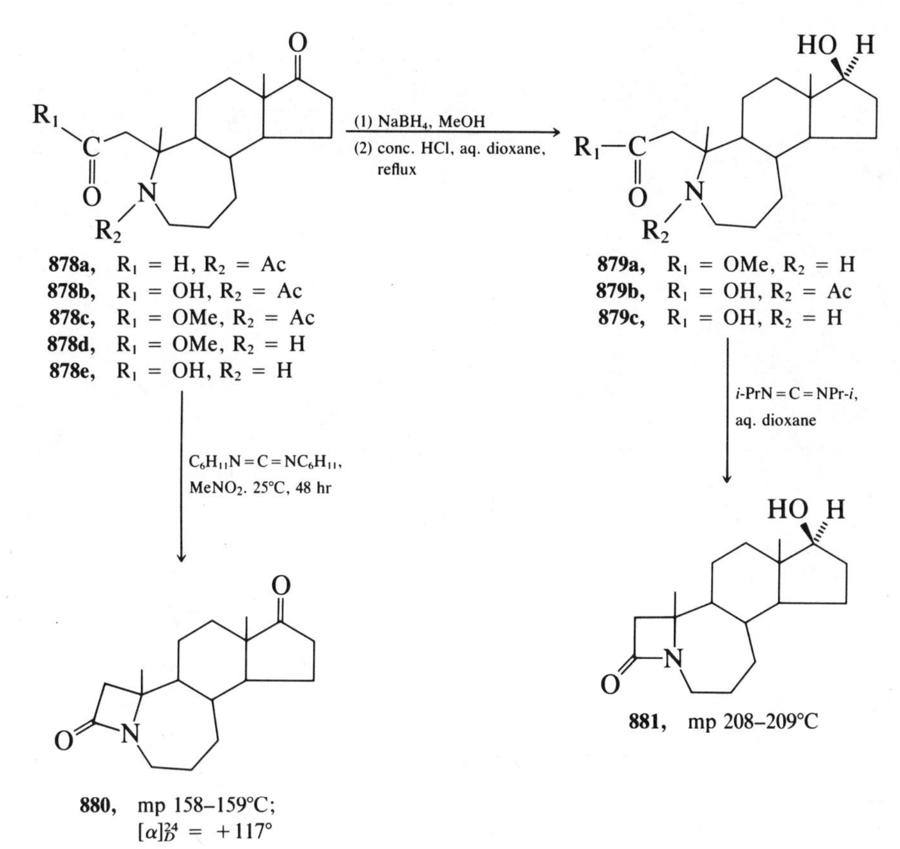

Figure 128 (*continued*)

Likewise, 17α-methyltestosterone (**869b**) was converted to the indeno[5,4-*c*]azepines **870c–f,** and **870f** was cyclized to the azaandrostane analog **871b** (256).

Indeno[5,4-*c*]azepines derived from steroids have been utilized by Levine and co-workers as intermediates in a synthesis of steroidal β-lactams (257–263) (Fig. 128). Oxidative cleavage of the conjugated double bond in A-nortestosterone (**872**) was achieved by sequential oxidation with osmium tetroxide and periodic acid to afford **873a.** On a large scale, periodate–permanganate proved more convenient for the preparation of the lactol **873a.** Selective acetylation at C_{17} with acetic anhydride in pyridine yielded **873b,** which in turn was esterified with diazomethane to obtain **874a.** Hydroxylamine hydrochloride in pyridine formed the oxime **874b,** which underwent Beckmann rearrangement with thionyl chloride in dioxane to give the lactam **875a** plus a trace of the abnormal Beckmann product **876.** The methyl ester **875b** was reduced with lithium aluminum hydride to the amino diol **877a.**

TABLE 49. INDENO[5,4-*c*]AZEPINES

R_1	R_2	R_3	R_4	X	mp (°C)	Additional Data	Refs.
CH_2CO_2H	H	H	6-Methyl-2-heptyl	O	210	$[\alpha]_D^{21} = +59.6°$, ir	253,254
CH_2CO_2Me	H	H	6-Methyl-2-heptyl	O	132.5	$[\alpha]_D^{21} = -3.2°$, ir	253,254
$(CH_2)_2OH$	H	H	6-Methyl-2-heptyl	O	151	$[\alpha]_D^{14} = -45.2°$, ir	253
$(CH_2)_2OAc$	H	H	6-Methyl-2-heptyl	O	163–164	$[\alpha]_D^{18} = +78.3°$, ir	253
$(CH_2)_2CO_2H$	H	H	6-Methyl-2-heptyl	O			264
$(CH_2)_2CO_2Me$	H	H	6-Methyl-2-heptyl	O			264
CH_2CO_2H	H	H	OH	O	218–219	$[\alpha]_D^{20} = +10.8°$, ir	255
CH_2CO_2Me	H	H	OAc	O	138–139	$[\alpha]_D^{20} = +42.0°$, ir	255
CH_2CO_2H	H	Me	OH	O	226–229	$[\alpha]_D^{20} = -1.7°$, ir	256
CH_2CO_2H	H	Me	OAc	O	214–216	$[\alpha]_D^{20} = +80.6$, ir	256
CH_2CO_2Me	H	Me	OH	O	Oil	ir	256
CH_2CO_2Me	H	Me	OAc	O	159.5–161.2	$[\alpha]_D^{20} = +38.8°$, ir	256
CO_2H	H	H	OH	O	275–276	$[\alpha]_D^{23} = +30.0°$, ir	258
CO_2Me	H	H	OH	O	Oil	ir,nmr	258
CH_2OH	H	H	OH	H_2	170.5–171	$[\alpha]_D^{25} = -16.0°$, pmr	258

CH_2OAc	Ac	H	OAc	H_2	139–140	$[\alpha]_D^{26} = -41.0°$, ir, pmr	258
CH_2OH	Ac	H	OH	H_2	172–172.5	$[\alpha]_D^{23} = -47.0°$, ir, pmr	258
CHO	Ac	—CO—		H_2	172–173	$[\alpha]_D^{24} = +60.0°$, ir, pmr	258
CO_2H	Ac	—CO—		H_2	180.5–181.5	$[\alpha]_D^{23} = -2.0°$, ir, pmr	258
CO_2Me	Ac	—CO—		H_2	131.5–132.5	ir, nmr	258
CO_2Me	H	—CO—		H_2	Oil	ir,nmr	258
CO_2H	H	—CO—		H_2	Oil	ir,nmr	258
CO_2Me	H	H	OH	H_2	Oil	nmr	258
CO_2H	Ac	H	OH	H_2	188–188.5	$[\alpha]_D^{24} = -71.0°$. ir	258
CO_2H	H	H	OH	H_2	233.5–234.5	$[\alpha]_D^{23} = 0°$, ir, pmr	258
CO_2H	H	H	CHOHMe	O	270–271.5	ir	260
CO_2Me	H	H	CHOHMe	O	154–155	$[\alpha]_D = +22.0°$, ir, pmr	260
CH_2OH	H	H	CHOHMe	H_2	159–160.5	$[\alpha]_D = -20.0°$, ir, pmr	260
CH_2OAc	Ac	H	CHOAcMe	H_2	Oil	nmr	260
CH_2OH	Ac	H	CHOAcMe	H_2	169.5–170.5	$[\alpha]_D = -9.0°$, ir, pmr	260
CHO	Ac	H	CHOAcMe	H_2	159.5–160.5	$[\alpha]_D = +19.0°$, ir, pmr	260
CO_2H	Ac	H	CHOAcMe	H_2	177–177.5	$[\alpha]_D = -21.0°$, ir, pmr	260
CH_2OH	Ac	H	CHOHMe	H_2	184.5–185	$[\alpha]_D = -51.0°$, ir, pmr	260
CHO	Ac	H	COMe	H_2	Oil	nmr	260
CO_2H	Ac	H	COMe	H_2	176–177	$[\alpha]_D = +10$, ir, pmr	260
$(CH_2)_2CO_2H$	H	H	6-Methyl-2-heptyl	O	165–167	$[\alpha]]_D^{25} = +29.3°$, ir, pmr	264
$(CH_2)_2CO_2Me$	H	H	6-Methyl-2-heptyl	O	135–137	$[\alpha]]_D^{25} = +46.3°$, ir, pmr	264

Acetylation gave **877b,** which could be selectively converted to the *N*-acetyl derivative **877c** with methanolic potassium carbonate. Chromic acid oxidation in acetone gave **878a,** and further oxidation with silver oxide led to the amino acid **878b.** Treatment with diazomethane afforded the methyl ester **878c,** which on treatment with refluxing 10% methanolic hydrogen chloride yielded the amino ester **878d.** Reduction with sodium borohydride afforded the 17β-hydroxy compound **879a.** Hydrolysis of **878d** with ethanolic sodium hydroxide gave the amino acid **878e.** Finally, cyclization of **878e** to the desired lactam **880** was effected by dicyclohexylcarbodiimide in nitromethane. A similar series of reactions was used to prepare the ring-D alcohol **881.** Thus, sodium borohydride reduction of **878b** gave the 17β-hydroxy compound **879b,** which on hydrolysis yielded the amino acid **879c.** Because of its insolubility in nitromethane, **879c** had to be cyclized in aqueous dioxane using diisopropylcarbodiimide, whereupon **881** was formed in low yield.

Subsequent to the preparation of **880** and **881,** a similar synthesis of a steroidal β-lactam bearing a pregnane side chain at C_{17} was reported (260–262). A-Norprogesterone (**882**) was converted to the β-lactam **883** by reactions closely analogous to those outlined in Fig. 128. Rodewald recently described the conversion of cholest-4-ene to β-homo-5-azacholestane (264). The reader is referred to Table 49 for these and other steroidal derivatives of the indeno[5,4-*c*]azepine ring system.

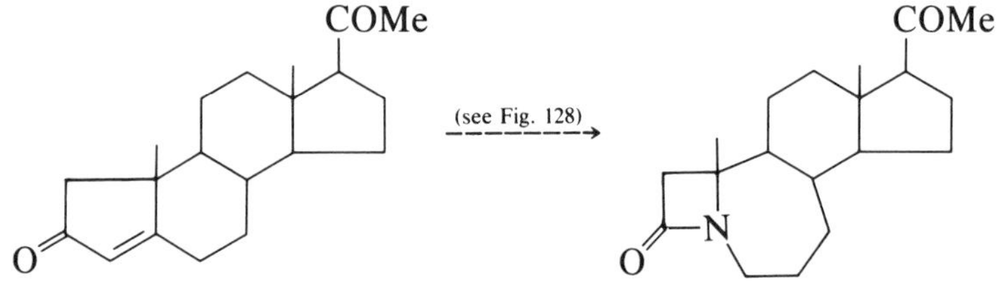

COMe

O

882

(see Fig. 128)

COMe

N

O

883

79. Indeno[7,1-*bc*]azepines

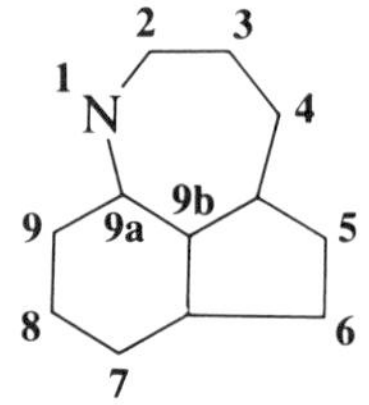

1 N 2 3 4 5 6 7 8 9 9a 9b

80. Indeno[7,1-*cd*]azepines

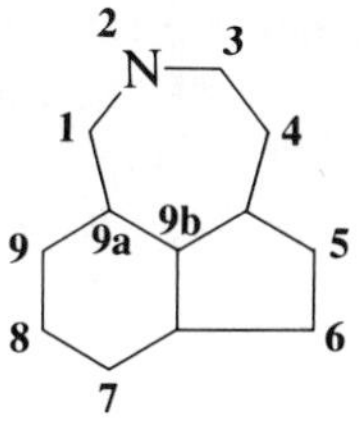

The reaction of 3,4-dihydroacenaphthen-5(2aH)-one (**884a**) with sodium azide in polyphosphoric acid (Fig. 129) afforded the indeno[7,1-*bc*]azepine **885a** and the isomeric indeno[7,1-*cd*]azepine **886a** in an approximate ratio of 2:3. When the corresponding oxime **884b** or the oxime tosylate **884c** were subjected to Beckmann rearrangement, only **886a** was obtained (87% and 64% yields, respectively). Reduction of **885a** and **876a** with lithium aluminum hydride led to the amines **885b** and **886b** (265).

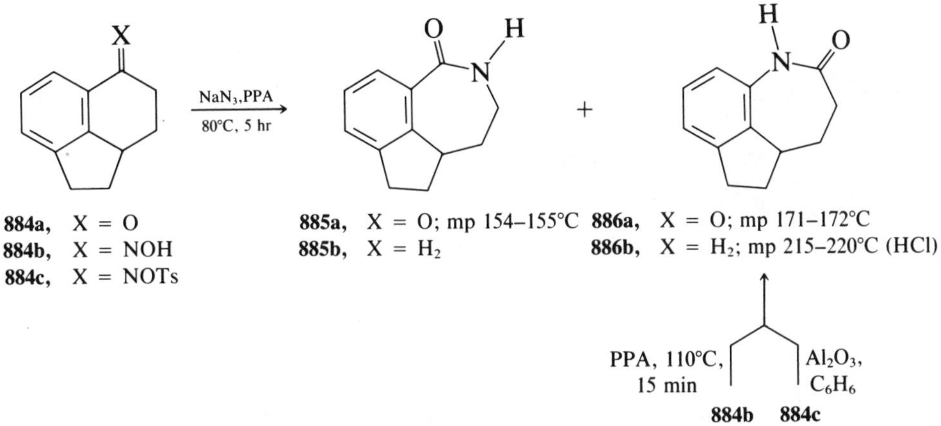

Figure 129

V. 5,7,7-SYSTEMS

1. Cyclohepta[4,5]thieno[3,2-*b*]azepines

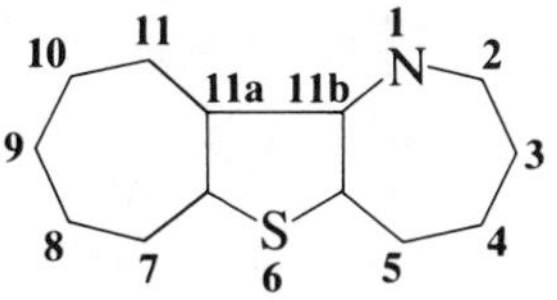

The only known member of the cyclopenta[4,5]thieno[3,2-*b*]azepine ring system is lactam **888,** which was prepared from the oxime tosylate **887** by Beckmann rearrangement (238,266). Desulfurization of **888** with Raney nickel produced the caprolactam **889** (Eq. 99). Lactam **888** was also hydrolyzed to the corresponding amino acid **890.**

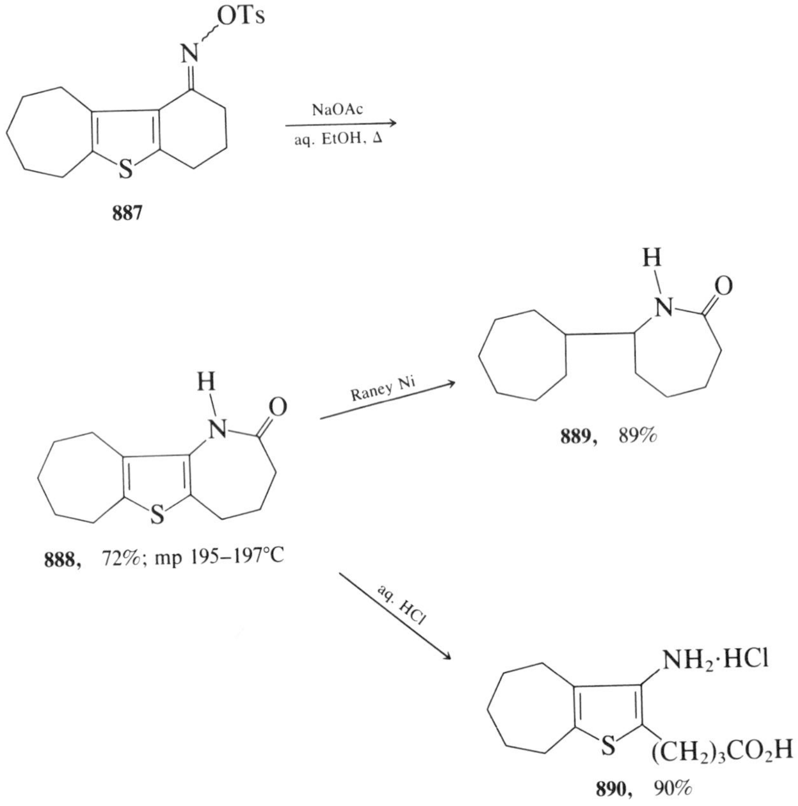

OTs
N
S
887

NaOAc
aq. EtOH, Δ

H
N
O
S
888, 72%; mp 195–197°C

Raney Ni

H
N
O
889, 89%

(99a)

aq. HCl

$NH_2 \cdot HCl$
S
$(CH_2)_3CO_2H$
890, 90%

(99b)

2. Azuleno[1,8-*cd*]azepines

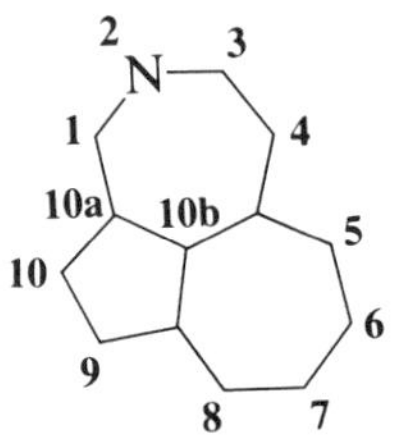

2 N 3
1 4
10a 10b 5
10 6
9 8 7

The only known example of the azuleno[1,8-*cd*]azepine ring system was formed inadvertantly, when, during the synthesis of new heterocyclic systems iso-π-electronic with the nonbenzenoid aromatic compound cyclopenta-[*e,f*]heptalene (**891**), the sodium salt **892** was accidentally treated with a large excess of benzonitrile. In addition to the expected ketone **893**, the π-equivalent compound **894** was obtained in about 10% yield (Eq. 100) (267,268).

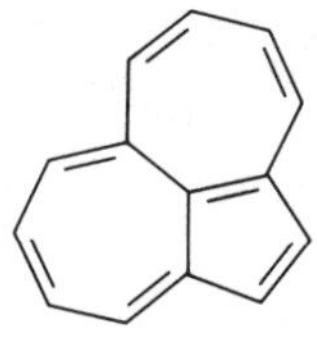

891

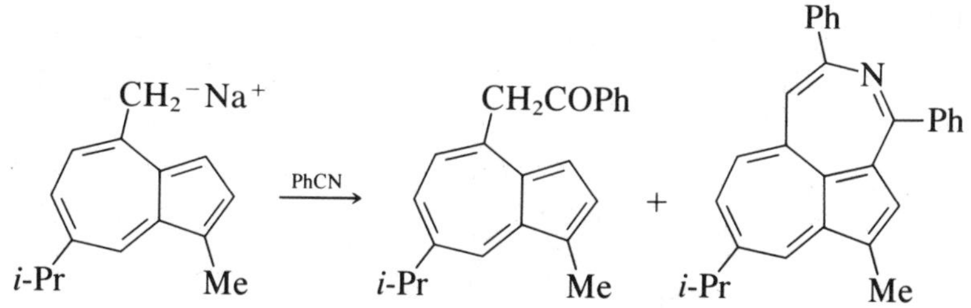

$CH_2^-Na^+$ — PhCN → CH_2COPh + Ph, N, Ph

i-Pr, Me — i-Pr, Me — i-Pr, Me (100)

892 **893** **894**, mp 152.5–153.5°C

VI. 6,6,7-SYSTEMS

1. Azepino[1,2-*b*][1,2,4]benzothiadiazine

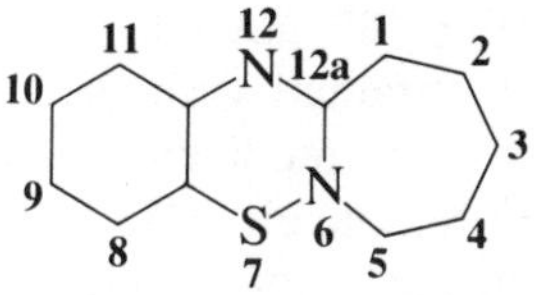

A single example of azepino[1,2-*b*][1,2,4]benzothiadiazines, which is claimed to have sedative and anticonvulsant activity, has appeared in a patent (269). The aminosulfonic acid **895** reacted with *O*-methylcaprolactim (**896**) at room temperature in methanol overnight to give a precipitate of **897**, which was cyclized with phosphoryl chloride to the hexahydroazepino[1,2-*b*][1,2,4]benzothiadiazine-7,7-dioxide **898** (Eq. 101). Variation of the ether **896** was claimed to yield a series of analogs of **898**, but no descriptive details were provided.

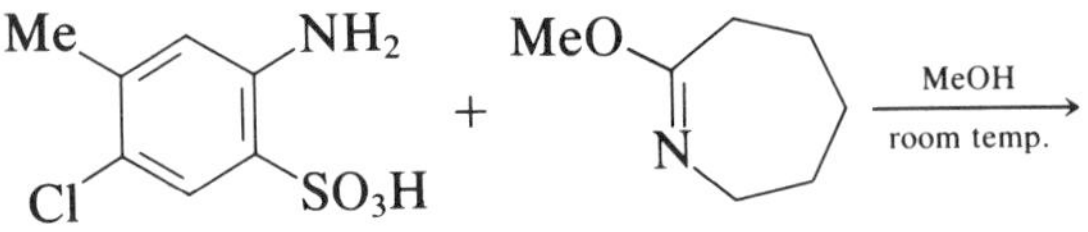

Me, NH_2, Cl, SO_3H **895** + MeO, N **896** $\xrightarrow[\text{room temp.}]{\text{MeOH}}$

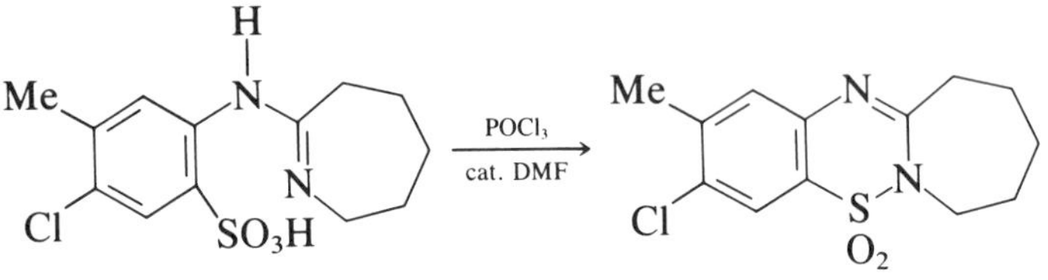

Me, H, N, Cl, SO_3H, N **897** $\xrightarrow[\text{cat. DMF}]{\text{POCl}_3}$ Me, Cl, N, S, N, O_2 **898**, mp 114–116.5°C (101)

2. Azepino[2,1-*c*][1,2,4]benzothiadiazines

1, 2, 3, 4, 12a, 12, N, 11, 10, 9, 8, 7, N 6, S 5

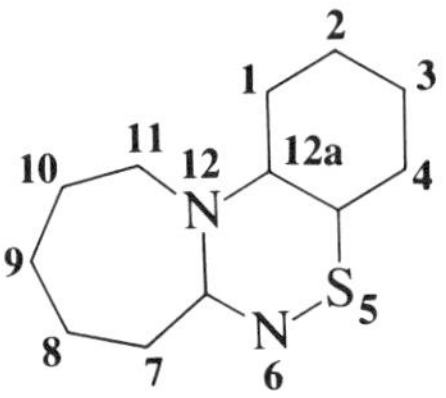

Two reports (270,271) concerning the azepino[2,1-*c*][1,2,4]benzothiadiazine system originate from the laboratories of Suschitzky and Meth-Cohn, who have explored nitrene chemistry as a route to various heterocycles (see Section IV. 20). Thermal decomposition of the perhydroazepinyl azide **899** in refluxing chlorobenzene (Fig. 130) gave the Cope elimination-type product **902** in 76% yield. Evidence obtained with the related piperidine and pyrrolidine analogs suggested the intermediacy of a sulfonyl nitrene (**900**) which formed the mesoionic benzothiadiazole **901.** The ylide **901** was not isolated in the decomposition of **899,** but could be isolated from the corresponding dimethylamino azide. The mesoionic intermediate **901,** however, could be intercepted in two ways. First the azide **899** could be converted in unspecified yield to the azepino[2,1-*c*][1,2,4]benzothiadiazine **903** by addition of piperidine hydrochloride. Second, thermal decomposition in the presence of copper, a widely used catalyst in such reactions, gave **903** in 81% yield, together with a trace of **902.** The authors suggested the possibility of a mesoionic benzothiadiazole–copper complex, which encourages N—N bond cleavage and inhibits Cope-type β-elimination to **902.**

Photolysis of **899** in dimethyl sulfoxide (Fig. 131) occurred rapidly to give

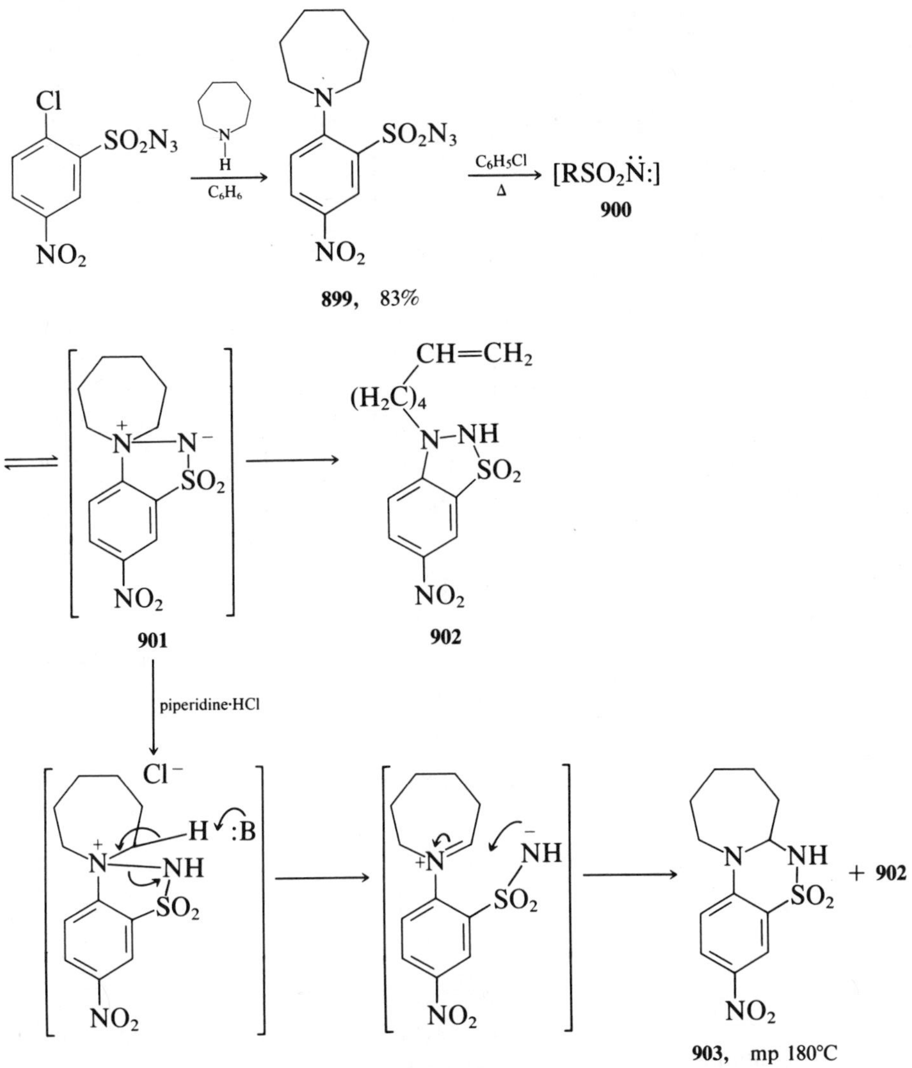

$[RSO_2\ddot{N}:]$

900

899, 83%

901

902

piperidine·HCl

903, mp 180°C

+ 902

Figure 130

a 42% yield of the yellow substance **905.** No reaction occurred in chlorobenzene. These data suggest the intermediacy of the sulfoximide **904,** particularly since photolysis of such intermediates is known to generate nitrenes. With the corresponding piperidine derivative, neither the mesoionic intermediate to **902** nor the tricyclic product analogous to **903** was affected by irradiation in DMSO. This confirmed that the extra double bond in **905,** as opposed to **903,** is the result of elimination rather than oxidation. Reduction of **905** with *bis*(methoxyethoxy)aluminum hydride gave **903.**

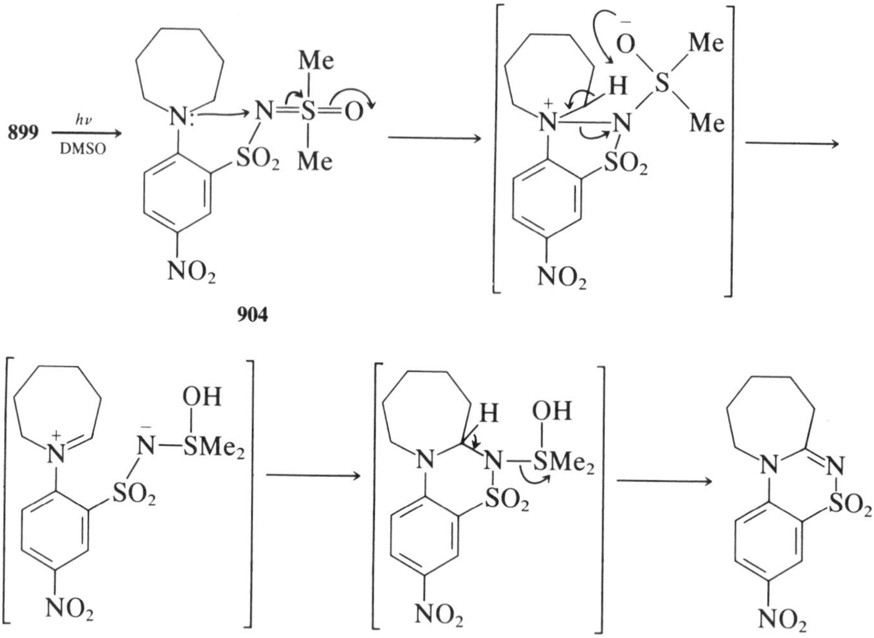

899 $\xrightarrow[\text{DMSO}]{h\nu}$

904

905, mp 194°C

Figure 131

3. Azepino[2,1-*b*][1,3]benzoxazines

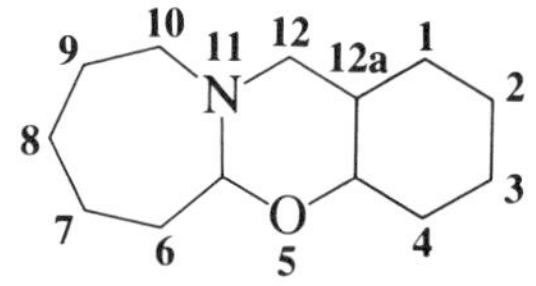

Structures which exemplify the azepino[2,1-*b*][1,3]benzoxazine system are also unique examples of stable, isolable orthoacid amides known as cyclols. The lactam **906** (Eq. 102) (272), prepared from caprolactam and 2-benzyloxybenzoyl chloride, on catalytic hydrogenolysis of the benzyl ether, gives the phenolic acylamide **907,** which immediately forms the stable cyclol **908** in 84% yield. Normally, the tetrahedral intermediates formed on addition to amide carbonyls are unstable, but the additional carbonyl attached to the amide nitrogen and the incorporation of this moiety into the six-membered 1,3-oxazinone ring make **908** a stable structure. In general, stable cyclols form from *N*-salicyloyl lactams with 10 or 11 atoms in the bicyclic system (there are 11 in compound **908**). The aromatic ring, which flattens the molecule, also imparts stability. No ring expansion to the cyclodepsipeptide **909** occurs in this system as sometimes happens in related structures (see Section V. 5 for related thiacyclol chemistry).

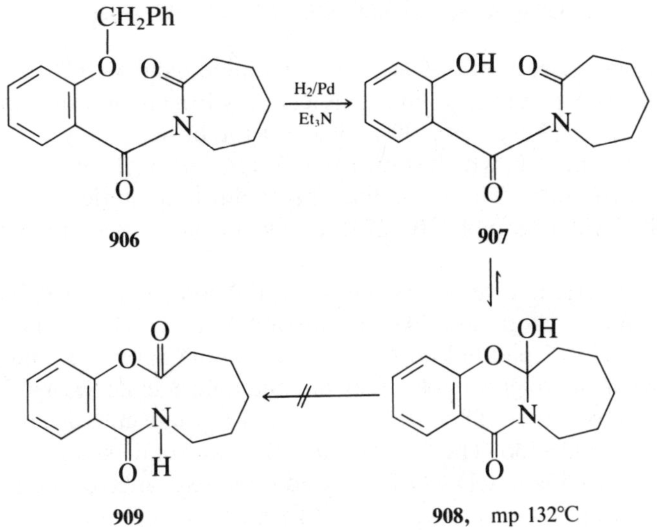

CH_2Ph ... H_2/Pd, Et_3N

906 **907**

(102)

909 **908,** mp 132°C

906

H_2/Pd, H^+ H_2/Pd, Et_3N H_2/Pd H^+ MeI Ag_2O Ac_2O HBr, HOAc H_2O $BF_3 \cdot Et_2O$

910, mp 79°C

908

X^-

913a, X = Br (93%)
913b, X = BF_4 (60%); mp 202°C

912, mp 117°C

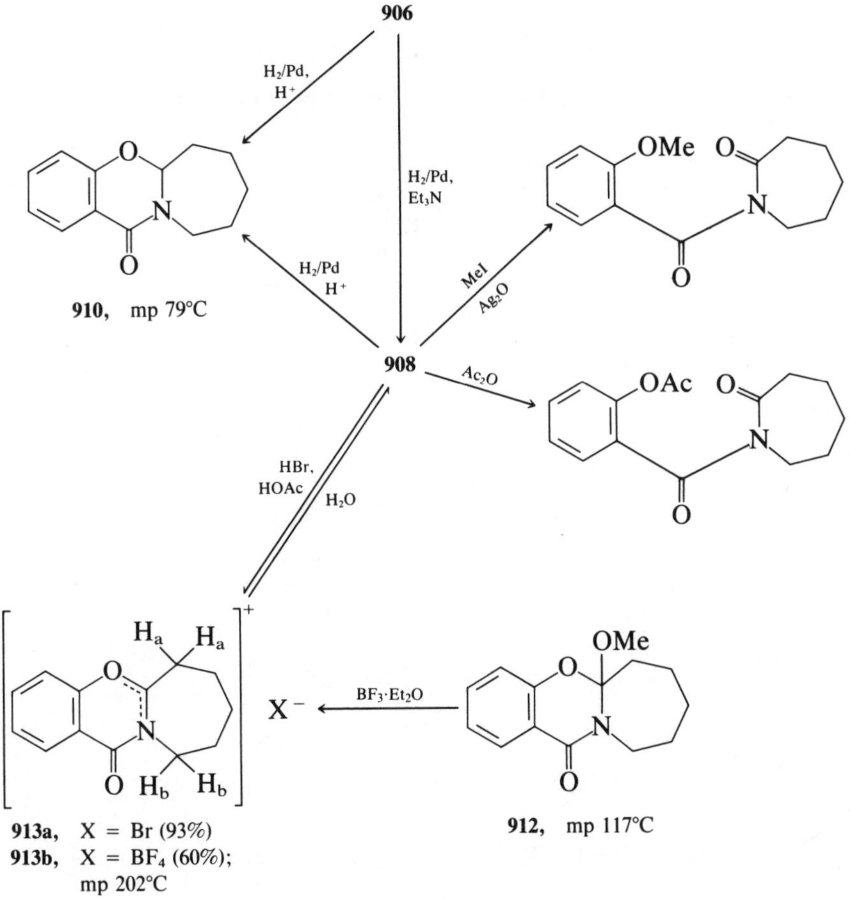

Figure 132

The infrared spectrum of **908** shows an amide band at 1676 cm^{-1} and no absorptions due to the acylamide moiety. Hydrogenation of either **906** or **908** in acid (Fig. 132) gives >90% yields of the basic, noncyclol, azepinobenzoxazin-12-one **910,** which shows an infrared carbonyl band similar to that of **908** at 1670 cm^{-1}. The *N*-acyllactam–cyclol tautomeric equilibrium has been studied polarographically (273), and **908** exists in solution primarily in the cyclol form.

The tautomeric nature of **908** and its partial conversion to **907** in solution is evident in the reaction of **908** with methyl iodide, which yields a mixture of **911a** and **912** (separated on thick-layer chromatography), and by acetylation (274) of **908** to give exclusively the phenolic acetate **911b** in 73% yield.

Treatment of either **908** or **912** with the appropriate anhydrous acids gave the salts **913a** and **913b.** The fluoroborate **913b** was quite stable. Nmr spectra of the bromide **913a** in CD_3CO_2D showed rapid exchange of the H_a protons, probably via an anhydro intermediate formed by deprotonation–reprotonation. Infrared, ultraviolet, and nmr data were provided for all compounds. The mass spectrum of **912** shows a base peak due to loss of methanol (275).

4. [1,3]Oxazino[2,3-*a*][2] benzazepines

1 O, 2, 3, 4, N 5, 6, 7, 8, 9, 10, 11, 12, 12a, 12b

(1) Br_2 (2) Δ → $(CH_2)_3Br$, CHO

$HO(CH_2)_3NH_2$

$\overset{+}{N}$ Br^- $(CH_2)_3OH$ ← N, O (103)

914, bp 105–110°C/0.3 mm, mp 198–200°C (HBr)

An example of the [1,3]oxazino[2,3-*a*] [2]benzazepine ring system is compound **914,** which was prepared in 87% yield from 2-(3-bromopropyl)benzaldehyde (Eq. 103) (19). The same approach was used to synthesize oxazolo- and thiazolo[2,3-*a*][2]benzazepines (see Sections IV.5 II and IV.9).

5. Azepino[2,1-*b*][1,3]benzothiazines

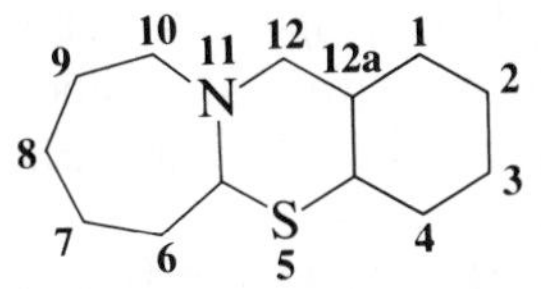

The first examples of the azepino[2,1-*b*][1,3]benzothiazine ring system (Fig. 133) were prepared via the thiacyclol **919** (276,277). This substance arises in a manner analogous to the cyclol (or oxocyclol) chemistry described for the azepino[2,1-*b*][1,3]benzoxazine system (Section VI. 3). Acylation of

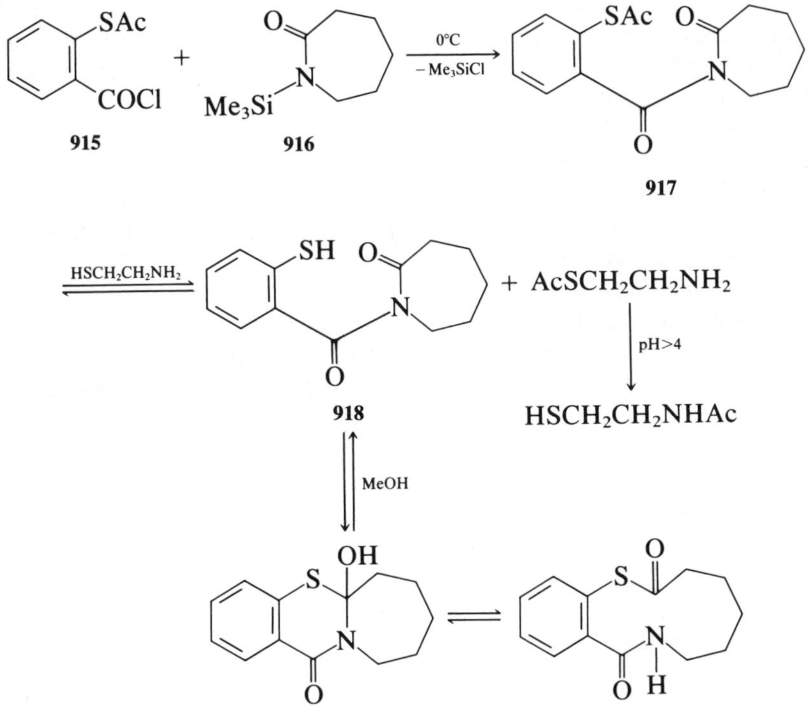

915 + **916** → (0°C, $-Me_3SiCl$) **917**

917 ⇌ ($HSCH_2CH_2NH_2$) **918** + $AcSCH_2CH_2NH_2$ → (pH>4) $HSCH_2CH_2NHAc$

918 ⇌ (MeOH) **919**, mp 122–123°C ⇌ **920**

Figure 133

the silylated lactam **916** with the *S*-protected acid chloride **915** gave the acylcaprolactam **917** in 70% yield. Cleavage of the *S*-acetyl group with cysteamine led initially to *S*-acetylcysteamine, which then rearranged irreversibly to the *N*-acetyl derivative. The concomitantly formed acylamide **918** cyclized at once to the thiacyclol **919** (64% yield). Infrared spectra (KBr) showed the absence of a thiol group (around 2250 cm^{-1}) and the acylamide (around 1730 and 1670 cm^{-1}) groups but indicated an OH group at 3236 cm^{-1} and an aromatic amide carbonyl at 1605 cm^{-1}. When the reaction time was prolonged, the cyclic thiodepsipeptide **920** began to form, as evidenced by infrared bands at 1700 cm^{-1} (thioester) and 1645 cm^{-1} (amide I band). An amide II band at 1550 cm^{-1} indicated *trans* geometry for the amide in **920.**

The mass spectrum of **919** showed only a small M^+ peak but a much larger peak at M^+-18 due to thermal loss of water. The infrared spectrum of **919** in methanol contained, in addition to an SH band at 2532 cm^{-1}, shoulders at 1709 and 1675 cm^{-1} due to the presence of small amounts of **918.** The relative proportions of **918, 919,** and **920** at equilibrium could be determined by obtaining infrared spectra in solvents of varying dielectric constants. Table 50 shows the percentage of **920** relative to **919** in various solvents.

Interconversion among structures **918, 919,** and **920** is also evident from their chemical transformation products (278). Methylation of **919** in aqueous base gave only the S-methylated lactam **921** (Fig. 134), which was also prepared by a variant of the synthesis of **917** using the corresponding thio ether in place of **915.** Treatment of **919** with a strong acid such as perchloric gave rise to the stable perchlorate salt **922** in 92% yield. The infrared spectrum of **922** showed a carbonyl absorption band (1721 cm^{-1}) and a band at 1536–1529 cm^{-1}, assigned to the iminium moiety. The marked displacement of the carbonyl band in **922** relative to **919** is due to the neighboring resonance-stabilized positive charge.

Salt **922** was a useful intermediate for the preparation of several other products. Hydrogenation of **922** with Pd/C gave the basic azepino[2,1-*b*][1,3]benzothiazin-12-one **923** in 80% yield, and treatment with sodium meth-

TABLE 50. PERCENT OF **920** IN EQUILIBRIUM WITH **919** AS A FUNCTION OF THE DIELECTRIC CONSTANT

Solvent	Dielectric Constant	% **920**
Acetone	20.5	0.4
Acetone/H_2O (80% w/w)	29.6	1.0
Methanol	32.6	2.3
Methanol/H_2O (80% w/w)	42.7	10.1
Acetone/H_2O (50% w/w)	48.2	20.5
Methanol/H_2O (60% w/w)	52.7	28.9

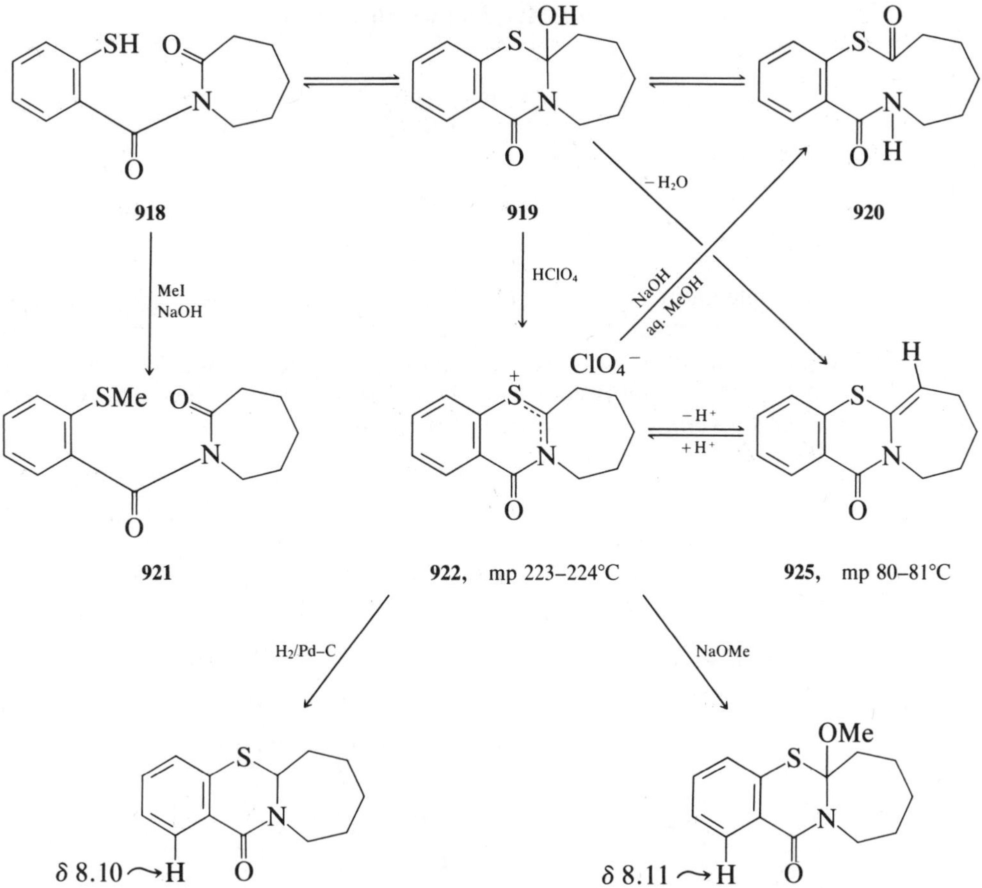

918 919 920

MeI NaOH $HClO_4$ $-H_2O$ NaOH aq. MeOH

ClO_4^- $-H^+$ $+H^+$

921   922, mp 223–224°C   925, mp 80–81°C

H_2/Pd–C   NaOMe

δ 8.10 → H   δ 8.11 → H

923, mp 91.5–92.5°C   924, mp 88–89°C

Figure 134

oxide gave the somewhat unstable ether **924** in 69% yield. The downfield shift of the aromatic proton adjacent to the carbonyl group in **923** and **924** was taken as evidence of coplanarity of the benzene and thiazinone rings in these structures.

Treatment of **922** with aqueous methanolic sodium hydroxide gave, via the thiacyclol **919**, the cyclothiodepsipeptide **920** in 91% yield. Unfortunately, no physical data, except for infrared spectral data, were given in support of structure **920**. Although the dehydrated product **925** could be obtained from **922** by heating in various solvents, a better yield (77%) was produced on treatment of the thiacyclol **919** with acid-washed alumina. The nmr spectrum of **925** ($CDCl_3$), clearly shows the presence of the vinyl proton at δ 5.5. (t, J = 5.5 Hz). This air-unstable material was converted back to the perchlorate **922** in 91% yield.

6. Azepino[2,1-*b*]pteridines

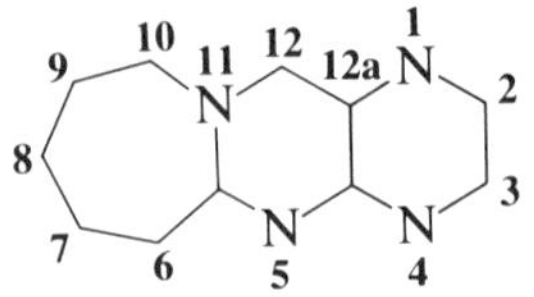

10 11 12 12a 1
9 N N 2
8 3
7 6 N 5 N 4

The single known example of the azepino[2,1-*b*]pteridine ring system was prepared by thermal cyclization of the pyrazine amino ester **926** with *O*-methylcaprolactim to give **927** (Eq. 104) in 52% yield (mp 217–218°C). The product showed an infrared carbonyl band at 1687 cm^{-1}. The ultraviolet and nmr spectra were also supportive of the structure.

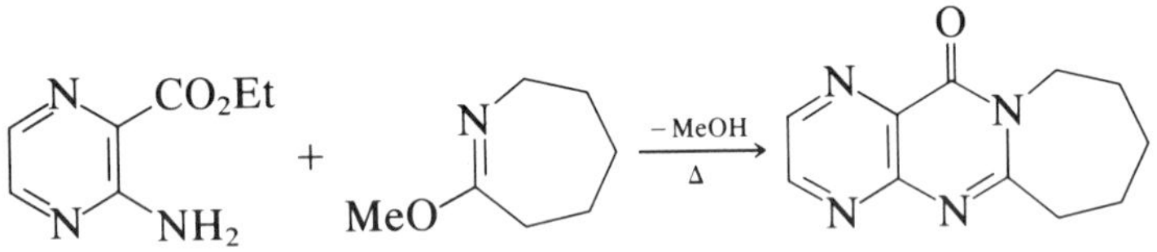

N CO_2Et N NH_2 + N MeO $\xrightarrow[\Delta]{-MeOH}$ O N N N N (104)

926 **927**

7. Dipyrimido[5,4-*c*:4′,5′-*e*]azepines

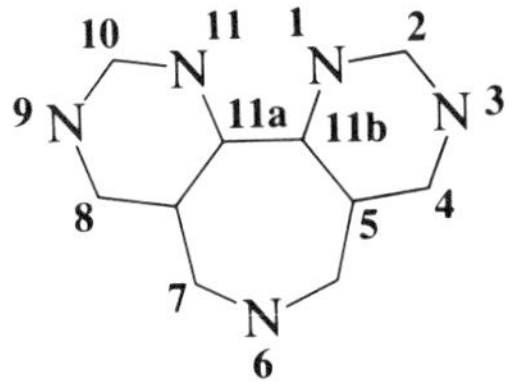

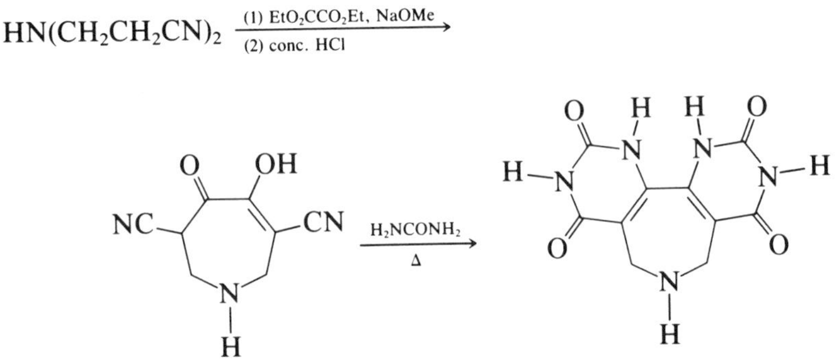

10 11 1 2
9 N N 11a 11b N N 3
8 5 4
7 N 6

$HN(CH_2CH_2CN)_2$ $\xrightarrow[\text{(2) conc. HCl}]{\text{(1) } EtO_2CCO_2Et\text{, NaOMe}}$

O OH NC CN N H $\xrightarrow[\Delta]{H_2NCONH_2}$ O H H O H—N N N N—H O O N H (105)

928 **929**, mp 147–148°C

When the azacycloheptane dione **928,** prepared in 63% yield from iminodipropionitrile, was heated with urea in ethanol (Eq. 105), compound **929** was obtained in unspecified yield as colorless crystals. The substance gave a red ferric chloride test (280).

8. Pyrimido[4′,5′:4,5]pyrimido[1,2-*a*]azepines

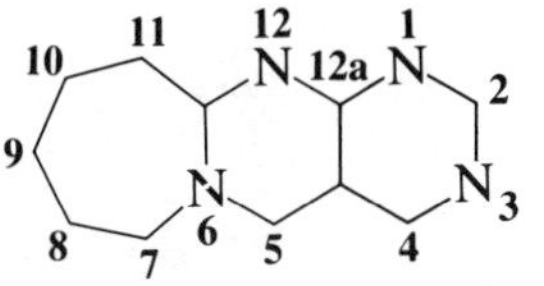

Two examples of the pyrimido[4′,5′:4,5]pyrimido[1,2-*a*]azepine system have been reported. Both were prepared by thermal cyclization of the pyrazine derivatives **930** and **931,** with *O*-alkylcaprolactims (Eq. 106). Compound **932** was obtained in 93% yield (281), but the methyl substituted product **933** was obtained in a much lower yield (21%) (279). In a study of the Dimroth rearrangement (281), heating of **932** with 1 *M* NaOH led only to hydrolysis of the imino group. As expected, the five-membered methylene

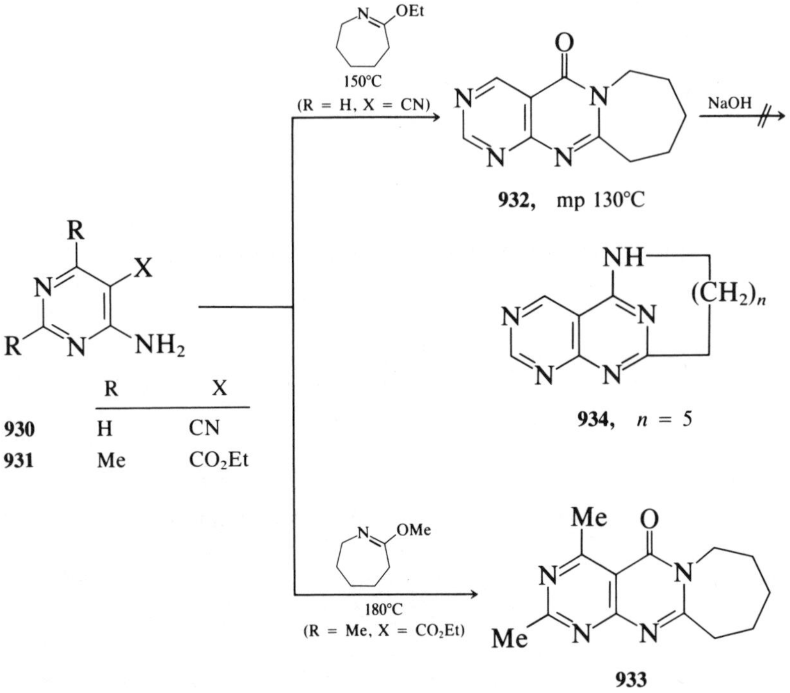

	R	X
930	H	CN
931	Me	CO_2Et

150°C (R = H, X = CN)

932, mp 130°C

NaOH

934, n = 5

(106a)

180°C (R = Me, X = CO_2Et)

933

(106b)

chain was too short to form strained structure **934.** However, with a methylene chain of seven carbons, the Dimroth rearrangement was found to readily give **934** ($n = 7$).

9. 2,3a,6a-Triazacyclohepta[*de*]naphthalenes

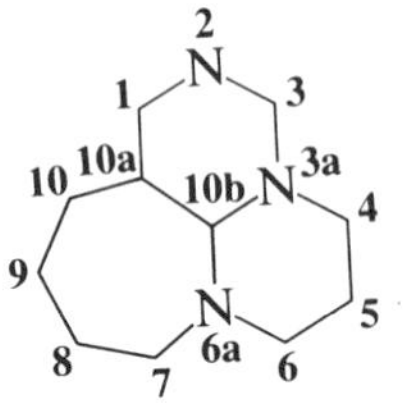

One example of the 2,3a,6a-triazacyclohepta[*de*]naphthalene system has been cited without detail in a patent (36). Reaction of the diazabicyclo compound **935** with phenoxycarbonyl isocyanate was suggested to give the tricyclic imide **936** via loss of phenol (Eq. 107). The reaction is analogous to one reported in the same patent for the preparation of triazabenz[*c*,*d*]azulenes (Section IV. 17).

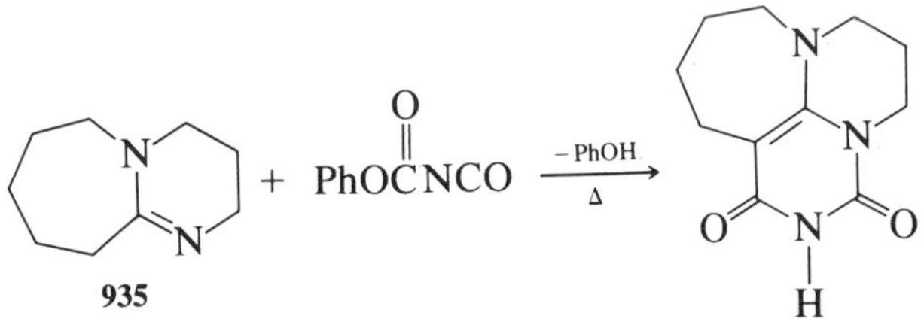

935 + PhOC(O)NCO $\xrightarrow[\Delta]{-\text{PhOH}}$ **936**, mp 210°C (107)

10. Pyrido[2′,3′:4,5]pyrimido[1,2-*a*]azepines

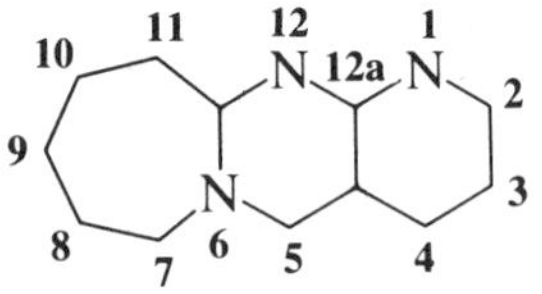

11. Pyrido[3′,2′:4,5]pyrimido[1,2-*a*]azepines

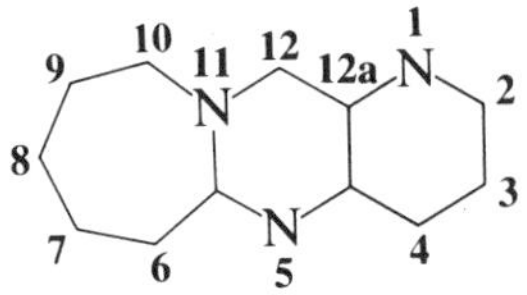

Examples of these isomeric systems were prepared by condensation of *O*-methylcaprolactim (**937**) with aminopyridine carboxylic acids (282,283) (Eq. 108). Apparently, the condensation of 2-aminonicotinic acid (**938**) with **937** is slow and required heating in refluxing glyme. On the other hand, 3-amino-2-picolinic acid (**940**) reacted smoothly with **937** in refluxing acetone. No yields or spectral data were reported. Compounds **939** and **941** could be vacuum distilled and are water soluble.

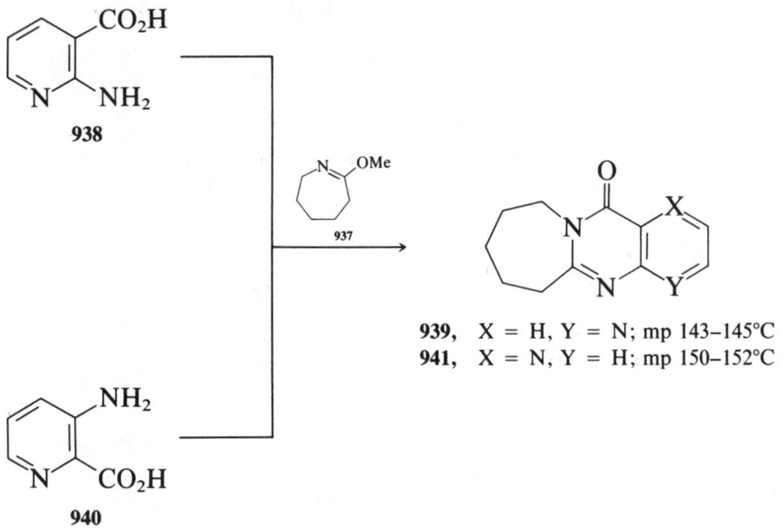

CO_2H NH_2 **938**

OMe **937**

(108)

939, X = H, Y = N; mp 143–145°C
941, X = N, Y = H; mp 150–152°C

NH_2 CO_2H **940**

12. Azepino[1,2-*a*]quinazolines

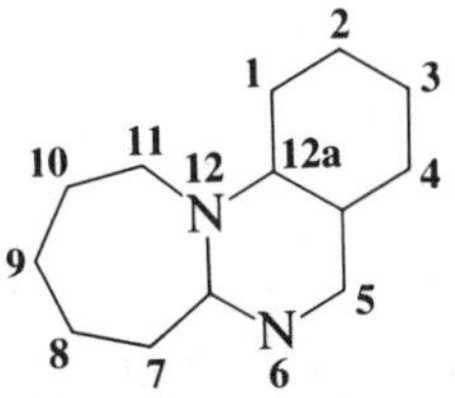

Because of biological interest in the hypnotic agent methaqualone (**942**), its reaction with dimethyl acetylenedicarboxylate (DMAD) was investigated as a source of new structures containing the 2-methyl-3-*o*-tolylquinazolone nucleus (284). Reaction of **942** with DMAD in refluxing acetonitrile for 30 hr gave, after chromatography, a 28% yield of a yellow substance as the only product (Fig. 135). The mass spectrum showed an M^+ of *m*/*e* 534, indicating that two moles of DMAD had been incorporated; the base peak of M^+-86 was quite characteristic for the loss of methyl acrylate, as is known to be the case for other DMAD adducts (see Sections VI. 13, VI. 15, VI. 23, and VI. 24). The authors suggested that the nmr spectrum of the yellow

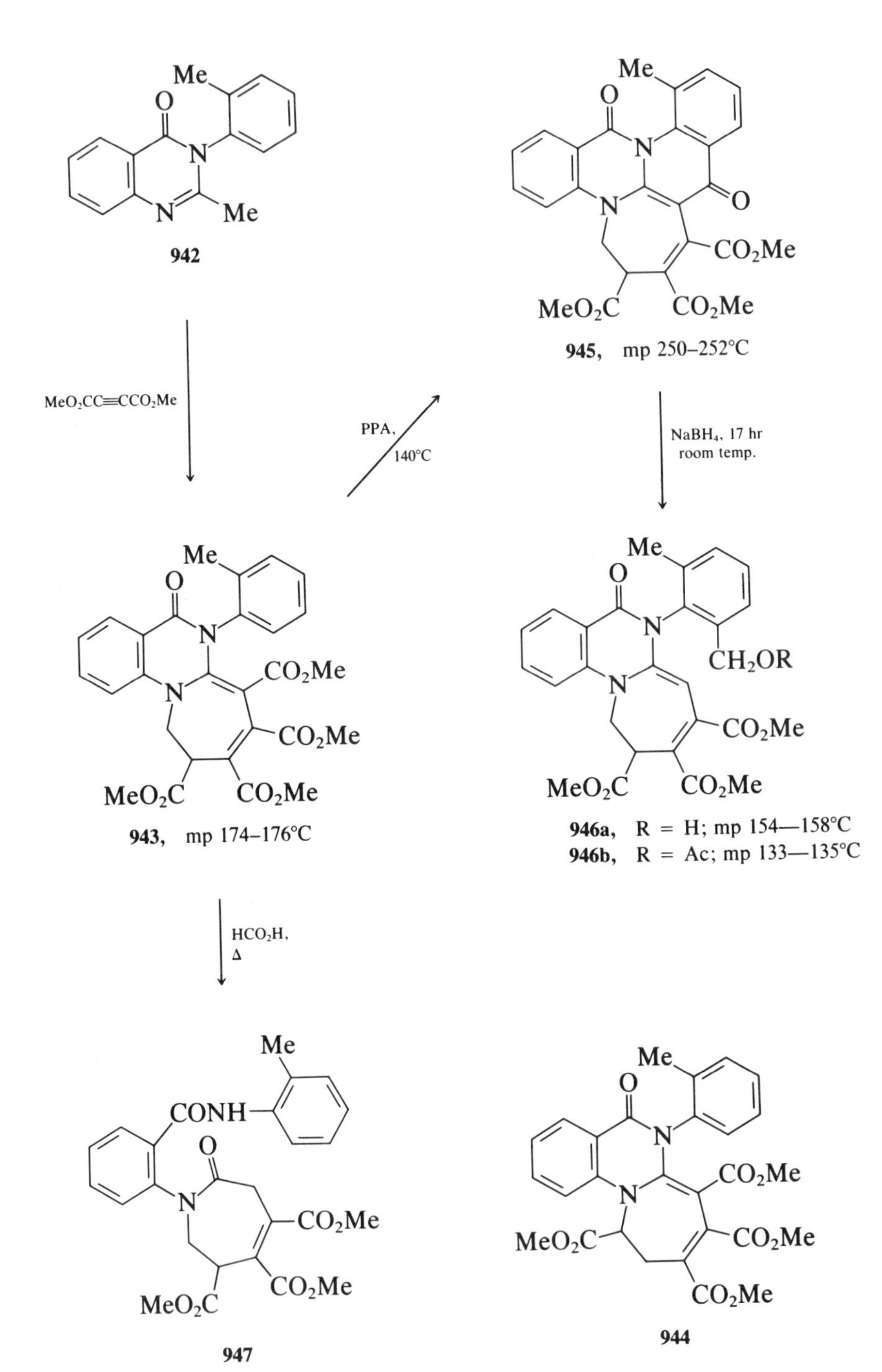
Me
O
N
N
Me
942
MeO2CC≡CCO2Me
PPA,
140°C
Me
O
N
N
O
CO2Me
MeO2C
CO2Me
945, mp 250–252°C
NaBH4, 17 hr
room temp.
Me
O
N
N
CO2Me
CO2Me
MeO2C
CO2Me
943, mp 174–176°C
Me
O
N
CH2OR
N
CO2Me
MeO2C
CO2Me
946a, R = H; mp 154—158°C
946b, R = Ac; mp 133—135°C
HCO2H,
Δ
Me
CONH
O
N
CO2Me
CO2Me
MeO2C
947
Me
O
N
N
CO2Me
MeO2C
CO2Me
CO2Me
944

Figure 135

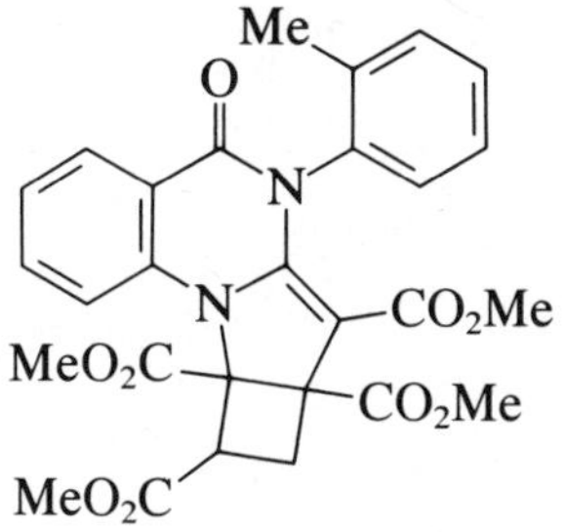

Me
O
N
N
CO_2Me
MeO_2C
CO_2Me
MeO_2C

948

product was more compatible with **943** than with **944,** by analogy, again, to related DMAD adducts.

The nmr spectrum of **943** is of interest since it shows at room temperature a pair of diastereomeric rotamers exhibiting double resonances for the *o*-tolyl methyl and the four ester signals due to slow rotation about the aryl bond. The relative proportions of the two conformers are 70:30 at 100°C and become equal at 150°C. The aryl ring also causes shielding of the proximate ester group, so that their doubled resonances are displaced upfield to δ 3.19 and δ 3.17.

Cyclization of **943** in PPA at 140°C gave a 57% yield of the ketone **945,** whose nmr spectrum showed two low-field aromatic protons (δ 8.10, δ 8.25) due to the hydrogens *peri* to the two ketone carbonyls. Reduction with sodium borohydride reportedly gave a 62% yield of the alcohol **946a** (R = H). It was suggested that **946a** could arise by reduction of the ketone, followed by protonation of the adjacent carbon, which is part of an enamine system. The aldehyde formed on retroaldol cleavage is reduced to **946a.** The presence of the hydroxymethyl group was established by an *AB* quartet in the nmr spectrum, which moved downfield in the acetylated derivative **946b** (50% yield). Nonequivalence of the methylene protons was observed again.

Attempted reduction of **943** with refluxing formic acid gave the crystalline azepinone **947** in 90% yield. This could conceivably occur by ester exchange–decarboxylation, followed by addition of water (possibly on workup), and ring opening to **947.** A small amount of **947** was also formed in the PPA cyclization of **943** to **945.**

It must be noted that recent reassignments of the structures of various adducts of acetylenedicarboxylic esters have been required (see Section VI. 24 and Ref. 285), and it is very possible that **948,** rather than **943,** is the correct structure for the reaction product of **942** with DMAD.

13. Azepino[1,2-*c*]quinazolines

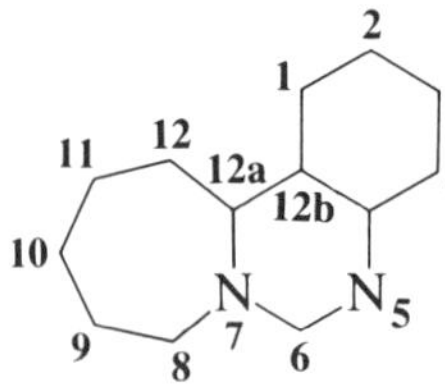

A tetracarbomethoxy derivative initially thought to be **950** was obtained (286) in very low yield by reaction of 4-methylquinazoline (**949**) with dimethyl acetylenedicarboxylate (Eq. 109). The nmr spectrum of the pale yellow 1:2 adduct showed a low-field singlet (δ 9.37), consistent with an aromatic proton deshielded by a neighboring CO_2Me group. The mass spectrum (26) showed a loss of methyl acrylate, which likewise was consistant with the proposed structure. However, because of structural reassignments that have been made for closely related adducts, this substance must now be reformulated as **951** (see Section VI. 24 and Ref. 285).

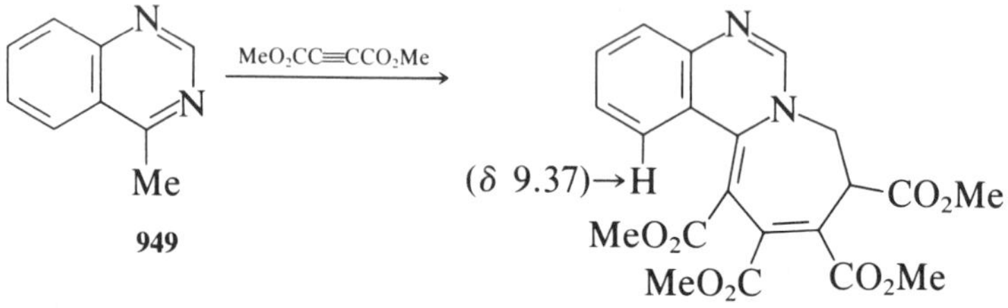

$MeO_2CC{\equiv}CCO_2Me$

949

(δ 9.37)→H

950, mp 187–188°C (109)

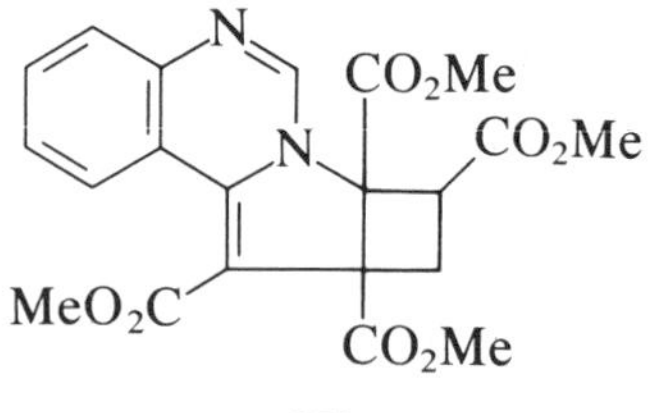

951

14. Azepino[2,1-*b*]quinazolines

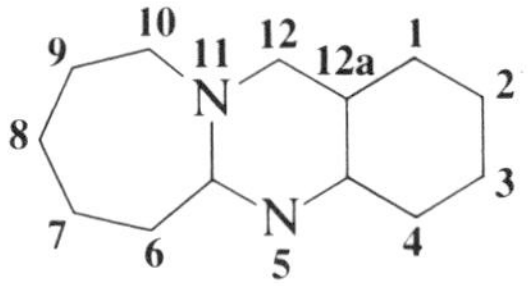

A number of azepino[2,1-*b*]quinazolin-12-ones (**954**) have been prepared by the facile condensation of anthranilic acids with caprolactam or its α-chloro (289) or *O*-lactim ether derivatives (282,290) (Table 51, Eq. 110). The temperature rises spontaneously to 85°C on mixing equal amounts of the acids **952** and the lactim ether **953**, but if the condensation is carried out in MeOH the intermediate **955** can be isolated (282).

Reaction of the azepinoquinazolones in Table 51 with zinc–HCl gave the corresponding azepino[2,1-*b*]quinazolines listed in Table 52 in 33–44% yields (289). This reduction removed both the 12-carbonyl function and, if present, the 6-chloro substituent.

Two oxidative cyclization procedures that may be used to obtain azepino[2,1-*b*]quinazolines or the corresponding quinazolones are illustrated in Eq. 111. Mercuric acetate oxidation of **956** followed by alkaline workup gives a 79% yield of **957** (291), which can be oxidized in 85% yield to the azepinoquinazolone **958** with manganese dioxide. Compound **958** can also be obtained in 71% yield directly from **956** by manganese dioxide oxidation (292). Although both **957** and **958** form addition salts, the greater basicity of **957** relative to **958** is evident by comparison of the tlc mobility of these compounds on silica gel (R_f 0.18 and 0.65, respectively). The infrared spectrum (291) of **958** shows bands at 1660 cm^{-1} (C═O) and 1610 cm^{-1} (C═N), whereas that of **957** shows a band at 1610 cm^{-1} only.

Another approach to azepino[2,1-*b*]quinazolines (Eq. 112) involves reduction of the nitro lactam **959** with stannous chloride in HCl (50% yield), followed by ring closure with loss of water, to give **960** (44% yield) (293). However, formation of the starting material **959** via amidomethylation of caprolactam proceeds only in very low yield.

The basic azepinoquinazoline **957** has also been isolated in 52% yield from the enzymatic oxidation of hexamethylenediamine with diamine oxidase and oxygen (Eq. 113). The azepine **962**, formed on cyclization of the oxidation product **961**, is "trapped" by condensation with *o*-aminobenzadehyde to give **957** (294).

A modification of the process shown in Eq. 110 is the condensation of anthranilamide and caprolactone (295). However, only a 7% yield of **958** was obtained, along with a 32% yield of 2-(5-hydroxypentyl)-4-quinazolone, which could be converted to **958** in unstated yield by heating at 300°C.

When the β-lactam **963** was heated sequentially with **964** at 130°C, and then at 180–200°C (Eq. 114), the 1,12a-decahydroazepinoquinazolone **965** was formed in 81% yield with loss of methanol. The infrared absorption band for the acylamidine functionality was seen at 1688 cm^{-1} (296).

On treatment with aqueous KOH, the methiodide (**966**) of **958** was converted into the cyclodipeptide **967** in good yield (Eq. 115) (297). The intermediate cyclol **968** could not be isolated as had been done in the synthesis of the analogous azepino[2,1-*b*][1,3]benzoxazine (Section VI. 3) and benzothiazine (Section VI. 5) systems. Infrared and mass spectral data were used to assign the structure of the product as **967** rather than **968** or **969.**

TABLE 51. AZEPINO[2,1-*b*]QUINAZOLIN-12(1H)-ONES

R_1	R_2	R_3	R_4	R_5	mp (°C)	Yield (%)	Refs.
H	H	H	H	H	96–98	56	289
					219 (HCl)		289
					96	62	282
					96–98	48	290
					96–97	67	274
					221–224 (HCl)		274
					231–233 (CH_3I)	63	274,299
					95–96.5	100	300
					209–211 (HCl)		300
					95	85	291
					95	71	292
					176 (*N*-oxide)		56
H	Me	H	H	H	78–80	46	290
H	MeO	H	H	H	116–117	54	290
H	H	Cl	H	H	118–119	50	290
					114	40–50	300

H	Cl	H	H	H	122–124	52	290
					105–106.5	56	289
					107	—	282
H	H	NO_2	H	H	150	72	282
					145–146	60	289
H	Br	H	H	H	103–104	58	289
H	I	H	H	H	122–124	66	289
H	NO_2	H	H	H	164–165	57	289
H	H	H	Me	H	105	50	289
H	H	H	MeO	H	138–139	50	289
H	NH_2	H	H	H	171–173	52	289
H	NO_2	H	H	H	164–165	57	289
H	H	H	Me	H	105	50	289
H	H	H	MeO	H	138–139	50	289
H	NH_2	H	H	H	171–173	52	289
H	H	NH_2	H	H	195–196	50	289
H	H	H	H	Cl	101–103	51	289
H	Br	H	H	Cl	127–129	53	289
H	Cl	H	H	Cl	126–127	54	289
H	I	H	H	Cl	129–130	56	289
H	NO_2	H	H	Cl	206–207	56	289
H	H	NO_2	H	Cl	169–170	56	289
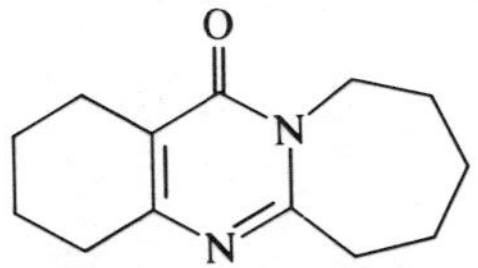					80–81	81	296

R_1 CO_2H NH_2 + H—N (O, R_2) $\xrightarrow[\Delta]{POCl_3}$ R_1 (O, N, N, R_2)

952 954 (110)

953 (EtO, N) MeOH, Δ → [R_1 CO_2H H N N] 955

The quinazolone **958** could not be converted directly to its *N*-oxide **970** by a variety of oxidation methods, but gave instead the nitrocaprolactam **971** (Eq. 116) (56). However, catalytic reduction of **971** in the presence of one equivalent of HCl provided the desired *N*-oxide **970.** When **970** was photolyzed in MeOH it underwent rapid deoxygenation to **958.** No reaction similar to that discussed with azepino[1,2-*a*]benzimidazoles (Section IV. 20) to give the methylene-bridged quinazolinedione **972** was observed. Attempted Dimroth rearrangement of the 12-imino analog of **958** with KOH gave only **958** by simple hydrolysis and none of the methylene-bridged quinazoline **973** (281).

Reduction of **958** with lithium aluminum hydride (Eq. 117) gave an excellent yield of the quinazoline **974,** which was isolated as its hydrochloride salt (298). Mass spectra, nmr (CF_3CO_2H), infrared, and uv data suggest the presence in samples of **974** of the ring-opened tautomer **976a.** Nmr spectra showed a vinyl triplet at δ 8.9 due to —N—CH=CH_2. Compound **974** could be converted via base treatment of its methiodide **(975)** to the stable *N*-methyl derivative **976b,** isolated as a monohydrochloride salt. Nmr spectra ($CDCl_3$) of the free base showed the vinyl triplet at δ 7.68. On hydrogenation, **976b** consumed one mole of hydrogen to give the octahydro-1H-[1,8]benzodiazacycloundecane **977.**

Some of the compounds listed in Table 51 were bioassayed because of their relationship to vascicine and peganine alkaloids (289,290), but were found to be devoid of pharmacological activity. Section VI. 25 should be consulted for additional information on this system as it relates to the preparation of azepino[2,3-*b*]quinolines.

TABLE 52. HEXAHYDROAZEPINO[2,1-*b*]QUINAZOLINES

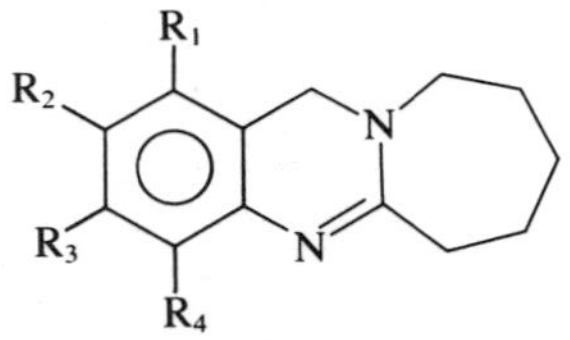

R_1	R_2	R_3	R_4	mp (°C)	mp (°C) (HCl)	Yield (%)	Refs.
H	H	H	H	94–96	202–204	44	289
				96–97		79	291
					148–149 (picrate)		294
H	Br	H	H	118–120	275	41	289
H	Cl	H	H	180	237–240	45	289
H	H	NO_2	H	150–152	260	48	289
H	H	H	Me	156–157	257–258	43	289
H	H	H	MeO	100–102	—	33	289
H	NH_2	H	H	97–98	255	33	289
H	MeO	MeO	H	124–125		50	293

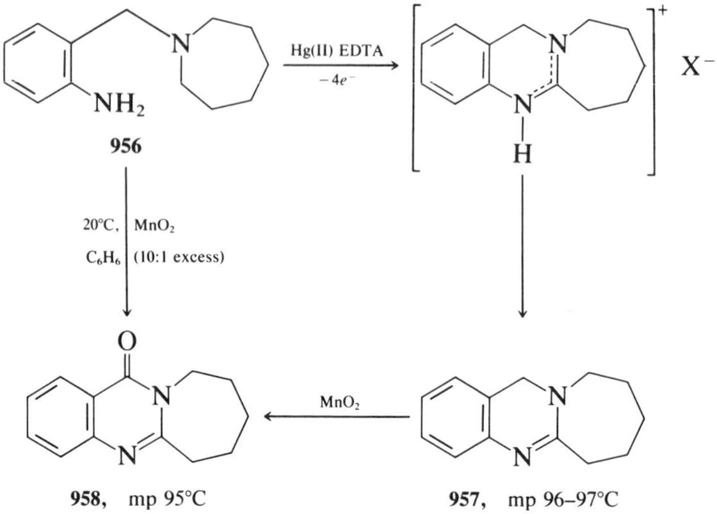

$$\text{956} \xrightarrow[-4e^-]{\text{Hg(II) EDTA}} [\ ]^+ X^- \longrightarrow \text{957, mp 96–97°C}$$

$$\text{956} \xrightarrow[\text{C}_6\text{H}_6,\ (10:1\ \text{excess})]{20°\text{C},\ \text{MnO}_2} \text{958, mp 95°C}$$

$$\text{957} \xrightarrow{\text{MnO}_2} \text{958} \quad (111)$$

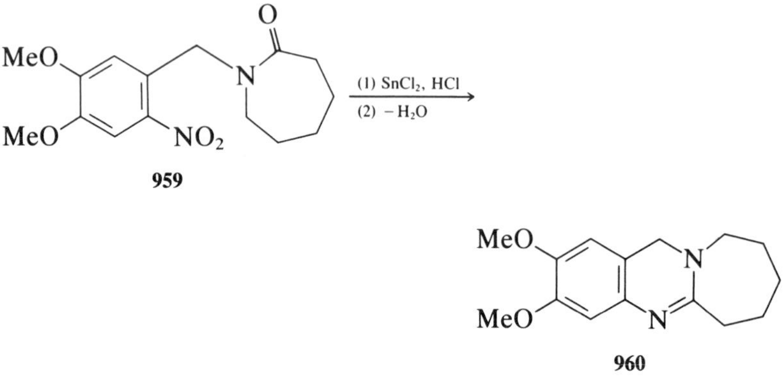

$$\text{959} \xrightarrow[(2)\ -H_2O]{(1)\ SnCl_2,\ HCl} \text{960} \quad (112)$$

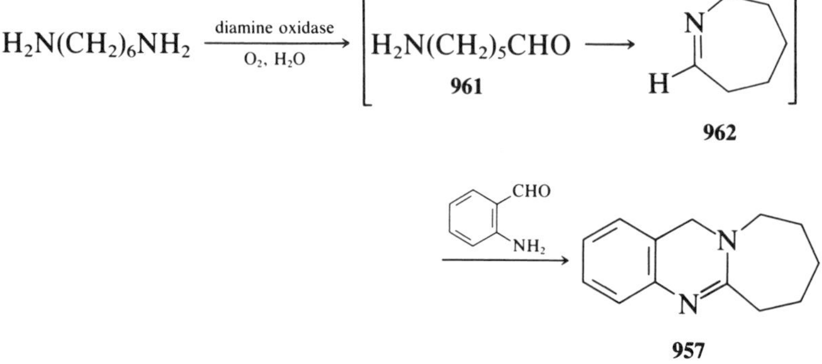

$$H_2N(CH_2)_6NH_2 \xrightarrow[O_2,\ H_2O]{\text{diamine oxidase}} \left[\underset{\textbf{961}}{H_2N(CH_2)_5CHO} \longrightarrow \textbf{962} \right] \xrightarrow{\text{o-aminobenzaldehyde (CHO, NH}_2)} \textbf{957} \quad (113)$$

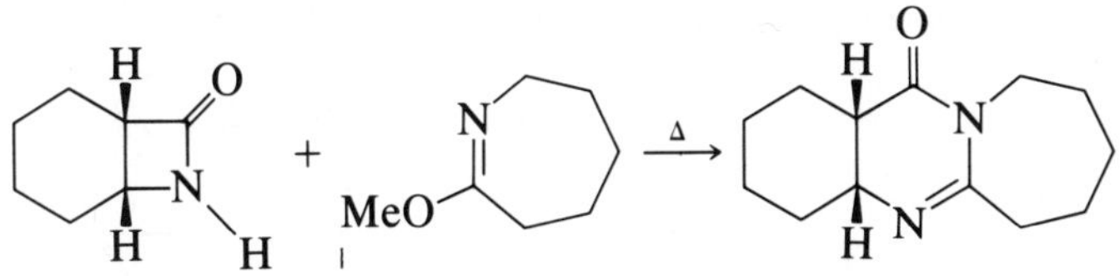

963 + 964 $\xrightarrow{\Delta}$ 965, mp 80–81°C (114)

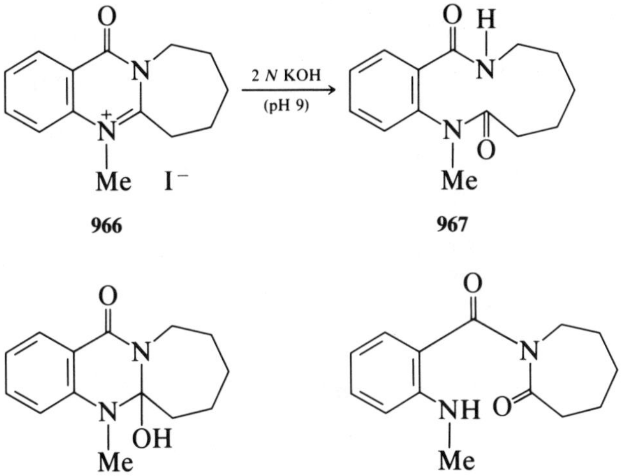

966 $\xrightarrow[\text{(pH 9)}]{2\,N\ \text{KOH}}$ 967

968 969 (115)

958 $\underset{h\nu,\ \text{MeOH}}{\overset{[\text{O}]}{\nrightarrow}}$ 970

958 $\xrightarrow{[\text{O}]}$ 971 $\xrightarrow{\text{H}_2\text{/Pd, 1 eq. HCl}}$ 970 (116)

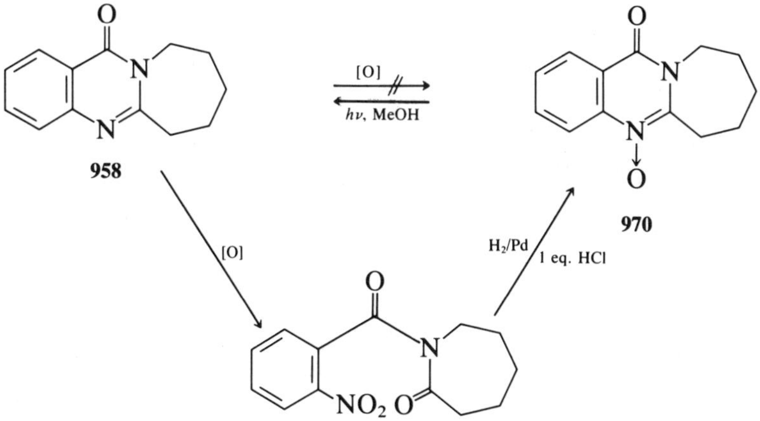

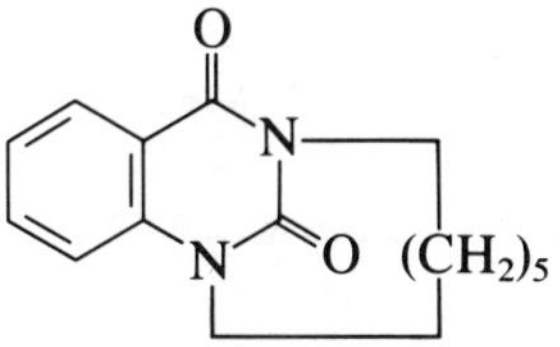

972 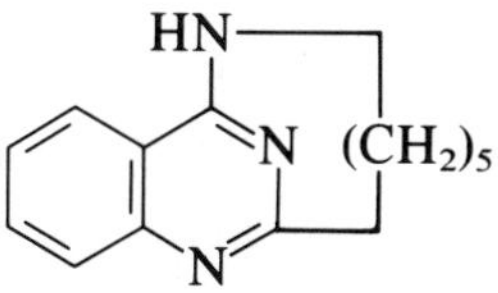 973

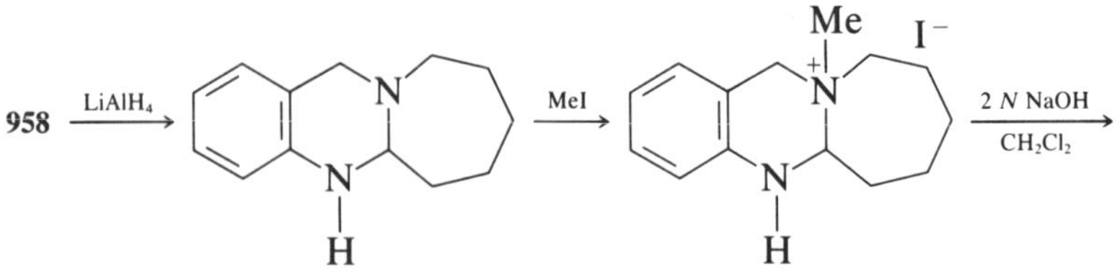

958 $\xrightarrow{LiAlH_4}$ [N, N–H] $\xrightarrow{MeI}$ [Me, I^-, $\overset{+}{N}$, N–H] $\xrightarrow[CH_2Cl_2]{2\ N\ NaOH}$

974, 96%; mp 199—200°C (HCl) **975**, mp 218–220°C

(117)

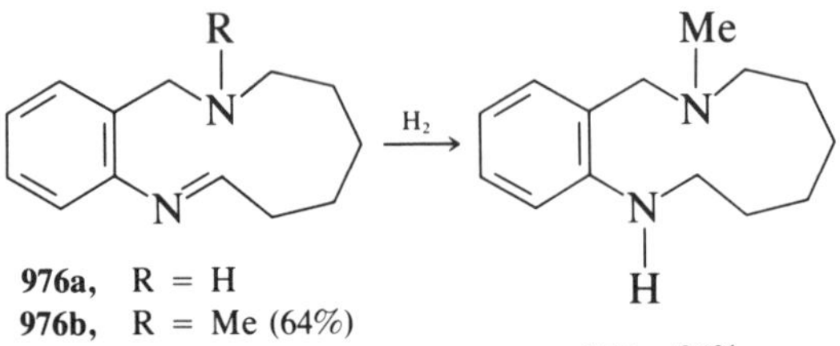

[R, N, N] $\xrightarrow{H_2}$ [Me, N, N–H]

976a, R = H
976b, R = Me (64%)

977, 86%

15. Azepino[1,2-*a*]quinoxalines

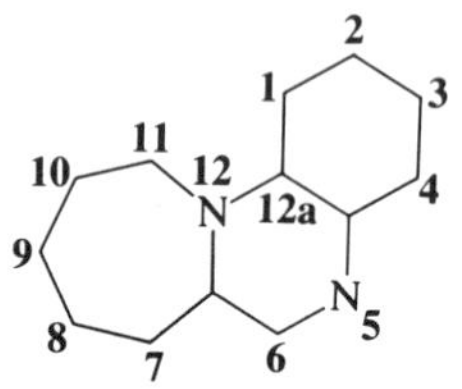

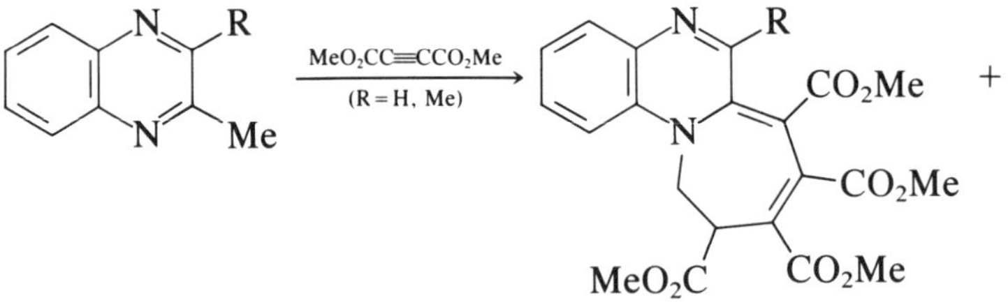

1, 2, 3, 4, N_5, 6, 7, 8, 9, 10, 11, 12 (N), 12a

[N, R, N, Me] $\xrightarrow[(R = H, Me)]{MeO_2CC{\equiv}CCO_2Me}$ [N, R, CO_2Me, N, CO_2Me, MeO_2C, CO_2Me] +

978

(118)

[N, R, CO_2Me, N, CO_2Me, MeO_2C, CO_2Me]

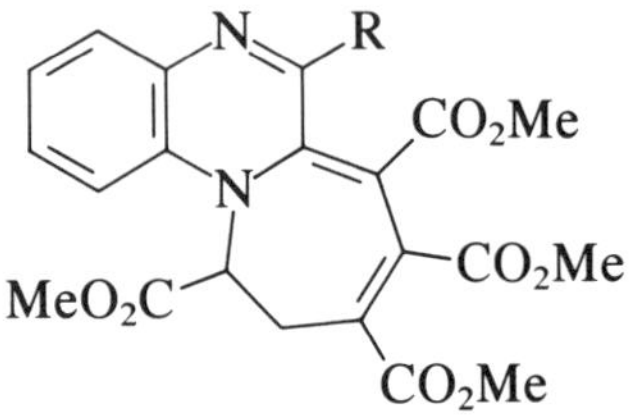

979

TABLE 53. AZEPINO[1,2-*a*]QUINOXALINES[a]

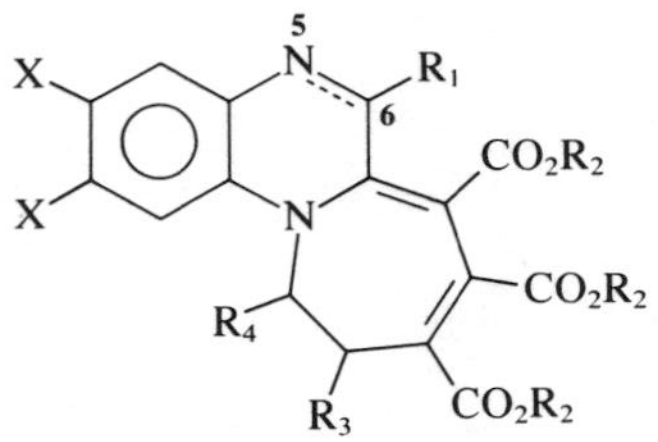

R_1	R_2	R_3	R_4	X	Yield (%)	mp (°C)
H	Me	CO_2Me	H	H	2.0	138–139
Me	Me	CO_2Me	H	H	8.7	170–171
H	Me	H	CO_2Me	H	16	157–158
H	Et	H	CO_2Et	H	5	127–128
H	Me	H	CO_2Me	Br	93	207–209
CBr_3	Me	CO_2Me	H	Br	90	320
CBr_3	Me	CO_2Me	H	H		182–183
CH=CHPh	Me	CO_2Me	H	H		209–210
			5,6-Dihydro			
Me	Me	CO_2Me	H	H	64	216–217
H	Me	H	CO_2Me	H	73	220–221

[a]Ref. 287.

The Acheson group, which has explored the chemistry of acetylenedicarboxylic esters for the synthesis of other heterocyclic systems (see Sections VI.24, VI.12, VI.13, and VI.23) reported (287) the preparation of the azepino[1,2-*a*]quinoxalines **978** and **979** (Eq. 118). As with the related systems mentioned above, ultraviolet, nmr, and mass spectra (26) were cited in support of the assigned structures. Hydrogenation of **978** or **979** reportedly gave dihydro derivatives **980**, an bromination gave **981**. Examples reported are given in Table 53.

However, a structural reassignment was made recently (288) on the basis of work done on the analogous azepino[1,2-*a*]quinolines (Section VI.24) and the interpretation of ^{13}C-nmr spectra. Structures **978** and **979** are now formulated as the *cis* and *trans* isomers of **982a,b** and **982c,d.**

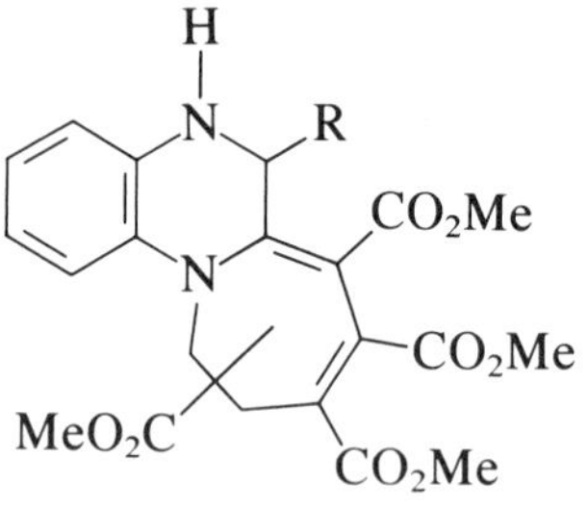

980 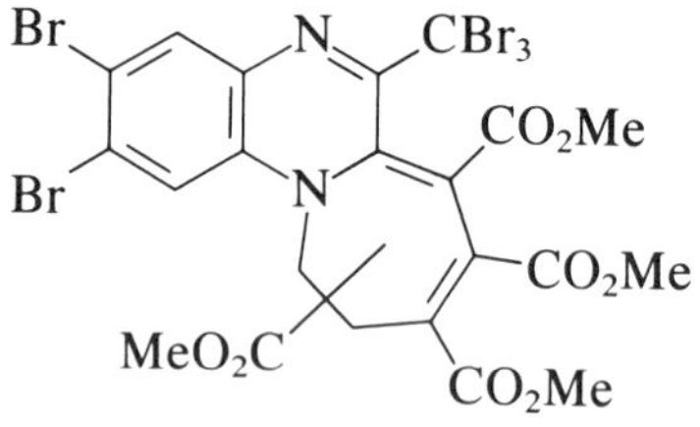 **981**

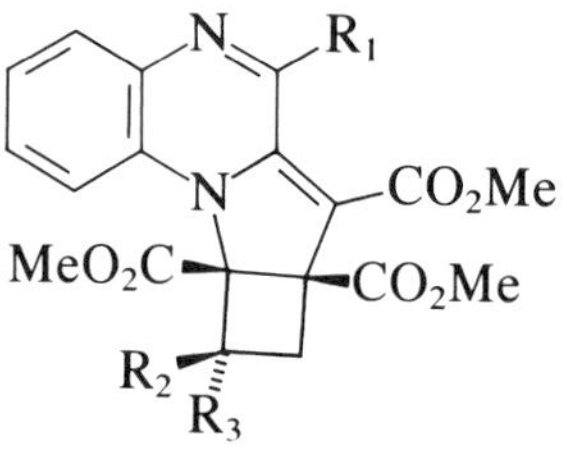

982a, R_1 = H, R_2 = CO_2Me, R_3 = H
982b, R_1 = H, R_2 = H, R_3 = CO_2Me
982c, R_1 = Me, R_2 = CO_2Me, R_3 = H
982d, R_1 = Me, R_2 = H, R_3 = CO_2Me

16. Azepino[2,3-*b*]quinoxalines

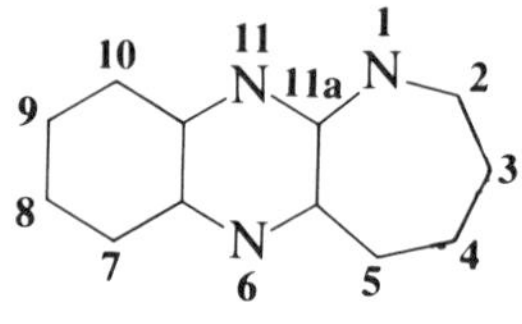

1 2 3 4 5 6 7 8 9 10 11 11a

Treatment of the keto lactam **983** with triethyloxonium tetrafluoroborate followed by alkaline workup provided the keto lactim ether **984** in excellent yield (ir, 1715 cm^{-1} for C=O, 1660 cm^{-1} for C=N). Condensation of **984** with *o*-phenylenediamine in refluxing ethanol (Eq. 119) afforded a 35% yield of **985,** the only know example of this system. The product showed infrared absorption at 3240 cm^{-1} (N—H) and ultraviolet absorption maxima at 255 nm (log ϵ 4.31) and 360 nm (log ϵ 3.93), was soluble in most organic solvents, and formed a yellow hydrochloride salt (mp 195–196°C). The direct condensation of **983** with *o*-phenylenediamine failed to yield **985** (301).

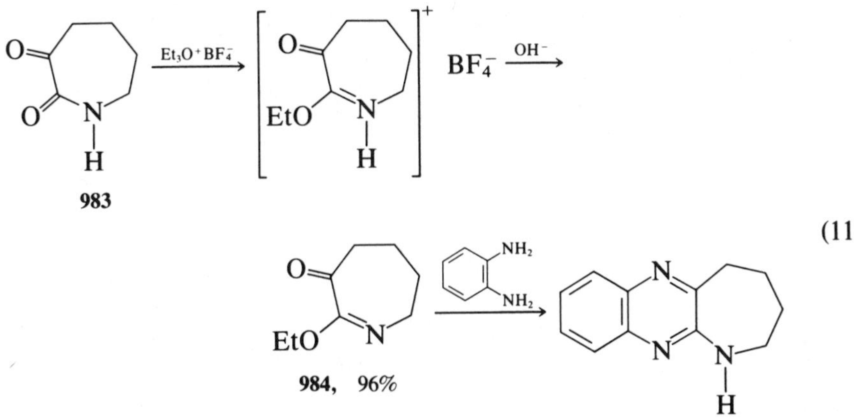

$Et_3O^+BF_4^-$; BF_4^- ; OH^-

983

(119)

984, 96%

985, 35%; mp 103.5–104.5°C

17. Azepino[4,5-*b*]quinoxalines

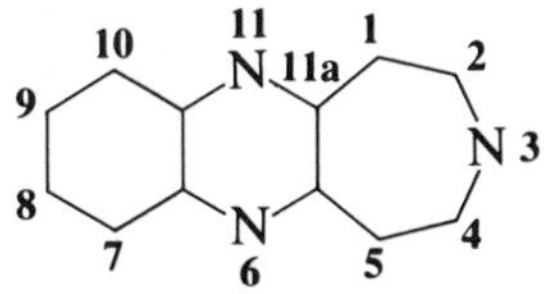

Numerous examples of 2,3,4,5-tetrahydro-1H derivatives (**987**) of the azepino[4,5-*d*]quinoxaline ring system, claimed to have appetite suppressant and antibacterial activity, have been reported, most of them in a patent. The most important synthetic route to compounds of this type (5,303,304) is the condensation of an *o*-phenylenediamine with a 4,5-diketohexahydroazepine (**986**), as shown in Fig. 136. The diketoazepine **986** is prepared via an acyloin condensation followed by oxidation of the acyloin or its trimethylsilyl derivative with cupric acetate, lead tetraacetate, or bromine. A second, less effective route involves condensation of the diaminobenzene with an α-

bromohexahydroazepinone (**988**) followed by oxidative workup (e.g., air oxidation).

The condensation products **987** can be converted to a variety of other substances (Table 54) by alkylation, acylation, Michael reactions of the basic nitrogen in **989,** and by rearrangement of the mono- or di-*N*-oxides **990.** Reported yields vary from moderate to good.

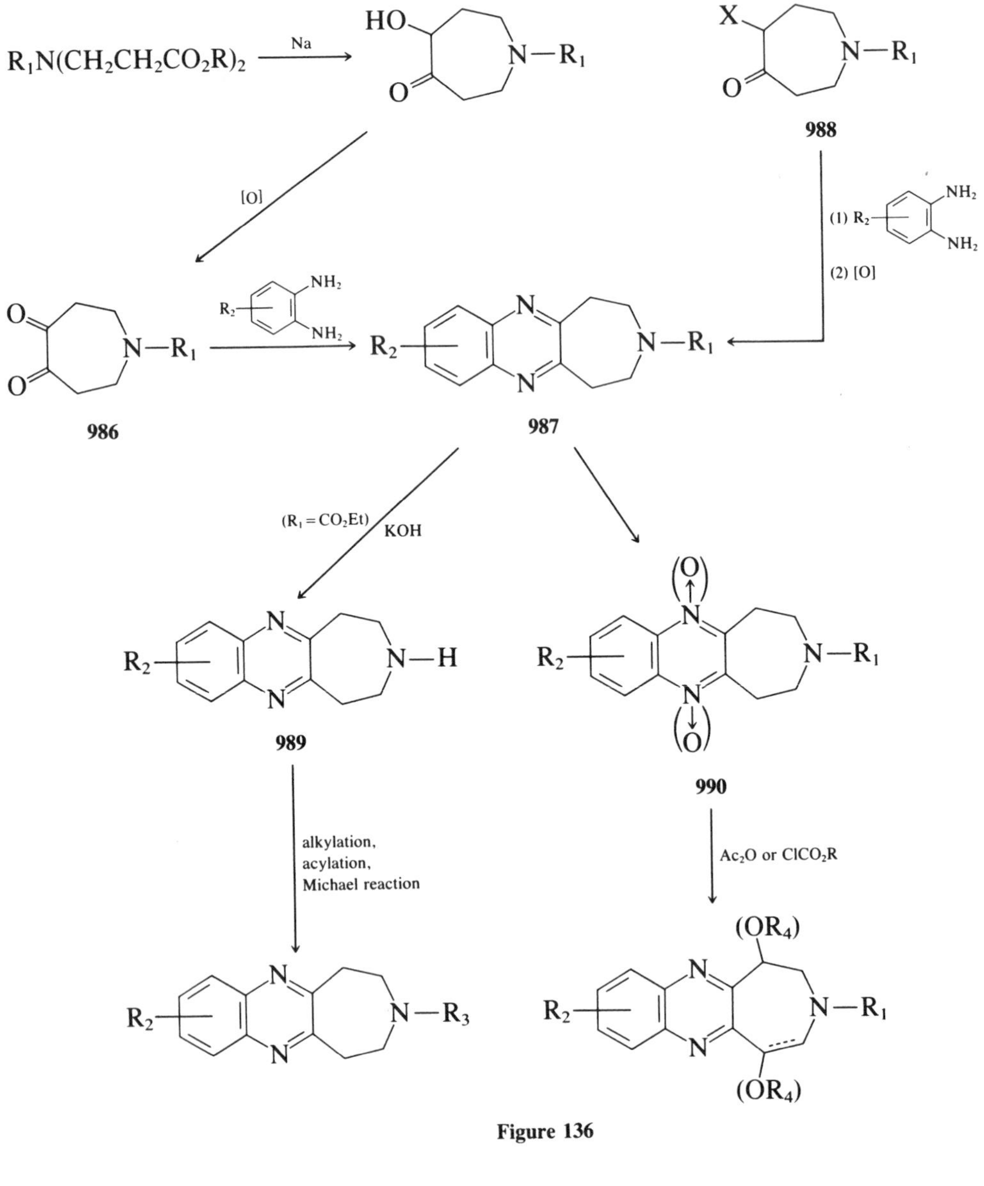

Figure 136

TABLE 54. DERIVATIVES OF 1H-AZEPINO[4,5-*b*]QUINOXALINE

R_1	R_2	R_3	R_4	R_5	R_6	R_7	mp (°C)	Refs.[a]
2,3,4,5-Tetrahydro Derivatives								
CH_2Ph	H	H	H	H	H	H	238–241 (HCl)	
CH_2Ph	H	H	7-NO_2	H	H	H	133	
CH_2Ph	H	H	8-NO_2	H	H	H	145–147	
CH_2Ph	H	H	8-Me	H	H	H	91–93	
CH_2Ph	H	H	8-MeO	H	H	H	215–220 (HCl)	
CH_2Ph	H	H	8-Cl	H	H	H	126–127	
CH_2Ph	H	H	8-Me	9-Me	H	H	146–147	
CH_2Ph	H	H	8-CO_2Et	H	H	H	110–111	
$CH_2C_6H_4Br$-4	H	H	H	H	H	H	187–189	
Ph	H	H	H	H	H	H	130–131	303
Ph	H	H	8-NO_2	H	H	H	236–238	
Ph	H	H	8-Me	9-Me	H	H	179–181	
Ph	H	H	8-Cl	H	H	H	170–172	
Ph	H	H	8-CO_2H	H	H	H	253–255 (HBr)	
Et	H	H	H	H	H	H	259–261 (HCl)	
							260–262 (HCl)	

TABLE 54. *(Continued)*

R_1	R_2	R_3	R_4	R_5	R_6	R_7	mp (°C)	Refs.[a]
Me	H	H	H	H	H	H	245–248 (HCl)	
n-Pr	H	H	H	H	H	H	239–241 (HCl)	
n-Bu	H	H	H	H	H	H	228–231 (HCl)	
$CH_2CH{=}CH_2$	H	H	H	H	H	H	236–237 (HCl)	
t-Bu	$(Me)_2$	H	H	H	H	$(Me)_2$	109–110	304
$CH_2CH_2CO_2Et$	H	H	H	H	H	H	207–209 (HCl)	304
CH_2CH_2CN	H	H	H	H	H	H	225–229 (HCl)	
CH_2CO_2Et	H	H	H	H	H	H	192–193 (HCl)	
CH_2CN	H	H	H	H	H	H	310–315 (HCl)	
CH_2CO_2H	H	H	H	H	H	H	272 (HCl)	
CH_2CONH_2	H	H	H	H	H	H	280 (HCl)	
CH_2CON(morpholino)	H	H	H	H	H	H	187–189	
CH_2CONEt_2	H	H	H	H	H	H	103–104	
CO_2Et	H	H	H	H	H	H	123	
CO_2Et	H	H	7-NO_2	H	H	H	163	
CO_2Et	H	H	8-NO_2	H	H	H	132	
CO_2Et	H	H	8-Cl	H	H	H	144–145	
CO_2Et	H	H	8-MeO	H	H	H	140–142	
CO_2Et	H	H	8-Me	H	H	H	141–143	
CO_2Et	H	H	8-Me	9-Me	H	H	160–161	
CO_2Ph	H	H	H	H	H	H	184–186	
$CO_2C_6H_{11}$-*c*	H	H	H	H	H	H	118–120	
CO_2Bu-*n*	H	H	H	H	H	H	86–87	

COPh	H	H	H	H	H	H	186–188	
COMe	H	H	H	H	H	H	179–180	
$CONHC_6H_{11}$-*c*	H	H	H	H	H	H	180–181	
CONHPr-*n*	H	H	H	H	H	H	185–186	
CSNHEt	H	H	H	H	H	H	220–233	
H	H	H	H	H	H	H	316 (HCl)	
H	H	H	8-Me	9-Me	H	H	340–345 (HCl)	
H	H	H	8-Me	H	H	H	296 (HCl)	
H	H	H	8-Cl	H	H	H	297–299 (HCl)	
H	H	H	8-MeO	H	H	H	276–278 (HCl)	
H	CN	H	H	H	H	CN	214–216 (HCl)	5
CO_2Et	H	O	H	H	H	H	122	
CO_2Et	H	O	H	H	O	H	159–160	
CO_2Et	OAc	H	H	H	H	H	95–97	
CO_2Et	OAc	H	H	H	H	OAc	144–146[b]	
CO_2Et	OCO_2Et	H	H	H	H	H	Oil[c]	
CO_2Et	OCO_2Et	H	H	H	O	H	104	
CO_2Et	OCO_2Et	H	H	H	H	OCO_2Et	103	
CO_2Et	OH	H	H	H	H	H	113	
CO_2Et	OH	H	H	H	O	H	143–145	
CO_2Et	OH	H	H	H	H	OH	132–133	
CO_2Et	Cl	H	H	H	H	H	115–117	
CO_2Et	Cl	H	H	H	O	H	144–145	
				1,2-Dihydro Derivatives				
CO_2Et	H	H	H	H	H	H	134	
CO_2Et	H	H	H	H	H	OAc	129	

[a]Ref. 302, unless noted otherwise.
[b]Two diastereomers, each with mp 144–146°C.
[c]*m/e* = 359.

18. Pyrimido[4,5-*b*][1]benzazepines

Several examples of the pyrimido[4,5-*b*]benzazepine system have been prepared via the reaction of amides with formamide in phosphoryl chloride, which produces pyrimidines in low yields (305,306). Figure 137 shows the conversion of the tetrahydrobenzazepin-2-one **991** to the tricyclic pyrimidine **992** in low to moderate yields by means of this one-step synthesis. The unsubstituted compound **992a** was *N*-alkylated with dimethylaminopropyl chloride to give **993a,** an analog of the antidepressant imipramine. Alkylation of **991** followed by condensation with formamide also gave **993** in poor yields. Compounds **993a** and **993b** possessed weak antidepressant properties, but no details were provided. Table 55 lists the known derivatives of this system.

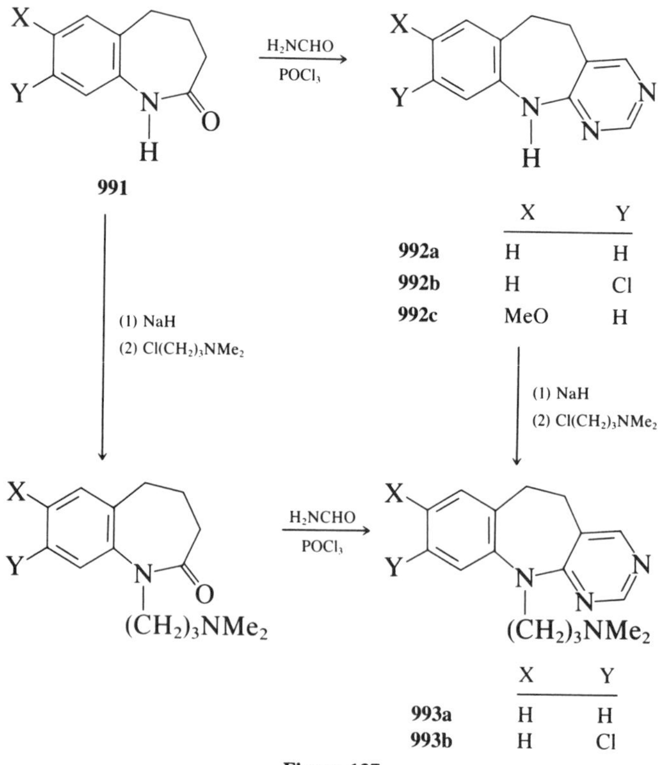

Figure 137

TABLE 55. DERIVATIVES OF PYRIMIDO[4,5-*b*][1]BENZAZEPINE

X	Y	R	mp (°C)	Yield (%)	Refs.
H	H	H	168–170.5	29	305,306
H	Cl	H	177–178	42	305,306
MeO	H	H	145–146	3.5	306
H	H	$(CH_2)_3NMe_2$	233–234 (2HCl)	5	305,306
H	Cl	$(CH_2)_3NMe_2$	93–94 234–235 (2HCl)	18	305,306

19. Pyrimido[4,5-*c*][2]benzazepines

In a study of reagents that could specifically modify the heterocyclic bases in nucleic acids (307), cytosine (**994**) was found to react with ninhydrin (Fig. 138) to give a stable adduct (**995**) in which a new carbon–carbon bond had been formed at the 5-position. As part of the structure determination, sodium periodate was used to cleave **995** between the hydroxyl and C═N bonds and generate the 2H-pyrimido[4,5-*c*]benzazepine 10-carboxylic acid (**997a**), presumably via the intermediate **996.** The alternate structure **997b** for the periodate oxidation product was rejected by comparing its physical constants (uv, ir, nmr, pK_a) and chemical properties with those of the model compound *N*-acetylcytosine. Unlike *N*-acetylcytosine, **997a** was attacked by nitrous acid and gave a bathochromic shift in the ultraviolet spectrum at pH 11. The acid **997a** was more acidic (pK_a 9.5) than cytosine (pK_a 12.2).

The acid **997a** reacted easily at 5°C with aqueous ammonia or methylamine to give, with decarboxylation, the 5-(3-phthalimidyl)cytosines **998a,b.** When **997a** was heated in refluxing acetic anhydride, 5-(3-phthalidyl)uracil (**999**) was obtained in low yield. Reaction with amines again yielded the cytosines **998a** and **998b.** Possible mechanisms for these conversions were suggested.

994

995

$NaIO_4$

996

997a, 83%; mp 300°C (sintering at 218–220°C)

997b

(1) Ac_2O
(2) H_2O

RNH_2

999, 12%

RNH_2

	R	Yield (%)
998a	H	54
998b	Me	88

Figure 138

20. Dipyrido[2,3-*b*:4′,3′-*e*]azepines

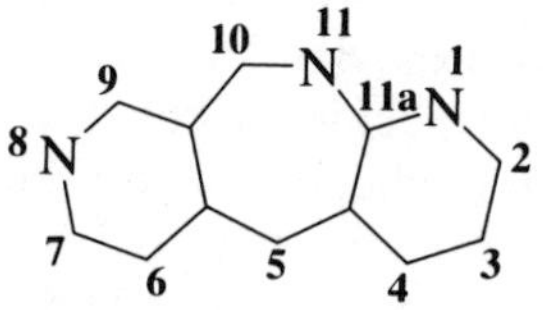

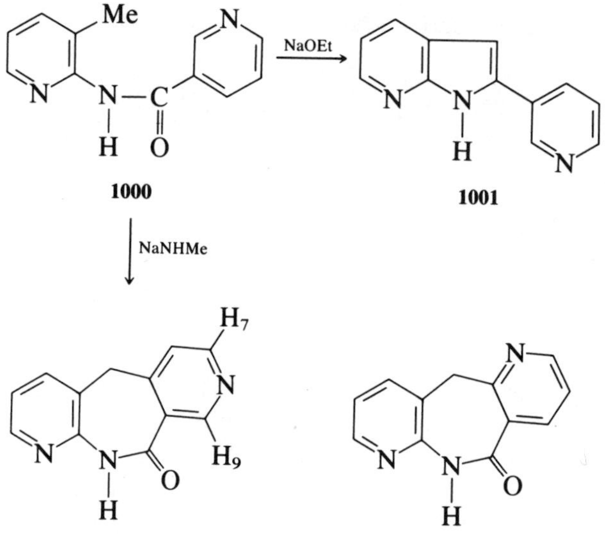

10 11 1 9 11a 8 2 7 3 6 5 4

Me N NaOEt N N—C N N H O H N

1000 1001

NaNHMe

(120)

H_7 N N N N N—O N—O H H H_9

1002, mp 276–279°C 1003

CH_2 1000 $-H^+$ N N—C N H O

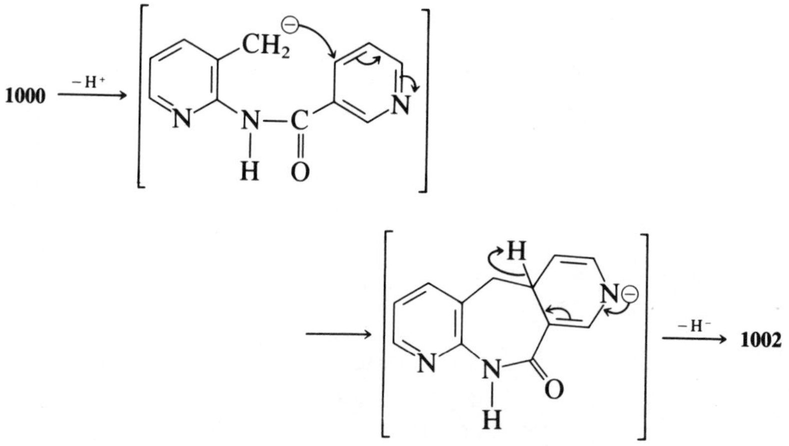

(121)

H N N—O H $-H^-$ 1002

In a study of the preparation and properties of 1H-pyrrolo[2,3-*b*]pyridines, 2-nicotinamido-3-picoline (**1000**) was subjected to base treatment (Eq. 120). With sodium ethoxide, the expected pyrrolo[2,3-*b*]pyridine **1001** was obtained, but when sodium *N*-methylamide was employed as the base, another product was obtained in unspecified yield. Infrared (3150 and 1650 cm^{-1}) and nmr spectra (CF_3CO_2H) indicated the product to be either **1002** or **1003.** However, the appearance of the most strongly deshielded proton as a singlet (H_9) and the next most deshielded proton as a doublet (H_7) favored structure **1002.** The authors suggested that the formation of **1002,** as depicted in Eq. 121, might be the first example of a cyclization involving intramolecular nucleophilic attack of a carbanion at the 4-position of the pyridine ring (308).

21. Dipyrido[4,3-*b*:3′,4′-*f*]azepines

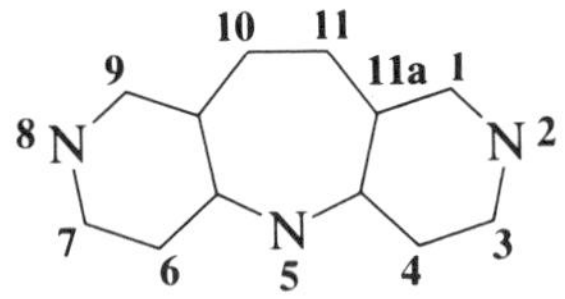

10 11 9 11a 1 8 N N 2 N 7 6 5 4 3

Two examples of the dipyrido[4,3-*b*:3′,4′-*f*]azepine ring system were prepared for pharmacological comparison with the antidepressant imipramine (309) (see Section VI.39 for the chemistry of this important medicinal agent). Pyrolysis of the diphosphate of the *bis*-pyridyl compound **1004** gave **1005** Eq. 122), which was alkylated to the diazaimipramine analog **1006.** This compound was inactive in several biological tests when compared to imipramine, particularly as an antagonist of tetrabenazine-induced sedation in

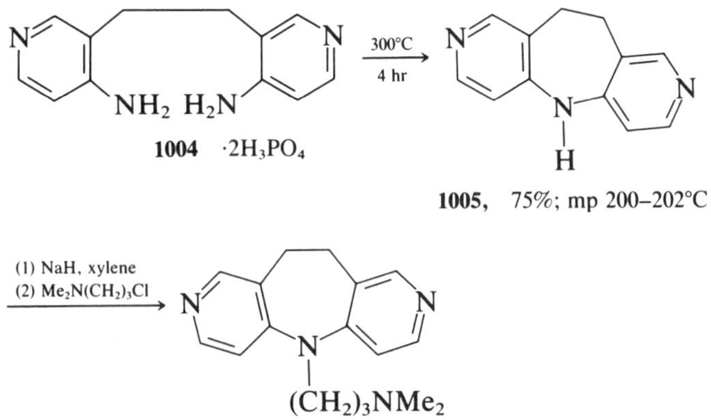

NH₂ H₂N → 300°C, 4 hr

1004 $\cdot 2H_3PO_4$

1005, 75%; mp 200–202°C (122)

(1) NaH, xylene
(2) $Me_2N(CH_2)_3Cl$

$(CH_2)_3NMe_2$

1006, 48%; mp 156–157°C (dimaleate)
mp 223–224°C (4HCl)

mice. Pyrolysis of a phosphate salt has been employed similarly in the synthesis of the pyrido[2,3-*b*][1]benzazepine system (Section VI.32).

22. Azepino[1,2-*b*]isoquinolines

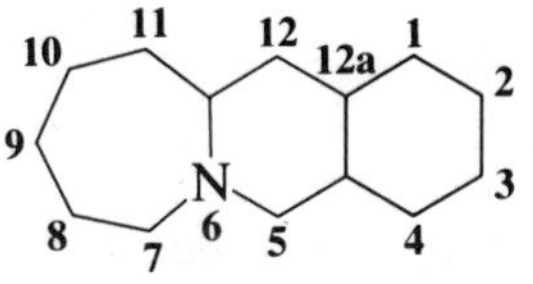
11
12
12a
1
10
2
9
N
3
8
7
6
5
4

Dieckmann cyclization of the diester **1007** provided the hexahydro-11-keto derivative **1008** (ir, 1708 cm^{-1}). Electrolytic reduction of this substance using a lead cathode–30% H_2SO_4 system gave the rearranged pyridobenzazepine **1009** and the 11-membered-ring amino alcohol **1010,** both isolated in crystalline form as picrates (Fig. 139). Compound **1009** was the sole isolated product (as a picrate salt) of the Clemmensen reduction of **1008.** Similar rearrangements have been observed in the Clemmensen reduction of other related systems. These results were obtained as part of a larger study of the effects of ring size on the products of electrolytic reduction of α-amino ketones (219).

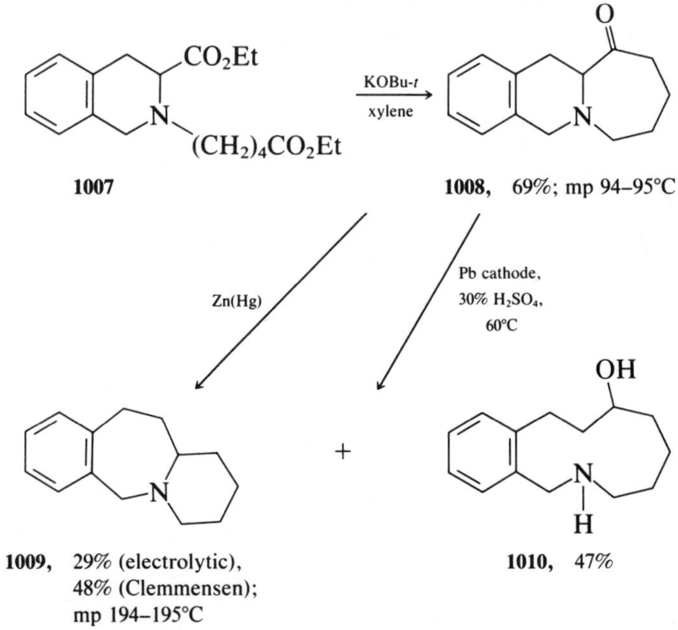
CO2Et
N
(CH2)4CO2Et
1007
KOBu-t
xylene
O
N
1008, 69%; mp 94–95°C
Zn(Hg)
Pb cathode,
30% H2SO4,
60°C
N
+
OH
N
H
1009, 29% (electrolytic),
48% (Clemmensen);
mp 194–195°C
1010, 47%

Figure 139

23. Azepino[2,1-*a*]isoquinolines

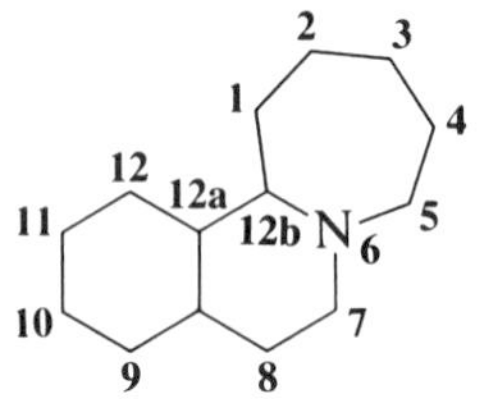

The first reported examples of the azepino[2,1-*a*]isoquinoline system can be seen in Eq. 123. Cyclization of the isoquinoline **1011** gave the quaternary bromide **1012** in modest yield, along with 40% recovered starting material (310). Reduction with tin and HCl gave the base **1013** in very low yield.

The ring system has also been prepared, again in low yield, by reaction of 1-methylisoquinoline with acetylenedicarboxylic esters (311) (Fig. 140). The reaction produced a single stereoisomer of the yellow 1:2 adduct **1014** as well as the orange isomeric benzo[*a*]quinolizine **1015** in the case of dimethyl acetylenedicarboxylate. The nmr signals for the C_{12} aromatic protons in **1014a** and **1014b** appear as low-field quartets due to the deshielding effect of the proximate ester group. Extensive nmr, infrared, and ultraviolet spectral data were reported in support of these structures. The mass spectrum of **1014a** contains a base peak of *m/e* 341 arising from loss of methyl acrylate (26). Catalytic hydrogenation of **1014a** led to a new tetrahydro derivative **1016a.** However, hydrogenation of **1014b** gave a 1:3 ratio of the stereoisomers **1016b** and **1016c.** The exact stereochemical configuration of compounds **1016a–**

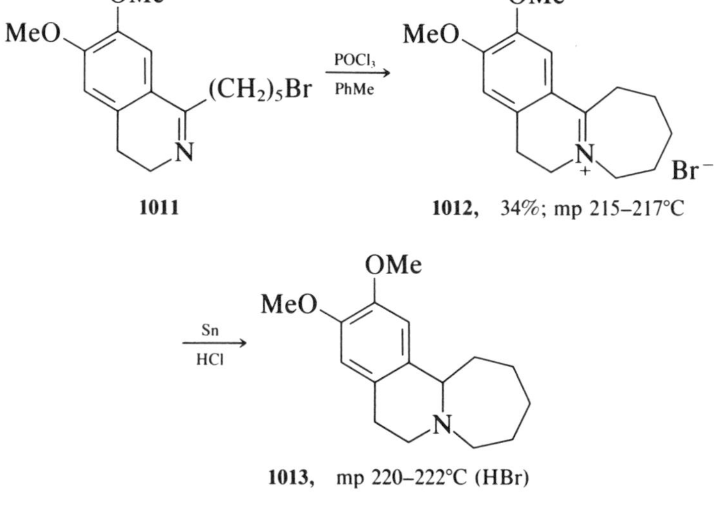

1011 $\xrightarrow[\text{PhMe}]{\text{POCl}_3}$ **1012,** 34%; mp 215–217°C $\xrightarrow[\text{HCl}]{\text{Sn}}$ **1013,** mp 220–222°C (HBr) (123)

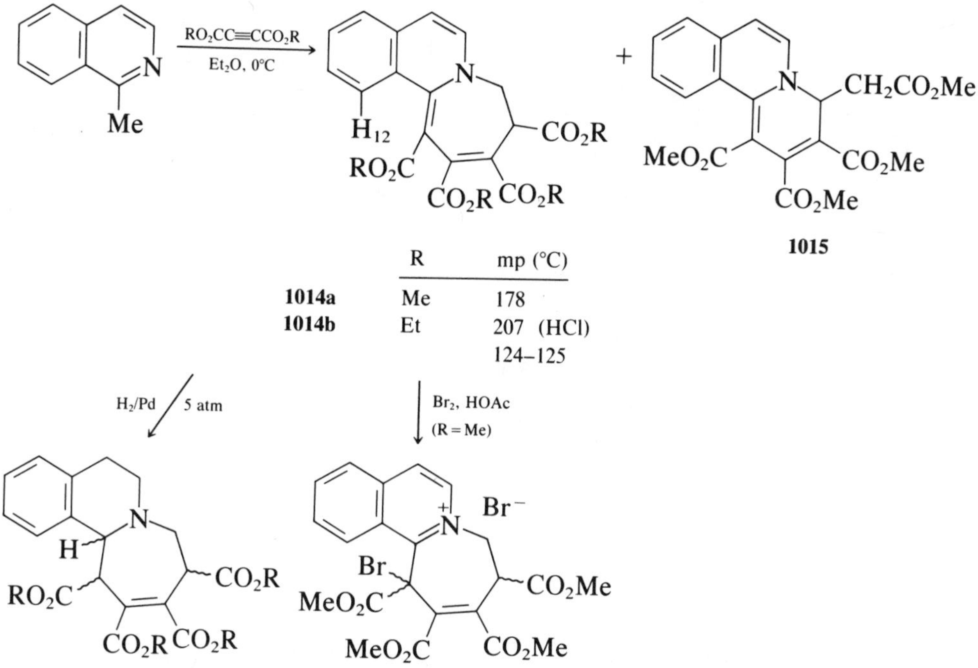

$RO_2CC\equiv CCO_2R$, Et_2O, 0°C

	R	mp (°C)
1014a	Me	178
1014b	Et	207 (HCl)
		124–125

1015

H_2/Pd, 5 atm

Br_2, HOAc (R = Me)

	R	mp (°C)
1016a	Me	200
1016b,c	Et	128, 154 (stereoisomers)

1017, mp 142°C

Figure 140

1016c was not determined. Addition of bromine to **1014a** gave a crystalline bromide for which the structure **1017** was suggested on the basis of a comparison of its ultraviolet and nmr spectra with those of isoquinoline model compounds. Treatment of **1014a** with hot nitric acid gave a good yield of mononitro product, mp 248°C, which could be reduced catalytically to an amine.

Reassignment of the structure of products **1014** may well be required in view of the recent correction of the structure of similar products from the reaction of 2-methylquinolines and acetylenedicarboxylic esters (see Section VI.24).

In a study of the competition between alkylation and rearrangement of Reissert anions, deprotonation of the isoquinoline **1018** with sodium hydride in DMF gave the 5-oxoazepino[2,1-*a*]isoquinoline-12b-carbonitrile **1019** (Eq. 124). The infrared spectrum (KBr) of **1019** showed bands at 2230 and 1673 cm^{-1}. On being heated in refluxing ethanol, this product generated smoothly the isoquinoline **1020** (312).

H C≡N N C(CH$_2$)$_4$Cl O

1018

NaH / DMF →

NC N O

1019, 72%; mp 91–92°C

(124)

EtOH / reflux →

N (CH$_2$)$_4$CO$_2$Et

1020, 68%

24. Azepino[1,2-*a*]quinolines

1 2 3 4 5 6 7 8 9 10 11 12 N 12a

Cyclization of the diester **1021** with potassium *t*-butoxide did not give the desired product **1023**, a potential intermediate to azasteroids, but produced instead the azepino[1,2-*a*]quinoline **1022** in good yield (313) (Eq. 125). The acyloin was obtained as a potassium salt. The free base was unstable but

Me CH$_2$CH$_2$CHCO$_2$Et N CO$_2$Et

1021

KOBu-*t*

Me OH OH N

1022, mp 96–97°C (HCl)

(125)

Me CO$_2$Et O N

1023

yielded an insoluble low-melting hydrochloride salt. Infrared bands were seen at 3350, 2500 (NH), and 1725 cm^{-1}.

All other chemistry reported for this ring system derives from investigations of products obtained from reactions of 2-substituted quinolines with acetylenedicarboxylic esters (Eq. 126). Jackman and co-workers (314) suggested the ylide structure **1025** for the "red adduct" isolated by Diels along with the benzo[*c*]quinolizine **1024.** However, Acheson and coworkers (26,311,315,316), on reinvestigation of this reaction with quinolines and related heterocyclic substrates, proposed azepino[1,2-*a*]quinoline structures **1026** and **1027** for these low-yield products. Their assignments of structure were based on careful interpretation of nmr spectra and rigorous chemical characterization. Although a plausible scheme was put forward to explain the formation of **1027,** the only route that could be devised to account for **1026** required the unlikely formation of a nonstabilized carbanion as an

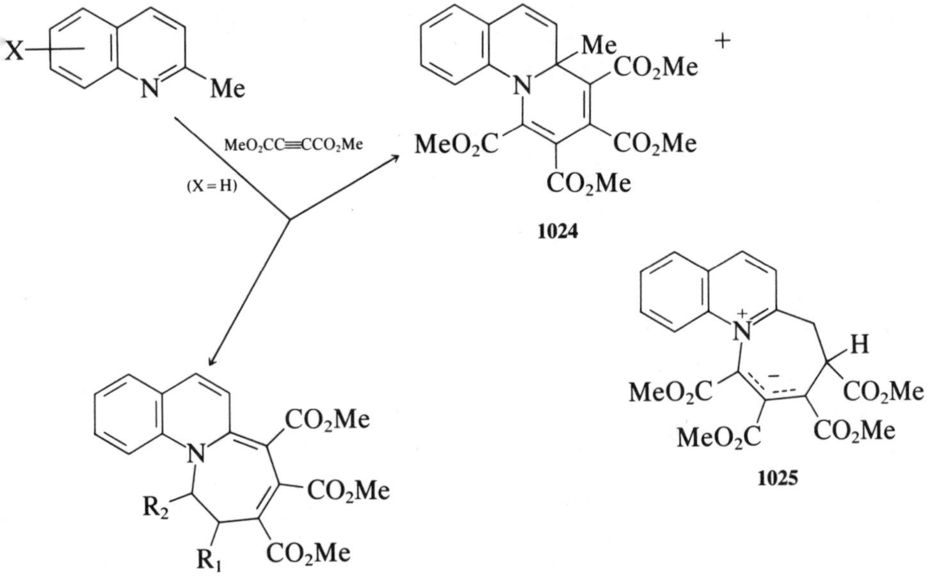

X— Me N $MeO_2CC{\equiv}CCO_2Me$ (X = H)

1024

1025

(126)

1026, R_1 = H, R_2 = CO_2Me
1027, R_1 = CO_2Me, R_2 = H

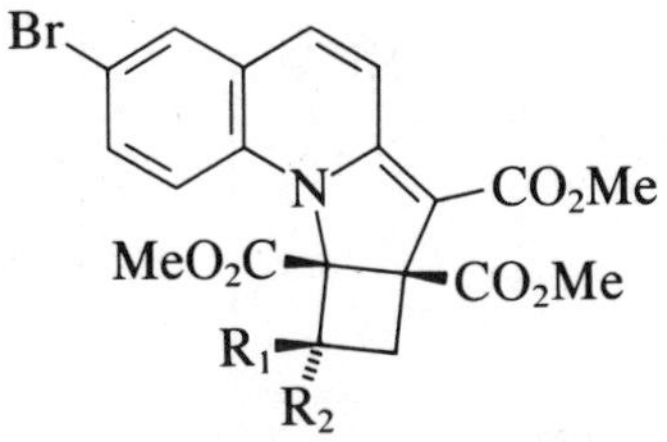

1028, R_1 = H, R_2 = CO_2Me
1029, R_1 = CO_2Me, R_2 = H

intermediate. X-ray crystallographic analysis of the product from 6-bromo-2-methylquinoline showed its structure to be **1028** (317,318). Thus the isomeric substance, whose ultraviolet and mass spectra were almost identical to those of **1028,** had to be **1029.** This finding necessitates a reappraisal of the structures of all products previously formulated as azepino[1,2-*a*]quinolines, since the physical data for these products are equally consistent with structures analogous to **1028** and **1029.** The ^{13}C-nmr spectra of these substances have been interpreted in accordance with these new assignments (288).

A review of the reactions of acetylenecarboxylic esters and nitrogen heterocycles by Acheson and Elmore (285), in which reassignments of structure are disclosed for several previously proposed azepine-containing ring systems, has recently appeared. The tricyclic systems affected, in addition to azepino[1,2-*a*]quinolines, are: azepino[1,2-*a*]quinazolines (Section VI.12) and azepino[1,2-*c*]quinazolines (Section VI.13), azepino[1,2-*a*]quinoxalines (Section VI.15), azepino[2,1-*b*]benzoxazoles (Section IV.4), azepino[1,2-*a*]-benzimidazoles (Section IV.20) and quite possibly azepino[2,1-*b*]benzothiazoles (Section IV.7) and azepino[2,1-*b*]benzoselenazoles (Section IV.13).

25. Azepino[2,3-*b*]quinolines

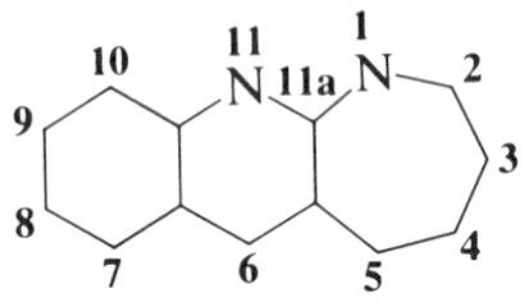

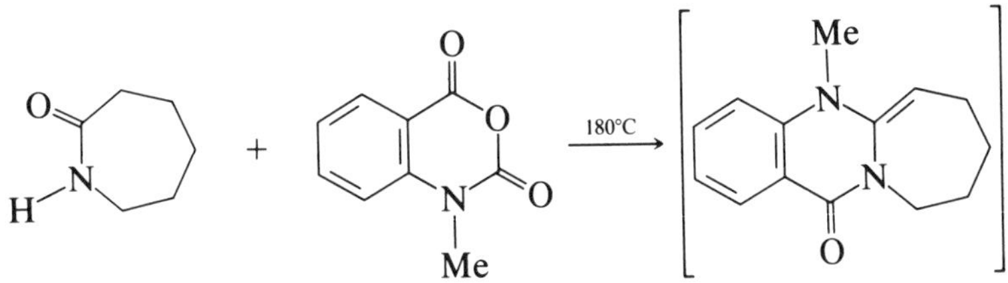

1030 **1031**

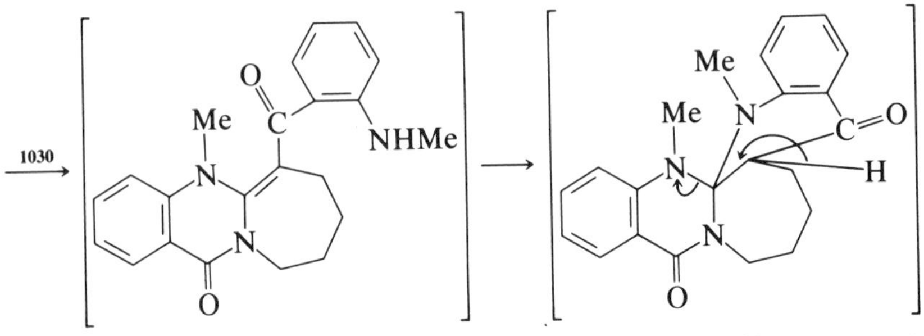

1032 **1033**

Figure 141

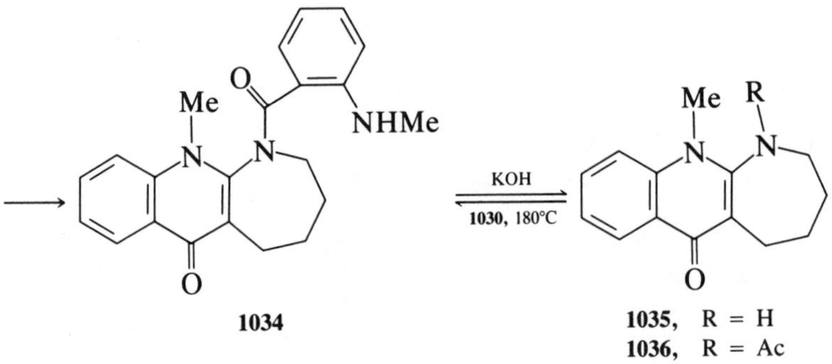

Figure 141 (*continued*)

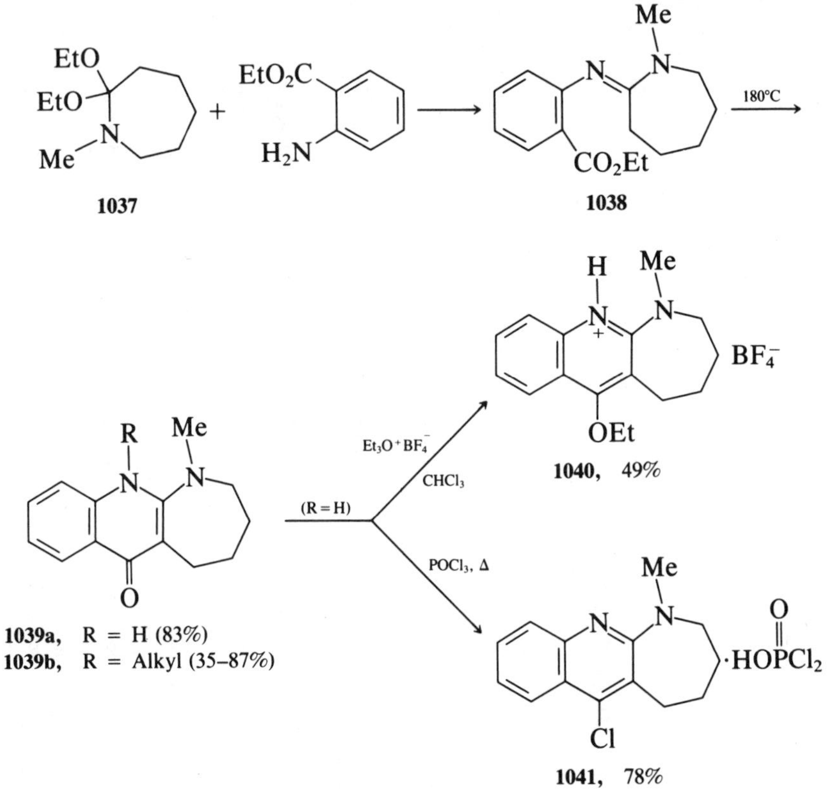

Figure 142

Russian workers have provided all the existing literature on azepino[2,3-*b*]quinolines (Fig. 141). Reaction of caprolactam with *N*-methylisatoic anhydride (**1030**) gave the azepino[2,1-*b*]quinazoline **1031** (299) (see Section VI.14 for the chemistry of this isomeric system). The product reacted immediately with a second equivalent of the anhydride **1030** to give **1032**, which underwent rearrangement via the azacyclol **1033** to give the azepino-

TABLE 56 AZEPINO[2,3-*b*]QUINOLINES

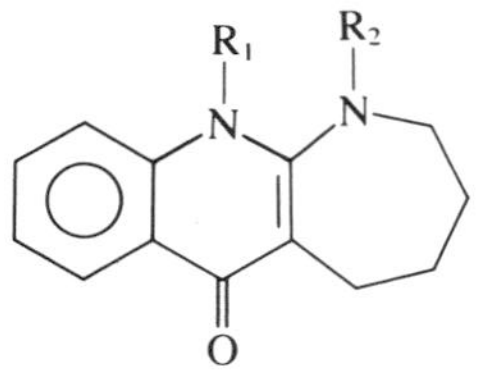

R_1	R_2	Yield (%)	mp (°C)	Spectra	Refs.
Me	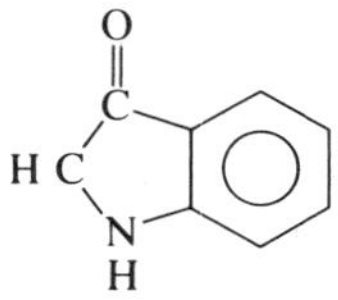 (3-oxo-indolin-2-ylidene: O=C, HC, NH)		178–180	ir	299
			257–259 (HCl)	ir	319
Me	H		248–249	ir	299
Me	Ac		142–144	ir	319
H	Me	83	178–179 (HBF_4)	ir,nmr,uv	320
			293–295	ir,nmr,uv	320
CH_2Ph	Me	60	152–153		321
$(CH_2)_3NMe_2$	Me	87	187–188		321
Me	Me	51	155–156		321
$(CH_2)_2NMe_2$	Me	60	174–176		321

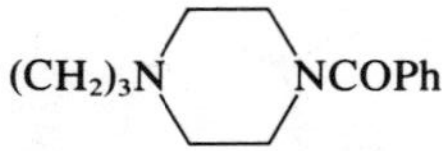	Me	35	163–165		321
$CH_2C_6H_4NO_2$-4	Me	13	164–165		321
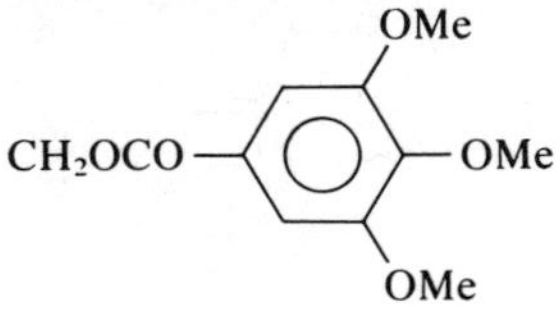	Me	45	156–157		321

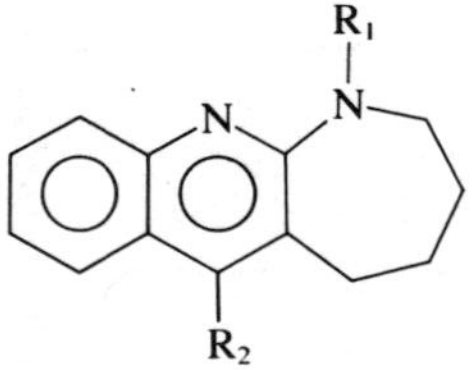

Me	EtO	49	195–197 (HBF_4)	ms	320
Me	Cl	78	131–133 (HO_2PCl_2) 173–175/3[a]		320

[a]Boiling point (°C/torr).

[2,3-*b*]quinoline **1034.** Hydrolysis of **1034** with aqueous KOH (319) gave **1035.** Reaction of **1035** with acetic anhydride gave the *N*-acetyl derivative **1036** (319).

A variant of this synthetic process is the reaction of the lactam ketal **1037** (Fig. 142) with ethyl anthranilate (320,321) to give the intermediate **1038,** which on thermal cyclization afforded a 44% yield of **1039a.** When a catalytic amount of *p*-toluenesulfonic acid was added, the yield increased to 83%. Compound **1039a** could be *O*-alkylated with Meerwein's reagent to give **1040,** and was converted on phosphoryl chloride treatment to the chloride **1041,** which was isolated as the dichlorophosphate salt. *N*-Alkylation of **1039a** with the aid of sodium hydride in DMF (322,333) provided a series of *N*-alkyl derivatives (**1039b;** see Table 56). These substances were evaluated as antidepressants but were essentially inactive in amphetamine interaction and antireserpine activity tests.

26. Azepino[3,2-*b*]quinolines

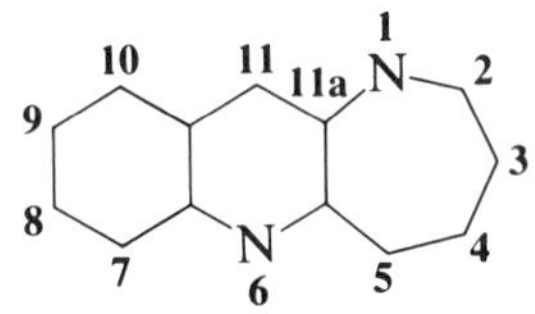
1
10
11
11a
N
2
9
3
8
N
4
7
6
5

Only a limited amount of information concerning the azepino[3,2-*b*]quinoline ring system is available. A communication (324) briefly describes the prep-

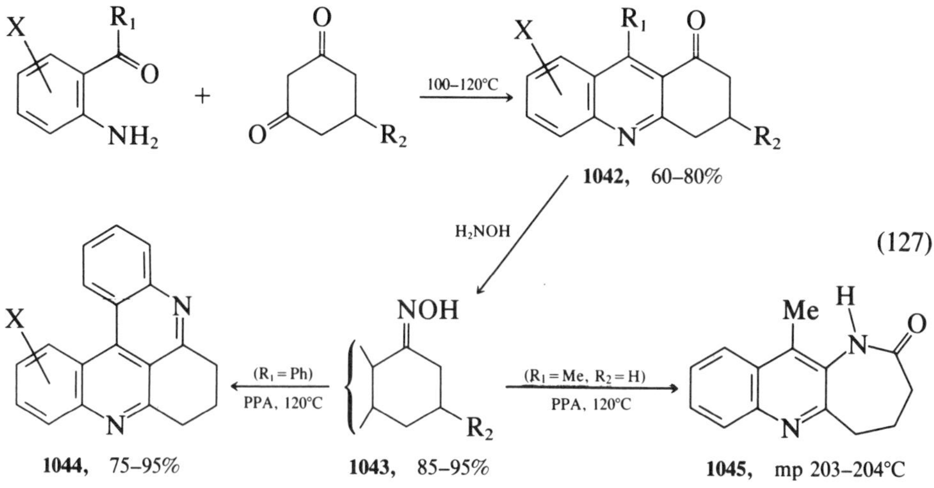
X
R₁
O
O
+
NH₂
O
R₂
100–120°C
X
R₁
O
N
R₂
1042, 60–80%
H₂NOH
X
N
N
1044, 75–95%
(R₁ = Ph)
PPA, 120°C
NOH
R₂
1043, 85–95%
(R₁ = Me, R₂ = H)
PPA, 120°C
Me
H
N
O
N
1045, mp 203–204°C

(127)

aration of a series of tetrahydroacridinones (**1042**) by condensation of *o*-aminobenzophenones or acetophenones with 1,3-cyclohexanediones (Eq. 127). The oximes (**1043**) of **1042,** when heated to 120°C in polyphosphoric acid (PPA), gave the pentacyclic products **1044** in good yield. However, in the single example described where $R_1 = CH_3$, PPA treatment gave **1045** in 85% yield via a Beckmann-type rearrangement.

A Russian reference (325) reported the formation of **1047** from **1046** (Eq. 128) by a Schmidt reaction, but details are not available.

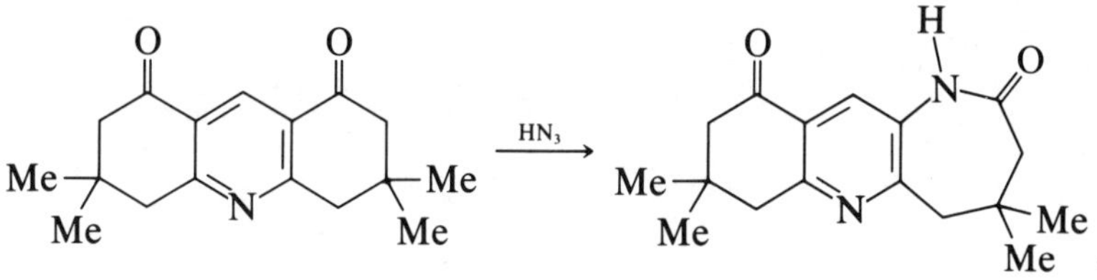

$$\mathbf{1046} \xrightarrow{HN_3} \mathbf{1047} \qquad (128)$$

27. Azepino[4,3-*b*]quinolines

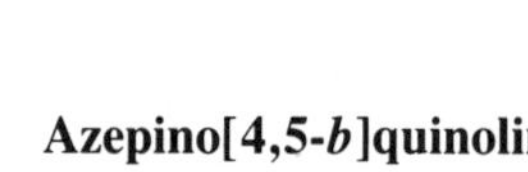

28. Azepino[4,5-*b*]quinolines

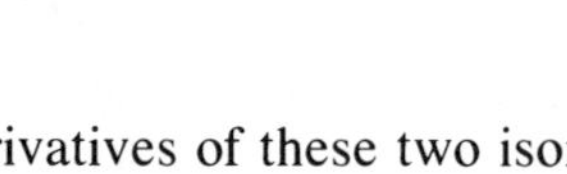

A large number of derivatives of these two isomeric systems have been patented (326,327) as anorexigenic agents and are listed in Tables 57–60. Two synthetic methods that provide both isomers were described. Condensation of **1048** (R_1 = alkyl, aryl, OH) with tetrahydroazepin-4-ones **1049** (R_2 = alkyl, acyl) gave a 4:6 mixture of the [4,5-*b*] and [4,3-*b*] isomers **1050** and **1051,** respectively (Eq. 129). Schmidt ring expansion of the ketal **1052**

TABLE 57. AZEPINO[4,5-*b*]QUINOLINES AND AZEPINO[4,3-*b*]QUINOLINES

[4,5-*b*] [4,3-*b*]

R_1	R_2	R_3	R_4	R_5	R_6	[4,5-*b*] Yield (%)	[4,5-*b*] mp (°C)[a]	[4,3-*b*] Yield (%)	[4,3-*b*] mp (°C)[a]
H	H	H	H	H	Me	30	(103) 296	33	(123) 284
H	H	H	H	H	H	10	249	24	270
H	H	H	H	H	Et	22	270	28	252
H	H	H	H	H	*n*-Pr	19	264	19	282
H	H	H	H	H	*i*-Pr	15	265	18	270
H	H	H	H	H	*sec*-Bu	12	280	15	255
H	H	H	H	H	*i*-Bu	8	264	13	298
H	H	H	H	H	*c*-C_6H_{11}	6	225	9	286
H	H	H	H	H	Ph	31	276	31	335
H	H	H	H	H	CO_2H	6	(233)	10	(269)
H	H	H	H	H	Cl[b]			4	(127) 260
CH_2Ph	H	H	H	H	Me	24.5	(110) 255	36.8	(126) 244

Me	H	H	H	H	Me	19	282	39	280
Et	H	H	H	H	Me	8	274	10	278
n-Pr	H	H	H	H	Me	10	246	18	232
i-Pr	H	H	H	H	Me	33	281	60	273
n-Bu	H	H	H	H	Me	8	251	12	245
i-Bu	H	H	H	H	Me	12	242	8	235
sec-Bu	H	H	H	H	Me	30	260	56	280
tert-Bu	H	H	H	H	Me	11	301	10	270
CH_2Ph	H	H	H	H	H	10	252	21	(134)
$CH_2C_6H_4Me$-4	H	H	H	H	H	7	263	19	209
Et	H	H	H	H	Ph	28	280	50	273
CH_2Ph	H	H	H	H	Ph	26	250	49	248
Et	H	H	H	H	*c*-C_6H_{11}	15	(104)	45	255
Et	H	H	H	H	Cl[b]	16	233–236	45	267
CH_2Ph	H	H	H	H	Cl	31	220	33	(148)
Et	H	H	H	H	OH	13	295	6	294
CH_2Ph	H	H	H	H	OH	10.4	274	20	258
H	H	H	H	H	Me	43	296	22	(122)
H	H	H	Br	H	Me	21	313	22	307
H	H	H	Cl	H	Me	19	300	30	288
H	H	H	NO_2	H	Me	30	308	15	308
H	H	NO_2	H	H	Me	20	284	22	300
H	NO_2	H	H	H	Me	15	(130)	18	(177)
H	H	H	H	NO_2	Me	18	190	19	210
Me	H	H	OH	H	Me	10	230	20	250
Me	OH	H	H	H	Me	19.5	300	23.5	317
CH_2Ph	H	H	OH	H	Me	10	278	20	210
H	H	H	MeO	H	Me	10	284	29	274
H	MeO	H	H	H	Me	19	260	35	224

(continued)

TABLE 57. *(Continued)*

						[4,5-*b*]		[4,3-*b*]	
R_1	R_2	R_3	R_4	R_5	R_6	Yield (%)	mp (°C)[a]	Yield (%)	mp (°C)[a]
CO_2Et	H	H	H	H	Me	43.5	(124) 247		
CO_2Et	H	OH	H	H	Me	12	263	8	254
H	H	H	Me	H	Me	27	(68–70) 300	42	(150) 293–295
H	Ph	H	Me	H	Me	10	239	10	196
$CH_2CH=CH_2$	H	H	H	H	Me	3	278	3	270
H	H	Me	H	H	Me	17	290–292	32	305–307
H	H	Cl	H	H	Me	24	287	16	3O4
H	H	H	F	H	Me	18	278	24	285
CO_2Et	H	MeO	MeO	H	Me	15	246	22	202
CO_2Et	H	OH	MeO	H	Me	5	325	4	294
CO_2Et	H	—OCH_2O—		H	Me	8	300	10	>300
H	H	CF_3	H	H	Me	38.2	275	38.2	270
H	H	H	CF_3	H	Me	24	283	23	274
H	CF_3	H	H	H	Me	28	(78)	27	(94)
Et	H	H	Cl	H	Ph	26	280	50	273
CH_2Ph	H	H	H	H	NH_2[c]	13	138	13	149
H	H	H	H	H	Me[d]	36	250		

[a]Mp's are given for the dihydrochloride. Mp's of free bases or acids are given in parentheses.
[b]Obtained from anthranilic acid on reaction with phosphoryl chloride.
[c]Obtained from *o*-aminobenzonitrile.
[d]Methyl group at position 5 also. Obtained from 3-methylhexahydroazepin-4-one.

TABLE 58. AZEPINO[4,5-*b*]QUINOLINES[a]

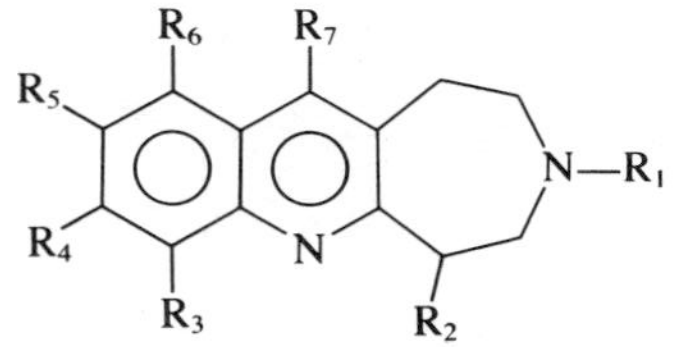

R_1	R_2	R_3	R_4	R_5	R_6	R_7	Yield (%)	mp (°C)
H	H	H	H	H	Cl	Me	30	285 (2HCl)
H	H	Cl	H	H	H	Me	18	245 (2HCl)
Ac	H	NO_2	H	H	H	Me	34.6	187 (HCl)
CO_2Et	H	NO_2	H	H	H	Me	36	175 (HCl)
Ac	H	H	H	H	NO_2	Me	18.2	155 (HCl1)
CO_2Et	H	H	H	H	NO_2	Me	20	149 (HCl)
Ac	H	H	NO_2	H	H	Me	10.7	173 (HCl)
CO_2Et	H	H	NO_2	H	H	Me	8	168 (HCl)
H	H	H	H	NH_2	H	Me		320 (3HCl)
Ac	H	NH_2	H	H	H	Me	73	129
H	H	NH_2	H	H	H	Me	83	285 (2HCl)
Ac	H	H	H	H	NH_2	Me	70	150
H	H	H	H	H	NH_2	Me	13	300 (2HCl)
CO_2Ph	H	H	H	H	H	Me	63	150
CO_2Et	H	H	H	H	H	Me	10	125
								247 (HCl)
CO_2Me	H	H	H	H	H	Me	91	245 (HCl)
CO_2Pr-*n*	H	H	H	H	H	Me	76	243–245 (HCl)
CO_2Pr-*i*	H	H	H	H	H	Me	83	227 (HCl)
CO_2Bu-*n*	H	H	H	H	H	Me	77	203 (HCl)

(continued)

TABLE 58. *(Continued)*

R_1	R_2	R_3	R_4	R_5	R_6	R_7	Yield (%)	mp (°C)
CO_2Bu-*i*	H	H	H	H	H	Me	83	244 (HCl)
CO_2Bu-*sec*	H	H	H	H	H	Me	87	212 (HCl)
$CO_2C_6H_{11}$-*c*	H	H	H	H	H	Me	67	233 (HCl)
CO_2Ph	H	H	H	H	H	Me	71	246 (HCl)
CO_2Bz	H	H	H	H	H	Me	79	199 (HCl)
CO_2Et	H	H	H	H	H	H	64	264 (HCl)
CO_2Et	H	H	H	H	H	Et	72	212 (HCl)
CO_2Et	H	H	H	H	H	*n*-Pr	84	201 (HCl)
CO_2Et	H	H	H	Cl	H	Ph	65	238 (HCl)
CO_2Et	H	Cl	H	H	H	Me	70	169 (HCl)
CO_2Et	H	OH	H	H	H	Me	22	129 (HCl)
CO_2Et	H	MeO	H	H	H	Me	49	128 (HCl)
CO_2Et	H	NO_2	H	H	H	Me	70	242 (HCl)
CO_2Et	H	NH_2	H	H	H	Me	38	168 (HCl)
CO_2Et	H	H	H	Br	H	Me	62	160 (HCl)
CO_2Et	H	H	H	Cl	H	Me	65	238 (HCl)
CO_2Et	H	H	H	NO_2	H	Me	51	234 (HCl)
CO_2Et	H	H	H	MeO	H	Me	66	142 (HCl)
CO_2Et	H	H	H	OH	H	Me	16	284 (HCl)
CO_2Et	Cl	H	H	H	H	Cl	84	184 (HCl)
CO_2Et	OCO_2Et	H	H	H	H	Me	73	128 (HCl)
CO_2Et	OH	H	H	H	H	Me	70	152 (HCl)
CO_2Et	H	H	H	Cl	H	Ph	88	216 (HCl)
CO_2Et	H	H	H	H	NO_2	Me	78	200 (HCl)
CO_2Et	H	H	H	H	NH_2	Me	76	182 4HCl)
CO_2Ph	H	H	H	H	H	Cl	35.7	157 (HCl)
CO_2Et	H	H	H	H	H	Cl	97	104 (HCl)
CO_2Ph	H	H	H	H	H	OH	44	232 (HCl)

CO_2Et	H	H	H	H	H	OH	20	252 (HCl)
Ac	H	H	H	H	H	Me	60	210 (HCl)
Ac	H	H	H	H	H	Ph	90	210 (HCl)
COC_6H_4Cl-4	H	H	H	H	H	Me	65	163 (HCl)
CO_2Et	OCO_2Et	H	H	H	H	Me	63	128 (HCl)
$HOCH_2CH_2$	H	H	H	H	H	Me	33	252 (HCl)
$HOCHCH_2$ CH_3	H	H	H	H	H	Me	58	273 (HCl)
CH_2CH_2OMe	H	H	H	H	H	Me	87	250 (HCl)
n-Hexyl	H	H	H	H	H	Me	83	240 (HCl)
n-Bu	H	H	H	H	H	Me	85	278 (HCl)
CO_2Et	H	H	H	CN	H	Me	35	217 (HCl)
CO_2Et	H	CN	H	H	H	Me	42	208 (HCl)
CO_2Et	H	H	H	Ac	H	Me	20	155 (HCl)
H	H	H	OH	OH	H	Me	90	>330
Ac	H	H	H	CN	H	Me	28	245 (HCl)
H	H	H	H	CO_2H	H	Me	75	300
H	H	H	H	CO_2Et	H	Me	90	250 (HCl)
CO_2Et	H	OH	Br	H	Br	Me	36	179 (HCl)
CO_2Et	H	H	H	NH_2	H	Me	95	214–216 (HCl)
CO_2Et	H	H	H	Me	H	Me	85	212 (HCl)
CO_2Et	H	H	Cl	H	H	Me	70	230 (HCl)
CO_2Et	H	H	H	H	OH	Me	29	303 (HCl)
CO_2Et	H	H	H	H	Cl	Me	37	138
CO_2Et	H	H	H	H	H	CO_2H	57	247
CO_2Et	H	H	H	H	H	CO_2Et	10	94
CO_2Et	H	H	H	H	H	CH_2OH	12	158
CO_2Et	OH	H	H	H	H	CO_2H	1	>300
CO_2Et	OH	H	H	H	H	CH_2OH	35	171 (HCl)
CO_2Et	Cl	H	H	H	H	Me	52	131
CO_2Et	MeO	H	H	H	H	Me	62	Oil

(continued)

TABLE 58. (*Continued*)

R_1	R_2	R_3	R_4	R_5	R_6	R_7	Yield (%)	mp (°C)
CO_2Et	NH_2	H	H	H	H	Me	10	236 (2HCl)
CO_2Et	NMe_2	H	H	H	H	Me	77	173 (2HCl)
CO_2Et	morpholino (N–O ring)	H	H	H	H	Me	51	222 (2HCl)
n-PrCO	H	H	H	H	H	Me	67	130 (HCl)
EtCO	H	H	H	H	H	Me	72	128–130 (HCl)
i-BuCO	H	H	H	H	H	Me	50	107 (HCl)
Caproyl	H	H	H	H	H	Me	82	109 (HCl)
Lauroyl	H	H	H	H	H	Me	86	100 (HCl)
$MeOCH_2CO$	H	H	H	H	H	Me	79	116 (HCl)
CF_3CO	H	H	H	H	H	Me	66	147 (HCl)
CH_3SO_2	H	H	H	H	H	Me	73	265 (HCl)
Ts	H	H	H	H	H	Me	50	236 (HCl)
$ClC_6H_4SO_2$	H	H	H	H	H	Me	75	189 (HCl)
CH_2CO_2Et	H	H	H	H	H	Me	61	225 (2HCl)
CH_2CO_2H	H	H	H	H	H	Me	72	287 (2HCl)
$CH_2CH_2CO_2Et$	H	H	H	H	H	Me	72.5	190 (2HCl)
$CH_2CH_2CO_2H$	H	H	H	H	H	Me	76	270 (2HCl)
CH_2CH_2CON (morpholino N–O ring)	H	H	H	H	H	Me	75	189 (HCl)
$CH_2CH_2CONMe_2$	H	H	H	H	H	Me	62	264 (2HCl)
CH_2CN	H	H	H	H	H	Me	25	250 (2HCl)
CH_2CH_2CN	H	H	H	H	H	Me	86	266 (2HCl)
CH_2CONMe_2	H	H	H	H	H	Me	68	261 (2HCl)

CH_2CON O	H	H	H	H	H	Me	59	253 (2HCl)
CO_2Et	H	H	CF_3	H	H	Me	98	142 (HCl)
CO_2Et	H	H	H	CF_3	H	Me	90	152 (HCl)
CO_2Et	H	H	H	H	H	PhO	58.2	112
CO_2Et	H	H	H	H	H	N	18	111
COPh	H	H	H	H	H	MeO	56.6	Oil
CO_2Et	H	H	H	H	H	MeO	62	122
H	H	H	H	H	H	MeO	7	112
COPh	H	H	H	H	H	OCH_2CH_2Ph	73	Oil
CO_2Et	H	H	H	H	H	OCH_2CH_2Ph	51	153 (HCl)
Et	H	H	H	H	H	$OCH(CH_3)CO_2Et$	22.4	20
CO_2Et	H	H	H	H	H	$OCH(CH_3)CO_2Et$	39.5	133 (HCl)
CO_2Et	H	H	Me	H	H	Me	79	121
H	H	H	H	H	OH	Me	72	>300
H	H	H	OH	H	H	Me	34	245
CO_2Et	H	CF_3	H	H	H	Me	92	136
$(C{=}NH)NH_2$	H	H	H	H	H	Me	10	140
$CONH_2$	H	H	H	H	H	Me	34	215
$CSNH_2$	H	H	H	H	H	Me	34	215
H	H[b]	H	H	H	H	Me	95	153
CO_2Et	H[b]	H	H	H	H	Me	42	218 (HCl)

[a]Ref. 326.
[b]Double bond located between C_3 and C_4.

TABLE 59. AZEPINO[4,3-*b*]QUINOLINES

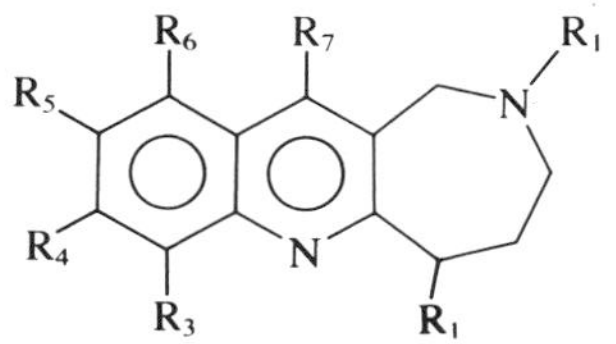

R_1	R_2	R_3	R_4	R_5	R_6	R_7	Yield (%)	mp (°C)
CO_2Et	H	H	H	H	H	Me	58	219 (HCl)
CO_2Et	H	H	H	H	H	H	81	214 (HCl)
CO_2Et	OCO_2Et	H	H	H	H	H	80	<40
CO_2Et	H	H	H	Cl	H	Ph	89	186 (HCl)
CO_2Ph	H	H	H	H	H	Cl	18.5	137 (HCl)
CO_2Ph	H	H	H	H	H	OH	51	299 (HCl)
$CO_2C_6H_4Cl$-4	H	H	H	H	H	Me	75	275 (HCl)
Ac	OH	H	H	H	H	Me	79	150 (HCl)
Ac	H	H	H	H	H	Ph	92	275 (HCl)
Ac	H	H	H	H	H	Cl	27	235 (HCl)
CO_2Et	OCO_2Et	H	H	H	H	Me	63	128 (HCl)

Ac	AcO	H	H	H	H	Me	70	40
H	OH	H	H	H	H	Me	70	240 (2HCl)
CH_2CH_2OH	H	H	H	H	H	Me	24	268 (HCl)
$CH_2CH(CH_3)OH$	H	H	H	H	H	Me	28	268 (HCl)
Ac	H	H	H	CN	H	Me	36	230 (HCl)
H	H	H	H	CO_2H	H	Me	80	305
Ac	H	H	H	OH	Br	Me	62	190
H	H	H	H	OH	Br	Me	80	> 300 (HCl)
H	H	OH	H	H	Br	Me	22	> 300 (2HCl)
H	H	H	H	NH_2	Br	Me	60	312 (2HCl)
CO_2Et	H	H	Me	H	H	Me	42	132 (HCl)
Ac	Cl	H	H	H	H	Me	77	Oil
Ac	morpholino (N-linked)	H	H	H	H	Me	13	179
Et	H	H	H	H	H	PhO	31	246 (2HCl)
CO_2Et	H	H	H	H	H	OH	9	285

[a]Ref. 326.

TABLE 60. 6-*N*-OXIDES OF AZEPINO[4,5-*b*] AND AZEPINO[4,3-*b*]QUINOLINES[a]

[4,5-*b*] [4,3-*b*]

Ring	R_1	R_2	R_3	Yield (%)	mp (°C)
[4,5-*b*]	CO_2Et	H	Me	66	164 (HCl)
	Ac	H	Me	83.5	128 (HCl)
	H	H	Me	82	249 (HCl)
	CO_2Et	H	CO_2H	3.7	202 (HCl)
	CO_2Et	OAc	Me	23	130 (HCl)
	CO_2Et	OH	Me	17	199 (HCl)
	CO_2Et	H	CH_2OH	21	207
	H	H	CO_2H	25	240 (2HCl)
	COPh	H	MeO	15	135
	CO_2Et	H	Cl	29	170
[4,3-*b*]	Ac	H	Me	83	168 (HCl)
	CO_2Et	H	Me	60	129 (HCl)
	H	H	Me	91	236 (HCl)

[a]Ref. 326.

1048 **1049**

(129)

1050, [4,5-*b*] **1051,** [4,3-*b*]

gave a mixture of **1053** and **1054,** with the [4,5-*b*] isomer predominating (Eq. 130). Reduction of the lactams gave the amines **1055** and **1056.**

The products of the azepinone condensation, **1050** and **1051,** could be separated by column chromatography or fractional crystallization. Each isomer was then subjected to various standard reactions (e.g., *N*-acylation, *N*-oxide formation, nitration, reduction, and Sandmeyer replacements), as indicated in Tables 58–60.

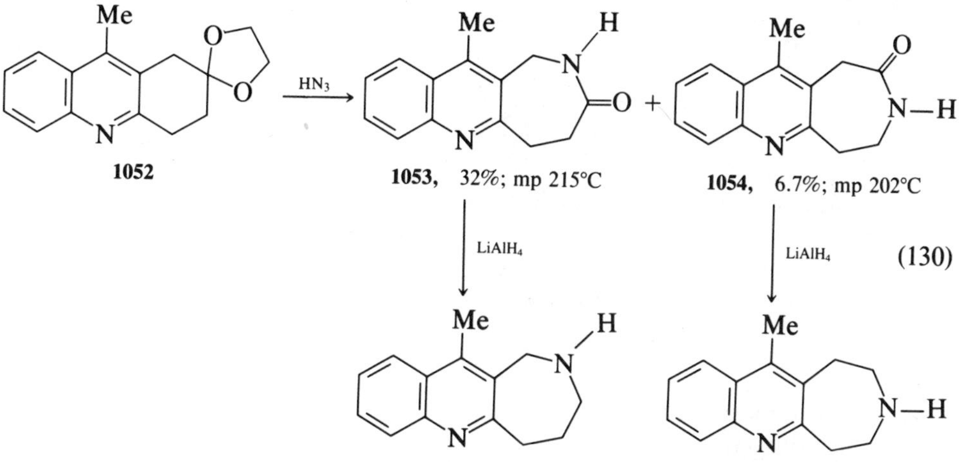
Me
O
O
N
1052
HN3
Me
H
N
=O +
N
1053, 32%; mp 215°C
Me
O
N—H
N
1054, 6.7%; mp 202°C
LiAlH4
LiAlH4
Me
H
N
N
1055, 12%
Me
N—H
N
1056, 10%

(130)

29. Pyrido[1,2-*b*][2]benzazepines

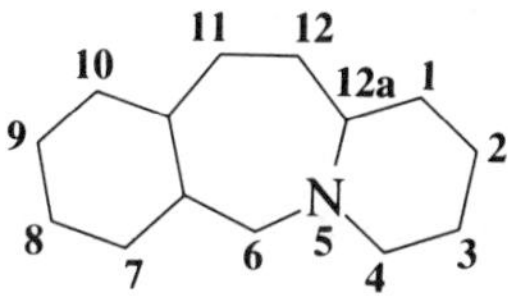
11 12
10
12a 1
9
2
N
8
7
6 5
4
3

An example of the pyrido[1,2-*b*][2]benzazepine system was prepared initially in the course of electrolytic reduction studies on α-amino ketones. The tricyclic ketone **1008** (Fig. 139; see Section VI. 22) gave two new products on electrolytic reduction using the lead–H_2SO_4 cathode (219). The products were identified as **1009** (29%) and **1010** (47%). Clemmensen reduction of **1008** likewise gave **1009** (47%) via a rearrangement process frequently observed in this reaction.

In an extension of similar chemistry in the pyrido[2,1-*b*][3]benzazepine system (see Section VI. 31), the 2-acetonylpyridines **1057** were quaternized with *m*-methoxybenzyl bromide (Eq. 131), and the resulting salts **1058** were cyclized to the 6H-pyrido[1,2-*b*][2]benzazepinium perchlorates **1059** (329). Oxidation of the 10-methyl salt **1059** (R = Me) with potassium permanganate

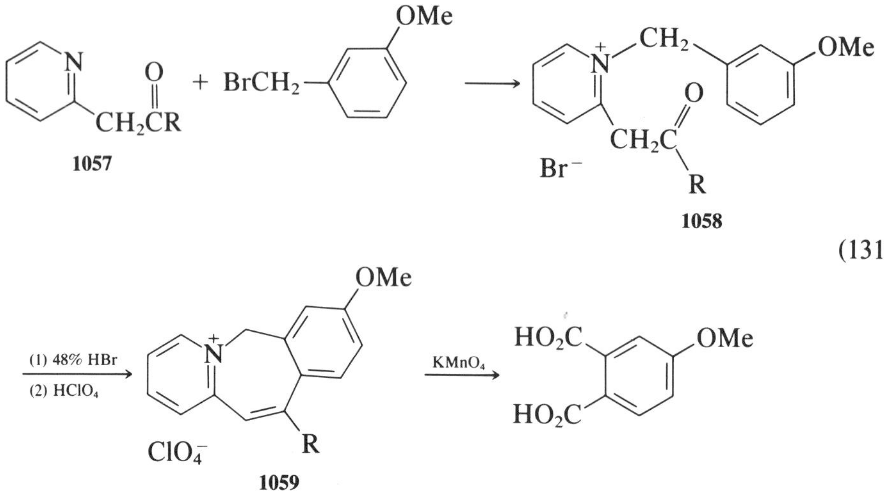

(131)

gave 4-methoxyphthalic acid. As can be seen in Table 61, an activating methoxy group is necessary for cyclization to occur. In general, cyclization to **1059** appears to be more difficult than the formation of the isomeric [2,1-*b*]-[3] system, which occurs in 75% yield from 1-acetonyl-2-benzylpyridium bromide.

A single additional example of this system was reported in a patent claiming major tranquilizer activity (330,331). Reaction of the dilithio salt (**1061**) of **1060** with 2-benzoylpyridine (Eq. 132) gave **1062,** which, on catalytic reduction followed by heating, cyclized to the lactam **1063.**

TABLE 61. 6H–PYRIDO[1,2-*b*][2]BENZAZEPINIUM PERCHLORATES[a]

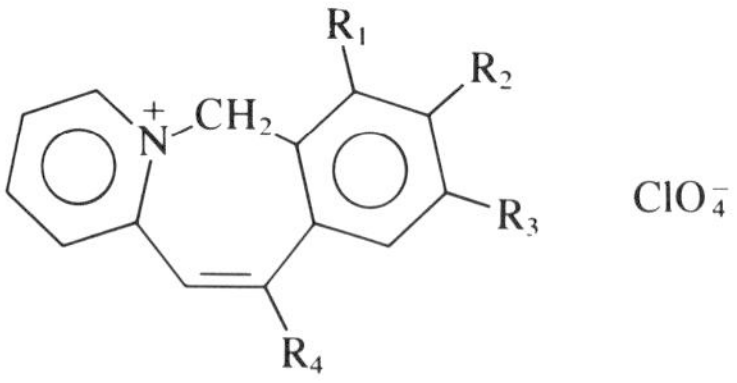

R_1	R_2	R_3	R_4	Reaction Time (hr)	Yield (%)	Spectra
H	H	H	Me	117	0	—
H	MeO	H	Me	24	33	uv
MeO	MeO	H	Me	2	18	uv
H	—OCH₂O—		Me	1.5	19	uv
H	MeO	H	Ph	2.6	8	uv

[a]Ref. 329.

1060 → (n-BuLi (2 eq.)) → [1061] + 2-benzoylpyridine

→ 1062 → ((1) H_2/Pt, HOAc; (2) Δ, o-$C_6H_4Cl_2$, 18 hr) → 1063, mp 184.5–188°C (132)

30. Pyrido[2,1-*a*][2]benzazepines

The major alkaloid astrocasine, isolated from an *Astrocasia phyllanthroides* species, was reported to have structure **1064** on the basis of degradation studies and spectral data (**1046**). This compound and its degradation products (Fig. 143) constitute the only reported examples of the pyrido-[2,1-*a*][2]benzazepine ring system. Astrocasine could be reduced catalytically to dihydroastrocasine (**1066**) or, with lithium aluminum hydride, to the diamine **1065.** Catalytic hydrogenation of **1065** or chemical reduction ($LiAlH_4$) of **1066** yielded the same saturated diamine **1067**. Hofmann elimination of dihydroastrocasine produced the olefin **1068.** Nmr spectra of **1064** and **1065** were consistent with a disubstituted double bond. The loss of four carbons and the NMe_2 moiety during conversion of **1068** to **1069** could be attributed to the presence of either an α-*N*-methylpiperidine or an α-*N*-methylpyrrolidylmethyl moiety in **1066.** However, the appearance of an intense ion with *m/e* 98.0968 in the mass spectra of both **1064** and **1065** was ascribed unequivocally to the *N*-methylpiperidine cation.

The subsequent isolation of astrophylline from the same plant and its characterization as *N*-*cis*-cinnamoyl-3(S)-[2′(R)-piperidyl]piperidine (**1070**)

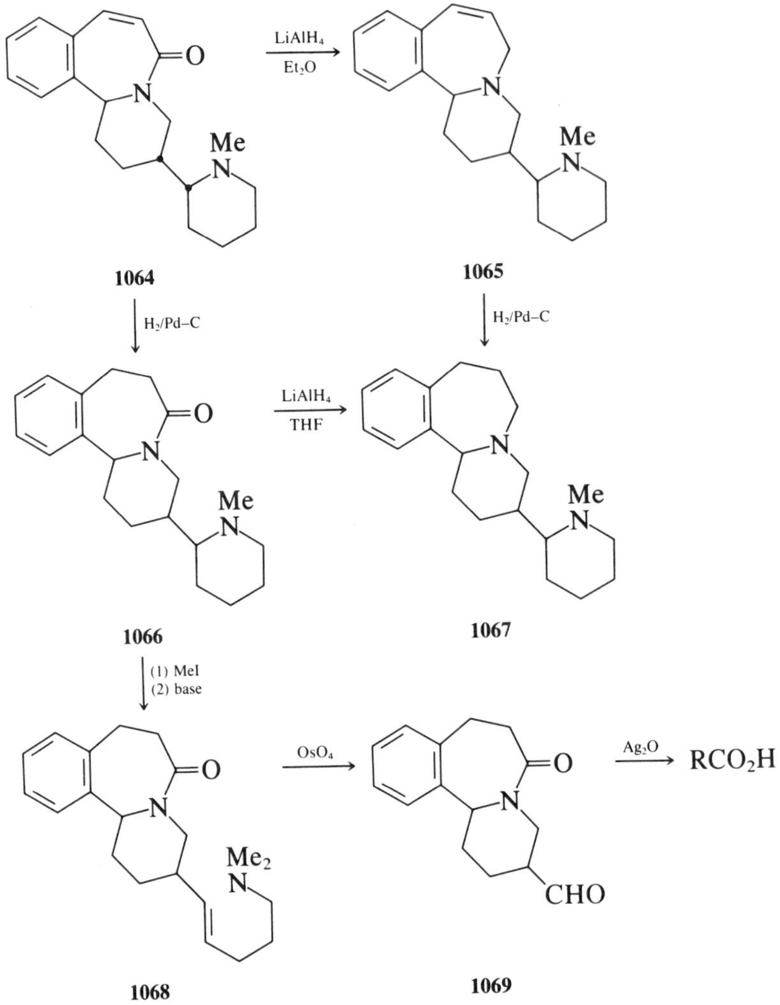

LiAlH$_4$ / Et$_2$O

1064 → 1065

H$_2$/Pd–C

LiAlH$_4$ / THF

1066 → 1067

(1) MeI
(2) base

OsO$_4$

Ag$_2$O → RCO$_2$H

1068 → 1069

Figure 143

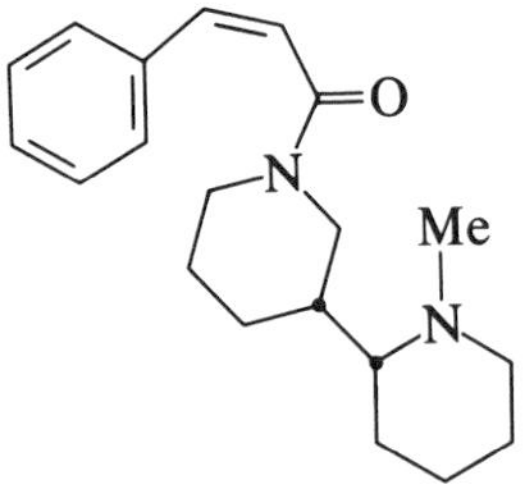

1070

TABLE 62. ASTROCASINE AND DEGRADATION PRODUCTS[a]

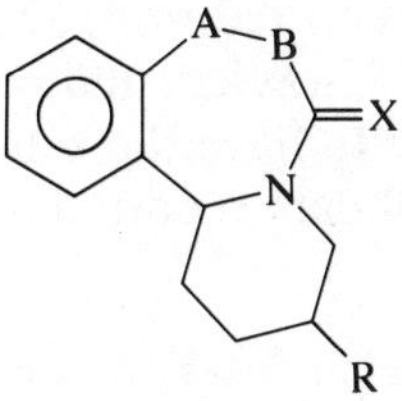

A—B	R	X	mp (°C)	Salt	Spectra
CH=CH	Me–N piperidin-2-yl 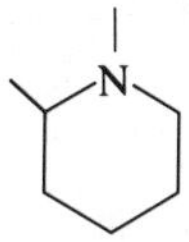	O	171–172		$[\alpha]_D^{24} = -270°$ (EtOH)
CH=CH	Me–N piperidin-2-yl 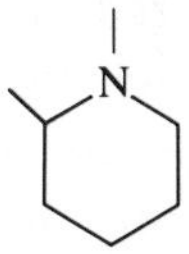	H_2	220 (dec)	$2HCl \cdot H_2O$	uv, nmr
CH_2CH_2	Me–N piperidin-2-yl 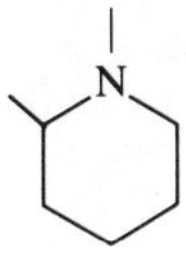	O	221–223	2HCl	
CH_2CH_2	Me–N piperidin-2-yl 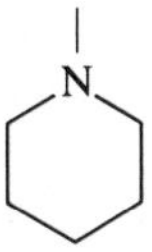	O	121–122 278–280	 MeI	$[\alpha]_D^{24} = +235°$ uv, ir
CH_2CH_2	Me–N piperidin-2-yl 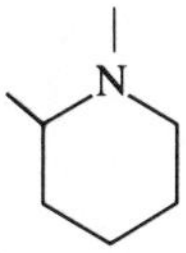	H_2	278–280 270	MeI 2MeI	 $[\alpha]_D^{24} = +82$ nmr, ir
CH_2CH_2	Me–N tetrahydropyridin-2-yl 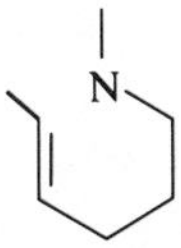	O			
Ch_2CH_2	CHO	O	190	2,4-DNP	ir
CH_2CH_2	CO_2H	O	224–225		ir

[a]Ref. 332.

provided further support for the structure of **1064**. Astrophylline was the first *cis*-cinnamoyl alkaloid reported in nature, and astrocasine appears to be biogenetically related. An X-ray crystallographic analysis (**1047**) showed astrocasine and astrophylline to have the same absolute configuration.

Table 62 provides physical data reported for the compounds in Fig. 143.

31. Pyrido[2,1-*b*][3]benzazepines

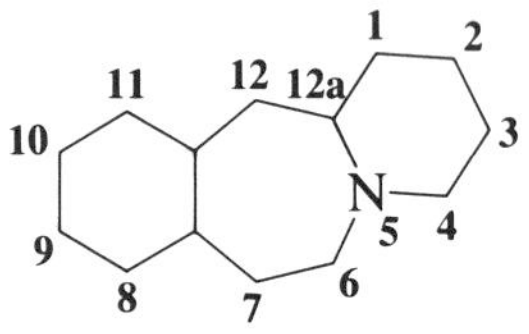

Bradsher and Moser (332) reported the acid-catalyzed cyclization of 1-acetonyl-2-benzylpyridinium salts such as **1072** to 7-substituted 12H-pyr-

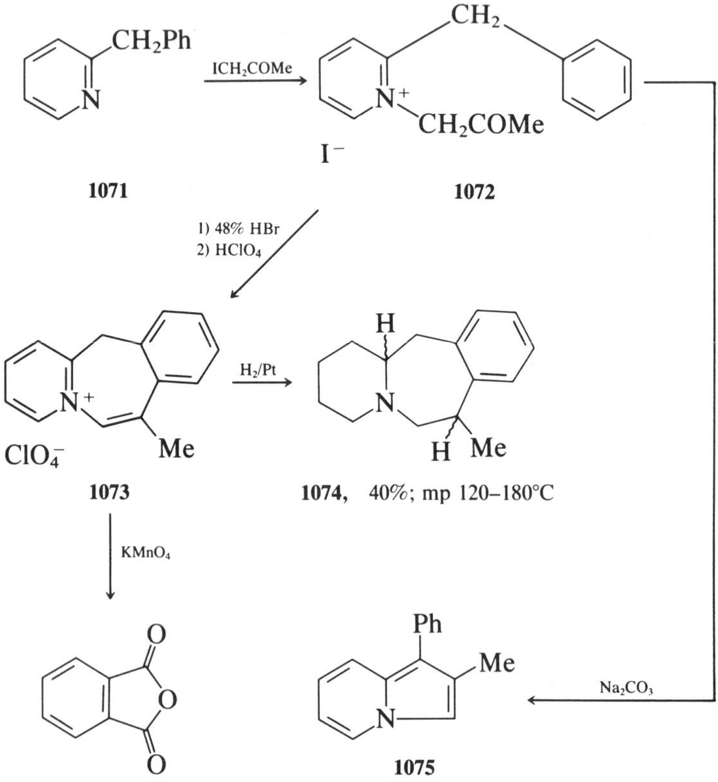

1071 → (ICH_2COMe) → **1072** (I^-)

1072 → (1) 48% HBr; 2) $HClO_4$) → **1073** (ClO_4^-)

1073 → (H_2/Pt) → **1074**, 40%; mp 120–180°C

1073 → ($KMnO_4$)

1072 → (Na_2CO_3) → **1075**

Figure 144

TABLE 63. 12H-PYRIDO[2,1-*b*][3]BENZAZEPINES

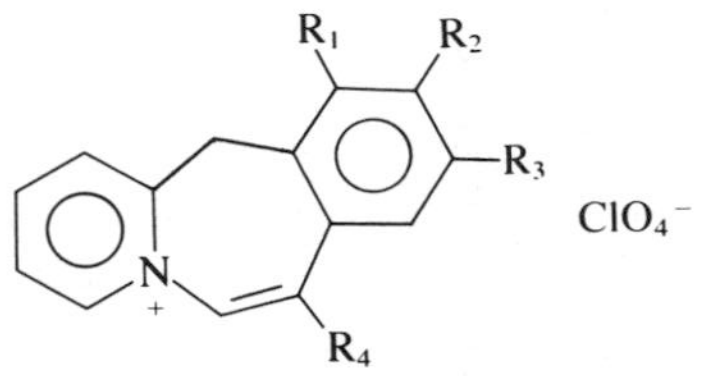

R_1	R_2	R_3	R_4	Quaternization (% Yield)	Cyclization Agent	Reaction Time (hr)	mp (°C)	Color	Spectra	Yield (%)	Refs.
H	H	H	H	71	HBr[a]	40	182–183	Colorless		81	333
H	H	H	Me	80	HBr	119	156.6–158 200.5–202.5 (picrate)	Tan	uv	75	332
H	H	H	Ph	66	HBr	156	—	—		—[c]	332
H	MeO	MeO	Me	68	HCl[b]	1.5	268–270 (HCl)	Yellow	uv	82	332
H	MeO	MeO	Ph	84	HCl	2.5	222.5–225	Yellow	uv	51	332
Me	H	H	Me	35	HBr	116	234.5–236	Tan	uv	78	332

[a] 48% HBr.
[b] Concentrated HCl.
[c] Starting material was recovered (92%).

ido[2,1-*b*][3]benzazepines in good yield (Fig. 144). The pyrido[1,2-*b*]-[2]benzazepine system has also been prepared by a similar condensation (see Section VI. 29). The 7-methyl compound **1073** was shown to be distinct from the pyrrocoline **1075,** a possible alternative structure for the cyclization product, by comparison of melting point and ultraviolet spectra of a legitimate sample. The pyrrocoline was formed from **1072** by cyclization in the presence of sodium carbonate. Oxidation of **1073** gave phthalic anhydride, further ruling out structure **1075.**

Hydrogenation of **1073** resulted in the uptake of four moles of hydrogen and yielded a crystalline product **1074**, which was a mixture of stereoisomers (mp 170–180°C). From this mixture only the higher-melting form was isolated in pure state (mp 253–255°C). On the basis of these observations, the authors eliminated the possibility that acid-catalyzed reaction had given rise to some type of bimolecular condensation product.

12H-Pyrido[2,1-*b*][3]benzazepines prepared by the synthetic route given in Fig. 144 are listed in Table 63. It should be noted that the 1-phenacyl salt did not cyclize, reflecting the low order of activity previously observed for phenyl ketones in related cyclizations.

Interestingly, while **1071** could not be quaternized with bromo- or chloroacetals or the corresponding aldehydes, α-chloroacetaldoxime was found to readily form **1076,** which on acid treatment cyclized to **1077** (333) (Eq. 133).

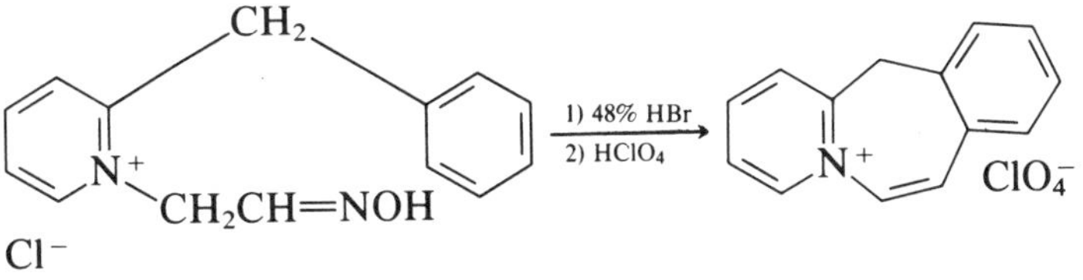

CH_2 ; N^+ ; $CH_2CH{=}NOH$; Cl^- — 1) 48% HBr, 2) $HClO_4$ → N^+ ; ClO_4^- (133)

1076 **1077**

32. Pyrido[2,3-*b*][1]benzazepines

1, 2, 3, 4, 5, 6, 7, 8, 9, 10, 11 N, 11a

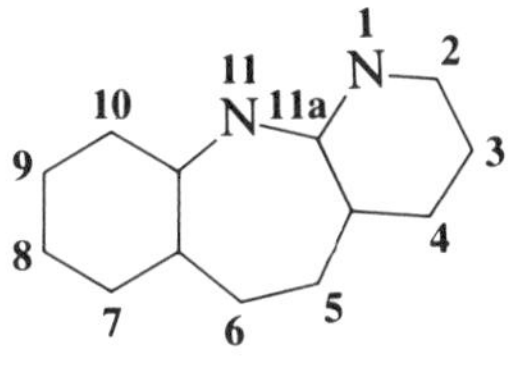

Two examples of the 6,11-dihydro-5H form of pyrido[2,3-*b*][1]benzazepines are known from the work of Villani and Mann (334). Reaction of 2-aminonicotinaldehyde (**1078**) with the Wadsworth–Emmons reagent **1079** led to the stilbazole **1080** (Fig. 145). This was hydrogenated to the diamine **1081,** which on pyrolysis of the phosphate salt, a procedure previously reported from their laboratory for the preparation of related heterocycles (309) (see Section VI. 21), gave **1082.** For pharmacological evaluation and comparison with the antidepressant imipramine, the *N*-dimethylaminopropyl analog **1083** was prepared.

Compound **1083** at oral doses of 1 and 3 mg/kg did not antagonize tetrabenazine-induced sedation in mice. The approximate ED_{50} for **1083** was between 5 and 10 mg/kg, whereas imipramine has an ED_{50} in the 1–3 mg/kg range. These and previously reported data from this group demonstrate that substitution of a pyridine ring for one or both benzene rings in the dibenz[*b,f*]azepine system of imipramine (see Section VI. 39) results in compounds with greater toxicity and low antidepressant activity.

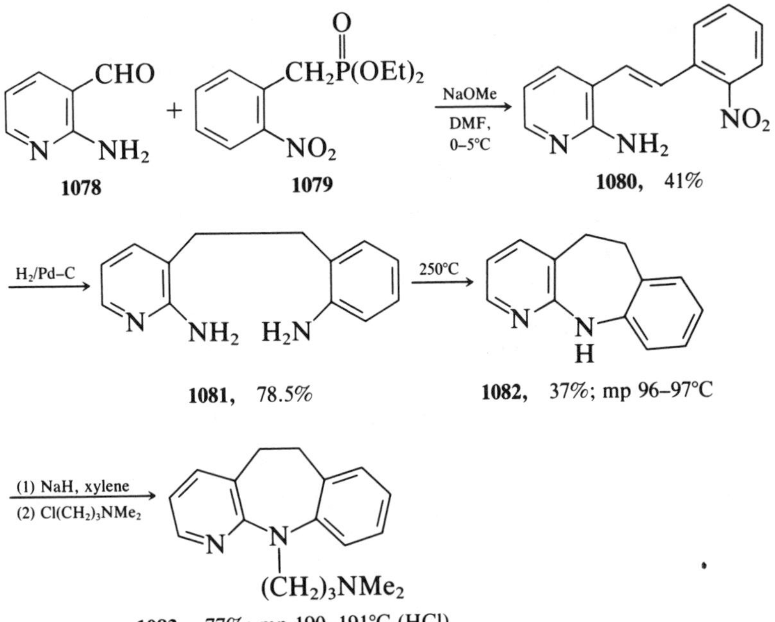

CHO NH2 N **1078** + $CH_2P(O)(OEt)_2$ NO_2 **1079** —NaOMe, DMF, 0–5°C→ **1080**, 41%

—H_2/Pd–C→ **1081**, 78.5% —250°C→ **1082**, 37%; mp 96–97°C

—(1) NaH, xylene (2) $Cl(CH_2)_3NMe_2$→ $(CH_2)_3NMe_2$ **1083**, 77%; mp 190–191°C (HCl) mp 89–90°C (maleate)

Figure 145

33. Pyrido[2,3-*c*][2]benzazepines

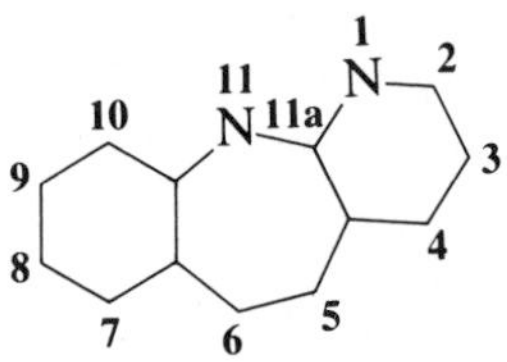

1 2 3 4 5 6 7 8 9 10 11 11a

CO_2H, Cl, N + $HNCH_2Ph$ (R_1) $\xrightarrow[110°C]{KI}$ CO_2H, N, $N-CH_2$, R_1

R_1	Yield (%)
H	70
Me	75

$\xrightarrow[\text{room temp.}]{SOCl_2}$ COCl, N, N·HCl, R_1 $\xrightarrow[ClCH_2CH_2Cl]{AlCl_3}$ O, N, N, R_1

	R_1	Yield (%)
1084a	H	20
1084b	Me	82

	R_1	Yield (%)
1085a	H	20
1085b	Me	40

$\xrightarrow{R_2MgX}$ R_2 OH, N, N, R_1 $\xrightarrow[EtOH]{HCl}$ H, R_2, N, N, R_1

1086 **1087**

Figure 146

1085b $\xrightarrow{NaBH_4}$ H OH, N, N, R

1088

(134)

$\xrightarrow{SOCl_2}$ H Cl, N, N, R $\xrightarrow{HN\;\;X}$ X, N, N, N, R

1089 **1090**

Two patents (335,336) describe the preparation of pyrido[2,3-*c*][2]-benzazepines and their utility as antidepressant agents. Figure 146 shows the conversion of 2-chloronicotinic acid via cyclization of the acid chlorides **1084a** and **1084b** to the pyridobenzazepin-5-ones **1085a** and **1085b**. These ketones were converted to basic tertiary alcohols **1086** by addition of aminopropyl Grignard reagents. Dehydration of **1086** produced the corresponding aminoalkylidene compounds **1087.**

The ketone **1085b** was also converted via the alcohol **1088** and the 5-chloro intermediate **1089** to the 5-amino derivatives **1090** (Eq. 134). Compounds prepared in this manner are listed in Table 64.

TABLE 64. PYRIDO[2,3-*c*][2]BENZAZEPINES

R_1	R_2	R_3	Yield (%)	mp (°C)	Refs.
H	C=O		20	173–175	335
Me	C=O		40	103–106	335
Me	$(CH_2)_3NMe_2$	OH	92	108–110	335
Me	$(CH_2)_3N$	OH	31	134–135	335
Me	$(CH_2)_3N$ N—Me	OH	70	109–110	335
Me	H	OH	89	132–133	335
Me	H	Cl	96	> 200	336
Me	H	N N—Me	75	146–147	336
Me	H	N	40	159	336
Me	H	N O	50	127	336
Me	$=CHCH_2NMe_2$		64	92–95	336
Me	$=CHCH_2N$		40	232–236	335
Me	$=CHCH_2N$ N—Me		68	177	335

34. Pyrido[3,2-*c*][2]benzazepines

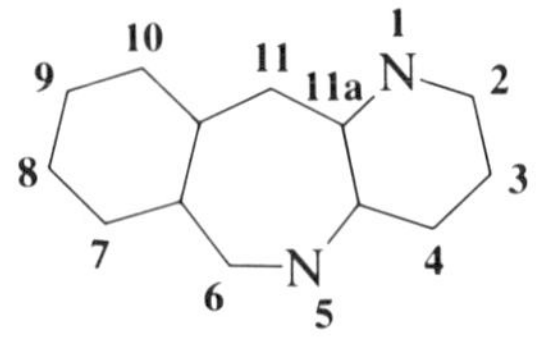

35. Pyrido[3,4-*c*][2]benzazepines

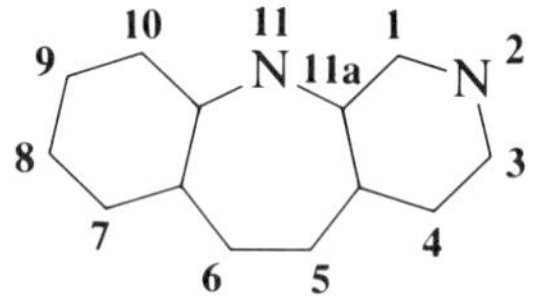

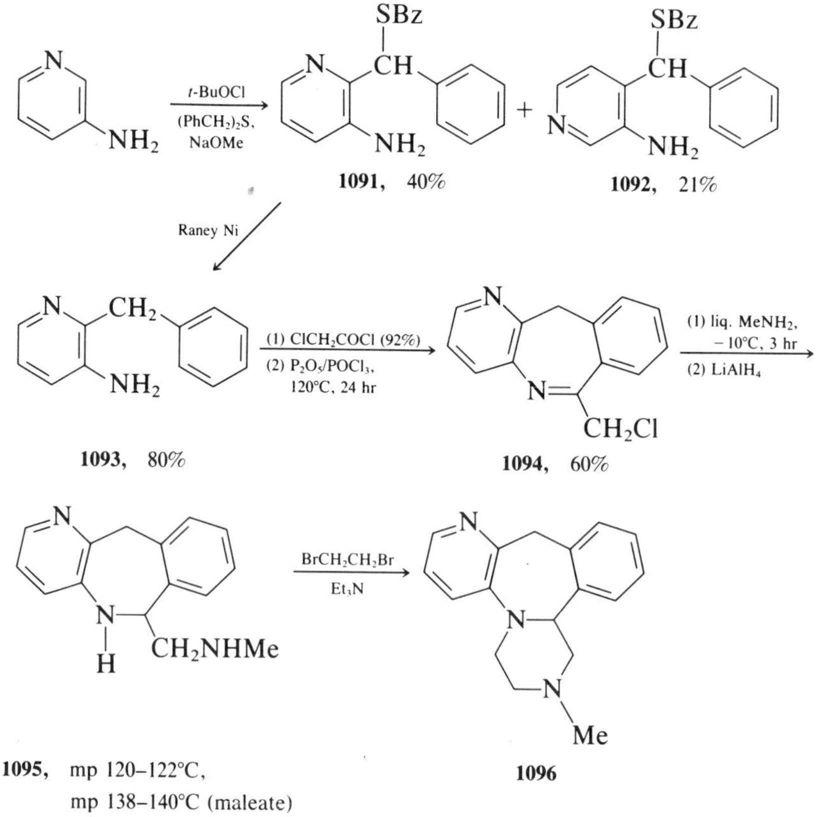

t-BuOCl, $(PhCH_2)_2S$, NaOMe

1091, 40%

1092, 21%

Raney Ni

1093, 80%

(1) $ClCH_2COCl$ (92%)
(2) $P_2O_5/POCl_3$, 120°C, 24 hr

1094, 60%

(1) liq. $MeNH_2$, −10°C, 3 hr
(2) $LiAlH_4$

$BrCH_2CH_2Br$, Et_3N

1095, mp 120–122°C,
mp 138–140°C (maleate)

1096

Figure 147

The synthesis of pyrido[3,2-*c*][2]benzazepine derivatives was accomplished as a part of a structure–activity study involving analogs of the interesting antidepressant mianserin (see Section VI. 38). 3-Amino-2-benzylpyridine (**1093**) was prepared by rearrangement of the sulfilamine salt generated by the action of *tert*-butyl hypochlorite and dibenzyl sulfide on 3-aminopyridine (Fig. 147). The intermediates **1091** and **1092** were separated by column chromatography. Desulfurization with Raney nickel provided **1093.** The seven-membered ring was formed by cyclization of the α-chloroacetamide of **1093** with a phosphorus pentoxide–phosphoryl chloride mixture. The oily product **1094** was converted to the diamine **1095** in good yield by amination in liquid methylamine followed, without isolation, by lithium aluminum hydride reduction. Further reaction of **1095** with ethylene dibromide yielded the azamianserin analog **1096.** The patent indicates the synthesis of the isomeric pyrido[3,4-*c*][2]benzazepines from **1092,** but gives no details beyond describing the formation of the isomeric azamianserin **1098** from **1097** (Eq. 135). No pharmacological or spectral data are reported (337).

1092 → **1097** → **1098** (135)

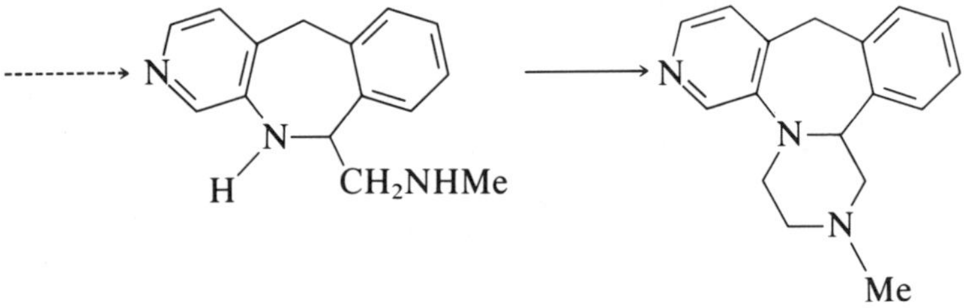

36. Pyrido[3,2,1-*jk*][1]benzazepines

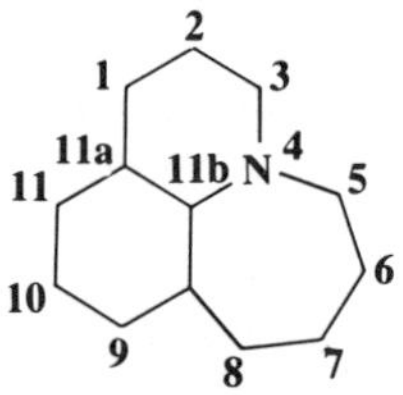

Reactions of quinisatin (**1099**) with various aliphatic diazo compounds have been studied (337). The trione **1103** could be prepared in several steps from tetrahydroisoquinoline (Fig. 148). Ring enlargement was postulated to involve the zwitterion **1100** and subsequently the aldol-type addition product **1102.** In contrast to other quinisatin derivatives, which yielded spiroepoxides as substantial by-products, **1099** did not form the epoxide **1101.** This was thought to be due to inhibition of amide resonance by the trimethylene bridge,

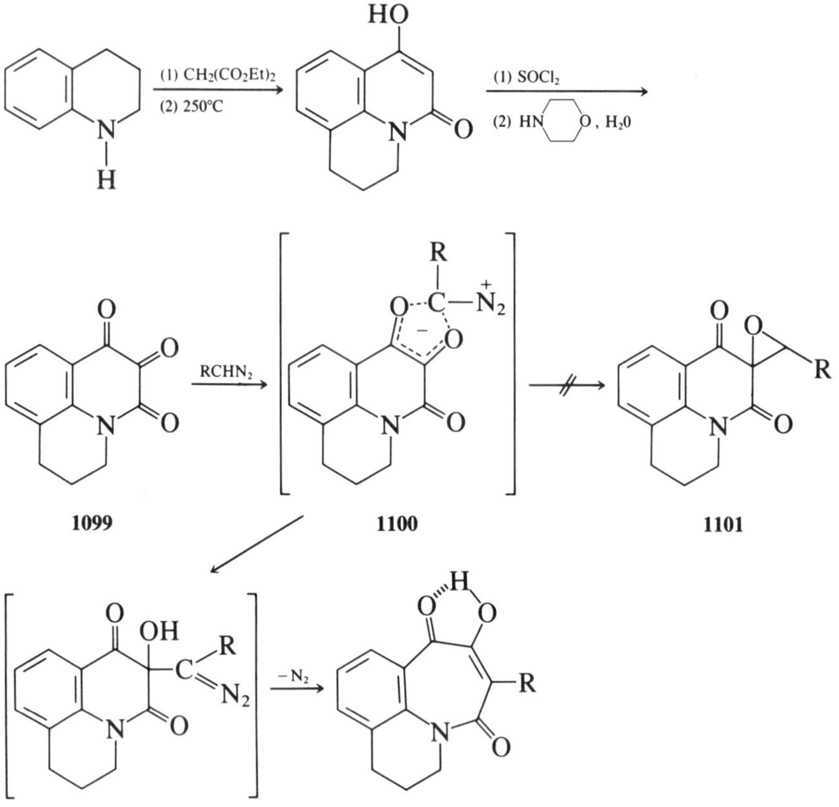

(1) $CH_2(CO_2Et)_2$
(2) 250°C

(1) $SOCl_2$
(2) HN O, H_2O

RCHN$_2$

1099 **1100** **1101**

$-N_2$

1102 **1103**

Figure 148

TABLE 65. 1H-PYRIDO[3,2,1-*jk*][1]BENZAZEPIN-5,8-DIONES[a]

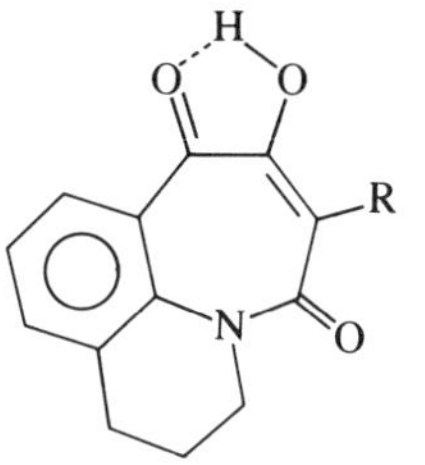

R	Yield (%)	mp (°C)	Color	$FeCl_3$ Test	Spectra
H	75	174	Yellow	Blue-green	
Me	~100	184	Colorless	Blue-green	
CO_2Et	73	137	Colorless		ir,nmr
CO_2Et	78	83[b]	Colorless		
COPh	0				

[a]Ref. 338.
[b]Enol methyl ether of the preceding compound.

which would have the effect of making the adjacent ketone carbonyl more reactive to addition of the diazo compound. Adduct **1102** was isolated and characterized only from the ethyl diazoacetate reaction, and readily lost nitrogen on warming to give the ring-expanded benzazepine **1103.** Compounds prepared in this way are the only reported representatives of the pyrido[3,2,1-*jk*][1]benzazepine ring system and are listed in Table 65.

37. Dibenz[*b,d*]azepines

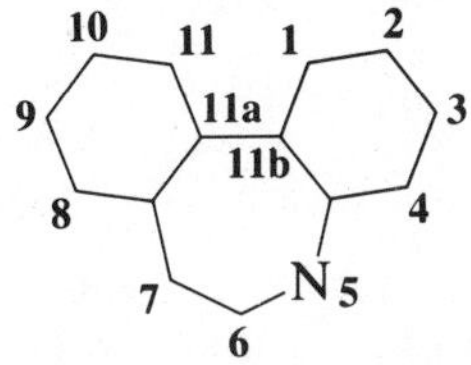

The synthesis of dibenz[*b,d*]azepin-6(5H)-one (**1106**) was accomplished from **1104** as shown in Eq. 136. Catalytic reduction of the nitro group in **1105** was accompanied by spontaneous lactamization (338). The active methylene group of **1106,** unlike that of oxindole, failed to condense with benzaldehyde or nitrosobenzene. This observation led to formulation of the product of the Stolle condensation of **1107** as **1106,** rather than the previously assigned structure **1108** (339).

The same synthetic approach described above was used by Wiesner and Valenta (340) in their synthesis of erythrina alkaloids. Figure 149 shows the

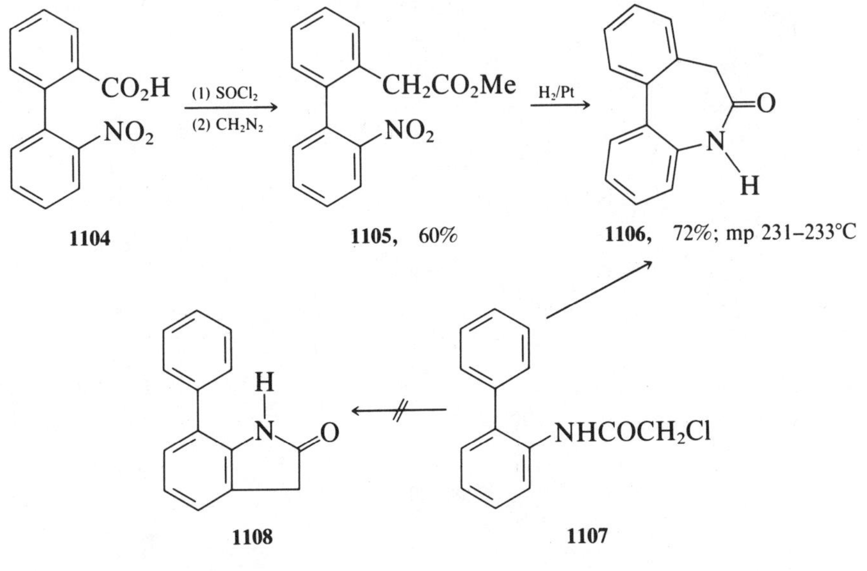

1104 → (1) $SOCl_2$, (2) CH_2N_2 → **1105,** 60% → H_2/Pt → **1106,** 72%; mp 231–233°C (136)

1107 → **1106**; **1107** ⇸ **1108**

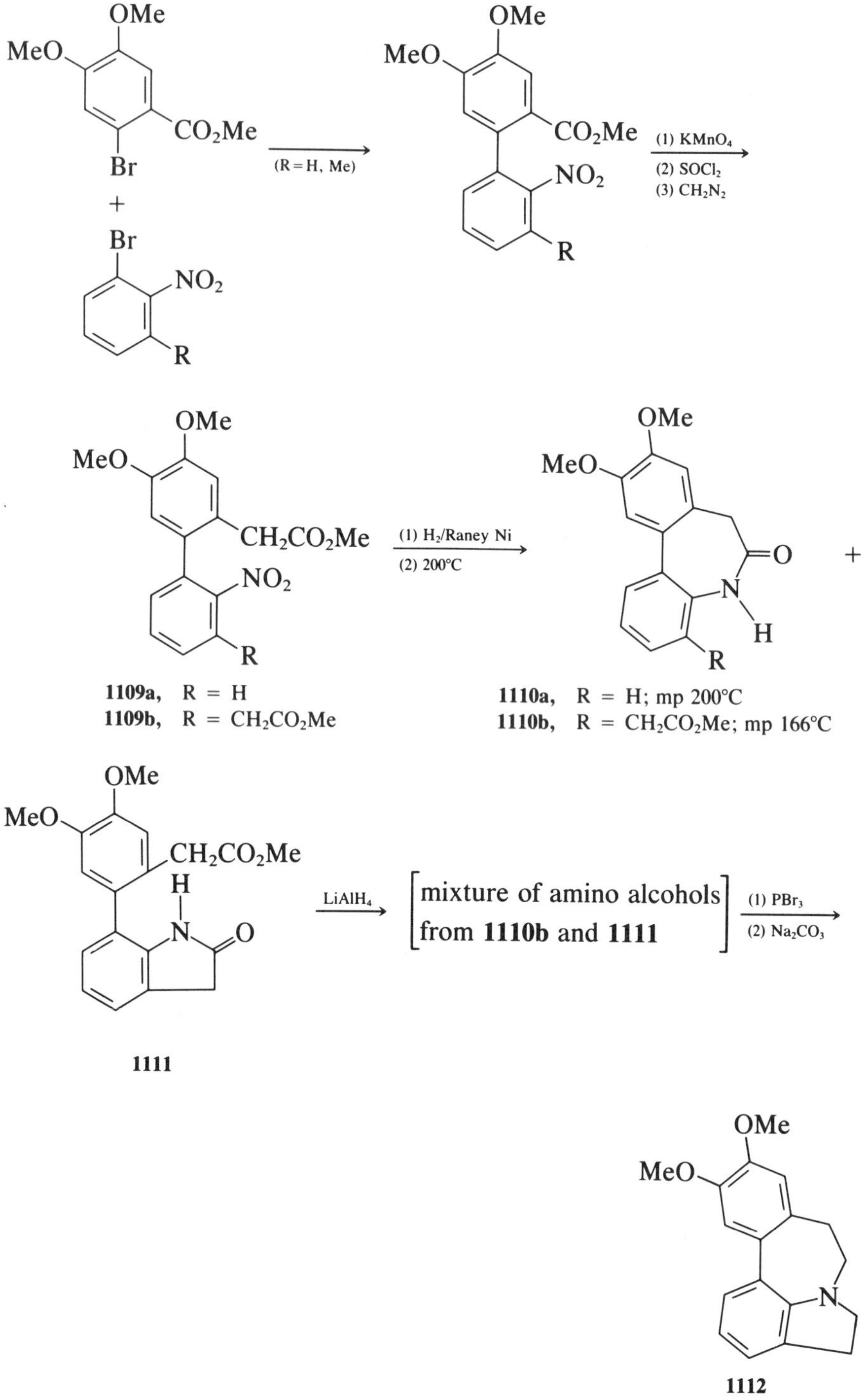

OMe
MeO
CO_2Me
Br
+
Br
NO_2
R

(R = H, Me)

OMe
MeO
CO_2Me
NO_2
R

(1) $KMnO_4$
(2) $SOCl_2$
(3) CH_2N_2

OMe
MeO
CH_2CO_2Me
NO_2
R

1109a, R = H
1109b, R = CH_2CO_2Me

(1) H_2/Raney Ni
(2) 200°C

OMe
MeO
O
N
H
R

1110a, R = H; mp 200°C
1110b, R = CH_2CO_2Me; mp 166°C

+

OMe
MeO
CH_2CO_2Me
H
N
O

1111

$LiAlH_4$

[mixture of amino alcohols from **1110b** and **1111**]

(1) PBr_3
(2) Na_2CO_3

OMe
MeO
N

1112

Figure 149

preparation of the dimethoxy compound **1110a** and the acetic acid derivative **1110b** via the nitro esters **1109a** and **1109b,** respectively. Ring closure of **1109b** to **1110b** was accompanied by formation of some of the oxindole **1111.** Although **1110b** could be obtained pure by fractional crystallization, the mixture of **1110b–1111** was reduced to a mixture of amino alcohols which could be converted to dimethylapoerysopine (**1112**). The lactam **1110a** was reduced with lithium aluminum hydride to the corresponding tricyclic amine (mp 147–148°C).

Proctor and Paterson (341,342), and also Raboman (343), have prepared dibenz[*b,d*]azepin-7(5H)-ones (**1113**) in high yield by Friedel–Crafts cyclization of appropriate acid chloride precursors with aluminum chloride at low temperature (Fig. 150). Other variants employing stannic chloride, PPA, or aluminum chloride at higher temperatures gave phenanthridine derivatives. The ketones **1113** (carbonyl absorption at 1685 cm^{-1}) could be reduced with lithium aluminum hydride to the alcohols **1114,** and could be converted to 2,4-DNP derivatives. Base-promoted elimination of *p*-toluenesulfinic acid from **1113** (R = H) was reported initially to give the azatropone structure **1115** (341). However, Proctor and Peaston (344), and also Cooke and Russell (345), reported the correct structure of the purple product of this reaction as the dimeric structure **1116.** Oxidation of **1116** with manganese dioxide led to the yellow *bis*-azatropone system **1117,** which could be quantitatively reconverted to **1116** by hydrogenation. Ozonolysis of **1116** gave the keto lactam **1118,** in agreement with the structure of **1116.** When the conversion of **1113** to **1116** was attempted with sodium methoxide in toluene saturated with oxygen, there was obtained a 35% yield of phenanthridone and a 42% yield of **1119.** The latter also formed in 7% yield in the reaction producing **1116.** It was reported, in addition, that heating **1116** (X = H) in a sealed tube at 190°C and treatment with lithium aluminum hydride in THF gave phenanthridone in 43% and 30% yields, respectively.

Treatment of **1113** with sodium ethoxide in ethanol (Fig. 151) gave **1120** (343,344). This arises presumably by conversion of **1113** to the dibenzazatropone **1115** which immediately adds ethoxide across the imine double bond to give **1120.** This substance was converted in high yield to the stable, colorless dibenz[*b,d*]azepin-7-ones **1121** by oxidation with manganese dioxide. It is interesting to note that the methylene protons of the ethyl group in **1120** show diastereotopic coupling, whereas those of **1121** (R = Et) give rise to a simple quartet. Reaction of **1113** with bromine gave, presumably via *N*-tosylated imminium intermediates, the solvent addition products **1122,** along with some phenanthridinecarboxylic acid (**1123,** R = H). Treatment of **1122** (X = H) with sodium hydride in benzene afforded **1121** (X = H, R = Me) in unspecified yield. No seven-membered ring derivatives of **1121** were obtained; reaction of **1121** with methyl iodide, 2,4-dinitrophenylhydrazine, or dry HBr gave **1123** (R = Et). The mass spectra of **1113** (X = R = H) and **1114** have been published (346). Tables 66A,B list the dibenz[*b,d*]azepin-7-ones prepared as shown in Figs. 150 and 151.

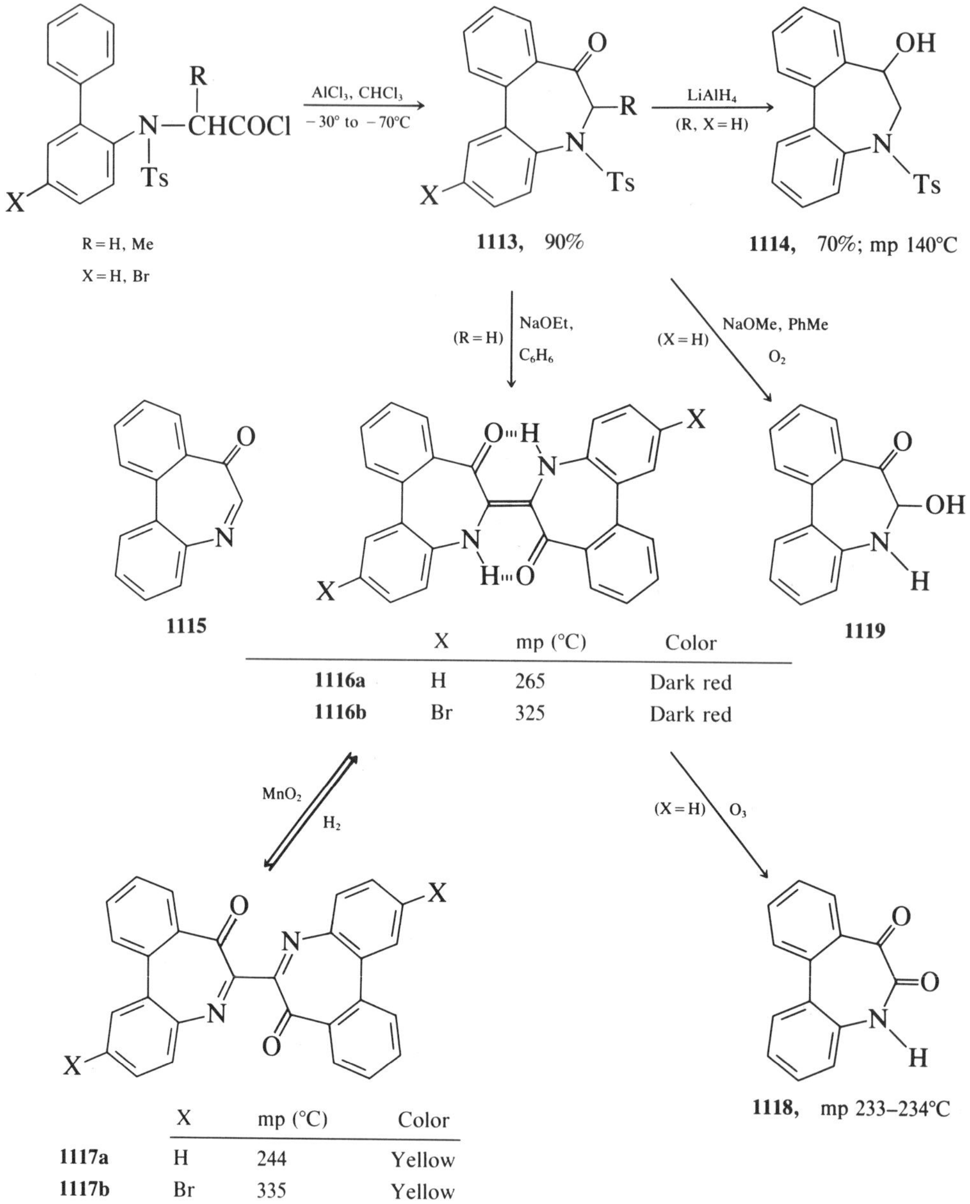

R = H, Me

X = H, Br

$AlCl_3$, $CHCl_3$, −30° to −70°C

1113, 90%

$LiAlH_4$ (R, X = H)

1114, 70%; mp 140°C

(R = H) NaOEt, C_6H_6

(X = H) NaOMe, PhMe, O_2

1115

	X	mp (°C)	Color
1116a	H	265	Dark red
1116b	Br	325	Dark red

1119

MnO_2 / H_2

(X = H) O_3

	X	mp (°C)	Color
1117a	H	244	Yellow
1117b	Br	335	Yellow

1118, mp 233–234°C

Figure 150

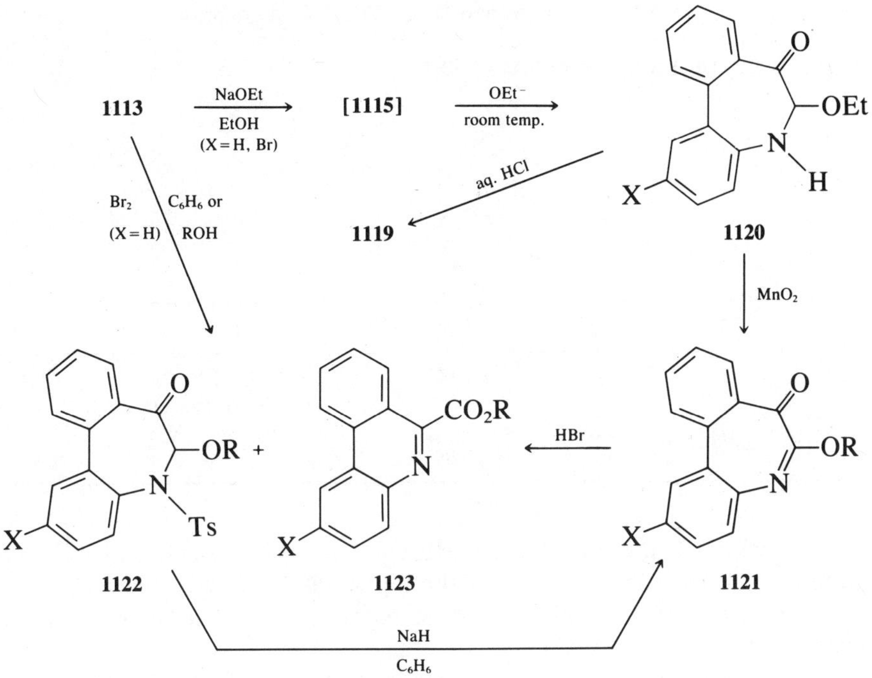

Figure 151

TABLE 66A. DIBENZ[*b*,*d*]AZEPIN-7-ONES

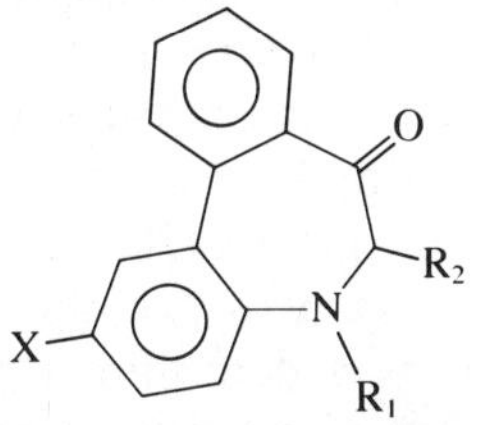

R_1	R_2	X	mp (°C)	Yield (%)	Spectra	Refs
p-MeC_6H_4	H	H	137			
			212 (2,4-DNP)	90	uv,ms	341,346
H	EtO	H	134		ms,ir,nmr	344
H	OH	H	239–240	42	nmr,ir,uv	341,344
H	AcO	H	267			344
p-MeC_6H_4	OH	H	207		nmr	342
pMeC_6H_4	EtO	H	103		ir,nmr	342
p-MeC_6H_4	MeO	H	117			
p-MeC_6H_4	H	Br	145		ir,nmr	
H	EtO	Br	151	93	ir	342
H	OH	Br	257			342
p-MeC_6H_4	Me	H	131–132	65	ir,nmr	343,344
			192–195 (2,4-DNP)			

TABLE 66B. DIBENZ[*b*,*d*]AZEPIN-7-ONES

R	X	mp (°C)	Yield (%)	Spectra	Refs
EtO	H	118	90	ir,nmr, dipole moment	344
MeO	H	112		ir,nmr	342
EtO	Br	103	90	ir	342

A very low yield of the dibenz[*b*,*d*]azepinone-*N*-oxide **1125** was reported from the thermal isomerization of the biphenyltolane derivative **1124** (Eq. 137). Infrared, uv, and mass spectral data were cited in support of the isatogen-like structure (347).

Examples of the dibenz[*b*,*d*]azepine system arise from Beckman rearrangements in various studies dealing with the structure and configuration of terpenes and related substances. The reactions shown in Fig. 152 were carried out (348) to determine the stereochemistry of the product generated from the cyclization of **1126.** A *cis*/*trans* mixture of **1127** was obtained and converted to the mixture of the oximes **1128,** which could be separated by fractional crystallization. Each oxime was carried through a Beckmann rearrangement to give **1129.** Acid hydrolysis of **1129** failed; base hydrolysis at 200°C (KOH, ethylene glycol) gave **1130,** but acidification resulted in immediate regeneration of the lactams. Oxidation of the salts **1130,** however, gave low yields of **1131.** It was determined by mixed melting point and infrared spectral comparisons with known isomers of **1131** that the major isomer of **1127** had the *cis* stereochemistry and gave rise to *cis*-**1129** and *cis*-**1131.** Similarly, the minor isomer of **1127** had a *trans* ring fusion and generated the *trans* isomer of **1129** and of **1131.** The lactams **1129** were reduced with lithium aluminum hydride to the *cis*- and *trans*-amines **1132.**

NO_2, I $\xrightarrow[C_5H_5N,\ 72\ hr]{CuC\equiv CPh}$ NO_2, C≡CPh $\xrightarrow{\Delta}$ O, Ph, N, O (137)

1124 **1125,** 2.6%; mp 239–240°C

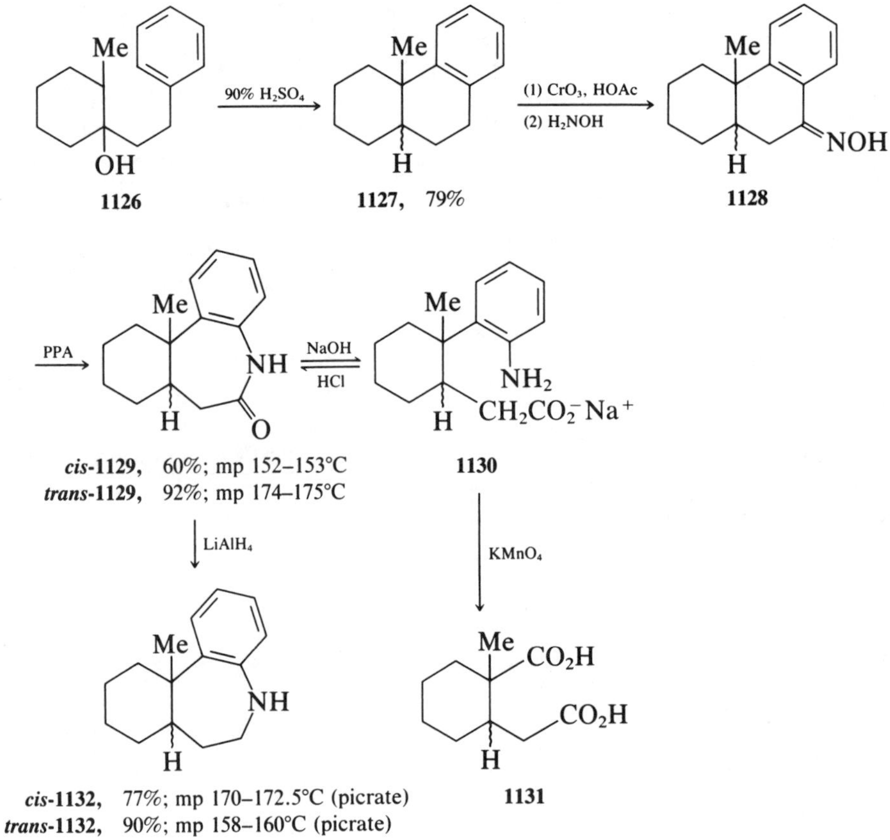

1126 → (90% H_2SO_4) → 1127, 79% → ((1) CrO_3, HOAc; (2) H_2NOH) → 1128

→ (PPA) → *cis*-1129, 60%; mp 152–153°C
trans-1129, 92%; mp 174–175°C

1129 ⇌ (NaOH / HCl) 1130 (NH_2, $CH_2CO_2^-Na^+$)

1129 → ($LiAlH_4$) → *cis*-1132, 77%; mp 170–172.5°C (picrate)
trans-1132, 90%; mp 158–160°C (picrate)

1130 → ($KMnO_4$) → 1131

Figure 152

Very similar chemistry is shown in Fig. 153. Arigoni and Jeger (349) have shown the absolute configurational relationship between abietic acid **(1133)** and α-onocerin by obtaining the diacid **1137** from the degradation of both natural products. The ketone **1134** (R = Me), obtained from abietic acid, yielded the lactam **1136** when subjected to the Schmidt reaction. Ozonolysis followed by oxidative workup gave **1137.** Japanese workers (350) have converted methyl 7-ketodehydroabietate (**1134,** R = CO_2Me) to the oxime **1135a,** and reported the preparation of the lactam **1138** in 76% yield from the oxime tosylate **1135b** and in 39% yield from the oxime **1135a** by Beckmann rearrangement. Reduction of **1138** with lithium aluminum hydride gave the amino alcohol **1139.** However, the physical properties, melting point, and rotations of **1138** and **1139** are not in agreement with those reported by a Swedish group (351).

Figure 154 shows the formation of the preferred lactam **1142** in 57% yield from Beckmann rearrangement of the oxime tosylate **1141,** and in 95% yield

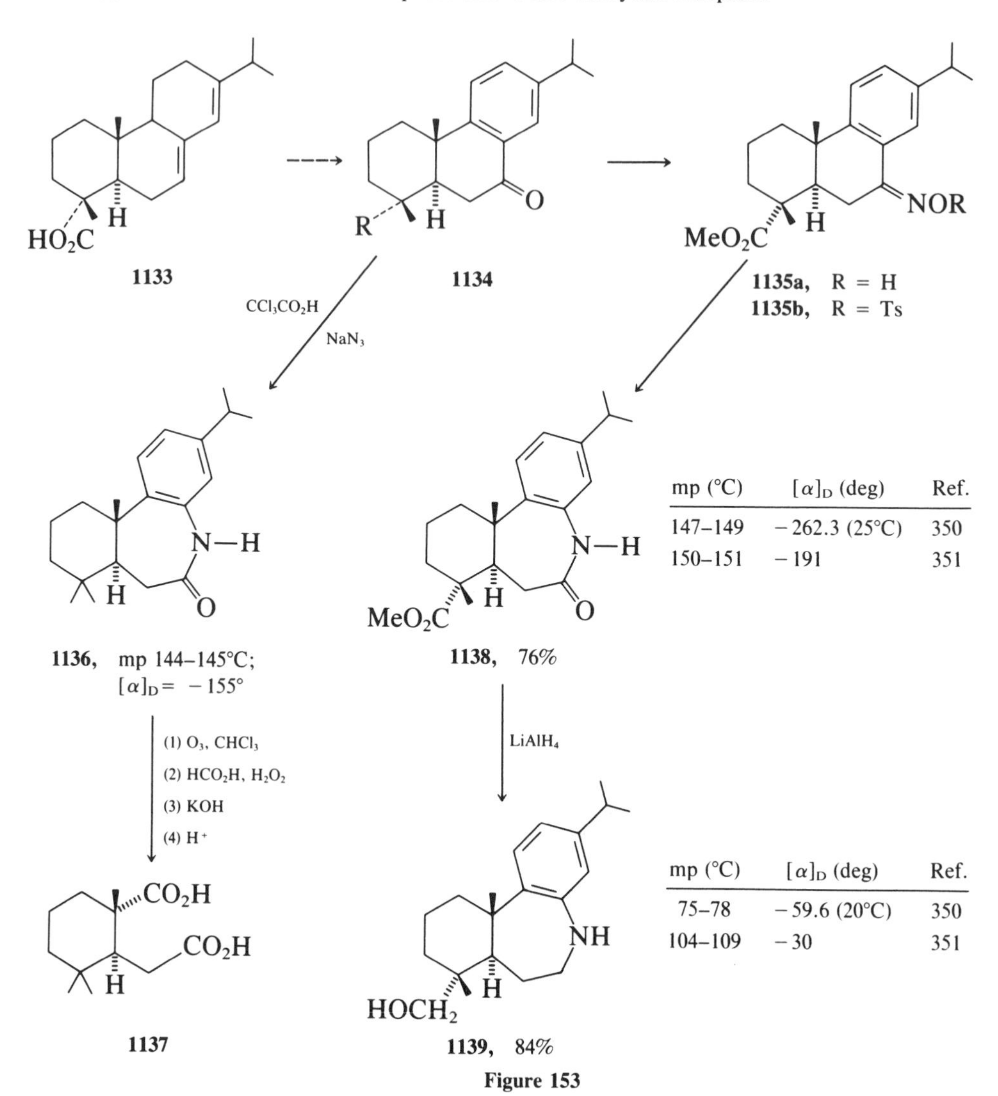

mp (°C)	[α]$_D$ (deg)	Ref.
147–149	−262.3 (25°C)	350
150–151	−191	351

mp (°C)	[α]$_D$ (deg)	Ref.
75–78	−59.6 (20°C)	350
104–109	−30	351

Figure 153

from the Schmidt reaction of the ketone **1140.** A trace amount of the isomeric dibenz[*c*,*e*]azepine lactam **1143** was isolated from the Schmidt rearrangement (352). Also reported in unspecified yield was the cyclization of the hydroxamic acid **1144** to **1142** in hot PPA. Although the crude yield in this reaction was good, extensive purification was required to obtain material with the correct melting point. The *anti* stereochemistry of the oxime **1141** was assigned from nmr spectra. Nmr, infrared, and ultraviolet spectra were used to characterize **1142** and **1143**.

A different synthetic approach (353) to dibenz[*b*,*d*]azepin-6-ones such as

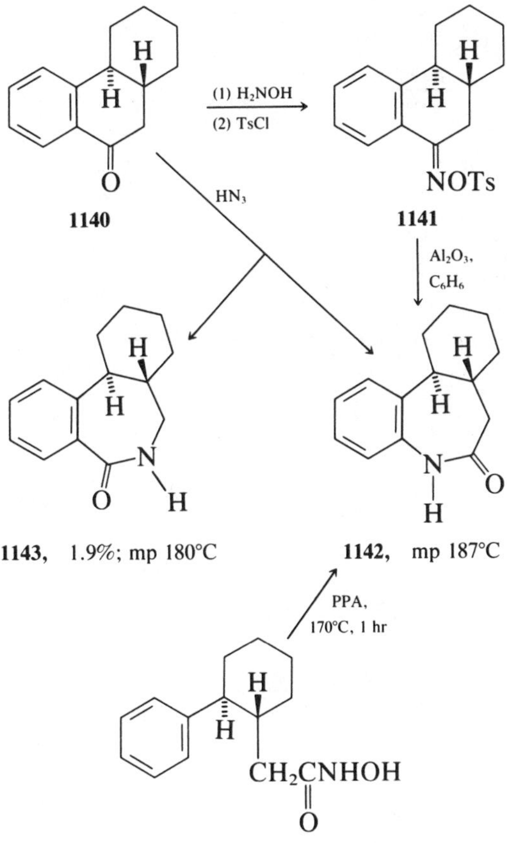

(1) H_2NOH
(2) TsCl

NOTs

HN_3

1140 **1141**

Al_2O_3,
C_6H_6

1143, 1.9%; mp 180°C **1142**, mp 187°C

PPA,
170°C, 1 hr

CH_2CNHOH

1144

Figure 154

1129 and **1142** combines the Diels–Alder reaction and the Wolff rearrangement. The phenylcyclohexanecarboxylic acid **1146** (Eq. 138) has *trans* stereochemistry from the Diels–Alder reaction of the *trans*-cinnamic acid **1145**. Chain lengthening and Wolff rearrangement of **1146** to **1147** proceeded in good yields. Reduction of the nitro group was accomplished with Raney nickel–hydrazine to avoid reduction of the double bond. The resulting amino ester **1148** had to be heated to 180°C before lactamization to **1149** occurred. The difficult lactamization of **1148** is in sharp contrast to the ease of cyclization of **1130** to **1129** (Fig. 152).

A patented process (354,355) claims the formation of **1151** from **1150** by Friedel–Crafts cyclization (Eq. 139). Although no yield was given, the process appears to be a valid one, judging from the good yields of related transformations, and would provide access to dibenz[*b*,*d*]azepines not available by the other routes discussed above.

1145

$H_2C=C(Me)-C(Me)=CH_2$, xylene, 130°C

1146, 70%

(1) $SOCl_2$
(2) CH_2N_2
(3) AgOBz, MeOH

(138)

1147

Raney Ni, N_2H_4

Amino ester **1148**

180°C

1149, mp 219°C

1150

$AlCl_3$

(139)

1151, bp 150–155°C/0.01 mm; $[n]_D^{20} = 1.4670$

38. Dibenz[*b,e*]azepines

The broad spectrum of pharmacologic activity exhibited by the dibenz[*b,e*]azepines, or morphanthridines, has generated an extensive patent literature. The name "morphanthridine" was first used by Scholl (356) because of the homologous relationship of these compounds to phenanthridine and morphine.

The three most important groups of compounds in this system are the 5,6-dihydrodibenz[*b,e*]azepin-6,11-diones (morphanthridine-6,11-diones), the 5,6-dihydrodibenz[*b,e*]azepin-6(11H)-ones (6-morphanthridones), and the 5,6-dihydromorphanthridines. These substances have been employed as key starting materials for further structural elaboration, including, for example, the attachment of aminoalkyl side chains at the 5-,6-, and 11-positions to generate a variety of products with antidepressant, antipsychotic, and antispasmodic activity.

An important route to 5,6-dihydrodibenz[*b,e*]azepin-6,11-diones involves the Schmidt reaction on anthraquinones (357–361). This reaction (Eq. 140) provides the parent unsubstituted compound in good to excellent yield. However, when the anthraquinone is substituted, isomeric lactams such as **1152** and **1153** are formed. The directing influence of such groups as chloro, methyl, and methoxy was found not to be very specific, and mixtures of isomeric lactams were obtained (361). The structures of the products were determined by hydrolysis with dilute sodium hydroxide to amino acids. The amino acids were then subjected to diazotization, followed by treatment with hypophosphoric acid, to obtain the corresponding benzoylbenzoic acids. The position of the substituent was determined either by comparison with an authentic sample or from nmr and infrared spectra. The results of Werner and co-workers (361) differ somewhat from those of Coronna and Palazzo (363,364), who indicated one product from monosubstituted anthraquinones but did not provide physical data or proof of structure. Drukker and Judd have determined the structure of the 2-chloro-6,11-dione, one of three products from 2-chloroanthraquinone, by synthesis from *N*-(4-chlorophenyl) phthalimide (365). 6,11-Diones prepared by the Schmidt reaction are listed in Table 67.

Another synthesis of 6,11-diones (**1155**) is the reported rearrangement (366,367,368) of *N*-arylphthalimides (**1154**) at elevated temperatures (Eq. 141). Werner and Rossi (361), however, found this method inferior to the Schmidt reaction for their purposes.

HN_3 → **1152** + **1153** (140)

TABLE 67. 5,6-DIHYDRO-11H-DIBENZ[*b*,*e*]AZEPIN-6,11-DIONES PREPARED FROM ANTHRAQUINONES BY THE SCHMIDT REACTION

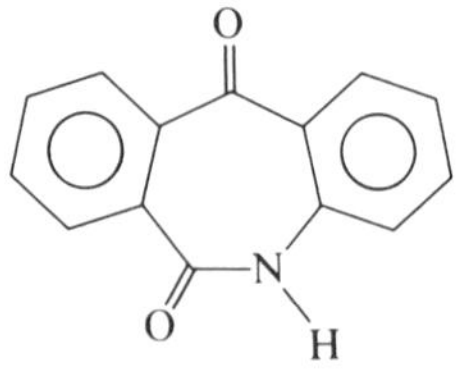

Anthraquinone	[*b*,*e*]	mp (°C)	Yield (%)	Refs
H	H	245	Excellent	357,363
		243–244		360
		250–251	84	359,392
		246–248	77	361,388
		280 (2,4-DNP)		388
1-Cl	4-Cl (?)	211–213		358
		280		364
	1,4- or 10-Cl	210–212	3.7	361
	7-Cl (?)	177–179		358
		271–273	10	361
2-Cl	3-Cl or 8-Cl	222–224		358
	8-Cl	234–235	7.5	361
	2-, 3-, or 9-Cl	216–220	7.5	361
	3-Cl	293		364
1-Me	10-Me	230–232	19	361
	1-, 4-, or 7-Me	217–219	23	361
2-Me	3-Me	260–264	20	361
	3-Me or 8-Me	253–256		358
	8-Me	204–207	17.5	361
1,4-Me_2	1,4-Me_2	225–226	97	436
2-*t*-Bu	isomer I	165–166	v28.9366	
	isomer II	214–215		366
2-Et	3-Et	178–180	22	361
1-MeO	1-MeO	254–255	63	361
2-MeO	3-MeO	258–259	22	361
	2,8- or 9-MeO	235–236	>5	361
1-NO_2	4-NO_2[a]	320		364
1-NH_2	10-NH_2[a]	248		363,373
2-NH_2	9-NH_2	230–231		363

[a]No physical data or structure proof provided.

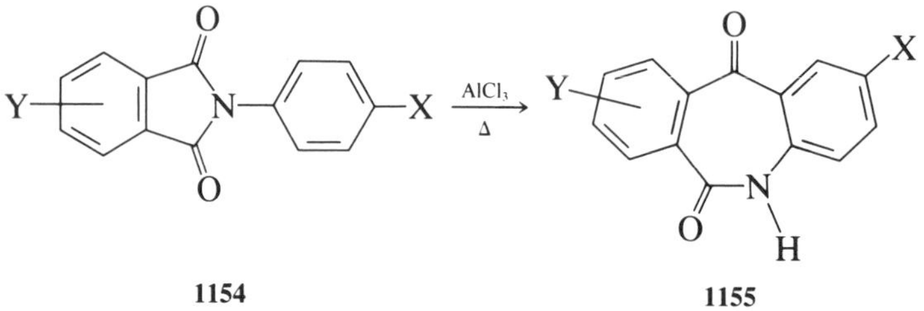

$$\mathbf{1154} \xrightarrow[\Delta]{AlCl_3} \mathbf{1155} \tag{141}$$

In a somewhat related process (369), phthalic anhydride was condensed with the acetylated anilines **1156** in the presence of aluminum chloride (Fig. 155), and the resulting benzophenones **1157a** and **1157b** were heated in dilute HC1 to form the 6,11-diones **1159a** and **1159b**, respectively. In the reaction of phthalic anhydride with **1156** (R = H), the products **1157b** and **1159b** were accompanied by some **1158**. Dilute HC1 treatment of **1157b** gave a quantitative yield of **1159b**. Other examples of this lactamization, which give 6,11-diones, have been reported (370–372). It is interesting to note that the action of concentrated sulfuric acid on **1157b** gave a different type of cyclization product, namely the anthraquinone **1160**.

Figure 156 depicts further chemical transformation of the 6,11-diones listed in Tables 67 and 68. Electrophilic substitution reactions have been carried out with chlorosulfonic acid (374), chlorine (375), and bromine (376) to obtain the 2-substituted derivatives **1162**. Numerous *N*-alkylations of the amide nitrogen have been carried out (361,375,377,378, 379) to give **1163**. Ring-substituted amines have been converted by diazotization to phenols

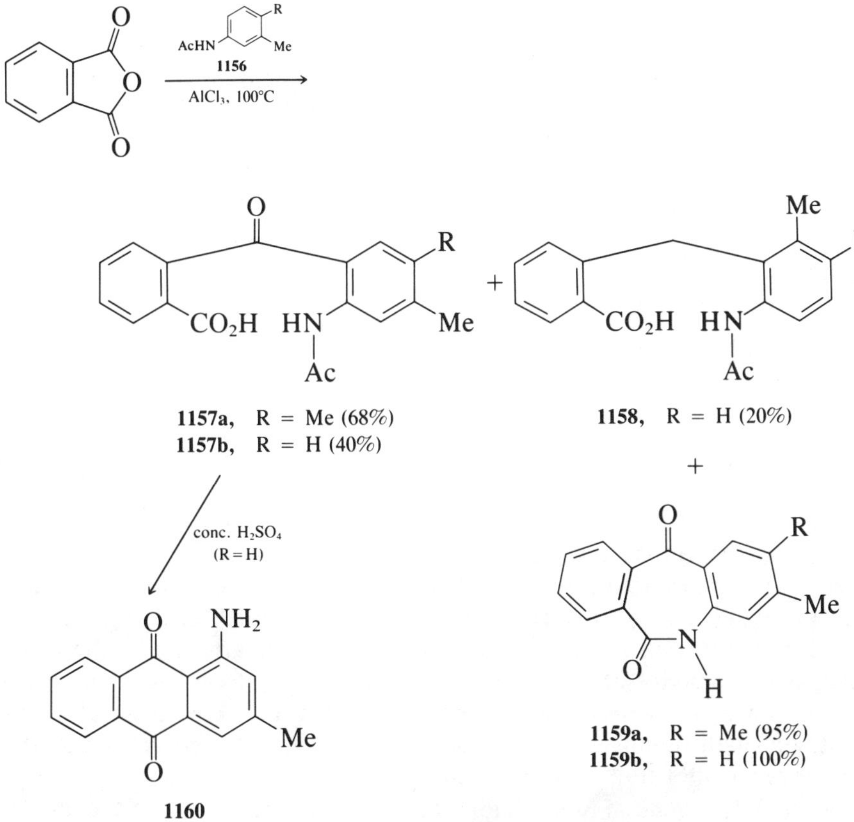

1156, $AlCl_3$, 100°C

1157a, R = Me (68%)
1157b, R = H (40%)

1158, R = H (20%)

conc. H_2SO_4 (R = H)

1159a, R = Me (95%)
1159b, R = H (100%)

1160

Figure 155

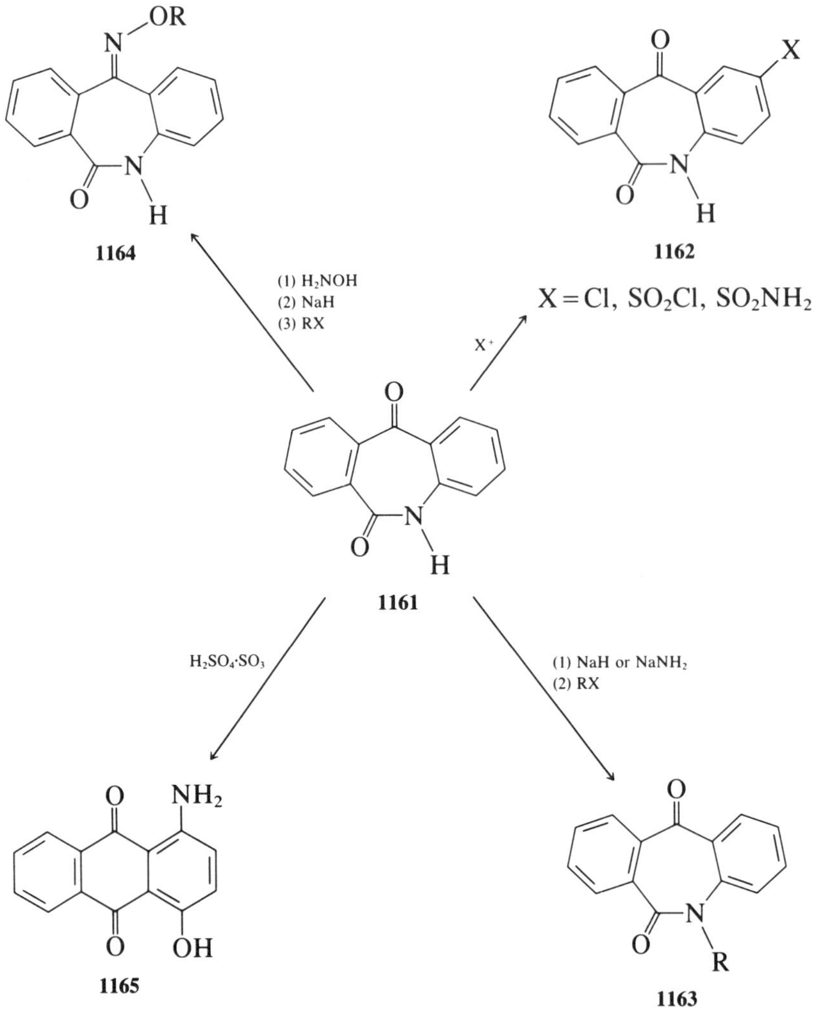

1164 **1162**

(1) H_2NOH
(2) NaH
(3) RX

X^+

X = Cl, SO_2Cl, SO_2NH_2

1161

$H_2SO_4 \cdot SO_3$

(1) NaH or $NaNH_2$
(2) RX

1165 **1163**

Figure 156

(380), and the ketone function has been converted to oximino derivatives **1164** (381,387), or reduced to alcohols (375 and later references). *O*-Aminoalkyl oxime derivatives (Table 69) have been patented for antidepressant activity, and this effort has been summarized in comparison with the pharmacological activity of related ring systems (378). Interestingly, the 6,11-diones can be converted back to anthraquinones (e.g. **1165** from **1161**) by treatment with fuming sulfuric acid, a process that may be viewed as the formal reversal of the Schmidt reaction (382).

A much less general preparation of a hexahydro derivative of the unsubstituted 6,11-diones is shown in Eq. 142. In an investigation of the photo-

TABLE 68. DERIVATIVES OF 5,6-DIHYDRO-11H-DIBENZ[*b,e*]AZEPIN-6,11-DIONE

R_1	R_2	mp (°C)	Yield (%)	Spectra	Refs.
H	H		60		368
2-Cl	H	291–294			367
		295–298	65–85		367,375
		300			366,368
2-Cl	Me	147–152	66		375
		154–155	79	ir,uv	377
H	Me	98–99	74	ir,uv	377
		93–95	71		447
H	CH_2Ph	106–107	78	ir,uv	377
		109–110	64	ir,uv	462
H	*n*-Pr	138			384
H	$CH_2C{\equiv}CH$	112–114	68	ir,nmr	462
H	CH_2CMe_3	186–187	28	ir,nmr	462
H	1-Naphthylmethyl	172–174	71	ir,nmr	462
H	9-Anthrylmethyl	218–220	62	ir,nmr	462
H	CH_2CN	172–173	33	ir,ur	377
H	NH_2	152–156	Good		409
2-Cl	Et	135–137			375
2-Cl-3-Me	H	303.5–304.5	100		372
2-Br	H	308–310	Good		376
		285			366
3-Cl	H	293–295			412,366
3-F	H	282–285			412,366
7-, 8-, 9-, or 10-Cl	H	281			366
2-Me	H	238–240			366
8- or 9-Me	H	236–237			366
3,4-Me_2	H	197–199	33.5		383
3-*t*-Bu	H	182.5–183.5			370
2-OH	H	267–269	75	uv,nmr	380
		200/0.005[a] (sublimes)			
2-MeO	H	244–245	68–72	uv,nmr	380
8-MeO	H	271–273			389
7-NH_2	H	298–299		ir,uv,nmr	371
7-NHAc	H	262–263		ir,uv,nmr	371
10-$NHCOCH_2Cl$	H	252–257	87		373
10-$NHCOCH_2I$	H	203–204	95		373
10-$NHCOCH_2NH_2$	H	250	42		373
10-$NHCOCH_2NMe_2$	H	224–226	82		373
2-SO_2NH_2	H	298–300			374
2-SO_2NH_2-3-Me	H	279–281			374

TABLE 68. *(Continued)*

R_1	R_2	mp (°C)	Yield (%)	Spectra	Refs.
2-SO_2NH_2-8-Me	H	238			374
2-SO_2NH_2-8-Cl	H	255			374
H	$(CH_2)_2NMe_2$	115–117	55		378
H	$(CH_2)_3NMe_2$	193–194/0.1[a]			358
		98–101 (HCl)			
		170–172 (HCl)			361
H	$CH_2CHMeNMe_2$	184–186/0.15[a]			358
		112–113 (HCl)			
H	$(CH_2)_2NEt_2$	98–100 (bitartrate)			358
H	$(CH_2)_2N(i\text{-}Pr)_2$	133 (HCl)			358
H	$(CH_2)_3NEt_2$	112 (HCl)			358
H	$CH_2CHMeNMe_2$	210–212 (HCl)			358
?-Cl	$CH_2CHMeNMe_2$	216–218 (HCl)			358
?-Cl	$(CH_2)_3NMe_2$	165–168 (HCl)			358
?-Cl	$(CH_2)_2NEt_2$	150 (HCl)			358
3-Me(?)	$(CH_2)_2NEt_2$	135–142 (bitartrate)			358
3-Me(?)	$(CH_2)_3NMe_2$	98–102 (tartrate)			358
3-Me(?)	$(CH_2)_3NEt_2$	143–145 (HCl)			358
8-Me(?)	$(CH_2)_3NMe_2$	191–193 (HCl)			358
H	$(CH_2)_2N$	210–212 (HCl)			361
H	$CH_2CHMeNEt_2$	213–215 (HCl)			361
H	CH_2	>300 (HCl)			361
		235	95	ir,nmr,ms	385

[a]Boiling point (°C/torr).

chemical reactivity of ketoimino ethers (385), the photocycloaddition of cyclohexene to **1166** gave **1167**. This [2 + 2] cycloaddition occurred in high quantum yield from a triplet state. The adduct was found by nmr spectra to have *cis-anti-cis* geometry. In 3*N* HCl, **1167** rearranged to the dibenz[*b*,*e*]azepinedione **1168** with no change in the geometry at the ring juncture.

Reaction with phosphoryl chloride in *N*,*N*-diethylaniline converted **1161** to 6-chlorodibenz[*b*,*e*]azepin-11-one (**1169**) in 60% yield (Fig. 157). From **1169** a series of amino acid derivatives (**1170**) have been prepared and patented as general central nervous system agents (386). Cooke and Russell (343,387) reduced **1169** by the Sonn–Muller method to the parent di-

TABLE 69. 11-OXIMINO DERIVATIVES OF 5,6-DIHYDRO-11H-DIBENZ[*b*,*e*]AZEPIN-6,11-DIONES

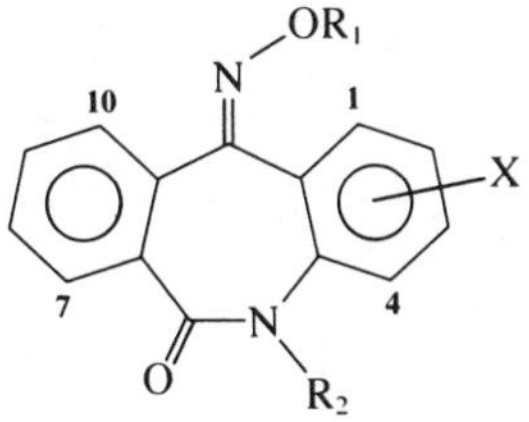

R_1	R_2	X	mp (°C)	Yield (%)	Refs.
H	H	H	206–208		381
H	Me	H	90–91		381
H	H	3-Me	273		378
H	H	1-MeO	271–273		378
$(CH_2)_2NHMe$	H	H	237–239 (HCl)		378
$(CH_2)_2NMe_2$	H	H	191–192		381
			160		384
			186–189 (HCl)		
			192–194	*syn*/*anti* isomers	378
			215–217 (HCl)		378
			192–194		
			206–208 (HCl)		
$(CH_2)_2NMe_2$	Me	H	206–207 (oxalate)		381
$(CH_2)_2NEt_2$	H	H	155–156	62	378
			230–232 (HCl)		
$(CH_2)_2NEt_2$	H	H	143	*syn*/*anti* isomers 384	
			230–232 (HCl)		
			151		
			233–234 (HCl)		
$(CH_2)_3NMe_2$	H	H	250–253 (HCl)		387
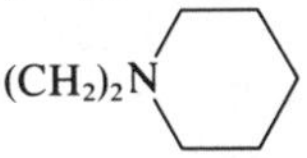 $(CH_2)_2N$ (piperidino ring)	H	H	161–162		384
			244–247 (HCl)		378
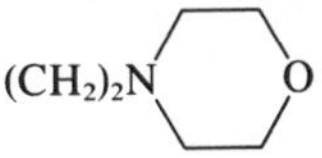 $(CH_2)_2N$ O (morpholino ring)	H	H	190		384
			265–267 (HCl)		
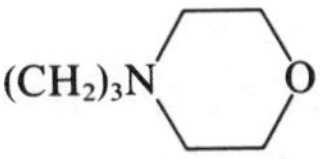 $(CH_2)_3N$ O (morpholino ring)	H	H	145		378
			254–257 (HCl)		
$(CH_2)_2NMe_2$	H	2-Cl	260–262 (HCl)		378
$(CH_2)_2NEt_2$	H	2-Cl	196–198		384
			258–260 (HCl)		
$(CH_2)_2NEt_2$	H	4-Cl	208–212 (HCl)		378
$(CH_2)_2NEt_2$	H	3-Me	160		384
			221–225 (HCl)		
$(CH_2)_2NEt_2$	H	1-MeO	174		384
			222–224 (HCl)		
$(CH_2)_2NMe_2$	$(CH_2)_2NMe_2$	H	220–224/0.3[a]	72	378
$(CH_2)_2NEt_2$	$(CH_2)_2NEt_2$	H	230–234/0.3[a]		378
H	$(CH_2)_2NEt_2$	H	155–156		378
Me	$(CH_2)_2NEt_2$	H	216–219/0.4[a]	85	378
$(CH_2)_2NEt_2$	*n*-Pr	H	206–208/0.3[a]		384
			185–188 (HCl)		

[a]Boiling point (°C/torr).

benz[*b*,*e*]azepin-11-one (**1171a**) in about 50% yield. Solvolysis of **1169** in ethanol gave **1171b**, which was identical to the product obtained from **1161** and triethyloxonium tetrafluoroborate.

This observation requires correction of the report by Moriconi and Maniscalco (388), who argued, on the basis of spectral and chemical evidence, that the Meerwein adduct of **1163** had the azatropone structure **1172**. Cooke and Russell (387) showed that **1171b** has ultraviolet and infrared absorptions very similar to those of the carbocyclic analog. The latter molecule is known to be nonplanar, with the seven-membered ring in a boat conformation. The principal chemical argument advanced by Moriconi and Maniscalco in support of structure **1172** was based on the two products obtained by reduction with sodium borohydride (**1173**) and subsequent hydrolysis (**1174**). Cooke and Russell suggested that **1173** and **1174** were unacceptable on general chemical grounds and were also inconsistent with reported spectral data.

An alternative interpretation of these reactions, based on structure **1171b**, is that the borohydride reduction product is **1175**, and the product of subsequent acid hydrolysis is **1176**. This conclusion was reinforced by the unequivocal preparation of **1176** from *o*-(*o*-aminobenzoyl)benzoic acid. The properties of **1176** were very similar to those reported for **1174**. Simple imino ethers are known to be reduced to amines by borohydride. However, in compounds such as **1171b** it is probable that the carbonyl group would be preferentially attacked.

Another preparation of dibenz[*b*,*e*]azepin-11-ones (**1178**) involves oxidation of the corresponding 5,6-dihydro compounds **1177** with manganese dioxide (389,390) (Eq. 143). The C=N bond in **1178** appears to be quite labile. Thus MacDonald and Proctor (389) observed that **1178** (X = 8-OMe, Y = H) could be converted to **1179** merely by recrystallization in ethanol.

Photolysis of the azide **1180** (X = N_3), prepared from the chloride **1180** (X = Cl), afforded the 11-ketone **1182** in good yield (391) (Eq. 144). The

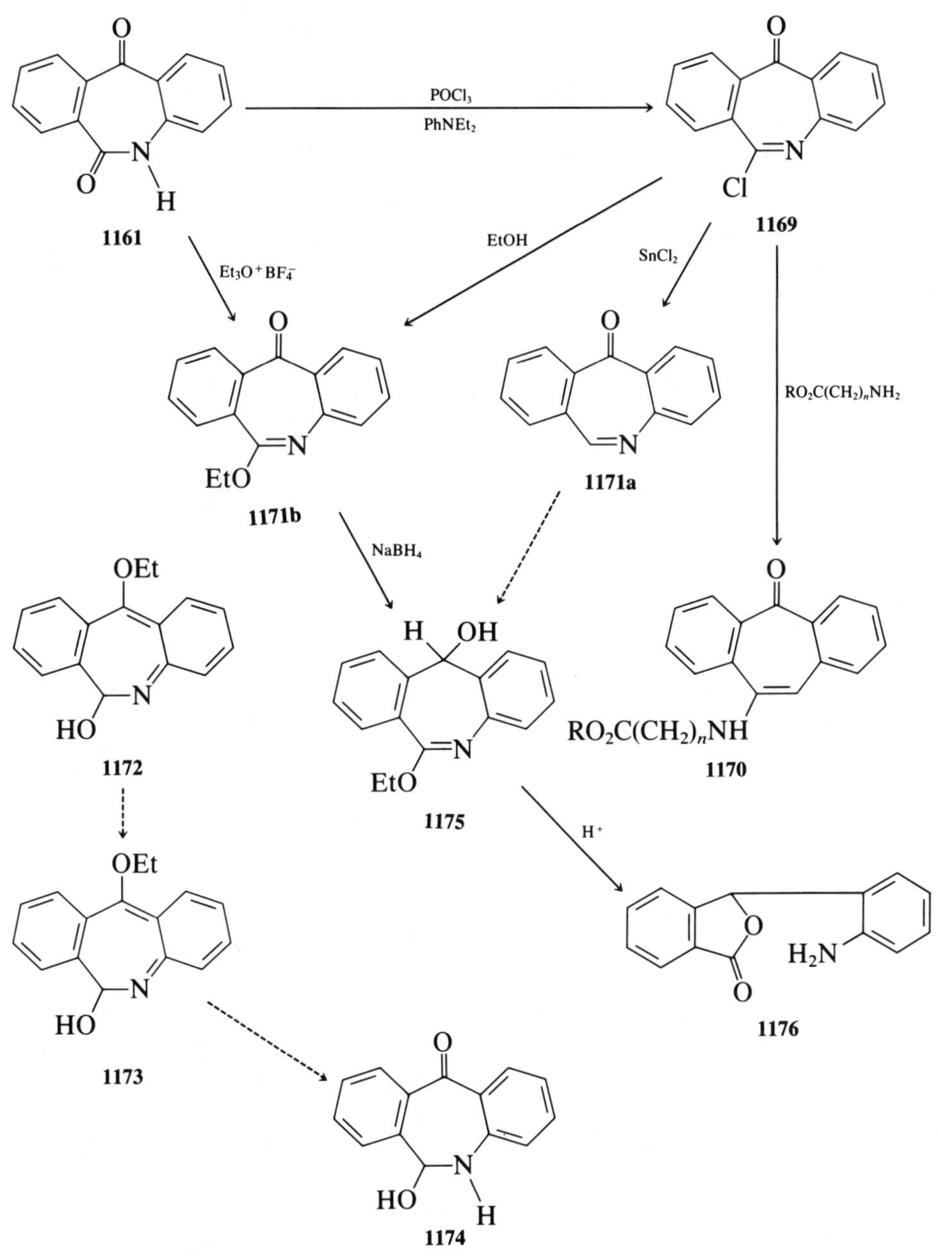

$POCl_3$
$PhNEt_2$
1161
1169
$Et_3O^+BF_4^-$
EtOH
$SnCl_2$
$RO_2C(CH_2)_nNH_2$
EtO
1171b
1171a
$NaBH_4$
OEt
HO
1172
H OH
1175
$RO_2C(CH_2)_nNH$
1170
H^+
1173
H_2N
1176
1174

Figure 157

X, Y; MnO_2; 1177; 1178; (143)

EtOH (air) (X = 8-MeO, Y = H); MeO; 1179

Ph X; $h\nu$ or Δ; 1180, X = Cl,N_3; 1181; 1182 (144)

Method	**1181:1182**
Heat	35:65
$h\nu$	15:85

photolysis reaction was more selective than the thermolysis of **1180**, which gave a 1:2 mixture of **1181** and **1182**. Dibenz[*b*,*e*]azepin-11-ones prepared as indicated in Fig. 157 and Eqs. 143,144 are listed in Table 70.

The synthesis of 5,6-dihydrodibenz[*b*,*e*]azepin-6-ones (6-morphanthridones, Table 71) has been accomplished in several ways (Eq. 145). Werner and Rossi (361) reduced the 6,11-diones **1183** to **1184** with palladium in acetic acid as well as with copper chromite and zinc–ammonium hydroxide. Protiva and co-workers (392) performed the reduction with zinc–acetic acid or with hydrogen in the presence of palladium catalyst and perchloric acid. Truce and Emrick (360) reduced the *o*-(*o*-aminobenzoyl)benzoic acid **1185**, obtained from hydrolysis of the 6,11-dione, with zinc–ammonium hydroxide to *o*-(*o*-aminobenzyl)benzoic acid, which on warming cyclized readily to **1186**. The nonreduced amino acid **1185** did not undergo ring closure. However, lac-

TABLE 70. 11H-DIBENZ[*b*,*e*]AZEPIN-11-ONES

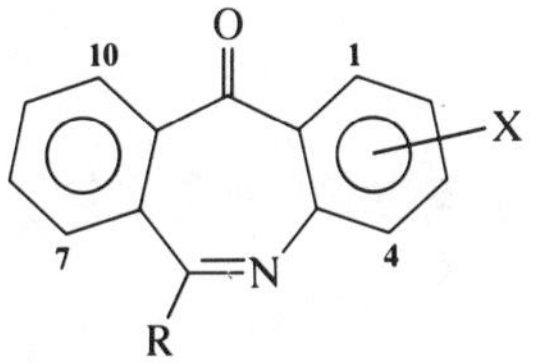

R	X	mp (°C)	Salt	Yield (%)	Spectra	Refs.
H	H	93–94		70	ir,nmr,ms	343,389
H	8-MeO	129–133		79	ms,ir	389
		142–145		98	ir,nmr,uv	387
H	3-Cl, 8-MeO	163		70	ir,uv	390
Ph	H	116			ir,nmr,ms	391
Cl	H	109–110		60	uv	386,389
		129–130		72	ir	427
$NH(CH_2)_5CO_2Et$	H	100–101				386
$NH(CH_2)_5CO_2H$	H	134–136	HCl			386
$NH(CH_2)_5CO_2Et$	2-Cl	114–118	$HCl\cdot 0.5H_2O$			386
$NH(CH_2)_6CO_2H$	H	116–118				386
$NH(CH_2)_6CO_2Et$	2-Cl	120	HCl			386
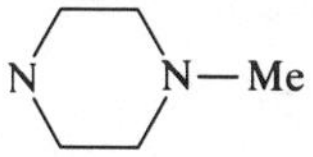 N N—Me	H	130–132		51	uv,ir	427
EtO	H	103		50	uv,nmr,ir	389
$O(CH_2)_2NMe_2$	H	184	Oxalate			440

tamization of **1185** to **1183** could be carried out via the action of sodamide on the *N*-acetyl derivative (393).

Beckmann rearrangement of anthrone oximes **1187** has been reported to give good yields of the 6-morphanthridones **1188** (394–396) (Eq. 146). The rearrangement was brought about by the action of polyphosphoric acid or phosphorus pentachloride. The starting oximes **1187**, with $R_1 = CH(CO_2Et)_2$ or CMe_2NO_2, were obtained by conjugate addition of the appropriate carbanions to 9-nitroanthracene, followed by reaction of the Meisenheimer complex with a benzyl halide to form the oxime and benzaldehyde via a Hass–Bender reaction.

An important synthesis of 6-morphanthridones (**1190**) which is described in a broad series of patents (397–401), involves intramolecular Friedel–Crafts–type cyclization of the isocyanates **1189** (Eq. 147), which are conveniently prepared from the corresponding anilines. A related cyclization of thiourea and thiocarbamate derivatives of similar anilines has likewise been patented (402).

A number of *N*-aminoalkyl derivatives (**1192**) of 6-morphanthridones have been patented (403–406) for antidepressant activity and are listed in Table

TABLE 71. DERIVATIVES OF 5,6-DIHYDRO-11H-DIBENZ[*b,e*]AZEPIN-6-ONE (6-MORPHANTHRIDONES)

R_1	R_2	R_3	R_4	mp (°C)	Yield (%)	Spectra	Refs.
H	H	H	H	193–194			360
				198–199	77		400
				193–196	96		401
				202–204			361
				200–202			392
				201–203	80–95,96		397,398
				221–233[a]			402
				218–219[a]			410
Me	H	H	H	106–107	61	ir,nmr	462
$CH_2C{\equiv}CH$	H	H	H	110–111	72	ir,nmr	462
CH_2Ph	H	H	H	148–150	75	ir,nmr	462
				150–151	74		447
H	H	H	2-Cl	261–262	93		397,399
				245–246	93		410
				259–261			400
H	H	H	3-Cl	273–275	95		397,398,400
				258–259.5			410
H	H	H	8-Cl	239–240	89		397,399
				237–239			400
				236–237			410
H	Me	H	H	203–206	87		397,398
				202–205	84		405
				204–205			394

H	Et	H	H	198–200	77		397,399
Et	Et	H	H	67–68	11	ir,uv,nmr	388
H	Me	H	10-Me	235	67		436
H	Me	H	2-Cl	235–236	87		397,399
H	Me	H	3-Cl	196–198			397
H	Me	H	8-Cl	258–260			397
				254–259	25		399
H	Me	Me	H	252–253	80		394
					73		396
H	CH_2CO_2Et	H	H	150–151			417
H	CH_2CO_2H	H	H	267–268			417
H	CH_2COCl	H	H	102–103			417
H	CH_2CONH_2	H	H	254–255	72		417
H	$CH(CO_2Et)_2$	H	H	197–201	85	ir,nmr	395
H	$CMe(CO_2Et)_2$	H	H	153–154.5	49	ir,nmr	395
H	CMe_2NO_2	H	H	225–226	63	ir,nmr	395
H	CHO	H	H	243–245	68–84		379
H	CH=NOH	H	H	242–243	80		379
Ac	CN	H	H	172–174	80		379
H	CN	H	H	262–263	76		379
H	CO_2H	H	H	221–223	51		379
Me	CHO	H	H	159–160	87–92		379
Me	CH=NOH	H	H	181–182	85		379
Me	CN	H	H	171–172	77		379
Me	CO_2H	H	H	197–198	53–65		379
Me	CO_2Me	H	H	186–188	80–88		379
H	H	OH	H	247–248	92	nmr	388
				245–248	27	ir	414
				228–230			392
				198–200			376
H	H	OMe	H	215–219	80	ir	392
H	H	Cl	H	226			376
H	H	OH	2-Cl	273–274			407,408

(*continued*)

TABLE 71. (*Continued*)

R_1	R_2	R_3	R_4	mp (°C)	Yield (%)	Spectra	Refs.
H	H	Cl	2-Cl	254–257			407,408,431
H	H	OH	2-Br	270–272			376
Me	H	OH	H	195–197			408
Me	H	Cl	H	162–167			408
				164–165			376
Me	H	OH	2-Cl	195–198	93		375
Me	H	Cl	2-Cl	188–190	93		375
H	Me	OH	H	179–180	80.5		411
				170			436
Me	CH_2CO_2Et	OH	2-Cl	162–165	72	ir,nmr	418
H	H	NH_2	2-Cl	240–242			407
H	H	NEt_2	2-Cl	178–180			431
H	H	NMe_2	2-Br	216–217			376
H	H	NEt_2	H	159–160	66		376
H	H	N-succinimido (structure)	H	279–280			376
H	H	N-phthalimido (structure)	H	244–245			376
H	H	N(Ac)Et	H	221–222			376
H	H	4-methylpiperazin-1-yl (structure: N NMe)	2-Cl	230–232 279–281 (MeI) 278 (EtI)			407

Ac	H	N NMe	2-Cl	193–195	407
H	H	N NMe	H	231–233	407
Ac	H	N NMe	H	223–225	407
Me	H	N NMe	H	156–158	407
Me	H	N NMe	2-Cl	285–286 (HI)	407
H	H	N NPh	2-Cl	248–250	407
H	H	N NCH_2Ph	H	208–209	407
H	H	N NCH_2Ph	2-Cl	150–155	407
Me	H	N NCH_2Ph	H	255–257 (MeI)	407
Me	H	N NCH_2Ph	2-Cl	214–216 (MeI)	407
H	H	N NCH_2CH_2OH	2-Cl	241–243	407

(continued)

TABLE 71. *(Continued)*

R_1	R_2	R_3	R_4	mp (°C)	Yield (%)	Spectra	Refs.
H	H	N O	H	206.5–208.5			407
H	H	N O	2-Cl	238–239			407
H	H	N NCO_2Et	2-Cl	212–213			407
Me	H	N NCO_2Et	2-Cl	164–167			407
Ac	H	N NCO_2Et	2-Cl	196–198			407
H	H	N NCH_2CH_2OH	H	210–212			407
Me	H	N NCH_2CH_2OH	H	274–276 (HCl)			407
H	H	N NCH_2CH_2OH	2-Cl	133–136 ($2.5H_2O$)			407
H	H	N NCH_2CH_2Cl	H	195–196			407
H	H	N NCH_2CH_2OMe	H	169–172 (HCl)			407

H	H	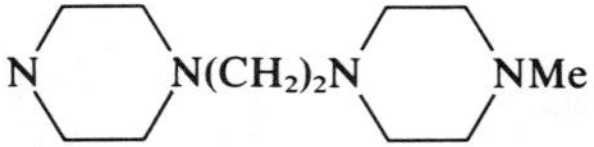 N N$(CH_2)_2$N NMe	2-Cl	153–156	407
H	H	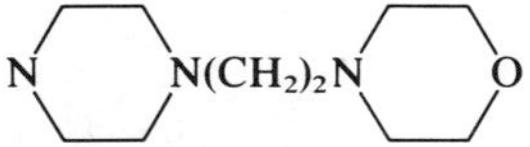 N N$(CH_2)_2$N O	H	227–228	407
H	H	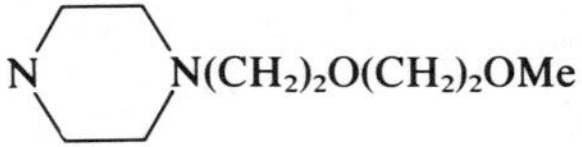 N N$(CH_2)_2O(CH_2)_2OMe$	H	155	407
H	H	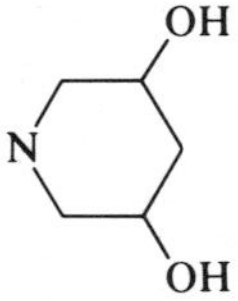 N (OH, OH)	H	180–188	407
H	H	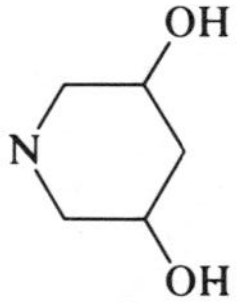 N (OH, OH)	2-Cl	172–173	407
H	H	$O(CH_2)_2NMe_2$	2-Cl	164–166	408
H	H	$O(CH_2)_2NMe_2$	H	179–180	408
H	H	$O(CH_2)_3NMe_2$	2-Cl	131–132	408
H	H	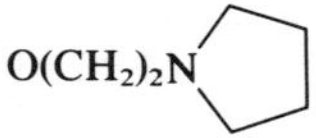 $O(CH_2)_2N$	2-Cl	135–136 222–233 ($HCl \cdot 0.5H_2O$)	408
H	H	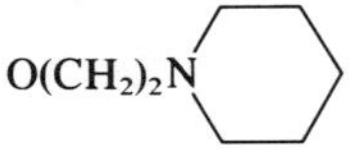 $O(CH_2)_2N$	2-Cl	142–143	408

(continued)

TABLE 71. (*Continued*)

R_1	R_2	R_3	R_4	mp (°C)	Yield (%)	Spectra	Refs.
H	H	OCH_2–(1-methyl-2-piperidinyl)	2-Cl	151–153 193–195 (MeI)			408
H	H	OCH_2–(1-methyl-3-piperidinyl)	H	126–130			408
H	H	OCH_2–(1-ethyl-3-piperidinyl)	H	254–255 (HCl)			408
H	H	$O(CH_2)_2N$(morpholino)	H	135–137			408
H	H	O–(1-methyl-4-piperidinyl)	H	148–150			408
H	H	O–(1-methyl-3-piperidinyl)	H	160 270–271 (HCl)			408
H	H	$S(CH_2)_2NMe_2$	2-Cl	141–143 246–248 (picrate) 262–264 (MeI)			408
$(CH_2)_3NMe_2$	H	H	H	236–239 (HCl)			361
$(CH_2)_3NMe_2$	H	H	H	166–170 ($MeCl{\cdot}H_2O$)			361
$(CH_2)_2N(i\text{-}Pr)_2$	H	H	H	202–210 (HCl)			361

CH_2–(2-imidazolinyl)	H	H	H	180 (maleate)			361
$(CH_2)_2NMe_2$	H	H	2-Cl	264–268 (HCl)	63		403
$(CH_2)_2NMe_2$	H	H	3-Cl	85–88			403
$(CH_2)_2NMe_2$	H	H	8-Cl	71–74			403
$(CH_2)_2NMe_2$	Me	H	H	239–241 (HCl)	73		404,405
$(CH_2)_2NMe_2$	Et	H	H	85–88			404
$(CH_2)_3NMe_2$	Me	H	H	172–174 (HCl)	74		404,405
$(CH_2)_2NMe_2$	Me	Me	H	209–210 (HCl)			406
$(CH_2)_3NMe_2$	Me	H	2-Cl	206–209 (HCl)			393
$(CH_2)_2NMe_2$	Me	H	2-Cl	175–178/0.05[b]	68		405
				246–252 (HCl)			393
$(CH_2)_2NMe_2$	Me	H	3-Cl	75–79			393
				260–264 (HCl)			
$(CH_2)_2NMe_2$	CMe_2NO_2	H	H	160–162	44	ir,nmr	395
$(CH_2)_2NMe_2$	$CH(CO_2Et)_2$	H	H	142–146	22	ir,nmr,ms	376
				162.5–165 (HBr)			
H	$(CH_2)_3NMe_2$	OH	H	188–189	42	ir	414
				152–153	24		365
				148.5–149.5	29		415
H	$(CH_2)_3NMe_2$	H	H	131–133[c]	40		414
				220–222 (HCl)			
				232–234 (HI)			
Me	$(CH_2)_3NMe_2$	H	H	150–152	68		365
H	$(CH_2)_3NMe_2$	OH	2-Br	190–192			413
H	$(CH_3)_3NEt_2$	OH	2-Br	186–188			413
H	$(CH_3)_3N$(pyrrolidinyl)	OH	2-Br	216			413
H	CH_2–(N-methylpyrrolidin-3-yl) NMe	OH	H	88–110		ir	416

(*continued*)

TABLE 71. (*Continued*)

R_1	R_2	R_3	R_4	mp (°C)	Yield (%)	Spectra	Refs.
H	CH_2— NMe	H	H	141–145 179–180 (maleate)			416
NH_2	Me	H	H	142–145	Good		409
NH_2	Me	Me	H	223–225	Good		409
H	$=CHCONH_2$		H	315–318	60		417
Me	$=CHCONH_2$		H	250–252	70		417
H	$=CHCONH_2$		H	154–155	75		417
H	$=CH_2$		H	209–211	95		411
H	$=CH_2$		3-F	220–230			412
$(CH_2)_2NMe_2$	$=CH_2$		H	105–106 205–207 (HCl)	62		411
$(CH_2)_3NMe_2$	CH_2		H	155–157 (HCl)	53		411
$(CH_2)_2N$	$=CH_2$		H	116–118 (maleate)	66		411
$(CH_2)_3N$	$=CH_2$		H	147–148 (succinate)	70		411
$(CH_2)_2N$ —OMe	$=CH_2$		H	201–203 (HCl)	77		411
$(CH_2)_3N$ OMe	$=CH_2$		H	112–114 (succinate)	36		411

H	$=CH(CH_2)_2NMe_2$	2-Br	135–142[d]			413
H	$CH(CH_2)_2NEt_2$	2-Br	114–119[d]			413
H	$=CH(CH_2)_2N$(pyrrolidine)	2-Br	243–247[e] (oxalate) 254–255[e] (oxalate)			413
H	$=C_5H_8NMe$ (4-piperidylidene, NMe)	3-Cl	183–186			421
H	spiro-epoxide (O)	H	222–224	74–86		379
H	spiro-epoxide (O)	2-Cl	204–207	60	ir,nmr	418
Me	spiro-epoxide (O)	H	116–117	74–80		379
Me	spiro-epoxide (O)	2-Cl	—[f]			418
H	$—OCH_2CH_2O—$	H	223–224	53		365
H	$—OCH_2CH_2O—$	2-Cl	269–273	53		365
Me	$—OCH_2CH_2O—$	H	190–191	64		365
Me	$—OCH_2CH_2O—$	2-Cl	173–176	82		365

[a]6-Thione compound.
[b]Boiling point (°C/torr).
[c]Impure free base.
[d]*Cis*/*trans* mixture?
[e]*Cis* and *trans* isomers?
[f]No details, but epoxide was opened to form described derivatives.

H_2/Pd, HOAc or H_2/copper chromite

1183 → 1184, 50%

H_3O^+

(145)

CO_2H H_2N

1185

(1) Zn, NH_4OH, Na_2CO_3 (2) Δ

1186, 75%

R_1 R_2

1187

PPA or PCl_5

(146)

	R_1	R_2	Yield (%)
1188a	H	Me	52
1188b	Me	Me	50
1188c	H	$CH(CO_2Et)_2$	85
1188d	H	$CMe(CO_2Et)_2$	49
1188e	H	$CMeNO_2$	63

$AlCl_3$

(147)

1189, R = H, Me, Et;
X = Y = H, Cl, Me

1190, 80–95%

71. These substances were prepared generally by the action of sodamide on **1191**, followed by treatment with a dialkylaminoalkyl halide (Eq. 148). A series of 11-substituted aminoalkylamines **1193** (X = NR_3) (407) and aminoalkyl ethers (X = O) (408) have been patented as antiulcer or antisecretory agents, or anticonvulsants (376). These substances were prepared by displacement of chloride **1191**, which was accessible from the 11-hydroxy compound by reaction with thionyl chloride. The *N*-amino derivative **1194** was obtained from **1191** (A = H) by reaction with *O*-(2,4-dinitrophenyl)-hydroxylamine, and was patented as an anti-inflammatory agent and anticoagulant (409,419).

6-Morphanthridones **1195** have been converted to the corresponding thiones **1196** in excellent yield by the action of phosphorus pentasulfide (Eq. 149),

A

Alk—NR_1R_2

1192

(148a)

(1) $NaNH_2$
(2) Cl-Alk—NR_1R_2
(A = H, alkyl)

X—Alk—NR_1R_2

A

H

1191

$HX—Alk—NR_1R_2$
(A = Cl)

H

1193, X = O, NR_3

(148b)

(1) $NaNH_2$
(2) H_2NO—C_6H_3—NO_2, O_2N
(A = H)

NH_2

1194

(148c)

(149a)

1195 → (P_2S_5) → **1196**, 93%

(149b)

1195 → (H_2NNHCO_2Et) → **1197**, X = H (44%), X = Cl (59%)

and have been used as intermediates for the synthesis of other heterocyclic systems, such as **1197** (410), by annelation with ethyl carbazate.

6,11-Morphanthridinediones are transformed into 6-morphanthridones by addition of hydride or organometallic agents to the 11-keto function. Thus (Fig. 158) reaction of **1198** with either sodium borohydride (375) or aluminum isopropoxide (388) affords the alcohol **1199** in 90–95% yield; a similar high yield is obtained by catalytic hydrogenation with platinum–acetic acid (392). Additions of alkyl Grignard reagents (411,412) and dialkylaminoalkyl Grignard reagents (365,413–416) have also been reported. The latter reaction gives products (**1200b**) that have been patented as antidepressant and antispasmodic agents. The tertiary OH group of **1200** can be removed by hydrogenolysis (414) to give **1201**. Where R = Me, dehydration of **1200a** has been carried out to obtain 11-*exo*-methylene compounds (411,412), which on *N*-alkylation gave **1202**, patented as antidepressant agents. Drukker and Judd were unable to dehydrate **1200b** ($n = 3$, $R_1 = R_2 =$ Me, X = H) (365,414), but a related patent (413) claimed that **1203** could be prepared from **1200b** by dehydration with sulfuric acid. Addition of phosphonate anions to the 11-keto group of **1198** gave **1204**, of unspecified stereochemistry, in good yield (417). Catalytic hydrogenation of **1204** yielded **1205**. These

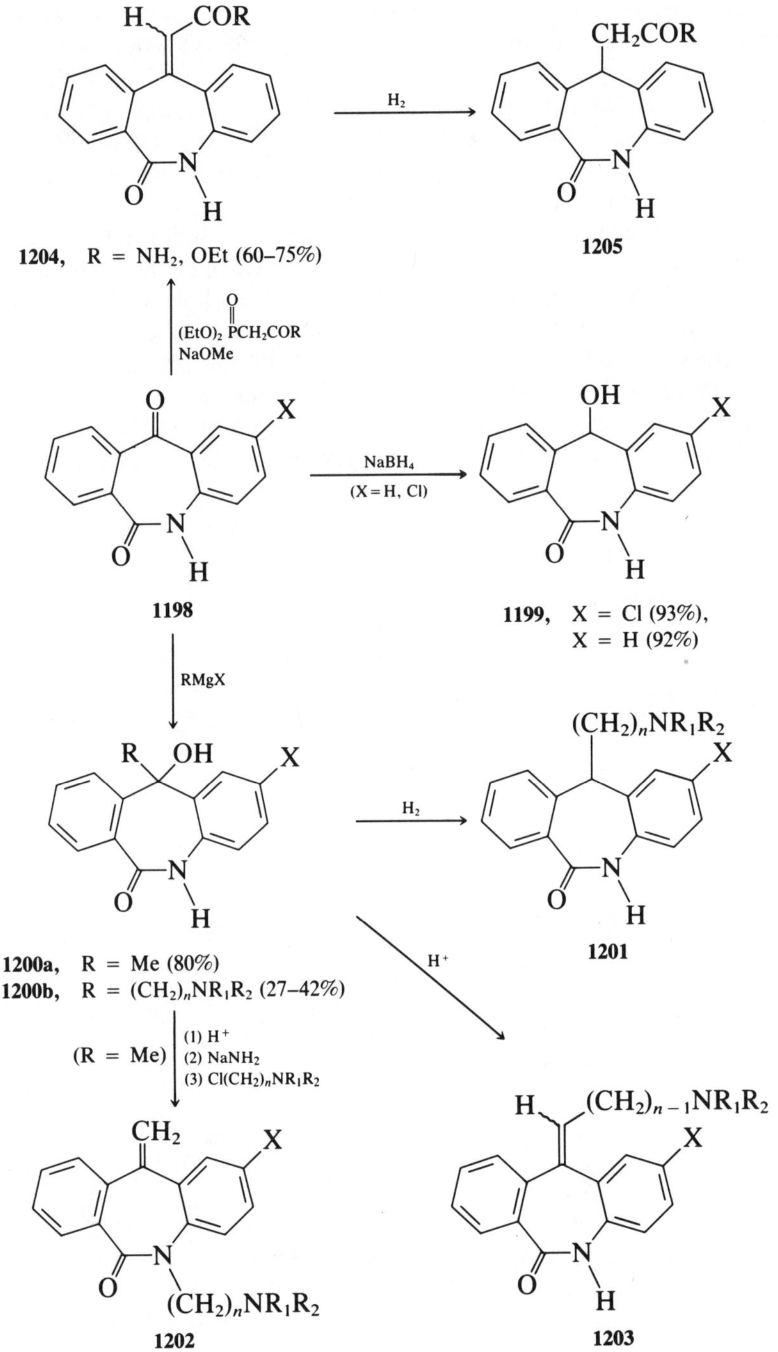

H COR

CH$_2$COR

H_2

O N H

1204, R = NH$_2$, OEt (60–75%)

1205

(EtO)$_2$ P(O)CH$_2$COR
NaOMe

O X

OH X

NaBH$_4$
(X = H, Cl)

1198

1199, X = Cl (93%),
X = H (92%)

RMgX

(CH$_2$)$_n$NR$_1$R$_2$
X

R OH X

H_2

1200a, R = Me (80%)
1200b, R = (CH$_2$)$_n$NR$_1$R$_2$ (27–42%)

H^+

1201

(R = Me) (1) H^+
(2) NaNH$_2$
(3) Cl(CH$_2$)$_n$NR$_1$R$_2$

H (CH$_2$)$_{n-1}$NR$_1$R$_2$
X

CH$_2$ X

(CH$_2$)$_n$NR$_1$R$_2$

1202

1203

Figure 158

substances have been patented for muscle relaxant, tranquilizing, and anticonvulsant activity. Compounds given in Fig. 158 are listed in Table 71.

The spiroepoxides **1206** have been obtained from the diones **1161** by the Corey procedure using dimethylsulfoxonium methylide (379), and have been rearranged to the aldehydes **1207** with boron trifluoride (Fig. 159). The aldehydes have been converted into various 11-substituted 6-morphanthridones (**1208**), generally in good yields (Table 71).

The spiroepoxides **1209** (R = H, Me) have been similarly prepared (Fig. 160). Hardtmann and co-workers (418) noted that opening of the epoxides with ammonia gave, presumably via the aminoalcohols **1210**, the azepine ring-opened products **1211**. When hydrazine was employed in place of ammonia, the corresponding *N*-amino product **1212** was produced.

Hardtmann (418) also reported a thermal rearrangement of the 11-hydroxy ester **1213**, which was prepared in 72% yield from the 6,11-dione by a Reformatsky reaction using zinc and ethyl bromoacetate. It was proposed that

R	Yield (%)
H	74–86
Me	74–80

R	Yield (%)
H	64–84
Me	87–92

R	X	Yield (%)
H	CH=NOH	80
Ac	CN	80
H	CO_2H	51
Me	CH=NOH	85
Me	CN	77
Me	CO_2H	53–65

Figure 159

NH_3, 100°C

1209, 60% **1210**

N_2H_4 (R = H)

1212 **1211**

R	Yield (%)
H	30
Me	71

Figure 160

the reaction proceeds via the transannular intermediate **1214**, which on ring opening gives rise to the intermediate phthalide **1215** (Fig. 161). Lactam formation with concomitant loss of ethanol then yields the spirophthalide **1216**. Reaction of **1213** with hydrazine gave the carbostyril **1217**. A similar rearrangement had been reported earlier (375) by Hardtmann and Ott. Treatment of the 6,11-dione **1218** with ammonia at 110°C generated **1219** (Fig. 162). Amination of **1220** under similar conditions gave **1221**, but in lower yield than was obtained from borohydride reduction of **1219**. The preparation of **1220** was achieved conveniently from **1218** by reduction to the alcohol followed by halogenation with thionyl chloride. The conversions of **1218** to **1219** and of **1220** to **1221** presumably occur via a transannular intermediate similar to **1214**.

The preparation of a 5,6-dihydro-11-morphanthridone (**1222**; R = Me, X = H) was first reported by Wittig and co-workers from the reaction of *o*-methoxymethylbenzonitrile with *o*-lithiodimethylaniline (420). However, the low yields in this reaction prompted Drukker and Judd (365) to devise alternative routes (Fig. 163) to these synthetically useful ketones. One route utilized the 6,11-diones **1198** as starting materials, and required ketalization

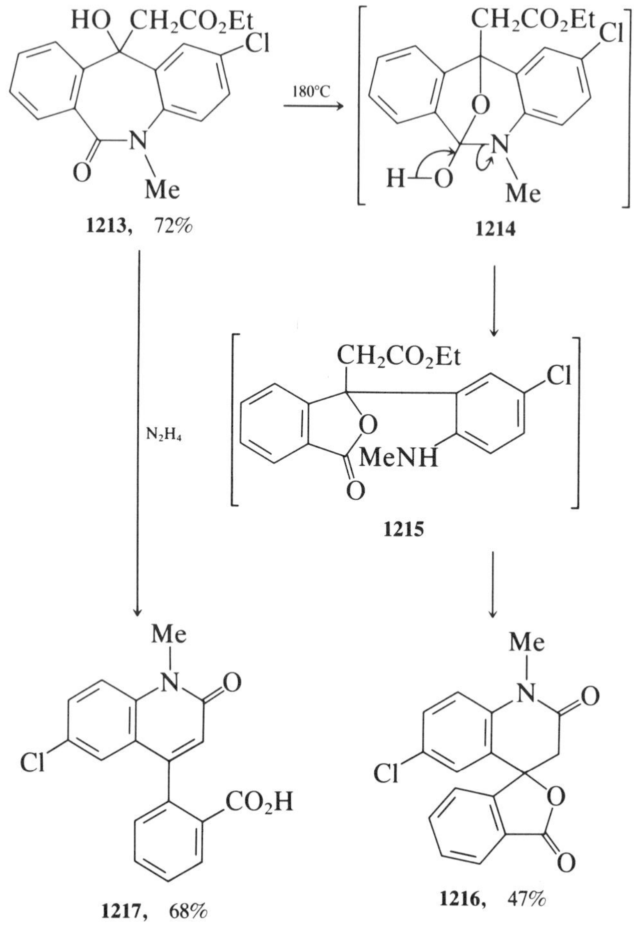

1213, 72%

1214

1215

1217, 68%

1216, 47%

Figure 161

of the 11-keto group. Although this proved troublesome, ketalization could be accomplished satisfactorily by heating **1198** in a large excess of ethylene glycol with a catalytic amount of *p*-TsOH, while distilling off glycol and water. *N*-Methylation by successive treatment with sodamide and methyl iodide then gave **1223a**, which on lithium aluminum hydride reduction yielded the amine **1223b**. Acidolysis of the ketal protecting group in **1223b** yielded **1222**.

The second route had as its key step the direct oxidation of the 11-position of *N*-acetylhomoacridan (**1224**) with chromium trioxide in glacial acetic acid to give **1225**. Saponification of **1225** afforded **1222** (R = H). Acylation of the nitrogen is obviously important in this reaction, since Protiva and co-workers reported that similar oxidation of *N*-methylhomoacridan gave (in about 30% yield) 5,6-dihydro-5-methyl-6-morphanthridone (ir band at 1640

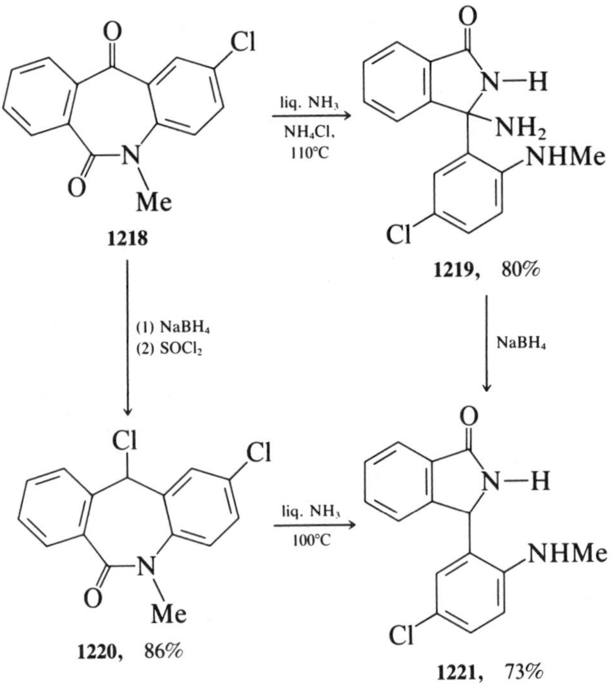

liq. NH_3, NH_4Cl, 110°C

1218

1219, 80%

(1) $NaBH_4$
(2) $SOCl_2$

$NaBH_4$

liq. NH_3, 100°C

1220, 86%

1221, 73%

Figure 162

cm^{-1}) (392). The ketone **1225** was also converted in low yield to **1226a**, and subsequently to **1226b**, via a Grignard reaction. The stereochemistry of the products was not discussed. Specific examples of compounds prepared as shown in Fig. 163 are listed in Tables 72 and 74.

Proctor and co-workers (389,390) devised the alternative route shown in Fig. 164 to obtain **1229**. *N*-Benzylanthranilic acid derivatives had been reported to be unsuitable substrates for cyclization to 5,5-dihydrobenz[*b*,*e*]azepin-11-ones (361,392). However, the Proctor group's experiences with tetrahydrobenz[*d*]azepin-1-ones indicated the route merited further exploration. Low-temperature Friedel–Crafts cyclization of the acid chloride of **1227** (R_1 = MeO, R_2 = H) have a 73% yield of **1228** (Fig. 164). However, with R_1 = R_2 = H or OMe, extensive cleavage of the benzylic C—N bond occurred in preference to cyclization. Proctor reasoned that the *m*-methoxy group should faciliate cyclization while not contributing additionally to stabilization of the incipient benzylic carbonium ion, and his prediction was borne out experiementally. Hot polyphosphoric acid was found to be an effective means of accomplishing *N*-detosylation in the presence of an electron-withdrawing group in the *ortho* or *para* positions. The ketone **1229a** (X = H) was brominated to the dibromide **1229b**, whereas the deactivated *N*-sulfonyl derivative **1228** failed to react. The infrared stretching frequency

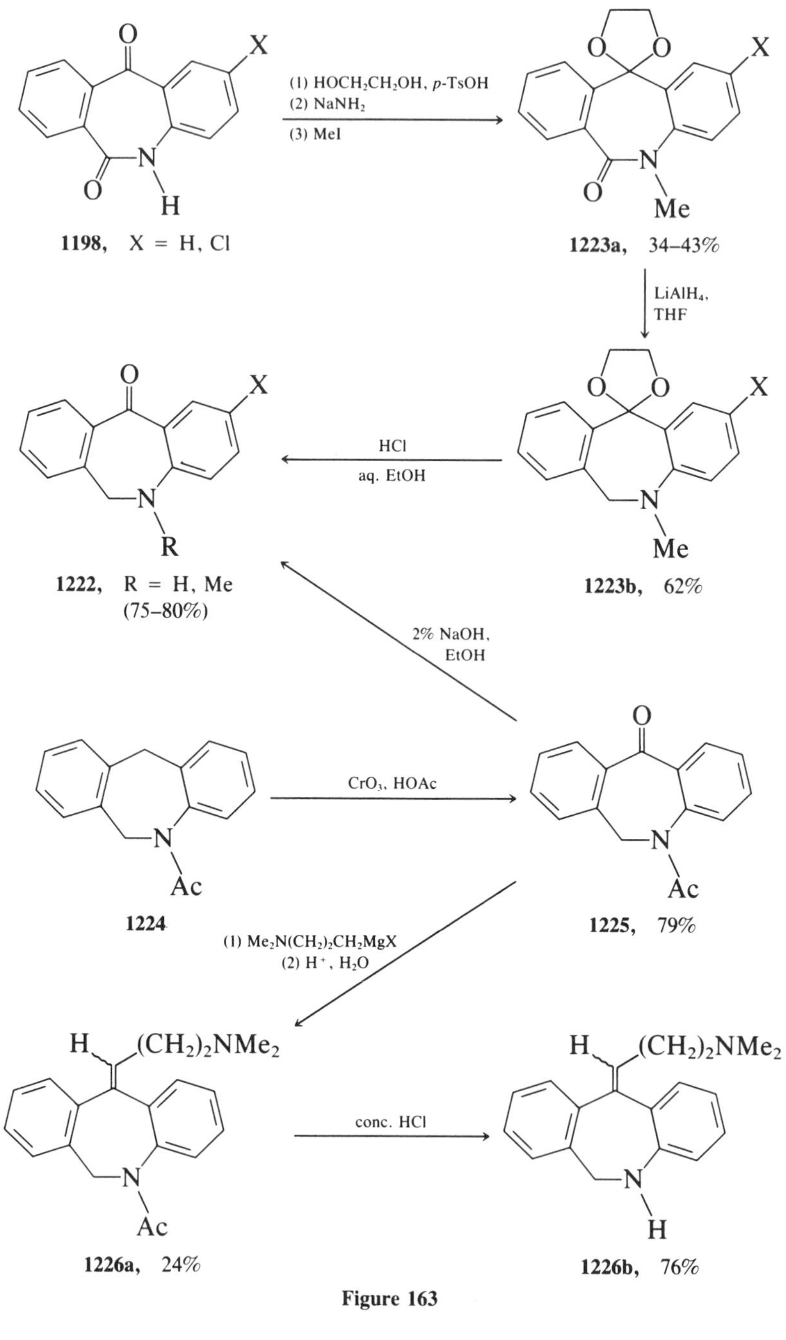

1198, X = H, Cl

(1) $HOCH_2CH_2OH$, *p*-TsOH
(2) $NaNH_2$
(3) MeI

1223a, 34–43%

$LiAlH_4$, THF

1223b, 62%

HCl
aq. EtOH

1222, R = H, Me
(75–80%)

2% NaOH, EtOH

1224

CrO_3, HOAc

1225, 79%

(1) $Me_2N(CH_2)_2CH_2MgX$
(2) H^+, H_2O

1226a, 24%

conc. HCl

1226b, 76%

Figure 163

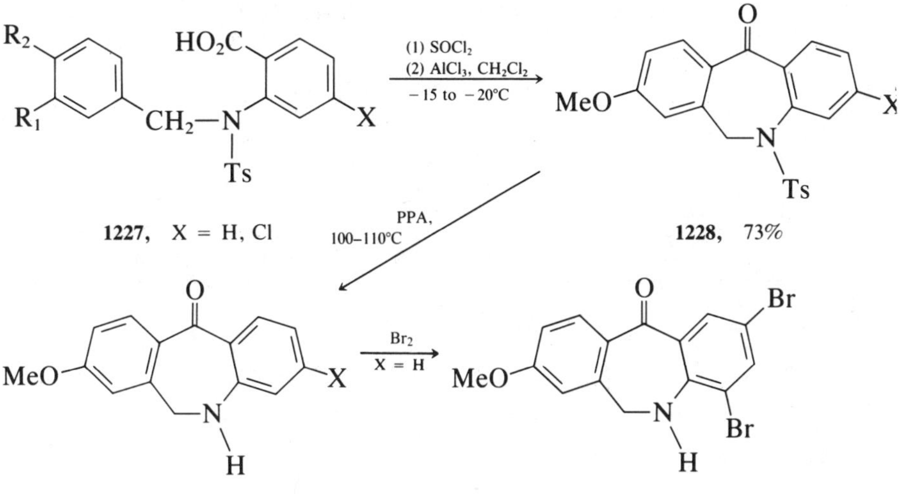

Figure 164

for **1228** was at 1640 cm^{-1}, while that for **1229a** was at 1605 cm^{-1}. Although an oxime could be prepared from **1228** and **1225**, none could be prepared from **1222**. These results suggest considerable loss of carbonyl group character in these substances, presumably as a result of enolization into the seven-membered ring. As indicated earlier (Eq. 143), oxidation of 5,6-dihydrodibenz[*b*,*e*]azepine-11-ones with manganese dioxide leads to the corresponding diben[*b*,*e*]azepin-11-ones. Known examples of 5,6-dihydrodibenz[*b*,*e*]azepin-11-ones are listed in Table 72.

Drukker and Judd (365) condensed the ketones **1230** with aminoalkyl Grignard reagents (Fig. 165) to obtain **1231**. In contrast to the difficulties they reported for the dehydration of **1200b** to **1203** (Fig. 158), the adducts **1231** were easily converted to **1232**, patented for antidepressant and other CNS activities. A second route employed the 11-lithio species **1233**, which reacted with *N*-methyl-4-piperidone to form **1234a**. A large amount of **1233** (Li = H) was recovered, possibly indicating incomplete lithiation. Addition of the corresponding piperidyl Grignard reagent to **1230** (R = Me) gave **1234b**. Dehydration of **1234b** by pyrolysis of the acetate ester produced **1235** in quantitative yield.

Other workers have prepared **1232** from the 6-morphanthridone derivatives **1236** (Eq. 150) (413,421). The "*cis* isomers" were claimed to be active as anti-inflammatory agents (413), but evidence for the *cis* stereochemistry was not provided. Antipsychotic and antiemetic claims were also made (421). Compounds of structure **1238**, possessing antihypertensive and antiparkinsonian activity (422–424), have been prepared from **1237** by Grignard addition (and quite possibly hydride reduction). Oxidation of **1238** with manganese

TABLE 72. 5,6-DIHYDRO-11H-DIBENZ[*b,e*]AZEPIN-11-ONES

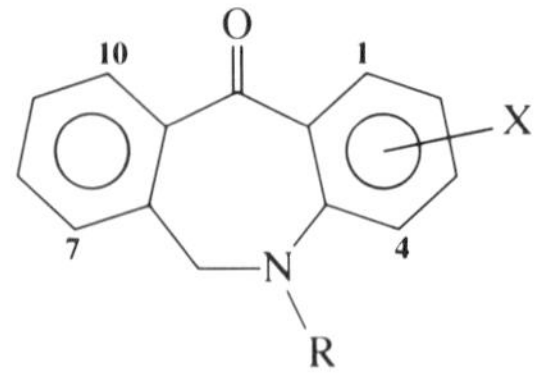

R	X	mp (°C)	Yield (%)	Spectra	Refs.
H	H	124–125	75	ir,uv	365
Me	H	108–110	52.5	ir,uv	365
		111–115	80		365
		115[a]			
		112–114	78		420
		169–170 (MeI)	29		420
CH_2Ph	H	210–0.025[b]	83	ir,uv	365
Ac	H	97–100	83		365
p-Ts	H	171	Low	ir,nmr,ms	389
p-Ts	8-MeO	198–200	73	ir,nmr	389
H	8-MeO	140–141	86	ms,ir,nmr	389
H	2,4-diBr-8-MeO	208–209	35	ms,ir	389
Ac	8-MeO	146–147	37	ir,nmr	389
p-Ts	3-Cl-8-MeO	238	~50	ir	390
H	3-Cl-8-MeO	148			390

[a]Sublimed.
[b]Boiling point (°C/torr).

dioxide gave **1237** (Eq. 151). Pure *cis* and *trans* isomers were obtained in the preparation of **1237**, but no supporting evidence of geometry was reported. Finally, polyphosphoric acid cyclization of the anilides **1239** gave **1240** (Eq. 152), patented as psychotropic and antiemetic agents (425,426). Compounds represented in Fig. 165 and Eqs. 150–152 are listed in Tables 73–75.

11H-Dibenz[*b,e*]azepines (morphanthridines) have been prepared by two basic methods. The first of these (Fig. 166) involved conversion of a 5,6-dihydro-6-morphanthridone **1241** to the corresponding chloroimine **1242** with phosphoryl chloride or phosphorus pentachloride. When R_1 in **1242** is alkyl, hydrogen, or *exo*-methylene, reaction of the chloroimine with amines generates **1243**. These compounds have been widely patented as antipsychotic or neuroleptic agents, chiefly by Schmutz and co-workers (427) and by Drukker and Judd (428). The chemistry and structure–activity relationships of perlapine (**1243**; R_1 = H, R_2R_3N = *N*-methylpiperazinyl) have been summarized and compared by Schmutz with loxapine, clozapine, and related drugs (429). The physical and chemical properties and stability of this drug have also been reported (430).

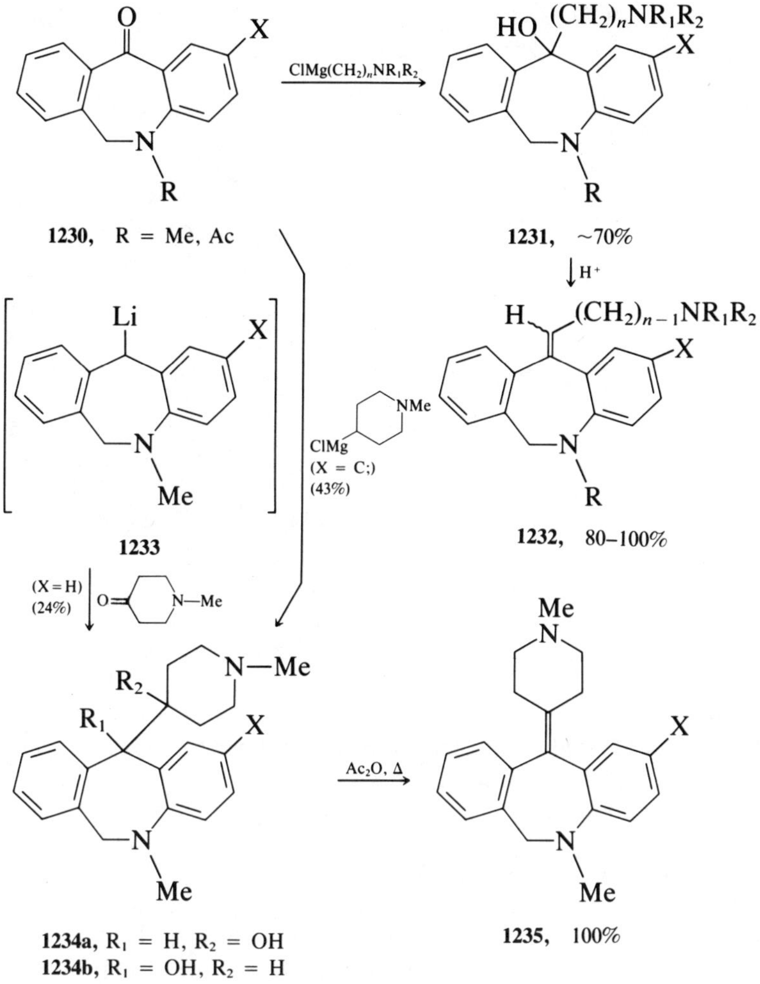

1230, R = Me, Ac

$ClMg(CH_2)_nNR_1R_2$ →

1231, ~70%

H^+ ↓

1232, 80–100%

1233

ClMg–(1-methyl-4-piperidyl) (X = Cl) (43%)

(X = H) (24%) 1-methyl-4-piperidone

1234a, R_1 = H, R_2 = OH
1234b, R_1 = OH, R_2 = H

Ac_2O, Δ →

1235, 100%

Figure 165

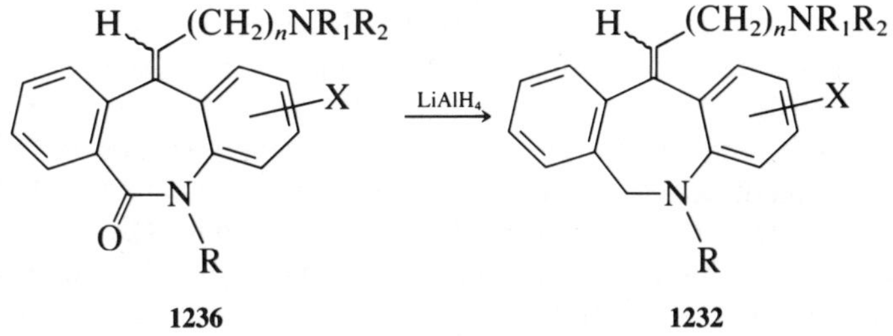

1236 —$LiAlH_4$→ **1232** (150)

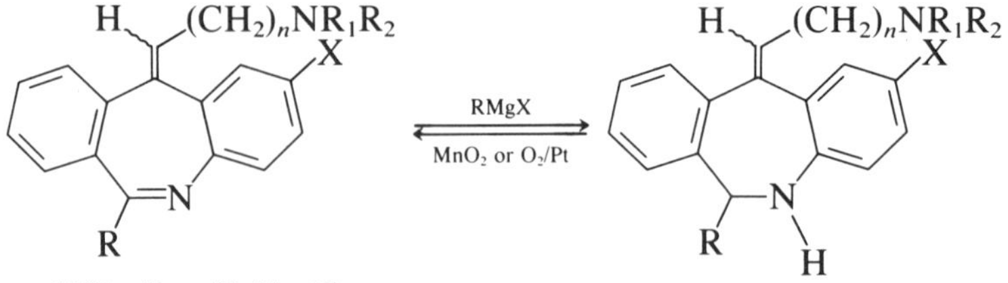

1237, R = H, Me, Ph

RMgX; MnO_2 or O_2/Pt

1238, 40–50%

(151)

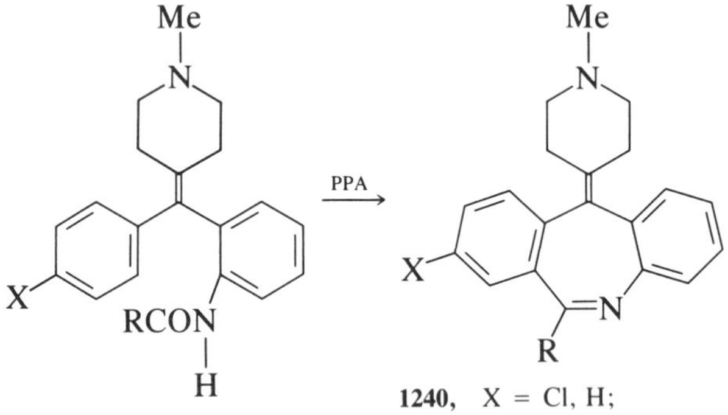

1239

PPA

1240, X = Cl, H; R = H, Me (20–40%)

(152)

When R_1 in **1242** is also chlorine, both halogens can be displaced by amines to give **1244**. These substances have been patented as antispasmodic, anti-inflammatory, and diuretic agents (407,431). A recent patent (432) claims the process of direct conversion of amide **1241** to amidine **1243** by employing titanium or zirconium tetrachloride.

A second approach to morphanthridines (Eq. 153) is the Bischler–Napieralski cyclization (392, 433–436). Ring closure of **1245** (X = CH_2) produces **1246** (X = CH_2) in yields of 25–80%. Reduction of **1246** by chemical or catalytic means gives the homoacridans **1247** (X = CH_2) in good yields (392,433). Inch and co-workers (436) have dehydrogenated a variety of ring-substituted homoacridans with sulfur to obtain the corresponding morphanthridines in yields of 49–79%. Specific compounds represented by the general structures in Fig. 166 and Eq. 153 are listed in Table 76. Schmutz and co-workers (434) have examined the Bischler–Napieralski cyclization of **1245** and the base strengths of **1246** and **1247** as a function of the heteroatom X (X = C, O, N, S).

6-Chloromorphanthridine (**1248**, X = C1) has been a useful starting point for the synthesis of new tetracyclic structures such as **1249** (396), **1250** (441), and **1251** (428) (Fig. 167). 6-Chloromethylmorphanthridine (**1248**, X = CH_2C1) is an important intermediate in the synthesis of the interesting antidepressant mianserin (**1252**) (435,442–444).

TABLE 73. DERIVATIVES OF 5,6-DIHYDRO-11-HYDROXY-11H-DIBENZ-[*b*,*e*]AZEPINE

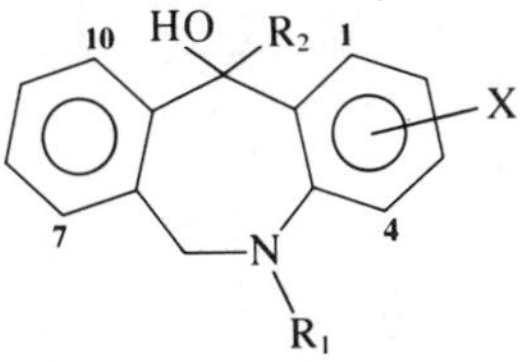

R_1	R_2	X	mp (°C)	Refs.
Me	$(CH_2)_3NMe_2$	H	133.5–134	365
Me	$(CH_2)_3NMe_2$	2-Cl	158–160	365
Me	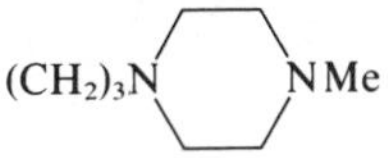 $(CH_2)_3N$ NMe	H	143–146.5	365
Me	 NMe	H	192–196	365
			152–153	465
Me	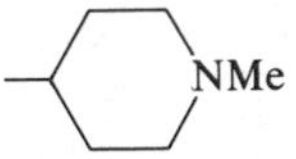 NMe	2-Cl	210–212	365

Decomposition of the diazonium tetrafluoroborate salts obtained from *o*-benzylanilines in nitrile solvents (Eq. 154) has been reported to yield various substituted morphanthridines (**1253**) (439) in yields of 39–58%, but this reaction does not seem to offer any synthetic advantage over other routes previously discussed. Table 77 lists the products of these decompositions.

Hydrogen peroxide oxidation of the parent morphanthridine **1253** ($X = R_1 = R_2 =$ H) gave predominantly **1254** along with a small amount of 6-morphanthridone (**1255**) (445). The suggested mechanistic pathway for this oxidation (Fig. 168) was through the oxaziridine **1256**, which opened to give the species **1257**, capable of undergoing phenyl migration (pathway *b*) to **1254**, or hydride migration (pathway *a*) to **1255**. Dibenz[*b*,*f*][1,4]oxazepines were more extensively studied in this report.

The first preparation of 5,6-dihydro-11H-dibenz[*b*,*e*]azepine (5,6-dihydromorphanthridine) was reported by Wittig and co-workers (420) (Fig. 169). *N*-Methyl-*N*-phenylisoindolium iodide (**1258**), on treatment with phenyllithium in refluxing ether, gave 5-methyl-5,6-diydromorphanthridine (**1259**) in 49% yield. This product was identical to that obtained by reduction of the 6,11-dione **1260** to **1261** (46% yield), followed by *N*-alkylation with methyl iodide in the presence of methyllithium (73% yield). Sodium–potassium alloy cleavage of **1259** gave **1262** (84% yield), identical to a sample prepared independently.

The mechanism suggested by the authors for the formation of **1259** is given in Fig. 170. Structures **1264** and **1265** also can be written in mesomeric

TABLE 74A. DERIVATIVES OF 5,6-DIHYDRO-11-AMINOALKYLIDENE-11H-DIBENZ[*b*,*e*]AZEPINES

R	A	X	Stereo chemistry	mp (°C) or bp (°C/torr)	Salt	Refs.
Me	$=CH(CH_2)_2NMe_2$	H		165/0.08		365
				137	Cyclohexyl sulfamate	
Me	$=CH(CH_2)_2NMe_2$	2-Cl		200–210	(0.35 mm)	365
				191–193	2HCl	
Me	$=CHCH(Me)CH_2NMe_2$	2-Cl		180–185/0.6		365
Me	$=CH(CH_2)_2NHMe$	H		163–167/0.02		365
Me	$=CH(CH_2)_2N$(piperazine)NMe	H		182–186	Dimaleate	365
Me	$=CHCH(Me)CH_2N(Me)CH_2Ph$	H		214–225/1.8		365
Me	$=CHCH(Me)CH_2NHMe$	H		170–175/0.2		365
Me	$=CH(CH_2)_2N$(piperazine)NCH_2Ph	H		220/0.7		365
				193–194	2 Maleate	365
Me	=(piperidin-4-ylidene)NMe	H		120–123		365
				145–147	Maleate	

CH_2Ph	= (ring) NMe	H		112–113		365
H	$=CH(CH_2)_2NMe_2$	H		200–210/1.3		365
H	= (ring) NMe	H		185	Maleate	365
Ac	$=CH(CH_2)_2NMe_2$	H		200/210/0.17		365
H	$=CH(CH_2)_2NMe_2$	2-Cl	*Trans*	102–104		413
				154–156		466
CHO	$=CH(CH_2)_2NMe_2$	2-Cl		91–93		314
Ac	$=CH(CH_2)_2NMe_2$	2-Cl		130–140/0.2		413
H	$=CH(CH_2)_2NMe_2$	2-Br	*Cis*	158–160		413
			Trans	103–107		
Ac	$=CH(CH_2)_2NMe_2$	2-Br	*Trans*	145–150/0.2		413
COEt	$=CH(CH_2)_2NMe_2$	2-Br	*Trans*	145–155/0.2		413
H	$=CH(CH_2)_2NEt_2$	2-Br	*Trans*	140–145/0.2		413
			Cis	180–183	Oxalate	
Ac	$=CH(CH_2)_2NEt_2$	2-Br		145–155/0.2		413
H	$=CH(CH_2)_2N$ (ring)	2-Br	*Trans*	138–139.5		413
Ac	$=CH(CH_2)_2N$ (ring)	2-Br	*Trans*	160–165/0.15		413
H	$=CH(CH_2)_2NMe$	3-Cl		192–195/0.02		421
				177–180	Maleate	
H	= (ring) NMe	3-Cl		184–185		421
				186–189	Maleate	
Me	$=CPH_2$	H		167–168		420

TABLE 74B. DERIVATIVES OF 5,6-DIHYDRO-11-AMINOALKYLIDENE-11H-DIBENZ[*b*,*e*]AZEPINES[a]

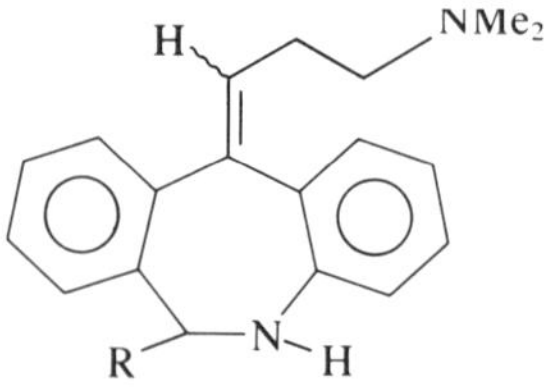

R	Stereochemistry	mp (°C) or bp (°C/torr)
Ph	*Trans*	140–150/0.02
	Cis	144–146
Me	*Trans*	130–135/0.2
	Cis	143–146

[a]Ref. 422.

TABLE 75. DERIVATIVES OF 11-AMINOALKYLIDENE-11H-DIBENZ-[*b*,*e*]AZEPINES

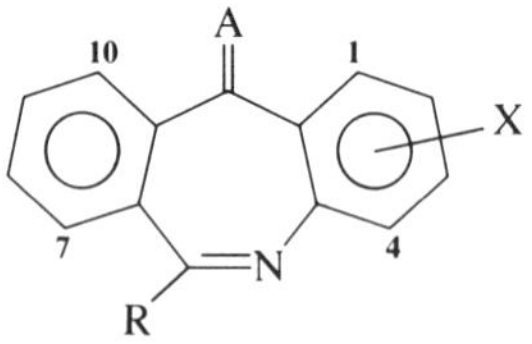

R	A[a]	X	mp (°C) or bp (°C/torr)	Yield (%)	Refs.
H	$=CH(CH_2)_2NMe_2$	H	188–200/0.05 116–117 (cyclohexyl-sulfamate)		424
H	$=CH(CH_2)_2NMe_2$	2-Cl	92–93		424
H	$=CH(CH_2)_2NMe_2$	3-Cl	119–120		426
Me	$=CH(CH_2)_2NMe_2$	2-Cl	125–133/2 120–125/0.7	Fair	423
Ph	$=CH(CH_2)_2NMe_2$	2-Cl	111–114 145–150/0.2	40–50	423
H	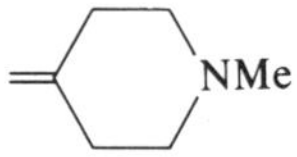 =(piperidylidene)NMe	H	101–103 277 (HCl)	Low	425 426
Me	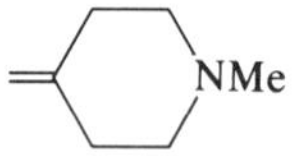 =(piperidylidene)NMe	H	146–147	30–40	426
Me	 =(piperidylidene)NMe	2-Cl	122–123		426
Me	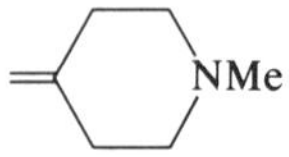 =(piperidylidene)NMe	3-Cl	109–111 229–231 (HCl)		

[a]*Cis*/*trans* geometry of double bond not specified.

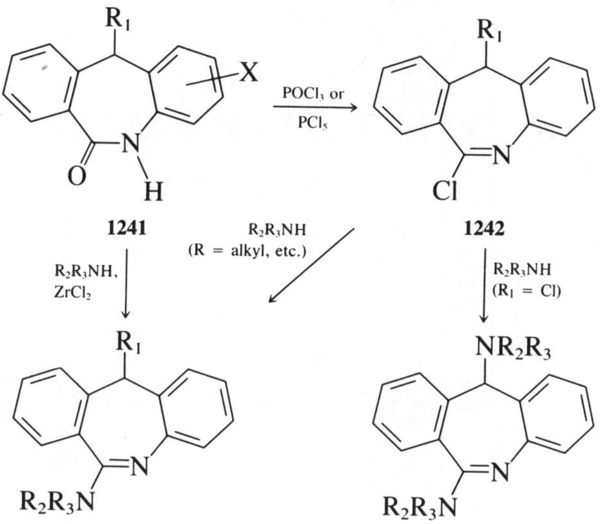

R_1 ... X, POCl$_3$ or PCl$_5$ → 1242; 1241 → R_2R_3NH, $ZrCl_2$ → 1243; 1242 → R_2R_3NH (R = alkyl, etc.) → 1243; 1242 → R_2R_3NH (R_1 = Cl) → 1244

1241 1242 1243 1244

Figure 166

1245 → PPA → **1246,** R = H, Me, Ph, R_1R_2N, $ClCH_2$

1246 → [H] (R = H) → 1247; 1247 → S, 130–140°C → 1246

1245 1247

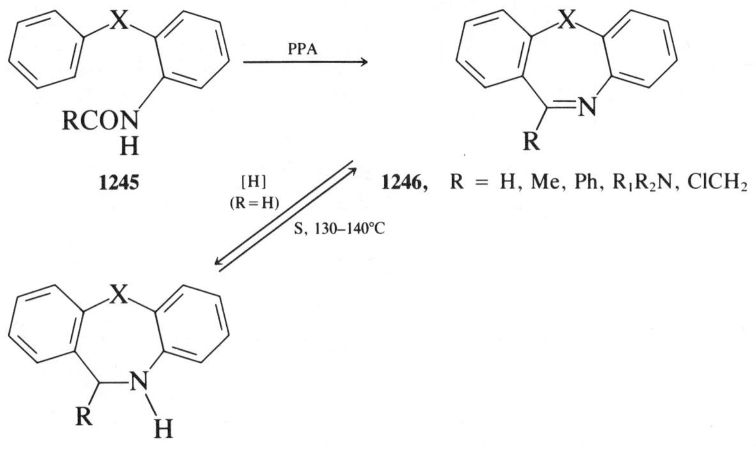

(153)

o-xylylene form. Carpenter (446) studied the mechanism of the conversion of **1258** to **1259** with the benzene ring fully deuterated. Deuterium distribution in **1259** indicated that the rearrangement proceeds via the *o*-xylylene–type intermediate rather than via a carbene or carbenoid species following collapse of the initially formed ylid **1263**.

When the rearrangement was carried out with phenyllithium and **1258** in a 1:1 ratio, one-half of **1258** was recovered. This suggested that **1259** undergoes lithiation as soon as it forms. Indeed, when **1259** was treated sequen-

TABLE 76. DERIVATIVES OF 11H-DIBENZ[*b*,*e*]AZEPINES (MORPHANTHRIDINES)

R_1	R_2	R_3	X	mp (°C)	Yield (%)	Spectra	Refs.
H	H	H	H	63	70	ir	392
				62	65		436
				63–64			434
				211–215 (picrate)			
H	H	H	1-Me	45	25		436
H	H	H	4-Me	68	43		436
H	H	H	10-Me	102/2[a]	50		436
H	Me	H	H	69	79		436
H	H	H	1,4-Me_2	80	50		436
H	H	H	7,10-Me_2	91	85		436
H	Me	H	10-Me	75	49		436
H	Me	H	1,4-Me_2	81	50		436
H	Et	H	H	61	69		436
Me	H	H	H	100	~50		392
				99–100	~80		434
				197–202 (picrate)			
Ph	H	H	H	159	~50		392
CH_2Cl	H	H	H	138–139	Good		391
				136–137	75		435
CH_2NHMe	H	H	H	154–157	89.5		442
$(CH_2)_3NMe_2$	H	H	H	161 (oxalate)	20	uv	428
$(CH_2)_3N(Me)CH_2Ph$	H	H	H	189–192 (oxalate)		uv	428

Cl	H	H	H	149–151	83		437
				162/2[a]	90		428
				66–69			441
Cl	H	Cl	H	166–169			407,431
Cl	Me	H	H	80–82			427
Cl	H	Cl	2-Cl	164–166			431
Cl	CMe_2NO_2	H	H	124–129	41	ir,nmr	395
NH_2	H	H	H	166.5–167.5			395
				204–206 (HCl)	59	uv	428
$NHCONHCO_2Et$	H	H	H	180.5–182	29		395
$NH(CH_2)_2CH(OMe)_2$	H	H	H	120–121	44		441
$NHCH_2CH_2OH$	H	H	H	205–207	45	uv	428
$NHCH_2CH_2Cl$	H	H	H	184–186 (HCl)	79		428
$NHNH_2$	H	H	H	203 (oxalate)	100	uv	428
NHPh	H	H	H	155.5–156			433
 N NMe (piperazine ring)	H	H	H	138–138.5	88		427,432,437
$NH(CH_2)_2NMe_2$	H	H	H	92–94			427,437
$NH(CH_2)_3NMe_2$	H	H	H	110–111			427
$N(Me)(CH_2)_3NMe_2$	H	H	H	223–225 (HCl)			427
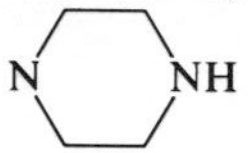 N NH (piperazine ring)	H	H	H	110–111			427
 N NH (piperazine ring)	H	H	8-Me	182–183			438
 N $N(CH_2)_2OH$ (piperazine ring)	H		H	143–144			427
 N $N(CH_2)_2OH$ (piperazine ring)	H	H	8-Me	157–161			438

(continued)

TABLE 76 (*Continued*)

R_1	R_2	R_3	X	mp (°C)	Yield (%)	Spectra	Refs.
4-[$(CH_2)_2OAc$]piperazin-1-yl (ring drawn: N⟨ ⟩N$(CH_2)_2OAc$)	H	H	H	105–107			467
4-methylpiperazin-1-yl (ring drawn: N⟨ ⟩NMe)	Me	H	H	106–107			427
4-methylpiperazin-1-yl (ring drawn: N⟨ ⟩NMe)	Et	H	H	136–140			427
4-methylpiperazin-1-yl (ring drawn: N⟨ ⟩NMe)	H	H	2-Cl	163–165			427
4-methylpiperazin-1-yl (ring drawn: N⟨ ⟩NMe)	Me	H	2-Cl	152–154			427
4-methylpiperazin-1-yl (ring drawn: N⟨ ⟩NMe)	H	H	3-Cl	202–204			427
4-methylpiperazin-1-yl (ring drawn: N⟨ ⟩NMe)	Me	H	3-Cl	162–164			427
4-methylpiperazin-1-yl (ring drawn: N⟨ ⟩NMe)	H	H	8-Cl	135–137			427
4-methylpiperazin-1-yl (ring drawn: N⟨ ⟩NMe)	H	H	8-Me	113–115			438

N N-Bu-*t*	H	H	H	138–141 (maleate)	432
N NMe	H	N NMe	H	95–99	407,431
NMe_2	H	NMe_2	H	157–160	431
N	H	N	H	125–127	431
N	H	N	H	128–131	431
N O	H	N O	H	182–184	431
NMe_2	H	N	H	124–125	431
NMe_2	H	N NMe	H	154–156	431
NMe_2	H	N O	H	182–184	431
N O	H	NMe_2	H	154–156	431
N O	H	N	H	183–185	431
N	H	N O	H	128.5–129.5	431

(*continued*)

TABLE 76. *(Continued)*

R_1	R_2	R_3	X	mp (°C)	Yield (%)	Spectra	Refs.
NEt_2	H	NMe_2	H	98–99			431
NEt_2	H	NEt_2	H	99–100			431
[morpholino]	H	NEt_2	H	142–144			431
$N(Me)CH_2Ph$	H	H	H	99–100			431
[4-methylpiperazino]	H	NEt_2	H	112–114			431
[piperidino]	H	NMe_2	H	141–142			431
$N(Me)(CH_2)_2OH$	H	NMe_2	H	106–107			431
[azepan-1-yl]	H	NMe_2	H	187–188 (2HCl,0.5MeOH)			431
NMe_2	H	NMe_2	2-Cl	182–184			431
[morpholino]	H	[morpholino]	2-Cl	213–217			431
[pyrrolidino]	H	[pyrrolidino]	2-Cl	113–116			431

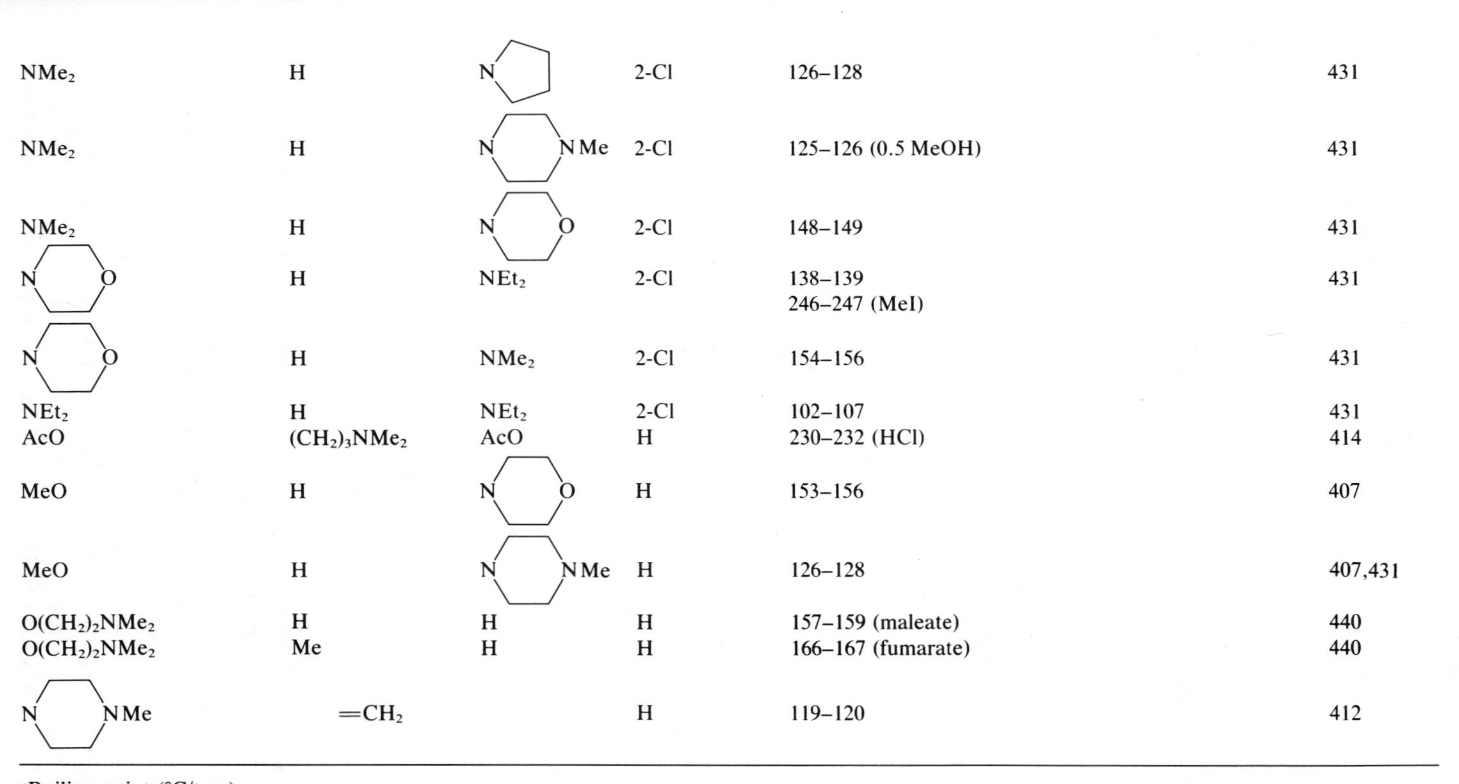

NMe_2	H	pyrrolidin-1-yl	2-Cl	126–128	431
NMe_2	H	4-methylpiperazin-1-yl	2-Cl	125–126 (0.5 MeOH)	431
NMe_2	H	morpholino	2-Cl	148–149	431
morpholino	H	NEt_2	2-Cl	138–139 246–247 (MeI)	431
morpholino	H	NMe_2	2-Cl	154–156	431
NEt_2	H	NEt_2	2-Cl	102–107	431
AcO	$(CH_2)_3NMe_2$	AcO	H	230–232 (HCl)	414
MeO	H	morpholino	H	153–156	407
MeO	H	4-methylpiperazin-1-yl	H	126–128	407,431
$O(CH_2)_2NMe_2$	H	H	H	157–159 (maleate)	440
$O(CH_2)_2NMe_2$	Me	H	H	166–167 (fumarate)	440
4-methylpiperazin-1-yl	$=CH_2$		H	119–120	412

[a]Boiling point (°C/torr).

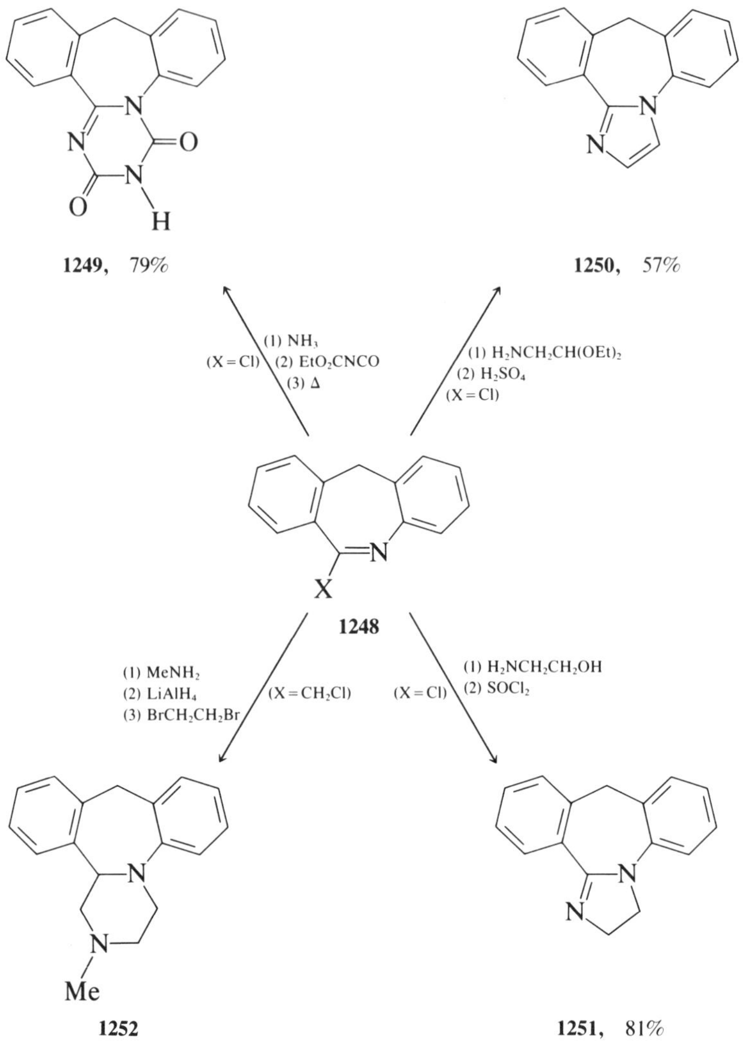

1249, 79% **1250**, 57%

(X = Cl) (1) NH_3 (2) EtO_2CNCO (3) Δ

(1) $H_2NCH_2CH(OEt)_2$ (2) H_2SO_4 (X = Cl)

1248

(1) $MeNH_2$ (2) $LiAlH_4$ (3) $BrCH_2CH_2Br$ (X = CH_2Cl)

(X = Cl) (1) $H_2NCH_2CH_2OH$ (2) $SOCl_2$

1252 **1251**, 81%

Figure 167

tially with phenyllithium and benzophenone, **1268** could be obtained in 65% yield, presumably via the 11-lithio species **1267** (Fig. 171). The methiodide salt **1269** was cleaved by Emde degradation to **1270**. Alternatively, reaction of **1269** with phenyllithium yielded **1271**. It is of interest to note that in this reaction phenyllithium does not metallate positions 6 or 11 but instead attacks the 6-position, with displacement of the dimethylammonium moiety. Drukker and Judd (447) later showed that **1259** could be cleaved to **1270** by hydrogenolysis.

Thermal fragmentation of **1272** under Hofmann conditions has been found

TABLE 77. 11H-DIBENZ[*b,e*]AZEPINES FROM DECOMPOSITION OF o-BENZYLDIAZONIUM SALTS IN NITRILE SOLVENTS[a]

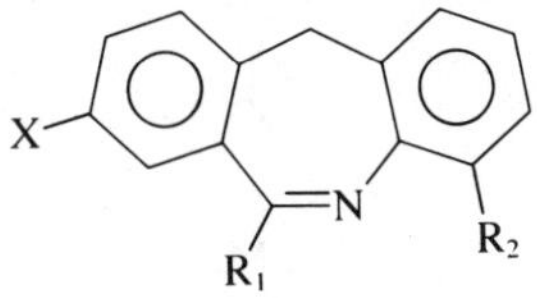

R_1	R_2	X	Solvent	mp (°C)	Yield (%)	Spectra
Ph	Me	H	PhCN	152	39	nmr
Me	H	H	MeCN	199 (picrate)	23	nmr
MeS	H	H	MeSCN	140–160/0.01[b] 121	51	nmr
MeS	Me	H	MeSCN	135/0.01[b] 111	58	nmr
Me	H	MeO	MeCN	185 (picrate)	—	—
Me	H	Me	MeCN	175–178 (picrate)	—	—

[a]Ref. 439.
[b]Boiling point (°C/torr).

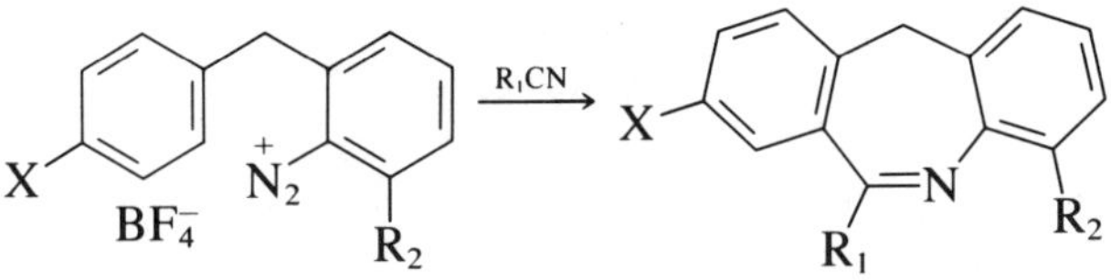

R_1CN (154)

1253

H_2O_2, HOAc, room temp., 3 days

1253 1254, 57% 1255, 18%

1256 H^+ 1257

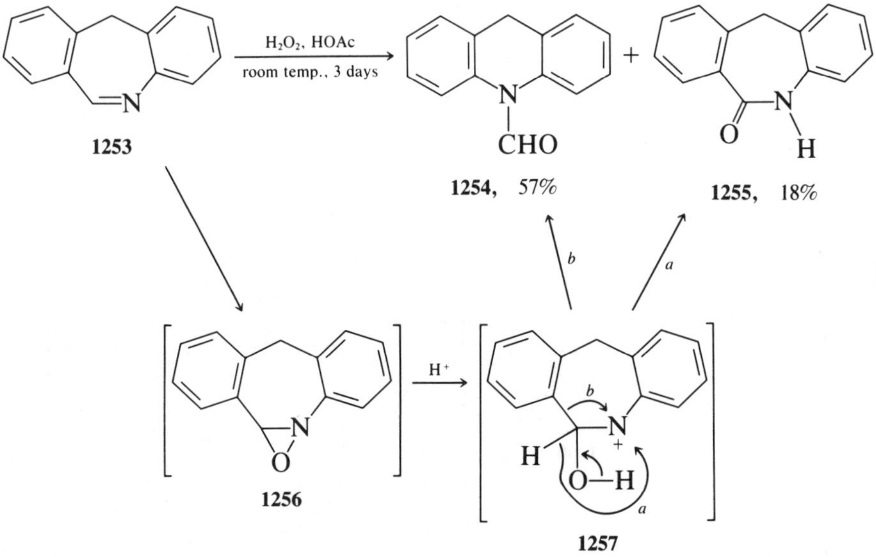

Figure 168

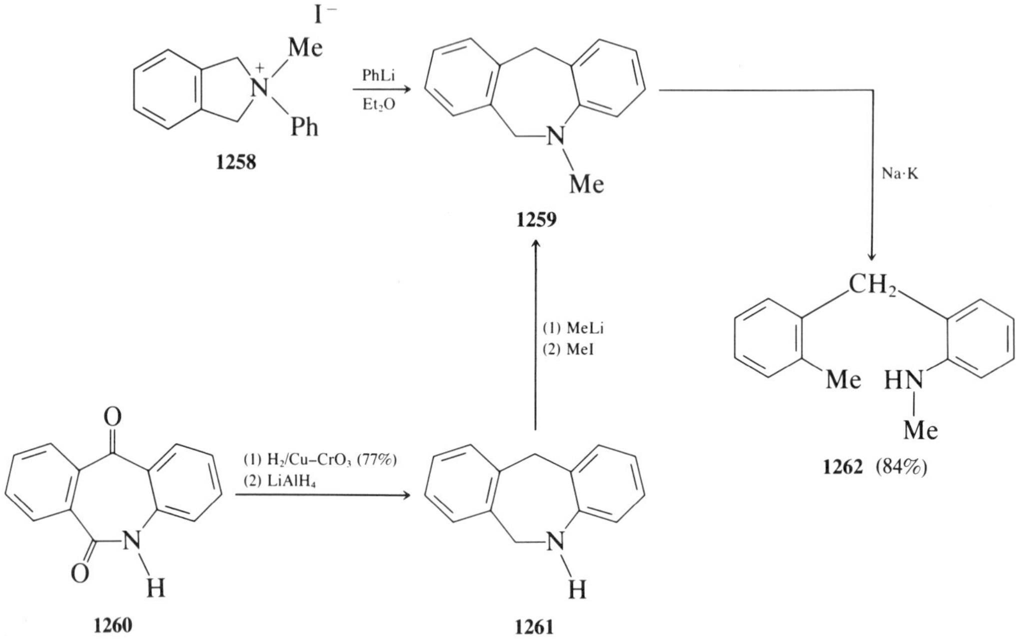

I^- Me N^+ Ph

1258

PhLi / Et_2O →

N Me

1259

Na·K

CH_2

Me HN Me

1262 (84%)

(1) MeLi
(2) MeI

O

N H O

1260

(1) H_2/Cu–CrO_3 (77%)
(2) $LiAlH_4$

N H

1261

Figure 169

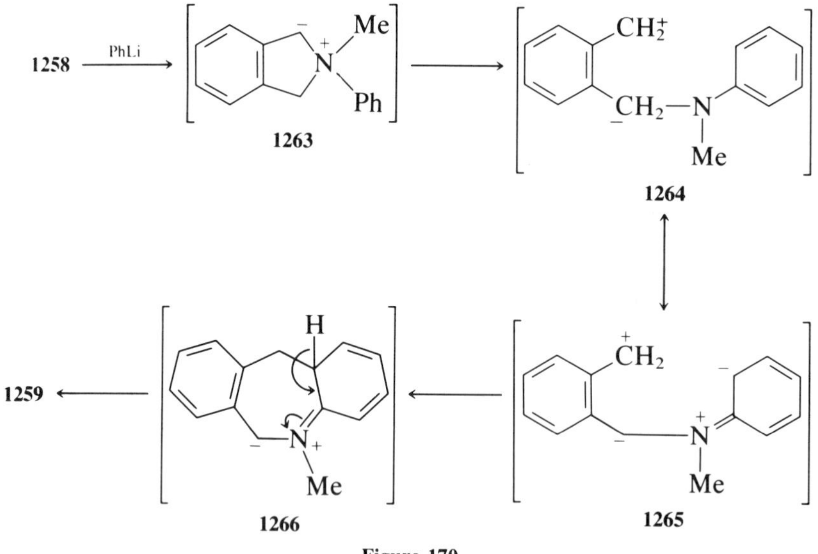

1258 —PhLi→ [N^+ Me Ph] 1263 → [CH_2^+ CH_2—N Me] 1264

↕

1259 ← [H N Me] 1266 ← [$\overset{+}{C}H_2$ N^+ Me] 1265

Figure 170

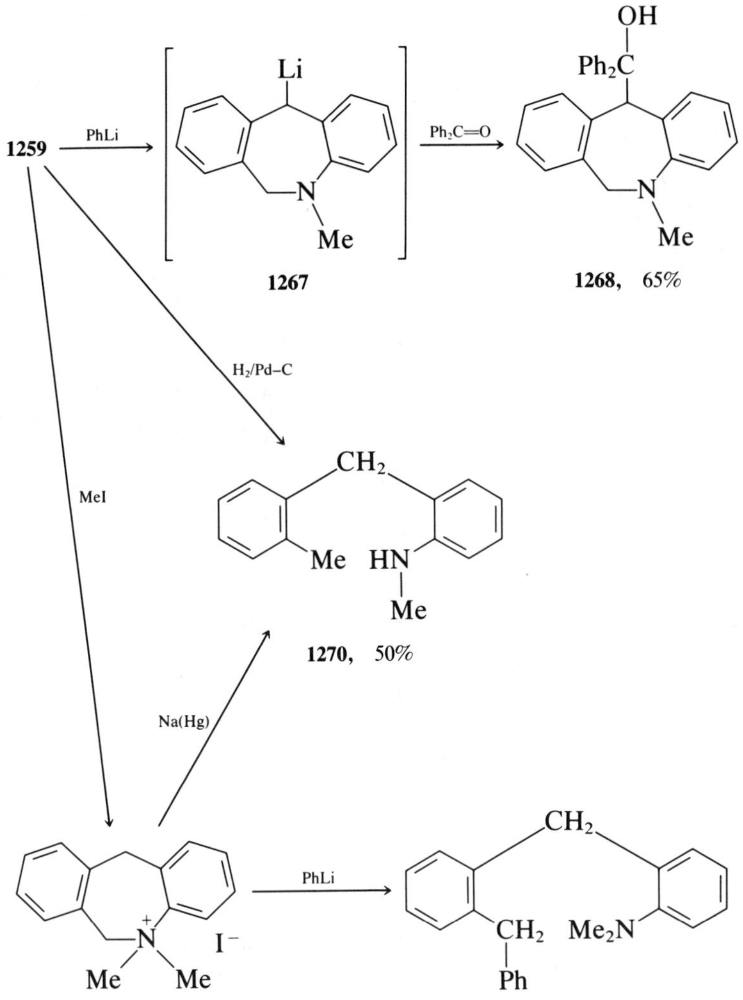

Figure 171

to give a 24% yield of the unusual compound **1274**, possibly via dimerization of species represented by **1273a-1273c** (Fig. 172). Emde degradation of **1274** gave **1275**, which was identical to material prepared independently. A trace of anthraquinone was also obtained in this reaction.

When the same preparation of the quaternary hydroxide ion **1272** was carried out in methanol in the presence of oxygen, no **1274** was obtained and the yield of anthraquinone rose to 95%. Careful investigation revealed the pathway for this oxidation to be as shown in Fig. 173. It was proposed that oxygenation occurs at the 11-position of **1276**, giving rise (via **1277**) to

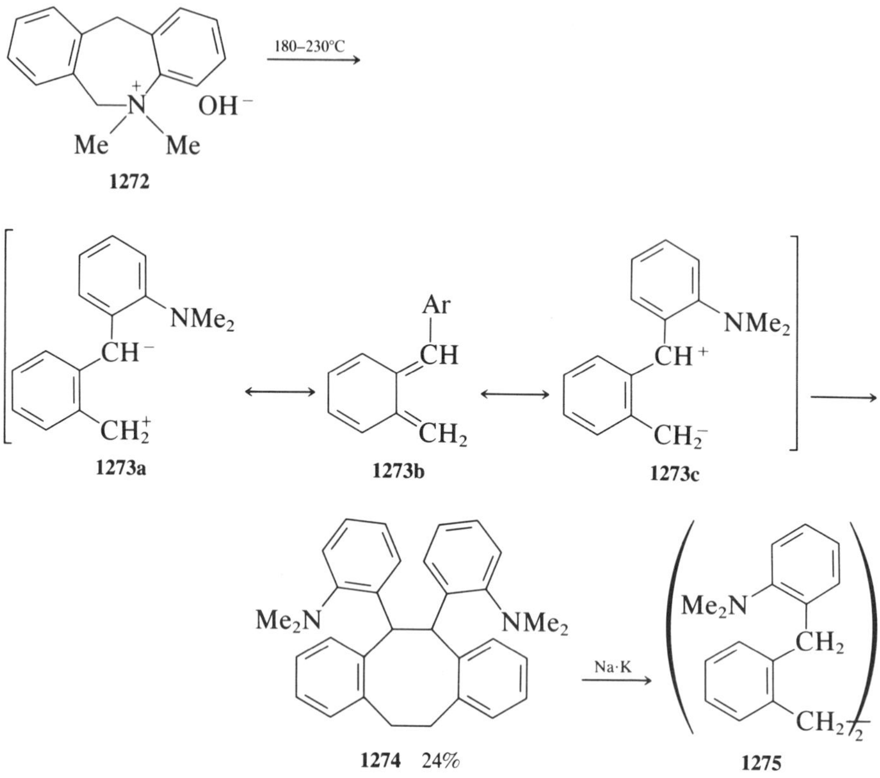

180–230°C

1272

1273a 1273b 1273c

Na·K

1274 24%

1275

Figure 172

1278. This substance was prepared synthetically, and was shown to be easily converted to **1279** in base in a Stevens-like rearrangement. In the presence of oxygen, this aminoanthrone forms anthraquinone in high yield.

Protiva and co-workers (392) have prepared 5,6-dihydromorphanthridines in high yields by chemical and catalytic reduction of morphanthridines. Their interest in this chemistry centered on an improved synthesis of the antidepressant agent propazepin, the *N*-dimethylaminopropyl derivative of **1261** (Fig. 169) (448). Related structures have been patented with similar claims of pharmacological activity (361,411,449–453) (see Table 78).

Drukker and Judd (447) used the Wittig method to prepare the 2-chloro derivative of **1259** in 39% yield. However, a more versatile procedure (Eq. 155) involved sequential alkylation of the dione **1260** to **1280** and reduction with lithium aluminum hydride (392,448).

The same approach could also be used with 6-morphanthridone in place of **1260**. Once in hand, the 5-substituted morphanthridines **1281** were lithiated by *n*-BuLi to give **1282** and alkylated at the 11-position with various aminoalkyl halides to give **1283**. When R was benzyl, hydrogenolysis provided the free N—H compound for further elaboration. It should be noted, however, that since **1259** can be cleaved catalytically to **1270** (Fig. 171), hydro-

1276 $\xrightarrow{O_2}$ [1277] ⟶

1278 $\xrightarrow{-H_2O}$ [ ] ⟶

1279 $\xrightarrow[-HNMe_2]{H_2O_2}$

Figure 173

1260 $\xrightarrow[(2)\ RX]{(1)\ NaNH_2}$ 1280 $\xrightarrow{LiAlH_4}$ 1281 $\xrightarrow{n\text{-BuLi}}$

[1282] $\xrightarrow{X\text{—Alk—}NR_1R_2}$ 1283 (155)

TABLE 78. 5,6-DIHYDRO-11H–DIBENZ[*b*,*e*]AZEPINES (5,6-DIHYDROMORPHANTHRIDINES)

R_1	R_2	R_3	X	mp (°C) or bp (°C/torr)	Yield (%)	Spectra	Refs.
H	H	H	H	129–131/0.01	46		420
				130–131	43,74,91		392,414,420
				129–130			361,434
				131–132	73		448
				188–190 (HCl)			
Me	H	H	H	78–78.5	48.5		420
				80	90		392
				78–80	84		447
				182–183 (MeI)	96		420
Et	H	H	H	150–155/2	Good		392
				114–117 (picrate)			
CHO	H	H	H	85–86	95		392
Ac	H	H	H	149–150	96		392
CH_2Ph	H	H	H	180–190/0.12	68		447
n-Bu	H	H	H	150/0.13	—		447
H	Me	H	H	73	70	ir	392
				71–72			434
H	Ph	H	H	151	Good		392
H	H	Me	H	80	91		436
				59–61			456
H	H	Me_2	H	79–80			456
H	H	H	3-Me	93–95			361

H	H	H	3-Et	Oil			361
H	H	H	1,4-Me_2	64–65	76		436
H	H	Me	10-Me	80–82	68		436
H	H	H	8-Cl	136–138			361
H	H	H	1-MeO	104–105			361
Me	H	H	2-Cl	57.5–60	39		447
H	H	H	8-MeO	113–115			460
Me	H	Ph_2COH	H	176–178	65		420
CH_2Ph	H	CO_2H	H	149–150			454
CH_2Ph	H	CO_2CH_2C	H	126–128			454
CH_2Ph	H	$CONH_2$	H	176			454
Me	H	CON(piperazine ring)NMe	H	196–197	25		454
Me	H	CONHMe	H	144–145			454
CH_2Ph	H	CON(piperidine ring)	H	250/0.06			454
Me	H	CO_2H	H	130			455
Me	H	CO_2CH_2CN	H	184.5			455
Me	H	$CONH_2$	H	146–147			455
H	H	$(CH_2)_3NMe_2$	H	146–148	40–52	ir	414
Ac	H	$(CH_2)_3NMe_2$	H	235–238	22	ir	414
Me	H	$(CH_2)_3NMeCO_2Et$	H	210/0.5	92		447
Me	H	$(CH_2)_3NHMe$	H	180/0.1 124–126 (maleate)	80		447
Me	H	$(CH_2)_3NMe_2$	H	156–159/0.125 148–149 (C)[a]			447
Me	H	$(CH_2)_3N(Me)CH_2Ph$	H	225/0.2 119 (2C)[a]			447
Me	H	$(CH_2)_3N$(piperidine ring)	H	190/0.08 184–186 (fumarate)			447

(*continued*)

TABLE 78. *(Continued)*

R_1	R_2	R_3	X	mp (°C) or bp (°C/torr)	Yield (%)	Spectra	Refs.
Me	H	$(CH_2)_3N$(piperazine)NMe	H	195/0.1 69–72			447
Me	H	$(CH_2)_3NHMe$	H	170/0.05 176 (maleate)			447
Me	H	$CH(Me)(CH_2)_2NMe_2$	H	145–147/0.12 117 (2C)[a]			447
Me	H	$(CH_2)_2N(Me)CH_2Ph$	2-Cl	163–170/0.35 122–124 (2C)[a]			447
Me	H	$(CH_2)_3NHMe$	2-Cl	200–204 (2HCl)			447
Me	H	$(CH_2)_3N$(piperidine)	2-Cl	223/0.2			447
Me	H	$(CH_2)_3N$(piperazine)NCH_2Ph	H	260/0.8			447
CH_2Ph	H	$(CH_2)_3N$(piperidine)	H	290/0.5			447
n-Bu	H	$(CH_2)_3N$(piperidine)	H	250/3.5 138 (maleate)			447
Me	H	$(CH_2)_2N$(piperidine)	H	169/0.05			447

Me	H	4-methylpiperidino (N–ring–Me)	H	162–167/0.1		447
H	CH_2NHMe	H	H	92–94	78	442
				176–178 (dibenzoyl-(+)-tartrate)[b]		444
				117–118[c] $[\alpha]_D^{25} = +166°$		444
				116–118 $[\alpha]_D^{25} = -165°$		444
H	CH_2CH_2NHMe	H	H	124–127		443
H	$CH_2CH_2NH_2$	H	H	152–153		443
$(CH_2)_2NMe_2$	H	H	H	150–154/0.5		448
				181–183 (HCl)		361
				195–198 (HCl)		450
				198–200 (HCl)		448
				171–172 (picrate)	63	448,450
				221–222 (MeI)		450
				209–210 (MeI)		448
				215–217 (EtI)		361
$(CH_2)_3NEt_2$	H	H	H	140–145/0.01		450
				175/0.9	63	448
				162–164 (HCl)		448,450
				209–210 (MeI)		448,450
$(CH_2)_2N$ (piperidino ring)	H	H	H	180–185/0.5	63	448
				208–210 (HCl)		450
				207–209 (HCl)		448
				200–201 (MeI)		448,450
$(CH_2)_2N$ (morpholino ring, O)	H	H	H	161–164/0.03		450

(*continued*)

TABLE 78. (*Continued*)

R_1	R_2	R_3	X	mp (°C) or bp (°C/torr)	Yield (%)	Spectra	Refs.
				225 (MeI)	62		448,450
$CH_2CH(Me)NMe_2$	H	H	H	153/0.4			448
				207–209 (MeI)	57		
$(CH_2)_2NHMe$	H	H	H	162–166/0.1			449
				138–142 (fumarate)			
				183–184 (oxalate)			
				133 (citrate)			
				176–178 (HCl)			450
				189–190 (MeI)			
				180/0.4	65		448
$(CH_2)_2NMe_2$	H	H	H	161–162/0.015			450
				157–158 (picrate)			448
				182–183 (HCl)			448
				185–187 (2MeI)			361
				189–190 (2MeI)			448
$(CH_2)_3NMe_2$	Ph	H	H	107–108 ($HCl \cdot H_2O$)	Good		392
$(CH_2)_3NHEt$	H	H	H	170–171 (fumarate)			449
$(CH_2)_3NHPr$	H	H	H	170–172 (fumarate)			449
$CH_2CH(Me)CH_2NHMe$	H	H	H	185–187 (fumarate)			449
$CH_2(Me)CH_2N$	H	H	H	152–153/0.01			449
$(CH_2)_3N$	H	H	H	151–153/0.01			450
$(CH_2)_3N$	H	H	H	165–166/0.02			450
				199–201 (HCl)			

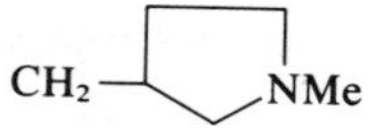 CH_2 NMe	H	H	H	152–154/0.02		450
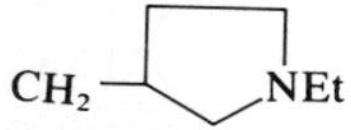 CH_2 NEt	H	H	H	165–168/0.01		450
CH_2CN	H	H	H	96–98		361
CH_2CN	H	H	1-OMe	117–118		361
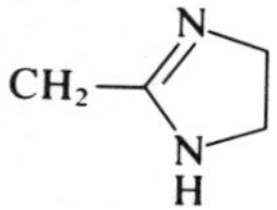 CH_2 N, N, H	H	H	H	245–247 (HCl)	nmr,ir	361,362
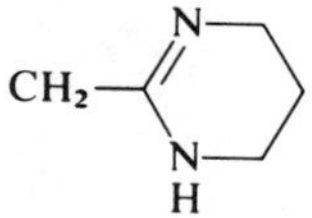 CH_2 N, N, H	H	H	H	169		361
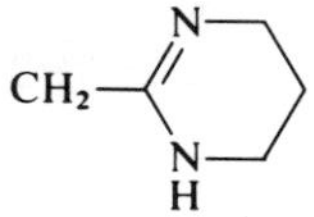 CH_2 N, N, H	H	H	H	190		361
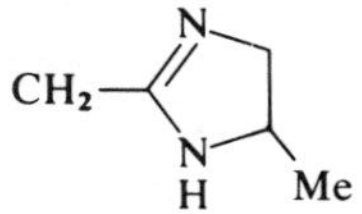 CH_2 N, N, H, Me	H	H	H	193–195 (HCl)		361
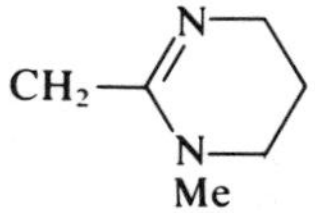 CH_2 N, N, Me	H	H	H	206–207 (MeI)		361

(*continued*)

TABLE 78. *(Continued)*

R_1	R_2	R_3	X	mp (°C) or bp (°C/torr)	Yield (%)	Spectra	Refs.
CH_2–(2-imidazolinyl)	H	H	3-Me	234–236 (maleate)			361
CH_2–(2-imidazolinyl)	H	H	1-, 4-, or 7-Me	239–241 (HCl)			361
CH_2–(2-imidazolinyl)	H	H	8-Cl	250 (HCl)			361
CH_2–(2-imidazolinyl)	H	H	1-OMe	254–255 (HCl)			361
$COCH_2Cl$	H	H	H	137–138			452
COCl	H	H	H	98–102			457
$COCH_2CH_2Cl$	H	H	H	77–79	63		448
$COCH_2NEt_2$	H	H	H	180–181/0.03 100 (HCl)	74		451
$CO(CH_2)_2NEt_2$	H	H	H	181–186/0.35 210–212 (MeI)			448

$CONHNH_2$	H	H	H	171–178		458
				171–175		459
$CONH(CH_2)_2NEt_2$	H	H	H	141–143 (oxalate)		457
$COCH_2S(CH_2)_2NMe_2$	H	H	H	200–205/0.1		452
$(CH_2)_2S(CH_2)_2NMe_2$	H	H	H	190–193/0.01		452
				165 (HCl)		452
				110 (citrate)		452
$(CH_2)_2S(CH_2)_2N$(pyrrolidine ring)	H	H	H	195–197/0.01		452
				106–108 (citrate)		452
$COCH(Me)S(CH_2)_2NMe_2$	H	H	H	210/0.9		452
$CH_2CH(Me)S(CH_2)_2NMe_2$	H	H	H	192–195/0.1		452
				125–126 (2HCl)		452
				112–114 (citrate)		452
$CO(CH_2)_3SMe$	H	H	H	204–206/0.4		448
				123–125 (MeI)		
$(CH_2)_3OH$	H	H	H	162–170/0.03		449
H	H	$=CH_2$	H	98–99	75	411
$(CH_2)_2NMe_2$	H	$=CH_2$	H	166–167 (maleate)	77	411
$(CH_2)_3NMe_2$	H	$=CH_2$	H	158–160 (maleate)	Good	411
$(CH_2)_2N$(morpholine ring, O)	H	$=CH_2$	H	162–164 (maleate)	76	411
$(CH_2)_3N$(piperidine ring)	H	$=CH_2$	H	128–130 (C)[a]	30	411

[a]C = Cyclohexyl sulfamate.
[b]Diastereomeric mixture (?).
[c]Melting point and rotations are of pure enantiomers.

genolysis has to be performed under stoichiometrically controlled conditions. Products that have been obtained by this route and tested as antidepressants are listed in Table 78. 5,6-Dihydromorphanthridines have also been prepared by lithium aluminum hydride reduction of 6-chloromorphanthridines (456) and 11-hydroxy-6-morphanthridones (414).

A series of octahydromorphanthridines (**1285**) have been prepared from 5,6-dihydromorphanthridines (**1284**) by catalytic reduction (Eq. 156). The stereochemistry of ring fusion, from the method of preparation, is presumably *cis*. Attachment of *N*-aminoalkanoyl side-chains gave products exemplified by **1286**, which were evaluated as antihypertensive agents (460,461). Examples are listed in Table 79.

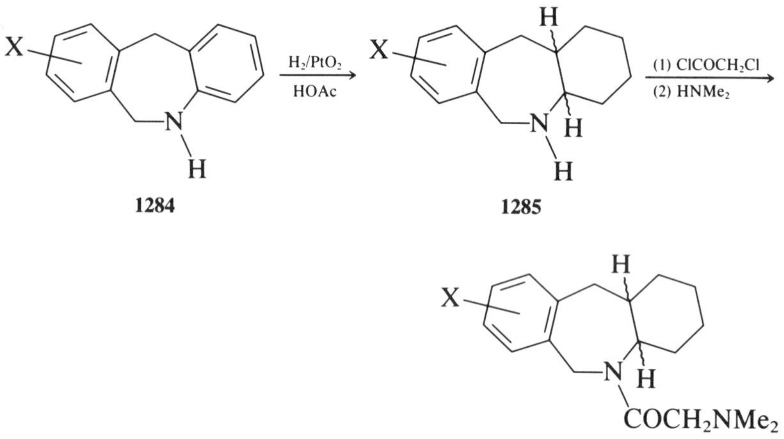

X H N H H/PtO₂ HOAc (1) ClCOCH₂Cl (2) HNMe₂ COCH₂NMe₂

1284 **1285** **1286**

(156)

Barriers to ring inversion in the dibenz[*b*,*e*]azepine series have been observed by several workers using nmr spectrometry. Proctor and MacDonald (389) proposed that the quartet seen in the nmr spectrum (40 mHz) of **1287** at δ 5.00 (J_1 = 16 Hz, J_2 = 1 Hz) arose from the nonequivalent protons of the C_6-methylene group. A similar quartet (δ 4.97; J_1 = 15 Hz, J_2 = 1 Hz) was observed for the ethyl ether **1288**. If ring inversion had been more rapid on the nmr time scale, the coupling constant might have been smaller.

Horning and Muchowski (379) reported that in the nmr spectra of a series of 6,11-disubstituted 6-morphanthridones, two well-separated resonances for the C_{11} proton, and in most cases for the 6-substituent as well, were observed at 60 mHz. The relative intensity of the absorptions was markedly dependent on solvent polarity as well as on the nature of the C_{11} substituent. This was interpreted as slow interconversion of diastereomeric boat conformers in which the C_{11} substituent is either quasi-axial or quasi-equatorial.

A particularly interesting spectrum was observed for 5-methyl-11-methoxy-6-dibenz[*b*,*e*]azepinone (**1289-1290**) in d_6-DMSO. The least abundant species was assigned structure **1290** on the basis that an axially oriented

TABLE 79. OCTAHYDRO DERIVATIVES OF 11H-DIBENZ[*b,e*]AZEPINE

R	X	mp (°C)	Refs.
H	H	51–52.3	460
		213–215 (HCl)	461
Ac	H	82–84	461
Me	H	242–243 (HCl)	461
Et	H	230–231 (HCl)	461
n-Bu	H	218–220 (HCl)	461
H	9-Cl	235–240 (HCl)	461
Me	9-Cl	262 (HCl)	461
H	7,9-Cl_2	284–285 (HCl)	461
H	8-Me	130–132 (HOAc)	461
$COCH_2Cl$	H	142–144	460
$COCH_2CH_2Cl$	H	129–130	460
COCH(Me)Br	H	147–149	460
$COCH_2CH(Me)Cl$	H	117–119	460
$COCH_2Cl$	9-Cl	148–150	460
$COCH_2Cl$	7,9-Cl_2	146–148	460
H	8-MeO	130–132	460
$COCH_2NHMe$	H	287–289 (HCl)	460
$COCH_2NHPr$	H	223–225 (HCl)	460
$COCH_2NMe_2$	H	110–112	460
		233–235 (HCl)	
$COCH_2NMe_2$	9-Cl	242–245 (HCl)	460
$COCH_2NEt_2$	H	85–87	460
$COCH_2NHBu$	H	224–226 (HCl)	460
$COCH_2N$(piperidino)	H	129–130	460
$COCH_2N$(morpholino, O)	H	152–158	460
$COCH_2N$(piperazino, NMe)	H	97–99	460
$COCH_2NMe_2$	8-Me	133–135	460
$COCH_2CH_2NMe_2$	H	74–76	460
$COCH(Me)NMe_2$	H	256–258 (HCl)	460
$COCH_2NMe_2$	7,9-Cl_2	284–286 (HCl)	460
$COCH_2NH_2$	H	182–192 (HCl)	460
$COCH_2CH(Me)NMe_2$	H	236–239 (HCl)	460

methoxyl group would lie in the positive shielding region above the plane of the aromatic rings. When the C_{11} substituent was chlorine, the quasi-axial configuration was predominant.

The barriers to ring inversion for several 6-dibenz[*b*,*e*]azepinones have been calculated from variable-temperature nmr data. Calculations of ultraviolet spectra by the Pariser–Parr–Pople method were in fairly good agreement with experimental results (462). Semiempirical calculations by the SCFMO method have indicated that 5H-dibenz[*b*,*e*]azepine(**1291**) should be unstable (463). A refinement of the Hückel method had also predicted the nonaromaticity of **1291** (464).

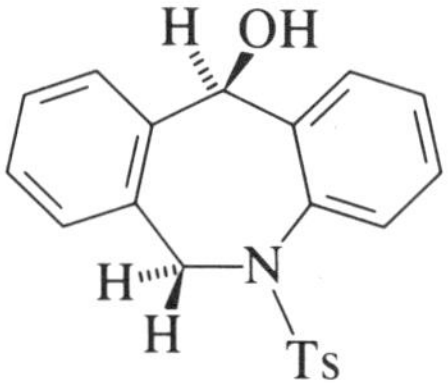

1287

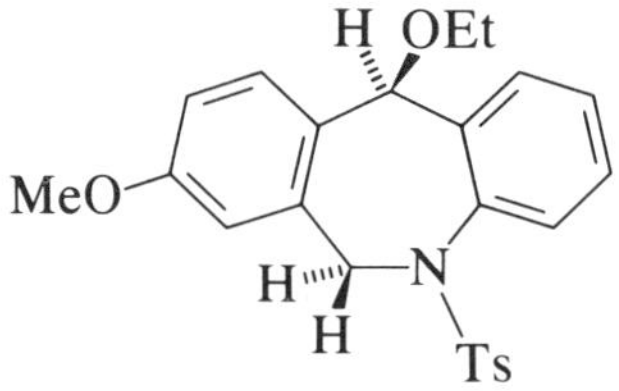

1288

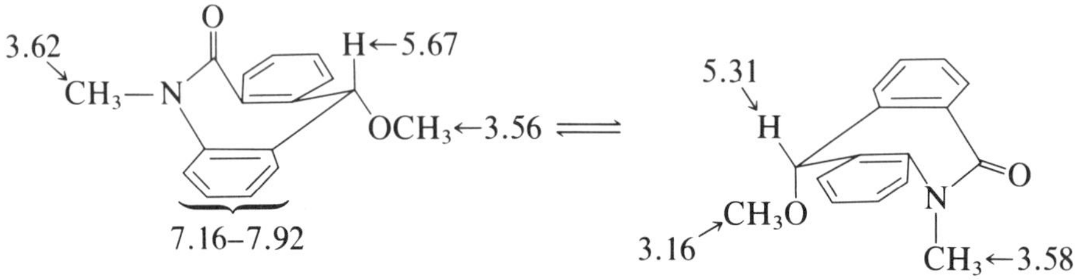

1289

1290

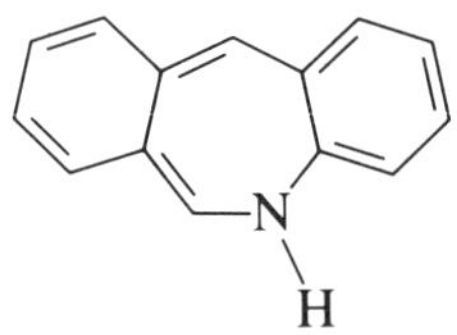

1291

39. Dibenz[*b*, *f*]azepines

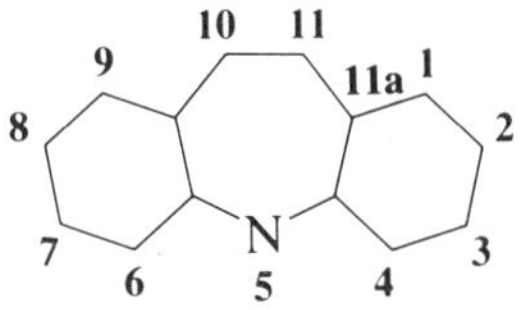

A. *Introduction*

The 5H-dibenz[*b,f*]azepine nucleus was first reported in 1899 by Thiele and Holzinger (468). Thermal cyclization of *o,o'*-diaminobibenzyl hydrochloride (**1292**) gave rise to 10,11-dihydrodibenz[*b,f*]azepine (**1293**), frequently referred to as iminobibenzyl, in 60% yield (Eq. 157). The system next appeared in 1938 when the reduced iminobibenzyl **1294** was described (469). Then, in a 1951 patent, Haefliger and Schindler described a series of aminoalkyliminobibenzyls (470). While the class of compounds encompassed by general structure **1295** was initially patented for antiallergic activity, it was subsequently found to also possess significant antidepressant activity. Today 5-(3-dimethylaminopropyl)iminobibenzyl (**1296a**, imipramine) can fairly be classified as a benchmark antidepressant agent.

The initial patent, and several that soon followed, prompted an extensive investigation of related compounds. Among the major dibenz[*b,f*]azepines that have grown out of these studies are imipramine (**1296a**, Tofranil), desipramine (**1296b**, Pertofrane), trimeprimine (**1296c**, Surmontil), opipramol (**1296d**, Ensidone), carpipramine (**1297a**) and carbamazepine (**1297b**, Tegretol). The biological literature associated with all of these compounds is vast, and no attempt has been made to cover it comprehensively in this review. The interested reader is directed toward several earlier reviews that address these topics (471–489).

Because of the widespread pharmaceutical interest in dibenz[*b,f*]azepines, much of the pertinent information on this ring system is recorded only in the patent literature, and many compounds are claimed without supporting physical data. Only those compounds for which minimal physical constants are available have been included in this survey. In contrast to *Chemical*

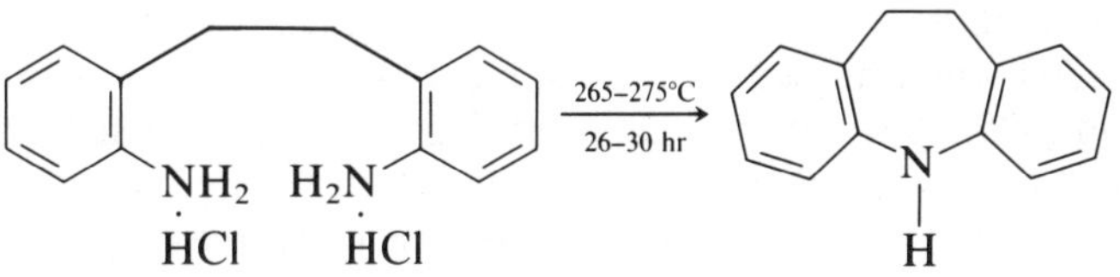

$NH_2 \cdot HCl$ $H_2N \cdot HCl$ —(265–275°C, 26–30 hr)→ N–H (157)

1292 **1293**

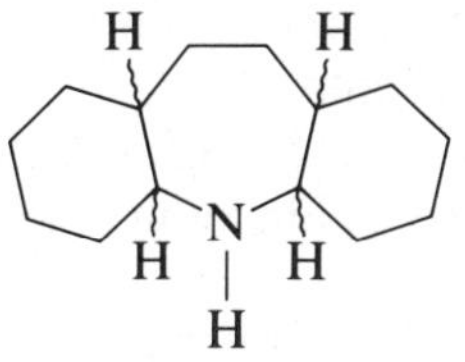

1294 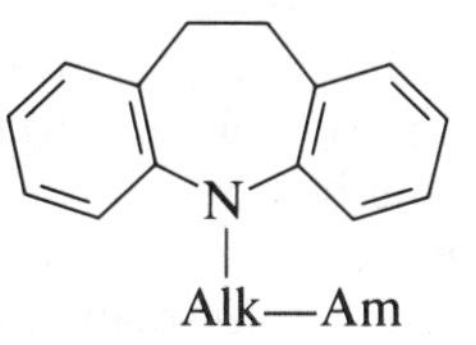 **1295** (N–Alk—Am)

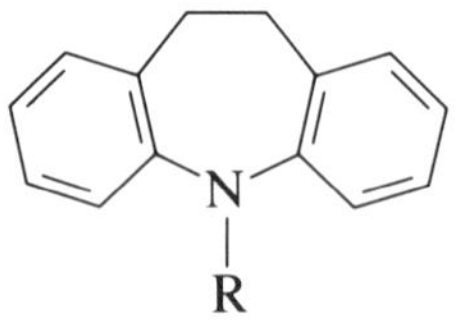

1296a, R = $(CH_2)_3NMe_2$
1296b, R = $(CH_2)_3NHMe$
1296c, R = $CH_2CHMeCH_2NMe_2$
1296d, R = $(CH_2)_3N$ (4-carbamoyl-4-piperidinopiperidino; $CONH_2$)

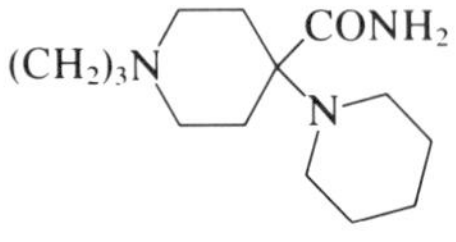

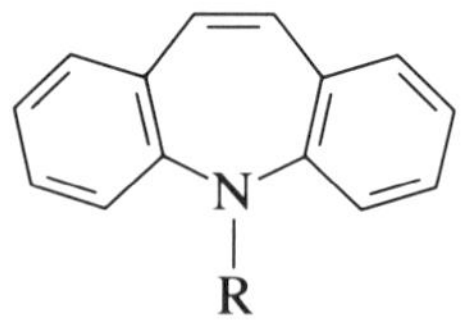

1297a, R = $(CH_2)_3N$ (piperazine) $N(CH_2)_2OH$
1297b, R = $CONH_2$

Abstracts nomenclature, no distinction is made among 1H-, 2H-, 5H-, and 10H-dibenz[*b,f*]azepines. The reader is referred to previous reviews by Kricka and Ledwith (490), Haefliger and Burckhard (491), and others (492,493) for discussions of the synthesis and pharmacology of 5H-dibenz[*b,f*]azepines.

B. *Synthesis*

a. 10,11–DIHYDRO–5H–DIBENZ[*b,f*]AZEPINES (See Table 80)

i. CYCLIZATION OF *o,o'*-DIAMINOBIBENZYL. Cyclization of *o,o'*-diaminobibenzyl (**1292**) was first reported in 1899 (468), and cyclization of various salts of **1292** is still the principal route to the iminobibenzyl nucleus (494–502) Small scale thermolysis of the dihydrochloride salt generally proceeds in high yield (497), but, unfortunately, the yield falls off as the scale of the reaction increases. Furthermore, with dihydrochloride salts of substituted diaminobibenzyls, thermolysis generally gives unacceptable results.

Because of the aforementioned difficulties, a number of other acid salts have been examined. The diphosphate salt of **1292** was found to give high yields of **1293** when heated without solvent (494, 498–501) or in polyphosphoric acid (495, 496). Sulfonic acid salts have also been used (500–502) as precursors to **1293**. In this method, the diamine is heated in the presence of an aliphatic or aromatic sulfonic acid, or the preformed salt alone can be heated to 295°C. A recent patent examined a number of acid catalysts in the cyclization reaction (500,501). Lewis acids that were found to bring about efficient cyclization of **1292** included aluminum chloride, ferric chloride, phosphorous tribromide, phosphoryl chloride, stannic chloride, titanium tetrachloride, iodine, cupric bromide, and cupric chloride. Also examined were the hydrochloride, hydrobromide, phosphate, and organic sulfonate salts. In general the best results were obtained using aluminum chloride, and the poorest with dihydrochloride salts.

The requisite *o,o'*-diaminobibenzyls for these cyclizations are readily available by reduction of *o,o'*-dinitrostilbenes (**1300**) (468,497,500), which

TABLE 80. 10,11-DIHYDRODIBENZ[*b, f*]AZEPINE AND ITS N-ALKYL AND N-ACYL DERIVATIVES

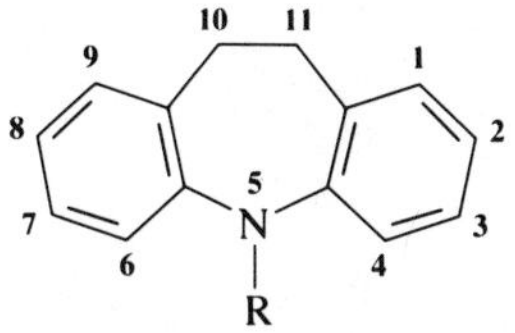

R	mp (°C)	Spectra	Refs.
H	92–93 (?)		814
	104–105		494
	107–108	uv	497
	107–108		784,809
	108	ir,uv	512
	108–109; 103–105		501
	108–110		498
	110		468,624
	—	pmr	584
	—	pmr	844
Me	106–107	pmr,^{13}C-nmr	584
	106–107	pmr,	844
	106–107		581,608
	106–107; 107–108	uv	497
Et	37–42		581
	52–53	ir,pmr,ms	584
$CH_2C{\equiv}CH$	105–106		620
CH_2Ph	66–68		581,582,608
$(CH_2)_3$-IB[a]	163–166	pmr,ms	586
CHO	134–135		659
CH=NPh	238 (HCl)		659
$CH{=}NSO_2Ph$	176–177		828
$CH{=}NSO_2Me$	197		828
COMe	85–86	pmr	584
	85–86	pmr,^{13}C-nmr	844
	—	pmr	843
	90		513
	96		509
	96–97		512
	98		513
$COCH_2Cl$	92–93		814
	96–98	pmr	584
	96–98	pmr^{13}C-nmr	844
	97–98		677
	97–100		734
COEt	68–70		679,680
	71–72	ir,pmr,ms	584
$COCHBrCH_3$	118–119		587,637,804
$CO(CH_2)_2Br$	110		587
$CO(CH_2)_2Cl$	10–106		634
COPr-*n*	—		608

TABLE 80. (*Continued*)

R	mp (°C)	Spectra	Refs.
COCHBrEt	100		637
$COCH(Me)CH_2Br$	118.5		634
COBu-*i*	66–67		679,680
COPh	126–127		679,680
	130–131	ir,pmr,ms	584
$COC_6H_2(OMe)_3$-3,4,5	145–147		734
$COCH{=}CH_2$	97–98		634
$COCH{=}CHC_6H_3(OCO_2Et)_2$-3,4	153		831–834
$COCH{=}CHC_6H_3(OH)_2$-3,4	255	ir	834
COCl	120–121		526,623
	122		826
CO_2Et	94–95	pmr,^{13}C-nmr	844
	95		512,523
$CONH_2$	200–202		851
	206–208		823,839
	210		826
CONHOH	181–182		836
CONHOMe	161–162		836
CONHCOR′			
R′ = Me	178–182		823,839
	185–187		835
$R' = CH_2Cl$	178–181		823,839
R′ = Et	150–152		835
$R' = CH_2Ph$	115–116		835
R′ = CH(Et)Ph	125–127		835
R′ = Ph	125–128		823,839
CON(piperidine-4-R_1,4-R_2)			
$R_1 = OH; R_2 = C_6H_4CF_3$-*m*	185–186		837
$R_1 = COMe; R_2 = Ph$	158–159		837

[a]IB = 5-Iminobibenzylyl≡10,11-dihydrodibenz[*b,f*]azepin-5-yl.

are most conveniently obtained by a Horner–Emmons reaction (Eq. 158) between the *o*-nitrobenzyl phosphonates **1298** and the *o*-nitrobenzaldehydes **1299** (500). Alternatively, *o,o′*-diaminobibenzyl can be prepared by reduction of the corresponding *o,o′*-dinitro compounds **1303**, which are readily obtained by base-catalyzed oxidative coupling of the *o*-nitrotoluenes **1301** (Eq. 159) (494–496, 498,499).

The base-catalyzed coupling of *o*-nitrotoluenes is an interesting reaction that has been carried out under a variety of conditions. The most frequently encountered procedure uses sodium ethoxide and isoamyl nitrite ester. The reaction has also been carried out in methanolic potassium hydroxide–air (503). The mechanism of coupling has been studied by Russell and co-workers (504), who postulated that an intermediate charge transfer complex **1302** collapses to **1303** in the presence of an electron acceptor.

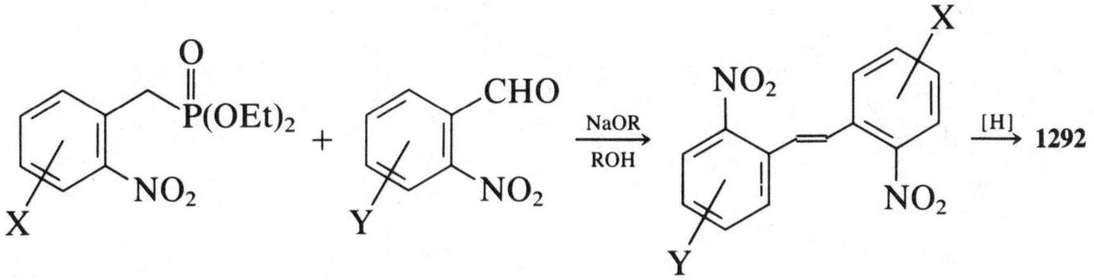

1298 + 1299 $\xrightarrow[ROH]{NaOR}$ 1300 $\xrightarrow{[H]}$ 1292 (158)

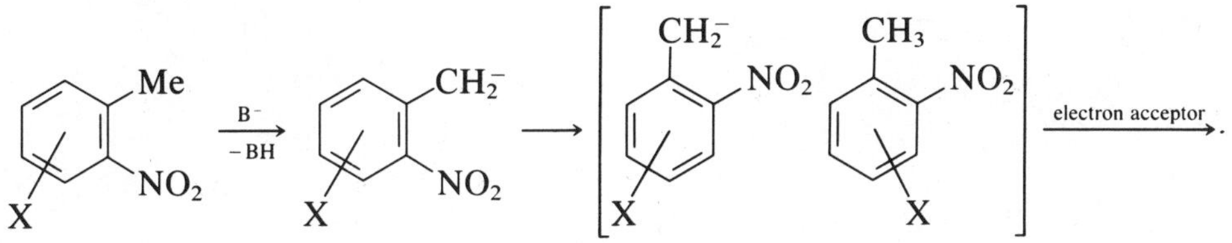

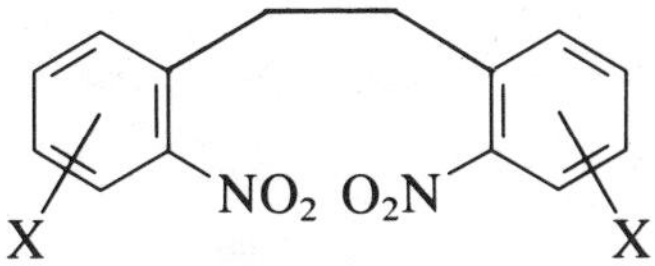

1301 $\xrightarrow[-BH]{B^-}$ → [1302] $\xrightarrow{\text{electron acceptor}}$ 1303 (159)

ii. Cyclization of *o*-Amino-*o*′-halobibenzyls. A method that is clearly related to that described in Section 39.B.a.i involves cyclization of the *o*-amino-*o*′-halobibenzyl **1304** to produce iminobibenzyls (Eq. 160) (505–507). Ring closure is said to occur when the precursors **1304** are heated in solvents such as dimethylformamide or dimethylacetamide, and is claimed to proceed at a faster rate in the presence of sodium or potassium carbonate and copper or copper bronze. A related patent describes the cyclization of the formamide **1305** under strongly alkaline conditions (508).

iii. Internal Coupling of *o,o*′-Di(halomethyl)diphenylamines. Bergmann has reported that treatment of the di(halomethyl)diphenylamine **1306** with phenyllithium (Eq. 161) gives a mixture of **1307** and **1308** in 66 and 3% yields, respectively (509,510). The sensitivity of the reaction to stoichiometry was demonstrated by treating the *o,o*′-di(halomethyl)diphenylamines with excess phenyllithium. Under these conditions, **1306** was converted to **1308** directly.

iv. Hydrogenation of 5H-Dibenz[*b,f*]azepines. In general, iminostilbenes are reduced readily to the corresponding iminobibenzyls (511–518). As a preparative source of iminobibenzyls, this methodology has been associated with iminostilbenes acquired via ring expansion of acridine methanols (see Section 39.B.b.iii). Reduction of the 10,11-double bond of imi-

DMF or DMA, Δ / K_2CO_3, Cu

NH X, R — **1304**

N, R — **1293**

$NaOCH_2CH_2ONa$ / $HOCH_2CH_2OH$

(160)

NH Br, CHO — **1305**

Br Br, CH_2 CH_2, N, Ac — **1306**

PhLi / Et_2O

N, Ac — **1307** + N, Ac — **1308**

(161)

nostilbenes can be achieved catalytically (511–515,517,518) or with sodium in ethanol (512,516). Catalytic reduction usually proceeds under mild conditions. The most commonly employed catalysts have been platinum oxide, palladium on carbon, and Raney nickel. The reduction of halogenated iminostilbenes can be complicated by competing hydrogenolysis. In this regard, platinum and palladium have generally given rise to a lesser incidence of hydrogenolysis when compared to Raney nickel. However, a Japanese patent has described the use of Raney nickel in chlorinated systems with good results (518).

b. 5H–DIBENZ[*b,f*]AZEPINES (See Table 80)

i. DEHYDROGENATION OF 10,11-DIHYDRODIBENZ[*b,f*]AZEPINES. Just as iminobibenzyls can be obtained by hydrogenation of iminostilbenes (Section 39.B.a.iv), the reverse reaction (i.e., catalytic dehydrogenation of iminobibenzyl) serves as a major source of iminostilbenes (497, 506,507,511,519-522). Yields are often high, especially when based on recovered starting material. A number of catalysts have been used to effect the reaction, with palladium on carbon being the most frequently encountered. In a typical experiment, the palladium on charcoal is sprinkled on a glass wool bed in an electrically heated glass column. After preheating the

TABLE 81. DIBENZ[b,f]AZEPINE AND ITS N-ALKYL AND N-ACYL DERIVATIVES

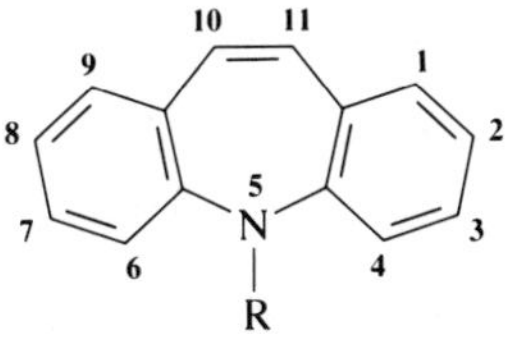

R	mp (°C)	Spectra	Refs.
H	189		510
	189–191	ir,uv	529
	193–195		497
	196		521
	195–196.5		514
	196.5–198		506,507,511
	197–198		519
	204–206	ir,uv	512
	206–208		523
	—	pmr	658
Me	141–143	pmr	583,658
	142–143		530
	143–144.5		497
	144–145		586
Et	102–103	ir,pmr,ms	583,609
	—	pmr	658
n-Pr	85–87	ir,pmr,ms	583
	—	pmr	658
CHO	133.5		838
COMe	99–101	uv	511
	112.1		838
	116–118	pmr	658
	117–119		523
	120		509
	120–121	ir,pmr,uv	629
	121–122	uv	512,628
$COCH_2Br$	136.8		838
$COCH_2Cl$	145–146	pmr	658
	147–148		636
$COCF_3$	86–99		630
	99.2		838
COEt	66–67	ir,pmr,ms	658
	75.7		838
$COCH_2CH_2Cl$	126.5–127.5	pmr	1041
	129.3		838
$CO(CH_2)_3Cl$	86.3		838
COPr-i	140.4		838
$COC(Me)_2Br$	86.3		838
$COCH(Me)CH_2Br$	94–95	pmr	1041
$COCH{=}CH_2$	120–121		638
	122.5–123.5	pmr	1041
$COC(Me){=}CH_2$	141	pmr	1041

TABLE 81. (*Continued*)

R	mp (°C)	Spectra	Refs.
COPh	131–132	ir,pmr,ms	658
$COCH_2CO$-IS[a]	214		829
$COCH(CH_2Ph)CO$-IS[a]	285		829
$CO(CH_2)_8CO$-IS[a]	137–138	ir,pmr,ms	658
COCl	156–157		528
	168–169		825
CO_2Et	124–126		658
	126–127		523
	126–128	ir,uv	512
	130–131	ir,pmr,uv	629
$CONH_2$	188–190		528
	188–192		823,839
	190–191		526,623
	191–192		851
	204–206		825
	—	ir,uv,ms	842
CN	109–110		840
CONHMe	202–204		841
$CONHCH_2CH_2Cl$	"Solid"		822
CONHPr-*n*	137–139		841
$CONHCH_2Ph$	137–140		841
$CONHC_6H_4Me$-*p*	192–195		841
$CONEt_2$	134–135		825
CONHCOMe	144–146		823,839
	145–146		835
$CONHCOCH_2Cl$	165–167		823,839
CONHCOEt	110–112		835
$CONHCOCH_2Ph$	160–162		835
CONHCOCH(Et)Ph	120–122		835
CONHPh	130–133		823,839
	144–146		835
Ts	166–167	ir,pmr,ms	658

[a]IS = 5-Dibenz[*b*, *f*]azepinyl≡5-iminostilbenyl.

column to 400–700°C, the iminobibenzyl is sublimed through the column under reduced pressure (0.2–2.0 mm Hg) and the volatiles are collected in a dry ice–cooled trap. The dehydrogenation reaction has also been carried out by refluxing the appropriate iminobibenzyl and catalyst in a high-boiling inert solvent such as dimethyl maleate (497), biphenyl (506), diphenyl oxide (506,519) or dimethylaniline (506).

Various catalysts have been employed in the dehydrogenation reaction. In addition to palladium, noble metals that have been claimed to be useful catalysts include ruthenium, iridium, rhodium, and osmium, preferably with a carrier such as charcoal, barium sulfate, or silica gel (520). Other catalysts include sulfur (497,506,519,520), selenium (507,520), nickel (521), iron oxide (521), chromium oxide (Cr_2O_3) (521), potassium dichromate (521), copper chromite (521), and ammonium vanadate (521). Combination catalysts have proven extremely effective for producing iminostilbenes in high yield and purity. For example, iron oxide–chromium trioxide–calcium oxide—potassium carbonate in a ratio of 54:3:10:33 at 410°C comprises the catalytic bed used for the commercial production of iminostilbene (522).

ii. BROMINATION–DEHYDROBROMINATION OF IMINOBIBENZYLS. Iminostilbenes have been prepared from iminobibenzyls not only by catalytic dehydrogenation, but also by a bromination–dehydrobromination procedure (499,512,523–528). The sequence is initiated by subjecting an *N*-acyliminobibenzyl, such as **1309a**, to a benzylic bromination to form the intermediate 10-bromoiminobibenzyl **1310a** (Eq. 162). The most frequently encountered brominating agent has been *N*-bromosuccinimide (NBS)–benzoyl peroxide in carbon tetrachloride (499,512). Other brominating agents that have been used include 1,3-dibromo-5,5-dimethylhydantoin **(1312)** (526) and molecular bromine (528). Several other halogenating agents have been claimed effective in the patent literature, including *N*-chlorosuccinimide, *N*-bromoacetamide, and *N*-bromophthalimide (523). Dehydrobromination of **1310a** has been carried out most often with alcoholic potassium hydroxide, the product being either *N*-acetyliminostilbene (**1311a**) or iminostilbene itself (**1311b**), depending on the stoichiometry of the base and the severity of the reaction conditions (512,523,525). Tertiary amines such as collidine have also been used for dehydrobromination. The 5-chlorocarbonyl derivative **1310b** was found to undergo dehydrobromination upon being heated in acetic anhydride, giving rise to 5-chlorocarbonyliminostilbene (**1311c**) (528). An interesting variant of this method involves bromination of **1309b** with molecular bromine to form the dibromide **1310c**, which on thermolysis gives **1311c** (528).

iii. RING EXPANSION OF 9-ACRIDINEMETHANOLS. The Wagner–Meerwein ring expansion of 9-(hydroxyalkyl)-9,10-dihydroacridines (9-acridanemethanols) affords a versatile route to substituted iminostilbenes (511,514,515,517,529–535). The method is especially useful for the preparation of 10- or 11-substituted iminostilbenes as well as 10,11-disubstituted derivatives. It also makes possible the synthesis of several ring-substituted 5H-dibenz[*b*, *f*]azepines that would otherwise be difficult to obtain. The ease

R$_2$ Br

bromination

−HBr

N R$_1$ N R$_1$ N R$_1$ (162)

1309a, R$_1$ = Ac
1309b, R$_1$ = COCl

1310a, R$_1$ = Ac, R$_2$ = H
1310b, R$_1$ = COCl, R$_2$ = H
1310c, R$_1$ = COCl, R$_2$ = Br

1311a, R$_1$ = Ac
1311b, R$_1$ = H
1311c, R$_1$ = COCl

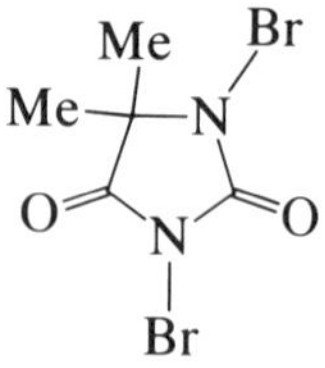

1312

with which iminostilbenes can be reduced (see Section 39.B.a.iv) makes this methodology valuable not only as a source of iminostilbenes, but also as a source of substituted iminobibenzyls. The simplest example of the reaction (Eq. 163) is the conversion of 9-acridanemethanol (**1313**) to iminostilbene (**1311b**) in hot polyphosphoric acid (529).

The requisite 9-acridanemethanols have been prepared by several different methods (511,514,515,529). In the most common approach (Fig. 174), an appropriately substituted 9-chloroacridine **1314** is converted to the corresponding cyanoacridine **1315** with methanolic cyanide. Hydrolysis of the nitrile **1315** to the corresponding carboxylic acid **1316a** has been carried out most frequently by acidic hydrolysis of **1315** to the primary amide, followed by diazotization. The acid **1316a** can be reduced with lithium aluminum hydride to give the corresponding 9-acridanemethanol **1317** directly (529). However, the more common procedure has been to convert **1316a** to its corresponding ester **1316b** prior to reduction. It should be noted that in either case, not only is the carboxylate group reduced, but the acridine ring is also reduced to the corresponding acridane.

Starting from **1314**, 9-acridanemethanols have been prepared by two other methods involving the common intermediate **1319a** (Fig. 175). The 9-methylacridine **1319a** is readily obtainable from **1314** by malonate displacement followed by acidic hydrolysis and decarboxylation. In the first route to **1317**, *N*-bromosuccinimide–benzoyl peroxide was used to convert **1319a** to the bromomethyl derivative **1320a** (514,515). The corresponding acetoxymethylacridine **1320** was obtained by the reaction of **1320a** with potassium acetate. Subsequent lithium aluminum hydride reduction of **1320b** gave the desired alcohol **1317**.

The alternate synthesis of **1317** involves base-catalyzed condensation of

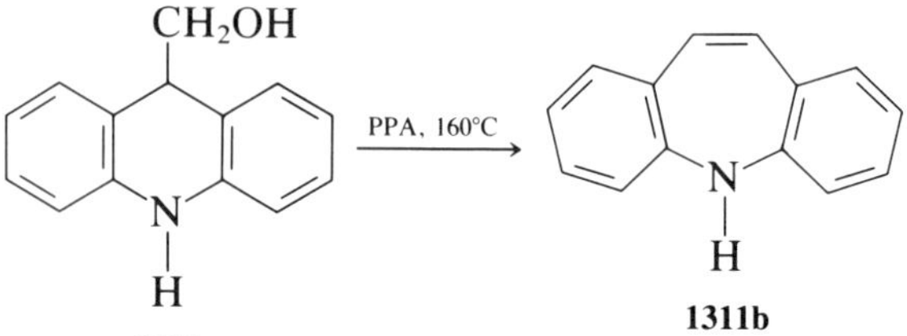

PPA, 160°C (163)

1313 **1311b**

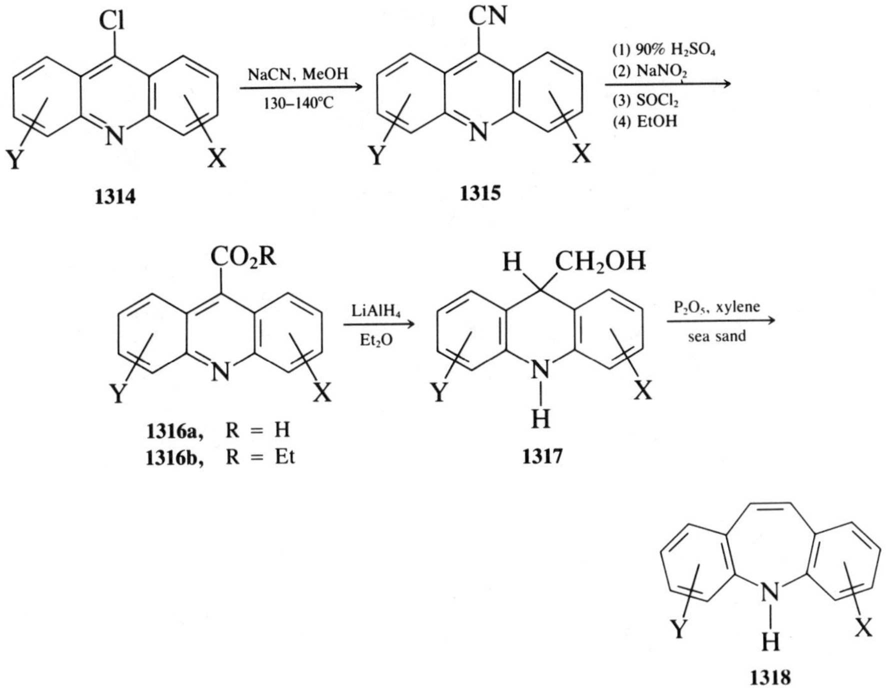

Cl — NaCN, MeOH, 130–140°C → CN — (1) 90% H_2SO_4 (2) $NaNO_2$ (3) $SOCl_2$ (4) EtOH →

1314 **1315**

CO_2R — $LiAlH_4$, Et_2O → H CH_2OH — P_2O_5, xylene, sea sand →

1316a, R = H
1316b, R = Et

1317

1318

Figure 174

1319a with *p*-nitroso-*N,N*-dimethylaniline to give the Schiff base **1321a**, which, without purification, was hydrolyzed to the aldehyde **1321b.** Reduction of **1321b** with lithium aluminum hydride yielded **1317** (507, 531).

The acridanemethanols thus far described have all been precursors to the 10,11-unsubstituted iminostilbenes **1318** (Fig. 174). Access to the 10- or 11-monosubstituted iminostilbenes **1322** (Fig. 176) would require as a pivotal intermediate either 9-(α-hydroxyalkyl)acridanes **1323** or 9-alkyl-9-hydroxymethylacridanes **1324**. Both types of intermediates have been prepared and ring expanded.

The preparation of **1323** was described by Bergmann and co-workers (517). According to their procedure, treatment of *N*-methylacridane (**1325**) with *n*-butyllithium in refluxing diethyl ether, followed by the addition of an appropriate aldehyde to the cooled solution, produced acceptable yields of the alcohol **1323** (R_1 = Me; R_2 = *n*-Pr, Ph). The preparation of **1324** is found in the patent literature (534,535) and involves conversion of a 9-carbomethoxyacridane (**1326**) to the corresponding 9-alkyl-9-carbomethoxyacridane (**1327**), which is then reduced with lithium aluminum hydride.

The requisite precursors for the preparation of 10,11-disubstituted iminostilbenes (**1331**), according to the methodology described in this section, would be 9-alkyl-9-(α-hydroxyalkyl)acridanes (**1330**) (Eq. 164). Such deriv-

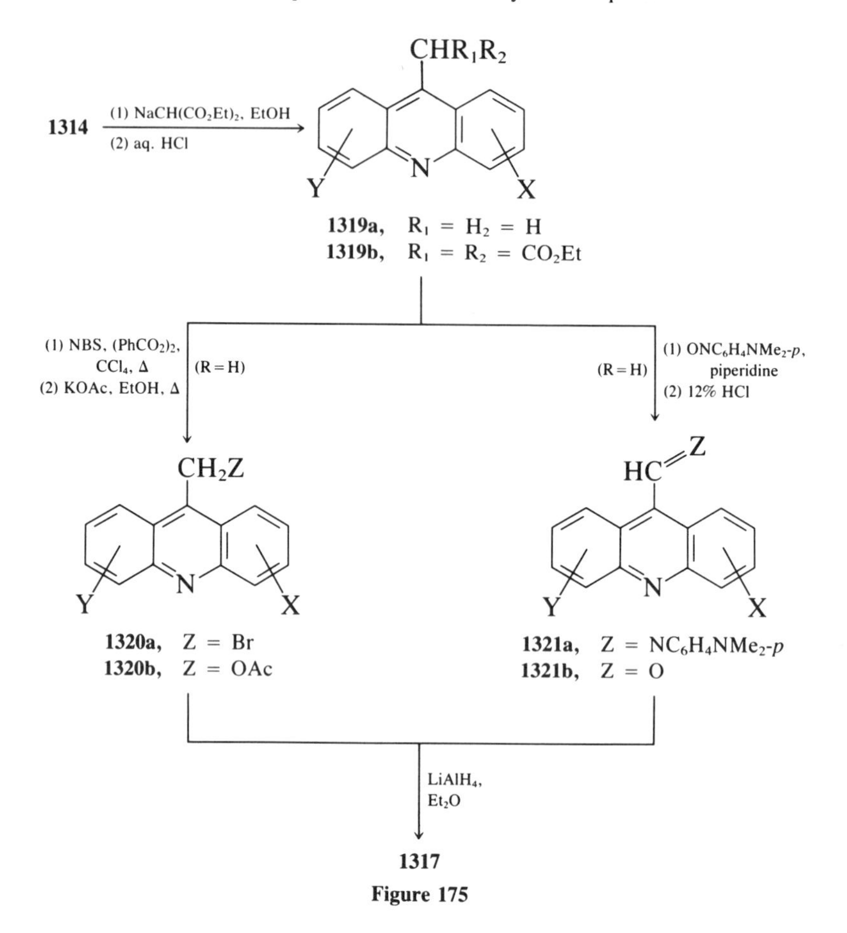

Figure 175

atives have been described in the patent literature (532,533). In an interesting reaction, when 9-methylacridine (**1328**) is dissolved in 2 *N* sulfuric acid and then treated sequentially with acetaldehyde, ferrous sulfate, and di-*tert*-butyl peroxide, 9-acridanyl methyl ketones (**1329**) are obtained. Sodium borohydride reduction then affords the desired carbinols **1330**.

Wagner–Meerwein rearrangements of 9-acridanemethanols are carried out under strongly acidic conditions. Specifically, rearrangement of **1317**, **1322**, and **1323** has been effected with polyphosphoric acid or, more commonly, phosphorus pentoxide in refluxing xylene, whereas rearrangement of **1327b** was carried out in 90–95% sulfuric acid. Unfortunately, reported yields have only been in the 25–50% range.

The rearrangement of **1317** to **1311b** can be visualized as proceeding through a series of intermediates (**1332–1334**, Fig. 177). An alternative competing process would be simple dehydration of **1317** (perhaps via **1332**) without rearrangement to give the 9-methyleneacridane **1335**, which would be ex-

1323

H^+, $-H_2O$

1322

H^+, $-H_2O$

1324

$LiAlH_4$

(1) n-BuLi, Et_2O
(2) R_2CHO

1326

(1) KOBu-t, HMPA, THF
(2) R_2X

1327

1325

Figure 176

1328

(1) 2 N H_2SO_4
(2) MeCHO
(3) $FeSO_4$, $(t\text{-}BuO)_2$

1329

$NaBH_4$, MeOH

(164)

1330

90–95% H_2SO_4, 30°C

1331

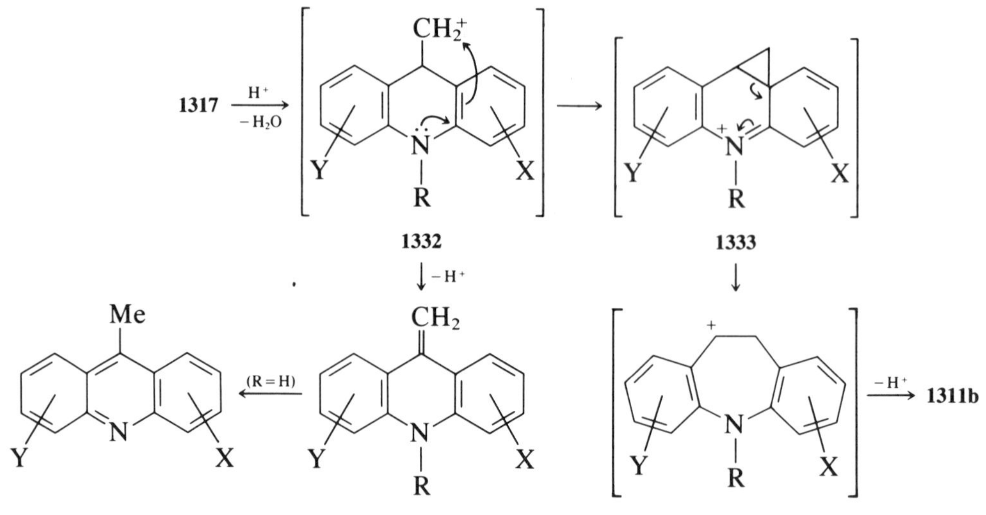

Figure 177

pected to rapidly isomerize to the 9-methylacridine **1328.** In at least one instance, the products of this process have been observed as minor impurities (517)

In a reaction closely related to the dehydrations described above, the action of ethereal diazomethane (Fig. 178) on 10-methylacridinium iodide (**1336**) afforded a low yield (less than 10%) of *N*-methyliminostilbene (**1337**) and a 50% yield of the 9-iodomethylacridane **1338** (530). When **1338** was treated with silver perchlorate, there was obtained a 29% yield of **1337** along with a 9% yield of the acridinium salt **1339.** The sequence is consistent with initial attack of diazomethane on **1336,** followed by loss of nitrogen, to give the intermediate **1332**.

iv. Dehydration of 10,11-Dihydro-10-hydroxydibenz[*b*,*f*]azepines. The addition of Grignard reagents to the dibenzazepinones **1340a** and **1340b** has been shown to give the corresponding 10-hydroxyiminobibenzyls **1341a** and **1341b,** respectively (Eq. 165). The alcohols **1341a** and **1341b** underwent ready dehydration, with formation of the substituted iminostilbenes **1322** and **1342,** respectively. Conditions that have been claimed to effect dehydration include dilute hydrochloric acid, polyphosphoric acid, vacuum distillation, potassium bisulfate, and *p*-toluenesulfonic acid in benzene or toluene. However, only the first two methods have been described in detail (516,536–541).

v. Internal Coupling of *o*,*o*′-Di(halomethyl)diphenylamines. This methodology has been previously discussed in Section 39.B.a.iii.

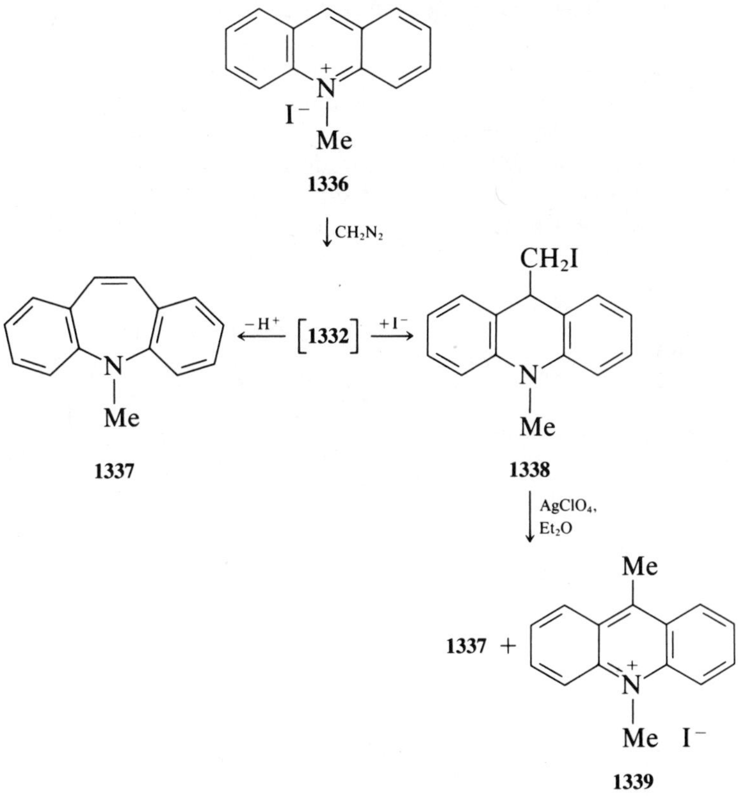

1336

CH_2N_2

$-H^+$ [1332] $+I^-$

1337

1338

$AgClO_4$, Et_2O

1337 +

1339

Figure 178

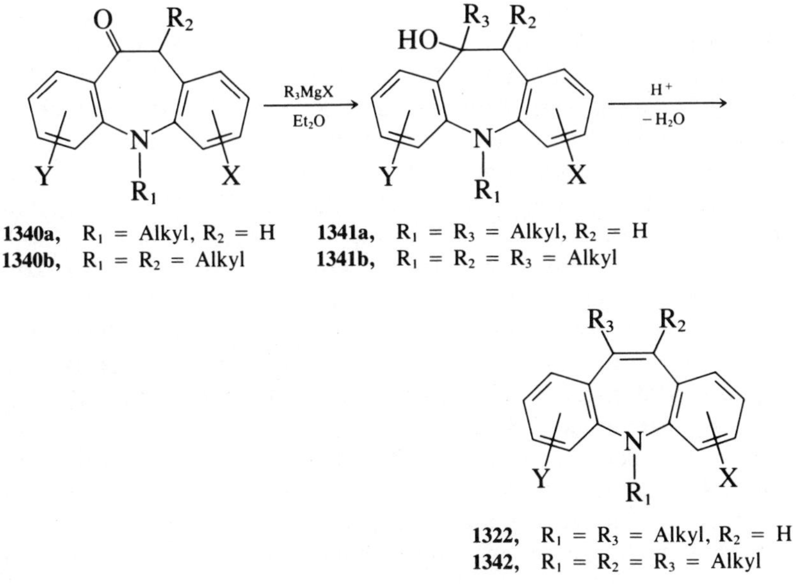

R_3MgX, Et_2O

H^+, $-H_2O$

1340a, R_1 = Alkyl, R_2 = H
1340b, $R_1 = R_2$ = Alkyl

1341a, $R_1 = R_3$ = Alkyl, R_2 = H
1341b, $R_1 = R_2 = R_3$ = Alkyl

(165)

1322, $R_1 = R_3$ = Alkyl, R_2 = H
1342, $R_1 = R_2 = R_3$ = Alkyl

vi. Cyclization of *o*-Amino-*o*′-halo-*cis*-stilbenes. When *cis*-o-(o′-bromostyryl)formanilide (**1344**) and sodium ethylene glycolate were heated to 100°C for 2 hr in ethylene glycol (Eq. 166), a 78% yield of iminostilbene (**1311b**) was obtained (508). The preparation of **1344** was achieved in four steps from *o*-bromobenzaldehyde and *o*-nitrophenylacetic acid by Knoevenagel condensation to the carboxylic acid **1343** followed by decarboxylation, reduction of the nitro group, and formylation.

CHO, Br + CH_2CO_2H, NO_2 —KOAc→ Br, CO_2H, O_2N **1343**

(166)

—(1) $-CO_2$ (2) [H] (3) HCO_2H (4) fractional crystallization→ Br HN, CHO **1344** —$NaOCH_2CH_2ONa$ / $HOCH_2CH_2OH$→ **1311b**

c. DIBENZ[*b,f*]AZEPINONES (See Table 82)

i. 5H-Dibenz[*b,f*]azepin-10-ones. The preparation of 5H-dibenz-[*b,f*]azepin-10-ones via *N*-acyliminostilbenes is by far the most common method in the literature (516,536,542–558). Bromination of *N*-acyliminostilbenes with the general structure **1345** leads smoothly to the corresponding dibromides **1346** (Fig. 179). Dehydrohalogenation of **1346** with potassium hydroxide in dioxane or dioxane–ethanol affords the vinyl bromide **1347** (542–544, 550,556). The vinyl bromide **1347** (R = H), in the presence of alcoholic alkoxide, is transformed into the vinyl ether **1348a.** Frequently, it has been found more convenient to avoid isolation of **1347,** and instead convert **1346** directly to **1348a** by using excess alkoxide. The *N*-unsubstituted vinyl ethers **1348a** can be hydrolyzed to the azepinones **1349a** with dilute acid (542–544,550,557). However, since the 11-position of dibenz[*b,f*]azepin-10-ones is readily alkylated, as shown in Eq. 167 (516,536–538), modification of the azepine nitrogen must be carried out at the enol ether stage. Thus in most of the literature in this field, alkylation of **1348a** is used to obtain the *N*-alkyl derivatives **1348b,** which upon hydrolysis with 2 *N* HC1 give the *N*-alkyl azepinones **1349b.**

As described in Section 39.B.b.iv, dibenz[*b,f*]azepinones can be converted easily to 10,11-substituted dibenz[b, f]azepines (Fig. 135). Thus it becomes apparent that the azepinones **1340a** and **1340b** (Eq. 167) are valuable intermediates for the conversion of iminostilbene to the substituted iminostilbenes **1322** and **1342** (Eq. 165).

TABLE 82A. DIBENZ[*b,f*]AZEPIN-10-ONES

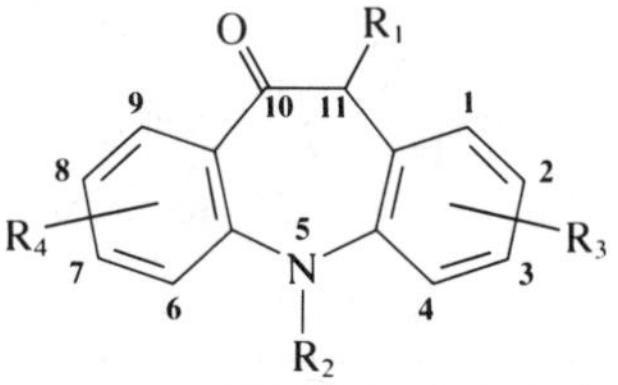

R_1	R_2	R_3	R_4	mp (°C)	Spectra	Refs.
H	H	H	H	141;168 (oxime)		548,549,557
				145–146		542–544
				102–103		565
H	Me	H	H	104		516,536,547 550,557,577,578
				106		562–564
				196 (oxime)		548,549,557
				120;207 (oxime)		548,549,557
H	Et	H	H	126–128		577,578
H	Pr-*n*	H	H	60–62;100 (oxime)		548,549
H	CH_2Ph	H	H	147;202 (oxime)		557
				152		516,536,547, 550,577,578
H	$CONH_2$	H	H	215–216		546
H	Me	1-Cl	H	129;193 (oxime)		559–564
H	Me	2-Br	H	128–129		625
H	Me	2-Cl	H	146;180–182 (oxime)		559–564
H	Me	3-Cl	H	130;209 (oxime)		559–564
H	CH_2Ph	3-Cl	H	122–125		556,631
H	Me	4-Cl	H	140;211 (oxime)		559–564
H	Me	H	6-Cl	80;182–183 (oxime)		559–564
H	Me	H	7-Cl	162;202 (oxime)		559–564
H	CH_2Ph	H	7-Cl	152–154		556
H	Me	H	8-Cl	106;163 (oxime)		560,562–564
H	H	3-Cl	7-Cl	318–320		516,542–544
H	CH_2Ph	3-Cl	7-Cl	152–154		631
H	Me	2-SEt	H	54–55		625
H	H	2-COMe	8-COMe	238–240	ir,pmr,ms	658
Me	Me	H	H	128–130		537,541
Me	CH_2Ph	H	H	178		516,536,550

Other less direct methods have been developed for the synthesis of the 5H-dibenz[*b,f*]azepin-10-one system. One of these involves cyclization of the dinitrile **1350** to the enaminonitrile **1351**, followed by hydrolysis with acetic acid–phosphoric acid, to form the azepinone **1340a** (Eq. 168). Unfortunately, 10 steps were required for the synthesis of the precursor **1350** (559–561).

The same patents (559–561) also described another potentially useful syn-

TABLE 82B. DIBENZ[*b, f*]AZEPIN-2-ONES

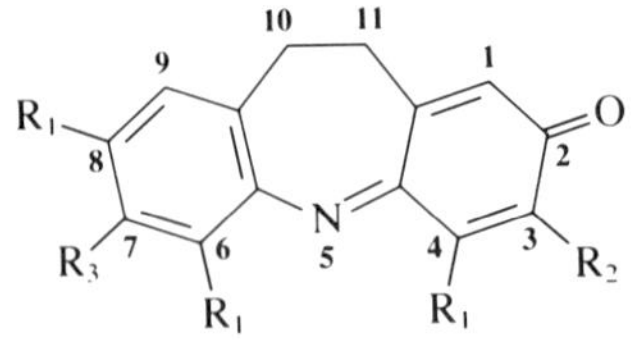

R_1	R_2	R_3	mp (°C)	Spectra	Refs.
H	H	H	101		624
			105–106		649–652
Br	H	H	147–149; 152–153	ir,uv	624
H	Cl	H	134–136		653
H	H	Cl	126–128		653

TABLE 82C. DIBENZ[*b, f*]AZEPIN-2-ONES

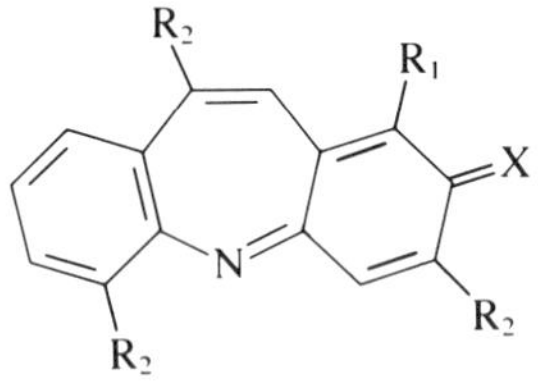

R_1	R_2	X	mp (°C)	Spectra	Refs.
H	H	O	135–136	ir,pmr	654
			135–136	ir,pmr,uv,ms	655
Br	H	O	170–172	pmr,ms	655
Cl	H	O	175–177	pmr,ms	655
NO_2	H	O	200–201 (dec.)	pmr,ms	655
H	OH	O	280–283 (dec.)	uv	572
H	OH	$NNHCONH_2$	260–310 (dec.)	uv	572
COMe	OH	O	220–227	uv	572
COMe	OH	NH	237–239	uv	572

TABLE 82D. 5H-1,2,3,4-TETRAHYDRO DIBENZ[*b, f*]AZEPIN-11-ONES

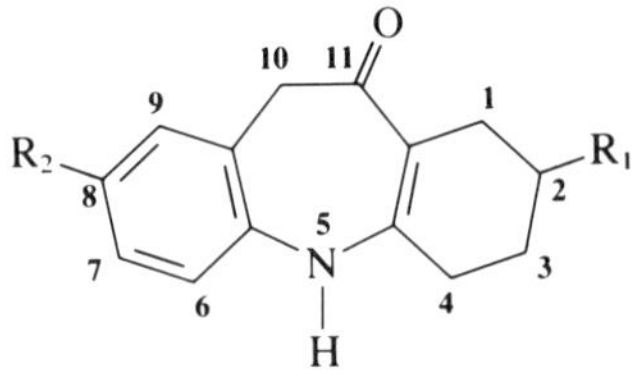

R_1	R_2	mp (°C)	Spectra	Refs.
H	H	241	ir,pmr,uv	566
H	MeO	209		567
CO_2Et	H	205		568
CO_2Et	MeO	244	ms	568

TABLE 82E. DIBENZ[b,f]AZEPIN-10,11-DIONES

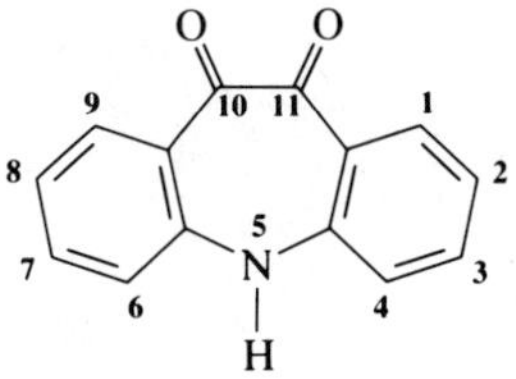

O O
9 10 11 1
8 2
5
7 N 3
6 4
H

mp (°C)	Spectra	Refs.
360 (dec.)	ir,pmr,ms	854
—	X-ray	855

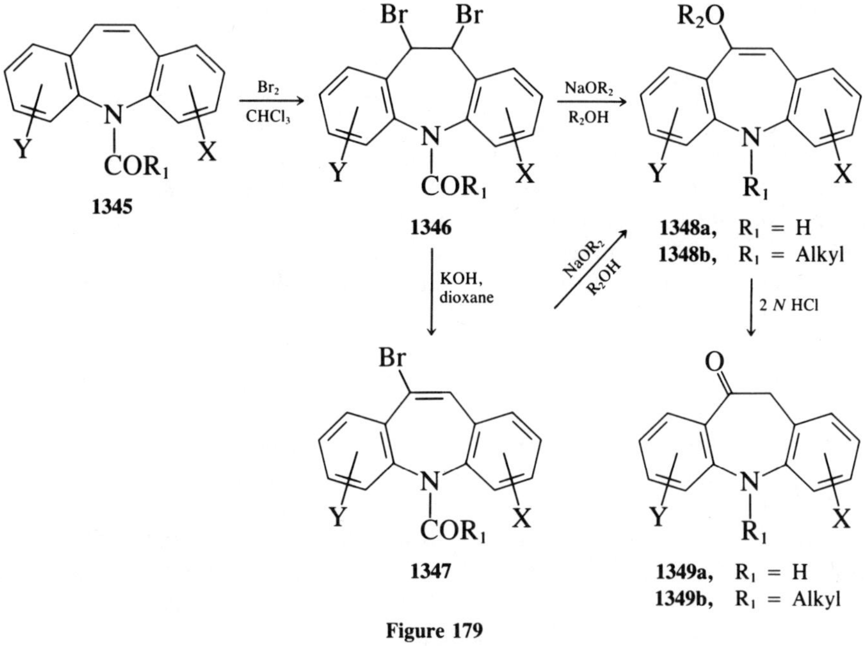

Y N X
COR$_1$
1345

Br_2
$CHCl_3$

Br Br
Y N X
COR$_1$
1346

$NaOR_2$
R_2OH

R_2O
Y N X
R$_1$
1348a, R_1 = H
1348b, R_1 = Alkyl

KOH,
dioxane

$NaOR_2$
R_2OH

2 N HCl

Br
Y N X
COR$_1$
1347

O
Y N X
R$_1$
1349a, R_1 = H
1349b, R_1 = Alkyl

Figure 179

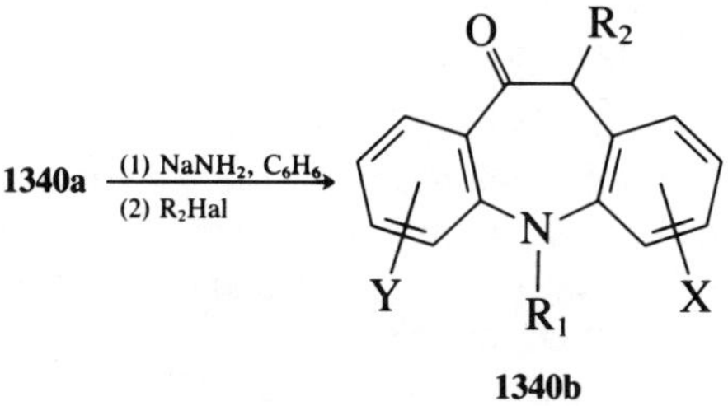

O R_2

1340a (1) $NaNH_2$, C_6H_6
(2) R_2Hal

Y N X
R$_1$
1340b

(167)

CN
CH_2 NC
NaOPr
n-PrOH
NC NH₂
Y N X
1350
Y N X
Me
1351
H_3PO_4, HOAc
Δ

(168)

O
Y N X
Me
1340a

thesis of dibenz[*b,f*]azepin-10-ones via intramolecular Friedel–Crafts acylation of the *o*-(phenylamino)arylacetic acids **1352a.** Thus, on being heated in polyphosphoric acid, **1352a** underwent ring closure to the tricyclic ketone **1340a** (Eq. 169). The preparation of the necessary arylacetic acids **1352a** is shown in Eq. 170. In a related reaction, the analogous nitrile **1352b** was cyclized by the action of anhydrous hydrogen chloride in toluene–ethanol (Eq. 169). The resulting imine salt **1353** was formed in good yield (565).

Base-catalyzed cyclizations of *N*-(*o*-tolyl)anthranilic esters **1354** have been achieved successfully by the action of lithium dimethylamide in benzene–hexamethylphosphoric triamide (Eq. 171) (562–564). The resulting azepinones **1340a** were formed in approximately 80% yield.

Finally, in the context of a synthesis of Iboga alkaloids, Rosenmund and co-workers obtained several tetrahydrodibenz[*b,f*]azepinones **1359** as by-products from a new indole synthesis (566–568). As shown in Fig. 180, acylation of the enamines **1355** with the *o*-nitrophenacyl chlorides **1356** afforded the *β*-diketones **1357.** Zinc–hydrochloric acid reduction of the nitro group of **1357** generally gave good yields of the desired indoles **1358**, but in all cases these were contaminated with 6 to 15% of the dibenz[*b,f*]azepinones **1359.** Spectral considerations supported structure **1359,** as opposed to the tautomers **1360** and **1361**.

ii. DIBENZ[*b,f*]AZEPIN-1-ONES. The only report of dibenz[*b,f*]azepin-1-ones is found in the French literature (569). When indole and dimedone were combined in methanolic hydrogen chloride, the condensation products **1362** and **1363** were formed in low yield (Eq. 172). Both products can be readily explained as arising from the known dimeric and trimeric adducts obtained from indole and hydrogen chloride (570).

(169)

1352a, Z = CO_2H
1352b, Z = CN

1340a, Z = O
1353, Z = NH·HCl

PPA, Δ

(1) MeOH, H_2SO_4 (2) NaH (3) MeI

(1) $LiAlH_4$ (2) $SOCl_2$

(1) NaCN, DMSO (2) KOH, aq. EtOH → **1352a**

(170)

LiNMe$_2$, C_6H_6, HMPA

1354 → **1340a**

(171)

iii. DIBENZ[*b, f*]AZEPIN-2-ONES. Most dibenz[*b, f*]azepin-2-ones have been prepared by the oxidation of iminobibenzyls and iminostilbenes, and this chemistry is discussed in detail in Section 39.C.c.ii. Known examples of this class of compounds are given in Table 82.

In another study, alkaline cleavage of the phenoxazone **1364** gave the new phenoxazone **1365** (Eq. 173) (571). As part of an investigation of the generality of this reaction, the phenoxazones **1366a** and **1366b** were treated with dilute methanolic alkali (572). Under these conditions, **1366a** and **1366b** did not give the expected product **1366c,** but instead gave the trihydroxydibenz[*b, f*]azepin-2-one **1367a** (Fig. 181). The expected product **1366c** was also converted to **1367a.** The phenylquinonimine **1368** was invoked as the likely intermediate in this transormation. That the product **1367a**

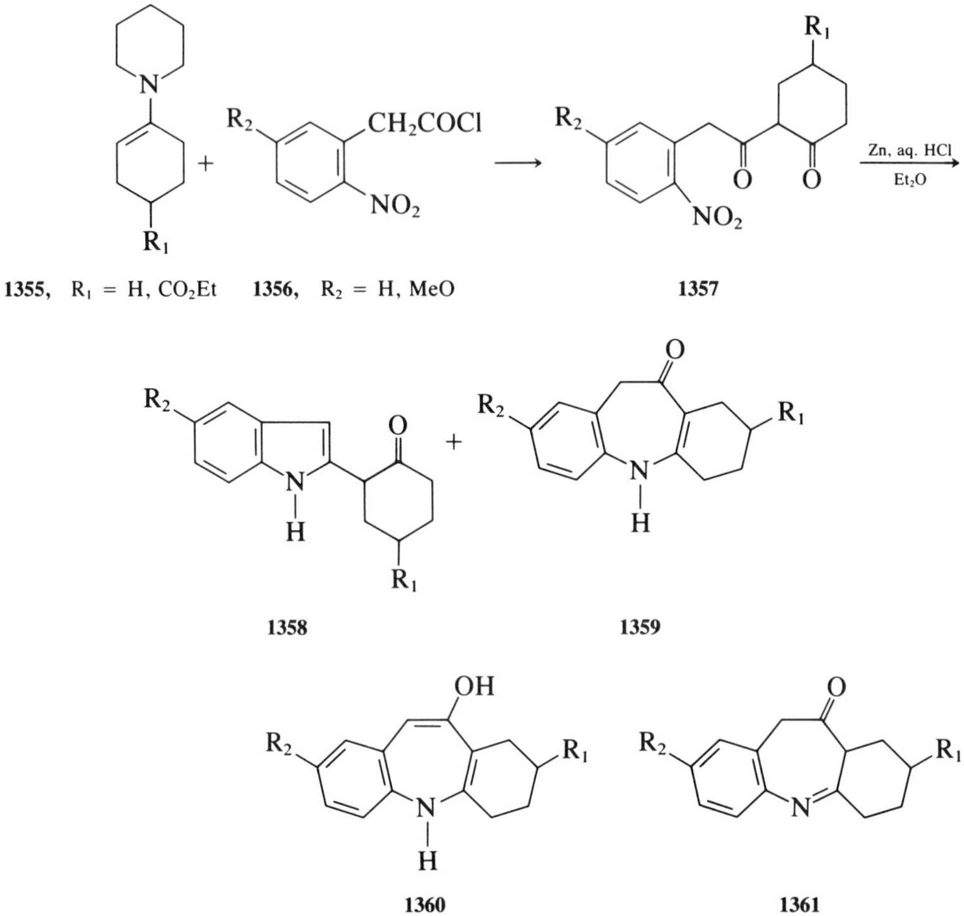

1355, R_1 = H, CO_2Et **1356,** R_2 = H, MeO **1357**

1358 **1359**

1360 **1361**

Figure 180

contained but one carbonyl group was supported by the formation of the monosemicarbazone **1367b.** Condensation of **1367a** with *o*-phenylenediamine produced the crystalline quinoxaline **1369a**.

In the same paper, some additional derivatives of this series were prepared (Eq. 174). Air oxidation of 2-amino-3-hydroxyacetophenone (**1370**) in an aqueous solution buffered to pH 8.5–9.0 with ammonia and ammonium acetate gave a new compound, **1371a.** Dichromate oxidation of **1370** in 8 *N* sulfuric acid gave a different compound, **1371b.** The ultraviolet spectra of **1371a** and **1371b** in acid as well as base were identical. Compound **1371b** was postulated to be a tautomer of **1371a** containing a mole of hydration. On treatment with methanolic hydrochloric acid, the red solid **1371b** was in fact converted to the yellow solid **1371a**.

Alkaline hydrolysis of either **1371a** or **1371b** gave the new compound **1371c**

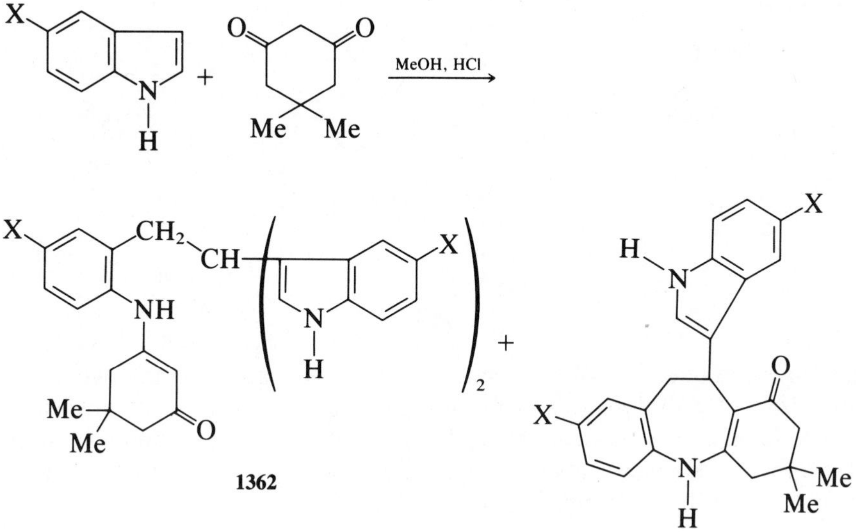

MeOH, HCl

1362

1363

(172)

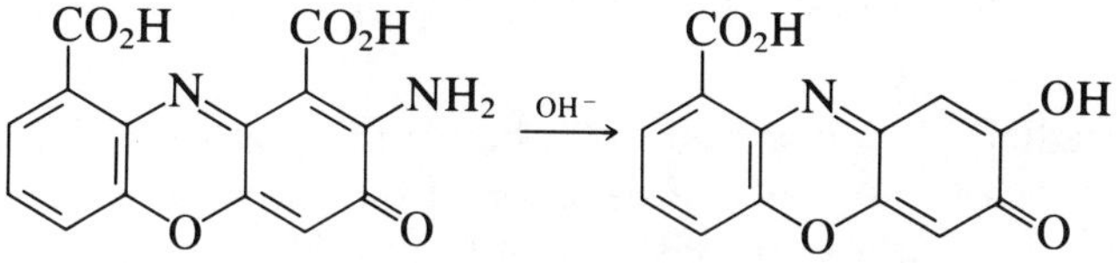

1364

OH^-

1365

(173)

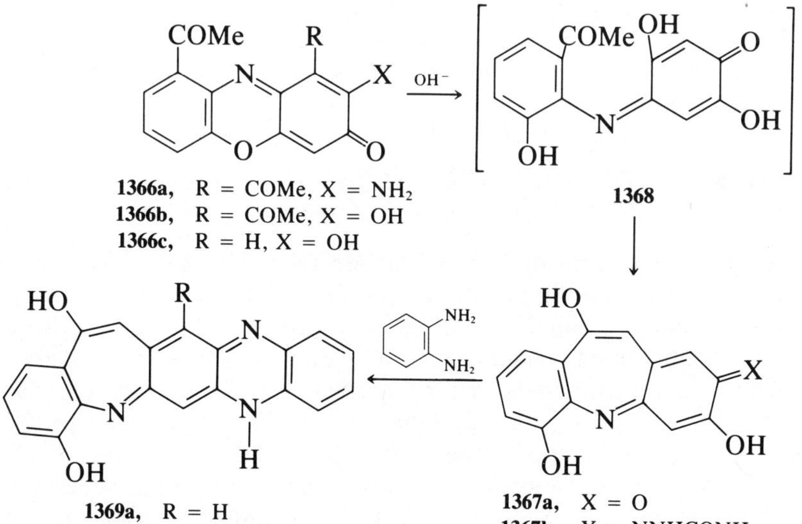

1366a, R = COMe, X = NH_2
1366b, R = COMe, X = OH
1366c, R = H, X = OH

OH^-

1368

1367a, X = O
1367b, X = $NNHCONH_2$

1369a, R = H
1369b, R = COMe

Figure 181

2 **1370** $\xrightarrow{[OX]}$ **1371** $\xrightarrow{o\text{-}C_6H_4(NH_2)_2}$ **1369b** (174)

1371a, X = NH
1371b, X = $NH{\cdot}H_2O$
1371c, X = O

as orange-red needles. Microchemical analysis indicated that the two structures differed only by the loss of an amino or imino group. In contrast to the phenoxazones **1366a** and **1366b,** the acetyl group on the quinonoid system of **1371c** was stable to alkaline cleavage. Condensation of **1371c** with *o*-phenylenediamine afforded the quinoxaline **1369b**.

C. *Functional Derivatives of Dibenz[b,f]azepines*

The preceding section served to introduce the principal approaches that have been developed for the construction of the parent dibenz[*b,f*]azepine nucleus. It is the purpose of this section to describe the preparation of the major functional derivatives of the parent system. The vast majority of the dibenz[*b,f*]azepine literature has dealt with derivatives, especially those with a basic nitrogen substituent. Since most of the chemistry is classical, much of the literature can be summarized in a few general equations. For specific compounds in any functional class, the reader should consult the tables cited below.

a. ALKYL–, ALKENYL–, and ALKYNYLDIBENZ[*b,f*]AZEPINES (See Tables 83 and 84)

Alkyl derivatives, for the sake of convenience, have been combined into three subgroups according to the position of the alkyl substituent: (1) 10- and 11-substituted compounds (bridge alkyl groups), (2) 1- to 4-substituted and 6- to 9-substituted compounds (alkyl groups on the benzene rings), and (3) 5-substituted compounds (*N*-alkyl derivatives). The *N*-alkyl derivatives comprise by far the largest subgroup, with 5-(aminoalkyl)dibenz[*b,f*]azepines such as imipramine (**1296a**) being the subject of literally scores of patents. Accordingly, aminoalkyl derivatives are covered in a separate section of this review (see Section 39.C.i.ii). The available methods of preparation of 10-alkyl and 10,11-dialkyl iminobibenzyls and iminostilbenes have all been previously described in Sections 39.B.a.iv, 39.B.b.iii, and 39.B.b.iv. A listing of these compounds is given in Tables 83 and 84.

Dibenz[*b,f*]azepines with alkyl substitutents on the benzene rings have been prepared via three basic strategies. The first of these has been to prepare the desired iminobibenzyl or iminostilbene according to methodology described earlier (Sections 39.B.a.i and 39.B.a.iii), using appropriately alkylated precursors. For example, cyclization of alkylated *o,o'*-diaminobibenzyls (Section 39.B.a.i) has been used to obtain monoalkyliminobibenzyls (500) as well as symmetrically substituted dialkyl iminobibenzyls (500,501,506). The internal coupling of the diphenylamine **1372** (see Section 39.B.a.iii for other examples) gave the 2,8-dimethyliminostilbene **1373** in 30% yield (Eq. 175) (510).

1372 $\xrightarrow[Et_2O]{PhLi}$ **1373** (175)

The second strategy for the preparation of the alkyl-substituted dibenz[*b,f*]azepines has been via reduction of the corresponding acyl derivatives. Thus the Huang–Minlon modification of the Wolff–Kishner reduction was used (Eq. 176) to convert the ketones **1374,** which are readily available by Friedel–Crafts acylation and subsequent hydrolysis, to the alkyl derivatives **1375** (512,527,573). Using methods already described in Section 39.B.b.ii, **1375** was converted to the corresponding alkyl iminostilbenes **1376.** A slightly different reductive approach (Eq. 177) allowed the conversion of iminobibenzyl-4-carboxylic acid (**1377**) to 4-methyliminobibenzyl(**1379**) (574). The reduction was achieved with alane and most likely proceeded via the azaquinone methide **1378.**

1307 $\xrightarrow[(2)\ KOH,\ aq.\ EtOH]{(1)\ RCOCl,\ AlCl_3,\ CS_2}$ **1374** $\xrightarrow[KOH,\ HOCH_2CH_2OH,\ \Delta]{H_2NNH_2 \cdot H_2O}$

1375 $\xrightarrow{(1)\ AcCl;\ (2)\ NBS,\ CCl_4,\ h\nu;\ (3)\ KOH,\ aq.\ EtOH}$ **1376** (176)

TABLE 83. ALKYL-, ALKENYL-, AND ARYLDIBENZ[*b*, *f*]AZEPINES

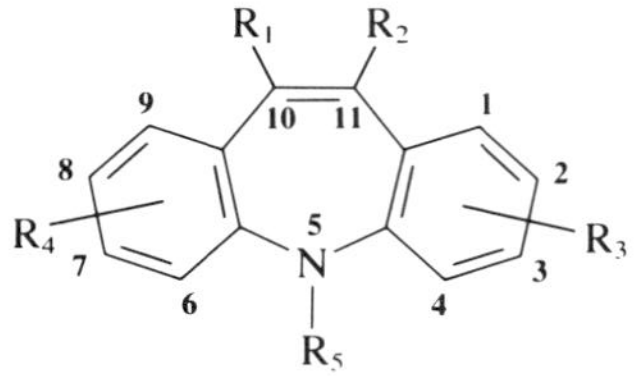

R_1	R_2	R_3	R_4	R_5	mp (°C)	Spectra	Refs.
Me	H	H	H	H	133–134		516,536,550
Me	H	H	H	Me	96–97		516,536,550
Me	H	H	H	Et	143–145		547,577,578
Me	H	H	H	COMe	97–98		556,631
Me	H	H	H	COCl	116–117		852
Me	H	H	H	$CONH_2$	180–181		852
Et	H	H	H	H	61–63		516,536,550
Et	H	H	H	Me	148–150/0.01[a]		516,536,550
Et	H	H	H	CH_2Ph	Oil		516,536,550
Pr	H	H	H	Me	Oil		517
CH_2Ph	H	H	H	H	148		516,536,550
Ph	H	H	H	Me	88–93		516,536,550
					200–200.5	uv,ir	517

Me	H	3-Cl	H	H	—		556,631
Me	H	3-Cl	H	COMe	170–172		556,631
H	Me	3-Cl	H	H	145–146		556,631
H	Me	3-Cl	H	COMe	126–127		556,631
H	H	3-Et	H	H	186	uv,ir	512,527
H	H	3-Et	H	COMe	Oil		512
H	H	2-Ph	H	H	198–200 (dec.)		655
Me	Me	H	H	H	130–131		533
					131–132		537–541
					134–135		516,536,550
Me	Me	H	H	Me	109–111		537–541
					113		533
Me	Me	H	H	CH_2Ph	130		516,536,550
Me	Me	H	H	COMe	109–111		537–541
Me	Me	H	H	$CONH_2$	107–108		852
Me	Et	H	H	H	164/0.005[a]		516,536,550
Me	Me	2-Cl	H	H	137–138		533
Me	Me	3-Cl	H	H	137–139		533
Me	Me	$3-CF_3$	H	H	153–155		533
Ph	Ph	2-Me	8-Me	Me	145–147	pmr,ms	510

[a]Boiling point (°C/torr).

TABLE 84. ALKYL-, ALKENYL-, AND ARYL-10,11-DIHYDRODIBENZ[*b,f*]AZEPINES

R_1	R_2	R_3	R_4	R_5	mp (°C)	Spectra	Refs.
Me	H	H	H	H	72–73		516,536,550
Me	H	H	H	Me	79–81		516,536,550
Me	H	H	H	COCl	135–136		615,852
Me	H	H	H	$CONH_2$	182–183		852
Et	H	H	H	H	57–58		516,536,550
Et	H	H	H	Me	137–139/0.005[a]		516,536,550
CH_2Ph	H	H	H	$CONH_2$	144–145		852
Ph	H	H	H	H	154		516,536,550
Ph	H	H	H	Me	Oil		516,536,550
Ph	H	H	H	$CONH_2$	148–152		852
H	H	2-*t*-Bu	H	H	106		575,576
H	H	2-*t*-Octyl	H	H	106		575,576
H	H	3-Me	H	H	103.5–104		500
H	H	3-Et	H	H	120–130/0.005[a]	uv,ir	512
					90–92		527
					93–94		
H	H	3-Et	H	COMe	84–85		512,527
H	H	3-*n*-Pr	H	H	74–75		527

H	H	4-Me	H	H	52–53		574,635
H	H	3-Cl	7-Me	H	130–131		500
Me	Me	H	H	H	96–97		516,536,550
H	H	2-Me	8-Me	COCl	102–105		827
H	H	2-Me	8-Me	$CONH_2$	144		827
H	H	2-*t*-Bu	8-*t*-Bu	H	160		575,576
H	H	3-Me	7-Me	H	162–163		501
H	H	3-Me	7-Me	COCl	125		827
H	H	3-Me	7-Me	$CONH_2$	200–201		827
H	H	2-CH=CH_2	H	Me	69–70	ir,pmr,ms	618
H	H	2-CH=C(NMe_2)$PO(OEt)_2$	H	Me	Oil		581
H	H	2-CH=C(NMe_2)$PO(OEt)_2$	H	Et	Oil		581
H	H	2-CH=C(NMe_2)$PO(OEt)_2$	H	CH_2Ph	Oil		581
H	H		H	Me	175–180		617
H	H	2-CH= (2-phenyl-5-oxo-oxazol-4-ylidene; O, O, N, Ph)	H	Me	171–172		581

[a]Boiling point (°C/torr).

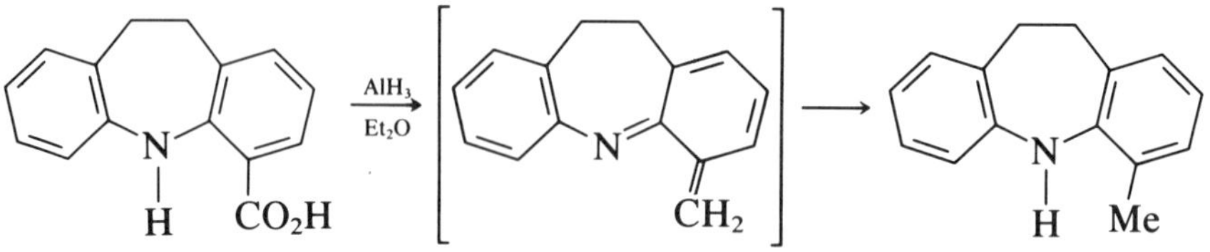

$\xrightarrow[Et_2O]{AlH_3}$ (177)

1377 **1378** **1379**

Finally, Friedel–Crafts alkylation of iminobibenzyl (**1293**) with diisobutylene in the presence of aluminum chloride at 108–146°C (Eq. 178) gave a mixture of mono- and di-(*tert*-octyl)iminobibenzyl **1380a** and **1380b,** as well as unreacted starting material (575,576). When the reaction was carried out at 180–190°C, the products were the corresponding mono- and di(*tert*-butyl)iminobibenzyls **1381a** and **1381b**.

The *N*-alkyldibenz[*b,f*]azepines have been the subject of many publications. *N*-Alkylations of iminobibenzyls and iminostilbenes (497,516,536, 547–549,577–616) have usually been carried out by reaction of the free base with an alkyl halide (chloride, bromide, or iodide) or tosylate in the presence of a strong base as a proton acceptor.

By far the most frequently used bases for these alkylations have been the alkali amides, especially sodium amide. The reactions have been carried out in refluxing benzene, toluene, or xylene (516,536,547,580,582,587–596). Alkyl and aryl lithiums have also been used, usually with ether as the solvent (497,584,600–602). Sodium hydride in dioxane (579), DMF (581,598,599), and DMSO (dimsyl anion) (586,603,604) have likewise yielded excellent results. Sodium carbonate has been used in conjunction with dimethyl sulfate (497), although an excellent yield of *N*-methyldibenz[*b,f*]azepine (**1337**) was obtained using dimethylsulfate alone (586). In addition, thallous ethoxide in a mixture of diethyl ether and DMF has been shown to be an effective base for alkylation of iminostilbene under extremely mild conditions (583).

The relative merits of each *N*-alkylation procedure must be weighed on an individual basis. Obviously, the conduct of a laboratory-scale reaction, where optimum yield is often a determinant of the choice of conditions, would allow a significant degree of freedom. This latitude may not be available on pilot-plant or larger scales where expense, safety, and ease of handling become of paramount importance. However, if optimum yield is the sole criterion for choosing one of the methods, some instructive comparisons are possible.

The *N*-methylation of iminobibenzyl (**1293**) has been carried out under a variety of conditions. For example, sodium carbonate and dimethyl sulfate converted **1293** to *N*-methyliminobibenzyl (**1382a**) in 56% yield (497). Phenyllithium–methyl iodide in ether effected the same transformation in 63% yield (497), and *n*-butyllithium–methyl iodide gave **1382a** in 70% yield (584).

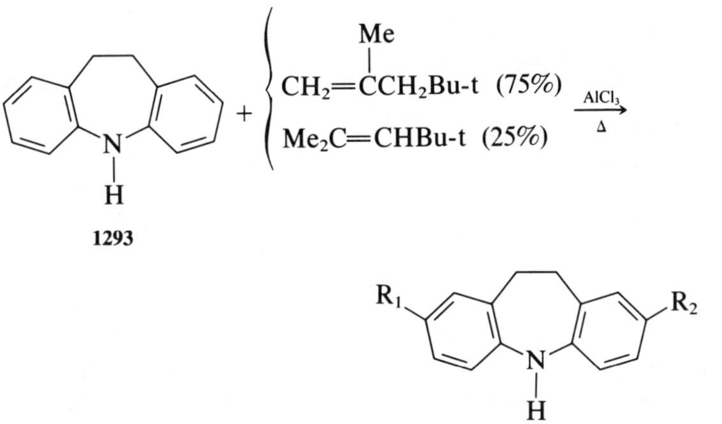

$$\mathbf{1293} + \left\{ \begin{array}{l} CH_2{=}C(Me)CH_2Bu\text{-}t\ (75\%) \\ Me_2C{=}CHBu\text{-}t\ (25\%) \end{array} \right. \xrightarrow[\Delta]{AlCl_3} \tag{178}$$

1293

1380a, R_1 = H, R_2 = CMe_2CH_2Bu-*t*
1380b, R_1 = R_2 = CMe_2CH_2Bu-*t*
1381a, R_1 = H, R_2 = *t*-Bu
1381b, R_1 = R_2 = *t*-Bu

The last method represents the best yield thus far reported for this conversion. The analogous *N*-ethyl derivative **1322b**, was also formed from *n*-butyllithium–ethyl iodide and **1293.** These conditions failed when extended to the *n*-propyl derivative **1382c**, suggesting the existence of steric hindrance associated with *N*-alkylation of **1293.** Thallous ethoxide–methyl iodide (or bromide) and **1293** failed to react.

The *N*-methylation of iminostilbene (**1311b**) has also been investigated. Thallous ethoxide–methyl iodide and **1311b** gave **1337** in 74% yield (583). Neat dimethyl sulfate and **1311b** gave **1337** in 84% yield, whereas sodium hydride–methyl iodide in DMSO afforded **1337** (see Fig. 178) quantitatively (586).

An interesting *N*-alkylation of **1293** with the cyclopropane derivative **1383** has been reported. Reaction of **1293** with either *cis*- or *trans*-**1383** in dimethyl sulfoxide–sodium hydride gave the *trans*-*N*-cyclopropyliminobibenzyl **1384** (Eq. 179). The reaction product and stereochemistry are readily explained by invoking *in situ* formation of the cyclopropene **1385** (603,604).

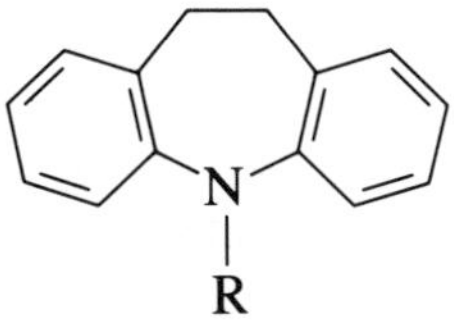

1382a, R = Me
1382b, R = Et
1382c, R = *n*-Pr

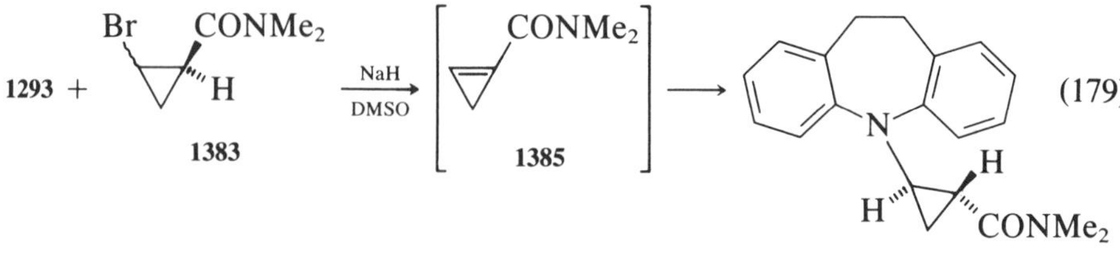

1293 + 1383 $\xrightarrow[\text{DMSO}]{\text{NaH}}$ [1385] ⟶ 1384 (179)

The preparation of N-alkyldibenz[$b,f,$]azepines has been accomplished by reduction of the corresponding N-acyl derivatives. Borane has been used with good results for the reduction of N-formyl-, N-acetyl-, and N-butyryl-iminobibenzyl to the corresponding N-alkyl derivatives, namely **1345** → **1386** (Eq. 180) (605,608). The same results have been obtained using alane as the reducing agent (606,607).

The amination of aldehydes, either reductively or via a Strecker synthesis, has been used for the alkylation of dibenz[b,f]azepines. When iminostilbene (**1311b**) was treated with sodium borohydride in glacial acetic acid (Eq. 181), the N-ethyl derivative **1387**, was obtained in 72% yield (609). The reaction was postulated to proceed by (1) reduction of acetic acid to acetaldehyde, perhaps via an acetyloxyborohydride species; (2) formation of an imminium ion (**1388**); and (3) hydride reduction of **1388**.

The principle of reductive amination having been demonstrated, it should be possible to prepare a number of N-alkyl derivatives by standard routes. That the reduction above did not involve N-acetyldibenz[b,f]azepine (**1345**) was demonstrated when this compound was shown to be stable to the reaction conditions. The alkylation of iminobibenzyl **1293** with aldehydes in acetic acid–potassium cyanide (Strecker amino acid synthesis) has been demonstrated (610,611) to form the adducts **1389**. In this reaction (Eq. 182),

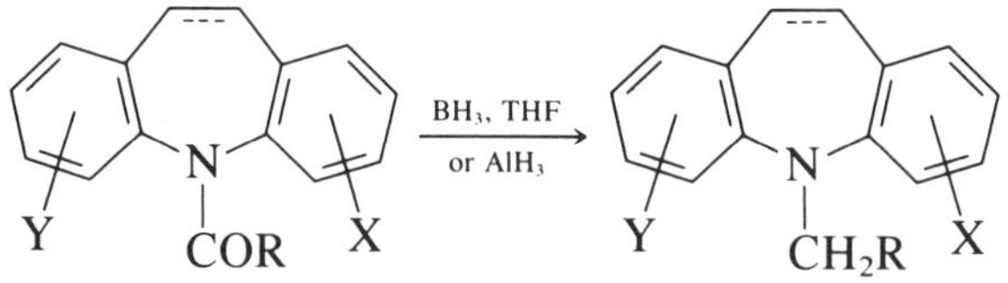

1345 $\xrightarrow[\text{or } AlH_3]{BH_3\text{, THF}}$ 1386 (180)

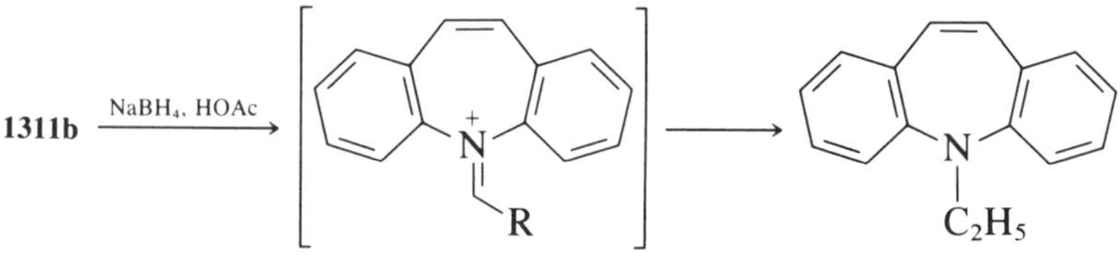

1311b $\xrightarrow{NaBH_4\text{, HOAc}}$ [1388] ⟶ 1387 (181)

1293 + RCHO $\xrightarrow{\text{KCN, HOAc}}$ [structure 1389: N-RCHCN iminobibenzyl] (182)

RCHCN

1389

the intermediate imminium ion is captured by the cyanide ion instead of by hydride.

Finally, *N*-aminoalkyl derivatives **1392** of iminobibenzyl and iminostilbene have been obtained by the thermal decarboxylation of the urethanes **1391** (612–616). The reaction has been carried out by heating the urethanes either without solvent, under reduced pressure, or in the presence of copper powder. The requisite urethanes **1391** are readily available from the *N*-chlorocarbonyl derivatives **1390.** While the reaction has been described only for *N*-(aminoalkyl)dibenz[*b, f*]azepines, it could be a general procedure for the introduction of simpler alkyl groups. (Eq. 183)

In spite of the voluminous literature dealing with the dibenz[*b, f*]azepine system, there have been few examples of alkenyl and alkynyl derivatives. The iminobibenzyl-2-carboxaldehydes **1393** have been converted to 2-vinyliminobibenzyls (Fig. 182). Wittig olefination of **1393** (R = Me) gave **1394** in 45% yield (618). The olefin **1394** could be hompolymerized by radical, cationic, or anionic initiators, and could also be copolymerized with styrene and maleic anhydride. Condensation of **1393** (R = Me) with nitromethane in benzene in the presence of ammonium acetate produced the nitroolefin **1395** (617). In a related condensation, **1393** was condensed with the sodium salt of tetraethyl dimethylaminomethylenediphosphonate to produce the diethylalkenylphosphonates **1396** (581).

Y COCl X → Y X O O—Alk-NR$_2$ $\xrightarrow[-CO_2]{\Delta}$

1390 **1391**

(183)

Y X Alk-NR$_2$

1392

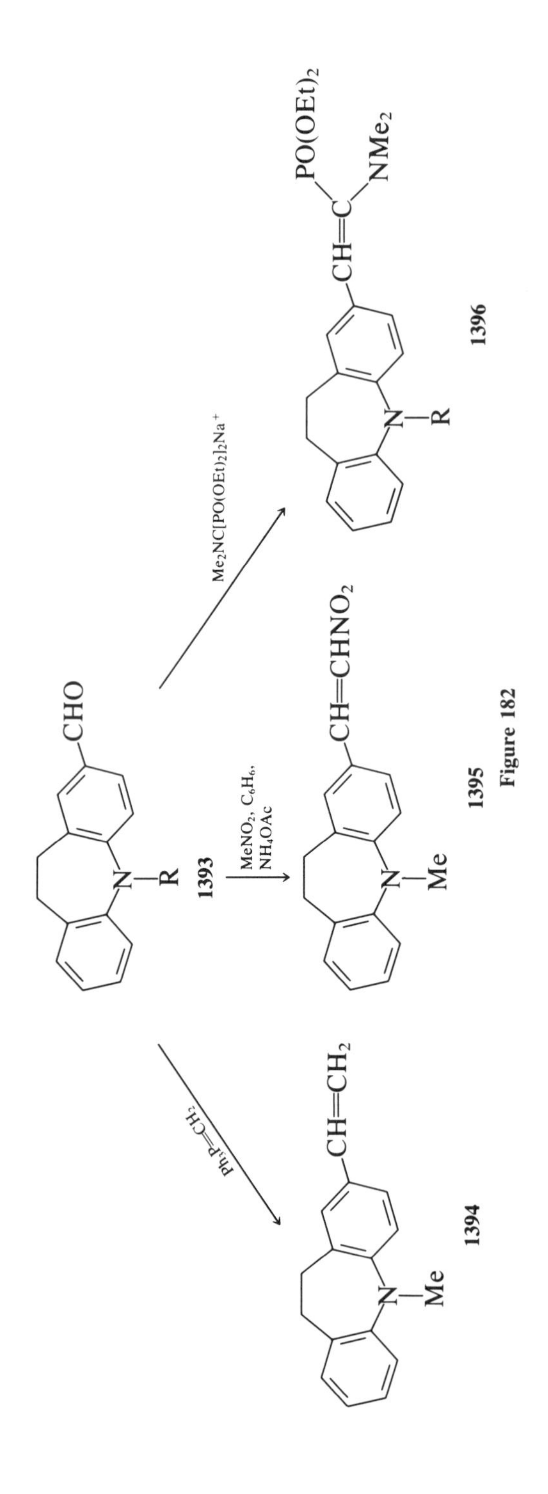
CHO
N—R
1393
MeNO$_2$, C$_6$H$_6$, NH$_4$OAc
CH=CHNO$_2$
N—Me
1395
Ph$_3$P=CH$_2$
CH=CH$_2$
N—Me
1394
Me$_2$NC[PO(OEt)$_2$]$_2$Na$^+$
PO(OEt)$_2$
CH=C
NMe$_2$
N—R
1396

Figure 182

The only *N*-vinyl compound reported to date is the 3-chloro-5-vinyl derivative **1398**, which was prepared by Hofmann degradation of the aminoalkyliminobibenzyl **1397** (Eq. 184). Since this is the only member of the *N*-vinyl series, it has been included in this review despite the lack of any supporting evidence for its structure (603).

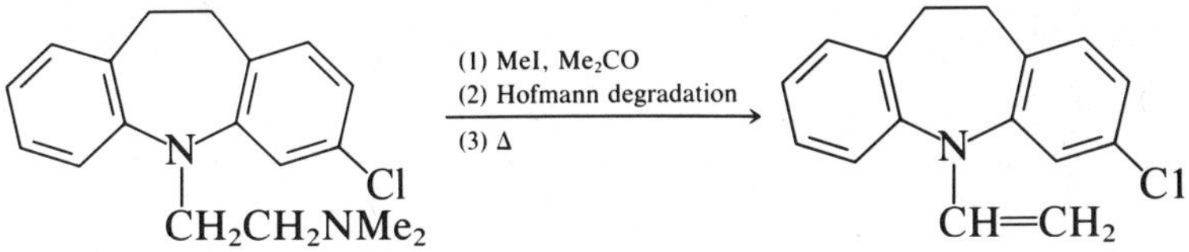

$$\text{1397 } (N\text{-}CH_2CH_2NMe_2\text{, 3-Cl}) \xrightarrow[\text{(3) }\Delta]{\text{(1) MeI, }Me_2CO\text{; (2) Hofmann degradation}} \text{1398 } (N\text{-}CH{=}CH_2\text{, 3-Cl}) \qquad (184)$$

No acetylenic derivatives with the alkyne function in conjugation with the ring have been described. However, two *N*-propargyl derivatives, **1399a** and **1399b,** have been prepared (619,620) via the alkylation reaction of the type described above.

b. HALODIBENZ[*b,f*]AZEPINES AND ORGANOMETALLIC DERIVATIVES OF DIBENZ[*b,f*]AZEPINES

i. Benzenoid Substitution. Almost all of the dibenz[*b,f*]azepines with aromatic halogen substitution (especially chlorine) have been prepared using the methods described in Section 39.B. For example, cyclization of chlorinated *o,o'*-diaminobibenzyls (495,499–501,621) (see Section 39.B.a.i) yields iminobibenzyls that can be converted to the corresponding chlorinated iminostilbenes by catalytic dehydrogenation (507) (see Section 39.B.b.i) or by bromination–dehydrobromination (512,523,525,526,622,623) (see Section 39.B.b.ii). A frequently encountered route to chlorinated iminobibenzyls involves ring expansion of the corresponding acridane carbinols (511,514,515,517,531,533) (see Section 39.B.b.iii). The preparative methods outlined in Section 39.B.c.i have been used to prepare ring-chlorinated dibenz[*b,f*]azepin-10-ones (542–544,559,562,564). Known examples of these compounds are listed in Tables 85 and 86.

Other halogenated derivatives have been prepared by chemical manipulation of iminobibenzyls. For example, 3-chloroiminobibenzyl (**1401a**) has

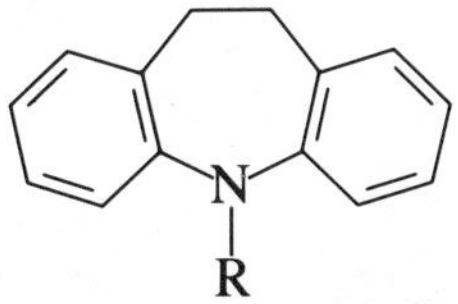

1399a, R = $CH_2C{\equiv}CH$
1399b, R = $CH_2C{\equiv}CCH_2NEt_2$

TABLE 85. HALO- AND HALOALKYLDIBENZ[*b*, *f*]AZEPINES

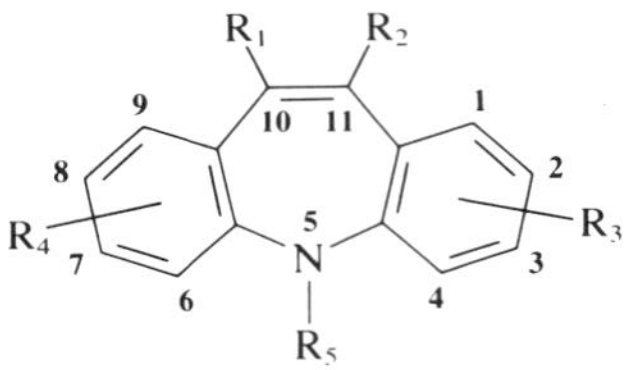

R_1	R_2	R_3	R_4	R_5	mp (°C)	Spectra	Refs.
Br	H	H	H	H	65–90		630
Br	H	H	H	Et	147.5–148.5	uv	606,607
Br	H	H	H	COMe	100–102		606
					108–109		628
					109–110		542,544
Br	H	H	H	COBr	125–127		643
Br	H	H	H	$CONH_2$	168–170		630
Br	H	3-Cl	H	COMe	148–149		556,631
H	Br	3-Cl	H	COMe	198–200		556,631
Cl	H	H	H	H	78–80		630
Cl	H	H	H	COMe	85–88		630
Cl	H	H	H	$CONH_2$	183–185		630
Cl	Cl	H	H	$CONH_2$	265–270		630
F	H	H	H	H	69–72		630
					74–76	ir,pmr	629
F	H	H	H	CH_2Ph	Oil		630
F	H	H	H	COCl	—		630
F	H	H	H	$CONH_2$	—		630
H	H	2-Cl	H	H	168–170		531
H	H	2-F	H	H	175–177		531
H	H	3-Br	H	H	217.5–218.5		622
H	H	3-Cl	H	H	208–209	ir,uv	511,514,515
					213–214		622
Me	H	3-Cl	H	H	—		556,631
Me	H	3-Cl	H	COMe	170–172		556,631

H	Me	3-Cl	H	H	145–146		556,631
H	Me	3-Cl	H	COMe	126–127		556,631
Me	Me	2-Cl	H	H	137–138		532,533,632
Me	Me	2-Cl	H	COMe	—		532,632
Me	Me	3-Cl	H	H	137–139		532,533,632
Me	Me	3-Cl	H	COMe	—		532,632
OMe	H	3-Cl	H	H	128–129	ir,uv	631
OEt	H	2-Br	H	Me	131		625
H	OMe	3-Cl	H	H	143–144	ir,uv	631
H	H	2-OMe	7-Cl	H	176	ir,pmr,uv	517
					173–175		531
H	H	3-Cl	7-Cl	H	282	ir,uv	512
					284–286		523
H	H	3-Cl	7-Cl	COMe	184–185		512,523
H	H	3-CF_3	H	H	180–181.5		514
Me	Me	3-CF_3	H	H	153–155		532,533,632
Me	Me	3-CF_3	H	COMe	Amorphous		532,632
CH_2Br	H	H	H	Me	109–111		577,578
CH_2Br	H	H	H	COMe	133–135		547,556,577, 578,631
CH(Me)Br	H	H	H	COMe	123–124		556,631
CH_2Br	H	3-Cl	H	COMe	—		556,631
H	CH_2Br	3-Cl	H	COMe	197–200		556,631
CH_2Br	CH_2Br	H	H	Me	127–130		537–541
CH_2Br	CH_2Br	H	H	COMe	175–176		537–541
CH_2Br	CH_2Br	2-Cl	H	COMe	Amorphous		532,632
CH_2Br	CH_2Br	3-Cl	H	COMe	Amorphous		532,632
CH_2Br	CH_2Br	3-CF_3	H	COMe	Yellow foam		532,632
OCH_2CH_2Br	H	H	H	Me	144–146		698
H	H	H	H	$(CH_2)_3Cl$	67		591
H	H	H	H	$(CH_2)_3Cl$			
H	H	H	H	$(CH_2)_3Br$	46–61[a]	ir,pmr,ms	586
H	H	H	H	$CH_2CCl{=}CH_2$	80–81		163

[a]60:40 mixture.

TABLE 86. HALO, HALOALKYL, AND ORGANOMETALLIC DERIVATIVES OF 10,11-DIHYDRODIBENZ[*b*, *f*]AZEPINE

R_1	R_2	R_3	R_4	R_5	mp (°C) or bp (°C/torr)	Spectra	Refs.
Br	H	H	H	COMe	118–119	ir,uv	512
Br	H	H	H	COCl	125–126		528
					128–129		526,623
					—	pmr	627
Br	Br	H	H	COMe	136–138		512,516,536, 542–544, 577,578,630
Br	*trans*-Br	H	H	$COCF_3$	186–189		630
Br	Br	H	H	COCl	169–172		528
Br	*trans*-Br	H	H	$CONH_2$	144.5		630
Cl	*cis*/*trans*-Cl	H	H	COMe	137–152		630
Cl	*trans*-Cl	H	H	$CONH_2$	140–142		630
F	*cis*-MeO	H	H	CO_2Et	101–102	ir,pmr, ^{19}F-nmr,uv	629
F	*cis*-CF_3O	H	H	COMe	114–116	ir,pmr,ms	629
F	*trans*-CF_3O	H	H	COMe	100–103	ir,pmr,ms	629
F	*cis*-CF_3O	H	H	CO_2Et	133–134	ir,pmr, ^{19}F-nmr,uv	629

F	*trans*-CF_3O	H	H	CO_2Et	107–107.5	ir,pmr, ^{19}F-nmr,uv	629
H	Br	3-Et	H	COMe	—		512
H	H	1-Cl	H	H	108–111		499
H	H	3-Cl	H	H	80–83		518
					84.5–86	ir,uv	511,514,515
					86–88		808
					87.5–88		500
					87–89		513,622
					88–89		621
H	H	3-Cl	H	Me	56–58		608
H	Me	3-Cl	H	COMe	119–120		608
					126		513,622
H	H	3-Br	H	COMe	141–143		622
Br(H)	H(Br)	3-Cl	H	COMe	135–136.5		622
Br	Br	3-Cl	H	COMe	122–124		556,631
H	H	3-Cl	7-Me	H	130–131		500
H	H	2-MeO	7-Cl	H	70		517
H	H	1-Cl	9-Cl	H	108–111		499
					114–115		495
H	H	2-Br	8-Br	H	167	ir,ms	624
					173–174		584
H	H	2-Br	8-Br	Me	108–109	ir,pmr,ms	584
H	H	3-Cl	7-Cl	H	112–114		501
					114		512
					113–114		499
					114–115		495
H	H	3-Cl	7-Cl	COMe	147–148		523
					148–149		512
H	H	3-Cl	7-Cl	$COCH_2Cl$	142		496
H	H	3-Cl	7-Cl	COCH(Me)Br	164		496

(*continued*)

TABLE 86. (*Continued*)

R_1	R_2	R_3	R_4	R_5	mp (°C) or bp (°C/torr)	Spectra	Refs.
H	H	3-Cl	7-Cl	COCl	145–152		827
H	H	3-Cl	7-Cl	$CONH_2$	220–225; 238–240		827
H	H	3-Cl	7-Cl	$CONMe_2$	180		827
Br	H	3-Cl	7-Cl	COMe	154–156		499,512
H	H	2,4-Br_2	6,8-Br_2	H	153	uv	624
H	H	2-CH_2Br	H	Me	Oil		608
H	H	2-CH_2Cl	H	Me	Oil		617
H	H	2-CH_2Cl	H	CH_2Ph	Oil		582,608,617
H	H	2-CH_2Cl	H	Et	Oil		617
H	H	2-CH_2Cl	7-Cl	Me	Oil		608
H	H	2-$(CH_2)_2Cl$	H	H	Oil		617
H	H	2-$(CH_2)_2Cl$	H	Me	Oil		617
H	H	2-$(CH_2)_2Cl$	H	Et	Oil		617
H	H	3-CH_2Br	H	COMe	106–107		608
H	H	3-CH(Me)Br	H	COMe	Oil		608
H	H	4-CH_2Br	6-CH_2Br	Me	—		579
H	H	H	H	$(CH_2)_2Cl$	84.5–85.5		597
H	H	H	H	$(CH_2)_2Br$	90–91	uv	749
H	H	H	H	$(CH_2)_3Cl$	150–153/0.03		601,602
					150–160/0.2–0.3		597
					160–166/0.3		749
H	H	3-MeO	H	$CH_2CH(Me)CH_2Cl$	103		600,633
BH_2	H	H	H	Me	127	ir	642
H	H	H	H	BCl_2	Pink solid		641
H	H	H	H	BHMe	120–150/0.1		641
H	H	H	H	BMe_2	110–120/0.1		641

been prepared (Eq. 185) by hydrolysis of the acetanilide **1400a,** diazotization, reaction with cuprous chloride to form **1401b,** and alkaline hydrolysis of **1401b** (622). The free amino iminobibenzyl **1400b** was likewise converted to the 3-bromo derivative **1401c** by diazotization and decomposition with cuprous bromide.

Ring-brominated derivatives of iminobibenzyl (Table 86) have also been obtained by electrophilic aromatic substitution reactions. In 1960 Teuber and Schmidtke reported that bromination of iminobibenzyl **1293** in acetic acid, chloroform, or carbon disulfide (Eq. 186) gave the tetrabromo derivative **1403a** (624). They also claimed that **1293** was converted to the dibromide **1403b** with *N*-bromosuccinimide and benzoyl peroxide in refluxing carbon tetrachloride. The latter result seems unlikely in view of the known propensity of **1293** for benzylic bromination under such conditions (see Sections 39.B.b.ii and 39.C.b.ii). Later, Kricka and Ledwith demonstrated that the reaction of *N*-methyliminobibenzyl **(1382a)** with molecular bromine in acetic acid produced the dibromo compound **1403c,** whereas under the same conditions the corresponding *N*-acetyl derivative **1307** was unreactive (584).

N, Ac, NHR

(1) 2 *N* HCl
(2) NaOH

1400b

(1) 2 *N* HCl
(2) NaNO$_2$
(3) CuCl$_2$, HCl
(4) KOH, MeOH

(1) 48% HBr
(2) NaNO$_2$
(3) CuBr$_2$

1400a, R = Ac
1400b, R = H

(185)

N, R, X

1401a, X = Cl, R = H
1401b, X = Cl, R = Ac
1401c, X = Br, R = Ac

X, N, R, X

Br$_2$, HOAc
H$_2$/cat.

X, X, N, Y, R, Y

(186)

1293, R = H, X = H
1307, R = Ac, X = H
1382a, R = Me, X = H
1402, R = H, X = Cl

1403a, R = H, X = Y = Br
1403b, R = Y = H, X = Br
1403c, R = Me, X = Br, Y = H

There are examples in the literature that claim successful ring bromination of dibenz[*b,f*]azepin-10-ones (625) and annelated iminobibenzyls (626) using *N*-bromosuccinimide.

The tetrabromo derivative **1403a** could be smoothly converted back to **1293** by catalytic hydrogenolysis (624). Catalytic hydrogenolysis of 3,7-dichloroiminobibenzyl **1402** with Raney nickel as the catalyst has been reported to yield a mixture of **1402, 1401a,** and **1293** (513). Palladium on carbon produced acceptable amounts of **1401a.** Raney nickel has been claimed to reduce chlorinated iminostilbenes with minimal hydrogenolysis (518).

ii. 10,11-HALOGEN SUBSTITUTION. The preparation of 10- or 11-monobromoiminobibenzyls by benzylic bromination of the parent iminobibenzyl with *N*-bromosuccinimide, 1,3-dibromo-5,5-dimethylhydantoin, *N*-bromoacetamide, or *N*-bromophthalimide has been described in Section 39.B.b.ii (499,512,523–528,623). The addition of molecular bromine to iminostilbenes to produce the corresponding 10,11-dibromoiminobibenzyls (516,536,542–558,577,578,606,628,629) has been treated in Section 39.B.c.i. The addition of molecular chlorine to the iminostilbenes **1297b** and **1298** (Eq. 187) gave the corresponding dichlorides **1404b** and **1404c** (630). The detailed pmr spectrum of **1404a** has been published (627).

Dehydrohalogenation of the dibromide **1346** (Eq. 187) with alcoholic potassium hydroxide (542–544), refluxing *n*-butylamine (606,628), or 1,5-diazabicyclo[4.3.0]non-5-ene (630) resulted in the formation of the monobromoiminostilbene **1347.** The dichlorides **1404b** and **1404c** were converted

COR

1297b, R = NH_2
1298, R = Me

$\xrightarrow{X_2}$

X Y

COR

1346, R = Me, NH_2; X = Y = Br
1404a, R = Cl, X = Br, Y = H
1404b, R = NH_2, X = Y = Cl
1404c, R = Me, X = Y = Cl

$\xrightarrow{B:}$

(187)

X

COR

1347, R = Me, NH_2; X = Br
1405a, R = NH_2, X = Cl
1405b, R = Me, X = Cl

to the vinyl chlorides **1405a** and **1405b** with 1,5-diazabicyclo[4.3.0]non-5-ene and ethyldiisopropylamine, respectively (630).

The corresponding vinyl fluoride **1407a** was prepared as a light yellow oil by treatment of the azepinone **1406** with hydrogen fluoride–sulfur tetrafluoride in methylene chloride solution (Eq. 188) (630). Hydrogenolysis over palladium on carbon gave **1407b** as a low-melting solid. This substance has also been obtained as a very minor impurity when the adducts **1408a** and **1408b,** obtained from the corresponding *N*-acyliminostilbenes and perfluoromethanol, were treated with ethanolic potassium hydroxide (629). The major product from **1408a** or **1408b** was **1407c**.

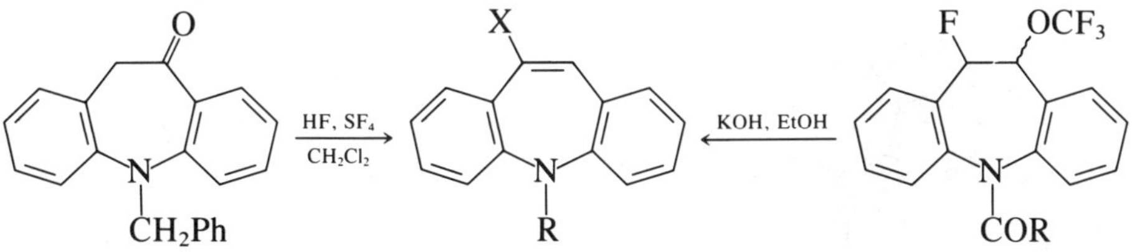

HF, SF_4 / CH_2Cl_2 ; KOH, EtOH (188)

1406 (N-CH_2Ph)

1407a, R = CH_2Ph, X = F
1407b, R = H, X = F
1407c, R = H, X = CF_3O

1408a, R = Me
1408b, R = OEt

iii. Halogen in an Alkyl Side-Chain. Haloalkyldibenz[*b,f*]azepines (Tables 85 and 86) have been used as intermediates in the preparation of several different types of derivatives and are accessible via classical methods. For example, *N*-bromosuccinimide converted the 10-alkyl derivatives **1409a** to the corresponding α-bromoalkyl derivatives **1409b** (556,577,578,631). The same conditions have been used to convert **1410a** to the *bis*(bromomethyl) compounds **1410b** (537–541,632). The reaction of **1410b** with primary amines produced the new heterocycles **1411,** which have been claimed to possess central nervous system depressant activity.

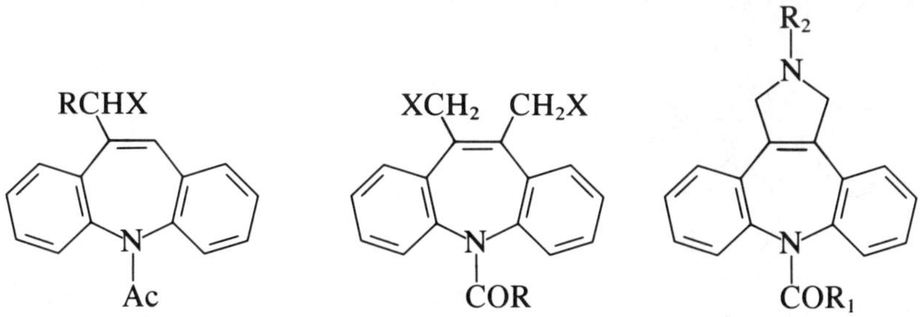

1409a, R = H, Me; X = H
1409b, R = H, Me; X = Br

1410a, X = H
1410b, X = Br

1411

Other haloalkyl derivatives have been obtained via the corresponding alcohols. The conversion of hydroxyalkyl derivatives of general structure **1412a** to the corresponding halides **1412b** has been achieved using thionyl chloride (582,592,595,596,617), hydrogen bromide in chloroform (608), and phosphorus tribromide (579,608).

A frequently encountered class of haloalkyl derivatives consists of the 5-(haloalkyl)dibenz[*b,f*]azepines **1413a**, which serve as precursors to analogs of imipramine (**1296a**). The haloalkyl derivatives **1413a** have been prepared from the corresponding alcohols **1413b** and thionyl chloride (592,595,596), and also by alkylation of the requisite dibenzazepines with either 1-bromo-3-chloropropane (591,595,601,602) or the equivalent chloro-ω-tosyloxy-alkanes (597,600,633) (Section 39.C.a).

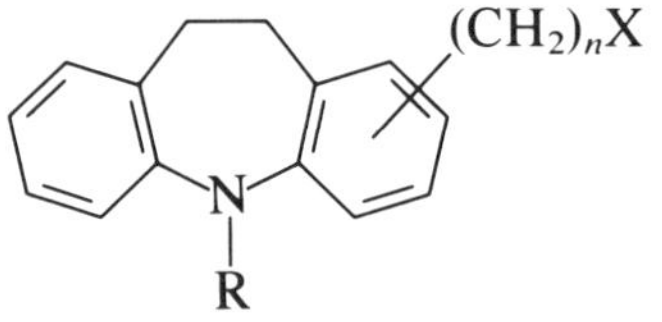

$(CH_2)_nX$ ... N—R

1412a, X = OH; n = 1, 2
1412b, X = Br, Cl; n = 1, 2

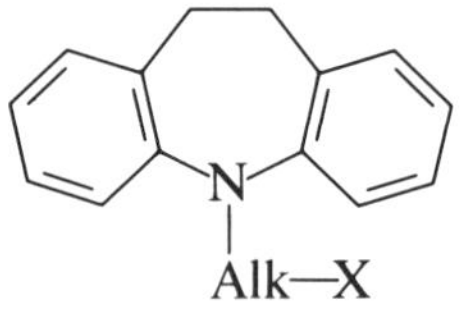

N—Alk—X

1413a, X = Br, Cl
1413b, X = OH

The 5-(ω-haloacyl)dibenz[*b,f*]azepines (**1414a**) represent yet another class of halogenated side-chain derivatives. They are readily prepared by acylation of the appropriate dibenzazepine with the desired ω-haloacyl chloride or bromide (496,587,605,634–637). When **1414b** and **1414c** were treated with 1,5-diazabicyclo[5.4.0]undec-5-ene in DMSO, the corresponding *N*-acryloyl and *N*-methacryloyl derivatives **1415a** and **1415b,** respectively, were obtained (635,638). Borane reduction of **1416a** gave the chloroalkyl derivative **1416b** (605).

iv. ORGANOMETALLIC DERIVATIVES. The alkali metal amides of iminobibenzyl (**1296a**) and iminostilbene (**1311b**) are prepared by reacting the appropriate azepine with a strong base such as an alkali amide or organolithium. The formation and subsequent alkylation of the resulting alkali metal

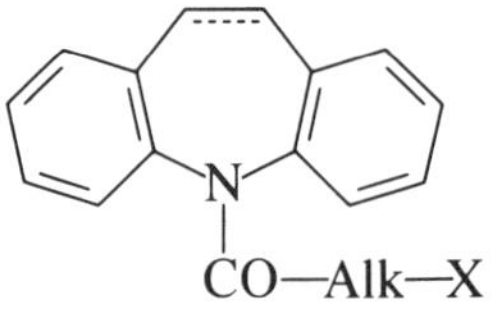

N—CO—Alk—X

1414a, X = Br, Cl
1414b, Alk = $(CH_2)_2$, X = Cl
1414c, Alk = $CHMeCH_2$, X = Br

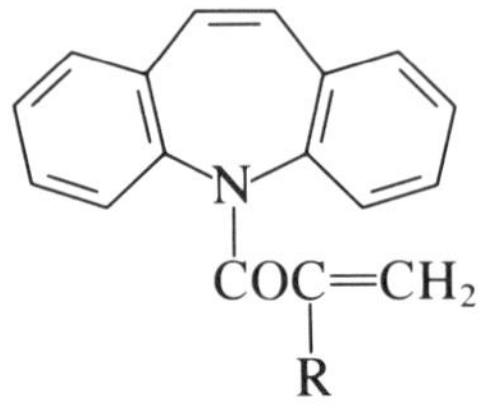

N—COC(R)=CH_2

1415a, R = H
1415b, R = Me

N—C(=X)—$(CH_2)_2Cl$; C(=X)—NHMe

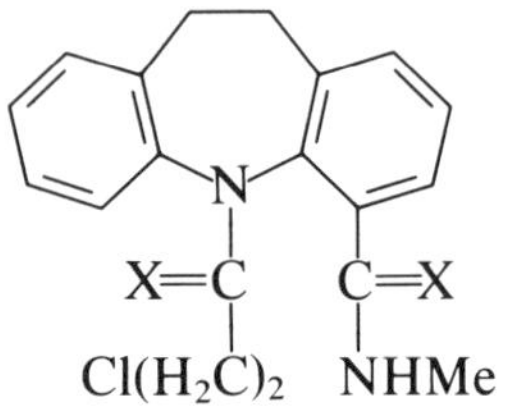

1416a, X = O
1416b, X = H_2

amides **1417** was discussed in depth in Section 39.C.a. None of these alkali metal amides have been isolated and physically characterized. However, the pmr spectrum of the sodium salt of iminostilbene (**1418**) has been analyzed in detail (639). A strong paramagnetic ring current was indicated in the seven-membered ring, although not as large as that observed for the carbocyclic analog **1419**.

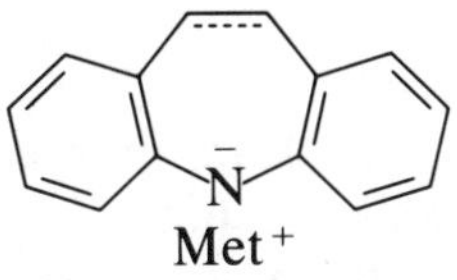

Met$^+$ Na$^+$

1417, Met = Li, Na, K 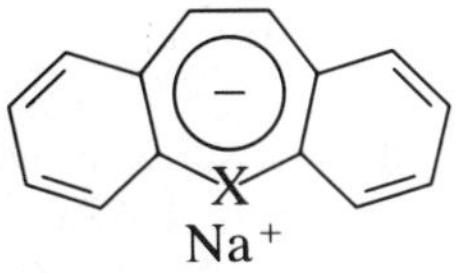 **1418**, X = N **1311b**
1419, X = CH

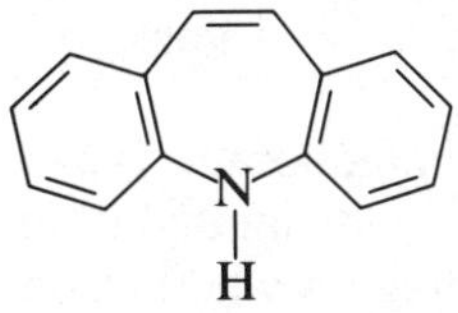

The thallium salt **1420** was prepared from **1311b** and thallous ethoxide and was shown to undergo alkylation under extremely mild conditions (583). However, in contrast to the alkylation of the alkali metal amides, alkylation of **1420** was restricted to primary alkyl halides (Eq. 189). In order to accommodate the observed steric limitation, a four-center transition state was proposed. It was suggested that branching at the α- or β-carbon of the alkyl halide might hinder close approach to the thallium salt.

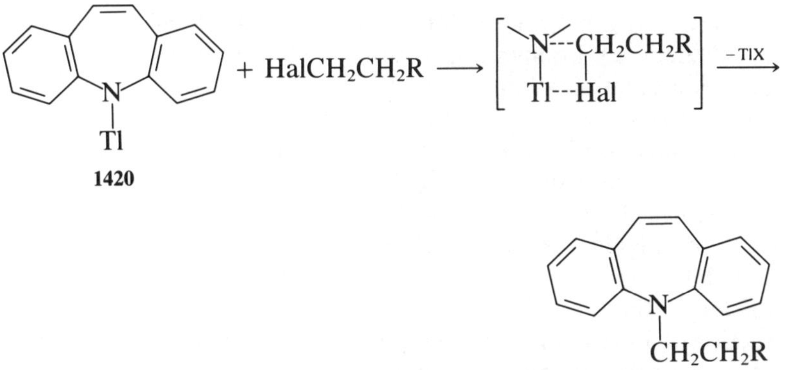

+ $HalCH_2CH_2R$ ⟶ [N---CH_2CH_2R / Tl---Hal] $\xrightarrow{-TlX}$

1420

(189)

CH_2CH_2R

1421

The reaction of iminobibenzyl with excess butyllithium has been shown to give the 4,5-dilithio species **1422a.** When treated with carbon dioxide, **1422a** afforded the corresponding carboxylic acid **1422b** (574,605,635). The 10-lithiodibenzazepine **1423** has also been reported, but without experimental details (640).

Finally, several boron derivatives of iminobibenzyl (**1293**) have been prepared (see Table 86). Reaction with boron trichloride (Eq. 190) produced

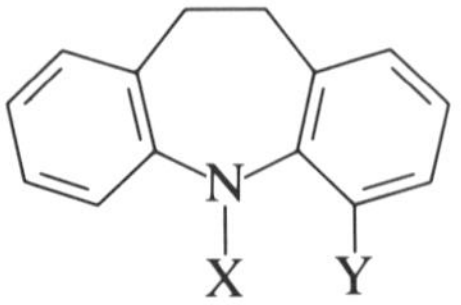

1422a, X = Y = Li
1422b, X = H, Y = CO_2H

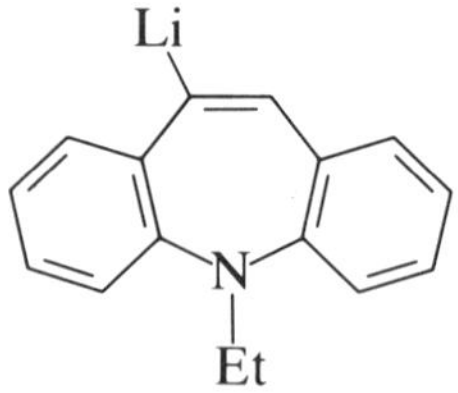

1423

the aminoborane **1424** as a light pink solid (641). When **1424** was allowed to react with methyllithium, the adducts **1425a** or **1425b** were obtained, depending on whether one or two equivalents of methyllithium were used.

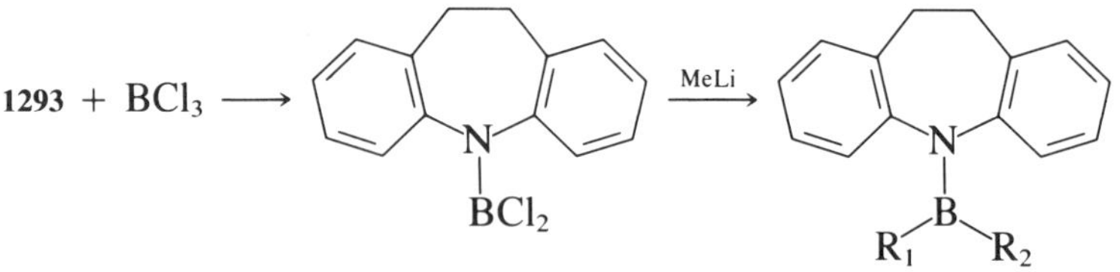

1293 + BCl_3 ⟶ **1424** $\xrightarrow{\text{MeLi}}$ **1425** (190)

1425a, R_1 = Me, R_2 = Cl
1425b, R_1 = R_2 = Me

An examination of the pmr pattern of the ethano bridge of **1425a** and **1425b** provided a useful probe of the ring conformations of these compounds. In the completely rigid molecules all four bridge protons would experience a different magnetic environment, generating an ABCD spectrum. Ring inversion or pseudorotation about the ethano bridge would reduce the spectrum to an AA′BB′ pattern. The rapid occurrence of both processes would result in a single absorption. The observed pattern, over a temperature range of −70 to + 50°C, was a simple AA′BB′ pattern, suggesting a large energy difference between the two processes. The authors argued in favor of ring inversion as the higher-energy process and suggested the conformations **1426a** and **1426b** as the two equilibrating forms. The boat confirmation **1427** was postulated as the transition state in this process.

The reaction of 5-methyliminostilbene (**1337**) with *N,N*-diethylamine–borane (Eq. 191) gave the 10-borohydro derivative **1428a.** Sequential treatment of **1428a** with caustic and chloramine produced the 10-amino derivative **1428b** (642).

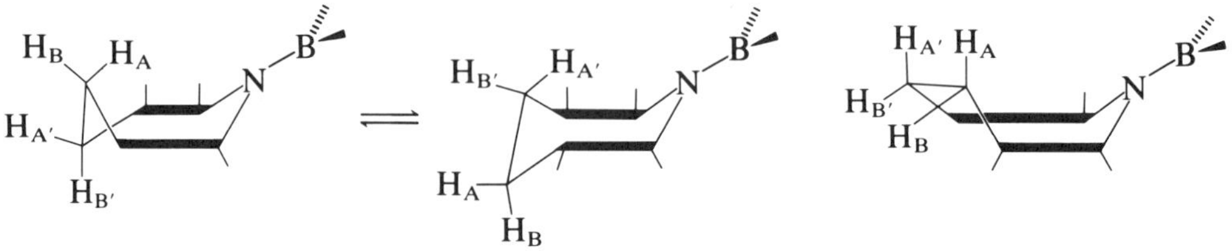

1426a ⇌ **1426b**

1427

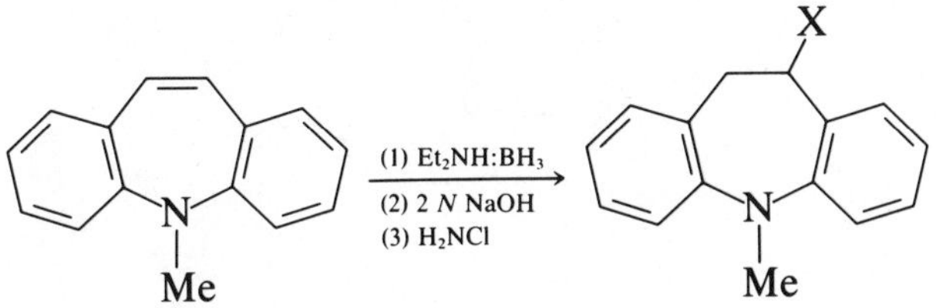

(1) $Et_2NH{:}BH_3$
(2) 2 *N* NaOH
(3) H_2NCl

Me

1337

X

Me

1428a, X = BH_2
1428b, X = NH_2

(191)

c. DIBENZ[*b,f*]AZEPINOLS AND THEIR ETHERS

i. 10- AND/OR 11-SUBSTITUTION. In an earlier discussion of the synthesis of 10- and 11-hydroxydibenz[*b,f*]azepines, the addition of Grignard reagents to dibenz[*b,f*]azepine-10-ones to produce the corresponding tertiary alcohols was described in detail (Section 39.B.b.iv, Eq. 165). A related reaction has been described (Eq. 192) wherein the anthranilic acid derivative **1429** gave the tertiary alcohol **1431** on treatment with methyllithium (645). The observed product strongly implicates the *in situ* formation of azepinone **1430,** as described in Section 39.B.c. (Eq. 171).

The reduction of dibenz[*b,f*]azepinones with the general structure **1432** to the corresponding secondary alcohols **1433** (Eq. 193) has been achieved with sodium borohydride (499) or lithium aluminum hydride (646) and also by catalytic reduction over copper chromite (546). Catalytic reduction of **1434** over copper chromite, was used to prepare **1436a,** one of the metabolites

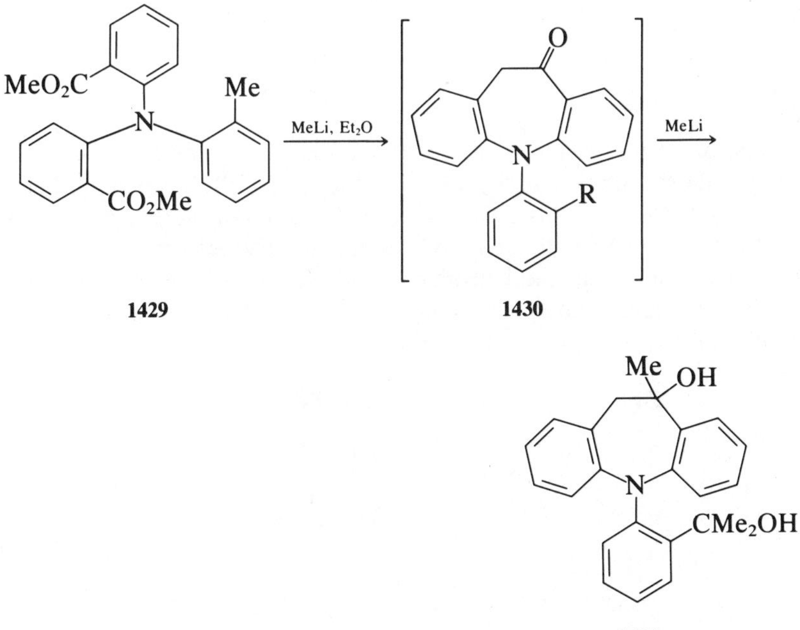

MeO_2C Me

CO_2Me

MeLi, Et_2O

O

R

MeLi

1429 **1430**

(192)

Me OH

CMe_2OH

1431

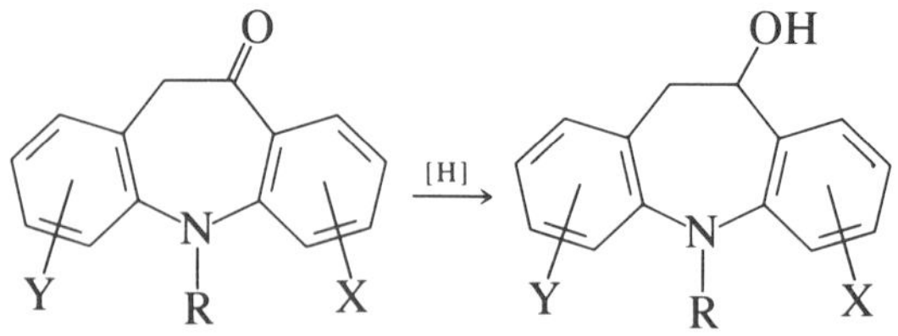

(193)

1432
1434, R = $CONH_2$, X = Y = H

1433

of Tegretol (**1297b**) (546). However, there are reported instances where catalytic reduction over copper chromite converted a dibenz[*b,f*]azepin-10-one directly to the corresponding iminobibenzyl (547,577,578). The Tegretol metabolite **1436a** has also been obtained by catalytic reduction of the epoxide **1435** (647). The reaction (Eq. 194) proceeded at ambient temperature and pressure in the presence of 10% Pd/C to produce **1436a** in 73% yield. The same paper described the preparation of the dihydroxymetabolite of Tegretol, **1436b.** The synthesis of **1436b** was achieved in 56% yield by treating a pyridine solution of Tegretol with osmium tetroxide followed by hydrolysis of the intermediate osmate ester.

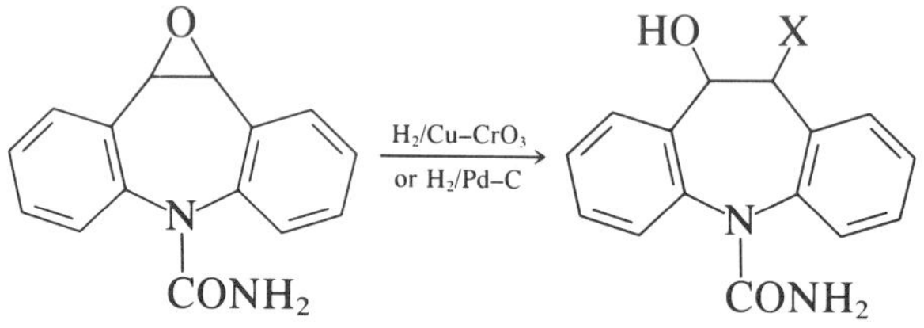

H_2/Cu–CrO_3 or H_2/Pd–C (194)

1435

1436a, X = H
1436b, X = OH

The preparation of 10- and 11-alkoxydibenz[*b,f*]azepines from the corresponding iminostilbenes, namely **1345** → **1348** (Fig. 179), has been utilized extensively as a route to dibenz[*b,f*]azepin-10-ones and was discussed in Section 39.B.c.i. However, there have been no publications dealing with the mechanism by which the vinyl bromides **1347** are converted to the enol ethers **1348.** It has been noted that the replaceability of the bromine in **1347** by alkoxide is somewhat surprising (542–544).

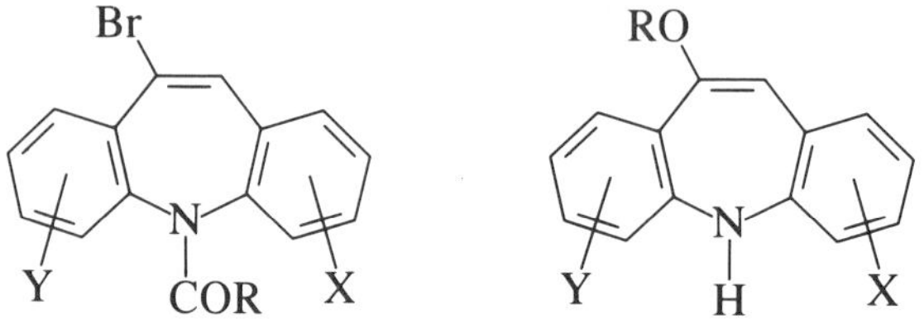

1347

1348a

Of the possible mechanistic paths for nucleophilic vinylic substitution (648), the most likely would be addition–elimination and elimination–addition. The lack of vinylic activation would tend to make the former route unlikely. The latter route (Eq. 195) would generate the highly reactive heterocyclic aryne intermediate **1438,** an intermediate for which there is precedent (628). Specifically, the reaction of **1437** with potassium *tert*-butoxide gave the furan cycloaddition adduct **1439.** The isolation of **1439** argues convincingly for the intermediacy of the hetaryne **1438** and supports elimination–addition as the most likely mechanistic path by which the enol ethers **1348** are formed.

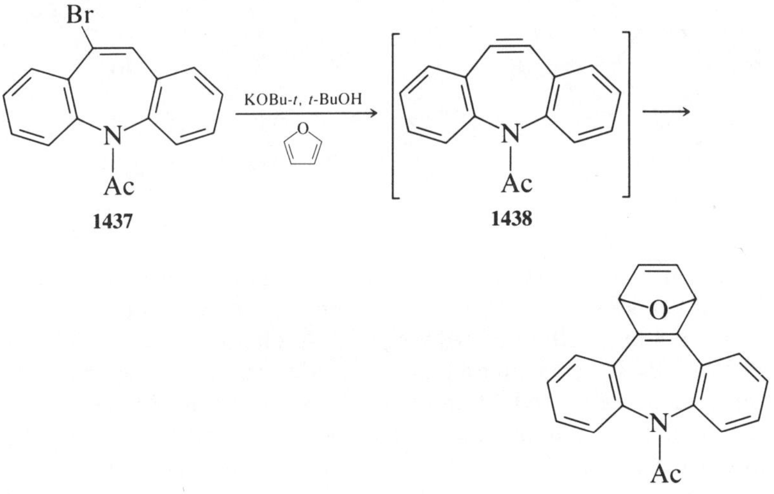

Br N Ac **1437** —KOBu-*t*, *t*-BuOH, furan→ [N Ac **1438**] → O N Ac **1439** (195)

A report in the patent literature claimed that, in the presence of alcoholic alkoxide, **1440a** gave exclusively **1440b,** and **1440c** gave exclusively **1440d** (556). If the hetaryne **1441** were an intermediate, it would be expected to give a mixture of the vinyl ethers **1440b** and **1440d.** Since both **1440a** and

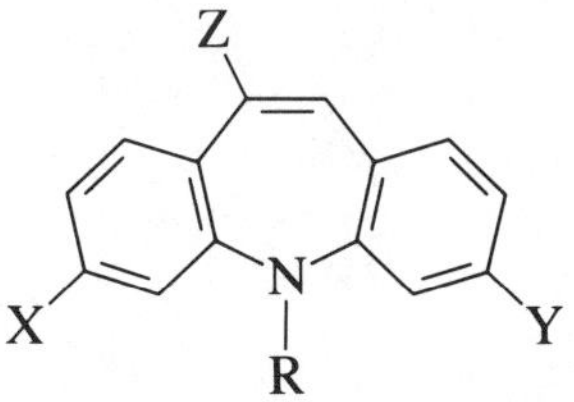

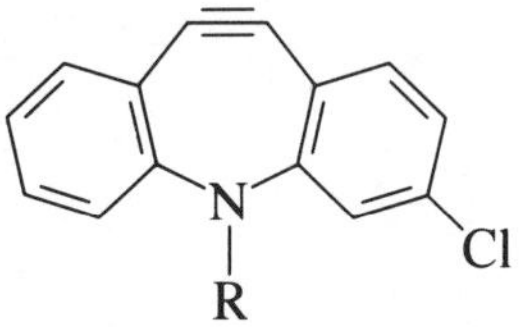

Z X N R Y

	X	Y	Z	R
1440a	H	Cl	Br	Ac
1440b	H	Cl	OMe	H
1440c	Cl	H	Br	Ac
1440d	Cl	H	OMe	H

N R Cl

1441

1440c would give the same intermediate **1441**, each would be expected to yield a mixture of isomers (**1440b** and **1440d**). That only a single isomer was formed from each vinyl bromide argues against the elimination–addition mechanism. Clearly, this is a problem deserving further investigation.

The action of triethyl orthoformate on the azepinone **1442** (Eq. 196) afforded the corresponding vinyl ether **1443** (625). The addition of perfluoromethanol to *N*-acyliminostilbenes to give the adducts **1408** (Eq. 190), and the subsequent conversion of these adducts to the vinyl ether **1407c** was described in Section 39.C.b.ii (629).

O Br N Me **1442** $\xrightarrow[EtOH]{HC(OEt)_3}$ EtO Br N Me **1443** (196)

ii. Benzenoid Substitution. Ring-hydroxylated dibenz[*b,f*]azepines (Tables 87 and 88) have been of considerable interest since they are formed as metabolites of the dibenz[*b,f*]azepine pharmaceuticals. The 2-hydroxy derivatives have been prepared by oxidation of the appropriate dibenzazepine followed by reduction of the resulting dibenz[*b,f*]azepin-2-ones. For example, the quinoneimine **1444a** was prepared by treating the tetrabromo derivative **1403** with nitrous acid in concentrated sulfuric acid (Eq. 197). Reduction of **1444a** with methanolic sodium borohydride afforded the 2-hydroxy derivative **1445a.** The aminophenol **1445a** could be converted back to **1444a** with ferric chloride in alcoholic hydrogen chloride (624).

X X N H X X

1403, X = Br
1293, X = H

(1) H_2SO_4, $NaNO_2$
(2) Fremy's salt

X O N X X

1444a, X = Br
1444b, X = H

[H] ⇄ $FeCl_3$, HCl, EtOH

(197)

X OH N H X X

1445a, X = Br
1445b, X = H

The oxidation of iminobibenzyl (**1293**) to 10,11-dihydro-2H-dibenz[*b,f*]azepin-2-one (**1444b**) has been effected with potassium nitrosodisulfonate (Fremy's salt) in acetone (624,649–652). The reduction of **1444b** to give 2-hydroxyiminobibenzyl (**1445b**) has been accomplished with sodium borohydride in methanol (624), by catalytic hydrogenation (649–652), and by chemical reduction using either sodium dithionite (624,649) or sodium bisulfite (650–652). The monochloro derivative **1446,** on oxidation with Fremy's salt (Eq. 198), gave a 1:4 mixture of **1447a** and **1447b** (653). Dithionite reduction of the mixture produced the alcohols **1448a** and **1448b** in 12% and 32% yields, respectively.

N H Cl

1446

Fremy's salt / Me_2CO

Y N O X

1447a, X = Cl, Y = H
1447b, X = H, Y = Cl

$Na_2S_2O_4$ / H_2O

(198)

OH Y N H X

1448a, X = Cl, Y = H
1448b, X = H, Y = Cl

While the formation of **1444b** from iminobibenzyl has been shown to be a direct, high-yield process, the same oxidation of iminostilbene (**1311b**) was found to be somewhat more complicated (Eq. 199). Fremy's salt oxidation of **1311b** in buffered acetone solution resulted in the formation of the acridine **1449** (37% yield), 2H-dibenz[*b,f*]azepin-2-one **1450** (53%), and recovered **1311b** (5%) (654,655). The azepinone **1450** was isolated as deep red, crystalline solid whose structure was assigned on the basis of spectral and chemical evidence. The yield of **1450** could be increased at the expense of **1449** by raising the pH of the reaction above 8.0. Catalytic reduction of **1450** yielded **1445b,** the same compound as the one obtained on reduction of **1444b.**

One of the major metabolites of imipramine is the 2-hydroxy derivative **1453.** This metabolite has been prepared synthetically as outlined in Fig. 183 (649–652). Benzylation of **1445b** gave the ether **1451,** which underwent a standard alkylation to produce the *N*-aminoalkyl derivative **1452.** Reductive debenzylation of **1452** gave the imipramine metabolite **1453.** The active proton of 2-hydroxyiminobibenzyls has also been protected as the tetrahydropyranyl ether (653). The metabolite **1453** has been obtained directly from

TABLE 87. DIBENZ[*b*, *f*]AZEPINOLS, HYDROXYALKYLDIBENZ[*b*, *f*]AZEPINES, AND THEIR ETHERS

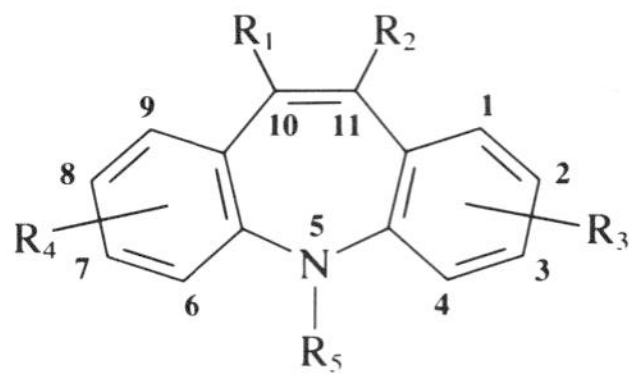

R_1	R_2	R_3	R_4	R_5	mp (°C)	Spectra	Refs.
MeO	H	H	H	H	124		516,536,542, 544,547,550, 577,578
					125		557
MeO	H	H	H	Me	145–146		516,536,547, 550,577,578
MeO	H	H	H	Et	180		548,549,557
					186–188		577,578
MeO	H	H	H	*n*-Pr	100		548,549
MeO	H	H	H	CH_2Ph	121		516,536,547 550,577,578
MeO	H	H	H	COCl	138		546

MeO	H	H	H	$CONH_2$	181		546,547
CF_3O	H	H	H	H	61–62	ir,pmr,uv	629
EtO	H	H	H	H	132–133		542–544
n-BuO	H	H	H	H	113–114		542–544
MeO	H	3-Cl	H	H	128–129	ir,uv	556,631
H	MeO	3-Cl	H	H	143–144	ir,uv	556,631
$O(CH_2)_2Br$	H	H	H	H	144–146		698
MeO	H	3-Cl	7-Cl	H	182–183		542–544
EtO	H	2-Br	H	Me	131		625
H	H	2-MeO	7-Cl	H	173–175	ir,pmr,uv	531
					176		517
$C(Me_2)_2OH$	H	H	H	Et	—		640
H	H	H	H	$(CH_2)_2OH$	88–89		519,783
H	H	H	H	$(CH_2)_2OMs$	148–151		783
H	H	H	H	$(CH_2)_2OTs$	104–108		519
H	H	H	H	$(CH_2)_3OH$	185–190/0.35[a]		519
H	H	H	H	$(CH_2)_3OTs$	128–129		519
H	H	H	H	$CH_2CH(Me)OH$	108–110		783
H	H	H	H	$CH_2CH(Me)OMs$	145		783
H	H	H	H	$CH_2CH—CH_2$ (epoxide, O bridging CH and CH_2)	150–160/.02[a]		589,593

[a]Boiling point (°C/torr).

TABLE 88. 10,11-DIHYDRODIBENZ[*b,f*]AZEPINOLS, HYDROXYALKYL-10,11-DIHYDRODIBENZ[*b,f*]AZEPINES, AND THEIR DERIVATIVES

R_1	R_2	R_3	R_4	R_5	mp (°C) or bp (°C/torr)	Spectra	Refs.
OH	H	H	H	H	106–107		499
OH	H	H	H	Me	78–79		499
OH	H	H	H	COMe	139–140		499
OH	H	H	H	$CONH_2$	189–192	ir,pmr	647
					195–196		546
OH	OH	H	H	$CONH_2$	242–245	ir,pmr	647
Me,OH	H	H	H	Me	138		516,536,550
Me,OH	Me	H	H	Me	102–104		537–541
Me,OH	H	H	H	$C_6H_4[C(Me)_2OH]$-2	198–199	pmr	645
Et,OH	H	H	H	Me	Oil		516,536
F	*cis*-CF_3O	H	H	COMe	114–116	ir,pmr,ms	629
F	*trans*-CF_3O	H	H	COMe	100–103	ir,pmr,ms	629
F	*cis*-CF_3O	H	H	CO_2Et	133–134	ir,pmr,^{19}F-nmr,uv	629
F	*trans*-CF_3O	H	H	CO_2Et	107–107.5	ir,pmr,^{19}F-nmr,uv	629
H	H	2-OH	H	H	155 (dec.)		655
					159		624
					169–171		649
					168–169		650–652

H	H	2-OH	H	COMe	184–186		649
H	H	2-OCH_2Ph	H	H	92–94		650–652
					96–97		649
H	H	2-OAc	H	COMe	118–120		649
H	H	2-OH	3-Cl	H	150–151		653
H	H	2-OTHP	3-Cl	H	104–105		653
H	H	2-OH	7-Cl	H	139–140		653
H	H	2-OTHP	7-Cl	H	121–122		653
H	H	2-OH	4,6,8-Br_3	H	144–145	uv	624
H	H	2-MeO	7-Cl	H	70		517
H	H	3-OH	H	COMe	78–82		650–652
H	H	3-MeO	H	H	94		650–652
					95–98		574,635
H	H	3-OCH_2Ph	H	H	111		650–652
H	H	3-OCH_2Ph	H	COMe	91		650–652
H	H	4-OH	H	COPh	171		650–652
H	H	2-CH_2OH	H	Me	78–79		608,617
H	H	2-CH_2OH	H	Et	70–72		617
H	H	2-CH_2OH	H	CH_2Ph	190–200/0.01		582,608,617
H	H	2-CH_2OH	3-Cl	Me	Oil		608
H	H	2-CH_2OH	7-Cl	Me	Oil		608
H	H	2-$(CH_2)_2OH$	H	H	107–109		617
H	H	2-$(CH_2)_2OH$	H	Me	155/0.001		617
					66–68		
H	H	2-$(CH_2)_2OH$	H	Et	150/0.001		617
H	H	3-CH_2OH	H	Me	Oil		608
H	H	3-CH_2OH	H	*n*-Bu	Oil		608
H	H	3-CH_2OH	H	COMe	118–120		608
H	H	3-CH_2OH	7-Cl	COMe	130–132		608
H	H	3-CH(Me)OH	H	Me	Oil		608
H	H	3-CH(Me)OH	H	CHO	111–113		608

(*continued*)

TABLE 88. *(Continued)*

R_1	R_2	R_3	R_4	R_5	mp (°C) or bp (°C/torr)	Spectra	Refs.
H	H	3-CH(Me)OH	H	COMe	Oil		608
H	H	4-CH_2OH	6-CH_2OH	Me	138–139	ir,pmr,uv	579
H	H	H	H	$(CH_2)_2OH$	185–200/0.6		592
					115–117		783
H	H	H	H	$(CH_2)_2OMs$	136–138		783
H	H	H	H	$(CH_2)_2OTs$	126–128		592
Me	H	H	H	$(CH_2)_2OH$	168/0.003		595,596
Me	H	H	H	$(CH_2)_2OTHP$	190/0.05		595,596
H	H	3-Cl	7-Cl	$(CH_2)_2OH$	222–230/0.5		592
H	H	3-Cl	7-Cl	$(CH_2)_2OTs$	110–112		592
H	H	H	H	$(CH_2)_3OH$	165–175/0.02		592
					185–200/0.08		594
H	H	H	H	$(CH_2)_3OTs$	100–102		592,594
H	H	3-Cl	7-Cl	$(CH_2)_3OH$	215–230/0.6		592
H	H	H	H	$CH_2CH(Me)OH$	188–198/0.6		590,592
H	H	H	H	$(CH_2)_4OH$	215/0.4		590,592
H	H	H	H	$CH_2CH(Me)CH_2OH$	190–200/0.6		592
H	H	H	H	$CH_2CH\text{-}CH_2$ (epoxide, O bridging CH and CH_2)	165/0.2		587,588
					73–74		

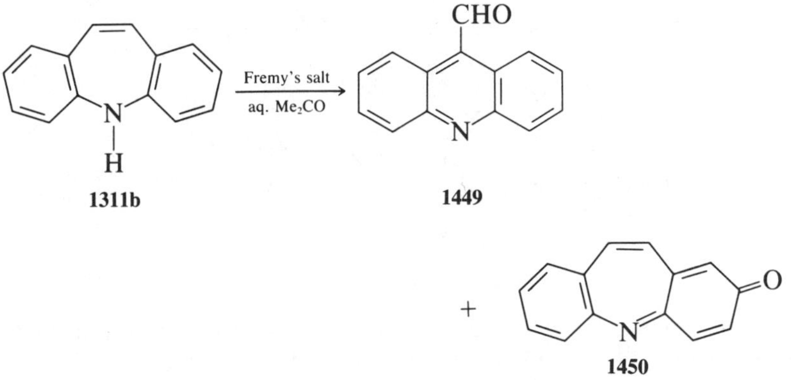

CHO

Fremy's salt, aq. Me_2CO

1311b → 1449 + 1450 (199)

imipramine (**1296a**) by oxidation with ferrous sulfate, disodium ethylenediaminetetraacetic acid, and ascorbic acid in a phosphate buffer solution (650–652).

The 3- and 4-hydroxydibenz[*b*, *f*]azepines **1454c** and **1455** have also been synthesized chemically (650–652). Compound **1454c** was obtained by diazotization of 3-amino-5-acetyliminobibenzyl (**1454a**) followed by aqueous thermolysis of the diazonium salt **1454b.** The 4-hydroxy derivative **1455** was obtained by treating iminobibenzyl with benzoyl peroxide in chloroform.

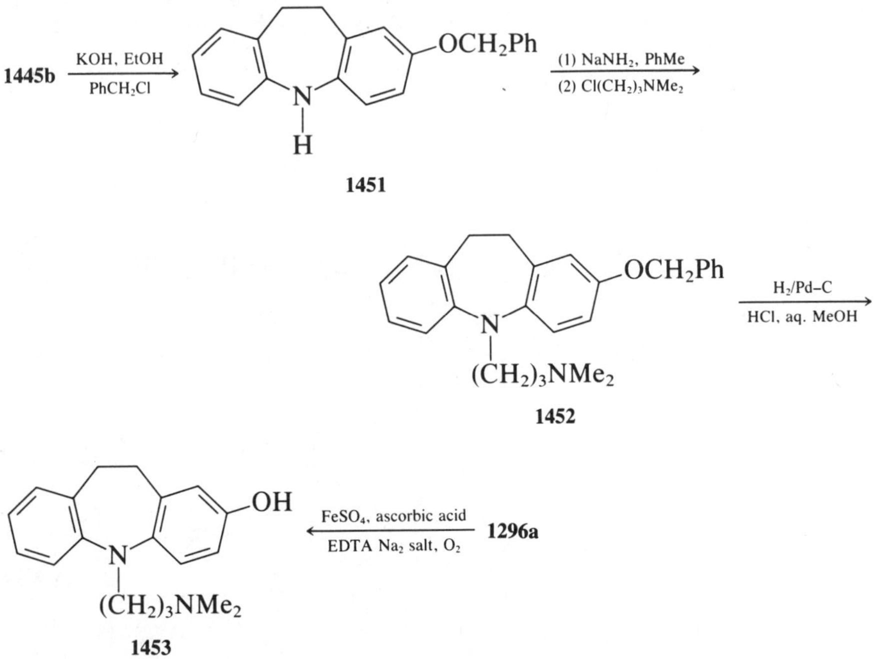

1445b —KOH, EtOH / $PhCH_2Cl$→ **1451** (OCH_2Ph) —(1) $NaNH_2$, PhMe / (2) $Cl(CH_2)_3NMe_2$→

1452 (OCH_2Ph, $(CH_2)_3NMe_2$) —H_2/Pd–C / HCl, aq. MeOH→

1453 (OH, $(CH_2)_3NMe_2$) ←$FeSO_4$, ascorbic acid / EDTA Na_2 salt, O_2— **1296a**

Figure 183

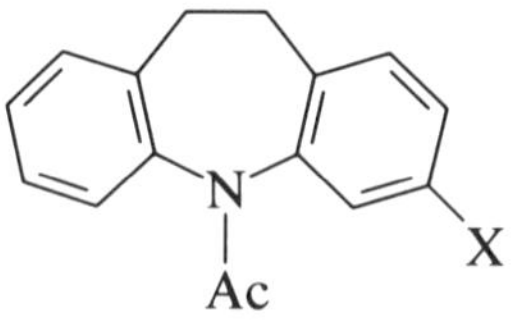

1454a, X = NH_2
1454b, X = N_2^+
1454c, X = OH

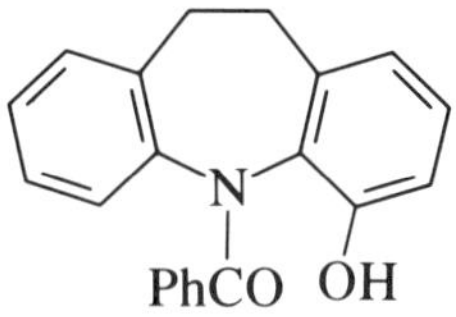

1455

Alkoxy-substituted dibenz[*b,f*]azepines have been prepared via the methods outlined in Section 39.B.a.i (506) and Section 39.B.b.iii (517,531).

iii. HYDROXYALKYLDIBENZ[*b,f*]AZEPINES. The principal synthetic route to hydroxyalkyldibenz[*b,f*]azepines of the general formula **1457** (Eq. 200) has been the reduction of carbonyl compounds **1456.** Reductions of dibenz[*b,f*]azepine carboxaldehydes have been effected with lithium aluminum hydride (582,608,617), while the analogous ketones have been reduced with sodium borohydride or lithium aluminum hydride, and also with borane (608). Dibenz[*b,f*]azepine carboxylic acids have been reduced with borane and with lithium aluminum hydride (608,617), while the corresponding carboxylic esters have been reduced to alcohols with lithium aluminum hydride (579,608). The reaction of organolithium compounds with acetone was used to prepare an α-hydroxyalkyl compound, e.g, **1423** → **1458** (Eq. 201) (640). Known examples of these hydroxyalkyl derivatives are given in Tables 87 and 88.

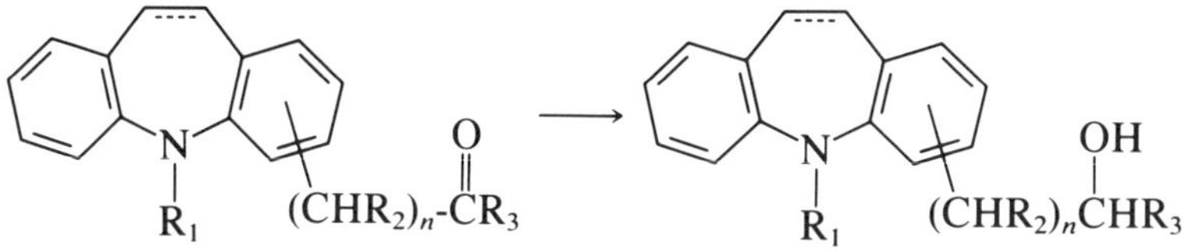

(200)

1456, n = 0, 1; R_2 = Alkyl; R_3 = H, Alkyl, OH, *O*-Alkyl

1457

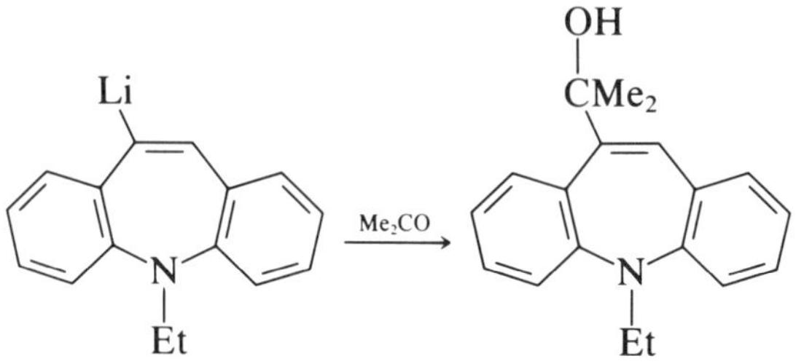

(201)

1423 **1458**

5-(Hydroxyalkyl)dibenz[*b,f*]azepines (**1459**) have been prepared by alkylation of the parent tricyclic compounds (Eq. 202). However, α-halo-ω-hydroxyalkyl groups cannot be introduced directly. Instead, the hydroxyl group must first be protected as a tetrahydropyranyl ether, which can be subsequently removed by acidic hydrolysis (592,594–596).

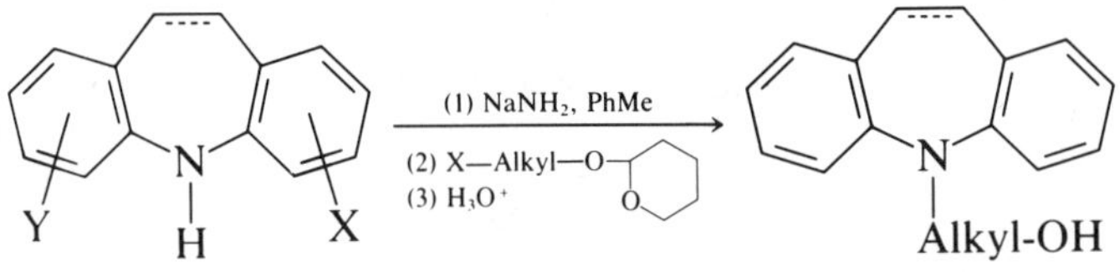

(1) $NaNH_2$, PhMe; (2) X—Alkyl—O—(tetrahydropyranyl); (3) H_3O^+ → **1459** (Alkyl-OH) (202)

d. DIBENZ[*b,f*]AZEPINE CARBOXALDEHYDES AND KETONES

The principal routes to dibenz[*b,f*]azepine carboxaldehydes and ketones (see Table 89) have been via classical electrophilic aromatic substitution reactions. For example, Friedel–Crafts acetylation (Eq. 203) of 5-acetyli-

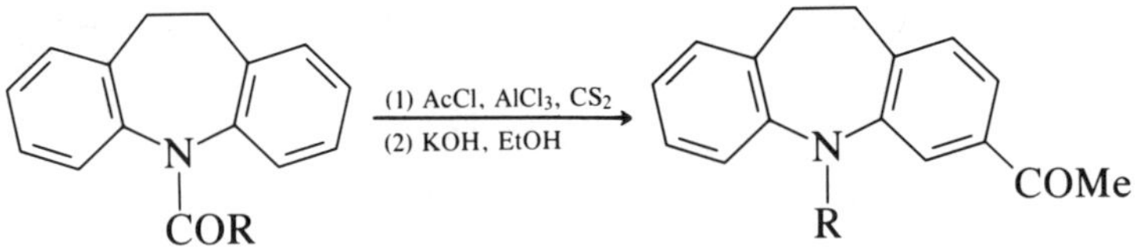

(1) AcCl, $AlCl_3$, CS_2; (2) KOH, EtOH (203)

1307, R = Me
1460, R = H

1461a, R = COMe
1461b, R = H

minobibenzyl (**1307**) gave 3,5-diacetyliminobibenyl (**1461a**) (512,527, 584,622,657). Alkaline hydrolysis of **1461a** afforded the parent ketone, 3-acetyliminobibenzyl (**1461b**) (512,527).

The aforementioned acetylation has most frequently been carried out by treating **1307** with acetyl chloride–aluminum chloride in carbon disulfide. The reaction is not limited to acetylation, as other acid chlorides have been used successfully (527,657). Catalysts such as boron trifluoride, aluminum tribromide, and ferric chloride have been claimed to be effective (527,622), although the only ones for which experimental documentation has been supplied are aluminum chloride and aluminum trichloride–iodine (608). In addition to carbon disulfide, nitrobenzene and chlorobenzene have been claimed to be useful reaction solvents. Interestingly, acetylation of **1307** with acetylium perchlorate was unsuccessful (584).

The aluminum chloride–catalyzed acetylation of several *N*-acyliminobibenzyls has been examined in carbon disulfide (584). The directing effect of the *N*-acetyl group to the 3-position was also displayed by the *N*-propionyl,

TABLE 89A. DIBENZ[b,f]AZEPINE CARBOXALDEHYDES AND KETONES

COR_1	R_2	R_3	R_4	mp (°C)	Spectra	Refs.
2-CHO	Me	H	H	90–93; 94–95		581,608,617, 618,659
				148–150 (oxime)		608
2-CHO	Me	3-Cl	H	Mixture as oil		608
2-CHO	Me	H	3-Cl			
2-CHO	Et	H	H	114–116		581,617
2-CHO	CH_2Ph	H	H	99.5–101		581,582,608,617
2-CHO	$(CH_2)_3NMe_2$	H	H	200–210/0.8–1.0[a]		659
2-COMe	Me	H	H	80–83		608
2-COMe	Me	H	8-COMe	124–125	ir,pmr,ms	584
3-COMe	H	H	H	156–157; 154–156		512,527,584,657
				240 (2,4-DNP)		657
				134–136 (ethylene ketal)		657
3-COMe	CHO	H	H	111–113		608
3-COMe	COMe	H	H	140–142; 143–144		512,527,584, 622,657,844
				170–171 (phenylhydrazone)		512
3-COMe	COEt	H	H	93–94	ir,pmr,ms	584
3-COMe	COPr-*n*	H	H	Oil		608
3-COMe	$COCH_2Cl$	H	H	145–146	ir,pmr,ms	584
3-COMe	COPh	H	H	155–156	ir,pmr,ms	584
3-COMe	COMe	H	7-Cl	Oil		608
3-COMe	Me	H	H	Oil		608
3-COMe	$(CH_2)_3NMe_2$	H	H	191 (HCl)		657
3-COEt	H	H	H	140		527
3-COEt	COMe	H	H	132		527
4-COMe	COPh	H	H	146–148		626
4-COMe	COPh	1-Br	H	163–165	pmr	626
4-COMe	COPh	2-Br	8-Br	179–181		626
4-COMe	COPh	3-Br	H	191–192		626
4-COMe	H	H	H	108–110	pmr	626
4-COMe	H	1-Br	H	79–81		626
4-COMe	H	2-Br	H	102–104	pmr	626
4-COMe	H	3-Br	H	Oil		626
4-COMe	H	2-Br	8-Br	143–145		626
4-COMe	H	H	8-Br	114–115	pmr	626
4-COMe	$COCH_2Br$	H	H	143–144		626
4-COMe	$COCH_2Br$	1-Br	H	141–142		626
4-COMe	$COCH_2Br$	2-Br	H	161–163		626
4-COMe	$COCH_2Br$	3-Br	H	—		626
4-COMe	$COCH_2Br$	2-Br	8-Br	156–158		626
4-COMe	$COCH_2Br$	H	8-Br	—		626
4-COPh	COMe	H	H	Oil		626

TABLE 89A. (*Continued*)

COR_1	R_2	R_3	R_4	mp (°C)	Spectra	Refs.
4-COPh	COMe	2-Br	8-Br	205–207		626
4-COPh	H	H	H	Oil		626
4-COPh	H	2-Br	H	146–148		626
4-COPh	H	2-Br	8-Br	106–108		626
4-COPh	H	H	8-Br	117–119		626
4-COPh	$COCH_2Br$	H	H	154–156		626
4-COPh	$COCH_2Br$	H	8-Br	203–205		626
4-COPh	$COCH_2Br$	2-Br	H	—		626
4-COPh	$COCH_2Br$	2-Br	8-Br	175–177		626

[a]Boiling point (°C/torr).

TABLE 89B. DIBENZ[*b,f*]AZEPINE CARBOXALDEHYDES AND KETONES[a]

10 11 9 1 R_4 8 2 R_3 5 N 7 4 3 6 R_1 COR_2

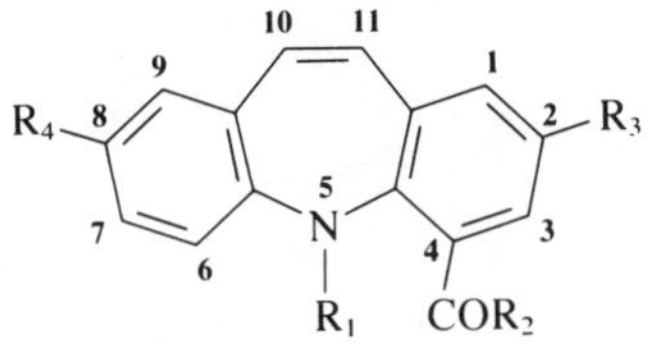

R_1	R_2	R_3	R_4	mp (°C)	Spectra
H	Me	H	H	98–99	
H	Me	Br	Br	165–167	pmr
$COCH_2Br$	Me	H	H	144–146	
$COCH_2Br$	Me	Br	Br	—	
H	Ph	H	H	157–158	
H	Ph	Br	Br	Oil	
$COCH_2Br$	Ph	H	H	172–174	
$COCH_2Br$	Ph	Br	Br	224–225	

[a]Ref. 626.

TABLE 89C. DIBENZ[*b,f*]AZEPINE CARBOXALDEHYDES AND KETONES[a]

O Me—C 9 10 11 1 8 2 5 N 7 3 6 4 COR

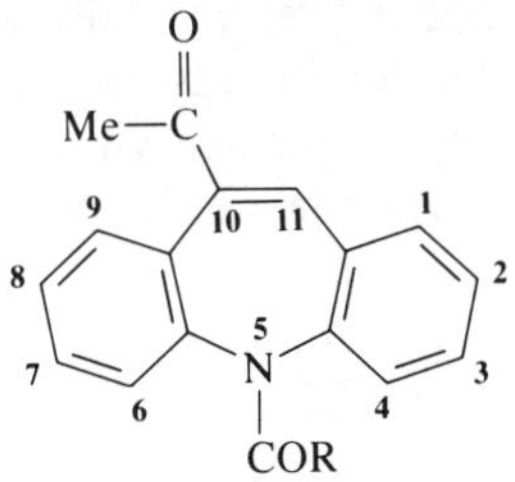

R	mp (°C)	Spectra
Me	163–164	ir,ms,pmr
Et	167–168	ir,ms,pmr
CH_2Cl	172–174	ir,ms,pmr

[a]Ref. 658.

N-chloroacetyl, *N*-benzoyl, and *N*-butyryl substituents (584,608). However, it is of interest to note that, unlike the *N*-acetyl derivative **1307**, other *N*-acyliminobibenzyls would undergo acetylation only after a prolonged reaction time with a large excess of acetyl chloride–aluminum chloride.

There have been no reported examples of diacylated products arising from an *N*-acyliminobibenzyl. A possible explanation has been advanced on the basis of the molecular geometry of 3,5-diacetyliminobibenzyl (**1461a**) (584). Since the iminobibenzyl ring is puckered, the most reasonable conformation for **1461a** would have the *N*-acetyl group coplanar and conjugated with the unsubstituted ring. This conformation would serve to deactivate the unsubstituted ring toward subsequent acetylation. Unlike the *N*-acetyl derivative **1307,** Friedel–Crafts acetylation of *N*-methyliminobibenzyl (**1382a**) afforded the 2,8-diacetyl compound **1462** and none of the monoacetyl derivative (584).

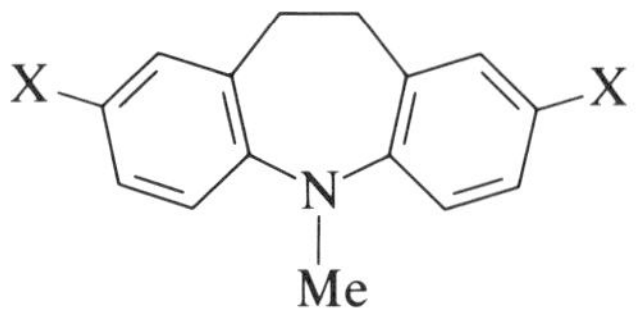

X X N Me

1382a, X = H
1462, X = COMe

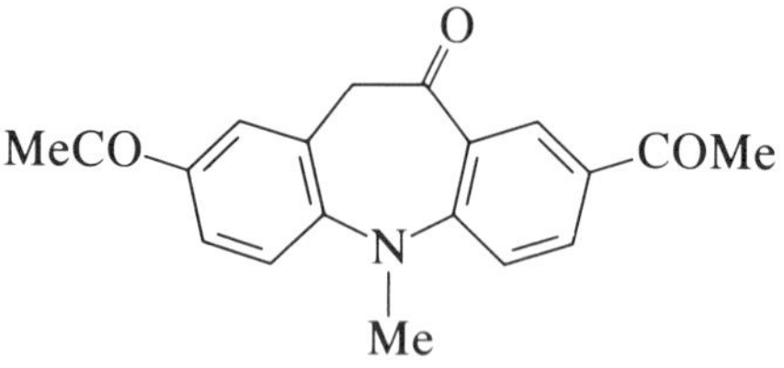

O MeCO COMe N Me

1463

The Friedel–Crafts acetylation of 5-acyliminostilbenes (Eq. 204) took a different course from that of the 10,11-dihydrobenz[*b,f*]azepines. Thus acetylation of **1308** with acetyl chloride–aluminum trichloride in carbon disulfide did not give a 3,5-diacetyl derivative, but instead gave the 5,10-disubstituted compound **1464** (658). The observed reactivity of the iminostilbene ring (i.e., the deactivation of the aromatic rings by *N*-acylation) was found to be consistent with π-electron density (θ) calculations, which had predicted that electrophilic attack should occur preferentially at the 10-position. Interestingly, the corresponding *N*-methyliminostilbene (**1337**) was essentially unreactive toward acetylation by either acetylium perchlorate or acetyl chloride–aluminum chloride in nitrobenzene (homogeneous) or carbon disulfide

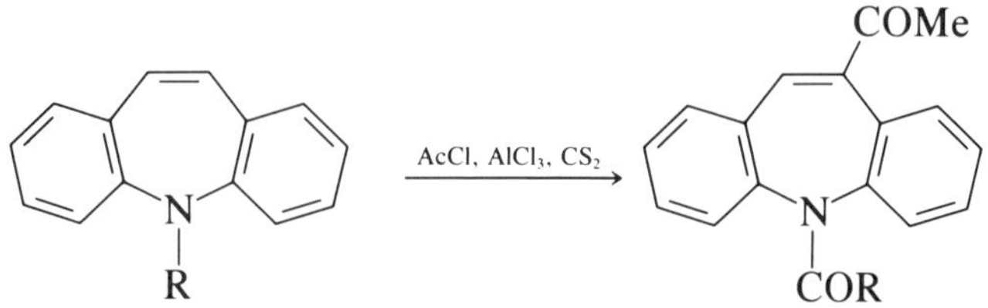

N R $\xrightarrow{\text{AcCl, AlCl}_3\text{, CS}_2}$ COMe N COR (204)

1308 R = COMe
1337, R = Me (unreactive)

1464

(heterogeneous) (658). The latter conditions ($AcCl–AlCl_3–CS_2$) did give a trace product that was tentatively identified as 2,8-diacetyldibenz[*b,f*]azepin-10-one (**1463**) on the basis of spectral data.

Vilsmeier–Haack formylation (Eq. 205) has been successfully employed to convert *N*-alkyliminobibenzyls such as **1382a** to the corresponding 2-carboxaldehydes (**1465**) (581,582,608,617,618,659). These reactions have typically been carried out by adding a DMF solution of the appropriate *N*-alkyliminobibenzyl to a cooled mixture of DMF and phosphoryl chloride. After about 2 hr at 60–70°C, the products can be isolated in good yield by dilution with water. Iminobibenzyl (**1293**) under the same conditions gave only the *N*-formyl compound **1460** (see Eq. 203).

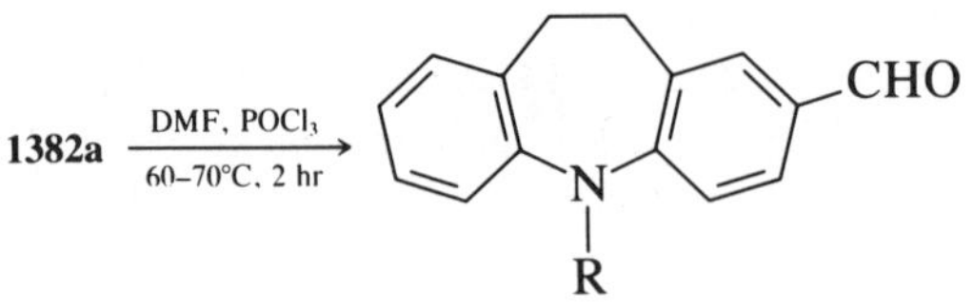

1382a $\xrightarrow[\text{60–70°C, 2 hr}]{\text{DMF, POCl}_3}$ **1465** (CHO; N–R) (205)

A series of 4-acyliminobibenzyls and iminostilbenes has been prepared according to the sequence shown in Fig. 184 (626). Application of the Fischer indole synthesis to 5-amino-10,11-dihydrodibenz[*b,f*]azepine (**1466**) produced the benzazepinoindoles **1467.** Oxidative cleavage afforded the 4,5-diacyl derivatives **1468,** which on acidic hydrolysis were converted to the 4-acyl derivatives **1469**.

The 2-cyanoderivative **1470** underwent a standard Grignard reaction (Eq. 206) to afford, after hydrolysis, the 2-acetyliminobibenzyl **1471** (608).

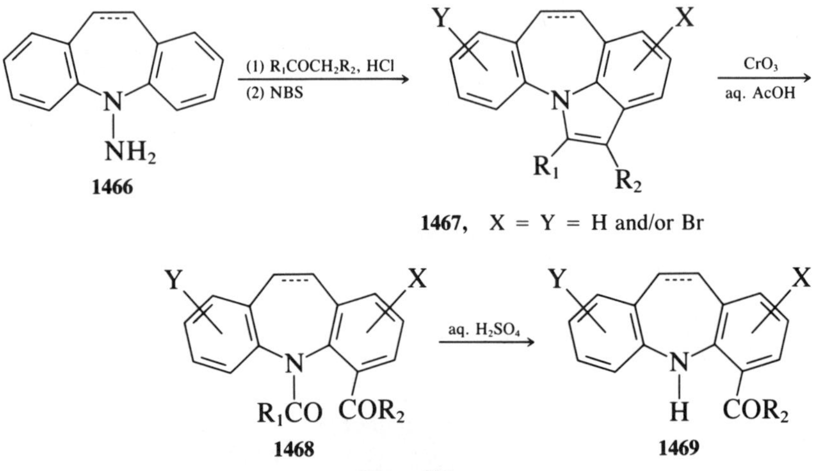

1466 (NH_2) $\xrightarrow[\text{(2) NBS}]{\text{(1) } R_1COCH_2R_2\text{, HCl}}$ **1467** (R_1, R_2, X, Y) $\xrightarrow[\text{aq. AcOH}]{CrO_3}$

1467, X = Y = H and/or Br

1468 (R_1CO, COR_2, X, Y) $\xrightarrow{\text{aq. } H_2SO_4}$ **1469** (H, COR_2, X, Y)

Figure 184

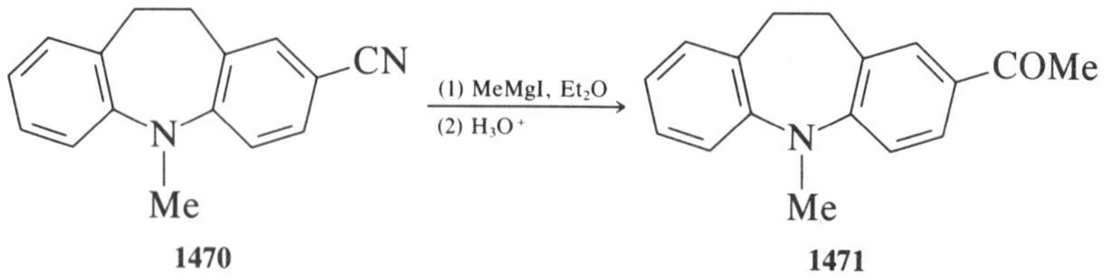

(206)

e. DIBENZ[*b,f*]AZEPINE CARBOXYLIC ACIDS

Several preparative routes to the dibenz[*b,f*]azepine carboxylic acids are available. For example (Fig. 185), cyanation of the bromo derivative **1472** with cuprous cyanide in DMF afforded the 10-cyanoiminostilbene **1473** in 91% yield (606,607,685). Ethanolic sodium borohydride gave a 77% yield of the corresponding iminobibenzyl **1474,** which was alkylated, using sodium hydride as the base, to give the α-aminoalkyl nitriles **1475.** Hydrolysis of either **1474** or **1475** to the primary amides **1476** or **1477,** respectively, was

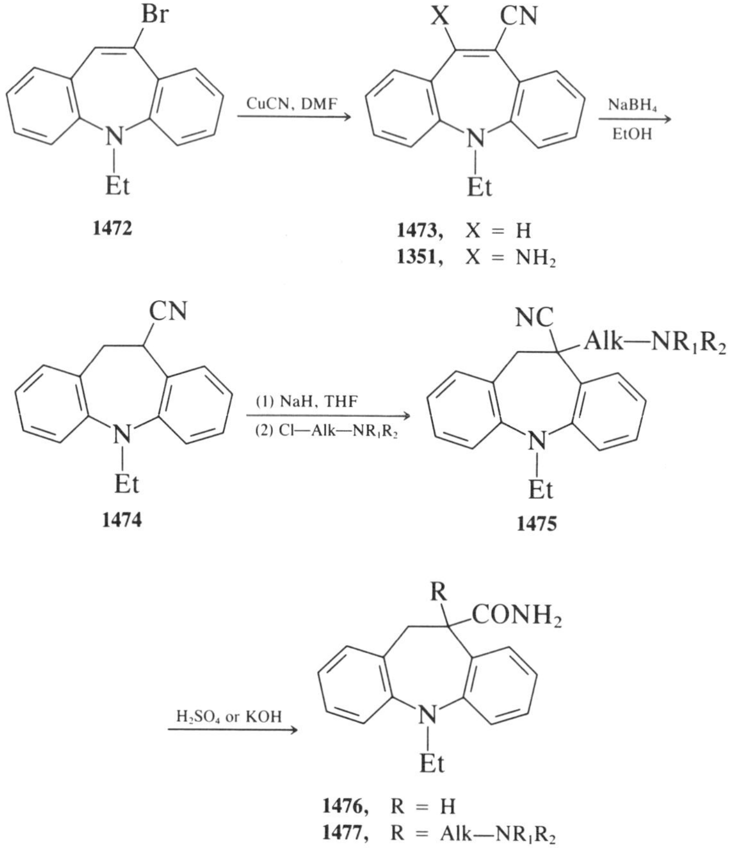

Figure 185

effected by either concentrated sulfuric acid or aqueous ethanolic potassium hydroxide.

The enaminonitrile **1351,** which was obtained by Thorpe–Ziegler cyclization of the diphenylamine **1350** (Eq. 168) was described in Section 39.B.c.i (559,560).

The dibenz[*b,f*]azepine carboxaldehydes and ketones whose synthesis was described in Section 39.C.d have found use in the preparation of dibenz[*b,f*]azepine carboxylic and alkanoic acid derivatives. Both the 2- and 3-dibenz[*b,f*]azepine carboxylic acids have been prepared via these valuable intermediates. Permanganate oxidation of the aldehyde **1478a** (Eq. 207) gave a 76% yield of the corresponding acid **1479a** (659). Alternatively, the same aldehyde was converted to the nitrile **1479b** via the intermediate aldoxime **1478b** (608). Haloform oxidation of the 3-acetyliminobibenzyls **1480** with aqueous sodium hypochlorite in dioxane (Eq. 208) gave the *N*-substituted acids **1481a**, which on basic hydrolysis produced the parent acids **1481b** (608).

(a) $KMnO_4$; (b) Ac_2O (207)

1478a, X = O
1478b, X = NOH

1479a, Z = CO_2H
1479b, Z = CN

(1) NaOCl, aq. dioxane; (2) KOH, EtOH (208)

1480

1481a, R′ = COR
1481b, R′ = H

Introduction of the carboxyl group into the 4-position of the dibenz[*b,f*]azepine nucleus has been achieved by electrophilic as well as nucleophilic substitution. For example, the reaction of the iminostilbenes (**1311b**) and iminobibenzyls (**1482**) with oxalyl chloride in ether followed by treatment with aluminum chloride in carbon disulfide (Fig. 186) produced the annelated derivatives **1483** and **1484**, respectively (574,579,635). Cleavage of the α-ketoamides **1483** and **1484** with alkaline hydrogen peroxide produced the acids **1485a** and **1485b**. In one instance, the acid **1485b** was not isolated, but was converted directly to the methyl ester **1485c** with ethereal diazomethane. When the entire process was then repeated, the diester **1485d** was formed

1311b, X = CH=CH
1482, X = CH_2CH_2

(1) $ClCOCOCl$, Et_2O
(2) $AlCl_3$, CS_2

1483, X = CH=CH
1484, X = CH_2CH_2

(1) OH^-, H_2O_2
(2) CH_2N_2

(1) *n*-BuLi
(2) CO_2

1485a + **1485b**

	X	R_1	R_2	Z
1485a	CH=CH	H	OH	H
1485b	CH_2CH_2	H	OH	H
1485c	CH_2CH_2	H	MeO	H
1485d	CH_2CH_2	H	MeO	6-CO_2Me
1485e	CH_2CH_2	H	NH_2	H
1485f	CH_2CH_2	Me	MeO	6-CO_2Me

Figure 186

(579). In general the yields for each of the steps in this process were good, often exceeding 80%. The methyl esters of these acids were also obtained with methanol–sulfuric acid and by the sequential action of thionyl chloride and methanol (574,635).

Treatment of the ester **1485c** with ammonia gave the primary amide **1485e** (574,635). Interestingly, attempts to *N*-methylate the diester **1485d** under standard conditions gave only unchanged starting material. However, the use of sodium hydride followed by prolonged heating with methyl iodide gave the desired *N*-methyl compound **1485f** in 88% yield (579). That the difficulty of alkylation was due to steric crowding was supported by the spectral properties of **1485d** and **1485f**. Whereas **1485d** was yellow in color, with a long-wavelength absorption band at 363 nm (ϵ 10,780), the *N*-methyl derivative **1485f** was colorless and showed only end absorption in the ultraviolet region. Thus it would appear that the coplanar, conjugated carbomethoxy groups of **1485d** were forced out of planarity by the introduction of the *N*-methyl group.

An alternative approach to the acids **1485a** and **1485b**, involving nucleophilic substitution, has also been described (574,605,635). In this method, treatment of either **1311b** or **1482** with butyllithium and subsequent treatment with carbon dioxide gave the 4-carboxylic acids **1485a** and **1485b**.

f. DIBENZ[*b,f*]AZEPINE ALKANOIC ACIDS

Just as the dibenz[*b,f*]azepine carboxaldehydes and alkyl ketones have been shown to be useful in the synthesis of dibenz[*b,f*]azepine carboxylic acids, they have also been demonstrated to be valuable intermediates in the preparation of the corresponding alkanoic acids. An illustrative example is outlined in Figs. 187A and 187B (582,608,617). Friedel–Crafts acetylation or Vilsmeier formylation of the iminobibenzyl **1486** gave the ketone **1480** and the aldehyde **1465**, respectively. Reduction of the aldehyde **1465** with lithium aluminum hydride gave the primary alcohol **1488a**, and reduction of **1480** with sodium borohydride afforded the secondary alcohol **1489a.** Conversion of the alcohols **1488a** and **1489a** to the corresponding alkyl halides **1488b** and **1489b,** followed by reaction with sodium cyanide in DMSO, produced the alkyl nitriles **1491a** and **1493a,** respectively. Alkaline hydrolysis then gave the arylacetic acid **1491b** and the arylpropanoic acid **1493b.** Hydrolysis of **1491a** with alkaline hydrogen peroxide provided the primary amide **1491c,** while anhydrous hydrogen chloride in ethanol produced the ethyl ester **1491d.**

Other processes for the conversion of **1480** and **1465** to alkanoic acid derivatives were also described. For example (Fig. 187A), conversion of **1465** to the nitrile **1470,** (Section 39.C.e, Eq. 207) followed by the addition of methylmagnesium iodide gave the methyl ketone **1471.** The methyl ketones **1480** and **1471** have been converted to the corresponding arylacetic acids in two ways. Hypochlorite oxidation of **1480** gave the acid **1481a** (Eq. 208). Following esterification, low-temperature (−78°C) lithium aluminum hydride reduction to the ester **1487** gave the alcohol **1489c,** which was converted stepwise to the bromide **1489d,** nitrile **1493c,** and the acid **1493d.** Alternatively, ketones **1471** and **1480** were transformed to the arylacetic acids **1491b** and **1493d** via the Willgerodt reaction (608,660). Treatment of **1471** and **1480** with sulfur in refluxing morpholine gave the thioamides **1492** and **1490,** and alkaline hydrolysis of the thioamides produced the acids **1491b** and **1493d,** respectively.

Another approach to the arylacetic acid **1491b** from the aldehyde **1465** involves condensation of **1465** with the sodium salt of tetraethyl dimethylaminomethylenediphosphonate to give the diethyl alkenylphosphonate **1396** (see Section 39.C.a, Fig. 182), which on acidic hydrolysis yields the acid **1491b** (581) (Eq. 209).

Alkylation of the ethyl ester **1491d** has been accomplished in two ways (Fig. 188). Direct α-alkylation to form **1494a** was carried out with ethyl iodide in hexamethylphosphoric triamide, using sodium hydride to generate the anion (608). Alternatively, carboethoxylation of **1491d** with diethyl carbonate in the presence of sodium ethoxide gave the malonate ester **1495.** Standard malonate alkylation afforded **1496,** which on alkaline hydrolysis provided the corresponding alkanoic acids **1494b** and **1494c.**

Other chemical transformations similar to those outlined above are summarized in Eq. 210. Treatment of the 10-methyliminostilbene **1409a** with *N*-bromosuccinimide gave the bromide **1409b,** which was converted to the

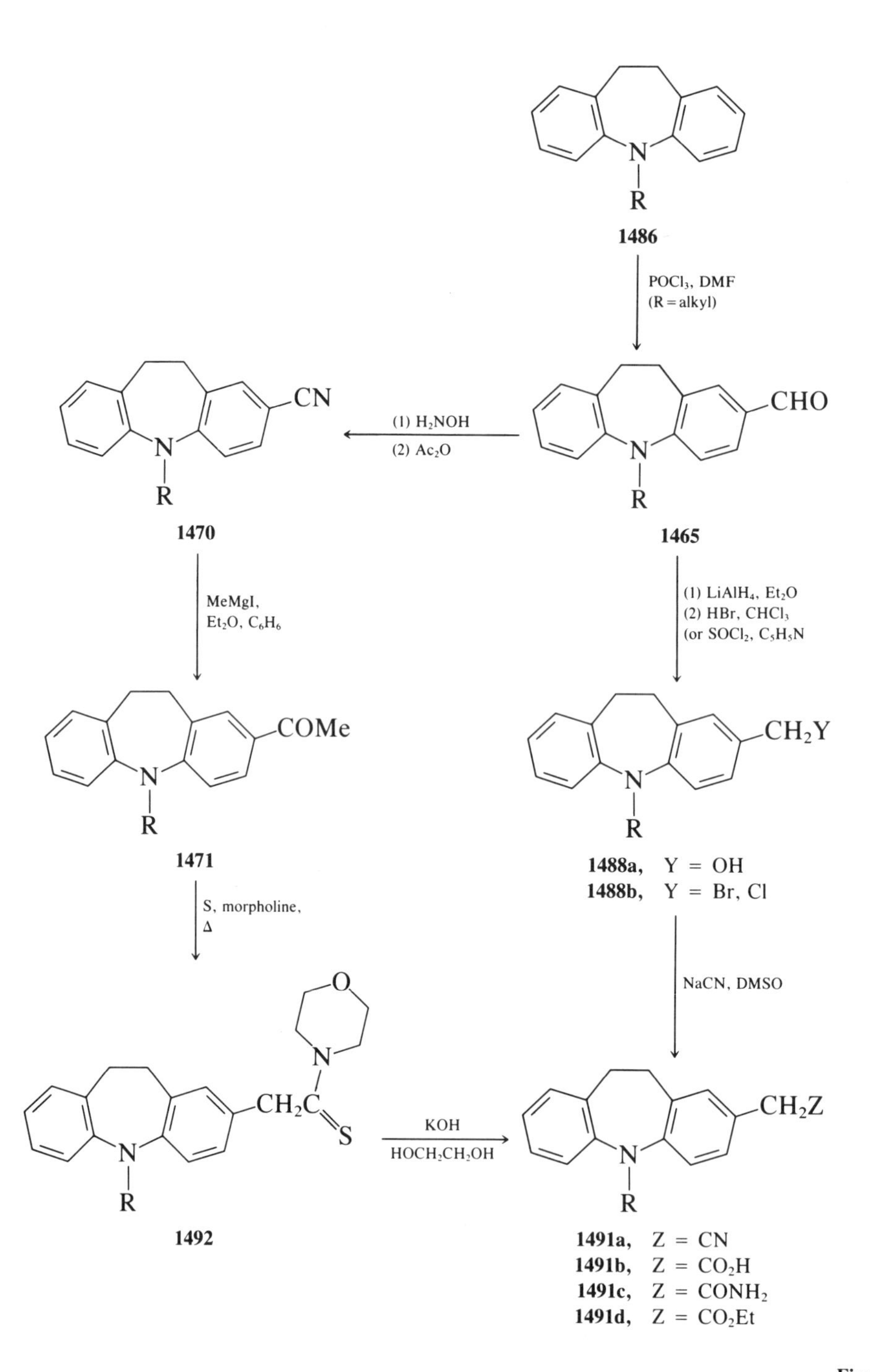
1486
$POCl_3$, DMF
(R = alkyl)
CHO
1465
(1) H_2NOH
(2) Ac_2O
CN
1470
MeMgI,
Et_2O, C_6H_6
COMe
1471
S, morpholine,
Δ
1492
CH_2C
S
O
N
KOH
$HOCH_2CH_2OH$
(1) $LiAlH_4$, Et_2O
(2) HBr, $CHCl_3$
(or $SOCl_2$, C_5H_5N
CH_2Y
1488a, Y = OH
1488b, Y = Br, Cl
NaCN, DMSO
CH_2Z
1491a, Z = CN
1491b, Z = CO_2H
1491c, Z = $CONH_2$
1491d, Z = CO_2Et
R
N

Figure 187A

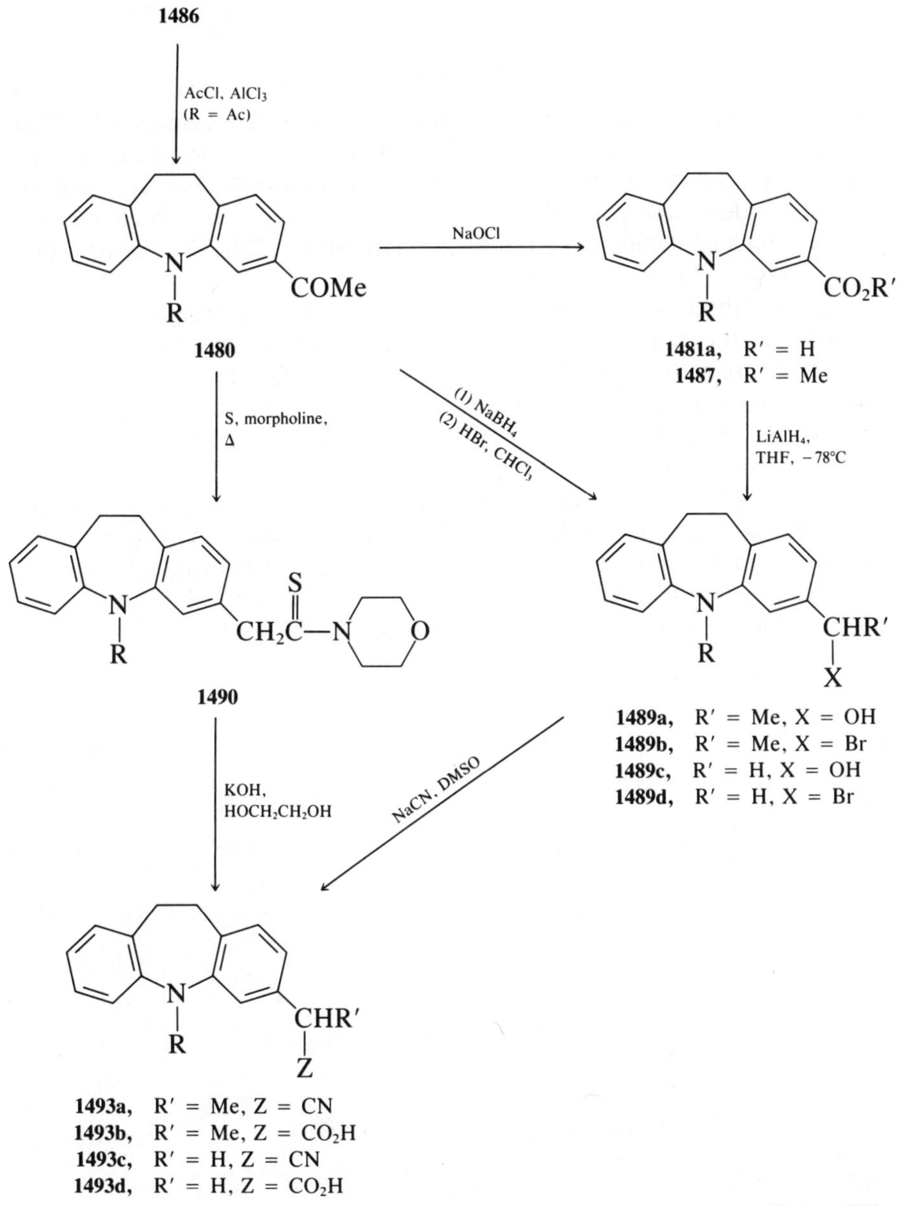

1486
AcCl, AlCl3
(R = Ac)
N
R
COMe
1480
NaOCl
N
R
CO2R′
1481a, R′ = H
1487, R′ = Me
S, morpholine,
Δ
(1) NaBH4
(2) HBr, CHCl3
LiAlH4,
THF, −78°C
N
R
S
CH2C—N
O
1490
N
R
CHR′
X
1489a, R′ = Me, X = OH
1489b, R′ = Me, X = Br
1489c, R′ = H, X = OH
1489d, R′ = H, X = Br
KOH,
HOCH2CH2OH
NaCN, DMSO
N
R
CHR′
Z
1493a, R′ = Me, Z = CN
1493b, R′ = Me, Z = CO2H
1493c, R′ = H, Z = CN
1493d, R′ = H, Z = CO2H

Figure 187B

$$\mathbf{1465} \xrightarrow[\text{(see Fig. 182)}]{Me_2N\bar{C}[PO(OEt)_2]_2Na^+} \mathbf{1396} \xrightarrow{\text{conc. HCl}} \mathbf{1491b} \quad (209)$$

NMe$_2$ CH=C P(OEt)$_2$ O N R

1396

acetonitrile **1497a** by reaction with sodium cyanide in aqueous ethanol (577,578). Basic hydrolysis produced the acetic acid **1497b.** The corresponding propionic acids **1498b** and **1498c** were obtained in two steps from the bromide **1409b.** Condensation with ethyl cyanoacetate produced the expected adduct **1498a,** which could be converted to either **1498b** or **1498c,** depending on the hydrolysis conditions.

The 10-iminostilbenylacetic acid **1502** (Eq. 211) was prepared in three steps from the azepinone **1499** (625). A Reformatsky reaction with ethyl bromoacetate converted **1499** to the hydroxy ester **1500.** Dehydration of **1500** with ethanolic hydrogen chloride produced a mixture of the isomeric (E)- and (Z)-esters **1501.** Alkaline hydrolysis gave, after acidfication, the acetic acid derivative **1502.**

Only a limited number of methods have been reported for the preparation of dibenz[*b,f*]azepine-5-alkanoic acids, and two of them have already been discussed in Section 39.C.a (Eqs. 179 and 182). The 5-(cyanoalkyl)dibenz-[*b,f*]azepines **1503a** have been prepared by cyanide displacement of the corresponding 5-(haloalkyl) derivative **1503b** or the analogous 5-(tosyloxy-

1491d —(1) NaH, HMPA; (2) EtI→ **1494**

R$_1$ CHCO$_2$R$_2$ N Me

1494a, R$_1$ = Et, R$_2$ = Et
1494b, R$_1$ = Me, R$_2$ = H
1494c, R$_1$ = Et, R$_2$ = H

1491d —(1) NaOEt, EtOH; (2) (EtO)$_2$CO→ **1495**

CH(CO$_2$Et)$_2$ N Me

1495

1495 —(1) NaOEt, EtOH; (2) RI→ **1496**

R$_1$ C(CO$_2$Et)$_2$ N Me

1496

1496 —(1) KOH, aq. *n*-BuOH; (2) H$^+$→ **1494**

Figure 188

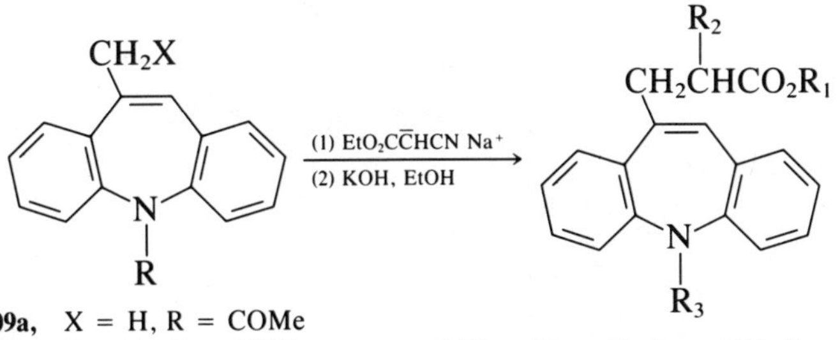

(1) $EtO_2C\bar{C}HCN\ Na^+$
(2) KOH, EtOH

(210)

1409a, X = H, R = COMe
1409b, X = Br, R = COMe
1497a, X = CN, R = COMe
1497b, X = CO_2H, R = H

1498a, R_1 = Et, R_2 = CN, R_3 = COMe
1498b, R_1 = R_2 = H, R_3 = COMe
1498c, R_1 = R_2 = R_3 = H

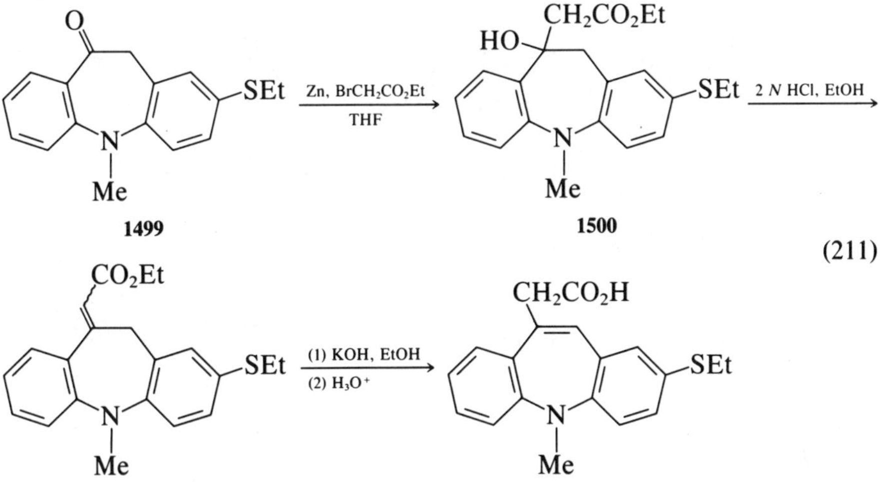

Zn, $BrCH_2CO_2Et$, THF; 2 *N* HCl, EtOH

1499 **1500**

(211)

(1) KOH, EtOH
(2) H_3O^+

1501 **1502**

alkyl) derivatives **1503c** (595,661–663). Direct alkylation of iminobibenzyl with *N,N*-dimethyl-α-bromopropionamide afforded the amide **1504** (664). Hydrolysis of the nitrile group in **1503a** and the amide group in **1504** constitutes a viable route to 5-alkanoic acids.

The acids and nitriles described above have been routinely converted to amides and esters by classical methods. Tables 90 and 91 should be consulted for specific examples.

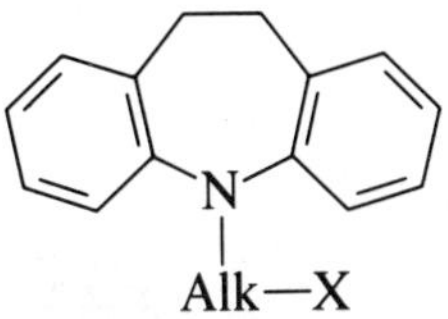

Alk—X

1503a, X = CN
1503b, X = Cl
1503c, X = OTs

$CHMeCONMe_2$

1504

TABLE 90. DIBENZ[*b*,*f*]AZEPINE CARBOXYLIC ACIDS AND DERIVATIVES

X	Y	Z	R	mp (°C)	Spectra	Refs.
CH_2CH_2	H	2-CO_2H	Me	195–197		659
CH_2CH_2	H	2-CN	Me	120–122		608
CH_2CH_2	H	3-CO_2H	H	196–197		608
CH_2CH_2	H	3-CO_2H	COMe	197–198		608
CH_2CH_2	H	3-CO_2Me	COMe	122–124		608
CH_2CH_2	H	3-CO_2H	CO Pr-*n*	108–110		608
CH_2CH_2	7-Cl	3-CO_2H	COMe	264–266		608
CH_2CH_2	7-Cl	3-CO_2Me	COMe	130–132		608
CH_2CH_2	H	4-CO_2H	H	226–228		574,635
CH_2CH_2	H	4-CO_2Me	H	145–148/0.1[a] 63–64	ir,uv,pmr	574,579, 635
CH_2CH_2	H	4-$CONH_2$	H	183–184		574,635
CH_2CH_2	H	4-CONHMe	H	163–165		605
CH_2CH_2	H	4-CONHMe	$CO(CH_2)_2Cl$	177–179.5		605
CH_2CH_2	H	4-CO_2H	COMe	200–203		574,635
CH_2CH_2	H	4-$CONH_2$	COMe	190–191		574,635
CH_2CH_2	H	4-CO_2H	COC_6H_4Cl-*p*	208–210		574,635
CH_2CH_2	H	4-CO_2Me	COC_6H_4Cl-*p*	173–175		574,635
CH_2CH_2	H	4-$CO_2H \cdot Et_2N(CH_2)_2Cl$	H	177.5–178.5		574,635

CH_2CH_2	H	4-$CO_2(CH_2)_2NEt_2$	H	178–180 (HCl)		574,635
CH_2CH_2	H	4-$CO_2(CH_2)_2NMe_2$	H	209–210 (HCl)		574,635
CH_2CH_2	3-MeO	4-CO_2H	H	158–160		574,635
CH_2CH_2	6-Me	4-CO_2H	H	188–190		574,635
CH_2CH_2	7-Cl	4-CO_2H	H	220–222		574,635
CH_2CH_2	6-CO_2Me	4-CO_2Me	H	114–115	ir,uv,pmr	579
CH_2CH_2	6-CO_2Me	4-CO_2Me	Me	122–124	ir,uv,pmr	579
CH=CH	H	4-CO_2H	H	243–245		574,635
CH=CH	H	4-$CO_2H \cdot Et_3N$	H	126–129		574,635
CH=CH	7-Cl	4-CO_2H	H	249–250		574,635
CH=CH	3,7-Cl_2	4-CO_2H	H	166–168		574-635
$_{10}CH=C_{11}(CN)$	H	H	Et	144–146.5	ir,uv	606,607,685
$_{10}CH_2CH_{11}(CN)$	H	H	Et	127–129	ir,uv	606,607,685
$_{10}CH_2\text{-}CH_{11}(CONH_2)$	H	H	Et	70–95 (MeOH)	ir	607
$H_2N\text{-}_{10}C=C_{11}\text{-}CN$	H	3-Cl	Me	162		559,560
$H_2N\text{-}_{10}C=C_{11}\text{-}CN$	H	2-Cl	Me	154		559,560

[a]Boiling point (°C/torr).

TABLE 91A. DIBENZ[*b,f*]AZEPINE ALKANOIC ACIDS AND DERIVATIVES

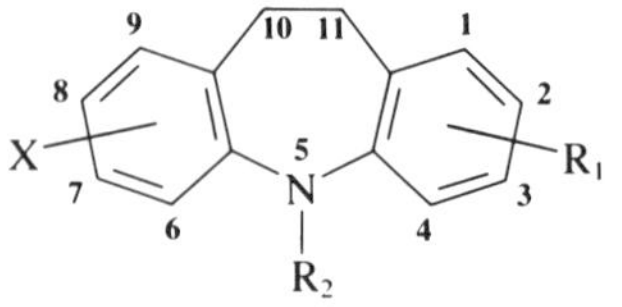

X	R_1	R_2	mp (°C) or bp (°C/torr)	Refs.
H	2-CH_2CO_2H	H	155–158	581,608,617
H	2-CH_2CO_2Et	H	77–79	617
H	2-CH_2CONH_2	H	245–250	617
H	2-$CH_2CONHMe$	H	57–58	617
H	2-CH_2CN	H	100–102	617
H	2-CH_2–(tetrazol-5-yl, N–H) 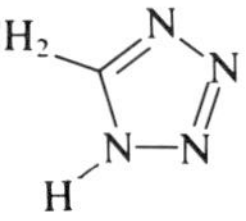	H	186–187	582
H	2-CH_2CO_2H	Me	121–123	581,608,617
H	2-$CH_2CO_2^-Na^+$	Me	192–194	608
H	2-CH_2CO_2Et	Me	170/0.001	608,617
H	2-CH_2CONH_2	Me	140–142	608,617
H	2-$CH_2CONHMe$	Me	129–131	617
H	2-CH_2CN	Me	70–71	608,617
7-Cl	2-CH_2CO_2H	Me	175–178	617
7-Cl	2-CH_2CN	Me	117–118	617
H	2-CH_2CO_2H	Et	113–115	581,617
H	2-CH_2CO_2Et	Et	175/0.001	617
H	2-CH_2CONH_2	Et	129–131	617
H	2-$CH_2CONHMe$	Et	—	617
H	2-CH_2CN	Et	77–78	617
H	2-CH_2CO_2H	CH_2Ph	138–139	608
H	2-CH_2CN	CH_2Ph	96–98	582,608,617
H	2-CH_2–(tetrazol-5-yl, N–H) 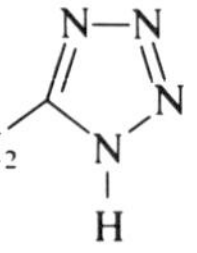	CH_2Ph	187–189	582
H	2-CH_2CO_2H	CHO	187–188	617
H	2-CH_2CO_2Et	CHO	200/0.001	617
H	2-CH_2CONH_2	CHO	208–210	617
H	2-$CH_2CONHMe$	CHO	127–129	617
H	2-CH_2CN	CHO	99–101	617
H	2-$CHMeCO_2H$	Me	153–157	608,684
H	2-$CHEtCO_2H$	Me	108–113	608,684
H	2-$CH(CO_2Et)_2$	Me	190–195/0.001	608,684
H	2-$CMe(CO_2Et)_2$	Me	Oil	608,684
H	2-$CEt(CO_2Et)_2$	Me	Oil	608,684
H	3-CH_2CO_2H	H	133–135 138	608,660
H	3-CH_2CO_2Me	H	Oil	608

TABLE 91A. (*Continued*)

X	R_1	R_2	mp (°C) or bp (°C/torr)	Refs.
H	$3\text{-}CH_2CO_2Al(OH)_2$	H	275 (foaming)	660
H	$3\text{-}CH_2CO_2H$	Me	140–141 143	580,608
7-Cl	$3\text{-}CH_2CO_2H$	Me	156–158	608
H	$3\text{-}CH_2CO_2Me$	Me	Oil	608
H	$3\text{-}CH_2CN$	Me	78–81	608
H	$3\text{-}CH_2CO_2Me$	*n*-Bu	Oil	608
H	$3\text{-}CH_2CN$	*n*-Bu	Oil	
H	$3\text{-}CH_2CO_2Me$	CHO	85–87	608
H	$3\text{-}CH_2CO_2H$	COMe	163–165	608
7-Cl	$3\text{-}CH_2CO_2H$	COMe	128–129	608
H	$3\text{-}CH_2CO_2Me$	COMe	—	608
H	$3\text{-}CH_2CN$	COMe	97–100	608
7-Cl	$3\text{-}CH_2CN$	COMe	112–114	608
H	$3\text{-}CH_2\overset{\overset{S}{\|}}{C}$—N⟨ ⟩O	COMe	116	660
H	$3\text{-}CH_2CO_2H$	COPr-*n*	Oil	608
H	$3\text{-}CH_2CO_2Me$	COPr-*n*	Oil	608
H	$3\text{-}CHMeCO_2H$	H	129–131	608
7-Cl	$3\text{-}CHMeCO_2H$	H	155–157	608
H	$3\text{-}CHMeCO_2Me$	H	Oil	608
H	$3\text{-}CHMeCO_2H$	Me	138–140	608
H	3-CHMeCN	Me	Oil	608
H	$3\text{-}CHMeCO_2Me$	CHO	Oil	608
H	$3\text{-}CHMeCO_2H$	COMe	153–154	608
7-Cl	$3\text{-}CHMeCO_2H$	COMe	128–129	608
H	3-CHMeCN	COMe	Oil	608
H	H	CH_2CN	90	611
H	H	CH(CN)Me	80–82	610,611
H	H	$CHMeCONMe_2$	114–116	644
H	H	CHEtCN	87	611
H	H	CH(CN)Pr-*n*	101	611
H	H	CH(CN)Pr-*i*	130	611
H	H	CH(CN)Bu-*n*	Amorphous	611
H	H	CH(CN)Me	113	611
H	H	$CHMeCONMe_2$	114–116	664
H	H	$(CH_2)_2CN$	—	595,661
H	H	$(CH_2)_2CO_2H$	148–149	661
H	H	$(CH_2)_2COCl$	Oil	661
H	H	$(CH_2)_2CONHC_5H_5N$	144–145	661
H	H	$CH_2CH(CN)Me$	133–134	662,663
H	H	$CH_2CH(Me)\underset{\underset{NMe}{\|}}{C}OEt$	Oil (HCl)	663
H	3-Cl	(cyclopropane ring) CO_2Et, H, H	—	603

TABLE 91B. DIBENZ[*b,f*]AZEPINE ALKANOIC ACIDS AND DERIVATIVES

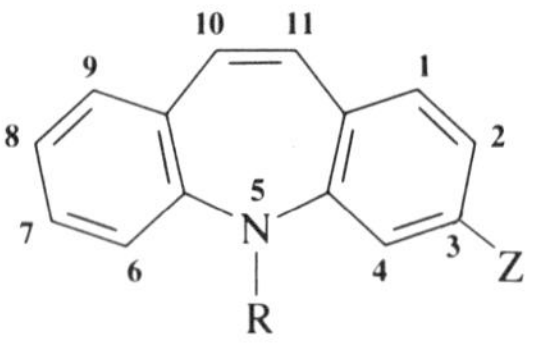

R	Z	mp (°C)	Refs.
H	CH_2CO_2H	208	660
Me	CH_2CO_2H	185	580
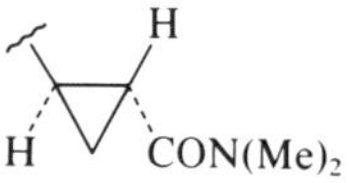 2-(H)-cyclopropyl with H and $CON(Me)_2$	Cl	Oil	603,604

TABLE 91C. DIBENZ[*b,f*]AZEPINE ALKANOIC ACIDS AND DERIVATIVES

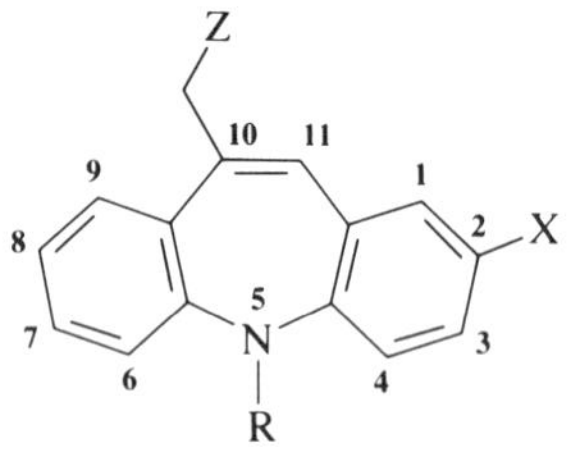

X	R	Z	mp (°C)	Spectra	Refs.
H	H	CO_2H	167		547,577,578
H	H	$CO_2C_6H_4NO_2$-*p*	124–127		547,577,578
H	H	$CONMe_2$	196–198		547,577,578
H	COMe	CN	137–138		547,577,578
2-SEt	Me	CO_2H	164		625
2-SEt	Me	$CO_2C_6H_4NO_2$-*p*	Oil	ir,uv	625
2-SEt	Me	$CONMe_2$	Oil	ir,uv	625
H	H	CH_2CO_2H	193–195		547,577,578
H	COMe	CH_2CO_2H	169–170		547,577,578
H	H	$CH_2CO_2C_6H_4NO_2$-*p*	—		547,577,578
H	H	CH_2CONMe_2	187–190		547,577,578
H	COMe	$CH(CN)CO_2Et$	Oil		547,577,578

TABLE 91D. DIBENZ[*b,f*]AZEPINE ALKANOIC ACIDS AND DERIVATIVES

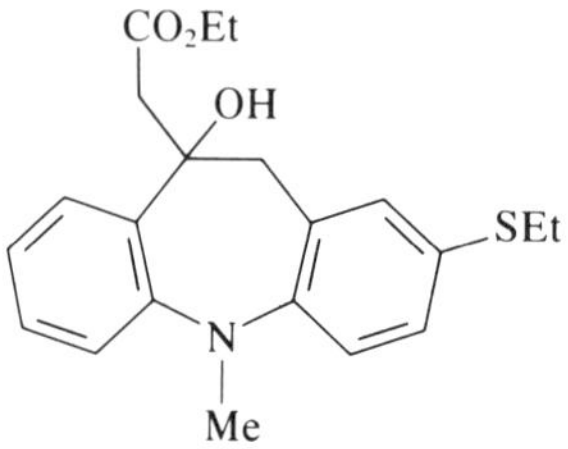

CO_2Et, OH, SEt, N, Me

ir,uv (Ref. 625)

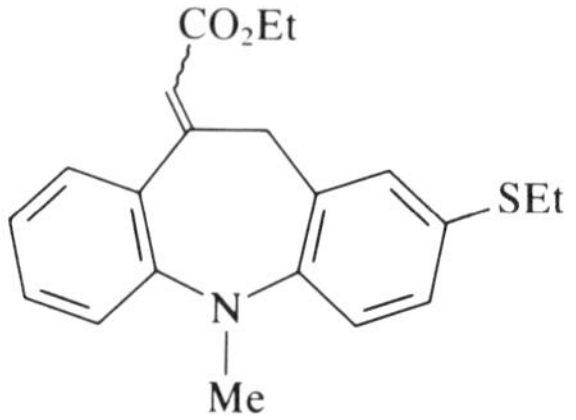

CO_2Et, SEt, N, Me

ir,uv (Ref. 625)

g. DIBENZ[*b,f*]AZEPINE SULFONAMIDES AND OTHER SULFUR–CONTAINING DERIVATIVES (see Table 92A–C)

The sulfur-containing derivatives of the dibenz[*b,f*]azepines have received only a modest amount of attention, with much of the chemistry being devoted to the 5H-dibenz[*b,f*]azepine-3-sulfonamides **1507a**, (Fig. 189) and the corresponding 5-(aminoalkyl) derivatives **1513**. The latter, since they are readily obtainable from **1507a** by standard methods, will not be discussed in this section. The interested reader should refer to Table 92A–C for specific examples.

TABLE 92A. SULFUR-CONTAINING DERIVATIVES OF THE DIBENZ[*b,f*]-AZEPINES

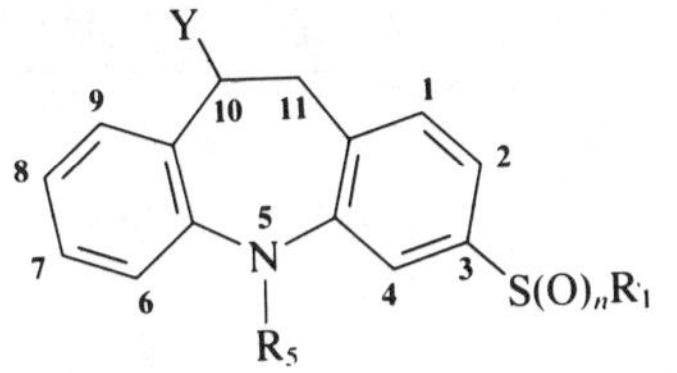

R_1	R_2	Y	*n*	mp (°C) or bp (°C/torr)	Refs.
[a]	COMe	H	0	110 (dec.)	669–671
Me	H	H	0	64	669–671
Me	COMe	H	0	160/0.003	669–671
Et	H	H	0	150/0.001	669–671c
Et	COMe	H	0	102	669–671
i-Pr	H	H	0	79	669–671
i-Pr	COMe	H	0	89	669–671
Ph	H	H	0	101	669,670
Ph	COMe	H	0	110–111	669
$C_6H_4NO_2$-*p*	COMe	H	0	126	669
Me	COMe	H	1	—	671
Me	H	H	1	—	671
Et	COMe	H	1	135	671
Et	H	H	1	128	671
OH	COMe	H	1	154	668
Me	COMe	H	2	153–154	668
Me	H	H	2	175–176	668
Cl	COMe	H	2	173.5–174	525,665,666
NMe_2	H	H	2	140	525,665,666
NMe_2	COMe	H	2	151	525,665,666
NMe_2	COMe	Br	2	174–176	525,666
NEt_2	H	H	2	174	665
NEt_2	COMe	H	2	136	665,666
Piperidino	H	H	2	244	665
Piperidino	COMe	H	2	145–146	665,666
Morpholino	H	H	2	211–212	665
Morpholino	COMe	H	2	196	665,666

[a] R_1 = 3-(5-Acetyl-10.11-dihydrodibenz[*b,f*]azepinyl)thio

TABLE 92B. SULFUR-CONTAINING DERIVATIVES OF DIBENZ[*b,f*]AZEPINE

R_1	R_2	*n*	mp (°C)	Refs.
Me	H	0	168	669,670
Et	H	0	143	669
Ph	H	0	—	669
Cl	COMe	2	Amorphous	669
NMe_2	H	2	186	525,666
NMe_2	COMe	2	180–181	525,666

TABLE 92C. SULFUR-CONTAINING DERIVATIVES OF DIBENZ[*b,f*]AZEPINE

mp 54–55°C (Ref. 625)

mp 230–231 °C, ir (Refs. 658,672)

The dibenz[*b,f*]azepines in Fig. 189 share the common feature of being substituted at the 3-position; they have all been prepared from the same readily available precursor **1505a** and its diazonium salt **1505b**. Saturation of an aqueous solution of **1505b** with sulfur dioxide and subsequent addition of cupric chloride gave the sulfonyl chloride **1506** (525,665–667). The sulfonamides **1507b** were prepared by the addition of amines to **1506,** and on alkaline hydrolysis yielded **1507a.** The corresponding iminostilbenes **1508a** and **1508b** were obtained from **1507b** by the bromination–dehydrobromination technique described in Section 39.B.b.ii (525,666).

When copper powder was added to a sulfur dioxide-saturated solution of **1505b,** the sulfinic acid **1509** was produced. Alkylation of **1509** under alkaline conditions produced the alkyl arylsulfones **1510a** (668). The corresponding sulfides and sulfoxides were prepared from the sulfonyl chloride **1506** (669–671). Hydriodic acid reduction of **1506** gave the disulfide **1511,** which, on sequential reductive cleavage and S-alkylation, afforded the sulfides **1512a.** The diaryl sulfide **1512b** was obtained on treatment of the diazonium salt **1505b** with sodium thiophenoxide (669,670). Finally, sodium metaperiodate oxidation of **1512a** gave the expected sulfoxides **1510b** (671).

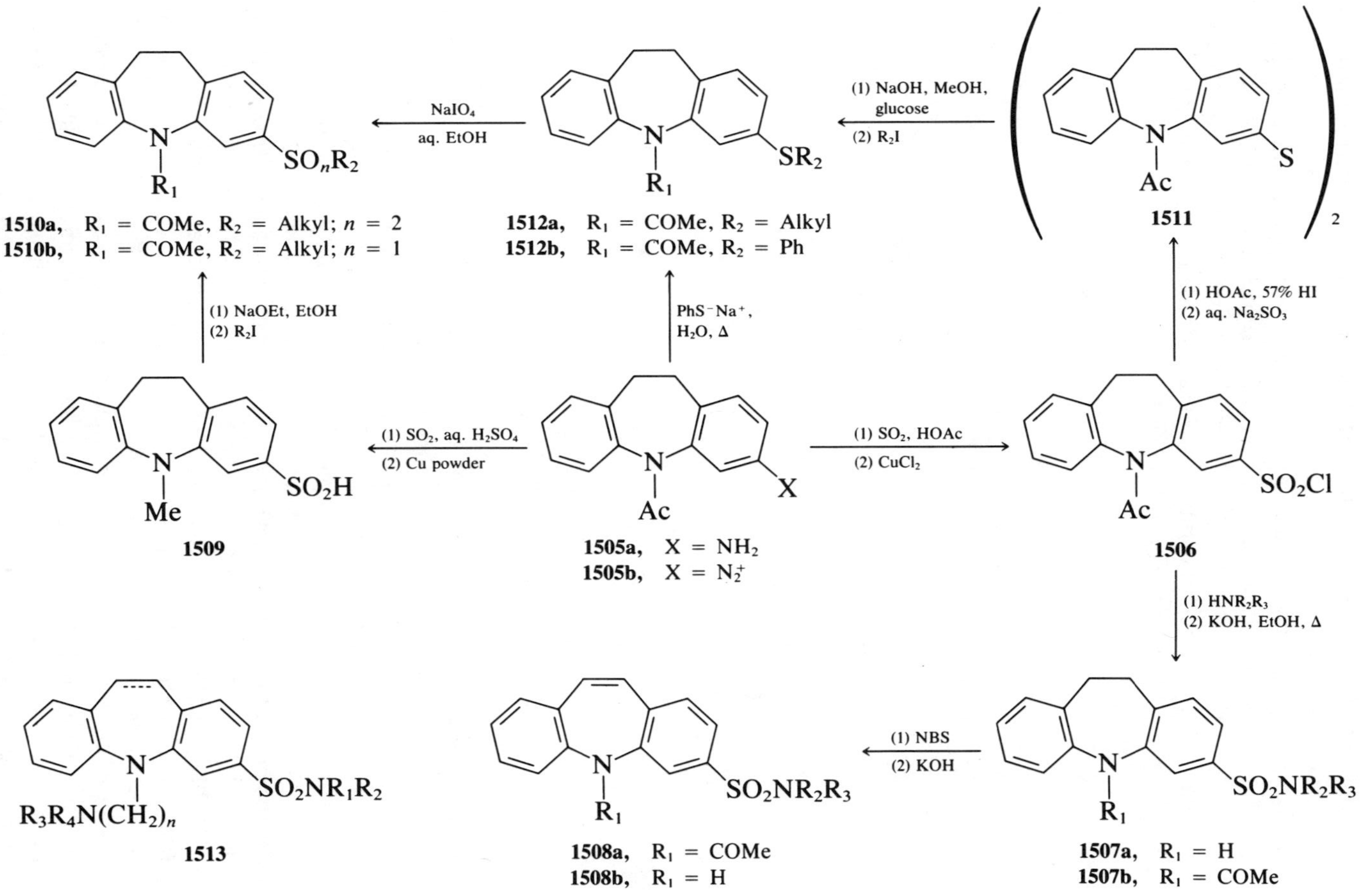

SO_nR_2

R_1

1510a, R_1 = COMe, R_2 = Alkyl; n = 2
1510b, R_1 = COMe, R_2 = Alkyl; n = 1

$NaIO_4$
aq. EtOH

SR_2

R_1

1512a, R_1 = COMe, R_2 = Alkyl
1512b, R_1 = COMe, R_2 = Ph

(1) NaOH, MeOH, glucose
(2) R_2I

S

Ac

1511

2

(1) NaOEt, EtOH
(2) R_2I

PhS^-Na^+,
H_2O, Δ

(1) HOAc, 57% HI
(2) aq. Na_2SO_3

SO_2H

Me

1509

(1) SO_2, aq. H_2SO_4
(2) Cu powder

X

Ac

1505a, X = NH_2
1505b, X = N_2^+

(1) SO_2, HOAc
(2) $CuCl_2$

SO_2Cl

Ac

1506

(1) HNR_2R_3
(2) KOH, EtOH, Δ

$SO_2NR_1R_2$

$R_3R_4N(CH_2)_n$

1513

$SO_2NR_2R_3$

R_1

1508a, R_1 = COMe
1508b, R_1 = H

(1) NBS
(2) KOH

$SO_2NR_2R_3$

R_1

1507a, R_1 = H
1507b, R_1 = COMe

Figure 189

Sulfur has been introduced into the 2-position of the dibenz[*b,f*]azepine system in two ways. The 2-bromodibenz[*b,f*]azepinone **1514** was converted in high yield to the corresponding sulfide **1499** with cuprous thioethoxide in pyridine–quinoline (Eq. 212) (625). Further conversions of the sulfide **1499** were described in Section 39.C.f (Eq. 211). Sensitized and unsensitized irradiation of 5-tosyliminostilbene (**1515a**) afforded the 2-tosyl derivative **1515b** in low conversion (Eq. 213) (672).

1514 $\xrightarrow[\text{quinoline, }\Delta]{\text{CuSEt, }C_5H_5N}$ **1499** (212)

1515a $\xrightarrow{h\nu}$ **1515b** (213)

h. NITRO– AND NITROSODIBENZ[*b,f*]AZEPINES (see Table 93)

The preparation of 5-nitroso-10,11-dihydrodibenz[*b,f*]azepine (**1516a,** Fig. 190) was first reported in 1899 by Thiele and Holzinger at the same time that they described the original preparation of iminobibenzyl (**1293**) (468). In spite of their 80-year history, the nitroso and even the nitrodibenz[*b,f*]azepines have not been widely examined.

There are now several reports describing the preparation of 5-nitrosodibenz[*b,f*]azepines by the action of nitrous acid on the appropriate 5H-dibenz[*b,f*]azepine. Thiele and Holzinger used sodium nitrite in acetic acid (468), while nitrosations have also been conducted with HCl in ethanol (673) or cold DMF (674–676,677). Alternatively, **1516a** has been prepared from **1293** by the action of amyl nitrite in diethyl ether (678). In general, all these conditions gave very high yields.

The chemistry of **1516a** has been studied only sparingly. Russian workers found that in methanol–ether, anhydrous hydrogen chloride caused quantitative rearrangement of **1516a** to the isomeric nitroso compound **1516b** (678). Sodium nitrite in methanol–thionyl chloride gave **1516b** directly, presumably via the intermediacy of **1516a.** Nitrous acid and imipramine (**1296a**) gave the 2-nitroso derivative **1516c** (678).

TABLE 93A. NITRO- AND NITROSODIBENZ[*b,f*]AZEPINES

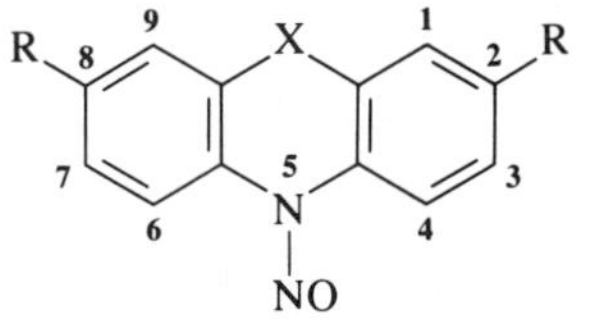

X	R	mp (°C)	Spectra	Refs.
CH=CH	H	116–118	ir,pmr,ms,uv	658,673
CH_2CH_2	H	120	468	
			uv	678
		112–113		674
		113–115		676
		113		677
CH_2CH_2	Me	127–129		675

TABLE 93B. NITRO- AND NITROSODIBENZ[*b,f*]AZEPINES

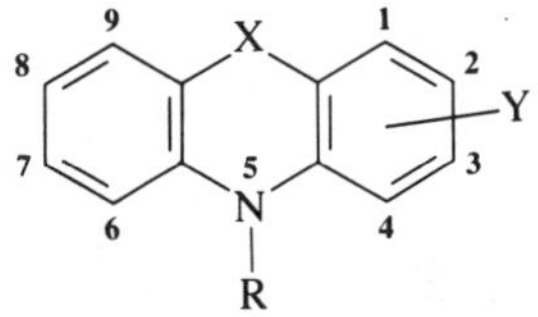

X	Y	R	mp (°C)	Spectra	Refs.
CH_2CH_2	2-NO	H		uv	678
CH=CH	2-NO_2	H	176.5–178	uv,ms	673
CH=CH	2-NO_2	COMe	195–196	ir,pmr,ms	673
CH_2CH_2	2-NO_2	H	179–181	ir,pmr,uv,ms	673
CH_2CH_2	4-NO_2	H	64–65	ir,pmr,uv,ms	673
CH=CH	3-NO_2	COMe	160–162	ir,pmr,uv,ms	524
CH_2CH_2	3-NO_2	COMe	157–158		679–683
CH_2CH_2	3-NO_2	COEt	165–166		679,680
CH_2CH_2	3-NO_2	COC_5H_{11}-*i*	115–116		679,680
CH_2CH_2	3-NO_2	COPh	178–180		679,680
CH_2CH_2	2-CH=$CHNO_2$	Me	175–180		617

Reduction of **1516a** to the corresponding 5-aminoiminobibenzyl (**1516d**) has been carried out. Although it was originally reported that all attempts to reduce **1516a** led only to iminobibenzyl (678), several reducing agents have since been found to work with varying degrees of success. The best of these is lithium aluminum hydride in ether (674–676,677). Yields of 75–90% were realized, and there were only trace amounts of cleavage. Sodium aluminum hydride was likewise effective, but sodium *bis*(2-methoxyethoxy)aluminum hydride failed (677). Zinc in acetic acid and aluminum and

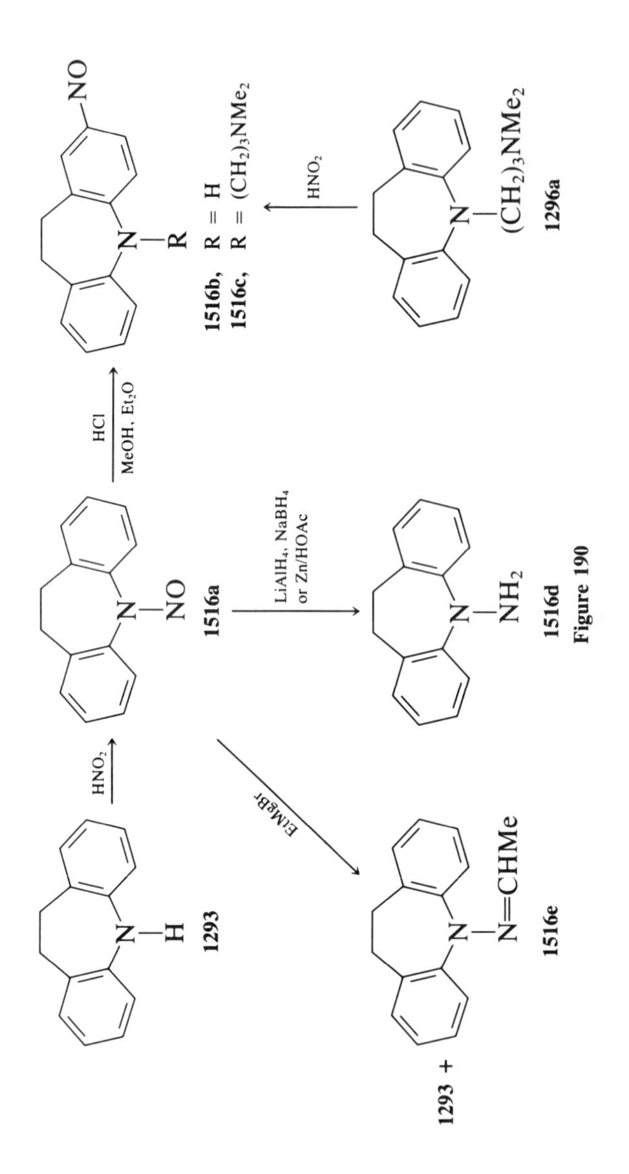
HNO$_2$
1293
1516a
HCl
MeOH, Et$_2$O
NO
1516b, R = H
1516c, R = (CH$_2$)$_3$NMe$_2$
HNO$_2$
(CH$_2$)$_3$NMe$_2$
1296a
LiAlH$_4$, NaBH$_4$
or Zn/HOAc
NH$_2$
1516d
EtMgBr
1293 +
N=CHMe
1516e

Figure 190

iodine in ethanol–tetrahydrofuran both gave erratic results (678), but the former method has been used successfully to generate **1516d** *in situ* (691–694,863,866). Catalytic reductions led exclusively to denitrosation (677).

In accord with the ease with which it was cleaved under reductive conditions, **1516a** readily nitrosated other amines. For example, **1516a** and aniline interacted in aqueous HCl to give **1293** and phenyldiazonium chloride (678). In the presence of ethyl magnesium bromide, **1516a** gave predominantly the cleavage product **1293,** but there was also obtained a 13% yield of 5-(ethylidenamino)iminobibenzyl (**1516e**) (677).

The thermal, photochemical, and acid-catalyzed decompositions of *N*-nitrosoiminostilbene (**1517a**) were investigated by Bremmer (673). The major products from these reactions (Eq. 214) were the acridines **1518,** iminostilbene (**1311b**), and 2-nitroiminostilbene (**1517b**). The exact composition of the product mixture was very dependent on the reaction conditions (Table 94) Aqueous acid gave primarily **1517b,** especially in the presence of oxygen. In contrast to the earlier Russian work with **1516a,** Bremmer found that **1517a** under anhydrous acidic conditions gave primarily the acridines **1518.** Thermal decomposition of **1517a** in several solvents generally gave either the acridines **1518** as the major products or the recovered starting material. Photochemical decomposition of **1517a** gave much iminostilbene, as well as **1517b,** and—in methanol—the acridines **1518.** Mechanisms for these transformations were proposed.

Nitration of the 5-acyliminobibenzyls (**1519**) to the corresponding 3-nitro derivatives **1520** has been described in the patent literature (Eq. 215) (679,680). When **1519** was treated with HNO_3 in acetic acid, a nitrate complex was isolated in 95% yield. On being added to cold sulfuric acid, the complex gave a 58% yield of **1520.** Reduction of **1520** to the corresponding amino

NO₂ — N — H — **1517b** (214a)

N — NO — **1517a** — O_2, H_3O^+

R — N — (214b)

1518a, R = H
1518b, R = CHO

TABLE 94. ACID–CATALYZED REACTIONS OF **1311b** AND **1517a**

		Products (%)				
Substrate	Conditions	**1559**	**1518b**	**1517b**	**1311b**	Others
1517a	HCl, MeOH, 20°C, 2 hr	59	Trace	3	—	HCO_2Me, MeONO, N_2O, NO_2^- NO_3^-
1517a	HCl, MeOH, 20°C, Argon, 2 hr	7	—	—	—	**1517a** and **1311b**
1517a	H_2SO_4, MeOH, 20°C, 24 hr	36	Trace	4	6	N_2O
1517a	HCl, EtOH, 20°C, 6 hr	39	Trace	4	—	HCO_2Et, EtONO, N_2O, and mixture of **1311b** and **1517a**
1517a	HCl, Me_2CO, 20°C, 24 hr	Trace	57	—	3	**1319a**
1517a	HCl (gas), EtOH–Et_2O (1:1)	14	—	—	24	—
1311b	5 *N* HCl, 100°C 1 hr	Trace	Trace	—	48	**1319a** (24%)
1311b	EtOH (100 ml), 0.15 *N* HCl (100 ml), 40–50°C, 30 hr	49	<1	—	<1	—

H_2SO_4, HNO_3, 0–5°C

N–COR **1519** → N–COR, NO_2 **1520** (215)

derivative is described in some detail in the following section. Irradiation of 5-nitrosoiminobibenzyl (**1516a**) in oxygen-enriched benzene gave the 2- and 4-nitro derivatives **1521a** and **1521b** in 28 and 21% yields, respectively (673).

i. DIBENZ[*b,f*]AZEPINES WITH BASIC NITROGEN SUBSTITUTION

i. Aminodibenz[*b,f*]azepines

(*a*) *10- or 11-Substitution* (*see Tables 95 and 96*). Most of the reported synthetic approaches to the 10-aminoiminostilbenes and 10-aminoiminobibenzyls have been via dibenz[*b,f*]azepin-10-ones or intermediates involved in their preparation. One general method of synthesis of the 10-aminoiminobibenzyls **1523** is to reduce the oximes **1522**, which are themselves readily

1516a $\xrightarrow{h\nu,\ C_6H_6,\ O_2}$ (216)

1521a, R_1 = H, R_2 = NO_2
1521b, R_1 = NO_2, R_2 = H

available via the corresponding azepinones **1340a** (Eq. 217) (548,549,557–561,759). Reductions of the oximes have been carried out with sodium or sodium amalgam in ethanol or *n*-butanol, generally at or near the reflux temperature.

$\xrightarrow[(R_1 = H,\ Me)]{Na,\ ROH,\ \Delta}$ (217)

1340a, X = O
1522, X = NOH

1523

Base-promoted cyclization of 2-(*N*-methyl-*N*-*o*-tolyl)aminobenzonitrile (**1524**) has been claimed in the patent literature to produce the tricyclic imine **1353** (Eq. 218) (681). A related cyclization was discussed earlier in Section 39.B.c.i (Eq. 171) (562–564) as an approach to the dibenz[*b,f*]azepin-10-ones, and tricyclic imine **1353** itself was cited (Eq. 166) (565). The base-promoted cyclization of **1524** was carried out in tetrahydrofuran using lithium dimethylamide. Quenching with ethanol, followed by saturation of the system with ammonia and reduction over Raney nickel, gave **1523** in 87% yield.

1524 $\xrightarrow[(3)\ NH_3\quad(4)\ H_2/\text{Raney Ni}]{(1)\ LiNMe_2,\ THF\quad(2)\ EtOH}$ [**1353**] ⟶ **1523** (218)

TABLE 95. 10-AMINO-10,11-DIHYDRODIBENZ[*b, f*]AZEPINES

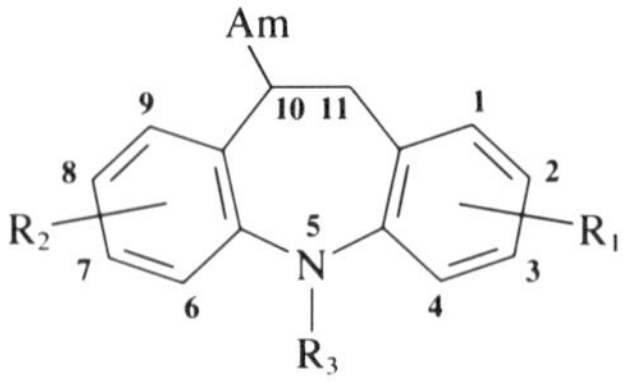

Am	R_1	R_2	R_3	mp (°C)	Salt	Refs.
NH_2	H	H	H	123		548,549
				194	Oxalate	557,759
$NHCO_2R$						
R = Me	H	H	H			548,549
R = Et	H	H	H	Oil		548,549
R = *n*-Pr	H	H	H			548,549
R = *i*-Pr	H	H	H	106		548,549
R = *n*-Bu	H	H	H			548,549
R = *i*-Bu	H	H	H			548,549
R = *n*-C_5H_{11}	H	H	H			548,549
R = *n*-C_6H_{13}	H	H	H			548,549
NH_2	H	H	COMe	94–98		499
N_3	H	H	COMe	131–132 (dec.)		499
NH_2	1-Cl	7-Cl	COMe	216–217		499
NH_2	3-Cl	7-Cl	COMe	108–112		499
N_3	3-Cl	7-Cl	COMe	99–100 (dec.)		499
NH_2	H	H	Me	90–92		
				270	HCl	681
				96		548,549,557
				98		642
NH_2	1-Cl	H	Me	73		559–561
NH_2	2-Cl	H	Me	102–103		559–561
NH_2	3-Cl	H	Me	93		559–561
NH_2	4-Cl	H	Me	275–278	HCl	559–561
NH_2	H	6-Cl	Me	108		559–561
NH_2	H	7-Cl	Me	247	Fumarate	559–561
NH_2	H	8-Cl	M	105		560
NH_2	H	H	Et	89		548,549
				90		557,759
				200	Fumarate	
NH_2	H	H	*n*-Pr	212–215	Fumarate	548,549
NH_2	H	H	CH_2Ph	205		557,759
NHCHO	H	H	Me	142		549
NHCHO	1-Cl	H	Me	174		559–561
NHCHO	2-Cl	H	Me	159–160		559–561
NHCHO	3-Cl	H	Me	148		559–561
NHCHO	4-Cl	H	Me	124		559–561
NHCHO	H	6-Cl	Me	180–182		559–561
NHCHO	H	7-Cl	Me	146		559–561
NHCHO	H	8-Cl	Me	188–190		560
NHCOMe	H	H	Me	187		557,759
NHTs	H	H	Me	158		557,759

TABLE 95. (*Continued*)

Am	R_1	R_2	R_3	mp (°C)	Salt	Refs.
$NHCO_2R$						
R = Me	H	H	Me	118–120		548,549
R = Et	H	H	Me	95		548,549
R = Et	H	H	Et	192–194/0.9[a]		548,549
R = Et	H	H	*n*-Pr			548,549
R = Et	2-Cl	H	Me	126–128		559–561
R = *n*-Pr	H	H	Me	70–72		548,549
R = *i*-Pr	H	H	Me	91		548,549
R = *n*-Bu	H	H	Me	77		548,549
R = *i*-Bu	H	H	Me	86		548,549
R = *n*-C_7H_{15}	H	H	Me	78		548,549
NHMe	H	H	Me	238–240	HCl	557,759
NHMe	1-Cl	H	Me	218	Fumarate	559–561
NHMe	2-Cl	H	Me	232–234	HCl	559–561
NHMe	3-Cl	H	Me	138–140	Maleate	559–561
NHMe	4-Cl	H	Me	240–242	HCl	559–561
NHMe	H	6-Cl	Me	115–116		559–561
NHMe	H	7-Cl	Me	184	Fumarate	559–561
NHMe	H	8-Cl	Me	90		560
NMeCHO	H	H	Me	95		557,759
NMeTs	H	H	Me	182–183		557,759
$NMeCO_2Et$	H	H	Me	172–175/0.1[a]		557,759
NMeCN	H	H	Me	85–87		557,759
NHEt	H	H	Me	135–138	Fumarate	557,759
				196	Methanesulfonate	557,759
NEtCHO	H	H	Me	Oil		557,759
NEtCOMe	H	H	Me	Oil		557,759
NMe_2	H	H	Me	65		557,759
NMeEt	H	H	Me	135	Mlleate	557,759
NEt_2	H	H	Me	70		557,759

[a]Boiling point (°C/torr).

Both *N*-acetyliminobibenzyls (**1309a**) and *N*-methyliminostilbene (**1337**) have been used as precursors of **1523.** In the former case (Eq. 219), *N*-bromosuccinimide treatment of **1309a** (to produce the bromide **1310a**) was

(1) NBS, CCl_4, $h\nu$; (2) NaN_3, H_2O, Et_2O; (3) $NaBH_4$, *i*-PrOH, Δ (219)

1309a → **1310a**, X = Br; **1525**, X = N_3; **1523a**, X = NH_2

TABLE 96. 10-AMINODIBENZ[*b,f*]AZEPINES

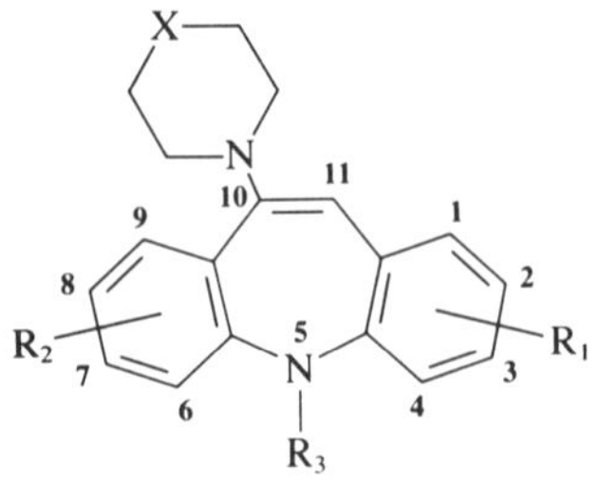

X	R_1	R_2	R_3	mp (°C)	Spectra	Refs.
CH_2	H	H	H	134–135	pmr	628
CH_2	H	H	Me	138–139	pmr	628
CH_2	H	H	COMe	148–149	pmr	628
CH_2	3-Cl	7-Cl	COMe	203–204		628
O	H	H	COMe	221–222		628
NMe	H	H	H	170–171		628
NMe	H	H	COMe	158–160	uv	687
				163–164		628
NMe	H	H	Me	165 (maleate)		688
NMe	H	H	Et	122–125	uv	687
				213 (maleate)		
NMe	2-Cl	H	Me	164–166		689
				200–204 (maleate)		
NMe	H	7-Cl	Me	—		688
NMe	3-Cl	7-Cl	H	176–177		628
NMe	3-Cl	7-Cl	Me	240–242		628
NPr-*n*	H	H	COMe	120–123		687
				168–170 (maleate)		
NCH_2Ph	H	H	COMe	180–182	uv	687
NCH_2CH_2OH	2-Cl	H	Me	130–135		689
				174–176 (maleate)		
NPh	H	H	COMe	207–208		628
NCO_2Et	H	H	COMe	156–157		628

followed by azide displacement to give the corresponding 10-azidoiminobibenzyl **1525** (499). Sodium borohydride in 2-propanol reduced **1525** to the amine **1523a** in acceptable yield. In the latter case (Eq. 220), reaction of **1337** and *N,N*-diethylaminoborane for 22 hr in refluxing toluene afforded a moderate yield of a compound that was characterized as the organoborane **1526a** (642). Treatment of a toluene solution of **1526a** with 2 *N* NaOH and either chloramine or *N*-chloromethylamine gave rise to the amines **1523b** and **1526b,** respectively. The yields in the second step were again unimpressive.

The primary amine **1523** can be converted easily to secondary or tertiary amines (548,549,557–561,759). The chemistry required for these transfor-

(1) Et_2NBH_2, PhMe
(2) 2 *N* NaOH
(3) ClNHR, H_2O

(220)

1337

1526a, X = BH_2
1526b, X = NHMe
1523b, X = NH_2

mations is unexceptional and not unique to this series of compounds. A summary is presented in Fig. 191.

Since 10-aminoiminostilbenes **1527** are enamines, it is not surprising that one of the methods by which they can be prepared is by the action of a secondary amine on a dibenz[*b,f*]azepin-10-one (Eq. 221). Standard enamine-forming conditions have been used, such as refluxing in toluene with a catalytic amount of *p*-toluenesulfonic acid and azeotropic removal of water (689). Titanium tetrachloride in benzene has also been used successfully to generate the enamines **1527** (688,689). In one instance, the ketone **1340a** and an excess of the secondary amine were heated directly in the presence of a catalytic amount of anhydrous hydrogen chloride (687).

An alternative method of synthesis of the enamines **1527** has been to start

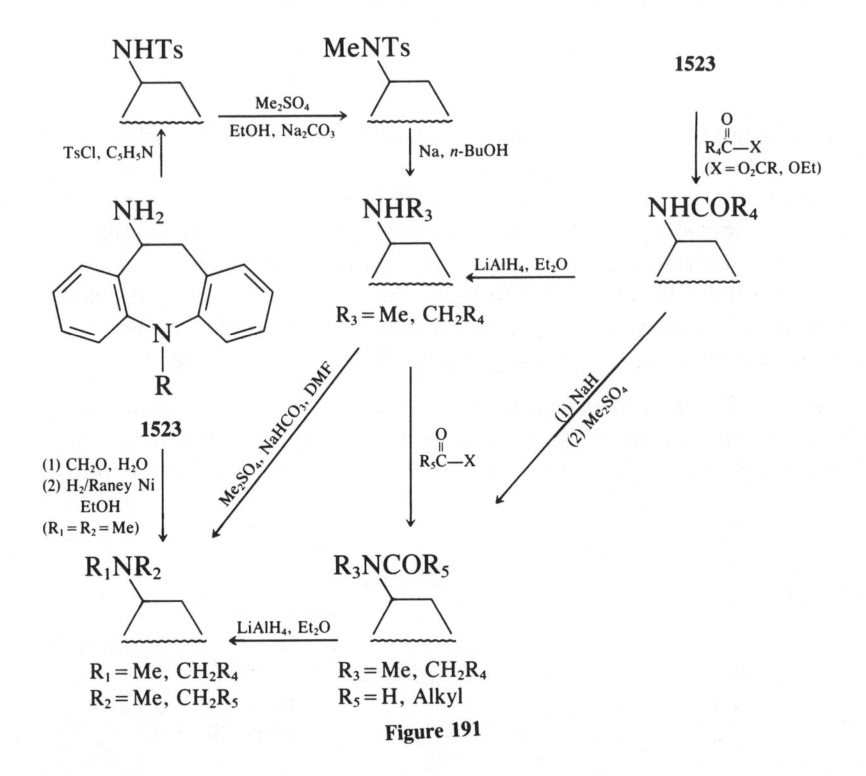

Figure 191

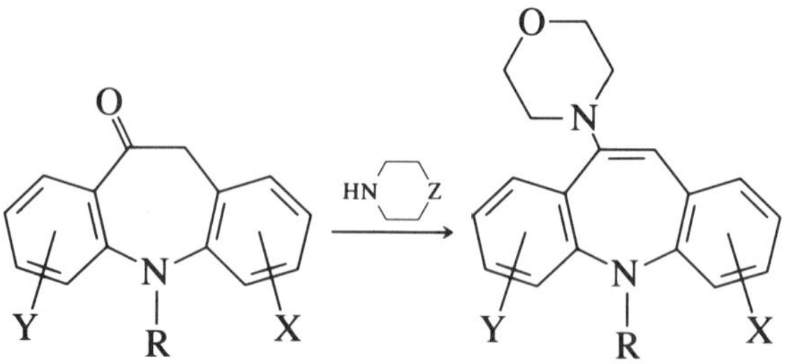

(221)

1340a → **1527**

with the vinyl bromides **1437** (Eq. 222) (628,687). Simply heating **1437** in the presence of a secondary amine gave no reaction. However, when the reaction was carried out in the presence of potassium *tert*-butoxide, the desired 5-acetyl-10-aminoiminostilbenes **1528a** were formed in fair to good yield. Evi-

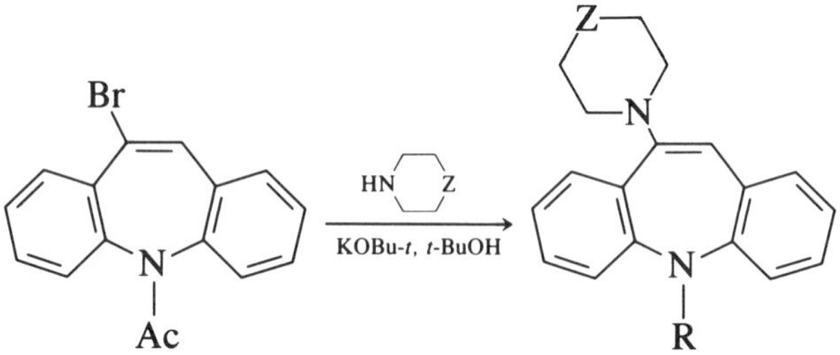

(222)

1437 → (KOBu-*t*, *t*-BuOH) **1528a**, R = COMe; **1528b**, R = H

dence for an elimination–addition mechanism was obtained when the vinyl bromide **1437** was treated with potassium butoxide in the presence of furan. Isolation of the Diels–Alder adduct **1439** (see Eq. 195) argued convincingly in favor of the hetaryne **1438.** Both alcoholic potassium hydroxide and lithium aluminum hydride cleaved the 5-acetyl group of **1528a** to produce the *N*-unsubstituted derivatives **1528b** (628). Some of these enamines displayed antimalarial activity. 10-Aminoiminobibenzyls are listed in Table 95 and 10-aminoiminostilbenes are listed in Table 96.

(*b*) *2-, 3-, and 5-Substitution* (*see Table 97*). A 1962 patent described the conversion of the readily available 3,5-diacetyliminobibenzyl (**1461a**) to 3-amino-5-acetyliminobibenzyl (**1400b**, Eq. 223) (622). The ketone **1461a** (see

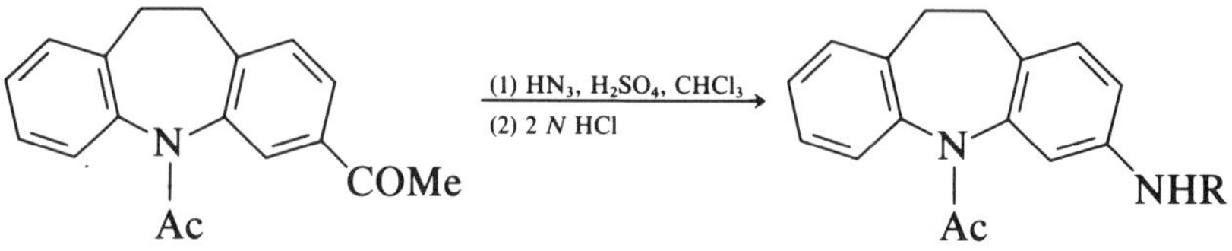

(223)

1461a → (1) HN_3, H_2SO_4, $CHCl_3$; (2) 2 *N* HCl → **1440a**, R = COMe; **1400b**, R = H

TABLE 97. 2-, 3-, AND 5-AMINODIBENZ[*b,f*]AZEPINES

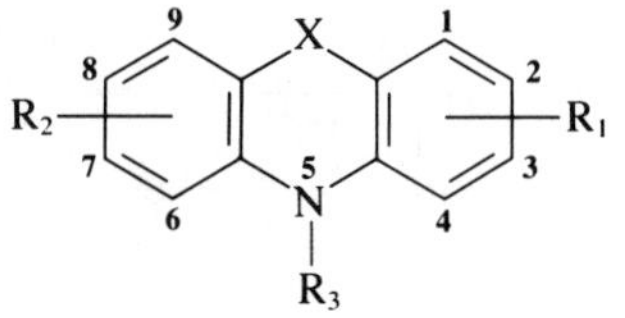

X	R_1	R_2	R_3	mp (°C)	Refs.
CH_2CH_2	2-NH_2	H	H	—	678
CH_2CH_2	3-NH_2	H	H	98–100	524,683
CH_2CH_2	3-NH_2	H	COMe	149–150	622
				151–153	679,680,683
				159	682
CH_2CH_2	3-NHCHO	H	H	121–123	524,683
CH_2CH_2	3-NHCOMe	H	H	170–172	524
CH_2CH_2	3-NHCOMe	H	COMe	220–221	622
CH_2CH_2	3-NHMe	H	H	87–89	524,683
CH_2CH_2	3-N(Me)COMe	H	H	182–183	678
CH_2CH_2	3-N(Me)COEt	H	H	162–163	678
CH_2CH_2	3-NHEt	H	H	142–144	524
CH_2CH_2	3-NHC_8H_{17}-*n*	H	H	134	524
CH_2CH_2	3-$NHCH_2Ph$	H	H	115–116	524
CH_2CH_2	3-NMe_2	H	H	103–105	682,683
CH_2CH_2	3-NMe_2	H	COMe	130–132	682,683
CH_2CH_2	3-NMeEt	H	H	60–62	678
CH_2CH_2	3-NMePr	H	H	170–175/0.01[a]	678
CH_2CH_2	3-$\overset{+}{N}Me_3I^-$	H	COMe	185–187	682,683
CH═CH	3-NH_2	H	H	217–218	524
CH═CH	3-NH_2	H	COMe	158–160	524
CH═CH	3-NHCHO	H	H	202–203	524
CH═CH	3-NHMe	H	H	182–184	524
CH═CH	3-NMe_2	H	H	170–172	524,682
CH_2CH_2	H	H	NH_2	52–53	676
				54–55	677
				140(HCl)	674,677
CH_2CH_2	2-Me	8-Me	NH_2	131–133	675
CH_2CH_2	H	H	N=CHMe	65	677,686
				68	
CH_2CH_2	H	H	N=CMe_2	88–89	677
CH_2CH_2	H	H	N=$CMeCO_2H$	113–114	677
CH_2CH_2	H	H	N=(cyclohexylidene) 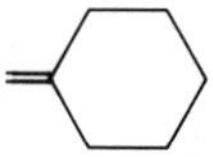	100	677
CH_2CH_2	H	H	N=CH–(2,5-furandiyl)–NO_2 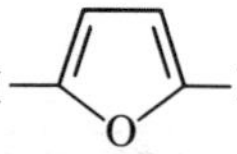	145–147	676
CH_2CH_2	H	H	N=CHC_6H_4OMe-*p*	143–144	676
CH_2CH_2	H	H	N=$CHC_6H_3(OCH_2O)$-3,4	127–129	676
CH_2CH_2	H	H	N=$CHC_6H_2(OMe)_3$-3,4,5	150–151	676
CH_2CH_2	H	H	N=$CHC_6H_3Cl_2$-2,4	107–109	676

TABLE 97. (*Continued*)

X	R_1	R_2	R_3	mp (°C)	Refs.
CH_2CH_2	H	H	N=(ring)N—Me	93–95	674
CH_2CH_2	H	H	NHCOMe	268	677
CH_2CH_2	H	H	$NHCOC_6H_4Cl$-*p*	281–282	676
CH_2CH_2	H	H	$NHCOC_6H_3Cl_2$-3,4	271–272	676
CH_2CH_2	H	H	$NHCOC_6H_2(OMe)_3$-3,4,5	150–151	676
CH_2CH_2	H	H	$NHCOC_6H_4(SO_2F)$-*m*	213–215 (dec.)	676

[a]Boiling point (°C/torr).

Section 39.C.d) was converted to the 3-acetamido derivative **1400a** by the action of hydrazoic acid in a classical Schmidt reaction. Hydrolysis of **1400a** with aqueous HCl gave the desired amine **1400b**.

Subsequently, the preparation of **1400b** was achieved by reduction of 5-acetyl-3-nitroiminobibenzyl (**1520**, see Section 39.C.h) with iron in acetic acid (679,680). Catalytic reduction of **1520** in the presence of formaldehyde (Fig. 192) gave the tertiary amine **1529a.** Treatment of **1400b** with methyl iodide in acetone afforded the quaternary ammonium iodide **1529b.** When either **1529a** or **1529b** was treated with potassium hydroxide in diethylene glycol monomethyl ether, the same tertiary amine **1529c** was produced (682,683).

Reduction of 5-acetyl-3-nitroiminostilbene (**1529d**) with iron in acetic acid (Fig. 193) produced the corresponding 5-acetyl-3-aminoiminostilbene (**1529e**).

Figure 192

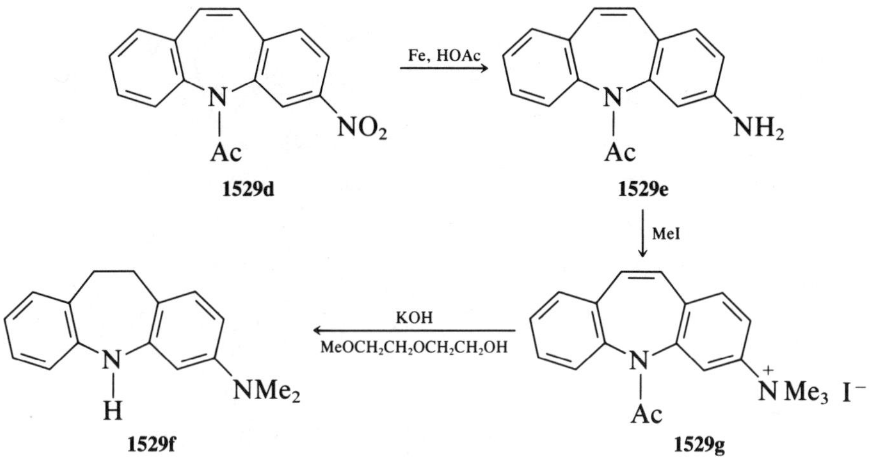

1529d → (Fe, HOAc) → **1529e** → (MeI) → **1529g** → (KOH, $MeOCH_2CH_2OCH_2CH_2OH$) → **1529f**

Figure 193

Conversion of **1529e** to the dimethylamino analog **1529f** was accomplished via the quaternary salt **1529g** (682).

The conversion of **1400b** to several secondary and tertiary derivatives by classical amine chemistry (Fig. 191) has been described in the literature (524,682,683), and the interested reader is referred to Table 97 for specific examples.

Reduction of 2-nitrosoiminobibenzyl (**1516b**) has been reported in the Russian literature to give an easily air-oxidized product. While such chemical behavior would be consistent with the existence of 2-aminoiminobibenzyl (**1530**), no other data were supplied (678).

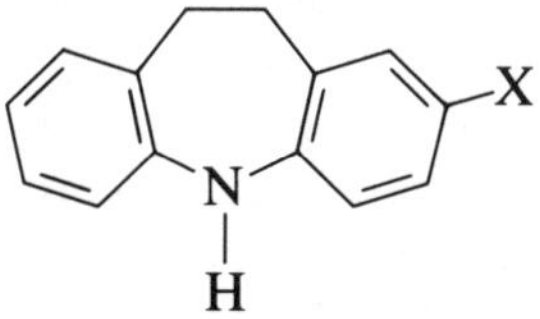

1516b, X = NO
1530, X = NH_2

The principal route to the 5-aminoiminobibenzyls (**1531a**) has been through reduction of the corresponding 5-nitroso derivatives (**1531b**), as discussed in Section 39.C.h. An alternative mode of access to **1531a** involves treatment of the sodium salts of the iminobibenzyls **1531c** with chloramine (Eq. 224) (677,690).

(224)

ii. AMINOALKYLDIBENZ[*b,f*]AZEPINES

(a) 10-Substitution (see Tables 98 and 99). Three approaches to the 10-aminoalkyliminostilbenes have been described. The first reported members of this class were the 10-(α-aminoalkyl)iminostilbenes **1532a** (556,577,578,631), which were produced by the reaction of secondary amines with the 10-(α-bromoalkyl)iminostilbene **1409b** (see Section 39.C.b.iii). Conversion of the tertiary amines to the corresponding secondary amines **1532b** was achieved (Fig. 194) by reaction with ethyl chloroformate and hydrolysis of the resulting carbamates **1532c**.

A second preparative method has been via lithium aluminum hydride reduction of the 10-iminostilbenylalkanoic amides **1533a** and **1533b** (577,578,625) to give the expected tertiary amines **1533c** and **1533d,** respectively (Eq. 225). The action of ethyl chloroformate on **1533c** has been used to prepare the secondary amine **1533e** by way of the carbamate **1533f** (625). The preparation of the parent alkanoic acids **1497b** and **1498c** was described in Section 39.C.f.

The last approach to the synthesis of 10-(β-dialkylaminoethyl)iminostilbenes **1537** involves ring expansion of the 9-hydroxymethylacridanes **1536** (see Section 39.B.b.iii for a complete discussion of the ring expansion of 9-hydroxymethylacridanes) (534,535,696). Dimethyl sulfate was used to convert 9-carbomethoxyacridine (**1534**) to a quaternary salt (Fig. 195), which was in turn reduced with sodium borohydride to the corresponding acridane **1535a.** Potassium metal in toluene or potassium *tert*-butoxide in a mixture of tetrahydrofuran and hexamethylphosphoric triamide (HMPT) converted **1535a** to its anion, which on alkylation with β-dimethylaminoethyl chloride gave **1535b.** Lithium aluminum hydride reduction of **1535b** afforded **1536,** which underwent ring enlargement to **1537** in about 70% yield on heating in xylene in the presence of P_2O_5. The amines were patented as antidepressants. Compounds discussed in this section are listed in Table 98.

TABLE 98. 10-AMINOALKYL-10,11-DIHYDRODIBENZ[*b, f*]AZEPINES AND AZEPIN-11-ONES

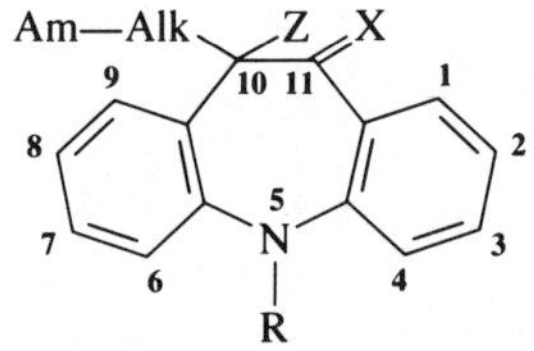

Alk	Am	X	Z	R	mp (°C) or bp (°C/torr)	Refs.
CH_2CH_2	NMe_2	H_2	H	H	97	695
CH_2CH_2	NMe_2	H_2	H	Me	165/0.005	695
CH_2CH_2	NMe_2	H_2	H	CH_2Ph	200/0.02	695
CH_2CH_2	NMe_2	O	H	Me	116–117	695
CH_2CH_2	Pyrrolidino	H_2	H	H	186/0.005; 160 (HCl)	695
CH_2CH_2	Pyrrolidino	H_2	H	Me	187/0.07	695
CH_2CH_2	NMe_2	H_2	CN	Et	243–246 (HCl)	606,607
CH_2CH_2	NMe_2	H_2	$CONH_2$	Et	148.5–151	606,607
CH_2CH_2	NMe_2	H_2	COEt	Et	182–183 ($HCl\cdot H_2O$)	607
CH_2CH_2	NMe_2	H_2	COPh	Et	213–216 ($HCl\cdot H_2O$)	607
CH_2CH_2	NEt_2	H_2	CN	Et	211–214 (HCl)	606,607
CH_2CH_2	NEt_2	H_2	$CONH_2$	Et	95–97	606,607
CH_2CH_2	NEt_2	H_2	CH_2NH_2	Et	234–236 (2HBr)	606
CH_2CH_2	$N(Pr\text{-}i)_2$	H_2	CN	Et	230–233 (HCl)	606,607
CH_2CH_2	$N(Pr\text{-}i)_2$	H_2	$CONH_2$	Et	101.5–105.5 103.5–105.5	606 607
CH_2CH_2	$N(Pr\text{-}i)_2$	H_2	CH_2NH_2	Et	239–240 (2HBr)	606
CH_2CH_2	$NMeCH_2Ph$	H_2	$CONH_2$	Et	153–156 (oxalate)	607
$(CH_2)_3$	NHMe	H_2	H	H	186/0.05	547,577, 578,695
$(CH_2)_3$	NHMe	H_2	H	Me	177/0.03	695
$(CH_2)_3$	NMe_2	H_2	H	H	160/0.004 276–278 (fumarate)	547,577 578,695
$(CH_2)_3$	NMe_2	H_2	H	Me	172–176/0.008	547,577, 578,695
$(CH_2)_3$	NMe_2	H_2	H	COMe	187/0.015	547,577,578
$(CH_2)_3$	NMe_2	O	H	Me	236–238 (HCl)	547,577,578
$(CH_2)_3$	NMe_2	H_2	CN	Et	217–218.5 220.5–222 (HCl)	607 606
$(CH_2)_3$	NMe_2	H_2	$CONH_2$	Et	118–120 (oxalate)	607

10-Aminoalkyliminobibenzyls **1538a** have been prepared primarily by catalytic reduction of the corresponding 10-aminoalkyldibenz[*b, f*]azepine-11-ones **1538b**, (Fig. 196) (547,577,578,695). The catalytic deoxygenations were carried out in dioxane under fairly severe conditions (e.g., hydrogen over 10% (w/w) copper chromite at 150 atm and 180–200°C).

The precursors **1538b** were synthesized readily by alkylation of the sodium

TABLE 99. 10-AMINOALKYLDIBENZ[*b*, *f*]AZEPINES

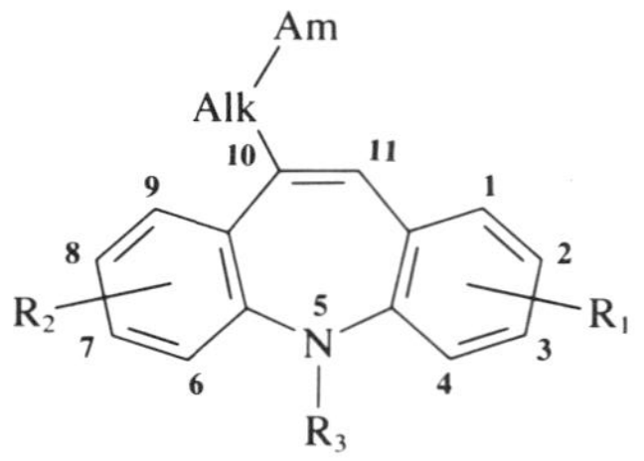

Alk	Am	R_1	R_2	R_3	mp (°C) or bp (°C/torr)	Spectra	Refs.
CH_2	NHMe	H	H	H	110		556,631
CH_2	NHMe	H	H	Me	147–149/0.09		547,577,
					175–177 (HCl)		578
CH_2	NHMe	2-SEt	H	Me	135 (fumarate)		625
CH_2	NHMe	2-SO_2Et	H	Me	Yellow resin	ir,uv	625
CH_2	NHEt	H	H	H	153–155/0.1		556,631
CH_2	NMeCO_2Et	2-SEt	H	Me	Oil		625
CH_2	NMeCO_2Et	2-SO_2Et	H	Me	Yellow resin	ir,uv	625
CH_2	NHEt	H	H	H	153–155/0.1		556,631
CH_2	NMeCO_2Et	2-SEt	H	Me	Oil		625
CH_2	NMeCO_2Et	2-SO_2Et	H	Me	Yellow resin	ir,uv	625
CH_2	NMe_2	H	H	H	127–128		547,556,577,578,631
					180 (fumarate)		535
CH_2	NMe_2	H	H	Me	140–144/0.01		547,577,578
					225–228 (HCl)		

CH_2	NMe_2	H	H	Et	150–152/0.04 247–249 (HCl)	547,577,578
CH_2	NMe_2	H	H	COMe	108–109	547,556,577, 578,631
CH_2	NMe_2	2-SEt	H	Me	176 (HCl)	625
CH_2	NMe_2	2-SO_2Et	H	Me	209	625
CH_2	NMe_2	3-Cl	H	H	124–125	556,631
CH_2	NMe_2	H	7-Cl	H	161–162	556,631
CH_2	NEt_2	H	H	Me	147–150/0.004 148–149 (fumarate)	547,577,578
CH_2	Pyrrolidino	H	H	H	114–116	556,631
CH_2	Pyrrolidino	H	H	Me	160–164/0.01 130–132 (HCl)	547,577, 578
CH_2	Pyrrolidino	H	H	Et	92;170–172 (HCl)	547,577,578
CH_2	Piperidino	H	H	H	238 (HCl)	556,631
CH_2	Piperidino	H	H	Me	172–175/0.01 171–174 (HCl)	547,577,578
CH_2	Piperazino	H	H	Me	184–189 (HCl)	547,577,578
CH_2	4-Methyl piperazino	H	H	H	199–204 (2HCl)	556,631
CH_2	4-Methyl piperazino	H	H	Me	224–229 (2HCl)	547,577,578
CH_2	4-(2-Hydroxyethyl) piperazino	H	H	H	177–179 (fumarate)	556,631
CH_2	4-(2-Hydroxyethyl) piperazino	H	H	Me	214–217 (2HCl)	547,577,578
CHMe	NMe_2	H	H	H	182–184/0.005	556,631
CHMe	NMe_2	H	H	Me	145–149/0.04 156–160 (HCl)	547,577, 578

(*continued*)

TABLE 99. (*Continued*)

Alk	Am	R_1	R_2	R_3	mp (°C) or bp (°C/torr)	Spectra	Refs.
CHMe	NEt_2	H	H	H	173–175/0.01		556,631
CHMe	Pyrrolidino	H	H	H	188–195/0.02		556,631
CHMe	Pyrrolidino	H	H	Me	168–172/0.3 193–196 (HCl)		547,577, 578
CHMe	4-(2-Hydroxyethyl) piperazino	H	H	H	140–144		556,631
CH_2CH_2	NHMe	H	H	Me	180 (HCl) 170 (pamoate)		535 696
CH_2CH_2	$NMeCO_2Et$	H	H	Me		ir	535
CH_2CH_2	NMe_2	H	H	H	200–204 (fumarate)		547,577,578
CH_2CH_2	NMe_2	H	H	Me	78	ir,uv	535
CH_2CH_2	NMe_2	2-Cl	H	Me	79	ir,uv	534,535
CH_2CH_2	NMe_2	H	7-Cl	H	Amorphous		535
CH_2CH_2	NMe_2	H	8-Cl	Me	82 150–164 (fumarate)	ir,uv	534,535
CH_2CH_2	NNe_2	3- C(H)[a]	H(7-Cl)[a]	Me	180 (fumarate)		534,535
OCH_2CH_2	NMe_2	H	H	Me	150.5–152 (maleate)		656,697
$(CH_2)_3$	NMe_2	H	H	H	162—163/0.001		577,578

[a]An isomeric mixture.

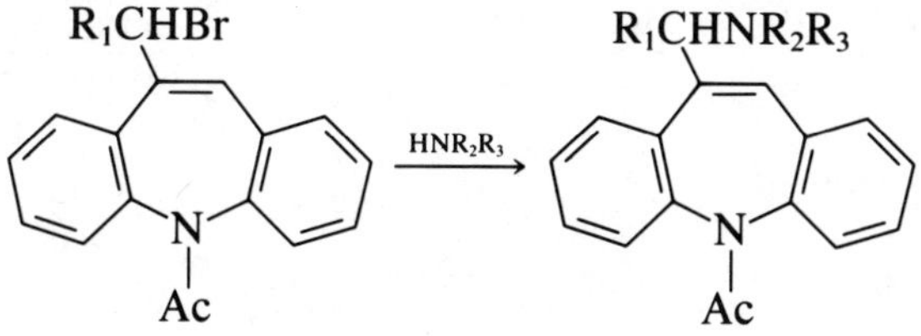

R_1CHBr → (HNR_2R_3) → $R_1CHNR_2R_3$

1409b, R_1 = H, Me

1532a, R_1 = H, Me

↓ EtO_2CCl

R_1CHNHR_2 ← (OH^-) ← $R_1CHNR_2CO_2Et$

1532b, R_1 = H, Me

1532c, R_1 = H, Me

Figure 194

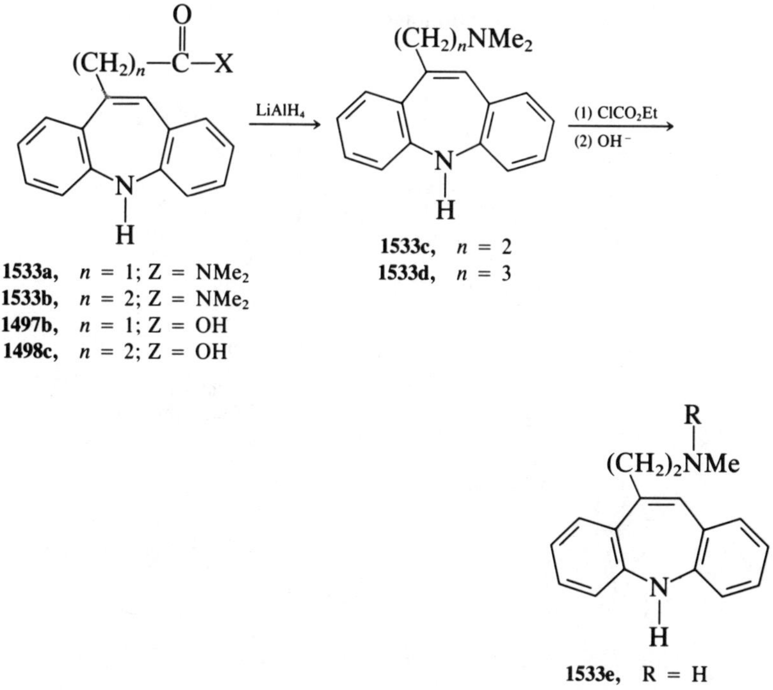

$(CH_2)_n$—C(=O)—X → ($LiAlH_4$) → $(CH_2)_nNMe_2$ → ((1) $ClCO_2Et$; (2) OH^-)

1533a, n = 1; Z = NMe_2
1533b, n = 2; Z = NMe_2
1497b, n = 1; Z = OH
1498c, n = 2; Z = OH

1533c, n = 2
1533d, n = 3

(225)

$(CH_2)_2N(R)Me$

1533e, R = H
1533f, R = CO_2Et

CO_2Me

1534

Me_2SO_4

CO_2Me

Me

$MeSO_4^-$

(1) $NaBH_4$
(2) K, PhMe
(3) $Cl(CH_2)_2NMe_2$

R CO_2Me

Me

1535a, R = H
1535b, R = $CH_2CH_2NMe_2$

$LiAlH_4$
THF

$HOCH_2$ $CH_2CH_2NMe_2$

Me

1536

P_2O_5, xylene
Δ

$CH_2CH_2NMe_2$

Me

1537

Figure 195

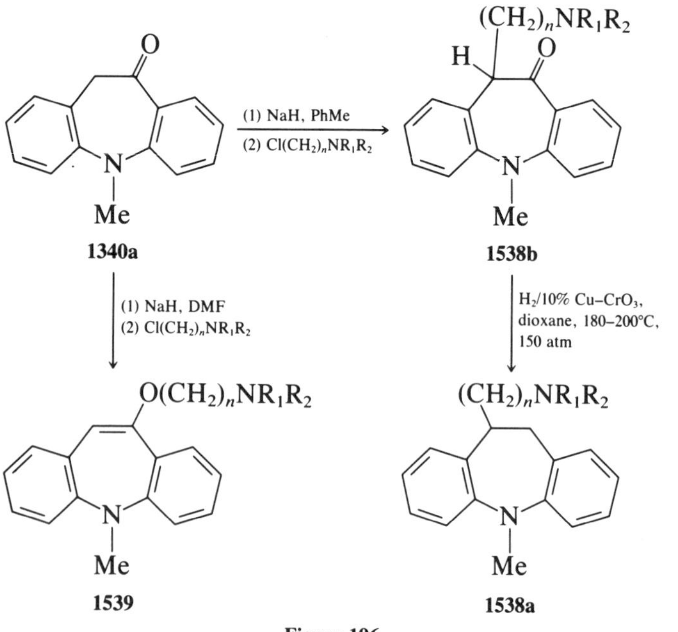

Figure 196

enolate of the azepinone **1340a** with the appropriate aminoalkyl halide. The enolate anion is an ambident nucleophile and, not surprisingly, the course of the alkylation has been found to be solvent dependent. For example, in toluene the sodium enolate gave exclusively *C*-alkylation (547,577,578), whereas in DMF the *O*-alkylated product **1539** was produced (656,697). Preference for alkylation at the more electronegative atom of an ambident nucleophile (i.e., oxygen) in a polar solvent is a recognized phenomenon (1043).

Elsewhere in this review (see Section 39.C.e), the preparation of several 10-aminoalkyl-10-cyanoiminobibenzyls **1475** by base-promoted alkylation of the corresponding 10-cyanoiminobibenzyl **1474** was described (606,607). Lithium aluminum hydride reduction of **1475** gave the diamines **1540** (Eq. 226), whereas hydrolysis of **1475** produced the amides **1477.** Several compounds in this series were claimed to possess antiarrhythmic properties. Table 99 should be consulted for a listing of compounds discussed in this section.

NC, R; N, Et

1475, R = $(CH_2)_nNR_1R_2$
1474, R = H

$LiAlH_4$ →

H_2NCH_2, $(CH_2)_nNR_1R_2$; N, Et

1540

(226a)

OH^- →

H_2NCO, $(CH_2)_nNR_1R_2$; N, Et

1477

(226b)

(b) 2- and 4-Substitution (see Table 100). Aromatic ring-substituted aminoalkyliminobibenzyls have been obtained by metal hydride and catalytic reduction of various nitrogen-containing functional groups. The nitroolefins **1395** were reduced with lithium aluminum hydride to give the amines **1541a** (Eq. 227). Catalytic reduction of the nitriles **1491a** afforded the same amines (617). Borane has been employed to reduce the amides **1541b** to the amines **1541c** (Eq. 228) (605,617).

(c) 5-Substitution. An enormous amount of work has been devoted to the synthesis, analysis, and biological evaluation of 5-aminoalkyldibenz-[*b,f*]azepines. Haefliger and Schindler first disclosed the preparation of several examples of this system in a 1951 patent (470). Although the 5-

TABLE 100. 2- AND 4-AMINOALKYL-10,11-DIHYDRODIBENZ[*b*, *f*]AZEPINES

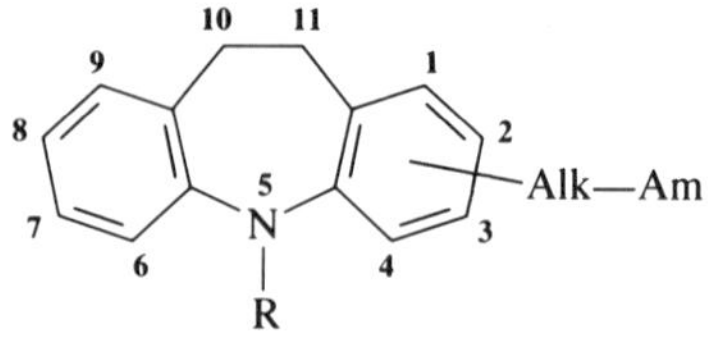

Alk	Am	R	mp (°C)	Refs.
2-CH_2CH_2	NH_2	H	175–180/0.001[a] 231–232 (fumarate)	617
2-CH_2CH_2	NH_2	Me	209–211 (HCl)	617
2-CH_2CH_2	NH_2	Et	172–176	617
2-CH_2CH_2	NH_2	CH_2Ph	93–94 (HCl)	617
2-CH_2CH_2	Phthalimido	Me	122–125	617
2-CH_2CH_2	NHMe	H	220–225 (HCl)	617
2-CH_2CH_2	NHMe	Me	172–174 (HCl)	617
2-CH_2CH_2	NMe_2	Me	202–204 (HCl·H_2O)	617
4-CH_2	$\overset{+}{N}HMe\,^{-}BH_3$	$(CH_2)_3Cl$	Oily solid	605

[a]Boiling point (°C/torr)

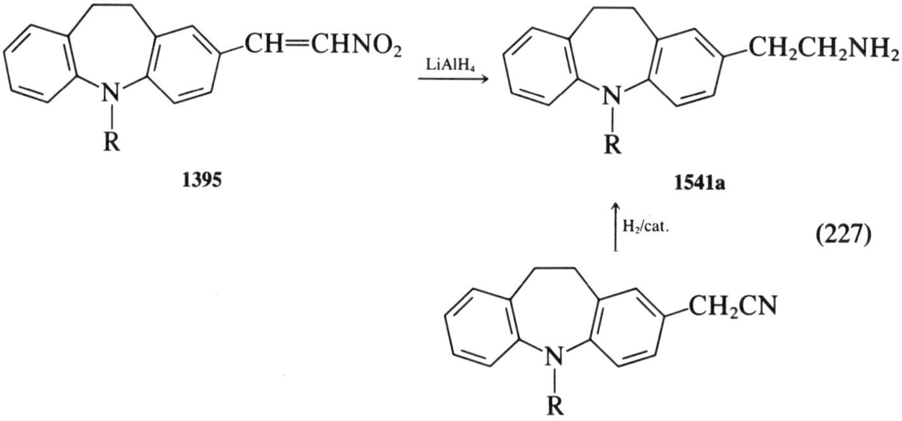

1395 → ($LiAlH_4$) → **1541a**; **1491a** → (H_2/cat.) → **1541a** (227)

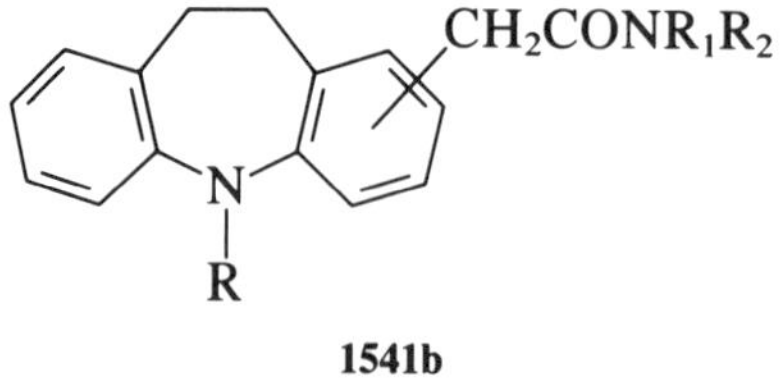

1541b → (BH_3 THF) → **1541c** (228)

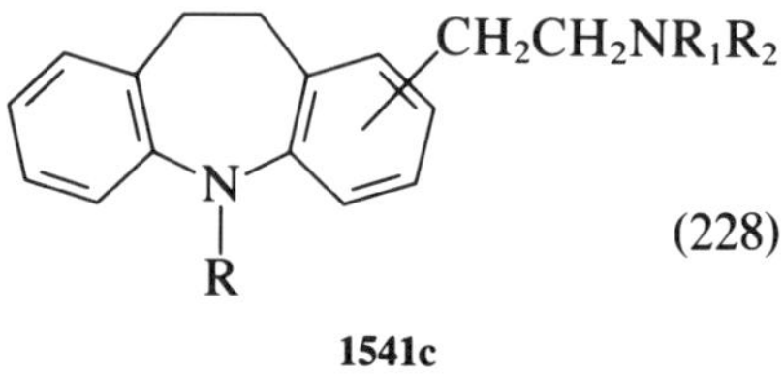

TABLE 101. 5-(AMINOMETHYL)DIBENZ[*b*, *f*]AZEPINES

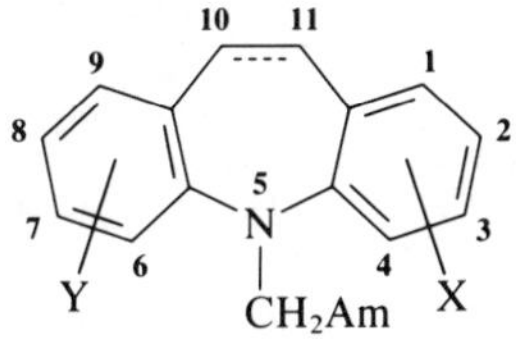

Am	Ring System[a]	X	Y	mp (°C) or bp (°C/torr)	Spectra	Refs.
1-Methylpyrrolidin-2-yl	IBB	H	H	145–147/0.01		703,745
1-Methylpiperidin-2-yl	IS	H	H	169–172/0.05		703,745
1-Methylpiperidin-2-yl	IBB	H	H	165–167/0.02		703,745
1-Methylpiperidin-2-yl	IBB	3-Cl	7-Cl	195–197/0.01		703
1-Methylpiperidin-3-yl	IS	H	H	166–169/0.01		703,745
1-Methylpiperidin-3-yl	IBB	H	H	157–160/0.01		703,745
1-Methylpiperidin-4-yl	IS	H	H	176–177/0.02		703,745
1-Methylpiperidin-4-yl	IBB	H	H	172–174/0.02		703,745
1-Methylpiperidin-4-yl	IBB	3-Cl	7-Cl	190–192/0.01		703,745
Quinuclidin-2-yl	IBB	H	H	123 204–205 (fumarate)		764
Quinuclidin-3-yl	IBB	H	H	135 196–198 ($MeSO_3H$)		764
	IS	H	H	148.5–150		722
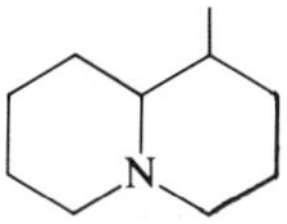	IBB	H	H	143–145		722
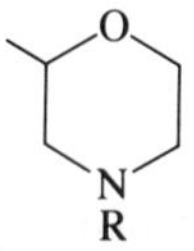 R						
R = H	IS	H	H	133–135		806
R = H	IBB	H	H	194–195 (oxalate)		806
R = H	IBB	3-Et	H	146–150 (oxalate)		806
R = Me	IS	H	H	207–209 (oxalate)		806
R = Me	IBB	H	H	210–211 (oxalate)		806
R = Me	IBB	3-Et	H	168–170 (oxalate)		806
R = Et	IS	H	H	99.5–100.5 (oxalate)		806
R = Et	IBB	H	H	210–211 (oxalate)		806
R = *i*-Pr	IS	H	H	146–147 (oxalate)		806
R = *i*-Pr	IBB	H	H	203.5–204.5 (oxalate)		806
R = CH_2CF_3	IS	H	H	99–100		806
R = CH_2Pr-*i*	IBB	H	H	137–138 (oxalate)		806
R = Cyclopropylmethyl	IS	H	H	153.5–154.5		806
R = Cyclopropylmethyl	IBB	H	H	161.5–162.5 (oxalate)		806
R = Cyclopropylmethyl	IBB	H	H	161.5–162.5 (oxalate)		806

TABLE 101. (*Continued*)

Am	Ring System[a]	X	Y	mp (°C) or bp (°C/torr)	Spectra	Refs.
R = $CH_2CH{=}CMe_2$	IS	H	H	99–100		806
R = $CH_2CH{=}CMe_2$	IBB	H	H	94–95 (oxalate)		806
R = CH_2Ph	IS	H	H	122–123		806
R = CH_2Ph	IBB	H	H	139–140 (oxalate)		806
R = $CH_2C{\equiv}CH$	IS	H	H	145–147		806
R = $(CH_2)_2OH$	IS	H	H	169.5–170.5 (oxalate)		806
R = $(CH_2)_2OH$	IBB	H	H	186–188 (oxalate)		806
R = $(CH_2)_2OMe$	IS	H	H		ir	806
R = $(CH_2)_2OMe$	IBB	H	H	148–149 (oxalate)		806
R = $(CH_2)_3OH$	IBB	H	H		ir	806
R = $(CH_2)_3OMe$	IBB	H	H	128.5–129.5 (oxalate)		806
R = $(CH_2)_3OPr$-*i*	IBB	H	H	133–135 (oxalate)		806
R = 2-(N, NH ring: 4,5-dihydroimidazol-2-yl)	IBB	H	H	268–270 (HCl) 270–272 (HCl)		806 750
R = 2-(N, NH ring fused to cyclohexane)	IBB	H	H	192–193 298–300 (HCl)		806

[a]IS = Iminostilbene or dibenz[*b*, *f*]azepine; IBB = iminobibenzyl or 10,11-dihydrodibenz-[*b*, *f*]azepine.

aminoalkyliminobibenzyls were initially claimed to possess antiallergic properties, they were soon recognized to be effective in the treatment of depression. Imipramine (**1296a**) was developed as a benchmark antidepressant, which for years has been the preeminent agent for the chemotherapeutic treatment of clinical depression.

The development of imipramine (Tofranil) triggered a dedicated effort to uncover new analogs that would equal or surpass the parent drug in terms of potency or spectrum of activity while showing fewer undesirable side effects. Several such analogs were introduced in Section 39.A. Amitriptyline and clomipramine may be cited as especially important examples of compounds arising from the search for novel imipramine analogs. Although this search began to abate somewhat with the arrival of newer-generation antidepressants, work in this area is by no means at an end.

While hundreds, or perhaps even thousands, of imipramine analogs have been prepared, the scope of the synthetic chemistry has been rather limited, and most of the work has been quite classical. Using imipramine itself as a model, most of the principal routes to 5-aminoalkyldibenz[*b*, *f*]azepines have been brought together in Eqs. 229–230 and Fig. 197. Because there have been so many examples of each of these routes, no attempt has been made to discuss them exhaustively in this review. Tables 101–117 should be consulted for specific examples.

The most direct route to imipramine analogs (Eq. 229) is one that can be followed only when the requisite dialkylaminoalkyl halide precursors are readily available. Alkylation of dibenz[*b*,*f*]azepines has almost always been carried out by treatment of **1542** with sodium amide in benzene or toluene (occasionally xylene) and addition of the appropriate dialkylaminoalkyl halide (470,506,511,514,517,531). In one instance phenyllithium was employed

TABLE 102. 5-(β-AMINOETHYL)DIBENZ[*b*,*f*]AZEPINES

Z; 10, 11; 9, 1; 8, 2; 5 N; 7, 3; 6, 4; CH_2CH_2Am

Am	X	Y	Z	mp (°C) or bp (°C/torr)	Refs.
NMe_2	H	H	H	95–96	612,711
NMe_2	H	H	10-MeO	90	544,545,714
NEt_2	H	H	H	53	523,612
NEt_2	3-Cl	7-Cl	H	95–96	523
Pyrrolidino	H	H	H	74–74.5 195–196 (HCl)	523,612,709
Pyrrolidino	H	H	10,11-Me_2	152–155/0.01 198–199 (HCl)	812
Pyrrolidino	H	H	10-EtO	162/0.001	544,545,714
1-Methylpiperidin-2-yl	H	H	H	170–172/0.004	703,745
1-Methylpiperidin-2-yl	3-SMe	H	H	200/0.001	669,670
N (ring) $C_6H_4CF_3$-*m*, OH	H	H	H	191 (HCl)	761
N (ring) $N(CH_2)_2OH$	H	H	H	159	519
N (ring) $N(CH_2)_2OCOPh$	H	H	H	203–206 (HCl)	748
N (ring) $N(CH_2)_2O(CH_2)_2OH$	H	H	H	145 (2HCl)	519
N (bicyclic ring) N	H	H	H	221–223 (2HBr,H_2O)	773,776

TABLE 103. 5-(β-AMINOETHYL)-10,11-DIHYDRODIBENZ[*b*, *f*]AZEPINES

Am	X	Y	Z	mp (°C) or bp (°C/torr)	Refs.
NH_2	H	H	H	256 (HCl)	721
NH_2	H	H	10-Me	—	595
NHMe	H	H	H	167–169/0.4[a] 214–216 (HCl)	758
NHMe	H	H	10-Me	155–158/0.005	595
NHMe	3-NHMe	H	H	223–225 (2HCl)	683
NHMe	3-SO_2NMe_2	H	H	256 (HCl)	707,784
$N(CO_2Et)Me$	H	H	H	—	758
NHEt	H	H	H	252–254 (HCl)	758
$N(CO_2Et)Et$	H	H	H	192–198/0.3	758
$NHCHMeCH_2Ph$	3-SO_2NMe_2	H	H	215–220/0.05 248 (HCl) 159–160 ($MeSO_3H$)	726
Piperidin-2-yl	H	H	H	200/0.05 221–224 (fumarate)	745
Piperidin-2-yl	3-SO_2N(piperidino)	H	H	165–167 ($HCl \cdot H_2O$)	573
1-Carbethoxypiperidin-2-yl	H	H	H	Oil	745
Piperidin-4-yl (N-ring, 4-R)					
R = H	H	H	H	180–181/0.1 278–280 (HCl)	470,587,699

R = H	3-Cl	7-Cl	H	191–193/0.007	
				238–240 (HCl)	700
R = H	3-COMe	H	H	206 (HCl)	657
R = H	$3-OCH_2Ph$	H	H	208–209 (HCl)	650
R = OH	H	H	H	92; 250 (HCl)	594
R = OCONHMe	H	H	H	152 (HCl)	594
R = $OCONMe_2$	H	H	H	174 (HCl)	594
R = $(CH_2)_2OH$	H	H	H	174 (HCl)	594
R = $(CH_2)_2OCONHMe$	H	H	H	~90 (HCl)	594
3-(3-Hydroxypropyl)piperidino	H	H	H	208.5–209 (HCl)	597
[piperidine ring: N; Ph; OR]					
R = H	H	H	H	135–136	749
R = CO_2Et	H	H	H	185–187 (HBr)	749
R = 1-Methylpiperidin-2-yl	H	H	H	178–181	703
R = 1-Methylpiperidin-2-yl	3-Cl	H	H	195/0.1	573,717,745
R = 1-Methylpiperidin-2-yl	3-Cl	7-Cl	H	195–197/0.01	745
R = 1-Methylpiperidin-2-yl	3-COMe	H	H	195/0.06	657,728
1-Methylpiperidin-2-yl	$3-NMe_2$	H	H	190–196/0.01	
				64–66 (HCl)	683
1-Methylpiperidin-2-yl	3-SMe	H	H	183 (HCl)	669,670
1-Methylpiperidin-2-yl	3-SEt	H	H	200/0.001	669,670
1-Methylpiperidin-2-yl	3-(=O)Me	H	H	—	671
1-Methylpiperidin-2-yl	$3-SO_2NMe_2$	H	H	125	616,667,707
1-Methylpiperidin-2-yl	$3-SO_2N$ [piperidine ring]	H	H	165–167 (HCl)	665,667,707
1-Methylpiperidin-2-yl	$3-SO_2N$ [morpholine ring, O]	H	H	165–167 (HCl)	665,667,707
1-Methylpiperidin-2-yl		H	H	211–212.5 ($HCl \cdot H_2O$)	665,667,707
1-Methylpiperidin-2-yl		H	10-Et	198/0.02	812
Quinuclidin-2-yl	H	H	H	96	764
1-Methylpiperidin-4-yl	H	H	H	125	785

(continued)

TABLE 103. *(Continued)*

Am	X	Y	Z	mp (°C) or bp (°C/torr)	Refs.
Morpholino	H	H	H	196–198/0.08	
				220–222 (HCl)	470,587,699
Morpholino	3-Cl	H	H	242 (HCl)	573,717
Morpholino	H	H	10-Me	175/0.005	
				216 (HCl)	812
3-Oxopiperazino	H	H	H	246–249 (HCl)	708
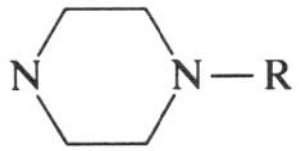 N(piperazine ring)N—R					
R = H	H	H	H	220–230/0.8	590
				220 (HCl)	
				220–222 (HCl)	748
R = H	3-Cl	7-Cl	H	162–164 (dimaleate)	590
R = Et	H	H	H	204–206/0.3	701
				239–241 (2HCl)	
R = $CONH_2$	H	H	H	157–159	590
R = $CONH_2$	3-Cl	7-Cl	H	198–200	590
R = CONHMe	H	H	H	145–148	590
R = $(CH_2)_2OH$	H	H	H	148	592
R = $(CH_2)_2OH$	3-Cl	7-Cl	H	250–252 (dimesylate)	592
R = $(CH_2)_2OCOPh$	H	H	H	155–158 (dimaleate)	748
R = $(CH_2)_2OCOC_6H_3(OCH_2O)$-3,4	H	H	H	230–232 (2HCl)	748
R = $(CH_2)_2OCOC_6H_2(OMe)_3$-3,4,5	H	H	H	230–233 (2HCl)	748
R = $(CH_2)_2O(CH_2)_2OH$	H	H	H	210 (2HCl)	592
R = $(CH_2)_2O(CH_2)_2OH$	3-Cl	7-Cl	H	220 (2HCl)	592
R 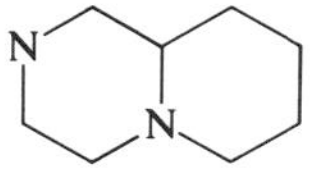 = (octahydro-2H-pyrido[1,2-a]pyrazin-2-yl, N ... N ring structure)	H	H	H	204.5–206 ($2HCl \cdot 2H_2O$)	772,773,776

NMe_2	H	H	H	138–140/0.08	
				216–217 (HCl)	
				163–166/0.6[b]	470,587,699
				217–219 (HCl)	758
NMe_2	3-Et	H	H	145–147/0.005	573
NMe_2	3-NHMe	H	H	151–160/0.007	683
NMe_2	3-NMe_2	H	H	214–216 (2HCl)	683
NNe_2	3-SPr-*i*	H	H	182 (HCl)	669,670
NMe_2	3-SO_2NMe_2	H	H	210 (HCl)	616,667,707, 744,785
NMe_2	H	H	10-Me	133–135/0.001	
				193–195 (HCl)	812
NMe_2	H	H	10-Et	142/0.005	812
NMe_2	H	H	10-OH	86–87	646
NMeEt	H	H	H	179–182/0.6	
				209–210 (HCl)	758
NEt_2	H	H	H	150–152/0.15	
				192–193 (HCl)	470,587,699
				170–178/0.5[c]	758
NEt_2	3-SPr-*i*	H	H	147 (HCl)	668,670
NMe($CHMeCH_2Ph$)	H	H	H	205–208/0.02	
				247–248 (HCl)	726
N(Ph)Pr-*i*	3-NMe_2	H	H	211–213 (2HCl)	683
Pyrrolidino	H	H	H	180/0.4	
				248–250 (HCl)	587
Pyrrolidino	3-Et	H	H	164/0.005	
				156–158 (HCl)	573
Pyrrolidino	3-SPr-*i*	H	H	170 (HCl)	669
Pyrrolidino	H	H	10-Me	155–157/0.001	
				217–219 (HCl)	812
Pyrrolidino	H	H	10,11-Me_2	155/0.01	812

[a] $[n]_D^{20} = 1.6109$.
[b] $[n]_D^{20} = 1.5961$.
[c] $[n]_D^{20} = 1.5809$.

TABLE 104. 5-(γ-AMINOPROPYL)DIBENZ[*b*, *f*]AZEPINES

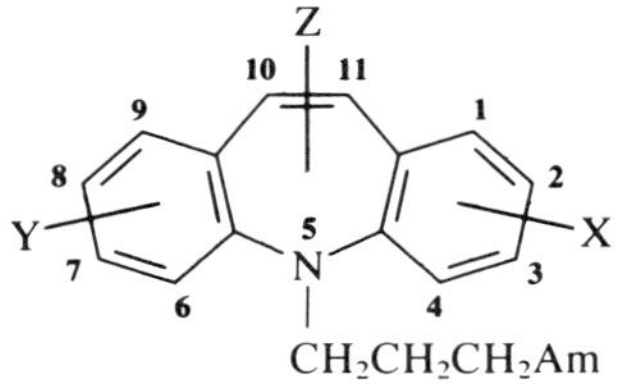

Am	X	Y	Z	mp (°C) or bp (°C/torr)	Salt	Other Data	Refs.
NHMe	H	H	10-Me	160/0.02			
				169–170	HCl		595,596,737
NHMe	H	H	10,11-Me_2	182–184/0.005			595,596,737
NHMe	3-NMe_2	H	H	164–166	Oxalate		683
NHMe	3-SO_2NMe_2	H	H	156–157	Oxalate		783
$NH(CH_2)_2NMe_2$	H	H	H	177/0.01			
				225–226	2HCl		727,753
$NH(CH_2)_2NEt_2$	H	H	H	192/0.01			
				213–215	2HCl		727,753
$NH(CH_2)_3NMe_2$	H	H	H	177/0.006			
				175–176	2HCl		
				151–153/0.02			
				56–57			727,753
				176–177	HCl		523,612,614
NMe_2	H	H	H	155–165/0.3			
				54–56			
				170–172	HCl		777
				146–148	Maleate		531
				148–149.5	Maleate		506,511
NMe_2	2-Cl	H	H	125–128	Maleate		531
NMe_2	2-F	H	H	133–134	Maleate		531

NMe_2	3-Cl	H	H	53			573
				124.5–125.5	Maleate		511,514
NMe_2	3-Cl	7-Cl	H	76–77			523,612
NMe_2	2-MeO	7-Cl	H	82–85		ir,uv	531
				88			517
NMe_2	3-NMe_2	H	H	174–176	2HCl		683
NMe_2	3-SMe	H	H	123	Oxalate		669,670
NMe_2	3-SEt	H	H	152/0.004			573
NMe_2	3-SO_2NMe_2	H	H	108			666
NMe_2	H	H	10-Me	159–160/0.05			812
				41–42			733
NMe_2	H	H	10-Et	161–162/0.05			812
NMe_2	H	H	10-CH_2Ph	185–190/0.005			812
NMe_2	H	H	10,11-Me_2	78–79			615,733,812
NMe_2	H	H	10-Et-11-Me	165–168/0.01			812
NMe_2	H	H	10-CH_2Ph-11-Me	210/0.003			812
NMe_2	H	H	10-Me	170/0.001			544,545,714
NMe_2	H	H	10-CF_3O	90/80(?)		ir,pmr,uv $[n]_D^{20}$ = 1.5525	629
NMe_2	H	H	10-EtO	160–161/0.001			
				166–169	HCl		544,545,714
NMe_2	H	H	10-*n*-BuO	173/0.003			544,545,714
NMe_2	3-Cl	H	11-MeO	96			544,545,714
$NMe(CH_2)_2NMe_2$	H	H	H	170/0.015			727,753
$NCH_2Ph(CH_2)_2NMe_2$	H	H	H	215/0.08	2HCl		727,753
				155–160			
Piperidino	H	H	10-MeO	191–193/0.01			544,545,714
4-Oxopiperidino	H	H	H	87–88	HCl		705
				94			

(*continued*)

TABLE 104. *(Continued)*

Am R₁	Am R₂	X	Y	Z	mp (°C) or bp (°C/torr)	Salt	Other Data	Refs.
—O$(CH_2)_2$O—		H	H	H	102–103			705
					206–208	HCl		
OH	H	H	H	H	208–210	HCl		705
MeO	H	H	H	H	181–182	HCl		705
OH	Ph	H	H	H	230	HCl·EtOH		768
OH	Ph	3-Cl	H	H	120–140	HCl		761,768
MeO	Ph	H	H	H	188	HCl·0.5H_2O		761,768
OH	C_6H_4Me-p	H	H	H	158	HCl·*i*-PrOH		761,768
					130–140	HCl		768
OH	C_6H_4Cl-*p*	H	H	H	199	HCl·*i*-PrOH		761
OH	$C_6H_4CF_3$-*m*	H	H	H	141–143	HCl·EtOH		761,765,768
OH	$C_6H_4CF_3$-*m*	3-Cl	H	H	95–105	HCl		768
OH	C_6H_4OMe-*p*	H	H	H	176–177	HCl		761,771
OH	CH_2Ph	H	H	H	100	HCl·H_2O		761,768
OH	2-Thienyl	H	H	H	134–136	Maleate		792,804
COMe	Ph	H	H	H	194–195	HCl		761,768
COMe	Ph	3-Cl	H	H	75	HCl		768
CN	Ph	H	H	H	209	HCl·H_2O		761,768
CN	Ph	3-Cl	H	H	201–203	HCl		761,768
$CONH_2$	Ph	H	H	H	260	HCl		761,768
$CONH_2$	NMe_2	H	H	H	225	2HCl·0.5H_2O		761,768
$CONH_2$	NMe_2	3-Cl	H	H	140–145	Maleate		761,768
$CONH_2$	Piperidino	H	H	H	235	2HCl·0.5H_2O		761,768
$CONH_2$	Piperidino	3-Cl	H	H	182–185	Maleate		761,768

—SCH_2CONH—	H	H	H	246	HCl	768,797
—$SCH_2CON(C_6H_4Me$-*p*)—	H	H	H	223	HCl	768
—OCHMeCONH—	H	H	H	261–262	HCl	775,781,796
—OCONMeCO—	H	H	H	218–219 (dec.)		757,788,804
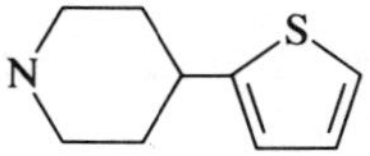 [structure: N-piperidin-4-yl–2-thienyl]	H	H	H	186–187	0.5 Fumarate $0.5H_2O$	792,793
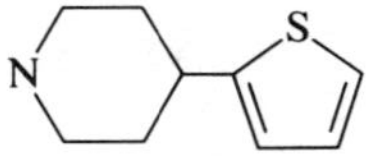 [structure: N-piperidin-4-yl–2-thienyl]	3-Cl	H	H	160–161	Fumarate	792
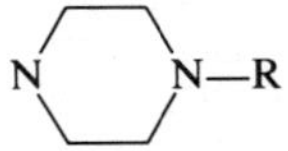 [structure: piperazine N—R]						
R = H	H	H	H	224–226	HCl	748
R = CO_2Pr-*i*	H	H	H	110–114	$HCl{\cdot}H_2O$	805
R = Me	H	H	H	86–88		591,613,706
				245–265	2HCl	511
R = Me	3-Cl	7-Cl	H	128–129		591,613,706
R = Me	3-SO_2NMe_2	H	H	197	Oxalate	666
R = Me	H	H	10-Me	224–245	2HCl	812
R = Me	H	H	10-Et	207–209	2HCl	812
R = Me	H	H	10-CH_2Ph	—		812
R = Me	H	H	$10,11$-Me_2	110–111		812
R = Me	H	H	10-MeO	195/0.03		544,545,714
R = C_6H_4Cl-*m*	H	H	H	250–270/1.0		790
				140	3HCl	
R = $C_6H_4CF_3$-*m*	H	H	H	260–280/1.0		790
				151		
R = $C_6H_4SCF_3$-*m*	H	H	H	260–280/1.0		790
				122		
R = 3-Pyridyl	H	H	H	120–122		782
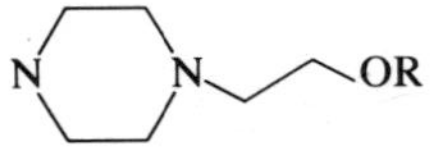 [structure: piperazine N—CH_2CH_2OR]						

(*continued*)

TABLE 104. *(Continued)*

Am	X	Y	Z	mp (°C) or bp (°C/torr)	Salt	Other Data	Refs.
R = H	H	H	H	100			591,613
				228–230	2HCl		
				210	2HCl		519,805
R = H	H	H	10-Me	81–83			812
R = H	3-NMe_2	H	H	138–140	Maleate		683
R = COMe	H	H	H	209–212	2HCl		591,613
R = COC_6H_{13}-*n*	H	H	H	161	2HCl		791
R = $COC_{10}H_{21}$-*n*	H	H	H	150–152	2HCl		791
R = $COCHPh_2$	H	H	H	164	Difumarate		734
R = COPh	H	H	H	74–75			748
				174–176	Dimaleate		
				208–210	2HCl		
R = COC_6H_4Me-*p*	H	H	H	195–196	2HCl		748
R = COC_6H_4Cl-*p*	H	H	H	202–204	2HCl		748
R = COC_6H_4OMe-*o*	H	H	H	157–158	Dimaleate		748
R = COC_6H_4OMe-*p*	H	H	H	175–176	Dimaleate		748
R = $COC_6H_3(OMe)_2$-3,4	H	H	H	145–147	2HCl		748
R = $COC_6H_2(OMe)_3$-3,4,5	H	H	H	170–171	Difumarate		734
				150–151	Dimaleate		
				196–198	2HCl		748
R = $COC_6H_2(OMe)_3$-3,4,5	3-Cl	H	H	159–160	Difumarate		748
				167–168	Dimaleate		
R = 2-(*n*-Butyloxy)3,4-dimethoxybenzoyl	H	H	H	163	Difumarate		734
R = Nicotinoyl	H	H	H	176	Difumarate		734
4-(2-Hydroxypropyl) piperazino	H	H	H	Oil			748

N(piperazine ring)NCH_2CHMeO-						
$COC_6H_2(OMe)_3$-3,4,5	H	H	H	195–198	2HCl	748
N(piperazine ring)$N(CH_2)_3O$-						
$COC_6H_2(OMe)_3$-3,4,5	H	H	H	212–214	2HCl	748
2-Methylpiperazino	H	H	H	143–145	2HCl	748
4-(2-Benzoyloxyethyl)-2-methylpiperazino	H	H	H	116–118	Dimaleate	748
N(homopiperazine ring)N—R						
R = H	H	H	H	216–222/0.5		770
R = Me	H	H	H	226–235/2.0		770
				173–174	Fumarate	
R = $(CH_2)_2OH$	H	H	H	219–224/0.3		770
				136–137	Fumarate	
R = $(CH_2)_2OAc$	H	H	H	142–145	Difumarate	770
R = $(CH_2)_2OCOC_6H_2$-$(OMe)_3$-3,4,5	H	H	H	208–212	2HCl	748

TABLE 105. 5-(γ-AMINOPROPYL)-10,11-DIHYDRODIBENZ[*b*, *f*]AZEPINES

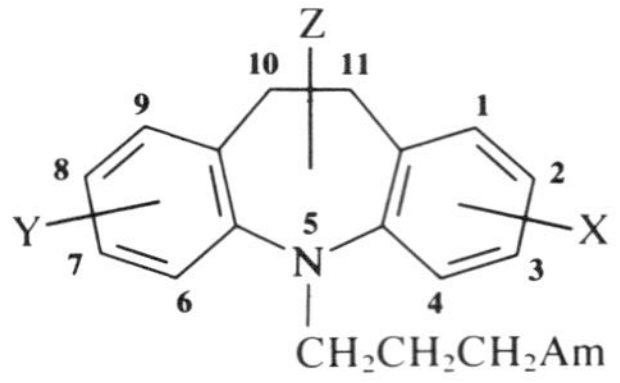

Am	X	Y	Z	mp (°C) or bp (°C/torr)	Salt	Other Data	Refs.
NH_2	H	H	H	258–260	HCl		732
NHMe	H	H	H	212–213	HCl		742
				179–186/0.15			758
				214–216	HCl		
				215–216	HCl		731,732
				125–150	Dinaponate		735,751
NHMe	2-OH	H	H	160			650–652
NHMe	2-OCH_2Ph	H	H	225/0.002			650–652
NHMe	3-COMe	H	H	179–183/0.04			657,728
NHMe	3-NMe_2	H	H	197–200	2HCl		683,752
NHMe	3-SMe	H	H	139	HCl		669,786
NHMe	3-SEt	H	H	210	Oxalate		669
NHMe	3-SPr-*i*	H	H	135	HCl		671
				185	Oxalate		669,786
				129–131	Fumarate		
NHMe	3-S(O)Pr	H	H	168	Oxalate		671
NHMe	3-SO_2NMe_2	H	H	133	HCl		616,707,744,784
NHMe	H	H	10-Me	155–158/0.005			595,596,737
				187–189	HCl		

$NMeCO_2Et$	H	H	H	200–210/0.6			758
$N(COCF_3)Me$	H	H	H	Oil		ir,uv,ms	778,813
N(COMe)Me	H	H	10-Me	Oil			737
$N(CO_2Et)Me$	H	H	10-Me	175–176/0.002			595,596,737
NHBu-*n*	H	H	H	180–183/0.03			721
$NHCH_2Ph$	H	H	H	175–205/0.015			732
				166–169	Maleate		
$NHCH_2R^c$	H	H	H	117–125	HCl		789
$NHC(=NH)NH_2$	H	H	H	133–135	AcOH		661
$NHC(=NH)NHNH_2$	H	H	H	163–164	HBr		661
$NH(CH_2)_2NMe_2$	H	H	H	184/0.005			727,753
				245–247	2HCl		
$NH(CH_2)_2NEt_2$	H	H	H	175–177/0.05			727,753
				245–247	2HCl		
$NH(CH_2)_3NMe_2$	H	H	H	172/0.008			727,753
				235–238	2HCl		
2-Pyridylamino	H	H	H	74–75			661
4-Pyridylamino	H	H	H	160–161			661
NMe_2	H	H	H	160/0.2			470,699
				161–168	HCl		
				155–160/0.1			809
				172–174	HCl		
				174–175	HCl		587,743,758
				172–175/0.8			498
				171–173	HCl[a]		775
				163–165	HCl[a]		775
				159	HCl[a]		775
				203	Pamoate		787
				125–150	Dinaponate		735,751
NMe_2	2-OH	H	H	132–133			735,751
				134–135			649–652
NMe_2	2-AcO	H	H	170–180/0.001			649

(*continued*)

TABLE 105. (*Continued*)

Am	X	Y	Z	mp (°C) or bp (°C/torr)	Salt	Other Data	Refs.
NMe_2	2-OCH_2Ph	H	H	212–215/0.002			649–652
NMe_2	3-Cl	H	H	53			573,717,758,808
				191–192	HCl		
				191–193	HCl		
				186–188	HCl		
				172–176	HCl[b]		875
NMe_2	3-*n*-Pr	H	H	142–144/0.001			573
NMe_2	3-COMe	H	H	175–176/0.01			657
				191	HCl		
NMe_2	3-COEt	H	H	180–184			657
NMe_2	3-NHMe	H	H	77–80			683
NMe_2	3-NMe_2	H	H	195–200	2HCl		683
NMe_2	3-MeO	H	H	140–142/0.001			650–652
				177	HCl		
NMe_2	3-OCH_2Ph	H	H	202/0.025			650–652
NMe_2	3-MeS	H	H	170	HCl		669,670
NMe_2	3-EtS	H	H	180/0.001			669,670
				180	Oxalate		
NMe_2	*i*-PrS	H	H	169	Oxalate		669,670
NMe_2	3-PhS	H	H	85			669,670
NMe_2	3-S(O)Me	H	H	157	Oxalate		671
NMe_2	3-SO_2NMe_2	H	H	66–68			616,665,667, 707,744,785
				189	HCl		
NMe_2	3-Pyrrolidinosulfonyl	H	H	211–212.5	$HCl \cdot H_2O$		616
NMe_2	3-Piperidinosulfonyl	H	H	109			616,665,667,707
NMe_2	3-Morpholinosulfonyl	H	H	132–133			665,667,707
				201–203	HCl		
NMe_2	H	H	10-Me	140/0.02			615,812
				167–168	HCl		

NMe_2	H	H	10-Et	152–154/0.005			615,812
				141–143	HCl		
NMe_2	H	H	10-*n*-Bu	180/0.1			812
NMe_2	H	H	10-Ph	187–189/0.007			812
NMe_2	H	H	10,11-Me_2	135–138/0.003			733,811
NMe_2	H	H	10-CH_2Ph-11-Me	210/0.003			733
NMe_2	H	H	10-OH	88–90			646
NMe_2	2-Me	8-Me	H	177–178	HCl		700
NMe_2	3-Cl	7-Cl	H	175–179/0.01			700
				174–175	HCl		
NMe_2	2-OH-3-Cl	H	H	108–112	Oxalate	uv	653
				145–147	Maleate		
NMe_2	2-OTHP-3-Cl	H	H	136–137	Oxalate		653
NMe_2	3-Cl	8-OH	H	167–168	Oxalate	uv	653
NMe_2	3-Cl	8-OTHP	H	125–126	Oxalate		653
NMe_2	3-Cl	H	11-OH	147–148			646
NMe_2	3-Cl	7-Cl	10-OH	—			646
NMe_2	2,4-D_2	6,8-D_2	H			ms	813
NMeEt	3-MeS	H	H	135	Oxalate		669,670
$NMeCH_2Ph$	H	H	H	195–201/0.04			731,732
$NMeCH_2Ph$	3-*i*-PrS	H	H	157	Oxalate		669,670
NEt_2	H	H	H	155–156/0.12			470,699
				168/0.1			587
				188–189	HCl		
N(Bu-*n*)$_2$	H	H	H	184/0.1			470,587,699
$N(CH_2Ph)_2$	H	H	H	Oil			732
N(*c*-C_5H_9)Me	H	H	H	173–175/0.02			704
				185–187	HCl		
$NEt(CH_2)_2NEt_2$	H	H	H	185/0.007			727,753
				177–179	2HCl		
$NMe(CH_2)_3NMe_2$	H	H	H	160–180/0.001			727,753
				241–243	2HCl		
N(*c*-C_6H_{11})$(CH_2)_2NMe_2$	H	H	H	206/0.01			727,753

(*continued*)

TABLE 105. (*Continued*)

Am	X	Y	Z	mp (°C) or bp (°C/torr)	Salt	Other Data	Refs.
$NMe(CH_2)_n-C(=O)-C_6H_4R$							
n R							
1 H	H	H	H	158–159	Oxalate		760
1 3-Cl	H	H	H	160	Oxalate		760
1 4-Cl	H	H	H	152–154	HCl		811
				154–156	HCl		760
1 4-F	H	H	H	151–153	Oxalate		760
1 2-MeO	H	H	H	139–141	Oxalate		760
1 3-MeO	H	H	H	156–157	Oxalate		760
1 4-MeO	H	H	H	212	HCl		760
1 4-EtO	H	H	H	166.5–167.5	Oxalate		760
1 4-*i*-PrO	H	H	H	168–169	Oxalate		760
1 4-*n*-BuO	H	H	H	166–167	Oxalate		760
1 3,4-$(OMe)_2$	H	H	H	187–189	Oxalate		760
1 3,4-OCH_2O	H	H	H	195	Oxalate		760
1 3,4-(1,1-Cyclo hexylidene)-dioxy	H	H	H	163.5–164.5	Oxalate		760
1 3,4,5-$(OMe)_3$	H	H	H	153–154	Oxalate		760
2 H	H	H	H	100–102			760
2 4-F	H	H	H	152–153	HCl		760
3 H	H	H	H	173–175	Oxalate		760
3 4-F	H	H	H	149–150	Oxalate		760
Piperidino	H	H	H	184–185/0.2			587
				214–215	HCl		

Piperidino	3-NMe_2	H	H	190–200/0.05		683
Piperidino	H	H	10-Me	168/0.001		615,812
				181	HCl	
Piperidino	H	H	10-CH_2Ph	215/0.009		812
Morpholino	H	H	H	185–188/0.1		587
				196–198	HCl	
4-Oxopiperidino	H	H	H	243–248/0.4		705
				91	HCl	
(4-piperidino ring with R_1, R_2 at C-4) R_1 R_2						
H OH	H	H	H	197–198	HCl	705
				206–209	HCl	597
				211–213	HCl	594
H MeO	H	H	H	172–174	HCl	705
H OCONHMe	H	H	H	102	HCl	594
H CH_2OH	H	H	H	112–113		597
				115		594
H $CH_2OCONHMe$	H	H	H	105		594
H $(CH_2)_2OH$	H	H	H	113–114.5		594,597
H $(CH_2)_2OCONHMe$	H	H	H	120		594
H $(CH_2)_3OH$	H	H	H	193–195	HCl	597
H COMe	H	H	H	214–217	HCl	597
H COC_6H_4F-*p*	H	H	H	156–159	Oxalate	601,602
—$O(CH_2)_2O$—	H	H	H	205–206	HCl	705
H Piperidino	H	H	H	300	2HCl	738,739,761
OH Ph	H	H	H	226	HCl	761
OH C_6H_4Cl-*p*	H	H	H	124–136	$HCl·0.5H_2O$	738,739, 741,761
OH $C_6H_4CF_3$-*m*	H	H	H	112–114	HCl·0.5MeOH	738,739, 741,761

(*continued*)

TABLE 105. (*Continued*)

Am		X	Y	Z	mp (°C) or bp (°C/torr)	Salt	Other Data	Refs.
OH	2-Thienyl	H	H	H	156–157	Maleate		792,793
OH	2-Thienyl	3-Me	H	H	—			793
OH	2-Thienyl	3-Cl	H	H	—			793
OH	2-Thienyl	3-MeO	H	H	Oil		ir,ms	792,793
OH	2-Furyl	3-Cl	H	H	86–87	Fumarate		792,793
CO_2Et	Ph	H	H	H	193–194	HCl		749
$CONH_2$	NMe_2	H	H	H	265–275	2HCl		738,739,761, 794,795
$CONH_2$	NMe_2	3-Cl	H	H	145–150	Maleate		756,761,766, 774,794, 795,801
$CONH_2$	Piperidino	H	H	H	278–279	2HCl·MeOH		738–740,761
$CONH_2$	Piperidino	3-Cl	H	H	181–183	Maleate		756,766,774,
					259	$2HCl\cdot 0.5H_2O$		761, 794,795,801
$CONH_2$	Piperidino	3-Me	H	H	260–263	2HCl		756,766,794, 795,801
$CONH_2$	Piperidino	3-COMe	H	H	240–245	$2HCl,H_2O$		756,766
$CONH_2$	Piperidino	3-MeO	H	H	186	Maleate		756,766,769, 794,795,801
$CONH_2$	Morpholino	3-Cl	H	H	259–260	$2HCl\cdot H_2O$		795,801
Ph	$CH_2NHCOMe$	H	H	H	75–80	$HCl\cdot 0.5H_2O$		738,739,761
3-Hydroxypiperidino		H	H	H	212–215/0.25			594
					214	HCl		
3-(*N*-Methylcarbonyloxy)piperidino		H	H	H	150			594
3-(Hydroxymethyl)piperidino		H	H	H	155	HCl		594

Structure									
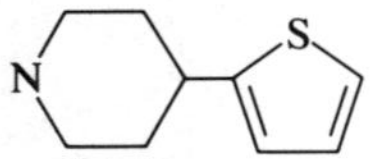 4-(2-thienyl)piperidine (structure)			3-Me	H	H	Oil		ir,ms	792,793
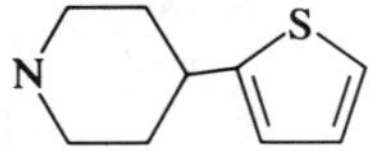 4-(2-thienyl)piperidine (structure)			3-Cl	H	H	160–161	Fumarate		793
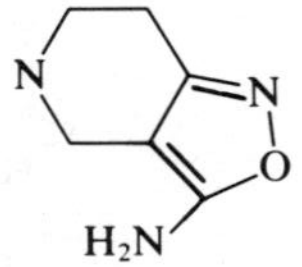 3-amino-4,5,6,7-tetrahydroisoxazolo[4,5-c]pyridine (structure, H_2N)			3-Cl	H	H	106–110 (dec.)			802
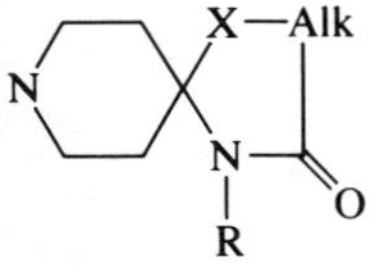 Spiro structure: X—Alk, N—R, O									
X	R	Alk							
O	H	CH_2	H	H	H	134–135	Oxalate		775,781,796
O	H	CHMe	H	H	H	251–252	HCl		775,781,796
O	H	CHPh	H	H	H	194–196	Maleate		775,781
S	H	CH_2	H	H	H	241–243	HCl		754,755
						263–264	HCl		797
S	H	CH_2	3-Cl	H	H	267–269	HCl		797
S	Me	CH_2	H	H	H	273–275	HCl		754,755,797
S	H	$(CH_2)_2$	H	H	H	249–252	HCl		754,755
S	Me	$(CH_2)_2$	H	H	H	236–239	HCl		754,755
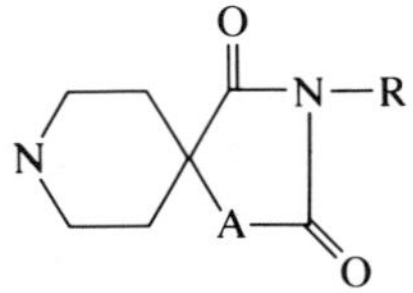 Spiro structure: O, N—R, A, O									

(*continued*)

TABLE 105. *(Continued)*

Am	X	Y	Z	mp (°C) or bp (°C/torr)	Salt	Other Data	Refs.
A R							
O Me	H	H	H	255–257			757,788,804
CH_2 Me	H	H	H	218–219	Fumarate		747
CH_2 Et	H	H	H	184–186	Fumarate		747
CH_2 CH_2Ph	H	H	H	112–115	Fumarate		747
CHMe Me	H	H	H	228–230	HCl		747
$N(CH_2)_6$	H	H	H	225–230/1.5			729,730
				125–127	$EtSO_3H$		
$N(CH_2)_7$	H	H	H	205–210/0.4			729,730
				130–132	$EtSO_3H$		
4-Oxopiperidino	H	H	H	227–230	HCl		705
N(piperazine)N—R, R							
H	H	H	H	215–230/0.25			590
				220	HCl		
H	3-Cl	7-Cl	H	185	Dimaleate		590
Me	H	H	H	84–87			810
				195–198/0.04			701
				245–248	2HCl		
Me	3-COMe	H	H	205–209/			657
				0.0005	2HCl		
				206–210			
Me	$3\text{-}NMe_2$	H	H	215–220	3HCl		683
Me	H	H	10-Me	185–189/0.005			812
Me	H	H	10-Bu-*n*	190–192/0.008			812

Me	H	H	10-Ph	220–222/0.005		812
Et	H	H	H	198/0.25		701
				234–238	2HCl	
CH_2Ph	H	H	H	111–114		810
2-Furylmethyl	H	H	H	199–202	Dimaleate	782
2-Furylmethyl	3-Cl	H	H	185–187	Dimaleate	782
2-Thienylmethyl	H	H	H	197–200	Dimaleate	782
2-Pyridylmethyl	H	H	H	188–190	Dimaleate	782
2-(4-Pyridyl)ethyl	H	H	H	156–168	Trimaleate	782
$(CH_2)_2OH$	H	H	H	191–193	Dimaleate	592
				225	2HCl	
				61		712,713
				236	2HCl	
$(CH_2)_2OH$	3-COMe	H	H	209–210	Oxalate	657,728
$(CH_2)_2OH$	3-Cl	7-Cl	H	205	2HCl	592
				100		712
				235–240	2HCl	
$(CH_2)_2OH$	H	H	10-CH_2Ph	159	Fumarate	812
$(CH_2)_2OAc$	H	H	H	198–199	Dimaleate	592
				218–221	2HCl	718
$(CH_2)_2OAc$	3-COMe	H	H	205–208	Oxalate	728
				207–209	Oxalate	657
$(CH_2)_2OCOC_6H_4Me$-*p*	3-MeO	H	H	153–155	Difumarate	748
$(CH_2)_2OCOC_6H_2(OMe)_3$-3,4,5	H	H	H	189–190	Difumarate	748
				185–189	Difumarate	734
$(CH_2)_2O(CH_2)_2OH$	H	H	H	215	2HCl	592
$(CH_2)_2O(CH_2)_2OH$	3-Cl	7-Cl	H	183–185	2HCl	592
C_6H_4Cl-*m*	H	H	H	250–260/0.2		790
				155–160	2HCl	
$C_6H_4CF_3$-*m*	H	H	H	250–260/0.4		790
				156	2HCl	
$C_6H_4SCF_3$-*m*	H	H	H	250–270/0.4		790
				192	2HCl	

(*continued*)

TABLE 105. (*Continued*)

Am	X	Y	Z	mp (°C) or bp (°C/torr)	Salt	Other Data	Refs.
$CONH_2$	H	H	H	134–136			590
$CONH_2$	3-Cl	7-Cl	H	118–120			590
CONHMe	H	H	H	140–142			590
CONHMe	3-Cl	7-Cl	H	190–192			590
$CONMe_2$	H	H	H	138–140	Maleate		590
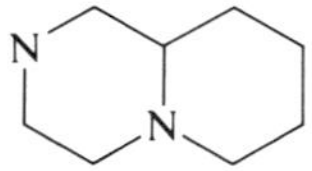 [octahydro-2H-pyrido[1,2-a]pyrazin-2-yl structure]	H	H	H	212–214	$2HCl \cdot H_2O$		773,776
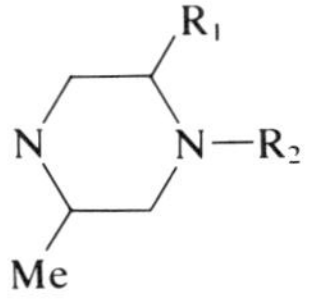 [piperazine structure: N; N—R_2; R_1; Me] R_1 R_2							
H H	H	H	H	156–158	Maleate		715
H Me	H	H	H	183–188/0.25			701
				244–247	HCl		

H	CH_2CO_2Et	H	H	H	206–208	2HCl	715
H	CHO	H	H	H	234–237/0.02		715
Me	H	H	H	H	190	2HCl	590
Me	$CONH_2$	H	H	H	168–170		590
N	N—R						
R							
H		H	H	H	215–220/1.0		770
Me		H	H	H	231–240/2.0		770
					188–191	Fumarate	
$(CH_2)_2OH$		H	H	H	225–230/1.0		770
					119–121	2HCl	
2-Aminopyridin-1-ium chloride		H	H	H	178–179		661
4-Aminopyridin-1-ium chloride		H	H	H	160–161		661

[a]Three polymorphic forms by differential scanning calorimetry.
[b]Three polymorphic forms.
[c]R = 2,3-dihydro-1,4-benzodioxin-2-yl

TABLE 106. 5-(β-AMINOPROPYL)DIBENZ[*b, f*]AZEPINES

[Structure: dibenz[b,f]azepine ring numbered 1–11 with X (ring positions 1–4) and Y (positions 6–9) substituents, 10,11 bond dashed; N-5 bears $CH_2—CH(CH_3)—Am$]

Am	Ring System[a]	X	Y	mp (°C) or bp (°C/torr)	Salt	Refs.
NMe_2	IS	H	H	151–153/0.2		612,710
NMe_2	IBB	H	H	145–146/0.05		
				199–200	HCl	470,692
				152–153/0.08		587
				200–201	HCl	
NMe_2	IBB	3-Cl	H	247	HCl	573
NMe_2	IBB	3-Cl	7-Cl	173–175/0.004		700
				254	HCl	
NMe_2	IBB	3-NMe_2	H	210–212	HCl	683
NMe_2	IBB	3-EtS	H	180/0.001		669
NEt_2	IBB	H	H	153–154/0.1		587
				196–197	HCl	
Piperidino	IBB	H	H	182–183/0.2		587
				241–242	HCl	
Morpholino	IBB	H	H	195–197/0.4		587
				252–253	HCl	
4-Hydroxypiperidino	IBB	H	H	215	HCl	594
3-(3-Hydroxypropyl) piperidino	IBB	H	H	199–208	HCl	597
4-(2-Hydroxy-ethyl)piperidino	IBB	H	H	208–210.5	HCl	597
4-(3-Hydroxy-propyl)piperidino	IBB	H	H	70–75		597
[piperazine ring] N N—R						
R = H	IBB	H	H	210	2HCl	590
R = $CONH_2$	IBB	H	H	141–142		590
R = $(CH_2)_2OH$	IBB	H	H	210	HCl	592
R = $(CH_2)_2O(CH_2)_2OH$	IBB	H	H	190	2HCl	592
R = $(CH_2)_2OC(=O)Ph$	IS	H	H	—		748
R = $(CH_2)_2OC(=O)C_6H_2$-$(OMe)_3$-3,4,5	IS	H	H	150–151	Dimaleate	748
R = C(=NH)NMe_2	IBB	H	H	195–196	Fumarate	663
R = C(=N—H)—N[piperazine]NMe	IBB	H	H	207–208	2HCl	663

[a] IBB = Iminobibenzyl; IS = Iminostilbene.

TABLE 107. 5-(γ-AMINO-β-METHYLPROPYL)DIBENZ[*b,f*]AZEPINES

(Structure: dibenz[b,f]azepine ring system, positions 1–11, with X on the ring bearing positions 1–4 and Y on the ring bearing positions 6–9; the ring N (position 5) carries CH_2CHCH_2Am with Me on the central carbon.)

Am	X	Y	mp (°C) or bp (°C/torr)	Salt	Refs.
NMe_2	H	H	158–161/0.05		523
			77–78		
			202–203	HCl	
NEt_2	H	H	164–165/0.03		523
Pyrrolidino	H	H	165/0.06		523
			108–115	HCl	
4-($C_6H_4CF_3$-*m*)-4-hydroxypiperidino	H	H	236	HCl	761
4-Phenyl-4-(NMe_2)piperidino	H	H	229	2HCl	761
4-R-piperazino					
R = H	H	H	77–79		762,763
			168–169	2HCl	
R = H	3-SO_2NMe_2	H	132	Oxalate	783
R = CO_2Et	H	H	90–92.5		762,763
R = Me	H	H	167–168/0.05		762,763
R = Me	3-SO_2NMe_2	H	115		666
R = 2-Pyridylmethyl	H	H	163–165	Dimaleate	782
R = $(CH_2)_2OC(=O)C_6H_2$-$(OMe)_3$-3,4,5	H	H	156–157	Dimaleate	748
R = $(CH_2)_2$-(3-methyl-2-oxoimidazolidin-1-yl)	H	H	107–109		762,763,779
			223–224	2HCl	
R = $(CH_2)_2$-(3-methyl-2-oxoimidazolidin-1-yl)	3-Cl	H	163–168	Maleate	762
R = $(CH_2)_3$-(3-methyl-2-oxoimidazolidin-1-yl)	H	H	132–135		762
			235–237	2HCl	

TABLE 108. 5-(γ-AMINO-β-METHYLPROPYL)-10,11-DIHYDRODIBEN[*b*, *f*]AZEPINES

Am	X	Y	Z	mp (°C) or bp (°C/torr)	Salt	Other Data	Refs.
NHMe	H	H	H	170–173/0.1			758
NHMe	3-NMe_2	H	H	162–165	Oxalate		683,752
NHMe	3-EtS	H	H	150	HCl		669
NHMe	3-SO_2NMe_2	H	H	216	Oxalate		707,744,784,785
NMe_2	H	H	H	160–165/0.06	HCl		743
				205–206			
				187–189/0.9			758
NMe_2	H	H	H	45		*d,l*	702
				142	Maleate		
				133–134	Picrate		
				Oil			
				132	Maleate	*d* : $[\alpha]_D^{23}(CHCl_3) = +25.5$	702
				143–145	Picrate		
				Oil		*l* : $[\alpha]_D^{23}(CHCl_3) = -25.3$	702
				132	Maleate		
				143–145	Picrate		
NMe_2	3-NMe_2	H	H	203–205	2HCl		683
NMe_2	3-MeS	H	H	148	Oxalate		669,670

NMe_2	3-EtS	H	H	155	Oxalate	669,670
				164	HCl	
NMe_2	*i*-PrS-3	H	H	156	HCl	669,670
NMe_2	S(O)Et	H	H	128–130	Oxalate	671
NMe_2	3-SO_2NMe_2	H	H	111		616,667,707, 744,785
NMe_2	3-SO_2N 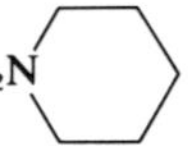	H	H	54–57		667,707
				119–128	HCl	
NMe_2	H	H	10-OH	138–140		646
Piperidino	3-SO_2NMe_2	H	H	168	Oxalate	744
3-Hydroxy-piperidino	3-MeO	H	H	134	Fumarate	600,633
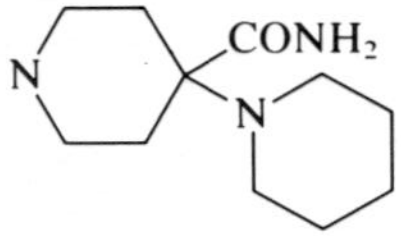 N, $CONH_2$, N	H	H	H	263	$2HCl \cdot H_2O$	794,795
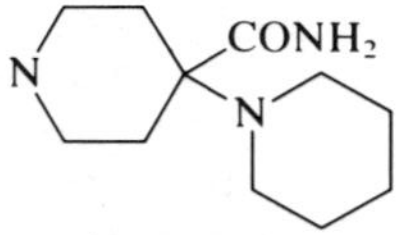 N, $CONH_2$, N	3-Cl	H	H	264	$2HCl \cdot H_2O$	794,795
Hexahydroazepino	3-COMe	H	H	211/0.003		657
				191–193	HCl	
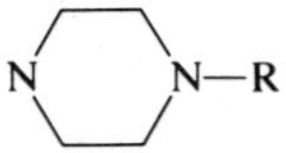 N, N—R						
R = H	H	H	H	228–265/0.8		590,592
				81–83		762,763
				268–270	2HCl	
R = H	3-Cl	H	H	Oil		762,763,780
R = H	3-SO_2NMe_2	H	H	168	Oxalate	616,667

(continued)

TABLE 108. (*Continued*)

Am	X	Y	Z	mp (°C) or bp (°C/torr)	Salt	Other Data	Refs.
R = Me	3-Cl	H	H	Oil			762,763,780
R = Me	3-SO_2NMe_2	H	H	193	Oxalate		616,667,744
R = Me	3-OCH_2Ph	H	H	260/0.008			650
				207–209	Oxalate		
R = CN	3-Cl	H	H	128.5–130			762,763,780
R = $CONH_2$	H	H	H	157–158			590
R = $(CH_2)_2OH$	H	H	H	102–103			592
R = $(CH_2)_2OH$	3-NMe_2	H	H	221–223	3HCl		683
R = $(CH_2)_2OH$	3-SO_2NMe_2	H	H	158	Oxalate		616,667,744
R = $(CH_2CH_2O)_2H$	H	H	H	102–103			592
R = $(CH_2)_2$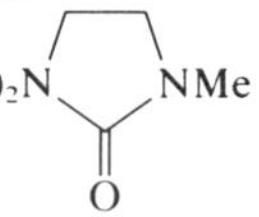	H	H	H	58–63			762,763,779
				231–234	2HCl		
R = $(CH_2)_2$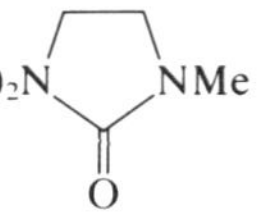	3-Cl	H	H	136.5–140			762,763, 779,780
				268–272	2HCl		
R = $(CH_2)_3$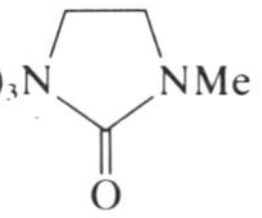	H	H	H	244–247	2HCl		762,763
R = $(CH_2)_3$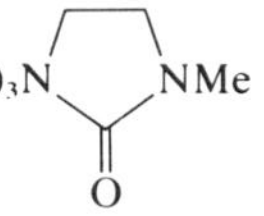	H	H	H	223–227	2HCl		762,763

TABLE 109. 5-(γ-AMINO-β-HYDROXYPROPYL)DIBENZ[*b, f*]AZEPINES

Am	Ring System[a]	X	Y	mp (°C) or bp (°C/torr)	Refs.
NHMe	IB	H	H	190–191 (HCl)	587,588
NMe_2	IB			212–213 (HCl)	587,588
NEt_2	IS	H	H	105–106	589,593
NEt_2	IS	3-Cl	7-Cl	268–270 (HCl)	593
NEt_2	IB	H	H	188–189 (HCl)	587,588
pyrrolidino	IS	H	H	185/0.002	593
pyrrolidino	IB	H	H	248–250 (HCl)	587,588
piperidino	IS	H	H	186–188/0.002	589,593
piperidino	IB	H	H	260 (HCl)	587,588
morpholino	IB	H	H	251–252 (HCl)	587
$NMe(CH_2)_2NMe_2$	IB	H	H	184–185/0.01	587
$N(Pr\text{-}n)(CH_2)_2NMe_2$	IB	H	H	171–172/0.002	587

[a]IS = Dibenz[*b, f*]azepine or iminostilbene; IB = 10,11-dihydrodibenz[*b, f*]azepine or iminobibenzyl.

as the base (498). When other base-labile functionalities are present in the dibenz[*b, f*]azepine, they must be protected prior to treatment with sodium amide. For example, the 3-acetyl group in **1461b** (Eq. 231) had to be converted to the ethylene ketal **1546** before it could be converted to **1547b** (657).

(1) $NaNH_2$, C_6H_6, Δ; (2) $Cl(CH_2)_3NMe_2$ (229)

1542 X = H, Cl → **1296a**, X = H; **1543**, X = Cl [N-$(CH_2)_3NMe_2$]

TABLE 110. 5-(β-AMINO-α-ALKYLETHYL)DIBENZ[*b,f*]AZEPINES

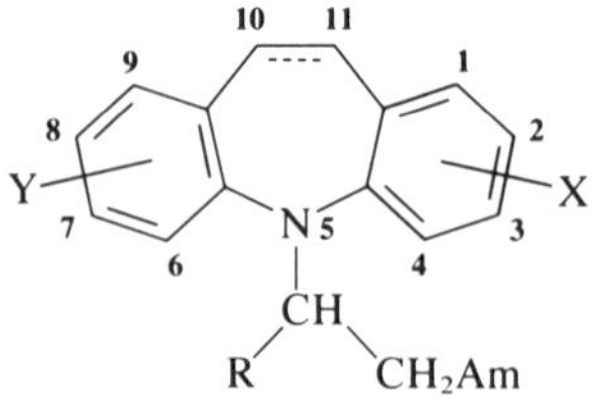

Am	R	Ring System[a]	X	Y	mp (°C) or bp (°C/torr)	Refs.
NH_2	Me	IS	H	H	266–268 (HCl)	721
NH_2	Me	IB	H	H	228–230 (HCl)	610,721
NH_2	Me	IB	3-Cl	H	250–252 (HCl)	721
NH_2	*i*-Pr	IB	H	H	248–250 (HCl)	719
NH_2	*n*-Bu	IB	H	H	209 (HCl)	719
NHCHO	Me	IB	H	H	114–115	610
NHCOMe	Me	IB	H	H	145	719
NHCOMe	*i*-Pr	IB	H	H	111–113	719
NHCOMe	*n*-Bu	IB	H	H	148	719
NHMe	Me	IB	H	H	226 (HCl)	720
					229–231(HCl)	610
NHEt	Me	IB	H	H	242–243 (HCl)	719
					224 (HCl)	721
NHEt	*i*-Pr	IB	H	H	223–224 (HCl)	719
NHEt	*n*-Bu	IB	H	H	160 (HCl)	719
NHPr-*i*	Me	IB	H	H	170–173/0.05	721
NMe_2	Me	IB	H	H	125–128/0.005	610,664
					185–187 (HCl)	
NMe_2	Me	IB	3-Cl	H	247 (HCl)	717

[a]IS = Dibenz[*b,f*]azepine or iminostilbene; IB = 10,11-dihydrodibenz[*b,f*]azepine or iminobibenzyl.

The methods summarized in Fig. 197 have found widespread application, no doubt because they afford maximum opportunity for variation of the aminoalkyl side chain. The preparation of alcohol **1413b** and the halide **1413a** was detailed in Sections 39.C.c.iii and 39.C.b.iii, respectively. The preparation of the tosylate **1544** from **1413b** has been achieved with tosyl chloride in pyridine (592,594). The conversion of **1413a** and **1544** to imipramine analogs has been carried out under standard conditions. Typically, **1413a** has been condensed with amines in methyl ethyl ketone in the presence of a catalytic amount of sodium iodide, with (507,597) or without (728) a carbonate buffer. Alternatively, refluxing ethanol or butanol has been used as the solvent (728,744,749). DMF has also been used widely as the reaction solvent, with no other additives (756,774), with a bicarbonate buffer (600,633), or with added sodium iodide and potassium carbonate (762,763,792).

Lithium aluminum hydride reduction of the 5-alkanoic amides (**1545**)

TABLE 111. MISCELLANEOUS 5-(AMINOALKYL)DIBENZ[*b*, *f*]AZEPINES

Am	Alk	Ring System[a]	X	Y	mp (°C) or bp (°C/torr)	Salt	Refs.
NMe	-	IB	H	H	165–167/0.02		703,745
NMe	-	IB	H	H	151–154/0.02		703,745
N	-	IB	H	H	150		764
N	-	IB	3-Cl	7-Cl	195–196		764
NMe_2	$CH_2CMe_2CH_2$	IB	H	H	152–154/0.2		587
					201–202	HCl	
NEt_2	$CH_2CMe_2CH_2$	IB	H	H	167/0.2		587
					177–178	HCl	

(*continued*)

TABLE 111. (*Continued*)

Am	Alk	Ring System[a]	X	Y	mp (°C) or bp (°C/torr)	Salt	Refs.
pyrrolidino	$CH_2CMe_2CH_2$	IB	H	H	169/0.02		523,612,709
pyrrolidino	$CH_2CMe_2CH_2$	IB	H	H	190–191/0.1		587
					188	HCl	
piperidino	$CH_2CMe_2CH_2$	IB	H	H	185–187/0.2		587
					218–220	HCl	
morpholino	$CH_2CMe_2CH_2$	IB	H	H	202–205/0.4		587
					222–224	HCl	
NMe_2	$CH_2CH(NMe_2)CH_2$	IB	H	H	230–232	HCl	505
NEt_2	$CH_2CH(NEt_2)CH_2$	IB	H	H	164–168/0.17		505
NCH_3COCF_3	$CH{=}C(COCF_3)CH_2$	IB	H	H	116–116.5		778,813
NMe_2	CH_2CH_2CHMe	IB	H	H	165/0.02		725
					146–148	HCl	
NMe_2	CH_2CH_2CHMe	IB	3-SEt	H	160/0.001		669
NMe_2	CH_2CH_2CHMe	IB	3-SO_2NMe_2	H	90		667
NEt_2	CH_2CH_2CHMe	IB	H	H	196/0.02		725
					165–166	HCl	
NMe_2	$(CH_2)_4$	IS	H	H	172–176/0.025		523,724
					165/0.02		716,723,767
					146–148	HCl	
NMe_2	$(CH_2)_4$	IB	H	H	200/2.5		619
					191.5–192.5	HCl	
NMe_2	$(CH_2)_4$	IB	3-SO_2NMe_2	H	90		616,707, 744,785

NEt_2	$(CH_2)_4$	IS	H	H	185–190/0.5		523
					193–195	HCl	
NEt_2	$(CH_2)_4$	IB	H	H	196–200/0.2		716,723,767
—N	$(CH_2)_4$	IS	H	H	206–208/0.35		523
—N N—R							
R = H	$(CH_2)_4$	IB	H	H	240/0.7		590
					200	2HCl	
$R = CONH_2$	$(CH_2)_4$	IB	H	H	~90		590
$R = (CH_2)_2OH$	$(CH_2)_4$	IB	H	H	225	2HCl	592
$R = (CH_2CH_2O)_2H$	$(CH_2)_4$	IB	H	H	200	2HCl	592
NMe_2	$CH_2C{\equiv}CCH_2$	IB	H	H	52–53		619
					171–171.5	HCl	
NMe_2	H, H, CH_2	IS	H	H	200–201	Maleate	603,604
NMe_2	H, H, CH_2	IB	H	H	149–151	Maleate	603,604
NEt_2	$(CH_2)_6$	IB	H	H	190–192/0.2		587
N O	$(CH_2)_6$	IB	H	H	206–208/0.05		587

[a]IS = Dibenz[*b, f*]azepine or iminostilbene; IB = 10,11-dihydrodibenz[*b, f*]azepine or iminobibenzyl.

TABLE 112. 5-(AMINOALKYL)DIBENZ[*b*, *f*]AZEPIN-10-ONES

Am	Alk	X	Y	mp (°C) or bp (°C/torr)	Salt	References
NHMe	$(CH_2)_3$	H	H	175/0.004		544,545,646
NMe_2	$(CH_2)_2$	H	H	174–175/0.001		544,545,646
				180	Fumarate	562–564
NMe_2	$(CH_2)_3$	H	H	174/0.005		544,545,646
				195–197/0.15		562–564
				208	HCl	
NMe_2	$(CH_2)_3$	H	7-Cl	186–188/0.03		646
				72–72.5		
NMe_2	$(CH_2)_3$	3-Cl	7-Cl	87		544,545,646
NMe_2	$(CH_2CHMeCH_2$	H	H	178–0.015		562–564
				165	Fumarate	
N	$(CH_2)_3$	H	H	188/0.005		544,545
N O	$(CH_2)_2$	H	H	202/0.01		544,545
Me N	$(CH_2)_2$	H	H	202/0.01		544,545
N NMe	$(CH_2)_3$	H	H	122		544,545

has likewise been used to prepare numerous imipramine analogs (505,603,604,661,664,743). The decarboxylation of carbamate esters such as **1391,** which was introduced earlier in Section 39.C.a (Eq. 183), represents a third major route to imipramine analogs (613–616,670). For example, acylation of **1542** with 3-(*N*,*N*-dimethylamino)propyl chloroformate led to a carbamate adduct, which on thermolysis at 200–275°C yielded **1296a** (Eq. 230). Alternatively, acylation of **1542** with phosgene produced the carbamoyl chloride **1390,** from which the urethane **1391** could be obtained by reaction with 3-(*N*,*N*-dimethylamino)propyl alcohol. Thermal decarboxylation of **1391** thereupon gave **1296a**.

Routes to 3-acyl analogs of imipramine have been developed. In one of these (Eq. 231), the ketal derivative **1546** of 3-acetyl-10,11-dihydro-

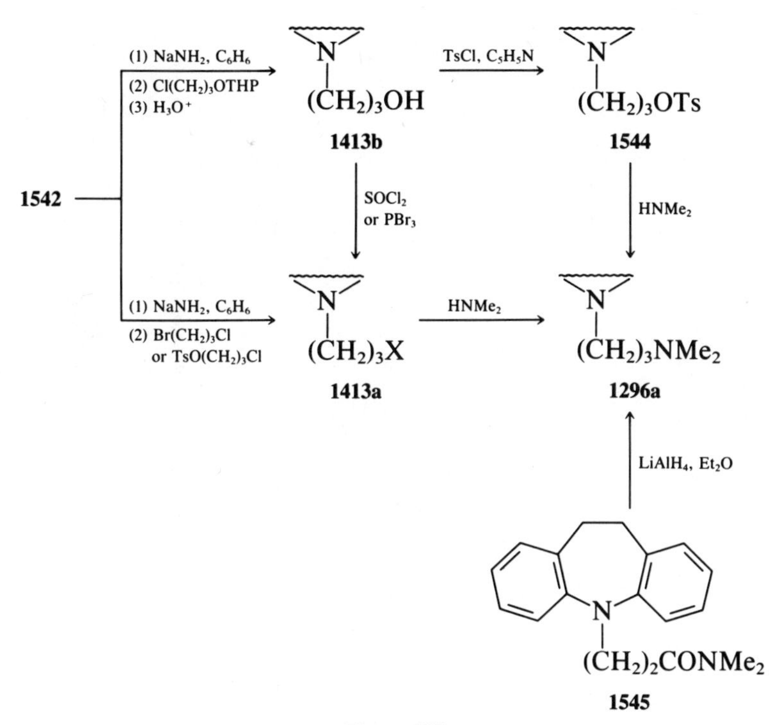

Figure 197

1542 — $COCl_2$, PhMe → **1390** (N–COCl) — $HO(CH_2)_3NMe_2$ → **1391** (N–$CO_2(CH_2)_3NMe_2$) — 220–240°C, $-CO_2$ → **1296a** (N–$(CH_2)_3NMe_2$)

1542 — (1) $XCO_2(CH_2)_3NMe_2$ (2) 200–275°C, KOAc → **1296a** (N–$(CH_2)_3NMe_2$)

(230)

1461b, X = O
1546, X = —OCH_2CH_2O—

(1) $NaNH_2$, C_6H_6
(2) $Cl(CH_2)_3NMe_2$
(3) H_3O^+

Me, N–H, X

Me, N–$(CH_2)_3NMe_2$, X

(231)

1547a, X = —OCH_2CH_2O—
1547b, X = O

TABLE 113. 5-(HYDROXYLAMINO)ALKYLDIBENZ[*b*, *f*]AZEPINES AND AMINE OXIDES

Am	Alk	Ring System[a]	X	Y	Z	mp (°C)	Refs.
ONH_2	$(CH_2)_2$	IB	H	H	H	188–189 (HCl)	598,599
$ON{=}C(Me)_2$	$(CH_2)_2$	IB	H	H	H	Oil	598,599
NMeOH	$(CH_2)_2$	IB	H	H	H	129–131	736
NMeOH	$(CH_2)_3$	IS	H	H	H	116–118	736
NMeOH	$(CH_2)_3$	IB	H	H	H	104–105	736
						163–165 (H_2SO_4)	
NMeOCOPh	$(CH_2)_2$	IB	H	H	H	130–132	736
NMeOCOPh	$(CH_2)_3$	IS	H	H	H	124–126	736
NMeOCOPh	$(CH_2)_3$	IB	H	H	H	120–122	736

HO—N (ring)	$(CH_2)_2$	IB	H	H	H	146–147	736
$PhCO_2$—N (ring)	$(CH_2)_2$	IB	H	H	H	132	736
$\overset{+}{N}Me_2—O^-$	$(CH_2)_3$	IS	H	H	H	72–75 ($1.5H_2O$)	746
$\overset{+}{N}Me_2—O^-$	$(CH_2)_3$	IS	3-Cl	H	H	115–118 (H_2O)	746
$\overset{+}{N}Me_2—O^-$	$(CH_2)_3$	IS	H	H	10-Me	97–102 ($2H_2O$)	746
$\overset{+}{N}Me_2—O^-$	$(CH_2)_3$	IB	H	H	H	75–79 (H_2O)	746
$\overset{+}{N}Me_2—O^-$	$(CH_2)_3$	IB	3-Cl	H	H	74–77 (H_2O)	746
$\overset{+}{N}Me_2—O^-$	$(CH_2)_3$	IB	3-Cl	7-Cl	H	115–119 (H_2O)	746
$\overset{+}{N}Me_2—O^-$	$(CH_2)_3$	IB	H	H	H	133–135 (H_2O)	746
$\overset{+}{N}Me_2—O^-$	$(CH_2)_3$	IB	H	H	H	122–125 ($2H_2O$)	746
$\overset{+}{N}Et_2—O^-$	$(CH_2)_3$	IB	H	H	H	51–60	746
$^-O—N^+$ (ring with O)	$(CH_2)_2$	IB	3-Cl	H	H	137.5 ($2H_2O$)	746

[a]IS = Dibenz[*b*, *f*]azepine or iminostilbene; IB = 10,11-dihydrodibenz[*b*, *f*]azepine or iminobibenzyl.

TABLE 114. 5-(ω-AMINOALKANOYL)DIBENZ[*b,f*]AZEPINES

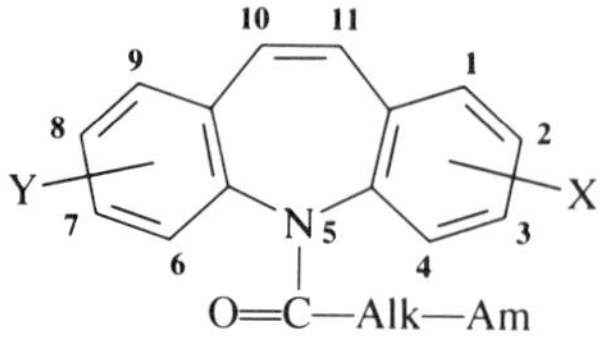

Y 7 8 9 10 11 1 2 3 4 6 X N5 O=C—Alk—Am

Alk	Am	X	Y	mp (°C)	Refs.
—	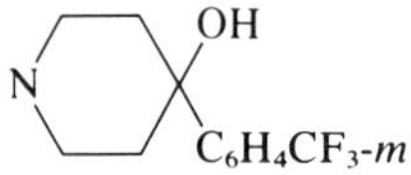 N, OH, $C_6H_4CF_3$-*m*	H	H	176–177	819
—	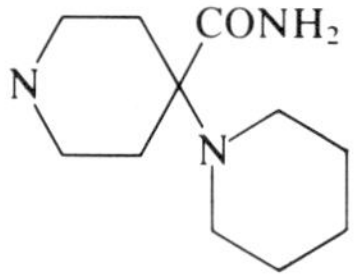 N, $CONH_2$, N	H	H	208–209	819
CH_2	NEt_2	H	H	217–220 215–216 (MeI)	636,816
CH_2	 N	H	H	104	636
CHMe	NEt_2	H	H	88–89 212–214 (MeI)	636
CHMe	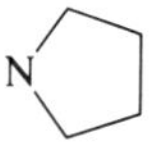 N	3-Cl	7-Cl	167–168	636
CHMe	 N	H	H	124–125	636
$(CH_2)_2$	 N, OH, Ph	H	H	189–191 (HCl)	821
$(CH_2)_2$	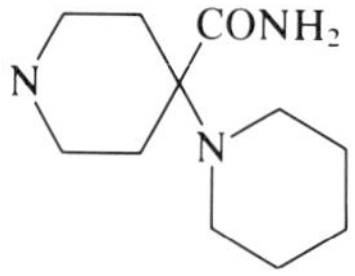 N, $CONH_2$, N	H	H	232–234 ($2HCl\cdot1.5H_2O$)	821

dibenz[*b,f*]azepine (**1461**) was alkylated with 3-(*N,N*-dimethylamino)propyl chloride, using sodium amide as the base. Removal of the ketal group from the resultant product **1547a** by acidolysis gave the ketone **1547b**.

Other routes to the preparation of 5-aminoalkyliminobibenzyls include catalytic or metal hydride reduction of cyanoalkyliminobibenzyls (**1389**, see Section 39.C.a) to give the primary amines **1548** (Eq. 232) (595,596,610,721). Amination of the epoxides **1549,** obtained by alkylation of **1542** with epichlorohydrin (Eq. 233), has been reported widely as a means of synthesis of the 5-(α-amino-β-hydroxypropyl)dibenz[*b,f*]azepines **1550** (587–589,593).

TABLE 115. 5-(ω-AMINOALKANOYL)-10,11-DIHYDRODIBENZ[*b,f*]AZEPINES

Alk	Am	X	Y	mp (°C) or bp (°C/Torr)	Salt	Refs.
—	3-pyridyl	H	H	125–127		
				201–203	HCl	703
—	1-Me-3-piperidyl	H	H	153–155	HCl	703
—	4-R_1-4-R_2-piperidino					
	R_1 / R_2					
	H / piperidino	H	H	132–133	$0.5H_2O$	819
	OH / CH_2Ph	H	H	140–142		819
	OH / $C_6H_4CF_3$-*m*	H	H	185–186		819
	CN / Ph	H	H	199–201		819
	COMe / Ph	H	H	158–159		819,837
	$CONH_2$ / piperidino	H	H	208–209		819,837

TABLE 115. *(Continued)*

Alk	Am	X	Y	mp (°C) or bp (°C/Torr)	Salt	Refs.
CH_2	NH-*i*-Pr	H	H	229 (dec.)	HCl	882
CH_2	NMe_2	H	H	71–72		
				236–237	HCl	587,814
CH_2	NEt_2	H	H	99–100		496
				228	HCl	587,814
CH_2	$N(CH_2CH{=}CH_2)_2$	H	H	189/0.05		814
CH_2	$NMe(CH_2)_2NMe_2$	H	H	236	2HCl	587
CH_2	$NEt(CH_2)_2NEt_2$	H	H	205–207/0.2		587,814
CH_2	N(piperidino ring)	3-Cl	7-Cl	105		496
CH_2	—N(piperazine ring)N-Alk-OR					
	Alk: CH_2CH_2, R: H	H	H	105–107		
				250–251	2HCl	734,882
	Alk: CH_2CH_2, R: $COC_6H_2(OMe)_3$ 3,4,5	H	H	190–192	Difumarate	734
	Alk: CH_2CHMe, R: H	H	H	175–180	Difumarate	734
	Alk: CH_2CHMe, R: $COC_6H_2(OMe)_3$-3,4,5	H	H	200–202	Difumarate	734
CHMe	NHMe	H	H	87		
				214–215	HCl	587
CHMe	NMe_2	H	H	110		587
CHMe	NEt_2	H	H	90		587,814
CHMe	—N(piperidino ring)	H	H	112–113		587,814
				166–167		496

CHMe	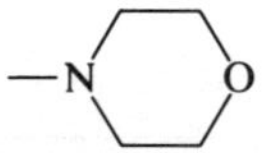 —N O (ring)	H	H	146		587
$(CH_2)_2$	NHMe	H	H	167	HCl	634
$(CH_2)_2$	NMe_2	H	H	165–167	HCl	634
				195–197/0.2		587,814
				168–170	HCl	
$(CH_2)_2$	NEt_2	H	H	192–194/0.1		587
				168–170	HCl	634
				179–180	HCl	814
$(CH_2)_2$	NMePh	H	H	172–174	HCl	634
$(CH_2)_2$	N(*i*-Pr)$_2$	H	H	187–188	HCl	634
$(CH_2)_2$	N(*n*-Bu)$_2$	H	H	126–127	Oxalate	634
$(CH_2)_2$	$N(CH_2Ph)(CH_2CH_2OH)$	H	H	101–103		
				170–172	Oxalate	634
$(CH_2)_2$	 N (ring)	H	H	158–159	HCl	634
$(CH_2)_2$	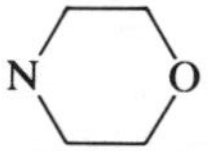 N O (ring)	H	H	205–206	HCl	634
$(CH_2)_2$	$N(CH_2)_2COIB$[a]	H	H	137 (dec.)		634
$(CH_2)_2$	 N (ring) R_1 R_2					
	R_1 / R_2					
	H / NMe_2	H	H	250–253	2HCl	821
	OH / Ph	H	H	203–205	HCl	821
	$CONH_2$ 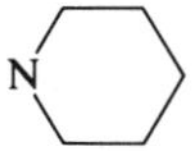 / N (ring)	H	H	237–238	$2HCl\cdot H_2O$	821

(*continued*)

TABLE 115. *(Continued)*

Alk	Am	X	Y	mp (°C) or bp (°C/Torr)	Salt	Refs.
	$CONH_2$ N(piperidino)	3-Cl	H	239–242	2HCl·0.5H_2O)	821
	-SCH_2CONH-	H	H	227–229	HCl	821
$(CH_2)_2$	N(piperazine)N—R					
	R					
	Me	H	H	227–232	HCl	634
	$(CH_2)_2OH$	H	H	130	2HCl	634
				173–175	Difumarate	734
		H	H	160–165	Difumarate	734
	$CH_2CHMeOH$	H	H	170–173	Difumarate	734
	$COC_6H_2(OMe)_3$-3,4,5					
	$(CH_2)_2O(CH_2)_2CO$-IB[a]	H	H	238–238.5		634
$CHMeCH_2$	NMe_2	H	H	240–241	HCl	634
$CHMeCH_2$	NEt_2	H	H	230–231	HCl	634
$CHMeCH_2$	N(piperidino)	H	H	248	HCl	634
$CHMeCH_2$	N(morpholino)O	H	H	250–251	HCl	634

[a]IB = 10,11-Dihydrodibenz[*b*, *f*] azepin-5-yl.

TABLE 116. ω-AMINOALKYL ESTERS, THIOESTERS, AND AMIDES OF THE DIBENZ [*b,f*]AZEPINE-5-CARBOXYLIC ACIDS

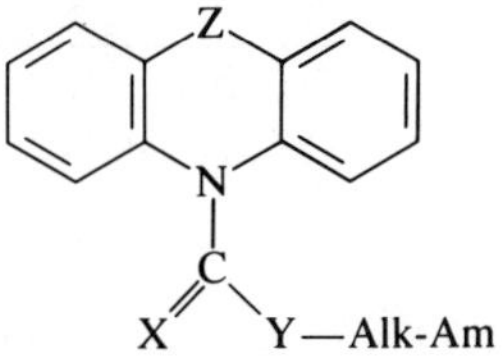

Z; N; C; X; Y—Alk-Am

X	Y	Z	Alk	Am	mp (°C)	Refs.
O	O	CH_2CH_2	$(CH_2)_3$	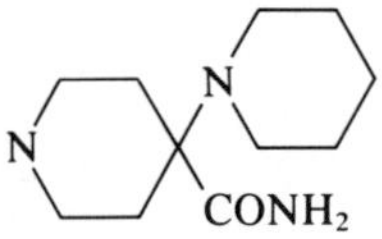 N; N; $CONH_2$	248–249	818
O	O	CH_2CH_2	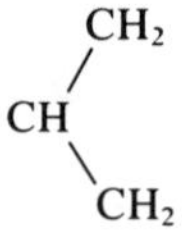 CH_2; CH; CH_2	*bis*-NMe_2	72 174–175 (picrate)	505
O	NH	CH_2CH_2	—	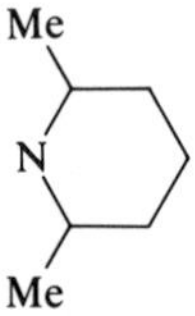 Me; N; Me	131–131	830
O	NH	CH_2CH_2	—	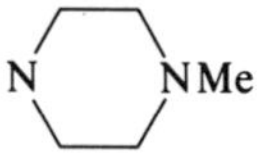 N; NMe	161–163	830
O	NH	CMe=CH	$(CH_2)_2$	NMe_2	145	817
O	NH	CMe=CH	$(CH_2)_2$	NEt_2	99	817
O	NH	CH=CH	$(CH_2)_2$	 N	157–159 207–209 (HCl)	822
O	NH	$CHMeCH_2$	$(CH_2)_2$	NMe_2	115	817
O	NH	$CH(CH_2Ph)CH_2$	$(CH_2)_2$	NEt_2	240/0.003[a]	817
O	NMe	CMe=CH	$(CH_2)_2$	NMe_2	145	817
O	NEt	$CHMeCH_2$	$(CH_2)_2$	NEt_2	166–168	817
O	NMe	$CH(CH_2Ph)CH_2$	$(CH_2)_2$	NMe_2	215–216 (HCl)	817
O	NEt	$CH(CH_2Ph)CH_2$	$(CH_2)_2$	NEt_2	166–167 (HCl)	817
O	NH	CH=CH	$(CH_2)_3$	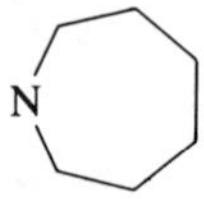 N	92–94 219–221 (MeI)	822
O	NH	$CMe=CH_2$	$(CH_2)_3$	NMe_2	110–111	817
O	NH	$CHMeCH_2$	$(CH_2)_3$	NEt_2	93	817
O	NH	CH_2CH_2	$CH_2CHOHCH_2$	NH-*i*-Pr	202 (HCl)	848
O	NH	CH_2CH_2	$CH_2CHOHCH_2$	NH-*t*-Bu	249–250 (HCl)	848
O	NH	CH=CH	$CH_2CHOHCH_2$	N-*i*-Pr(CH_2Ph)	190–191 (HCl)	848
O	NH	CH_2CH_2	$CH_2CHOHCH_2$	N-*i*-Pr(CH_2Ph)	180–181 (HCl)	848
O	S	CH_2CH_2	$(CH_2)_2$	NMe_2	230–231 (HCl)	824

TABLE 116. (*Continued*)

X	Y	Z	Alk	Am	mp (°C)	Refs.
O	S	CH_2CH_2	$(CH_2)_2$	NMe-*i*-Pr	215–217 (HCl)	820
O	S	CH_2CH_2	$(CH_2)_2$	NEt-*i*-Pr	76–78 224–227 (HCl) 204–205 (MeBr)	820
O	S	CH═CH	$(CH_2)_2$	N-*i*-Pr_2	203–205 (HCl) 225–227 (MeBr)	820
O	S	CH_2CH_2	$(CH_2)_2$	N-*i*-Pr_2	206–208 (HCl) 205–207 (MeBr)	820
S	S	CH_2CH_2	$(CH_2)_2$	NEt_2	240 (HCl)	815

[a]Boiling point (°C/torr)

TABLE 117. 5H-AMINOACETAMIDO-10,11-DIHYDRODIBENZ[*b,f*]AZEPINES[a]

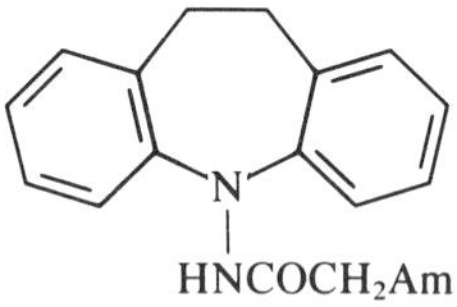

N
HNCOCH$_2$Am

Am	mp (°C)
NHMe	160–162
NMe_2	188–190
 N (piperidino)	207.5–209
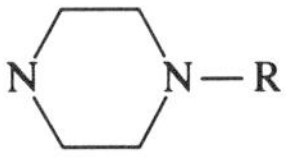 N N—R (piperazino)	
R = Me	202–204
R = *n*-Pr	197–200
R = CH_2Ph	199–200
R = CH_2CH_2OH	194–196
R = CH_2CH_2OAc	164–165
R = Ph	236–238
R = C_6H_4Cl-*p*	259–261
R = C_6H_4OMe-*p*	220–224
R = $C_6H_4CF_3$-*m*	264–266

[a]Ref. 803.

H_2/Ra-Ni (30–140 atm), EtOH, 60–100°C

RCHCN **1389** → $RCHCH_2NH_2$ **1548** (232)

(1) $NaNH_2$, C_6H_6; (2) $ClCH_2CH—CH_2$ (epoxide); then HNR_1R_2, ROH

1542 (X, N–H) → **1549** ($CH_2CH—CH_2$, epoxide O) → **1550** ($CH_2CHCH_2NR_1R_2$, OH) (233)

5-Aminoalkyl-10,11-dihydrodibenz[*b,f*]azepin-10-ones (**1552**) have been prepared by two different approaches (Fig. 198). In the first of these, the 10-alkoxyiminostilbenes **1348a,** were alkylated with the aid of sodium amide, and the resultant products **1551** were hydrolyzed with acid to the desired ketones **1552** (544,545). The alkylation of **1348a** represents an example of the need to protect the base-labile 10-keto group as an enol ether prior to alkylation. The second approach to **1552** has been via lithium dimethylamide–promoted cyclization of the *N*-(aminoalkyl)-*N,N*-diarylamine **1553** (562–564).

Several other classes of compounds covered most conveniently in this section (Fig. 199, Eq. 234) are the 5-(α- or 5-β-dialkylaminoalkanoyl) dibenz[*b,f*]azepines **1554,** and the dialkylaminoalkyl carbamates, ureas, thiolocarbamates, and dithiocarbamates **1391, 1555a, 1555b** and **1556.** The dialkylaminoacyl compounds **1554** have been prepared by condensing dialkylaminoacyl halides with iminobibenzyl (816) or by the action of an amine on an intermediate haloacyl derivative (**1414a**) (496,634,636,814). The carbamate **1391** (613–616, 818), the urea **1555a** (822), and the thiolocarbamate **1555b** (820,824) can be formed by treatment of the chlorocarbonyl derivative **1390** with the appropriate alcohol, amine, or thiol. The dithiocarbamates **1556** are accessible by sequential treatment of iminobibenzyl **1542** with phenyllithium, carbon disulfide, and a dialkylaminoalkyl halide (815).

Figure 198

Figure 199

HX—AlkNR$_1$R$_2$ / C_6H_6

COCl
1390

X—AlkNR$_1$R$_2$

1391, X = O
1555a, X = NH
1555b, X = S

(234)

Procedures that have been reported to convert tertiary aminoalkyldibenz[*b,f*]azepines to the corresponding secondary and primary amines, are detailed in Figs. 200 and 201. Conversion of side-chain secondary amines to tertiary *N*-methyl derivatives can be accomplished by reductive alkylation with formaldehyde and sodium borohydride (704,785), or by direct alkylation with dimethyl sulfate (785) (Eq. 235a). Reductive alkylation of primary side-chain amine groups has likewise been described (Fig. 202). Treatment of secondary amines with benzoyl peroxide, followed by alkaline hydrolysis, leads to side-chain *N*-hydroxy compounds (736) (Eq. 235b).

D. *Miscellaneous Chemistry of Dibenz[b,f]azepines*

a. REACTIONS WITH ELECTROPHILES

i. ACID-PROMOTED REACTIONS. Schindler and Blattner were the first to demonstrate that 5-acetyliminostilbene (**1311a**) and 5-acetyl-3-ethyliminostilbene (**1557**) gave the 9-methylacridines **1319a** and **1558,** respectively,

H_2/Pd–C or H_2/Raney Ni

$(CH_2)_3NCH_2Ph$ with R_1 → $(CH_2)_3NR_2R_3$

R_1	Conditions	R_2	R_3	Ref.
(*a*) Me	H_2/10% Pd–C, 60 psi. EtOH, 1 eq. aq. HCl	H	Me	731
(*b*) Me	Method *a* (except Raney Ni)	H	Me	784
(*c*) CH_2Ph	Method *a*, 40°C	H	CH_2Ph	732
(*d*) CH_2Ph	Method *a*, 64°C	H	H	

Figure 200

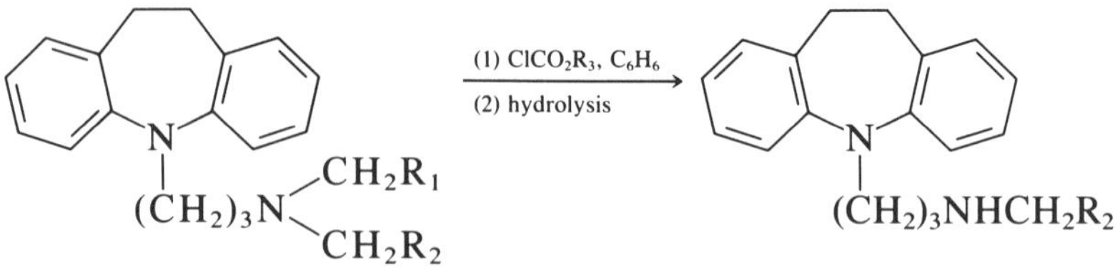

(1) $ClCO_2R_3$, C_6H_6
(2) hydrolysis

$(CH_2)_3N(CH_2R_1)(CH_2R_2)$ → $(CH_2)_3NHCH_2R_2$

R_3	Hydrolysis Conditions	Refs.
Et	KOH, $HO(CH_2)_2O(CH_2)_2OEt$, Δ	595, 737, 745, 752
CH_2Ph	30% HBr, HOAc, Δ	742
i-Pr	NaOMe, HMPA	805

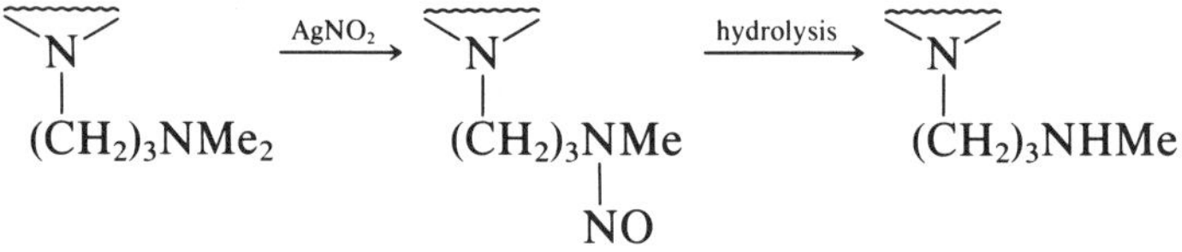

$(CH_2)_3NMe_2$ —$AgNO_2$→ $(CH_2)_3N(Me)NO$ —hydrolysis→ $(CH_2)_3NHMe$ (Ref. 798)

Figure 201

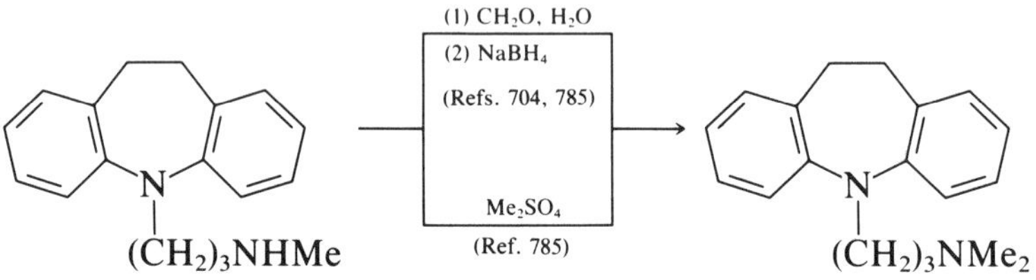

$(CH_2)_3NHMe$ → $(CH_2)_3NMe_2$ (235a)

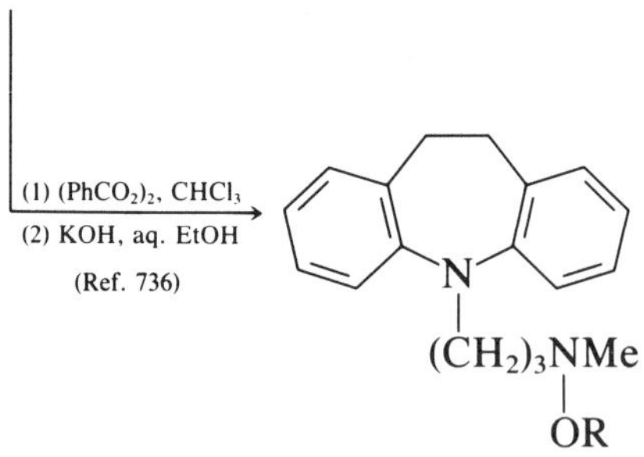

(1) CH_2O, H_2O
(2) $NaBH_4$
(Refs. 704, 785)

Me_2SO_4
(Ref. 785)

(1) $(PhCO_2)_2$, $CHCl_3$
(2) KOH, aq. EtOH
(Ref. 736)

$(CH_2)_3N(Me)OR$ (235b)

R = H, COPh

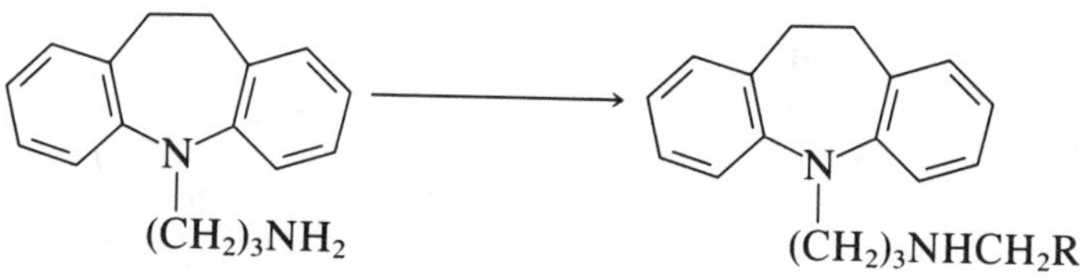

$(CH_2)_3NH_2$ → $(CH_2)_3NHCH_2R$

Reaction Conditions	R	Refs.
(1) AcCl (2) $LiAlH_4$	Me	719
(1) RCHO, dioxane, Δ (2) H_2/Raney Ni or H_2/Pd–C	H, Alkyl	595, 596, 610, 721
(1) HCO_2Me, Δ (2) $LiAlH_4$	H	610

Figure 202

in the presence of 48% HBr (Eq. 236) (512,532). The mechanism of the rearrangement was investigated subsequently by Rumpf and Reynaud (1042), who confirmed that under vigorous conditions (5 *N* HCl), the major product obtained from iminostilbene (**1311b**) was **1319a.** However, under mild conditions (0.15 *N* HCl), the major rearrangement product was not **1319a,** but instead acridine (**1559**) itself. Their proposed mechanism (Fig. 203) invoked azaquinonemethide **1560** as the initial protonated species in the rearrangement. Equilibration with the isomeric intermediate **1561** can be seen as leading to **1319a.** They suggested that acridine (**1559**) was formed by reaction of **1561** with molecular oxygen followed by the subsequent expulsion of a hydroperoxymethylene fragment.

A more thorough investigation of the rearrangement was published recently by Bendall and co-workers (673), who found that when an alcoholic solution of 5-nitrosoiminostilbene (**1517a**) was acidified with a few drops of 5% hydrochloric acid, acridine formed rapidly ($t_{1/2}$ = 10 min) as the major product (some of this work was discussed in Section 39.C.h). Several minor products, including **1311b,** 2-nitroiminostilbene (**1517b**), and acridine-9-carboxaldehyde (**1518b**), were also isolated. These results further confirmed the

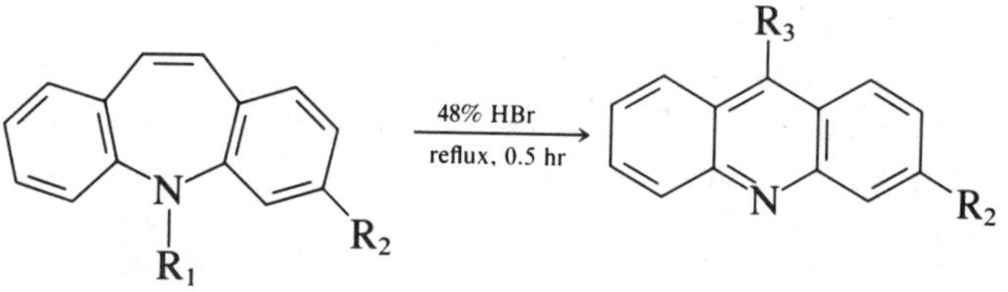

48% HBr, reflux, 0.5 hr → (236)

1311a, R_1 = COMe, R_2 = H
1311b, R_1 = R_2 = H
1557, R_1 = COMe, R_2 = Et

1319a, R_2 = H, R_3 = Me
1558, R_2 = Et, R_3 = Me
1559, R_2 = R_3 = H

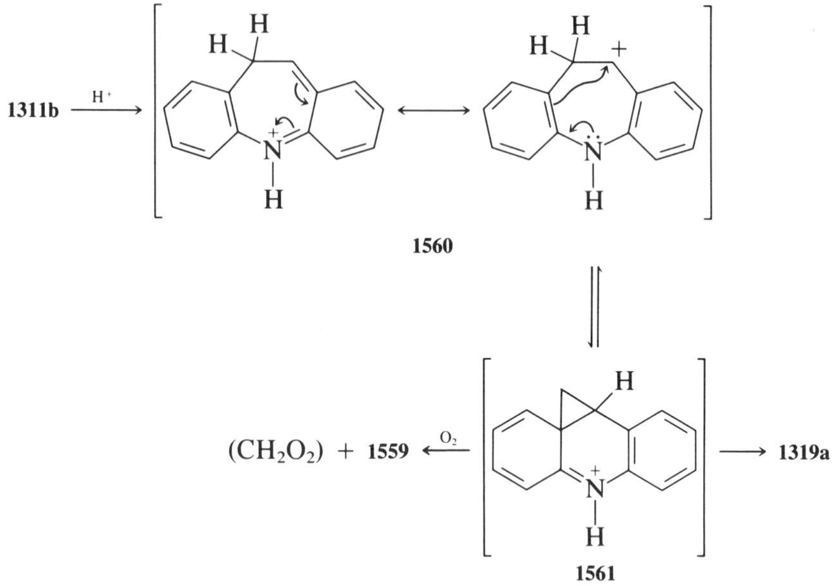

Figure 203

earlier suggestion (1042) that oxygen is involved in the formation of **1559.** They found oxygen to be equally crucial in the production of **1518b** and **1517b.** However, the isolation of alkyl formates prompted them to question the existence of the hydroperoxymethylene fragment proposed by Rumpf and Reynaud to explain the origin of **1559.** Table 94 summarizes the work of Bendall and co-workers, and Fig. 204 summarizes their mechanistic conclusions.

The rearrangement of **1311b** and **1517a** most probably proceeds via the cation **1562.** The radical cation **1563,** which can be formed directly from **1517a** or from **1311b** via **1562,** was postulated as the critical intermediate leading directly to **1517b.** Ring contraction of **1562** by way of the hydroperoxides **1564–1566** was proposed to be the source of **1559** and **1518b.** Under vigorous conditions, **1562** apparently undergoes rearrangement via the quinonemethide **1560** to give **1319a,** whereas under mild conditions it undergoes slow auto-oxidation to **1517b, 1518b,** and **1559.** This mechanistic interpretation also helps explain (1) the formation of the alkyl formates via **1566,** and (2) the preponderance of **1518b** when rearrangement was carried out in the nonhydroxylic medium acetone.

Interestingly, while **1311b** underwent rearrangement to 9-methylacridine (**1319a**) in 48% HBr (512), the corresponding 10,11-dialkyliminostilbenes were stable under these conditions (516,536). Thus when 5-benzyl-10,11-dimethyliminostilbene (**1567a**) was heated for 12 hr at 90–95°C in 48% HBr, the debenzylated iminostilbene **1567b** was isolated in good yield (Eq. 237).

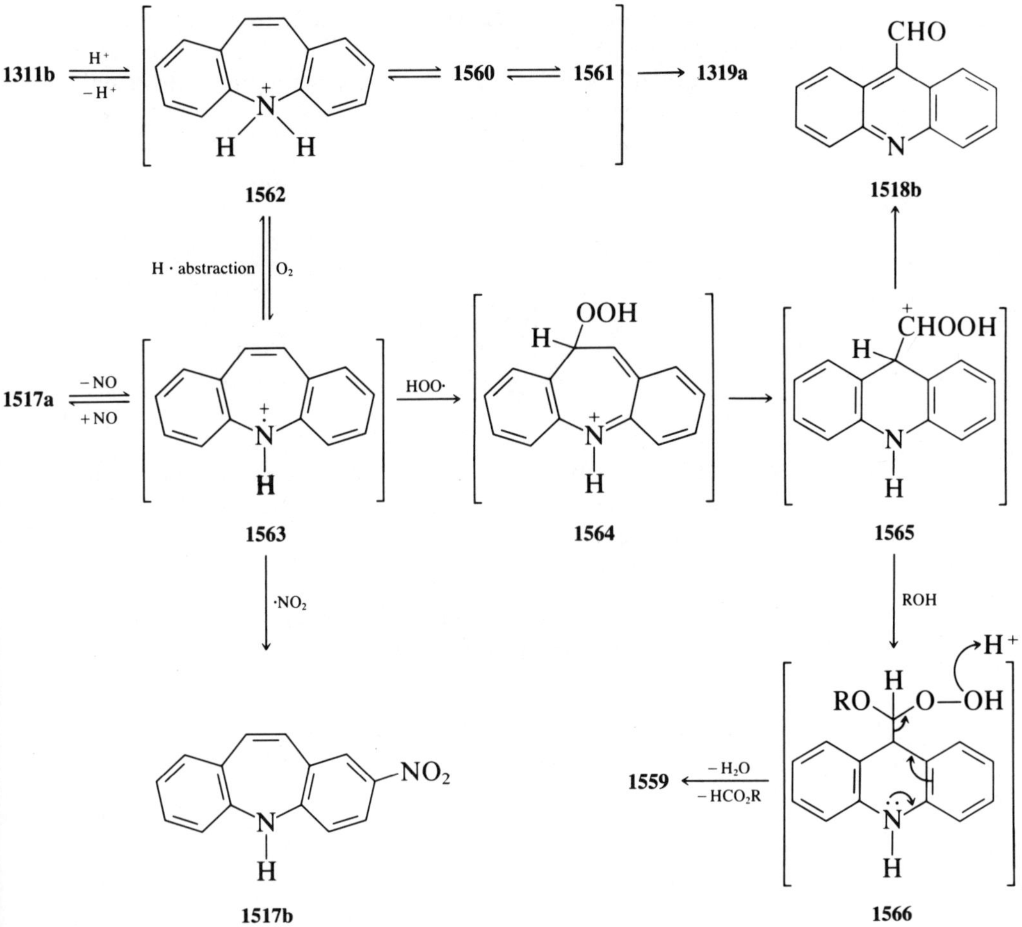

1311b ⇌ (H^+ / $-H^+$) [1562 ⇌ 1560 ⇌ 1561] → 1319a

1562: $\overset{+}{N}$, H, H

1518b: CHO, N

1562 ⇌ (H · abstraction, O_2) 1563

1517a ⇌ ($-NO$ / $+NO$) [1563: $\overset{+}{\dot{N}}$—H]

1563 → ($HOO\cdot$) [1564: H, OOH, $\overset{+}{N}$—H] → [1565: H, $\overset{+}{C}HOOH$, N—H] → 1518b

1563 → ($\cdot NO_2$) 1517b: NO_2, N—H

1565 → (ROH) [1566: RO, H, O—OH, H^+, $\ddot{N}$—H]

1559 ← ($-H_2O$, $-HCO_2R$) 1566

Figure 204

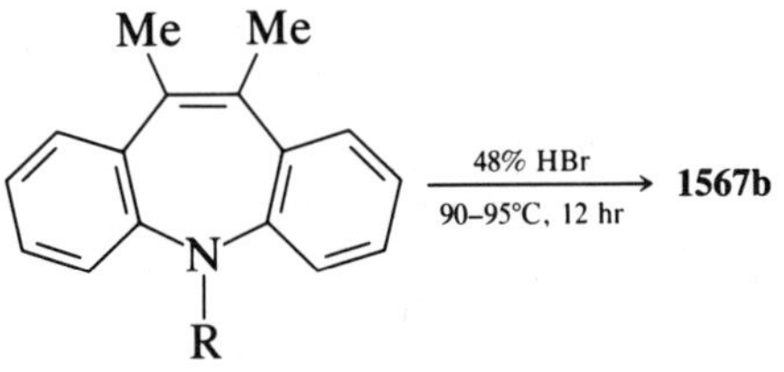

$$\text{1567a} \xrightarrow[\text{90–95°C, 12 hr}]{\text{48\% HBr}} \text{1567b} \qquad (237)$$

Me Me, N—R

1567a, R = CH_2Ph
1567b, R = H

As expected, iminobibenzyls are likewise quite stable to acidic conditions, allowing the use of 48% HBr as a reagent for the dealkylation of **1567c** or **1567d** to **1567e** (Eq. 238) (516,536). Alternatively, catalytic hydrogenolysis is used when the alkyl group is benzyl (608,630).

48% HBr, Δ

H_2/Pd–C

MeOH

(238)

1567c, R = Me
1567d, R = CH_2Ph

1567e

ii. Electrophilic Aromatic Substitution. Electrophilic aromatic substitution reactions were discussed in detail in Section 39.C; alkylations were covered in Section 39.C.a, halogenations in Section 39.C.b.i, acylations in Section 39.C.d, and nitrosations in Section 39.C.h.

iii. *N*-Acylation Reactions. One of the most common reactions of the dibenz[*b,f*]azepine family is *N*-acylation. Both iminostilbene (**1311b**)

TABLE 118. N-ACYLATIONS OF DIBENZ[*b,f*]AZEPINES

R	Method	Refs.
Me	Ac_2O–C_5H_5N	512,629
Me	Ac_2O–C_6H_6, Δ	511,523
Me	AcCl	499,513
Alkyl	RCOCl–C_6H_6	512,584 831–834,838
H	HCO_2H–Ac_2O	630
OEt	(1) NaH–PhMe (2) $ClCO_2Et$	512,523
Cl	Cl_2CO–PhMe	526,528,546,623, 608,825,826
	Cl_2CO–$C_6H_4Cl_2$-*o*	827
NHR_1 (R_1 = H, alkyl,acyl)	RNCO–C_6H_6 (or $CHCl_3$)	822,823, 835,839
NR_1R_2	$ClC(=O)NR_1R_2$–Xylene	825,837

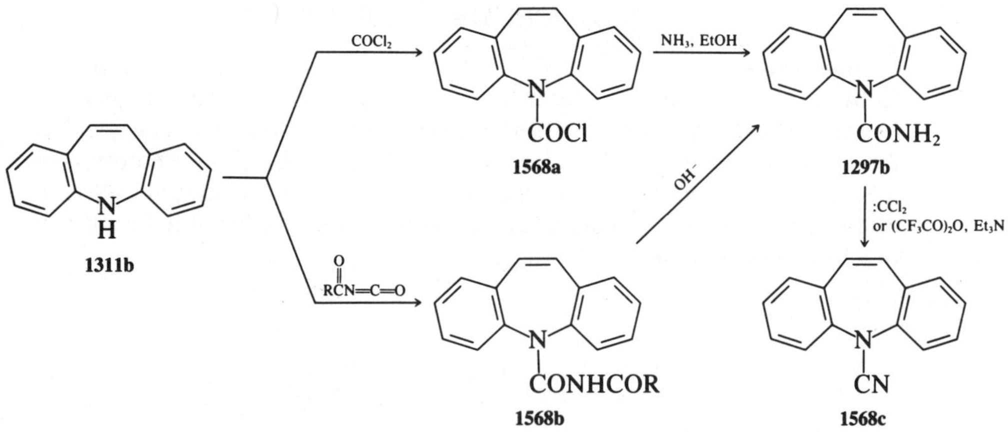

Figure 205

and iminobibenzyl (**1293**) undergo acylation in a normal manner to give *N*-acyl derivatives. A summary of various acylation conditions that have been employed is collected in Table 118.

Of considerable interest are *N*-carboxamidoiminostilbenes such as the anticonvulsant agent carbamazepine (Tegretol). The most frequently employed synthesis of **1297b** (Fig. 205) has been via the carbonyl chloride **1568a,** which can be prepared by the action of phosgene on iminostilbene (**1311b,** see Table 118). The conversion of **1568a** to **1297b** has been carried out with ethanolic ammonia (528,546,623,630,825–827). Benzene (817) has also been used as the reaction solvent. Secondary and tertiary analogs of **1297b** have been obtained in the same way from the corresponding primary and secondary amines (817,819,822). An alternate approach to **1297b** has been via the action of *N*-acyl isocyanates on **1311b** followed by hydrolysis of the intermediate acylureas **1568b** (823,827,835,839). Dichlorocarbene reacted with **1297b** to give the *N*-cyanoiminostilbene **1568c** (840). Trifluoroacetic anhydride–triethylamine also converted **1297b** to **1568c** (900).

An interesting nonelectrophilic approach to the preparation of an *N*-acyl derivative involved the permanganate oxidation of *N*-methyl-10,11-methanoiminobibenzyl (**1569a**) to the *N*-formyl compound **1569b** (586) (Eq. 239). The reaction, which proceeded in 68% yield, may be applicable to other *N*-aklyliminobibenzyls.

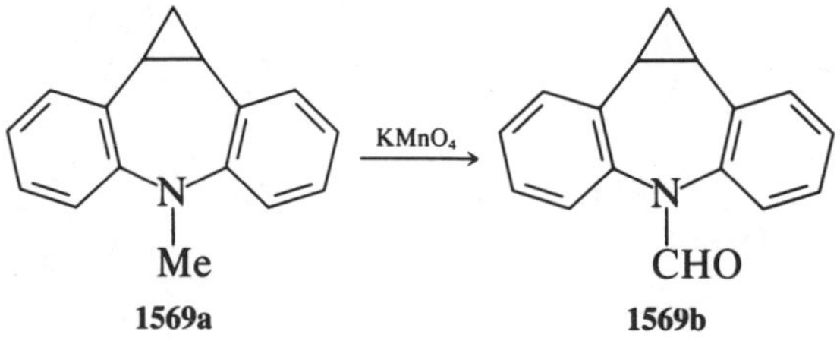

(239)

b. PHOTOCHEMICAL REACTIONS

Iminostilbene (**1311b**) and *N*-nitrosoiminostilbene (**1517a**) were reported to be stable to photolysis under argon (673). However, in the presence of oxygen, a benzene solution of **1517a** reacted readily to give a 23% yield of 2-nitroiminostilbene (**1517b**), along with **1311b** (21%, Eq. 240). Photolysis of a benzene solution of **1311b** in the presence of oxygen produced a considerable amount of a red, insoluble material that was speculated to be a dimer of acridine or acridane.

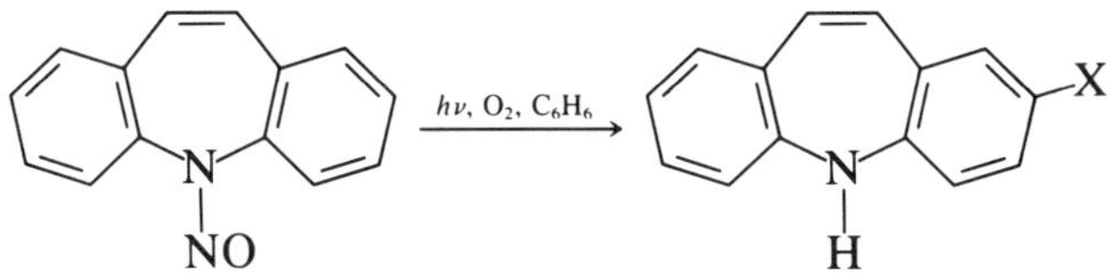

$h\nu$, O_2, C_6H_6 (240)

1517a

1517b, X = NO_2
1311b, X = H

In a subsequent publication, Kricka and co-workers confirmed that degassed solutions of iminostilbene and its *N*-alkyl derivatives were stable to both sensitized and unsensitized photolysis (658,672). In contrast, they found that benzophenone-sensitized irradiation of 5-acetyliminostilbene (**1311a**) gave a 77% yield of the dimer **1571** after only 5 min (Eq. 241). The quantum yield (ϕ) was 0.27. The unsensitized reaction was much less efficient (36% yield after 1 hr). In both instances oxygen retarded the reaction, further suggesting triplet intermediates. Michler's ketone (E_t = 61.0 kcal/mole), but not fluorenone (E_t = 53.3 kcal/mole), sensitized the dimerization reaction, suggesting an E_t between 53.3 and 61.0 kcal/mole. The sensitized irradiation of several other *N*-acyl derivatives also gave dimeric products. In every example, only one product was found, and since the reaction was presumed to proceed via triplet-state reactants, the more stable *trans* configuration was assigned, as shown in **1571**.

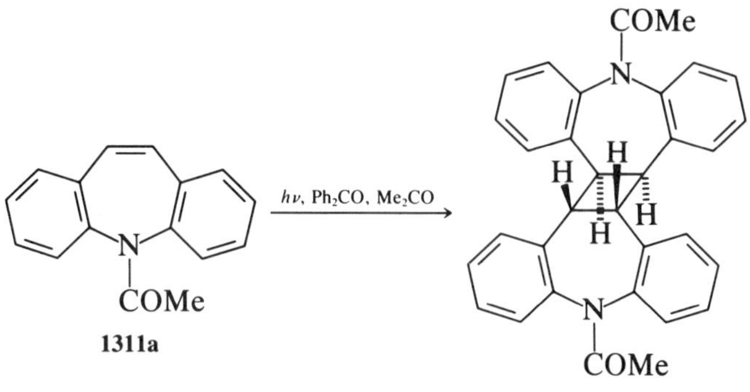

$h\nu$, Ph_2CO, Me_2CO (241)

1311a

1571

Irradiation of a benzophenone-sensitized solution of **1311a** in acetone containing either *N*-methyl- or *N*-phenylmaleimide (**1572**) gave the mixed dimers **1573a** along with the homodimer **1571** and a trace of the maleimide dimers **1574** (Eq. 242) (658). Again, only one type of mixed dimer was found, which was presumed to be the *trans* adduct. The photodimerization of *N*-formyliminostilbene (**1570**) and maleic anhydride (Eq. 243) was reported to give the adduct **1573b** (862). Treatment of **1573b** with lead tetraacetate was reported to give the tetracycle **1575,** although no supporting evidence was supplied. Dimethylmaleate and dichloromaleic anhydride also gave adducts with **1311a,** but the reaction mixtures were complex, and the adducts could be characterized only on the basis of mass spectral analysis (658). Several other olefins and acetylenes, covering a wide range of triplet energy values, failed to react. The unsensitized irradiation of 5-tosyliminostilbene (**1515a**) proceeded in low conversion to the 2-tosyl isomer **1515b** (658) (see Eq. 213).

1311a + **1572**, R = Me, Ph $\xrightarrow[\text{Ph}_2\text{CO, Me}_2\text{CO}]{h\nu}$ **1571** + **1573a**, R = Me, Ph + **1574**, R = Me, Ph (242)

1570 $\xrightarrow[\text{Ph}_2\text{CO, Me}_2\text{CO}]{\text{maleic anhydride}, h\nu}$ **1573b** $\xrightarrow[\text{C}_5\text{H}_5\text{N}]{\text{Pb(OAc)}_4}$ **1575** (243)

c. THERMAL REACTIONS

Only a limited amount of data has been reported on the thermolysis of dibenz[*b,f*]azepines. During the thermal dehydrogenation of iminobibenzyl (**1293**) over a ferric oxide catalyst, trace amounts of acridine (**1559**) and 5-methylacridine (**1319a**) were formed, presumably via decomposition of the dehydrogenation product iminostilbene (522) (see Section 39.B.b.i).

A more detailed account of the thermal decomposition of 5-nitrosoiminostilbene (**1517a**) has been given (Fig. 206) (673). In refluxing hydroxylic solvents (methanol, ethanol, or *n*-propanol), the major products were **1559** and **1311b,** together with minor amounts of **1517b.** When **1311b** was refluxed under similar conditions, no reaction occurred, thus ruling out **1311b** as the source of **1559** and **1517b.** In nonhydroxylic solvents (benzene and carbon tetrachloride), and major products were **1311b** and acridine-9-carboxaldehyde (**1518b**). In refluxing cumene, the reaction proceeded cleanly to afford **1518b** and **1559** in good yield. The formation of both products was shown to be oxygen dependent. Homolytic cleavage of the N—NO bond has been postulated as the first step in the process. The resonance-stabilized radical **1576** could abstract a proton to give **1311b** or react with oxygen to produce the hydroperoxides **1577** and **1564.** The fate of **1564,** leading to **1518b** and **1559,** has been discussed in Section 39.D.a.i (see Fig. 204).

The thermal degradation of carbamazepine (**1297b**) during gas chromatographic analysis has been reported (846,849,856,887,1044) to produce minor amounts of iminostilbene (**1311b**) and 9-methylacridine (**1319a**). The temperatures at the injection port were 250–280°C and the column (3% OV-17 on Gas Chrom Q) was maintained at 200–260°C. The epoxide **1435** underwent rearrangement to give exclusively 9-acridinecarboxaldehyde (**1518b**, Fig. 207). The dihydroxycarbamazepine **1436b** also gave **1518b.** Mechanist-

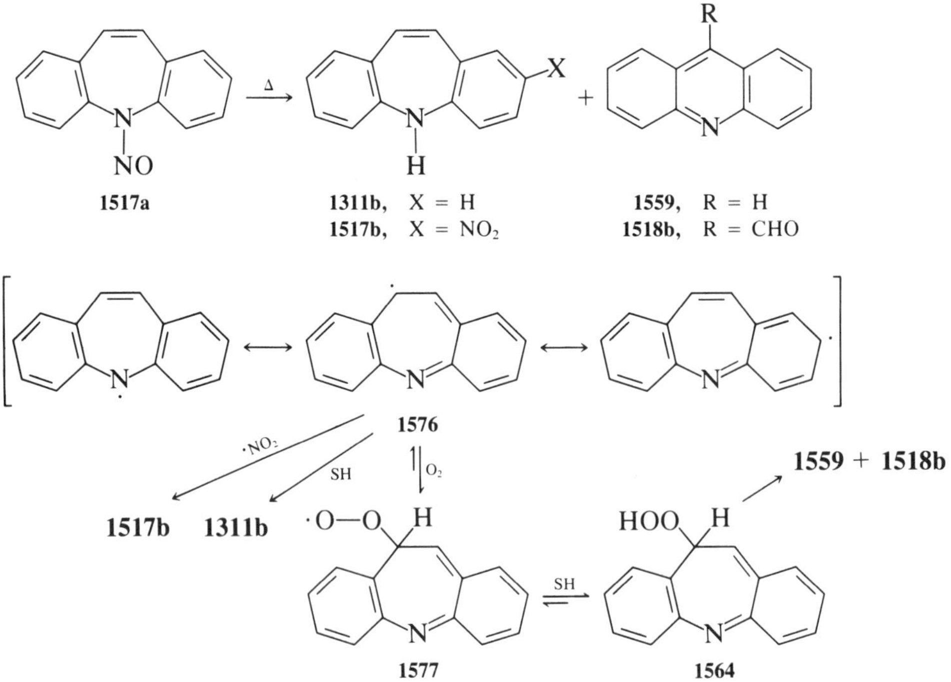

NO

1517a

H

1311b, X = H
1517b, X = NO_2

R

1559, R = H
1518b, R = CHO

1576

$\cdot NO_2$ SH O_2

1517b **1311b**

$\cdot$O—O H

HOO H

1559 + **1518b**

SH

1577 **1564**

Figure 206

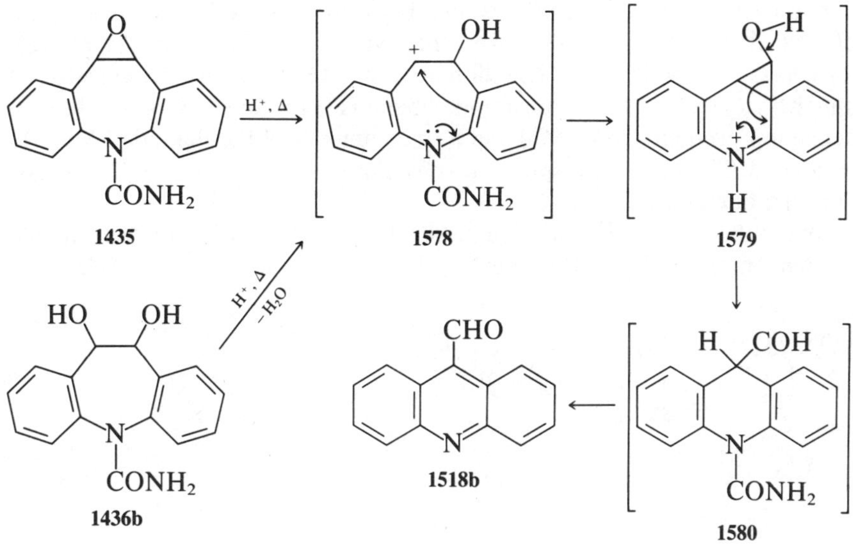

1435 1578 1579 1436b 1518b 1580

Figure 207

ically, all of the processes appear to be acid catalyzed, since the rearrangement of **1436b** did not take place on the less acidic SE-30 column, and the degradation of **1297b** occurred only when the compound was injected as a methanol solution. Thus the decomposition of **1297b** most likely occurred by methanolysis of the amide linkage to give **1311b** followed by the acid-catalyzed rearrangement of **1297b,** as described in Section 39.D.a.i (see Fig. 203). Rearrangement of **1435** and **1436b** no doubt proceeds via the pinacol-type intermediates **1578–1580**.

d. ANNELATION REACTIONS

Although the tetracyclic products obtained on annelation of dibenz[*b,f*]-azepines do not, strictly speaking, fall within the scope of this review, they merit brief consideration as a relevant aspect of dibenz[*b,f*]azepine chemistry. Most reported annelation reactions have resulted in the creation of new rings across the 4- and 5-positions of the dibenz[*b,f*]azepine system. Condensation of either iminobibenzyl or iminostilbene with oxalyl chloride (Eq. 244) gave the ketoamides **1483** and **1484,** a reaction already cited in

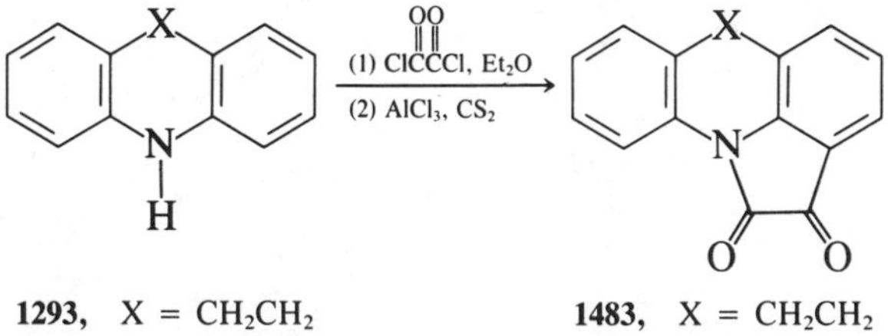

(1) ClCOCOCl, Et_2O; (2) $AlCl_3$, CS_2 (244)

1293, X = CH_2CH_2 **1483,** X = CH_2CH_2
1311b, X = CH=CH **1484,** X = CH=CH

Section 39.C.3 (574, 579, 635). When benzylmalonic acid *bis* 2,4-dichlorophenyl ester (**1581**) and **1293** were heated to 260–270°C (Eq. 245a), **1583** formed in 85% yield (829). The diphenyl ester **1582** likewise produced **1583,** but the yield was only 29%. Iminostilbene (**1311b**) and **1581** gave the condensation product **1584** (28%) and a trace amount of the diamide **1585.** Malonic acid, **1311b,** and phosphorous oxychloride in quinoline (Eq. 245b) gave only the diamide **1586.**

When the N-(α-haloalkanoyl)iminobibenzyls **1414a** were treated with aluminum chloride (Eq. 246), they cyclized to the tetracyclic amides **1587**

1293, 1311b + $PhCH_2CH(CO_2R)_2$ $\xrightarrow{260–270°C}$

1581, R = $C_6H_3Cl_2$-2,4
1582, R = Ph

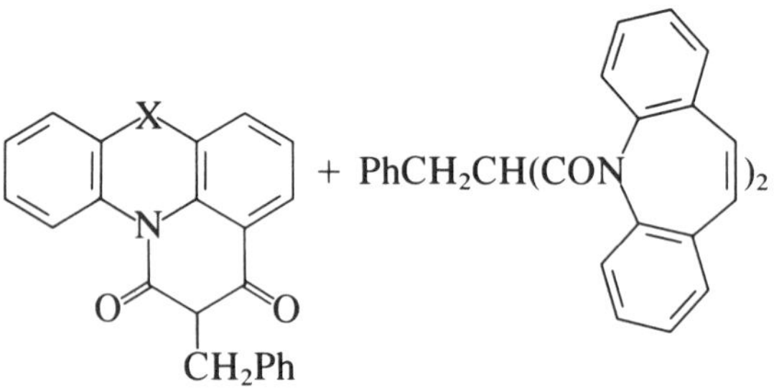

+ $PhCH_2CH(CON$ $)_2$ (245a)

CH_2Ph

1583, X = CH_2CH_2
1584, X = CH=CH

1585

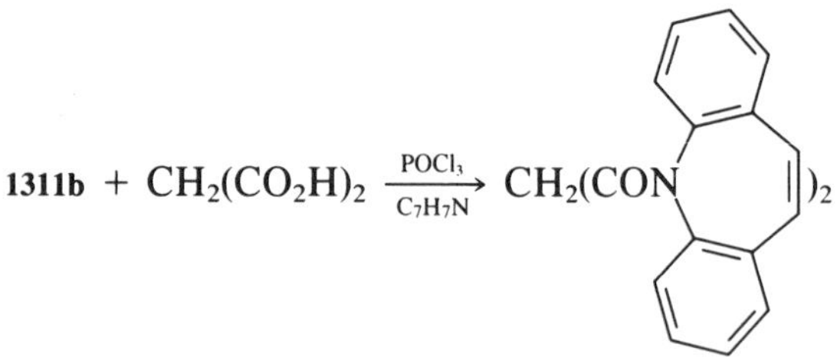

1311b + $CH_2(CO_2H)_2$ $\xrightarrow[C_7H_7N]{POCl_3}$ $CH_2(CON$ $)_2$ (245b)

1586

$\xrightarrow{AlCl_3}$ (246)

COCHX
R

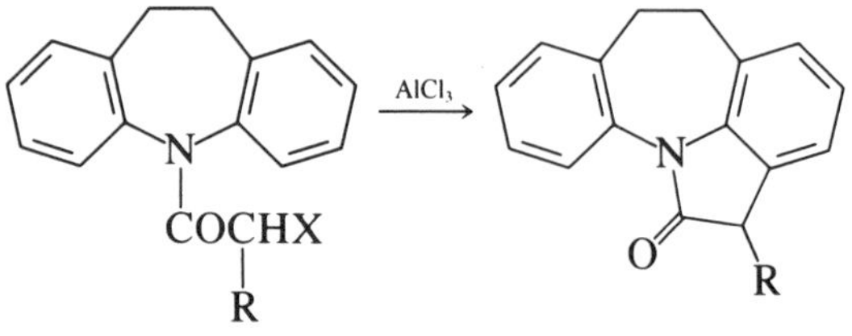

1414a **1587**

(637,677,844). Heating 5-iminobibenzylpropanoic acid (**1588**) and trifluoroacetic anhydride (Eq. 247) in refluxing benzene afforded the ketones **1589** and **1590** in 62 and 13% yields, respectively (867).

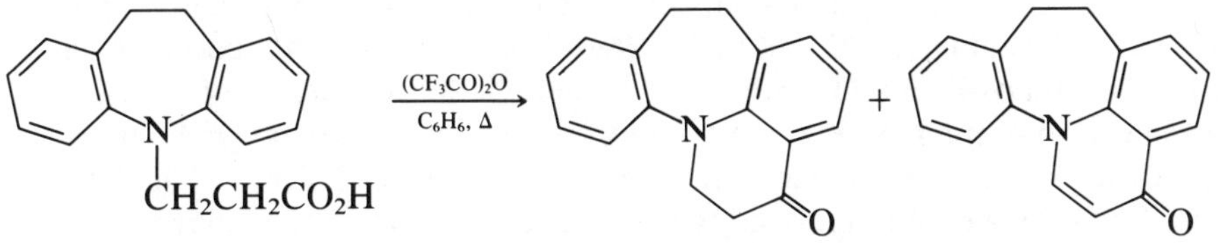

$CH_2CH_2CO_2H$ $\xrightarrow[C_6H_6,\ \Delta]{(CF_3CO)_2O}$ + (247)

1588 **1589** **1590**

A modified Fischer indole synthesis involving *N*-aminoiminobibenzyl (**1516d**) has been employed to prepare the indolo [1,7-*ab*][1]benzazepines **1594** (Fig. 208). The annelations were accomplished under three different sets of reaction conditions. Simply refluxing an ethanol solution of the hydrochloride salt of **1516d** with a ketone (**1591**) gave the indole **1594** (626, 674,675,677,864,865). Alternatively, when ethanol–glacial acetic acid solutions of *N*-nitrosoiminobibenzyl (**1516a**) and **1591** were treated with zinc, **1516a** was reduced *in situ* to **1516d,** and acceptable yields of **1594** resulted (691–694,863,866). Finally, **1594** was formed when *N*-ethylidenaminoiminobibenzyl (**1516e**) and **1591** were heated to 100°C in glacial acetic acid (677,686).

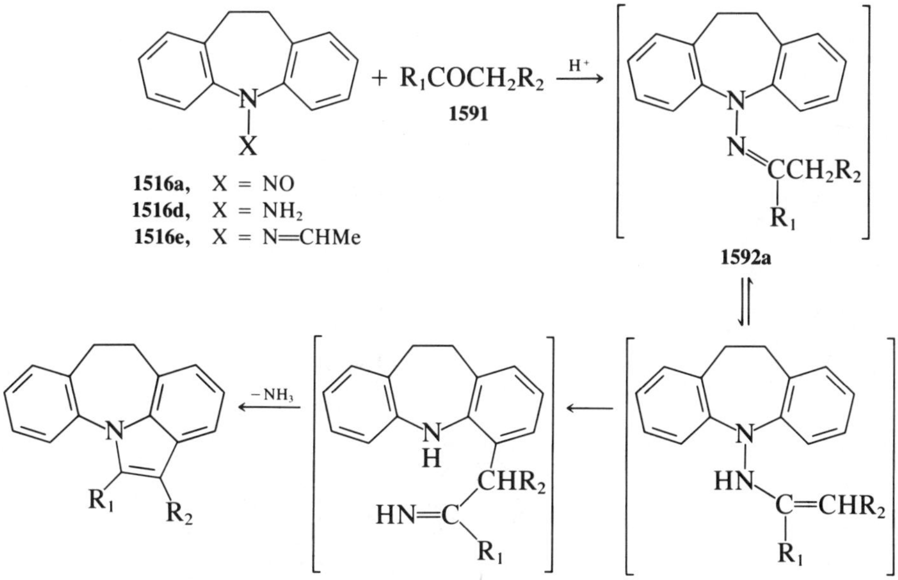

1516a, X = NO
1516d, X = NH_2
1516e, X = N=CHMe

+ $R_1COCH_2R_2$ $\xrightarrow{H^+}$ **1591**

1592a, **1592b**, **1593**, **1594** ($-NH_3$)

Figure 208

The mechanism of ring closure via the intermediates **1592** and **1593** is shown in Fig. 208. When 4-(*N*-methylaminomethyl)-5-(3-chloropropyl)iminobibenzyl (**1416b** see Section 39.C.b.iii) was treated with sodium hydride in diglyme, the tetracyclic amine **1595** was produced (605) (Eq. 248).

Simmons–Smith cyclopropanation of *N*-methyliminostilbene (**1412b**), using diiodomethane and zinc–copper couple, generated the dibenz[*b,f*]cycloprop[*d*]azepine **1597a** (586) (Eq. 249). When the same reaction was applied to iminostilbene (**1311b**), only *N*-methylation occured to give **1412b.** Application of the reaction to 5-(3-chloropropyl)iminostilbene (**1596**) did not give **1597b,** but gave instead a mixture of the products **1597c, 1597d, 1598a,** and **1598b.** In a reaction related to the one described above, dichlorocarbene and *N*-formyliminostilbene (**1570**) gave the tetracyclic product **1597e** (Eq. 250) (847).

The Robinson annelation reaction has been successfully applied to the azepinone **1340a** (Eq. 251). When an ethanolic solution of **1340a** and methyl vinyl ketone was treated with sodium ethoxide, the tetracyclic ketone **1599** was formed (868,869). Another type of annelation that has appeared several times in the patent literature (see Section 39.C.b.iii) is the reaction of the

$Cl(CH_2)_3$ CH_2NHMe

1416b

NaH, dioxane

Me

1595

(248)

CH_2I_2/Zn–Cu

1311b, R = H
1412b, R = Me
1596, R = $(CH_2)_3Cl$

1597a, R = Me
1597b, R = $(CH_2)_3Cl$
1597c, R = $CH_2CH{=}CH_2$
1597d, R = CH_2Pr-*c*

(249)

1598a **1598b**

:CCl_2 Cl Cl N CHO N CHO 1570 1597e (250)

O + CH_2=CHCOMe NaOEt, EtOH O N Me N Me 1340a 1599 (251)

$BrCH_2$ CH_2Br H_2NR_3 R_3 N N R_1 N R_1 1410b, R_1 = Me, COR_2 1411 (252)

1311b + Ph—$\overset{+}{C}$=N—$\overset{-}{N}$—Ph $h\nu$ Ph N N Ph N H 1600 1601 (253)

dihalide **1410b** with primary amines to give the dibenzo[*b,f*]pyrrolo[3,4-*d*] azepines **1411** (Eq. 252). These compounds have been studied actively because of their effects on the central nervous system (532,537–541,632). Annelation of iminostilbene (**1311b**) with the pbotochemically generated 1,3-dipolarophile **1600** was reported to give the 1:1 adduct **1601** (Eq. 253) (1045).

TABLE 119. OXIDIZED AND REDUCED DIBENZ[*b,f*]AZEPINES

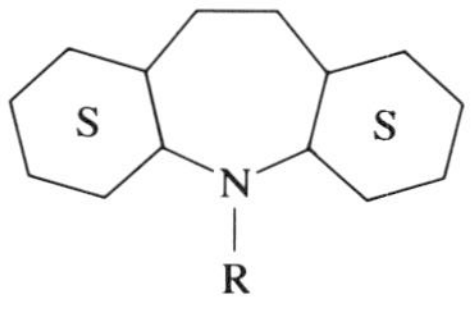

R	mp, °C(bp, °C/ torr) and other data	References
H	(107–110/2mm)	
	n_D = 1.5090	
	256–257 (HCl)	469
CONHPh	165–167	469
CONH-α-Naphthyl	153–154	469

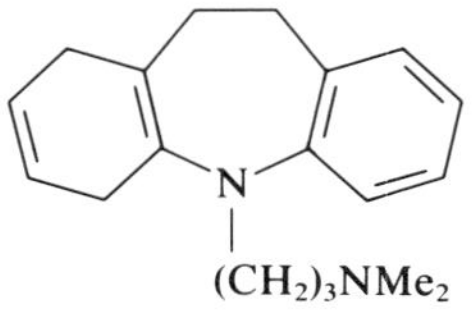

mp, °C	References
uv, pmr	853

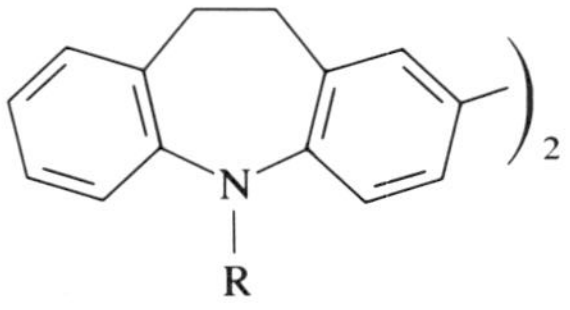

R	mp, °C	Spectra	References
H	300–302	ir, pmr	858
Me	169–170;		
	185–190	ir, pmr, ms	858

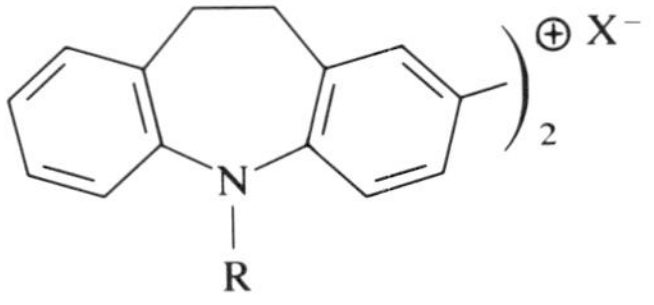

R	X	Data	References
H	SbF_6	ir	858
H	ClO_4	ir	858
Me	SbF_6	ir	858
Me	ClO_4	ir	858

e. OXIDIZED AND REDUCED DIBENZ[*b,f*]AZEPINES (see Table 119)

The oxidation of iminobibenzyl (**1293**) and iminostilbene (**1311b**) to give the azepinones **1444b** and **1450** was reviewed thoroughly in Section 39.C.c.ii (Eqs. 197–199). The preparation of the related azepinones **1367a** and **1371c** was detailed in Section 39.B.c.iii (Fig. 181 and Eq. 174). The azepinone **1450** was shown to undergo electrophilic substitution reactions. The action of cupric nirate in acetic anhydride on **1450** gave the 1-nitro derivative **1602,** whereas *N*-bromo- and *N*-chlorosuccinimide converted **1450** to **1603a** and **1603b**, respectively, (655). Pariser–Parr–Pople calculations predicted the most nucleophilic center to be C_1, followed by C_9.

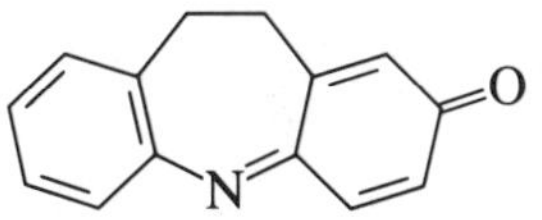

1444b

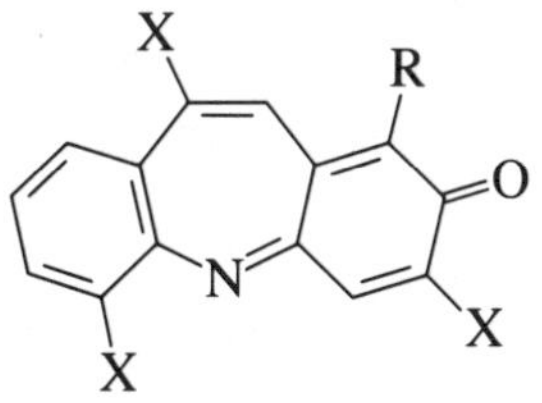

1450, R = X = H
1367a, R = H, X = OH
1371c, R = COMe, X = OH
1602, R = NO_2, X = H
1603a, R = Br, X = H
1603b, R = Cl, X = H

Selenium dioxide oxidation of the azepinone **1604** and subsequent hydrolysis gave dibenz[*b,f*]azepine-10,11-dione (**1605**) as an orange-red crystalline solid that decomposed without melting above 360°C (854). The dione **1605** underwent a facile benzylic acid rearrangement and dehydration to give acridine-9-carboxylic acid **1316a** (Eq. 254). On catalytic reduction, one carbonyl group of **1605** was reduced, and the resultant acyloin **1606** readily underwent acid-catalyzed rearrangement and dehydration to give 9-acridinecarboxaldehyde (**1518b**).

The epoxide **1435** (see Fig. 207), a metabolite of carbamazepine (**1297b**), was prepared by *m*-chloroperbenzoic acid oxidaton of **1297b** (841,856). Peracetic acid has also been used (841).

The electrochemical oxidation of iminobibenzyl (**1293**), 5-methyliminobibenzyl (**1382a**), imipramine (**1296a**), and 5-methyliminostilbene (**1337**) has been reported (Fig. 209 and Eq. 255) (857). The general oxidative behavior of all the compounds was found to be similar. The first step in the process was postulated to be the formation of radical–cation **1607**. Fast coupling of **1607** at the 2-position to form a dimeric compound **1608** was suggested to be the next step of the process. A final oxidation of **1608**, an overall two-electron process with a higher negative potential than is required for oxidation of the original compound, gave the dimeric species **1609**.

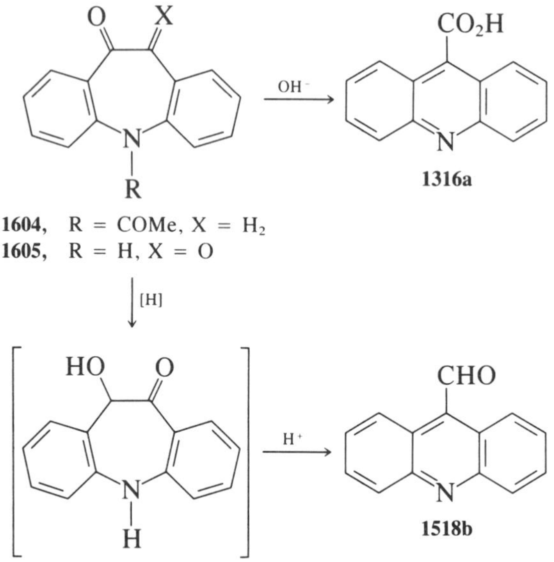

1604, R = COMe, X = H_2
1605, R = H, X = O

1316a

1606

1518b

(254)

With **1337**, coupling is possible at more than one position. In view of the extreme ease with which the 5-methyliminostilbene dimer was oxidized, the authors suggested the dimeric structure **1610**, which is coupled at the 10-position (Eq. 255). Such diaminobutadienic moieties are known to be exceedingly easy to oxidize.

The same coupling reactions have been carried out chemically. The ox-

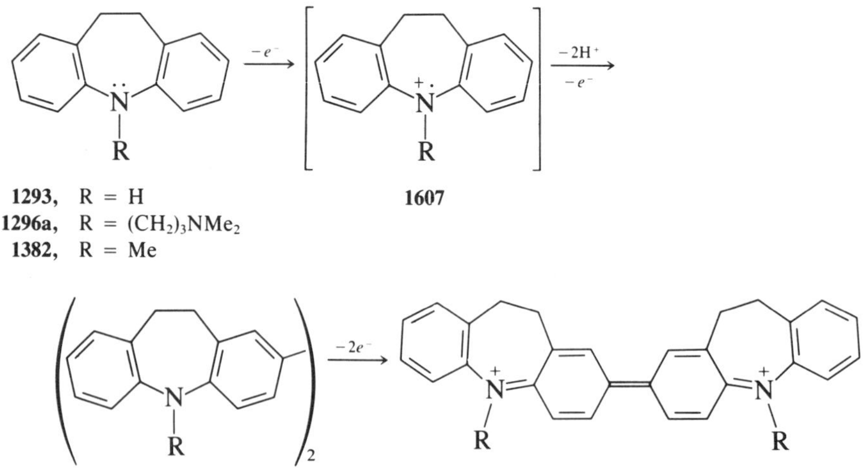

1293, R = H
1296a, R = $(CH_2)_3NMe_2$
1382, R = Me

1607

1608

1609

Figure 209

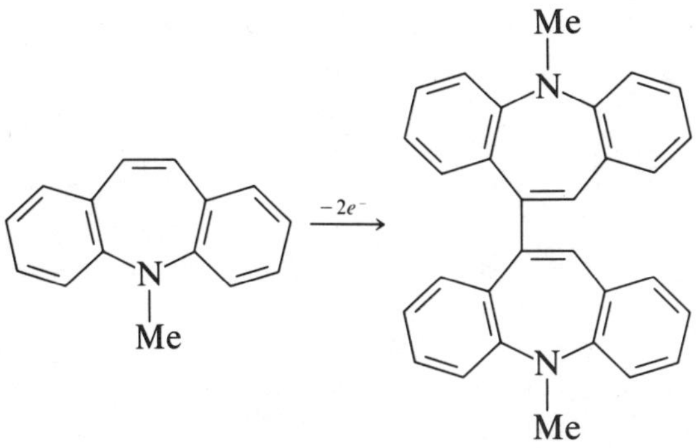

Me

1337 $\xrightarrow{-2e^-}$ **1610** (255)

idation of **1293** and **1382a** with 2,3-dichloro-5,6-dicyano-*p*-benzoquinone (DDQ) and perchloric acid or with *tris(p*-bromophenyl)ammonium hexachloroantimonate gave the cation radical perchlorates **1611a** and **1611b**, and the hexachloroantimonates **1611c** and **1611d**, respectively (858). Each salt was reduced by sodium dithionite to its respective neutral dimer **1612a** and **1612b**, (Eq. 256) Infrared spectra of the coupled products **1612a** and **1612b** were identical with those of authentic samples obtained by oxidation of **1293** and **1382a** with sodium dichromate in acetic acid. The related chemical oxidations of iminostilbene (**1311b**) and 5-methyliminostilbene (**1337**) gave complex mixtures of intensely colored products (858). Mass spectrometric analysis indicated the materials were dimeric.

In a more recent publication, the lead tetraacetate oxidation of **1293** and **1311b** in trifluoroacetic acid was reported to give the relatively stable radical cations **1613a** and **1613b** (Eq. 257b) (860). The radical cations were studied by esr spectroscopy. The nitroxides **1614a** and **1614b** were prepared by the action of lead dioxide (Eq. 257a) (859). Again, the esr spectra of **1614a** and **1614b** were studied.

The reduction of dibenz[*b*,*f*]azepine derivatives has received very little attention. Hydrogenation of a dioxane solution of *o*,*o*′-diaminobibenzyl (**1292**) over Raney nickel (220°C, 200–250 atm) gave the perhydroiminostilbene **1294**

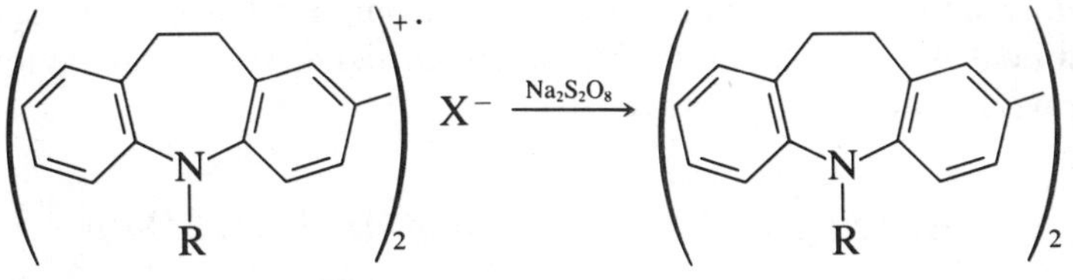

X^- $\xrightarrow{Na_2S_2O_8}$ (256)

1611a, R = H, X = ClO_4
1611b, R = Me, X = ClO_4
1611c, R = H, X = $SbCl_6$
1611d, R = Me, X = $SbCl_6$

1612a, R = H
1612b, R = Me

(257a)

1614a, X = CH_2CH_2
1614b, X = CH=CH

1293, X = CH_2CH_2
1311b, X = CH=CH

(257b)

1613a, X = CH_2CH_2
1613b, X = CH=CH

PbO_2; $Pb(OAc)_4$, CF_3CO_2H

1292 $\xrightarrow{H_2\text{/Raney Ni (200–250 atm), dioxane, 220°C}}$ **1294** (258)

1296a $\xrightarrow{\text{Li, } NH_3\text{, THF, MeOH}}$ **1615** (259)

$(CH_2)_3NMe_2$

(Eq. 258) (469). One can only speculate as to whether iminobibenzyl might have been an intermediate.

Birch reduction of imipramine (**1296a**) with lithium in a liquid ammonia–tetrahydrofuran–methanol solution (Eq. 259) was reported to give the 1,4,10,11-tetrahydrodibenz[*b,f*]azepine **1615** (853). Compound **1615** was sensitive to air and heat but was stable at 5°C under nitrogen. It was characterized by its ultraviolet and pmr spectrum.

E. *Physical Properties and Analytical Data*

a. INFRARED SPECTRA

The N—H stretching frequency of most iminostilbenes and iminobibenzyls occurs in the 3300–3340 cm^{-1} region (511,512,517,529,895). The band at 1316 cm^{-1} in the spectrum of iminostilbene has been attributed to the C—N stretching frequency, while a strong absorption at 1481 cm^{-1} has been assigned to a characteristic *o*-disubstituted benzene skeletal vibration (529). The aromatic C—H out-of-plane bending region of iminostilbene features strong bands at 802, 754, and 729 cm^{-1}, while the same region in the spectrum of iminobibenzyl shows a strong, broad absorption at 748 cm^{-1}, with a weaker shoulder at 763 cm^{-1} (511,512,529). The relatively low intensity of the N—H absorptions of iminostilbene when compared to indole, carbazole, diphenylamine, or even iminobibenzyl was attributed to a high electron density at nitrogen (895). The reader interested in more information about specific compounds in the dibenz[*b,f*]azepine family should refer to Tables 80–120 for references to the primary literature.

b. ELECTRONIC SPECTRA

Ultraviolet spectral data from several iminostilbenes and iminobibenzyls have been collected by Kricka and Ledwith (490). The ultraviolet spectrum of iminostilbene displays three characteristic bands at 365 nm (log ϵ 2.89), 293 nm (3.43), and 258 nm (4.65) (497,511,512,529,895). The band at 365 nm, which extends into the visible region, accounts for the yellow-orange color of iminostilbenes. The absorption has been attributed to the promotion of a π-electron from the highest occupied to the lowest unoccupied molecular orbital of iminostilbene (895). A significant degree of "charge transfer" character (i.e., a significant level of shift of negative charge away from the imino nitrogen toward the stilbene portion of the molecule) has been attributed to the 365-nm absorption. For example, the absorption in methanol, as compared to cyclohexane, was displaced by 10 nm toward the blue end of the sectrum, a shift attributed to hydrogen bonding of the ground state nitrogen electrons. Acetylation eliminates the band and the orange color.

The spectrum of iminobibenzyl is very similar to that of diphenylamine, with both compounds absorbing at 285–290 nm (497,512,624,870,874). Apparently, the ethylene bridge has little effect on the conjugation existing in diphenylamine.

N-alkylation of iminobibenzyl and iminostilbene has a hypsochromic effect, with an accompanying reduction in intensity (497,586,895). A similar hypsochromic shift is displayed by imipramine relative to iminobibenzyl (872,873,883). This effect has been attributed to the greater steric interactions of the *N*-alkyl group, as compared to N—H, in opposing coplanarity.

c. NUCLEAR MAGNETIC RESONANCE SPECTRA

The tables that accompany this review provide a complete listing of the published nuclear magnetic resonance data. In addition, Kricka and Ledwith have collated some of the data into tabular form in their earlier review (490).

One of the principal applications of proton magnetic resonance (pmr) spectroscopy has been in the study of the conformational equilibria associated with iminobibenzyl (**1293**). The ethano bridge protons in **1293** and its

TABLE 120. ^{13}C CHEMICAL SHIFTS (δ) OF *N*-SUBSTITUTED DIHYDRODIBENZ[*b*, *f*]*AZEPINES*[a]

R	C_1,C_9	C_2,C_8	C_3,C_7	C_4,C_6	C_{10},C_{11}	C_{9a},C_{11a}	C_{4a},C_{5a}
H	130.1	119.0	126.2	117.5	34.7	128.1	141.9
Me	129.2	118.3	125.9	121.3	32.8	132.6	148.2
CO_2Et[b]	129.3	126.9	126.1	127.9	30.6	135.5	140.3
$(CH_2)_3NMe_2$[c]	129.5	119.1	126.1	122.6	32.0	133.4	146.7
$(CH_2)_3NMe_2$[d]	129.5	119.7	126.5	122.6	31.8	133.6	147.3
$(CH_2)_3NMe_2$[e]	129.2	119.6	125.8	122.0	32.2	133.7	147.8

[a] Approximately 1 *M* solutions in $CDCl_3$ unless stated otherwise.
[b] $CO_2CH_2CH_3$: 154.5, 61.6, and 14.5, respectively.
[c] $CH_2CH_2CH_2\overset{+}{N}HMe_2$: δ 47.2, 22.2, 55.6, and 42.3, respectively.
[d] In D_2O solution $CH_2CH_2CH_2NHMe_2$: δ 46.8, 22.4, 55.4, and 42.5, respectively (standardized with $CDCl_3$ solution).
[e] Free base, $CH_2CH_2CH_2NMe_2$: δ 48.7, 26.1, 57.5, and 45.2, respectively.

N-alkyl derivatives appear as a sharp singlet at δ 2.83–3.10 down to −100°C (584,843). However, acylation of **1293** has a profound effect on the ethano bridge, and this phenomenon has been studied in some detail (843,844). At −60°C the pmr spectrum of *N*-acetyliminobibenzyl (**1309a**) exhibited an *ABCD* pattern, indicating total nonequivalence of the ethano bridge protons. At room temperature the pattern collapsed to an *AA′BB′* system, and at 112°C the signals eventually coalesced to a singlet.

These spectra have been interpreted in terms of restricted rotation about the amide C—N bond and inversion of the central seven-membered azepine ring. At low temperatures the fixed planar conformation of the amide moiety effectively "freezes" the conformation of the seven-membered ring. It was reasoned that ring flip would involve severe steric interactions between the planar amide group and the juxtapositioned 4- and 6-aromatic protons. As the temperature is raised, the amide rotational barrier is overcome, removing the steric restrictions imposed on the ring inversion process. At still higher temperature an increasingly rapid rate of ring flip results in collapse of the *AA′BB′* spectrum of **1309a** to an A_4 system. A full analysis of the 3 J_{HH} couplings in the AA′BB′ spectrum of **1309a** was performed. These values predicted a dihedral angle (ϕ) of approximately 45°. When the amide group is held in a rigidly planar conformation, as in **1587**, there is no longer a requirement for movement of the amide moiety, and the spectrum of **1587** shows the ethano bridge protons as a singlet at δ 3.0 down to −60°C (843,844).

Related phenomena have been reported for several other systems. The *N*-acyl derivatives **1414b** and **1414c** displayed restricted amide rotation similar to that described above for **1309a** (1041). The temperature profile of 5-acetyl-10-cyanoiminostilbene (**1616**) closely paralleled that of **1309a** (685). The vinyl proton of the enamine **1528a** appeared as a doublet, as did the *N*-acetyl signal. This result was again attributed to restricted rotation around the C—N amide bond (628). Evidence has been given for the stereochemical stability about the trivalent nitrogen of 5-methyl-4,6-*bis*(hydroxymethyl)-

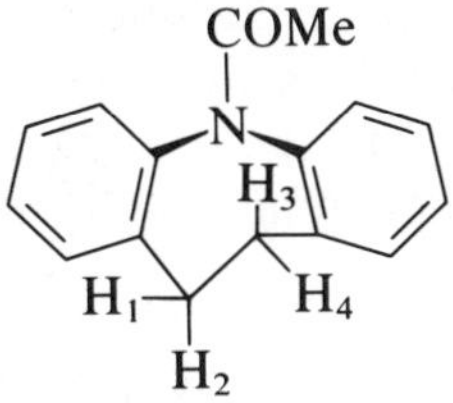

COMe
N
H_3
H_1
H_4
H_2
1309a

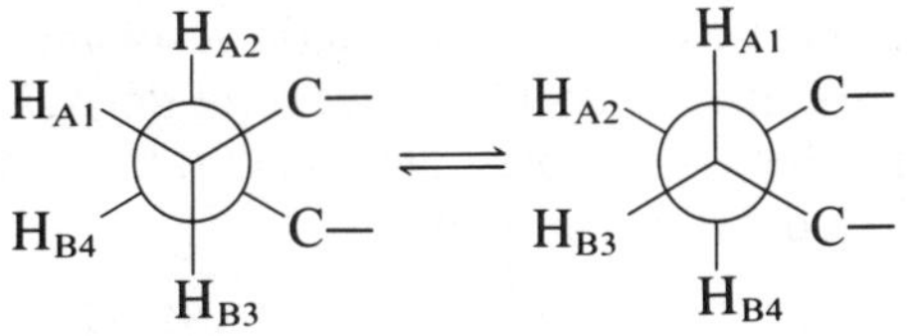

H_{A2}
H_{A1}
C—
H_{B4}
C—
H_{B3}
⇌
H_{A1}
H_{A2}
C—
H_{B3}
C—
H_{B4}

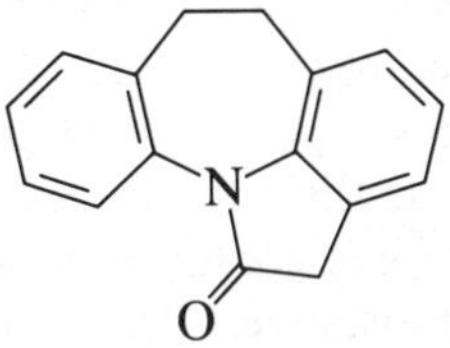

N
O
1587

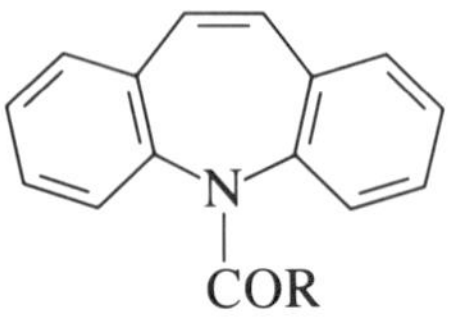

COR

1414b, R = $(CH_2)_2Cl$
1414c, R = $CBrMe_2$

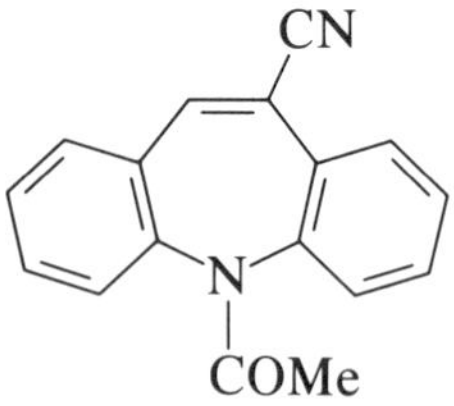

CN

COMe

1616

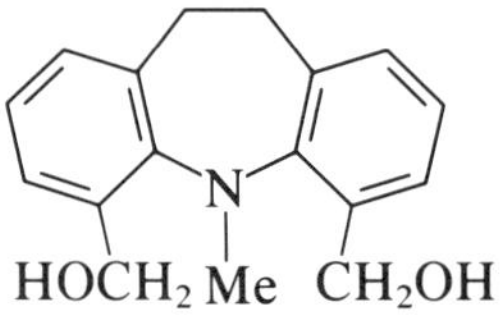

HOCH₂ Me CH₂OH

1617

R_1

R_2

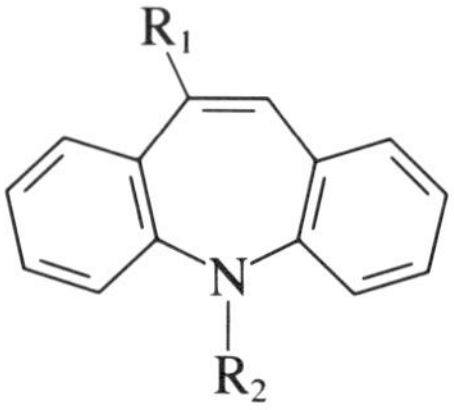

1458, R_1 = CMe_2OH, R_2 = Et
1528a, R_1 = N(piperidino), R_2 = COMe

10,11-dihydrobenz[*b*,*f*]azepine (**1617**) and related compounds (898). In the presence of a chiral shift reagent, **1617** exhibited diastereotopic resonances for the benzylic and ethano protons as well as a double resonance for the *N*-methyl signal. This phenomenon did not disappear at temperatures up to 150°C, suggesting a minimum energy barrier about the trivalent nitrogen of at least 22 kcal/mole. The methyl signals of the alcohol **1458a** were diastereotopic below 125°C, indicating a chiral conformation. Above 125°C the signals coalesced, and it was suggested that inversion of the seven-membered azepine ring proceeded via a planar, achiral transition state (640,897).

Proton magnetic resonance data have been reported for several of the psychotropic dibenz[*b*,*f*]azepine derivatives (844,876,885). Kricka and Ledwith carefully analyzed the conformational preference of the side chain in imipramine by nuclear magnetic resonance (844). In the same paper, they reported the ^{13}C spectra of several derivatives of iminobibenzyl, including imipramine (see Table 120). Finally, the sodionitranion of iminostilbene has been studied by pmr (639).

d. MASS SPECTRA

Isolated examples of mass spectral data have appeared throughout the dibenz[*b*,*f*]azepine literature. More recently, Ledwith and Kricka have published mass spectral data on several compounds, including *N*-ethyliminostilbene (**1387**) (583,584). The use of mass spectrometry in conjunction with gas chromatography as an analytical method for the detection of psychotherapeutic drugs and their metabolites has greatly expanded the amount of mass spectral information on these substances. In these papers the mass spectral patterns of carbamazepine (**1297b**) (647,842,845,846,849–851,856,

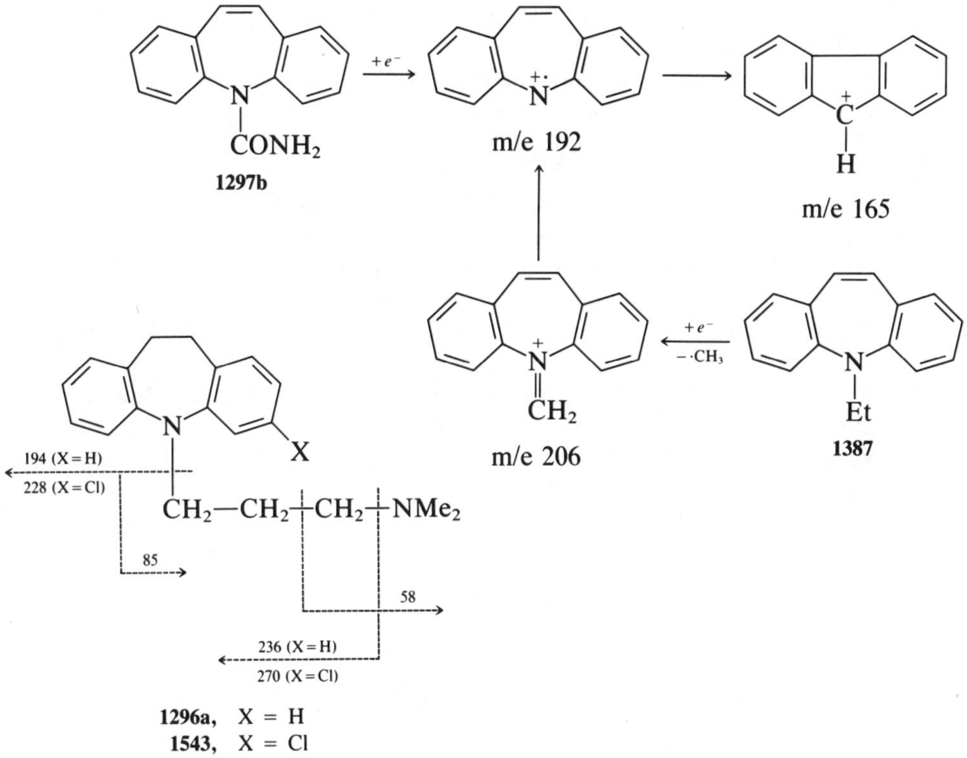

Figure 210

887,900), imipramine (**1296a**) (813,846,886,887), and clomipramine (**1543**) (807) have been thoroughly documented. A summary of *m/e* values for the fragmentation of imipramine and clopramine is given in Fig. 210.

e. X–RAY DATA

The structure of imipramine hydrochloride (**1296a**·HCl), has been determined by X-ray crystallography (799). The crystal was monoclinic, with space group P2l/*c* (a = 11.303(3), b = 29.227(8), c = 14.282(3) Å ; β = 130.91(1)°; Z = 8. The two molecules in the asymmetric unit displayed slightly different conformations of the azepine ring. The dimethylamino side chain exhibited a different conformation in each molecule. Interestingly, the dihedral angle (ϕ) for one of the two molecules was found to be 49.10°, a value in fairly good agreement with the value of 45° predicted by Kricka and Ledwith by nmr (843,844).

Denne and Mackay described their study of 5H-10,11-dioxo-10,11-dihydrodibenz[*b*,*f*]azepine (**1605**) (855) (Eq. 254). The crystals were found to belong to the orthorhombic space group *Pccn*, with a = 5.07, b = 14.50, c = 13.78 Å. The molecule was almost planar, and although the closeness to planarity might imply a certain level of aromaticity, a careful analysis of bond lengths suggested that **1605** is better viewed as a combination of two vinylogous amides.

X-Ray powder diffraction studies have been applied to the identification of some of the psychotherapeutic dibenz[*b*,*f*]azepines (890).

f. MOLECULAR ORBITAL CALCULATIONS

Simple Hückel LCAO–MO theoretical calculations were first performed on dibenz[*b*,*f*]azepine **1311b** by Schmid (895). However, the lack of correlation between Hückel MO delocalization energies and observed aromatic behavior for heteroaromatic systems led to subsequent semiempirical SCF–MO calculations by Dewar (463) and Hess (464). Dewar's calculations on iminostilbene led to a heat of atomization of 130.826 eV, a resonance energy of 38.40 kcal/mole, and an ionization potential of 8.19 eV. However, his calculated ionization potential is not in good agreement with experimental values (589,879). Dewar's estimated bond lengths and electron densities for iminostilbene are shown in **1618**.

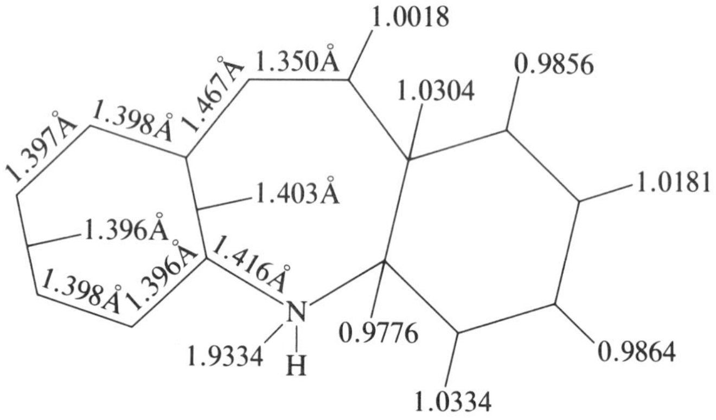

1618

g. MISCELLANEOUS DATA

A number of physicochemical and analytical studies have been performed on the dibenzazepine systems in addition to those already discussed. Several of these have been selected from the literature and are presented here, without discussion, for the interested reader.

Numerous chromatographic separations have been carried out on dibenz[*b*,*f*]azepines (see Section 39·E·d), including paper and thin-layer (902–908), liquid–solid adsorption (910,912), ion-pairing (909,912), and gas–liquid (902,911,913) chromatography. Partition coefficients have been determined for tricylic antidepressants such as imipramine (**1296a**) (888,889). A study of the charge transfer complex between iminostilbene and tetracyanoehtylene has been reported (878). Likewise, complexes with boron trifluoride, bromine, and sulfur dioxide have been described (884). Measurements of pK^a values have been made for iminostilbene, iminobibenzyl (894), and also for a variety of aminoalkyldibenz[*b,f*]azepines (891,892,893). Fluorescence and phosphorescence spectra of iminobibenzyl have been determined (877,880,881). Charge transfer spectroscopy has been used to determine the ioization energy of imipramine (**1296a**) and desmipramine (**1296b**) (871). The photoelectron spectra of **1293** and **1311b** have been determined, and their ionization potentials have been reported (658,879).

40. Dibenz[*c,e*]azepines

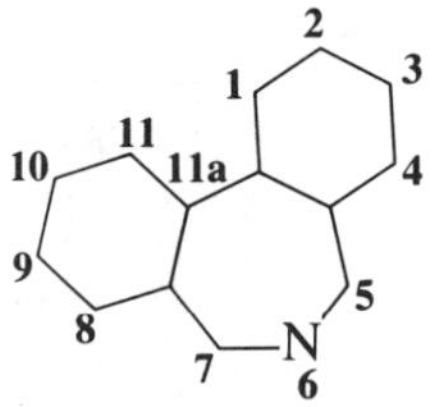

Considerable interest is attached to the chemistry of dibenz[*c,e*]azepines because of the biological activity shown by some of the derivatives of this ring system. The most widely employed route to these compounds is one that was first reported by Wenner (Fig. 211) (915,917,918). A 2,2′-*bis*(bromomethyl)biphenyl derivative (**1622**) is aminated with a primary or secondary amine to give **1623.** The dibromide **1622** can be prepared from

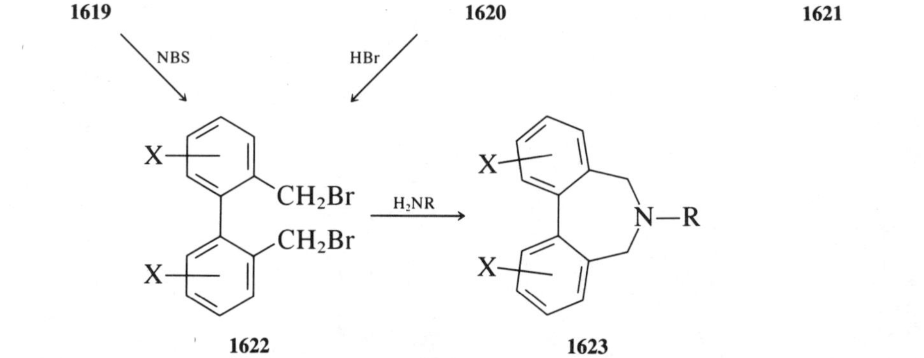

Figure 211

TABLE 121A. 6,7-DIHYDRO-5H-DIBENZ[*c,e*]AZEPINES PREPARED BY THE WENNER PROCEDURE

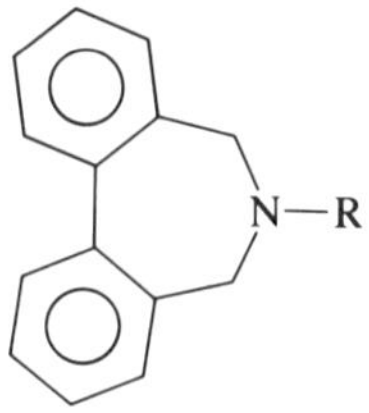

R	Salt	Yield (%)	mp (°C) or bp (°C/torr)	Other Data	Refs.
H	HBr		283–284		915
H	HCl		284–286		915,917,948
H	HCl		286–288		918
H	HCl		298–300		923
Me	—		175–178/10	Phosphorescense	915
Me	H_3PO_4		187		915
Me	HBr		223–225		915,948
			227–229		923
Me	Oxalate		173–174		915
Me	Methiodide		287–288		915
			294	uv	964
Me	Methobromide		276–277		915
			285–287		964
			260–262		917
Me	Picrate		162		915,964
Et	HBr		203–204		915
n-Pr	—	85	109/0.005	$[n]_D^{25}$ = 1.5590	934
n-Pr	$HBr\cdot0.5H_2O$		203–204	nmr	915,934
			248–250		918
n-Pr	Picrate		172–173		934
n-Pr	Perchlorate		198–199		934
i-Pr	HBr		238–240		915
i-Pr	—		195–205/20–24		918
n-Bu	—	77.5	121–122/0.01	nmr $[n]_D^{25}$ = 1.5893	934
n-Bu	$HBr\cdot0.5H_2O$		163–164		915
n-Bu	Methiodide		172–173		948
n-Bu	Methiodide		225–226		934
$CH_2CH=CH_2$	—		180–195/15		915
$CH_2CH=CH_2$	—		176–179/12		915
$CH_2CH=CH_2$	—		131–134/0.02		915
$CH_2CH=CH_2$	HCl	70	214–215		915
			208–210		918
$CH_2CH=CH_2$	HBr		212–213		915
$CH_2CH=CH_2$	H_3PO_4		210		915
$CH_2CH=CH_2$	H_3PO_4		210		915
$CH_2CH=CH_2$	H_3PO_4		205–206		918,948
$CH_2CH=CH_2$	H_3PO_4		206–209		920
$CH_2CH=CH_2$	H_3PO_4		211–215		915,918

TABLE 121A. (*Continued*)

R	Salt	Yield (%)	mp (°C)or bp (°C/Torr)	Other Data	Refs.
$CH_2CH{=}CH_2$	H_3PO_4		215–217		922
$CH_2CH{=}CH_2$	Methiodide·$0.5H_2O$		182–184		915
Ph	—	82	89.5–90.5		921
Ph	—		85.5–86.5	uv	948
					974
Ph	HBr		230–232		915
C_6H_4Me-*p*	—	88	104–105		921
C_6H_4Cl-*p*	—	98	147–149		921
C_6H_4Cl-*p*	—	78	147–148.5		940
C_6H_4OMe-*p*	—	94	148–149		921
2-Naphthyl	—	94	144–145		921
t-Bu	—	79	137–140/0.01	$[n]]_D^{20}$ = 1.5991	934
t-Bu	Picrate		206–208	nmr	
CH_2Ph	—	99	195–200/0.1	nmr	918
			101–102		934
CH_2Ph	Picrate		218–219		934
CH_2Ph	HCl·$0.5H_2O$		205		917,918
			201–202		948
CH_2Ph	HBr		196–197		918
			205–207		950
CH_2Ph	Methiodide		188–190		917
CH_2CH_2Ph	—	75	250–255/0.1	nmr	917
			66–67		934
CH_2CH_2Ph	Picrate		182–183		934
CH_2CH_2Ph	HCl		221		917
CH_2CH_2Ph	HBr·$0.5H_2O$		105–106		918
CH_2CH_2Ph	Methiodide		122		918
c-C_6H_{11}	HBr		264		918,948
CH_2CH_2OH	H_3PO_4		179–181		917
			185–186		948
$CH_2CH_2CH_2OMe$	—		195–200/10–15	nmr	918
			147–150/0.1		925
$CH_2CH_2CH_2OMe$	H_3PO_4		171–172		918
$CH_2CH_2CH_2OMe$	HCl		158–159		925
$CH_2CH_2CH_2OCHMe_2$	—		205–210/10		918
$CH_2CH_2CH_2OCHMe_2$	H_3PO_4		145–146		918
$CH_2CH_2CH_2OCHMe_2$	HCl		156		918
$(CH_2)_2NMe_2$	2MeI		244–245		922
$(CH_2)_3NMe_2$	2HCl·H_2O		235–237		922
$(CH_2)_2NEt_2$	$2H_3PO_4$		233–234		917
			238–239		948
$(CH_2)_3NEt_2$	H_3PO_4		223–225		948
NH_2	—	94	109–110		924
OH	—	67	168–170		924
p-Ts	—		164		941
CH_2CH_2SMe	—		150/0.005		925
$(CH_2)_2S(CH_2)_2NMe_2$	—		190–195/0.01		926

1619 by treatment with *N*-bromosuccinmide (919) or via the action of HBr on the diol **1620,** which is accessible in turn by reduction of the diphenic ester **1621.** A process patent (920) describes the preparation of the chloro analog of **1622** from **1619** by means of sulfuryl chloride.

The 6,7-dihydro-5H-dibenz[*c*,*e*]azepines prepared by the Wenner procedure are listed in Table 121A–E. Interest in this family of compounds was

TABLE 121B. 6,7-DIHYDRO-5H-DIBENZ[*c*,*e*]AZEPINES PREPARED BY THE WENNER PROCEDURE

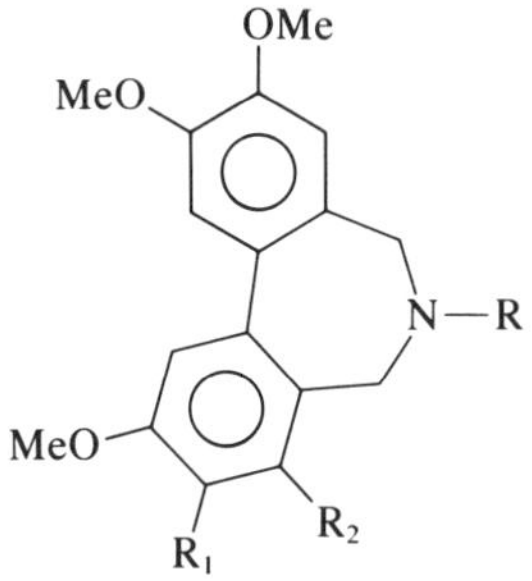

N—R	R_1	R_2	Yield (%)	Mp (°C)	Salt	Refs.
NCH_2CH_2OH	MeO	H	40	186–187 ($0.5H_2O$)		927
NCH_2CH_2Cl	MeO	H	64			927
				178–179	HCl	927
				272–273	Picrate	927
$N(c\text{-}C_6H_{11})$	MeO	H	66	92	Picrate	927
				244–246	Methiodide	927
				233–234		
NEt_2	MeO	H	50	285–286	Chloride (H_2O)	927
				234–236	Picrate	927
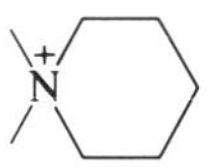 N^+ (piperidine ring)	MeO	H	65	315–316	Chloride	927
				225–226	Picrate	927
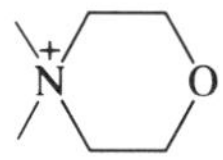 N^+ (ring with O)	MeO	H	60	168	Chloride ($0.5H_2O$)	927
				246	Picrate	927
NMe	H	H	53	160–162		928
				211–213	Picrate	928
				277–278	Methiodide	928
NMe	OH[a]	H		170–172		928
NCH_2Ph	H	MeO	81	147–149		929
				256–257	HCl	
NH	H	MeO	85	274–276	HCl	929
NMe	H	MeO	88	179–181	HCl	929
				247–249		
$NCH_2CH{=}CH_2$	H	MeO		241–242	HCl	930,931

[a]All MeO groups = OH.

TABLE 121C. 6,7-DIHYRO-5H-DIBENZ[*c,e*]AZEPINES PREPARED BY THE WENNER PROCEDURE

R_1	R_2	R_3	Yield (%)	mp (°C)	Spectra	Refs.
Me	H	NO_2	Good	105–107	nmr	939
CH_2Ph	H	NO_2	95	119–121	uv,nmr	939
Me	Ph	H	69	215–216	uv	953

generated by the potent hypotensive activity of the 6-allyl derivative, which was marketed under the name Ilidar. *N*-Oxides of several of these substances were prepared by hydrogen peroxide oxidation (see Table 122). Derivatives of **1623** with R = H are obtained in poor yield by amination of **1622** with ammonia; a quaternary spiro compound is formed (Table 121E). *N*-Unsubstituted analogs are obtained best by amination with benzylamine followed by *N*-debenzylation.

A second important synthesis of 6,7-dihydro-5H-dibenz[*c,e*]azepines, developed by Hawthorne, utilizes the dialdehyde **1624** (Fig. 212), which is available by ozonolysis of the corresponding phenanthrene (942–946). The dialdehyde **1624** may be converted to **1627** either via sodium hydrosulfite reduction of the *bis*-Schiff's base **1625** (method A) or directly by catalytic reductive amination (method B). Table 123A,B lists 6,7-dihydro-5H-dibenz[*c,e*]azepines that have bee prepared by this route. When the dialdehyde is allowed to react for a brief time with ammonia, the tricyclic carbinolamine **1626** is formed. This intermediate can be converted in high yield to **1627** by catalytic hydrogenation. Quaternary spiro compounds are not observed. A variant of this procedure utilizes formic acid at 100–120°C as the reducing agent (948). Table 123A,B should be consulted for specific examples and yields.

In another process (937) that yields spiro-quaternary salts of 6,7-dihydro-5H-dibenz[*c,e*]azepines, diphenic anhydride (**1628**) is allowed to react with a secondary amine to obtain the amide acid **1629** (Fig. 213). Reduction with lithium aluminum hydride gives **1630,** which, on treatment with *p*-toluenesulfonyl chloride under mild conditions, affords the quaternary salts **1631** in moderate to excellent yields.

TABLE 121D. 6,7-DIHYDRO-5H-DIBENZ[*c*,*e*]AZEPINES PREPARED BY THE WENNER PROCEDURE

A	R_1	R_2	R_3	Yield (%)	mp (°C)	X^-	Refs.
$\overset{+}{N}$ (piperidinium)	H	H	H	72	70–74 ($2H_2O$) 255	Br	956
$\overset{+}{N}$ (4-phenylpiperidinium)—Ph	H	H	H	Good	>300	Br	962
$\overset{+}{N}Et_2$	H	MeO	MeO	77		I	956
$\overset{+}{N}Me_2$	H	MeO	MeO	76	333 328	Br I	967 967
$\overset{+}{N}$ (piperidinium)	H	MeO	MeO	82	114	Br	956
$\overset{+}{N}Me_2$	H	Me	Me	85	259–261	Br	957

 $\overset{+}{N}$	H	NO_2	NO_2	65	306–308 302–303	I Br	961 961
 $\overset{+}{N}$	H	F	F	60	268–270 234–235	$Br \cdot 2H_2O$	961 963
 $\overset{+}{N}$	H	Cl	Cl		287–288 283–284	$Br \cdot 2H_2O$ $I \cdot 2H_2O$	963 963
 $\overset{+}{N}$	H	Br	Br		304–305 272–273	I $Br \cdot 2H_2O$	963 963
 $\overset{+}{N}$	H	I	I		315–317 300–301	I $Br \cdot 2H_2O$	963 963
$\overset{+}{N}Me_2$	H	NO_2	NO_2		253–254	I	961
 $\overset{+}{N}$	H	NO_2	H	68	137–140	$Br \cdot H_2O$	965
 $\overset{+}{N}$	NO_2	H	H		246–248	Br	965
 $\overset{+}{N}$	H	MeO	H	89	170–172 or 253–255[a]	Br	966

[a]Melting point depends greatly on the rate of heating.

TABLE 121E. 6,7-DIHYDRO-5H-DIBENZ[*c*,*e*]AZEPINES PREPARED BY THE WENNER PROCEDURE

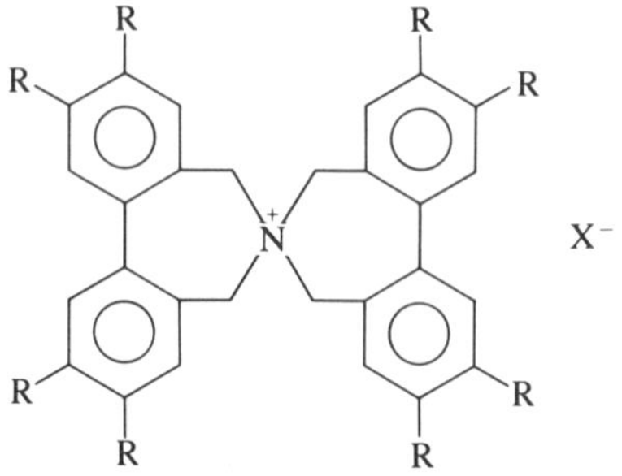

R	X^-	mp (°C)	Refs.
H	Br	>300 (0.5H_2O)	915,923
MeO	Cl	306–307 (picrate)	927

Photolysis of iodo-substituted dibenzylamines **1632** gave 6,7-dihydro-5H-dibenz[*c*,*e*]azepines **1633** in poor to moderate yields (Eq. 260) (949). Table 124 lists the results of photolysis experiments. The synthesis of the starting iodides is not difficult.

A chemical process very similar to the photolytic one just described is the Pschorr-type decomposition (951) of the diazonium salts of **1634** (Fig. 214). The 6-sulfonylmethyl-6,7-dihidro-5H-dibenz[*c*,*e*]azepines **1635** were obtained in poor to moderate yields, along with the side products **1636** and **1637.** Radical mechanisms were proposed for the formation of **1636** and **1637.** This process exemplifies only the second reported formation of a seven-membered ring by Pschorr cyclization. Table 125 lists compounds that have been synthesized via this approach.

In any discussion of the chemistry of 6,7-dihydro-5H-dibenz[*c*,*e*]azepines, stereochemstry is of pivotal importance. Substantial effort has been devoted to gaining an understanding of the effects of chirality and restricted rotation on the properties of the conjugated biphenyl system contained in this structure. The biphenyl moiety is forced to adopt a noncoplanar configuration by the flexible bridge of the azepine ring. However, ultraviolet absorption spectra suggest some conjugation between the aromatic rings. 1,11-Disubstituted derivatives have been prepared and resolved, and their optical stability has been studied.

Hall and co-workers have employed the Wenner procedure to prepare a number of quaternary salts of secondary amines (see Tables 121A–E and 126). Figure 215 summarizes the approaches employed by this group to obtain optically active 6,7-dihydro-5H-dibenz[*c*,*e*]azepines for racemization studies. Amination of a racemic dibromide (**1638**) with (−)-ephedrine

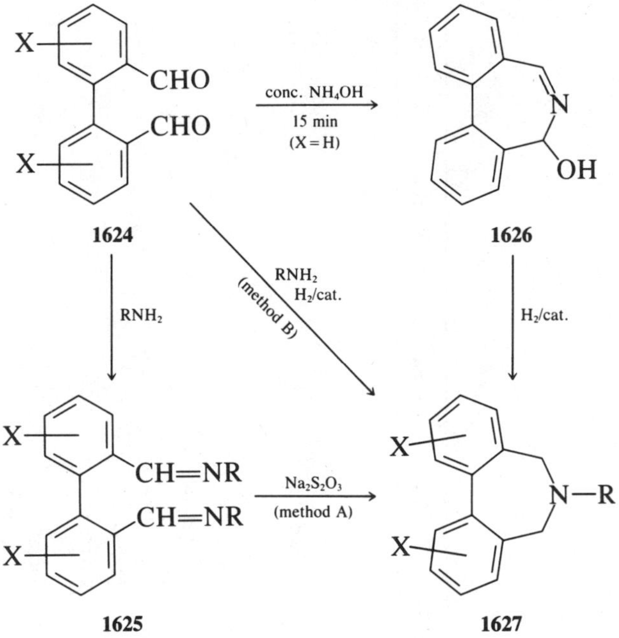

X X CHO CHO

1624

conc. NH_4OH
15 min
(X = H)

N OH

1626

RNH_2

RNH_2
H_2/cat.
(method B)

H_2/cat.

X X CH=NR CH=NR

1625

$Na_2S_2O_3$
(method A)

X X N—R

1627

Figure 212

O O O + HN X

1628

Δ

O C—N X CO_2H

1629, 80–85%

$LiAlH_4$

N X OH

1630, 84–97%

TsCl
Et_3N

$\overset{+}{N}$ X OTs^-

1631, 38–97%

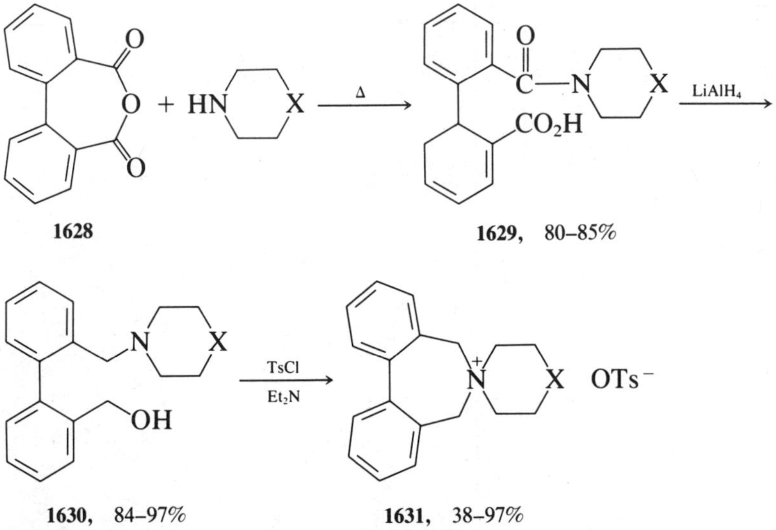

Figure 213

TABLE 122. DERIVATIVES OF 6,7-DIHYDRO-5H-DIBENZ[*c,e*]AZEPINES

A	B	R_1	R_2	mp (°C)	Salt	Yield (%)	Spectra	Refs.
NOH	Ph	H	H	171–172	HCl			933
				197–198				
NOAc	Ph	H	H	149–150				933
N(O)Pr-*n*	H	H	H	110–111	Picrate		nmr	934
				198–199	Perclorate			
				182–183	HCl			
				181–182				
N(O)Bu-*n*				179–180	Perchlorate			
				182–183	HCl			
N(O)Bu-*t*	H	H		94–95			nmr	934
				180–181	Picrate			
				164–165	Perchlorate			
N(O)CH_2CH_2Ph	H	H	H	102–103			nmr	934
				172–173	Picrate			
				188–189	Perchlorate			
				203–204	HCl			
N(O)CH_2Ph	H	H	H	167–168			nmr	934
				221–223	Picrate			
				204–205	Perchlorate			
				192–193	HCl			

NCH_2— (imidazoline ring: N, N–H)	H	H	H	260 (dec.)	HCl			935
$(CH_2)_4CN$	H	H	H	230–233	HCl			936
$(CH_2)_5CN$	H	H	H	175–176	HCl			936
Piperidino	H	H	H	203–204	Tosylate			937
Morpholino	H	H	H	224–225	Tosylate			937
Piperidino	H	H	H	237–238	Tosylate			937
4-Methylpiperazino	H	H	H	257–258	Ditosylate			937
NH	H	H	H	232–233	HCl			938
NHTs	H	H	H	149.6–151.5				938
NMe	H	PhCONH	H	176–178	Benzoate		nmr	939
$NN{=}NC_6H_4Cl$-*p*	H	H	H	114–115.5			ir,uv,nmr,ms	940
NNO	H	H	H	109–110.5			ir,uv,nmr,ms	940
				107–112				942
NCHO	H	H	H	179–180				948
$NCH_2C{\equiv}CH$	H	H	H	201–203	HCl			952
NH	H	Me	Me	272–273	HCl		nmr	954
NNO	H	Me	Me	123–123.5			nmr	954
NNO	Me	Me	Me	109–110			nmr	954
NNO	CO_2H	Me	Me				nmr	954
NMe	H	Me	Me	258–259	MeI		nmr	955
NPh	$-O\!+_{\frac{1}{2}}$	H	H	184–185		48, 35[a]		921
NC_6H_4Me-*p*	$-O\!+_{\frac{1}{2}}$	H	H	167–168		59, 99[a]		921
NC_6H_4Cl-*p*	$-O\!+_{\frac{1}{2}}$	H	H	156–158		55, 94[a]		921
N-2-Naphthyl	$-O\!+_{\frac{1}{2}}$	H	H	125–127		19, 49[a]		921
NC_6H_4OMe-*p*	$-O\!+_{\frac{1}{2}}$	H	H	121–123		— 87[a]		921

[a]Yields given are of the peroxygenation reaction followed by the hydrogen peroxide synthesis.

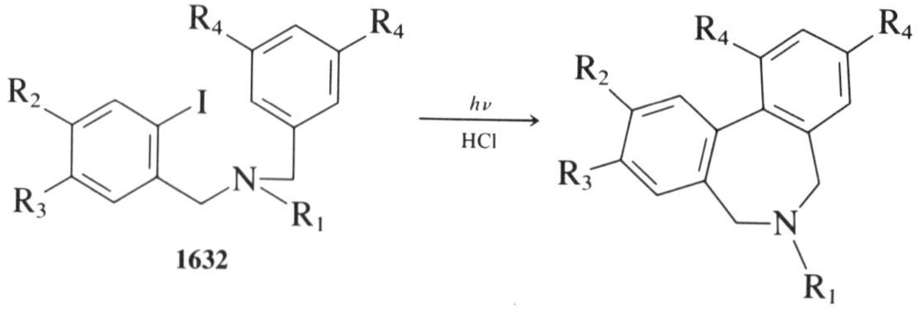

1632 → (hν, HCl) → 1633, 27–57% (260)

(956,961,962,965) gave a diastereomeric mixture of quaternary salts **1639,** which could be separated and individually converted to the pure enantiomeric amines **1640** by Hofmann degradation. Alterntively, reaction of the racemic dibromide with a secondary amine gave **1641,** which could be resolved with optically active acids such as camphorsulfonic acid (956) Finally, conversion of the (+)- and (−)-enantiomers of the diphenic acid **1642** into

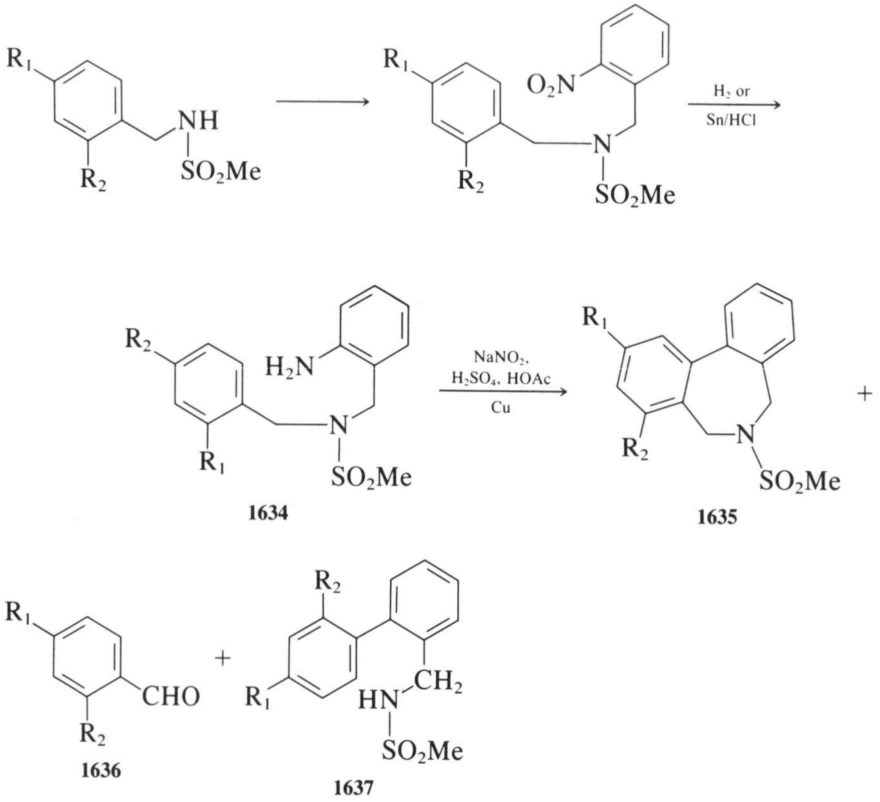

H_2 or Sn/HCl

1634 → ($NaNO_2$, H_2SO_4, HOAc, Cu) → 1635 + 1636 + 1637

Figure 214

TABLE 123A. 6,7-DIHYDRO-5H-DIBENZ[*c*,*e*]AZEPINES PREPARED BY THE HAWTHORNE PROCEDURE[a]

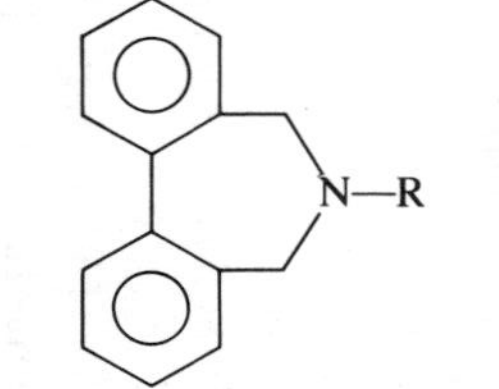

R	Method	Yield (%)	mp (°C)
H	A (from **1626**)	92	290–291 (HCl)
	A (from monoxime)	81	88–93
	B (Raney Ni, 850 psi, 75°C)	51	287–290 (HCl)
	B (5% Pt/C, 820 psi, 85°C)	84	284–290 (HCl)
CH_2Ph	A	87	203–205 (HCl)
	B (5% Pt/C, 80 psi, 50°C)	91	186–188 (MeI)
			202–208 (HCl)
Me	A	81	221–224 (HCl)
	B (5% Pt/C, 60 psi, 50°C)	97	287–289 (MeI)
			221–227 (HCl)
Ph	A	76	90–92
	B (Raney Ni, 50 psi, 25°C)	76	89.5–91.6
$CH_2CH{=}CH_2$	A	57	214–215 (HCl)
$(CH_2)_3NMe_2$	A	62	231–232 ($2HCl \cdot 2H_2O$)
$(CH_2)_3NEt_2$	A	70	222–223 ($H_3PO_4 \cdot H_2O$)
$(CH_2)_2NH_2$	A	63	268–270 (2HCl)

[a]Fig. 212, Ref. 942.

TABLE 123B. 6,7-DIHYDRO-5H-DIBENZ[*c*,*e*]AZEPINES PREPARED BY THE HAWTHORNE PROCEDURE[a]

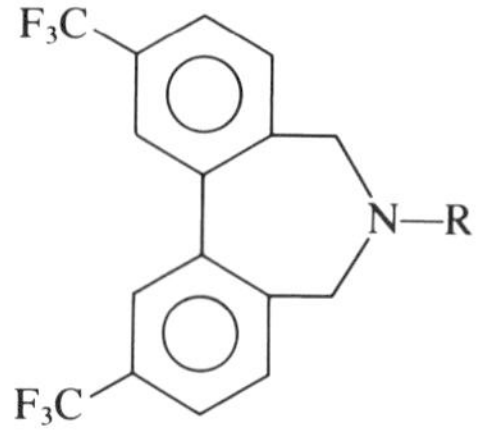

R	Method	Yield (%)	mp (°C)	Spectra
$(CH_2)_3NEt_2$	A	88	108 (HCl)	nmr
$CH_2CH{=}CH_2$	A	74	120	nmr
			185–187 (HCl)	
$CH(CH_2Ph)COOH$	A	79	207–207.5	nmr
$(CH_2)_4CH(NH_2)CO_2H$	A	17	230–232 (2HBr)	nmr

[a]Fig. 212, Ref. 946.

the optically active dibromide **1638** yielded optically active quaternary salts or amines **1641** directly (966–968).

Of the several 1,11-disubstituted quaternary salts (**1641**), only the 1,11-dimethoxy compound **1643** could be resolved into enantiomers by (+)- and (−)-camphorsulfonic acid (956). The (+)- and (−)-iodides of **1643** had rotations of + 3.0 ± 0.4 and −3.8 ± 0.2°. The salts showed considerable optical stability and were not racemized by heating at 100°C for 7 hr.

Enantiomers of 1,11-dinitro and 1,11-difluoro derivatives of **1640** were obtained by separation of the corresponding diastereomers **1639** (961,962). The nitro compounds were optically stable in benzene up to 100°C but were suprisingly less stable than the 1,11-dimethoxy compounds. Mononitro derivatives, likewise prepared by the (−)-ephedrine method, were studied. No resolution could be obtained with a 3-nitro derivative (965), but transient optical activity was seen with the 1-nitro isomer. However, mutarotation occurred rapidly ($t_{1/2}$ = 8.1 min).

The 1,11-dimethoxy quaternary salt **1643,** prepared from the resolved diphenic acid **1642,** was subsequently found to have a much larger rotation $[\alpha]_D$ = 46.0°), showing that the diastereomeric camphor sulfonates had been incompletely resolved (966). Table 126 lists the rotations of 6,7-dihydro-5H-dibenz[*c*,*e*]azepines obtained in these studies. Table 127 gives the E_{act} and $t_{1/2}$ values for these racemizations. The 1,11-dichloro analog of **1641,** prepared from dimethylamine, showed that the optical stability of these azepines was greater than that of the corresponding dibenzoxepins because of the larger dihedral angle between the rings of the biphenyl moiety in dibenz[*c*,*e*]azepines (470° versus 430°).

The ultraviolet absorption spectra of these compounds have been dis-

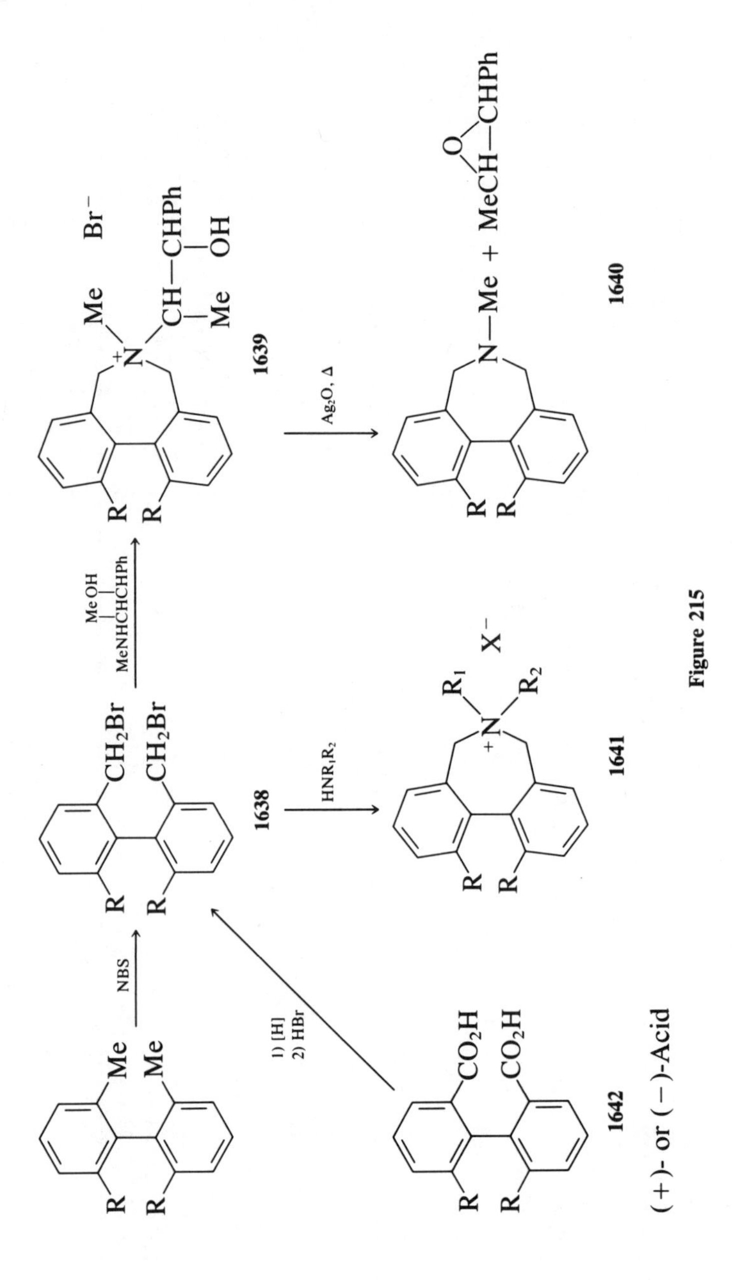

NBS
MeNHCHCHPh
Me OH
Ag2O, Δ
HNR1R2
1) [H]
2) HBr
1638
1639
1640
1641
1642
(+)- or (−)-Acid

Figure 215

TABLE 124. 6,7-DIHYDRO-5H-DIBENZ[*c*,*e*]AZEPINES FROM PHOTOLYSIS OF 2-IODOBENZYLAMINES[a]

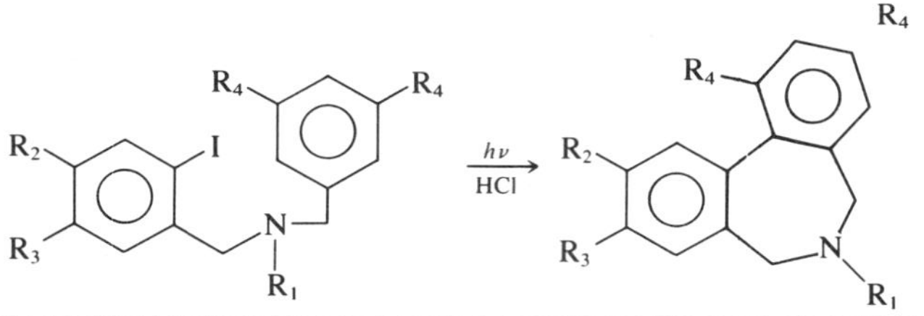

R_1	R_2	R_3	R_4	Irradiation Time (hr)	Starting Material (%)	Product (%)	Mp (°C)	Spectra
H	H	H	H	196	13	57	50 292 (HCl)	ir,uv,nmr
Me	H	H	H	96	10	44	225–227 (HBr) 193–195 ($HClO_4$)	nmr,ms
H	H	NO_2	H	192	8	32	320–321 (HCl)	uv,ir,nmr,ms
H	H	NO_2	MeO	480	26	27	281–282 (HCl)	uv,nmr,ms
H	H	H	MeO	192	0	0		

TABLE 125. PREPARATION OF 6,7-DIHYDRO-5H-DIBENZ[*c,e*]AZEPINES BY PSCHORR CYCLIZATION[a]

			Compounds (% Yield)			
R_1	R_2	**1635**[b]	**1636**	**1637**	Azepine mp (°C)	Spectra
H	H	55	11	—	147–148	nmr,ir
Me	H	40	19	8	145–146	nmr,ir
Cl	H	41	18	9	142–143	nmr,ir
H	Cl	28	17	11	150–151	nmr,ir
OMe	H	15[c]	18	—	108–109	nmr,ir

[a]Ref. 951.
[b]Products separated by column chromatography.
[c]Yield would have been higher but for difficulty in separation of *N*-benzylmethanesulfonamide by-product.

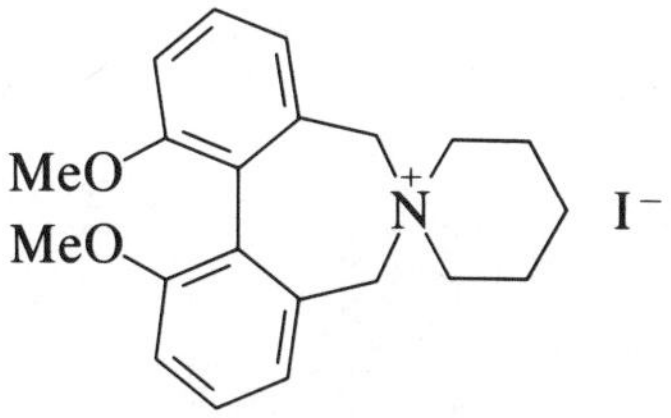

1643 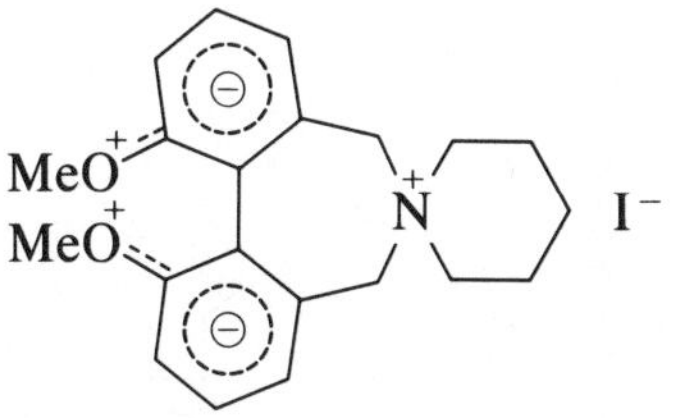 **1644**

cussed extensively (956,963,968). The spectrum of the 1,11-dimethoxy derivatives **1641** showed a long-wavelength band, assigned to the electronic interaction of the methoxy groups with the aromatic rings, as in structure **1644** (see above) To the extent that this would diminish the double bond character of the biphenyl bond in **1643,** the greater optical stability of the 1,11-dimethoxy compounds relative to the 1,11-dinitro derivatives (Table 126) might be explained (956). Some conjugation between the aromatic rings was evident even in 1,11-diiodo analogs, indicating that the steric effect of halogen atoms is smaller in dibenz[*c,e*]azepines than in the simpler 2,2′-biphenyl (963).

The ultraviolet and nmr spectra of 1,11-disubstituted 6,7-dihydro-5H-dibenz[*c,e*]azepines has been discussed comprehensively by Mislow and coworkers (960) in the context of the angle of torsion of various types of bridged and nonbridged biphenyl compounds. The dihedral angle for 6,7-dihydro-5H-dibenz[*c,e*]azepine was reported in this work to be 45.8°.

Fraser and co-workers (955) have analyzed the nmr spectrum of **1645** in trifluoroacetic acid (TFA). In this protonated amine, diaryl rotation is prevented by the 1,11-methyl groups, and deprotonation is slow on the nmr time scale. Nitrogen thus acts as a pseudoasymmetric center, and the spectrum of **1645** in TFA shows separate signals for all four protons adjacent to

TABLE 126. 6.7-DIHYDRO-5H-DIBENZ[*c*,*e*]AZEPINES: SPECIFIC ROTATION OF SALTS AND FREE BASES

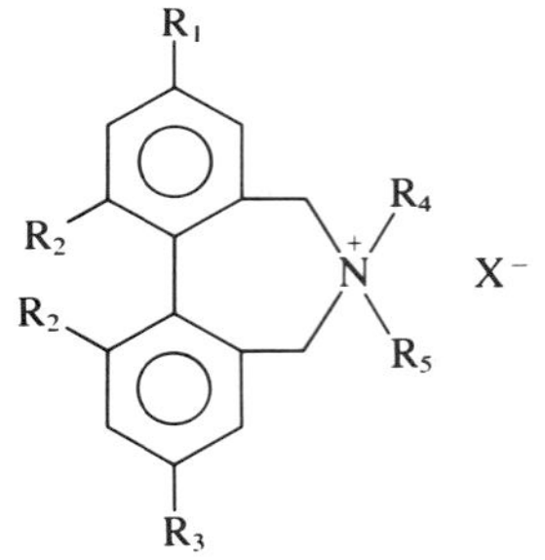

R_1	R_2	R_3	R_4	R_5	X^-	mp (°C)	$T_{[\alpha]}$ (°C)	$\lambda_{[\alpha]}$ (nm)	$[\alpha]$ (deg)	Refs.
H	H	H	H	$CH(Me)CH(OH)C_6H_4$	(+)-Bromide		25	546	+53.5	956
								579	+44.0	
H	H	H	H	$CH(Me)CH(OH)C_6H_4$	(+)-Iodide	226–228	23	546	+41.4	956
								579	+33.5	
H	H	H	—$(CH_2)_5$—		(+)-Camphorsulfonate		20	546	+14.0	956
					(+)-Bromocam-		20	579	+11.1	
					phor π-sulfonate			546	+60.4	956
								579	+50.7	
H	H	H	—$(CH_2)_2CH(Ph)(CH_2)_2$—		(+)-Camphorsulfonate		20	546	+21.6	956
								579	+18.2	
					(+)-Bromocam-		20	546	+57.0	956
					phor π-sulfonate			579	+47.8	
H	MeO	H	Et	Et	(+)-Camphor-					
					sulfonate	218	21	546	+22.3	956
H	MeO	H	—$(CH_2)_5$—		(+)-Camphor-					
					sulfonate	248–250	21	546	+19.0	956

								579	+15.8	
H	MeO	H	—$(CH_2)_5$—		(−)-Iodide		22	546	−3.8 ± 0.2	956
								579	−3.4 ± 0.2	
							21	546	− 46.0	966
					(−)-Camphor-			546	−14.5	966
					sulfonate	245–248	18	579	−12.0	
					(−)-Bromide	230–232	21	546	−46.9	966
								578	−43.2	
					(+)-Iodide		22	546	+4.0 ± 0.4	956
								579	+3.6 ± 0.4	
H	MeO	H	Me	Me	(+)-Bromide	182–184 (rapid) 256–258 (2nd melt) 330 (3rd melt)	24	546	+46.6	967
					(−)-Bromide[a]		20	546	−47.1	967
					(+)-Iodide	169–170	18	546	42.7	967
					(−)-Iodide	169–171	20	546	−41.6	967
H	Me	H	Me	Me	(+)-Bromide	278–288	20	546	+30.3	957 958
						287	26	546	+34.4	960
					(−)-Bromide	280–282	20	546	−30.3	957
					(−)-Iodide	265	26	546	−31.7	960
H	NO_2	H	—$(CH_2)_5$—		(−)-Bromide	168–169	25	546	−800	958
								578	−637	
							28	546	−527	958
H	Cl	H	—$(CH_2)_5$—		(−)-Bromide	297–298.5	27	546	−84	959
								578	−84	
H	Cl	H	Me	Me	(−)-Bromide		22	546	−159	968
H	Me	H	Me	—	Free base	50.5–52 105–110/0.08[b]	21	546	+87.8	960
H	NO_2	H	Me	CH(Me)CH(OH)Ph	(−)-Bromide	248	23	546	−709	961
					(+)-Bromide		18	546	+689	961
H	F	H	Me	CH(Me)CH(OH)Ph	(+)-Bromide		18.5	546	+50	961
					(−)-Bromide	230–232	18.5	546	−55	961

(*continued*)

TABLE 126. (*continued*)

R_1	R_2	R_3	R_4	R_5	X^-	mp (°C)	$T_{[\alpha]}$ (°C)	$\lambda_{[\alpha]}$ (nm)	$[\alpha]$ (deg)	Refs.
H	NO_2	H	Me	—	(−)-Free base	169–170	19	546	−1343	961
					(+)-Free base	169–170	19	546	+1333	961
					(−)-Hydroiodide	252–254	19	546	−813	961
H	F	H	Me	—	(+)-Hydrochloride		18.5	546	+45	961
					(−)-Hydrochloride		18.5	546	−41.5	961
H	NO_2 (mono)	H	Me	CH(Me)CH(OH)Ph	(−)-Bromide	223	24.2	546	−284[c]	965
					(−)-Iodide	175–190	22	546	−62.6[d]	965
					(−)-Bromide	190–195	21.3	546	−79.7[e]	965
					Bromocamphor-sulfonate	267–269	22	546	−240[f]	965
NO_2	H	H	Me	CH(Me)CH(OH)Ph	(+)-Bromide	229 (iodide)	22	546	+30.7	965
NO_2	H	H	Me	Me	(+)-Bromo-camphor-sulfonate	195–198	23	546	+63.4	965

[a]Same mp behavior as (+)-bromide.
[b]Boiling point (°C/torr).
[c]Mutarotated to −109; $t_{1/2}$ = 8.1 min.
[d]Mutarotated to +87.6.
[e]Mutarotated to +89.2.

nitrogen. Each proton is coupled to N—H and to its diastereotopic partner. Coupling constants were analyzed on the basis of an *ABCDX* spectrum, and there was excellent agreement between the calculated and observed values. Sutherland and Ramsay (969), in an nmr investigation of energy barriers to inversion in bridged biphenyls, reported the free energy barrier for conformational inversion ($\Delta F^{\neq}$) at -1°C to be 13.4 kcal/mole for **1646.** However, the *AB* system, corresponding to the H_A and H_B protons, was resolved at 60 mHz into only two broad resonances. In the corresponding protonated N—H compound, the benzylic protons were sharp singlets down to -60°C. According to Sutherland and co-workers (970), hindered rotation associated with the N—CO_2Me bond in **1647** causes the two methylene groups to be nonequivalent, but each methylene is represented by a single peak because

1645 **1646** **1647**

TABLE 127. 6,7-DIHYDRO-5H-DIBENZ[*c*,*e*]AZEPINES: ENERGY OF ACTIVATION AND HALF-LIFE DATA FOR RACEMIZATION

A	R_1	R_2	X	E_{act} (kcal/mole)	$t_{1/2}$ (°C)	Refs.
NMe	NO_2	NO_2	None	30	16 hr (125) 2.6 hr (145)	961,965
NMe	F	F	None	28	6.5 hr (80)	961,965
NMe_2	NO_2	H	Picrate	25.1	21.6 min (26.1)	965
$\overset{+}{N}$ (piperidinium)	MeO	MeO	Bromide	34	250 min (158)	966,967
$\overset{+}{N}Me_2$	MeO	MeO	Bromide	36	89 min (167.8)	967
$\overset{+}{N}Me_2$	MeO	MeO	Iodide	35.6	94 min (167.8)	967
$\overset{+}{N}Me_2$	Cl	Cl	Bromide	38.3		968

ring inversion is rapid at the temperatures at which the rotational process could be studied. The energy barrier ($\Delta G^{\neq}$) at $-3°C$ for rotation about the N—CO_2Me bond was calculated to be 15.8 kcal/mole.

Fraser and co-workers (954) reported that α-methylation of the *N*-nitrosamine **1648** gives **1649** in excellent yield (Fig. 216). The methyl group was determined to be *syn* and pseudoaxial to the nitroso group. The stereochemical relationships among the four bridge hydrogens are shown clearly in the projection formula **1651.** The formation of **1649** proceeds by removal of the pseudoaxial H_4 followed by stereospecific methylation of the resulting carbanion (971). No nmr signals for the *anti* isomer were seen for at least one hour. The carbanion could also be carboxylated to give **1650** (no yield given). Exchange studies with KOBu-*t* in *t*-BuOD revealed that each pseudoaxial proton (H_3, H_4) exchanged 100 times faster than its pseudoequatorial

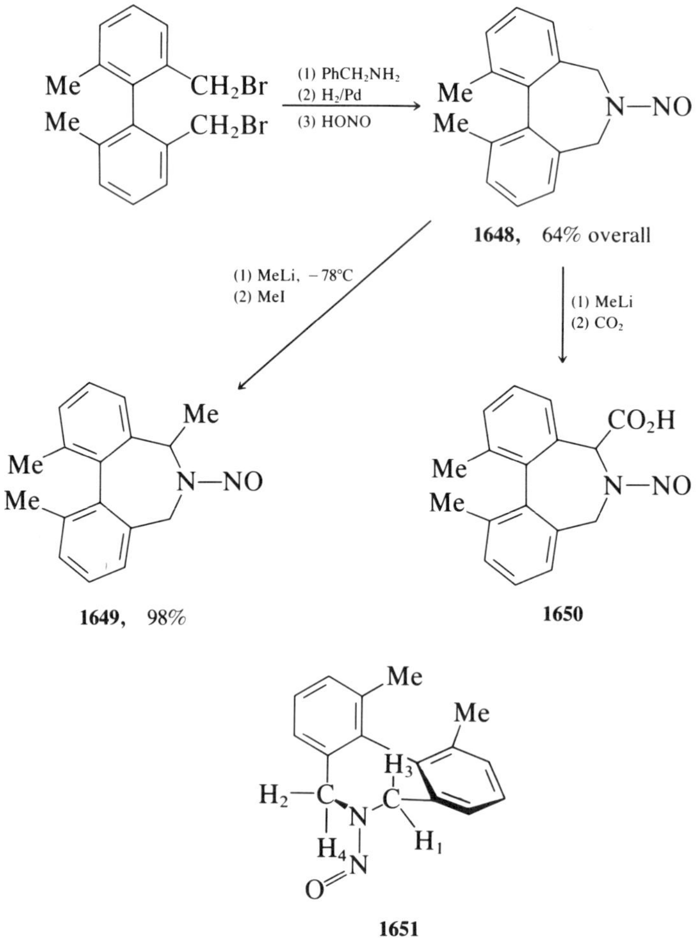

1651

Figure 216

partner (H_1, H_2), and each *syn* proton (H_2, H_4) exchanged 1000 times faster than its *anti* partner (H_1, H_3) (972). Interestingly, the 6,6-dimethyl iodide corresponding to **1648** did not exchange under these conditions, showing that ylide-type stabilization of the carbanion is not a factor.

An interesting study by Wittig and Zimmerman (957), later confirmed and extended by Mislow and co-workers (959) and by Hall and Poole (964), is summarized in Fig. 217. When the quaternary salt **1652** (R = Me) was treated with a strong base, rearrangement to the aminodihydrophenanthrene **1653** occurred without loss of optical activity. Hofmann degradation of **1653** gave inactive **1654.** Treatment of **1652** (R = H) with sodium bicarbonate or alkaline silver oxide gave predominately **1655** or **1656,** depending on which reagent was used. A related transformation reported by Carpino (938) was the rapid conversion of **1658** by aqueous NaOH at 80°C to 9,10-dihydrophenanthrene (**1659**) in high yield (Eq. 261). The starting material (**1657**) was prepared from *t*-butyl carbazate via the Wenner procedure.

The 1-nitro-6,7-dihydro-5H-dibenz[*c*,*e*]azepines **1660** were converted in good yield to the carbazole products **1661** by deoxygenation with triphenylphosphine (Fig. 218). Attempts to convert the benzamido derivative **1662** to **1663** with polyphosphoric acid met with failure (939). Compound **1662** was obtained from **1660** by catalytic reduction followed by acylation.

Another route to 5H-dibenz[*c*,*e*]azepines (Fig. 219) involves amination

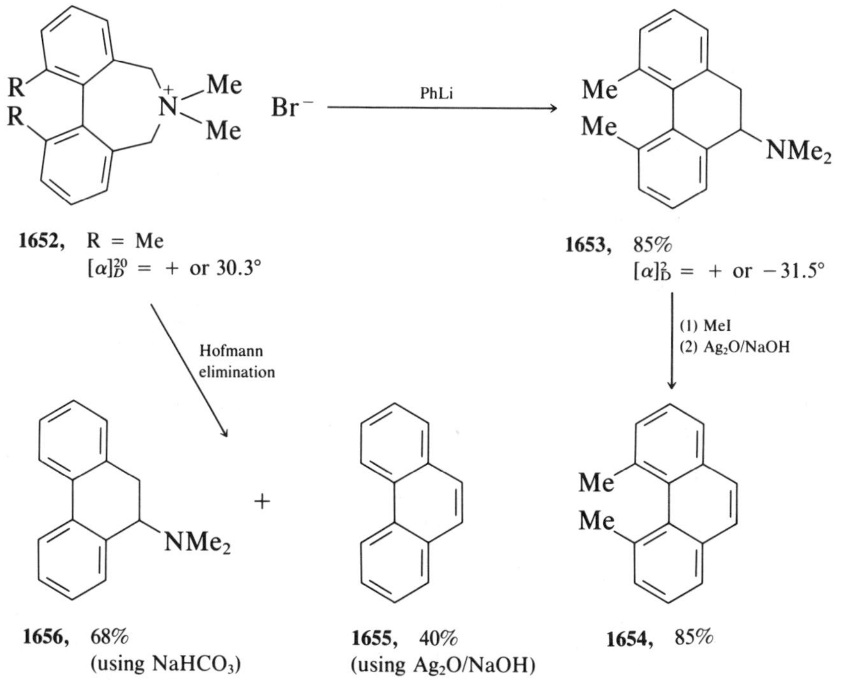

Figure 217

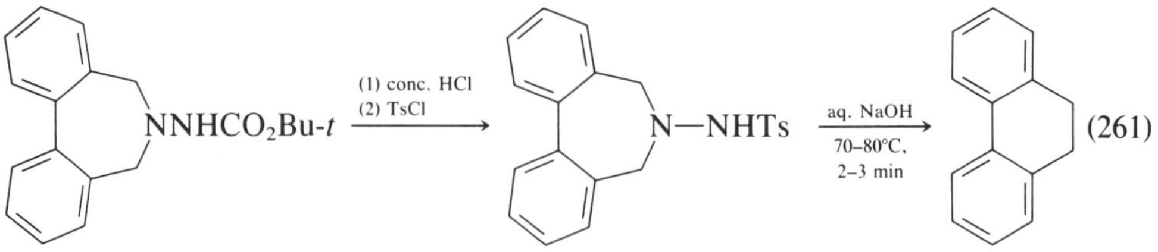

$NNHCO_2Bu$-*t* —(1) conc. HCl, (2) TsCl→ N—NHTs —aq. NaOH, 70–80°C, 2–3 min→ (261)

1657 **1658,** 25% overall **1659,** 95%

of **1665,** which is formed in 79% yield from **1664** by bromination (921). A series of 6-arylimmonium bromides (**1666,** see Table 128) were formed in good yield by this method and were converted to the corresponding peroxides **1667** (Table 122) by the action of hydrogen peroxide in methanol. The peroxides were identical to the products obtained by light-catalyzed peroxygenation of the 6,7-dihydroazepines **1668.** Hydrogenation of **1667** regenerated the dihydro compounds in good yield.

The 5-hydroxy-5H-dibenz[*c*,*e*]azepine **1626,** discussed earlier (see Fig. 212), could be converted to the nitrile **1669** and thence to the acid **1670** in good yield (Fig. 220) (942). Copper-catalyzed *bis*-ether formation with ethylene glycol gave **1671,** but an attempt to acetylate the hydroxy compound yielded the ring-opened triacetyl product **1673,** possibly indicating that **1626** is in tautomeric equilibrium with a small concentration of **1672.**

5H-Dibenz[*c*,*e*]azepine (**1676**) itself was first reported (941) as a product

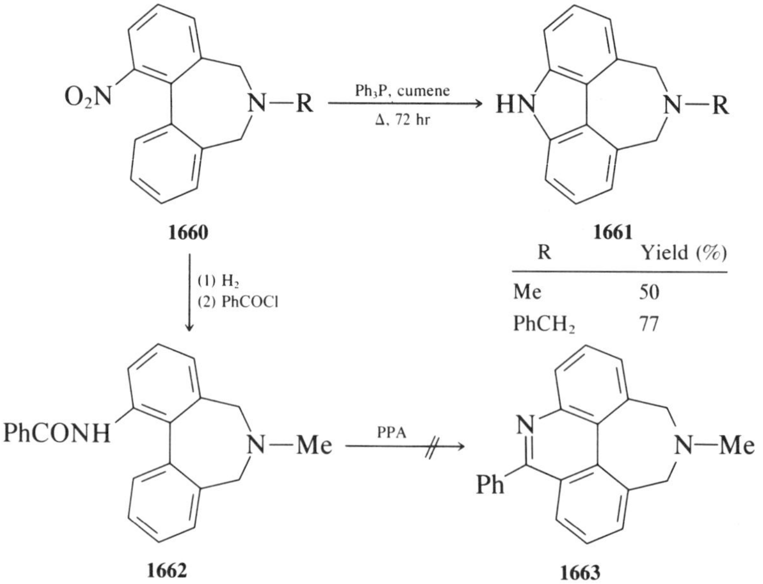

O_2N N—R —Ph_3P, cumene, Δ, 72 hr→ HN N—R

1660 **1661**

R	Yield (%)
Me	50
$PhCH_2$	77

↓ (1) H_2 (2) PhCOCl

PhCONH N—Me —PPA ⫽→ N Ph N—Me

1662 **1663**

Figure 218

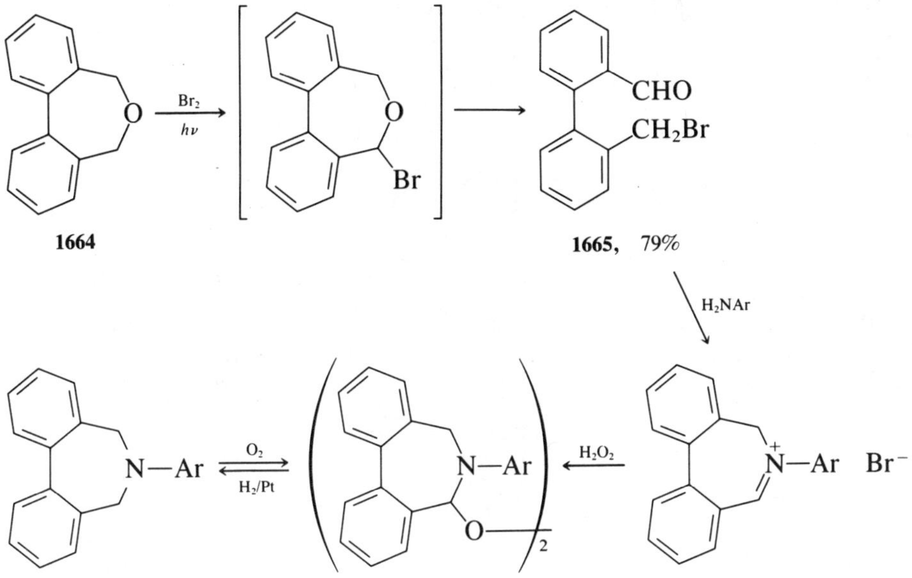

1664 → (Br_2, $h\nu$) → [bromo ether] → 1665, 79% (CHO, CH_2Br) → (H_2NAr) → 1666, 77–98% (Br^-) → (H_2O_2) → 1667 ⇌ (O_2 / H_2/Pt) 1668

Figure 219

of thermal decomposition of the *bis*-azide **1674** (Eq. 262). Thermolysis of **1674** in refluxing diphenyl ether gave a 29% yield of **1676** along with phenanthridine (**1675**). Small-scale vapor phase thermolysis gave **1676** (57%), but a larger-scale run gave only **1675** (77%). Phenanthridine was not formed from **1676.** The ultraviolet absorption spectrum of **1676** was similar to that of the carbocyclic analog. It was suggested that **1676** could arise either from the *bis*-carbene **1677,** or by intramolecular carbene attack on the azidomethyl group in **1678** with loss of nitrogen.

A better method of preparation (932) of **1676** involves base-promoted elimination of methanesulfinic acid from **1679,** which proceeds under very mild conditions in 85% yield (Fig. 221). Treatment of the immonium salts **1680,** derived from **1676,** with 2 *N* NaOH, failed to give the 6-methyl-6H-dibenz[*c*,*e*]azepine (**1681**), but instead yielded the carbinolamine **1682.** Had a strong nonnucleophilic base been employed, the formation of an ylide of **1680** might have been observed.

Cope elimination (934) converted the *N*-oxide **1683** to the *N*-hydroxyazepine **1684,** which could be oxidized in excellent yield to the nitrone **1685** (Fig. 222) (933). No detectable amount of the valence tautomer **1686** was observed by nmr spectrometry. On reaction with phenyl isocyanate and *N*-phenylmaleimide, **1685** gave high yields of 1:1 adducts, which are presumed to be **1687** and **1688.** Phenyllithium added to **1685** to give **1689,** which was easily oxidized with mercuric oxide to the nitrone **1690.**

From SCF–MO calculations, Dewar (975) has predicted a very small

Table 128. DERIVATIVES OF 5H-DIBENZ[*c*,*e*]AZEPINE

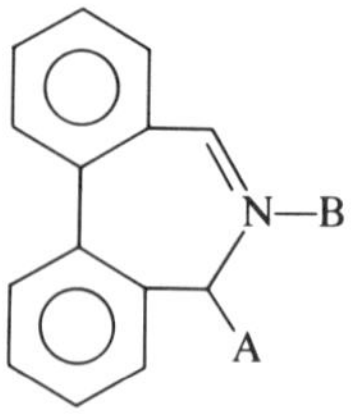

A	B	MP (°C)	Salt	Spectra	References
H	—	83–85		nmr	932
		217	MeI	ir,nmr	941
H	O	157–158		ir,uv,nmr	933
		122–123	Picrate		
		202–207	HCl		
Ph	O	185–189		uv	933
		216–217	HCl	nmr	
OH	—	126.8–128.8			942
		141–142			945,948
CN	—	154–156			942
		150–153			942
CO_2H	—	249–253			942
		221–222	Cu complex		942
OCH_2CH_2O (*bis*)	—	244–246			942
—	Ph	253–254	Br		921
		171–172	Picrate		
—	C_6H_4Me-*p*	253–254			921
		175–176	Picrate		
—	C_6H_4Cl-*p*	233–235			921
		170–172	Picrate		
—	2-Naphthyl	257–259			921
		206–207	Picrate		
—	C_6H_4OMe-*p*	236–237			921
		188–189	Picrate		

resonance energy for valence tautomers such as **1681** or **1686.** In structures of this kind, benzene rings could be aromatic only if there were extensive charge separation leading to unfavorable zwitterionic structures.

Hawthorne has used a modification of the previously described procedure for the synthesis of 6,7-dihydro-5H-dibenz[*c*,*e*]azepin-5-ones (976). The aldehyde acid **1691** (R = CHO) was subjected to reductive amination, either catalytically with Raney nickel or chemically with sodium bisulfite, to give the amino acids **1692** (Eq. 263). On being heated under vacuum, the amino acids underwent ring closure to the lactams **1693.** When an ester of **1691** was used, lactams were likewise obtained. A patented modification of this

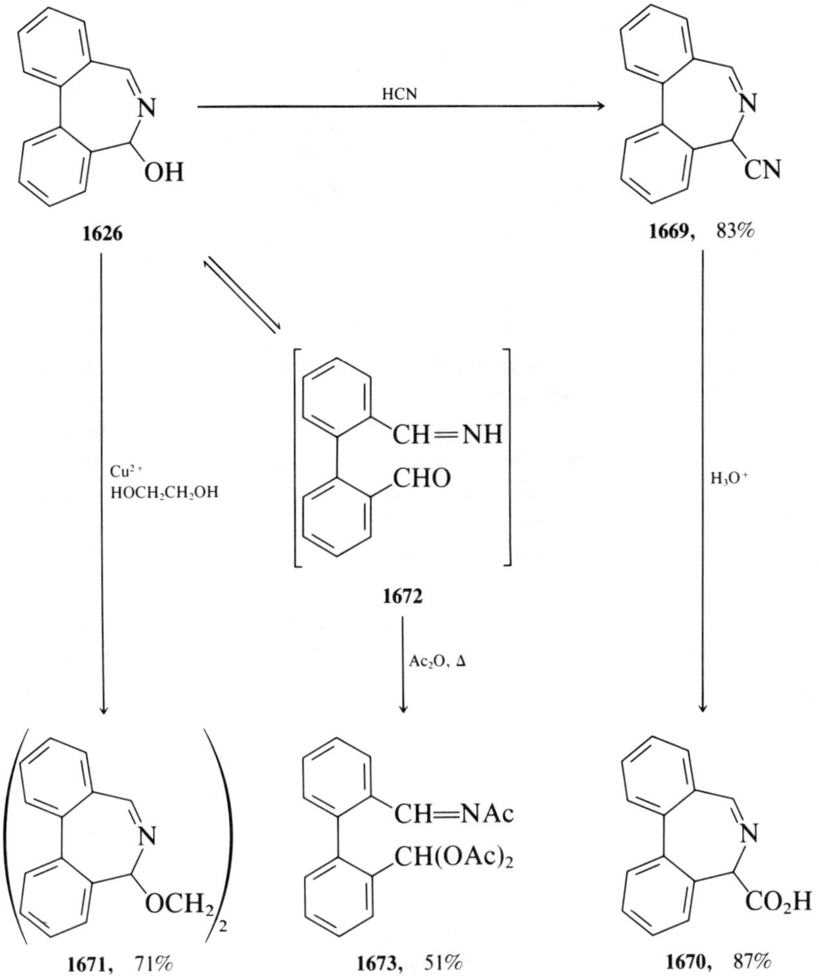

Figure 220

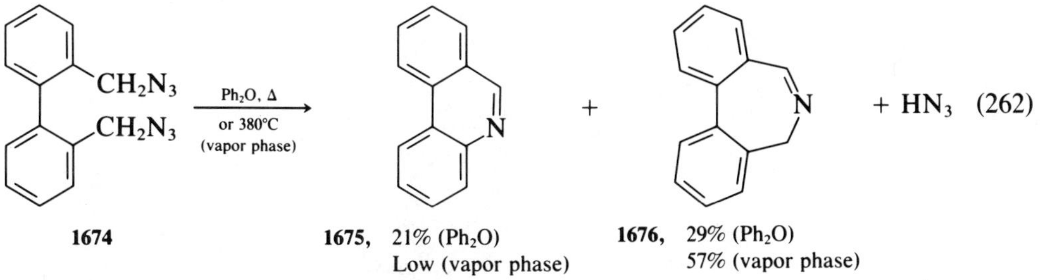

(262)

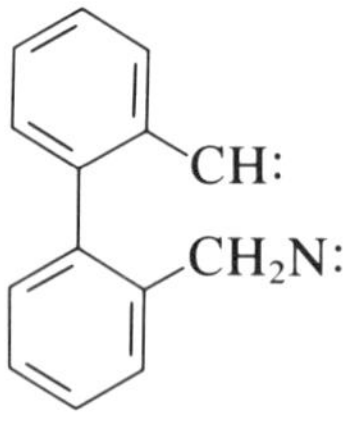

1677 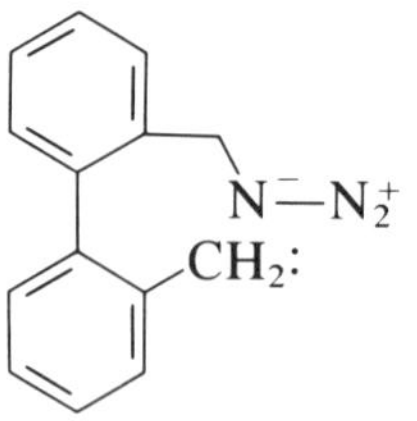 1678

process claims the use of the nitrile **1691** (R = CN) to form **1691** (R′ = H) directly (977).

In a study of the photochemistry of α-diketones (978), the 5-hydroxy-7-dibenz[*c*,*e*]azepinone **1695** was obtained in 53% yield from 9,10-phenanthrene quinone (**1694,** Fig. 223). For confirmation of structure, **1695** was synthesized from the amide ester **1696** by lithium borohydride reduction to the alcohol **1697** followed by oxidation with CrO_3. Yields in this sequence were unspecified.

The azepinone **1699** was obtained from the open-chain amine **1698** by anodic oxidation (979), but the yield was only 2% (Eq. 264). The corresponding azocine was obtained in 45% yield.

An X-ray crystallographic structure of 6,7-dihydro-6-methyl-5H-dibenz[*c*,*e*]azepine-5-one, the unexpected product of thermal rearrangement

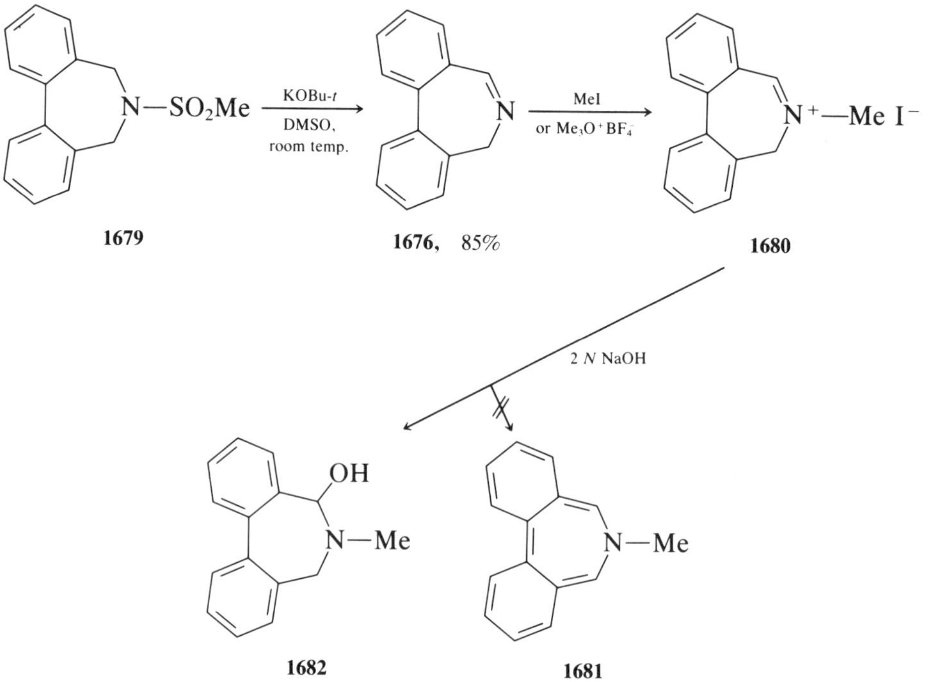

1679 1676, 85% 1680

1682 1681

Figure 221

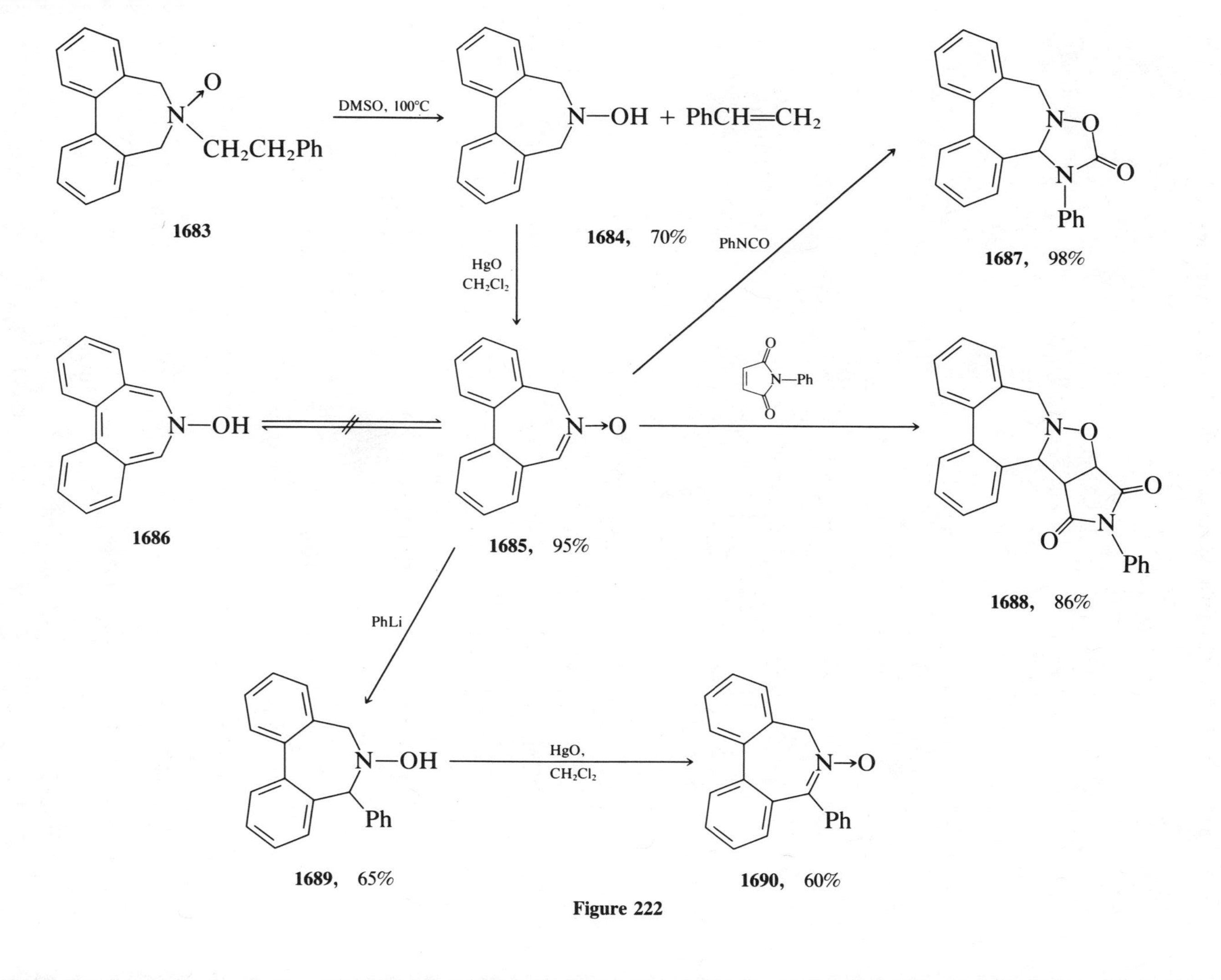
1683
DMSO, 100°C
N—OH + PhCH=CH2
1684, 70%
PhNCO
1687, 98%
HgO
CH2Cl2
1686
1685, 95%
1688, 86%
PhLi
HgO,
CH2Cl2
1689, 65%
1690, 60%

Figure 222

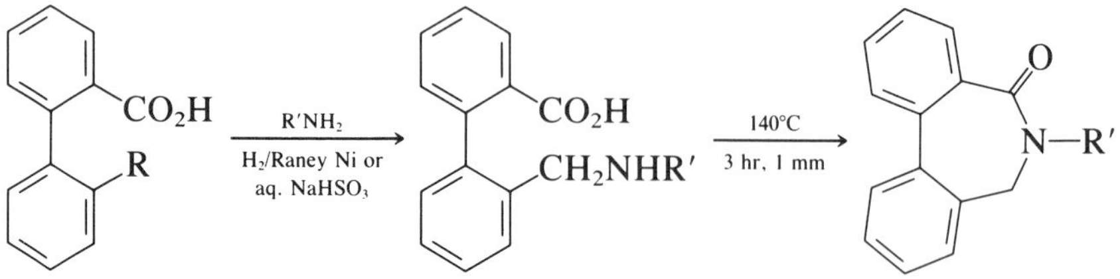

CO₂H
R
R'NH₂
H₂/Raney Ni or
aq. NaHSO₃
CO₂H
CH₂NHR'
140°C
3 hr, 1 mm
O
N—R'
(263)
1691
1692
1693

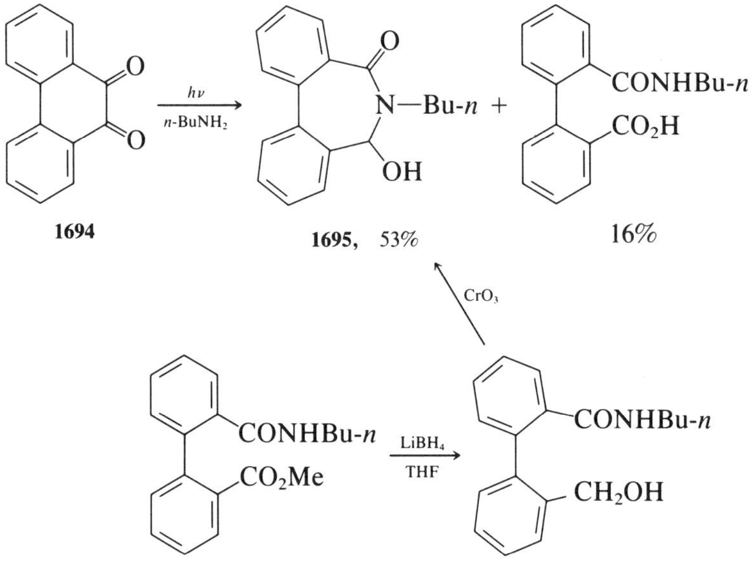

O
O
hν
n-BuNH₂
O
N—Bu-n +
OH
CONHBu-n
CO₂H
1694
1695, 53%
16%
CrO₃
CONHBu-n
CO₂Me
LiBH₄
THF
CONHBu-n
CH₂OH
1696
1697

Figure 223

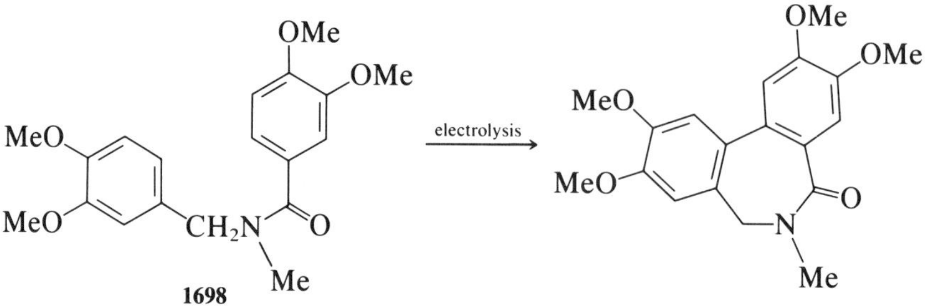

OMe
OMe
MeO
MeO
CH₂N
O
Me
1698
electrolysis
OMe
OMe
MeO
MeO
O
N
Me
(264)
1699, 2%

of 6,7-dihydro-6-methyldibenz[*c*,*e*]azepine-*N*-oxide, has been reported (980). Nmr, infrared, and mass spectral data were considered insufficient for firm structure proof. The biphenyl dihedral angle was found to be 40°, a somewhat small value for seven-membered-ring bridged biphenyls.

Nitration of the azepinone **1700** (Eq. 265) has been used to prepare the 9-nitro derivative **1701,** which could be converted to a series of 9-substituted

$$\mathbf{1700} \xrightarrow[H_2SO_4,\ 0\text{–}5°C]{KNO_3} \mathbf{1701}\ (NO_2) \xrightarrow{\text{standard reaction}} \mathbf{1702}\ (X) \tag{265}$$

TABLE 129A. DERIVATIVES OF 6,7-DIHYDRO-5H-DIBENZ[*c*,*e*]AZEPIN-5-ONES

R_1	R_2	R_3	MP (°C)	Yield (%)	Refs.
H	H	H	194–196		976
			190–191	59–61	977
Me	H	H	148–149		976
			115–116	10–17	950
CH_2Ph	H	H	114–115		976
n-Bu	OH	H	150.5–152.5	53	976
H	H	NO_2	250–252		978
H	H	NH_2	275–279		981
H	H	F	205–206		981
H	H	Br	228–230		981
H	H	CN	272–274		981
H	H	$CONH_2$	298		981
H	H	OH	290		981
H	H	MeO	183		981
H	*N*-Succinimide	H	281–283		981
H	CH_2CO_2H	H	242–244	22	982
H	CH_2CO_2Me	H	193–194		982

TABLE 129B. DERIVATIVES OF 6,7-DIHYDRO-5H-DIBENZ[*c*,*e*]AZEPIN-5-ONES

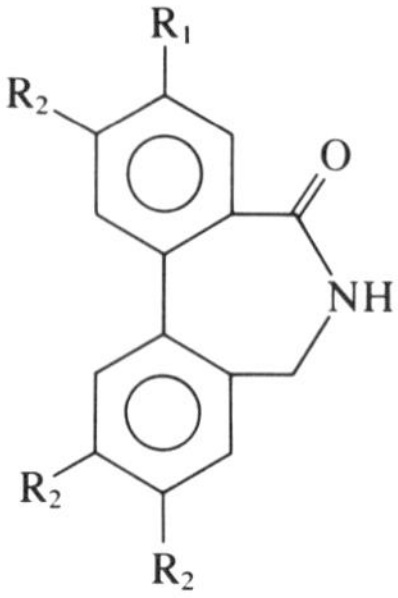

R_1
R_2
O
NH
R_2
R_2

R_1	R_2	mp (°C)	Yield (%)	Spectra	Refs.
NO_2	H	328			981
MeO	MeO	220–221	2	uv,ir,*m/e*	979

derivatives (**1702**) possessing anti-inflammatory, analgesic, and antipyretic properties (981). These substances, some of which were reported without yields, are listed in Table 129A,B.

Acid hydrolysis of the cinnamic acid derivative **1703** has been shown to proceed through the dibenz[*c*,*e*]azepinone **1705** and the amino acid **1706** to give the diacid **1707** (Fig 224) (982). Some loss of ammonia occurs rapidly from **1704** to give **1707** directly, but **1705** and **1706** can also be isolated and have been shown to be converted to **1707** by slow loss of ammonia.

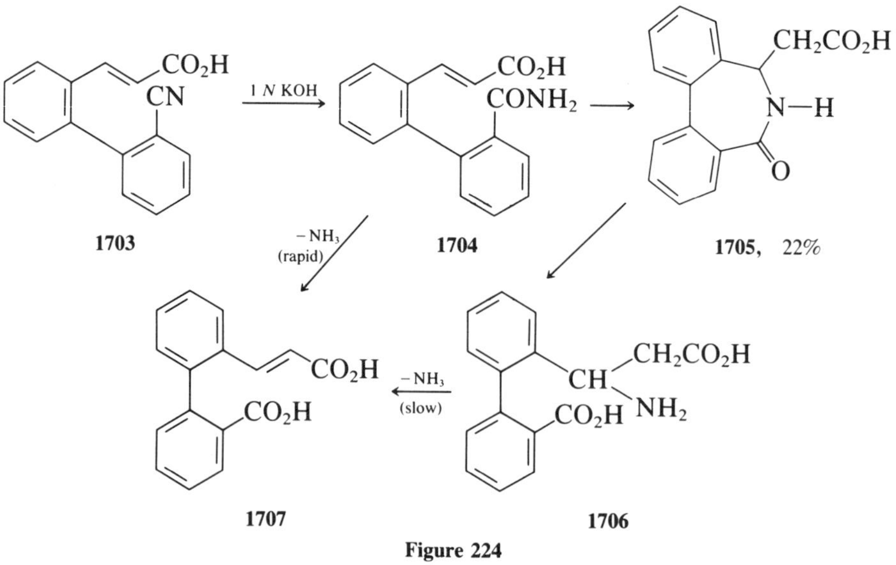

CO_2H
CN
1 *N* KOH
CO_2H
$CONH_2$
CH_2CO_2H
N—H
O
1703
$-NH_3$ (rapid)
1704
1705, 22%
CO_2H
CO_2H
$-NH_3$ (slow)
CH_2CO_2H
CH
CO_2H NH_2
1707
1706

Figure 224

Thermal decomposition of the diazonium salt derived from the *N*,*N*-dibenzylbenzamide **1708** has been reported to give the dibenz[*c*,*e*]azepinone **1709** as a minor product in 10–17% yield (Eq. 266) (950). This constituted the first example of seven-membered-ring formation by the Pschorr reaction. Evidence was provided to support a mechanism involving 1,5-hydride transfer from a benzyl methylene to a phenyl carbonium ion, as opposed to a free radical pathway. Final structure proof was provided by reduction of **1709** to an amine that was identical to the product of the Wenner reaction with benzylamine and 2,2′-*bis*-bromomethylbiphenyl.

$$\mathbf{1708} \xrightarrow[\text{HOAc, } H_2SO_4]{NaNO_2} \mathbf{1709},\ 10\text{–}17\% \qquad (266)$$

The preparation of the hexahydro-*trans*-(4a,10a) analog of **1693** (R = H) as a side product (1–2%) in a Schmidt reaction has been discussed previously (See Fig. 154).

Dibenz[*c*,*e*]azepin-5,7-diones (**1715**), also called diphenimides, are prepared in good to excellent yield by reaction of a diphenic anhydride (**1714**) with ammonia or amines to give an amide acid such as **1711** (Fig. 225). The amide acid can then be cyclized by fusion at elevated temperatures, by heating in acetic anhydride, thionyl chloride, and other dehydrating agent, or by the action of base. Diphenimide (**1712**) itself was first prepared in this manner (984). Substituted diphenic anhydrides (**1714**) are available generally from the appropriate phenanthrenes (**1713**) by oxidation (e.g., ozonolysis), followed by heating, to effect ring closure with loss of water. Reaction of diphenic anhydrides **1714** with a variety of amines and hydrazines and cyclization of the resulting amide acids by one of the aforementioned methods leads to *N*-substituted diphenimides. Table 130 lists these substances.

Cyclization of 2,2′-dicyanobiphenyl (**1716**) with 30% HBr in glacial acetic acid (Fig. 226) provides 6,7-dihydro-5-imino-5H-dibenz[*c*,*e*]azepin-7-one (**1717**) in 98% yield (997). If the dinitrile is treated with dry HBr in benzene, a 98% yield of 7-bromo-5-imino-5H-dibenz[*c*,*e*]azepine hydrobromide (**1718**) is obtained. The latter compound can be converted to **1712** in poor yield by aqueous acid hydrolysis. Although it seems likely that **1718** would be a good substrate for amines and other nucleophiles, such reactions have not been described to date.

Figure 225

Figure 226

Diphenimide was reported to form in 11% yield on treatment of 9,10-phenanthrenedione monoxime with thionyl chloride in liquid sulfur dioxide (988), but other workers were unable to repeat this work (999).

Like phthalimide, diphenimide can be alkylated with a variety of reagents. Reaction with diazomethane gave the *N*-methyl derivative, albeit in unspecified yield (1001). Because of the relationship between azapetine (Ilidar), an antiadrenergic agent, and pargyline, an MAO inhibitor, the *N*-propargyl analog of Ilidar (**1721**), was prepared (Fig. 227). Diphenimide (**1712**) was converted to its potassium salt (**1719**), which could be alkylated in 85% yield to **1720.** This substance was reduced to **1721** in 75% yield with lithium aluminum hydride (952). Alkylation of **1719** is not uniformly successful, however, as evidenced by the fact that an attempt to alkylate **1719** with 2-chloroethanol gave only ethyl diphenamate (2-carboxamido-2′-carbethoxybiphenyl) as the product (1002, 1003). A Japanese patent (1000) reports the thermal alkylation of **1712** with *N*,*N*-diethylaminoethanol at 210–230°C in 57% yield.

Monosubstituted and disubstituted diphenimides have been analyzed by the extended Hückel method, and correct predictions of dipole moments and pK_a values were reported (1004). An electrolytic process for the reduction of diphenimide to the amine has been patented (1005). At 15 A/10–15 V in ethanol–dioxane, a 44% yield of 6,7-dihydro-5H-dibenz[*c*,*e*]azepine was obtained, whereas a 67.5 % yield was realized at 20–25 A/10 V in glyme/MeOH–sulfuric acid.

N—H $\xrightarrow{KOH}$ N^- K^+

1712 **1719**

$BrCH_2C{\equiv}CH$

N—$CH_2C{\equiv}CH$ $\xleftarrow{LiAlH_4}$ N—$CH_2CH{\equiv}CH$

1721, 75% **1720**, 85%

Figure 227

TABLE 130. DERIVATIVES OF 5H,6H-DIBENZ[*c,e*]AZEPIN-5,7-DIONES (DIPHENIMIDES)

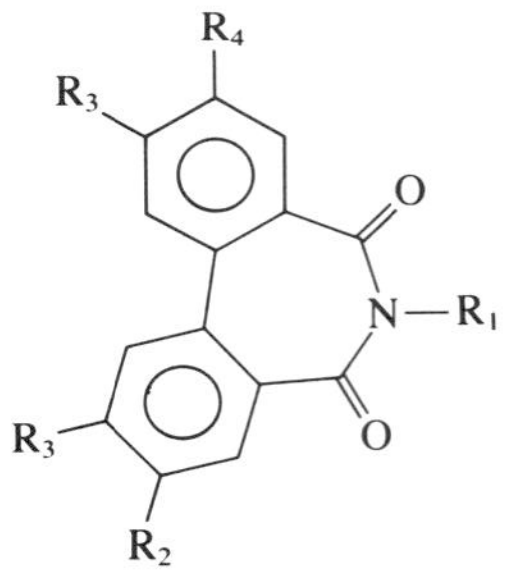

R_1	R_2	R_3	R_4	mp (°C)	Yield (%)	Spectra	Refs.
H	H	H	H	219–221	Good		947,984
				217–219	96		992
				223–224			237
Me	H	H	H	168			1001
CH_2CH_2OAc	H	H	H	123	86		997
Ph	H	H	H	199			994
				198–199	90		989
$C_6H_4NH_2$-*p*	H	H	H	194–195	93		994
$C_6H_3(OCH_2O)$-3,4	H	H	H	146–147	100		944
C_6H_4Me-*o*	H	H	H	189–190			983
C_6H_4Me-*p*	H	H	H	147–148			983
C_6H_4Cl-*p*	H	H	H	219–220			983
$C_6H_4NO_2$-*p*	H	H	H	261			983
$C_6H_4NO_2$-*o*	H	H	H	224.5			983
$C_6H_4NO_2$-*m*	H	H	H	257			983
C_6H_4NHAc-*o*	H	H	H	233			983
NH_2	H	H	H	180–181	48		985
NH_2	NO_2	H	H	319	50		987
NHAc	H	H	H	155		ir,nmr	986
NAc_2	H	H	H	162		ir,nmr	986
NHPh	H	H	H	194		ir,nmr	986
N(Ac)Ph	H	H	H	152		ir,nmr	986

NMe_2	H	H	H	98		ir,nmr	986
NAc_2	NO_2	H	H	188		ir,nmr	986
N(Ac)Ph	NO_2	H	H	256		ir,nmr	986
OH	H	H	H	286–287	52		985
$CH_2C{\equiv}CH$	H	H	H	175–176	85		952
$CH_2CH{=}CH_2$	H	H	H	170–190/3[a]	50		997
$(CH_2)_2NEt_2$	H	H	H	236–238/6[a]	57		1000
H	H	Br	H	287			990
H	H	Cl	H	234	86.7		991
2-Fluorenyl	H	CF_3	H	310–313	98		993
2-Fluorenyl	H	CH_3	H	322–325			993
1-Fluorenyl	H	H	H	146.5–150	80	ir	994
2-Fluorenyl	H	H	H	239–240	80	ir	994
2-Fluorenyl	Cl	H	Cl	311–312	94		995
2-Fluorenyl	NO_2	H	NO_2	302–303	100		995
2-(5,7-Dichlorofluorenyl)	Cl	H	Cl	298–299	100		995
2-(7-F-Fluorenyl)	H	H	H	276.5–277.5	60		994
2-(7-MeO-Fluoreny)	H	H	H	212.5–213.5	88		994
3-Fluorenyl	H	H	H	253.5–254.5	87		994
3-(2-MeO-Fluorenyl)	H	H	H	249–250	87		994
4-Fluorenyl	H	H	H	236–237	96		994
2,7-*bis*(Fluorenylene)	H	H	H	357–358.5	82		994
7-(3,4-Benzocoumarinyl)	H	H	H	315–316	91		994
6-(3,-Benzocoumarinyl)	H	H	H	353–354	82		994
6-Chrysenyl	H	H	H	247–248	97		994
3-(9-Et-Carbazolyl)	H	H	H	188.5–190.5	80		994
N—NH—C(=O)—N (bis-diphenimide structure)				186–187			996

[a]Boiling point (°C/torr).

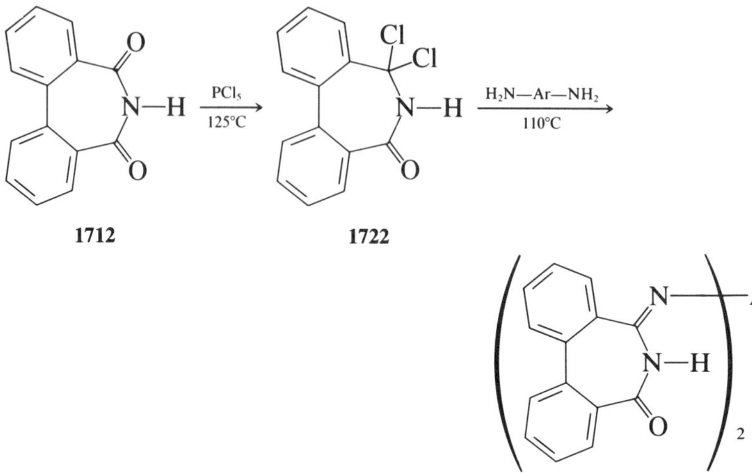

1712 $\xrightarrow[125°C]{PCl_5}$ 1722 $\xrightarrow[110°C]{H_2N—Ar—NH_2}$ (267)

1723, mp >300°C

Ar = aryl, diaryl

N-Bromodiphenimide has been studied as an allylic brominating agent (1006), but no details of its preparation or properties were given.

According to a series of patents (1007–1009), azomethine dyes of the general structure **1723** can be prepared from the dichloride **1722,** which was alleged to form when **1712** is heated with PCl_5 (Eq. 267). Although solubilities and absorption maxima for these dyes are cited, further structural characterization is lacking.

41. Naphth[1,2-*b*]azepines

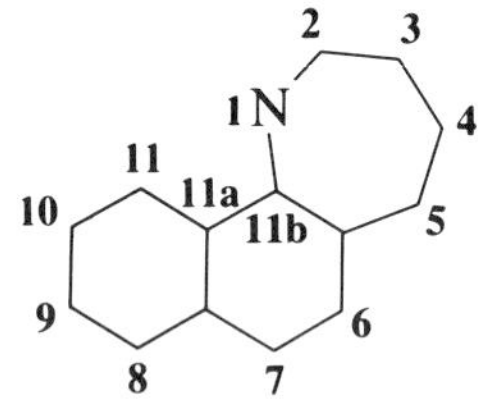

Napth[1,2-*b*]azepines were first described in 1930 when Beckmann rearrangement of the oxime acetates **1724a** and **1724b** gave the lactams **1725a** and **1725b,** respectively (Eq. 268) (1010). Chromic acid oxidation of 1,2,3,4,5,6,7,8-octahydrophenanthrene (**1724c**) gave the ketone **1724d** as one component of a mixture, while the methoxy ketone **1724e** was obtained by cyclization of 3-methoxy-2-tetralinbutyric acid. Ketones **1724d** and **1724e** were converted to their oximes **1724f** and **1724g** and subsequently to the oxime acetates **1724a** and **1724b**, respectively. The lactams **1725a** and **1725b** were converted with hot hydrochloric acid to the amino acid salts **1726a** and

converted with hot hydrochloric acid to the amino acid salts **1726a** and **1726b.** When heated above their melting points, **1726a** and **1726b** underwent recyclization to **1725.**

More recently, Sammes (1011) reported the preparation of the naphthazepine **1730** by a multistep synthesis from ketone **1724d** (Fig. 228). Methylmagnesium iodide converted **1724d** to the carbinol **1727a,** which on treatment with nitrosyl chloride gave the nitrite ester **1727b** in quantitative yield.

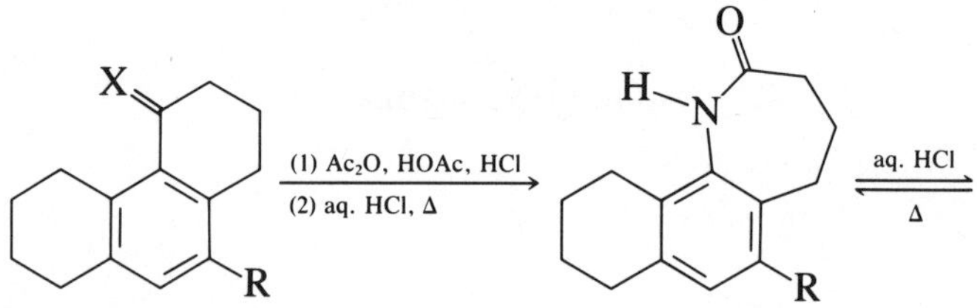

(1) Ac_2O, HOAc, HCl; (2) aq. HCl, Δ

aq. HCl; Δ

1724a, X = NOAc, R = H
1724b, X = NOAc, R = OMe
1724c, X = H_2, R = H
1724d, X = O, R = H
1724e, X = O, R = OMe
1724f, X = NOH, R = H
1724g, X = NOH, R = OMe

1725a, R = H; mp 156–158°C
1725b, R = OMe; mp 170–171°C

(268)

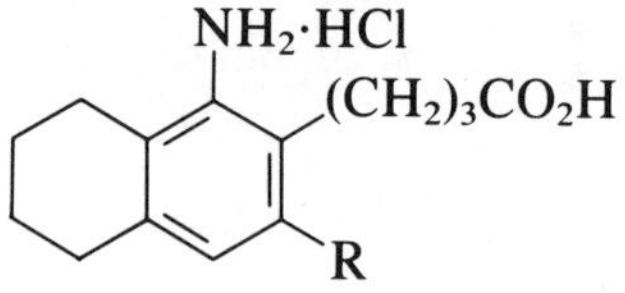

$NH_2 \cdot HCl$, $(CH_2)_3CO_2H$

1726a, R = H
1726b, R = OMe

1724d (1) MeMgI, Et_2O, 0°C; (2) NOCl, C_5H_5N → (1) *hν*, PhH; (2) Ac_2O, C_5H_5N

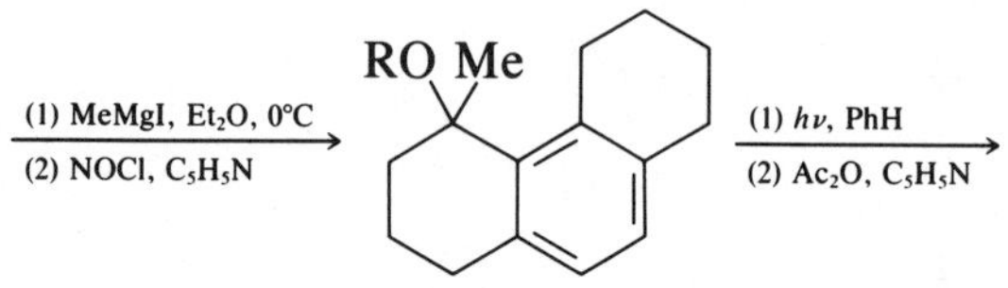

1727a, R = H
1727b, R = NO

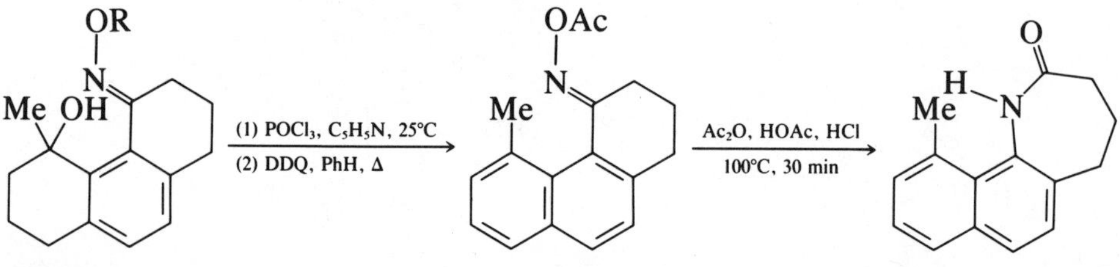

(1) $POCl_3$, C_5H_5N, 25°C; (2) DDQ, PhH, Δ

Ac_2O, HOAc, HCl; 100°C, 30 min

1728a, R = H
1728b, R = Ac

1729

1730, mp 198–199°C

Figure 228

The Barton reaction converted **1727b** to the hydroxy oxime **1728a** in 35–69% yield. Dehydration of the oxime acetate **1728b** with phosphoryl chloride and subsequent oxidation with 2,3-dichloro-5,6-dicyanoquinone (DDQ) gave the naphthalene derivative **1729.** The oxime **1729** was subjected to the action of acetic anhydride–acetic acid and dry hydrogen chloride at 100°C, whereupon Beckmann rearrangement afforded an 80% yield of **1730.** The same product was obtained when **1729** was treated with hot polyphosphoric acid or with acetyl bromide.

42. Naphth[1,2-*c*]azepines

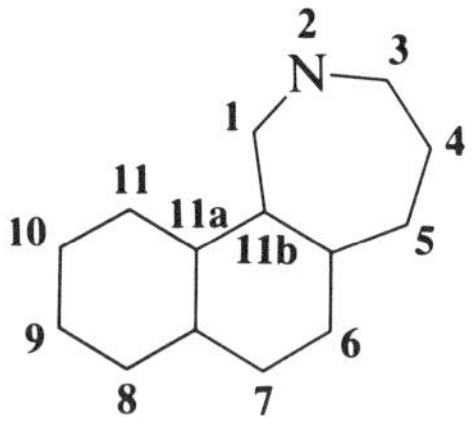

The napth[1,2-*c*]azepine ring system was first reported in a paper describing the Beckmann rearrangement of the oximes of dihydrothebainones (**1731**) and their Hofmann degradation products (Eq. 269) (1012). One of the

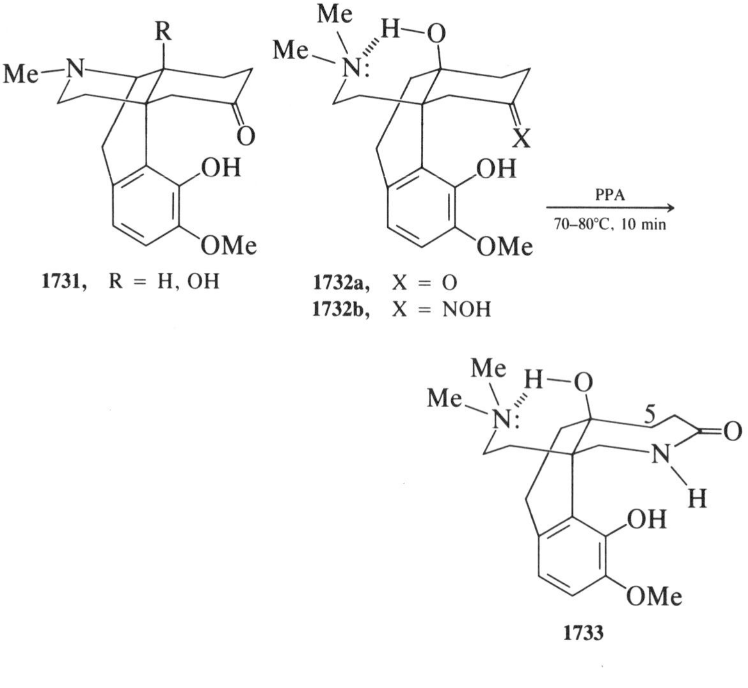

1731, R = H, OH

1732a, X = O
1732b, X = NOH

PPA
70–80°C, 10 min

(269)

1733

Hofmann degradation products of **1731** is the tricyclic ketone **1732a.** Treatment of the corresponding oxime **1732b** with polyphosphoric acid gave the lactam **1733** in 34% yield. The direction of rearrangement of the oxime **1732b** was deduced from the transannular coupling observed between the amide proton and the geminal protons at C_5. After deuterium oxide exchange, the C_5 geminal protons gave rise to a pair of one-proton doublets (J = 15 Hz) at δ 4.28 and δ 3.69 ppm.

In 1971, Oppolzer (1013) described the stereoselective synthesis of annelated heterocycles by the thermal rearrangement of the benzocyclobutenes **1734a–1734c** (Fig. 229). Thermolysis of **1734a** and **1734b** gave good yields of **1735a** and **1735b,** respectively. Similar treatment of **1734c** gave a 20% yield of the naphthazepine **1735c** along with 4% of the dimer **1737.** The results implicate *o*-quinonedimethanes, such as **1736,** as key intermediates in the reaction and suggest that the activation entropy for the intramolecular cycloaddition **1736**→**1735** depends strongly on the distance between the side-chain and quinonedimethane double bonds.

The Beckmann rearrangement has been used to prepare a series of naphth[1,2-*c*]azepines (1014–1017). The *cis*-fused phenanthrenone **1738a** was converted to its oxime **1738b**; when the latter was heated in polyphosphoric acid, the lactam **1739a** was obtained in 76–82% yield (Eq. 270) (1014,1016). Alkylation of **1739a** with the aid of sodium hydride as the base afforded moderate yields of **1739b–1739d.** Acetylation of **1739a** produced **1739e** in good yield. Lithium aluminum hydride reduction of the appropriate amides gave the corresponding amines **1740a–1740e** in good yield. The same se-

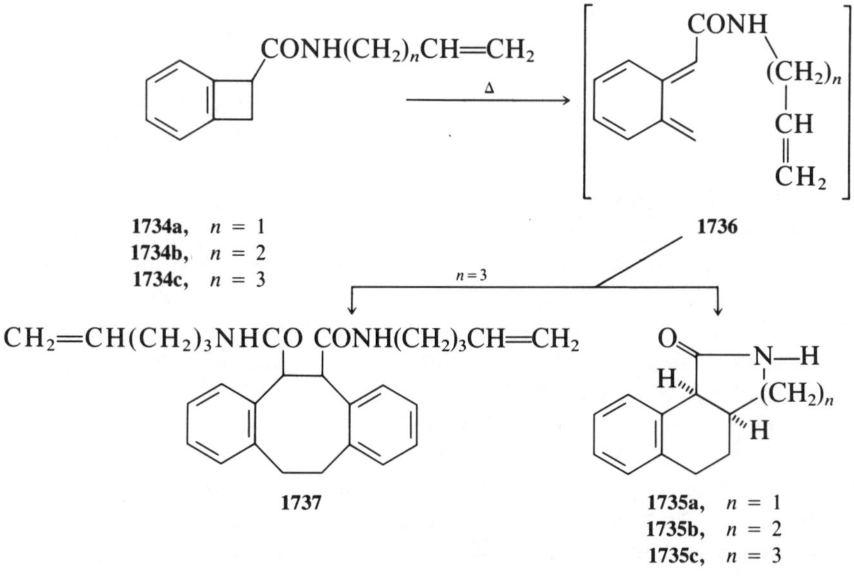

$CONH(CH_2)_nCH{=}CH_2$ $\xrightarrow{\Delta}$ [CONH, $(CH_2)_n$, CH, CH_2]

1734a, n = 1
1734b, n = 2
1734c, n = 3

1736

$n=3$

$CH_2{=}CH(CH_2)_3NHCO$ $CONH(CH_2)_3CH{=}CH_2$

1737

O, N—H, H, $(CH_2)_n$, H

1735a, n = 1
1735b, n = 2
1735c, n = 3

Figure 229

(1) PPA, 140°C 10–15 min (2) NaH, DMF (3) RX; $LiAlH_4$, Δ, 15 min dioxane (270)

1738a, X = O
1738b, X = NOH

1739a, R = H
1739b, R = Me
1739c, R = CH_2Ph
1739d, R = $(CH_2)_2NMe_2$
1739e, R = Ac

1740a, R = H
1740b, R = Me
1740c, R = CH_2Ph
1740d, R = $(CH_2)_2NMe_2$
1740e, R = Et

quence was applied to the *trans*-fused phenanthrenone **1741a,** giving the isomeric oxime **1741b,** the lactams **1742,** and the amines **1743** (Eq. 271) (1016). As in the Beckmann rearrangement of **1732b** (Eq. 269), isomeric naphth[1,2-*d*]azepines were not formed from either **1738b** or **1741b.** Physical constants for the naphth[1,2-*c*]azepines in Eqs. 270 and 271 are listed in Table 131.

(1) PPA, 140°C, 10–15 min (2) NaH, DMF (3) RX (271)

1741a, X = O
1741b, X = NOH

1742a, R = H
1742b, R = Me
1742c, R = Ph
1742d, R = $(CH_2)_2NMe_2$
1742e, R = Ac

1743a, R = H
1743b, R = Me
1743c, R = CH_2Ph
1743d, R = $(CH_2)_2NMe_2$
1743e, R = Et

At the time of this review, only one other example of the naphth[1,2-*c*] azepine system had been reported. Padwa found during his extensive investigations of the azirine system that irradiation of the *Z* and/or *E*-β-naphthylvinylazirine **1744** (Eq. 272) gave a single crystalline product **1745** in 85% yield (1018,1019). Similarly, photolysis of the *Z*- and/or *E*-α-naphthylvinylazirine **1746** gave an 80% yield of a single crystalline product identified as the naphth[2,3-*c*]azepine **1747** (Eq. 273).

It was of interest that both *Z*- and *E*-**1744** gave the single product **1745,** and that *Z*- and *E*-**1746** each gave **1747.** These results were in contrast to the behavior of the styryl azepines **1748,** where the *Z* isomer gave the benzazepine **1749,** and the *E* isomer afforded the pyrrole **1750** (Fig. 230). These results suggest that **1744** and **1746** isomerize much more rapidly than their

TABLE 131. NAPHTH[1,2-*c*]AZEPINES

X	R	mp (°C)	Other Data	Refs.
O	H	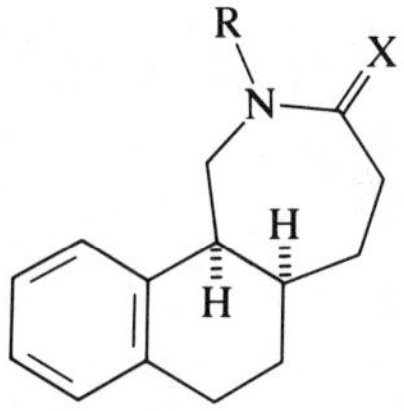 216–217		1014,1016
O	Me	—		1014,1016
O	CH_2Ph	—		1014,1016
O	$(CH_2)_2NMe_2$	—		1014,1016
O	Ac	—		1014,1016
H_2	H	171–173/3[a]		1016
H_2	Me	—		1016
H_2	CH_2Ph	—		1016
H_2	$(CH_2)_2NMe_2$	—		1016
H_2	Et	—		1016
O	H	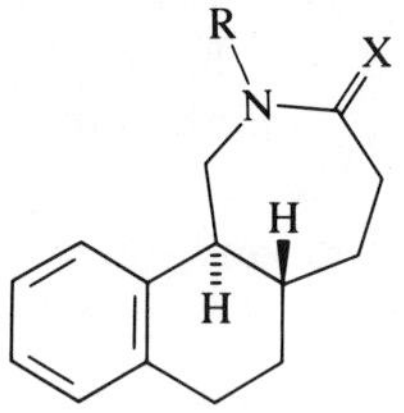 167	ir,pmr,R_f	1015
O	Me	143	ir,R_f	1015
O	CH_2Ph	173–175	ir,R_f	1015
O	$(CH_2)_3NMe_2$	126	ir,R_f	1015
O	Ac	82–83	ir,pmr,R_f	1015
H_2	H	126 (HCl)		1015
H_2	Me	56 (HCl)		1015
H_2	CH_2Ph	103 (HCl)		1015
H_2	$(CH_2)_3NMe_2$	68 (2HCl)		1015
H_2	Et	85 (HCl)		1015

[a]Boiling point (°C/torr).

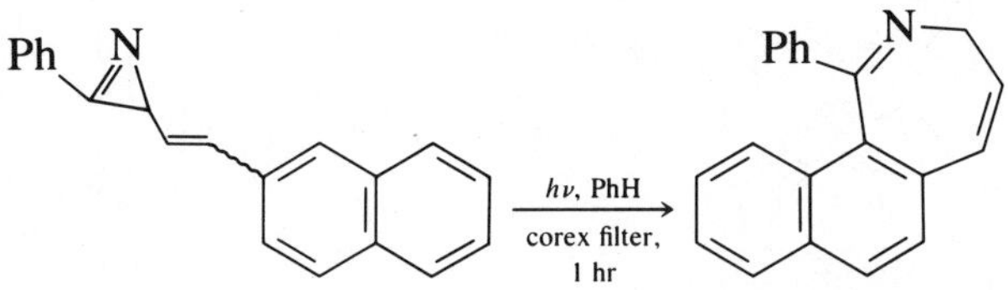

$h\nu$, PhH; corex filter, 1 hr

1744 → **1745** (272)

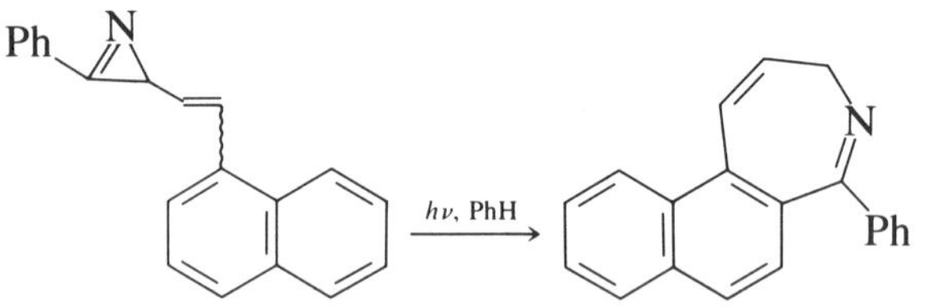

(273)

1746 **1747**

Z isomers undergo cyclization. The suggestion was verified by the observation that a photostationary state of the *Z*/*E*-naphthylazirine system **1744** was established prior to cyclization.

The results are rationalized by noting that the initial excitation energy is most probably localized in the arylvinyl moiety. However, it is the azirine end of the molecule that initiates the photoreaction, and, as a consequence, it is necessary to transfer excitation energy from the arylvinyl to the aziridinyl chromophore. Such an energy transfer is accomplished less easily with the lower-energy naphthyl systems (**1744** and **1746**) than with phenyl system (**1748**). Hence, **1744** and **1748** have the opportunity to undergo energy dissipation by rotation about the C=C bond. This rationale is consistent with the lower quantum yield observed for cyclization of the naphthylazirine (Φ (**1744**→**1745**) = 0.35) as compared to the styrylazirine system (Φ (**1748**→**1749**) = 0.82).

Mechanistically, the photocycloadditions were shown to proceed via the excited singlet state of the azirine. The observed products from **1744, 1746,**

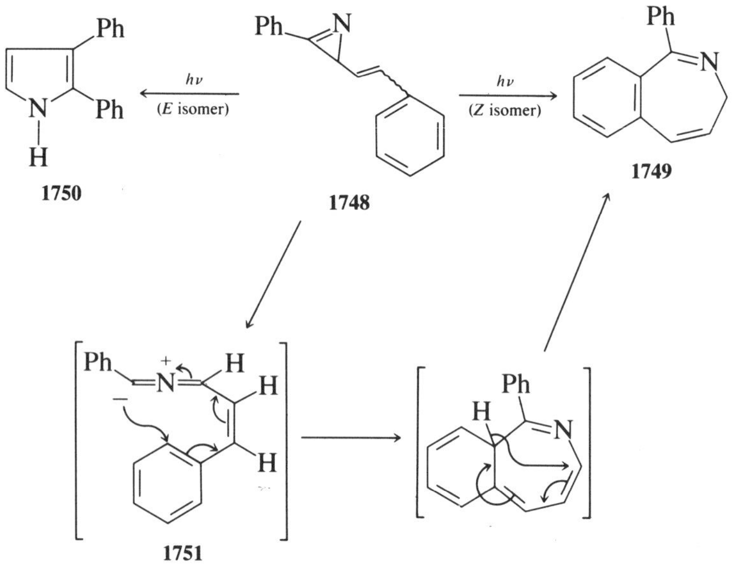

1750 **1748** **1749** **1751**

Figure 230

and *Z*-**1748** were explained on the basis of ring opening of azirine to give a nitrile ylide (**1748**→**1751**). Subsequent intramolecular reorganization to a seven-membered ring followed by a 1,5-sigmatropic shift would give the product **1749.**

43. Naphth[1,2-*d*]azepines

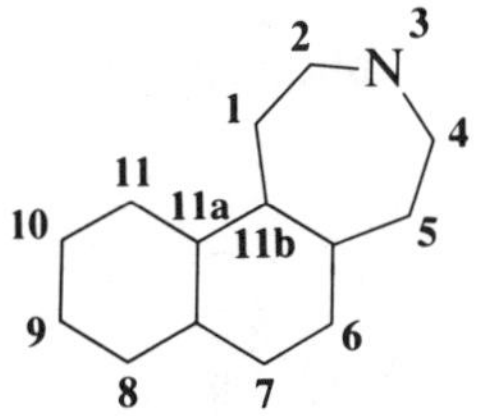
2
3
N
1
4
11
11a
10
11b
5
9
6
8
7

As part of a program exploring the scope of hydrogen halide–induced cyclization of α,ω-dinitriles, the *bis*(cyanomethyl)naphthalene **1752** (Eq. 274) was treated with 30% hydrogen bromide in acetic acid (237,1020). The precipitate that formed after 2 hr was a mixture of two salts. On neutralization, only one of the expected products was isolated (32% yield). The authors were unable to distinguish between the two possible isomers **1753a** and **1753b.**

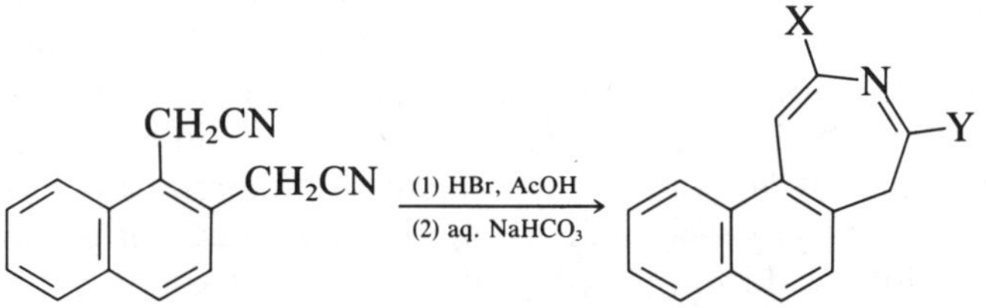
CH2CN
CH2CN
(1) HBr, AcOH
(2) aq. NaHCO3
X
N
Y

(274)

1752

1753a, X = Br, Y = NH_2 } mp 232–235°C
1753b, X = NH_2, Y = Br

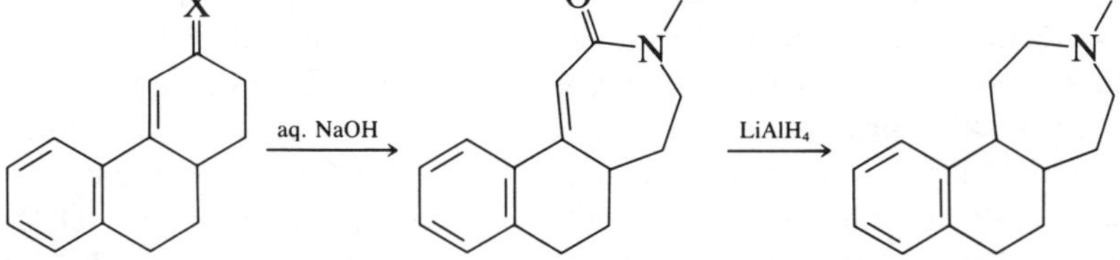
X
aq. NaOH
O
H
N
LiAlH4
H
N

(275)

1754a, X = O
1754b, X = *syn*-NOH
1754c, X = *anti*-NOH
1754d, X = *syn*-NOTs

1755 mp 155–156°C

1756, mp 73–75°C (HCl)

The naphth[1,2-*d*]azepine system has also been prepared via a Beckmann rearrangement (Eq. 275). Reaction of the phenanthrenone **1754a** with hydroxylamine in ethanol gave a 1:3 mixture of the *syn* and *anti* oximes **1754b** and **1754c,** respectively. The *syn* isomer **1754b** was converted to the corresponding tosylate **1754d,** which, on standing in aqueous sodium hydroxide, rearranged to the azepinone **1755** in 91% yield. Lithium aluminum hydride reduction of **1755** gave the saturated amine **1756** (1021).

44. Naphth[1,8,-*bc*]azepines

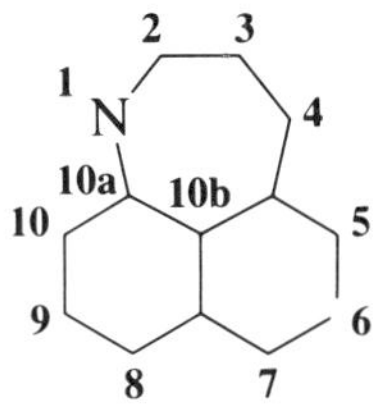

The ring enlargement of the phenalenones **1757a** and **1757b** (Fig. 231) has been examined as a route to the naphth[1,8-*bc*]azepine system. Treatment of **1757a** with sodium azide in glacial acetic acid–concentrated sulfuric acid gave a 9:1 mixture of the isomeric lactams **1758a** and **1759a** (1022). Similarly, **1757b** with hydrazoic acid in chloroform gave rise to a mixture of **1758b** and **1759b** in unspecified ratio (1023). Whereas the Schmidt reaction of **1757a** and **1757b** produced mixtures, Beckmann rearrangement of the oxime **1757c** gave exclusively **1758a** in 91% yield (1024). Reaction of the sodium salts of **1758a** and **1758b** with the appropriate alkyl halides afforded the tertiary amides **1758c–f** (1023,1024). Lithium aluminum hydride reduction gave the corresponding amines **1758g–k** (1022–1024). The structures of the lactams **1758a** and **1759a** followed from the pK_a values of the corresponding amines **1758g** and **1759c** (4.0 for the aromatic amine and 8.4 for the aliphatic amine) (1022).

The ring expansion of the phenalenone **1760a** was investigated as a possible route to the unsaturated azepines **1761** and **1762** (1023). However, attempted Beckmann rearrangement of the oxime **1760b** with phosphorus pentachloride resulted in the formation of an unstable material of undetermined structure. The tosylate **1760** was recovered unchanged under conditions that would normally be expected to cause rearrangement. Nitrous acid converted **1760d** to intractable materials, in addition to a small amount of **1760a.** Finally, treatment of **1758b** with dichlorodicyanoquinone (DDQ) afforded a 1:1 mixture of unchanged **1758b** and a material that may be **1763.** Although the latter was not isolated, its existence was supported by the appearance of a pair of doublets in the pmr spectrum at δ 5.70 and δ 6.65 ppm. In addition, the low-voltage mass spectrum of the mixture (**1758b** and

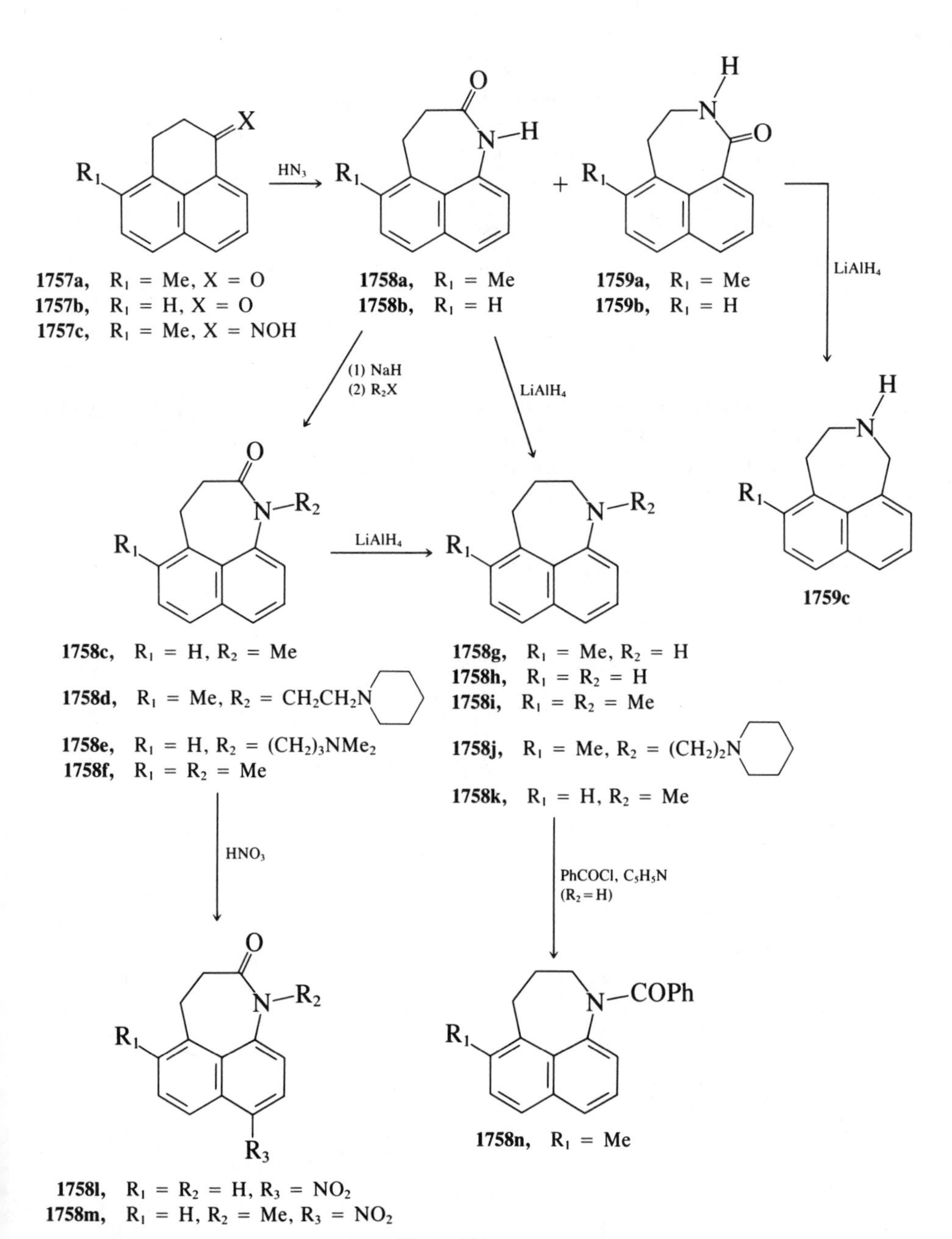
HN3
1757a, R1 = Me, X = O
1757b, R1 = H, X = O
1757c, R1 = Me, X = NOH
1758a, R1 = Me
1758b, R1 = H
1759a, R1 = Me
1759b, R1 = H
LiAlH4
(1) NaH
(2) R2X
LiAlH4
1759c
LiAlH4
1758c, R1 = H, R2 = Me
1758d, R1 = Me, R2 = CH2CH2N
1758e, R1 = H, R2 = (CH2)3NMe2
1758f, R1 = R2 = Me
1758g, R1 = Me, R2 = H
1758h, R1 = R2 = H
1758i, R1 = R2 = Me
1758j, R1 = Me, R2 = (CH2)2N
1758k, R1 = H, R2 = Me
HNO3
PhCOCl, C5H5N
(R2 = H)
N—COPh
1758l, R1 = R2 = H, R3 = NO2
1758m, R1 = H, R2 = Me, R3 = NO2
1758n, R1 = Me

Figure 231

1760a, X = O
1760b, X = NOH
1760c, X = NOTs
1760d, X = NNH_2

1761, X = H, OH

1762, X = H, OH

1763

1763) contained a major peak at *m/e* 195 (for **1763**), which was absent in the mass spectrum of **1758b.** Attempted oxidation of **1758b** or **1758f** with nitric acid yielded only the nitro compounds **1758l** and **1758m** (Fig. 231). See Table 132 for naphth[1,8-*bc*]azepines discussed in this section.

TABLE 132. NAPHTH[1,8-*bc*]AZEPINES

R_1	R_2	R_3	X	mp (°C)	Spectra	Refs.
Me	H	H	O	211–212 215–217	ir	1022,1024
H	H	H	O	193	pmr	1023
Me	Me	H	O	184–186		1024
Me	$(CH_2)_2N$(piperidino)	H	O	195–197 (HCl)		1024
Me	$(CH_2)_3NMe_2$	H	O	214–216 (HCl)		1024
H	Me	H	O	87	pmr	1023
Me	H	H	H_2	44–46 232–234 (HCl) 231–232		1022
H	H	H	H_2	73	pmr	1023
Me	Me	H	H_2	188–190 (HCl)		1023
Me	$(CH_2)_2N$(piperidino)	H	H_2	240–242 (HCl)		1023
H	H	NO_2	O	236	pmr,ms	1023
H	Me	NO_2	O	159	pmr,ms	1023
Me	COPh	H	H_2	194–195	pmr,ms	1022

45. Naphth[1,8-*cd*]azepines

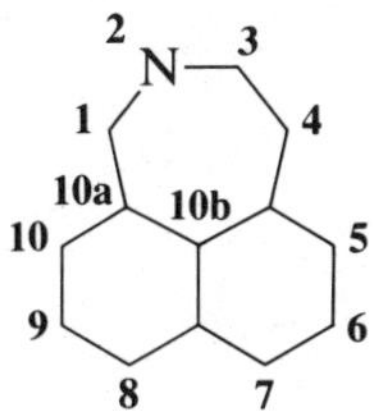
N
1
2
3
4
5
6
7
8
9
10
10a
10b

In 1960, Smith described a simple *o*-carboxylation procedure which he utilized to convert benzoic acids to phthalic acids and arylacetic acids to homophthalic acids (1025). This method could also be employed to prepare homonaphthalic acid (**1766**) from the naphth[1,8-*cd*]azepine **1765a** (Eq. 276). Reaction of α-naphthylacetyl chloride with lead thiocyanate gave a 72% yield of the acyl isothiocyanate **1764,** which on prolonged treatment with aluminum chloride in carbon disulfide led to **1765a** in 41% yield. Alkaline hydrolysis gave **1766.**

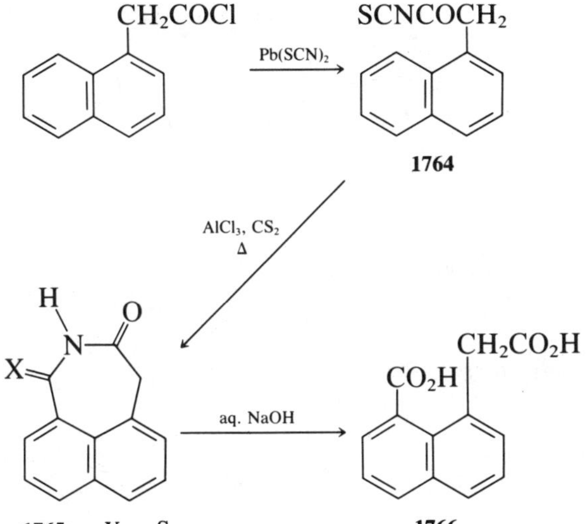
CH2COCl
Pb(SCN)2
SCNCOCH2
AlCl3, CS2
Δ
H
N
O
X
aq. NaOH
CO2H
CH2CO2H

1764

1765a, X = S
1769, X = O

1766

(276)

In an unrelated naphth[1,8-*cd*]azepine synthesis (Eq. 277), ethyl acetoacetate and 8-bromo-1-naphthoic acid (**1767**) were condensed in the presence of sodium ethoxide and copper power to give **1768** in 32% yield. Alkaline hydrolysis gave **1766,** which on sequential treatment with acetic anhydride, ammonia, and acetic anhydride–sodium acetate afforded the naphthazepine **1769** in 50% yield (1026).

1767: Br, CO_2H → ($MeCOCH_2CO_2Et$, NaOEt, EtOH / Cu powder) → 1768: $MeCOCHCO_2Et$, CO_2H (277)

1768 → (5% aq. NaOH) → 1766

1766 → ((1) Ac_2O, Δ (2) NH_4OH, Δ (3) NaOAc, Ac_2O, Δ) → 1769

The preparation of naphth[1,8-*cd*]azepines from phenalenones was described in Section VI. 44 (see Fig. 231). Table 133 lists known members of the naphth[1,8-*cd*]azepine family.

46. Naphth[2,1-*b*]azepines

The napth[2,1-*b*]azepine system was first reported, along with several isomeric naphthazepines in 1930. When the oxime tosylate **1770** was heated

TABLE 133. NAPHTH[1,8-*cd*]AZEPINES

R	X	Y	mp (°C)	Other Data	Refs.
Me	O	H_2	158–160	ir	1022
H	O	H_2	165	pmr	1023
Me	H_2	H_2	247–249 (HCl)	pK_a	1022
H	S	O	254–255		1025
H	O	O	151	ir,uv,pmr	1026

in methanol (Eq. 278), the amino ester **1771a** was formed in moderate yield. Warming with either aqueous acid or base gave the amino acid **1771b,** which on being heated above its melting point underwent cyclization to the lactam **1772** (1010).

NOTs

(1) MeOH, 95°C, 1hr
(2) 2 *N* NaOH or conc. HCl

1770

$(CH_2)_3CO_2R$
NH_2

1771a, R = Me
1771b, R = H

(278)

O
N
H

1772, mp 201–202°C

More recently, Hassner (1027) reported that when 4-azido-1,2-dihydronaphthalene (**1773**) was thermolyzed in the presence of the cyclopentadienone **1774,** the naphthazepine **1775** was formed, along with a large amount of polymeric material (Fig 232). The formation of **1775** supports the intermediacy of the 1-azirine **1776.** Interception of **1776** by the cyclopentadienone **1774** would give rise to the Diels–Alder adduct **1777.** The formation of **1775** from **1777** follows logically via the intermediates **1778** and **1779.**

47. Naphth[2,1-*c*]azepines

2 3 1 N 4 11 11a 10 11b 5 9 6 8 7

The synthesis of several azasteroids was described in Section IV.78. The indenoazepine **1780** (Eq. 279), was hydrolyzed in dioxane–hydrochloric acid, with removal of the *N*-acetyl group and formation of the amino acid **1781.** The amino acid was not characterized, but instead was cyclized with the aid

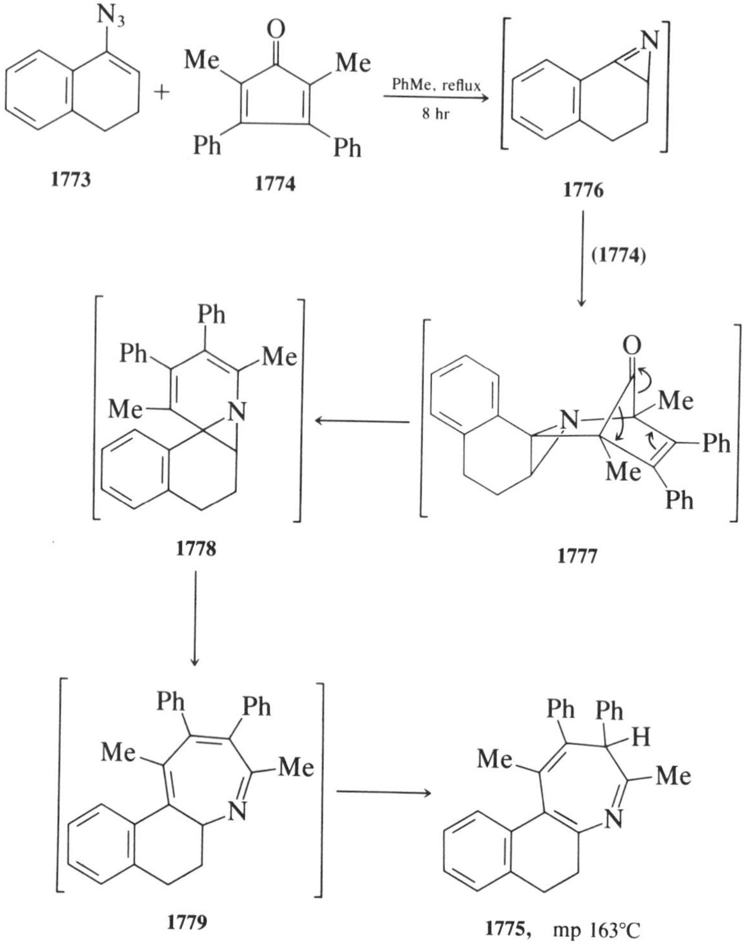

1773 1774 1776 1777 1778 1779

1775, mp 163°C

Figure 232

of dicyclohexylcarbodiimide to the β-lactam **1782** (261). The only other reported example of this system was disclosed by Padwa (1018,1019) and was discussed, along with the naphth[1,2-*c*]azepines, in Section V.42.

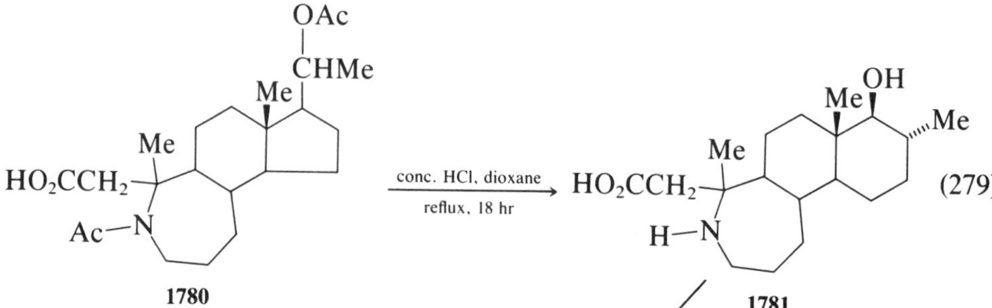

1780 1781 (279)

(see next page)

$MeNO_2$, DCC, 25°C, 3 days

(279 *cont'd*)

1782

48. Naphth[2,3-*b*]azepines

The naphth[2,3-*b*]azepine system was described in 1930, but has not been reported since (1010). The oxime tosylate **1783** gave the amino ester **1784a** on being heated in methanol. Alkaline hydrolysis afforded the amino acid **1784b,** which on further heating to a higher temperature (Fig. 233) underwent ring closure to the lactam **1785** in 87% yield. However, when **1783** was heated in phenol, the lactam ether **1786a** was isolated instead of the amino ester

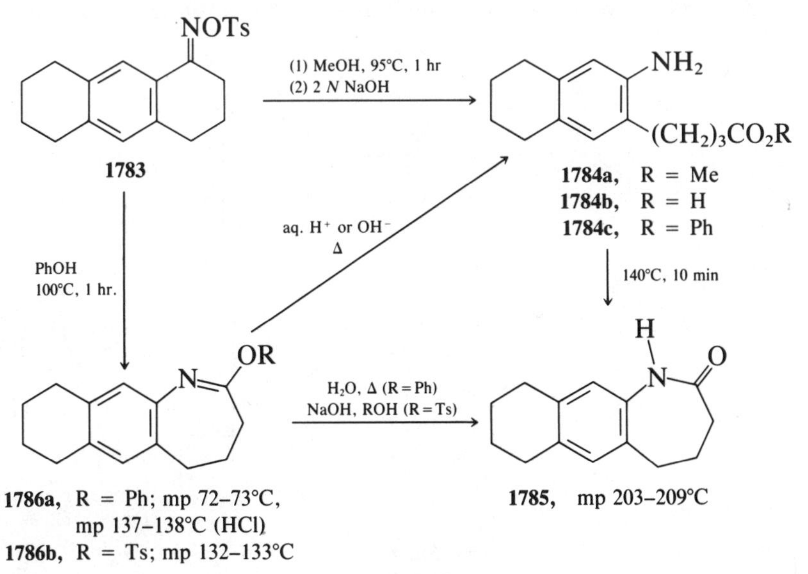

Figure 233

1784c. In addition to **1786a,** a minor product was obtained to which the authors assigned structure **1786b.** When the imino ether **1786a** was heated in water, it generated **1785,** but in aqueous acid or base **1786a** was hydrolyzed to the amino acid **1784b.** When **1786b** was warmed in alcoholic hydroxide, it reverted to **1785.**

49. Naphth[2,3-*c*]azepines

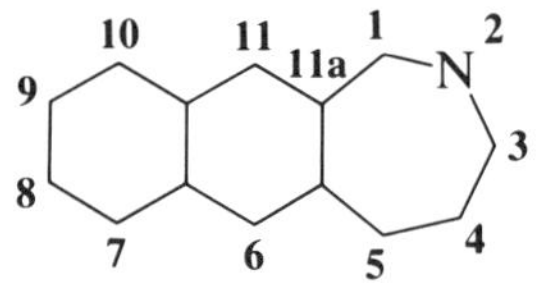

The photolytic decomposition of phenylazide **1787** in the presence of primary and secondary amines (Fig. 234) is known to generate the 2-amino-3H-azepines **1789a** (1028). The reaction is believed to proceed via the azirine **1788.** When the same photolysis is conducted in alcohols, which are less efficient than amines in trapping **1788,** only low yields of the corresponding 2-alkoxy-3H-azepines **1789b** are obtained (1029). However, it has been demonstrated recently that photolysis of 2-azidoanthracene (**1790**) in a 1:1 mixture of dioxane and 3 *N* potassium methoxide in methanol (Fig. 235) gives a nearly quantitative yield of 3-methoxy-1H-naphth[2,3-*c*]azepine (**1794a**), provided that the reaction mixture is refluxed for 15 min after irradiation (1030). In contrast, if the solution is neutralized immediately after irradiation, the main product (60% yield) is 1-amino-2-methoxyanthracene (**1792a**).

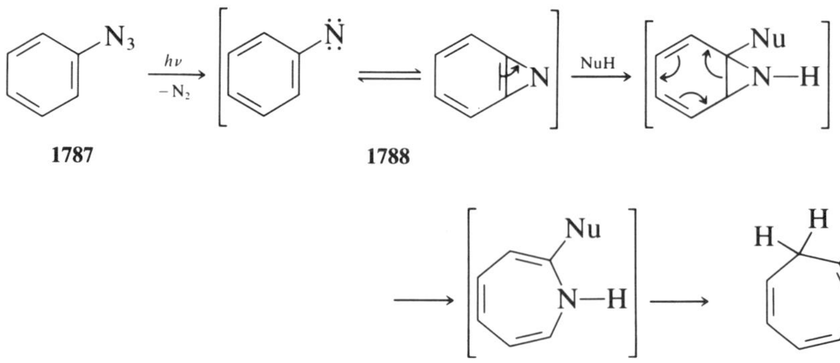

1787 **1788**

1789a, Nu = NR'_2
1789b, Nu = OR′

Figure 234

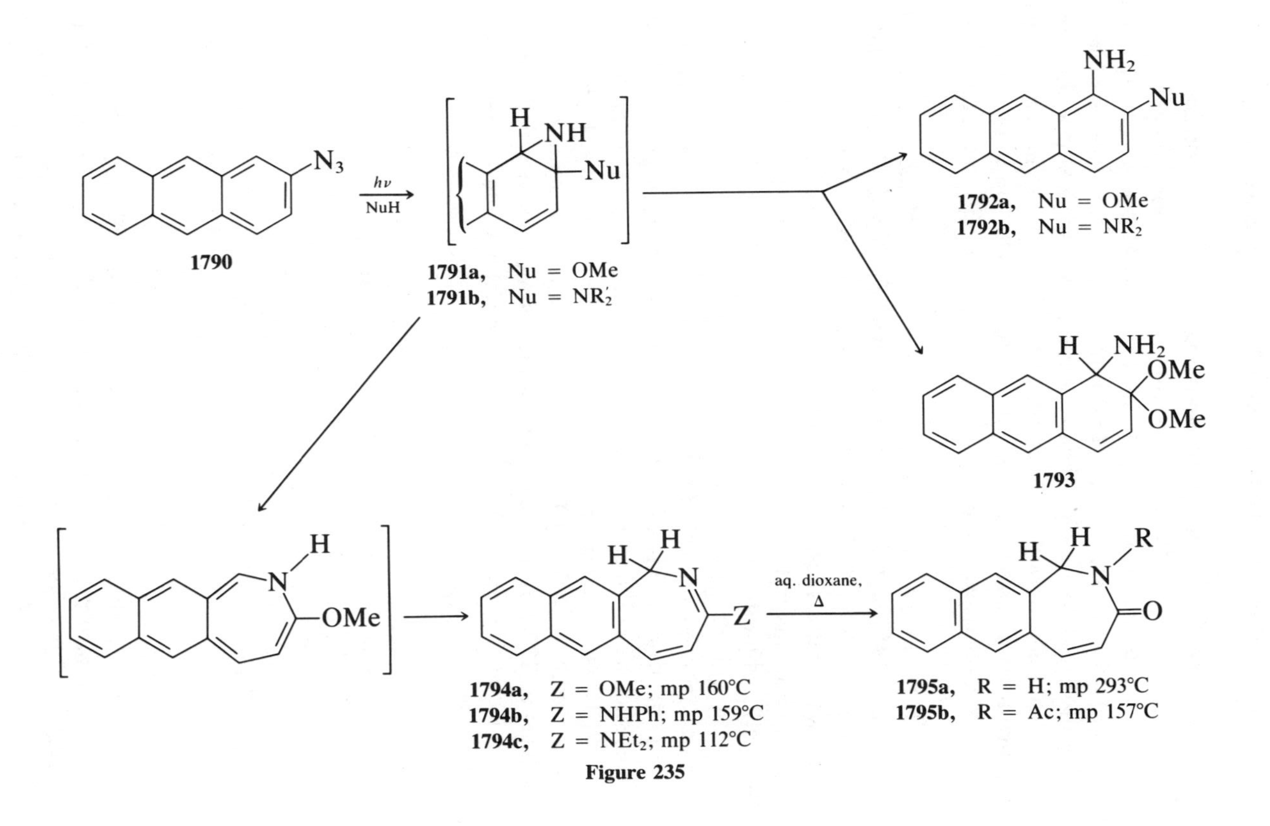
N₃
hν
NuH
1790
H
NH
Nu
1791a, Nu = OMe
1791b, Nu = NR′₂
NH₂
Nu
1792a, Nu = OMe
1792b, Nu = NR′₂
H
NH₂
OMe
OMe
1793
H
N
OMe
H
H
N
Z
aq. dioxane,
Δ
H
H
R
N
O
1794a, Z = OMe; mp 160°C
1794b, Z = NHPh; mp 159°C
1794c, Z = NEt₂; mp 112°C
1795a, R = H; mp 293°C
1795b, R = Ac; mp 157°C

Figure 235

The results were interpreted by invoking the methoxyaziridine **1791a** as an intermediate. The ultraviolet spectrum of the solution immediately after irradiation was distinctly different from that of **1794a** and was characteristic of a 1,2-dihydroanthracene. Thus **1791a** must be a fairly stable species, and its finite lifetime in strongly alkaline medium is attributable either to the acid-catalyzed nature of its isomerization to **1792** or to an increased energy requirement for electrocyclic opening in this series. The latter seems reasonable, since cleavage would result in loss of aromaticity in the naphthalene nucleus.

The high tendency for acid-catalyzed heterolysis of the N—C_2 bond in the aziridine **1791a** was further demonstrated when a new compound was observed on photolysis of **1790** in 1:1 dioxane–0.5 *N* potassium methoxide in methanol. Under these less strongly alkaline conditions, when the irradiated solution was left for four days at −15°C to limit thermal rearrangement, the 2,2-dimethoxy-1,2-dihydroanthracene **1793** was obtained in 40% yield, along with **1794a** and **1792a.** Heterolysis of the N—C_2 bond is apparently much more facile with an aminoaziridine such as **1791b,** since irradiation of **1790** in solutions of aliphatic secondary amines gave exclusively diamines such as **1792b** and no naphthazepine. Although the naphthazepines **1794b** and **1794c** were not directly obtainable by photolysis, they could be prepared readily by heating **1794a** for a long time in the appropriate amine. Hydrolysis of **1794a** in refluxing 50% aqueous dioxane afforded the corresponding lactam **1795a** in 74% yield, and the latter could be *N*-acylated to **1795b.**

50. Naphth[2,3-*d*]azepines

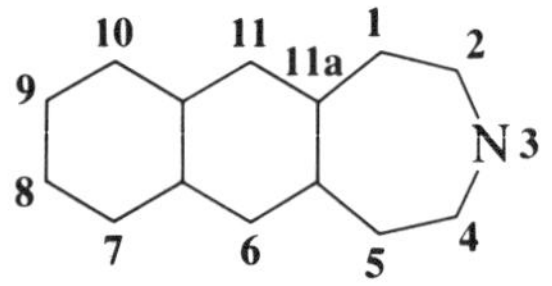

The hydrogen halide induced cyclization of α,ω-dinitriles, which was dicussed in several previous sections, was applied to the dinitrile **1796** (Eq. 280). Hydrogen bromide or hydrogen iodide in acetic acid gave the naphth[2,3-*d*]azepines **1797a** and **1797b** in 79 and 87% yields, respectively (237,1020).

In 1965, Dimroth and co-workers (1031) described the condensation of several *N,N-bis*(cyanomethyl)amines with phthalic dialdehydes (Eq. 281). As part of this work, 4-chloro-*N*,N-*bis*(cyanomethyl)aniline was condensed with 2,3-naphthalenedialdehyde to give the naphth[2,3-*d*]azepine **1798** in 11% yield.

Within a broad effort to gain insight into the possible 6π donor capabilities of lH-azepines, Paquette and co-workers (1032) described the addition of

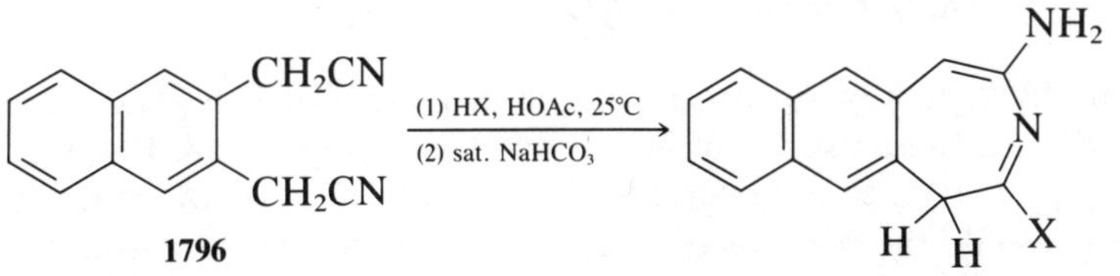

CH_2CN / CH_2CN (1796) $\xrightarrow[\text{(2) sat. } NaHCO_3]{\text{(1) HX, HOAc, 25°C}}$ NH_2, N, X, H H (280)

1796

1797a, X = Br; mp 240–242°C
1797b, X = I; mp 240–242°C

Cl, $N(CH_2CN)_2$ + CHO, CHO $\xrightarrow[\text{dioxane, }\Delta]{\text{KOMe, MeOH}}$ (281)

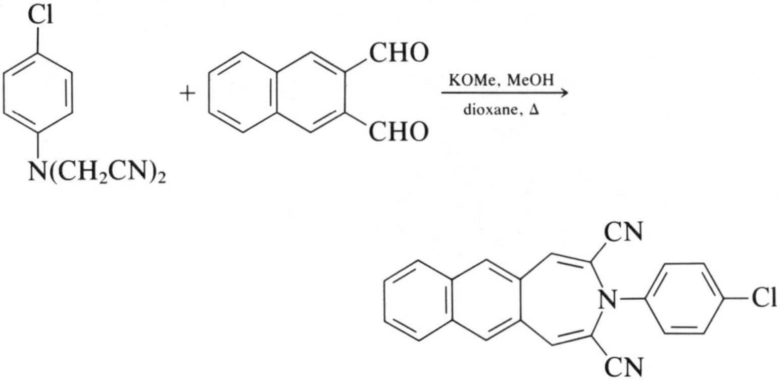

CN, N, Cl, CN

1798, mp 262-264°C

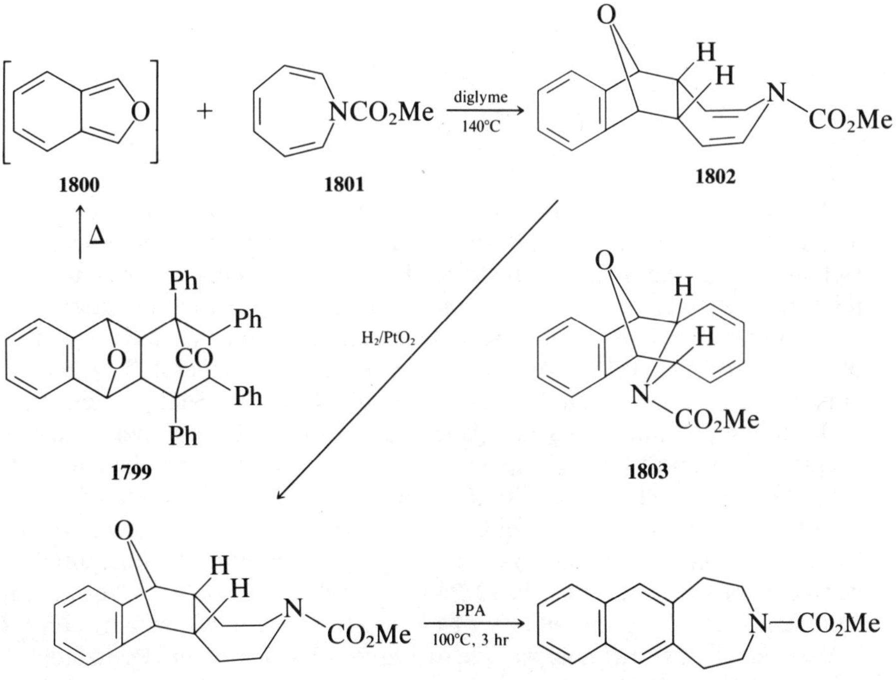

O (1800) + NCO_2Me (1801) $\xrightarrow[\text{140°C}]{\text{diglyme}}$ O, H, H, N, CO_2Me (1802)

1800 **1801** **1802**

↑ Δ

Ph, Ph, O, CO, Ph, Ph

1799

H_2/PtO_2

O, H, H, N, CO_2Me

1803

O, H, H, N, CO_2Me $\xrightarrow[\text{100°C, 3 hr}]{\text{PPA}}$ N—CO_2Me

1804 **1805,** mp 138.5–139°C

Figure 236

isobenzofuran (**1800**) to several *N*-substituted azepines. *In situ* generation of **1800** from **1799** in the presence of *N*-carbomethoxyazepine (**1801**) gave a 47% yield of the [4 + 2] π-adduct **1802** and none of the [4 + 6] π-adduct **1803** (Fig. 236). While the combined spectral data supported the formation of the [4 + 2] π-adduct **1802,** final confirmation was obtained chemically. When **1802** was reduced catalytically, two equivalents of hydrogen were absorbed to form the tetrahydro derivative **1804.** Heating **1804** in polyphosphoric acid led to the napth[2,3-*d*]azepine **1805** in 83% yield.

51. Pyrazino[1,2-*b*][2]benzazepines

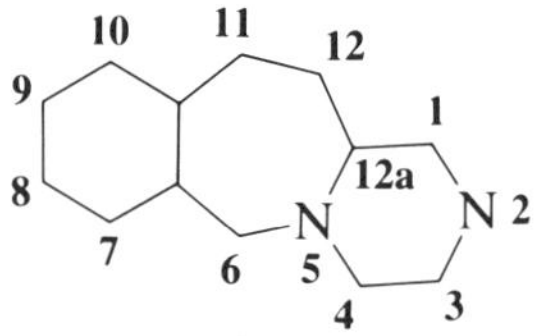

52. Pyrazino[2,1-*b*][3]benzazepines

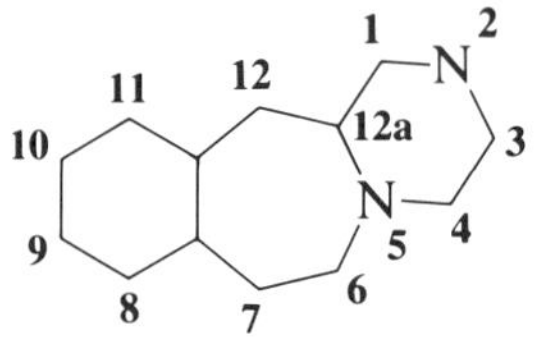

The known members of these ring systems were synthesized as shown in Fig. 237 (1033). When β-tetralone was treated with hydrazoic acid, the lactams **1806a** and **1806b** were formed in approximately a 1:1 ratio. The mixture was treated with Meerwein's reagent to give the isomeric imino ethers **1807a** and **1807b.** Condensation with nitromethane then gave a mixture of the nitroolefins **1808a** and **1808b,** which could be separated chromatographically. Individual catalytic reduction of **1808a** and **1808b,** or reduction with lithium aluminum hydride, gave good yields of the aminomethylbenzazepines **1809a** and **1809b,** respectively. Condensation of the diamines with diethyl oxalate, followed by lithium aluminum hydride reduction of the resultant tricyclic adducts **1810a** and **1810b,** afforded the pyrazinobenzazepines **1811a** and **1811b.** Both isomers were alkylated readily to *N*-substituted piperazines of the general types **1812a** and **1812b.** The existence of nitrogen inversion created the possibility of either *cis* or *trans* fusion between the hetero rings. Spectral analysis clearly favors the *cis* form. Pyrazino[1,2-*b*][2]- and [2,1-*b*][3]benzazepine depicted in Fig. 237 are listed in Table 134.

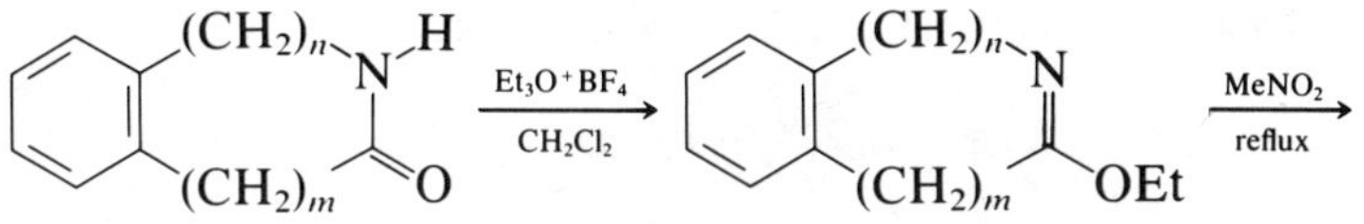

1806a, $n = 2, m = 1$
1806b, $n = 1, m = 2$

1807a, $n = 2, m = 1$
1807b, $n = 1, m = 2$

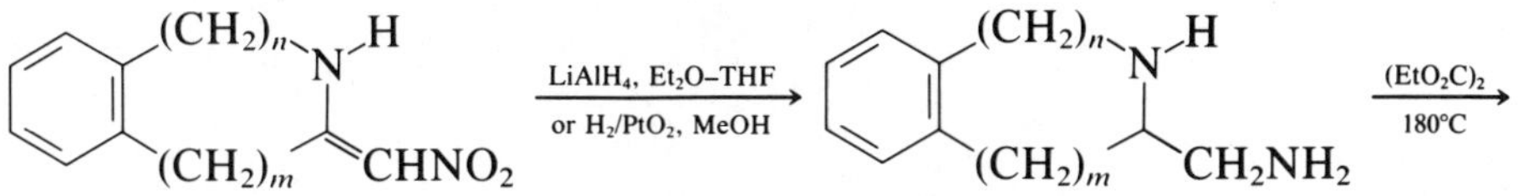

1808a, $n = 2, m = 1$
1808b, $n = 1, m = 2$

1809a, $n = 2, m = 1$
1809b, $n = 1, m = 2$

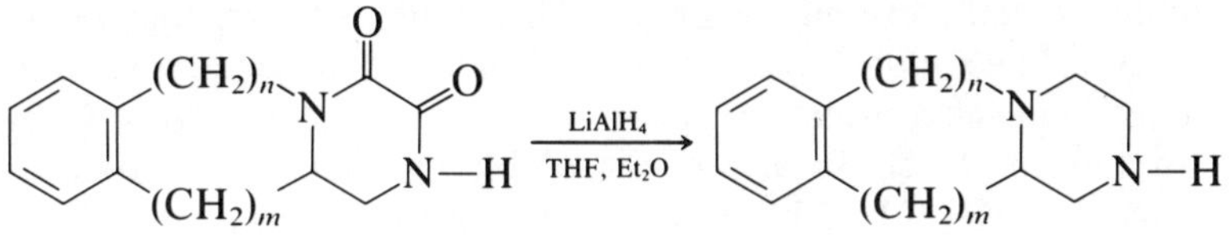

1810a, $n = 2, m = 1$ (66%)
1810b, $n = 1, m = 2$ (40%)

1811a, $n = 2, m = 1$ (85%)
1811b, $n = 1, m = 2$ (91%)

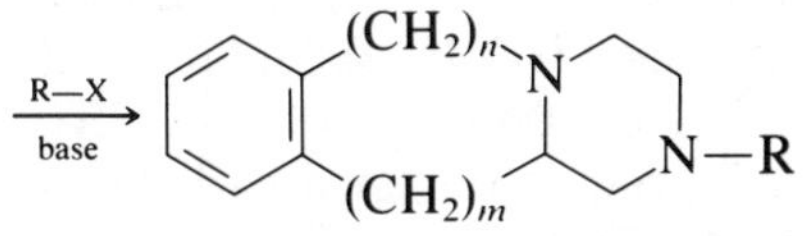

1812a, $n = 2, m = 1$
1812b, $n = 1, m = 2$

Figure 237

TABLE 134. PYRAZINO[1,2-*b*][2]- AND [2,1-*b*][3]BENZAZEPINES[a]

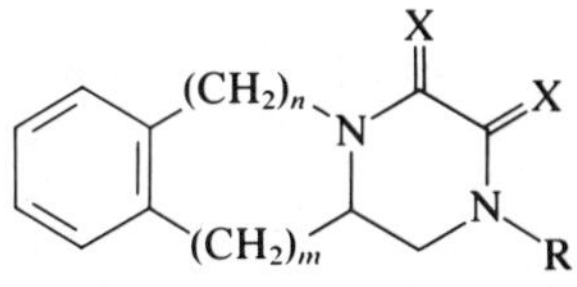

System	m	n	X	R	mp (°C)	Spectra
[1,2-*b*][2]	2	1	0	H	Hygroscopic	
	2	1	H_2	H	128–130 (fumarate)	pmr
	2	1	H_2	p-$FC_6H_4CO(CH_2)_3$	239–243 (2HCl)	
	2	1	H_2	$PhCH(OH)CH_2$	169–170 (2HCl)	
[2,1-*b*][3]	1	2	O	H	151–152	
	1	2	H_2	H	45–48	pmr
					195–197 (fumarate)	
	1	2	H_2	p-$FC_6H_4CO(CH_2)_3$	258–260 (2HCl)	pmr
	1	2	H_2	$Ph(CH_2)_2$	262–265 (2HCl)	
	1	2	H_2	$PhCH(OH)CH_2$	160–161 (fumarate)	
	1	2	H_2	CN	122–124	
	1	2	H_2	CH_2CN	84–86	

[a]Ref. 1033.

VII. 6,7,7-Systems

1. Pyrazino[1,2-*a*:4,5-*a*′]bisazepines

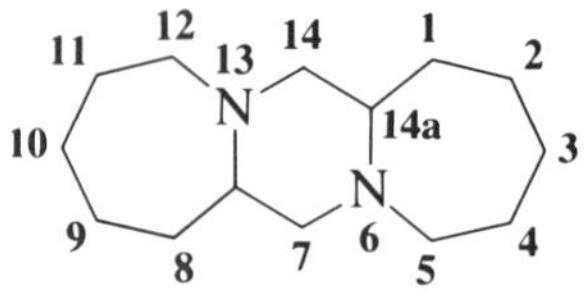

In two publications describing the preparation of the ring homologs **1814a** and **1814b** of L-proline, chlorolactam **1813** was converted by aqueous sodium hydroxide to the imino acid **1814b** in a Favorskii-like rearrangement (1034,1035) (Eq. 282). In addition to **1814b,** a water-insoluble product (6% yield) with the formula $C_{14}H_{22}N_2O_2$ (M^+ 250) was obtained; it displayed carbonyl absorption at 1658 cm^{-1} in its infrared spectrum, but no N—H absorption. Vigorous acid hydrolysis converted the latter material to **1814b.** Therefore, this compound was identified as racemic or meso-**1815.** When **1813** was treated with potassium *tert*-butoxide in refluxing THF, **1814b** and both isomers of **1815** (mp 118–119°C, 11%; mp 200–201°C, 26%) were obtained. The higher-melting material was identical to the product obtained earlier, and either isomer on hydrolysis generated **1814b.** The stereochemistry of the isomeric dimers was not determined.

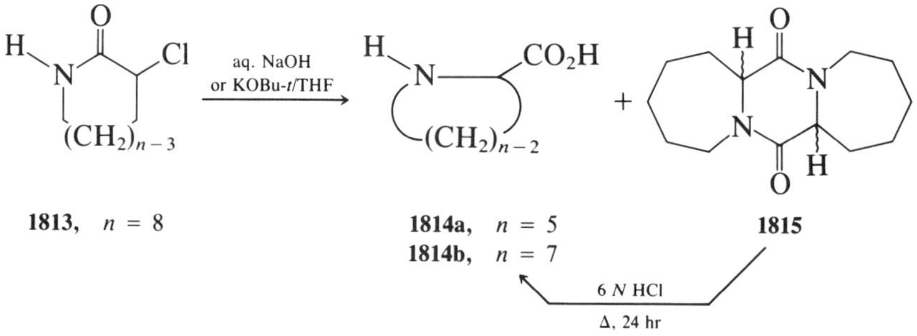

aq. NaOH or KOBu-*t*/THF (282)

1813, $n = 8$ **1814a,** $n = 5$ **1814b,** $n = 7$ **1815**

6 *N* HCl, Δ, 24 hr

2. 11b-Azabenzo[*e,f*]heptalenes

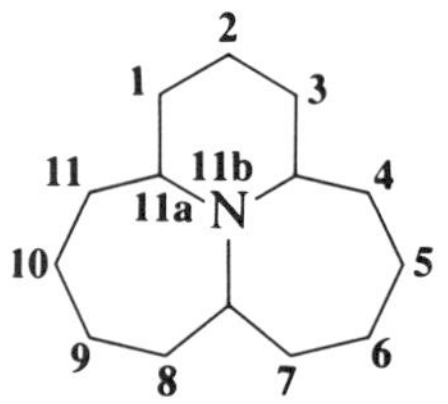

There are to date no reported examples of the 11b-azabenzo[*e*,*f*]heptalene ring system, although the parent heterocycle has been the subject of theoretical considerations. On the basis of either a modified Hückel molecular orbital calculation (464) or a topological resonance energy (TRE) calculation (1036) the parent system has been predicted to be essentially nonaromatic, in spite of its 14π periphery.

3. Azepino[3,2,1-*jk*][1,4]benzodiazepines

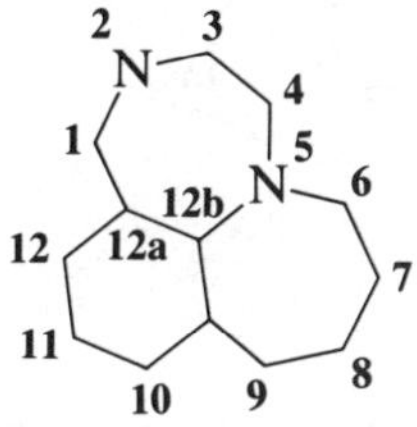

Two representatives of the azepino[3,2,1-*jk*][1,4]benzodiazepines have been claimed to exhibit anticonvulsant activity (149). Compounds **1817a** and

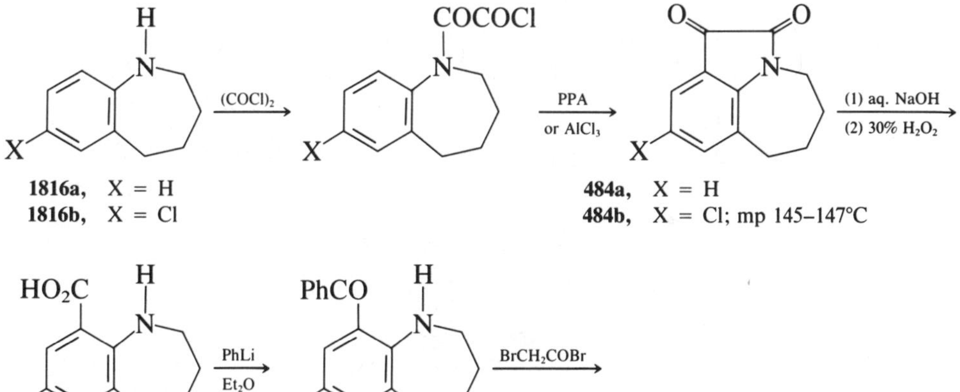

1816a, X = H
1816b, X = Cl

$(COCl)_2$

PPA or $AlCl_3$

484a, X = H
484b, X = Cl; mp 145–147°C

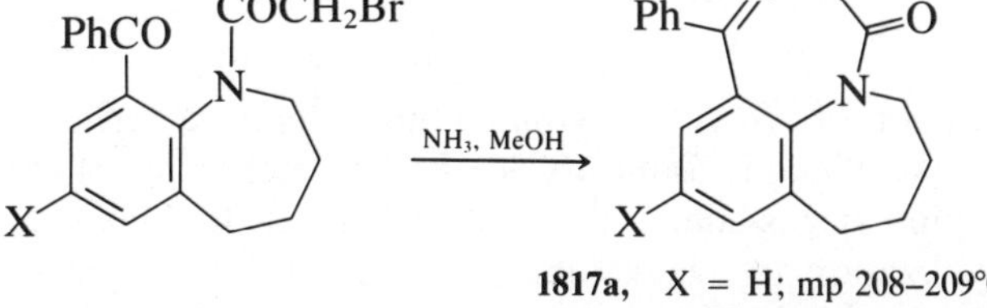

(1) aq. NaOH (2) 30% H_2O_2

PhLi Et_2O

$BrCH_2COBr$

NH_3, MeOH

1817a, X = H; mp 208–209°C
1817b, X = Cl; mp 165–166°C

Figure 238

1817b were synthesized in several steps from benzazepines **1816a** and **1816b,** respectively (Fig. 238). The tricyclic azepines **484a** and **484c,** which were intermediates in the sequence, have been discussed earlier (see Fig. 72, Section IV.39).

VIII. 6,7,8-Systems

1. Cycloocta[4,5]pyrimido[1,2-*a*]azepines

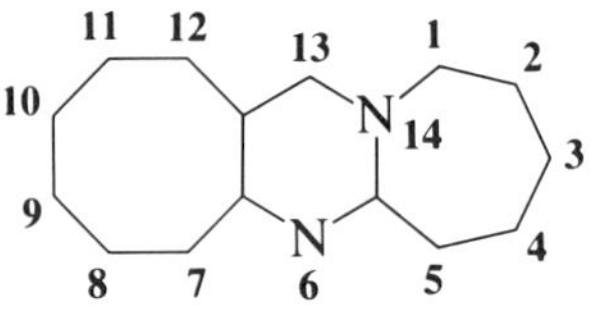

When equimolar amounts of the β-lactam **1818** and imino ether **1819** were heated for 3 hr at 130°C and then for 30 min at 200°C (Eq. 283), the pyrimidone derivative **1820,** the only known example of the cycloocta[4,5]pyrimido-[1,*a*]azepine system, was formed in 77% yield (1037).

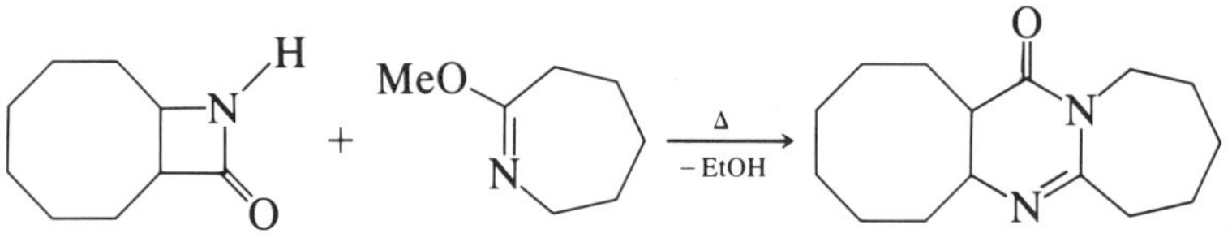

(283)

1818 **1819** **1820** bp 164-7°C (0.3 Torr)

2. Benzo[*e*]cyclooct[*b*]azepines

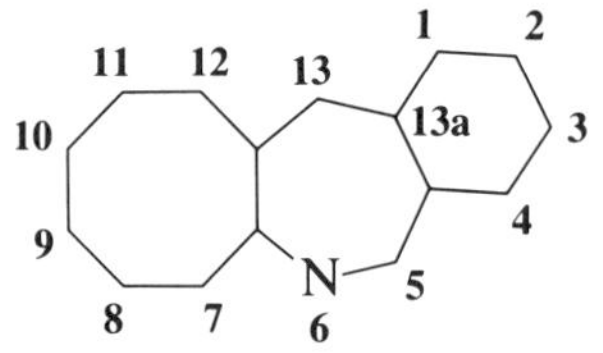

The photolytic cyclization of *N*-cycloalkylphthalimides gives a series of tricyclic azepinones, among which is found **1823c,** the single example of the benzo[*e*]cyclooct[*b*]azepine ring system (1038). *N*-Alkylphthalimides generally undergo photocyclization to azacyclobutanols in a Norrish type II process (1039), followed by retrotransannular ring opening to benzazepinones (1040). Irradiation of phthalimides **1821a–d** produced a variety of photoproducts, depending on the size of the cycloalkyl ring and the photolysis medium (Fig. 239). Phthalimides **1821a** and **1821d,** when irradiated in acetonitrile and *tert*-butanol, respectively, yielded the tricyclic structures **1823a** and **1823c.** Photolysis of **1821b** in either solvent afforded, in addition to **1823b,** a new product—the pentacyclic compound **1826a**—whose formation is consistant with a second Norrish type II process via **1825.** Compound **1823c** in *tert*-butanol yielded only **1826b.** When **1821b** and **1821c** were irradiated in ethanol, the sole products were the alcohols **1824a** and **1824b,** presumably formed by reduction of the intermediate ketones **1823a** and **1823b,** respectively.

Alcohols **1824a** and **1824b** were also obtained from lithium aluminum

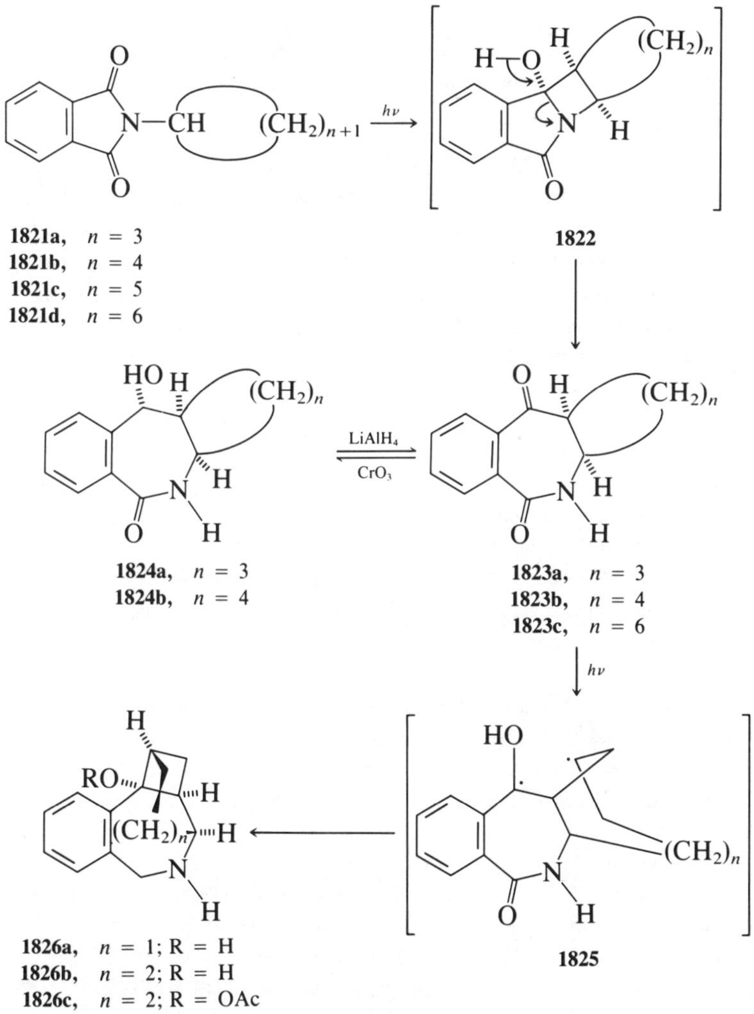

1821a, $n = 3$
1821b, $n = 4$
1821c, $n = 5$
1821d, $n = 6$

1822

1824a, $n = 3$
1824b, $n = 4$

1823a, $n = 3$
1823b, $n = 4$
1823c, $n = 6$

1826a, $n = 1$; R = H
1826b, $n = 2$; R = H
1826c, $n = 2$; R = OAc

1825

Figure 239

TABLE 135. SUMMARY OF THE PHOTOLYTIC REACTIONS OF *N*-CYCLOALKYLPHTHALIMIDES **1821a–d**

Starting Material	Solvent	Products	mp (°C)	Spectra
1821a	MeCN	**1823a** (48%)	210–212	ir,nmr,ms
	EtOH	**1824a**		
1821b	MeCN	**1823b** (10%)	230–234	
		1826a (14%)	244–246	ms,ir
	EtOH	**1824b**		
	t-BuOH	**1823b**		
		1826a (23%)		
1821c	*t*-BuOH	**1826b** (55%)	213–215[a]	ir,ms, X-ray[a]
1821d	*t*-BuOH	**1823c** (30%)		

[a]Melting point and x-ray analysis are on the O-acetyl derivative **1826c**.

hydride reduction of **1823a** and **1823b,** and could be reoxidized with chromium trioxide. That the ring junction of **1823a, 1823b,** and (by analogy) **1823c** was *cis* was confirmed by an independent synthesis. Therefore, **1824a** and **1824b** were also assigned *cis* stereochemistry. The stereochemistry of **1826b** was determined by X-ray crystallographic analysis of its *O*-acetate **1826b**, and the stereochemistry of **1826a** was assigned by analogy. Table 135 summarizes the photolysis data.

REFERENCES

1. A. Hassner and D. J. Anderson *J. Org. Chem.*, **39,** 2031 (1974).
2. W. Ross and G. Proctor, *J. Chem. Soc. Perkin I,* **1972,** 889.
3. K. Schloegl, M. Fried, and H. Falk, *Monatsh. Chem.*, **95,** 596 (1964).
4. J. C. Meslin and G. Duguay, *Bull. Soc. Chim. France,* **1976,** 1200.
5. W. Treibs and A. Lange, *J. Prakt. Chem.*, **14,** 208 (1961).
6. S. Rajappa and R. Sreenivasan, *Ind. J. Chem.*, **14B,** 400 (1976).
7. M. Choussy and M. Barbier, *Helv. Chim. Acta,* **58,** 2651 (1975).
8. M. Choussy and M. Barbier, *C. R. Acad. Sci. Paris (C),* **281,** 619 (1976).
9. D. Johnson and G. Jones, *J. Chem. Soc.*, **1972,** 844.
10. I. Hiroshi, H. Harada, K. Ohno, T. Mizutani, and S. Uyeo, *J. Chem. Soc. (D),* **1970,** 268.
11. H. Wynberg and M. Cabell, *J. Org. Chem.*, **38,** 2814 (1973).
12. N. Hjelte and T. Agback, *Acta Chem. Scand.*, **18,** 191 (1964).
13. C. V. Greco and R. P. Gray, *Tetrahedron,* **26,** 4329 (1970).
14. Japanese Kokai 76 100,095 (1976); *Chem. Abstr.*, **86,** 121341w (1977).
15. U.S. Patent 4,002,638 (1977); *Chem. Abstr.*, **83,** 58833d (1975).
16. R. M. Acheson, M. W. Foxton, P. Abbot, and K. R. Mills, *J. Chem. Soc. (C),* **1967,** 882.

17. R. M. Acheson and W. R. Tully, *J. Chem. Soc. (C),* **1968,** 1623.
18. D. W. Cameron and R. Giles, *J. Chem. Soc. (C),* **1968,** 1462.
19. V. Gomez Parra and R. Madronero, *An. Quim.,* **70,** 614 (1974).
20. S. Rajappa, B. Advani, and R. Sreenivasan, *Ind. J. Chem.,* **14B,** 391 (1976).
21. R. M. Acheson, M. W. Foxton, and G. Miller, *J. Chem. Soc.,* **1965,** 3200.
22. F. Babichev and V. Neplyuev, *Zh. Obshch. Khim.,* **32,** 857 (1962); *Chem. Abstr.,* **58,** 4535f (1961).
23. F. Babichev, N. Dzhigirei, and S Gukalov, *Zh. Obshch. Khim.,* **34,** 2433 (1964); *Chem. Abstr.,* **61,** 13454f (1964).
24. F. Babichev, N. Bubnovskaya, and L. Mukhina, *Ukr. Khim. Zh.,* **1968,** 1265; *Chem. Abstr.,* **70,** 115047h (1969).
25. I. McRobbie, O. Meth-Cohn, and H. Suschitzky, *J. Chem. Res. (S),* **1977,** 17.
26. R. M. Acheson, R. T. Aplin, and D. Harrison, *J. Chem. Soc. (C),* **1968,** 383.
27. N. Hassine-Coniac, G. Hazebroucq, and J. Gardent, *Bull. Soc. Chim. France,* **1971,** 3985.
28. V Khilya and I. Ya. Doroshko, *Khim. Str. Svoistva Reakt. Org. Soedin.,* **1969,** 48; *Chem. Abstr.,* **72,** 134122m (1970).
29. A. Kiprianov and V. Khilya, *Zh. Org. Khim.,* **1967,** 1091; *Chem. Abstr.,* **69,** 37080b (1968).
30. R. Glushkov and O. Magidson, *Zh. Obshch. Khim.,* **31,** 1173 (1961).
31. K. Schlogl and H. Mechtler, *Monatsh. Chem.,* **97,** 150 (1966).
32. H. Goldner, G. Dietz, and E. Carstens, *Justus Liebigs Ann. Chem.,* **692,** 134 (1966).
33. H. Goldner, G. Dietz, and E. Carstens, East German Patent 39,142; *Chem. Abstr.,* **63,** P13342c (1965).
34. H. Goldner, G. Dietz, and E. Carstens, *Z. Chem.,* **4,** 454 (1964).
35. H. Wamhoff, *Chem. Ber.,* **102,** 2739 (1969).
36. K. Ley, G. Aichinger, A. Botta, H. Hagemann, and E. Niemers, German Patent 2,126,148; *Chem. Abstr.,* **78,** 111306x (1973).
37. O. Meth-Cohn, R. K. Smalley, and H. Suschitzky, *J. Chem. Soc.,* **1963,** 1666.
38. R. K. Smalley, *J. Chem. Soc. (C),* **1966,** 80.
39. H. Suschitzky and M. E. Sutton, *Tetrahedron,* **24,** 4581 (1968).
40. K. H. Saunders, *J. Chem. Soc.,* **1955,** 3275.
41. M. D. Nair and R. Adams, *J. Am. Chem. Soc.,* **83,** 3518 (1961).
42. O. Meth-Cohn and H. Suschitzky, *J. Chem. Soc.,* **1963,** 4666.
43. D. Ainsworth and H. Suschitzky, *J. Chem. Soc.,* **1966,** 111.
44. R. De Selms, *J. Org. Chem.,* **27,** 2165 (1962).
45. A. Botta and C. Rasp, British Patent 1,463,258; Derwent Chemdoc 11422X-E; *Chem. Abstr.,* **84,** P164782s (1976).
46. R. Garner and H. Suschitzky, *J. Chem. Soc. (C),* **1967,** 74.
47. R. Smith and H. Suschitzky, *Tetrahedron,* **16,** 80 (1961).
48. D. Ainsworth, O. Meth-Cohn and H. Suschitzky, *J. Chem. Soc. (C),* **1968,** 923.
49. H. Suschitzky and M. Sutton, *Tetrahedron,* **24,** 4581 (1968).
50. R. Grantham and O. Meth-Cohn, *J. Chem. Soc. (C),* **1969,** 70.
51. R. Fielden, O. Meth-Cohn, D. Price, and H. Suschitzky, *J. Chem. Soc. (D),* **1969,** 772.
52. R. Fielden, O. Meth-Cohn, and H. Suschitzky, *J. Chem. Soc.* Perkin I, **1973,** 696.
53. R. Fielden, O. Meth-Cohn, and H. Suschitzky, *Tetrahedron Lett.,* **1970,** 1229.
54. R. Fielden, O. Meth-Cohn, and H. Suschitzky, *J. Chem. Soc. (D),* **1970,** 1658.

55. R. Fielden, O. Meth-Cohn, and H. Suschitzky, *J. Chem. Soc. Perkin I*, **1973,** 705.
56. R. Fielden, O. Meth-Cohn, and H. Suschitzky, *J. Chem. Soc. Perkin I*, **1973,** 702.
57. R. K. Grantham and O. Meth-Cohn, *J. Chem. Soc. (C)*, **1969,** 1444.
58. A. El'tsov *J. Org. Chem.* (USSR), **3,** 191 (1967).
59. R. K. Grantham, O. Meth-Cohn, and M. Naqui, *J. Chem. Soc. (C)*, **1969,** 1438.
60. R. Garner, G. V. Garner, and H. Suschitzky, *J. Chem. Soc. (C)*, **1970,** 826.
61. A. J. Boulton, A. Gripper-Gray, and A. Katritzky, *J. Chem. Soc.*, **1965,** 5958.
62. R. Perera and R. Smalley, *J. Chem. Soc. (D)*, **1970,** 1459.
63. R. Bacon and S. Hamilton, *J. Chem. Soc. Perkin I*, **1974,** 1975.
64. R. Hayward and O. Meth-Cohn, *J. Chem. Soc. Perkin I*, **1975,** 212.
65. A. Albini, G. Bettinetti, and S. Pietra, *Tetrahedron Lett.*, **1972,** 3657.
66. A. Albini, G Bettinetti, and S. Pietra, *Gazz. Chim. Ital.*, **105,** 15 (1975).
67. A. Albini, A. Barinotti, G. Bettinetti, and S. Pietra, *J. Chem. Soc. Perkin II*, **1977,** 238.
68. R. Acheson and M. Verlander, *J. Chem. Soc. Perkin I*, **1972,** 1577.
69. R. Acheson and M. Verlander, *J. Chem. Soc. Perkin I*, **1974,** 430.
70. R. C. Perera, R. K. Smalley, and C. Rogerson, *J. Chem. Soc. (C)*, **1971,** 1348.
71. U.S. Patent 3,652,570; *Chem. Abstr.*, **72,** 12601w (1970).
72. O. Meth-Cohn and H. Suschitzky, *J. Chem. Soc.*, **1964,** 2609.
73. R. Garner an H. Suschitzky, *J. Chem. Soc. (C)*, **1966,** 1572.
74. O. Aki and Y. Nakagawa, *Chem. Pharm. Bull.*, **20,** 1325 (1972).
75. P. A. S. Smith, Ed., *The Chemistry of Open-Chain Nitrogen Compounds*, Vol. 2, W. A. Benjamin, New York, 1966, pp. 126–127.
76. H. Teuber and L. Vogel, *Chem. Ber.*, **103,** 3302 (1970).
77. L. Swett and G. Aynilian, *J. Hetrocyclic Chem.*, **12,** 1135 (1975).
78. U.S. Patent 3,947,585; *Chem. Abstr.*, **84,** 121827n (1976).
79. U.S. Patent 3,939,271; *Chem. Abstr.*, **84,** 150623d (1976).
80. Netherlands Appl. 75 06,570; *Chem. Abstr.*, **85,** 160087s (1976).
81. F. Dallacker, D. Bernabei, R. Katzke, and P. Benders, *Chem. Ber.*, **104,** 2526 (1971).
82. F. Dallacker, K. W. Glombitza, and M. Lipp, *Justus Liebigs Ann. Chem.*, **643,** 82 (1961).
83. H. O. Bernhard and V. Sniekus, *Tetrahedron*, **27,** 2091 (1971).
84. B. Pecherer, R. Sundbury, and A. Brossi, *J. Hetrocyclic Chem.*, **1972,** 609.
85. U.S. Patent 3,795,683; *Chem. Abstr.*, **76,** 140768a (1972).
86. U.S. Patent 3,906,006; *Chem. Abstr.*, **84,** 59432e (1976).
87. B. Pecherer, R. Sundbury, and A. Brossi, *J. Heterocyclic Chem.*, **1972,** 617.
88. F. Santavy, J. Kaul, L. Hruban, L. Dolejs, V. Hanus, K. Blaha, and A. D. Cross, *Coll. Czech. Chem. Commun.*, **30,** 3479 (1965).
89. J. Hrbek, T. Hruban, V. Simaek, F. Santavy, and G. Snatzki, *Coll. Czech. Chem. Commun.*, **38,** 2799 (1973).
90. V. Simanek, A. Klasek, and F. Santavy, *Tetrahedron Lett.*, **1973,** 1779.
91. V. Simanek, A. Klasek, T. Hruban, V. Preininger, and F. Santavy, *Tetrahedron Lett.*, **1974,** 2171.
92. W. Klotzer, S. Teitel, J. Blount, and A. Brossi, *J. Am. Chem. Soc.*, **93,** 4321 (1971).
93. R. Hohlbrugger and W. Klotzer, *Chem. Ber.*, **107,** 3457 (1974).
94. W. Klotzer, S. Teitel, and A. Brossi, *Helv. Chim. Acta*, **55,** 2228 (1972).
95. W. Klotzer, S. Teitel, J. Blount, and A. Brossi, *Monatsh. Chem.*, **103,** 435 (1972).

96. H. Irie, S. Tani, and H. Yarmane, *J. Chem. Soc. Perkin I,* **1972,** 2986.
97. T. Kametani, S. Hirata, and K. Ogasawara, *J. Chem. Soc. Perkin I,* **1973,** 1466.
98. S. Teitel, W. Klotzer, J. Borgese, and A. Brossi, *Can. J. Chem.,* **50,** 2022 (1972).
99. T. Kametani, H. Nemoto, K. Suzuki, and K. Fukumoto, *J. Org. Chem.,* **41,** 2988 (1976).
100. F. Troxler, A. Stoll, and P. Nicklaus, *Helv. Chim. Acta,* **51,** 1870 (1968).
101. R. Windgassen, W. Saunders, and V. Boekelheide, *J. Am. Chem. Soc.,* **81,** 1459 (1959).
102. F. Gerson, E. Heilbronner, N. Joop, and H. Zimmermann, *Helv. Chim. Acta,* **46,** 1940 (1963).
103. O. Ceder and B. Beijer, *J. Heterocyclic Chem.,* **13,** 1029 (1976).
104. W. J. le Noble, *Highlights of Organic Chemistry,* Marcel Dekker, New York, 1974, pp. 348–353.
105. D. Farquhar and D. Lewis, *J. Chem. Soc., Chem. Commun.,* **1969,** 24.
106. F. Gerson, J. Joachimowicz, and D. Leaver, *J. Am. Chem. Soc.* **95,** 6702 (1973).
107. W. Flitsch and B. Muter, *Angew. Chem. Int. Ed., Eng.* **12,** 501 (1973).
108. W. Flitsch, A. Gurke, and B. Muter, *Chem. Ber.,* **108,** 2969 (1975).
109. R. Glushkov, T. Yakhontov, E. Pronina, and O. Magidson, *Khim. Geterotsikl. Soedin.,* **1969,** 564.
110. L. Yakhontov, R. Glushkov, E. Pronina, and V. Smirnova, *Dokl. Akad. Nauk SSSR,* **212,** 389 (1973).
111. C. Kaneko, Sa. Yamada, I. Yokoe, and M. Ishkawa, *Tetrahedron Lett.,* **1967,** 1873.
112. C. Kaneko, I. Yokoe, Sa. Yamada, and M. Ishikawa, *Chem. Pharm. Bull.,* **17,** 1290 (1969).
113. C. Kaneko, Sa. Yamada, and M. Ishikawa, *Chem. Pharm. Bull.,* **17,** 1294 (1969).
114. Sa. Yamada, M. Ishikawa, and C. Kaneko, *Tetrahedro Lett.,* **1972,** 977.
115. Sa. Yamada, M. Ishikawa, and C. Kaneko, *J. Chem. Soc., Chem. Commun.,* **1972,** 1093.
116. Sa. Yamada, M. Ishiawa, and C. Kaneko, *Chem. Pharm. Bull.,* **23,** 2818 (1975).
117. E. W. Collington and G. Jones, *Tetrahedron Lett.,* **1968,** 1935.
118. E. W. Collington and G. Jones, *J. Chem. Soc. (C),* **1969,** 1028.
119. G. R. Cliff, E. W. Collington, and G. Jones, *J. Chem. Soc. (C),* **1970,** 1490.
120. L. Krbechek and H. Takimoto, *J. Org. Chem.,* **33,** 4286 (1968).
121. G. R. Cliff and G. Jones, *J. Chem. Soc. (D),* **1970,** 1705.
122. G. R. Cliff and G. Jones, *J. Chem. Soc. (C),* **1971,** 3418.
123. G. V. Garner, D. B. Mobbs, H. Suschitzky, and J. S. Millership, *J. Chem. Soc. (C),* **1971,** 3693.
124. J. O. Madsen and S. O. Lawesson, *Bull. Soc. Chim. Belg.,* **85,** 819 (1976).
125. Hoang Nhu Mai, N. Langlois, B. C. Das, and P. Potier, *C. R. Acad Sci. Paris, (C),* **270,** 2154 (1970).
126. U.S. Patent 3,514,462; *Chem. Abstr.,* **72,** P90426 (1970).
127. J. B. Hester, Jr., *J. Org. Chem.,* **35,** 875 (1970); U.S. Patent 3,573,322 (1971); *Chem. Abstr.,* **75,** P49060g (1972); U.S. Patent 3,573,323-4 (1971); *Chem. Abstr.,* **75,** P5873v (1972); U.S. Patent 3,595,874 (1971); *Chem. Abstr.,* **75,** P98552c (1972).
128. J. B. Hester, Jr., *J. Org. Chem.,* **32,** 3804 (1967).
129. U.S. Patent 3,563,979; *Chem Abstr.,* **72,** P90425f (1970).
130. I. Saito, Y. Takahashi, M. Imuta, S. Matsugo, H. Kaguchi, and T. Matsura, *Heterocycles,* **5,** 53 (1976).
131. R. W. Jackson and R. M. Manske, *J. Am. Chem. Soc.,* **52,** 5029 (1930).

132. S. Morosawa, *Bull. Chem. Soc. Japan,* **33,** 1113 (1960).
133. U.S.S.R. Patent 221,708; *Chem. Abstr.,* **70,** P3859t (1969); U.S.S.R. Patent 242,175; *Chem. Abstr.,* **71,** P91344v (1969).
134. R. G. Glushkov, V. A. Volskova, V. G. Smirnova, and O. Yu. Magidson, *Dokl. Akad. Nauk SSSR,* **187,** 327 (1969).
135. H. J. Teuber, D. Cornelius, and U. Wolcke, *Justus Liebigs Ann. Chem.,* **696,** 116 (1966).
136. U.S.S.R. Patent 398,547; *Chem. Abstr.* **80,** P59942p (1974).
137. German Patent 2,357,320; *Chem. Abstr.,* **81,** P91594h (1974).
138. R. G. Glushkov, T. K. Trubitsyna O. Yu. Magidson, and M. D. Mashkovskii, *Khim.-Farm. Zh.,* **4,** 9 (1970).
139. R. G. Glushkov, V. A. Volskova, N. P. Kostyuchenko, Yu. N. Sheinker, and O. Yu. Magidson, *Khim.-Farm. Zh.,* **4,** 227 (1970).
140. British Patent 1,409,935; *Chem. Abstr.,* **84,** P31133j (1976).
141. V. Boekelheide, M. F. Grundon, and J. Weinstock, *J. Am. Chem. Soc.,* **74** 1866 (1952).
142. M. F. Grundon and V. Boekelheide, *J. Am. Chem Soc.,* **75,** 2537 (1953).
143. M. F. Grundon, G. L. Sauvage, and V. Boekelheide, *J. Am. Chem. Soc.,* **75,** 2541 (1953).
144. J. Weinstock and V. Boekelheide, *J. Am. Chem. Soc.,* **75,** 2546 (1953).
145. V. Boekelheide, J. Weinstock, M. F. Grundon, G. L. Sauvage, and E. J. Anello, *J. Am. Chem. Soc.,* **75,** 2550 (1953).
146. V. Boekelheide, A. E. Anderson, Jr., and G. L. Sauvage, *J. Am. Chem. Soc.,* **75,** 2558 (1953).
147. V. Boekelheide and M. F. Grundon, *J. Am. Chem. Soc.,* **75,** 2563 (1953).
148. B. D. Astill and V. Boekelheide, *J. Am. Chem. Soc.,* **77,** 4079 (1955).
149. British Patent 1,290,277; *Chem. Abstr.,* **76,** P59669x (1972).
150. J. Blake, J. R. Tretter, and H. Rapoport, *J. Am. Chem. Soc.,* **87,** 1397 (1965).
151. J. Blake, J. R. Tretter, G. J. Juhasz, W. Bonthrone, and H. Rapoport, *J. Am. Chem. Soc.* **88,** 4061 (1966).
152. M. Gotz, T. Bogri, and A. H. Gray, *Tetrahedron Lett.,* **1961,** 707.
153. H. Harda, H. Irie, N. Masaki, K. Osaki and S. Uyeo, *Chem. Commun.,* **1967,** 460.
154. M. Goetz and G. M Strunz in *MTP (Medical Technology Publishing Co.) International Review of Science: Organic Chemistry, Series 1,* Vol. 9 (K. Wiesner, Ed.), Butterworths, London, 1973, pp. 143–160.
155. T. Shingu, Y. Tsuda, S. Uyeo, Y. Yamato, and H. Harada, *Chem. Ind. (London),* **1962,** 1191.
156. S. Uyeo, T. Shingu, Y. Tsuda, and Y. Yamato, *Yakugaku Zasshi,* **84,** 548 (1964).
157. S. Uyeo and T. Shingu, *Yakugaku Zasshi,* **84,** 552, 555 (1964).
158. M. Gotz, T. Bogri, A. H. Gray, and G. M. Strunz, *Tetrahedron,* **24,** 2631 (1968).
159. G. M. Strunz, *Tetrahedron,* **24,** 2645 (1968).
160. S. Uyeo, H. Irie, and H. Harada, *Chem. Pharm. Bull.,* **15,** 768 (1967).
161. T. Kaneko, *ITSUU Kenkyusho Nempo,* No. 14, 49 (1965); *Chem. Abstr.,* **68,** 69167f (1968).
162. S. Uyeo, T. Shingu, and Y. Tsuda, *Yakugaku Zasshi,* **84,** 663 (1964).
163. B. G. McDonald and G. R. Proctor, *J. Chem. Soc., Perkin Trans. I,* **1975,** 1446.
164. German Patent 2,306,605; *Chem. Abstr.,* **79,** P137117u (1973).
165. U.S. Patent 3,906,000; *Chem. Abstr.,* **84,** P31033b (1976).
166. U.S. Patent 3,904,645; *Chem. Abstr.,* **84,** P44004t (1976).

167. S. Naruto and A. Terada, *Chem. Pharm. Bull.*, **23,** 3184 (1975).
168. Netherlands Appl. 65 09,222; *Chem. Abstr.*, **64,** PC 19565f (1966).
169. U.S. Patent 3,249,623 (1966); *Chem. Abstr.*, **64,** PC 19620b (1966).
170. J. B. Hester, Jr., *J. Org. Chem.*, **32,** 4095 (1967).
171. J. Harley-Mason and A. H. Jackson, *J. Chem. Soc.*, **1955,** 374.
172. T. S.-T. Wang, *Dissertation Abstr.*, **26,** 1361 (1965).
173. Netherlands Appl., 65 15,701; *Chem. Abstr.*, **65,** P13714b (1966).
174. J. B. Hester, A. H. Tang, H. H. Keasling, and W. Veldkamp, *J. Med. Chem.*, **11,** 101 (1968).
175. N. M. Sharkova, N. F. Kucherova, S. L. Portnova, and V. A. Zagorevskii, *Khim. Geterotsikl. Soedin.*, **4,** 131 (1968).
176. U.S. Patent 3,553,232; *Chem. Abstr.*, **71,** P91455g (1969).
177. French Patent 1,524,495; *Chem. Abstr.*, **72,** P55430g (1970).
178. R. K. Brown, in *The Chemistry of Heterocyclic Compounds,* Vol. 25, Part 1 (W. J. Houlihan, Ed.), Wiley, New York, 1972, pp. 260–273.
179. S. Naruto and O. Yonemitsu, *Tetrahedron Lett.*, **1975,** 3399.
180. K. S. Bhandari, J. A. Eenkhoorn, A. Wu, and V. Sniekus, *Synth. Commun.*, **5,** 79 (1975).
181. K. Freter, *Justus Liebigs Ann. Chem.*, **721,** 101 (1969).
182. L. S. Galstyan and G. L. Papayan, *Arm. Khim. Zh.*, **27,** 331 (1974); *Arm. Khim. Zh.*, **27,** 776 (1974).
183. L. S. Galstyan and G. L. Papayan, *Arm. Khim. Zh.*, **29,** 255 (1976).
184. A. N. Kost, V. I. Gorbunov, and S. M. Gorbunova, *Zh. Org. Khim.*, **12,** 1815 (1976).
185. M. Julia, J. Bagot, and O. Siffert, *Bull Soc. Chim. France,* **1973,** 1424.
186. U.S. Patent 3,652,588 (1972); *Chem. Abstr.*, **75,** 5876g (1971).
187. J. B. Hester, A. D. Rudzik, H. H. Keasling, and W. Veldkamp, *J. Med. Chem.*, **13,** 23 (1970); see also ref. 176.
188. Japanese Kokai 72 29,395; *Chem. Abstr.*, **78,** P29748w (1973).
189. U.S. Patent 3,676,558; *Chem. Abstr.*, **75,** 5876g (1971).
190. U.S. Patent 3,839,357; *Chem. Abstr.*, **75,** 5876g (1971).
191. U.S.S.R. Patent 215,218; *Chem. Abstr.*, **69,** 59212y (1968).
192. N. M. Sharkova, N. F. Kucherova, and V. A. Zagerevskii, *Khim. Geterotsikl. Soedin.*, **5,** 88 (1969).
193. U.S. Patent 3,525,750; *Chem. Abstr.*, **72,** 21675a (1970).
194. U.S. Patent 3,776,922; *Chem. Abstr.*, **80,** 82934w (1974).
195. D. G. Murray, A. Szakolcai, and S. McLean, *Can. J. Chem.*, **50,** 1486 (1972).
196. S. Mclean, G. I. Dmitrienko, and A. Szakolcai, *Can. J. Chem.*, **54,** 1262 (1976).
197. F. Ungemach and J. M. Cook, *Heterocycles,* **9,** 1089 (1978).
198. J. Shavel, Jr., M. von Strandtmann, and M. P. Cohen, *J. Am. Chem. Soc.*, **84,** 881 (1962).
199. M. von Strandtmann, M. P. Cohen, and J. Shavel, Jr., *J. Med. Chem.*, **6,** 719 (1963).
200. U.S. Patent 3,182,071; *Chem. Abstr.*, **63,** 11510g (1965).
201. R. E. Bowman, D. D. Evans, J. Guyett, H. Nagy, J. Weake, D. J. Weyell, and A. C. White, *J. Chem. Soc., Perkin Trans. I,* **1972,** 1926.
202. R. Littel and G. R. Allen, Jr., *J. Org. Chem.*, **38,** 11504 (1973).
203. G. S. King, P. G. Mantle, C. A. Szozyrbak, and E. S. Waight, *Tetrahedron Lett.*, **1973,** 215.

204. R. S. Bajwa, R. D. Kohler, M. S. Saini, M. Cheng, and J. A. Anderson, *Phytochemistry*, **14**, 735 (1975).
205. M. S. Saini, M. Cheng, and J. A. Anderson, *Phytochemistry*, **15**, 1497 (1976).
206. J. E. Robbers and H. G. Floss, *Tetrahedron Lett.*, **1969**, 1857.
207. M. von Strandtmann, M. P. Cohen and J. Shavel, Jr., *J. Med. Chem.*, **8**, 200 (1965).
208. Y. Kanaoka and M. Yoshihiro, *Tetrahedron Lett.*, **1974**, 3693.
209. Y. Kanaoka, *Acc. Chem. Res.*, **11**, 407 (1978).
210. Japanese Kokai 74 102,696; *Chem. Abstr.*, **82**, P171117r (1975).
211. Japanese Kokai 74 102,696; *Chem. Abstr.*, **82**, P170683k (1975).
212. Y Sato, H. Nakai, T. Mizoguchi, Y. Hatanaka, and Y. Kanaoka, *J. Am. Chem. Soc.*, **98**, 2349 (1976).
213. G. R. Clemo, J. G. Cook, and R. Raper, *J. Chem. Soc.*, **1938**, 1318.
214. V. Prelog and R. Seiwerth, *Chem. Ber.*, **72**, 1638 (1939).
215. G. R. Clemo and T. P. Metcalfe, *J. Chem. Soc.*, **1937**, 1518.
216. G. R. Clemo, J. G. Cook, and R. Raper, *J. Chem Soc.*, **1938**, 1183.
217. N. J. Leonard and W. C. Wildman, *J. Am. Chem. Soc.*, **71**, 3089 (1949).
218. R. M. Acheson and J. K. Stubbs, *J. Chem. Soc. (C)*, **1969**, 2316.
219. N. J. Leonard, S. Swann, Jr., and G. Fuller, *J. Am. Chem. Soc.*, **76**, 3194 (1954).
220. R. Huisgen and E. Laschtuvka, *Chem. Ber.*, **93**, 65 (1960).
221. British Patent 1,499,047; *Chem. Abstr.*, **86**, P171287d (1977).
222. H. Wamhoff, *Chem. Ber.*, **101**, 3377 (1968).
223. H. Wamhoff, *Chem. Ber.*, **102**, 2739 (1969).
224. U.S.S.R. Patent 461,930 (1975); *Chem. Abstr.*, **86**, P156253p (1975); V. G. Granik, A. M. Zhidkova, O. S. Anisimova, and R. G. Glushkov, *Khim. Geterotsikl. Soedin.*, **1975**, 716.
225. R. Royer, J. Guillaumel, P. Demerseman, and N. Platzer, *Bull. Soc. Chim. France*, **1973**, 793.
226. H. B. Becker and K. Gustafsson, *Tetrahedron Lett.*, **1976**, 1705.
227. H. Ogura, H. Takayanagi, and K. Furuhata, *Chem. Lett.*, **1973**, 387; *J. Chem. Soc., Perkin Trans. I*, **1976**, 665.
228. H. Ogura, H. Takayanagi, and C. Miyahara, *J. Org. Chem.*, **37**, 519 (1972).
229. H. Alper and E. C. H. Keung, *J. Heterocyclic Chem.*, **10**, 637 (1973).
230. R. M. Acheson and J. N. Bridson, *J. Chem. Soc. (D)*, **1971**, 1225.
231. R. M. Acheson, J. N. Bridson, and T. S. Cameron, *J. Chem. Soc., Perkin Trans. I*, **1972**, 968.
232. S. Rajappa and B. Advani, *Ind. J. Chem.*, **12**, 4 (1974).
233. R. Neidlein and M. Hoehle, *Pharm. Z.*, **119**, 1651 (1974).
234. R. Neidlein and M. Ziegler, *Arch. Pharm.*, **306**, 531 (1973).
235. U.S. Patent 3,714,193; *Chem. Abstr.*, **78**, 111281x (1973).
236. U.S. Patent 3,767,659; *Chem. Abstr.*, **80**, 27233s (1974).
237. F. Johnson and W. Nasutavicus, *J. Heterocyclic Chem.*, **2**, 26 (1965).
238. Y. Vikhlyaev, T. Klygul, E. Slyn'ko, Ya. Gol'dfarb, B. P. Fabrichnyi, I. Shalavina, and S. Kostrova, *Khim.-Farm. Zh.*, **8**, 8 (1974); *Chem. Abstr.*, **81**, 105342p (1974).
239. B. Fabrichnyi, I. Shalavina, Ya. Gol'dfarb, and S. Kostrova, *Zh. Org. Khim.*, **10**, 1956 (1974); *Chem. Abstr.*, **82**, 57495n (1975).
240. U.S Patent 3,959,282; *Chem. Abstr.*, **84**, 4934g (1976).

241. German Patent 2,310,012; *Chem. Abstr.*, **79,** 137115s (1973).
242. German Patent 2,246,706; *Chem. Abstr.*, **78,** 159399u (1973).
243. U.S. Patent 3,856,910; *Chem. Abstr.*, **77,** 152146e (1972).
244. French Patent 2,181,491; *Chem. Abstr.*, **80,** 120897w (1974).
245. German Patent 2,364,793; *Chem. Abstr.*, **82,** 4228w (1975).
246. U.S. Patent 3,842,082; *Chem. Abstr.*, **79,** 42549x (1973).
247. A. Bertho, *Chem. Ber.*, **90,** 29 (1957).
248. A. Bertho and H. Kurzmann, *Chem. Ber.*, **90,** 2319 (1957).
249. W. Treibs, H. M. Barchet, G. Bach, and W. Kirchhof, *Justus Liebigs Ann. Chem.*, **574,** 54 (1951).
250. A. P. Boyakhchyan, L. L. Oganesyan, and G. T. Tatevosyan, *Khim. Geterotsikl. Soedin.*, **1974,** 1129.
251. D. Berney and K. Schuh, *Helv. Chim. Acta,* **58,** 2228 (1975).
252. R. M. Pinder, *J. Chem. Soc. (C),* **1969,** 1690.
253. W. J. Rodewald and J. Wicha, *Bull. Acad. Pol. Sci. Ser. Sci. Chim.*, **11,** 437 (1963).
254. W. J. Rodewald and J. Wicha, *Rocz. Chem.*, **40,** 837 (1966).
255. W. J. Rodewald and T. Rotuska, *Rocz. Chem.*, **40,** 1255 (1966).
256. W. J. Rodewald and J. Jaszczynski, *Rocz. Chem.*, **42,** 625 (1968).
257. S. D. Levine, *Chem. Commun.*, **1968,** 580.
258. S. D. Levine, *J. Org. Chem.*, **35,** 1064 (1970).
259. U.S. Patent 3,590,031; *Chem. Abstr.*, **75,** P64109g (1971).
260. I. Harper, K. Tinsley, and S. D. Levine, *J. Org. Chem.*, **36,** 59 (1971).
261. U.S. Patent 3,652,544; *Chem. Abstr.*, **73,** P56318j (1970).
262. U.S. Patent 3,557,087; *Chem. Abstr.*, **75,** P20791k (1971).
263. U.S. Patent 3,696,109; *Chem. Abstr.*, **78,** P4421w (1973).
264. W. J. Rodewald and K. Olejniczak, *Rocz. Chem.*, **50,** 1089 (1976).
265. D. D. Evans, J. Weale, and D. J. Weyell, *Aust. J. Chem.*, **26,** 1333 (1973).
266. B. P. Fabrichnyi, I. F. Shalavina, Ya. L. Gol'dfarb and S. M. Kostrova, *Zh. Org. Khim.*, **10,** 1956 (1974).
267. L. L Replogle, K. Katsumoto, and T. C. Morrill, *Tetrahedron Lett.*, **1965,** 1877.
268. L. L. Replogle, K. Katsumoto, T. C. Morrill, and C. A. Minor, *J. Org. Chem.*, **33,** 823 (1968).
269. U.S. Patent 3,257,395; *Chem. Abstr.*, **65,** 7201e (1966).
270. J. Martin, O. Meth-Cohn, and H. Suschitzky, *J. Chem. Soc. Perkin I,* **1974,** 2451.
271. J. Martin, O. Meth-Cohn, and H. Suschitzky, *J. Chem. Soc. Chem. Commun.*, **1971,** 1319.
272. M. Shemyakin, V. Antonov, A. Shkrob, V. Shchelokov, and Z. Agadzhanyan, *Tetrahedron,* **21,** 3537 (1965).
273. G. Tsagareli, A. Shkrob, and A. Ershler, *Izv. Akad. Nauk SSSR, Ser. Khim.*, **1967,** 1952; *Chem. Abstr.*, **68,** 74625h (1968).
274. A. Shkrob, Yu. Krylova, V. K. Antonov, and M. Shemyakin, *Zh. Obshch. Khim.*, **38,** 2030 (1968); *Chem. Abstr.*, **70,** 47397 (1969).
275. V. Zaretsky, N Vulfson, V. Zaikin, A. Kisin, A. Shkrob, V. Antonov, and M. Shemyakin, *Izv. Akad. Nauk SSSR, Ser. Khim.*, **1964,** 2076; *Chem. Abstr.*, **62,** 8499g (1965).
276. M. Rothe and R. Steinberger, *Tetrahedron Lett.*, **1970** 649.
277. M. Rothe and R. Steinberger, *Angew. Chem. Int. Ed.*, **7,** 884 (1968).

278. M. Rothe and R. Steinberger, *Tetrahedron Lett.*, **1970,** 2467.
279. H. Wamhoff, *Chem. Ber.*, **105,** 743 (1972).
280. W. Treibs and A. Lange, *J. Prakt. Chem.*, **14,** 214 (1961).
281. D. J. Brown and K. Ienaga, *J. Chem. Soc. Perkin I*, **1975,** 2182.
282. S. Peterson and E. Tietze, *Justus Liebigs Ann. Chem.*, **623,** 166 (1959).
283. U.S. Patent 2,992,221; *Chem. Abstr.*, **55,** 27381d (1961).
284. J. B. Taylor, D. Harrison, and F. Fried, *J. Heterocyclic Chem.*, **9,** 1227 (1972).
285. R. M. Acheson and N. F. Elmore, in *Advances in Heterocyclic Chemistry*, Vol. 23, Academic Press, New York, 1978, pp. 263–482.
286. R. Acheson, M. Foxton, and J. Stubbs, *J. Chem. Soc. (C)*, **1968,** 926.
287. R. Acheson and M. Foxton, *J. Chem. Soc. (C)*, **1968,** 378.
288. R. Acheson and G. Proctor, *J. Chem. Soc. Perkin I*, **1977,** 1924.
289. K. Shakhidoyatov, A. Irisbaev, L. Yun, E. Oripov, and C. Kodyrov, *Khim. Geterotsikl. Soedin.*, **11** 1564 (1976); *Chem. Abstr.*, **86,** 106517q (1977).
290. G. Devi, R. Kapil, and S. Popli, *Ind. J. Chem.*, **14B,** 354 (1976).
291. H. Morhle and P. Grundlach, *Arch. Pharm.*, **306,** 541 (1973).
292. O. Meth-Cohn, H. Suschitzky, and M. Sutton, *J. Chem. Soc. (C)*, **1968,** 1722.
293. A. Irisbaev, K. Shakhidoyatov, and C. Kadyrov, *Khim. Priv. Soedin.*, **11,** 531 (1975); *Chem. Abstr.*, **84,** 4894b (1976).
294. K. Hasse and H. Maisack, *Biochem. Z.*, **328,** 429 (1957).
295. H. Morhle and C. Seidel, *Arch. Pharm.*, **309,** 542 (1976).
296. D. Borman, *Chem. Ber.*, **103,** 1797 (1970).
297. Y. Krylova, A. Shkrob, V. Antonov and M. Shemyakin, *Zh. Obshch. Khim.*, **38,** 2046 (1968); *Chem. Abstr.*, **70,** 20034n (1969).
298. T. Brzechffa, M Eberle, and G. Kahle, *J. Org. Chem.*, **40,** 3062 (1975).
299. A. Shkrob, Y. Krylova, V. Antonov, and M. Shemyakin, *Tetrahedron Lett.*, **1967,** 2701.
300. R. Glushkov and O. Magidson, *Zh. Obshch. Khim.*, **31,** 189 (1961); *Chem. Abstr.*, **55,** 22336e (1961).
301. V. Smirnova and R. Glushkov, *Khim. Farm. Zh.*, **7,** 3 (1973); *Chem. Abstr.* **80,** 47816n (1974).
302. German Patent 2,519,258; *Chem. Abstr.*, **86,** 89889t (1977).
303. N. J. Leonard, R. Fox, and M. Oki, *J. Am. Chem. Soc.*, **76,** 5708 (1954).
304. P. Johnson, I. Jacobs, and D. Kerkman, *J. Org. Chem.*, **40,** 2710 (1975).
305. K. Morita, S. Kobayashi, H. Shimadzu, and M. Ochiai, *Tetrahedron Lett.*, **1970,** 861.
306. S. Kobayashi, *Bull. Soc. Chem. Japan*, **46,** 2835 (1973).
307. R. Shapiro and S. Agarwal, *J. Am. Chem. Soc.*, **90,** 474 (1968).
308. R. Hebert and D. Wibberley, *J. Chem. Soc. (C)*, **1969,** 1505.
309. F. Villani, *J. Med. Chem.*, **10,** 497 (1967).
310. J. Osbond, *J. Chem. Soc.*, **1951,** 3464.
311. R. Acheson, J. Gagan, and D. Harrison, *J. Chem. Soc. (C)*, **1968,** 362.
312. H. Gibson, D. Chesney, and F. Popp, *J. Heterocyclic Chem.*, **9,** 541 (1972).
313. G. Jones and J. Wood, *Tetrahedron*, **21,** 2961 (1965).
314. A. Crabtree, I. Jackman, and A. Johnson, *J. Chem. Soc.*, **1962,** 4417.
315. R. Acheson and D. Nisbet, *J. Chem. Soc. (C)*, **1971,** 3291.
316. R. Acheson and D. Nisbet, *J. Chem. Soc. Perkin I*, **1973,** 1338.
317. R. Acheson, G. Proctor, and S. Critchley, *J Chem. Soc. Chem. Commun.*, **1972,** 692.

318. R. Acheson, G. Procter, and S. Critchley, *Acta Crystallogr.*, **B33,** 916 (1977).
319. A. Shkrob, Yu. Krylova, V. Antonov, and M. Shemyakin, *Tetrahedron Lett.*, **1967** 2701.
320. T. Vlasova, O. Anisimova, T. Gus'kova, and G. Pershin, *Khim. Geterotsikl. Soedin.*, **1974,** 670.
321. U.S.S.R. Patent 386,940; *Chem. Abstr.*, **79,** 1265045 (1973).
322. A. Zhidkova, V. Granik, R. Glushkov, A. Polezhaeva, and M. Mashkovskii, *Khim. Farm. Zh.*, **1976,** 18; *Chem. Abstr.*, **85,** 108567j (1976).
323. U.S.S.R. Patent 496,276; *Chem. Abstr.*, **84,** 121800d (1976).
324. G. Kempter and P. Klug, *Z. Chem.*, **11,** 61 (1971).
325. A. Strakov, D. Zicane, D. Brutane, and M. Opinane, "Nov. Issled. Obl. Khim. Tekhnol., Mater. Nauchnotech. Konf. Professorsko-Prepod. Sostav. Nauchn. Rab. Khim. Fak. RPI, 1972, Red.-Izd. Otd Rith. Politekh. Inst., Riga, USSR, 1973, p. 23; *Chem. Abstr.*, **82,** 4195h (1975).
326. U.S. Patent 3,987,047; *Chem. Abstr.*, **84,** 31038g (1976).
327. German Patent 2,357,253; *Chem. Abstr.*, **83,** 97254k (1975).
328. German Patent 2,521,544; *Chem. Abstr.*, **86,** 72619g (1977).
329. C. Bradsher, L. Quinn, and R. LeBleu, *J. Org. Chem.*, **26,** 3278 (1961).
330. U.S. Patent 3,824,244; *Chem. Abstr.* **81,** 105321f (1974).
331. U.S. Patent 3,892,752; *Chem. Abstr.*, **83,** 178856k (1975).
332. K. Moser and C. Bradsher, *J. Am. Chem. Soc.*, **81,** 2547 (1959).
333. R. Kimber and C. Bradsher, *J. Org. Chem.*, **28,** 3205 (1963).
334. F. Villani and T. Mann, *J. Med. Chem.*, **11,** 894 (1968).
335. East German Patent 78,584; *Chem. Abstr.*, **76,** 99712n (1972).
336. East German Patent 80,449; *Chem. Abstr.*, **76,** 85803f (1972).
337. B. Eistert and P. Donath, *Chem. Ber.*, **106,** 1537 (1973).
338. C. Muth, W. Sung, and Z. Papanastassiou, *J. Am. Chem. Soc.*, **77,** 3393 (1955).
339. R. Brown and M Butcher, *Tetrahedron Lett.*, **1971,** 667.
340. K. Wiesner, Z. Valenta, A. Manson, and F. Stonner, *J. Am. Chem. Soc.*, **77,** 675 (1955).
341. W. Paterson and G. Proctor, *J. Chem. Soc.*, **1962,** 3468.
342. G. Proctor and W. Peaston, *J. Chem. Soc. (C)*, **1969,** 2151.
343. M. Raboman, *J. Nat. Sci. Math.*, **1970,** 161; *Chem. Abstr.*, **75,** 129645s (1971).
344. W. Peaston and G. Proctor, *J. Chem. Soc. (C)*, **1968,** 2481.
345. R. Cooke and I. Russell, *Tetrahedron Lett.*, **1968,** 4587.
346. S. Aftalion and G. Proctor, *Org. Mass Spectrom.*, **1969,** 337.
347. C. Leznoff and R. Hayward, *Can. J. Chem.*, **49,** 3596 (1971).
348. R. Barnes and M. Beachem, *J. Am. Chem. Soc.*, **77,** 5388 (1955).
349. K. Schaffner, R. Viterbo, D. Arigoni, and O. Jeger, *Helv. Chim. Acta*, **34,** 174 (1956).
350. M. Ohta, *J. Pharm. Soc. Japan*, **75,** 289 (1955).
351. H. Erdtmann and L. Malmborg, *Acta Chem. Scand.*, **24,** 2252 (1970).
352. W. Nelson, D. Miller, and R. Wilson, *J. Heterocyclic Chem.*, **6,** 132 (1969).
353. M. Samimi, V. Kraatz, and F. Korte, *Heterocycles*, **5,** 73 (1976).
354. U.S. Patent 3,652,543; *Chem. Abstr.*, **71,** 61251v (1969).
355. Swiss Patent 500,194; *Chem. Abstr.*, **74,** 141586k (1971).
356. R. Scholl and J. Muller, *Chem. Ber.*, **64,** 639 (1931).
357. German Patent 837,537; *Chem. Abstr.*, **50,** 1897g (1956).

358. Swiss Patent 345,011; *Chem. Abstr.*, **55,** 2707a (1961).
359. M. M. Coombs, *J. Chem. Soc.*, **1958,** 4200.
360. D. Emrick and W. Truce, *J. Org. Chem.*, **26,** 1329 (1961).
361. L. Werner, S. Ricca, E. Mohacsi, A. Rossi, and V. Arya, *J. Med. Chem.*, **8,** 74 (1965).
362. F. Tishler, H. Hagman, and S. Brody, *Anal. Chem.*, **37,** 906 (1965).
363. G. Caronna and S. Palazzo, *Gazz. Chim. Ital.*, **83,** 533 (1953).
364. G. Caronna and S. Palazzo, *Gazz. Chim. Ital.*, **84,** 1135 (1954).
365. A. Drukker and C. Judd, *J. Heterocyclic Chem.*, **2,** 276 (1965).
366. German Patent 551,256; *Chem. Abstr.*, **26,** 4344 (1932).
367. U.S. Patent 3,475,449; *Chem. Abstr.*, **67,** 43829c (1967).
368. Japanese Kokai 75 112,387; *Chem. Abstr.*, **83,** 195241y (1975).
369. P. Kranzlein, *Chem. Ber.*, **70,** 1952 (1937).
370. W. Bradley and H. Nursten, *J. Chem. Soc.*, **1951,** 2170.
371. J. Ballantine, A. Johnson, and A. Katner, *J. Chem. Soc.*, **1964,** 3323.
372. J. Arient, J. Slosar, V. Sterba, and K. Obruba, *J. Chem. Soc. (C)*, **1967,** 1331.
373. German Patent 2,361,467; *Chem. Abstr.*, **83,** 164019c (1975).
374. U.S. Patent 3,354,147; *Chem. Abstr.*, **69,** 10385t (1968).
375. G. Hardtmann and H. Ott, *J. Org. Chem.*, **34,** 2244 (1969).
376. W. Waring and B. Whittle, *J. Pharm. Pharmacol.*, **21,** 520 (1969).
377. G. Walker, A. Engle, and R. Kempton, *J. Org. Chem.*, **37,** 3755 (1972).
378. G. Aichinger, O. Behner, F. Hoffmeister, and S. Schutz, *Artzneim. Forsch.*, **19,** 838 (1969).
379. K. Ackermann, J. Chapuis, D. Horning, G. Lacasse, and J. Muchowski, *Can. J. Chem.*, **47,** 4327 (1969).
380. D. Horning, D. Ross, and J. Muchowski, *Can. J. Chem.*, **51,** 2347 (1973).
381. British Patent 1,132,516; *Chem. Abstr.*, **70,** 77824z (1969).
382. German Patent 2,230,122; *Chem. Abstr.*, **81,** 14721v (1974).
383. M. Dokunikhin and T. Kurbyumova, *Sb. Stat. Obshch. Khim.*, **2,** 1411 (1953); *Chem. Abstr.*, **49,** 5491f (1955).
384. U.S. Patent 3,431,257; *Chem. Abstr.*, **67,** 82128t (1967).
385. K. Howard and T. Koch, *J. Am. Chem. Soc.*, **97,** 7288 (1975).
386. U.S. Patent 3,845,074; *Chem. Abstr.*, **74,** 100114m (1971).
387. R. Cooke and I. Russell, *Aust. J. Chem.*, **25,** 2421 (1972).
388. E. Moriconi and I. Maniscalco, *J. Org. Chem.*, **37,** 208 (1972).
389. I. MacDonald and G. Proctor, *J. Chem. Soc. (C)*, **1969,** 1321.
390. M. Fennon, A. McLean, I. McWatt, and G. Proctor, *J. Chem. Soc. Perkin I*, **1974,** 1828.
391. J. P. LeRoux, P. L. Desbene, and M. Seguin, *Tetrahedron Lett.*, **1976,** 3141.
392. J. Jilek, J. Pomykacek, E. Svatek, V. Seidlova, M. Rajsner, K. Pelz, B. Hoch, and M. Protiva, *Coll. Czech. Chem. Commun.*, **30,** 445 (1965).
393. Swiss Patent 464,933; *Chem. Abstr.*, **70,** 106405w (1969).
394. Japanese Patent 71 41,549; *Chem. Abstr.*, **76,** 59483g (1972).
395. R. Williams and H. Snyder, *J. Org. Chem.*, **38,** 809 (1973).
396. U.S. Patent 3,847,910; *Chem. Abstr.*, **82,** 730345 (1975).
397. J. Schmutz, F. Kunzle, F. Hunziker, and A. Burki, *Helv. Chim. Acta*, **48,** 336 (1965).
398. U.S. Patent 3,367,930; *Chem. Abstr.*, **69,** 27464r (1968).

399. Swiss Patent 436,303; *Chem. Abstr.*, **64,** 8223h (1966).
400. U.S. Patent 3,910,890; *Chem. Abstr.*, **84,** 59245w (1976).
401. U.S. Patent 4,058,561; *Chem. Abstr.*, **85,** 177100 (1976).
402. U.S. Patent 3,541,085; *Chem. Abstr.*, **74,** 76459e (1971).
403. Netherlands Appl. 66 00,293; *Chem. Abstr.*, **66,** 46346h (1967).
404. Swiss Patent 449,023; *Chem. Abstr.*, **69,** 67251a (1968).
405. U.S. Patent 3,454,561; *Chem. Abstr.*, **66,** 46354g (1967).
406. Japanese Patent 71 41,548; *Chem. Abstr.*, **76,** 59487m (1972).
407. Belgian Patent 707,523; *Chem. Abstr.*, **78,** 72213e (1973).
408. U.S. Patent 3,652,550; *Chem. Abstr.*, **77,** 48291n (1972).
409. U.S. Patent 3,663,533; *Chem. Abstr.*, **74,** 22906e (1971).
410. U.S. Patent 3,853,882; *Chem. Abstr.*, **79,** 105308r (1973).
411. U.S. Patent 3,580,907; *Chem. Abstr.*, **75,** 63648m (1971).
412. Belgian Patent 844,881; *Chem. Abstr.*, **86,** 171505g (1977).
413. U.S. Patent 3,993,757; *Chem. Abstr.*, **86,** 155531y (1977).
414. S. Winthrop, M. Davis, F. Herr, J. Stewart, and R. Gaudry, *J. Med. Pharm. Chem.*, **5,** 1199 (1962).
415. U.S. Patent 3,173,913; *Chem. Abstr.*, **62,** 537d (1965).
416. U.S. Patent 3,351,588; *Chem. Abstr.*, **68,** 105029y (1968).
417. British Patent 1,144,829; *Chem. Abstr.*, **70,** 96656n (1969).
418. G. Coppola, G. Hardtmann, and R. Mansukhani, *J. Org. Chem.*, **40,** 3602 (1975).
419. Japanese Patent 74 34,997; *Chem. Abstr.*, **82,** 170733b (1975).
420. G. Wittig, G. Closs, and F. Mindermann, *Justus Liebigs Ann. Chem.*, **594,** 7 (1955).
421. U.S. Patent 3,391,136; *Chem. Abstr.*, **69,** 96513t (1968).
422. U.S. Patent 3,931,158; *Chem. Abstr.*, **84,** 90027a (1976).
423. U.S. Patent 3,950,326; *Chem. Abstr.*, **85,** 46441c (1976).
424. U.S. Patent, 3,692,906; *Chem. Abstr.*, **78,** 16070u (1973).
425. German Patent 1,470,379; *Chem. Abstr.*, **85,** 177276c (1976).
426. Belgian Patent 732,405; *Chem. Abstr.*, **74,** 13023n (1971).
427. F. Hunziker, F. Kunzle, and J. Schmutz, *Helv. Chim. Acta*, **49,** 1433 (1966).
428. A. Drukker, C. Judd, and D. Dusterhoft, *J. Heterocyclic Chem.*, **3,** 206 (1966).
429. J. Schmutz, *Arzneim.-Forsch.*, **25,** 712 (1975).
430. K. Sano and T. Yoshiki, *J. Takeda Res. Lab.*, **29,** 602 (1970).
431. U.S. Patent 3,530,219; *Chem. Abstr.*, **70,** 57692b, 87604s (1969).
432. U.S. Patent 3,962,248; *Chem. Abstr.*, **80,** 14969m (1974).
433. British Patent 696,473; *Chem. Abstr.*, **49,** 385a (1955).
434. F. Hunziker, F. Kunzle, O. Schindler, and J. Schmutz, *Helv. Chim. Acta*, **47,** 1163 (1964).
435. U.S. Patent 3,534,041; *Chem. Abstr.*, **70,** 47499e (1969).
436. A. Wardrop, G. Sainbury, J. Harrison, and T. Inch, *J. Chem. Soc., Perkin Trans I*, **1976,** 1279.
437. S. African Patent 64/2568; *Chem. Abstr.*, **69,** 27277g, 27279j, 27280c (1968).
438. British Patent 1,142,596; *Chem. Abstr.*, **70,** 96821n (1969).
439. R. Schmidt, W. Schneider, J. Karg, and U. Burkent, *Chem. Ber.*, **105,** 1634 (1972).
440. British Patent 1,268,454; *Chem. Abstr.*, **73,** 131053z (1970).
441. U.S. Patent 3,917,633; *Chem. Abstr.*, **84,** 44062h (1976).

442. W. van der Burg, I. Bonta, J. Delobelle, C. Ramon, and B. Vargaftig, *J. Med. Chem.*, **13,** 35 (1970).
443. U.S. Patent 3,892,695; *Chem. Abstr.*, **81,** 3986j (1974).
444. U.S. Patent 3,850,956; *Chem. Abstr.*, **79,** 5339h (1973).
445. K. Brewster, R. Chittenden, J. Harrison, T. Inch, and C. Brown, *J. Chem. Soc., Perkin Trans. I*, **1976,** 1292.
446. T. Carpenter, *Diss. Abstr. Int. B*, **33,** 3546 (1973); *Chem. Abstr.*, **78,** 135389g (1973).
447. A. Drukker, C. Judd, J. Spoerl, and F. Kaminski, *J. Heterocyclic Chem.*, **2,** 283 (1965).
448. M. Borovicka and M. Protiva, *Coll. Czech. Chem. Commun.*, **23,** 1330 (1958).
449. French Patent 6171M; *Chem. Abstr.*, **72,** 43502u (1970).
450. Swiss Patent 341,829; *Chem. Abstr.*, **55,** 2707a (1961).
451. U.S. Patent 2,861,987; *Chem. Abstr.*, **53,** 7218a (1958).
452. German Patent 1,136,706; *Chem. Abstr.*, **58,** 4529c (1963).
453. Z. Votava and J. Metysova, *Farmakol. Neurotropn. Sredstv, Acad. Med. Nauk SSSR*, Sb. 43 (1963); *Chem. Abstr.*, **60,** 11241g (1964).
454. U.S. Patent 3,316,246; *Chem. Abstr.*, **67,** 90699v (1967).
455. U.S. Patent 3,316,245; *Chem. Abstr.*, **67,** 54044u (1967).
456. Japanese Patent 72 11,987; *Chem. Abstr.*, **77,** 34371b (1972).
457. W. Coyne and J. Cusic, *J. Med. Chem.*, **10,** 541 (1967).
458. W. Coyne and J. Cusic, *J. Med. Chem.*, **11,** 1158 (1968).
459. U.S. Patent 3,336,303; *Chem. Abstr.*, **68,** 78163d (1968).
460. U.S. Patent 3,862,135; *Chem. Abstr.*, **82,** 170736e (1975).
461. U.S. Patent 3,905,977; *Chem. Abstr.*, **83,** 193115m (1975).
462. A. Lindquist and J. Sandstrom, *Chem. Script.*, **5,** 52 (1974); *Chem. Abstr.*, **80,** 95136n (1974).
463. M. Dewar and N. Trinajstic, *Tetrahedron*, **26,** 4269 (1970).
464. B. Hess, Jr., L. Schaad, and C. Holyoke, Jr., *Tetrahedron*, **28,** 3657 (1972).
465. German Patent 1,670,334; *Chem. Abstr.*, **85,** 177276a (1976).
466. U.S. Patent 3,776,723; *Chem. Abstr.*, **74,** 141583g (1971).
467. Swiss Patent 436,308; *Chem. Abstr.*, **69,** 19212s (1968).
468. J. Thiele and O. Holzinger, *Justus Liebigs Ann. Chem.*, **305,** 96 (1899).
469. J. R. Durland and H. Adkins, *J. Am. Chem. Soc.*, **60,** 1501 (1938)
470. U.S. Patent 2,554,736; *Chem. Abstr.*, **46,** 3094d (1952).
471. G. L. Klerman and J. O. Cole, *Pharmacol. Rev.*, **17,** 101 (1965).
472. J. Meneses Hoyas, *Medicina* (Mexico City), **52,** 545 (1972).
473. G. Caille, C. Sauriol, J. M. Albert, J. A Mockle, and J. C. Panisset, *Contrib. Neurosci., Psychopharmacol., Mot. Syst., Int Symp.*, (J. Saint-Laurent, Ed.), Presses Univ. Montreal, Montreal, Quebec, 1972, p. 59.
474. H. Goldenberg and V. Fishman, in *Principles of Psychopharmacology: A Textbook for Physicians, Medical Students, and Behavioral Scientists* (W. J. Clark and J. Del Guidice, Eds.), Academic Press, New York, 1970, pp. 179–191 and p. 723.
475. E. B. Sigg and R. T. Hill, Proc. Int. Congr. Coll. Int. Neuro-Psycho-Pharmacol., 5th, Washington, D.C., 1966, p. 367 (publ. 1967).
476. I. P. Lapin, *Therapie*, **19,** 1107 (1964).
477. A. Todrick and A. C. Tait *J. Pharm. Pharmacol.*, **21,** 751 (1969).
478. J. Biel, in *Principles of Psychopharmacology: A Textbook for Physicians, Medical*

Students, and Behavioral Scientist (W. J. Clark and J. Del Guidice, Eds.), Academic Press, New York, 1970, p. 269.

479. W. Theobald, O. Buech, H. A. Kunz, C. Morpurgo, E. G. Stenger, and G. Wilhelmi, *Arch. Int. Pharmacodyn.*, **148,** 590 (1964).

480. A. P. Horovitz, A. R. Furgiuele, J. P. High, and J. C. Burke, *Arch. Int. Pharmacodyn.*, **151,** 180 (1964).

481. E. Kombos and L. Petocz, Conf. Hung. Ther. Invest. Pharmacol., 2, Budapest, 1962, p. 95.

482. M. Nakanishi Y. Kato, M. Setoguchi, A. Tsuda, and H. Yasudo, *Yakugaku Zasshi,* **90,** 197 (1970).

483. M. Nakanishi, T. Tsumagari, T. Okada, and Y. Kase, *Arzneim.-Forsch.*, **18,** 1435 (1965).

484. H. Meinarde, *Antiepileptic Drugs,* **1972,** 487.

485. B. Zetterstrom, *Sv. Farm. Tidskr.*, **74,** 520 (1970).

486. R. M. Julien and R. P. Hollister, *Adv. Neurol.*, **11,** 263 (1975).

487. A. Frigerio and P. L. Morselli, *Adv. Neurol.*, **11,** 295 (1975).

488. P. L. Morselli and A. Frigerio, *Drug Metab. Rev.*, **4,** 97 (1975).

489. L. J. Kricka and A. Ledwith, *Chem. Rev.*, **74,** 118–120 (1974).

490. L. J. Kricka and A. Ledwith, *Chem. Rev.*, **74,** 101 (1974).

491. F. Haefliger and V. Burckhardt, *Psychopharmacological Agents,* Vol. 1, Academic Press, New York, 1964, p. 35.

492. J. A. Moore and E. Mitchell, *Heterocyclic Compounds,* Vol. 9 (R. C. Elderfield, Ed.), Wiley, New York, 1967, p. 224.

493. J. Diamond, in ref. 492, p. 355.

494. U.S. Patent 2,764,580; *Chem. Abstr.*, **51,** 4447f (1957).

495. U.S. Patent 2,800,470; *Chem. Abstr.*, **52,** 4691a (1958).

496. U.S. Patent 2,809,200; *Chem. Abstr.*, **52,** 16388g (1958).

497. R. Huisgen, E. Laschtuvka, and F. Bayerlein, *Chem. Ber.*, **93,** 392 (1960).

498. V. G. Yashunskii, M. N. Shuhukina, V. G. Ermolaeva, and O. I. Samoilova, *Med. Prom. SSSR,* **15,** 10 (1961).

499. B. P. Das, R. W. Woodward, L. K. Whisenant, W. F. Winecoff III, and D. W. Boykin, Jr., *J. Med. Chem.*, **13,** 979 (1970).

500. British Patent 1,407,554; *Chem. Abstr.*, **80,** 120804p (1974).

501. British Patent 1,321,589; *Chem. Abstr.*, **75,** 151695e (1971).

502. French Patent 1,351,837; *Chem. Abstr.*, **60,** 15847d (1964).

503. H. O. House, *Organic Synthesis,* Coll. Vol. IV, Wiley, New York, 1963, p. 367.

504. G. A. Russell and W. C. Danen, *J. Am. Chem. Soc.*, **90,** 347 (1968) and references cited therein.

505. U.S. Patent 2,811,520; *Chem. Abstr.*, **52,** 5490e (1958).

506. U.S. Patent 3,074,931; *Chem. Abstr.*, **59,** 3900g (1963).

507. U.S. Patent 3,126,373; *Chem. Abstr.*, **60,** 13231f (1964).

508. U.S. Patent 3,591,604; *Chem. Abstr.*, **75,** 76855b (1971).

509. E. D. Bergmann, I. Shahak, and Z. Aizenshtat, *Tetrahedron Lett.*, **1968,** 3469.

510. E. D. Bergmann, Z. Aizenshtat, and I. Shahak, *Israel J. Chem.*, **6,** 507 (1968).

511. P. N. Craig, B. M. Lester, A. J. Saggiomo, C. Kaiser, and C. L. Zirkle, *J. Org. Chem.* **26,** 135 (1961).

512. W. Schindler and H. Blattner, *Helv. Chim. Acta,* **44,** 753 (1961).

513. U.S. Patent 3,056,776; *Chem. Abstr.*, **58,** 5646h (1963).
514. U.S. Patent 3,016,373; *Chem. Abstr.*, **57,** 2201f (1962).
515. U.S. Patent 3,033,866; *Chem. Abstr.*, **57,** 47c (1962).
516. U.S. Patent 3,144,440; *Chem. Abstr.*, **59,** 11454e (1963).
517. E. D. Bergmann, M. Rabinovitz, and A. Bromberg, *Tetrahedron*, **24,** 1289 (1968).
518. Japanese Kokai, 71 08,980; *Chem. Abstr.*, **75,** 35818y (1971).
519. British Patent 881,398; *Chem. Abstr.*, **56,** 14248h (1962).
520. U.S. Patent 3,068,222; *Chem. Abstr.*, **55,** 12437a (1961).
521. British Patent 1,077,648; *Chem. Abstr.*, **71,** 3292y (1969).
522. U.S. Patent 3,449,324; *Chem. Abstr.*, **69,** 35976z (1968).
523. U.S. Patent 3,130,192; *Chem. Abstr.*, **54,** 3466c (1960).
524. British Patent 1,040,737; *Chem. Abstr.*, **65,** 16952c (1966).
525. Swiss Patent 408,927; *Chem. Abstr.*, **66,** 10859p (1967).
526. British Patent 1,246,606; *Chem. Abstr.*, **75,** 140723m (1971).
527. U.S. Patent 3,056,775; *Chem. Abstr.*, **59,** 2783d (1963).
528. German Patent 2,238,904; *Chem. Abstr.*, **83,** 97062w (1975).
529. E. D. Bergmann and M. Rabinovitz, *J. Org. Chem.*, **25,** 827 (1960).
530. H. W. Whitlock, *Tetrahedron Lett.*, **1961,** 593.
531. R. S. Varma, L. K. Whisenant, and D. W. Boykin, *J. Med. Chem.*, **12,** 913 (1969).
532. U.S. Patent 3,772,348; *Chem. Abstr.*, **76,** 85721c (1972).
533. U.S. Patent 3,917,603; *Chem. Abstr.*, **76,** 72428n (1972).
534. French Patent 2,088,214; *Chem. Abstr.*, **77,** 126458p (1972).
535. U.S. Patent 3,979,515; *Chem. Abstr.*, **74,** 53577c (1971).
536. German Patent 1,142,870; *Chem. Abstr.*, **59,** 11454e (1963).
537. U.S. Paten 3,636,046; *Chem. Abstr.*, **74,** 53571w (1971).
538. Swis Patent 521,978; *Chem. Abstr.*, **77,** 126601e (1972).
539. Swiss Patent 521,981; *Chem. Abstr.*, **77,** 126603g (1972).
540. Swiss Patent 521,982; *Chem. Abstr.*, **77,** 126600d (1972).
541. U.S. Patent 3,749,790; *Chem. Abstr.*, **73,** 45488 (1970).
542. British Patent 943,277; *Chem. Abstr.*, **61,** 1815e (1964).
543. U.S. Patent 3,144,441; *Chem. Abstr.*, **64,** 3507b (1966).
544. U.S. Patent 3,144,442; *Chem. Abstr.*, **61,** 4328f (1964).
545. U.S. Patent 3,144,443; *Chem. Abstr.*, **61,** 1815e,g (1964).
546. U.S. Patent 3,637,661; *Chem. Abstr.*, **73,** 130908v (1970).
547. U.S. Reissue 27,622; *Chem. Abstr.*, **79,** 31944w (1973).
548. U.S. Patent 3,792,042; *Chem. Abstr.*, **74,** 76346r (1971).
549. U.S. Patent 3,882,235; *Chem. Abstr.*, **74,** 76346r (1971).
550. U.S. Patent 3,130,191; *Chem. Abstr.*, **59,** 11454e (1963).
551. Belgian Patent 618,411; *Chem. Abstr.*, **59,** 8715b (1963).
552. Belgian Patent 631,162; *Chem. Abstr.*, **60,** 14486a (1964)
553. British Patent 940,165; *Chem. Abstr.*, **61,** 1816a (1964).
554. Belgian Patent 631,161; *Chem. Abstr.*, **61,** 4327g (1964).
555. Swiss Patent 374,075; *Chem. Abstr.*, **61,** 5620a (1964).
556. French Patent 1,377,680; *Chem. Abstr.*, **62,** 9116d (1965).
557. U.S. Patent 3,622,565; *Chem. Astr.*, **70,** 57689f (1969).

558. Netherlands Appl. 66 09,781; *Chem. Abstr.*, **67**, 82124p (1967).
559. French Patent 2,092,877; *Chem. Abstr.*, **77**, 101405f (1972).
560. French Patent 2,069,831; *Chem. Abstr.*, **77**, 34370a (1972).
561. British Patent 1,301,366; *Chem. Abstr.*, **77**, 5379e (1972).
562. British Patent 1,301,367; *Chem. Abstr.*, **77**, 5379e (1972).
563. U.S. Patent 3,821,197; *Chem. Abstr.*, **77**, 152012h (1972).
564. French Patent 2,120,304; *Chem. Abstr.*, **78**, 111160g (1973).
565. Japanese Patent 71 09,340; *Chem. Abstr.*, **75** 35814u (1971).
566. P. Rosenmund and W. H. Haase, *Chem. Ber.*, **99**, 2504 (1966).
567. P. Rosenmund, J. Bauer, and D. Sauer, *Chem. Ber.*, **104**, 1379 (1971).
568. P. Rosenmund, W. H. Haase, J. Bauer, and R. Frische, *Chem. Ber.*, **106**, 1459, 1474 (1973).
569. E. Gonzalez and R. Saslin, *C. R. Acad. Sci. Paris (C)*, **275**, 965 (1972).
570. W. A. Remers and R. K. Brown, in *Indoles*, Part 1 (W J. Houlihan, Ed.), Wiley-Interscience, New York, pp. 66–67.
571. A. Butenhandt, E. Biekert, and G. Neubert, *Justus Liebigs Ann. Chem.*, **602**, 72 (1957).
572. A. Butenhandt, E. Biekert, and G. Neubert, *Justus Liebigs Ann. Chem.*, **603**, 200 (1957).
573. U.S. Patent 3,467,650; *Chem. Abstr.*, **60**, 5470b (1964).
574. U.S. Patent 3,624,072; *Chem. Abstr.*, **76**, 59479k (1972).
575. U.S. Patent 3,475,412; *Chem. Abstr.*, **64**, 739d (1966).
576. U.S. Patent 3,577,343; *Chem. Abstr.*, **64**, 739d (1966).
577. Swiss Patent 457,447; *Chem. Abstr.*, **70**, 19952d (1969).
578. U.S. Patent 3,501,459; *Chem. Abstr.*, **73**, 35241p (1970).
579. B. A Hess, Jr. and V. Boekelheide, *J. Am. Chem. Soc.*, **91**, 1672 (1969).
580. Japanese Patent 72 14,116; *Chem. Abstr.*, **77**, 34375f (1972).
581. Swiss Patent 509,317; *Chem. Abstr.*, **76**, 14371c (1972).
582. Swiss Patent 513,894; *Chem. Abstr.*, **76**, 113221k (1972).
583. L. J. Kricka and A. Ledwith, *J. Chem. Soc., Perkin Trans. I*, **1972**, 2292.
584. L. J. Kricka and A. Ledwith, *J. Chem. Soc., Perkin Trans. I*, **1973**, 859.
585. O. E. Edwards and G. Feniak, *Can. J. Chem.*, **40**, 2416 (1962).
586. K. Kawashima and Y. Kawano, *Chem. Pharm. Bull.*, **24**, 2751 (1976).
587. W. Schindler and F. Hafliger, *Helv. Chim. Acta*, **37**, 472 (1954).
588. U.S. Patent 2,674,596; *Chem. Abstr.*, **49**, 11030c (1955).
589. U.S. Patent 2,976,281; *Chem. Abstr.*, **55**, 17671c (1961).
590. U.S. Patent 2,981,736; *Chem. Abstr.*, **55**, 19972f (1961).
591. British Patent 862,297; *Chem. Abstr.*, **55**, 26001i (1961)
592. U.S. Patent 3,045,017; *Chem. Abstr.*, **55**, 14488d (1961).
593. British Patent 872,320; *Chem. Abstr.*, **56**, 1436h (1962).
594. U.S. Patent 3,102,888; *Chem. Abstr.*, **56**, 8695a (1962).
595. Belgian Patent 618,829; *Chem. Abstr.*, **59**, 5142e (1963).
596. British Patent 966,418; *Chem. Abstr.*, **59**, 9987f (1963).
597. U.S. Patent 3,123,610; *Chem. Abstr.*, **60**, 15847b (1964).
598. L. A. Paquette, *J. Org. Chem.*, **29**, 3545 (1964).
599. U.S. Patent 3,317,515; *Chem. Abstr.*, **64**, 11185f (1966).
600. French Med. 7329; *Chem. Abstr.*, **76**, 14368g (1972).

601. U.S. Patent 3,886,170; *Chem. Abstr.*, **83,** 79103f (1975).
602. U.S. Patent 3,968,217; *Chem. Abstr.*, **85,** 112759r (1976).
603. U.S. Patent 3,335,133; *Chem. Abstr.*, **68,** 68902e (1968).
604. C. Kaiser, D. H Tedeschi, P. J. Fowler, A. M. Pavloff, B. M. Lester, and C. L. Zirkle, *J. Med Chem.*, **14,** 179 (1971).
605. U.S. Patent 3,544,558; *Chem Abstr.*, **74,** 100132r (1971).
606. U.S. Publ. Patent Appl. B 520,256; *Chem. Abstr.*, **84,** 180094z (1976).
607. C. R. Ellefson and J. W. Cusio, *J. Med. Chem.*, **19,** 1345 (1976).
608. U.S. Patent 3,624,075; *Chem. Abstr.*, **72,** 43496v (1970).
609 G. W. Gribble, P. D. Lord, J. Skotnicki, S. E. Dietz, J. T. Eaton, and J. L. Johnson, *J. Am. Chem. Soc.*, **96,** 7812 (1974).
610. Swiss Patent 368,493; *Chem. Abstr.*, **60,** 510g (1964).
611. Swiss Patent 368,495; *Chem. Abstr.*, **60,** 2916a (1964).
612. British Patent 828,495; *Chem. Abstr.*, **54,** 13153f (1960).
613. Swis Patent 359,143; *Chem. Abstr.*, **58,** 10218f (1963).
614. Swiss Patent 363,029; *Chem. Abstr.*, **59,** 10011g (1963).
615. Swiss Patent 390,258; *Chem. Abstr.*, **63,** 14831d (1965).
616. Swiss Patent 404,671; *Chem. Abstr.*, **68,** 2878r (1968).
617. U.S. Patent 3,706,735; *Chem. Abstr.*, **74,** 99912m (1971).
618. P. Hyde, L. J. Kricka, A. Ledwith, and K. C. Smith, *Polymer,* **15,** 387 (1974).
619. N. M. Libman and S. G. Kuznetsov, *Zh. Org. Khim.*, **3,** 2012 (1967).
620. U.S. Patent 3,354,178; *Chem. Abstr.*, **68,** 114430z (1968).
621. Japanese Kokai 73 81,844; *Chem. Abstr.*, **80,** 108402y (1974).
622. U.S. Patent 3,056,774; *Chem. Abstr.*, **58,** 12526a (1963).
623. East German Patent 82,719 (1970); *Chem. Abstr.*, **77,** 101406g (1972).
624. H. J. Teuber and W. Schmidtke, *Chem. Ber.*, **93,** 1257 (1960).
625. U.S. Patent 3,931,151; *Chem. Abstr.*, **81,** 13403f (1974).
626. L. Toscano, E. Seghetti, and G. Fioriello, *J. Med. Chem.*, **18,** 976 (1975).
627. S. G. Rozenberg, K. F. Turchin, V. A. Zagorevskii, H. Wunderlich, A. Stark, and A. N. Gritsenko, *Khim.-Farm. Zh.*, **9,** 6 (1975)
628. B. P. Das and D. W. Boykin, Jr., *J. Med. Chem.*, **14,** 56 (1971).
629. German Patent 2,440,592; *Chem. Abstr.*, **82,** 156138e (1975).
630. U.S. Patent 4,076,812; *Chem. Abstr.*, **85,** 32888s (1976).
631. Swiss Patent 416,645; *Chem. Abstr.* **67,** 43689g (1967).
632. Belgian Patent 767,061; *Chem. Abstr.*, **80,** 12089v (1974).
633. French Patent 1,578,748; *Chem. Abstr.*, **72,** 111316p (1970).
634. V. N. Bagal, I. Ya Kvitko, I. P. Lapin, B. A. Porai-Koshits, and O. V. Favorskii, *Khim.-Farm. Zh.*, **1,** 21 (1967).
635. U.S. Patent 3,711,612; *Chem. Abstr.*, **78,** 101997w (1973).
636. U.S. Patent 3,025,228; *Chem. Abstr.*, **55,** 8436g (1961).
637. U.S. Patent 3,052,689-70; *Chem. Abstr.*, **57,** 13744c (1962).
638. P. Hyde, L. J. Kricka, and A. Ledwith, *J. Polymer Sci., Polymer Lett. Ed.*, **11,** 415 (1973).
639. H. W. Vos, Y. W. Bakker, C. MacLean, and N. H. Vetthorst, *Chem. Phys. Lett.*, **25,** 80 (1974).
640. M. Rogradi, W. D. Ollis, and I. O. Sutherland *J. Chem. Soc. (D),* **1970,** 158.

641. J. Casanova and M. Geisel, *Inorg. Chem.*, **13,** 2783 (1974).
642. British Patent 1,323,219; *Chem. Abstr.*, **77,** 88351p (1972).
643. East German Patent 102,149; *Chem. Abstr.*, **81,** 25576c (1974).
644. Swiss Patent 521,981; *Chem. Abstr.*, **77,** 126603g (1972).
645. D. Hellwinkel, M. Melan, W. Egan, and C. R. Degel, *Chem. Ber.*, **108,** 2219 (1975).
646. U.S. Patent 3,185,679; *Chem. Abstr.*, **63,** 9925e (1965).
647. K. M. Baker, J. Csetenyi, A. Frigerio, and P. L. Morselli, *J. Med. Chem.*, **16,** 703 (1973).
648. Zvi Rappoport, in *Advances in Physical Organic Chemistry*, Vol. 7 (V. Gold, Ed.), Academic Press, New York, 1969, pp. 1–114.
649. W. Schindler, *Helv. Chim. Acta*, **43,** 35 (1960).
650. British Patent 926,816; *Chem. Abstr.*, **59,** 11455f (1963).
651. Swiss Patent 383,981; *Chem. Abstr.*, **62,** 16218a (1965).
652. U.S. Patent 3,324,113; *Chem. Abstr.*, **62,** 16218a (1965).
653. A. Zirnis, F. F. Piszkiewicz, and A. A. Manian, *J. Heterocyclic Chem.*, **13,** 269 (1976).
654. K. E. Haque and G. R. Proctor, *Chem. Commun.*, **1968,** 1412.
655. K E. Haque, K. M. Hardie, and G. R. Proctor, *J. Chem. Soc., Perkin Trans. I*, **1972,** 539.
656. U.S. Patent 3,704,245; *Chem. Abstr.*, **72,** 55287r (1970).
657. U.S. Patent 3,040,031; *Chem. Abstr.*, **57,** 13785e (1962).
658. L. J. Kricka, M. C. Lambert, and A. Ledwith, *J. Chem. Soc., Perkin Trans. I*, **1974,** 52.
659. V. A. Porai-Koshits, I. Y. Kvitko, and O. V. Favorskii, *Zh. Org. Khim.*, **1,** 1516 (1965).
660. Japanese Kokai 72 14,115; *Chem. Abstr.*, **77,** 34604e (1972).
661. M. Thiel and K. Stach, *Monatsh. Chem.*, **93,** 1080 (1962).
662. Japanese Patent 11,874; *Chem. Abstr.*, **65,** 13671d (1966).
663. Japanese Patent 12,102; *Chem. Abstr.*, **65,** 16951c (1966).
664. German Patent 1,113,935; *Chem. Abstr.*, **56,** 11576f (1962).
665. U.S. Patent 3,192,197; *Chem. Abstr.*, **59,** 1606g (1963).
666. Swiss Patent 408,019; *Chem. Abstr.*, **66,** 18683x (1967).
667. Swiss Patent 403,770; *Chem. Abstr.*, **65,** 13670a (1966).
668. Swiss Patent 403,768; *Chem. Abstr.*, **65,** 13669e (1966).
669. U.S. Patent 3,446,798; *Chem. Abstr.*, **69,** 19040j (1968).
670. Swiss Patent 451,162; *Chem. Abstr.*, **69,** 96510q (1968).
671. British Patent 1,114,970; *Chem. Abstr.*, **69,** 86837n (1968).
672. L. J. Kricka, M. C. Lambert, and A. Ledwith, *J. Chem. Soc., Chem. Commun.*, **1973,** 244.
673. M. R. Bendall, J. B. Bremner, and J. F. W. Fay, *Aust. J. Chem.*, **25,** 2451 (1972).
674. U.S. Patent 3,373,168; *Chem. Abstr.*, **65,** 12180f (1966).
675. British Patent 1,149,508; *Chem. Abstr.*, **71,** 22040f (1969).
676. R. W. Woodward, K. Baldzer, and D. W. Boykin, Jr., *J. Med. Chem.*, **14,** 1131 (1971).
677. C. J. Cattanach, A. Cohen, and B. Heath-Brown, *J. Chem. Soc. Perkin Trans. I*, **1973,** 1041.
678. B. A. Porai-Koshits, E. N. Sofino, and I. Ya. Kvitko, *Zh. Obshch. Khim.*, **34,** 2094 (1964).
679. U.S. Patent 3,178,406; *Chem. Abstr.*, **57,** 15082h (1962).

680. U.S. Patent 3,196,148; *Chem. Abstr.*, **57**, 15082h (1962).
681. German Patent 2,345,972; *Chem. Abstr.*, **81**, 3785t (1974).
682. Swiss Patent 430,723; *Chem. Abstr.*, **68**, 39497w (1968).
683. French Patent 1,463,803; *Chem. Abstr.*, **69**, 67252b (1968).
684. Swiss Patent 506,529; *Chem. Abstr.*, **75**, 140717n (1971).
685. C. R. Ellefson, L. Swenton, R. H. Bible, Jr., and P. M. Green, *Tetrahedron*, **32**, 1081 (1976).
686. U.S. Patent 3,373,153; *Chem. Abstr.*, **69**, 43898r (1968).
687. U.S. Patent 3,600,391; *Chem. Abstr.*, **70**, 96823q (1969).
688. German Patent 1,801,523; *Chem. Abstr.*, **71**, 112976v (1970).
689. Belgian Patent 804,985; *Chem. Abstr.*, **81**, 13560e (1974).
690. U.S. Patent 3,373,154; *Chem. Abstr.*, **69**, 35977a (1968).
691. U.S. Patent 3,764,684; *Chem. Abstr.*, **78**, 136256j (1973).
692. U.S. Patent 3,790,675; *Chem. Abstr.*, **80**, 82930s (1974).
693. U.S. Patent 3,959,300; *Chem. Abstr.*, **85**, 160058h (1976).
694. British Patent 1,449,331; *Chem. Abstr.*, **81**, 136132z (1974).
695. Swiss Patent 408,021; *Chem. Abstr.*, **66**, 28688j (1967).
696. French Patent 2,278,341; *Chem. Abstr.*, **85**, 149124j (1976).
697. German Patent 1,907,670; *Chem. Abstr.*, **72**, 55287r (1970)
698. Japanese Patent 75 17,074; *Chem. Abstr.*, **84**, 59244v (1976).
699. British Patent 668,659; *Chem. Abstr.*, **47**, 5460e (1953).
700. British Patent 778,936; *Chem. Abstr.*, **51**, 18016h (1957).
701. British Patent 804,193; *Chem Abstr.*, **53**, 15111c (1959).
702. R. M. Jacob and M. Messer, *C. R. Acad. Sci. Paris*, **252**, 2117 (1961).
703. U.S. Patent 3,156,692; *Chem. Abstr.*, **56**, 1437b (1962).
704. U.S. Patent 3,036,064; *Chem. Abstr.*, **57**, 12446h (1962).
705. K. Stach, M. Thiel, and F. Bickelhaupt, *Monatsh. Chem.*, **93**, 1090 (1962).
706. Swiss Patent 366,542; *Chem. Abstr.*, **59**, 10082a (1963).
707. U.S. Patent 3,225,032; *Chem. Abstr.*, **59**, 11453g (1963).
708. U.S. Patent 3,091,615; *Chem. Abstr.*, **59**, 12825b (1963).
709. Swiss Patent 368,181; *Chem. Abstr.*, **60**, 1717f (1964).
710. Swiss Patent 368,182; *Chem. Abstr.*, **60**, 1717g (1964).
711. Swiss Patent 368,183; *Chem. Abstr.*, **60**, 1718a (1964).
712. Swiss Patent 369,453; *Chem. Abstr.*, **60**, 2976c (1964).
713. Swiss Patent 368,496; *Chem. Abstr.*, **60**, 8046c (1964).
714. British Patent 943,276; *Chem. Abstr.*, **61**, 1815e (1964).
715. U.S. Patent 3,125,576; *Chem. Abstr.*, **61**, 7027h (1964).
716. U.S. Patent 3,503,314; *Chem. Abstr.*, **61**, 8284d (1964).
717. Swiss Patent 371,800; *Chem. Abstr.*, **60**, 13232a (1964).
718. Swiss Patent 373,758; *Chem. Abstr.*, **61**, 8323a (1964).
719. Swiss Patent 374,680; *Chem. Abstr.*, **61**, 14647b (1964).
720. Swiss Patent 375,722; *Chem. Abstr.*, **61**, 6998e (1964).
721. Swiss Patent 376,508; *Chem. Abstr.*, **61**, 9478h (1964).
722. F. Sparatore and V. Boido, *Ann. Chim.* (Rome), **54**, 591 (1964).
723. U.S. Patent 3,320,129; *Chem. Abstr.*, **62**, 7738c (1965).

724. U.S. Patent 3,444,159; *Chem. Abstr.*, **62,** 7738c (1965).
725. U.S. Patent 3,505,314; *Chem. Abstr.*, **62,** 7740a (1965).
726. U.S. Patent 3,345,361; *Chem. Abstr.*, **62,** 14702e (1965).
727. Belgian Patent 642,903; *Chem. Abstr.*, **62,** 16216f (1965).
728. Swiss Patent 383,978; *Chem. Abstr.*, **62,** 16217b (1965).
729. French Patent M2523; *Chem. Abstr.*, **63,** 2982f (1965).
730. L. Toldy, J. Borsy, I. Toth, and M. Fekete, *Acta Chim. Acad. Sci. Hung.*, **43,** 253 (1965).
731. U.S. Patent 3,454,554; *Chem. Abstr.*, **63,** 14830e (1965).
732. U.S. Patent 3,454,698; *Chem. Abstr.*, **63** 14830e (1965).
733. Swiss Patent 390,259; *Chem. Abstr.*, **63,** 14831d (1965).
734. L. Toldy, I. Toth, M. Fekete, and J. Borsy, *Acta Chim. Acad. Sci. Hung.*, **44,** 301 (1965).
735. Netherland Appl. 65 00,845; *Chem. Abstr.*, **64,** 3506g (1966).
736. U.S. Patent 3,337,534; *Chem. Abstr.*, **64,** 9696d (1966).
737. Swiss Patent 395,103; *Chem. Abstr.*, **64,** 9698b (1966).
738. U.S. Patent 3,407,256; *Chem. Abstr.*, **64,** 17559c (1966).
739. U.S. Patent, 3,329,683; *Chem. Abstr.*, **64,** 17559c (1966).
740. German Patent 1,745,668; *Chem. Abstr.*, **65,** 8886e (1966).
741. Japanese Patent 66 7,587; *Chem. Abstr.*, **65,** 8886f (1966).
742. U.S. Patent 3,258,459; *Chem. Abstr.*, **65,** 13743b (1966).
743. Japanese Patent 66 11,986; *Chem. Abstr.*, **65,** 16951g (1966).
744. Swiss Patent 407,128; *Chem. Abstr.*, **65,** 20110e (1966).
745. Swiss Patent 407,127; *Chem. Abstr.*, **66,** 2476w (1967).
746. Swis Patent 408,018; *Chem. Abstr.*, **66,** 10857m (1967).
747. Netherlands Appl. 66 02,482; *Chem. Abstr.*, **66,** 85706v (1967).
748. U.S. Patent 3,511,839; *Chem. Abstr.*, **67,** 11503a (1967).
749. M. A. Davis, F. Herr, R. A. Thomas, and M. P. Charest, *J. Med. Chem.*, **10,** 627 (1967).
750. L. H. Werner, S. Ricca, A. Rossi, and G. De Stevens, *J. Med. Chem.*, **10,** 575 (1967).
751. U.S. Patent 3,326,896; *Chem. Abstr.*, **68,** 39498x (1968).
752. Swiss Patent 430,724; *Chem. Abstr.*, **68,** 95715c (1968).
753. U.S. Patent 3,458,499; *Chem. Abstr.*, **71,** 81221p (1969).
754. U.S. Patent 3,574,204; *Chem. Abstr.*, **72,** 12753x (1970).
755. U.S. Patent 3,651,052; *Chem. Abstr.*, **72,** 12753x (1970).
756. U.S. Patent 3,668,210; *Chem. Abstr.*, **72,** 43501t (1970).
757. U.S. Patent 3,654,270; *Chem. Abstr.*, **72,** 66954g (1970).
758. German Patent 1,920,170; *Chem. Abstr.*, **72,** 100722s (1970).
759. Netherlands Appl. 68, 00363; *Chem. Abstr.*, **72,** 111316p (1970).
760. U.S. Patent 3,637,660; *Chem. Abstr.*, **72,** 111489x (1970).
761. M. Nakanishi, C. Tashiro, T. Munakata, K. Araki, T. Tsumagari, and H. Imamura, *J. Med. Chem.*, **13,** 644 (1970).
762. U.S. Patent 3,646,037; *Chem. Abstr.*, **74,** 13179t (1971).
763. U.S. Patent 3,767,796; *Chem. Abstr.*, **74,** 13179t (1971).
764. British Patent 1,252,320; *Chem. Abstr.*, **74,** 141581e (1971).
765. Japanese Patent 70 41,391; *Chem. Abstr.*, **75,** 20232d (1971).

766. Japanese Patent 71 14,903; *Chem. Abstr.*, **75**, 48935 (1971).
767. U.S. Patent 3,591,691; *Chem. Abstr.*, **75**, 143994m (1971).
768. U.S. Patent 3,637,713; *Chem. Abstr.*, **76**, 153627d (1972).
769. Japanese Patent 71 39,710; *Chem. Abstr.*, **76**, 34135g (1972).
770. U.S. Patent 3,337,538; *Chem. Abstr.*, **76**, 25318s (1972).
771. Japanese Patent 72 04,069; *Chem. Abstr.*, **76**, 140577n (1972).
772. Japanese Patent 71 43,795; *Chem. Abstr.*, **76**, 59649r (1972).
773. British Patent 1,314,996; *Chem. Abstr.*, **76**, 113091 (1972).
774. U.S Patent 3,668,210; *Chem. Abstr.*, **77**, 87625u (1972).
775. British Patent 1,282,405; *Chem. Abstr.*, **77**, 114416k (1972).
776. U.S. Patent 3,755,315; *Chem. Abstr.*, **77**, 114430k (1972).
777. Danish Patent 121,167; *Chem. Abstr.*, **77**, 48481z (1972).
778. T. Walle, H. Ehrsson, and C. Bogentoft, *Acta Pharm. Suec.*, **9**, 509 (1972).
779. Swiss Patent 532,597; *Chem. Abstr.*, **78**, 136335f (1973).
780. Swiss Patent 527,832; *Chem. Abstr.*, **78**, 16224x (1973).
781. Belgian Patent 766,052; *Chem. Abstr.*, **78**, 29780a (1973).
782. U.S. Patent 3,780,035; *Chem. Abstr.*, **78**, 29823s (1973).
783. Swiss Patent 419,142; *Chem. Abstr.*, **67**, 21851b (1967).
784. Swiss Patent 415,643; *Chem. Abstr.*, **67**, 21853d (1967).
785. Swiss Patent 423,783; *Chem. Abstr.*, **68**, 78161b (1967).
786. Swiss Patent 441,333; *Chem. Abstr.*, **69**, 35986c (1968).
787. J. F. Magalhaes, *Rev. Farm. Bioquim. Univ. Sao Paulo*, **10**, 141 (1972).
788. Japanese Patent 73 21,941; *Chem. Abstr.*, **79**, 105237s (1973).
789. Canadian Patent 924,305; *Chem. Abstr.*, **79**, 18740g (1973).
790. French Patent 2,186,249; *Chem. Abstr.*, **81**, 3969f (1974).
791. Belgian Patent 808,496; *Chem. Abstr.*, **82**, 149529e (1975).
792. Japanese Kokai 73 72,173; *Chem. Abstr.*, **80**, 82727f (1974).
793. Japanese Kokai 73 72,175; *Chem. Abstr.*, **80**, 95768v (1974).
794. Japanese Kokai 74 36,692; *Chem. Abstr.*, **81**, 49587h (1974).
795. Japanese Kokai 74 36,691; *Chem. Abstr.*, **81**, 120507q (1974).
796. Japanese Patent 74 21,146; *Chem. Abstr.*, **82**, 140160v (1975).
797. Japanese Patent 74 31,994; *Chem. Abstr.* **82**, 140159b (1975).
798. L. Bernardi and G. Bosisio, *J. Chem. Soc. Chem. Commun.*, **1974**, 690.
799. M. L. Post and O. Kennard, *Acta Crystallogr., Sect. B*, **B31**, 1008 (1975).
800. Japanese Kokai 74 111,087; *Chem. Abstr.*, **84**, 135504x (1976).
801. Japanese Patent 74 25,949; *Chem. Abstr.*, **84**, 43885 (1976).
802. Japanese Kokai 76 63,192; *Chem. Abstr.*, **86**, 5442f (1977).
803. U.S. Patent 3,947,447; *Chem. Abstr.*, **85**, 32885p (1976).
804. Y. Nagai, A. Maki, H. Kanda, K. Natsuka, and S. Umemoto, *Chem. Pharm. Bull.*, **24**, 1179 (1976).
805. East German Patent 116,822; *Chem. Abstr.*, **85**, 21466b (1976).
806. U.S. Patent 4,017,621; *Chem. Abstr.*, **86**, 16683m (1977).
807. G. Alfredsson, F. A. Wiesel, B. Fyro, and G. Sedvall, *Psychopharmacology*, **52**, 25 (1977).
808. East German Patent 119,591; *Chem. Abstr.*, **86**, 139885 (1977).
809. East German Patent 119,592; *Chem. Abstr.*, **86**, 106656j (1977).

810. East German Patent 119,593; *Chem. Abstr.*, **86,** 139886n (1977).
811. German Patent 2,628,558; *Chem. Abstr.*, **86,** 139887p (1977).
812. British Patent 965,909; *Chem. Abstr.*, **62,** 1790e (1965).
813. M. Claeys, G. Muscettola, and S. P. Markey, *Biomed. Mass Spectrom.*, **3,** 110 (1976).
814. U.S. Patent 2,666,051; *Chem. Abstr.*, **49,** 9041f (1955).
815. British Patent 802,775; *Chem. Abstr.*, **53,** 16167d (1959).
816. Swiss Patent 358,806; *Chem. Abstr.*, **59,** 2784b (1963).
817. Swiss Patent 400,162; *Chem. Abstr.*, **65,** 5476 (1966).
818. Japanese Patent 70 28,991; *Chem. Abstr.*, **74,** 22859s (1971).
819. Japanese Patent 70 25,696; *Chem. Abstr.*, **73,** 131014n (1970).
820. German Patent 2,144,893; *Chem. Abstr.*, **77,** 164725f (1972).
821. Japanese Patent 73 43,743; *Chem. Abstr.*, **81,** 13404q (1974).
822. German Patent 2,533,738; *Chem. Abstr.*, **85,** 46442d (1976).
823. Japanese Kokai 74 126,689; *Chem. Abstr.*, **85,** 21150u (1976).
824. Swiss Patent 581,115; *Chem. Abstr.*, **86,** 72476 (1977).
825. U.S. Patent 2,948,718; *Chem. Abstr.*, **55,** 1671b (1961).
826. U.S. Patent 2,762,796; *Chem. Abstr.*, **51,** 4447f (1957).
827. Swiss Patent 340,232; *Chem. Abstr.*, **56,** 463a (1962).
828. G. Tosolini, *Chem. Ber.*, **94,** 2731 (1961).
829. E. Ziegler, H. Junek, E. Nolken, K. Gelfert, and R. Salvador, *Monatsh. Chem.*, **92,** 814 (1961).
830. Swiss Patent 433,329; *Chem. Abstr.*, **68,** 105032u (1968).
831. U.S. Patent 3,523,944; *Chem. Abstr.*, **70,** 3860m (1969).
832. French Patent 1,515,586; *Chem. Abstr.*, **70,** 106406x (1969).
833. U.S. Patent 3,670,081; *Chem. Abstr.*, **70,** 106406x (1969).
834. M. Flammang, C. G. Wermuth, J. Schreiber, J. Barth, M. Herold, and J. Cahn, *Chim. Ther.*, **4,** 120 (1969).
835. Japanese Patent 70 06810; *Chem. Abstr.*, **72,** 132569p (1970).
836. U.S. Patent 3,510,476; *Chem. Abstr.*, **73,** 14726k (1970).
837. Japanese Patent 70 28,992; *Chem. Abstr.*, **74,** 22718v (1971).
838. E. Gipstein, E. M. Barrall II, and K. E. Bredfelt, *Anal. Colorimetry, Proc. Symp., 2nd,* 1970, p. 127.
839. German Patent 2,307,174; *Chem. Abstr.*, **79,** 136999w (1973).
840. T. Saraie, T. Ishiguro, K. Kawashima, and K. Morita, *Tetrahedron Lett.*, **1973,** 2121.
841. U.S. Patent, 3,842,091; *Chem. Abstr.*, **79,** 5322x (1973).
842. J. L Brazier, D. Beruoz, . M. Rouzioux, and A. Badinand, *J. Eur. Toxicol.*, **6,** 24 (1973).
843. R. J. Abraham, L. J. Kricka, and A. Ledwith, *J. Chem. Soc., Chem. Commun.*, **1973,** 282.
844. R. J. Abraham, L. J. Kricka, and A. Ledwith, *J. Chem. Soc., Perkin Trans. II,* **1974,** 1648.
845. K. M. Baker and A. Frigerio, *J. Chem. Soc., Perkin Trans. II,* **1973,** 648.
846. A. Frigerio and P. L. Morselli, *Boll. Chim. Farm.*, **112,** 429 (1973).
847. British Patent 1,398,234; *Chem. Abstr.*, **81,** 135997y (1974).
848. U.S. Patent 3,925,369; *Chem. Abstr.*, **81** 63503j (1974).
849. A. Frigerio, K. M. Baker, and G. Belvedere, *Anal. Chem.*, **45,** 1846 (1973).
850. A. Frigerio, *Cron. Chim.*, **44,** 12 (1974).

851. R. J. Perchalski, B. D. Andresen, and B. J. Wilder, *Clin. Chem.* (Winston Salem, N.C.), **22,** 1229 (1976).
852. Swiss Patent 403,767; *Chem. Abstr.,* **59,** 6370h (1964).
853. W. A. Remers, E. N. Greenblatt, L. Ellenbogen, and M. J. Weiss, *J. Med. Chem.,* **14,** 331 (1971).
854. R. G. Cooke and I. M. Russell, *Tetrahedron Lett.,* **1968,** 4587.
855. W. A. Denne and M. F. Mackay, *Tetrahedron,* **26,** 4435 (1970).
856. K. M. Baker, A. Frigerio, P. L. Morselli, and G. Pifferi, *J. Pharm. Sci.,* **62,** 475 (1973).
857. S. N. Frank, A. J. Bard, and A. Ledwith, *J. Electrochem. Soc.,* **122,** 898 (1975).
858. P. Beresford, D. H. Iles, L. J. Kricka, and A. Ledwith, *J. Chem. Soc., Perkin Trans. I,* **1974,** 276.
859. A. Neugebauer and S. Bamberger, *Chem. Ber.,* **107,** 2362 (1974).
860. S. Bamberger, D. Hellwinkel, an F. A. Neugebauer, *Chem. Ber.,* **108,** 2416 (1975).
861. F. N. Pirnazarova, *Uzb. Khim. Zh.,* **14,** 62 (1970).
862. British Patent 1,428,481; *Chem. Abstr.,* **85,** 46267a (1976).
863. U.S. Patent 3,829,431; *Chem. Abstr.,* **79,** 126480f (1973).
864. L. Toscano, G. Grisanti, G. Fioriello, E. Seghetti, A. Bianchetti, G. Bossoni, and M. Riva, *J. Med. Chem.,* **19,** 208 (1976).
865. U.S. Patent 3,931,223; *Chem. Abstr.,* **84,** 90026z (1976).
866. U.S. Patent 3,983,123; *Chem. Abstr.,* **86,** 72613a (1977).
867. L. Toscano, E. Seghetti, and G. Fioriello, *J. Heterocyclic Chem.,* **13,** 475 (1976).
868. U.S. Patent 3,950,425; *Chem. Abstr.,* **82,** 156133z (1975).
869. Netherlands Appl. 73 06,069 (1973); *Chem. Abstr.,* **82,** 170732a (1975).
870. A. M. Monro, R. M. Quinton, and T. I. Wrigley, *J. Med. Chem.,* **6,** 255 (1963).
871. A. Fulton and L. E. Lyons, *Aust. J. Chem.,* **21,** 873 (1968).
872. J. Lagubeau, R. Crockett, and P. Mesnard, *Bull. Soc. Pharm. Bordeaux,* **110,** 10 (1971).
873. A. De Leenheer, *J. Assoc. Off. Anal. Chem.,* **56,** 105 (1973).
874. J. Itier and A. Casadevall, *Bull. Soc. Chim., France,* **1969,** 2355.
875. J. Masse, A. Chauvet, R. Malaviolle, and F. Sabon, *Trav. Soc. Pharm. Montpellier,* **36,** 81 (1976).
876. H. Auterhoff and J. Kuhn, *Arch. Pharm.* (Weinheim, Ger.), **306,** 241 (1973).
877. J. R. Huber and J. E. Adams, *Ber. Bunsenges Phys. Chem.,* **78,** 217 (1974).
878. G. Cauquis, O. Chalvet, and Y. Thibaud, *Bull. Soc. Chim, France,* **1971,** 1015.
879. H. Gusten, L. Klasino, T. Toth, and J. V. Knop, *J. Electron Spectrosc. Rel. Phenom.,* **8,** 417 (1976).
880. H. J Haink and J. R. Huber, *J. Mol. Spectrosc.,* **60,** 31 (1976).
881. J. E. Adams, W. W. Mantulin, and J. R. Huber, *J. Am. Chem. Soc.,* **95,** 5477 (1973).
882. V. N. Sharma, R. L. Mital, S. P. Banerjee, and H. L. Sharma, *J. Med. Chem.,* **14,** 68 (1971).
883. J. Gonzalez and J. I. Fernandez-Alonso, *An. Quim.,* **66,** 919 (1970).
884. S. Chan, C. M. Gooley, and H. Keyzer, *Tetrahedron Lett.,* **1975,** 1193.
885. J. Poirot-Lagubeau, P. Mesnard, P. Gerval, and E. Frainnet, *Ann. Pharm. France,* **33,** 279 (1975).
886. G. Eckhardt, K. J. Goebel, G. Tondorf, and S. Goenechea, *Org. Mass Spectrom.,* **9,** 1073 (1974).
887. A. Frigerio, K. M. Baker, and P. L. Morselli, *Mass Spectrom. Biochem. Med. Symp.,* **1974,** p. 65.

888. P. Seiter, *Eur. J. Med. Chem.-Chim. Ther.*, **9,** 663 (1974).

889. N. Nambu, S. Sakurai, and T. Nagai, *Chem. Pharm. Bull.*, **23,** 1404 (1975).

890. A. De Leenheer and A. Heyndrickx, *J. Assoc. Off. Anal. Chem.*, **51,** 633 (1971).

891. R. A. Maxwell, P. D. Keenan, E. Chaplin, B. Roth, and S. B. Eckhardt, *J. Pharmacol. Exp. Ther.*, **166,** 320 (1969).

892. I. P. Lapin, T. A. Ksenofontova, I. Ya Kvitko, and B. A. Poray-Koshits, *Farmakol. Toksikol. (Moscow)*, **33,** 8 (1970).

893. A. Pardo, S. Vivas, F. Espana, and J. I. Fernandez-Alonso, *Afinidad*, **29,** 640 (1972).

894. R. Reynaud and P. Rumpf, *Bull. Soc. Chim, France*, **1963,** 1805.

895. R. W. Schmid, *Helv. Chim. Acta*, **45,** 1982 (1962).

896. C. A. Ponce, A. Devia, and J. I. Fernandez-Alonso, *Experientia, Suppl.*, **23,** 185 (1976).

897. A. P. Downing, W. D. Ollis, M. Nogradi, and I. O. Sutherland, *Jerusalem Symp. Quantum Chem. Biochem.*, **3,** 296 (1971).

898. L. C. Keifer, *Diss. Abstr. Int. B*, **36,** 6172 (1976).

899. I. Floderer and V. Horvathy, *Act Pharm. Hung.*, **35,** 98 (1965).

900. A. Gerardin, F. Abadie, and J. Laffont, *J. Pharm. Sci.*, **64,** 1940 (1975).

901. P. O. Lagerstrom and A. Theodorsen, *Acta Pharm. Suec.*, **12,** 429 (1975).

902. B. Herrmann, W. Schindler, and R. Pulver, *Med. Exp.*, **1,** 381 (1959).

903. B. Herrmann and R. Pulver, *Arch. Int. Pharmacodyn. Ther.*, **126,** 454 (1960).

904. B. Herrmann, *Helv. Physiol. Acta*, **21,** 402 (1963).

905. K. Adank and W. Hammerschmidt, *Chimia*, **18,** 361 (1964).

906. J. J. Thomas and L. Dryon, *J. Pharm. Belg.*, **19,** 481 (1964).

907. J. J. Thomas and L. Dryon, *J. Pharm. Belg.*, **22,** 163 (1967).

908. L. Vignoli, B. Cristau, F. Gouezo, and J. M. Vassallo, *Bull. Trav. Soc. Pharm. Lyon*, **9,** 277 (1965).

909. H. J. Weder and M. H. Bickel, *J. Chromatogr.*, **37,** 181 (1968).

910. J. A Marca and H. Muhlemann, *Pharm. Acta Helv.*, **46,** 558 (1971).

911. A. De Leenheer, *J. Chromatogr.*, **74,** 35 (1972).

912. J. H. Knox and J. Jurand, *J. Chromatogr.*, **103,** 311 (1975).

913. P. Hartvig, N. O. Ahnfelt, and K. E. Karlsson, *Acta Pharm. Suec.*, **13,** 181 (1976).

914. French Patent 1,359,676 (1964); *Chem. Abstr.*, **62,** 9147a (1965).

915. W. Wenner, *J. Org. Chem.*, **16,** 1475 (1951).

916. O. Chapman and J. Meinwald, *J. Org. Chem.*, **23,** 162 (1958).

917. U.S. Patent 2,619,484; *Chem. Abstr.*, **48,** 4011b (1954).

918. W. Wenner, *J. Org. Chem.*, **17,** 1451 (1952).

919. W. Wenner, *J. Org. Chem.*, **16,** 523 (1951).

920. British Patent 765,617; *Chem. Abstr.*, **51,** 12158 (1957).

921. A. Rieche, E. Hoft, and H. Schultze, *Justus Liebigs Ann. Chem.*, **697,** 188 (1966).

922. K. Kotera, M. Motomura, S. Miyazaki, T. Okada, Y. Hamada, R. Kido, K. Hirose, M. Eigyo, H. Jyoyama, and H. Sato, *Ann. Rep. Shionogi Res. Lab.*, **17,** 88 (1967); *Chem. Abstr.*, **68,** 105398t (1968).

923. M. Kolesnikova, I. Redkin, and A. Tochilkin, *Zh. Vses. Khim. Obshchest.*, **16,** 99 (1971); *Chem. Abstr.*, **75,** 20093 (1971).

924. U.S.S.R. Patent 382,622; *Chem. Abstr.*, **79,** 92009s (1973).

925. Swiss Patent 367,829; *Chem. Abstr.*, **59,** 12771h (1963).

926. U.S. Patent 3,038,896; *Chem. Abstr.*, **59,** 2841b (1963).

927. E. Matarasso-Tchiroukhine and R. Ouelet, *Bull. Soc. Chim. France,* **1959,** 630.
928. R. Cromartie, J. Harley-Mason, and D. Wannigarna, *J. Chem. Soc.,* **1958,** 1982.
929. B. Pecherer, R. Sunbury, and A. Brossi, *J. Org. Chem.,* **12,** 149 (1969).
930. U.S. Patent 3,751,487; *Chem. Abstr.,* **79,** 104395e (1973).
931. U.S. Patent 3,888,907; *Chem. Abstr.,* **83,** 114004g (1975).
932. R. Kreher and W. Gerhadt, *Angew. Chem. Int. Ed.,* **14,** 265 (1975).
933. R. Kreher and H. Pawelczyk, *Z. Naturforsch.,* **29b,** 425 (1974).
934. R. Kreher, H. Pawelczyk, and W. Gerhardt, *Z. Naturforsch.,* **30b,** 926 (1975).
935. L. Werner, S. Ricca, A. Rossi, and G. DeStevens, *J. Org. Chem.,* **10,** 575 (1967).
936. K. Hoffmann, P. Stenberg, C. Ljunggren, U. Svensson, J. Nilsson, O. Eriksson, A. Hartroorn, and R. Lunden, *J. Med. Chem.,* **18,** 278 (1975).
937. N. Mehta, R. Brooks, J. Strelitz, and J. Horodniak, *J. Org. Chem.,* **28,** 2843 (1963).
938. L. Carpino, *J. Am. Chem. Soc.,* **79,** 4427 (1957).
939. D. Ames, K. Hansen, and N. Griffiths, *J. Chem. Soc. Perkin I,* **1973,** 2818.
940. M. Akhtar and A. Oehschlager, *Tetrahedron,* **26,** 3245 (1970).
941. B. Coffin and R. Robbins, *J. Chem. Soc.,* **1965,** 1252.
942. J. Hawthorne, E. Mihelic, M Morgan, and M. Wilt, *J. Org. Chem.,* **28,** 2831 (1963).
943. U.S. Patent 3,075,966; *Chem. Abstr.,* **58,** 13928e (1963).
944. U.S. Patent 3,023,199; *Chem. Abstr.,* **56,** 15683h (1962).
945. N. S. Dokunikhin, A. Poplavski, and G. Migachev, *Zh. Vses. Khim. Obshchest.,* **21,** 709 (1976); *Chem. Abstr.,* **86,** 139683a (1977).
946. A. Warner and J. Neumeyer, *J. Pharm. Sci.,* **65,** 928 (1976).
947. U.S. Patent 2,693,495; *Chem. Abstr.,* **49,** 12546e (1955).
948. U.S. Patent 3,116,283; *Chem. Abstr.,* **55,** 16581c (1961).
949. P. Jeffs, J. Hansen, and G. Brine, *J. Org. Chem.,* **40,** 2883 (1975).
950. T. Cohen, A. Dinwoodie, and D. McKeever, *J. Org. Chem.,* **27,** 3385 (1962).
951. J. Huppatz, *Aust. J. Chem.,* **26,** 1307 (1973).
952. R. Grunder, L. San, and P. Kaul, *J. Pharm. Sci.,* **62,** 1204 (1973).
953. R. Taber, G. Daub, N. Hayes, and D. Ott *J. Heterocyclic Chem.,* **2,** 181 (1965).
954. R. Fraser, G. Boussard, D. Postescu, J. Whiting, and Y. Wigfield, *Can. J. Chem.,* **51,** 1109 (1973).
955. R. Fraser, R. Renaud, J. Saunders, and Y. Wigfield, *Can. J. Chem.,* **51,** 2433 (1973).
956. G. Beaven, M. Hall, M. Leslie, and E. Turner, *J. Chem. Soc.,* **1952,** 854.
957. G. Wittig and H. Zimmermann, *Chem. Ber.,* **86,** 629 (1953).
958. H. Joshua, R. Gans, and K. Mislow, *J. Am. Chem. Soc.,* **90,** 4884 (1968).
959. D. Fitts, M. Siegel, and K. Mislow, *J. Am. Chem. Soc.,* **80,** 480 (1958).
960. K. Mislow, M. Glass, H. Hopps, E. Simon, and G. Wahl, *J. Am. Chem. Soc.,* **86,** 1710 (1964).
961. S. Ahmed and M. Hall, *J. Chem. Soc.,* **1958,** 3043.
962. M. Hall and M. Harris, *J. Chem. Soc.,* **1960,** 490.
963. S. Ahmed and M. Hall, *J. Chem. Soc.,* **1960,** 4165.
964. D. M. Hall and T. Poole, *J. Chem. Soc.,* **1963,** 268.
965. D. M. Hall and T. Poole, *J. Chem. Soc. (B),* **1966,** 1034.
966. D. M. Hall and J. Insole, *J. Chem. Soc. Perkin II,* **1972,** 1164.
967. J. Insole, *J. Chem. Soc. Perkin II,* **1972,** 1168.

968. P. Browne and D. M. Hall, *J. Chem. Soc. Perkin I,* **1972,** 2717.
969. I. Sutherland and V. Ramsey, *Tetrahedron,* **21,** 3401 (1965).
970. B. Price, R. Smallman, and I. Sutherland, *Chem. Commun.,* **1966,** 319.
971. R. Fraser and Y. Wigfield, *Tetrahedron Lett.,* **1971,** 2515.
972. R. Fraser and L. Ng, *J. Am. Chem. Soc.,* **98,** 5895 (1976).
973. R. Bonner, M. De Armond, and G. Wahl, Jr., *J. Am. Chem. Soc.,* **94,** 988 (1972).
974. E. Braude and W. Forbes, *J. Chem. Soc.,* **1955,** 3776.
975. M. Dewar and N. Trinajstic, *Tetrahedron,* **26,** 4269 (1970).
976. U.S. Patent 3,551,414; *Chem. Abstr.,* **74,** 125484v (1971).
977. U.S.S.R. Patent 261,390; *Chem. Abstr.,* **73,** 14725j (1970).
978. G. Gream, J. Paice, and B. Uszynski, *Chem. Commun.,* **1970,** 895.
979. M Sainsbury and J. Wyatt, *J. Chem. Soc. Perkin I,* **1976,** 661.
980. G. Wahl, Jr., K. Wildonger, and J. Bordner, *Crystallogr. Struct. Commun.,* **2,** 267 (1973).
981. U.S. Patent 3,821,201; *Chem. Abstr.,* **78,** 136115j (1973).
982. H. Rapoport, A. Williams, O. Lowe, and W. Spooncer, *J. Am. Chem. Soc.,* **75,** 1125 (1953).
983. T. Kulev and G. Stepnova, *Izv. Tomsk. Politekhn. Inst.,* **111,** 16 (1961); *Chem. Abstr.,* **58,** 4461d (1963).
984. H. Underwood and E. Kochmann, *J. Am. Chem. Soc.,* **46,** 2069 (1924).
985. A. Wasfi, *J. Ind. Chem Soc.,* **43,** 723 (1966).
986. N. Riggs and S. Verma, *Aust. J. Chem.,* **23,** 1913 (1970).
987. R. Labriola and A. Felitte, *J. Org. Chem.,* **8,** 536 (1943).
988. A. Bistrzycki and K. Fassler, *Helv. Chim. Acta,* **6,** 519 (1923).
989. W. Warren and R. Briggs, *Chem. Ber.,* **64,** 26 (1931).
990. Ch. Courtot and J. Kronstein, *Chem. Ind. (Paris),* **45,** 66 (1941); *Chem. Abstr.,* **37,** 2369 (1943).
991. C. Weis, *Helv. Chim. Acta,* **51,** 1582 (1968).
992. C. Muth and W. Sung., *West. Va. Univ. Bull., Ser. 56,* **1955,** 12, *Chem. Abstr.,* **51,** 5736a (1957).
993. E. Atkinson and F. Granchelli, *J. Med. Chem.,* **17,** 1009 (1974).
994. C. Cole, H. Pan, M. Namkung, and L. Fletcher, *J. Med. Chem.,* **13,** 565 (1970).
995. H. Pan and L. Fletcher, *J. Med. Chem.,* **13,** 567 (1970).
996. B. Banerjee, *Ind. J. Chem.,* **12,** 4 (1935).
997. V. Shabrov and A. Pechenkin, *Tr. Nauchn. Konf. Tomskoe Otd. Vses. Khim. Obshch.,* **1969,** 94; *Chem. Abstr.,* **74,** 111737t (1971).
998. D. Mukakami and N. Tokura, *Bull. Soc. Chem. Japan,* **31,** 1044 (1958).
999. N. Tokura and S. Anazawa, *Sci. Repts., Research Insts., Tohoku Univ., Ser. A,* **9,** 239 (1957); *Chem. Abstr.,* **52,** 1964g (1958).
1000. R. Goto and T. Ogata, *Nippon Kagaku Zasshi,* **84,** 653 (1963); *Chem. Abstr.,* **60,** 449b (1964).
1001. L. Irrera, *Gazz. Chim. Ital.,* **65,** 464 (1935).
1002. A. Knevel and J. Wells, *J. Pharm. Sci.,* **59,** 1845 (1970).
1003. F. Demers, Jr. and G. Jenkins, *J. Am. Pharm. Assoc.,* **61,** 41 (1952).
1004. A. Botrel and C. Guerillot, *J. Chim. Phys., Phys. Chim. Biol.,* **71,** 1293 (1974); *Chem. Abstr.,* **82,** 139039f (1975).

1005. U.S. Patent 3,017,337; *Chem. Abstr.*, **55**, 7450c (1961).

1006. N. Ghosh, J. Roy, and D. Chatterjee; *Sci. Cult.*, **38**, 42 (1972); *Chem. Abstr.*, **77**, 139180s (1972).

1007. T. Sekiguchi, *Kogyo Kagaku Zasshi*, **73**, 1853 (1970); *Chem. Abstr.*, **74**, 100575n (1971).

1008. Japanese Patent 73 33,613; *Chem. Abstr.*, **81**, 65187h (1974).

1009. Japanese Kokai 72 21,431; *Chem. Abstr.*, **78**, 99036m (1973).

1010. A. Gluschke, S. Gotzky, J. Huang, G. Irmisch, E. Laves, O. Schrader, and G. Stier, *Chem. Ber.*, **63B**, 1308 (1930).

1011. D. H. R. Barton, P. G. Sammes, and G. G. Weingarten, *J. Chem. Soc. (C)*, **1971**, 729.

1012. I. Seki, *Chem. Pharm. Bull.*, **18**, 1269 (1970).

1013. W. Oppolzer, *J. Am. Chem. Soc.*, **93**, 3833 (1971).

1014. A. P. Boyakhchyan, L. L. Oganesyan, and G. T. Tatevosyan, *Arm. Khim. Zh.*, **24**, 1000 (1971).

1015. A. P. Boyakhchyan and G. T. Tatevosyan, *Arm. Khim. Zh.*, **26**, 44 (1973).

1016. A. P. Boyakhchyan, G. T. Taevosyan, R. S. Gyuli-Kevkhyan, and L. L. Oganesyan, *Sin. Geterotsikl. Soedin.*, **1972**, 71.

1017. R. R. Safrazbekyan and N. M. Savel'eva, *Biol. Zh. Arm.*, **26**, 74 (1973).

1018. A. Padwa and J. Smolanoff, *Tetrahedron Lett.*, **1974**, 33.

1019. A. Padwa, J. Smolanoff, and A. Tremper, *J. Am. Chem. Soc.*, **97**, 4682 (1975).

1020. U.S. Patent 3,205,221; *Chem. Abstr.*, **64**, 2114f (1966).

1021. A. P. Boyakhchyan, L. V. Khazhakyan, K. S. Lusararyan, and G. T. Tatevosyan, *Arm. Khim Zh.*, **26**, 1026 (1973).

1022. D. H. Jones, *J. Chem. Soc. (C)*, **1969**, 1642.

1023. E. Doomes, *J. Heterocyclic Chem.*, **13**, 371 (1976).

1024. C. Evans and D. Waite, *J. Chem. Soc. (C)*, **1971**, 1607.

1025. P. A. S. Smith and R. O. Kan, *J. Am. Chem. Soc.*, **82**, 4753 (1960).

1026. K. A. Cirigottis, E. Ritchie, and W. C. Taylor, *Aust. J. Chem.*, **27**, 2209 (1974).

1027. D. J. Anderson and A. Hassner, *J. Org. Chem.*, **38**, 2565 (1973).

1028. W. von E. Doering and R. A. Odum, *Tetrahedron*, **22**, 81 (1966).

1029. R. J. Sundberg and R. H. Smith, Jr., *J. Org. Chem.*, **36**, 295 (1971).

1030 J. Rigaudy, C. Igier, and J. Barcelo, *Tetrahedron Lett.*, **1975**, 3845.

1031. K. Dimroth, D. Holzner, and H. G. Aurich, *Chem. Ber.*, **98**, 3907 (1965).

1032. L. A. Paquette, D. E. Kuhla, J. H. Barrett, and L. M. Leichter, *J. Org. Chem.*, **34**, 2888 (1969).

1033. V. M. Dixit, J. M. Khanna, and N. Anand, *Ind. J. Chem.*, **13**, 893 (1975).

1034. H. T. Nagasawa and J. A. Elberling, *Tetrahedron Lett.*, **1966**, 5393.

1035. H. T. Nagasawa, J. A. Elberling, and P. L. Fraser, *J. Med. Chem.*,**14**, 501 (1971).

1036. I. Gutman, M. Milun, and N. Trinajstic, *J. Am. Chem. Soc.*, **99**, 1692 (1977).

1037. German Patent 1,803,785; *Chem. Abstr.*, **73**, P35400g (1970).

1038. Y. Kanaoka, K. Koyama, T. L Flippen, I. L. Karle, and B. Witkop, *J. Am. Chem. Soc.*, **96**, 4719 (1974).

1039. Y. Kanaoka, Y. Migita, K. Koyama, Y. Sato, H. Nakai, and T. Mizoguchi, *Tetrahedron Lett.*, **1973**, 1193.

1040. P. J. Wagner, *Acc. Chem. Res.*, **4**, 168 (1971).

1041. E. Gipstein, W. Hewett, and O. Need, *J. Polymer Sci.*, A-1, **8**, 3285 (1970).

1042. P. Rumpf and R. Reynaud, *Bull, Soc. Chim. France*, **1962**, 2241.

1043. H. O. House, "Modern Synthetic Reactions", 2nd Ed., W. A. Benjamin, Inc., Menlo Park, Calif., 1972, pp. 520–530.

1044. J. C. Roger, G. Rodgers, Jr., and A. Soo, *Clin. Chem.,* **19,** 590 (1973).

1045. R. Hüisgen, M. Seidel, G. Wallbüllich, and H. Knüpfer, *Tetrahedron,* **17,** 3 (1962).

1046. H. A. Lloyd, *Tetrahedron Lett.,* **1965,** 1761.

1047. W. M. Bright, H. A. Lloyd, and J. V. Silverton, *J. Org. Chem.,* **41,** 2454 (1976).

CHAPTER II

Azepine Ring Systems Containing Two Rings

GEORGE R. PROCTOR

Department of Pure and Applied Chemistry
University of Strathclyde
Glasgow, Scotland

I. INTRODUCTION

Approximately 80 ring systems containing a seven-membered ring of six carbon atoms and one nitrogen atom (excluding the bridge head nitrogen) have been selected for inclusion in this chapter. A substantial number of these are mentioned only briefly in the literature and have not been studied systematically. The practice of discussing chemical reactions, methods of synthesis, and physical measurements under separate headings has been retained only where it seemed a useful way of presenting the data. It was felt that only four groups of compounds, namely the three benzazepine ring systems and the 3-azabicyclo[3.3.2]nonanes merited tabulation. Compounds for which no physical data were cited in the literature are excluded from the tables.

Literature was searched via *Chemical Abstracts* and was covered fairly exhaustively up to and including Volume 94 (mid-1981). Compounds appearing after Volume 84 were not tabulated. The nomenclature used is that currently used in *Chemical Abstracts*.

II. Fused Ring Systems

1. 3, 7-Systems

Representatives of the three azabicyclo[5.1.0]octane systems shown below have been reported. For the sake of convenience, they will be discussed in a single section.

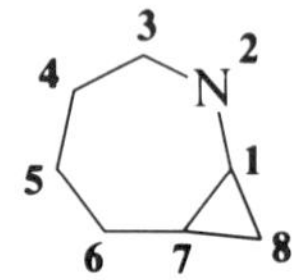

2-Azabicyclo[5.1.0]octanes

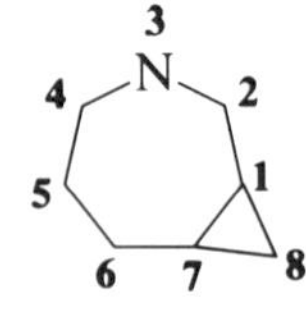

3-Azabicyclo[5.1.0]octanes

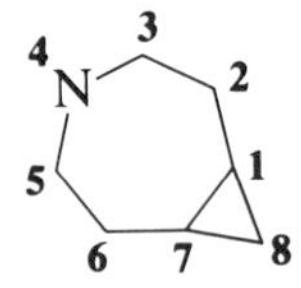

4-Azabicyclo[5.1.0]octanes

A. *Synthesis*

Azabicyclo[5.1.0]octane ring systems have been obtained by three approaches. In the first approach, reaction of cycloheptatriene with methyl or ethyl azidoformate gives mixtures of the 2-azabicyclo[5.1.0]octane **1** (R = Me or Et) and the 4-isomer **2** (R = Me or Et) in a 2:1 ratio, presumably by rearrangement of the initially formed aziridines (1–3). Of these, the ester **1** (R = Et) has been obtained as a pure crystalline solid (mp 45.5–46°C) (3). Dichlorocarbene addition to a caprolactim ether gave 23% of the 2-azabicyclo[5.1.0]octane derivative **3** (496).

The second approach gives 4-azabicyclo[5.1.0]octanes **4** (R = CO_2Me, COMe, Ts) by thermolysis (4) of the pyrazolo azepine derivatives **5**, which are themselves obtained by reaction of diazomethane with the appropriate azepines. Finally, Beckmann rearrangement of certain terpene oximes of the carane series (5) has yielded 4-azabicyclo[5.1.0]octane derivatives such as **6**. At present, this reaction provides the only known access to 3-azabicyclo[5.1.0]octanes such as **7**. Nmr and ir studies indicate chair comformations for several derivatives of **6** and **7** (497).

1 **2** **3** **4**

5 **6** **7**

B. *Reactions*

Thermal rearrangement of the 4-aza compound **2** (R = Et) at 220°C slowly gives **8,** whereas conversion of the 2-aza isomer **1** (R = Et) to **9** proceeds in only 30 min (50% yield) at this temperature. Dichlorocarbene insertion into **1** (R = Et) leads to the all-*trans* product (6), irradiation (7) gives **10,** and normal [4 + 2] cycloaddition with tetracyanoethylene and nitrosobenzene has been reported to yield both possible isomeric products (8). Reaction of bicyclodienes **1** and **2** (R = Et) with iron carbonyls has been shown to yield a series of bicyclononadienones (e.g., **11**) (498).

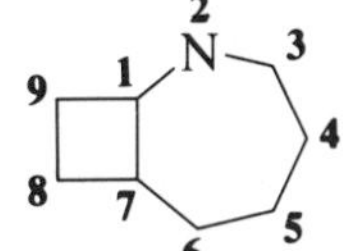

8

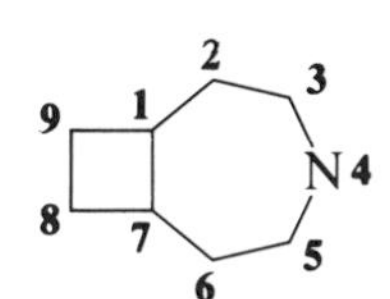

9

10

11

2. 4, 7-Systems

2-Azabicylo[5.2.0]nonane

4-Azabicyclo[5.2.0]nonane

3-Azabicyclo[5.2.0]nonane

Only members of 2-aza- and 4-azabicyclo[5.2.0]nonane ring systems are well known. Acylnitrene insertion into cyclooctatetraene gives the *N*-acyl-9-azabicyclo[6.1.0]nonatrienes **12** (R = CO_2Et, CO_2Me, $COCH_3$), which on gentle (56°C) thermolysis give the 4-azabicyclo[5.2.0]nonatrienes **13** (R = CO_2Et, CO_2Me, $COCH_3$) (9–11). At 76°C the 4-aza isomers **13** (R = CO_2Me, $COCH_3$) give, among other products, the isomeric 2-aza compounds **14** (R = CO_2Me, $COCH_3$) (10–12), whereas photolysis gives azonines (13). In **13**

12

13

14

($R = CO_2Et$) the ester group has been converted to CH_2OH, H, Me, $COCH_3$, and $CONMe_2$ by standard chemical methods (14).

Another approach to the 4-azabicyclo[5.2.0]nonane system involves photolytic conversion of **15** to **16** in diethylamine (15). 2-Azabicyclo-[5.2.0]nonane derivatives **17** (R = CN, CO_2Et) are available (16) by addition of acrylonitrile or ethyl acrylate to **18.** The 3-aza derivative **19** was recently (539) reported to arise by photolysis of compound **20.**

3. 5, 7-Systems

A. *Cyclopentazepines*

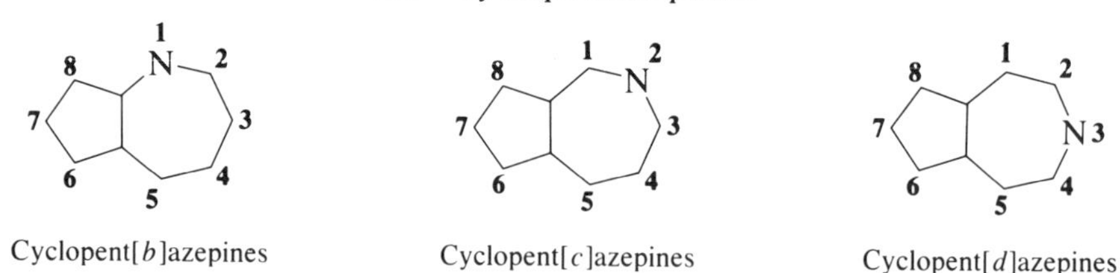

Cyclopent[*b*]azepines Cyclopent[*c*]azepines Cyclopent[*d*]azepines

Much of the interest in the cyclopentazepine ring systems stems from their relationship to azulenes. Chloramine reacts with the phenol **21** to give a 37% yield of a mixture of the cyclopent[*b*]azepine lactam **22** and the cyclopent[*c*]azepine isomer **23** in a 38:62 ratio. Nmr spectra of these compounds were particularly useful in their assignment of structure (17). Ring expansion of the azide **24** by photolysis in diethylamine (15) likewise gives a mixture of the isomeric products **25–27** in a 2:5:3 ratio.

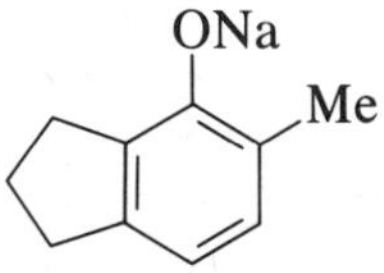

21

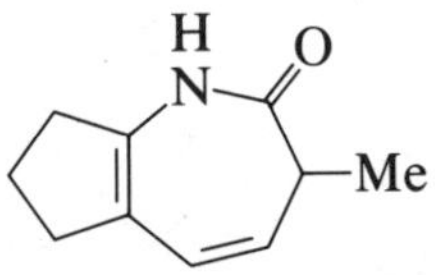

22

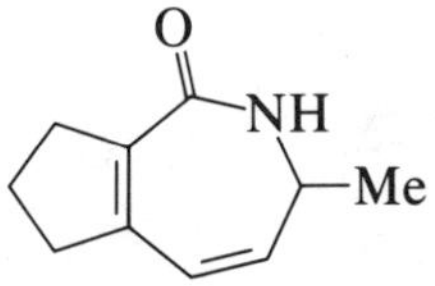

23

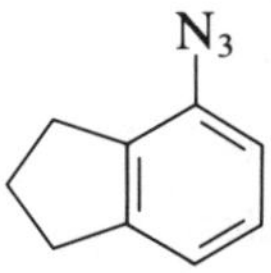

24

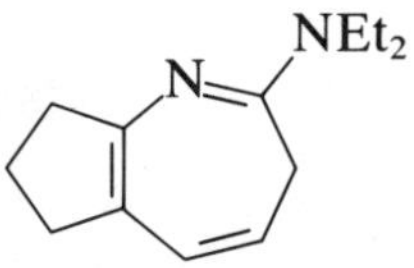

25

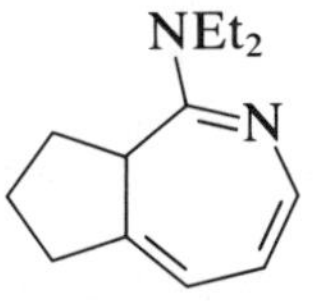

26

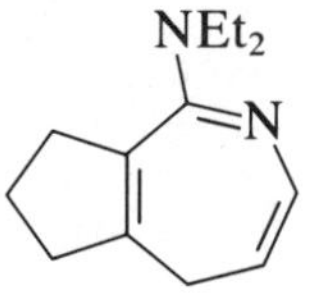

27

Another method of synthesis of cyclopent[*b*]azepines is illustrated by the thermal rearrangement of the tricyclic carbamate **28** at 180°C, which gives the azulenoid product **29.** Structure proof rests on the reduction of **29** to the perhydro compound, which was shown (18) to be identical to the reduction product of lactam **30.** The latter was formed, along with the isomeric lactam **31,** by Beckmann rearrangement (and methylation) of the oxime of ketone **32** (541). Recently (435), it has been shown that photolysis of *N*-cyclopentylsuccinimide gives a 45% yield of the keto lactam **33.**

Ferrocenyl derivative **34** was obtained by rearrangement and lithium aluminum hydride reduction of the appropriate *syn*-oxime (19). Although the *anti*-oxime failed to react analogously, the isomeric cyclopent[*c*]azepine structure was accessible (20) by reduction of the ferrocenyl lactam **35,** which can be prepared via the Schmidt reaction. Similar methods were used to obtain the cyclopent[*c*]azepine lactams **36** and **37** from the ketones **38** and **39,** respectively (21,22). In the latter instance, the isomeric lactam **40** was also found, along with several other products.

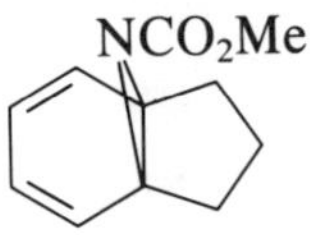

28

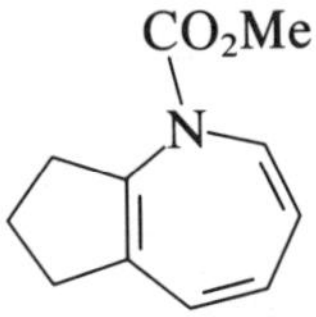

29

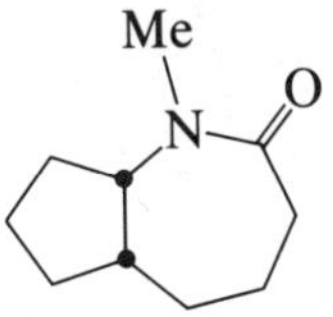

30

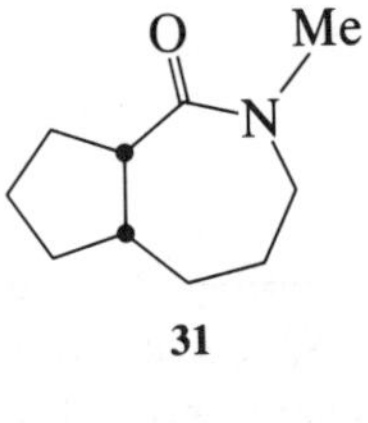

31

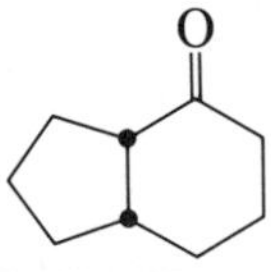

32

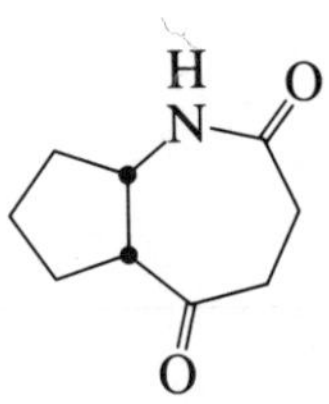

33

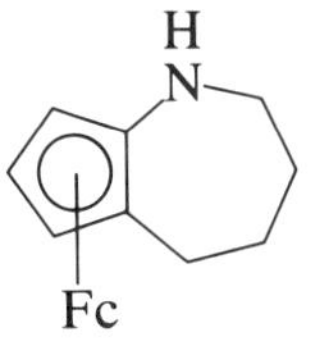

34 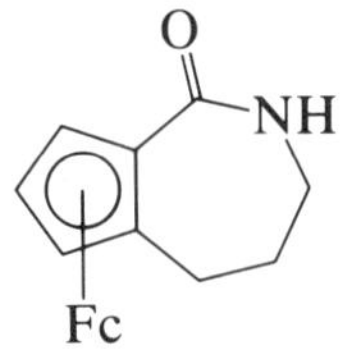 35 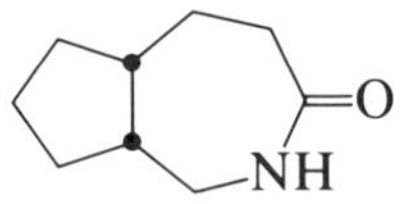 36 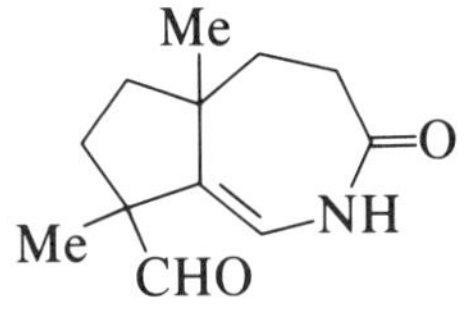 37

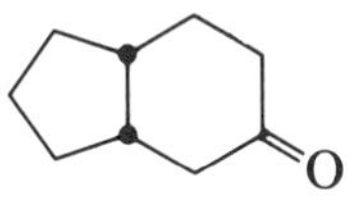

38 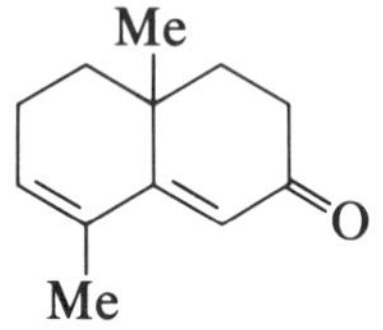 39 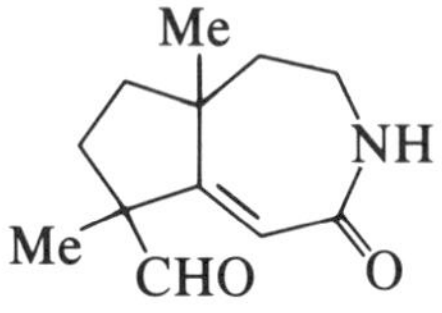 40

Only in the cyclopent[*c*]azepine series have azaazulenes been obtained consistently. Thus reaction of fulvene **41** with *N,N*-dimethylacetamidinium chloride (**42**) gave the 5-azaazulene **43** (R = NMe_2) (23). The phenyl analog (**43,** R = Ph) was obtained similarly (24). X-ray crystallographic (25) and mass spectral data (26) have been reported for compound **43,** as well as its reaction with hydrogen peroxide (27). Recently (499), it has been shown that the normal thermal cyclization of *N*-allylacrylamides to 3-vinyl-2-pyrrolidones can be diverted to yield azepin-2-one derivatives. In particular, compound **44** (R = H, Me) gives **45** (R = H, Me) in 65 and 70% yields, respectively, when heated in xylene with sodium hydride followed by hydrolysis.

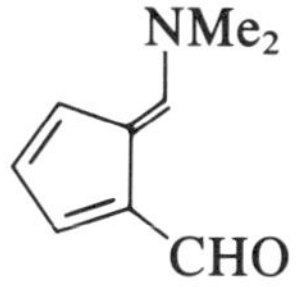

41 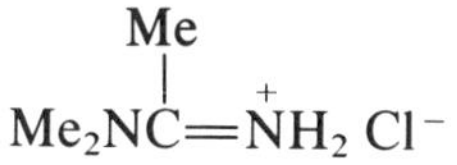 42 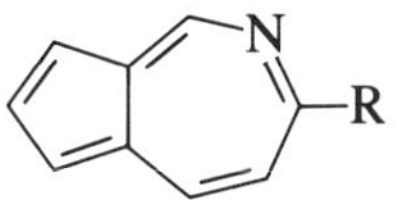 43

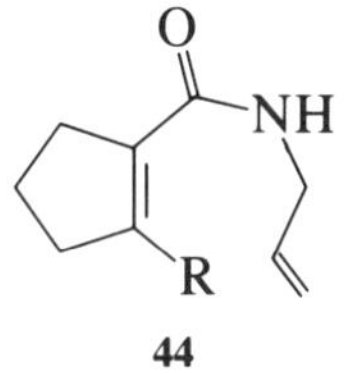

44 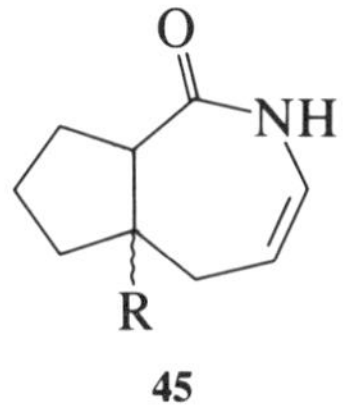 45

Several synthetic approaches to cyclopent[*d*]azepine derivatives are available. The earliest (28) involved Raney nickel reductive cyclization of dinitriles to the *cis* and *trans* isomers **46** and **47.** Subsequently, it was found that acid-induced cyclization (29) converts dinitrile **48** to the azepine **49.**

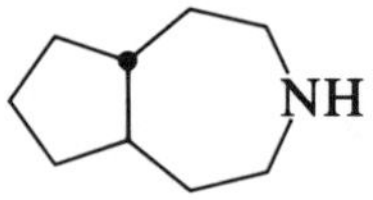

Reaction of the fulvene **50** with *N,N-bis*(carbethoxymethyl)aniline (**51**) yielded the azaazulene **52** (30). Photolysis of 5-azidoindane in diethylamine resulted in ring expansion to the azepines **53** and **54**, with the latter being the minor product. Cyclopent[*d*]azepine derivatives have been obtained by reaction of chlorosulfonyl isocyanate with certain vinylcyclopropanes (e.g., **55** [$n = 1$] → **56**) (500, 501) and by nitrene insertion into a bicyclo-[6.1.0]nonatriene derivative (502), but the latter method yielded by-products.

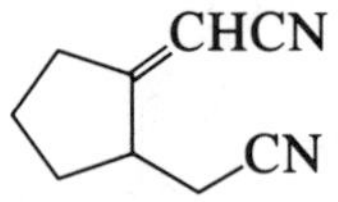

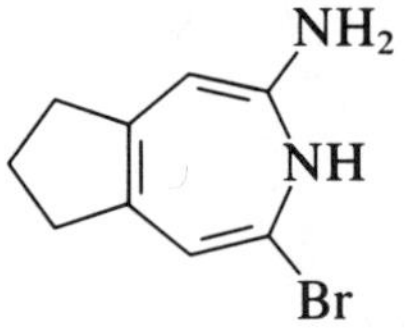

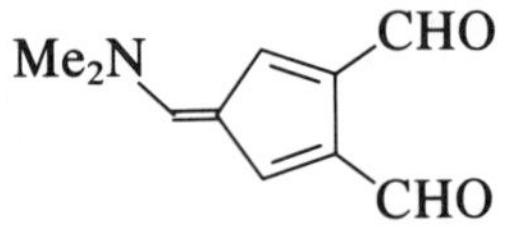

Although several studies (31–32) have predicted some degree of aromaticity for the 4-, 5-, and 6-azaazulenes, attempts to obtain these products by dehydrogenation have failed (21,28). X-ray crystallographic data for the 5-azaazulene **43** (R = NMe_2) reveal considerable C—C bond alternation (25). Two other 5-azaazulenes are known, and these were obtained by nondehydrogenative methods. The parent compound of the series (**43,** R = H) is a water-soluble violet solid (mp 34–35°C) (34); its ^{13}C-nmr spectrum has been reported (542). In a brief report (35a), 1,2,3-triphenyl-4-azaazulene (**57**) was obtained as a green solid in 4% yield by heating ketone **58** with pyrrole in the presence of *p*-toluenesulfonic acid. Analysis of the nmr spectrum shows

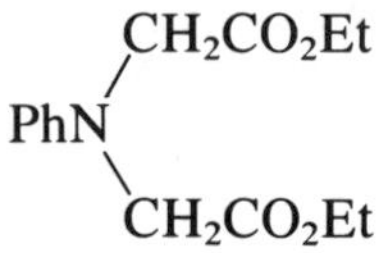

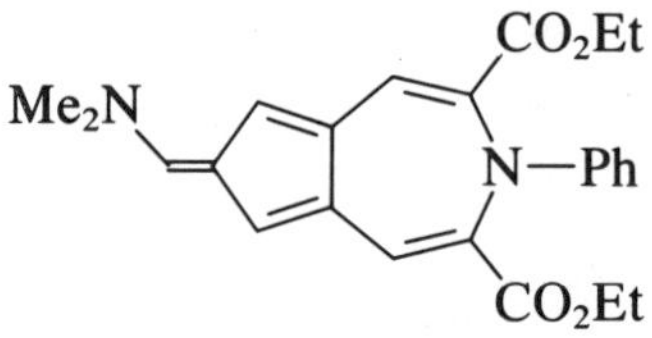

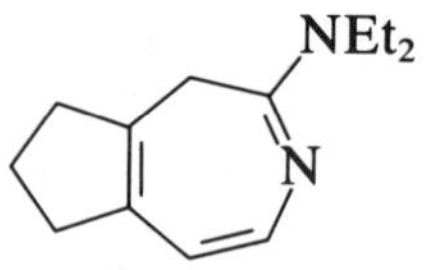

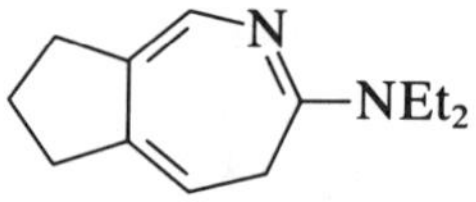

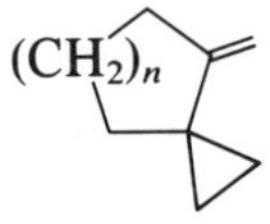

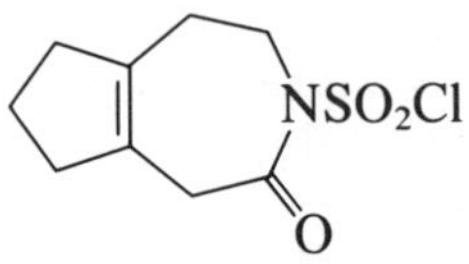

57 **58**

a four-spin system, and there seems little doubt about the correctness of the formulated structure (35b). Recently, synthesis of a 6-azaazulene derivative was reported (540), but the yield was only 4%.

B. *Furoazepines*

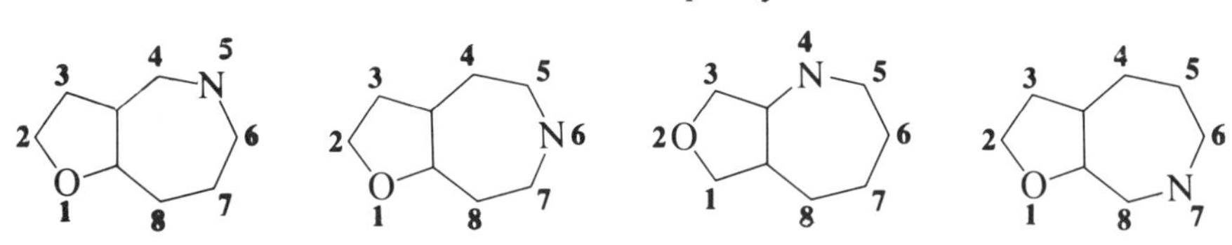

Of the ten possible furoazepine ring systems, only four have been reported. Furo[3,2-*c*]azepine derivatives have been obtained by the Schmidt reaction on furocyclohexanones, as exemplified by the preparation of lactam **59** from ketone **60**, and by Claisen rearrangement of certain azepine *O*-allyl ethers (543). The lactams have been reduced to amines, which have been substituted with various pharmacophoric groups (36,37). Furo[2,3-*d*]azepine **61** (R = $PhCH_2$) is said to be obtained from the silylated azepine **62** (R = $PhCH_2$) with malononitrile (38). The only known furo[3,4-*b*]azepine is compound **63**, which is formed in a simple lactone ring-closure reaction (39). Furo[2,3-*c*]azepine derivative **64** was obtained when ethyl acetoacetate reacted with lactam **65** (503).

59 **60** **61**

62 **63**

64

65

C. *Isoxazoloazepines*

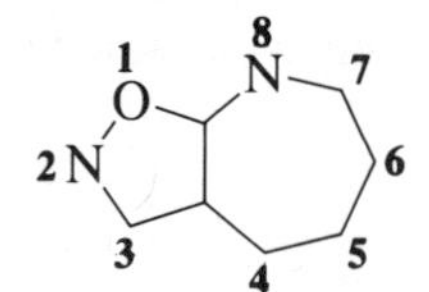

Isoxazolo[5,4-*b*]azepine

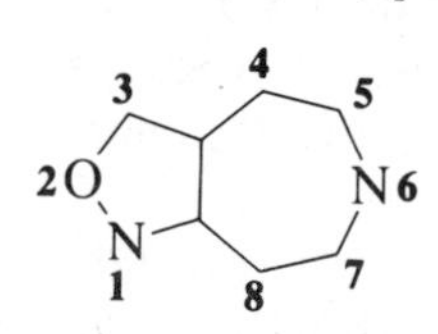

Isoxazolo[3,4-*d*]azepine

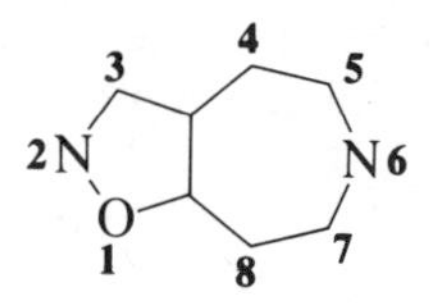

Isoxazolo[4,5-*d*]azepine

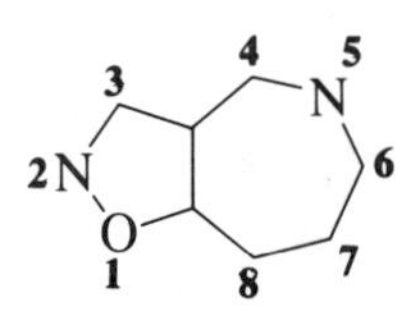

Isoxazolo[4,5-*c*]azepine

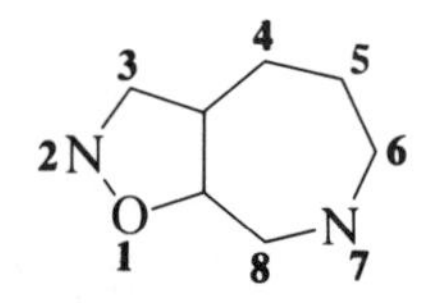

Isoxazolo[5,4-*c*]azepine

Isoxazolo[4,5-*b*]azepine

A member of the isoxazolo[5,4-*b*]azepine ring system known for 70 years is compound **66,** which was obtained by the successive treatment of carvone (**67**) with hydrogen cyanide and amyl nitrite (40,41). Isoxazoloazepines of [3,4-*d*] and [4,5-*d*] types were obtained from the β-keto esters **68** (R = Ac, CO_2Et, CH_2Ph). Thus treatment of **68** (R = CO_2Et) with benzylamine gave an enamine that was converted to **69** (R = CO_2Et). Decarboxylation of the latter compound yielded the zwitterion **70** as a hemihydrate (42). A ketal of

66

67

68

69

70

71

68 (R = CO_2Et) was treated with hydroxylamine, followed by hydrochloric acid, to obtain the isoxazolo[4,5-*d*]azepine derivative **71**, which was decarboxylated to furnish a zwitterion isomeric with **70**, whose structure has been confirmed by X-ray analysis (504).

The isomeric isoxazolo[4,5-*c*]- and [5,4-*c*]azepines **72** and **73** were obtained from appropriate hexahydroazepinone β-keto esters by reaction with hydroxylamine (505), while heating an oximino caprolactam derivative gave the isoxazolo[5,4-*c*]azepine **74** (506). Recently, isoxazolo[4,5-*b*]azepines **75** (R_1 and R_2 = various aryl groups) were obtained by stannous chloride treatment of suitably substituted 4-nitroisoxazoles (507), themselves obtained by condensation of chalcones with methyl nitroisoxazoles.

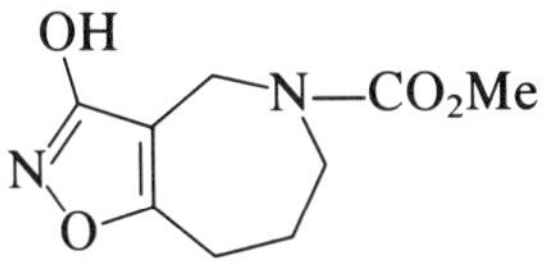

OH N—CO_2Me N O

72

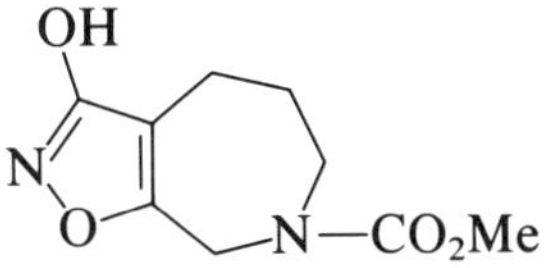

OH N O N—CO_2Me

73

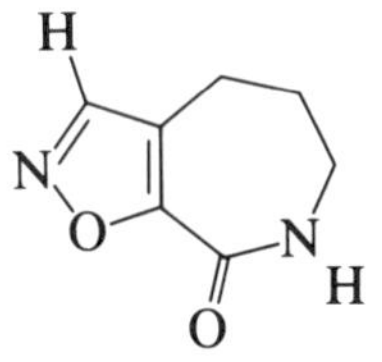

H N O N H O

74

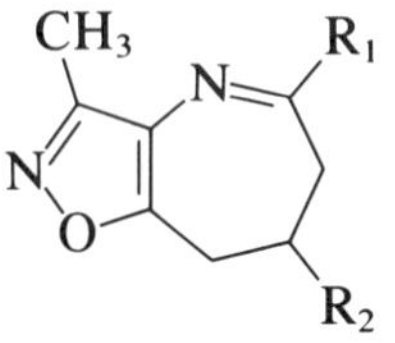

CH_3 N R_1 N O R_2

75

D. *Oxazoloazepines*

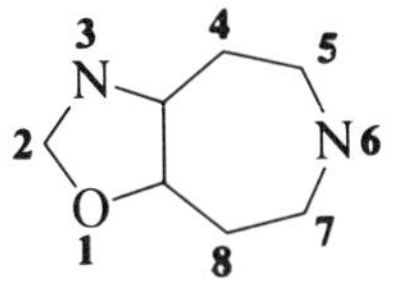

3 4 5 N N6 2 O 1 8 7

Oxazolo[4,5-*d*]azepine

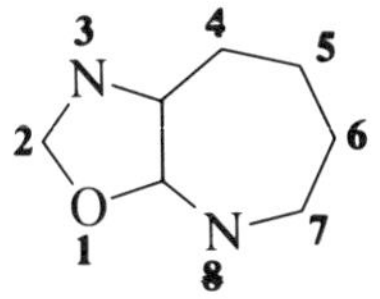

3 4 5 N 6 2 O N 7 1 8

Oxazolo[5,4-*b*]azepine

A method of synthesis of oxazolo[4,5-*d*]azepines has been reported (44). When α-bromoketones **76** were allowed to react with urea under fusion

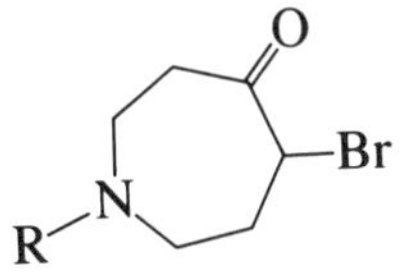

O Br R N

76

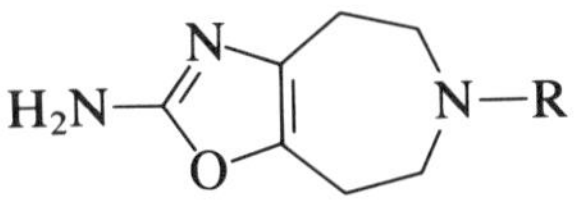

H_2N N O N—R

77

Me N O N COMe

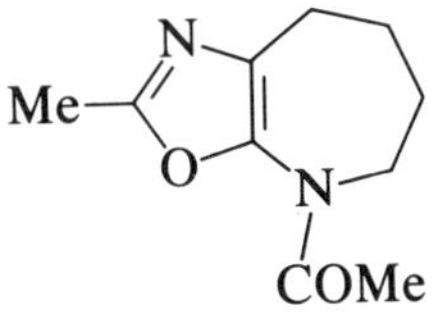

78

conditions, the products **77** were obtained. Although spectral data for the oxazolo[5,4-*b*]azepine **78** are mentioned (45), a synthesis does not appear.

E. *Pyrazoloazepines*

Pyrazolo[4,3-*c*]azepine

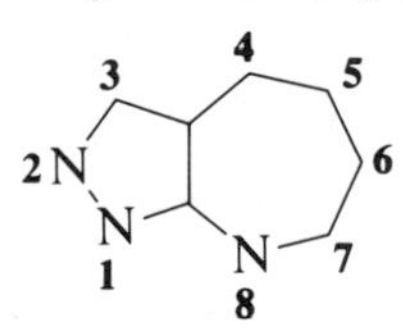

Pyrazolo[3,4-*b*]azepine

Pyrazolo[3,4-*d*]azepine

Pyrazolo[4,3-*b*]azepine

Pyrazolo[3,4-*c*]azepine

Examples of all the possible pyrazoloazepine ring systems are known. Pyrazolo[3,4-*b*]azepine (**79**) was obtained by reaction of the imino ether **80** with hydrazine (46). Several syntheses of pyrazolo[3,4-*d*]azepines have been reported. In the earliest of these (47), the hydroxymethylene ketone **81** (R_1 = CHOH, R_2 = CO_2Et) was condensed with hydrazine to give **82** (R_1 = H, R_2 = CO_2Et). In subsequent work, the azepine intermediates **81** (R_1 = CO_2Et; R_2 = PhCO, CH_3CO) were prepared by ring expansion of the corresponding piperidones and then treated with hydrazine to obtain the pyrazolone derivatives **82** (R_1 = OH; R_2 = PhCO, CH_3CO) (48,49). The isolation of compounds **5** has already been cited (4).

The only known pyrazolo[4,3-*c*]azepine is compound **83,** which was obtained by treatment of the amide ester **84** with thionyl chloride (50). Also from the same laboratory (51) is a report that pyrazolo[4,3-*b*]azepine **85** was obtained from ketone **86** via the Schmidt reaction. However, other authors (508) found that both possible isomers were obtained in this reaction. Pyrazolo [3,4-*c*]azepine derivative **87** was obtained by treatment of a 4-cyanocaprolactam derivative with hydrazine (506). 3-Aminopyrazoles react with ethyl levulinate to give pyrazolo[3,4-*b*]azepine derivatives (544).

79 **80** **81** **82**

83 **84**

85 **86** **87**

F. *Pyrroloazepines*

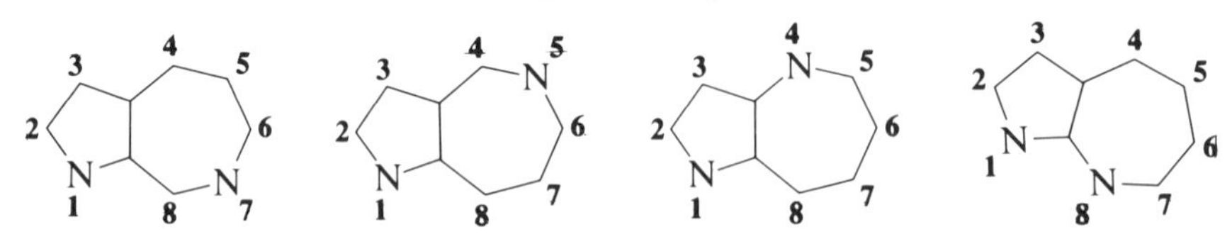

Pyrrolo[2,3-*c*]azepine Pyrrolo[3,2-*c*]azepine Pyrrolo[3,2-*b*]azepine Pyrrolo[2,3-*b*]azepine

Only four pyrroloazepine ring systems are known. Three groups of workers (52,53,509) have obtained pyrrolo[2,3-*c*]azepine derivatives by cyclization of the pyrrole amino esters, as exemplified by the preparation of **88** from **89.** Pyrrolo[3,2-*c*]azepines are available (54,36) by solvolysis of *syn*-4,5,6,7-tetrahydroindol-4-one oxime tosylate (**90**). The products (**91**) can be reduced and subjected to further substitution (36) or Mannich reactions (55). Solvolysis of the *anti* isomer of **90** yielded the pyrrolo[3,2-*b*]azepine **92,** which could likewise be reduced and converted to *N*-substituted derivatives (54,55). Pyrrolo[3,2-*b*]azepine derivatives can also be obtained by application of the Schmidt reaction, whereas Beckmann rearrangement gives pyrrolo[3,2-*c*]azepines (510). Some pyrrolo[2,3-*b*]azepine derivatives have recently been reported (545).

88 **89**

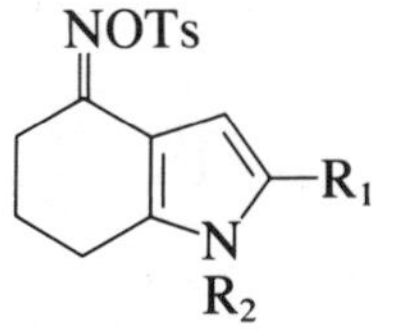

90 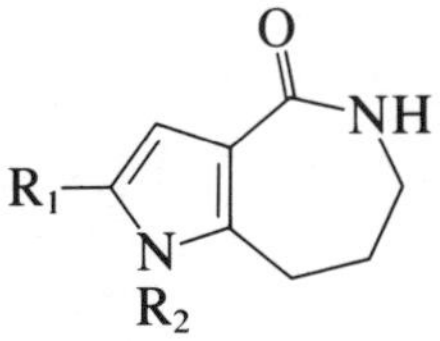 91 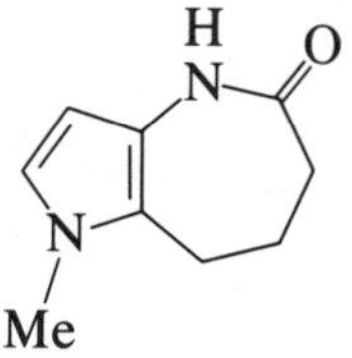 92

Pyrrolo[2,3-*c*]azepines (e.g., **93**) have recently been made by annelation of the five-membered ring onto azepine structures (546), and the dione **94** was demonstrated to arise by degradation of a substance obtained from the marine sponge *Phakellia flabellata* (547).

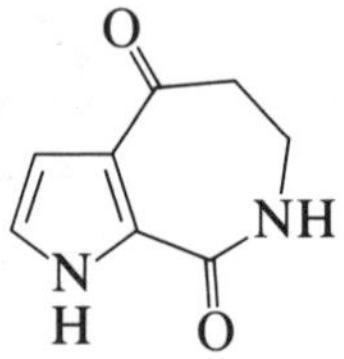

94 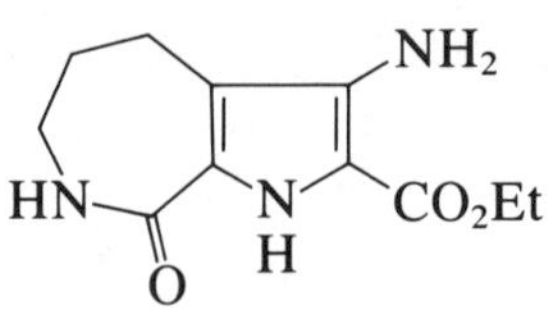 93

G. *Thiazoloazepines*

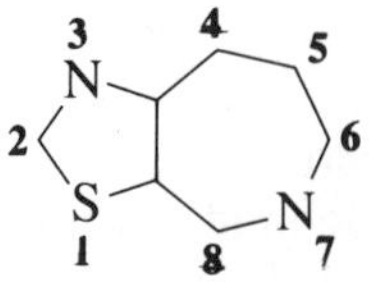

Thiazolo[5,4-*c*]azepine 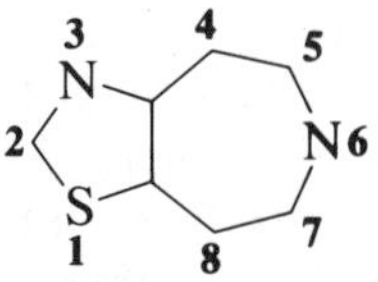 Thiazolo[4,5-*d*]azepine 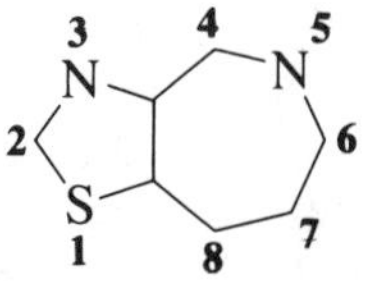 Thiazolo[4,5-*c*]azepine

Reaction of α-bromoketone **81** (R_1 = Br, R_2 = CO_2Et) with thioacetamide gave the thiazolo[5,4-*c*]azepine **95** (47,56), whereas condensation of the nonbrominated ketones **81** (R_1 = H; R_2 = $PhCH_2$, aryl) with thiourea yielded the thiazolo[4,5-*d*]azepines **96.** Thiazolo[4,5-*c*]azepine **97** was obtained from the reaction of thiourea with the bromolactam **65** (503). Suschitzky and coworkers (548) have succeeded in obtaining thiazolo[5,4-*c*]azepine derivatives (e.g., **98**) by photolysis of azidobenzothiazoles in diethylamine.

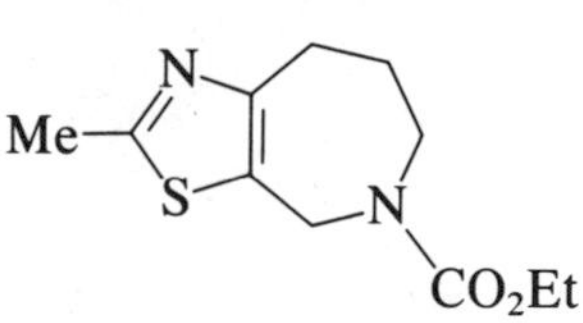

95 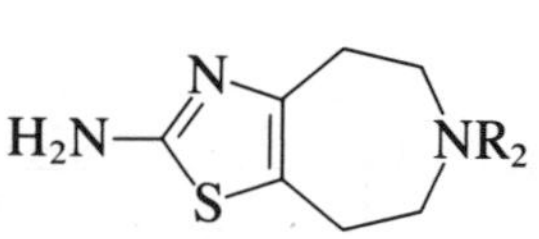 96

97 **98**

H. *Thienoazepines*

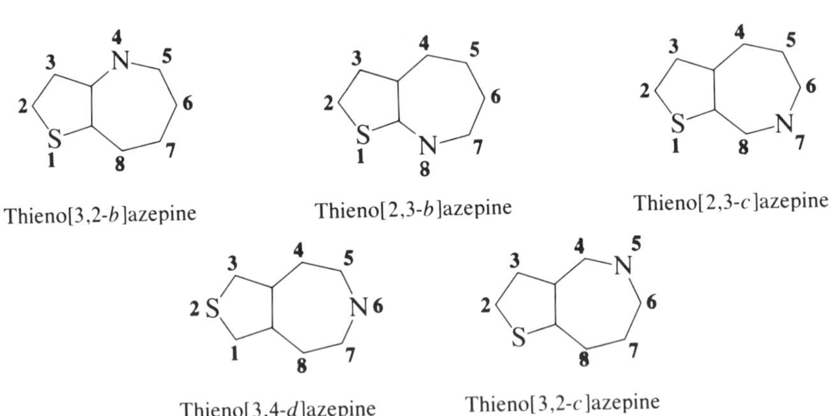
Thieno[3,2-*b*]azepine Thieno[2,3-*b*]azepine Thieno[2,3-*c*]azepine

Thieno[3,4-*d*]azepine Thieno[3,2-*c*]azepine

Of the ten possible thienoazepine ring systems, five are known. Thieno[2,3-*b*]azepines **99** (R = Ph, Me) were obtained by Dieckmann cyclization of the diesters **100** (R = Ph, Me) and were decarboxylated and detosylated by standard chemical methods (57). Thieno[2,3-*c*]azepine derivative **101** was prepared from the ketone precursor **102** via a Beckmann or Schmidt reaction, with none of the isomeric ring-enlargement product being reported (58).

99 **100**

101 **102** **103**

Photolysis of 6-azidobenzothiophene in diethylamine has given thieno[2,3-*c*]azepines such as **103** (R_1 = CO_2Me, R_2 = H; R_1 = R_2 = Br) (59).

Hydrogen bromide cyclization (29) of the appropriate dinitrile (60) gave the thieno[3,4-*d*]azepine **104.** A Schmidt reaction (36,37,60) was used to convert ketone **105** exclusively to thieno[3,2-*c*]azepine **106,** whereas Beckmann reaction on the oxime gave both **106** and the isomeric lactam **107** (R = Me) (61).

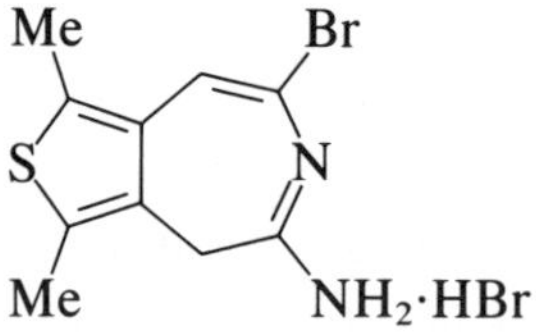

Me Br S N Me $NH_2 \cdot HBr$

104

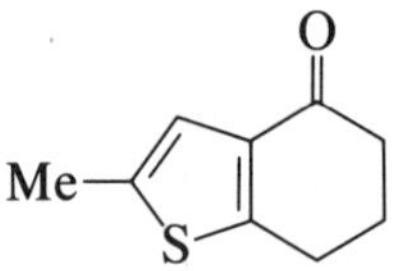

O Me S

105

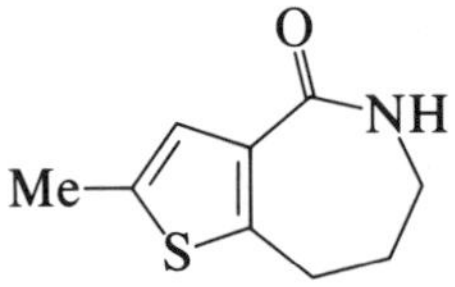

O NH Me S

106

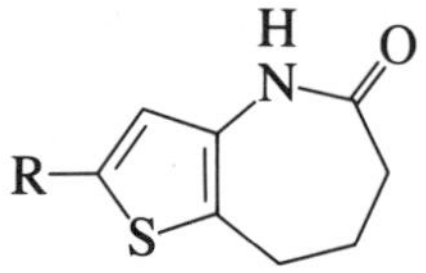

H N O R S

107

In the unsubstituted case, the thieno[3,2-*b*]azepine **107** (R = H) is the major product from the solvolysis of the corresponding oxime tosylate (62,63). Reduction of 3-nitro derivatives of this ring system gives the corresponding amines (61), and Raney nickel desulfurization of the latter affords a useful route to diamino acid derivatives. Infrared spectroscopy (64) and mass spectrometry (65) have been used to identify isomers in the [3,2-*c*] and [3,2-*b*] series. Apparently, the only example of intramolecular Friedel–Crafts cyclization in this series is preparation of the thieno[3,2-*c*]azepine **108** from the isothiocyanate **109** (511).

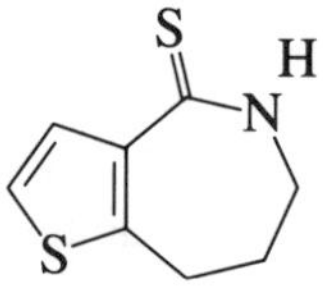

S H N S

108

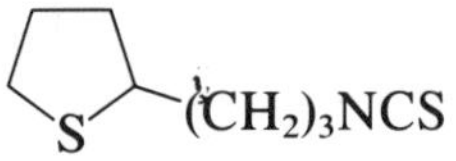

S $(CH_2)_3NCS$

109

Recently a thieno[3,2-*c*]azepine derivative (**110**) was reported to arise from the tricyclic molecule **111** (549). Thieno[3,2-*b*]azepine derivative **112** has been made by thermolysis of the appropriate *N*-thienovinylaziridine (**113,** R = thieno). This reaction has been used to build azepine rings onto other heterocycles (550).

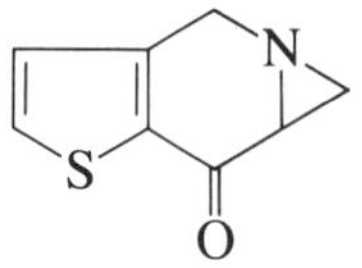

111 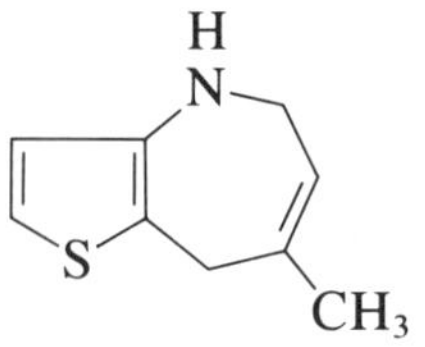 112 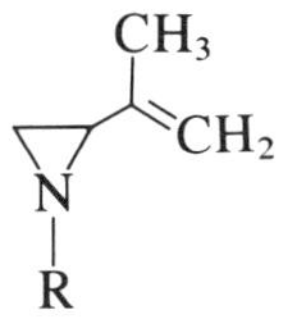 113 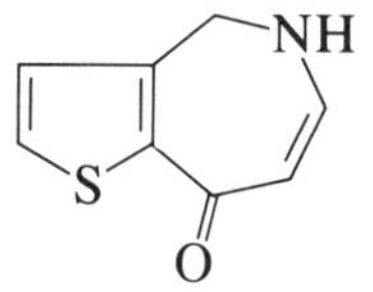 110

4. 6, 7-Systems

By far the most studied compounds of the 6,7-systems are the benzazepines. These have been the subject of a recent excellent review by Kasparek (66), which should be consulted by those particularly interested in this area. To avoid repetition, this chapter will concentrate on the salient points of benzazepine chemistry and present in tabular form the significant members of each series. In this connection, complex molecules in which the benzazepine moiety does not contribute the major fraction of the total molecular weight are excluded, especially if their synthesis involves predictable manipulations of preformed azepine derivatives.

A. *1-Benzazepines*

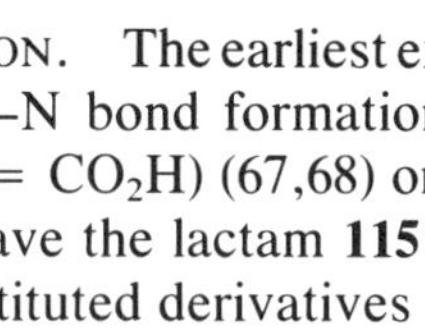

a. PREPARATION

i. C—N BOND FORMATION. The earliest examples of 1-benzazepine derivatives were made by C—N bond formation. Thus heating 4-(2-aminophenyl)butyric acid (**114,** R = CO_2H) (67,68) or treating the chloride **114** (R = CH_2Cl) with alkali (69) gave the lactam **115** (R = H) and the amine **116** (R = H), respectively. Substituted derivatives of **117** have been cyclized to the dione **118** (R_1 = 7-Me, R_2 = 8-Me) by heating the acid (70) or treating

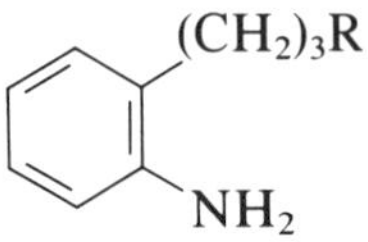

114 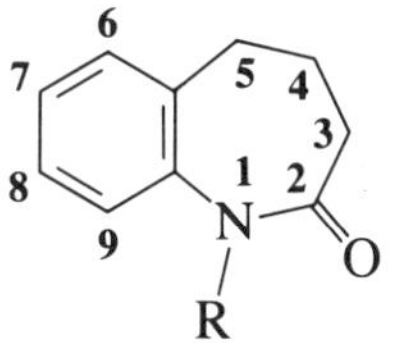 115 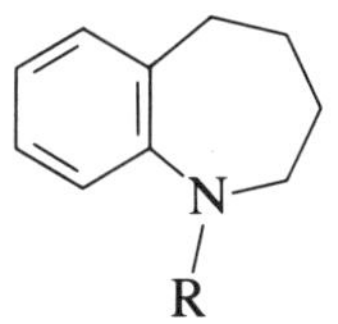 116

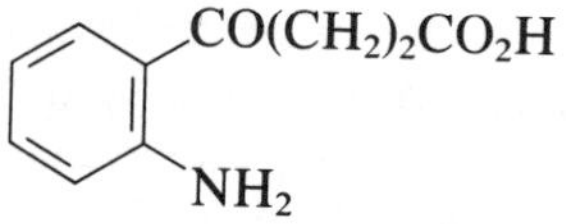

$CO(CH_2)_2CO_2H$ NH_2

117

R_1 O R_2 N H O

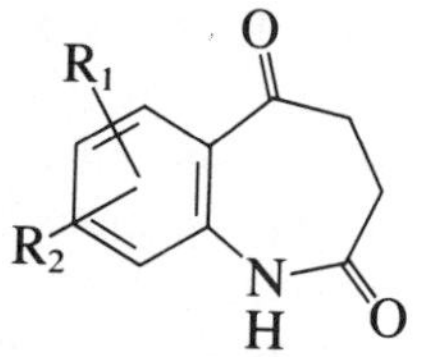

118

the corresponding ester with sodium hydride in dimethylformamide to obtain the keto lactams **118** (R_1 = Br, CO_2Me; R_2 = 9-Et) (71).

The principal limitation of this method is the labor involved in obtaining the amino acid precursors **117.** The ring opening of indoles (71) affords an ingenious and versatile means of achieving this end. During the reductive cyclization of the *o*-nitrobenzyl compound **119** to an indole, a minor product proved to be the dihydroazepinone **120.** However, this approach has not been developed further.

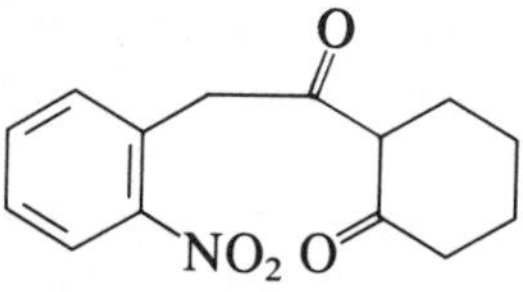

O NO_2 O

119

O N H

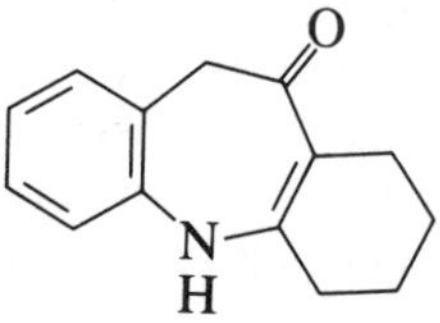

120

One of the most convenient syntheses of 1-benzazepine derivatives involves both C—N and C—C bond formation. In this route, anthranilic esters condense with succinate esters in the presence of base (73,74) to form the lactams **121** (R = Me, Et). A two-stage approach (C—N bond first) has been reported for compound **121** (R = Me) and the corresponding 8-methoxy derivative (75). *N*-Alkyl derivatives have also been cyclized (76). This reaction usually also gives quinolones such as **122,** as expected on the basis of the acidity of protons α to an amide carbonyl group (77).

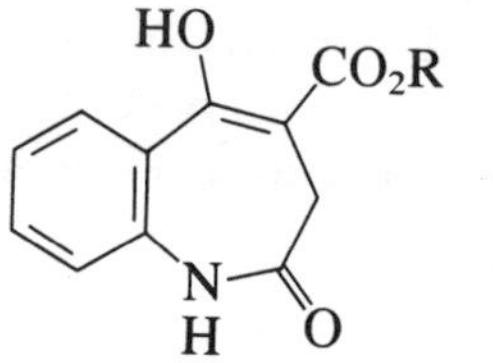

HO CO_2R N H O

121

OH CH_2CO_2Me N H O

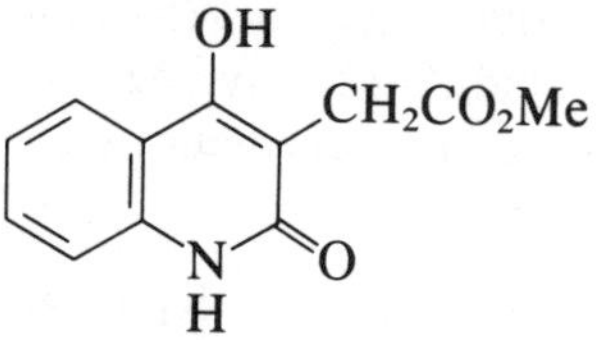

122

ii. C—C Bond Formation. Intramolecular electrophilic cyclization is successful with certain resonance-stabilized carbonium ions, as exemplified by the transformation of the allylic carbinols **123** (R_1 = Me, Et; R_2 = H,

Me; R_3 = Me, Et; R_4 = H, Me) to 2,5-dihydro-1-benzazepines **124** (78,79). It is interesting that this reaction does not invariably give the indolines (79b) that occasionally arise and are known to be produced by prolonged acid treatment, and that *N*-phenyl pyrrolines are reported to form only occasionally (126). In related work, certain vinyl-*N*-arylaziridines have been shown to yield tetrahydro-1-benzazepines (512). Cyclization of saturated alcohols such as **125** is also possible (88).

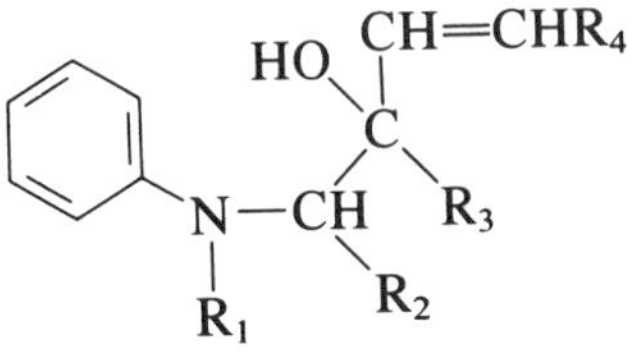

123 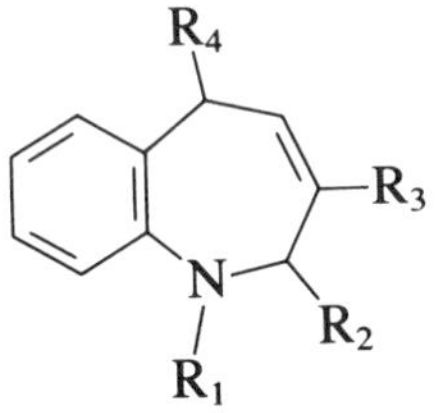 **124** 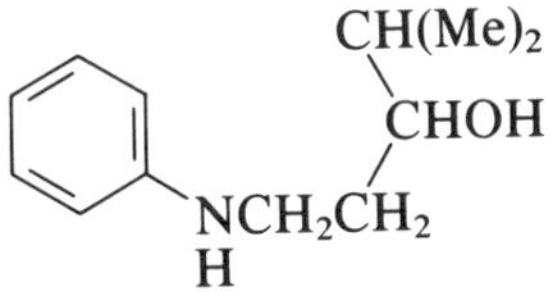 **125**

Intramolecular Friedel–Crafts acylation of *N*-alkyl and *N*-aryl amino acids with phosphoryl chloride to the tetrahydrobenzazepin-5-ones (**126,** R_1 = H, MeO; R_2 = Me, Ph) has been reported (80). However, this method does not work for *N*-tosyl amino acids because of the ease of formation of *N*-arylpyrrolid-2-ones, which is postulated to follow the course depicted in **127** (81–83).

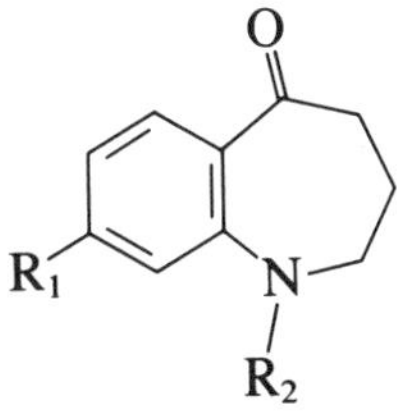

126 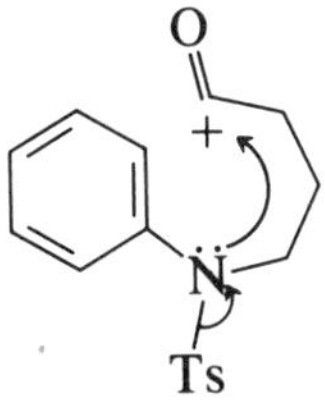 **127**

Since the claim (84) that 3-benzoylpropionanilide cyclizes to **128** (R = Ph) has been proved incorrect (85), the acid-induced condensation of aniline with levulinic acid (84) to give **128** (R = CH_3) deserves reexamination. *N*-(2-Chloroallyl)-*N*-tosyl derivatives of *o*-aminobenzyl alcohol, such as **129**

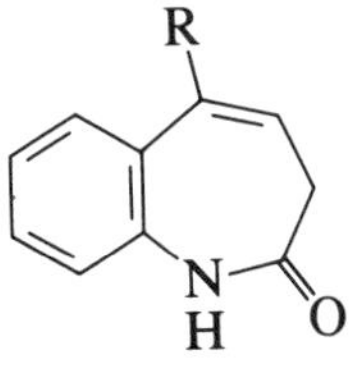

128 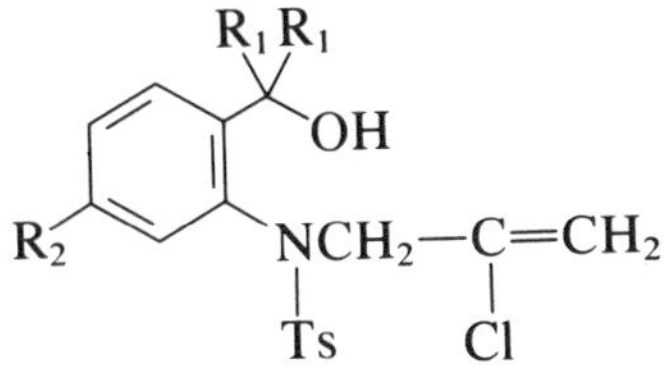 **129** 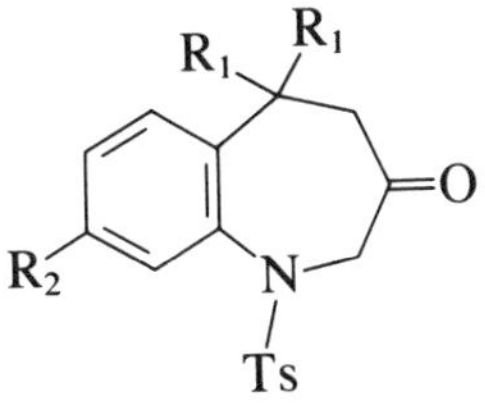 **130**

(R_1 = Me, H; R_2 = H, Cl), undergo ring closure to the ketones **130** on treatment at −5°C with 90% sulfuric acid, but other acidic reagents and *N*-protecting groups cannot be employed (86).

It has been stated that, in some cases, alkali treatment of the products from arylhydroxylamines and diketene yielded dihydroazepin-2-ones such as **131** (513). Reaction of the ketal **132** with 1-methoxy-2-buten-3-one affords an enamine adduct that can be cyclized to benzazepine **133** in aqueous alcoholic mineral acid (87).

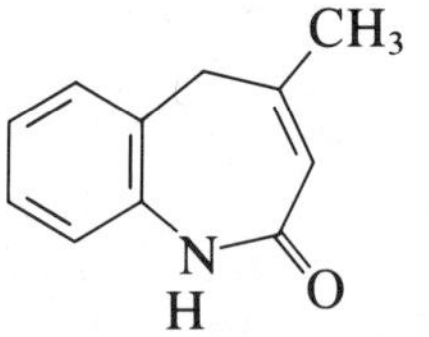

131

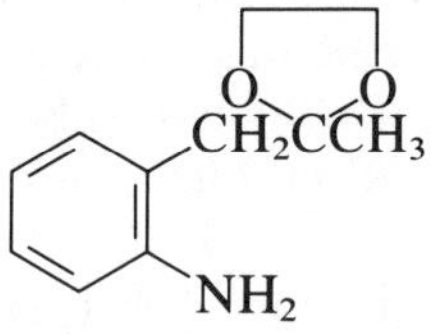

132

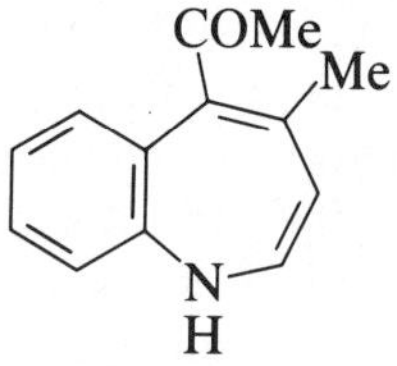

133

Dieckmann cyclization (80,81,89–91) of the diesters **134** (R_1 = tosyl, Me; R_2 = H, 4-Cl, 4,5-$(MeO)_2$, n = 3) probably is the most practical method of large-scale synthesis of tetrahydro-1-benzazepin-5-ones. The initially formed β-keto esters **135** can easily be decarboxylated with mineral acid to give the ketones **126** (R_1 = H, Cl; R_2 = Me, tosyl). The *N*-acetyl protecting group is of limited use, since it can participate in base-induced reactions yielding quinolones (77,93). Acyloin condensation of the diester **134** (R_1 = R_2 = H, n = 2) had not been of preparative value (92).

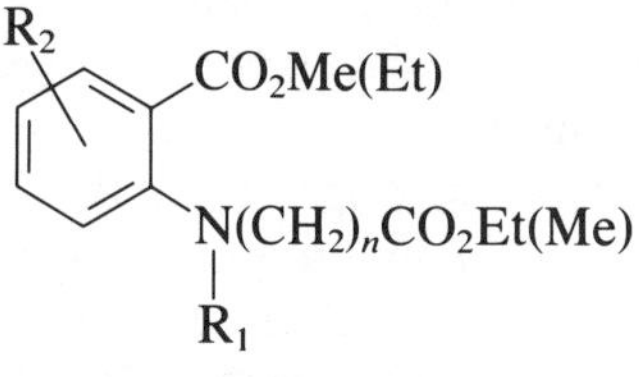

134

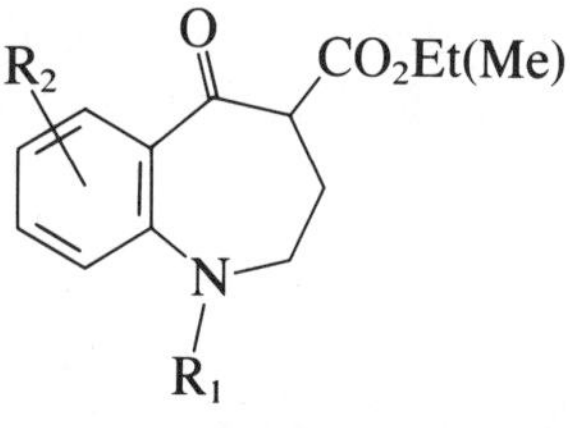

135

iii. Ring Expansion by Beckmann and Schmidt Reactions. The Beckmann and Schmidt reactions, which have been very fully reviewed (66), are widely applied to α-tetralones by medicinal chemists. The Beckmann rearrangement is critically sensitive to changes in reaction conditions and to substitution on the aryl ring and oxime oxygen. Commonly used reagents include "Beckmann's mixture" (acetic acid–acetic anhydride–hydrogen chloride)(68), polyphosphoric acid (94–104), potassium acetate on the *o*-benzenesulfonate ester of the oxime (105), and arylsulfonyl chlorides in pyridine (95,106). Aryl migration often takes preference over alkyl migration, and is explained on the basis of a bridged phenonium ion intermediate

(100,104,107). In some instances, both modes of migration occur (98,99), but this does not constitute a widely used route to 2-benzazepine derivatives (108).

The Schmidt reaction of α-tetralones is also an important source of 1-benzazepin-2-one derivatives. Forty years ago Briggs and De Ath (109) used hydrazoic acid, sulfuric acid, and chloroform to convert **136** (R = H) to the lactam **115** (R = H). Later workers have preferred to use sodium azide in the presence of acidic reagents such as sulfuric and acetic acid (110,111), polyphosphoric acid (112,113), methanesulfonic acid (114), or trichloroacetic acid (112), although sodium azide in sulfuric acid remains popular (108,112). The course of the reaction is sensitive to substituent changes. For example, 6-methoxy-1-tetralone (**136,** R = MeO) affords very little of the 1-benzazepin-2-one derivative **137** (R = 7-MeO) but gives a 90% yield of the isomeric 2-benzazepin-1-one product (108,112). Some of the apparently conflicting directional influences noted in the literature (66,112) for amino-substituted tetralones may be ascribed to the presence of protonated nitrogen species in concentrations that may vary as a function of the precise conditions.

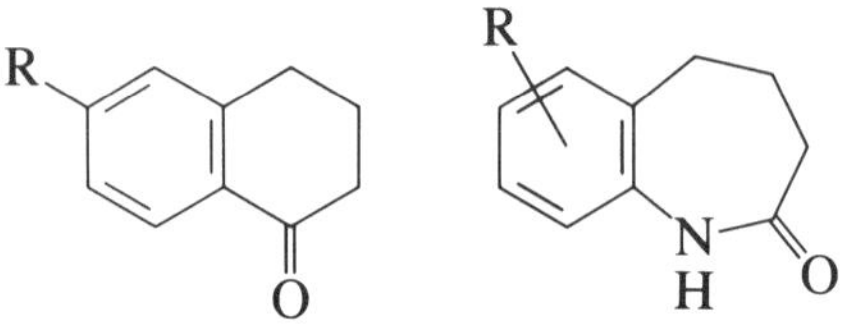

136 **137**

For the synthesis of 2,5-dihydro-1-benzazepine-2,5-diones, the Schmidt reaction with various 1,4-naphthoquinones has proved very convenient. The original, understandable misinterpretation of product structures (115) has been adequately corrected (116,117), and a wide range of products have been reported (118–121). Treatment of 4-azido-1,2-naphthoquinone with concentrated sulfuric acid led to the ring-enlarged hydroxydione **138** (R = H) (122), which subsequently could be converted to the azatropolone **139** (X = OH) (123). The corresponding azatropone (**139** X = H) has been synthesised by the same approach (124).

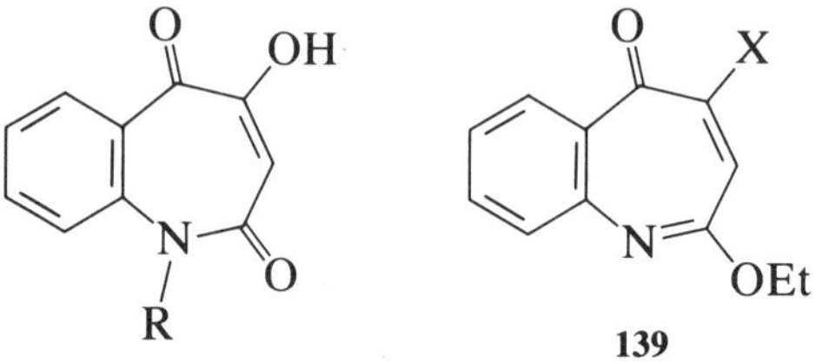

138 **139**

iv. Other Ring-Expansion Reactions. Sigmatropic rearrangement of the vinylarylaziridine **140** (R = Br) gave the 2,5-dihydro-1-benzazepine **141** (R = Br) in good yield (125). Other examples of this reaction have also been reported (126).

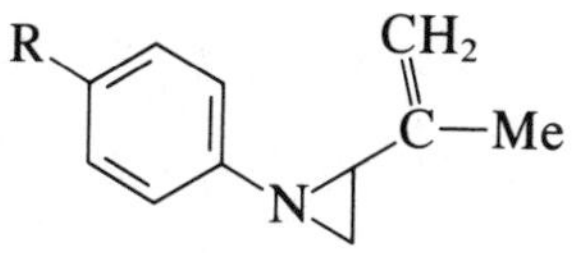

140

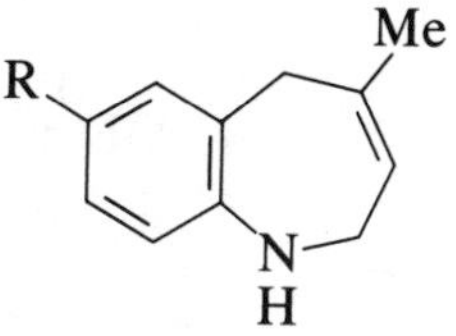

141

Ring expansion of indoles by reaction with dimethyl acetylenedicarboxylate has been examined thoroughly (127–131) and shown to give 1-benzazepines such as **142** (R_1 = Ac, R_2 = H, R_3 = piperidino; R_1 = Me, R_2 = CO_2Me, R_3 = H) (127,128). Tosylate esters of 2-hydroxyethylindoles have been observed to undergo ring enlargement on treatment with ethyl cyanoacetate, especially when dimethyl sulfoxide was used as the solvent. This type of rearrangement has been used to obtain the 1-benzazepine derivative **144,** presumably via the indolenine intermediate **143** (132). Indolenines themselves react with acetylenic amines, giving 5H-1-benzazepines. For example, addition of **146** to the indolenine **145** yields the adduct **147** (133).

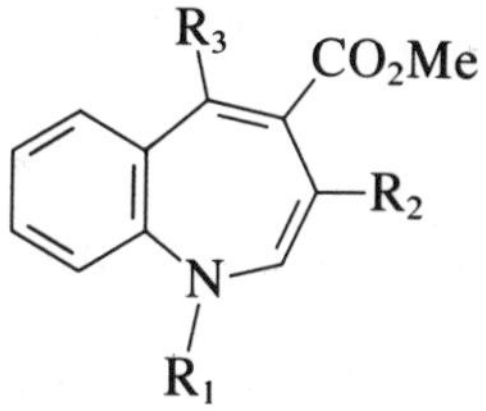

142

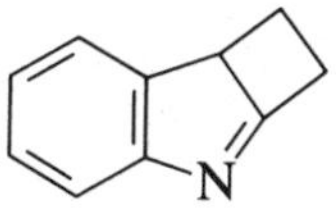

143

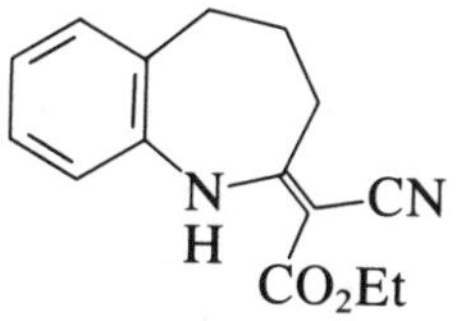

144

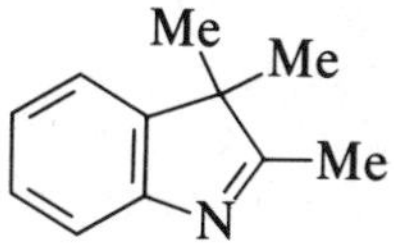

145

Ph—C≡C—NMe_2

146

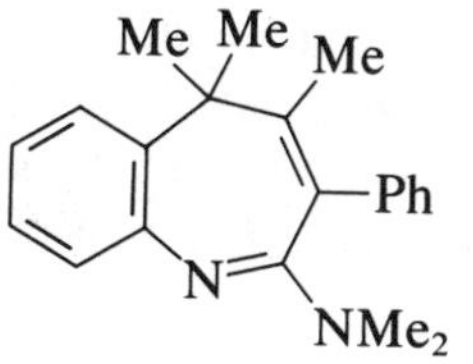

147

Ring expansion of quinoline enol ethers has been reported to follow dibromocarbene insertion in some instances (134). Whereas dibromocarbene adds to **148** (R_1 = tosyl, R_2 = H) to give **149** (R_1 = tosyl, R_2 = H), the same reaction fails with **148** when R_1 = tosyl and R_2 = Me. The blocking effect of the methyl group can be overcome by using the less bulky mesyl substituent in derivative **148** (R_1 = mesyl, R_2 = Me), which affords the adduct **149** (R_1 = mesyl, R_2 = Me). The adducts can be converted to 1-benzazepines such as **150** (R_1 = mesyl, R_2 = Me) or hydrolyzed to dihydro-1-benzazepin-5-ones such as **151** (R_1 = mesyl, R_2 = Me).

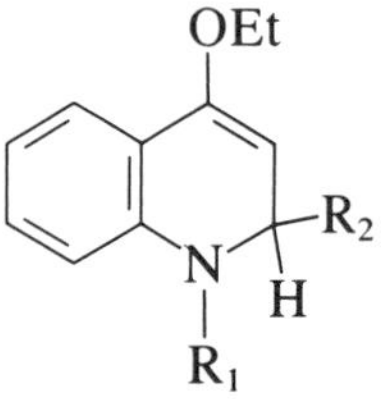

OEt, R_2, N, H, R_1 **148** | Br, OEt, Br, R_2, N, H, R_1 **149** | OEt, Br, N, R_2, R_1 **150** | O, Br, N, R_2, R_1 **151**

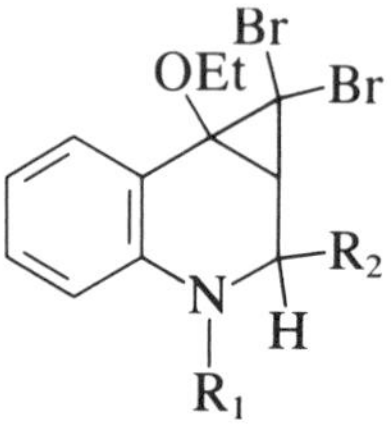

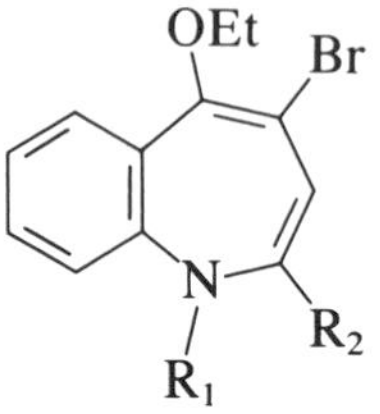

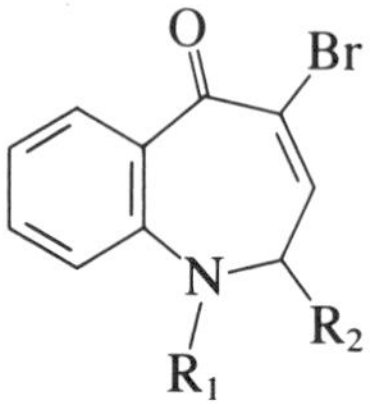

Thermal rearrangement of the 11-azabicycloundecane **152** yielded the 6,7,8,9-tetrahydrobenzazepine **153** (18). The claimed conversion of **154** to **138** (R = Ph) on treatment with alcohols has been disputed (136). ^{13}C-Nmr studies showed (325) that the product actually has the oxirane structure **155.**

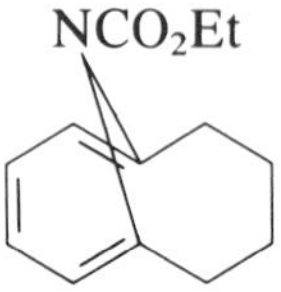

NCO_2Et **152** | N, CO_2Et **153** | O, CHN_2, OH, N, O, Ph **154** | O, O, N, O, Ph **155**

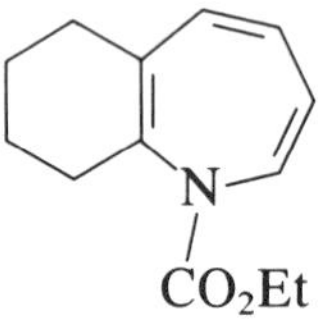

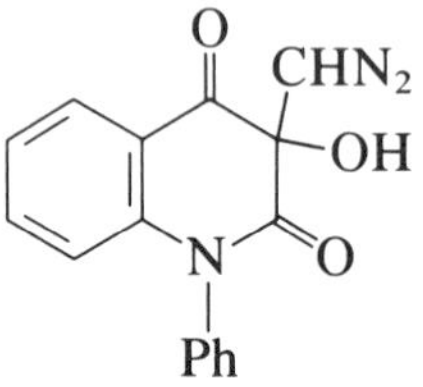

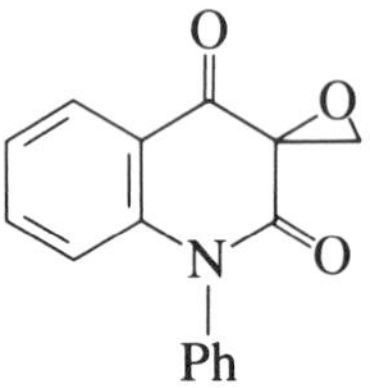

Recently, a nitrene-insertion rearrangement reaction was used to convert 1,2-dihydroquinolines to 2,3-dihydro-1-benzazepin-2-one derivatives (189). It was also claimed (192) that phenylhydroxylamines react with ketene dimer to form products that, on subsequent aldol condensation, give 1-benzazepin-2-ones such as **156.** 7-Methoxy-4-phenylquinoline *N*-oxide reacts with dimethylsulfoxonium methylide to give a low yield of the 1-benzazepine derivative **157** (R_1 = MeO, R_2 = Ph) in the presence of benzenesulfonyl

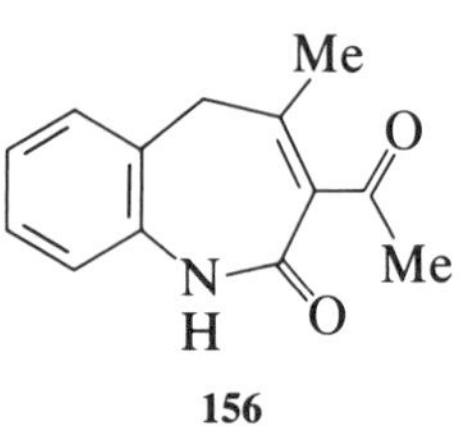

Me, O, Me, N, H, O **156** | R_2, R_1, N, H **157**

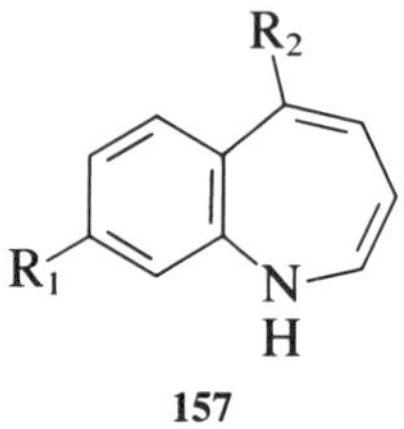

chloride. The parent benzazepine (**158,** $R_1 = R_2 = H$) cannot be prepared by this method (193).

b. PHYSICAL PROPERTIES

Several calculations predict that benzazepine systems will not be aromatic (137–139). Several variable-temperature nmr studies confirm that the 1-benzazepine system is not flat (127,134,140), with inversion barriers in the range of 15–17 kcal/mole (134,140). X-ray crystallographic information in this area is generally sparse (141). However, the azatropolone **139** (X = OH, Me, or Et) is known to be flat, albeit with considerable bond alternation. The greater acidity of 1-benzazepine **157** ($pK_a = 5.50$) as compared with that of 2,3,4,5-tetrahydro-1-benzazepine (**116,** R = H; $pK_a = 5.63$) suggests that there is little additional conjugation between the π electrons of the azepine ring in **157** ($R_1 = R_2 = H$) and the nitrogen atom. Since an even lower value has been reported for the pK_a of **116** (R = H) (144), it would appear that there is a need for additional data.

Quite apart from the questions of aromaticity, physical measurements have revealed peculiarities in the behavior of some reduced derivatives. For example, it was recognized over 30 years ago (145) that a steric effect inhibited the electronic interaction of the benzene ring with the C═O and NH groups in **126** ($R_1 = R_2 = H$). More recently, the ultraviolet spectrum of the lactam **115** (R = H) was shown to differ significantly from that of acetanilide (105). The infrared carbonyl stretching frequency of *N*-methyl-2,3,4,5-tetrahydro-1-benzazepin-5-ones such as **126** ($R_1 = H$, $R_2 = Me$) has been compared with that of the corresponding quinol-4-ones (91). The ultraviolet spectrum of **126** ($R_1 = H$, $R_2 = Me$) suggested less interaction between the carbonyl group and amino nitrogen than in the six-membered analog (91). This is supported by the difficulties encountered during attempts to detosylate *N*-tosyl ketones such as **126** ($R_1 = H$, $R_2 =$ tosyl) (146). Rates of quaternization of the amine **116** (R = Me) are abnormally low (147). The mass spectra of some *N*-tosyl di- and tetrahydro-1-benzazepinones have been reported (148), and azatropylium ions have been detected in the mass spectra of some substituted quinolines (149). Recently, it was observed (437) that electron impact causes alkyl elimination from 1-benzazepine derivatives, possibly by a combination of several different mechanisms.

c. CHEMICAL PROPERTIES

Ring contractions have been observed with several 1-benzazepine derivatives (77,98,119,436) and are generally attributable to the presence of carbonyl groups in the azepine ring and the fundamentally greater stability of five- and six-membered ring compounds. This topic has been reviewed in detail (66). The 1-benzazepine **142** ($R_1 = Me$, $R_2 = CO_2Me$, $R_3 = H$) undergoes cycloaddition with excess dimethyl acetylenedicarboxylate to give adduct **158** ($R = CO_2Me$), whose structure has been confirmed by X-ray

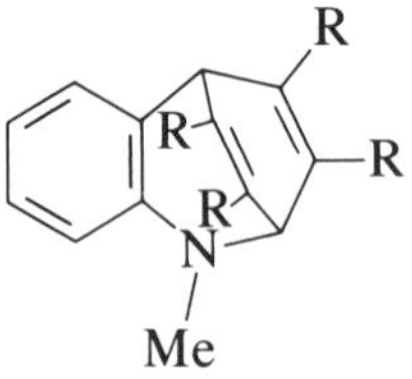

158 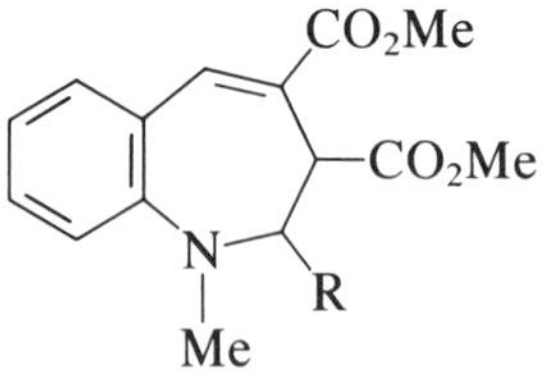 **159**

crystallography (128). The same 1-benzazepine reacted at position 2 with nucleophiles in the presence of acid. In this manner the diesters **159** (R = MeO, 3-indolyl, 3-*N*-methylindolyl) were obtained from methanol, indole, and *N*-methylindole, respectively (128).

N-Alkylation and *N*-acylation of partly reduced benzazepines have been studied by many groups. Amines (150,151,157) such as **116** (R = H) and lactams (152) such as **115** (R = H) were employed successfully. The cyclic amidine **160** (R = NHMe) (153) and thiolactim ether **160** (R = SMe) (154) have been obtained from lactam **115** (R = H). The lactams are more reactive with Grignard reagents than are the corresponding *O*-methyl enol ethers.

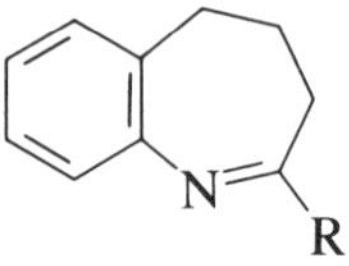

160

Intramolecular *N*-acylation reactions in reduced 7-carboxytetrahydro-1-benzazepine derivatives constitute an ingenious approach to the synthesis of the Ibogamine alkaloids (71,97). The lactam **137** (R = 8-Cl) reacted with phosphoryl chloride and formamide to give the β-chloroaldehyde **161** (135). When the same reactants were heated under pressure, however, the pyrimidobenzazepine **162** was obtained (156); several examples were given.

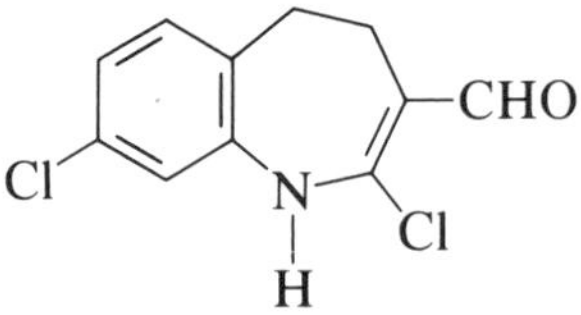

161 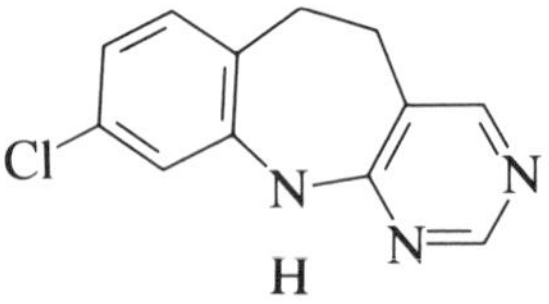 **162**

The reactivity of 5-keto derivatives is well documented. Normal carbonyl derivatives are obtained from **126** (R_1 = H, R_2 = tosyl) (158) and also from vinylogous amides (**126;** R_1 = H, R_2 = Me) (91). The α-methylene group in tetrahydro-1-benzazepin-5-ones likewise shows the expected ability to yield hydroxymethylene derivatives (81,158). These and the β-keto esters have been used to annelate several other heterocyclic ring systems (imidazole, pyrazole, oxazole, and pyrimidine) onto the 1-benzazepine skeleton (159). The Fischer method has been employed to prepare indole derivatives (91).

Substitution reactions involving the azepine ring of several 1-benzazepine derivatives have been discussed (161). The β-keto esters **163** (R_1 = H, Cl; R_2 = tosyl), as sodium salts, have been shown to undergo ring expansion on treatment with dimethyl acetylenedicarboxylate to compounds containing a nine-membered azonine ring (538). An X-ray crystallographic analysis has been performed on the chloro derivative (162).

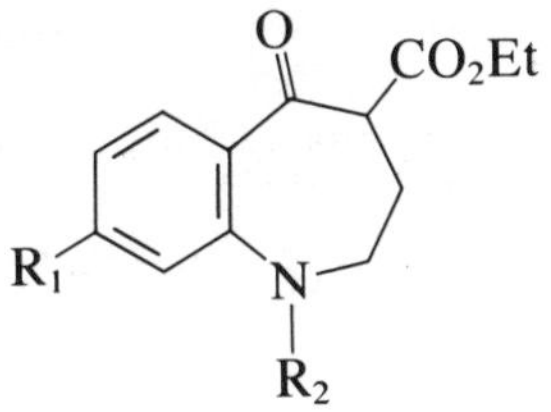

O CO_2Et R_1 N R_2

163

The Schmidt reaction transformed the ketone **126** (R_1 = R_2 = H) into several compounds, the predominant one being the indoline **164** (163,164). In contrast, enol ethers of the *N*-tosylated ketones (**126,** R_1 = H, Cl; R_2 = tosyl) gave exclusively ring-expanded products (165). The seven-membered ring in tetrahydro-1-benzazepine has an effect on the direction of electrophilic substitution in the benzene ring. Thus the compound **165** (R = H, n = 4) gave **165** (R = Br, n = 4), whereas the six-membered (**165;** R = H, n = 3) gave the 5- bromo (*peri*) compound (166). Known 1-benzazepine derivatives whose physical constants have been reported in the literature are listed Tables 1–15.

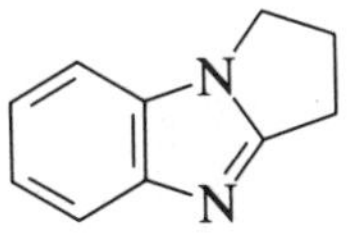

N N

164

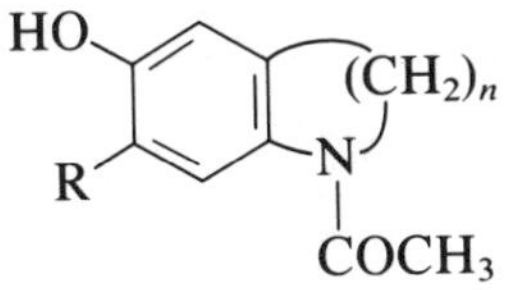

HO $(CH_2)_n$ R N $COCH_3$

165

More recent work in this area has produced some novel methods of preparation. Thus it has been shown that treatment of the compound **166** with methylmagnesium bromide in the presence of 3% $NiCl_2(PPh_3)_2$ gave the tetrahydro-1-benzazepine **167** in 60% yield (557), and the amino ketone **168** was obtained by heating allene with the *N*-oxide of benzalaniline in a sealed tube at 72°C (558).

Photolytic [2 + 2] cycloaddition of acetylenic esters to indoles gives the expected tricyclic products, whose thermolysis leads to 1-benzazepines (559).

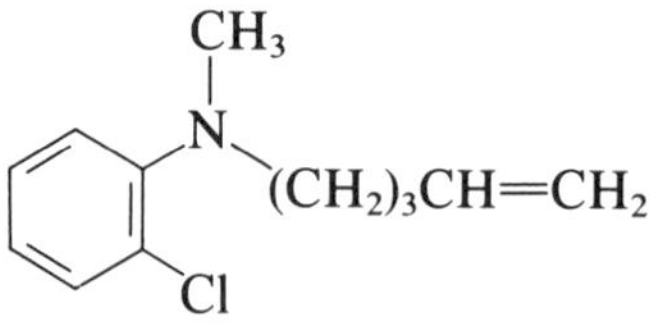

CH3 N (CH2)3CH=CH2 Cl

166

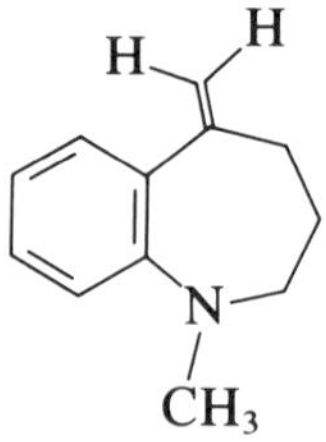

H H N CH3

167

O N H Ph

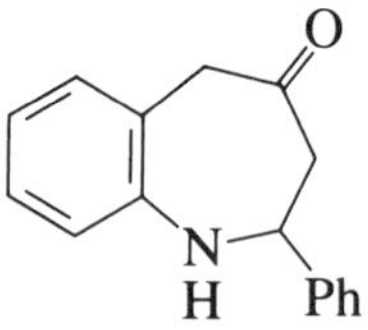

168

TABLE 1. REDUCED 1-BENZAZEPINES

Structure[a]		mp (°C) or bp (°C/torr)	Refs.
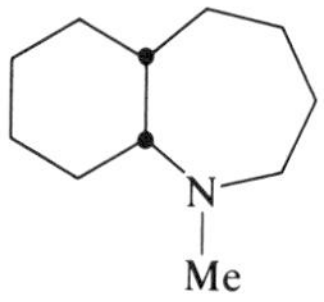 N-Me		72–74/2	18
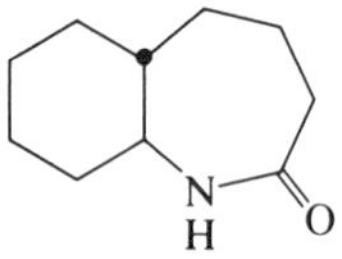 NH, O		170–171	181
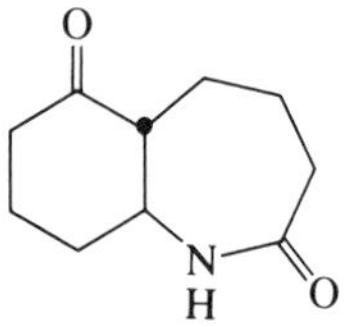 O; NH, O		199–200	181,176
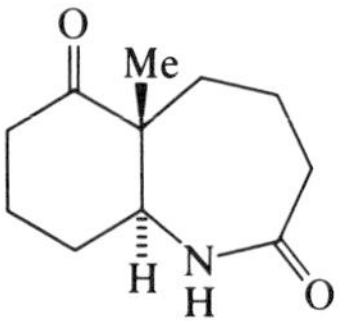 O, Me; H, NH, O		199–200	181
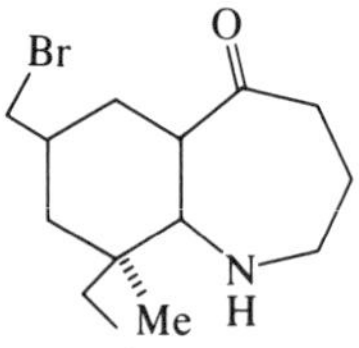 Br, O; Me, NH		150 (HBr)	106
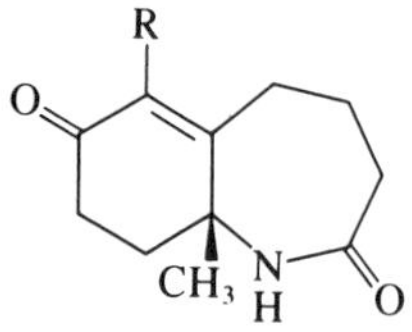 R, O; CH3, NH, O	R = H	234–236	182
	R = Me	192–194	182

[a]Stereochemistry of ring junction specified where known.

TABLE 2. 2,3,4,5-TETRAHYDRO-1H-1-BENZAZEPINES

R_1	R_2	R_3	R_4	R_5	mp (°C) or bp (°C/torr)	Derivatives (mp, °C)	Refs.
H	H	H	H	H	32–33	Picrate, 179–181	77,81,92
						HBr, 212–213	94
Ac	H	H	H	H	80		160,169
COPh	H	H	H	H	85–86		92
Tosyl	H	H	H	H	88		134
$SiMe_3$	H	H	H	H	60–61/0.1		177
Me	H	H	H	H	110–115/15	Picrate, 140–140.5	81
Et	H	H	H	H	89–90/1.7	HCl, 136–138	108
CH_2Ph	H	H	H	H	127/0.3		135
$CH_2CH_2NH_2$	H	H	H	H	60–65/0.025		170
CH_2CH_2N(morpholino)O	H	H	H	H	140–143/0.4	2HCl, 202–204	108
$CH_2CH(CH_3)NHMe$	H	H	H	H	92–93/0.001		171
$(CH_2)_3NMe_2$	H	H	H	H	78–80/0.001		171
$(CH_2)_3NH_2$	H	H	H	H	55–56		171
H	H	H	H	Me	36.5		84
H	Me	H	H	OH	115		77
Tosyl	Me	H	H	OH	97		77

(continued)

TABLE 2. *(Continued)*

R_1	R_2	R_3	R_4	R_5	mp (°C) or bp (°C/torr)	Derivatives (mp, °C)	Refs.
CH_2–(ring: N, N H) 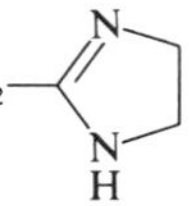	H	H	H	Ph		HCl, 230–232	110
$(CH_2)_3N$(ring) 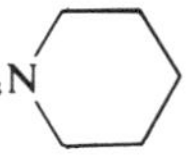	H	H	H	Ph		HCl, 208–209	110
H	H	H	Me	Me, Me	260–270/0.002		88
H	H	H	Me	Me, Me	220–230/0.07		88
CONHPr-*n*	H	H	H	H	64–65		94
H	H	H	H	Ph, CO_2Et	176–178		96
Me	H	H	H	Ph, CO_2Et	133–135		96
Me	H	H	H	Ph, CH_2OH		Picrate, 152–154	96
Me	H	Ph	H	H	64–66		110
CH_2–(ring: N, N Me) 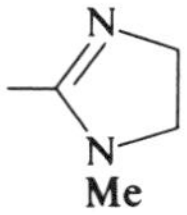	H	Ph	H	H		HI, 145 (dec.)	110
CH_2–(ring: N, N H, Me) 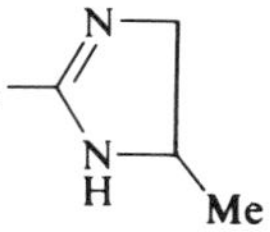	H	Ph	H	H		HCl, 218–220	110
CH_2–(ring: N, N H) 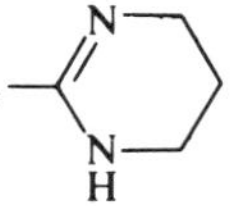	H	Ph	H	H		HCl, 256–259	110

CH_2—(2-imidazolinyl, NH)	H	m-ClC_6H_4	H	H		HCl, 241–242	110
CH_2—(2-imidazolinyl, NH)	H	p-ClC_6H_4	H	H		HCl, 239–241	152
CH_2—(2-imidazolinyl, NPh)	H	H	H	H		HCl, 280–283	152
CH_2—(2-imidazolinyl, NMe)	H	H	H	H	145 (dec.)		152
CH_2—(2-imidazolinyl, NMe)	H	p-MeC_6H_4	H	H	190–200		152
CH_2—(2-imidazolinyl, NH)	H	o-ClC_6H_4	H	H		HCl, 212	152
CH_2—(2-imidazolinyl, NH)	H	Ph	H	H		HCl, 238 (dec.)	110,152

(*continued*)

TABLE 2. (*Continued*)

R_1	R_2	R_3	R_4	R_5	mp (°C) or bp (°C/torr)	Derivatives (mp, °C)	Refs.
H	H	*o*-ClC_6H_4	H	H	138–139		152
H	H	*p*-ClC_6H_4	H	H	100–102		152
H	H	Ph	H	H	122–124		111,152
CH_2CN	H	*o*-ClC_6H_4	H	H	103–104		152
CH_2CN	H	*p*-ClC_6H_4	H	H	102–104		152
CH_2CN	H	Ph	H	H	144–155/0.1		152
CH_2CN	H	H	Ph	H		HCl, 154	180
CH_2-(2-imidazolinyl)	H	H	Ph	H	153–154		180
H	H	H	H	OH	88–90		
H	H	H	Me, Me	OH	101–102		161
Tosyl	H	H	OH	OH	106–108		191
Tosyl	H	H	H	OH	124–125		191
Me	H	H	CH_2OH	*trans*-Ph		HCl, 158 (dec.)	98
Me	H	H	CH_2NMe_2	*trans*-Ph		2HCl, 220 (dec.)	98

$(CH_2)_2NEt_2$	H	Ph	H	H		Oxalate, 138–140	111
$(CH_2)_3NEt_2$	H	Ph	H	H		Oxalate, 130–132	111
$CONH_2$	H	Ph	H	H	202		111
$COCH_2Cl$	H	Ph	H	H	155–157		111
Tosyl	H	H	H	OH, CH_2Ph	122.5–124		161
Tosyl		H	H	OH, $CH_2C_6H_4OMe$-*m*	116–118		161
H	$=C(CN)CO_2Et$	H	H	H	110–112		132
$CH(CH_3)CH_2NMe_2$	H	H	H	H	102–108/0.15	2HCl, 183–184	94
$CH_2CH{=}CH_2$	H	H	H	H	140–146/25	HCl, 134–135	94
CH_2CH_2CN	H	H	H	H	47–48		94
$(CH_2)_2N$(piperidino)	H	H	H	H		2HCl, 223–224	94
Me	H	H	H	H	58–60/0.1–0.2		18
Tosyl	H	=O	H	Me, Me	125–126		86
Tosyl	H	=O	H	H	117		86
H	H	*m*-ClC_6H_4	H	H	Oil		110
H	H	=O	H	H		Propylene ketal	86
H	=S	Ph	H	H	222–223		153
H	=S	H	H	H	158 (136)		167(183)

TABLE 3. ARYL-SUBSTITUTED 2,3,4,5-TETRAHYDRO-1H-1-BENZAZEPINES

R_1	R_2	Other Substituents	mp (°C) or bp (°C/torr)	Derivatives (mp, °C)	Refs.
H	7-Me		40–41		135
CH_2Ph	7-CH_2OH		225/0.6		135
CH_2Ph	7-CHO		78		135
Ac	9-NH_2		100		169
H	8-MeO		45–47	HCl, 225–227	94
H	8-NO_2		76–77		94
$(CH_2)_2N$(piperidino)	8-MeO			2HCl, 198–200	94
$(CH_2)_3NH(C{=}NH)NH_2$	8-Me		239–240	H_2SO_4, 239–240	94
$COCH_2N$(morpholino)	8-Et		96–97	HCl, 246–248	94
Ac	8-Ac		107–108.5		160
Ac	8-Et		58–58.5		160
Ac	8-CO_2H		233.5–234		160

H	8-CO_2H	186–187		160
2,4,6-$Me_3C_6H_4CO$	7-MeO	109–112		179
p-$ClC_6H_4SO_2$	7-MeO	132.5–134		179
p-$MeOC_6H_4SO_2$	7-MeO	128–130		179
p-$MeOC_6H_4SO_2$	7-MeO	92–94		179
$PhSO_2$	7-MeO	124–126		179
$MeSO_2$	7-MeO	84–85		179
CHO	7-MeO	60–62		179
PhNHCO	7-MeO	146.5–148.5		179
PhNHCS	7-MeO	139.5–141.5		179
Ac	7-MeO	80–83		179
t-BuCO	7-MeO	72.5–75		179
Cyclopropyl-CO	7-MeO	62–63		179
Cyclopentyl-CO	7-MeO	57–59		179
Cyclohexyl-CO	7-MeO	69–70		179
p-ClC_6H_6CO	7-MeO	70–73		179
p-MeC_6H_4CO	7-MeO	78–81		179
Cyclohexyl-NHCO	7-MeO	70–73		179
Ac	7,8-Me_2	90		178
H	7,8-Me_2	82		178
H	7-MeO		Oxalate, 166–171	19,173
$CO(CH_2)_3CH_3$	6,8-$(MeO)_2$, 9-CHO	76–78		172
$CO(CH_2)_3CH_3$	6,8-$(MeO)_2$, 9-CO_2H	158–159		172
H	6,8-$(EtO)_2$	140–180/20	Styphnate, 131–132	172

(*continued*)

TABLE 3. *(Continued)*

R_1	R_2	Other Substituents	mp (°C) or bp (°C/torr)	Derivatives (mp, °C)	Refs.
$CH_2CH(CH_3)CH_2NMe_2$	8-MeO		120–122/0.001		171
Me	7-NH_2		137–141		190
Me	7-Br		82–84		190
Ac	7-OH, 8-Br		222–224		166
Ac	7-OH		183–185		166
H	7-OH			HCl, 213–214	166
Ac	7-Et		120/0.4		97
Ac	7-Ac		84–86		97
Ac	7-CO_2H		200–201.5		97
H	7-CO_2H		173–174.5		97
H	7-CO_2H, 9-Ac		223–225.5		97
H	6-MeO			HCl, 256–260	188
H	6-MeO, 9-Cl			HCl, 221–225	188
Ac	9-NO_2		131		169
Ac	7,8-Me_2	4,5-Br_2	173		70
H	7,8-Me_2	5-OH	132–135		70
Ac	7,8-Me_2	5-OH	136		70
Ac	7-CO_2H	5-OH	185		97
H	8-Cl	3-Ph	73–75		152
H	7-MeO	2-(=C(CN)CO_2Et)	125–126		132
H	9-MeO	2-(=C(CN)CO_2Et)	162–163		132
Tosyl	8-Cl	3-Keto	107		86

1,3-Dimethylindole gave unusually complicated results (560). Several derivatives of the 7,8-dimethoxytetrahydrobenzazepine system **169** (R = Ts) have been made (561), and the bridged molecule **170** has been synthesised by intramolecular electrophilic cyclization of the amino acid **171** (562). The action of palladium acetate and triphenylphosphine on the aryl halide **172** yields the 1-benzazepine derivative **173** (567).

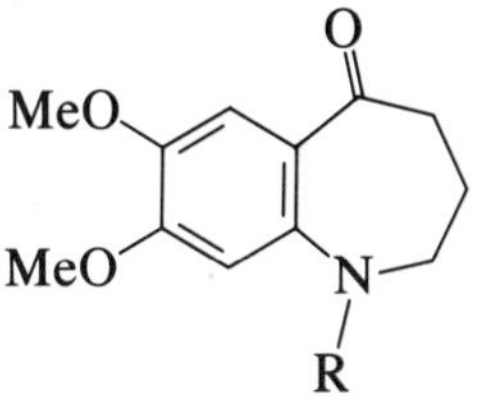

169 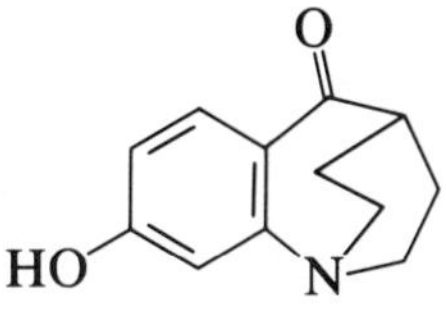 **170** 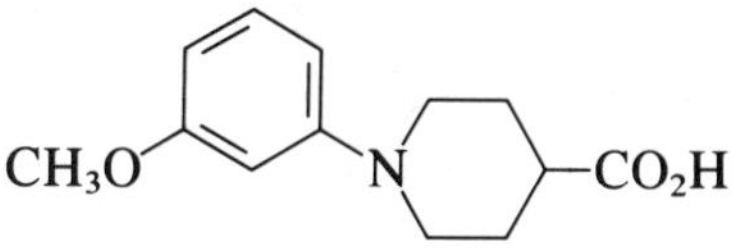 **171**

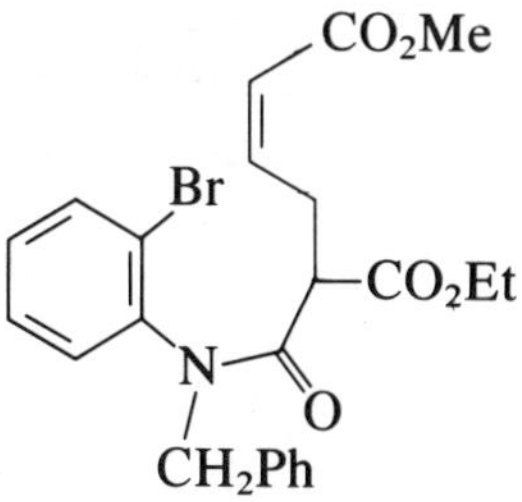

172 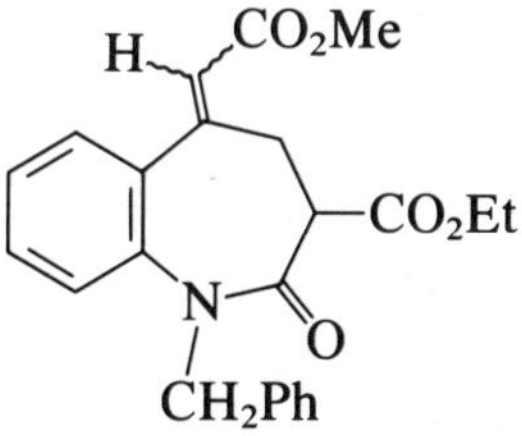 **173**

B. *2-Benzazepines*

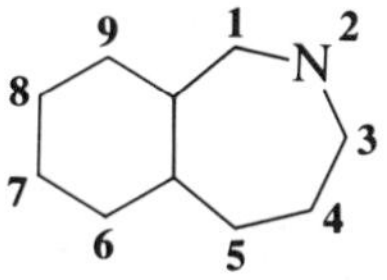

a. PREPARATION

i. C—N BOND FORMATION. A number of 2-benzazepine derivatives have been prepared by intramolecular alkylation of *o*-substituted benzylamines. Thus the halide **174** gave the tetrahydro-2-benzazepine **175** ($R_1 = R_2 = R_3 = H$) (194), and substituted analogs have yielded derivatives such as **175** (R_1 = Me, CH_2Ph; R_2 = 3-Me; R_3 = 5-Ph) (195). Cyclization of hydroxy-

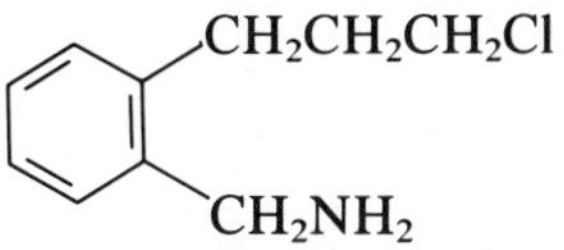

174 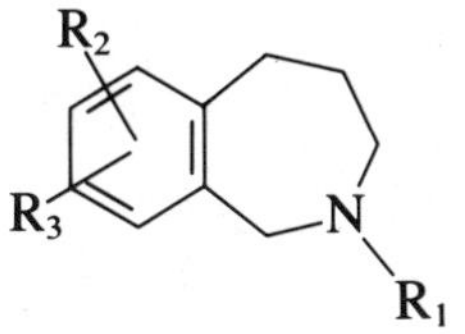 **175**

TABLE 4. 2,3,4,5-TETRAHYDRO-1H-1-BENZAZEPIN-5-ONES

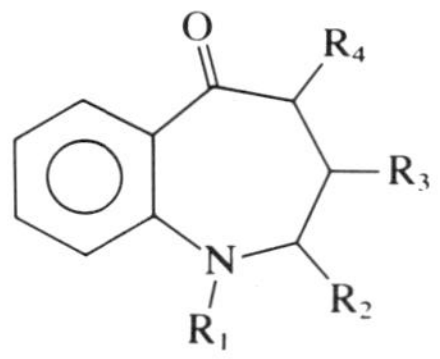

R_1	R_2	R_3	R_4	Other Substituents	mp (°C) or bp (°C/torr)	Derivatives (mp, °C)	Refs.
Tosyl	Me	H	Br, Br		135		77
Tosyl	H	H	CO_2Me	8-Cl	130		77
Tosyl	H	H	H	8-Cl	96–97		77
Tosyl	Me	H	H		122		77
Ac	Me	H	H		123		77
H	Me	H	H		75		111
Ph	H	H	H		65–66		80
$CH_2C(Cl)=CH_2$	H	H	H		81–82		161
$C(CO_2Me)=CHCO_2Me$	H	H	H		143–144		161
$(CH_2)_2NMe_2$	H	H	H		140/0.1		161
Tosyl	H	H	Me, Me		164		161
Tosyl	H	H	CO_2Et, $CH_2(Cl)=CH_2$		129–130		161
Tosyl	H	H	CO_2H, CH_2CO_2H		161–162	Hydrate	161
Tosyl	H	H	CO_2Et, CH_2CO_2H		166/0.15		161
Tosyl	H	H	CO_2Me, Me		118		161
Tosyl	H	H	Me		126		161
Tosyl	H	H	CO_2Et, Br		78–79		191
Ph	H	H	OH		150		92
Me	H	H	OH		Oil	Ferrotungstate, >300	92
Tosyl	H	H	OH		150–152		92
H	H	H	Br		114–115		77
Ac	H	H	H		122		92,127
PhCO	H	H	H		117		92

Me	H	H	H		120/0.15	HCl, 160–161	91,92
						2,4-DNP, 202–203	
						Semicarbazone, 198	
Tosyl	H	H	CO_2Et		120	2,4-DNP, 208	158
Tosyl	H	H	Br		132	Ethyleneketal, 147	158,191
Tosyl	H	H	=CHOH		126	2,4-DNP, 207	158
Tosyl	H	H	=CHOMe		197		158
Tosyl	H	H	Br, Br		173	Ethyleneketal, 162	158,191
Tosyl	H	H	H		126	2,4-DNP, 236	89,191
						Propyleneketal, 130–131	
Tosyl	H	H	$=CHC_6H_4NO_2$-*p*		249–250		89
Ac	H	CO_2Me	H		96–97		127
Me	H	H	CHO		57–57.5		81
Me	H	H	CH_2OH		Oil		81
Me	H	H	=CHOBz		108–109		81
Me	H	H	Me		90–110/0.5		81
H	H	H	Br, Br	7,9-Br_2, 8-Cl	142		77
H	H	H	H		69	2,4-DNP, 248	92
						Picrate, 228	
H	H	H	Br	7,9-Br_2	104		92
H	H	H	Br	7-Br	133		92
H	H	H	H	7-Br	107	2,4-DNP·HCl, 247	92
H	H	H	OH		70		92
H	H	H	Br, Br	7,9-Br_2	90–91	2,4-DNP, 274	191
						Bis(acetate), 157–158	
H	H	H	Cl	7,9-Br_2	117		191
H	H	H	=O	7,9-Br_2	—	Quinoxaline, 173	191
H	H	H	H	7,8-Me_2	106	2,4-DNP, 205	70
Ac	H	H	H	7,8-Me_2	105	Semicarbazone, 206	70
H	H	H	H	7-Br, 9-Et	91–92		71
H	H	H	H	7-CN, 9-Et	95–96		71
H	H	H	H	7-CO_2H	286–287.5	*N*-Acetyl, 187–188	97
Ac	H	H	H	7-Ac	134–138		97
Tosyl	H	H	=O		148	2,4-DNP, 252	158
						Ethyl enol ether	

TABLE 5. 2,3,4,5-TETRAHYDRO-1H-1-BENZAZEPIN-2-ONES

R_1	R_2	R_3	R_4	Other Substituents	mp (°C) or bp (°C/torr)	Derivatives (mp, °C)	Refs.
H	*m*-ClC_6H_4	H	H		204–207		110
Me	H	H	H		97–100/0.1		18
H	H	H	H	8-MeO, 9-NO_2	177.6–178.5		114
H	H	H	H	7-OH, 8-NO_2	205–207		114
H	H	H	H	6,9-Me_2	153–155 162–165		185,100
NO	H	H	H	6,9-Me_2	102		185
H	H	H	H	7-MeO, 9-Me	149–151		101
H	CH_2NMe_2	H	H	7-MeO	143–144		108
H	H	H	H	6-Et, 8-Pr-*i*	108–110		184
Me	H	H	H	7-Cl	59–61		85
$(CH_2)_2NMe_2$	H	H	H	7-Cl	51–53		85
$(CH_2)_3NMe_2$	H	H	H	7-Cl	127/0.01		85
H	H	H	Ph		180–181.5		84,85,110
H	H	H	Ph	7-Cl	185–186		85
Me	H	H	Ph	7-Cl	99–101		85
Me	H	H	Ph		101–103.5		85
$(CH_2)_3NMe_2$	H	H	Ph		142–143.5		85
$(CH_2)_3NMe_2$	H	H	Ph		66.5–68.5	HCl, 230–231	85,110

H	H	H	H	7-NO_2	220–222.5		112,114,190
H	H	H	H	8-NO_2	220–222		112,114
H	H	H	H	8-OMe	132–134		104,112
H	H	H	H	8-OH	227–228	8-OAc, 121–122	112,114
H	H	H	H	8-NH_2	229–231 (dec.)	8-NHAc, 216–217	112
H	H	H	H	8-Cl	157–158		112
H	H	H	H	7-NO_2, 8-OH	236–238		114
H	H	H	H	7-NH_2, 8-OH	229–230		114
H	H	H	H	7-OMe	141–142		112
H	H	H	H	7-OH	244–245	7-OAc, 140–141	112
H	H	H	H	7-NH_2	186–187	7-NHAc, 242–243	112,190
H	H	H	H	7-Cl	164–165		112,190
H	H	H	H	7-CO_2H, 9-Et	133–134	Me ester, 111–112	71
H	H	H	H	7,9-Me_2	156–157		108
$(CH_2)_2NMe_2$	Ph	H	H		60–62	HCl, 241–242	95,110, 174
$(CH_2)_3NMe_2$	Ph	H	H			HCl, 241–242	95
Et	H	H	H		105–106/0.5		108
$(CH_2)_2N$(morpholino)	H	H	H		178–180/0.9	Maleate, 146–147	108
$(CH_2)_2N$(piperidino)	H	H	H		156–158/0.6	Citrate, 154–156	108
$(CH_2)_3NMe_2$	H	H	H		140–142/0.4	Citrate	108
					110–112/0.001		171
$(CH_2)_2NMe_2$	H	Ph	H			HCl, 227–229	95
$(CH_2)_3NMe_2$	H	Ph	H			HCl, 197–199	95,174
H	H	CO_2Et	*trans*-Ph		159	Carboxylic acid, 257	98

(continued)

TABLE 5. (*Continued*)

R_1	R_2	R_3	R_4	Other Substituents	mp (°C) or bp (°C/torr)	Derivatives (mp, °C)	Refs.
Me	H	CO_2H	*trans*-Ph		257	Ethyl ester, 128	98
Me	H	CO_2Me	*trans*-Ph		162		98
Me	H	$CONMe_2$	*trans*-Ph		200		98
H	H	$CONMe_2$	*trans*-Ph		273		98
H	H	H	H	7-Br	165–166		190
H	H	H	H	7-I	182–184		190
H	H	H	H	7-F	191–193		190
H	H	H	H	7-Ac	164–166		97
H	H	H	H	7-Et	166.5–168		97
H	H	H	H	$7-CH_3CH(OH)$	177–179		97
H	Ph	H	H		196–198		152,153 111
H	H	H	H	$7,8-Me_2$	175		178
H	H	H	OH	$7,8-Me_2$	192		178
$(CH_2)_2N$(4-hydroxy-4-phenylpiperidino; ring drawn with OH, Ph)	H	H	H		117–119		187
$(CH_2)_3NMe_2$	H	H	H		165–166.5		187
$(CH_2)_2N$(4-phenyl-4-hydroxypiperidino; ring drawn with Ph, OH)	H	H	H		228–229		187
$(CH_2)_3NMe_2$	H	H	H	8-Cl	Oil	HCl, 160–161	156
$(CH_2)_2NEt_2$	H	H	H		Oil		111

$(CH_2)_2N$[piperidine ring]	H	H	Ph			HCl, 230–231	110
$CH_2CH(CH_3)NMe_2$	H	H	Ph			HCl, 248–249	110
CH_2–[2-imidazolinyl ring, N, NH]	Ph	H	H		>265		110
CH_3	Ph	H	H		113–114		110
H	H	CO_2H	H		141–142	Me ester, 121	103
H	Ph	H	H	8-Cl	224–227		152
H	*o*-ClC_6H_4	H	H		240–242		152
H	*p*-ClC_6H_4	H	H		193–195		152
H	H	H	H	6,8-Et_2	101–102		172
H	H	Ph	H		140–142		174
H	Me_2	H	H		139–140		175
H	H	H	H	6-Br	124–125		100
H	$(CH_2)_2$–[2-pyridyl ring, N]	H	H		130–132		99
H	$(CH_2)_2$–[4-pyridyl ring, N]	H	H		131–134		99
$(CH_2)_3Cl$	H	H	H		116–118/0.001		171
$(CH_2)_3NH_2$	H	H	H		112–144/0.001		171
H	H	H	CO_2H		141–144	Me ester, 121	102
H	H	H	H		141		168
H	H	H	Me		167–167.5		84

TABLE 6. 4,5-DIHYDRO-3H-1-BENZAZEPINES

R_1	R_2	R_3	R_4	Other Substituents	mp (°C)	Derivatives (mp, °C)	Refs.
SMe	Ph	H	H		66–70		153
NHMe	Ph	H	H			HCl, 260–272	153
NHEt	Ph	H	H			HCl, 264–270	153
NHPr	Ph	H	H			HCl, 196–197	153
NHPr	Ph	H	H	7-MeO		HCl, 197–198	153
NHPr	p-ClC_6H_4	H	H			HCl, 196–197	153
NH—cyclopropyl	Ph	H	H	7-MeO		HCl, 258–262	153
$NHCH_2Ph$	p-ClC_6H_4	H	H			HCl, 220–221	153
$NHCH_2Ph$	H	H	H			HCl, 248–250	153
$NHCH_2Ph$	H	H	Ph			HCl, 183–185	153
NHMe	H	CO_2Et	*trans*-Ph		136		98
NHMe	H	CO_2H	*trans*-Ph		278 (dec.)		98
$NHCH_2Ph$	H	CO_2H	*trans*-Ph		180 (dec.)		98
Me	H	H	OH		87		77
H	H	H	OH		88–89		161
H	H	Me_2	OH		90–91		161
NEt_2	=O	H	Me_2		105		133
NHPr-*n*	H	H	Ph			HCl, 210–212	153

TABLE 7. 4,5-DIHYDRO-1H-1-BENZAZEPINES

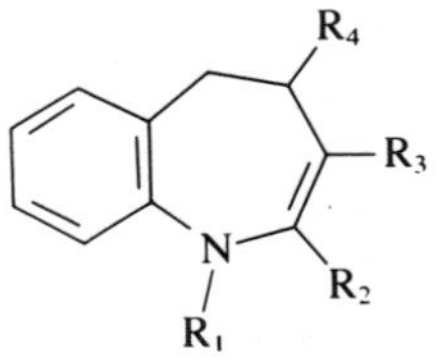

R_1	R_2	R_3	R_4	Other Substituents	mp (°C)	Refs.
Me	H	CO_2Me	CO_2Me		120–120.5	128
H	Cl	CHO	H	8-Cl	156–156.5	155

TABLE 8. 2,5-DIHYDRO-1H-1-BENZAZEPINES

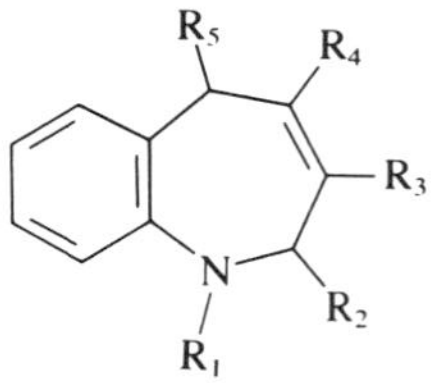

R_1	R_2	R_3	R_4	R_5	Other Substituents	mp (°C) or bp (°C/torr)	Other Data	Refs.
H	H	H	Me	H	7-Br	130–136/1.2	Picrate, mp 175–176°C	125
H	H	Et	Me	H		120/0.5		79b
Me	H	Et	Me	H		102/0.5		79b
Me	H	Et	Me	Me		110/1.25		79b
H	Me	Me	Me	Me		106/0.8		79b
H	H	H	H	H	7,9-Br_2	100–102		77
H	Me	H	Me	H			$[n]_D$ = 1.5680	126
H	Me	H	H	H			$[n]_D$ = 1.5687	126
H	Et	H	Me	H			$[n]_D$ = 1.5597	126

H	Et	H	H	H		$[n]_D$ = 1.5746	126
Tosyl	H	H	EtO	=O	135		92
H	=O	H	Me	OH	221–223		118
H	=O	H	Cl	OH	185–188		118
H	=O	Cl	Cl	OH	183–186		118
H	=O	H	Me	Cl	130–132		118
H	=O	H	Me	NMe_2	226–227		118
H	=O	H	Me	N	194–196		118
H	=O	Cl	Cl	N	202–204		118
H	=O	H	Cl	N	182–183		118
H	=O	H	Me	$CH(CO_2Et)_2$	191–192		118
Tosyl	H	H	Br	=O	131		134
Tosyl	H	H	Br	=O 7-MeO		Mass spectrum	134
Mesyl	Me	H	Br	=O	154		134

TABLE 9. 2,3-DIHYDRO-1H-1-BENZAZEPINES

R_1	R_2	R_3	R_4	R_5	Other Substituents	mp (°C) or bp (°C/torr)	Derivatives (mp, °C)	Refs.
Me	H	H	H	Cl		100–110/0.1	HBr, 160–161	80
Ph	H	H	H	Cl		140–150/0.4		80
Me	H	H	H	Cl	8-MeO	132–134/0.4	HBr, 157–160	80
Tosyl	H	H	CH_2Ph	H		79–80		161
H	H	H	H	H	7,8-Me_2	89	HCl, 196 Picrate, 174	178,70
Me	H	Me	Me	H		83–85/0.03		189
Me	$NHCO_2Et$	Me	Me	H		98–102/0.02		189
Me	H	H	H	Ph	7-Cl	104–106		85
Me	H	H	H	Ph		82–84		85
Tosyl	H	Br	H	H		128–129		191
Tosyl	H	=O	H	H		153–154		191
Tosyl	H	OH	H	H		114–115		191
H	H	H	H	H		48 95–100/0.4		92,191
Tosyl	H	H	H	H		108		191
PhCO	H	H	H	H		80		92

Ac	H	H	H	H		66		191
$CH_2C(Cl){=}CH_2$	H	H	H	H		55		161
Me	H	H	H	H			HBr, 176–177	161
$(CH_2)_2NMe_2$	H	H	H	H		125–130/0.1		161
Ac	H	H	H	H	7,8-Me_2	95		70
Ac	H	H	CO_2Me	1-Pyrrolidinyl		165–166		127
Ac	H	CO_2Me	CO_2Me	1-Pyrrolidinyl		150–152		127
Ac	H	H	CO_2Me	OH		93–97		127
Ac	H	CO_2Me	CO_2Me	OH		116–119		127
Me	H	CO_2Me	CO_2Me	H		289–290		128
Me	MeO	CO_2Me	CO_2Me	H		112–114		128
H	β-Indolyl	CO_2Me	CO_2Me	H		240–242		128
Me	β-Indolyl	CO_2Me	CO_2Me	H		228–229		128
Me	β-*N*-Methyl-indolyl	CO_2Me	CO_2Me	H		157–158		128
Me	H	H	Me	H		115–117/0.3	Picrate, 155–156 MeI, 159–161	81
Me	$=NCO_2Et$	Me	Me	H		130–132/0.03		189
Me	$=NPh$	Me	Me	H		152–156/0.03		189
Me	$=NCOPh$	Me	Me	H		158–162/0.03		189
Tosyl	H	$=O$	H	H		153–154	2,4-DNP, 209–211	191

TABLE 10. 2,3-DIHYDRO-1H-1-BENZAZEPIN-2-ONES

R_1	R_2	R_3	R_4	Other Substituents	mp (°C) or bp (°C/torr)	Refs.
Me	Me	Me	H		103–105/0.025	189
H	H	CO_2H	H		295–300	129
H	CO_2Me	CO_2Me	H		211–213	129
Me	CO_2Me	CO_2Me	H		127–128	129
H	H	Cl	Ph		270–272	85
Me	H	Cl	Ph		129–131	85
Me	H	Cl	Ph	7-Cl	137–139	85
Me	H	H	Ph		88–90	85
Me	H	H	Ph	7-Cl	132–133	85
H	H	H	Ph		200–201	84,85
H	H	H	Ph	7-Cl	211–214	85
$(CH_2)_2NMe_2$	H	H	Ph		99.5–101.5	85
$(CH_2)_3NMe_2$	H	H	Ph		93–95	85
H	H	CO_2H	OH		222.5–225	74,75
H	H	CO_2Me	OH	8-MeO	246	73
H	H	CO_2Et	OH		210–213	73
H	H	H	Me		158–158.5	84
H	H	Me	SCH_2CO_2H		84–85	118
H	=O	H	H		165–170	186

TABLE 11. 2,3,4,-TETRAHYDRO-1H-1-BENZAZEPINE-2,5-DIONES

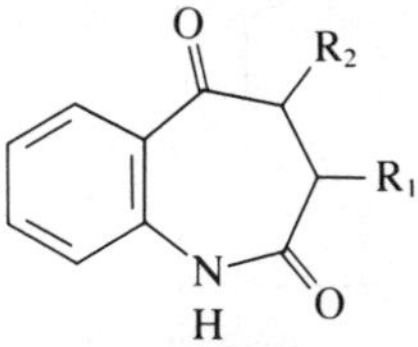

R_1	R_2	Other Substituents	mp (°C)	Derivatives (mp, °C)	Refs.
H	Me		149–150		118
H	H		191–192		118
H	H	7-CO_2Me, 9-Et	161–162		71
H	H	7-Br, 9-Et	92–93		71
H	H	7,8-Me_2	199	2,4-DNP, 300 Oxime, 226	70
H	Br	7,8-Me_2	135 (dec.)		70
H	=$NC_6H_4NMe_2$-*p*		310–312		124

amines has also been used, as in the synthesis of **175** (R_1 = Me, R_2 = 6-OBz, R_3 = 7-OMe) (196). The lactam **176** (R_1 = R_2 = R_3 = H) was obtained from the corresponding amino acid on heating strongly *in vacuo* (197). Imides **177** and **178** were obtained from the appropriate half-amides (198,199), whereas reaction of difunctional electrophiles with amines (200–202) has proved a versatile method of preparation for a variety of tetrahydro-2-benzazepines (138). Modifications of these methods are still popular, as illustrated, for example, by the conversion of the phthalides **179** (R = H, Me) in the absence of solvents (203). The products (**180;** R = H,Me) can be dehydrated and reduced to the lactams **181** (R = H,Me).

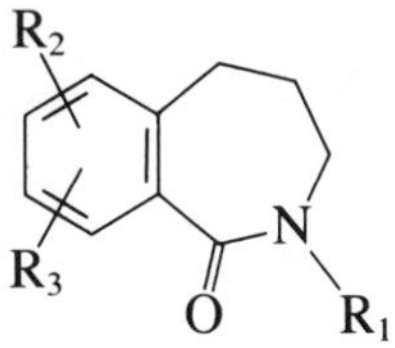

176 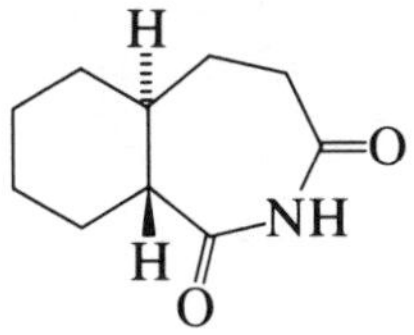 **177** 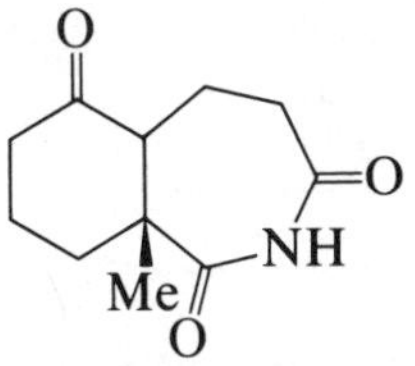 **178**

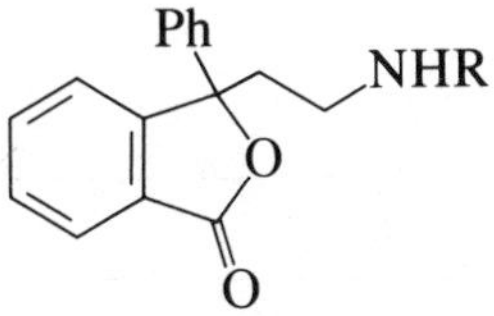

179 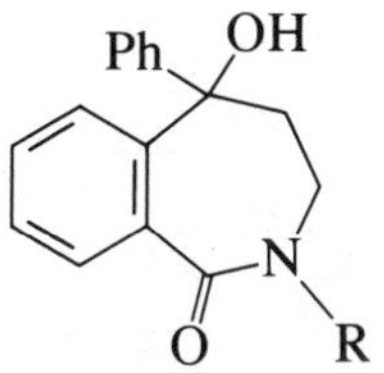 **180** 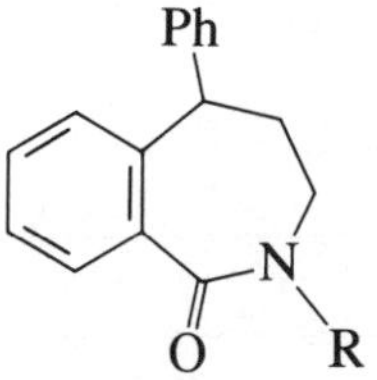 **181**

TABLE 12. 2,5-DIHYDRO-1H-1-BENZAZEPINE-2,5-DIONES

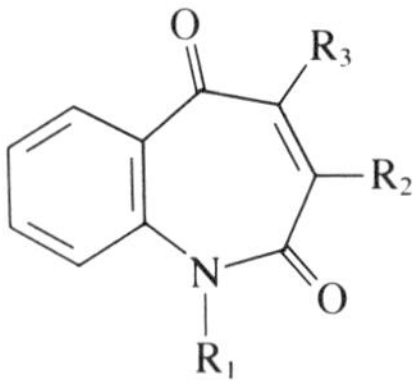

R_1	R_2	R_3	Other Substituents	mp (°C)	Refs.
Me	H	Me		84–86	118
H	H	H	7,8-Me_2	258	70
H	H	OH	7,8-Me_2	290	70
H	H	Br	7,9-Br_2	174	77
H	H	H	7,9-Br_2	149	77
H	H	Br	7,9-Br_2, 8-Cl	198	77
H	Me	OH		241	136
H	CO_2Et	OH		179	136
H	CO_2Et	MeO		164	136
Ph	Me	OH		206	136
Ph	CO_2Et	OH		176	136
Ph	CO_2Et	MeO		185	136
Ph	PhCO	OH		188	136
Ph	4-MeOPhCO	OH		165	136
H	H	H		222–223	116
H	H	Me		200–201	116,118
H	Me	H		217–218	116
Ph	H	OH		156	136
H	H	OH		260	122
H	H	Cl		205–206	118
H	Cl	Cl		260–262	118
H	Cl	Br		282–284	118
H	OH	H		245–248	119
H	OH	Br		260–265 (dec.)	119
H	AcO	H		201–203	119
H	MeO	H		255	119
H	MeO	H	7,8-Me_2	295–298	119
H	EtO	H	7,8-Me_2	256	119
H	OH	H	7,8-Me_2	288	119
H	Cl	H		204	119

TABLE 13. 3H-1-BENZAZEPINES

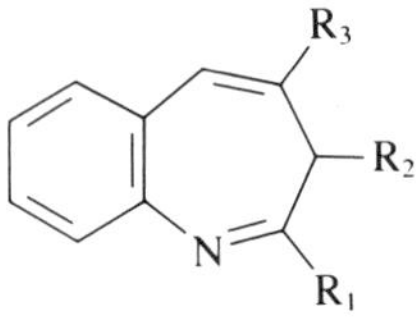

R_1	R_2	R_3	mp (°C)	Ref.
OEt	CO_2Me	CO_2Me	109–110	129

TABLE 14. 1H-1-BENZAZEPINES

R_1	R_2	R_3	R_4	R_5	Other Substituents	mp (°C)	Refs.
Me	H	CO_2Me	CO_2Me	Me		87–89	131
Me	EtO	CO_2Me	CO_2Me	H		138–139	129
H	H	H	Me	COMe		180	87
H	H	H	Me	COEt		176	87
H	H	H	Me	COPr-*i*		173	87
H	H	H	Me	$CO(CH_2)_3CH_3$		148	87
Ac	H	H	CO_2Me	N		150–151	127
Ac	H	CO_2Me	CO_2Me	N		171–172	127
Ac	H	H	CO_2Me	OH		137–139	127
Ac	H	CO_2Me	CO_2Me	OH		163–165	127
Me	H	CO_2Me	CO_2Me	H		105–107	128
H	H	H	H	Ph	8-MeO	82–83	193
Tosyl	H	H	Br	EtO		98	134
Tosyl	H	H	Br	EtO	7-MeO	136–137	134
Mesyl	Me	H	Br	EtO		126–127	134
CO_2Me	H	H	H	H	6,7,8,9-Tetrahydro	100–110/0.5[a]	18

[a]Boiling point (°C/torr).

TABLE 15. 5H-1-BENZAZEPINES

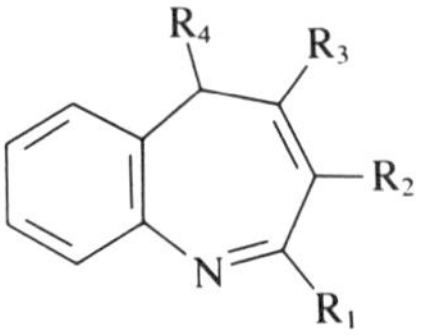

R_1	R_2	R_3	R_4	Other Substituents	mp (°C)	Refs.
NMe_2	Ph	H	Me_2		142	133
NMe_2	Ph	Me	Me_2		138	133
NEt_2	NEt_2	H	Me_2		81	133
H	H	Br	=O	7,9-Br_2	158	77
Cl	H	Br	=O	7,9-Br_2	142	77
H	H	Br	=O	7,9-Br_2, 8-Cl	158–160	77
EtO	H	OH	=O		81–82.5	123
EtO	H	H	=O		46	124
EtO	H	Me	=O		36–37	123
EtO	Me	H	=O		38–40	123

A variety of cyclizations involving nitriles have also been reported. Thus dinitrile **182** yielded the aminoazepine **183** on treatment with hydrogen bromide (29), the nitrile **184** gave the chloroazepine **185** in the presence of hydrogen chloride (204), and polyphosphoric acid converted the nitriles **186** and **187** to the 2-benzazepine-1,3-diones **188** (R = Ph) and **189** (R = Ph), respectively (205). The corresponding synthesis of **188** (R = H) and **189** (R = H) from the nitriles **186** (R = H) and **187** (R = H) has likewise been described (206).

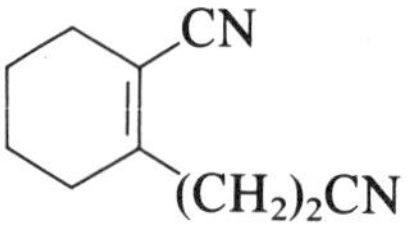

182

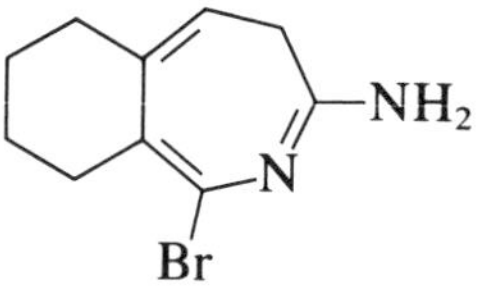

183

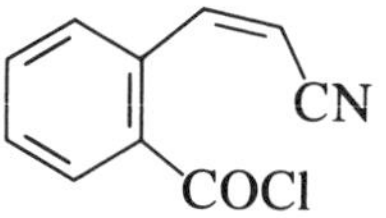

184

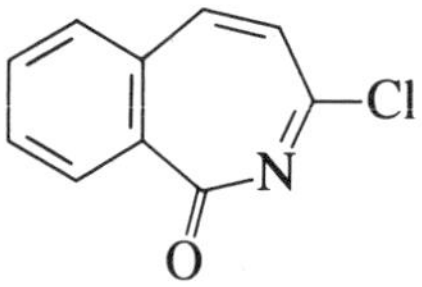

185

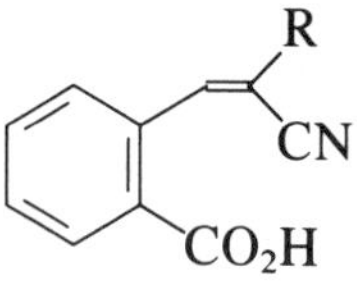

186

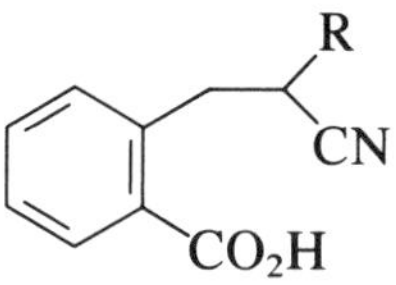

187

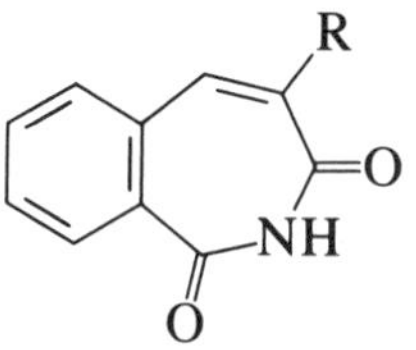

188

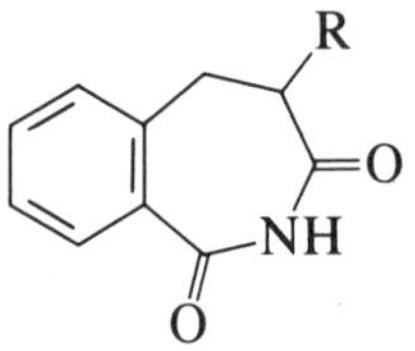

189

Saturated and unsaturated nitriles required for these syntheses have been obtained by ring opening of oximes such as **190** and nitroso compounds such as **191** (207–209). Photolysis of azidonaphthalene in the presence of potassium methoxide has been reported (438) to yield an aziridine that, on standing, was converted to the 2-benzazepine **192** (R = MeO). The latter could be converted to the 3-amino derivative **192** (R = NHPh, NEt_2) by heating with the appropriate amine, and to the 3-oxo compound **193.**

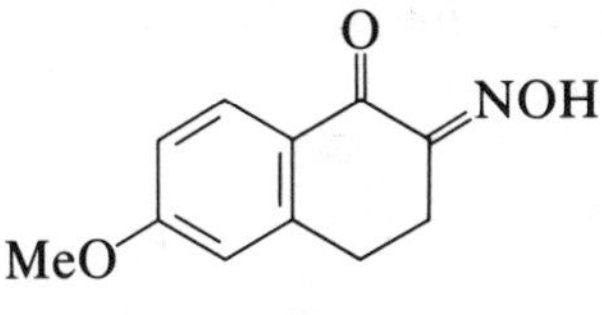

190

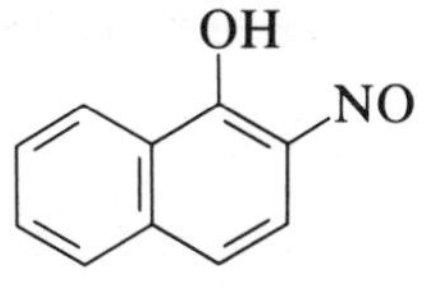

191

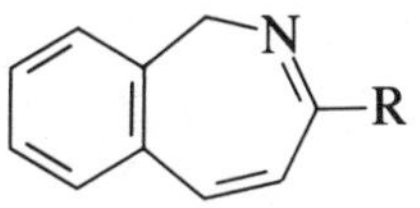

192

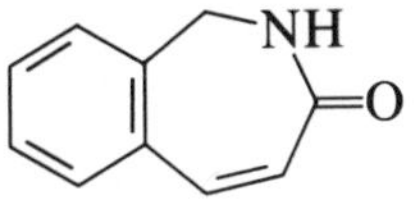

193

ii. C—C Bond Formation. The only satisfactory Friedel–Crafts cyclization of an *N*-benzyl-β-alanyl chloride proceeds at low temperature (−70 to 0°C) in the presence of alumium chloride to give ketone **194** (R = Ts) (210). In other instances (82), cleavage of the benzylic carbon–nitrogen bond intervenes. Mechanistically similar reactions are the cyclization of the *N*-benzyl-*N*-propargylamine derivative **195** to the 1,2-dihydro-2-benzazepine **196** (211) and the conversion of the bromides **197** (R = H, Me, Cl) to the tetrahydro-2-benzazepines **175** (R_1 = R_2 = H; R_3 = 7-H, 7-Me, 7-Cl) (212). Both processes require aluminum chloride as a catalyst.

Intramolecular benzene ring alkylations comparable to the Bischler–Napieralski and Pictet–Spengler isoquinoline syntheses are known, and many 3-

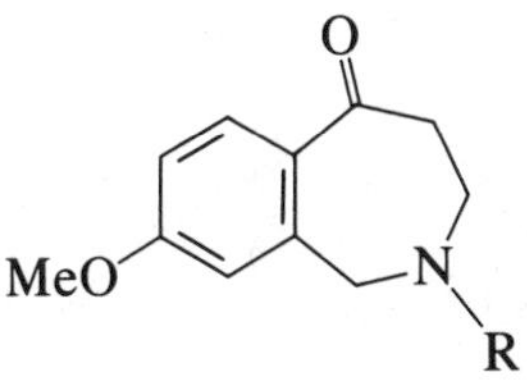

194

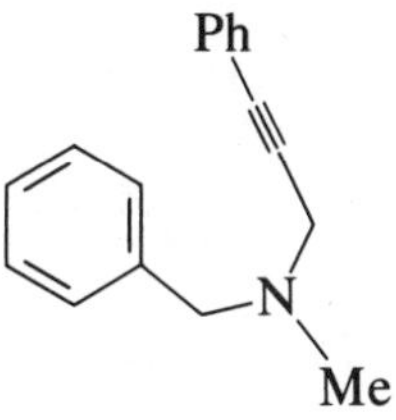

195

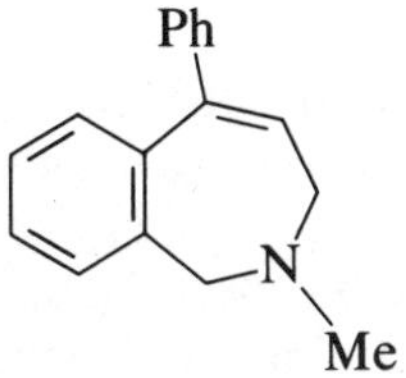

196

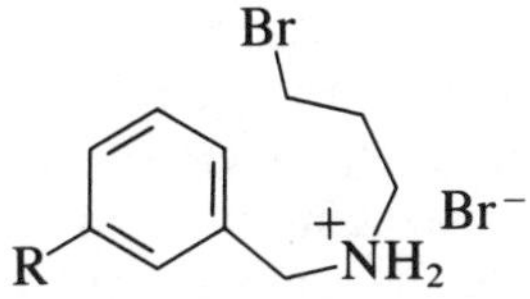

197

arylpropylamine derivatives have been cyclized. Dihydro-2-benzazepines such as **198** (R_1 = aryl, R_2 = CN) can be obtained from the corresponding amides (**198;** R_1 = aryl, R_2 = $CONH_2$) using phosphoryl chloride (213–216), polyphosphoric acid (217,218), phosphorus pentoxide (219, 225), and polyphosphoric ester (220). Dihydro-2-benzazepines such as **198** (R_1 = R_2 = H) are known to dimerize reversibly on standing (217–222).

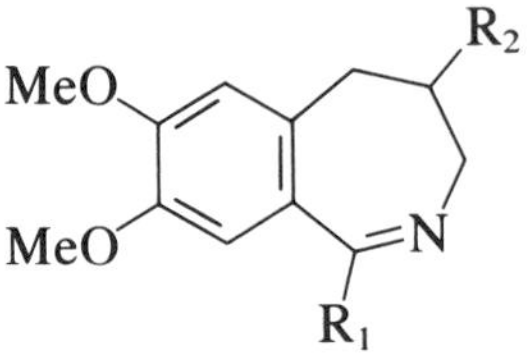

MeO, MeO, R_2, N, R_1

198

3-Phenylpropyl isothiocyanate has been found to cyclize to the thiolactam **199** in the presence of aluminum chloride (221). Pictet–Spengler-type cyclizations of the amide **200** and amine **201** with arylaldehydes have been observed to give the tetrahydro-2-benzazepines **202** and **203,** respectively (223,224). In the latter instance, the yields are reported to be poor (224). *N*-Tosyl-1,2,3,4-tetrahydro-2-benzazepines (**175;** R_1 = tosyl, R_2, R_3 = alkyl or H) were obtained by decarbonylative cyclization of appropriately substituted *N*-tosyl glycyl chlorides (232,233). [1H]-2-Benzazepines such as **192** (R_1 = NMe_2, R_2 = aryl) have been obtained by phosphoryl chloride treatment of ureas obtained from 2-(β-styryl)benzylamines (514).

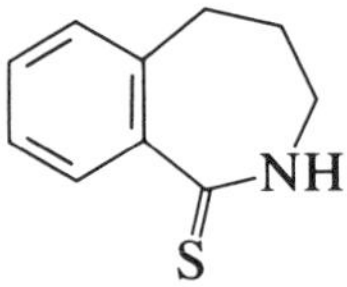

NH, S

199

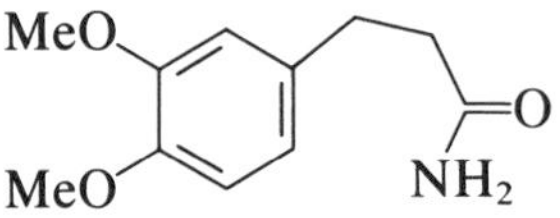

MeO, MeO, O, NH_2

200

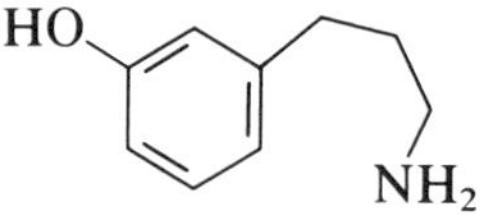

HO, NH_2

201

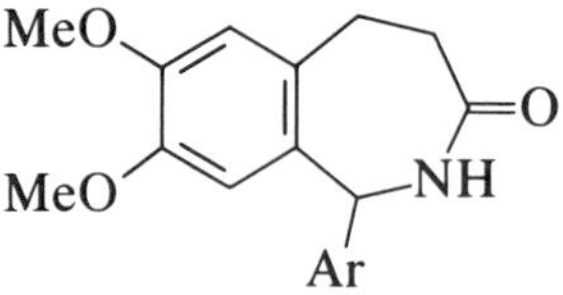

MeO, MeO, O, NH, Ar

202

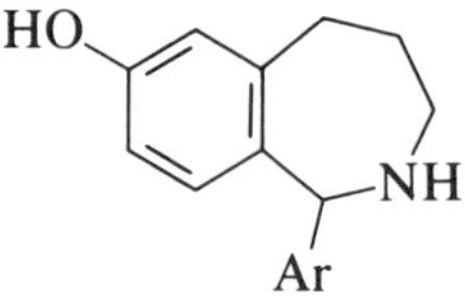

HO, NH, Ar

203

iii. Ring Expansion via Beckmann and Schmidt Reactions. As pointed out earlier, aryl migratory aptitudes generally favor 1-benzazepin-2-ones over 2-benzazepin-1-ones in the rearrangement of 1-tetralone oximes. Apart from the instances already cited (66,98,99,108,112), there are not many authentic examples of the conversion of 1-tetralones to 2-benzazepin-2-ones.

Production of 3-piperidinoethyl- and 3-piperidinomethyl-2-benzazepin-1-ones (226) represents a special situation (66). Also atypical is the preparation of the 2-benzazepin-1-one **176** ($R_1 = R_2 = R_3 = H$) by reduction of the tetrazole **204,** which was the product of reaction of 1-tetralone with sodium azide and hydrochloric acid (227). In general, one can expect to obtain either 1-benzazepin-2-ones or a mixture of both isomers (110).

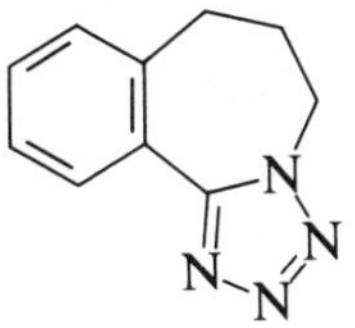

204

On the other hand, 2-tetralones can be a useful source of 2-benzazepine derivatives. Thus the oxime **205** (R = Me) gave **206** (R = Me) (228, 229), while Conley and Lange (230) demonstrated that Beckmann fragmentation of **205** (R = H) gives the nitrile **207**, which can re-close to either **206** (R = H) or **208**. From the saturated ring systems **209** (R = H, Me) the 2-benzazepine derivatives **210** (R = H, Me) were obtained (231), but the α,β-unsaturated ketone **211** (R = Me) reacted with HN_3 to give several products including the ring-enlarged lactams **212** (20% yield) (22). In a related study (182), ketone **213** (R = H) reacted with sodium azide in trichloroacetic acid to give **214.** However, under other conditions the ketones **213** (R = H, Me) gave either 1-benzazepine or 3-benzazepine derivatives.

iv. Other Ring-Expansion Reactions. The perchlorate salt **215** was converted to **216** in a two-stage process involving reaction with diazomethane

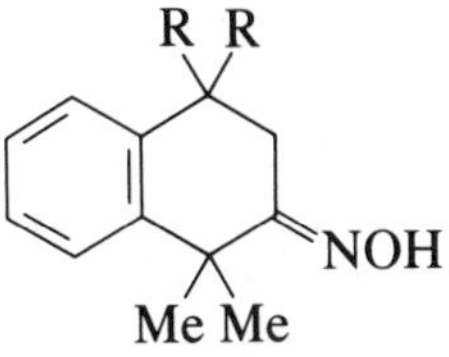

205 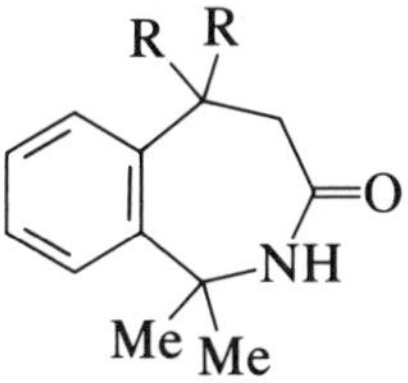 **206** 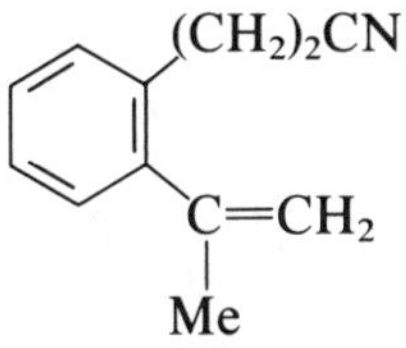 **207**

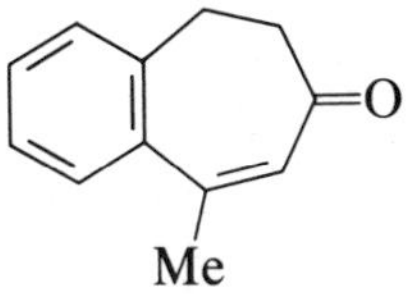

208 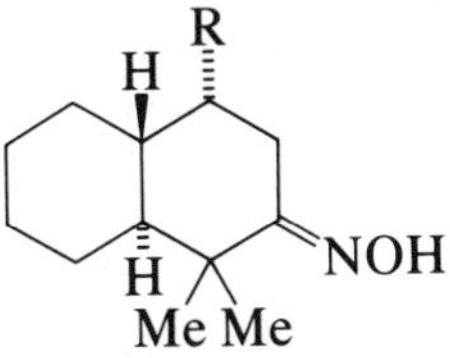 **209** 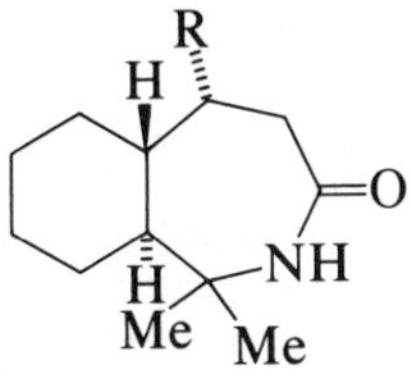 **210**

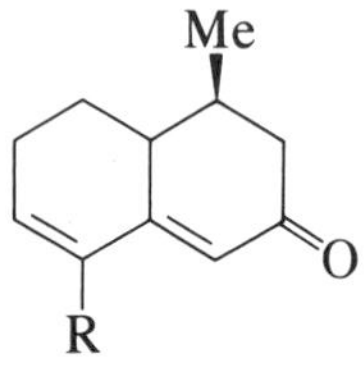

211

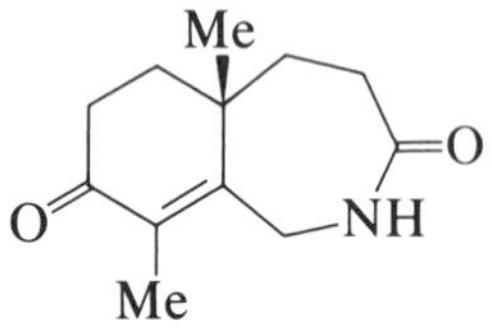

212

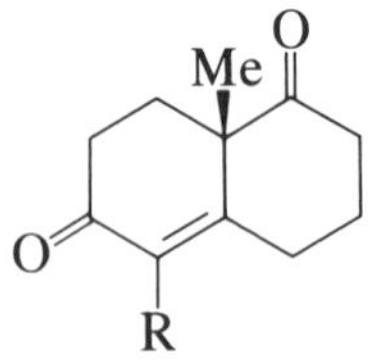

213

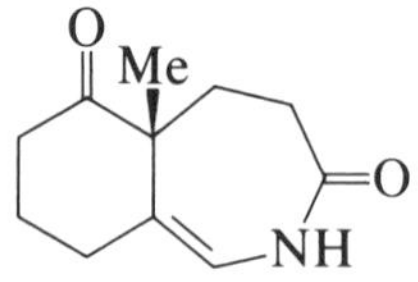

214

followed by treatment with methanolic sodium hydroxide (234). Thermal reaction of 1-aziridines with 1,3-diphenylinden-2-one gave the 3H-2-benzazepines **217** (R = H, Me, Ph) (235,236), while the addition of methylamine to the aminonaphthoquinones **218** (R = H, Cl) yielded substances that were formulated as the carbinolamines **219** (R′ = OH, NHMe) (237).

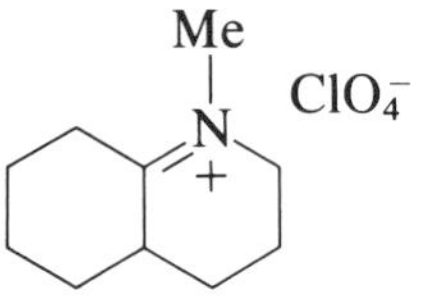

215

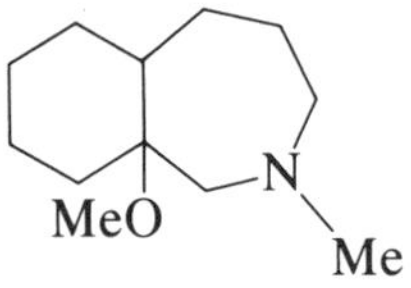

216

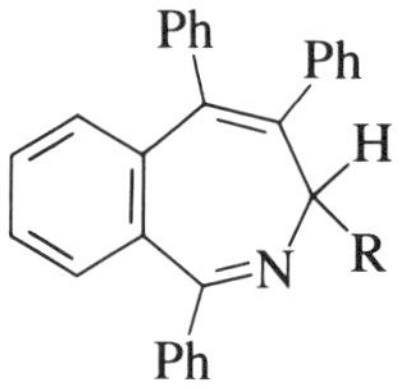

217

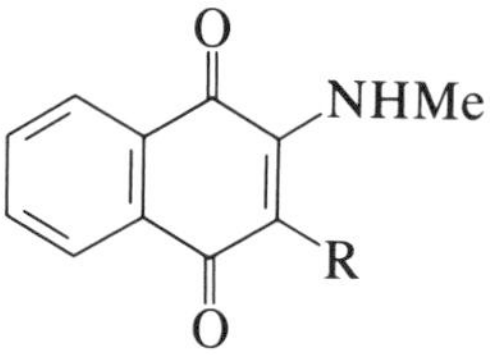

218

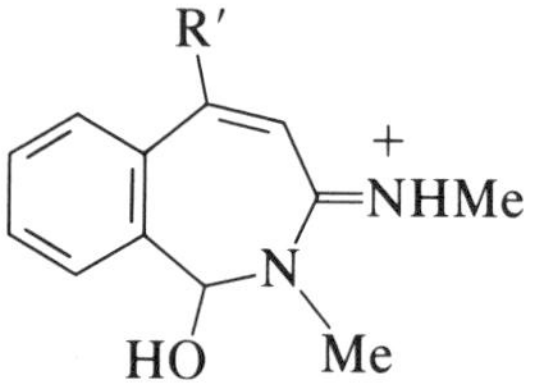

219

Photochemistry has played a significant part in the development of this area. Thus photolysis of *N*-alkylphthalimides led to the 2-benzazepine-1,5-diones **220** (R_1 = H, Me; R_2 = H, Me, R_3 = H) (238), while irradiation of *N*-methylphthalimide with buta-1,3-diene gave *E*- and *Z*-4-ethenyl compounds **220** (R_1 = CH_3CH═, R_2 = H, R_3 = Me) (515). The stereochemistry

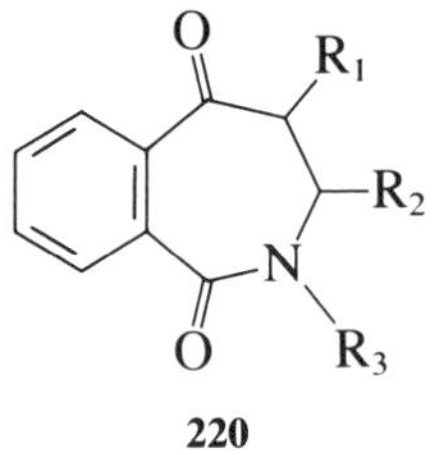

220

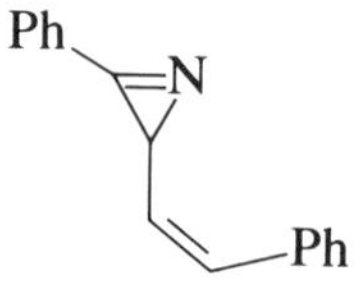

221

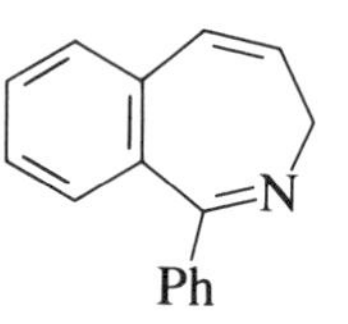

222

of this reaction was further explored using *E*- and *Z*-butene (516). Irradiation of either the *E* or *Z* isomer of 2-styryl-2H-azirine (**221**) gave 1-phenyl-3H-2-benzazepine **222** (239,240).

Alkenes add to the isoindolone **223** under photolytic conditions to form adducts of the general structure **224** (R_1, R_2, R_3, and R_4 are H, Me, or MeO), which on exposure to acid are cleaved to the corresponding 2-benzazepine-1,5-diones (241). Photolysis of the ylide **225** and the oxazirane **226** is said to have given the ring-enlarged products **227** and **228**, respectively (242,243), though the yield in the former instance was only 5%.

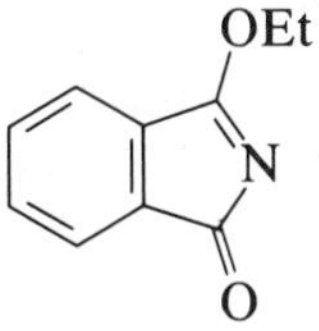

OEt, N, O — **223**
R_1, OEt, R_2, R_3, N, R_4, O — **224**
CO_2Et, N^{+}–C^{-}, CO_2Et — **225**
O–NCH_2Ph — **226**
CO_2Et, N, CO_2Et — **227**
N, O, CH_2Ph — **228** 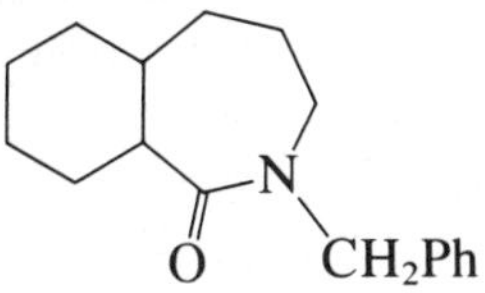

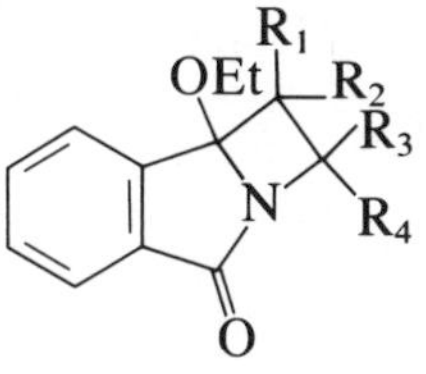

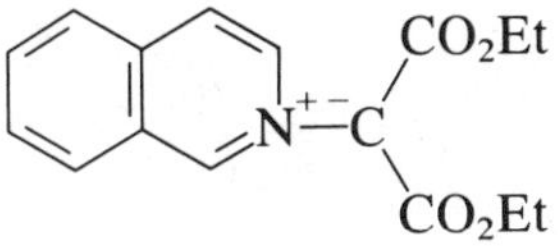

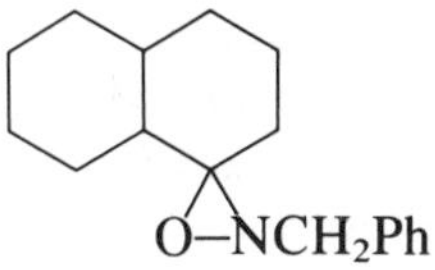

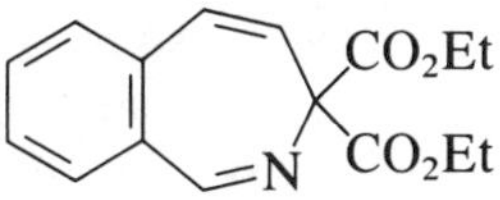

b. CHEMICAL PROPERTIES

Most partially reduced 2-benzazepines behave as simple cyclic amides or amines that are amenable to substitution on nitrogen by conventional pharmacophoric groups (e.g., *N,N*-dialkylaminoalkyl) (219,225,244–246). Reduction of the imine bond in 4,5-dihydro-2-benzazepines proceeds easily with $NaBH_4$ (219,240). The facile reversible dimerization of 4,5-dihydro-2-benzazepines was mentioned earlier (222).

Little is known about fully dehydrogenated 2-benzazepines (236,240), but it appears that they are stable in the 3H-2-benzazepine form (see, for example, structures **217** and **222**). Efforts to obtain azatropones (1H-2-benzazepin-5-ones) such as **229** were unsuccessful (247), although the existence of the 3-chloro derivatives **185** was noted (204). Simple cyclic Mannich bases (e.g., **194,** R = H, Me) are exceptionally unstable (247). However, *N*-tosyl derivatives of these compounds are more stable, and derivatives of **194** (R = tosyl) with bromine substituents at the 1- and 4- positions have been reported (210). Treatment of the 1-arylethyl-2-benzazepine **230** with dimsyl sodium resulted in the intramolecular displacement of bromine and a novel skeletal rearrangement, giving rise to the 10-membered ring product **231** (254,517). 2-Benzazepine derivatives for which physical constants are recorded in the literature are listed in Tables 16–25.

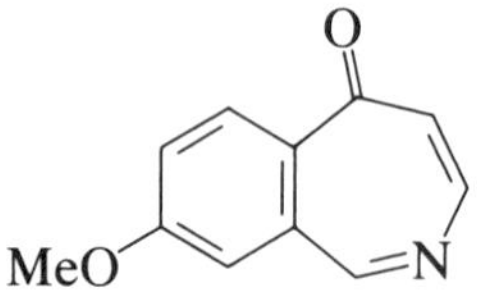

229

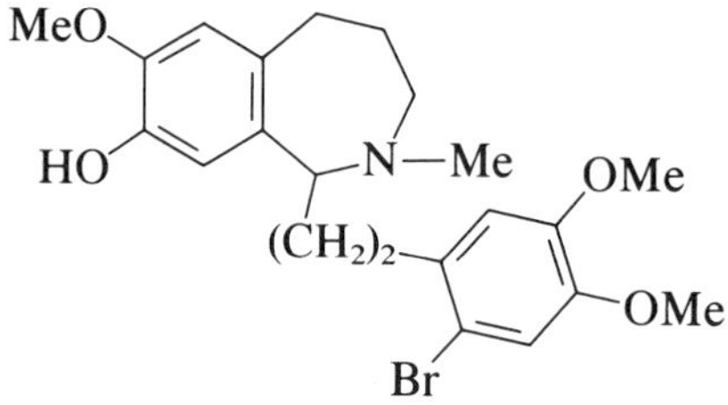

230

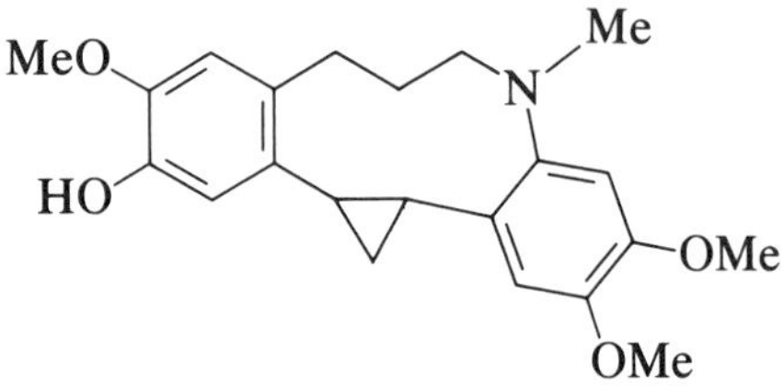

231

Some alkoxy-2-benzazepin-1-one derivatives (**232**) have been made by alcoholysis of dichlorocarbene adducts (**233**) of isoquinolones (563). Intramolecular cyclization of the isocyanate **234** gave the benzazepinone **235** in 80% yield (564) and, unexpectedly, treatment of the aminonaphthalenone **236** with aqueous sodium hydroxide gave the 2-benzazepinone **237** (565). A mechanism for the latter transformation has been proposed.

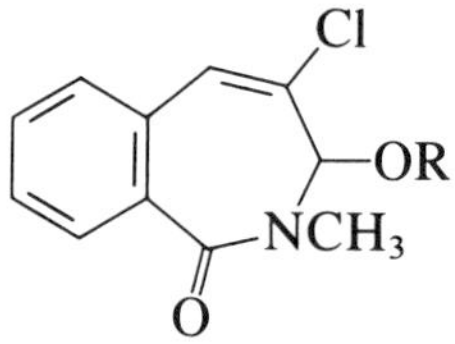

232

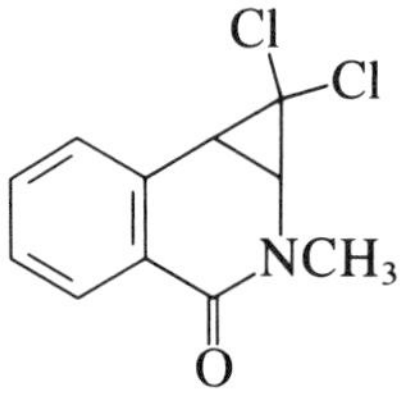

233

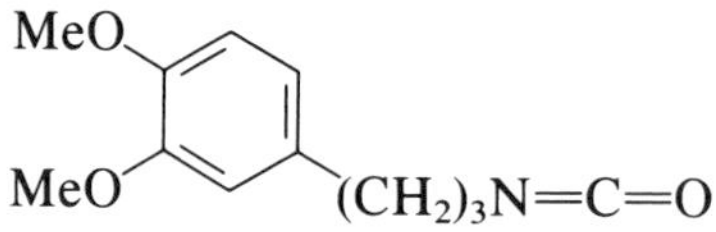

234

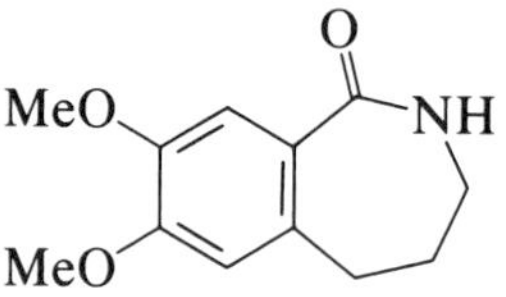

235

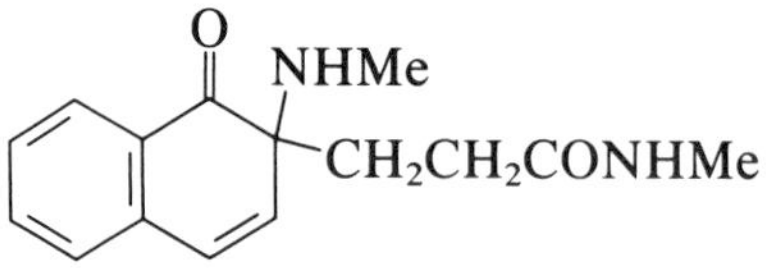

236

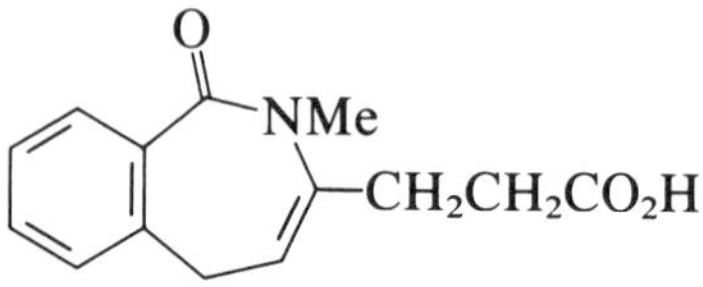

237

TABLE 16. OCTA- AND DECAHYDRO-2-BENZAZEPINES[a]

Structure	mp (°C)	Derivatives (mp, °C)	Refs.
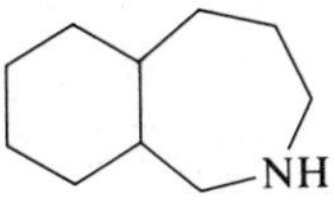	115/19[b]		250
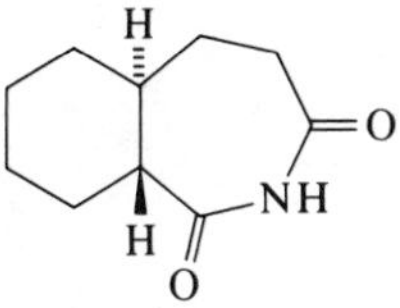	160–161		198
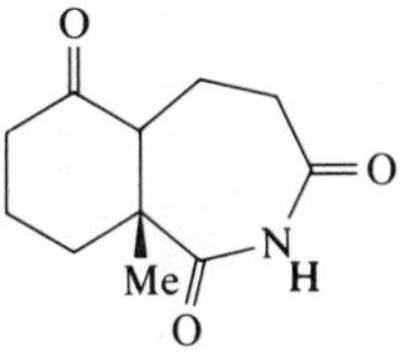	249–251		199
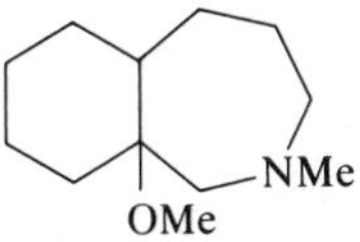		$HClO_4$, 174–175	234
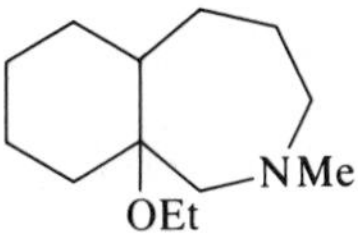		$HClO_4$, 163–163.5	234
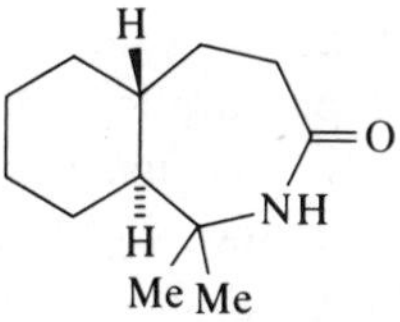	139–140		231
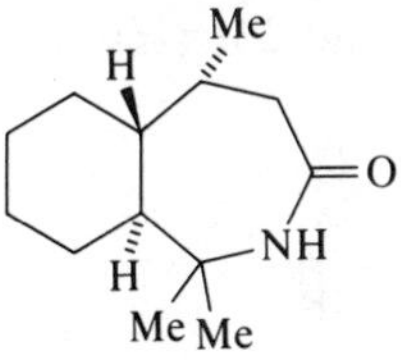	175–180		231
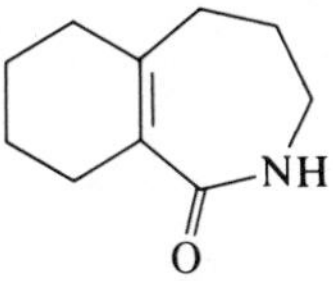	159–160		256
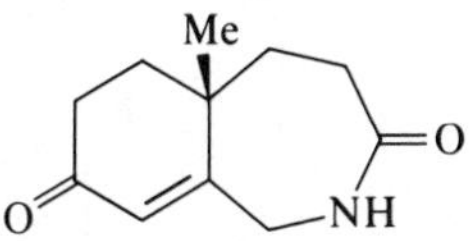	207–210		257

TABLE 16. (*Continued*)

Structure	mp (°C)	Derivatives (mp, °C)	Refs.
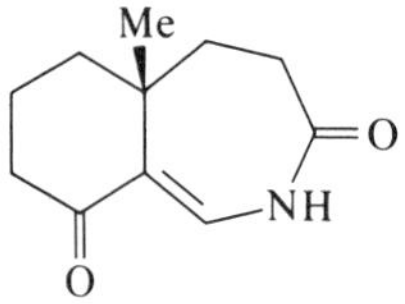 [structure]	163–165		257
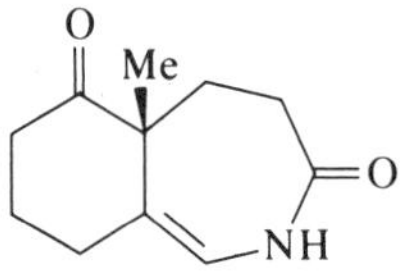 [structure]	142.4–144.5		182

[a]Stereochemistry of ring junctions is shown where known.
[b]Boiling point (°C/torr).

C. *3-Benzazepines*

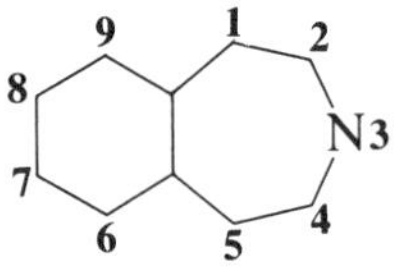

a. PREPARATION

i. C—N BOND FORMATION. Intramolecular dehydration of the amino acid **238** was the earliest reported method (197,258) of preparation of the benzazepinone **239** (R_1 = R_2 = R_3 = H), and this has been used again recently (259) to gain access to the substituted benzazepinones **239** (R_1 = Me, H; R_2 = H, CH_2Ph; R_3 = 7,8-$(MeO)_2$, 7,8-methylenedioxy). Other examples include the thermal cyclodehydration of **240** and **241** (260–262).

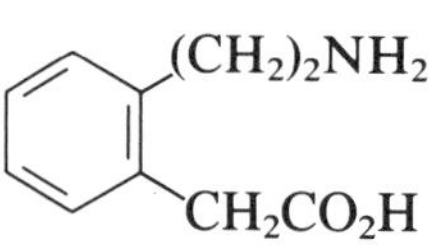

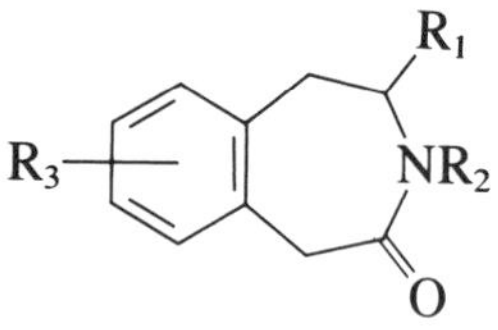

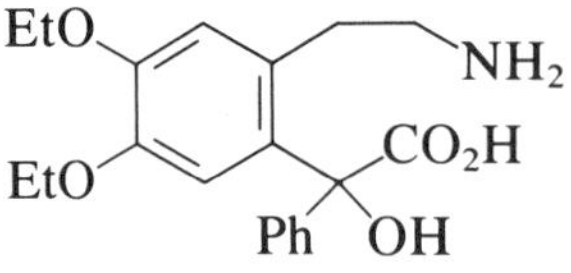

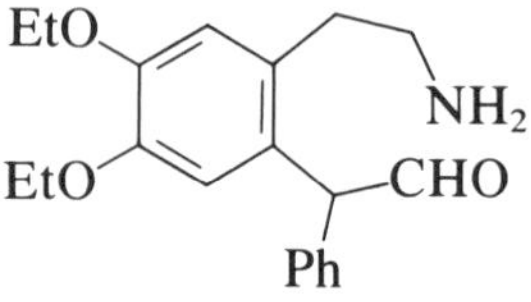

TABLE 17. 1,2,3,4-TETRAHYDRO-5H-2-BENZAZEPINES

R_1	R_2	R_3	R_4	R_5	Other Substituents	mp (°C) or bp (°C/torr)	Derivatives (mp, °C)	Refs.
H	H	H	H	H		130/30	HCl, 221–224	194
							HBr, 200–201	
							MeI, 180–181	245
H	H	H	H	H	7-Me	71–72 86–88/1	Picrate, 211–212	194
H	H	H	H	H	7-Cl	145–148/3	Picrate, 174–175	194
H	$PhCH_2$	H	H	H		47–48	Picrate, 149–150	201
H	Bu	H	H	H		148–151/13	Picrate, 126–128	201
H	Ph	H	H	H		136–147/ 0.005	Picrate, 170–172	201
H	$CH(Me)CH_2NMe_2$	H	H	H		130–132/0.8	diPicrate, 170	245
H	$(CH_2)_2CN$	H	H	H		130–140/19	HCl, 193–194	245
H	CONHPr	H	H	H		100–101/1		245
H	$(CH_2)_3N$(piperidino)	H	H	H	8-Cl	206–208/0.2	2HCl, 302–203	245
H	$CH_2CH(Me)OH$	H	H	H		136–139/3	HCl, 172–173	245
H	$CH_2CH(Me)Cl$	H	H	H			HCl, 169–171	245

(*continued*)

TABLE 17. (*Continued*)

R_1	R_2	R_3	R_4	R_5	Other Substituents	mp (°C) or bp (°C/torr)	Derivatives (mp, °C)	Refs.
H	$CH_2CH(Me)N$ 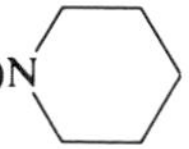	H	H	H		149–153/0.25	Oxalate, 140	245
$PhCH_2$	H	H	H	H			Maleate, 106–110	255
p-$ClC_6H_4CH_2$	H	H	H	H			HCl, 268–272	255
o-$NO_2C_6H_4CH_2$	H	H	H	H		62–67		255
3,4-$(MeO)_2C_6H_4CH_2$	H	H	H	H			Naphthalene disulfonate, 295–297	255
3,4-$(MeO)_2$-2-$NO_2C_6H_2CH_2$	H	H	H	H		107–110		255
H	Me	H	H	Ph		74–75	MeBr, 249–250	211,195
H	H	CH_2N 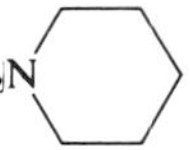	H	H		63–65		226
H	$C(=NH)NH_2$	H	H	H			H_2SO_4, 272 (dec.)	252
CN	Me	H	H	H		97–98		222
CN	Et	H	H	H		48		222
CH_2NH_2	Me	H	H	H		153/13		222
CH_2NH_2	Et	H	H	H		163/13		222
Ph	H	H	H	H	7-OH	261		224
3,4-$(MeO)_2C_6H_3$	H	H	H	H	7-OH	261		224
3,4-$(OCH_2O)C_6H_3$	H	H	H	H	7-OH	265		224
Ph	Me	H	H	H	7-OH	244–246		224

=S	H	H	H	H		223		221
H	H	=S	H	H		152		253
H	Me	H	Ph	H			HCl, 218–219 (dec.)	205
H	$PhCH_2$	H	Ph	H		130–131		205
H	$(CH_2)_2OH$	H	Ph	H			Picrate, 150–151 HCl, 186–188	205
H	$(CH_2)_2NMe_2$	H	Ph	H			Picrate, 220–221 (dec.)	205
H	$(CH_2)_3NMe_2$	H	Ph	H			Picrate, 203–205 (dec.) MeI, 261–263 (dec.)	205
H	$(CH_2)_3NMe_2$	H	p-MeC_6H_4	H			Picrate, 198–200 (dec.)	205
H	$(CH_2)_3NMe_2$	H	p-$MeOC_6H_4$	H			Picrate, 198–200 (dec.)	205
Ph	CN	H	H	H	7,8-$(MeO)_2$	114–115		223
H	$(CH_2)_2Me_2$	H	H	Me_2	7-OH		HCl, 247	244
H	$(CH_2)_2NMe_2$	H	H	Me_2	7-MeO		HCl, 253	244
H	$(CH_2)_3NMe_2$	H	H	Me_2	7-MeO		HCl, 262	244
H	Tosyl	H	H	Me_2	7-MeO	128		209
H	H	H	H	H	7-MeO	208–210		209
H	$CH_2CH(CH_2)_2$	H	H	H	7-MeO		HCl, 163–165	209
H	H	H	H	Me	7-MeO		HCl, low-melting	209
H	$CH_2CH(CH_2)_2$	H	H	Me	7-MeO		HCl, 184–186	209
H	H	H	H	Me_2	7-MeO		HCl, 164–166 HBr, 154	209 225
H	$CH_2CH(CH_2)_2$	H	H	Me_2	7-MeO		HCl, 162–166	209

(*continued*)

TABLE 17. *(Continued)*

R_1	R_2	R_3	R_4	R_5	Other Substituents	mp (°C) or bp (°C/torr)	Derivatives (mp, °C)	Refs.
H	$(CH_2)_2Ph$	H	H	Me_2	7-MeO		HCl, 176–178	209
3,4-$(MeO)_2C_6H_3$	CHO	H	H	H	7,8-$(MeO)_2$	235–245/0.05		223
3,4-$(MeO)_2C_6H_3$	Me	H	H	H	7,8-$(MeO)_2$	185–195/0.05	HCl, 210–211	223
H	$(CH_2)_2Ph$	H	H	Me_2	7-MeO		HBr, 186	225
H	$(CH_2)_2Ph$	H	H	Me_2	7-OH		HBr, 246	225
H	$(CH_2)_2Ph$	H	H	Me, Et	7-MeO		HBr, 208	225
H	$(CH_2)_2Ph$	H	H	Me, Pr	7-MeO		Picrate, 129	225
H	$(CH_2)_2Ph$	H	H	Me, Et	7-OH		HBr, 269	225
H	$(CH_2)_2C_6H_4Cl$-*p*	H	H	Me_2	7-MeO		HBr, 221	225
H	H	H	H	Me_2	7-OH		HBr, 259	225
H	Dimethylallyl	H	H	Me_2	7-OH		HCl, 102–103	225
H	Dimethylallyl	H	H	Me_2	7-MeO	130–132/3		225
H	$CH_2CH<(CH_2)_2$ (cyclopropylmethyl, drawn as ring)	H	H	Me_2	7-OH		HCl, 104–106	225
H	$CH_3C{\equiv}C$—	H	H	Me_2	7-OH		HCl, 106	225
H	$CH_3C{\equiv}C$—	H	H	Me_2	7-MeO	141/3		225
H	Me	H	H	Me_2	7-MeO	143/5		219,225
H	Allyl	H	H	Me_2	7-MeO	147/3		225
H	Me	H	H	Me_2	7-OH		HBr, 250	225
Me	H	H	H	H	7,8-$(MeO)_2$	95–100/3	HBr, 195–196	223
Ph	H	H	H	H	7,8-$(MeO)_2$	116–118	HCl, 247–249	220,223
3,4-$(MeO)_2C_6H_4$	H	H	H	H	7,8-$(MeO)_2$	200–205/0.1	HBr, 220–221	223
Ph	H	H	$CONH_2$	H	7,8-$(MeO)_2$	200–202		214

Ph	H	H	NMe_2	H	7,8-$(MeO)_2$		Fumarate, 239–241	214
$PhCH_2$	H	H	H	H	7,8-$(MeO)_2$		Maleate, 156–157	215
p-$ClC_6H_4CH_2$	H	H	H	H	7,8-$(MeO)_2$		Naphthalene disulfonate, 280–282	215
3,4-$(MeO)_2C_6H_3CH_2$	H	H	H	H	7,8-$(MeO)_2$		Naphthalene disulfonate, 262–266	215
CH_2CH_2Ph	H	H	H	H	7,8-$(MeO)_2$		Naphthalene disulfonate, 288–290	215
$(CH_2)_2C_6H_4Cl$-p	H	H	H	H	7,8-$(MeO)_2$		Naphthalene disulfonate, 257–260	215
MeO	Me	H	H	H	7,8-(OCH_2O)	79–80		216
NHOH	Me	H	H	H	7,8-(OCH_2O)	145–146		216
H	H	H	H	H	7,8-$(MeO)_2$	172–173	Dimer(?)	220
Me	H	H	H	H	7,8-$(MeO)_2$	115–120/3	Picrate, 163.5	220
H	Me	H	H	H	6-OH, 7-MeO	194	OBz ether, bp 130°C/0.007 mm	196
H	Tosyl	H	H	H		134–136		227
H	Me	Me	H	Ph			MeCl, 246–248	195
H	CH_2OH	H	H	H		138–142/2		249
H	$PhCH_2$	H	H	Ph		195–225/2.5		195
H	H	H	H	Ph			HCl, 201–202	195
H	Me	H	H	Ph, CO_2H		227–228	Et ester HCl, 110–112	195
H	Me	H	H	OH	8-MeO	180/0.3	MeI, 183–184	247
H	Ac	H	H	OH	8-MeO	127–128		247
H	H	H	H	OH	8-MeO	137–139		247
H	Tosyl	H	H	OH	8-MeO	109		247

TABLE 18. 1,2-DIHYDRO-3H-2-BENZAZEPINES

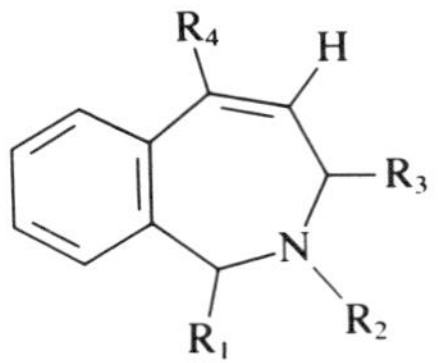

R_1	R_2	R_3	R_4	Other Substituents	mp (°C) or bp (°C/torr)	Derivatives (mp, °C)	Refs.
H	Tosyl	H	H	8-MeO	117–119		210
H	Me	H	Ph		146–148/0.55	Picrate, 209–210	211
Ph	H	H	H		Oil	Picrate, 215–218	239
H	H	=O	H		185		438
=O	H	H	Ph		233–235		203
=O	H	H	Ph		136.5–138		203

Sodium borohydride treatment of the amino ester **242** unexpectedly gave the *trans*-hydroxy lactam **243** (276), but it is not clear whether reduction preceded cylization or vice versa. The imides **244** (R_1 = Me; R_2 = H, $CH_2CH_2NMe_2$) have been obtained by heating the appropriate diacid with ammonia or *N,N*-dimethylaminoethylamine (263), while the imide **244** (R_1 = R_2 = H) was obtained by heating the corresponding diamide *in vacuo* (264).

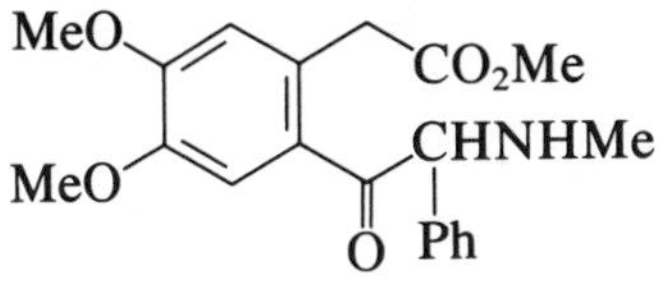

242

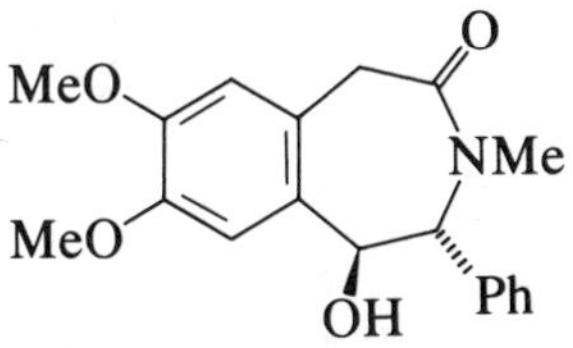

243 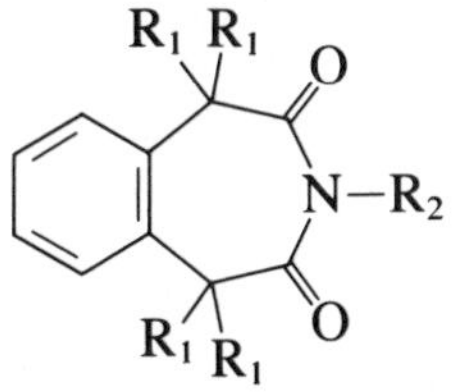 **244**

An interesting synthesis of 1,4-methano-3-benzazepine-2,5-diones involves cyclization of suitably constructed α-haloketocarboxamides, as illustrated by the conversion of **245** to **246** (326). Treatment of **247** (X = Br) with amines yielded the *N*-substituted tetrahydro-3-benzazepines **248** (265), whereas hydrogenation of the dinitrile **247** (X = CN) over Raney nickel (266,267,280) or rhodium on alumina (268) gave the unsubstituted compound

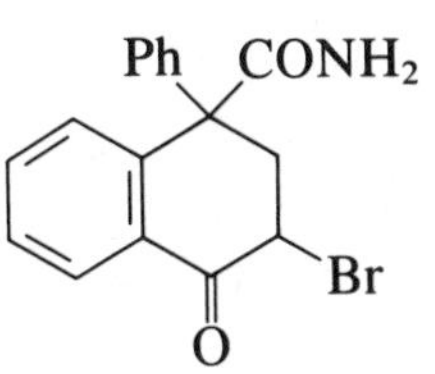

245 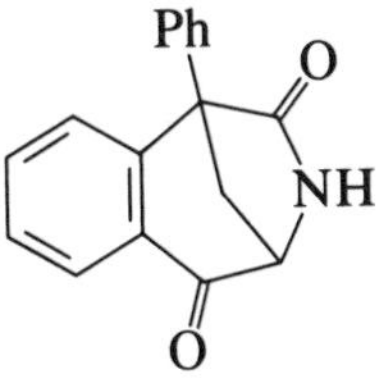 **246** 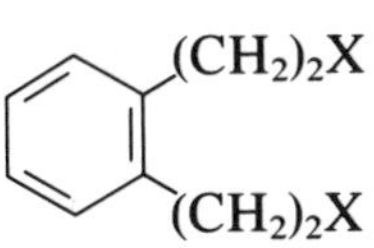 **247** 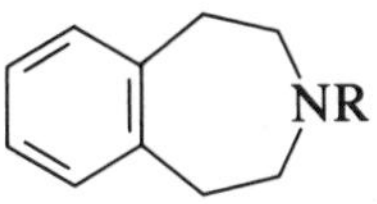 **248**

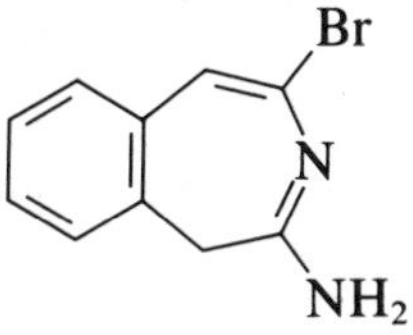

249 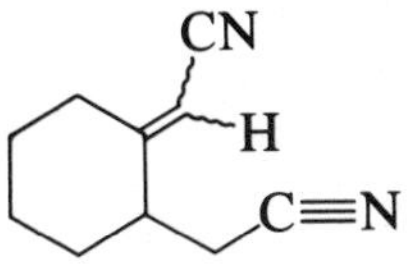 **250** 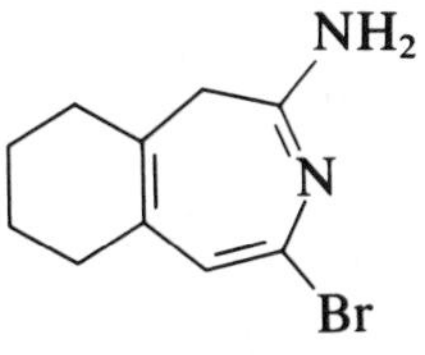 **251**

TABLE 19. 4,5-DIHYDRO-3H-2-BENZAZEPINES

R_1	R_2	R_3	R_4	Other Substituents	mp (°C) or bp (°C/torr)	Derivatives (mp, °C)	Refs.
p-ClC_6H_4	H	CO_2Et	H	7-MeO	94.5–96.5		214
Me	H	CO_2Et	H	7,8,9-$(MeO)_3$		HCl, 165–166	214
H	H	H	Et, Me	7-MeO		MeI, 189	219
H	H	H	H	7,8-$(MeO)_2$	165–166	Dimer (?)	220
Me	H	H	H	7,8-$(MeO)_2$	140–143/3	Picrate, 183–184	22
						MeI, 190–191	
Ph	H	H	H	7,8-$(MeO)_2$	84.5–85.5	Picrate, 222	220
					Oil	257–259	223
						MeI, 206–206.5	220

Ph	H	CN	H	7,8-$(MeO)_2$	145–147		214
Ph	H	CO_2Et	H	7,8-$(MeO)_2$	101–103	Acid + H_2O, 206.5–208.5	214
Ph	H	$CONH_2$	H	7,8-$(MeO)_2$	170–172		214
Ph	H	$CONMe_2$	H	7,8-$(MeO)_2$	152–154		214
p-ClC_6H_4	H	CO_2Et	H	7,8-$(MeO)_2$	122–124		214
Ph	H	CH_2NH_2	H	7,8-$(MeO)_2$		$2HClO_4 \cdot H_2O$, 189–191°C	214
MeS	H	H	H		102/1		221
NH—CH(CH_2—CH_2) (cyclopropyl)	H	H	H			HCl, 243–245	221
H	H	H	H			MeI, 120–122 EtI, 145	222
H	H	H	Me_2	7-MeO	139/3	HCl, 193 Picrate, 196	219
H	H	H	Et, Me			Picrate, 154	219

TABLE 20. 2,3,4,5-TETRAHYDRO-1H-2-BENZAZEPIN-1-ONES

R_1	R_2	R_3	R_4	Other Substituents	mp (°C)	Comments	Refs.
H	H	H	Ph, OH		145–147		203
Me	H	H	Ph, OH		140–141		203
H	H	H	Ph		224–225		110,203
H	H	H	Me_2	7-MeO	145–146		209
H	H	H	H	7-MeO	159–160 (117–118)		112,209 (108)
H	H	H	Me	7-MeO	158–159.5		209
H	$(CH_2)_2N$(piperidino)	H	H		115		226
H	CH_2N(piperidino)	H	Ph		230–235		226

H	$CH_2N(CH_2)_5$	H	H		118–120		226
$(CH_2)_3NMe_2$	H	H	Ph		114–115		85
H	H	Ph	H		103–105		110
CH_2Ph	H	H	H	7-MeO	219–221		207
H	H	H	H		109–110		108,245
H	CH_2CH_2-(2-pyridyl)	H	H		124–127		99
H	CH_2CH_2-(4-pyridyl)	H	H		167–170		99
H	H	CO_2Et	*trans*-3,4-$(MeO)_2C_6H_3$	7,8-$(MeO)_2$	148		98
H	H	H	H	7-OH	238–239	7-OAc, mp 131–132°C	112
H	H	H	H	7-NH_2	165–166	7-NHAc, mp 192–193°C	112
H	H	H	H	7-Cl	183–184		112
H	H	H	H	8-MeO	100–101		112
H	H	H	H	8-OAc	120–121		112
H	H	H	H	8-NHAc	240–241		112
H	H	H	H	8-Cl	160–161		112

TABLE 21. 2,3,4,5-TETRAHYDRO-1H-2-BENZAZEPIN-3-ONES

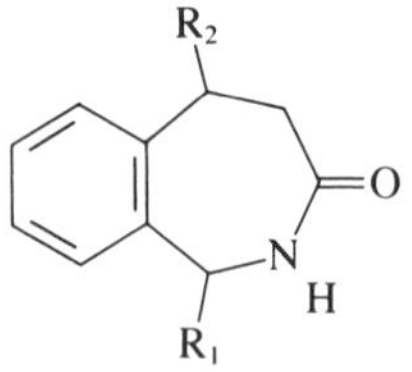

R_1	R_2	Other Substituents	mp (°C)	Refs.
Me_2	Me_2		144–145	228
H	H		109–110	248
Me_2	H		113–114.5	230
$PhCH_2$	H		153–155	255
o-$ClC_6H_4CH_2$	H		156–158	255
p-$NO_2C_6H_4CH_2$	H		200–209	255
2,3-$(MeO)_2C_6H_3CH_2$	H		207–210	255
2,3-$(MeO)_2$-4-$NO_2C_6H_2CH_2$	H		153–158	255
H	H	7,8-$(MeO)_2$	170–171	223
Me	H	7,8-$(MeO)_2$	165–166	223
Ph	H	7,8-$(MeO)_2$	191–192	223
3,4-$(MeO)_2C_6H_3$	H	7,8-$(MeO)_2$	183–184	223

248 (R = H). On the other hand, reaction of the dinitrile **247** (X = CN) with hydrogen bromide followed by mild base treatment afforded 2-amino-4-bromo-1H-3-benzazepine (**249**) (269,270,278). Under the same conditions, cyclohexylidine dinitrile (**250**) gave the 6,7,8,9-tetrahydro-3-benzazepine **251** (29).

ii. C—C Bond Formation. Cyclization of β-phenylethylglycyl derivatives has been well studied and can give good yields of tetrahydro-3-benzazepin-1-ones of the general structure **252.** For example, the ketone **252** (R_1 = tosyl, R_2 = H) can be obtained from the glycyl chloride in the presence of 3 molar equivalents of anhydrous aluminum chloride in methylene chloride at −15°C (271–273). For the preparation of the alkoxy-substituted derivatives (**252;** R_1 = tosyl, mesyl; R_2 = EtO, MeO), β-phenylethylglycinamides were treated with moist $POCl_3$ (274, 275). Prolonged treatment of the corresponding free acid with polyphosphoric acid at room temperature gave **252** (R_1 = tosyl, R_2 = MeO) directly (277).

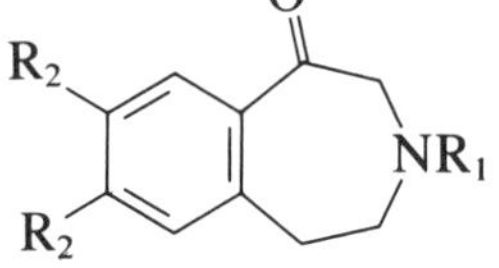

252

Other electrophilic intramolecular cyclizations are known to give 3-benzazepine derivatives. Thus polyphosphoric acid caused hydroxyamide **253** (R = H, EtO) to cyclize to the lactam **254** (R = H, EtO) (279), whereas aluminum chloride is preferred for the ring closure of α-haloamides such as **255** which yields **239** ($R_1 = R_2 = R_3 =$ H). The latter type of cyclization can be induced photochemically when an activating *ortho*- or *para*-hydroxy substituent is present. An example of such a photolytic cyclization is the formation of **239** ($R_1 = CO_2Me$, $R_2 =$ H, $R_3 =$ 7-OH) (282,283). Cyclization *meta* to a hydroxyl group proceeds in very low yield (284). Cyclizations *para* (285) as well as *ortho* (286) to a methoxy substituent have likewise been observed. Displacement of a methoxy group has been noted (287), as well as the formation of 4-azabicyclo[5.3.1]undecane derivatives, when methoxy groups were placed unfavorably for other *ortho* or *para* intramolecular electrophilic attacks (290).

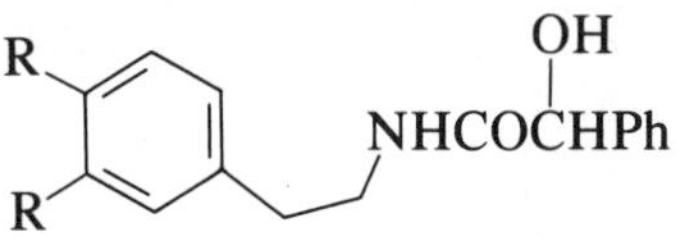

253 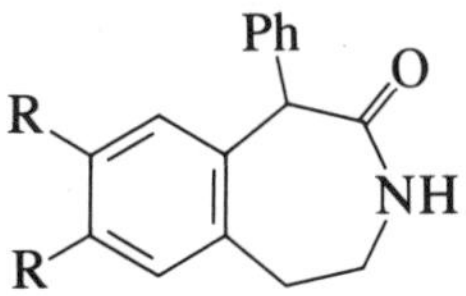 **254** 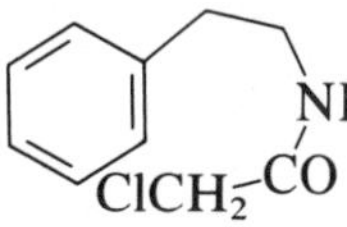 **255**

Electrophilic cyclization of halides such as **256** (R = H, X = Br) proceeds readily in the presence of $AlCl_3$ at 140°C (212). A patent to Hoegerle and Habicht (288) describes a number of examples of this kind. A review of the

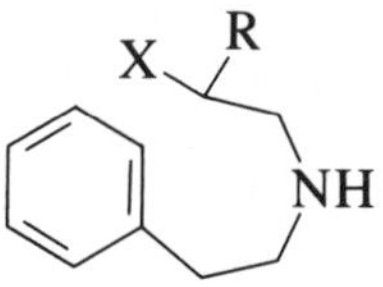

256 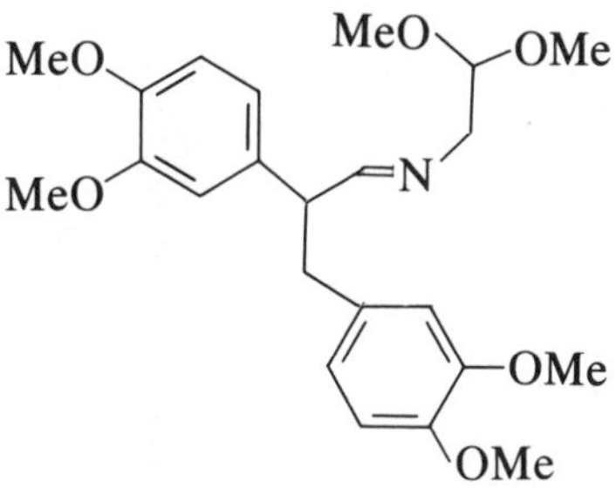 **257**

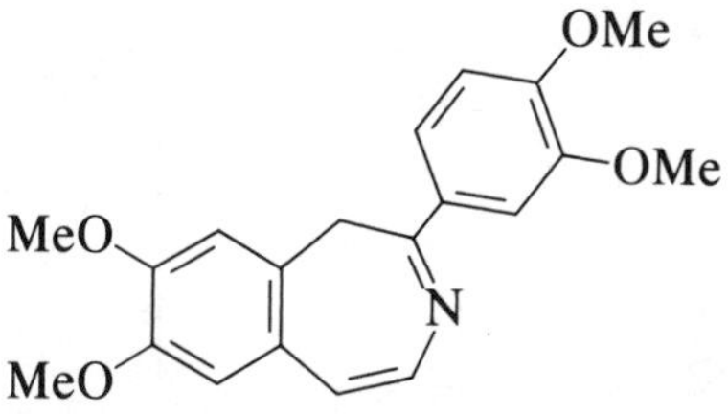

258 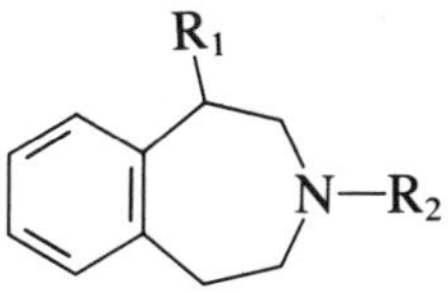 **259**

TABLE 22. 1,2,3,4-TETRAHYDRO-5H-2-BENZAZEPIN-5-ONES

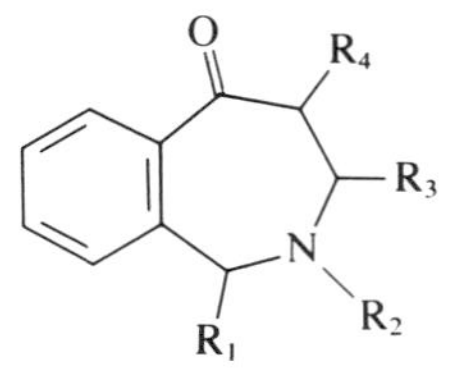

R_1	R_2	R_3	R_4	Other Substituents	mp (°C) or bp (°C/torr)	Derivatives (mp, °C)	Refs.
H	Ac	H	H	8-MeO	87–88		247
H	$PhCH_2OCO$	H	H	8-MeO	73–74		247
H	H	H	H	8-MeO	180/0.05		247
H	Tosyl	H	H	8-MeO	154–155	Oxime, 181–183	210
H	Tosyl	H	Br	8-MeO	134–136		210
H	Tosyl	H	Br_2	8-MeO	143–145		210
EtO	Tosyl	H	H	8-MeO	135–136		210

N-succinimidyl	Tosyl	H	Br	8-MeO	210–215	210
H	Ac	$\Delta^{3,4}$		8-MeO	121–122	247
H	Tosyl	$\Delta^{3,4}$		8-MeO	160–162	210
=O	H	Me_2	Me_2		165	241
=O	H	H	Me_2		167	241
=O	H	H	H		161–163	209
=O	H	H	Me		160–161	209
=O	H	Me	H		227–228	209
=O	H	Me_2	H		159–160	209
=O	H	H	Me_2		183–185	209
=O	H	H	=cyclohexylidene		169–171	209
=O	Me	3-MeNOH, $\Delta^{3,4}$			283–284	237
=O	H	3-Me, $\Delta^{3,4}$			245–250	237

TABLE 23. 1,2,3,4-TETRAHYDRO-5H-2-BENZAZEPINE-1,3-DIONES

R_1	R_2	mp (°C)	Derivatives (mp, °C)	Refs.
H	H	118.5–120.5		206
H	$CONH_2$	219–221		206
H	Ph	180–182		205
H	*p*-MeC_6H_4	206–208		205
H	*p*-$MeOC_6H_4$	158–160		205
H	*p*-FC_6H_4	180–181		205
H	3,4-$(MeO)_2C_6H_3$	130.5–132.5		205
H	3-Pyridyl	220–222		205
CH_2Ph	Ph	99–100		205
$(CH_2)_2Ph$	Ph	96–97		205
CH_2CN	Ph	123–124		205
$(CH_2)_2NMe_2$	Ph		MeI, 191–195	205
Me	*p*-MeC_6H_4	140–141		205
Me	*p*-FC_6H_4	122–123		205
Me	3,4-$(MeO)_2C_6H_3$	116–118		205
Me	3-Pyridyl	171–172		205

TABLE 24. 2,3-DIHYDRO-1H-2-BENZAZEPINE-1,3-DIONES

R_1	R_2	mp (°C)	Refs.
H	H	142–143	206
H	$CONH_2$	268–270	206
H	Ph	211–213	205
H	*p*-MeC_6H_4	198–200	205
H	*p*-$MeOC_6H_4$	208–210	205
H	*p*-FC_6H_4	196–197	205
H	3,4-$(MeO)_2C_6H_3$	223.5–226	205
H	3-Pyridyl	249–251	205
$PhCH_2$	Ph	111–112	205
CH_2CO_2Et	Ph	104–106	205
Me	*p*-MeC_6H_4	74–76	205
Me	*p*-$MeOC_6H_4$	122–124	205
Me	3,4-$(MeO)_2C_6H_3$	149–151	205

TABLE 25. 1H- AND 3H-2BENZAZEPINES

Structure	Physical Constants	Refs.
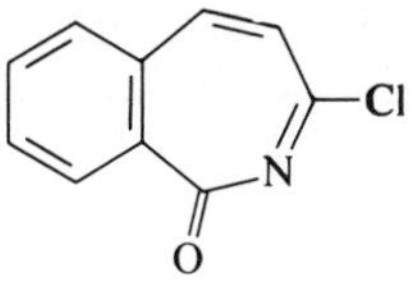 [structure: 3-Cl, 1-oxo]	mp 117°C	204
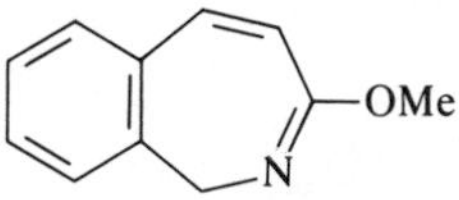 [structure: 3-OMe]	Liquid; picrate, mp 146°C	438
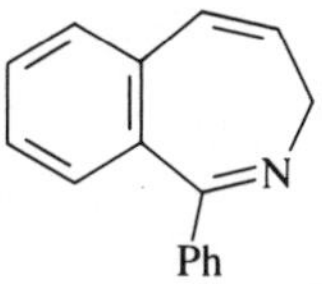 [structure: 1-Ph]	Oil; picrate, mp 208–211°C	240
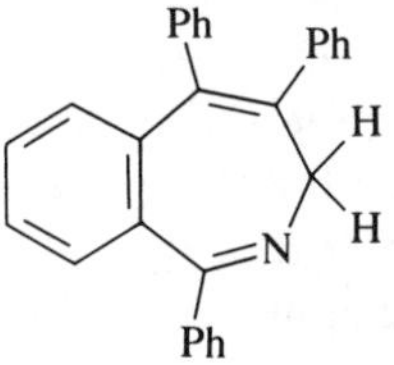 [structure: 1,4,5-Ph, 3-H, H]	mp 157°C	236
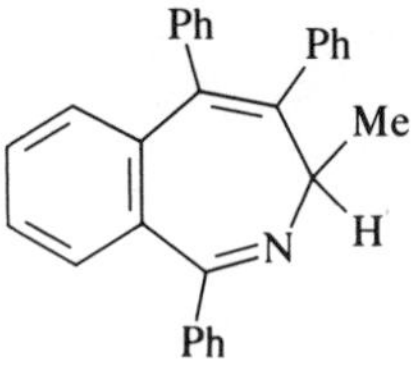 [structure: 1,4,5-Ph, 3-Me, H]	mp 201°C	236
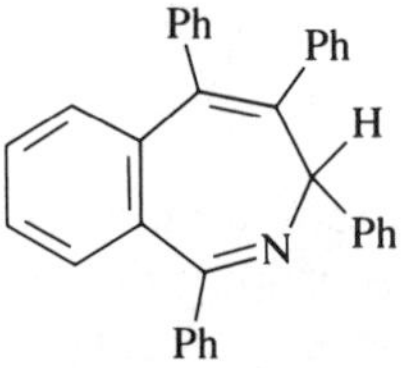 [structure: 1,4,5-Ph, 3-H, Ph]	mp 185°C	236

various available methods of preparation of the starting halides has appeared (66). In contrast, cyclization of hydroxy compounds (**256;** R = Ph, X = OH) is carried out most effectively in sulfuric acid (289) or PPA (518). All these reactions proceed via a carbonium ion intermediate. This is presumably also true in the cyclizations of the acetal **257** to **258** under the influence of hydrochloric or sulfuric acid (291) and of the simpler acetal **256** (X = R = MeO) to **259** (R_1 = MeO, EtO;R_2 = H) in the presence of boron trifluoride (292). The latter method is based on well-known isoquinoline syntheses (293), but experimental details are scant.

It has been shown that the β-phenylethylamines **260** (R_1 = H, Me, *i*-Pr;

R_2 = OH) react with hydrated glyoxylic acid under mild conditions to yield the dihydroxy lactams **261** (R_1 = H, Me, *i*-Pr; R_2 = OH), but it should be noted that when R_2 = H this reaction gives tetrahydroisoquinoline derivatives (294,295). The reason for this difference is not immediately apparent, since the additional hydroxyl group that causes cyclization to **261** is *meta* to the point of attack.

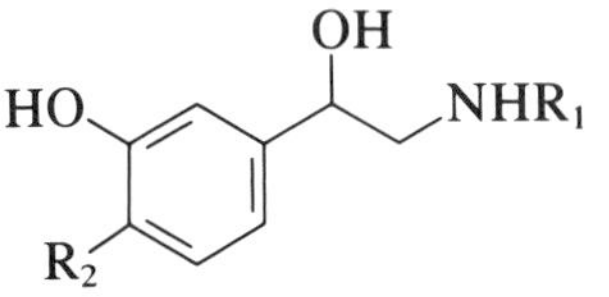

260

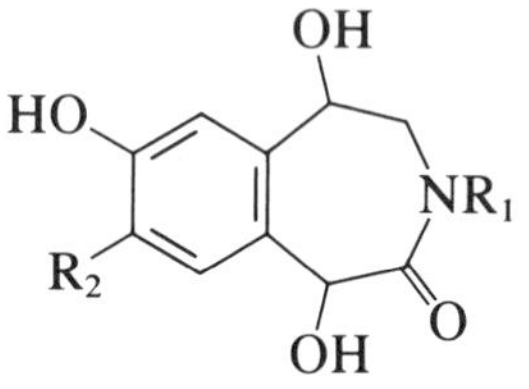

261

Formation of 3H-3-benzazepines (**262**) has been achieved by several groups. Esters of iminodiacetic acid react with *o*-phthalaldehyde in the presence of sodium methoxide, potassium methoxide, or potassium *tert*-butoxide (296–298). The products (**262;** R_1 = Me; R_2 = CO_2H, CO_2Et, CO_2Me; R_3 = H) were never obtained in more than 40% yield. It has been claimed (297) that

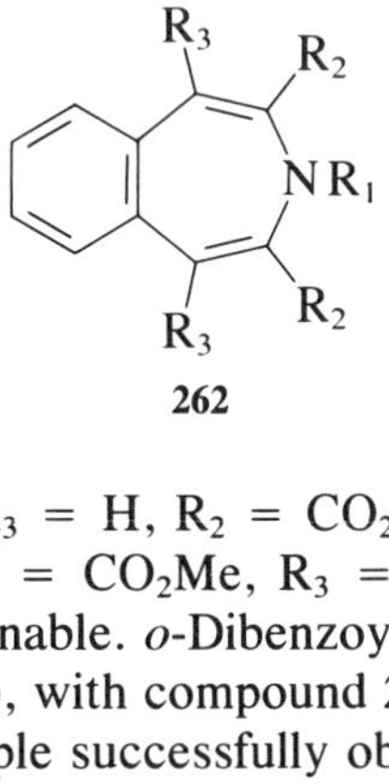

262

bromination of (**262** R_1 = R_3 = H, R_2 = CO_2Me) yields the *N*-bromobenzazepine **262** (R_1 = Br, R_2 = CO_2Me, R_3 = H), but the validity of this observation appears questionable. *o*-Dibenzoylbenzene was less successful in this type of reaction (299), with compound **262** (R_1 = *i*-Pr, R_2 = CN, R_3 = Ph) being the only example successfully obtained by condensation with the appropriate amino dinitrile. On the other hand, several dinitriles did condense satisfactorily with *o*-phthalaldehyde in the presence of sodium methoxide (299). The *N*-phenyl-3-benzazepine **262** (R_1 = Ph, R_2 = CO_2Me, R_3 = H) has been made (300), and 4,5-methylenedioxyphthalaldehyde has been utilized in a condensation reaction with dimethyl-*N*-methylimino diacetate (302). The latter diester and other *N*-substituted analogs have been shown to undergo Claisen condensation with dimethyl phthalate (301) to give **262** (R_1 = Me, Et, Ph; R_2 = CO_2Me; R_3 = MeO, AcO) along with the corresponding cyclic β-keto esters.

iii. Ring Expansion via Beckmann and Schmidt Reactions. In contrast to 1-benzazepine and 2-benzazepine derivates, 3-benzazepines are not readily accessible by means of Beckmann or Schmidt reactions. The lactam **239** ($R_1 = R_2 = R_3 = H$) was obtained on solvolysis of the tosylate of β-tetralone oxime. Under more traditional Beckmann conditions, both possible isomers were obtained (248).

More recent Russian work (305) confirms the generality of the *o*-tosyl oxime ring expansion to 3-benzazepin-2-one derivatives. The 3-benzazepine derivative **263** was obtained by irradiation of the 1,1-dimethylnaphthalen-2(1H)-one oxime (303). The corresponding dihydro compound **264** was prepared by conventional Beckmann rearrangement of β-tetralone oxime (304). However, this is contrary to the general tendency of bicyclic ketones to yield a diversity of products under the conditions of the Schmidt reaction (see earlier discussion in Section A.a.iii). Additional examples of unidirectional ring expansion include the formation of **265** (R = H, Me) from **213** (R = H, Me) (182), **266** (R = Me) from **39** (22), **267** from **268** (256), and both **266** (R = H) and **269** from the dienone **211** (R = H) (257).

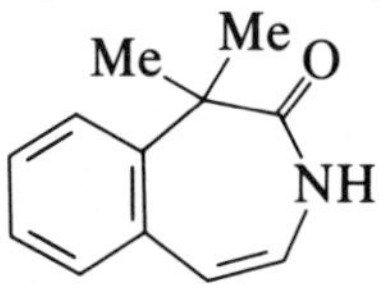

263 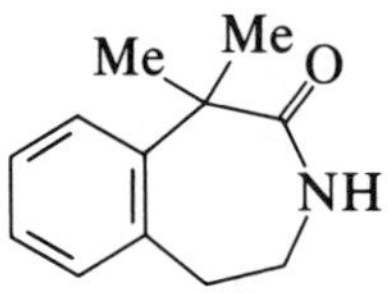 **264**

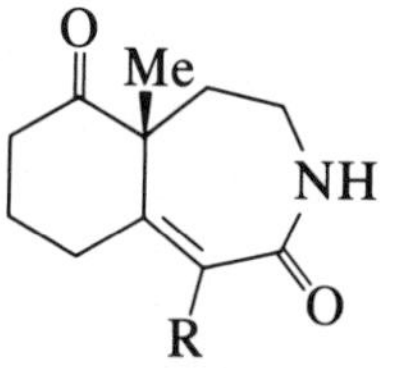

265 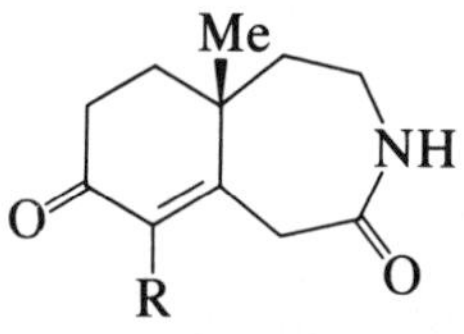 **266**

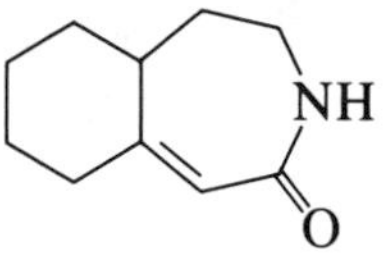

267 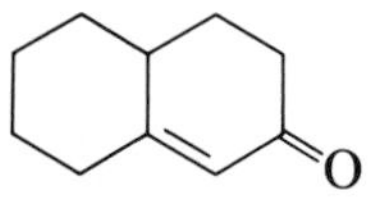 **268** 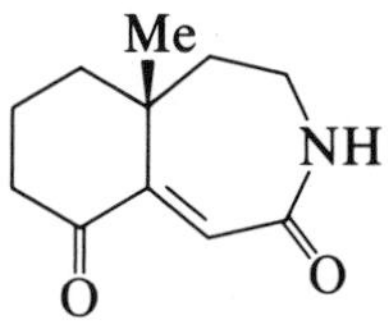 **269**

iv. Other Ring-Expansion Reactions. The most intensively studied approach concerns the application of ring-expansion techniques to isoquinoline derivatives. The additional carbon atom required for the 3-benzazepine skeleton is provided either from a substituent at C_1 of the isoquinoline ring or by addition of a carbon atom from an external source such as diazomethane. In reactions of the first type, a ring-opened intermediate may be isolated (261,262,279,306). Ring expansion may take place without isolation of an intermediate, as in the conversion of the salts **270** or **271** to **272** with zinc in acetic acid (307,308). Other examples of direct ring expansion of isoquinolines include the formation of **273** (R_1 = 3,4-diMeOC_6H_3, R_2 = Me) in 3% yield on treatment of β-hydroxylaudanosine with tosyl chloride and triethylamine (309), of **274** from the spiro compound **275** under similar reaction conditions (310), and of **276** by solvolysis of the mesylate **277** (311). Compound **274** could be oxidized to simpler 3-benzazepines (310).

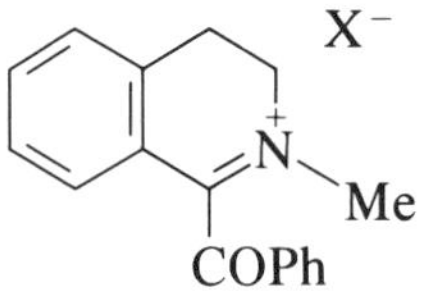

X^- COPh Me **270**

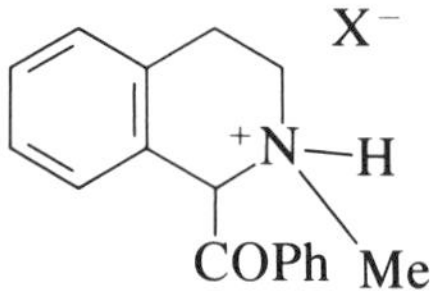

X^- N—H COPh Me **271**

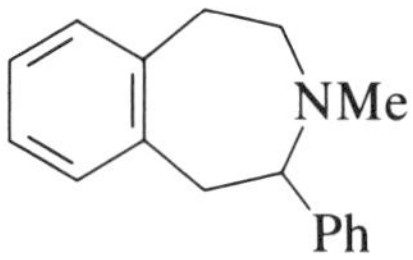

NMe Ph **272**

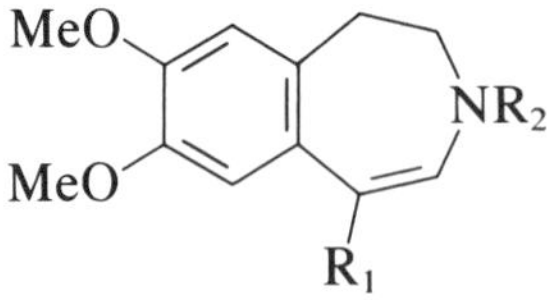

MeO MeO NR_2 R_1 **273**

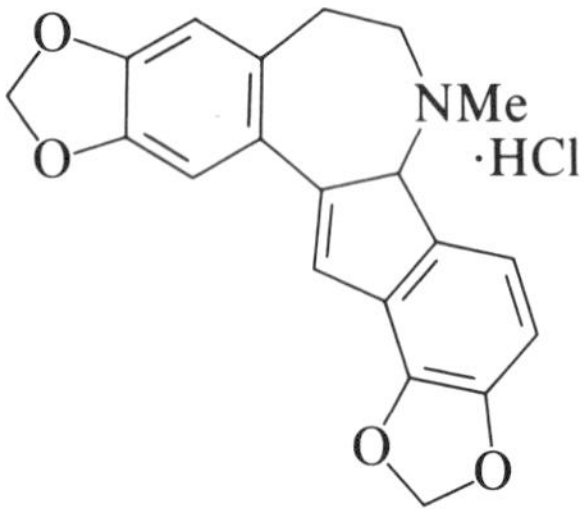

NMe ·HCl **274**

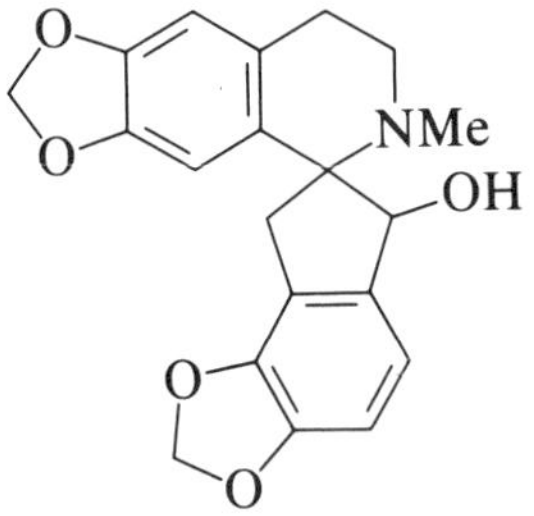

NMe OH **275**

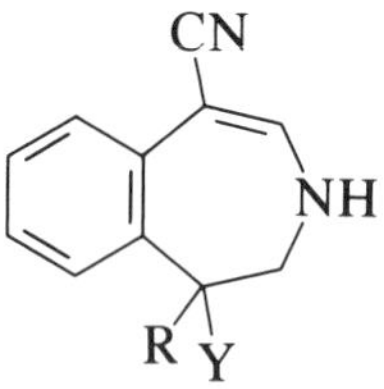

CN NH R Y **276**

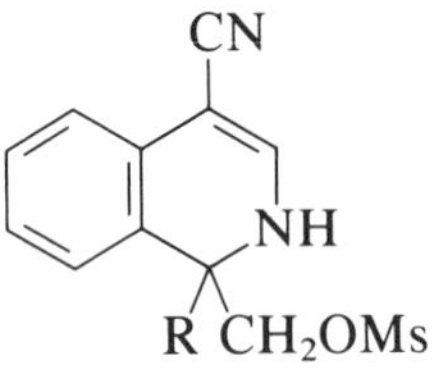

CN NH R CH_2OMs **277**

Other examples in which a ring-opened intermediate has been isolated include the preparation of the dihydro-3-benzazepines **273** (R_1 = 3,4-diMeOC_6H_3, R_2 = Me) (312) and **278** (313). In one instance (345), the intermediate **279** was isolated from 1-halomethyl-6,7-dimethoxytetrahydroisoquinoline. Hydrogenation of **279** over Raney nickel led to opening of the aziridine ring and formation of the 7,8-dimethoxy derivative of **248** (R = H).

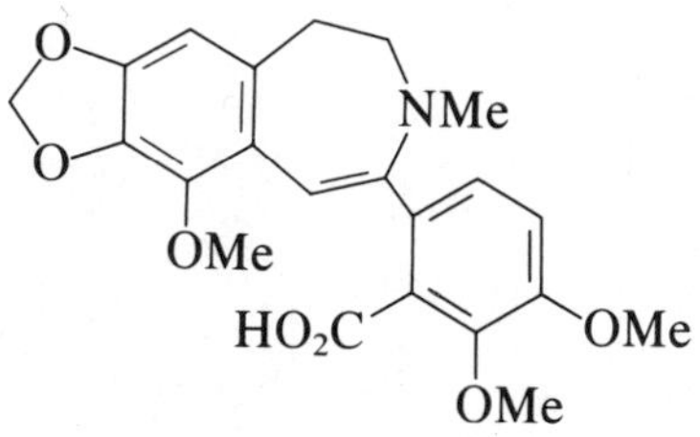

278

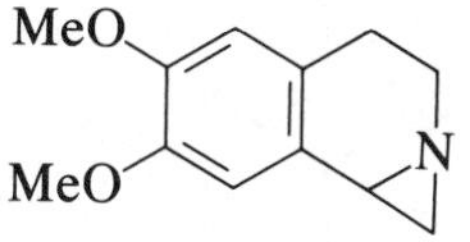

279

Diazoalkanes are the most commonly employed sources of an extra carbon atom in ring expansions of the second kind. Thus addition of diazomethane to the dihydroisoquinoline salt **280** (R = H) yielded a mixture of products, one of which was the fused aziridinium ion **281.** On treatment with methanol compound **281** underwent rearrangement to the 7,8-methylenedioxy derivative of **259** (R_1 = MeO, R_2 = Me) (314). Similar examples (315) employed phenyldiazomethane in methanol. In the reaction of certain 1-aroyldihydroisoquinolinium salts such as **280** (R = 3,4-diMeOC_6H_3CO), a diazonium ion intermediate (**282**) was postulated (316). These reactions led to dihydro-3-benzazepines (e.g., **273** R_1 = 3,4-diMeO$C_6H_3CH_2$, R_2 = Me) in moderate yield. Quite recently (317) it was reported that isoquinoline *N*-oxide reacts with dimsyl sodium to give the 3H-3-benzazepine **262** (R_1 = $MeSO_2$, R_2 = R_3 = H) along with other products. A mechanism for this ring-enlargement process was proposed.

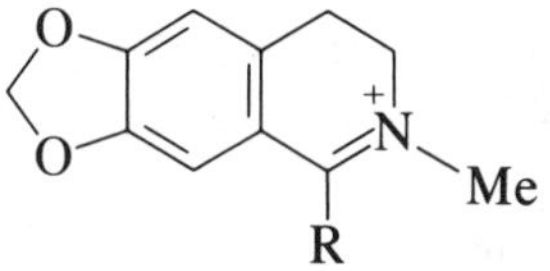

280

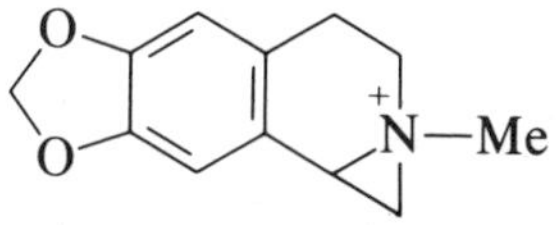

281

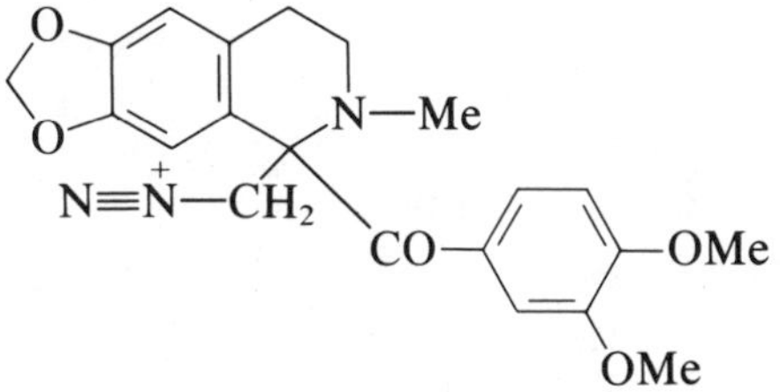

282

Other types of ring expansion are known. Thus the spiropyridine compounds **283** and **284** yielded the 3-benzazepines **285** and **239** (R_1 = H, R_2 = CH_2Ph, R_3 = 7-OH), respectively. In the former instance potassium *tert*-butoxide was the reagent (318), whereas in the latter ultraviolet irradiation was employed (319).

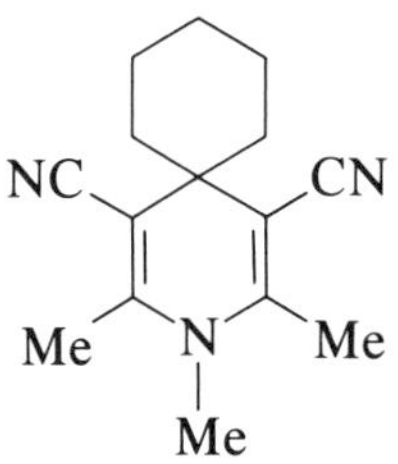

283 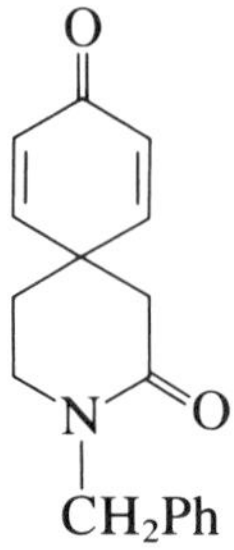 **284** 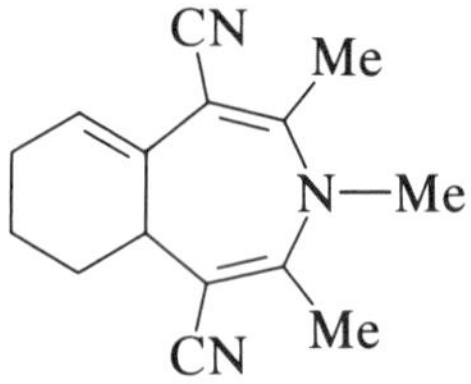 **285**

There is one example (320) of the introduction of a nitrogen atom from outside the ring, namely the conversion of the triketone **286** to the diketo lactam **287** on treatment with cold ammonium hydroxide. It is also relevant to cite at this point the interesting three-stage process commencing with the reaction of 1,2,3,4,5,8-hexahydronaphthalene with iodine isocyanate (323) and ending ultimately with the cleavage of aziridine **288** to the tetrahydro-3-benzazepine **289.** Finally, certain bridged systems can be rearranged to 3-benzazepines by photochemical means. This is illustrated by the conversion of the bridged structure **290** (R_1 = R_2 = CO_2Me) to **262** (R_1 = R_2 = CO_2Me, R_3 = H) (321) and of **290** (R_1 = CO_2Bu-*tert*, R_2 = H) to **262** (R_1 = CO_2Me, R_2 = R_3 = H) (322).

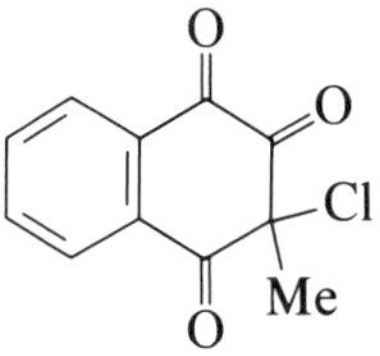

286 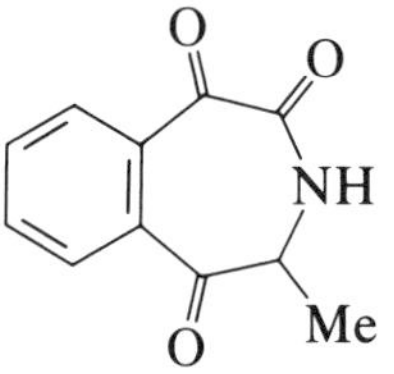 **287**  **288**

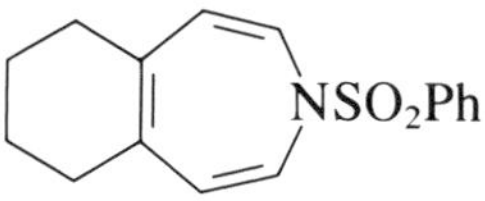

289 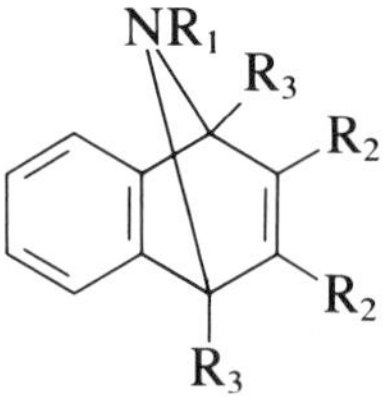 **290**

b. PHYSICAL PROPERTIES

Since only a few 3-benzazepines have been available for study, information on the physical properties of this ring system is sparse. Though X-ray crystallographic data are lacking, complete infrared, ultraviolet and nmr spectra have been reported (269) for 1H-3-benzazepines (199). There is some evidence from ultraviolet spectra (322,323) that the fused benzene ring in 3H-3-benzazepines (**262**) is in resonance with the π-system of the seven-membered ring (321–323), but more definitive studies are required in this regard. A polarographic reduction potential of -1.55 volts has been recorded for the 3-benzazepine **262** (R_1 = Ph, R_2 = CO_2Me, R_3 = H) (327), and mass spectral data for various 3-benzazepines have been presented (282,321,328,329).

c. CHEMICAL REACTIONS

3H-3-Benzazepines (**262**) are stable to acids under moderate conditions (298). However, strong mineral acid has been observed to convert **262** (R_1 = Me, R_2 = CO_2H, R_3 = H) to the indane carboxylic acid **291** (297). On the other hand, the *N*-phenyl analog **262** (R_1 = Ph, R_2 = CO_2H, R_3 = H) was not affected by acid. 1H-3-Benzazepines such as **249** are comparatively stable. Compound **249** is unchanged in the presence of triethylamine or sodium thiocyanate, for example, and does not undergo catalytic hydrogenolysis (269).

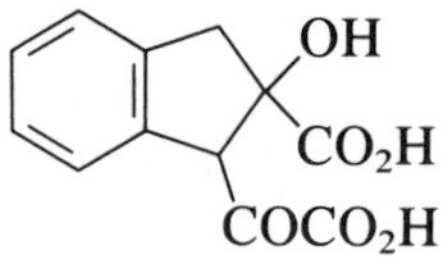

OH
CO_2H
$COCO_2H$

291

Among the partly reduced 3-benzazepines may be found numerous examples of reactivity attributable to specific structural aspects of the seven-membered ring. Thus lactams such as **239** and **246** may be hydrolyzed, and imines are similarly vulnerable (269). 4,5-Dihydro-3*H*-3-benzazepines are particularly interesting since they are enamines. Even when the nitrogen atom is substituted by a sulfonyl group, these compounds continue to behave like enamines. Thus Vilsmeier formylation of the enamines **273** (R_1 = H; R_2 = mesyl, tosyl) gave the 1-formyl derivatives (R_1 = CHO, R_2 = mesyl, tosyl) (330), whereas irradiation caused rearrangement (R_1 = tosyl, R_2 = H) (331). The enamine properties of a cyclopentano-3-benzazepine (**292**) have been utilized in the synthesis of cephalotaxine (332). Ring contraction has been reported to take place when **273** (R_1 = Ph; R_2 = PhCO, EtO, or MeO)

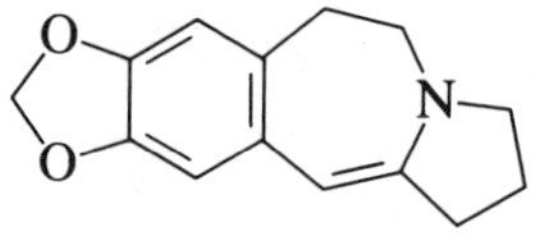

O
O
N

292

is treated with 85% H_3PO_4 (306), but the mechanism proposed for this reaction is open to debate.

Cyclic amino ketones such as **252** have been studied intensively as potential precursors to benzazatropones. In this connection the substitution reactions of the seven-membered ring were examined. Halogenation of **252** (R_1 = tosyl, R_2 = H) at C_2 and C_5 was carried out (271), along with nucleophilic displacement of halogen and oxidation. Condensation of **252** (R_1 = mesyl, R_2 = MeO) with benzaldehyde yielded a 2-benzylidene derivative, which was then subjected to reduction and base treatment (275,333,334). Hydride reduction has been explored (329), and a base-induced ring contraction of the 2-benzyl derivative of **252** (R_1 = mesyl, R_2 = MeO) has been shown to give the aminotetralones **293** (335) rather than the isomer products originally formulated (334). In several instances, difficulty was encountered on attempted direct removal of the tosyl or mesyl groups; however, the desired basic ketones **252** (R_1 = H, Me; R_2 = H) could be prepared by alternative, less direct routes (277).

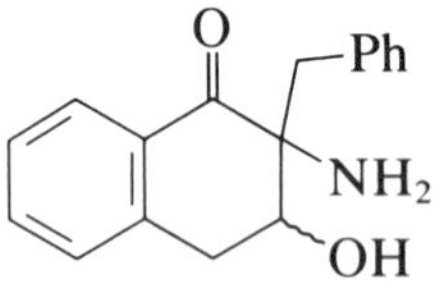

293

2,3,4,5-Tetrahydro-1H-3-benzazepines such as **259** are chemically unexceptional. Many straightforward alkylations and acylations have been reported (336–341). Nitration of the tetrahydro-3-benzazepines **248** (R = H, Me, $COCH_3$) gave the corresponding 7-nitro derivatives, which were further converted to 7-chloro and 7-methoxy derivatives by conventional means (325).

Elaboration of groups attached to the azepine ring of the 3-benzazepine skeleton has been reported. Thus the hydrazino compound **294** reacted with *N*-methylformamide to give the fused triazinone **295** (356). An intramolecular aldol-type condensation, which made use of the acidity of the CH_3SO_2 group to give cyclic structures such as **296,** has also been reported (330). 3-Benzazepine derivatives for which physical constants are reported in the literature are listed in Tables 26–32.

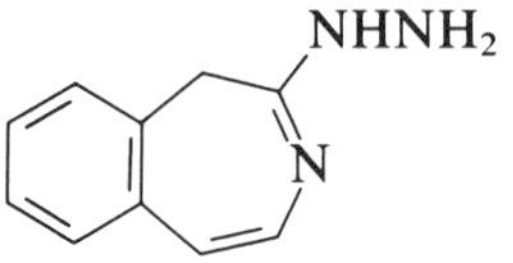

294 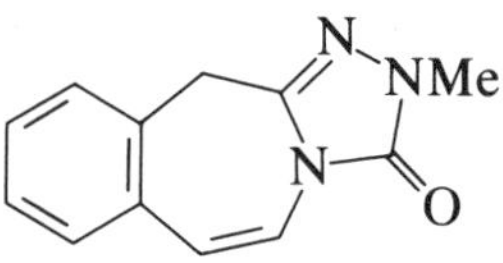 **295** 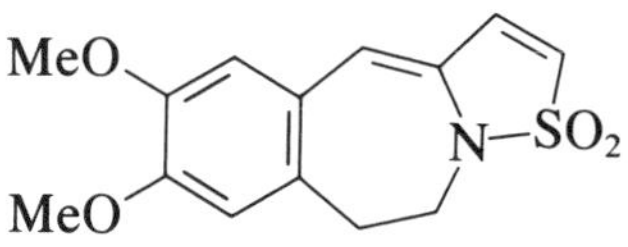 **296**

TABLE 26. OCTA-AND DECAHYDRO-3-BENZAZEPINES

Structure	Physical Constants	Refs.
 NH	bp 36°C/0.05 mm; HCl, mp 222°C	346
 NAc	bp 93–97°C/0.05 mm	346
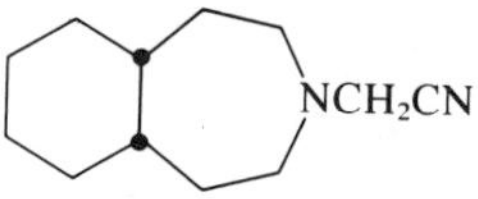 NCH_2CN	bp 160–162°C/14 mm	346
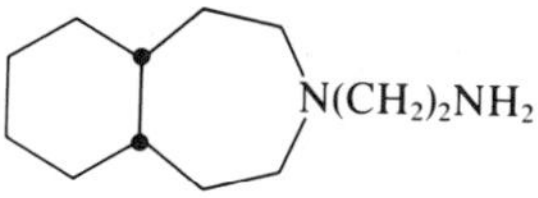 $N(CH_2)_2NH_2$	bp 60–65°C/0.025 mm	346
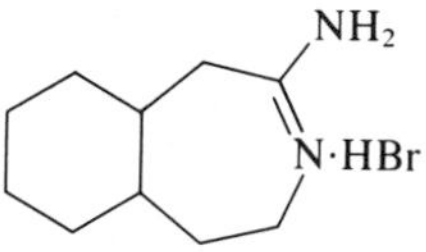 NH_2; N·HBr	mp 204–205°C	346
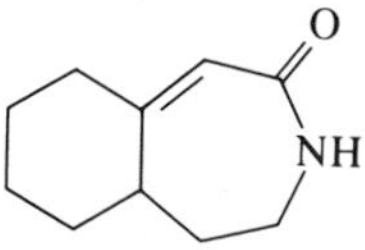 O; NH	mp 111–112°C	256
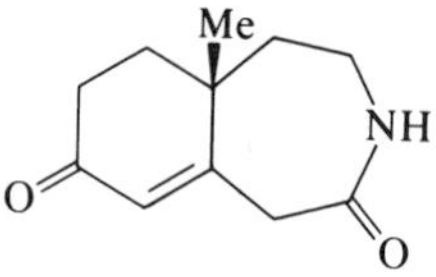 Me; NH; O; O	mp 185–195°C	257
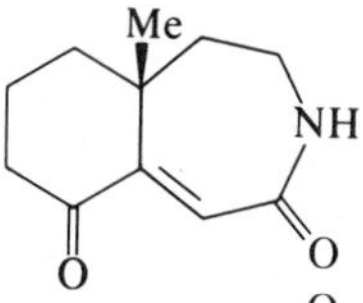 Me; NH; O; O	mp 135–137°C	257
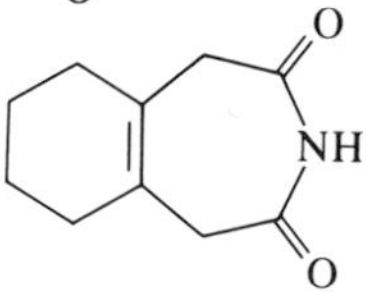 O; NH; O	mp 188–189°C; mono-*O*-methyl ether, mp 174–176°C	29

TABLE 27. 1,2,4,5-TETRAHYDRO-1H-3-BENZAZEPINES

[Structure: 1,2,4,5-tetrahydro-1H-3-benzazepine ring with substituents R_1 (C-1), R_2 (C-2), R_3 (N-3), R_4 (C-4), R_5 (C-5)]

R_1	R_2	R_3	R_4	R_5	Other Substituents	mp (°C) or bp (°C/torr)	Derivatives (mp, °C)	Refs.
H	H	H	H	H		126–127/18	HCl, 249–250 (132–133) Picrate, 218–219	266 (280) 212
H	H	Tosyl	H	H		130–131		267,271
H	H	$Me_2N(CH_2)_3$	H	H			2HCl, 287–290	339
H	H	$NHCONHC(=NH)NH_2$	H	H			HCl, 230–232	339
H	H	SO_2NMe_2	H	H		108–110		339
H	H	SO_2NH_2	H	H		170–173		339
H	H	NO	H	H		64–67		339
H	H	NH_2	H	H			HCl, 197–199	339
H	H	$(CH_2)_2NH_2$	H	H		117–140/0.05		347
H	H	CH_2CN	H	H		113–116		228,247
H	H	CH_2CH_2CN	H	H		61–64 160–162/7		265
H	H	$(CH_2)_2C(NH_2)=NOH$	H	H			2HCl, 197–198 (231–235)	265 (228)
H	H	$(CH_2)_3NMe_2$	H	H		160–165/12	2HCl, 301–302 (283–285)	265 (280)

H	H	$(CH)(Me)CH_2NMe_2$	H	H		140–146/6	2HCl, 244–255	265
H	H	$(CH_2)_3OH$	H	H		66–67 155–157/0.75		265
H	H	$(CH_2)_3Cl$	H	H			HCl, 208–210	265
H	H	$(CH_2)_3N$[ring] 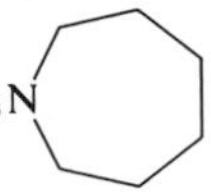	H	H		196–200/1.75	2HCl, 290–291	265
H	H	$CH_2C\equiv CH$	H	H		70–72	HCl, 206–207	280
H	CO_2H	Me	CO_2H	H		81–82	MeI, 133–134	297
Ph	H	H	H	H	7,8-$(EtO)_2$		HCl, 110°C *N*-Benzoyl, 129–130	262
OH	*cis*— 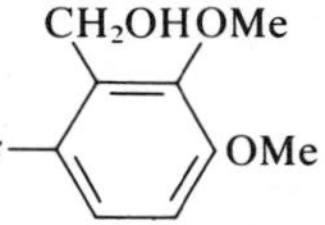 CH_2OH OMe OMe	Me	H	H	7,8-$(MeO)_2$		Picrate, 208–209	310
OH	*cis*-*p*-$MeOC_6H_4$	Me	H	H	7-MeO, 8-OH	204		328
OH	*trans*— 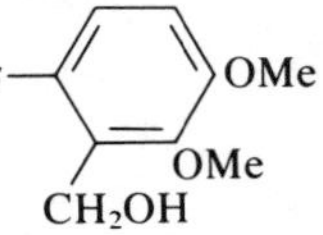 OMe OMe CH_2OH	Me	H	H	7,8-$(MeO)_2$		Picrate, 189–191	310
OH	*cis*— 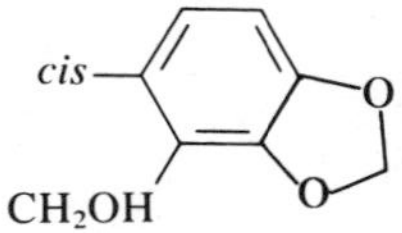 O O CH_2OH	Me	H	H	7,8-(OCH_2O)	142–143		310
H	H	Tosyl	H	OH		118–119		271
H	H	Tosyl	=O	=O		196–198		271
H	H	Tosyl	Br_2	*N*-Succini-mido		227–229		271
H	H	H	H	H	7-Cl		HCl, 175–176	212
H	Me	Me	H	H	7,8-(OCH_2O)		MeI, 268–269.5	314

(*continued*)

TABLE 27. (*Continued*)

R_1	R_2	R_3	R_4	R_5	Other Substituents	mp (°C) or bp (°C/torr)	Derivatives (mp, °C)	Refs.
H	H	Me	H	H	7,8-(OCH_2O)		MeI, 258–260	314
							HCl, 277 (dec.)	259
H	Me	CO_2Et	H	H	7,8-(OCH_2O)	135–140/0.15		314
H	H	CO_2Et	H	H	7,8-(OCH_2O)	96–97		314
OH	H	Tosyl	H	H	7,8-$(MeO)_2$	100		274
H	H	H	H	H	7,8-$(MeO)_2$		HCl, 247 (280 d)	259,331
H	H	Ac	H	H	7,8-$(MeO)_2$	128		331
H	H	Tosyl	H	H	7,8-$(MeO)_2$	114		331
OH	H	H	H	H	7,8-$(MeO)_2$	161		333
OH	H	Mesyl	H	H	7,8-$(MeO)_2$	183		333
OH	=CHPh	Mesyl	H	H	7,8-$(MeO)_2$	150		333
OH	CH_2Ph	Mesyl	H	H	7,8-$(MeO)_2$	217 or 161?		333
CH_2OH	H	Mesyl	H	H	7,8-$(MeO)_2$	120		330
H	CHO	Mesyl	H	H	7,8-$(MeO)_2$	165		330
H	H	H	H	H	7-NO_2	76.5–77.5	*N*-Acetyl, 126–127	235
H	H	Me	H	H	7-NO_2		HCl, 235–236	235
H	H	H	H	H	7-NH_2		HCl, 283–286	235
							N-Acetyl, 180–183	
H	H	Me	H	H	7-NH_2		H_2SO_4, 261 (dec.)	235
H	H	H	H	H	7-Cl	110–115/0.1	HCl, 177–178	235
							N-Acetyl, 85–87	337
H	H	H	H	H	7-OH		$0.5H_2SO_4$, 317–318 (dec.)	
							N-Acetyl, 181–181.5	235
							HCl, 191–193	336
							HBr, 248–249	
H	H	Me	H	H	7-OH	142–146	HCl, 232–235 (dec.) (244–248)	235 (336)

H	H	H	H	H	7-MeO	90–93/0.05	HCl, 231–234 *N*-Acetyl, 94–96	235,336
							HCl, 194–195	235,336
H	H	Me	H	H	7,8-$(MeO)_2$		HCl, 262–263	259,325
H	H	H	H	H	7-OH, 8-MeO		HCl, 240–242 (dec.)	325
H	H	H	H	H	7,8-$(OH)_2$		HCl, mp, 263–270°C (dec.)	325
AcO	*cis*-$PhCH_2$	Ac	H	H	7,8-$(MeO)_2$	179		329
OH	*trans*-$PhCH_2$	Tosyl	H	H	7,8-$(MeO)_2$	~80		329
OH	*cis*-$PhCH_2$	Tosyl	H	H	7,8-$(MeO)_2$	~80		329
AcO	*trans*-$PhCH_2$	Tosyl	H	H	7,8-$(MeO)_2$	134		329
AcO	*cis*-$PhCH_2$	Tosyl	H	H	7,8-$(MeO)_2$	145		329
OH	*cis*-$PhCH_2$	H	H	H	7,8-$(MeO)_2$	181		329
OH	*cis*-$MeOC_6H_4CH_2$	H	H	H	7,8-$(MeO)_2$	180		329
OH	*cis*-*p*-$ClC_6H_4CH_2$	H	H	H	7,8-$(MeO)_2$	189		329
OH	*cis*-$PhCH_2$	Me	H	H	7,8-$(MeO)_2$	139		329
H	H	H	H	H	7,8-(OCH_2O)	82–84	HCl, 282–284	357
							N-Ac, 178–180	359
H	H	Ac	H	H	6-CH_2Cl, 7,8-(OCH_2O)	133–135		357
H	H	Ac	H	H	6,9-$(CH_2Cl)_2$, 7,8-(OCH_2O)	202–202.5		357
H	H	H	H	H	6-Me, 7,8-(OCH_2O)		HCl, 264.5–266 *N*-Ac, 103–104	357
H	H	Ac	H	H	6-CH_2OH, 7,8-(OCH_2O)	145–147	*O*-Ac, 167–169	357
H	H	Ac	H	H	6-CHO, 7,8-(OCH_2O)	178–179		357
H	H	Ac	H	H	6-CO_2H, 7,8-(OCH_2O)	204–206		357
H	H	Ac	H	H	6-CH_2CN, 7,8-(OCH_2O)	179–182		357

(continued)

TABLE 27. (*Continued*)

R_1	R_2	R_3	R_4	R_5	Other Substituents	mp (°C) or bp (°C/torr)	Derivatives (mp, °C)	Refs.
H	H	H	H	H	6-CH_2OH, 7,8-(OCH_2O)		HCl, 250–251	357
H	H	H	H	H	6-CO_2H, 7,8-(OCH_2O)		HCl, 273–275	357
H	H	H	H	H	6-CH_2CO_2H, 7,8-(OCH_2O)		HCl, 265–266	357
H	H	Ac	H	H	6,9-$(CH_2OAc)_2$; 7,8-(OCH_2O)			357
H	H	H	H	H	6,9-Me_2, 7,8-(OCH_2O)		HCl, 290–292 *N*-Ac, 177–178	357
H	H	Ac	H	H	6,9-$(CH_2CN)_2$, 7,9-(OCH_2O)	188.5–190		357
H	H	$PhCH_2$	H	H	7,8-$(MeO)_2$	69–71	HCl, 228–229.5	259
H	H	H	H	H	6,7,8-$(MeO)_3$	64–66	HCl, 191–192 (dec.)	259
H	H	$PhCH_2$	H	H	7,8-(OCH_2O)	98–100	HCl, 242–244 (dec.)	259
H	Me	Me	H	H	7,8-(OCH_2O)		HCl, 271.5–272.5	259
H	H	H	H	H	6,9-$(CH_2OH)_2$, 7,8-(OCH_2O)	221–222		357
H	H	H	H	H	6,9-$(CH_2CO_2H)_2$ 7,8-(OCH_2O)	>330		357
H	H	Me	H	H	6,9-Me_2, 7,8-(OCH_2O)		HCl, 284.5–286 (dec.)	357
OH	H	H	H	H		131–132	*N*-Ac, 100–101 *N*-PhCO, 71	277
OH	H	Me	H	H		180/0.3	MeI, 183–184	277
H	Ph	Me	H	H			HCl	307
H	3,4-$(MeO)_2C_6H_3$	Me	H	H	7,8-$(MeO)_2$		HCl, 255–260 MeI, 95–99	307 309
H	3,4-$(MeO)_2C_6H_3$	Et	H	H	7,8-$(MeO)_2$	84		307

H	p-ClC$_6$H$_4$	Me	H	H	7,8-(MeO)$_2$	125	HCl, 230	307
H	p-ClC$_6$H$_4$	Et	H	H	7,8-(MeO)$_2$		HCl, 180	307
H	m-MeC$_6$H$_4$	Me	H	H	7,8-(MeO)$_2$		HCl, 190	307
H	p-ClC$_6$H$_4$	H	H	H	7,8-(MeO)$_2$		HCl, 260	263
Me$_2$	H	H	H	Me$_2$		77–85	HCl, 275	263
Me$_2$	H	(CH$_2$)$_3$NMe$_2$	H	Me$_2$			2HCl, 240–241	263
Me$_2$	H	CHO	H	Me$_2$		116–117		263
Me$_2$	H	CH$_2$CO$_2$Bu-t	H	Me$_2$		50–52		263
Me$_2$	H	NO	H	Me$_2$			HCl, 109–110	263
Me$_2$	H	PhCH$_2$	H	Me$_2$		135–137	HCl, 245	263
Me$_2$	H	Allyl	H	Me$_2$			HCl, 224	263
Me$_2$	H	C≡CMe	H	Me$_2$			HCl, 226–227	263
Me$_2$	H	PhCH$_2$	H	Me$_2$	7-NO$_2$	168–171		263
Me$_2$	H	PhCH$_2$	H	Me$_2$	7-NH$_2$	90–92	HCl, 243–245	263
Me$_2$	H	NH$_2$	H	Me$_2$			HCl, 218–220	263
Me$_2$	H	Me	H	Me$_2$			HCl, 216–218	263
H	H	CONHSO$_2$Ph	H	H		185–187		350
						(176–179)		(344)
H	H	CONHSO$_2$C$_6$H$_4$Me-p	H	H		167.5–169		350
						(160)		(344)
H	H	CONHSO$_2$C$_6$H$_4$F-p	H	H		154–156		350
						(101–103)		(344)
H	H	CONHSO$_2$C$_6$H$_4$NH$_2$-p	H	H		216–218		350
						(182–183)		(344)
Ph	H	Me	H	H	7,8-(MeO)$_2$	82–84	MeI, 246–249	348,349
Ph	H	H	H	H	7,8-(MeO)$_2$	198–200/2	Maleate, 198–200	343,349
Ph	H	CH$_2$CH=CH$_2$	H	H		65–68	HCl, 203–205	349
Ph	H	CH$_2$CH$_2$OH	H	H		95–97		349
Ph	H	H	H	H	7,8-(OH)$_2$		HBr, 283–285	348
H	H	Me	H	H		74–75/2	HCl, 244–245	340
H	H	Et	H	H		90–100/3	HCl, 238–239	270,340
H	H	i-Pr	H	H		95–105/2	HCl, 260–261	340
H	H	Allyl	H	H		97–102/1	HCl, 210–212	340

(continued)

TABLE 27. *(Continued)*

R_1	R_2	R_3	R_4	R_5	Other Substituents	mp (°C) or bp (°C/torr)	Derivatives (mp, °C)	Refs.
H	H	*n*-Pr	H	H		95–105/2	HCl, 236–237	340
H	H	Bu	H	H		107–110/2–3	HCl, 226–227	340
H	H	$PhCH_2$	H	H			HCl, 248–256	340
H	H	$PhCH_2CH_2$	H	H			HCl, 260–263	340
H	H	$Ph(CH_2)_3$	H	H			HCl, 211.5–213	340
Ph	H	H	H	H	7-MeO		Maleate, 196–197	348
H	H	CH_2CH_2OH	H	H		87–88	AcO·HCl, 197–199	341
H	H	$CH_2CH(Me)OH$	H	H		120–121	HCl, 187–189	341
H	H	$(CH_2)_2OEt$	H	H			HCl, 196–198	341
H	H	$CH_2CH(Me)OMe$	H	H			HCl, 194–195	341
Ph	H	H	H	H		140–150/0.01		337
Me	H	H	H	H		72/0.6		337
Me	H	H	H	H	7-Pr-*i*	71–72/0.2		337
H	Me	H	H	H		60/0.2		337
Ph	H	CH_2CN	H	H	7,8-$(MeO)_2$	108–109 (152–153)		342 (338)
Ph	H	$(CH_2)_2NMe_2$	H	H	7,8-$(MeO)_2$		HCl, 266–267	338
Ph	H	$(CH_2)_2C(=NH)NHOH$	H	H	7,8-$(MeO)_2$	155–157	$0.5H_2SO_4$, 75	338
p-ClC_6H_4	H	$(CH_2)_2N$(pyrrolidinyl)	H	H	7-Cl	180–190/0.09		338
Ph	H	CH_2CH(imidazolidin-2-yl, NH/NH ring)	H	H	7,8-$(MeO)_2$		2HCl, 250–251	338

Ph	H	CH_2CH_2Cl	H	H	7,8-$(MeO)_2$		HCl, 90	343
Ph	H	$(CH_2)_2SC(=NH)NH_2$	H	H	7,8-$(MeO)_2$		2HCl, 185–190	343
Ph	H	$(CH_2)_2CN$	H	H	7,8-$(MeO)_2$	119–120		342
Ph	H	$(CH_2)_3NH_2$	H	H	7,8-$(MeO)_2$		Dimaleate, 171–172	342
Ph	H	$CH_2CH(Me)CN$	H	H	7,8-$(MeO)_2$	96–98		342
Ph	H	$CH_2CH(Me)CH_2NH_2$	H	H	7,8-$(MeO)_2$		Difumarate, 176.5–177.5	342
Ph	H	$(CH_2)_2NH_2$	H	H	7,8-$(MeO)_2$		Dimaleate, 182–130	342
Ph	H	$(CH_2)_3NH_2$	H	H	7,8-$(MeO)_2$		Dimaleate, 167.5–168.5	342
Ph	H	H	H	H	8-Cl	185/0.05		343
Ph	H	Me	H	H	8-Cl		HCl, 157–159	343
p-C_6H_4Cl	H	H	H	H	8-Cl	107–110		343
H	H	Dimethylallyl	H	H	7-MeO		HCl, 204–206.5	336
H	H	Cyclopropylmethyl	H	H	7-MeO		HCl, 222–223	336
H	H	Cyclopropylcarbonyl	H	H	7-MeO	58–60		336
H	H	Allyl	H	H	7-MeO		HCl, 196–199	336
H	H	$CH_2C\equiv CH$	H	H	7-MeO		HCl, 194–195	336
H	H	Et	H	H	7-MeO		HCl, 219–221	336
H	H	Ac	H	H	7-MeO	90–91		336
H	H	*n*-Pr	H	H	7-MeO		HCl, 208–210	336
H	H	Dimethylallyl	H	H	7-OH		HCl, 254.5–256	336
H	H	Cyclopropylmethyl	H	H	7-OH		HCl, 220–222	336
H	H	Cyclobutylmethyl	H	H	7-OH		HCl, 252–254	336
H	H	Allyl	H	H	7-OH		HCl, 176–178	336
H	H	2-Methylallyl	H	H	7-OH		HCl, 219–221	336
H	H	$CH_2C\equiv CH$	H	H	7-OH		HCl, 201–202	336
H	H	Et	H	H	7-OH	168–171	HCl, 247–250	336
H	=S	H	H	H		154		253

TABLE 28. 1,2,4,5-TETRAHYDRO-3H-3-BENZAZEPIN-2-ONES

R_1	R_2	R_3	R_4	Other Substituents	mp (°C)	Derivatives (mp, °C)	Refs.
H	H	H	H		159–160		248,270
H	Et	H	H		107		270
Ph	H	H	H		189		279
H	H	=O	H		191–192		264
H	$(CH_2)_3NMe_2$	H	H		158/0.07[a]	HCl, 193–195	280
$(CH_2)_3NMe_2$	$(CH_2)_3NMe_2$	H	H			2HCl. H_2O, 215–218	280
Ph_2	$(CH_2)_2NMe_2$	H	H		91–93	H_2SO_4, 84–86; MeI, 213–215	353
OH	H	H	OH	7,8-$(OH)_2$	205		295
OH	Me	H	OH	7,8-$(OH)_2$	185–186		295
3,4-$(MeO)_2C_6H_3$	H	H	H	7,8-$(MeOH)_2$	180–181		309
Ph	H	H	H	7-MeOH	169–171		348
Ph	H	H	H	7,8-$(MeO)_2$	199–201		342
Ph	CH_2CH_2CN	H	H	7,8-$(MeO)_2$	228–230		342

Ph	$(CH_2)_3NH_2$	H	H	7,8-$(MeO)_2$	189–190.5	342
Ph	COPh	H	H	7,8-$(MeO)_2$	175–178	343
Ph	H	H	H	8-Cl	220–222	343
p-ClC_6H_4	H	H	H	8-Cl	231–232	343
Ph_2	H	Me	H		203–204	353
Ph_2	CH_2CO_2Et	H	H		136–137	354
Ph_2	$CH_2CONHPh$	H	H		162–163	354
H	H	H	H	8-OH	201–203	284,319
H	Me	H	H	8-OH	206–209	284,319
H	CH_2Ph	H	H	8-OH	213–215	284,319
H	H	H	H	6,7,8-$(MeO_3$	166–168	259
H	H	H	H	8-MeO	163–164	284
H	H	H	H	9-MeO	163–164.5	351
H	H	H	H	7-MeO	156–157.5	287,351
H	H	H	H	7,9-$(MeO)_2$	183–184.5	351
H	H	H	H	7,8,9-$(MeO)_3$	166.5	285,352
H	Me	Ph	*trans*-OH	7,8-$(MeO)_2$	211–213	276
H	Me	Ph	*cis*-OH	7,8-$(MeO)_2$	175–177	276
H	H	H	H	6,9-$(MeO)_2$	221	287
H	H	H	H	7-OH	256	287
H	CH_2Ph	H	H	7,8-$(MeO)_2$	141–142	259
H	CH_2Ph	H	H	7,8-(OCH_2O)	141–142	259
H	H	Me	H	7,8-(OCH_2O)	175–175.5	259
Me_2	H	=O	Me_2		142–144	263

[a]Boiling point (°C/torr).

TABLE 29. 1,2,4,5-TETRAHYDRO-3H-3-BENZAZEPIN-1-ONES

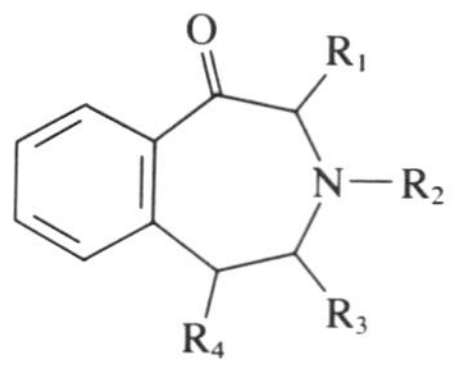

R_1	R_2	R_3	R_4	Other Substituents	mp (°C) or bp (°C/torr	Derivatives (mp, °C)	Refs.
H	Tosyl	H	H		156–157	Ethylene ketal, 113	82,271
						2,4-DNP, 213°	
						Oxime, 174–175	277
Br	Tosyl	H	H		95–96		271
MeO	Tosyl	H	H		103–104		271
AcO	Tosyl	H	H		133–134		271
OH	Tosyl	H	H		169–170		271
Me	Tosyl	H	H		132	2,4-DNP, 283–285	271

H	H	H	H			HCl, 150 (dec.)	277
H	Me	H	H		126/0.3 (dec.)		277
H	Mesyl	H	H		161		330
CO_2Me	Me	CO_2Me	=O		144–145		301
CO_2Me	Et	CO_2Me	=O		159–161		301
CO_2Me	Ph	CO_2Me	=O		190–191		301
=O	H	Me	=O		189–190		320
H	Tosyl	H	H	7,8-$(MeO)_2$	211	2,4-DNP, 235–238	274
H	Tosyl	H	H	7,8-$(EtO)_2$	127		274
=O	Tosyl	H	H	7,8-$(MeO)_2$	210		333
=O	Mesyl	H	H	7,8-$(MeO)_2$	235		333
OH	Mesyl	H	H	7,8-$(MeO)_2$	164		331
=CHPh	Mesyl	H	H	7,8-$(MeO)_2$	231		331
CH_2Ph	Mesyl	H	H	7,8-$(MeO)_2$	173		331
H	Mesyl	H	H	7,8-$(MeO)_2$	219	2,4-DNP, 235–238; Ethylene ketal, 143	333
H	$PhCH_2SO_2$	H	H	7,8-$(MeO)_2$	172		333
Ph	Me	=O	H	7,8-$(MeO)_2$	132–137		276

TABLE 30. 4,5-DIHYDRO-3H-3-BENZAZEPINES

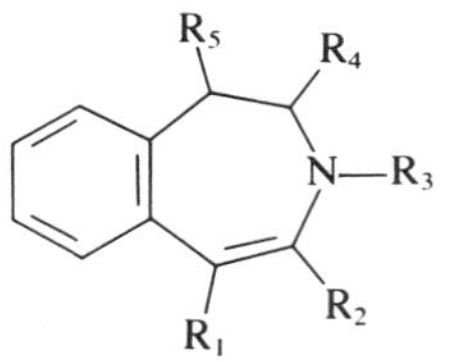

R_1	R_2	R_3	R_4	R_5	Other Substituents	mp (°C) or bp (°C/torr)	Derivatives (mp, °C)	Refs.
Ph	H	H	H	H		94		262
Ph	H	Ac	H	H		115		262
H	H	Tosyl	H	H		138		271
AcO	AcO	Tosyl	H	H		135		271
H	Br	Tosyl	H	H		102.5–103		271
N-Succinimido	Br	Tosyl	H	H		157–158		271
H	NHAc	Ac	H	H		142–143		269
Ph	H	Ac	H	H	7,8-$(EtO)_2$	108		262
Ph	H	COPh	H	H	7,8-$(EtO)_2$	91		261
Ph	H	H	H	H	7,8-$(EtO)_2$	148		260
3,4-$(MeO)_2C_6H_3$	H	Me	H	H	7,8-$(MeO)_2$	95–96		355
3,4-$(OCH_2O)C_6H_3$	H	Me	H	H	7,8-$(MeO)_2$	179–181		316
H	H	Tosyl	AcO	=O		196–197		271
3,$(MeO)_2C_6H_3$	Me	H	H	H	7,8-(OCH_2O)	204–205		316
3,4-$(OCH_2O)C_6H_3$	Me	H	H	H	7,8-(OCH_2O)	165		316
H	H	Tosyl	H	H	7,8-$(MeO)_2$	143		274,333
CHO	H	Mesyl	H	H	7,8-$(MeO)_2$	166	Oxime, 213	330
CHO	H	Tosyl	H	H	7,8-$(MeO)_2$	120		330
CHO	H	Mesyl	H	H		139		330
Cl	H	Tosyl	H	H	7,8-$(MeO)_2$	127		330

Cl	CHO	Tosyl	H	H	7,8-$(MeO)_2$	169		330
Cl	H	Mesyl	H	H		122		330
Cl	H	Mesyl	H	H	7-OH, 8-MeO	162		330
Cl	CHO	Mesyl	H	H	7-OH, 8-MeO	279		330
Cl	H	Mesyl	H	H	7,8-$(MeO)_2$	165		330
Cl	CHO	Mesyl	H	H	7,8-$(MeO)_2$	145	Oxime, 120	330
Cl	CH_2OH	Mesyl	H	H	7,8-$(MeO)_2$	155		330
H	H	Mesyl	H	H	7,8-$(MeO)_2$	174		330
H	H	Mesyl	H	H		163		330
CN	H	Mesyl	H	H	7,8-$(MO)_2$	224		330
H	CH_2OH	Mesyl	H	H	7,8-$(MeO)_2$	110		330
H	CHO	Mesyl	H	H	7,8-$(MeO)_2$	194		330
H	CHO	Tosyl	H	H	7,8-$(MeO)_2$	169		330
OH	CHO	Mesyl	H	H	7,8-$(MeO)_2$	210		330
MeO	CHO	Tosyl	H	H	7,8-$(MeO)_2$	188		330
EtO	CHO	Tosyl	H	H	7,8-$(MeO)_2$	139		330
H	H	H	CH_2Ph	=O	7,8-$(MeO)_2$	185		329,333
CN	H	H	H	MeO		160–161		311
CN	H	H	H	MeO, Me		141–143		311
CN	H	H	H	NEt_2			Picrate, 204–205	311
CN	H	H	H	Me, NEt_2			Picrate, 129.5–131	311
CN	H	H	H	Me_2		174–175		311
CN	H	H	H	Me, Et		156–157.5		311
CN	H	H	Ph	H		185–186		311
3,4-$(MeO)_2C_6H_3$	H	Me	H	H	7,8-$(MeO)_2$	115–116		309
H	H	H	=O	H		161	*N*-Acetyl, 81	270
Br	H	H	=O	H		184	*N*-Ethyl, 78	270
H	H	Et	=O	H		127–128/0.4		270
Br	H	H	=O	Br		173		270
H	H	Et	H	H		103–105/0.25		270
Ph	H	COPh	H	H	7,8-$(MeO)_2$	128		306

TABLE 31. 3H-3-BENZAZEPINES

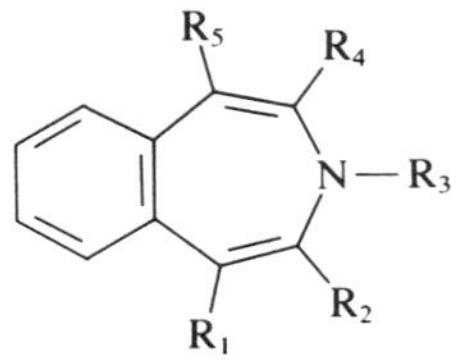

R_1	R_2	R_3	R_4	R_5	Other Substituents	mp (°C)	Refs.
H	CN	Me	CN	H		195–196	299
H	$CONH_2$	p-MeC_6H_4	$CONH_2$	H		324–329 (dec.)	299
H	$CONH_2$	p-BrC_6H_4	CO_2H	H		262–268	299
H	CO_2Me	p-ClC_6H_4	CO_2Me	H		201–203	299
H	CO_2Me	p-$NO_2C_6H_4$	CO_2Me	H		198–201	299
H	CO_2Me	Me	CO_2Me	H		146–147	297
H	CO_2Et	Me	CO_2Et	H		145–146	297
H	H	SO_2Ph	H	H	6,7,8,9-Tetrahydro	105–107	323
Ph	CN	i-Pr	CN	Ph		205–207	299
H	CO_2Me	Me	CO_2Me	H	7,8-(OCH_2O)	167.5	302
H	CO_2Me	Ph	CO_2Me	H		177–178	300
H	CO_2H	Ph	CO_2H	H		203–205	298,300
H	CO_2Et	Ph	CO_2Et	H		136–137	298,300
H	CO_2Me	Br	CO_2Me	H		150–151	297
CN	Me	Me	Me	CN		180–182	318
CN	Me	Me	Me	CN	6,7,8,9-Tetrahydro	216–217	318
H	NHAc	Ac	Br	H		160	269
H	NHAc	Ac	H	H		154	269
H	H	CO_2Bu-t	H	H		95.5–97.0	322
MeO	CO_2Me	Me	CO_2Me	MeO		108–110	301
MeO	CO_2Me	Ph	CO_2Me	MeO		195–197	301
AcO	CO_2Me	Me	CO_2Me	MeO		112.5–114	301
AcO	CO_2Me	Ph	CO_2Me	AcO		182–184	301

TABLE 32. 1H-3-BENZAZEPINES

R_1	R_2	Other Substituents	mp (°C)	Derivatives (mp, °C)	Refs.
NHCOMe	Br		172–173		269,270
NH_2	H		165 (dec.)	MeI, 205 EtI, 218–220 HBr, 234	269,270
NH_2	Br	6,7,8,9-Tetrahydro	160–165	HBr, 221–222	29
NH_2	SCN	6,7,8,9-Tetrahydro	170–174		29
3,4-$(MeO)_2C_6H_3$	H	7,8-$(MeO)_2$	161–162		291
NH_2	Br		194–196 (dec.)	HBr, 256–257	269
NH_2	I		191–193		269
N-Morpholino	H		70–71		269
CONHPh	Br		245–246		269
CONHPh	H		202–205		269
CSNHPh	H		173–175		269
NH_2	SCN		179–181		269
$NHCH_2OH$	H		136–139		269
NH_2	$\overset{+}{N}C_5H_5$ Br^-		262–264		269
			144–148		269
$NHCHOHCCl_3$	H	144–148	269		
NHAc	=O (4,5-dihydro)		261–263		269

D. *Pyranoazepines*

Pyrano[2,3-*b*]azepine   Pyrano[3,2-*c*]azepine

The pyrano[2,3-*b*]azepine salt **297** was made (358) during a study of the electrophilic substitution of ketene *S,N*-acetals. In this particular example the reactants were α-chloro-β-benzoylethene and the azepine thioacetal **298.** Quite recently the pyrano[3,2-*c*]azepine derivative **299** was reported to arise from a product of Claisen rearrangement of an *O*-allyl azepinone (543).

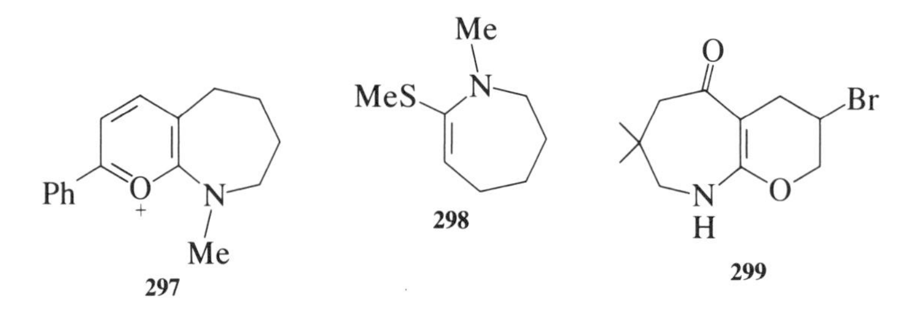

E. *Pyridoazepines*

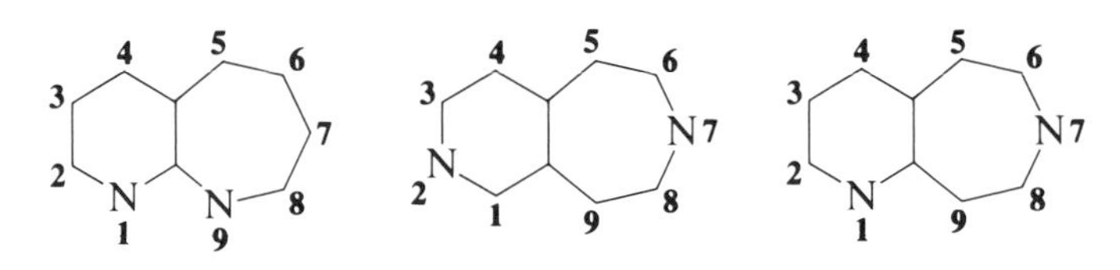

Pyrido[2,3-*b*]azepine   Pyrido[3,4-*d*]azepine   Pyrido[2,3-*d*]azepine

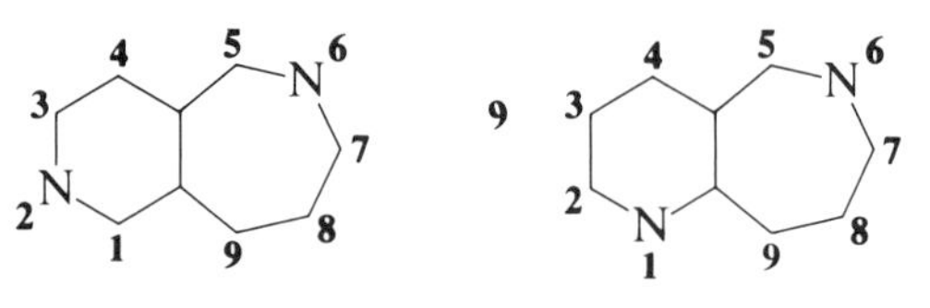

Pyrido[4,3-*c*]azepine   Pyrido[3,2-*c*]azepine

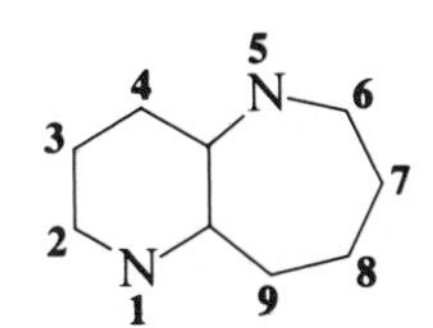

Pyrido[4,3-*b*]azepine Pyrido[3,2-*b*]azepine Pyrido[3,4-*b*]azepine

Eight of the ten possible pyridoazepine ring systems have been described. The pyrido[2,3-*b*]azepine **300** was obtained by treatment of the epoxy ketone **301** with boron trifluoride etherate followed by acetylation (359), and the tetrahydro derivative **302** ($R_1 = R_2 = H$) was the product when 4-(3′-pyridyl)butylamine was subjected to the action of strong base (360).

Both oximes **303** and **304** (R = Me, H) reacted with polyphosphoric acid (361); the first gave the pyrido[2,3-*d*]azepines **305** (R = Me, H), whereas the second gave the pyrido[3,2-*c*]azepines **306** (R = Me, H). Both pyrido[4,3-*c*]azepine **307** and pyrido[3,4-*d*]-azepine **308** were obtained from the Schmidt reaction on the α,β-unsaturated ketone **309** (362,519); the former

300 **301** **302**

303 **304** **305**

306 **307** **308**

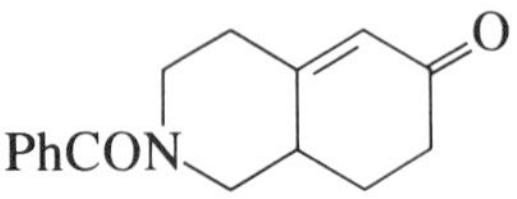

PhCON 309 O; PhCON 310 H H NH O; PhCON 311 H H O NH

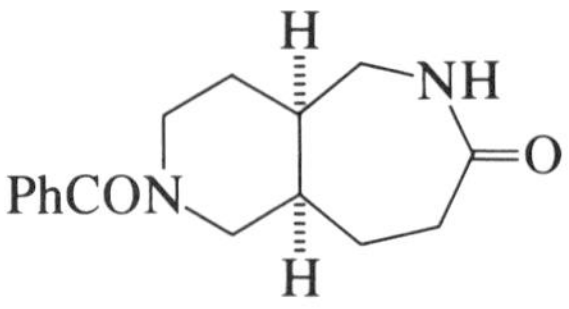

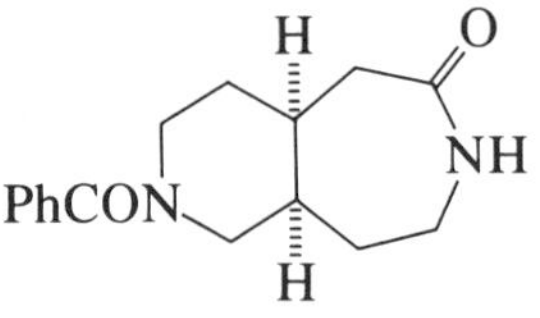

compound (**307**) was only a minor product in this reaction. In contrast, the dihydro derivative of **309** reacted with sodium azide in polyphosphoric acid to give **310** and **311** in a 1:2 ratio.

Beckman and Schmidt reactions on pyridylcyclanone derivatives have been fruitful. Thus the tosylates **312** and **313** reacted with potassium acetate in ethanol to give the lactams **302** ($R_1 + R_2 = O$) and **314** respectively (520), while Schmidt reaction on the ketone corresponding to **313** gave a mixture of the pyrido[3,2-*b*]azepine derivative **314** and the pyrido[3,2-*c*]azepine isomer **315** (521).

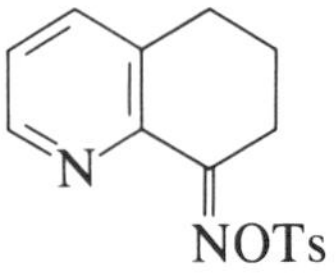

N NOTs 312; NOTs N 313; H N O N 314; O NH N 315

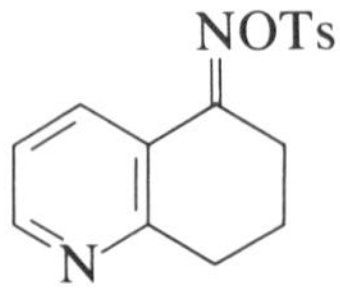

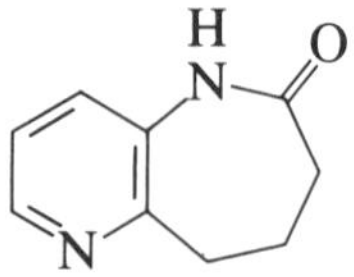

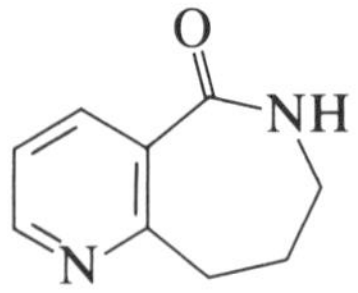

Reductive cyclization of compounds **316** (R = H, Me) gave the corresponding lactams of the pyrido[3,2-*c*]azepine series (522). A more versatile approach favored by workers in Japan and the U.S.S.R. involves fusion of the pyridine ring onto a preformed azepine nucleus. Thus β-keto ester **317** was converted by conventional condensations to the pyrido[2,3-*d*]azepine **318,** which underwent further, predictable transformations (523). On the

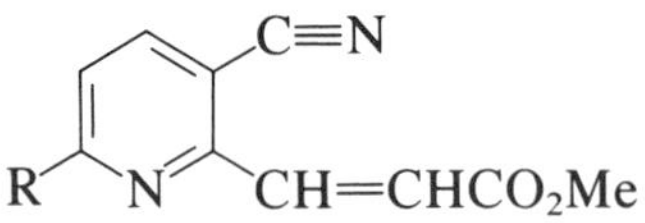

C≡N R N CH=CHCO$_2$Me 316; PhCH$_2$N CO$_2$Et O 317

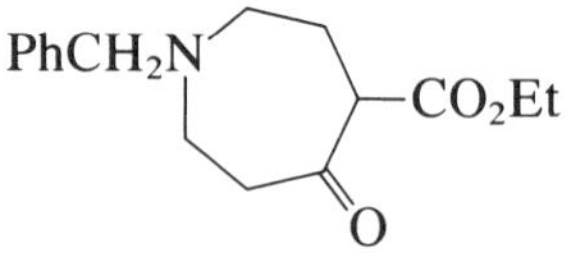

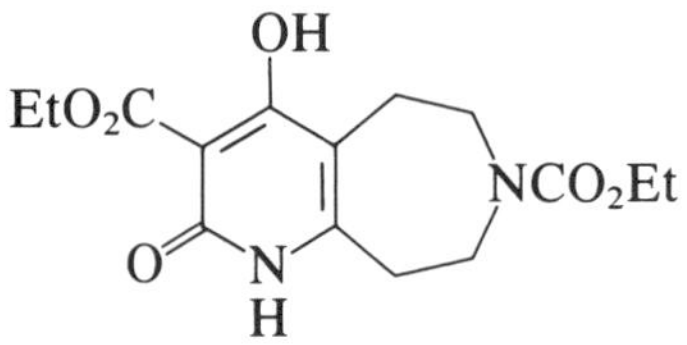

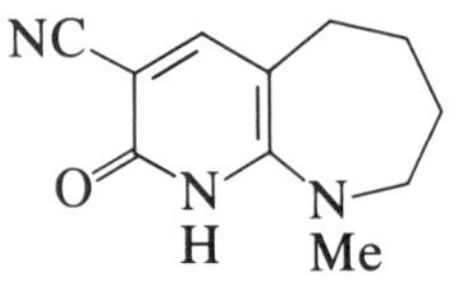

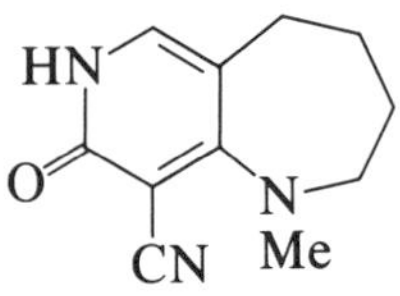

OH EtO$_2$C NCO$_2$Et O N H 318; NC O N H N Me 319; HN O CN N Me 320

other hand, caprolactam derivatives have been converted to both pyrido[2,5-*b*]azepines **319** (524) and pyrido[4,3-*b*]azepines **320** (525).

Finally, it has been claimed that an intramolecular benzyne reaction carried out on the bromopyridine **321** gave a mixture that contained the pyrido[4,3-*b*]azepine **322** as one of the products (363), and the previously mentioned vinylaziridine rearrangement (550) has been applied to two *N*-pyridyl compounds, giving the pyrido[4,3-*b*]azepine derivative **323** and the pyrido[3,4-*b*]azepine **324**. Apparently, this is the first example of that ring system.

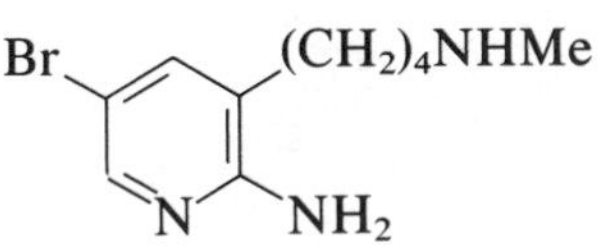

321 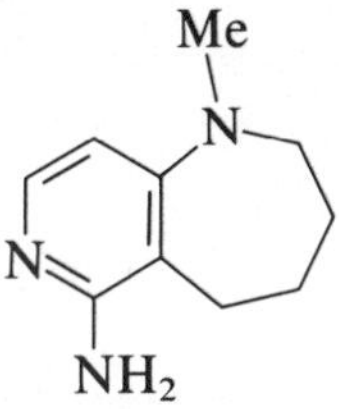 **322**

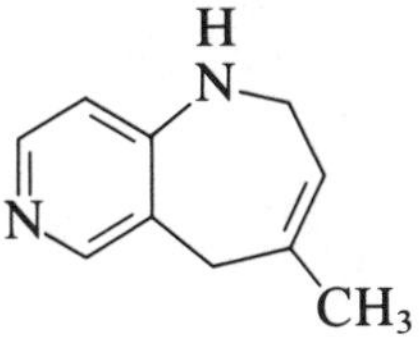

323 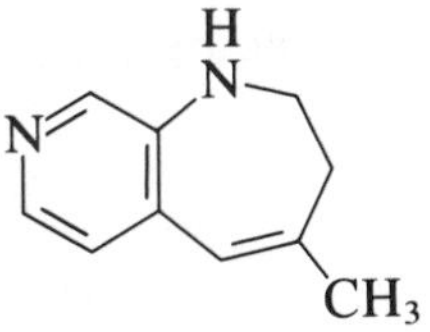 **324**

F. *Pyrimidoazepines*

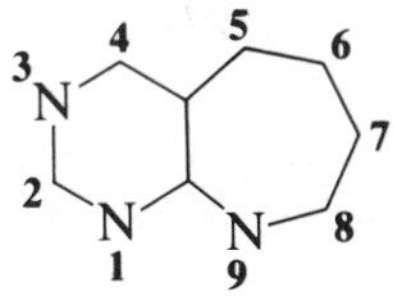

Pyrimido[4,5-*b*]azepine

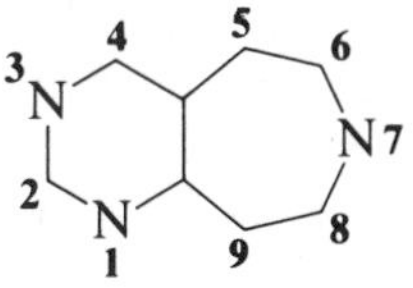

Pyrimido[4,5-*d*]azepine

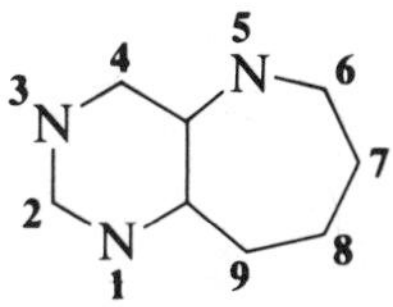

Pyrimido[5,4-*b*]azepine

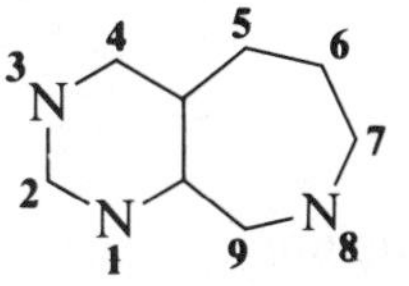

Pyrimido[4,5-*c*]azepine

Two approaches to pyrimido[4,5-*b*]azepines have been explored. In one, caprolactam reacted with phosphoryl chloride in formamide under pressure to yield **325** (R_1 = R_2 = H) (156); in the other, the enol lactim ethers **326** (R = CN, CO_2Et) were condensed with guanidine, acetamidine, or thiourea to obtain the various tetrahydropyrimido[4,5-*b*]azepines **325** (R_1 = NHMe, SH; R_2 = NH_2, OH) (46).

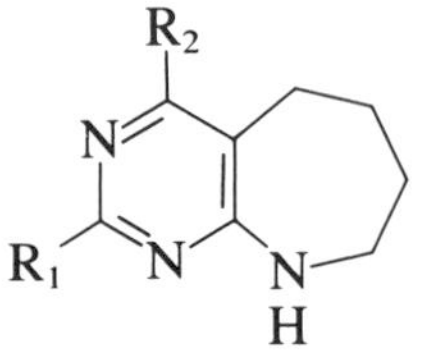

325

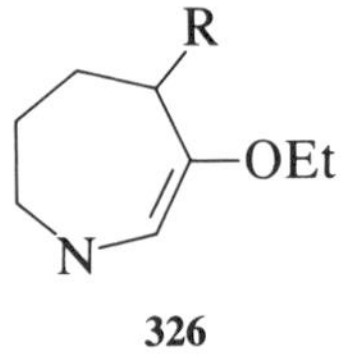

326

Pyrimido[5,4-*b*]azepine derivatives have been made either by a Schmidt reaction on the appropriate ketone (364) or by solvolysis of the corresponding oxime tosylate (365). The product reported in each instance was the lactam **327,** but the latter procedure was markedly more efficient (365).

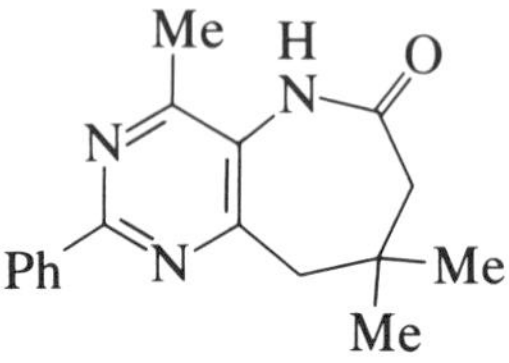

327

Condensation of the hexahydroazepinone **317** with guanidine, thiourea, or urea provided the pyrimido[4,5-*d*]azepines **328** (R = NH_2, SH, OH), which could be elaborated further by conventional methods (49,526). The starting material **317** could be made by cyclization of the appropriate amino diester or by ring expansion of *N*-benzyl-4-piperidone with ethyl diazoacetate (48).

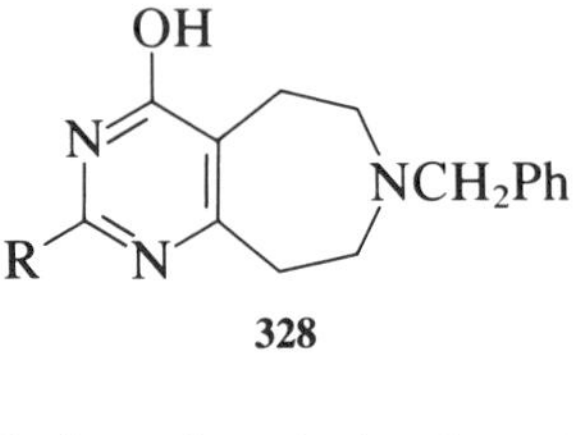

328

Recently, pyrimido[4,5-*c*]azepine derivatives **329** (R = NH_2, Me, Ph, SH) were obtained by converting caprolactam dione to 4-dimethylamino-

329 330 331 332

methylene-3-ketocaprolactam **330,** which reacted with various amidines (527). Also, the lactam acetal **331** reacted with phenyl isothiocyanate to give the pyrimido[4,5-*b*]azepine derivative **332** (551).

G. *Pyridazinoazepines*

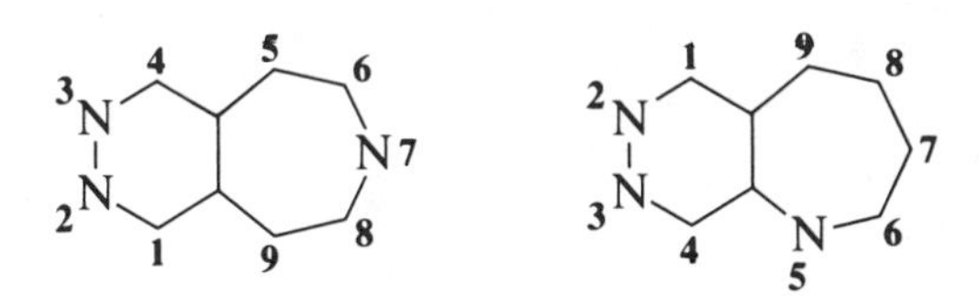

Pyridazino[4,5-*d*]azepine Pyridazino[4,5-*b*]azepine

Heating of *N*-ethoxycarbonylazepine with the tetrazine **333** afforded the triester **334,** which on treatment with chloranil was dehydrogenated to the pyridazino[4,5-*d*]azepine **335** (439,528). The same tetrazine **333** reacted with the lactim ethers **326** (R = Me, Ph) to yield the pyridazino[4,5-*b*]azepines **336** (R = Me, Ph) (529).

333 334

335 336

5. 7, 7 and Higher Ring Systems

The synthesis of 7,7 and higher ring systems is a little-explored area. A general method consists of treating certain dinitriles with hydrogen bromide, as illustrated by the preparation of the cyclooctanoazepine **337** (n = 4) (29). The method was extended to other homologous cycloalkanoazepines (**337;** n = 3, 5, 6) (366). Recently, another general method has been reported, which involves photolysis of *N*-cycloalkylsuccinimides (435). Thus irradiation of *N*-cycloheptylsuccinimide gave the keto lactam **338** as a mixture of *cis* and *trans* isomers. Reaction of chlorosulfonyl isocyanate with compound **55** (n = 3) and analogues, followed by treatment with thiophenol in pyridine, gave cyclohept[*d*]azepines **339** (R = H, CN) (501).

337 **338** **339**

III. Bridged Bicyclic Ring Systems

1. Azabicyclooctanes

2-Azabicyclo[4.1.1]octane 3-Azabicyclo[4.1.1]octane

Both 2- and 3-azabicyclo[4.1.1]octanes have been obtained. In every instance, the starting materials were oximes of naturally occurring terpenes. Thus Beckmann rearrangement of nopinone oxime (367) and 4-pinanone oxime (368) yielded the lactams **340** (R = H) and **340** (R = Me), respectively. Rearrangements of oxime **341** (R = H) led to lactam **342** (R = H) (369), whereas **341** (R = Me) yielded **342** (R = Me) (370). In the latter reaction rearrangement was effected by treating the oxime tosylate with sodium hydroxide, and the reaction apparently proceeded via the more stable ion, a phenomenon observed generally under these conditions. Recently, it has been claimed (440) that the *anti*-oxime of nopinone reacts with phosphorus pentachloride to form a compound formulated as the 3-azabicyclooctane

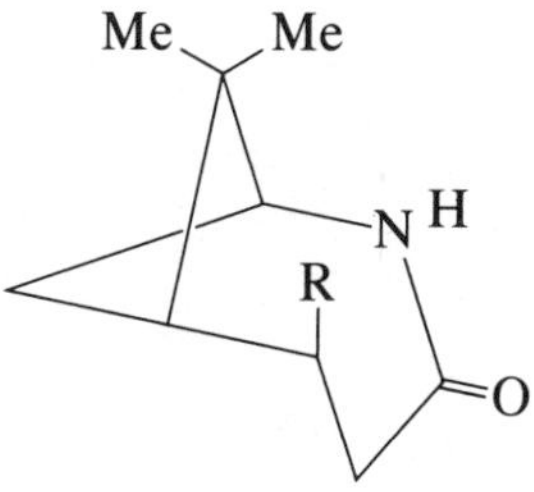

340

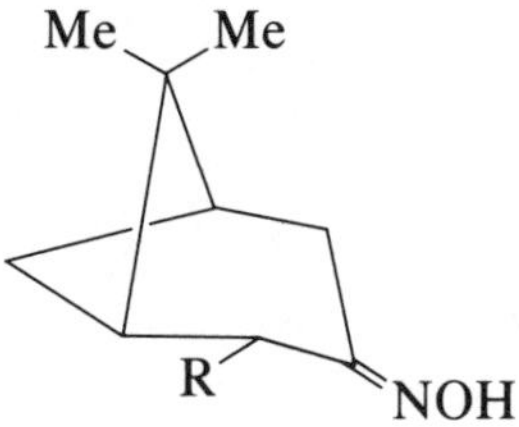

341

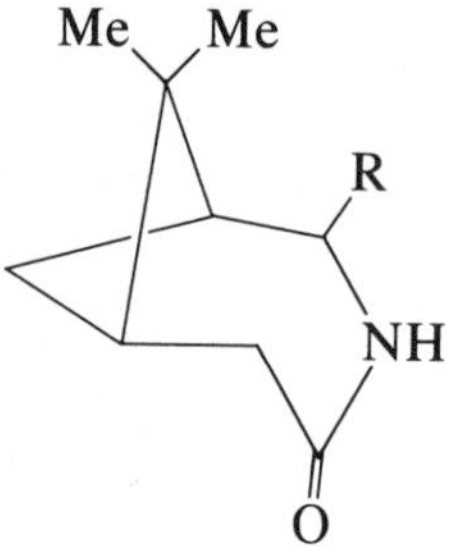

342

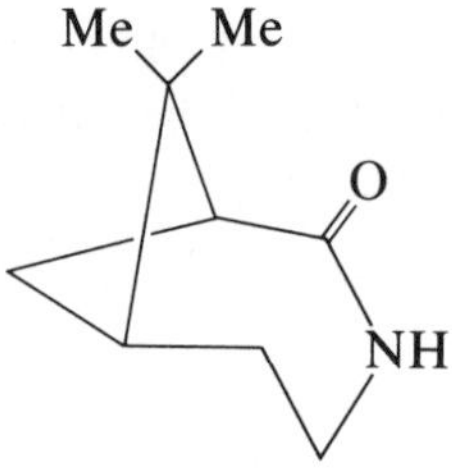

343

lactam **343.** Since the melting point of this product is virtually identical to that reported for **340** (R = H), it is possible that one of these structural assignments is in error.

2. Azabicyclononanes

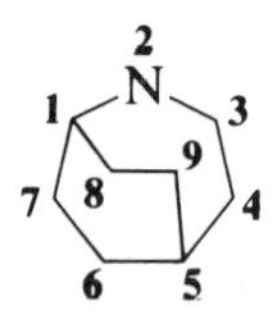

A. *2-Azabicyclo[3.2.2]nonanes*

The 2-azabicyclo[3.2.2]nonane ring system is well represented in the literature. Predictably, Beckmann ring expansion of the oxime of bicyclo[2.2.2]octan-2-one gave lactam **344** (R_1 = R_2 = H, mp 166°C) of the 2-azabicyclo[3.2.2]nonane series (371,377), whereas Schmidt reaction on the corresponding ketone gave the 3-azabicyclo[3.2.2]nonan-2-one **345** (R = H, mp 152°C). Brief reference to this reaction had appeared previously (372). Lactam **344** (R_1 = MeO, R_2 = H) was obtained in similar fashion (373). Recently it was revealed that bicyclo[2.2.2]octan-2-iminonitrile (374) reacts with hydrogen peroxide in the presence of sodium carbonate to give a mixture of the amide **344** (R_1 =, R_2 = $COHN_2$; 83%) along with 17% of the isomeric 3-azabicyclo[3.2.2]nonan-2-one.

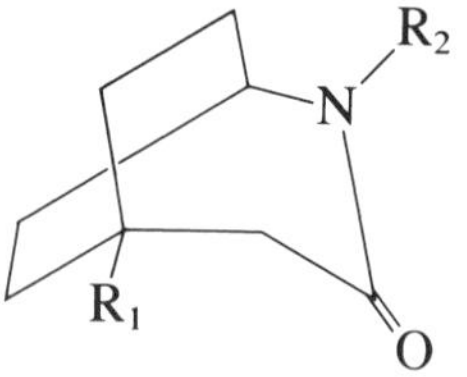

344

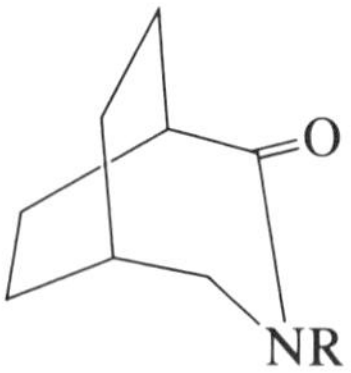

345

An interesting ring expansion with diazomethane converted the isoquinuclidinium salt **346** to the fused aziridinium salt **347,** and thence to **348,** on refluxing with isopropanol (378). The lactam **344** ($R_1 = R_2 = H$) was obtained by heating *cis*-4-aminocyclohexaneacetic acid, but the yield was poor (375). The infrared spectrum of this compound has been discussed (376).

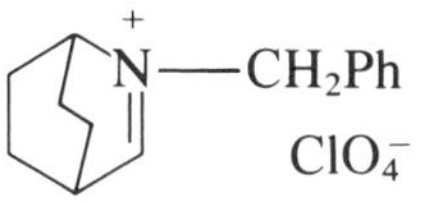

346

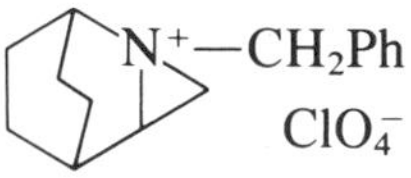

347

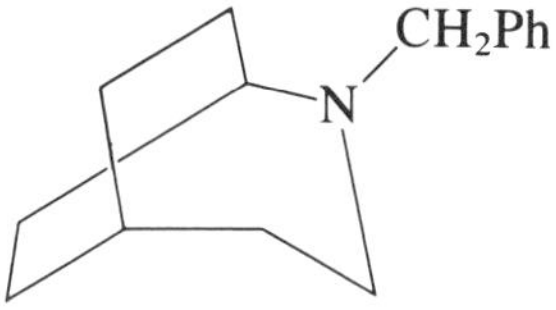

348

The other major method of access to the 2-azabicyclo[3.2.2]nonane system involves the symmetry-allowed [4 + 2] cycloaddition of azepine derivatives to dienophiles. The first example (379) of this type of conversion involved the reaction of the lactams **349** (R = H, Me, Et) with tetracyanoethylene to yield the adducts **350** (R = H, Me, Et), of which the first was obtained in virtually quantitative yield. On the other hand, dimethyl

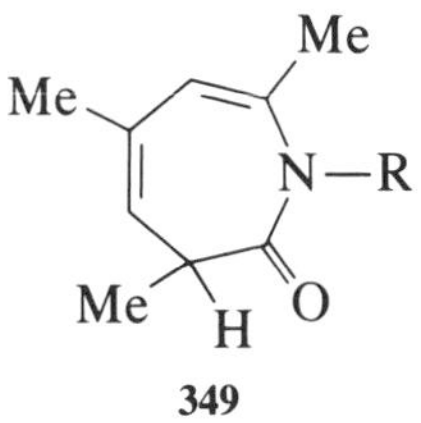

349

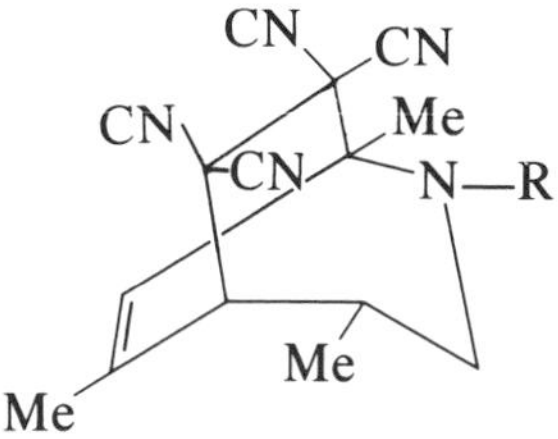

350

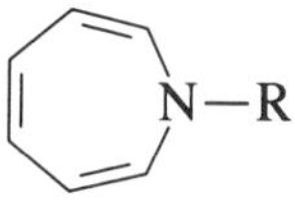

351

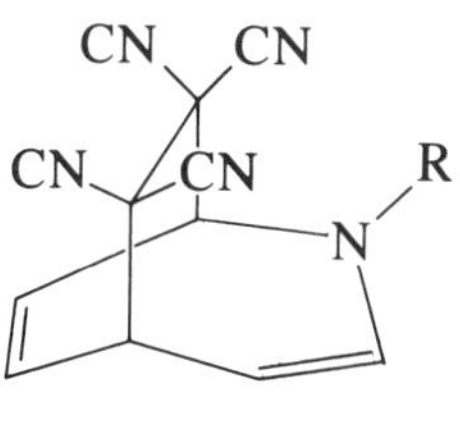

352

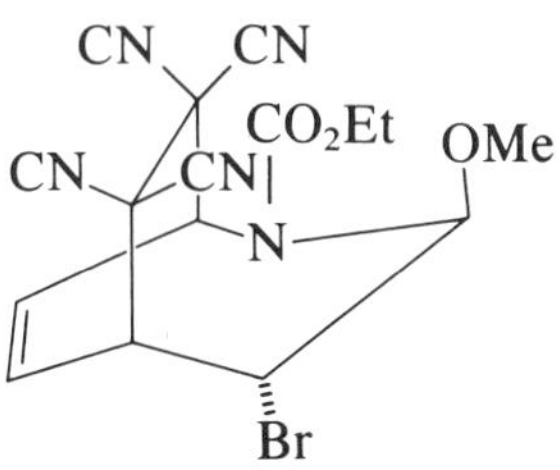

353

acetylenedicarboxylate gave only 19.6% of the corresponding adduct, and in most of the subsequent work tetracyanoethylene (TCE) has been the dienophile of choice. The azepine **351** (R = CO_2Et) reacted with TCE (380,381) to form the adduct **352** (R = CO_2Et), which on bromination in methanol afforded a mixture of stereoisomers, of which the major one was shown by X-ray diffraction to be **353.**

Addition of TCE to monomethylazepines has been used to prepare the four possible monomethyl derivatives of **352** (R = CO_2Me) (383,384). Thus the 2-methylazepine gave a 2.5% yield of the 3-methyl derivative of **352** (R = CO_2Me), the 4-methylazepine gave a 65% yield of the corresponding 6-methyl derivatives, and the 3-methylazepine gave a mixture of the 4-methyl and 7-methyl derivatives.

Some dimethyl azepines have also been condensed with TCE (385). In the four examples studied, the 2,4-, 2,5-, 3,4-, and 3,5-dimethyl derivatives of **351** (R = CO_2Me) each yielded only a single adduct. The adducts were determined to be the 1,6-, 3,6-, 6,7-, and 4,6-dimethyl derivatives of **352** (R = CO_2Me). It should be noted that in the sterically hindered 3,6-di-*tert*-butyl derivatives of **351** (R = CO_2Et), a 1,6-addition product was formed (386). Low yields were obtained in the reaction of TCE with the azepines **351** (R = pentafluorophenyl and mesyl) (387,388).

Cycloadditions have also been carried out with the dihydroazepines **354** and **355,** with formation of the expected 3,4-dihydro derivatives of **352** (389,390,552). In the first example (389), dimethyl acetylenedicarboxylate proved to be a satisfactory dienophile. In **352** (R = CO_2Et), nmr spectroscopy revealed a long-range coupling of 1–2 Hz between H_1 and H_3 or H_5 and H_7 (381).

Me, Me, N—Me, Me

354

CO_2Me, SMe, MeO_2C, N—Me

355

B. *3-Azabicyclo[3.2.2]nonanes*

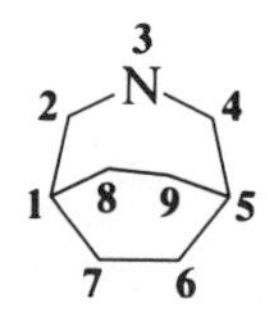

a. PREPARATION

The parent substance, 3-azabicyclo[3.2.2]nonane (**356,** R = H), became commercially available in the early 1960s by processes involving the passage

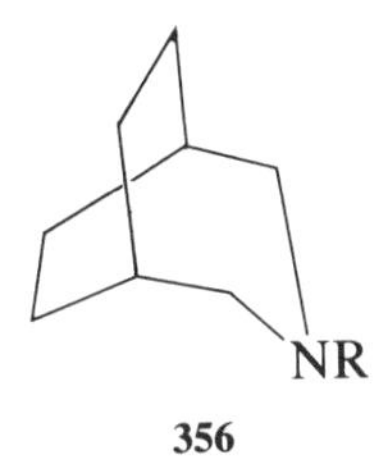

356

of 1,4-*bis*(aminomethyl)cyclohexane over Al_2O_3 at about 400°C (391–393), or by reaction of 1,4-*bis*(hydroxymethyl)cyclohexane with ammonia in an autoclave at 225°C (394). Ring expansion of bicyclo[2.2.2]octane derivatives is the only other synthetic approach that has been explored. As reported in the preceding section, Schmidt reaction on bicyclo[2.2.2]octan-2-one gives the lactam **345** (R = H) in 56% yield (371, 395).

An alternative method (396) subjected bicyclo[2.2.2]octane-2-carboxylic acid, prepared from 1,3-cyclohexadiene and acrylic acid, to the action of nitrosylsulfuric acid and fuming sulfuric acid at 70°C. As stated previously, the *N*-cyano imine of the bicyclo[2.2.2]octane series reacted with hydrogen peroxide in alkaline medium to give the lactam **345** (R = $CONH_2$) as a minor product.

b. PHYSICAL PROPERTIES

The basicity of 3-azabicyclo[3.2.2]nonane (**356,** R = H) was examined, and the pK_a was found to be 10.58 at 25°C in water (397). In studies of the thermodynamic properties of this molecule (398–400), the high-temperature solid phase was deduced to be a plastic crystal (398), and the enthalpy of combustion was shown to be 1240.51 kcal/mole (400). An X-ray crystallographic structure determination on the low-temperature solid phase at 291°K showed that the crystal was orthorhombic with space group *Aba*2 (401).

The physical properties of the 3-nitroso compound **356** (R = NO) have likewise been of interest. The nmr spectrum was interpreted as being indicative of significant contributions from a dipolar form ($N^+{=}N{-}O^-$), and the mass spectrum showed a significant M − OH peak, which arose via a McLafferty type of rearrangement (403). In an esr study of the action of peracids on the amine **356** (R = H), nitroxide radicals were detected (404), and the 3-nitro derivative (**356,** R = NO_2) was compared with other *N*-nitro compounds by differential thermal analysis (405).

With regard to the energy barrier to inversion of the 3-nitrogen atom in 3-azabicyclo[3.2.2]nonanes, it was shown that the barrier for the *N*-methyl compound **356** (R = Me) is greater than that for the *N*-chloro compound (**356,** R = Cl) (406), and that the barrier for all 3-azabicyclo[3.2.2]nonanes studied (including the N—CD_3 derivative) is greater than in the corresponding nonbridged hexahydroazepines. The barrier to rotation about the N—N bond of **356** (R = NO) has been estimated from nmr measurements (407), and the ^{13}C-nmr spectrum was discussed in relation to that of similar molecules (408). More recently, further temperature-dependent nmr studies have been reported (553).

c. CHEMICAL PROPERTIES

The availability of 3-azabicyclo[3.2.2]nonane as a starting material has made possible a large number of chemical modifications involving alkylation and acylation of the nitrogen atom. As there are over 100 papers and patents covering several hundred compounds, we shall consider only those reactions which add to our basic knowledge of the 3-azabicyclo[3.2.2]nonane system and will include in the table only a representative number of the more significant derivatives. A comprehensive list of simple alkyl, acyl, and sulfonyl derivatives has been published (392), and several studies on Mannich bases (409,410) and enamines (411,412) have appeared.

Photolysis of the *N*-nitroso compound **356** (R = NO) in acid solution gave the 2-oximino derivative of **345** (R = H). A mechanism has been proposed for this reaction which assumes that the cyclic imine is one of the intermediates and that recombination of this imine with NOH in the acid medium leads to the observed product (413).

Acid hydrolysis of the lactam **345** (R = H) gave *cis*-4-aminomethylcyclohexane-1-carboxylic acid (395), whereas alkaline hydrolysis yielded the thermodynamically favored isomer (414). Mercuric acetate treatment of the amines **356** (R = H, Me, $PhCH_2$) gave the corresponding 4-alkylaminocyclohexane-1-carboxaldehydes (415), which were isolated as dimedone derivatives. The benzylamino analog was demonstrated to possess *trans* geometry.

When the amide **356** (R = COPh) was incubated with *Sporotrichum sulfurescens* (416), a mixture of the ketone **357** (R = COPh) and the *endo*-alcohol **358** was obtained, with the latter predominating. Sodium borohydride reduction of ketone **357** (R = CH_2Ph) gave the *exo*-alcohol **359**.

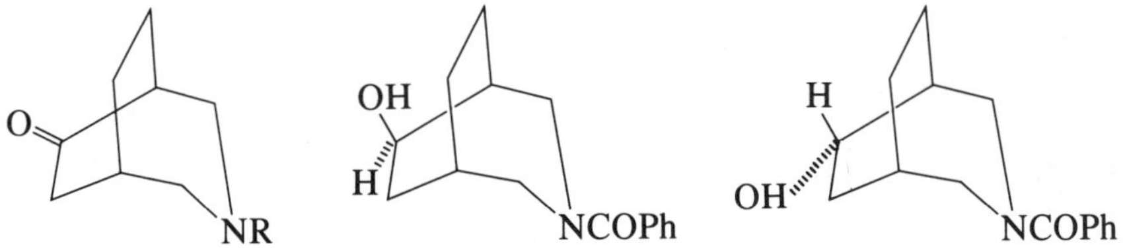

O NR **357**

OH H NCOPh **358**

H OH NCOPh **359**

Catalytic hydrogenolysis of ketone **357** (R = COPh) gave the amino ketone **357** (R = H), which was isolated as a hydrochloride salt (416). The lithium salt **356** (R = Li) was used to study nucleophilic attack on epoxides (427). Representative 3-azabicyclo[3.2.2]nonanes are listed in Table 33.

TABLE 33. 3-AZABICYCLO[3.2.2]NONANES

R	Other Substituents	mp (°C) or bp (°C/torr)	Derivatives (mp, °C)	Refs.
H		186	MeI, 245 (dec.)	392,395
			HCl, 281–283	
			Picrate, 234–235	
Me		37–38/2	Picrate, 249–251	415
			$HClO_4$, 250	
Et		53–56/4	$HClO_4$, 226–228	415
n-Bu		148–153		
		56–57/0.1–0.2		392
Ph		136/0.7	$HClO_4$, 236–238	415
			Picrate, 156–158	
$PhCH_2$		111–113/0.5	$HClO_4$, 142–143	415
CO_2Et		98/0.7		415
CH_2CH_2CN		26		
		84–86/0.1		392
CH_2CN		51–53		392
$CH_2CH_2NH_2$		97–98/1.5		392
CH_2CH_2OH		94/1.8		392
CH_2CH_2Cl			HCl, 198–225	420
$CH{=}CMe_2$		70–74/1.5		392,422
$CH{=}C_6H_{10}$ (cyclohexylidene)		168–174/2.7		392,422

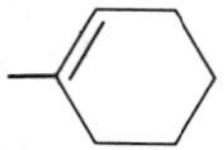 1-cyclohexenyl (structure)	88–92/0.7		392,422
Ac	87–91		392
$COCH_2Cl$	73–76		392,424
$COCH_2I$	61–61.5		424
p-$NO_2C_6H_4$	148–153		392,418
$CO(CH_2)_{16}Me$	64–65		392
$COCH_2Ph$	70–72		392
$COCH_2OPh$	101–103		392
$COCH{=}CHPh$	104–105		392
Cl	88–90/10		415
NH_2	75–77		425
NO	161–164		425
NO_2	110–111		405
CO–cyclohexyl (structure) 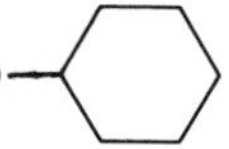	80–83		392
PhCO	93–94		392
α-Furoyl	75–78		392
3,5-$(NO_2)_2C_6H_3$	158–163		392
p-$MeOC_6H_4$	107–113		392
p-$NH_2C_6H_4$		HCl, 190–200	418
SO_2Me	115–116		392
$PhSO_2$	96–98		392
p-$NO_2C_6H_4SO_2$	193–195		392
p-$NH_2C_6H_4SO_2$	149–151		392
SO_2Et	63–66		419
SO_2NMe_2	90–91		419
SO_2NH_2	161–162.5		419
SCl	93–95		426
$CS_2 \cdot Na \cdot H_2O$	271–273		421

(continued)

TABLE 33. (*Continued*)

R	Other Substituents	mp (°C) or bp (°C/torr)	Derivatives (mp, °C)	Refs.
CONHPh		196–198		392
CSNHPh		173–174		392
α-Naphthyl-NHCO		242–248		392
CS_2CH_2Ph		98–99		423
$2,6\text{-}Cl_2C_6H_4CS_2$		98–99		423
$2,3,4\text{-}Cl_3PhCS_2$		111–113		423
CH_2CH_2NC		104–106/0.1		418
$CH_2CH_2NMe_2$		74–76/0.06		417
$CH_2CH_2NEt_2$		84–87/0.05		417
$CH(Me)CH_2NMe_2$		88–93/0.06		417
CH_2CH_2N(piperidino)		118–123/0.06		417
H	2-Keto	152 154–155/6	HCl, 208.5–210	395
Ac	2-Keto	96–97/0.2		395
H	6α-Ph, 6β-OH		HCl, 218–220 (dec.)	416
COPh	6β-OH	73–75		416
H	6-Keto		HCl, 227–229 (dec.)	416
CH_2Ph	6-Keto	222 (dec.)		416
COPh	6-Keto	59–62	Semicarbazone, 197–200 Oxime, 156–158 2,4-DNP, 198–201	416
COPh	6α-OH	135–137		416
H	6α-*O*-Benzoate		HCl, 205–208	416
H	6α-OH, 6β-Ph		HCl, 238–240 (dec.)	416
H	6α-OH, 6β-Me		HCl, 230–232	416
$PhCH_2$	6α-OH		HCl, 185–187	416

C. *6-Azabicyclo[3.2.2]nonanes*

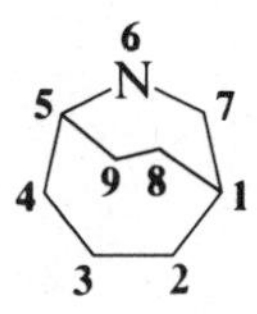

Derivatives of the 6-azabicyclo[3.2.2]nonane system were first made (428) by boron trifluoride–catalyzed addition of a series of *N*-ethoxycarbonylimines to 1,3-cycloheptadiene, giving adducts **360** (R_1 = H, Ph; R_2 = H, Ph). Later, it was shown that compound **361** cyclized to **362** either by radical (530) or by chromous chloride initiation (531).

360 **361** **362**

D. *2-Azabicyclo[4.2.1]nonanes*

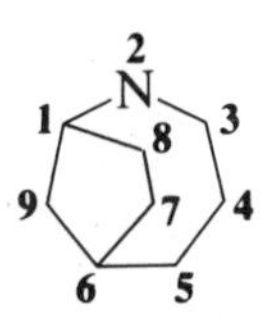

Beckmann rearrangement of bicyclo[3.2.1]octan-2-one oxime sulfonate esters gives lactams of the 2-azabicyclo[4.2.1]nonane series. Thus the benzenesulfonate ester of bicyclo[3.2.1]octan-2-one oxime yielded lactam **363** on treatment with sodium hydroxide (429) (but this reaction has recently been shown to give **364** as well (566)), and the unsaturated oxime tosylate **365** (R = Me) gave lactam **366** (R_1 = Me, R_2 = H) (430). The desmethyl

363 **364**

365 **366**

analog **365** (R = H) gave **366** (R_1 = R_2 = H) in analogous fashion (431). The *N*-nitroso derivatives were shown to undergo fragmentation in the presence of sodium methoxide to give variously substituted cyclopentanes.

Complexation of *N*-methoxycarbonylazepines with transition metals alters the normal course of TCE cycloaddition. Under these circumstances, either [2 + 2] and [2 + 6] cycloadditions are predominant, leading to complexes such as **367** (M = Fe, Ru) and **368** (432). With iron as the transition metal, compound **368** was the major product and **367** (M = Fe) was the minor one. X-ray crystallographic analyses have been performed with these complexes (432).

367 **368**

E. *3-Azabicyclo*[*4.2.1*]*nonanes*

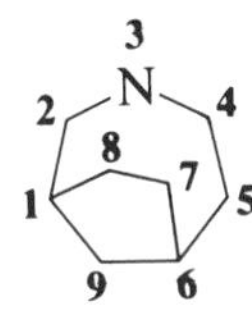

Predictably, polyphosphoric acid treatment of the oxime of bicyclo[3.2.1]octan-3-one gave the lactam **364**, which could be reduced with lithium aluminum hydride to the parent amine (369,433). Polyphosphoric acid with the oxime corresponding to **365** (R = Me) afforded the lactam **369** (430). It has been claimed (434) that 4-chlorobenzenediazonium sulfate causes ring expansion of **370** to **371**.

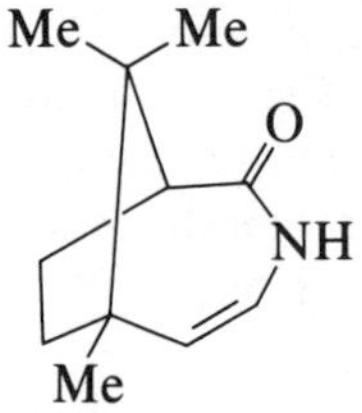

369

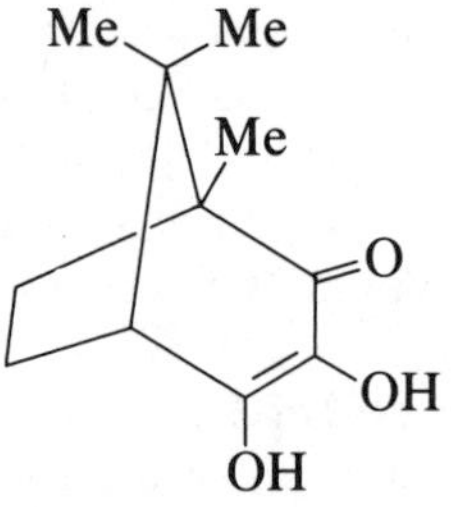

370

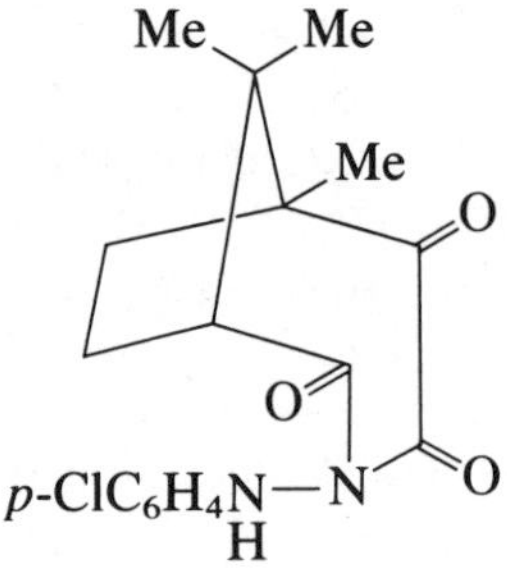

371

F. *9-Azabicyclo[4.2.1]nonanes*

a. PREPARATION

The classical work on 9-azabicyclo[4.2.1]nonanes (441) involved Tiffeneau ring expansion of the reduced cyanohydrin of tropinone to the ketone **372,** which was converted to several other 9-azabicyclo[4.2.1]nonane derivatives. Subsequent work has generally used the approach of adding a one-nitrogen source to a suitable cyclooctane derivative. For example, a patented procedure (442) involves heating 1,4-cyclooctanediol with ammonia and hydrogen in an autoclave to give the parent substance 9-azabicyclo[4.2.1]nonane (mp 62°C). Reaction of cyanogen azide with cyclooctatetraene at 78°C yielded a mixture that included compound **373** (R = CN) as one of the products (443,444). It was proposed that this compound arises from the addition of triplet cyanonitrene to the cyclooctatetraene system.

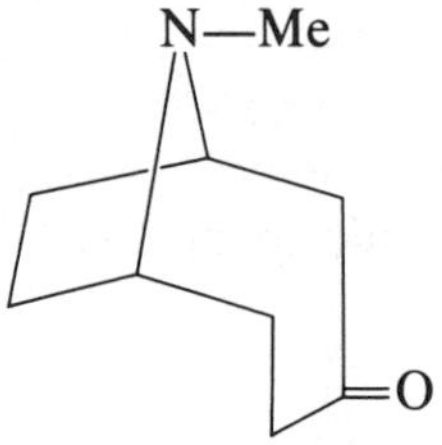

372

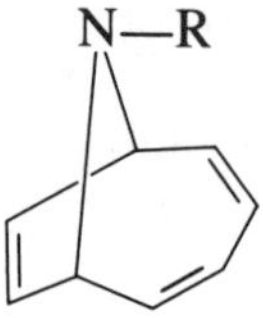

373

The potassium salt of cyclooctatetraenide dianion reacted with isoamyl nitrite in tetrahydrofuran to give a 74% yield of the bicyclic product **373** (R = OH), which was converted to a number of other derivatives (**373;** R =

H, NH_2, $PhCH_2$) (445). It was also demonstrated that compound **373** (R = NH_2) could be obtained by treating the phthalimide **374** successively with sodium hydroxide and hydrogen chloride. Finally, homolytic decomposition of an *N*-chloro derivative in the presence of silver perchlorate in acetone has been reported to give compound **375** along with the isomeric 2-chloro-9-azabicyclo[3.3.1]nonane (446).

The proof of structure of **375** and of **373** (R = CN) rested on comparisons with authentic material prepared by alternate routes (441). Similar product distributions were noted in aminomercuration reactions of *N*-phenyl-9-azabicyclo[6.1.0]non-4-ene (533), and the authors have used ^{13}C-nmr to distinguish the various azabicyclic systems (534).

374 **375**

b. REACTIONS

Transformations of **373** (R = CN) included modification of the nitrile group to give **373** (R = $CONH_2$, CHO) and reduction of the 7,9-double bond (447). Photolytic rearrangement of **373** in the presence of a sensitizer yielded the cage compound **376,** which displayed a temperature-dependent nmr spectrum (448); without the use of a sensitizer, a dimer was obtained (449). The *N*-nitroso compound **373** (R = NO) was quite stable and did not decompose thermally to cyclooctatetraene and nitrous oxide (450), though under the influence of basic sodium hydrosulfite it did yield cyclooctatetraene by what was regarded as a nonlinear chelotropic reaction (451). The latter decomposition mode was also observed with the *N*-nitroso compound **377** (R = NO), but not with **378** (R = NO). Compound **377** (R = H) was obtained by reacting 3,8-dibromocyclooctane with disodium cyanamide. Analysis of their photoelectronic spectra (452) reveals that trienes such as **373** (R = H) are

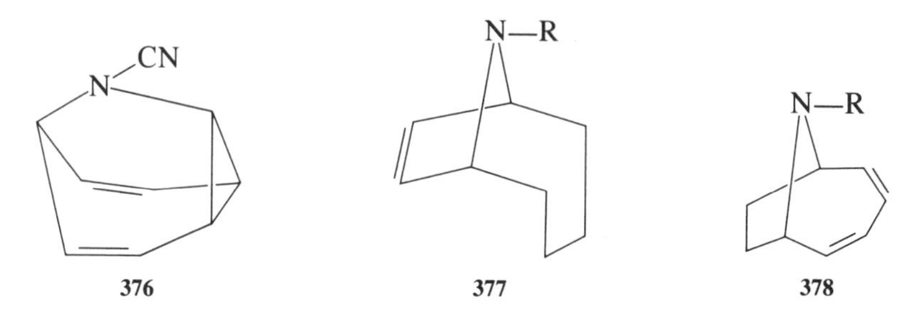

376 **377** **378**

"bicyclo-conjugated" molecules, which are unlike the partly reduced derivatives **377** and **378** and related carbocyclic analogs.

The iron complex **368** has been mentioned previously. X-ray crystallographic data (453) were also obtained for **379,** the *N*-acetyl derivative of a toxin produced by the blue-green freshwater algae *Anabaena flos-aquae*. Recently, another 9-azabicyclo[4.2.1]nonane derivative, called anatoxin, was isolated from freshwater algae (494) and synthesized from cocaine by a carbene-mediated ring expansion (495) and later by an interesting application of the cyclization of an iminium salt (532).

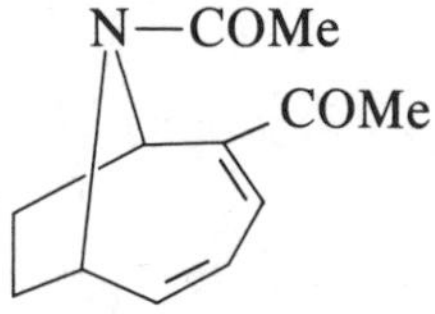

379

G. *6-Oxa-3-azabicyclo[3.2.2]nonanes*

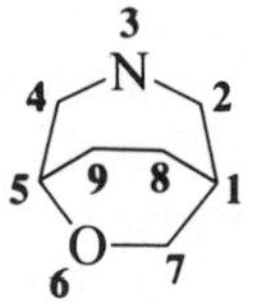

Treatment of the cineolic acid derivative **380** with sodium methoxide in toluene gave the lactam **381** (454). Reduction of such imides gives amines with biological activity (554).

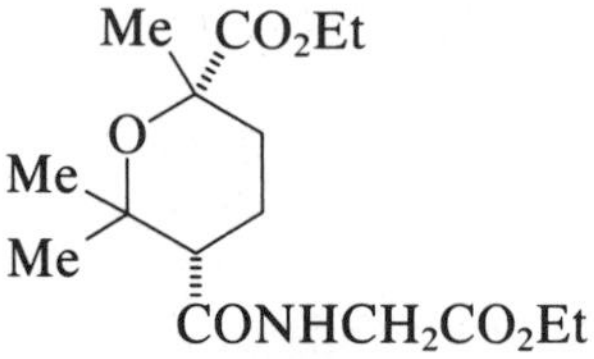

380 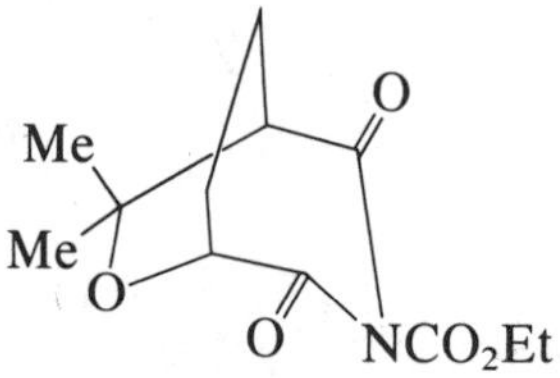 **381**

H. *8-Oxa-3-azabicyclo[4.2.1]nonanes*

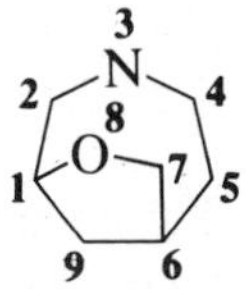

A double nucleophilic substitution by methylamine on the bromo mesylate **382** gave the bicyclic compound **383** (455).

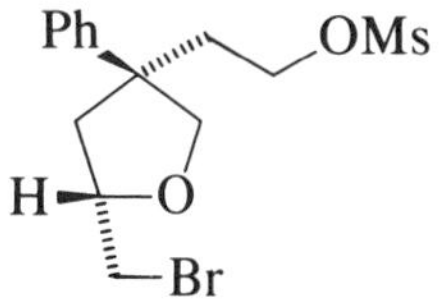

382

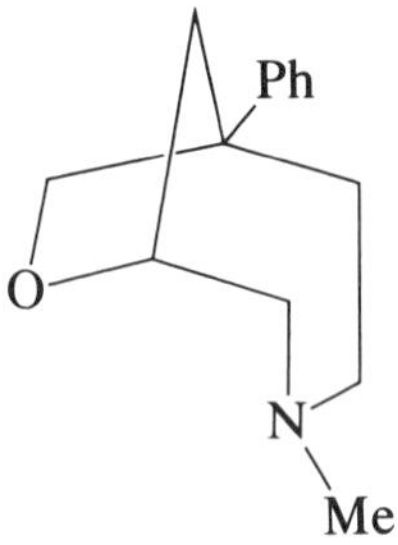

383

I. *7-Oxa-8,9-diazabicyclo*[*4.2.1*]*nonanes*

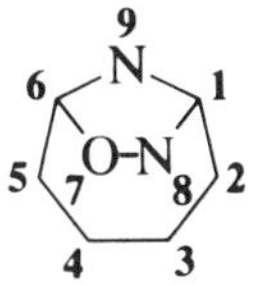

As was noted earlier (see Section 2.A), tetracyanoethylene (TCE) undergoes [4 + 2] cycloaddition with azepines to give 2-azabicyclo[3.2.2]nonane derivatives. On the other hand, nitrosobenzene reacted by [6 + 2] addition with 3,6-dimethyl-1-ethoxycarbonylazepine (385) to form the bicyclic adduct **384.** This structure has been confirmed by X-ray analysis (535).

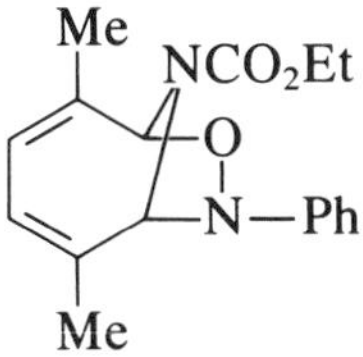

384

J. *7,8,9-Triazabicyclo*[*4.2.1*]*nonanes*

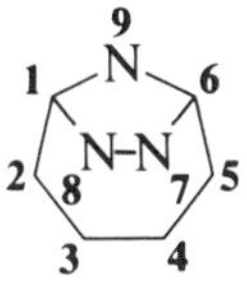

A "*trans*-dienophile" such as diethyl diazodicarboxylate was shown (456) to react with *N*-ethoxycarbonylazepine to produce compound **385** by a [6 + 2] process. In contrast, "*cis*-dienophiles" react (456) with *N*-ethoxycarbonylazepine via [4 + 2] cycloaddition, as illustrated by the formation of compound **386** from *N*-phenyl-1,3,4-triazine-2,5-dione.

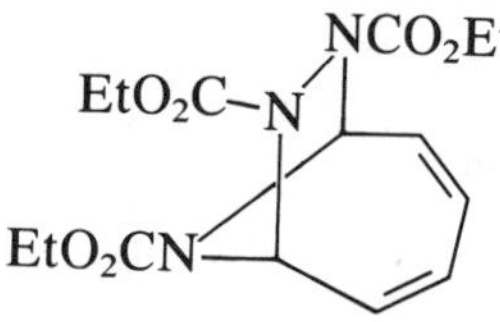

385

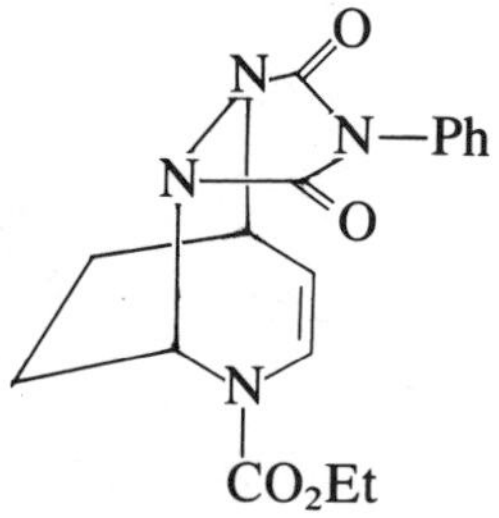

386

3. Azabicyclodecanes

A. *3-Azabicyclo[3.3.2]decanes*

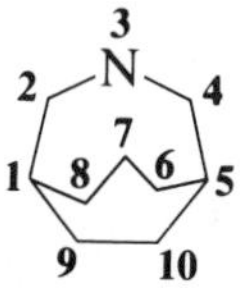

Members of the 3-azabicyclo[3.3.2]decane ring system have been obtained from azabullvalene derivatives. The diene **387** (R = H) arose when the lactam **388** was reduced with lithium aluminium hydride (457), and diene **387** (R = CH_2Ph) was the product from the subsequent treatment of **389** with mercuric acetate in methanol followed by sodium borohydride (458). Exhaustive methylation of **387** (R = H) led to 1,4-dimethylcycloheptane (457), whereas thermolysis at 250°C caused 90% conversion to the 8-azabicyclo[4.3.1]decane derivative **390** (459).

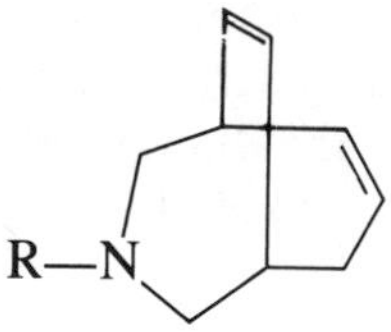

387

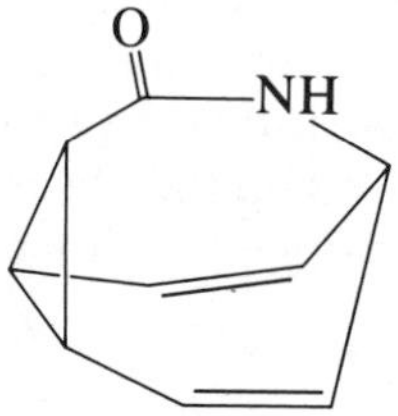

388

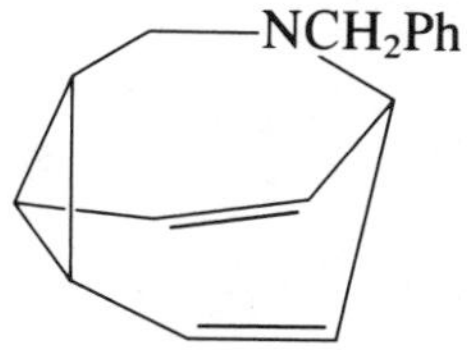

389

390

B. *9-Azabicyclo[3.3.2]decanes*

Members of the 9-azabicyclo[3.3.2]decane ring system have all been made by Beckmann or Schmidt reactions starting with the accessible bicyclo[3.3.1]nonan-9-ones. Thus the oxime **391,** derived from the adduct of cyclohexanone and benzalacetophenone, yielded lactam **392,** which was stable to dilute acid or base but could be hydrolyzed to a cyclooctane derivative on prolonged acid hydrolysis (460). Strong alkaline treatment caused skeletal rearrangement (461.)

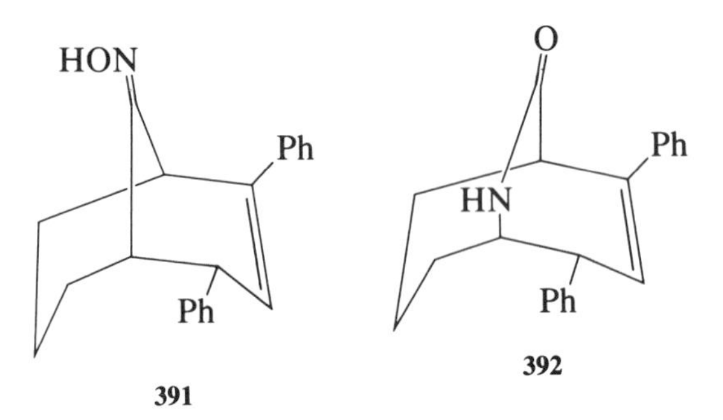

391 **392**

More recently, the synthetic part of this work has been extended to chalcones other than benzalacetophenone (462). The parent lactam **393** (R = H) was obtained in similar fashion (457) and reduced to the parent amine (457). Interestingly, 2-*N*-pyrrolidinylbicyclo[3.3.1]nonan-9-one underwent Schmidt reaction to give two products, which proved to be the α and β isomers of **376** (R = pyrrolidine) rather than positional insertion isomers. This structural assignment was based on the observation that both products yielded ketone **393** (R = O) on treatment with mercuric acetate in aqueous acetic acid.

393

C. *2-Azabicyclo[4.3.1]decanes*

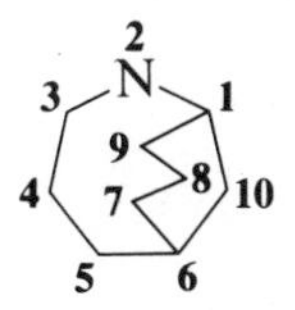

Heating the amino bromo compound **394** in propanol caused ring closure to the corresponding quaternary salt, which on thermolysis gave **395** (464). The carbonyl group in **395** was reduced successfully by the Wolff–Kishner method.

Br O $(CH_2)_3NMe_2$ OMe

394

O Me N OMe

395

D. *3-Azabicyclo[4.3.1]decanes*

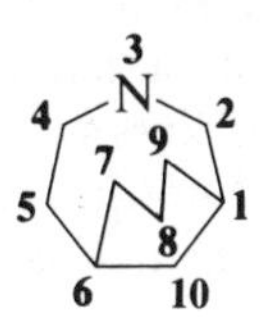

Until recently 3-azabicyclo[4.3.1]decanes were obtained exclusively via Beckmann reactions on oximes of bicyclo[3.3.1]nonan-3-ones. The unsubstituted lactam **396** (R = H) was obtained (367) by this method, as was the 2-phenyl derivative (**396,** R = Ph) (465). Another example (466) involved mixtures of ketones from natural sources. As expected, lithium aluminum hydride reduction of the amides afforded the corresponding bicyclic amines. Another route to 3-azabicyclo[4.3.1]decanes (467) is illustrated by the conversion of the amino alcohol **397** to the bicyclic amine **398.**

R
NH
O
396

$HOCH_2$—CPh_2
—CH_2NHMe
397

Ph
Ph
NMe
398

E. *10-Azabicyclo[4.3.1]Decanes*

The trivial name "homogranatanine" was employed in the literature for the parent member of the 10-azabicyclo[4.3.1]decane system prior to 1950. The classical Robinson tropinone synthesis offered a practical route to the 8-ketoamine **399** starting from adipaldehyde (468). The same method has been employed by subsequent workers who studied the course of reduction of the carbonyl group to α- and β-hydroxy compounds (469,470). Exhaustive degradation to bicyclo[4.3.0]nonane derivatives with barium hydroxide was

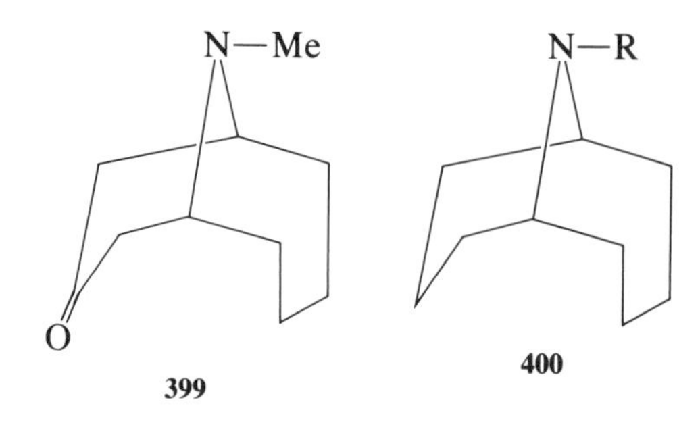

399 **400**

also investigated (471,472). Demethylation of **400** (R = Me) was effected with mercuric acetate (473); in acetic acid solution, the *N*-formyl compound (**400,** R = CHO) accompanied the secondary amine (**400,** R = H). The steric requirements for a successful demethylation have been discussed (473); the postulated intermediate is a methyleneiminium salt.

4. Azabicycloundecanes

A. *10-Azabicyclo[4.3.2]undecanes*

9 1 3 8 5 10 N2 7 4 11 6

The only example of 10-azabicyclo[4.3.2]undecanes is compound **401,** which appears as a by-product in 5% yield in the reaction of the bicyclo[6.1.0]nonane **402** with chlorosulfonyl isocyanate (474).

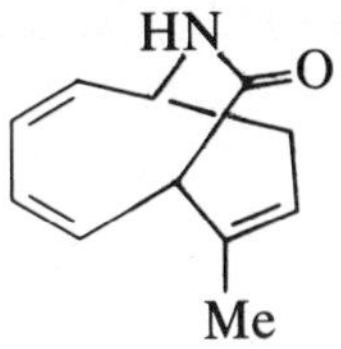

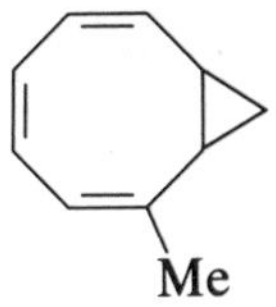

HN O Me Me **402**

B. *11-Azabicyclo[4.4.1]undecanes*

6 N 1 8 5 7 9 10 2 4 3

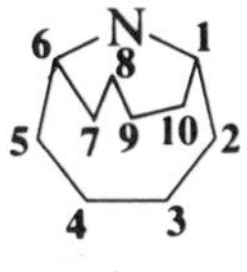

a. SYNTHESIS

Much of the synthetic activity relating to 11-azabicyclo[4.4.1]undecanes has concerned the "nonbenzenoid" aromatic system iminocyclodecapentaene (**403,** R = H). Basically, the synthesis of **403** (R = Ac) depended on the addition of nitrosyl chloride to "isotetralin," followed by acetylation, bromination, and dehydrobromination (475). This method has been used to make potential antiviral compounds of related structure (476).

A number of less extensively unsaturated 11-azabicyclo[4.4.1]undecanes have also been prepared. 6-Hydroxycyclodecanone oxime on reduction with sodium in butanol gave enamine **404** (R_1 = R_2 = H), which was reduced to the parent amine **405** (477). Hydrazoic acid reacted with Δ^5-octalin (478) to furnish another route to the enamine **404** (R_1 = R_2 = H). Both of these reactions proceeded in about 25% yield. Dimeric azepines have been found to undergo Hofmann degradation to bicyclic products such as **404** (R_1 = Me; R_2 = $CH_2CH_2NMe_2$, CH = CH_2). The preparation of compound **152** (323) has been mentioned elsewhere in this chapter (see Section II.4.A.i)

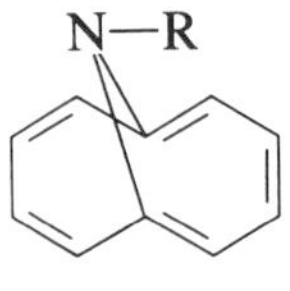

N—R N—R_1 NH

403 R_2 **405**

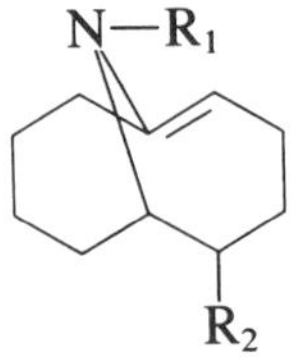

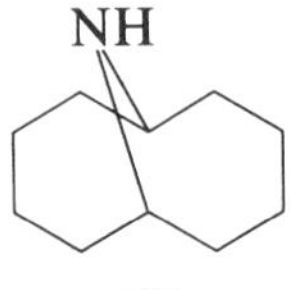

404

b. PHYSICAL PROPERTIES

Nmr spectral data for **403** (R = H, Me, Ac) indicate that a diamagnetic ring current is present in the molecule (480–482). Ultraviolet absorption spectra (480,481) support this conclusion, as does measurement of dipole moments (483). Circular dichroism studies (484) indicate that there is some interaction between the *p*-electrons on the nitrogen atom and the π-system.

c. CHEMICAL PROPERTIES

The chemical properties of iminocyclodecapentaene systems (328) have been less well studied than one would expect. The partly reduced compound **152** was converted to **153** at 180°C (18). The nitroso derivative **404** (R_1 = NO, R_2 = H) has been demonstrated (485) to consist of *E* and *Z* isomers by nmr spectroscopy, indicating a lack of free rotation about the N—N bond.

C. *2-Azabicyclo*[*4.4.1*]*undecanes*

2 N 1 3 10 9 4 8 11 7 5 6

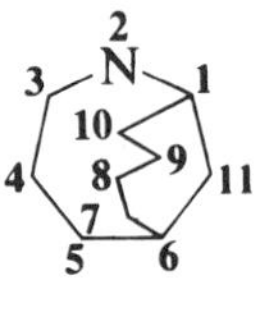

The sole example of 2-azabicyclo[4.4.1]undecanes, **406,** was obtained by classical elaborations of a suitably 1,6-disubstituted cycloheptatriene (536).

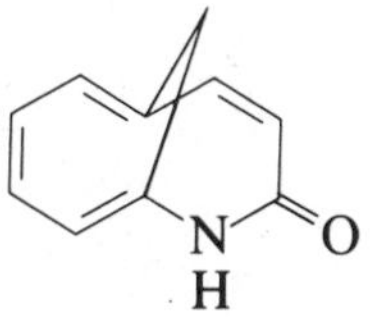

406

D. *3-Azabicyclo[4.4.1]undecanes*

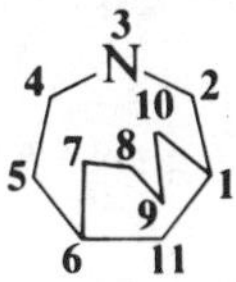

Here also, one recent example exists. Thus the azamethanoannulene **407** was obtained by bromination–dehydrobromination of compound **408** followed by oxidation (537).

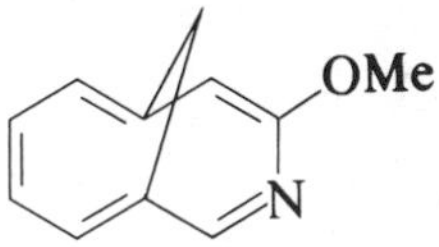

407

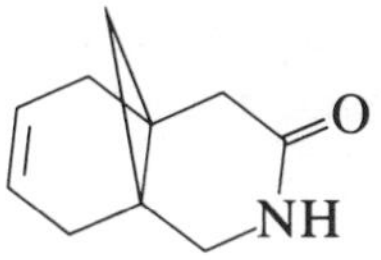

408

E. *3,11-Diazabicyclo[4.4.1]undecanes*

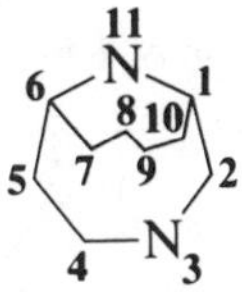

Ring enlargement of ketone **399** via the Schmidt reaction gave the lactam **409** (n = 4), which underwent Hofmann elimination to give the ten-membered ring product **410** (n = 4) (486). Perchlorate treatment of **410** (n = 4) did not lead to reclosure of the bridge, although similar treatment of **410** (n = 2, 3) did regenerate the salts of **409** (n = 2, 3). Mass spectra of **409** (n = 4) and other related lactams have been reported (487).

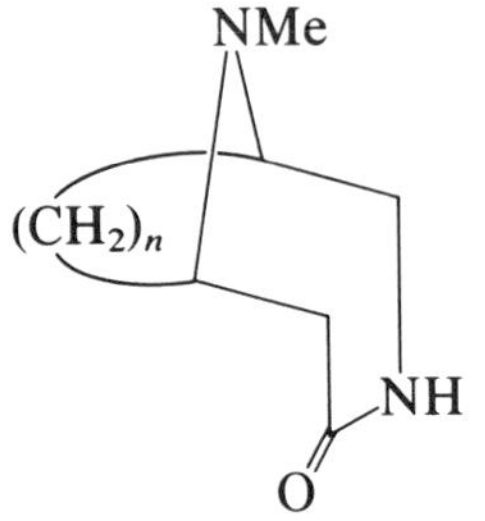

409

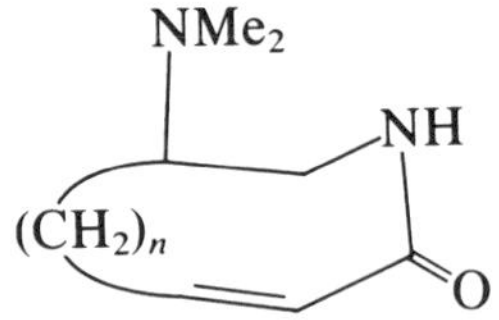

410

IV. Spiro Systems

1. 6-Azaspiro[3.6]decanes

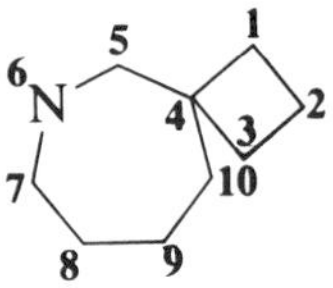

The only reference to the 6-azaspiro[3.6]decane ring system reveals that the *N*-methyl lactam **411** can be obtained in 49% yield by base treatment of the chloro compound **412** (556).

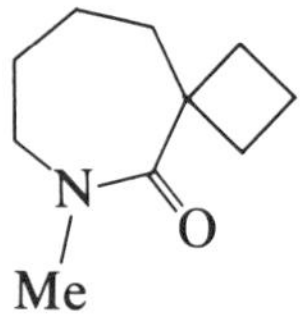

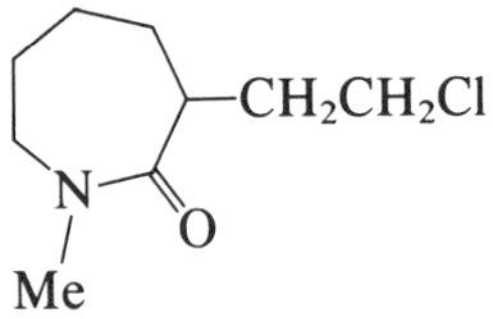

411

412

2. 6-Azazpiro[4.6]undecanes

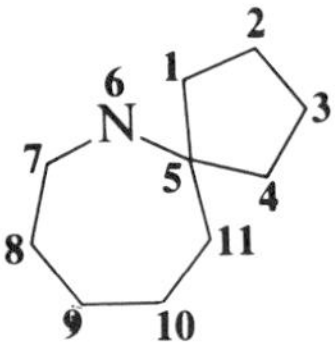

Reduction of the nitro compound **413** and heating of the resultant amino ester gave the lactam **414** (488), which was later also obtained by treatment of oxime **415** with benzenesulfonyl chloride and sodium hydroxide (489) or with phosphorus pentoxide (490). Alternatively, compound **414** could be prepared by polyphosphoric acid treatment of the unsaturated amide **416** (491) at 70–73°C. This ring system has also been made by ring expansion of the appropriate azido thioether **417** (555).

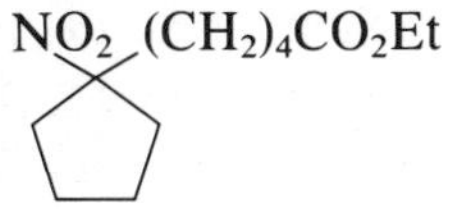

NO_2 $(CH_2)_4CO_2Et$

413

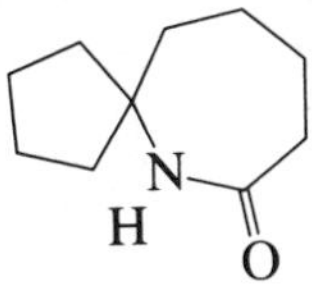

N H O

414

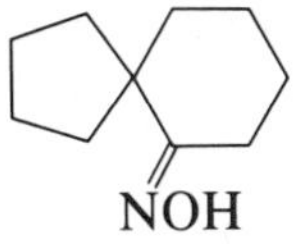

NOH

415

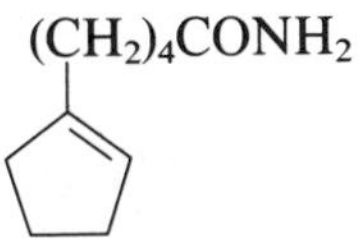

$(CH_2)_4CONH_2$

416

N_3 SMe

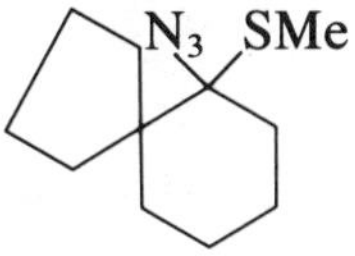

417

3. 7-Azaspiro[4.6]undecanes

7 N 8 6 1 9 5 2 10 11 4 3

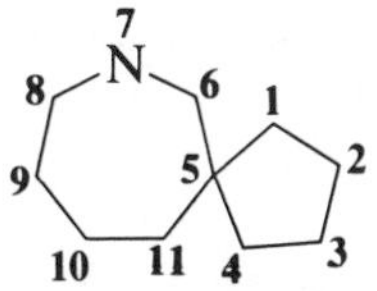

The only known example of 7-azaspiro[4.6]undecanes is the lactam **418**, which was formed in 6% yield when oxime **415** was subjected to the action of polyphosphoric acid at 120–125°C (489). More recently, a much better yield of the *N*-methyl derivative was obtained by intramolecular cyclization of compound **419** (556).

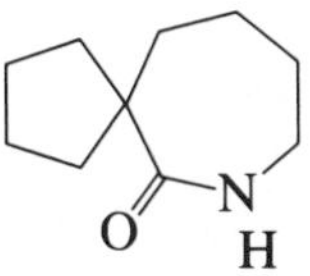

O N H

418

H $CH_2CH_2CH_2Cl$ N O Me

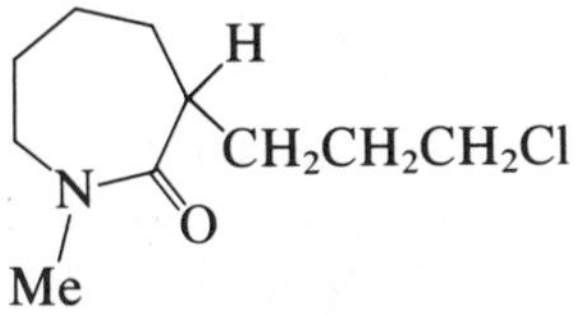

419

4. 9-Azaspiro[5.6]dodecanes

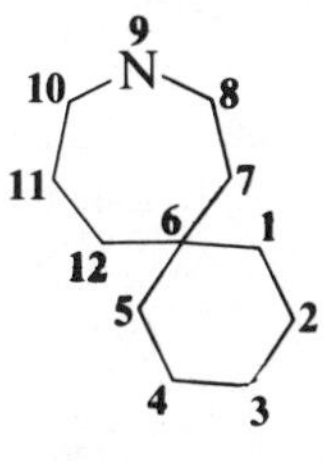

9 N 10 8 11 7 6 1 12 5 2 4 3

Ring expansion of the spiro ketone **420** via a Schmidt reaction gave lactam **421,** from which several derivatives were made by acylation and reduction (492).

420  **421**

5. 1-Oxa-7-azaspiro[4.6]undecanes

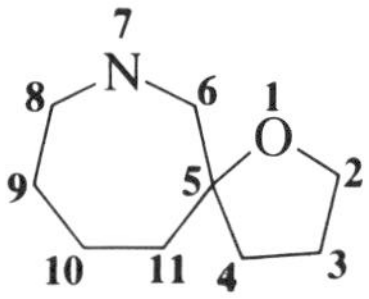

Reduction of the cyano ester **422** in methanol over Raney cobalt led to the isolation of the lactam **423.** The amino ester that presumably is formed in this reaction was not isolated (493).

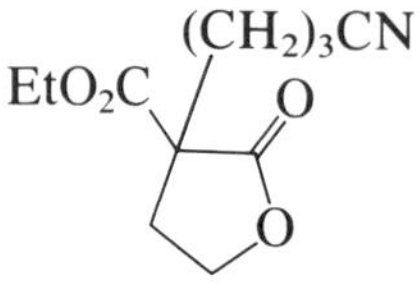

422 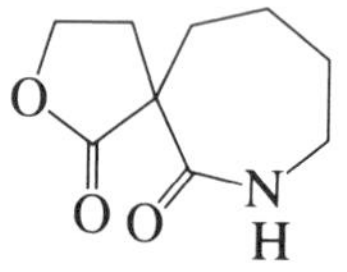 **423**

References

1. L. A. Paquette and R. J. Haluska, *J. Chem. Soc. (D)*, **1968,** 1370.
2. L. A. Paquette and R. J. Haluska, *J. Org. Chem.*, **35,** 132 (1970).
3. W. H. Okamura, W. H. Snider, and T. J. Katz, *Tetrahedron Lett.*, **1968,** 3367.
4. H. Prinzbach, D. Stusche, J. Markert, and H. H. Limbach, *Chem. Ber.*, **109,** 3505 (1976).
5. A. Zabza, C. Wawrzenczyk, and H. Kuczynski, *Bull. Acad. Pol. Sci., Ser. Sci. Chim.* **20,** 521 (1972); *Chem. Abstr.*, **78,** 111524d (1973).
6. T. Sasaki, K. Kanematsu, and Y. Yukimoto, *J. Org. Chem.*, **39,** 455 (1974).
7. K. Shudo and T. Okamoto, *Chem. Pharm. Bull.* (Tokyo) **22,** 1204 (1974).
8. T. Sasaki, K. Kanematsu, and A. Kakeni, *J. Chem. Soc (D)*, **1970,** 1030.
9. S. Masamune and N. J. Catellucci, *Angew. Chem.*, **76,** 569 (1964).
10. A. G. Anastassiou, R. L. Elliott, and A. Lichtenfeld, *Tetrahedron Lett.*, **1972,** 4569.
11. A. G. Anastassiou, S. W. Eachus, R. L. Elliott, and E. Yakali, *J. Chem. Soc (D).*, **1972,** 531.

12. A. G. Anastassiou, R. L. Elliott, H. W. Wright, and J. Clardy, *J. Org. Chem.*, **38**, 1959 (1973).
13. S. Masamune, K. Hojo, and S. Takada, *J. Chem. Soc. (D)*, **1969**, 1204.
14. A. G. Anastassiou and R. L. Elliott, *J. Chem. Soc. (D)*, **1973**, 601.
15. R. N. Carde and G. Jones, *J. Chem. Soc. Perkin I*, **1975**, 519.
16. V. G. Granik, N. S. Kuryatov, V. P. Pakhomov, O. Anisimova, and R. G. Glushkov, U.S.S.R. Patent 389,094; *Chem. Abstr.*, **79** 126337q (1973).
17. L. A. Paquette and W. C. Farley, *J. Am. Chem. Soc.*, **89**, 3595 (1967).
18. L. A. Paquette, D. E. Kuhla, and J. H. Barrett, *J. Org. Chem.*, **34**, 2879 (1969).
19. K. Schloegel and H. Mechtler, *Monatsh. Chem.*, **97**, 150 (1966).
20. K. Schloegel, M. Fried, and M. Falk, *Monatsh. Chem.*, **95**, 576 (1964).
21. M. Pailer and I. Allmer, *Monatsh. Chem.*, **86**, 819 (1955).
22. K. Nomura, J. Adachi, and Y. Hamatani, *Chem. Pharm. Bull.* (Tokyo), **23**, 701 (1975).
23. U. Mueller-Westerhoff and K. Hafner, *Tetrahedron Lett.*, **1967**, 4341.
24. K. Hafner, J. Haering, and W. Jaekel, *Angew, Chem. Int. Ed. Engl.*, **9**, 159 (1970).
25. H. J. Lindner, *Chem. Ber.*, **102**, 2464 (1969).
26. N. Neuner-Jehle, *Tetrahedron Lett.*, **1968**, 2047.
27. G. Schaden, *Chem. Ber.*, **105**, 3128 (1972).
28. J. O. Halford and B. Weissmann, *J. Org. Chem.*, **17**, 1276 (1952).
29. W. A. Nasutavicus and F. Johnson, *J. Org. Chem.*, **32**, 2367 (1967).
30. K. Hafner and K. H. Vopel, *Angew. Chem.*, **71**, 672 (1959).
31. H. Kon, *Sci. Rep. Tohoku Univ., Ser. 1*, **38**, 67 (1954); *Chem. Abstr.*, **49**, 10046f (1955).
32. M. Godfrey and J. N. Murrell, *Proc. R. Soc. London Ser. A*, **64**, 278 (1964); *Chem. Abstr.*, **60**, 7575h (1964).
33. J. P. H. Boyer, R. J. P. Corriu, R. J. M. Perz, and C. G. Reye, *Tetrahedron*, **31**, 377 (1975).
34. K. Hafner and M. Kreuder, *Angew, Chem. Int. Ed. Engl.*, **1963**, 132.
35. (a) M. K. Conner and E. Legoff, *Tetrahedron Lett.*, **1970**, 2687; (b) M. K. Conner, Ph.D. Thesis, Michigan State University, 1969.
36. M. J. Weiss, G. J. Gibbs, J. F. Poletto, and W. A. Remers, U.S. Patent 3,758,501; *Chem. Abstr.*, **79** 115550c (1973).
37. R. Royer, G. Lamotte, J. P. Bachelet, P. Demerseman, R. Cavier, and J. Lemoine, *Eur J. Med. Chem., Chim. Ther.*, **11**, 221 (1976).
38. K. Ruehlmann, A. Kokkali, H. Becker, H. Seefluth, and U. Friedenberger, *J. Prakt, Chem.*, **311** 844 (1969).
39. M. Anderson and A. W. Johnson, *J. Chem. Soc. (C)*, **1966**, 1075.
40. A. Lapworth and E. Wechsler, *J. Chem. Soc.*, **91**, 1920 (1907).
41. M. Elliott, N. F. Janes, and K. A. Jeffs, *Chem. Ind.* (London), **1967**, 1175.
42. P. Krogsgaard-Larsen, H. Hjeds, S. B. Christensen, and L. Brehm, *Acta Chem. Scand., Ser. B*, **27**, 3251 (1973).
43. P. Krogsgaard-Larsen and H. Hjeds, *Acta Chem. Scand., Ser. B.* **28**, 533 (1974).
44. G. Griss, M. Kleeman, W. Grell, and H. Ballhouse, German Offen. 2,040,510 and 2,206,385; *Chem. Abstr.*, **76**, 140925z (1972) and **79**, 146503u (1973).
45. P. Bassignana, C. Cogrossi, and M. Gandino, *Spectrochim. Acta*, **19**, 1885 (1963).
46. V. G. Granik and R. G. Glushkov, *Khim. Farm. Zh.*, **1**, 21 (1967); *Chem. Abstr.*, **68**, 12942a (1968).

47. S. Morosawa, *Bull. Chem. Soc. Japan,* **33,** 1108 (1960); *Chem. Abstr.* **55,** 27363b (1961).
48. T. Moriya, T. Oki, S. Yamaguchi, S. Morosawa, and A. Yokoo, *Bull. Chem. Soc. Japan,* **41,** 230 (1968); *Chem. Abstr.,* **69,** 2851b (1968).
49. H. Yamamoto, M. Nakata, S. Morosawa, and A. Yokoo, *Bull. Chem. Soc. Japan,* **44,** 153 (1971).
50. I. A. Strakova, A. Ya. Strakov, L. R. Yapina, D. Zicane, and E. Gudriniece; *Latv. PSR Zinat, Akad. Vestis, Kim. Ser. 100,* (1975); *Chem. Abstr.,* **83,** 28146q (1975).
51. A. Ya. Strakov, D. Zicane, D. Brutane, and M. Opmane, Nov. Issled. Obl. Khim. Khim. Tekhnol., Mater. Nauchno-Tekh. Konf. Professorkto-Prepod. Sostava Nauchn. Rob. Khim. Fak. RPI, 1972, p. 23 (1973); *Chem. Abstr.,* **82,** 4195h (1975).
52. A. Valasinas, E. S. Levy, and B. Frydman, *J. Org. Chem.,* **39,** 2872 (1974).
53. A. R. Battersby, J. F. Beck, and E. McDonald, *J. Chem. Soc. Perkin I,* **1974,** 160.
54. F. Troxler, A. P. Stoll and P. Niklaus, *Helv. Chim. Acta,* **51,** 1870 (1968).
55. A. P. Stoll and F. Troxler, *Helv. Chim. Acta,* **51,** 1864 (1968).
56. A. Yokoo and S. Morosawa, *Bull. Chem. Soc. Japan,* **36,** 599 (1963); *Chem. Abstr.,* **60,** 9386c (1964).
57. R. F. Koebel, L. L. Needham, and De W. C. Blanton, Jr., *J. Med. Chem.,* **18,** 192 (1975).
58. B. P. Fabrichnyi, I. F. Shalavina, S. E. Zurabyan, Ya. L. Gol'dfarb, and S. M. Kostrova, *Zh. Org. Khim.,* **4,** 680 (1968); *Chem. Abstr.,* **69,** 18565 (1968).
59. B. Iddon, M. W. Pickering, H. Suschitzky, and D. S. Taylor, *J. Chem. Soc. Perkin I,* **1975,** 1686.
60. C. Hoogzand, J. Nielsen, and E. H. Braye, *J. Chem. Soc. (D),* **1971,** 1520.
61. B. P. Fabrichnyi, I. F. Shalavina, and Ya. L. Gol'dfarb, *Zh. Org. Khim.,* **1,** 1507 (1965); *Chem. Abstr.,* **64,** 586d (1966).
62. (a) B. P. Fabrichnyi, I. F. Shalavina, and Ya. L. Gol'dfarb, *Zh. Obshch. Khim.,* **31,** 1244 (1961); *Chem. Abstr.,* **55,** 23488e (1961); (b) *Zh. Org. Khim.,* **5,** 361 (1969); *Chem. Abstr.,* **70,** 114910x (1969); (c)B. B. Fabrichnyi, I. F. Shalavina, S. M. Kostrova, and Ya. L. Gol'dfarb, *Zh. Org. Khim.,* **6,** 1091 (1970); *Chem. Abstr.,* **73,** 35280a (1970).
63. K. Aparajithan, A. C. Thompson, and J. Sam, *J. Heterocyclic Chem.,* **3,** 466 (1966).
64. Ya. L. Gol'dfarb, B. M. Zolotarev, V. I. Kadentsev, and O. S. Chizhov, *Izv. Akad. Nauk SSSR, Ser. Khim.,* **1970,** 1014; *Chem. Abstr.,* **73,** 76370y (1970).
65. Ya. L. Gol'dfarb, I. P. Yakovlev, and O. S. Chizhov, *Izv. Akad. Nauk SSSR, Ser. Khim.,* **1970,** 1011; *Chem. Abstr.,* **73,** 76348x (1970).
66. S. Kasparek, *Adv. Heterocyclic Chem.,* **17,** 45 (1974).
67. J. Von Braun, *Chem. Ber.,* **40,** 1834 (1907).
68. G. Schroeter, A. Gluschke, S. Goetzky, J. Huang, G. Irmisch, E. Laves, O. Schrader, and G. Stier, *Chem. Ber.,* **63,** 1308 (1930).
69. J. Von Braun and B. Bartsch, *Chem. Ber.,* **45,** 3376 (1912).
70. A. H. Rees, *J. Chem. Soc.,* **1959,** 3111.
71. J. Witte and V. Boekelheide, *J. Org. Chem.,* **37,** 2850 (1972).
72. P. Rosenmund and W. H. Haase *Chem. Ber.,* **99,** 2504 (1966).
73. H. B. MacPhillamy, R. L. Dziemian, R. A. Lucas, and M. E. Kuehne, *J. Am. Chem. Soc.,* **80,** 2172 (1958).
74. T. A. Geissman and A. K. Cho, *J. Org. Chem.,* **24,** 41 (1959).
75. R. G. Cooke and H. F. Haynes, *Aust. J. Chem.,* **11,** 225 (1958).

76. U. Hoerlein and W. Geiger, *Arch. Pharm.*, **304,** 167 (1971).
77. A. Cromarty, G. R. Proctor, and M. Shabbir, *J. Chem. Soc. Perkin I,* **1972,** 2012.
78. R. Tillais, A. Lattes, H. Bouget, J. Huet, and J. Bonnic, *C. R. Acad. Sci. Paris, Ser. C,* **267,** 1350 (1968).
79. (a) J. Bonnic, J. Huet, A. Lattes, and H. Bouget, *C. R. Acad. Sci., Ser. C.,* **272,** 672 (1971); (b) J. Bonnic, J. Huet, H. Bouget, and A. Lattes, *C. R. Acad. Sci., Ser. C,* **278,** 1461 (1974); (c) J. Bonnic, P. Uriac, H. Bouget, and J. Huet, *C. R. Acad. Sci., Ser. C,* **286,** 83 (1978).
80. G. R. Proctor, W. I. Ross, and A. Tapia, *J. Chem. Soc. Perkin I,* **1972,** 1803.
81. B. D. Astill and V. Boekelheide, *J. Am. Chem. Soc.,* **77,** 4079 (1955).
82. G. R. Proctor and R. H. Thomson, *J. Chem. Soc.,* **1957,** 2302.
83. J. T. Braunholtz and F. G. Mann, *J. Chem. Soc.,* **1957,** 4174.
84. A. Bertho, *Chem. Ber.,* **90,** 29 (1957).
85. B. Loev, R. B. Greenwald, M. M. Goodman, and C. L. Zirkle, *J. Med. Chem.,* **14,** 849 (1971).
86. D. N. Gupta, I. McCall, A. McLean, and G. R. Proctor, *J. Chem. Soc. (C),* **1970,** 2191.
87. M. J. Teuber and G. Emmerich, *Tetrahedron Lett.,* **1970,** 4069.
88. B. D. Tilak, V. N. Gogte, and T. Ravindranathan, *Ind. J. Chem.,* **7,** 24 (1969).
89. G. R. Proctor and R. H. Thomson, *J. Chem. Soc.,* **1957,** 2312.
90. I. McCall, G. R. Proctor, and L. Purdie, *J. Chem. Soc. (C),* **1970,** 1126.
91. J. T. Braunholtz and F. G. Mann, *J. Chem. Soc.,* **1958,** 3377.
92. W. H. Bell, E. D. Hannah, and G. R. Proctor, *J. Chem. Soc.,* **1964,** 4926.
93. B. G. McDonald and G. R. Proctor, *J. Chem. Soc. Perkin I,* **1975,** 1446.
94. P. E. Reyl and J. L. A. Rollet, French Patent 1,473,839; *Chem. Abstr.,* **68,** 78164e (1968).
95. J. Bernstein, E. J. Pribyl, and J. Krapcho, U.S. Patent 3,330,823; *Chem. Abstr.* **68,** 95713a (1968).
96. G. N. Walker, D. Alkalay, and R. T. Smith, *J. Org. Chem.,* **30,** 2973 (1965).
97. R. L. Augustine and W. G. Pierson, *J. Org. Chem.,* **34,** 1070 (1969).
98. A. Vogel, F. Troxler, and A. Lindenmann, *Helv. Chim. Acta,* **52,** 1929 (1969).
99. L. Bauer and R. E. Hewitson, *J. Org. Chem.,* **27,** 3982 (1962).
100. P. T. Lansbury and W. R. Mancuso, *J. Am. Chem. Soc.,* **88,** 1205 (1966).
101. R. Futaki, *Tetrahedron Lett.,* **1967,** 2455.
102. H. A. Lloyd and E. C. Horning, *J. Am. Chem. Soc.,* **76,** 3651 (1954).
103. H. A. Lloyd, L. U. Matternas, and E. C. Horning, *J. Am. Chem. Soc.,* **77,** 5932 (1955).
104. S. Nizamuddin and D. N. Chaudhury, *J. Ind. Chem. Soc.,* **40,** 960 (1963).
105. R. Huisgen, I. Ugi, H. Brade, and E. Rauenbusch, *Justus Liebigs Ann. Chem.,* **586,** 30 (1954).
106. S. I. Sallay, *J. Am. Chem. Soc.,* **89,** 6762 (1967).
107. R. Huisgen, J. Witte, and I. Ugi, *Chem. Ber.,* **90,** 1844 (1957).
108. D. Evans and I. M. Lockhart, *J. Chem. Soc.,* **1965,** 4806.
109. L. M. Briggs and G. C. De Ath, *J. Chem. Soc.,* **1937,** 456.
110. L. H. Werner, S. Ricca, A. Rossi, and G. De Stevens, *J. Med. Chem.,* **10,** 575 (1967).
111. J. M. Khanna and N. Anand, *J. Med. Chem.,* **10,** 944 (1967).
112. M. Tomita, S. Minami, and S. Uyeo, *J. Chem. Soc (C),* **1969,** 183; see also earlier papers.
113. R. T. Conley, *J. Org. Chem.,* **23,** 1330 (1958).

114. P. A. S. Smith and W. L. Berry, *J. Org. Chem.*, **26,** 27 (1961).
115. D. Misiti, H. W. Moore, and K. Folkers, *Tetrahedron,* **22,** 1201 (1966); *Tetrahedron Lett.*, **1965,** 1071.
116. R. W. Rickards and R. M. Smith, *Tetrahedron Lett.*, **1966,** 2361.
117. G. R. Bedford, G. Jones, and B. R. Webster, *Tetrahedron Lett.*, **1966,** 2367.
118. G. Jones, *J. Chem. Soc. (C),* **1967,** 1808.
119. G. R. Birchall and A. H. Rees, *Can. J. Chem.*, **52,** 610 (1974).
120. H. Uno, Y. Nakano, and Y. Takase, Japanese Patent 74 28,753; *Chem. Abstr.*, **82,** 139978m (1975).
121. H. Uno and Y Nakano, Japanese Patent 74 28,754; *Chem. Abstr.*, **82,** 139981g (1975).
122. H. W. Moore, H. R. Shelden, and W. Weyler, *Tetrahedron Lett.*, **1969,** 1243.
123. R. G. Cooke and I. M. Russel, *Aust. J. Chem.*, **25,** 2421 (1972).
124. E. J. Moriconi and I. A. Maniscalco, *J. Org. Chem.*, **37,** 208 (1972).
125. P. Scheiner, *J. Org. Chem.*, **32,** 2628 (1967).
126. A. Sauleau, J. Sauleau, H. Bourget, and J. Huet, *C. R. Acad. Sci., Paris Ser. C,* **279,** 473 (1974).
127. M.-S. Lin and V. Snieckus, *J. Org. Chem.*, **36,** 645 (1971).
128. R. M. Acheson, J. N. Bridson, and T. S. Cameron, *J. Chem. Soc Perkin I,* **1972,** 968.
129. H. Plieninger and H. Wild, *Chem. Ber.*, **99,** 3070 (1966).
130. R. M. Acheson, J. N. Bridson, T. R. Cecil, and A. R. Hands, *J. Chem. Soc. Perkin I,* **1972,** 569.
131. F. Fried, J. B. Taylor, and R. Westwood, *J. Chem. Soc. (D),* **1971,** 1226.
132. T. Sakan, S. Matsubara, H. Takagi, Y. Tokunaga, and T. Miwa, *Tetrahedron Lett.*, **1968,** 4925.
133. R. Fuks and H. G. Viehe, *Chem. Ber.*, **103,** 573 (1970).
134. A. Cromarty, K. E. Haque, and G. R. Proctor, *J. Chem. Soc. (C),* **1971,** 3536.
135. D. Thon and W. Schneider, *Chem. Ber.*, **109,** 2743 (1976).
136. B. Eistert and P. Donath, *Chem. Ber.*, **103,** 993 (1970).
137. R. W. Schmid, *Helv. Chim. Acta,* **45,** 1982 (1962).
138. M. J. S. Dewar, and N. Trinajstic, *Tetrahedron,* **26,** 4269 (1970).
139. B. A. Hess, Jr., L. J. Schaad, and C. W. Holyoke, Jr., *Tetrahedron,* **28,** 3657 (1972).
140. A. Mannschreck, G. Rissmann, F. Vogtle, and D. Wild, *Chem. Ber.*, **100,** 335 (1967).
141. R. M. Acheson and J. N. Bridson, *J. Chem. Soc. (D),* **1971,** 1225.
142. W. A. Denne and M. F. MacKay, *Tetrahedron,* **28,** 1795 (1972).
143. R. Reynaud and P. Rumpf, *Bull. Soc. Chim. France,* **1963,** 1805.
144. G. Baddeley, J. Chadwick, and H. T. Taylor, *J. Chem. Soc.*, **1956,** 448.
145. W. R. Remington, *J. Am. Chem. Soc.*, **67,** 1838 (1945).
146. P. C. Carpenter and M. Lennon, *J. Chem. Soc. (D),* **1973,** 664.
147. W. G. Brown and S. Fried, *J. Am. Chem. Soc.*, **65,** 1841 (1943).
148. S. Aftalion and G. R. Proctor, *Org. Mass Spectrom.*, **2,** 337 (1969).
149. S. D. Sample, D. A. Lightner, O. Buchardt, and C. Djerassi, *J. Org. Chem.*, **32,** 997 (1967).
150. C. M. C. Koo, T. W. Pattison, and D. R. Herbst, U.S. Patent 3,458,498; *Chem. Abstr.*, **71,** 81225t (1969).
151. J. Huet, M. Sado-Odeye, M. Martin, P. Guibet, Ph. Linee P. Lacroix, P. Quiniou, and J. Laurent, *Eur. J. Med. Chem., Chim. Ther.*, **9,** 376 (1974).

152. L. H. Werner, W. L. Bencze, and G. De Stevens, Belgian Patent 641,254; *Chem. Abstr.*, **63,** 9956g (1965).
153. W. L. Bencze, U.S. Patent 3,509,130; *Chem. Abstr.*, **73,** 98832c (1970).
154. K. Joshi, V. A. Rao, and N. Anand, *Ind. J. Chem.*, **11,** 1222 (1973).
155. O. Aki and Y. Nakagawa, *Chem. Pharm. Bull.* (Tokyo), **20,** 1325 (1972).
156. S. Kobayashi, *Bull. Chem. Soc. Japan,* **46,** 2835 (1973).
157. J. Krapcho, U.S. Patent 3,748,321; *Chem. Abstr.*, **79,** 92303h (1973)
158. G. R. Proctor, *J. Chem. Soc.*, **1961,** 3989.
159. G. R. Proctor and B. M. L. Smith, *J. Chem. Soc. Perkin I,* **1978,** 862.
160. R. L. Augustine and W. G. Pierson, *J. Org. Chem.*, **34,** 2235 (1969).
161. M. Lennon, A. McLean, I. McWatt, and G. R. Proctor, *J. Chem. Soc. Perkin I,* **1974,** 1828.
162. F. D. Sancilio and J. F. Blount, *Acta Crystallogr., Sect. B,* **B72,** 2123 (1976).
163. D. Misiti, V. Rimatori, and F. Gatta, *J. Heterocyclic Chem.*, **10,** 689 (1973).
164. H. Kanamoto, T. Matsuo, S. Morosawa, and A. Yokoo, *Bull. Chem. Soc. Japan,* **46,** 3898 (1973).
165. W. I. Ross and G. R. Proctor, *J. Chem. Soc. Perkin I,* **1972,** 889.
166. K. G. Svensson, H. Selander, M. Karlsson, and J. L. G. Nilsson, *Tetrahedron,* **29,** 1115 (1973).
167. (a) H. Behringer and H. Meier, *Justus Liebigs Ann. Chem.*, **607,** 67 (1957); (b) H. Behringer and G. F. Grunwald, German Patent 943,227; *Chem. Abstr.*, **53,** 6262b (1959).
168. R. Huisgen and I. Ugi, *Justus Liebigs Ann. Chem.*, **610,** 57 (1957).
169. O. Meth-Cohn and H. Suschitzky, *J. Chem. Soc.*, **1964,** 2609.
170. British Patent 910,428; *Chem. Abstr.*, **58,** 7914b (1963).
171. U.C.B. Societe Anon., Netherlands Appl. 6,514,240; *Chem Abstr.*, **65,** 12182d (1966).
172. S. Uyeo and T. Shingu; *Yakugaku Zasshi,* **84,** 555 (1964); *Chem. Abstr.*, **61,** 8353c (1964).
173. C. M. C. Koo, T. W. Pattison, and D. R. Herbst, U.S. Patent 3,542,760; *Chem. Abstr.*, **74,** 22717u (1971).
174. Netherlands Appl. 6,516,320; *Chem. Abstr.*, **65,** 15354g (1966).
175. R. T. Conley and L. J. Frainier, *J. Org. Chem.*, **27,** 3844 (1962).
176. W. Eisele, C. A. Grob, and E. Renk, *Tetrahedron Lett.*, **1963,** 75.
177. J. F. Klebe, H. Finkbeiner, and D. M. White, *J. Am. Chem. Soc.*, **88,** 3390 (1966).
178. A. H. Rees and K. Simon, *Can. J. Chem.*, **47,** 1227 (1969).
179. L. H. Werner, U.S. Patent 3,312,691; *Chem. Abstr.*, **68,** 49473z (1968).
180. K. G. Artz and C. A. Grob, *Helv. Chim. Acta,* **51,** 807 (1968).
181. W. Eisele, C. A. Grob, E. Renk, and H. Von Tschammer, *Helv. Chim. Acta,* **51,** 816 (1968).
182. C. V. Greco and R. P. Gray, *Tetrahedron,* **26,** 4329 (1970).
183. A. I. Kiprianov and V. P. Khilya, *Zh. Org. Khim.*, **3,** 1091 (1967); *Chem. Abstr.*, **69,** 37080b (1968).
184. G. M. Strunz, *Tetrahedron,* **24,** 2645 (1968).
185. D. Gordon, L. Frye, and H. Sheffer, *Acta Chem. Scand.*, **23,** 3577 (1969).
186. C. G. Hughes and A. H. Rees, *Chem. Ind.* (London), **1971,** 1439.
187. H. J. Havera, J. W. Van Dyke, Jr., T. M. H. Liu, and L. F. Sancilio, *J. Med. Chem.*, **12,** 580 (1969).

188. R. Sarges, J. R. Tretter, S. S. Tenen, and A. Weissman, *J. Med. Chem.*, **16,** 1003 (1973).
189. Y. Sato, H. Kojima, and H. Shirai, *J. Org. Chem.*, **41,** 195, 3325 (1976).
190. British Patent 359,285; *Chem. Abstr.*, **81,** 152031w (1974).
191. E. D. Hannah, W. C. Peaston, and G. R. Proctor, *J. Chem. Soc. (C)*, **1968,** 1280.
192. T. Kato, K. Tabei, and E. Kawashima, *Chem. Pharm. Bull.* (Tokyo), **24,** 1544 (1976).
193. V. N. Gogte, K. M. More, and B. D. Tilak, *Ind. J. Chem.*, **12,** 1238 (1974).
194. J. Von Braun and F. Zobel, *Chem. Ber.*, **56B,** 690 (1923).
195. M. H. Sherlock, U.S. Patent, 3,225,031; *Chem. Abstr.*, **64,** 9696g, 19577d (1966).
196. F. Ceasar and A. Mondon, *Chem. Ber.*, **101,** 990 (1968).
197. J. Von Braun and H. Reich, *Justus Liebigs Ann. Chem.*, **445,** 225 (1925).
198. J. L. Charlish and W. H. Davies, *J. Chem. Soc.*, **1950,** 1385.
199. H. H. Inhoffen and E. Prinz, *Chem. Ber.*, **87,** 684 (1954).
200. J. Von Braun and W. Kaiser, *Chem. Ber.*, **58,** 2162 (1925).
201. A. Rieche and E. Hoeft, *J. Prakt. Chem.*, **17,** 293 (1962).
202. V. Gomez Parra and R. Madronero, *An. Quim.*, **70,** 614 (1974); *Chem. Abstr.*, **81,** 151970q (1974).
203. K. Ackerman, D. E. Horning, and J. M. Muchowski, *Can. J. Chem.*, **50,** 3886 (1972).
204. G. Simchen, *Angew. Chem. Int. Ed. Engl.*, **7,** 464 (1968).
205. G. N. Walker and D. Alkalay, *J. Org. Chem.*, **36,** 461 (1971).
206. G. N. Walker *J. Org. Chem.*, **37,** 3955 (1972).
207. B. Pecherer, F. Humiec, and A. Brossi, *Helv. Chim. Acta*, **54,** 743 (1971).
208. (a) B. R. Vogt, U.S. Patent 3,887,544; *Chem. Abstr.*, **83,** 131649e (1975); (b) U.S. Patent 3,985,731; *Chem. Abstr.*, **86,** 55304h (1977).
209. Y. Sawa, T. Kato, T. Matsuda, M. Hori, and H. Fujimura, *Chem. Pharm. Bull.* (Tokyo), **23,** 1917 (1975).
210. I. MacDonald and G. R. Proctor, *J. Chem. Soc. (C)*, **1970,** 1461.
211. J. R. Brooks and D. N. Harcourt, *J. Chem. Soc. (C)*, **1969,** 625.
212. L. W. Deady, N. H. Pirzada, and R. D. Topsom, *J. Chem. Soc. Perkin I*, **1973,** 782.
213. F. J. McCarty, L. J. Lendvay, A. J. Vazakas, W. W. Bennetts, F. P. Palopoli, R. Orzechowski, and S. Goldstein, *J. Med. Chem.*, **13,** 814 (1970).
214. J. A. Meschino, U.S. Patents 3,483,186, 3,894,072, and 3,828,096; *Chem. Abstr.*, **72,** 121383x (1970); **81,** 136001f (1974); and **83,** 114031e (1975).
215. D. Berney and T. Jauner, *Helv. Chim. Acta*, **59,** 623 (1976).
216. F. Dallacker, D. Bernabei, B. Katzke, and P. H. Benders, *Chem. Ber.*, **104,** 2526 (1971).
217. I. M. Goldman, J. K. Larson, J. R. Tretter, and E. G. Andrews, *J. Am. Chem. Soc.*, **91,** 4941 (1969).
218. M. Tomita and S. Minami, *Yakugaku Zasshi*, **83,** 1022 (1963); *Chem. Abstr.*, **60,** 7998c (1964).
219. H. Fujimura and M. Hori, U.S. Patent 3,409,607; *Chem. Abstr.*, **70,** 77827c (1969).
220. Y. Kanaoka, E. Sato, O. Yonemitsu, and Y. Ban, *Tetrahedron Lett.*, **1964,** 2419.
221. M. W. Gittos, J. W. James, and J. P. Verge, German Offen. 1,911,519; *Chem. Abstr.*, **72,** 12601w (1970).
222. H. Bohme and K. P. Stöcker, *Arch. Pharm.*, **306,** 271 (1973).
223. R. R. Wittekind and S. Lazarus, *J. Heterocyclic Chem.*, **8** 495 (1971).
224. T. Kametani, K. Kigasawa, M. Hiiragi, H. Ishimaru, and S. Haga, *J. Chem. Soc. Perkin I*, **1974,** 2602.

225. M. Hori, H. Fujimura, T. Matsuda, and Y. Sawa, *J. Pharm. Soc. Japan,* **95,** 131 (1975).
226. H. J. Schmidt, A. Hunger, and K. Hoffmann, *Helv. Chim. Acta,* **39,** 607 (1956).
227. N. S. Hjelte and T. Agback, *Acta Chem. Scand.,* **18,** 191 (1964).
228. R. P. Mull, P. Schmidt, M. R. Dapero, J. Higgins, and M. J. Weisbach, *J. Am. Chem. Soc.,* **80,** 3769 (1958).
229. H. A. Bruson, F. W. Grant, and E. Bobko, *J. Am. Chem. Soc.,* **80,** 3633 (1958).
230. R. T. Conley and R. J. Lange, *J. Org. Chem.,* **28,** 210 (1963).
231. J. A. Marshall, N. H. Anderson, and J. W. Schlicher, *J. Org. Chem.,* **35,** 858 (1970).
232. J. Von Braun and K. Wirz, *Chem. Ber.,* **60,** 102 (1927).
233. J. Von Braun and O. Bayer, *Chem. Ber.,* **60,** 1257 (1927).
234. N. J. Leonard, K. Jann, J. V. Paukstelis, and C. K. Steinhardt, *J. Org. Chem.,* **28,** 1499 (1963).
235. A. Hassner and D. J. Anderson, *J. Org. Chem.,* **39,** 2031 (1974).
236. A. Hassner and D. J. Anderson, *J. Org. Chem.,* **39,** 3070 (1974).
237. V. S. Kuznetsov, E. A. Korkhova, Yu. A. Ignat'ev, and A. B. El'tsov, *Zh. Org. Khim.,* **11,** 808 (1975); *Chem. Abstr.,* **83,** 8724y (1975).
238. Y. Kanaoka, Y. Migita, K. Koyama, Y. Sato, H. Nakai, and T. Mizoguchi, *Tetrahedron Lett.,* **1973,** 1193.
239. A. Padwa and J. Smolanoff, *Tetrahedron Lett.,* **1974,** 33.
240. A. Padwa, J. Smolanoff, and A. Tremper, *J. Am. Chem. Soc.,* **97,** 4682 (1975).
241. K. A. Howard and T. H. Koch, *J. Am. Chem. Soc.,* **97,** 7288 (1975).
242. A. Lablachecombier and G. Surpateanu, *Tetrahedron Lett.,* **1976,** 3081.
243. E. Desherces, M. Riviere, J. Parello, and A. Lattes, *C. R. Acad. Sci. Paris, Ser C,* **275,** 581 (1972).
244. T. Matsuda, H. Fujimura, M. Hori, Y. Sawa, I. Mikami, and T. Kato, Japanese Patent 70 21,715; *Chem. Abstr.,* **73,** 87806w (1970).
245. P. E. Reyl and J. L. A. Rollet, French Patent 1,472,930; *Chem. Abstr.,* **68,** 59453g (1968).
246. B. Loev, U.S. Patent, 3,686,165; *Chem. Abstr.,* **77,** 152005h (1972).
247. A. McLean and G. R. Proctor, *J. Chem. Soc. Perkin I,* **1973,** 1084.
248. I. L. Knunyants and B. P. Fabrichnyi, *Dokl. Akad. Nauk SSSR,* **68,** 523 (1949); *Chem. Abstr.* **44,** 1469f (1950).
249. B. Belleau, *J. Med. Pharm. Chem.,* **1,** 343 (1959).
250. British Patent 956,613; *Chem. Abstr.,* **61,** 8284h (1964).
251. W. L. Bencze and L. I. Barsky, *J. Med. Chem.,* **14,** 40 (1971).
252. A. Stankevicius, A. Kost, and V. Vizas, *Khim. Farm. Zh.,* **3,** 21 (1969); *Chem. Abstr.,* **72,** 66776a (1970).
253. V. P Khilya and I. Ya. Doroshko, *Khim. Str. Svoistva, Reakt. Org. Soedin.,* **1969,** 48; *Chem. Abstr.,* **72,** 134122m (1970).
254. S. Kano, T. Ogawa, T. Yokomatsu, Y. Takamagi, E. Komiyama, and S. Shibuya, *Heterocycles,* **3,** 129 (1975).
255. D. Berney and K. Schuh, *Helv. Chim. Acta,* **59,** 2059 (1976).
256. K. Mitsuhashi, K. Nomura, N. Minami, and M. Matsuyama, *Chem. Pharm. Bull.* (*Tokyo*), **17,** 1578 (1969).
257. K. Mitsuhashi, K. Nomura, and F. Miyoshi, *Chem. Pharm. Bull.* (*Tokyo*), **19,** 1983 (1971).
258. J. Von Braun and H. Reich, *Chem. Ber.,* **58,** 2765 (1925).
259. B. Pecherer, R. C. Sunbury, and A. Brossi, *J. Heterocyclic Chem.,* **9,** 609 (1972).
260. J. Chazerain, *Ann. Chim.* (*Paris*), **8,** 255 (1963); *Chem. Abstr.,* **59,** 8703a (1963).

261. M. Hamon, *C. R. Acad. Sci. Paris,* **225,** 1519 (1962); *Chem. Abstr.,* **58,** 2436h (1963).
262. M. Hamon, *Ann. Chim.* (Paris), **10,** 213 (1965); *Chem. Abstr.,* **63,** 18027g (1965).
263. J. P. Yardley, H. Smith, and R. W. Rees, German Offen. 1,921,861; *Chem. Abstr.,* **72,** 31646f (1970).
264. J. O. Halford and B. Weissman, *J. Org. Chem.,* **17,** 1646 (1952).
265. P. E. Reyl and J. L. A. Rollet, French Patent 1,437,840; *Chem. Abstr.,* **68,** 78160a (1968).
266. P. Ruggli, B. B. Bussemaker, W. Müller, and A. Staub, *Helv. Chim. Acta,* **18,** 1388 (1935).
267. J. H. Wood, M. A. Perry, and C. C. Tung, *J. Am. Chem. Soc.,* **73,** 4689 (1951).
268. K. Hoegerle and E. Habicht, Swiss Patent 498,123; *Chem. Abstr.,* **74,** 125489a (1971).
269. F. Johnson and W. A. Nasutavicus, *J. Heterocyclic Chem.,* **2,** 26 (1965).
270. J. Gardent and G. Hazebroucq, *Bull. Soc. Chim. France,* **1968,** 600.
271. M. A. Rehman and G. R. Proctor, *J. Chem. Soc. (C),* **1967,** 58.
272. J. Schlademan and R. Partch, *J. Chem. Soc. (C),* **1972,** 213.
273. I. MacDonald and G. R. Proctor, *J. Chem. Soc. (C),* **1969,** 1321.
274. G. Hazebroucq and J. Gardent, *C. R. Acad. Sci. Paris, Ser. C,* **257,** 923 (1963).
275. G. Hazebroucq, *Ann. Chim.* (Paris), **1,** 221 (1966).
276. Y. Inubushi T. Harayama, and K. Takeshima, *Chem. Pharm. Bull.* (Tokyo), **20,** 689 (1972).
277. M. Lennon, A. McLean, G. R. Proctor, and I. W. Sinclair, *J. Chem. Soc. Perkin I,* **1975,** 622.
278. B. V. Shetty, German Offen. 2,207,430; *Chem. Abstr.,* **79,** 126338r (1973).
279. M. Hamon, *C. R. Acad. Sci. Paris, Ser. C.,* **255,** 1619 (1962).
280. M. Pelz, M. Rajsner, J. O. Jilek, and M. Protiva, *Coll. Czech. Chem. Commun.,* **33,** 2111 (1968).
281. M. D. Nair and P. A. Malik, *Ind. J. Chem.,* **5,** 169 (1967).
282. O. Yonemitsu, T. Tokuyama, M. Chaykovsky, and B. Witkop, *J. Am. Chem. Soc.,* **90,** 776 (1968).
283. O. Yonemitsu, S. Naruto, N. Kanamaru, and K. Kimura, *J. Am. Chem. Soc.,* **93,** 4053 (1971).
284. T. Iwakuma, H. Nakai, O. Yonemitsu, and B. Witkop, *J. Am. Chem. Soc.,* **96,** 2564 (1974).
285. O. Yonemitsu, Y. Okuno, Y. Kanaoka, and B. Witkop, *J. Am. Chem. Soc.,* **92,** 5686 (1970).
286. Y. Okuno, and O. Yonemitsu, *Chem. Pharm. Bull.* (Tokyo), **23,** 1039 (1975).
287. Y. Okuno, M. Kawamori, and O. Yonemitsu, *Tetrahedron Lett.,* **1973,** 3009.
288. K. Hoegerle and E. Habicht, Swiss Patent 500,194; *Chem. Abstr.,* **74,** 141586k (1971).
289. L. A. Walter and W. K. Chang, Swiss Patent 555,831; *Chem. Abstr.,* **82,** 72813h (1975).
290. Y. Okuno and O. Yonemitsu, *Heterocycles,* **4,** 1371 (1976).
291. D. W. Brown, S. F. Dyke, G. Hardy, and M. Sainsbury, *Tetrahedron,* **25,** 1881 (1969).
292. J. Likforman and J. Gardent, *C. R. Acad. Sci. Paris, Ser. C,* **268,** 2340 (1969).
293. J. M. Bobbitt, *Adv. Heterocyclic Chem.,* **15,** 99 (1973).
294. J. P. Fourneau, C. Gaignault, R. Jacquier, O. Stoven, and M. Davy, *Chim. Ther.,* **4,** 67 (1969); *Chem. Abstr.,* **71,** 81108g (1969).
295. J. P. Fourneau and J. Delourme, German Offen. 1,944,121; *Chem. Abstr.,* **72,** 111311h (1970).
296. K. Dimroth and H. Freyschlag, *Angew. Chem.,* **68,** 518 (1956).

297. K. Dimroth and H. Freyschlag, *Chem. Ber.*, **89,** 2602 (1956).
298. K. Dimroth and H. Freyschlag, *Chem. Ber.*, **90,** 1628 (1957).
299. K. Dimroth, D. Holzner, and H. G. Aurich, *Chem. Ber.*, **98,** 3907 (1965).
300. R. Huisgen, E. Laschtuvka, I. Ugi, and A. Kammermeier, *Justus Liebigs Ann. Chem.*, **630,** 128 (1960).
301. W. E. Hahn, J. Epsztajn, and Z. Madeja-Kotkowska, *Rocz. Chem.*, **39,** 1423 (1965); *Chem. Abstr.*, **64,** 17540 (1966).
302. F. Dallacker, K. W. Glombitza, and M. Lipp, *Justus Liebigs Ann. Chem.*, **643,** 82 (1961).
303. T. Oine and T. Mukai, *Tetrahedron Lett.*, **1969,** 157.
304. P. A. Petyunin, A. K. Sukhomlinov, and N. G. Panferova, *Khim. Geterotsikl. Soedin.*, **1968,** 1033; *Chem. Abstr.*, **70,** 68045u (1969).
305. A. K. Durgaryan, S. G. Chshmarityan, and G. T. Tatevoysan, *Ann. Khim. Zh.*, **27,** 510 (1974); *Chem. Abstr.*, **81,** 120432m (1974).
306. G. Mahuzier and M. Hamon, *Bull Soc. Chim. France,* **1969,** 687.
307. C. Reby and J. Gardent, *Bull. Soc. Chim. France,* **1972,** 1574.
308. R. Grewe and G. Winter, *Chem. Ber.*, **92,** 1092 (1959).
309. T. Kametani, S. Hirata, S. Shibuya, and K. Fukumoto, *J. Chem. Soc. (C),* **1971,** 1927.
310. H. Irie, S. Tani, and H. Yamane, *J. Chem. Soc. Perkin I,* **1972,** 2986.
311. M. Natsume and M. Wada, *Chem. Pharm. Bull.* (Tokyo), **20,** 1836 (1972).
312. T. Kametani, H. Nemoto, K. Suzuki, and K. Fukumoto, *J. Org. Chem.*, **41,** 2988 (1976).
313. W. Klotzer, S. Teitel, J. F. Blount, and A. Brossi, *J. Am. Chem. Soc.*, **93,** 4321 (1971).
314. H. O. Bernhard and V. Sniekus, *Tetrahedron,* **27,** 2091 (1971).
315. B. Goeber and G. Engelhardt, *Pharmazie,* **24,** 423 (1969).
316. T. Kametani, M. S. Premila, S. Hirata, H. Seto, H. Nemoto, and K. Fukumoto, *Can. J. Chem.*, **53,** 3824 (1975).
317. I. Takeuchi, I. Ozawa, Y. Hamada, H. Masuda, and M. Hirota, *Chem. Lett.*, **1976,** 519.
318. A. W. Johnson and M. Mahendran, *J. Chem. Soc. (C),* **1971,** 1237.
319. T. Iwakuma, H. Nakai, O. Yonemitsu, D. S. Jones. I. L. Karle, and B. Witkop, *J. Am. Chem. Soc.,* **94,** 5136 (1972).
320. Yu. B. Shetsov, I. A. Red'kin, and M. M. Shemyakin, *Zh. Obshch. Khim.*, **21,** 339 (1951); *Chem. Abstr.*, **45,** 7556f (1951).
321. G. Kaupp, J. Perreten, R. Leute, and H. Prinzbach, *Chem. Ber.*, **103,** 2288 (1970).
322. J. S. Swenton, J. Oberdier, and P. D. Rosso, *J. Org. Chem.*, **39,** 1038 (1974).
323. L. A. Paquette, D. E. Kuhla, J. H. Barrett, and R. J. Haluska, *J. Org. Chem.*, **34,** 2866 (1969).
324. B. Pecherer, R. C. Sunbury, and A. Brossi, *J. Heterocyclic Chem.*, **8,** 779 (1971).
325. H. Wittmann, E. Ehrlich, H. Siegel, and H. Sterk, *Z. Naturforsch., Sect. B,* **31,** 1716 (1976).
326. G. N. Walker and K. Schenker, U.S. Patent 3,291,806; *Chem. Abstr.*, **66,** 75922y (1967).
327. F. Sparatore, *Ann. Chim.* (Rome), **49,** 2162 (1959); *Chem. Abstr.*, **54,** 16446e (1960).
328. T. Ibuka, T. Konoshima, and Y. Inubushi, *Chem. Pharm. Bull.* (Tokyo), **23,** 114 (1975).
329. A. Graftieaux, G. Hazebroucq, and J. Gardent, *Bull. Soc. Chim. France,* **1976,** 455.
330. N. Ben Hassine-Coniac, G. Hazebroucq, and J. Gardent, *Bull. Soc. Chim. France,* **1971,** 3985.
331. A. Graftieaux and J. Gardent, *Tetrahedron Lett.*, **1972,** 3321.
332. J. Auerbach and S. M. Weinreb, *J. Am. Chem. Soc.*, **94,** 7172 (1972).
333. G. Hazebroucq, Ph. D. Thesis, University of Paris, 1966.

334. J. Gardent, G. Hazebroucq, and G. Cormier, *Bull. Soc. Chim. France,* **1969,** 4001.
335. J. Henin, G. Hazebroucq, and J. Gardent, *Bull. Soc. Chim. France,* **1976,** 771.
336. British Patent, 1,268,243; *Chem. Abstr.,* **76,** 153628e (1972).
337. K. Hoegerle and E. Habicht, S. African Patent 68 01,019; *Chem. Abstr.,* **71,** 61251v (1969).
338. R. P. Mull and G. De Stevens, U.S. Patent 3,496,166; *Chem. Abstr.,* **73,** 14724h (1970).
339. J. Tokolics, G. A. Hughes, and H. Smith, French Patent 1,535,085; *Chem. Abstr.,* **71,** 81224s (1969).
340. L. A. Walter, U.S. Patent, 2,520,264; *Chem. Abstr.,* **45,** 675c (1951).
341. L. A. Walter, U.S. Patent, 2,684,962; *Chem. Abstr.,* **49,** 11030g (1955).
342. L. A. Walter, U.S. Patent 3,743,731; *Chem. Abstr.,* **79,** 78638q (1973).
343. R. P. Mull and G. De Stevens, U.S. Patent 3,609,138; *Chem. Abstr.,* **75,** 140722k (1971).
344. H. Dietrich, S. African Patent 67 05,527; *Chem. Abstr.,* **70,** 96654k (1969).
345. J. Kobor and K. Koczka, *Chem. Abstr.,* **77,** 151861 (1972).
346. R. P. Mull, U.S. Patent 3,252,972; *Chem. Abstr.,* **65,** 15354d (1966).
347. R. P. Mull, U.S. Patent 3,093,632; *Chem. Abstr.,* **59,** 12771g (1963).
348. L. A. Walter and W. K. Chang, British Patent 1,118,688; *Chem. Abstr.,* **69,** 106576q (1968).
349. L. A. Walter and W. K. Chang, U.S. Patent 3,393,192; *Chem. Abstr.,* **69,** 96507u (1968).
350. H. Dietrich, S. African Patent 67 05,100; *Chem. Abstr.,* **70,** 106409a (1969).
351. Y. Okuno, K. Hemmi, and O. Yonemitsu, *Chem. Pharm. Bull.* (Tokyo), **20,** 1164 (1972).
352. O. Yonemitsu, H. Nakai, Y. Kanaoka, I. L. Karle, and B. Witkop, *J. Am. Chem. Soc.,* **92,** 5691 (1970).
353. P. A. Petyunin and P. A. Bezuglyi, *Khim. Geterotsikl. Soedin.,* **1970,** 954.
354. P. A. Petyunin, V. V. Bolotov, and A. F. Soldatova, *Zh. Org. Khim.,* **7,** 1069 (1970).
355. T. Kametani, S. Hirata, F. Satoh, and K. Fukumoto, *J. Chem. Soc. Perkin I,* **1974,** 2509.
356. Y. Kuwada, H. Tawada, and K. Meguro, German Offen. 2,442,987; *Chem. Abstr.,* **83,** 58833d (1975).
357. B. Pecherer, R. C. Sunbury, and A. Brossi, *J. Heterocyclic Chem.,* **9,** 617 (1972).
358. R. Gompper and W. Elser, *Justus Liebigs Ann. Chem.,* **725,** 73, (1969).
359. S. Carboni, A. De Settimo, D. Bertini, P. L. Ferrarini, O. Livi, and I. Tonetti, *J. Heterocyclic Chem.,* **12,** 743 (1975).
360. E. M. Hawes and H. L. Davis, *J. Heterocyclic Chem.,* **10,** 39 (1973).
361. Y. Tamura, Y. Ki'a, and J. Uraoka, *Chem. Pharm. Bull.* (Tokyo), **20,** 876 (1972).
362. M. R. Uskokovic, J. Gutzwiller, and T. Henderson, *J. Am. Chem. Soc.,* **92,** 203 (1970).
363. F. M. Stoyanovich, V. G. Glimenko, and Ya. L. Gol'dfarb, *Izv. Akad. Nauk SSSR, Ser. Khim.,* **1970,** 2585; *Chem. Abstr.,* **74,** 111870f (1971).
364. Ya. A. Strakov and D. Brutane, *Latv. PSR Zinat. Acad. Vestis, Kim. Ser.,* **1973,** 225; *Chem. Abstr.,* **79,** 126438y (1973).
365. A. Ya. Strakov, D. Zicane, D. Brutane, and M. Opmane, Nov. Issled. Obl. Khim. Khim. Tekhnol., Mater. Nauchno-Tekh. Konf. Professorsko-Prepod. Sostava Nauchn. Rab. Khim. Fak. RPI, Riga, USSR, **1972,** p. 23; *Chem. Abstr.,* **82,** 4195h (1975).
366. F. Johnson and W. A. Nasutavicus, U.S. Patent 3,321,466; *Chem. Abstr.,* **68,** 21857d (1968).
367. H. K. Hall, Jr., *J. Org. Chem.,* **28,** 3213 (1963).
368. H. Erdtman and S. Thoren, *Acta Chem. Scand.,* **24,** 87 (1970).
369. P. Brun, R Furstoss, P. Teissier, W. Tubiana, and B. Waegell, *C. R. Acad. Sci. Paris, Ser. C.,* **269,** 427 (1969).

370. A. Zabza, C. Wawrzenczyk, and H. Kuczynski, *Bull. Acad. Pol. Sci., Ser. Sci. Chim.*, **22,** 855 (1974); *Chem. Abstr.*, **82,** 73196w (1975).
371. H. Bara and H. Klare, *Chem. Ber.*, **99,** 856 (1966).
372. Swiss Patent 280,367; *Chem. Abstr.*, **47,** 351c (1953).
373. M. Takshashi and S. Suzuki, *Bull. Chem. Soc. Japan*, **41,** 264 (1968).
374. F. D. Marsh, U.S. Patent 3,845,086; *Chem. Abstr.*, **82,** 86097q (1975).
375. H. K. Hall, Jr., *J. Am. Chem. Soc.*, **80,** 6412 (1958).
376. H. K. Hall, Jr., and R. Zbinden, *J. Am. Chem. Soc.*, **80,** 6428 (1958).
377. R. K. Hill, R. T. Conley, and O. T. Chortyk, *J. Am. Chem. Soc.*, **87,** 5646 (1965).
378. W. Schneider, R. Dillman, and H. J. Dechow, *Arch. Pharm.*, **299,** 397 (1966).
379. L. A. Paquette, *J. Org. Chem.*, **29,** 3447 (1964).
380. J. H. Van Den Hende and A. S. Kende, *J. Chem. Soc. (D)*, **1965,** 384.
381. J. E. Baldwin and R. A. Smith, *J. Am. Chem. Soc.*, **87,** 4819 (1965).
382. A. S. Kende, P. T. Izzo, and J. E. Lancaster, *J. Am. Chem. Soc.*, **87,** 5044 (1965).
383. L. A. Paquette, D. E. Kuhla, J. H. Barrett, and L. M. Leichter, *J. Org. Chem.*, **34,** 2888 (1969).
384. T. Sasaki, K. Kanematsu, and A. Kakehi, *Bull. Chem. Soc. Japan*, **43,** 2893 (1970).
385. J. M. Photis, *J. Heterocyclic Chem.*, **7,** 1249 (1970).
386. J. M. Photis, *J. Heterocyclic Chem.*, **8,** 729 (1971).
387. R. A. Abramovitch, S. R. Challand, and E. F. V. Scriven, *J. Am. Chem. Soc.*, **94,** 1374 (1972).
388. R. A. Abramovitch, T. D. Bailey, T. Takaya, and V. Uma, *J. Org. Chem.*, **39,** 340 (1974).
389. L. A. Paquette, U.S. Patent, 3,267,093; *Chem. Abstr.*, **67,** 3004y (1967).
390. T. Oishi, S. Murakami, and Y. Ban, *Chem. Pharm. Bull.* (Tokyo), **20,** 1740 (1972).
391. H. Wollweber, R. Hiltmann, H. G. Kroneberg, and H. Wilms, Belgian Patent 608,905; *Chem. Abstr.*, **57,** 16561g (1961).
392. V. L. Brown, Jr. and T. E. Stanin, *Ind. Eng. Chem. Prod. Res. Develop.*, **4,** 40 (1965).
393. V. L. Brown, Jr., J. G. Smith, and T. E. Stanin, French Patent 1,575,505; *Chem. Abstr.*, **72,** 100547p (1970).
394. V. L. Brown, Jr. and T. E. Stanin, British Patent 1,057,113; *Chem. Abstr.*, **66,** 115618n (1967).
395. G. Reinisch, M. Bara, and H. Klare, *Chem. Ber.*, **99,** 856 (1966).
396. S. Gomi, S. Suzuki, H. Takita, M. Takahashi, and K. Asano, German Patent 1,817,699; *Chem. Abstr.*, **73,** 66462r (1976).
397. H. K. Hall, Jr., *J. Org. Chem.*, **29,** 3135 (1964).
398. C. A. Wulff and E. F. Westrum, Jr., *J. Phys. Chem.*, **68,** 430 (1964).
399. C. M. Barber and E. F. Westrum, Jr., *J. Phys. Chem.*, **67,** 2373 (1963).
400. S.-W. S. Wong and E. F. Westrum, Jr., *J. Am. Chem. Soc.*, **93,** 5317 (1971).
401. L. M. Amzel, S. Baggio, R. F. Baggio, and L. N. Becka, *Acta Crystallogr., Sect. B*, **30,** 2494 (1974).
402. J. G. Traynham and M. T. Yang, *Tetrahedron Lett.*, **1965,** 575.
403. J. W. Ap Simon and J. D. Cooney, *Can. J. Chem.*, **49,** 1367 (1971).
404. G. Chapelet-Letourneux, M. Lemaire, and A. Rassat, *Bull. Chim. Soc. France*, **1965,** 3283.
405. Y. P. Carignan and D. R. Satriana, *J. Org. Chem.*, **32,** 285 (1967).
406. J. M. Lehn and J. Wagner, *J. Chem. Soc. (D)*, **1970,** 414.
407. J. D. Cooney, S. K. Brownstein, and J. W. Ap Simon, *Can. J. Chem.*, **52,** 3028 (1974).

408. G. E. Ellis, R. G. Jones, and M. G. Papadopoulos, *J. Chem. Soc. Perkin II*, **1974,** 1381.
409. W. L. Nobles and N. D. Potti, *J. Pharm. Sci.*, **57,** 1097 (1968).
410. R. A. Magarian and W. L. Nobles, *J. Pharm. Sci.*, **56,** 987 (1967).
411. M. E. Herr and R. B. Moffet, U.S. Patent 3,385,846; *Chem. Abstr.*, **69,** 59122u (1968).
412. N. J. Harper, G. B. A. Veitch, and D. G. Wibberley, *J. Med. Chem.*, **17,** 1188 (1974).
413. Y. L. Chow, *Can. J. Chem.*, **45,** 53 (1967).
414. M. Fukumi, M. Okubo, and G. Inoue, Japan Kokai 73 52,746; *Chem. Abstr.*, **80,** 59576x (1974).
415. W. Schneider an- D. K. Pomorin, *Chem. Ber.*, **105,** 1553 (1972).
416. R. A. Johnson, M. E. Herr, H. C. Murray, L. M. Reineke, and G. S. Fonken, *J. Org. Chem.*, **33,** 3195 (1968).
417. H. Kreiger, Belgian Patent 636,766; *Chem. Abstr.*, **61,** 16053g (1964).
418. Netherlands Patent 6,508,905; *Chem. Abstr.*, **64,** 19507c (1966).
419. W. J. Houlihan, U.S. Patent 3,198,785; *Chem. Abstr.*, **63,** 14830f (1965).
420. V. L. Brown, Jr., and T. E. Stanin, U.S. Patent 3,173,909; *Chem. Abstr.*, **62,** 13132d (1965).
421. Netherlands Patent 6,409,516; *Chem. Abstr.*, **64,** 2074b (1966).
422. V. L. Brown, Jr., and T. E. Stanin, U.S. Patent 3,282,925; *Chem. Abstr.*, **66,** 37789c (1967).
423. J. J. D'Amico, French Patent 1,463,732; *Chem. Abstr.*, **67,** 54045r (1967).
424. J. J. D'Amico, French Patent 1,478,974; *Chem. Abstr.*, **67,** 107584d (1967).
425. E. Jucker, A. Lindenmann, E. Schenker, and F. Gadient, French Patent 1,497,450; *Chem. Abstr.*, **69,** 77125b (1968).
426. E. Morita and J. J. D'Amico, *Rub. Chem. Technol.*, **44,** 881 (1971); *Chem. Abstr.*, **76,** 15537y (1972).
427. C. L. Kissel and B. Rickborn, *J. Org. Chem.*, **37,** 2060 (1972).
428. G. R. Krow, R. Rodebaugh, M. Grippi, G. Devicaris, C. Hyndman, and J. Marakowski, *J. Org. Chem.*, **38,** 3094 (1973).
429. H. K. Hall, Jr., *J. Am. Chem. Soc.*, **82,** 1209 (1960).
430. I. Fleming and R. B. Woodward, *J. Chem. Soc. Perkin I*. **1973,** 1653.
431. E. Billett, I. Fleming, and S. W. Hanson, *J. Chem. Soc. Perkin I*, **1973,** 1661.
432. M. Green, S. Tolson, J. Weaver, D. C. Wood, and P. Woodward, *J. Chem. Soc. (D)*, **1971,** 222.
433. V. Arya and S. J. Shenoy, *Ind. J. Chem.*, **10,** 815 (1972).
434. B. Eistert, D. Greiber, and I. Caspari, *Justus Liebigs Ann. Chem.*, **659,** 64 (1962).
435. Y. Kanaoka and Y. Hatanaka, *J. Org. Chem.*, **41,** 400 (1976).
436. G. J. B. Cajipe, G. Landen, B. Semler, and H. W. Moore, *J. Org. Chem.*, **40,** 3874 (1975).
437. H. Budzikiewicz and U. Lenz. *Org. Mass Spectrom.*, **10,** 992 (1976).
438. J. Rigaudy, C. Igier, and J. Barcelo, *Tetrahedron Lett.*, **1975,** 3845.
439. G. Seitz and T. Kaepchen, *Chem. Z.*, **99,** 503 (1975); *Chem. Abstr.*, **84,** 120840e (1976).
440 H. H. Quon and Y. L. Chow, *Tetrahedron*, **31,** 2349 (1975).
441. A. C. Cope, H. R. Nace, and L. L. Estes, Jr., *J. Am. Chem. Soc.*, **72,** 1123 (1950).
442. D. Anker, Y. Colleuille, and P. Perras, French Patent 1,364,067; *Chem. Abstr.*, **62,** 1637a (1965).
443. A. G. Anastassiou, *J. Am. Chem. Soc.*, **87,** 5512 (1965).

444. A. G. Anastassiou, *J. Am. Chem. Soc.*, **90,** 1527 (1968).
445. G. C. Tustin, C. E. Monken, and W. H. Okamura, *J. Am. Chem. Soc.*, **94,** 5112 (1972).
446. J. W. Bastable, J. D. Hobson, and W. D. Riddell, *J. Chem. Soc. Perkin I,* **1972,** 2205.
447. A. G. Anastassiou and R. P. Cellura, *J. Org. Chem.*, **37,** 3126 (1972).
448. A. G. Anastassiou, A. E. Winston, and E. Reichmanis, *J. Chem. Soc. (D),* **1973,** 779.
449. A. G. Anastassiou and R. M. Lazarus, *J. Chem. Soc. (D),* **1970,** 373.
450. W. L. Mock and P. A. H. Isaac, *J. Am. Chem. Soc.*, **94,** 2749 (1972).
451. A. G. Anastassiou and H. Yamamoto, *J. Chem. Soc. (D),* **1973,** 840.
452. H. Schmidt, A. Schweig, A. G. Anastassiou, and H. Yamamoto, *J. Chem. Soc. (D),* **1974,** 218.
453. C. S. Huber, *Acta Crystallogr., Sect. B,* **28,** 2577 (1972).
454. V. M. Thuy, *Ann. Chim.* (Paris), **2,** 183 (1967).
455. H. Wong, J. Chapuis, and I. Monkovic, *J. Org. Chem.*, **39,** 1042 (1974).
456. W. S. Murphy and J. P. McCarthy, *J. Chem. Soc. (D),* **1970,** 1129.
457. L. A. Paquette, J. R. Malpass, G. R. Krow, and T. J. Barton, *J. Am. Chem. Soc.*, **91,** 5296 (1969).
458. G. R. Krow and J Reilly, *J. Org. Chem.*, **40,** 136 (1975).
459. G. R. Krow and J. Reilly, *J. Am. Chem. Soc.*, **97,** 3837 (1975).
460. A. C. Cope, F. S. Fawcett, and G. Munn, *J. Am. Chem. Soc.*, **72,** 3399 (1950).
461. A. C. Cope E. L. Wick, and F. S. Fawcett, *J. Am. Chem. Soc.*, **76,** 6156 (1954).
462. J. R. Merchant, A. P. Moghe, and J. R. Patell, *J. Ind. Chem. Soc.*, **48,** 483 (1971).
463. J. Plostnieks, *J. Org. Chem.*, **31,** 634 (1966).
464. E. L. May, *J. Org. Chem.*, **21,** 223 (1956).
465. L. Baiocchi, M. Giannangeli, and G. Palazzo, *Gazz. Chim. Ital.*, **102,** 36 (1972).
466. F. Derichs, W. Schade, O. Glosauer, and W. Franke, *Justus Liebigs Ann. Chem.*, **687,** 116 (1965).
467. D. Schill and W. Schneider, *Arch. Pharm.*, **308,** 925 (1975).
468. B. C. Blount and R. Robinson, *J. Chem. Soc.*, **1932,** 1429.
469. K. Alder and H. A. Dortman, *Chem. Ber.*, **87,** 1905 (1954).
470. K. Alder, H. Wirtz, and H. Koppelberg, *Justus Liebigs Ann. Chem.*, **601,** 138 (1956).
471. J. Meinwald and M. Koskenkyla, *Chem. Ind.* (London), **1955,** 476.
472. O. L. Chapman and J. Meinwald, *J. Org. Chem.*, **23,** 162 (1958).
473. N. J. Leonard and D. F. Morrow, *J. Am. Chem. Soc.*, **80,** 371 (1958).
474. L. A. Paquette and M. J. Broadhurst, *J. Am. Chem. Soc.*, **94,** 632 (1972).
475. E. Vogel, M. Biskup, W. Pretzer, and W. A. Boell, *Angew. Chem.*, **76,** 785 (1964).
476. G. L. Grunewald, A. M. Warner, S. J. Hays, R. H. Bussell, and M. K. Seals, *J. Med. Chem.*, **15,** 747 (1972).
477. A. C. Cope, R. J. Cotter, and G. G. Roller, *J. Am. Chem. Soc.*, **77,** 3590 (1955).
478. K. Biemann, G. Büchi, and B. H. Walker, *J. Am. Chem. Soc.*, **79,** 5558 (1957).
479. L. A. Paquette, J. H. Barrett, and D. E. Kuhla, *J. Am. Chem. Soc.*, **91,** 3616 (1969).
480. H. R. Blattmann, W. Boll, E. Heilbronner, G. Hohlneicher, E. Vogel, and J. P. Weber, *Helv. Chim. Acta,* **49,** 2017 (1966).
481. E. Vogel, *Chimia,* **22,** 21 (1968).
482. H. Guenther and H. H. Hinrichs, *Tetrahedron,* **24,** 7033 (1968).
483. W. Bremser, H. T. Grunder, E. Heilbronner, and E. Vogel, *Helv. Chim. Acta,* **50,** 84 (1967).

484. B. Briat, D. A. Schooley, R. Records, E. Bunnenberg, C. Djerassi, and E. Vogel, *J. Am. Chem. Soc.*, **90,** 4691 (1968).
485. D. R. Battiste and J. G. Traynham, *J. Org. Chem.*, **40,** 1239 (1975).
486. L. A. Paquette and L. D. Wise, *J. Am. Chem. Soc.*, **87,** 1561 (1965).
487. A. M. Duffield, C. Djerassi, L. D. Wise, and L. A. Paquette, *J. Org. Chem.*, **31,** 1599 (1966).
488. R. K. Hill, *J. Org. Chem.*, **22,** 830 (1957).
489. R. K. Hill and R. T. Conley, *J. Am. Chem. Soc.*, **82,** 645 (1960).
490. R. Lukes and J. Hofman, *Coll. Czech. Chem. Commun.*, **26,** 523 (1961).
491. R. T. Conley and B. E. Nowak, *J. Org. Chem.*, **26,** 692 (1961).
492. L. Toscano, A. Calvi di Bergolo, and A. Bianchetti, British Patent 1,343,916; *Chem. Abstr.*, **80,** 120807 (1974).
493. W. A. W. Cummings and A. C. Davis, *J. Chem. Soc.*, **1964,** 4591.
494. J. P. Devlin, O. E. Edwards, P. R. Gorham, N. R. Hunter, R. K. Pike, and B. Stavric, *Can. J. Chem.*, **55,** 1367 (1977).
495. (a) H. F. Campbell, O. E. Edwards, and R. Kolt, *Can. J. Chem.*, **55,** 1372 (1977); (b) H. F. Campbell, O. E. Edwards, J. Elder, and R. J. Kolt, *Pol. J. Chem.*, **53,** 27 (1979); *Chem. Abstr.*, **91,** 5097a (1979).
496. T. Oishi, M. Fukui, R. Kenkyusho, and Ban, *Heterocycles*, **5,** 281 (1976).
497. C. Wawrzenczyk and A. Zabza, *Bull. Acad. Pol. Sci., Ser. Sci. Chim.*, **24,** 939, 951 (1976).
498. R. Aumann and J. Knecht, *Chem. Ber.*, **111,** 3429, 3927 (1978).
499. M. Bortolussi, R. Bloch, and J. M. Conia, *Tetrahedron Lett.*, **1977,** 2289.
500. S. Sarel, A. Felzenstein, and J. Yovell, *Tetrahedron Lett.*, **1976,** 451.
501. M. Langbeheim and S. Shalom, *Tetrahedron Lett.*, **1978,** 1219.
502. A. G. Anastassiou and R. L. Mahaffey, *J. Chem. Soc., Chem. Commun.*, **1978,** 915.
503. R. G. Glushkov, V. G. Smirnova, I. M. Zasosova, T. Stezhko, I. M. Oucharova, and T. F. Vlasova, *Khim. Geterotsikl. Soedin.*, **1978,** 374; *Chem. Abstr.*, **89,** 43306j (1978).
504. L. Brehm and A. L. N. Larsen, *Acta Crystallogr., Sect. B*, **1332,** 3336 (1976).
505. P. Krogsgaard-Larsen, *Acta Chem. Scand., Ser. B*, **B31,** 584 (1977).
506. R. G. Glushkov and T. V. Stezhko, *Khim. Geterotsikl. Soedin.*, **1978,** 1252; *Chem. Abstr.* **90,** 22784b (1979).
507. C. J. Rao and K. Murthy, *Ind. J. Chem., Sect. B*, **16B,** 636 (1978).
508. V. Bardakos and W. Sucrow, *Chem. Ber.*, **109,** 1898 (1976).
509. G. Ufer, S. S. Tjoa, and S. F. MacDonald, *Can. J. Chem.*, **56,** 2437 (1978).
510. V. Bardakos and W. Sucrow, *Chem. Ber.*, **111,** 1780 (1978).
511. R. V. Davies, B. Iddon, M. W. Pickering, H. Suschitzky, P. T. Gallacher, M. W. Gittos, and M. D. Robinson, *J. Chem. Soc., Perkin Trans. I*, **1977,** 2357.
512. J. Sauleau, A. Sauleau, and J. Huet, *Bull. Soc. Chim. France*, **1978,** 97.
513. T. Kato, K. Tabei, and E. Kawashima, *Chem. Pharm. Bull.*, **24,** 1544 (1976).
514. G. Gast, J. Schmutz, and D. Sorg. *Helv. Chim. Acta.*, **60,** 1644 (1977).
515. P. H. Mazzocchi, M. J. Bowen, and N. K. Narain, *J. Am. Chem. Soc.*, **99,** 7063 (1977).
516. P. H. Mazzocchi, S. Minamikawa, and P. Wilson, *J. Org. Chem.*, **44,** 1186 (1979).
517. S. Kano, T. Yokomatsu, and S. Shibuya, *Heterocycles*, **4,** 933 (1976).
518. M. Kumari, J. M. Khanna, and N. Anand, *Ind. J. Chem., Sect. B*, **16B,** 129 (1978).
519. M. R. Uskokovic, T. Henderson, C. Reese, H. L. Lee, G. Grethe, and J. Gutzwiller, *J. Am. Chem. Soc.*, **100,** 571 (1978).

520. A. Jossang-Yanagida and C. Gansser, *J. Heterocyclic Chem.*, **15,** 249 (1978).
521. H. Klar, *Arch. Pharm.*, **309,** 550 (1976).
522. L. Zukauskaite, A. Stankevicius, and A. N. Kost, *Khim. Geterotsikl. Soedin.*, **1978,** 63; *Chem. Abstr.*, **88,** 17005q (1978).
523. H. Yamamoto, H. Kawamoto, S. Morosawa, and A. Yokoo, *Heterocycles,* **11,** 267 (1978).
524. V. G. Granik, O. Ya. Belyaeva, R. B., Glushkov, T. F. Vlasova, and O. S. Anisimova, *Khim. Geterotsikl. Soedin.*, **1977,** 1106; *Chem. Abstr.*, **88,** 6779z (1978).
525. V. G. Granik, N. B. Marchenko, T. F. Vlasova, and R. G. Glushkov, *Khim. Geterotsikl. Soedin.*, **1976,** 1509; *Chem. Abstr.*, **86,** 139783b (1977).
526. H. Yamamoto, T. Komazawa, K. Nakave, and A. Yokoo, *Heterocycles,* **11,** 275 (1978).
527. R. G. Glushkov, O. Ya. Belyaeva, V. G. Granik, M. K. Polievktov, A. B. Grigor'ev, V. E. Serokhvostova, and T. F. Vlasova, *Khim. Geterotsikl. Soedin.*, **1976,** 1640; *Chem. Abstr.*, **86,** 121287h (1977).
528. G. Seitz, T. Kaempchen, and W. Overheu, *Arch. Pharm.*, **311,** 786 (1978).
529. G. Seitz and W. Overheu, *Arch. Pharm.*, **310,** 936 (1977).
530. P. Mackiewicz, R. Furstoss, B. Waegell, R. Cote, and J. Lessard, *J. Org. Chem.*, **43,** 3746 (1978).
531. J. Lessard, R. Cote, P. Mackiewicz, R. Furstoss, and B. Waegell, *J. Org. Chem.*, **43,** 3750 (1978).
532. H. A. Bates and H. Rapoport, *J. Am. Chem. Soc.*, **101,** 1259 (1979).
533. M. Barrelle and M. Apparu, *Tetrahedron,* **33,** 1309 (1977).
534. M. Barrelle and M. Apparu, *Can. J. Chem.*, **56,** 85 (1978).
535. W. S. Murphy, K. P. Raman, and B. J. Hathaway, *J. Chem. Soc., Perkin Trans. I,* **1977,** 2521.
536. M. Schaeffer-Ridder, A. Wagner, M. Schwamborn, H. Schreiner, E. Devrout, and E. Vogel, *Angew. Chem.*, **90,** 894 (1978).
537. H. J. Goelz, J. M. Muchowski, and M. L. Maddox, *Angew. Chem.*, **90,** 896 (1978).
538. A. J. Frew, G. R. Proctor, and J. V. Silverton, *J. Chem. Soc., Perkin Trans. I,* **1980,** 1251.
539. Y. Kanaoka, O. Haruo, and Y. Hatanaka, *Kokagaku Toronkai Koen Yoshishu,* **1979,** 16; *Chem. Abstr.*, **93,** 45533n (1980).
540. M. Kimura and S. Tai, *J. Chem. Soc.*, **1980,** 974.
541. M. R. Roberts and R. H. Schlessinger, *J. Am. Chem. Soc.*, **101,** 762 (1979).
542. S. Braun, J. Kinkeldei, and L. Walther, *Tetrahedron,* **36,** 1353 (1980).
543. B. A. Mooney, R. H. Prager, and D. A. Ward, *Aust. J. Chem.*, **33,** 2717 (1980).
544. S. C. Sharma and B. M. Lynch, *Can. J. Chem.*, **57,** 3034 (1979).
545. M. M. Vora, C. S. Yi, and C. De. W. Blanton, *Heterocycles,* **16,** 399 (1981).
546. R. G. Glushkov and T. V. Stezhko, *Khim. Geterotsikl. Soedin.*, **1980,** 1097; *Chem. Abstr.*, **94,** 47260r (1981).
547. G. M. Sharma, J. S. Buyer, and M. W. Pomerantz, *J. Chem. Soc. (D),* **1980,** 435.
548. P. T. Gallacher, B. Iddon, and H. Suschitzky, *J. Chem. Soc. Perkin I,* **1980,** 2362.
549. K. Satake, M. Kimura, and S. Morosawa, *Chem. Lett,* **1980,** 1389; *Chem. Abstr.*, **94,** 83985q (1981).
550. H. P. Figeys and R. Jammar, *Tetrahedron Lett.*, **21,** 2995 (1980).
551. J. Singh, V. Virmani, P. C. Jain, and N. Anand, *Ind. J. Chem., Sect. B,* **19B,** 195 (1980).
552. W. Eberbach and J.-C. Carre, *Tetrahedron Lett.*, **21,** 1145 (1980).
553. I. Yavari, *J. Mol. Struct.*, **67,** 293 (1980); *Chem. Abstr.*, **94,** 3369d (1981).

554. F. Bondavalli, P. Schenone, S. Lanteri, M. Longobardi, F. Rossi, F Rosatti, R. Ottave, and E. Marmo, *Farmaco, Ed. Sci.*, **35,** 380 (1980).

555. B. M. Trost, M. Vaultier, and M. L. Santiago, *J. Am. Chem. Soc.*, **102,** 7929 (1980).

556. T. Cuvigny, P. Hullot, P. Mulot, M. Larcheveque, and H. Normant, *Can. J. Chem.*, **57,** 1201 (1979).

557. M. Mori, S. Kudo, and Y. Ban, *J. Chem. Soc., Perkin Trans. I,* **1979,** 771.

558. J. J. Tufariello, S. A. Ali, and H. O. Klingele, *J. Org. Chem.*, **44,** 4213 (1979).

559. P. D. Davis and D. C. Neckers, *J. Org. Chem.*, **45,** 456 (1980).

560. P. D. Davis, D. C. Neckers, and J. R. Blount, *J. Org. Chem.*, **45,** 462 (1980).

561. P. D. Carpenter, V. Peesapati, and G. R. Proctor, *J. Chem. Soc., Perkin Trans I,* **1979,** 103.

562. M. Lennon and G. R. Proctor, *J. Chem. Soc., Perkin Trans I,* **1979,** 2009.

563. H. P. Soetens and U. K. Pandit, *Heterocycles,* **11,** 75 (1978).

564. T. Fushima, H. Ikuta, H. Irie, K. Nakadachi, and S. Uyeo, *Heterocycles,* **12,** 1311 (1979).

565. D. Berney and K. Schuh, *Helv. Chim. Acta,* **63,** 924 (1980).

566. G. R. Krow and S. Szczepanski, *J. Org. Chem.*, **47,** 1153 (1982).

567. M. Mori and Y. Ban, *Tetrahedron Lett.*, **20,** 1133 (1979).

Author Index

Numbers in parentheses are reference numbers and indicate that author's work is referred to although his name is not mentioned in the text. Numbers in *italics* show the pages on which the complete references are listed.

Subject Index